Behavioral Biology of Laboratory Animals

Behavioral Biology of Laboratory Animals

edited by
Kristine Coleman and Steven J. Schapiro

CRC Press
Taylor & Francis Group
Boca Raton London New York

CRC Press is an imprint of the
Taylor & Francis Group, an **informa** business

First edition published 2022
by CRC Press
6000 Broken Sound Parkway NW, Suite 300, Boca Raton, FL 33487-2742

and by CRC Press
2 Park Square, Milton Park, Abingdon, Oxon, OX14 4RN

© 2022 Taylor & Francis Group, LLC

CRC Press is an imprint of Taylor & Francis Group, LLC

Reasonable efforts have been made to publish reliable data and information, but the author and publisher cannot assume responsibility for the validity of all materials or the consequences of their use. The authors and publishers have attempted to trace the copyright holders of all material reproduced in this publication and apologize to copyright holders if permission to publish in this form has not been obtained. If any copyright material has not been acknowledged please write and let us know so we may rectify in any future reprint.

Except as permitted under U.S. Copyright Law, no part of this book may be reprinted, reproduced, transmitted, or utilized in any form by any electronic, mechanical, or other means, now known or hereafter invented, including photocopying, microfilming, and recording, or in any information storage or retrieval system, without written permission from the publishers.

For permission to photocopy or use material electronically from this work, access www.copyright.com or contact the Copyright Clearance Center, Inc. (CCC), 222 Rosewood Drive, Danvers, MA 01923, 978-750-8400. For works that are not available on CCC please contact mpkbookspermissions@tandf.co.uk

Trademark notice: Product or corporate names may be trademarks or registered trademarks and are used only for identification and explanation without intent to infringe.

ISBN: 978-0-367-02923-4 (hbk)
ISBN: 978-1-032-03435-5 (pbk)
ISBN: 978-0-429-01951-7 (ebk)

DOI: 10.1201/9780429019517

Typeset in Times
by codeMantra

Contents

Preface vii
Editors ix
Contributors xi

Part 1

1. Introduction to the *Behavioral Biology of Laboratory Animals* 3
Steven J. Schapiro and Kristine Coleman

2. Animal Behavior: An Introduction 7
Kristine Coleman and Melinda A. Novak

3. Abnormal Behavior of Animals in Research Settings 27
Melinda A. Novak and Jerrold S. Meyer

4. Utilizing Behavior to Assess Welfare 51
Daniel Gottlieb and Ori Pomerantz

5. An Overview of Behavioral Management for Laboratory Animals 65
Steven J. Schapiro

Part 2

6. Behavioral Biology of Mice 89
Aileen MacLellan, Aimée Adcock, and Georgia Mason

7. Behavioral Biology of Rats 113
Sylvie Cloutier

8. Behavioral Biology of Guinea Pigs 131
Gale A. Kleven

9. Behavioral Biology of Deer and White-Footed Mice, Mongolian Gerbils, and Prairie and Meadow Voles 147
Kathleen R. Pritchett-Corning and Christina Winnicker

10. Behavioral Biology of Hamsters 165
Christina Winnicker and Kathleen R. Pritchett-Corning

11. Behavioral Biology of Rabbits 173
Lena Lidfors and Kristina Dahlborn

12. Behavioral Biology of Ferrets 191
Claudia M. Vinke, Nico J. Schoemaker, and Yvonne R. A. van Zeeland

13. Behavioral Biology of Dogs 205
Laura Scullion Hall and Mark J. Prescott

14. Behavioral Biology of the Domestic Cat 223
Judith Stella

15. Behavioral Biology of Pigs and Minipigs 243
Sandra Edwards and Nanna Grand

16. Behavioral Biology of Sheep 261
Cathy M. Dwyer

17. Behavioral Biology of Cattle 273
Clive Phillips

18. Behavioral Biology of Horses 285
Janne Winther Christensen

19. Behavioral Biology of Chickens and Quail 299
Laura Dixon and Sarah Lambton

20. Behavioral Biology of the Zebra Finch 315
Samantha R. Friedrich and Claudio V. Mello

21. Behavioral Biology of Zebrafish 331
Christine Powell, Isabel Fife-Cook, and Becca Franks

22. Behavioral Biology of Amphibians 345
Charlotte A. Hosie and Tessa E. Smith

23. Behavioral Biology of Reptiles 361
Dale F. DeNardo

24. Behavioral Biology of Marmosets 377
Arianna Manciocco, Sarah J. Neal Webb, and Michele M. Mulholland

25. Behavioral Biology of Squirrel Monkeys 395
Anita I. Stone and Lawrence Williams

26. Behavioral Biology of Owl Monkeys 409
Alba García de la Chica, Eduardo Fernandez-Duque, and Lawrence Williams

27. Behavioral Biology of Capuchin Monkeys 421
Marcela Eugenia Benítez, Sarah F. Brosnan, and Dorothy Munkenbeck Fragaszy

28. Behavioral Biology of Macaques 437
Paul E. Honess

29. Behavioral Biology of Vervets/African Green Monkeys 475
Matthew J. Jorgensen

30. Behavioral Biology of Baboons 495
Corrine K. Lutz

Part 3 Selected Ethograms 513

Index 529

Preface

The Behavioral Biology of Laboratory Animals is designed to provide readers with knowledge about the behavioral biology of animals that are common subjects in research settings. Knowing the natural behavior of animals is a critical part of modern animal care practices. Each species evolved under specific environmental conditions resulting in unique behavioral patterns, many of which are maintained in captivity even after generations of breeding. Thus, factors such as how individuals obtain food and avoid predation, whether they maintain territories, how they interact and communicate with conspecifics, and how they care for their young are important considerations when providing appropriate housing options for laboratory animals. Knowing the behavioral biology allows facilities to provide conditions that functionally simulate their environment and encourage the expression of natural behavior.

In addition, knowing behaviors that are normal for the species can help those caring for them recognize behaviors of concern. Certain vocalizations or behavioral patterns, such as scratching in nonhuman primates and dogs, can indicate that the animal is feeling anxious. Abnormal behaviors such as excessive stereotypy can indicate reduced welfare. For this reason, the *Guide for the Care and Use of Laboratory Animals* suggests that "Personnel responsible for animal care and husbandry should receive training in the behavioral biology of the species they work with to appropriately monitor the effects of enrichment as well as identify the development of adverse or abnormal behaviors" (National Resource Council 2011, p. 53).

This book provides behavioral information for many, although not all, common laboratory animals. Because not every reader has a background in animal behavior, the book is broken into three parts to cover both broad aspects of animal behavior and how to maintain animals in captivity, as well as the behavioral biology of specific laboratory animals.

Part 1 of this book covers some of the basics of animal behavior and how behavior can be used to assess welfare. The heart of the book is the next part, which focuses on the behavioral biology of specific laboratory animal species, or group of species. Each chapter discusses the natural behavior of the animal, and because the behaviors animals express in captivity can be somewhat different than those found in their wild counterparts, each includes a section on common captive behavior, including those considered both normal and abnormal. Additionally, the chapters in this part of the book provide information on ways to maintain the behavioral health of the animals. The third part of the book contains ethograms (lists of behaviors and their definitions) for many of the species. These ethograms will provide a quick reference for care staff, researchers, and casual observers alike.

As you look through this book, you will read about the fascinating and unique behaviors displayed by each of the animals. Working with these animals is a great privilege, accompanied, of course, by a great responsibility. We are so grateful to the authors of this book for sharing their knowledge of, and love for, the animals with which they work. We hope that not only does this book help you to learn more about the species in your care, but also to better *honor the contributions made by the animals* to research endeavors.

Editors

Kristine Coleman, PhD, is an associate professor in the Division of Comparative Medicine, and Head of the Behavioral Services Unit at the Oregon National Primate Research Center (ONPRC), Oregon Health & Science University. Dr. Coleman received her PhD in behavioral ecology from Binghamton University, where she studied individual differences in temperament in pumpkinseed sunfish. She went to the Oregon Regional (now National) Primate Research Center for her postdoctoral training and never left. Since 2001, she has overseen the ONPRC behavioral management program, where she studies ways to improve the psychological well-being of laboratory macaques.

Dr. Coleman has coauthored over 150 papers, book chapters, and abstracts on various topics in the field of animal behavior and management. She regularly teaches courses and workshops on behavioral management topics. For the past 5 years, she has taught a course on the behavioral biology of laboratory animals in the Eastern Virginia Medical School's Master of Laboratory Animal Science program. Dr. Coleman is a member of a number of animal behavior societies and is currently the treasurer and co-chair of the Member and Finance Committee of the American Society of Primatologists. In addition, Dr. Coleman is vice-chair of the ONPRC IACUC, and an ad hoc specialist with AAALAC International.

Steven J. Schapiro, PhD, is an associate professor of comparative medicine in the Department of Comparative Medicine at the Michale E. Keeling Center for Comparative Medicine and Research of The University of Texas MD Anderson Cancer Center. Dr. Schapiro earned his PhD from the University of California at Davis in 1985 after receiving his BA in behavioral biology from Johns Hopkins University. He completed a postdoctoral research fellowship at the Caribbean Primate Research Center of the University of Puerto Rico.

In 1989, he joined the Department of Comparative Medicine at MD Anderson's Keeling Center and has been there ever since. In 2009, Dr. Schapiro was designated an honorary professor in the Department of Experimental Medicine at the University of Copenhagen, Denmark. He is a founding faculty member of both the Primate Training and Enrichment Workshop and the Primate Behavioral Management Conference, educational programs conducted at the Keeling Center that have reached over 1,000 individuals from primate facilities around the globe. He is the founder and organizer of Project Monkey Island, a rescue and restoration mission for the monkeys and people associated with Cayo Santiago in Puerto Rico.

Dr. Schapiro has coauthored almost 200 peer-reviewed papers and book chapters examining various aspects of nonhuman primate behavior, management, and research. He has also edited the three volumes of the third edition, and the fourth edition of the *Handbook of Laboratory Animal Science* along with Jann Hau, and served as the sole editor of the *Handbook of Primate Behavioral Management*. He has also coedited one issue of the *ILAR Journal*.

Dr. Schapiro has participated in international meetings and courses on primatology, behavioral management, and laboratory animal science in North America, Europe, Asia, and Africa. He is a member of a number of primatology and animal behavior societies and served as the treasurer and vice president for membership of the International Primatological Society for many years. He is also a past president, former treasurer, and former meeting coordinator of the American Society of Primatologists, as well as an honorary member of the Association of Primate Veterinarians. Dr. Schapiro is an advisor or consultant for a number of primate facilities that produce, manage, and conduct research with nonhuman primates in the United States and abroad.

Contributors

Aimée Adcock
Formerly Department of Animal Bioscience
University of Guelph
Guelph, Ontario, Canada

Marcela Eugenia Benítez
Department of Anthropology
Emory University
Atlanta, Georgia, United States

Sarah F. Brosnan
Departments of Psychology and Philosophy
Language Research Center
Center for Behavioral Neuroscience and
  Neuroscience Institute
Georgia State University
Atlanta, Georgia, United States

Janne Winther Christensen
Department of Animal Science
Aarhus University
Aarhus, Denmark

Sylvie Cloutier
Animal Behavior Scientist
Ottawa, Ontario, Canada

Kristine Coleman
Oregon National Primate Research Center
Oregon Health & Science University
Beaverton, Oregon, United States

Kristina Dahlborn
Department of Anatomy Physiology and Biochemistry
Swedish University of Agricultural Sciences
Uppsala, Sweden

Alba García de la Chica
Universidad de Barcelona
Barcelona, Spain

Dale F. DeNardo
Department of Animal Care and Technologies
School of Life Sciences
Arizona State University
Tempe, Arizona, United States

Laura Dixon
Animal Behaviour and Welfare
Scotland's Rural College
Penicuik, Midlothian, United Kingdom

Cathy M. Dwyer
Animal Behaviour and Welfare
Scotland's Rural College
Penicuik, Midlothian, United Kingdom

Sandra Edwards
School of Natural and Environmental Sciences
Newcastle University
Newcastle, United Kingdom

Eduardo Fernandez-Duque
Department of Anthropology
Yale University
New Haven, Connecticut, United States

Isabel Fife-Cook
Department of Environmental Studies
New York University
New York, New York, United States

Dorothy Munkenbeck Fragaszy
Department of Psychology
University of Georgia
Athens, Georgia, United States

Becca Franks
Department of Environmental Studies
New York University
New York, New York, United States

Samantha R. Friedrich
Department of Behavioral Neuroscience
Oregon Health & Science University
Portland, Oregon, United States

Daniel Gottlieb
Oregon National Primate Research Center
Oregon Heath & Science University
Beaverton, Oregon, United States

Nanna Grand
Scantox A/S
Lille Skensved, Denmark

Laura Scullion Hall
Refining Dog Care and
University of Stirling
Stirling, United Kingdom

Paul E. Honess
School of Veterinary Medicine and Science
University of Nottingham
Nottingham, United Kingdom

Charlotte A. Hosie
Department of Biological Sciences
University of Chester
Chester, United Kingdom

Matthew J. Jorgensen
Department of Pathology
Section on Comparative Medicine
Wake Forest School of Medicine
Winston-Salem, North Carolina, United States

Gale A. Kleven
Department of Psychology
Behavioral Neuroscience Area
Wright State University
Dayton, Ohio, United States

Sarah Lambton
Bristol Veterinary School
University of Bristol
Langford, North Somerset, United Kingdom

Lena Lidfors
Department of Animal Environment and Health
Swedish University of Agricultural Sciences
Skara, Sweden

Corrine K. Lutz
Animal Behavior Scientist
San Antonio, Texas, United States

Aileen MacLellan
Department of Integrative Biology
University of Guelph
Guelph, Ontario, Canada

Arianna Manciocco
Istituto di Scienze e Tecnologie della Cognizione
Consiglio Nazionale delle Ricerche
Rome, Italy

Georgia Mason
Department of Integrative Biology
University of Guelph
Guelph, Ontario, Canada

Claudio V. Mello
Department of Behavioral Neuroscience
Oregon Health & Science University
Portland, Oregon, United States

Jerrold S. Meyer
Department of Psychological and Brain Sciences
University of Massachusetts-Amherst
Amherst, Massachusetts, United States

Michele M. Mulholland
Department of Comparative Medicine
The University of Texas MD Anderson Cancer Center
Bastrop, Texas, United States

Sarah J. Neal Webb
Department of Comparative Medicine
The University of Texas MD Anderson Cancer Center
Bastrop, Texas, United States

Melinda A. Novak
Department of Psychological and Brain Sciences
University of Massachusetts, Amherst
Amherst, Massachusetts, United States

Clive Phillips
Curtin University Sustainability Policy Institute
Curtin University
Perth, Western Australia, Australia

Ori Pomerantz
California National Primate Research Center
University of California, Davis
Davis, California, United States

Christine Powell
Department of Environmental Studies
New York University
New York, New York, United States

Mark J. Prescott
National Centre for the Replacement, Refinement and
    Reduction of Animals in Research (NC3Rs)
London, United Kingdom

Kathleen R. Pritchett-Corning
Office of Animal Resources
Harvard University Faculty of Arts and Sciences
Department of Comparative Medicine
Cambridge, Massachusetts, United States
and
University of Washington
Seattle, Washington, United States

Steven J. Schapiro
Department of Comparative Medicine
The University of Texas MD Anderson Cancer Center
Bastrop, Texas, United States
and
Department of Experimental Medicine
University of Copenhagen
Copenhagen, Denmark

Nico J. Schoemaker
Division of Zoological Medicine
Department of Clinical Sciences
Faculty of Veterinary Medicine
University of Utrecht
Utrecht, The Netherlands

Tessa E. Smith
Department of Biological Sciences
University of Chester
Chester, United Kingdom

Judith Stella
Good Dog, Inc.
New York, New York, United States

Anita I. Stone
Department of Biology
California Lutheran University
Thousand Oaks, California, United States

Yvonne R.A. van Zeeland
Division of Zoological Medicine
Department of Clinical Sciences
Faculty of Veterinary Medicine
University of Utrecht
Utrecht, The Netherlands

Claudia M. Vinke
Division of Animals in Science and Society
Department of Population Health Sciences
Faculty of Veterinary Medicine
University of Utrecht
Utrecht, The Netherlands

Lawrence Williams
Department of Comparative Medicine
The University of Texas MD Anderson Cancer Center
Bastrop, Texas, United States

Christina Winnicker
Research Office of Animal Welfare Ethics Strategy & Risk
GlaxoSmithKline
Collegeville, Pennsylvania, United States

Part 1

1

Introduction to the Behavioral Biology of Laboratory Animals

Steven J. Schapiro
The University of Texas MD Anderson Cancer Center
University of Copenhagen

Kristine Coleman
Oregon National Primate Research Center

CONTENT

References 5

Welcome to the *Behavioral Biology of Laboratory Animals* (*BBLA*). This volume contains 30 chapters divided into three sections, which all focus on aspects of the behavioral biology of animals that are often/frequently/sometimes/occasionally found in laboratories and related research settings. The overall goal of the *BBLA* is to provide you, someone working in a maintenance, care, and/or research program that involves laboratory animals, with information about the way the animals live in the wild, and the way that they should live in captive research settings. We hope that the contents of the chapters (descriptions, data, guidance, resources, and recommendations) will help you understand your animals better, allow you to refine the care and treatment that your animals receive, and improve the well-being, welfare, and wellness of your animals. If the data show that when intact adult male New Zealand White (NZW) rabbits are housed together in captivity (a social situation that is extremely unlikely to be observed in nature), they are quite likely to fight, then, unfortunately, captive, intact adult male NZWs should probably be housed individually (Lidfors and Dahlborn, 2021). At the opposite end of the spectrum, if the data show that zebra finches that are housed singly in captivity (again, a social situation that is extremely unlikely to be observed in the wild) fail to prosper, then (fortunately) captive zebra finches should be housed socially (in pairs or flocks; Friedrich and Mello, 2021).

The *BBLA* includes an entire section (Part 2; 25 chapters) on the behavioral biology of specific taxonomic groups (e.g., mice, zebrafish, zebra finches, reptiles, macaques), preceded by five introductory chapters that focus on the science of studying behavior as a discipline and its impact on our understanding of the species with which we work. The book concludes with a section that includes ethograms for some/many of the species discussed. The ethograms in Part 3 should provide a relatively centralized resource for those interested in understanding, and potentially quantifying, baboon, bird, or bunny behavior.

Most of the chapter authors in this volume study and/or work with animals in "research" settings; therefore, the information contained in their chapters is likely to be most relevant to animals living in such settings. However, many of the guidelines and recommendations contained in this book will also be valuable to those managing and working with animals in other environments, including zoological parks, aquaria, and sanctuaries.

The *BBLA* evolved as we each attempted to teach courses on the behavior of laboratory animals (SJS at the University of Copenhagen and KC at Eastern Virginia Medical School). Neither of us could find a single resource that addressed all of the issues that we felt were important to impress upon our students. KC, being a person of action, decided that we could fill this resource void by putting together a volume that addressed the issues that we thought would be most important to those working with laboratory animals. SJS came along for the ride, as he has certain organizational skills and connections that helped this project come to fruition. So, here you are, reading the result of a slightly delayed (partially due to COVID-19), but hopefully, successful, collaboration among the publishers, editors, and the authors. We are both reasonably seasoned observers of animal behavior, but we each learned quite a bit from working on these chapters. Some of our favorites include popcorning and rumblestrutting in guinea pigs (Kleven, 2021), the weasel war dance in ferrets (Vinke et al., 2021), and clumping in zebra finches (Friedrich and Mello, 2021). We hope that you will learn a few things and discover a few favorites as well.

Part 1 of the *BBLA* includes five chapters by Schapiro and Coleman; Coleman and Novak; Novak and Meyer; Gottlieb and Pomerantz; and Schapiro; respectively. Schapiro's and Coleman's chapter, which you are currently reading, establishes the motivation for the book and describes the chapters herein. The chapter by Coleman and Novak explains the role that animal behavior research plays in the selection and utilization of animals in laboratory research programs. Their chapter establishes the fundamental role that studying and understanding animal behavior plays in scientific research that involves animals. Pay particular attention to the discussion of the differences between animal and human sensation/perception.

This chapter has much to offer those responsible for studying and caring for laboratory animals. In Chapter 3, Novak and Meyer provide a brief review of abnormal behaviors in laboratory animals, with an emphasis on nonhuman primates (NHPs). While colony managers, veterinarians, researchers, and behaviorists make every attempt to establish captive environments that discourage the performance of abnormal activities, we are still learning how to prevent abnormal behavior. It is important that those of us working with animals in the laboratory have some understanding of behavioral abnormalities, the conditions that may yield them, and potential treatments to address them. Gottlieb and Pomerantz, in Chapter 4, then provide several behaviorally oriented perspectives that can be used to assess welfare in laboratory animals. They describe three approaches that should be helpful in determining whether strategies employed to maintain and enhance the welfare of laboratory animals are successful. Part 1 closes with a discussion of behavioral management by Schapiro, again with an emphasis on NHPs. Strategies to stimulate species-typical behavior and to minimize abnormal behavior are outlined in general terms, as each of the taxon-specific chapters in Part 2 contains several sections intended to address the ways that questions related to behavioral management are being handled in the animals included in the chapter. All five of these chapters help to establish a foundation for the taxon-specific information contained in the chapters that follow.

Part 2 of the *BBLA* comprises 25 chapters describing the behavioral biology of different taxonomic groups of animals. It should be obvious that we could not include every animal species that is involved in laboratory research. It should also be obvious that not every author we asked to write a chapter was able to complete the task. We are missing a few chapters that we would have liked to include (e.g., cephalopods, additional fish and avian species), but overall, we feel as though we included many of the most important animal species involved in research.

Some chapters cover a relatively small amount of taxonomic real estate (e.g., Chapter 8 by Kleven on guinea pigs), while others cover a considerable range of genera (e.g., Chapter 22 by Hosie and Smith on amphibians, Chapter 23 by DeNardo on reptiles) and/or species (Chapter 28 by Honess on macaques). We asked the authors of the chapters in this section to adhere to a particular format, beginning with a discussion of the typical research involvement of the taxa, followed by the natural history (e.g., ecology, social organization, feeding behavior, communication) of the animals and their common behaviors in captivity (normal and abnormal). Authors then present sections on ways to maintain behavioral health and stimulate species-typical behavior in captive settings. After brief discussions of situations that are special to that taxon, the chapters conclude with recommendations, resources, and, of course, references. The authors of these chapters have quite a bit of relevant experience, and reading their work should provide you with a wealth of useful information on the behavioral biology of these taxonomic groups.

Part 2 starts with a set of six chapters on rodents and lagomorphs, and begins with a fantastic chapter (Chapter 6) by MacLellan and colleagues on the behavioral biology of mice. As mice comprise the taxonomic group that is most frequently involved in research laboratories, it makes sense to start with them. There is much to be learned in this chapter. Chapter 7, by Cloutier, addresses the behavioral biology of rats, another popular and well-utilized taxon. A great deal is known about rat behavior in captive settings, but relatively little is known about the way they function in natural habitats. Chapter 8 by Kleven on guinea pigs comes next, with "popcorning" and the "rumblestrut", new behaviors for SJS and KC to add to our internal ethograms of animal behaviors. Guinea pigs are also known to "stampede", which creates a fun visual image. Pritchett-Corning and Winnicker describe gerbils, voles, and deer mice in Chapter 9, including arboreal *Peromyscus* and monogamous *Microtus*. Winnicker and Pritchett-Corning change author order for Chapter 10, in which they discuss hamsters. As those who have worked with hamsters know, they can be a bit grumpy, when rudely awakened. This is the final rodent chapter in this part. Lidfors and Dahlborn (Chapter 11) then address the intricacies of socially housing intact, adult male rabbits in the context of their natural history, among other topics.

Three chapters focusing on the behavioral biology of carnivores (ferrets by Vinke and colleagues, Chapter 12; Hall and Prescott on dogs, Chapter 13; and Stella on cats, Chapter 14) follow the rodent/lagomorph chapters, and one must pay particular attention to the weasel war dance performed by ferrets. Among "humans' best friends", which have olfactory systems that are a million to a billion times more sensitive than those of humans, selectively and purpose-bred beagles are especially important. It is also interesting to note that the cat is one of only a few types of animals in which the wild ancestor/relative is solitary, and the "domesticated" members of the species are quite social.

A series of five chapters on farm-type animals comes next, beginning with pigs and minipigs by Edwards and Grand (Chapter 15). The behavior of domestic pigs and mini-pigs differs little from that of their wild relatives, and considerable attention must be paid to weaning-related transitions and the provision of nest-building opportunities for pigs in research. Also, it should be noted that mother pigs do not lick their infants, a bit of an anomaly among mammals. Dwyer discusses the behavioral biology of sheep in Chapter 16, explaining that they are highly social, engage in many types of allomimetic behavior, and are, surprisingly, highly trainable. The behavioral biology of cattle that participate in research is described by Phillips (Chapter 17), followed by horses in Chapter 18 (Christensen). Horses are interesting, in that (1) negative reinforcement training (rather than positive reinforcement) is typically employed in practical situations, and (2) the mare-foal bond is extremely important for normal development. This subsection concludes with Chapter 19 on the behavioral biology of chickens by Dixon and Lambton, the first of only two chapters on avian taxa. The second avian contribution is a wonderful chapter (Chapter 20) by Friedrich and Mello that beautifully describes the behavioral biology of zebra finches (affectionately referred to by some as "zebbies"), including their vocalizations and their tendency to clump.

The next three chapters continue the departure from mammalian species, with chapters (21 by Powell and colleagues) on zebrafish, (22 by Hosie and Smith) on amphibians, and (23 by

DeNardo) on reptiles. Not only is the zebrafish chapter highly informative, all of the impressive illustrations were drawn by one of the authors. While Powell and colleagues were asked to focus their efforts on only the increasingly research-relevant zebrafish, Hosie and Smith and DeNardo were given the monumental tasks of addressing the behavioral biology of the multileveled taxonomic groups of amphibians and reptiles, respectively. They have each done a fantastic job of presenting important information on the behavioral biology of frogs, toads, and salamanders, and of snakes, lizards, crocodilians, and turtles/tortoises, respectively.

Part 2 concludes with seven chapters on various "species" of NHPs; four New World monkeys and three Old World monkeys. Chapter 24, by Manciocco and colleagues, discusses marmosets, with an emphasis on the common marmoset, an increasingly important NHP species in laboratory research, given its amenability to CRISPR-Cas9 technology. Stone and Williams, in Chapter 25, report on squirrel monkeys, a taxon with a long history of studies in the wild and the laboratory. In Chapter 26, Garcia de la Chica and colleagues discuss owl (night) monkeys, primates that are primarily nocturnal (as their common name implies), making them quite difficult to observe in the wild. Studies of owl monkeys in captive settings (making use of altered light cycles) have truly informed our understanding of their natural behavioral biology. This chapter is the only one in the book to discuss nocturnal NHPs, but not the only chapter to discuss nocturnal animals (e.g., mice, rats). Benitez and colleagues discuss the so-called "capuchin" monkeys (their taxonomy is somewhat in flux) in Chapter 27. These extremely interesting and intelligent animals have made important contributions toward answering a variety of psychological and anthropological research questions. Chapter 28 (by Honess) is another chapter devoted to the behavioral biology of a very large taxonomic group, the macaques. This chapter does an excellent job of providing information on the behavioral biology of a wide range of macaque species, while emphasizing the two most frequently encountered NHPs in laboratory research (*Macaca mulatta* and *M. fascicularis*). The penultimate chapter (Chapter 29 by Jorgensen) focuses on another group that is in a bit of taxonomic flux, "vervet monkeys". Vervets are involved less frequently in laboratory research than macaques, and Jorgensen clearly explains that the behavior in nature of African vervets differs considerably from the natural behavior of the primarily Asian macaques. Therefore, vervet monkeys cannot be managed in the same way as macaques. Part 2 ends with a chapter about baboons (Chapter 30 by Lutz), another NHP species with a long research history in the wild and the laboratory, and a recent history that involves taxonomic questions. This excellent chapter is a fitting way to end Part 2. In an odd twist of fate, Lutz's chapter was the first one received by us, but is the last one out to you.

Part 3 of the *BBLA* is simply a collection of ethograms for a selection of the taxa included in Part 2. We felt that it would be extremely useful for you to have a centralized resource that contains verified ethograms for many of the species that most frequently live in laboratory settings and participate in research projects. Over the years, we have searched for this type of resource, without much luck. Hopefully, you will find Part 3 to be a useful addition to this volume.

There are just three more things to quickly mention before you venture off into the biological landscape of the *BBLA*.

1. You will encounter some common themes, techniques, and terminology as you read the chapters in this volume. These include in no particular order: temperament/personality, behavioral assessments, functional simulations, functionally appropriate captive environments, prevention vs. cure, refinements, research–management synergisms, positive reinforcement training, enhancing the definition of animal models, minimizing confounds, and better scientific data. Please pay attention to these points when you encounter them.

2. In a similar vein, you will read about similar things that are, at times, labeled in slightly different ways in different parts of the book. We tried to establish a consistent framework across chapters, but sometimes it did not make sense to change what the authors had written. Keep your eyes open for similarities and differences in terms used. For instance, the definitions for abnormal behaviors or types of environmental enrichment may differ slightly across chapters, but there are really many more similarities than differences in these cases.

3. And finally, do not forget that the goal of *Behavioral Biology of Laboratory Animals* is to provide you with research-based information, from experts, that you can use to better understand, and to benefit, the animals that you care for, and work with, in your "laboratories". Understanding their behavioral biology is good for the animals and good for science, contributing to the optimization of the welfare of the animals, and the reliability and validity of the data. Behavioral biology as a science is constantly evolving, as new methods, techniques, and care strategies are designed and implemented. We can always learn more about our animals and do more to enhance their lives in research settings.

REFERENCES

Friedrich SR, Mello CV. Behavioral biology of zebra finches (*Taeniopygia guttata*). In Coleman K, Schapiro SJ, eds. *Behavioral Biology of Laboratory Animals*. Boca Raton, FL: CRC Press, Taylor & Francis Group; 2021:315–329.

Kleven GA. Behavioral biology of guinea pigs. In Coleman K, Schapiro SJ, eds. *Behavioral Biology of Laboratory Animals*. Boca Raton, FL: CRC Press, Taylor & Francis Group; 2021:131–146.

Lidfors L, Dahlborn K. Behavioral biology of rabbits. In Coleman K, Schapiro SJ, eds. *Behavioral Biology of Laboratory Animals*. Boca Raton, FL: CRC Press, Taylor & Francis Group; 2021:173–190.

Vinke CM, Schoemaker NJ, van Zeeland YRA. Behavioral biology of ferrets. In Coleman K, Schapiro SJ, eds. *Behavioral Biology of Laboratory Animals*. Boca Raton, FL: CRC Press, Taylor & Francis Group; 2021:191–204.

2

Animal Behavior: An Introduction

Kristine Coleman
Oregon National Primate Research Center

Melinda A. Novak
University of Massachusetts, Amherst

CONTENTS

Introduction 7
The Definition and Challenge of Animal Behavior 8
    Description Phase 8
    Quantification Phase 9
    The Explanation Phase 10
Biologically Relevant Questions about Behavior 10
    Proximate Questions (Mechanism and Development) 11
    Ultimate Questions (Function and Evolutionary History) 11
    Bird Song as an Example of a Levels of Analysis Perspective 11
Using the Scientific Method to Study Behavior 12
Concepts from Animal Behavior That Impact Laboratory Animal Care and Welfare 13
    Ecology 13
        Spatial Distributions 13
        Predation and the Landscape of Fear 13
    Social Interactions and Organization 14
        Social Organization 14
        Dominance Hierarchies 15
        Sexual Selection and Mating 15
    Foraging and Feeding Behavior 17
    Communication 17
    Learning 18
    Individual Variation 20
Cautions: Our Limitations as Observers 21
    Sensory Motor Capabilities 21
    Anthropomorphism 21
Summary 21
Acknowledgments 22
References 22

Introduction

If you were fortunate enough to have had time to take a walk in your neighborhood today, you may have heard robins sing. Or, perhaps you witnessed a squirrel run across a field to retrieve a previously buried nut. You may have caught a rabbit off guard and watched as he remained motionless before running away. All of these behaviors, as well as others we may not notice, help the animals survive in their world. Many bird species sing to attract mates, squirrels hide (cache) nuts to consume later, and rabbits freeze and/or flee to escape predation. Animals evolved to perform specific behaviors, such as the ones listed, that help them survive and reproduce. As such, animals have a strong drive to engage in these behaviors, even when they are living in other environments, such as research facilities. Even genetically altered animals maintain behavioral instincts of their wild counterparts. In order to properly care for these animals, it is important that we understand their behavioral biology, a point that is emphasized in regulatory and accreditatory documents, including the eighth edition of the *Guide for the Care and Use of Laboratory Animals* (National Resource Council 2011) and the

European Convention for the Protection of Vertebrate Animals Used for Experimental and Other Scientific Purposes, Council of Europe (ETS 123, Appendix A).

There are several reasons why understanding the behavioral biology of animals is important to their care. As mentioned above, animals evolved specific behavioral repertoires that help them survive in their natural environment. When animals are not able to perform these behaviors, it can cause stress and anxiety, as well as the development of undesired behavior, which can negatively impact their welfare (see Gottlieb and Pomerantz 2021; Novak and Meyer 2021). Knowing the behavioral biology of the animals is key to being able to provide them the features of their environment that can best promote these behaviors. For example, species that build nests, such as mice, are generally provided with appropriate nest making material. In addition, an individual's behavior provides information about its welfare state. As you will read in this book (e.g., Gottlieb and Pomerantz 2021), the absence of normal behavior, as well as the presence of abnormal behavior, can indicate pain and/or distress in animals. Thus, understanding normative behavior is key to assessing well-being. Lastly, while we are perhaps biased, having chosen to devote our careers to the study of animal behavior, we believe it is one of the most fascinating scientific fields. As you read through the chapters in this book, you will see that animals engage in some pretty amazing behaviors.

This book examines the behavioral biology of animals that participate in research. Each chapter in the main part of the book, Part 2, examines how a specific animal – or group of animals – behaves in its natural environment. What you will find, if you read all of those chapters, is that while specifics vary, there are a number of similarities as well. In order to survive, animals, regardless of their size or speed, have to be able to find and process food, while simultaneously avoiding being a food source for someone else. Most of them try to find mates and many care for offspring. These aspects of their behavior are often interconnected and are related to the environment in which they live.

Equally important to understanding environmental factors is understanding how animals perceive their environment. An animal's perceptual world, or umwelt, is closely tied to its senses (von Uexküll 2010; Burnett 2011). Humans have five senses, but we perceive our world largely through hearing and vision. We hear sounds in a frequency range from about 20 Hz to 20 kHz, and see three colors (blue, green, and red). However, this is not the same way other animals perceive their world. Many animals hear sounds in ultrasonic ranges that we cannot. Birds are tetrachromatic and see UV in addition to the colors we see. A dog's sense of smell can be over 10,000 times as sensitive as ours and is one of the main ways in which they perceive their environment. Bats and other animals use echolocation, a sense we do not possess. These senses are central to an animal's behavior; for example, they influence how animals migrate, how they find food, and how they communicate with one another. Therefore, understanding how animals perceive their world is key to understanding their behavior.

As you read through the chapters in this book, it will become clear why understanding behavioral biology is important to animal care. However, it can also be helpful to have a basic understanding of animal behavior as an academic discipline. Having some knowledge about the way in which behavioral scientists address behavioral questions can help guide your own observations. In addition, as with any scientific discipline, there are some terms that are common in animal behavior that have somewhat different meanings outside of the field. For example, people often use the terms "territory" and "home range" interchangeably, but the former implies active defense, while the latter does not. This distinction may not seem critically important, but it has implications for the animal. Vervet monkeys are territorial in nature; rhesus macaques, on the other hand, maintain home ranges but are not territorial. This seemingly subtle difference is important to those responsible for socially housing these species. A common method of pair housing rhesus macaques involves moving one potential partner next to the cage of the other (Truelove et al. 2017). Because vervets are territorial in their natural environment, this method is not recommended for them (see Jorgensen 2021).

Our goal for this chapter is to introduce you, the reader, to the field of animal behavior. In the first section, we describe steps involved in studying behavior, including biologically relevant questions used by ethologists to explain and understand behavior. We then discuss some behavioral concepts important to the care of animals in the laboratory, which are addressed by the authors in the taxa-specific part of the book. Clearly, a single chapter cannot cover the entire field of animal behavior. However, this chapter will, hopefully, provide an overview that will help put into context the specific information presented in the other chapters.

The Definition and Challenge of Animal Behavior

The field of animal behavior, or ethology, broadly described, is the study of everything that animals are observed or inferred to do, which includes, but is not limited to, movements, foraging activities, breeding and parental behavior, predator-prey interactions, communication, migration, learning, cognitive capabilities, and in testing situations, reactions to a variety of stimuli.

Below we describe three phases for the study of animal behavior.

Description Phase

The study of animal behavior starts first with description, a basic, but detailed, characterization of the behavior of a species. A number of steps are involved in this process. The first step is to acquire knowledge about the general behavior of the species in question. Many sources are available online to assist in this process, including general scientific journals (e.g., *Animal Behaviour, Institute for Laboratory Animal Welfare Journal, Journal of Neuroscience, Applied Animal Behaviour Science*), species-specific journals (e.g., *American Journal of Primatology, Mammalogy, Journal of Ornithology*), book chapters (such as those in Part 2 of this volume), monographs, and online video clips, as well as other reputable sources, such as The Audubon Society, National Geographic, The Cornell Lab of Ornithology, the National Centre for the Replacement,

Refinement & Reduction of Animals in Research (NC3Rs), and the National Research Council's *Guide for the Care and Use of Laboratory Animals*.

Because the above information is either in the form of written descriptions or short video clips that are selected only to highlight some activity, the second step involves the live observation of representative members of the species. Live observation is crucial in understanding the natural ebb and flow of behavior over time. Finally, this effort culminates in the development of an ethogram. Ethograms describe and define behaviors of interest. They can include species-typical behaviors, as well as abnormal, idiosyncratic behaviors that may not be displayed much in the animals' natural habitat. Additionally, because scientists often use apparatuses and other testing situations to assess behavior (e.g., Skinner box, open field, radial arm maze, and Novel Object Test), it is important to add behaviors or behavioral constructs derived from these situations to the ethogram. For example, in the open field test for rodents, one would include not only locomotion but also proximity to walls (to measure thigmotaxis, or preferences for walls).

Ethograms ensure that all observers are using a common language or terminology. While many behaviors are rather straightforward (e.g., drinking water from a water bottle), others may be open to interpretation. For example, you will read about "motor stereotypies" in the later chapters, which are often defined as repetitive behaviors, such as pacing and route tracing. Operationally defining these behaviors is not always straightforward. If you were to witness a dog walk across his pen one time, you likely would not consider that behavior to be pacing. On the other hand, if that same dog were to walk back and forth, in the same pattern, 50 times, you would probably (and rightly) call that pacing. But, what if the dog walks back and forth twice? Would you consider that to be pacing behavior? Effective ethograms clearly define each behavior, reducing this kind of ambiguity. For example, Jorgensen (2021) defines pacing as when an animal (a vervet monkey in this case) walks back and forth at least three times. Many of the chapters in Part 2 of this book contain ethograms for particular species. Additional ethograms are presented in Part 3.

Quantification Phase

Following the development of an ethogram, careful consideration must be given to how, where, and when the behavioral categories will be assessed and measured. Actual observations are at the heart of most behavioral research. Depending on the circumstance, these observations may be collected directly by the observer standing in front of the animal, or more remotely, by an observer scoring videotapes or live web cams. Additionally, some behaviors can be quantified through automated systems. For example, jumping and somersaulting in deer mice can be quantified using a photo beam array, in which the beams are set high enough so that all four paws must leave the ground to break the beam (Bechard and Lewis 2016). Standard measuring units include latencies (i.e., amount of time between a starting point and the occurrence of a particular behavior), frequencies (i.e., the number of times a behavior occurs), and durations (i.e., the length of time).

Observations can occur in many different situations. Normative or baseline observations often take place either in an animal's home cage environment or in unmanipulated free-ranging populations. Laboratory animals are often studied in response to some challenge, either using a testing arena (such as an elevated T-maze) or in their home cage (e.g., Figure 2.1). Free-ranging animals can also be tested in areas in which manipulanda are set out in the field and various capabilities assessed (Figure 2.2; see van de Waal et al. 2013 for more details).

Because animals may be studied in social groups, issues of how to observe multiple subjects arise. One can use focal sampling, in which individuals are observed one at a time for a certain amount of time (e.g., 15 min a day for several days), and behaviors of interest are recorded. Conversely, one can use scan sampling, in which a group of animals is observed, and the behavior of each individual is recorded at set intervals (e.g.,

FIGURE 2.1 Example of a rhesus macaque inspecting (a) and avoiding (b) a brightly colored bird toy placed on the home cage as part of a novel object test. (Republished by permission of Taylor and Francis Group, LLC, a division of Informa plc, from the *The Handbook of Primate Behavioral Management*, S.J. Schapiro (Ed), 2017.)

FIGURE 2.2 Example of testing in free-ranging animals. This study (van de Waal et al. 2013) examined social learning of food preference in wild vervet monkeys. Monkeys that immigrate into new troops adopt the local traditions and eat food even if that food was avoided in their previous group. (Photo copyright Erica van de Waal.)

one would record the behavior of all animals in a group every 60s for a 30min time period). Subject sampling procedures are determined by the research question (see Altmann 1974). However, to be effective, observation periods must be long enough to measure the behavior of interest, and frequent enough across days and perhaps weeks to ensure that unforeseen or uncommon events (e.g., illness, change in care taker, etc.) do not skew the data. For example, if someone were to take observations of you at this moment, they would observe you reading this book (and hopefully learning a great deal from it!). If that were the only observation period, the observer might come away thinking that you spend all of your time reading this book. While we think that would be a great use of your time, it is probably not the only thing you will do today or this week. Thus, taking shorter, more frequent observations over several days, as opposed to longer, less frequent observations, generally provides a more complete activity budget of the individual. Detailed descriptions methodologies and observational sampling techniques are beyond the scope of this chapter; however, there are very good resources available on these topics, including Jeanne Altmann's classic paper, Observational study of behavior: Sampling methods (Altmann 1974) and *Measuring Behavior: An Introductory Guide* (Martin and Bateson 2007).

The Explanation Phase

The field of animal behavior is not merely about description and quantification. Another important phase is determining why animals behave the way they do. However, developing accurate explanations of behavior is extraordinarily complicated. Many factors contribute to the production of behavior. In fact, the expression of behavior can be considered the ultimate end product of a complex series of interactions between the organism and the environment.

The sources of these interactions can be divided into two general types: internal variables (e.g., neurotransmitters, hormones, gene expression) and external variables (e.g., experience, food availability, predation risk, environmental complexity). Complicating the picture further is the fact that organisms develop and change over time, which in turn can yield changes in how they respond to these variables. Thus, any attempt to explain behavior must take into account the developmental stage of the organism.

Here we present two examples to illustrate the points raised above. Meadow voles and prairie voles are closely related rodent species within the genus, *Microtus*. However, whereas meadow voles are promiscuous breeders and females almost exclusively care for pups, many prairie voles show social monogamy, forming pair bonds with their partners and engaging in bi-parental care of pups (see reviews, Gobrogge and Wang 2016; Rogers et al. 2018). One question that has been considered extensively is why male prairie voles care for pups. Although this process has been described numerous times and various patterns of behavior measured (e.g., time spent crouching over pups, number of pup retrievals), the explanation of why this occurs is considerably more challenging. Many variables contribute to paternal behavior, including internal factors, such as vasopressin and oxytocin, neuropeptides implicated in social bonding (Bamshad et al. 1994; Bosch and Young 2018; Gobrogge and Wang 2015), and changes in gene expression (Barrett et al. 2013). Environmental pressures include mate guarding as a means to ensure paternity (Getz et al. 2003; Wolff et al. 2002) and changes in population density (Getz and McGuire 2008). From a developmental perspective, the quality of paternal care is dependent, in part, on previous experience with pups (Stone et al. 2010), allowing adolescent males to perfect basic parenting skills.

In the second example, African Thomson's gazelles engage in a rather startling behavior when pursued by a cheetah. Instead of running as fast as possible in the opposite direction, they perform an unusual behavior called stotting. The gazelles run a short distance in the opposite direction and then paradoxically leap high in the air, run again and leap high in the air, and continue to do this as long as the cheetah maintains its pursuit. It is not difficult to describe this phenomenon. In fact, one can film it and obtain various descriptive measures of average leap height and number of leaps per unit time. The challenge is to explain why gazelles stott. Very young gazelles do not stott, thus there is a developmental component. Additionally, both external variables (e.g., type of predator, whether the gazelle is alone or in the middle of a herd) and internal variables (e.g., activation of neurotransmitters, increased secretion of adrenaline) play a role in explaining this behavior. However, the interactions between these variables are not fully understood. Furthermore, many other variables remain to be identified.

Biologically Relevant Questions about Behavior

As demonstrated by the examples above, one of the challenges facing behavioral scientists is that a variety of factors internal and external to the organism contribute to the expression of behavior. Many years ago, Niko Tinbergen, an internationally recognized ethologist, created a structure for explaining behavior by identifying four biologically relevant questions (Tinbergen 1963). Two of these were termed proximate questions, proximate because they depended on understanding behavior in individual animals at the present moment in time. The remaining two were termed ultimate questions, because they relied on the process of natural selection, which frequently required many generations. While most often employed by behavioral scientists, these questions can help guide the ways in which we look at the behavior of the animals in our care.

Proximate Questions (Mechanism and Development)

The first proximate question is concerned with the underlying mechanisms that explain how behavior is expressed. These include, but are not limited to, the role of neurotransmitters, hormones, genes, epigenetic processes, neural circuitry, etc. The second proximate question involves an exploration of the role of development, and focuses both on the need for critical internal and external input as young organisms mature, as well as on the timing of such input.

Ultimate Questions (Function and Evolutionary History)

The first ultimate question explores the functional significance of behavior from the standpoint of the ways in which the behavior affects reproductive success. This question also involves a consideration of the costs and benefits of such behavior, to explain why it persists in a population. The second question is concerned with how the behavior may have developed within various taxonomic groups, and involves a comparison of existing species using both morphological and molecular techniques. The output of these comparisons is a putative history of how various species groups are related to one another and how traits may have arisen over evolutionary time.

Each question formulated by Tinbergen can be viewed as representing a different, but equally important, level of analysis. Traditionally, scientists focus on only one level of analysis; e.g., behavioral ecologists look at ultimate causes for behavior, while neuroscientists typically focus on the role of neurotransmitters, etc. However, there may be significant value to scientific creativity through awareness of findings at all interpretational levels. Similarly, interpreting behavior at multiple levels can provide a framework for understanding why animals in our care behave the way they do. Much of the material about behavior presented in the chapters of this book is provided in a functional and/or evolutionary context. Authors describe behaviors that have been shaped by natural selection and provide information about how to promote those behaviors in the laboratory environment. Unless you are specifically studying the underlying neural mechanisms of behavior, you may not think too much about the proximate mechanisms involved. However, to really understand why animals are engaging in behavior, it is important to consider both proximate and ultimate causes. As stated above, experience, particularly early in life, is one proximate mechanism that underlies behavior of animals. Early interactions with the mother can influence later behavior of young animals of a variety of species. For example, mouse pups raised by mothers that licked and groomed them a great deal go on to show a decreased startle response and increased exploration in an open field compared to pups that received less maternal licking and grooming (Caldji et al. 1998). In addition, certain drugs or scientific treatments can affect hormones and/or neurotransmitters, which can alter behavior. Genome editing technologies (such as CRISPR) alter the genetic code and may have unexpected downstream behavioral effects. As an example, Qiu et al. (2019) used CRISPR/Cas9 editing to knock out a gene involved in circadian rhythm, BMAL1, in cynomolgus macaque embryos. As expected, these monkeys, as juveniles, had altered sleep patterns, but they also showed decreased exploration and increased fear and anxiety toward caregivers, compared to control animals. Any time new technologies that can alter neural function are introduced, there is the chance that they can also affect the behavior of the animals. Below, we look at a single example and examine it from both a proximate and ultimate perspective.

Bird Song as an Example of a Levels of Analysis Perspective

Some birds produce melodious sounds referred to as song. Song learning is present only in three distinct avian groups, constituting somewhat less than 50% of all avian species. These are the oscine passerine birds, hummingbirds, and parrots. Song is distinct from the species-typical calls found in all avian species (e.g., begging calls, alarm calls, food location calls), in terms of its structure, complexity, and its sex specificity (only male birds sing). Although females do not sing, they nonetheless have to recognize the songs of males of their species. Bird song has been studied extensively and considerable information is available relating to the four biologically relevant questions (levels of analysis).

1. *Mechanism*: Although many aspects of bird song have been explored in this domain, we highlight the research on the neuroanatomical structures involved in song (Wild 2004). Birds that sing must first learn the species song before reliably producing it. Extensive evidence suggests that these two processes (learning and producing) are controlled by different neuroanatomical structures. Pathways that connect areas of the brain, specifically the lateral magnocellular nucleus of the anterior nidopallium with area X, appear to be activated during the learning of song. In contrast, the ongoing production of song involves a complex circuit that includes the robust nucleus of the arcopallidum, the high vocal center, and the hypoglossal nucleus that connects to muscles of the syrinx or voice box. The distinction between learning song and producing it is explored further under the next section "Development".

2. *Development*: Male white-crowned sparrows do not begin to sing until about 200 days of age, and even then, the song is quite different from the species-typical song. Extensive research on the white-crowned sparrow has shown that song learning is dependent on a combination of innate cues and crucial experiences across development (Brainard and Doupe 2002). Without previous song exposure, young white-crowned sparrows are genetically

attuned to respond to conspecific song with increased heart rate and begging responses (Nelson and Marler 1993). Subsequently, male white-crowned sparrows acquire the full species song in stages. During the sensory stage, nestlings hear their father sing the species song and encode it into their memory. For maximal memory encoding, exposure should occur before 100 days of life (Nelson et al. 1997). They then fledge and begin to sing at around 200 days of age. Initially, the songs produced are imperfect versions of the species song. During the sensory-motor stage, male birds appear to compare their vocal output with their encoded memory, improving their song through successive approximations. Males enter the final stage of crystallization when a match between the two sources of information is achieved (Nelson and Marler 1994). At that point, the song remains relatively invariant throughout the remainder of the bird's life, and thus, the white-crowned sparrow is a closed-end learner. It should be noted that the process of song learning as described in the white-crowned sparrow is also observed in zebra finches. However, it is by no means universal. Many other bird species are open-ended learners. For example, canaries retain some old songs from the previous year, while also learning new songs in the new year.

3. *Function*: Singing has some very clear costs. Consider that adult males perch on high branches and loudly sing their song. What better way to attract predators. Additionally, singing is energetically costly and takes time away from foraging for food. Given those clear costs, the benefits must be powerful enough to exceed those costs. There are obviously many proposed explanations for bird song that include, but are not limited to, attracting mates and defending territories from intruder males.

   Here we examine one function, namely, that songs aid in territorial defense and act to repel intruder males. Testing the territorial defense hypothesis is not as easy as it seems. Males not only sing, they also engage in aggressive behavior against intruders. How does one separate the behavior from the song? In an ingenious experiment, this hypothesis was tested by replacing song sparrow males with speakers playing their song and examining the rate of territorial intrusion (Nowicki et al. 1998). The authors first audiotaped the songs of 22 males. Then 11 matched pairs of males were removed temporarily from their territories. One male of each pair was replaced with a loudspeaker producing its song whereas the other male's territory remained silent. The mean number of intrusions per hour was significantly lower on territories in which the male's song was played, compared to the silent territories, thus providing support for the idea that song alone can temporarily deter intrusion by strange males.

4. *Evolutionary History*: Three groups of birds learn songs; the oscine passerine birds, hummingbirds, and parrots. However, the actual ancestral state reconstruction of this pattern across avian species is not well established. Initial research involving morphological and molecular comparisons suggested that these three groups were unrelated and song must have evolved independently three separate times (Sibley et al. 1988). A subsequent comparison suggested that parrots may be closely related to oscine passerine birds, and thus, song learning may only have evolved independently twice (Hackett et al. 2008). New information now suggests that some suboscine passerine bird species may also learn their song (Kroodsma et al. 2013), all of which indicates that the phylogeny of bird song remains uncertain (Jarvis et al. 2014).

Using the Scientific Method to Study Behavior

We can generate endless questions about animal behavior and develop numerous explanations. However, use of the scientific method is essential in determining whether a particular explanation is a viable answer to a question. For example, Tim Caro (1986) proposed and tested an explanation for stotting in gazelles. He started with a simple **question**; why do gazelles stott? He then **hypothesized** that gazelles leap high in the air to get a better view of the actual location of the cheetah. He termed this the terrain hypothesis. If the terrain hypothesis were true, he **predicted** that the environment would matter, namely, that in tall grass situations, where cheetah movements might be difficult to detect, gazelles should stott. In contrast, in short grass situations, where cheetahs could be easily seen, stotting would be unnecessary. He then **tested** this prediction by observing cheetahs and gazelles in both tall and short grass situations, and discovered that gazelles were equally likely to stott in both situations. Thus, he ruled out the terrain hypothesis. Frequently, multiple explanations (hypotheses) are possible. In some cases, two competing hypotheses that yield different predictions can be tested simultaneously. Such hypotheses are considered mutually exclusive.

Caro went on to test four **mutually exclusive hypotheses** of stotting in gazelles. Two of these were based on the idea that stotting was a form of communication to other gazelles, either to warn them of the presence of a predator or to cause them to group together to dilute the risk of predation. The remaining two were based on potential interactions with the predator, either to confuse the predator when many gazelles were stotting at the same time or to alert the predator that it had been detected and further pursuit was futile (cheetahs rely on stealth to sneak up on prey). Without going into the details of the predictions and testing, Caro found support for only one hypothesis related to communication between predator and prey, namely, detection. However, several years later, Fitzgibbon and Fanshawe (1988) proposed that gazelles stotted to indicate their fitness to the predator. Unfortunately, detection and fitness are *not* mutually exclusive hypotheses, because both lead to the same prediction. Gazelles should stott to alert the cheetah that it has been *detected*, enabling them to outrun the predator, or they should stott as a *demonstration of superior* *strength and stamina* to the cheetah, indicating that they can outrun the predator. However, a further examination of the stotting behavior combined with the success or failure of the hunt provided support for the fitness hypothesis. Gazelles that were taken by a predator stotted less frequently per unit time and did not attain the higher leaps of the gazelles that escaped.

The story does not end here. Recent evidence indicates that females are more likely to stott than males, and stotting may be used by mothers to warn fawns of dangerous situations (Blank 2018). These varied findings suggest that stotting serves more than one purpose and that its use is context dependent, a not uncommon finding in much of animal behavior research.

This example illustrates several points. First, it demonstrates the rigor with which behavioral science is performed. It also shows that, as with other fields of science, explanations may change over time, as new information is uncovered. Much of the information about the behavior of the species discussed in this book was learned through rigorous behavioral studies.

Concepts from Animal Behavior That Impact Laboratory Animal Care and Welfare

There are a number of concepts derived from studying animals in nature that may have some application to laboratory animal care and welfare. Both regulatory documents overseeing animal care (e.g., United States Animal Welfare Act) and the *Guide for the Care and Use of Laboratory Animals* (National Research Council 2011) emphasize the importance of promoting species-typical behavior for research animals. In the sections below, we highlight various evolutionary selection pressures and features of the environment that act to produce species-typical behavior. Additionally, we provide some suggestions for using this information to improve the captive environment.

Ecology

Spatial Distributions

Animals are distributed across space and time in various social or nonsocial configurations. Two concepts are used to characterize spatial distribution: **home range** and **territory**. Home range refers to an area that an animal or group traverses in their daily search for food, water, mates, resting sites, and sleeping sites (first defined by Burt 1943). Territories are areas within the home range that are actively defended from intrusion by conspecifics. There are benefits to territoriality for animals, including exclusive access to food and mates, as well as safety from predation. However, these benefits come at a cost (associated with defense), and territoriality may only "make sense", if the benefits outweigh the costs. All animals have home ranges, but not all animals defend territories. Further complicating the picture is that spatial distributions can vary across time (breeding vs. nonbreeding season) and within a species by sex.

A comparison of vole species is useful in characterizing this variability. Although population-level differences exist, in general, prairie vole males are socially monogamous, occupy and defend a territory (an area near the nesting burrow), and help care for the pups. Males show intense aggressive behavior toward male conspecifics, particularly after mating (Winslow et al. 1993). Prairie vole home ranges are much larger than those of meadow voles, and encompass areas of dry grasslands and hay fields. During the nonbreeding season, prairie voles are often found in communal groups consisting of the breeding pair and offspring, as well as other males (Getz et al. 1993).

In contrast, promiscuous meadow voles have a different spatial pattern (Pritchett-Corning and Winnicker 2021). Females tend to exist alone in territories independent of other females, whereas males have larger home ranges that overlap the territories of several females and the home ranges of other males (Madison 1980). Males and females interact at the time of mating, and typically only females care for the pups (McGuire and Novak 1984). During the nonbreeding season, meadow voles overwinter in larger communal groups, consisting of females and their offspring and males (Madison and McShea 1987).

In this example, studies of home range and territoriality can be used to inform the establishment of laboratory breeding colonies of prairie voles and meadow voles from wild stock. Whereas prairie voles can ideally be maintained as breeding pairs, maintaining meadow voles as breeding pairs can be more difficult, and other options may be necessary. Additionally, the above findings suggest that sociality may be driven in part by day length, potentially necessitating light cycle management in captivity.

Predation and the Landscape of Fear

Another factor that can influence habitat use is the risk of predation. Predation risk in free-ranging animals is known to vary across different environments, presumably eliciting different perceptions of risk. This spatial variation in the perception of risk by prey species is referred to as the **landscape of fear** (Laundré 2010). As risk increases, prey species respond with increased vigilance behavior and with changes in habitat use (Blanchard et al. 2018). For example, a shift in habitat preference from aspen stands to conifer forests was observed in elk after the reintroduction of wolves to Yellowstone Park (Fortin et al. 2005).

Clearly, changing habitats is not the only way that animals respond to predation threat. There are many ways in which animals may respond to a potential predator. One common response is to remain motionless, or freeze, to avoid detection; some animals, such as various frog species (e.g., Ramalho et al. 2019), even feign death by adopting a rigid posture in addition to the immobility (see Figure 2.3). Many animals flee and/or hide when detected by a predator, while others fight back, with chemical (e.g., skunk spray) or mechanical (e.g., teeth) defenses. Individuals do not always have to avoid predators on their own; individuals within a group may work together to "mob" a potential predator. The chapters in Part 2 of this book describe anti-predator behaviors for the various species.

Do such animals perceive the laboratory environment as a landscape of fear? Obviously, laboratory animals are not exposed to predators or predator scent, unless such exposure is part of a research protocol (or, if carestaff inadvertently introduce predator scent by, for example, working with a predator

FIGURE 2.3 South African mongoose (left) feigns death (right) when approached by a human. (Photos by M. Novak.)

species, such as a cat, before working with a prey species, such as a mouse). However, it is reasonable to consider that species-typical reactions to predators honed over evolutionary time by natural selection can impact behavior in the laboratory. For example, research tamarins (small New World primates, closely related to the marmosets discussed in Chapter 24) that lived in a laboratory environment their entire lives, were given a choice of sleeping boxes with differing levels of concealment. The tamarins preferred the boxes with the highest level of cover, even though they had never seen a predator. Further, when forced to sleep in boxes with relatively little concealment, they increased their rates of vigilant scanning (Caine et al. 1992). They maintained this antipredator behavior, despite the lack of predators in their environment.

There are, of course, events that occur across the day that can increase the perception of risk for laboratory animals. Human observers and/or caretakers may be viewed as potential threats (Caine 2017, Waiblinger et al. 2006), particularly if they are involved in a negative interaction (e.g., handling for injections). There are other events that can be perceived as high risk, including being transferred to new rooms, new locations, or new partners. It is likely that some of the natural fear responses (increased vigilance and movements indicative of flight) can be mitigated by desensitization and positive reinforcement training techniques (e.g., blood glucose testing in chimpanzees: Reamer et al. 2014; reducing abnormal behavior: Baker et al. 2009; Coleman and Maier 2010) and by providing safe havens when possible (e.g., transferring marmosets to new cages by carrying them in their nest boxes).

Social Interactions and Organization

Social Organization

Social organization refers to the structural features of social relationships that include, but are not limited to, sexual composition, age composition, cohesiveness, division of labor, and patterns of communication. Social organization varies widely, ranging from solitary females, whose primary social interaction is with offspring, to large multimale, multifemale social groups consisting of every age class (infants, juveniles, adolescents, and adults). An example of such widespread variability can be observed in the primate order. Five distinct organizations are recognized: (1) **Solitary living**: Only a very few primate species are solitary (e.g., lorises, galagos, orangutans), with the social unit being the mother-infant dyad. In this form of social organization, solitary males form overlapping home ranges around solitary females. (2) **Monogamy**: Relatively few primate species live as breeding pairs surrounded by immature offspring that disperse prior to sexual maturity (e.g., gibbons, siamangs, owl monkeys, titi monkeys). These species tend to be territorial and monomorphic. (3) **Polyandry**: In polyandrous groups, one resident female lives with several males. This social structure is also quite rare, but has been observed in some tamarins. (4) **Polygyny**: In polygynous groups, one resident male lives with several females (e.g., gelada and hamadryas baboons, langurs, patas monkeys) in a "harem" group. Coexisting with these harem groups are all-male groups. Males from these all-male groups frequently attempt to oust harem males, and the tenure of individual harem males is often short. (5) **Polygynandry**: Polygynandrous groups consist of multiple adult males and multiple adult females plus offspring of varying ages. This is a common form of primate social structure typical of many macaques, most baboon species, vervet monkeys, capuchins, ringtailed lemurs, and some colobines.

The problem with creating the above classification is that within-species variation is not adequately represented. For example, marmosets are considered monogamous primates that live in extended family groups, and most studies of free-ranging marmosets are consistent with this view (Arruda et al. 2005; Yamamoto et al. 2014). However, other marmoset social groupings have been reported, including multimale and multifemale groups of individuals that are not necessarily closely related (Faulkes et al. 2003), as well as groups containing two breeding females (Digby and Ferrari 1994). Marmosets are not the only primate species to show such variability. There are other factors, such as resource availability and predation pressure, that are not taken into account in the above scheme, yet do influence social groupings.

For many species, it may not be possible to recreate complex social organizations in the laboratory environment (e.g., multimale, multifemale groups of rhesus macaques). However, a detailed understanding of the natural social organization of a species may help guide the way that animals are grouped in a laboratory environment. It may also explain other aspects of their basic social structure. For example, laboratory rhesus macaques are often housed in pairs, with ten or more pairs per room. Although care staff often focus their attention on the individual pairs, the larger social unit is the group (all the animals in a room and their interactions with one another). In this broader context, moving animals to different places in the same room or removing individuals to other locations, may disrupt the social organization at the level of the room.

Dominance Hierarchies

Dominance hierarchies are a salient feature in the social organization of many species, including fish, chickens, and nonhuman primates. In their simplest form, dominance hierarchies establish a rank order, often linear, for the members of a group. Dominance confers access to resources such as food, sleeping locations, mating opportunities, etc. Thus, in a linear hierarchy, the dominant individual gets access to all of the resources s/he wants, the second dominant animal gets access to whatever s/he wants (unless it is desired by the dominant individual) and so on. Rank within the dominance hierarchy can be relatively stable (e.g., females in many macaque species) or can fluctuate (e.g., males of many macaque species), and is often maintained with sensory signals (e.g., visual cues, threatening facial expressions; auditory cues, roars; ritualized displays, head butting in bighorn sheep), rather than actual fighting, although the latter may also occur. Some dominance styles (e.g., despotic) are marked by frequent conflict, unequal resource distribution, and infrequent postconflict resolution, but that is not the case with all social hierarchies. In egalitarian dominance systems, subordinates often help the dominant attain status and dominance can be maintained with relatively little overt aggression (Vehrencamp 1983). Even within a genus, there can be vast differences in dominance styles; rhesus macaques have a despotic dominance style, while stumptailed macaques are known to be egalitarian (Honess 2021; Matsumura 1999).

It is often assumed that being a subordinate member of a group is more stressful than being dominant, and that is often the case. However, the social context of the dominance rank is as important as the rank itself (Sapolsky 2005). Dominant animals that have to fight to maintain their rank (i.e., they are not certain of their rank) often show more signs of stress than subordinate conspecifics. On the other hand, dominants that are secure in their rank and that use facial expressions, as opposed to overt aggression, to intimidate lower ranking individuals, often show fewer signs of stress (Sapolsky 2005). Because stress can impact many physiological systems, having to maintain a "stressful" rank is associated with various negative health outcomes, including immunosuppression, compromised cardiovascular function, and decreased reproductive output.

It is easy to see why understanding the dominance system of research animals is important. Disrupting (e.g., disbanding and reforming) groups, which often happens in research environments, can increase fighting for a variety of species, as individuals try to establish dominance relationships in the "new" group. Even removing only one animal can destabilize the group in certain circumstances. For example, in rhesus macaques, removing a high ranking individual can lead to group instability.

Sexual Selection and Mating

a. *Mating Strategies*: Male and female mating strategies are very different. Males typically attempt to mate with as many females as possible, whereas females typically attempt to choose the single fittest male. These differences are, in part, the result of gamete size and rarity. Males invest in and produce many small gametes, whereas females invest in and produce a limited number of very large gametes. For example, male rhesus monkeys generate large quantities of sperm, enabling them to breed with many females at any time during the breeding season. Rhesus females, on the other hand, produce one egg per month when cycling, resulting in one infant per year. A second consequence of these differences, barring any health-related issues, is that *most females* will usually produce offspring, whereas *only some males* will sire offspring. This is the basis of Bateman's Principle, wherein variability in reproductive success is significantly greater for males than females. There is a notable exception. In species with nonbreeding alloparents (e.g., naked mole rats), variability in reproductive success may be greater for females (Hauber and Lacey 2005). The mating strategies described above result in sexual selection, a special case of natural selection, in which selection typically acts on males to increase their competitive ability to access females, and on females to discriminate among males to select the fittest male (e.g., peahens "think" that if a peacock can survive to breeding age with that beautiful, but very costly tail, then he must have very strong genes).

b. *Male Competition*: Although agonistic behavior and dominance interactions are thought of as the primary way in which males compete for access to females, other strategies include, but are not limited to, mate guarding (e.g., baboons, Setchell 2016), sealing the vagina with a sperm plug after mating (e.g., house mice, Sutter and Lindholm 2016), impersonating a female to cause males to waste their gametes (e.g., bluegill sunfish, Fu et al. 2001), and killing young infants to bring females into breeding condition (e.g., during male takeovers in langurs, Borries et al. 1999). Perhaps the best example of competition and how sexual selection can lead to genetically diverse male phenotypes is seen in the ruff, a European sandpiper. The ruff is a lekking species, wherein males vigorously defend small, mating-only territories contained within a large ancestral breeding ground (lek) for the opportunity to breed with females, who wander around the lek and eventually select a mate. Three genetically different types of males, each representing very different mating strategies, have been identified (Ekblom et al. 2012; Küpper et al. 2016; Lank et al. 2013). *Independent* males defend a small breeding territory by fighting with other independent males. Location (being in the centermost area of the lek surrounded by other males) is important, in as much as females spend most of their time in the center of the lek. *Satellite* males passively exist on the territories of independent males (one per territory) and neither fight with males nor defend the territory. However, independent males benefit because territories containing satellite males receive more visits by females. Satellite males also benefit by occasionally breeding with females when independent males fight. The third genetic variant is a *female mimic* that resembles females and can blend in with them. When a real female selects a male, the mimic can sometimes rush in before the independent male mates with the female. The female mimic then mounts the real female.

c. *Female Choice*: The primary evolutionary goal of females is to assess male fitness. Several models have been used in the past to characterize the process of female choice. In the direct benefits model, males provide necessary resources to females, such as nest sites or food during courtship. Females then compare the quality of the relevant features before breeding with a particular male. However, not all males provide resources. In indirect benefits models, males provide only sperm (see Jones and Ratterman 2009 for discussion). Choice is then based on females being able to discriminate males that possess genes that will increase the fitness of their offspring. Two challenges to this model are: (1) identifying the relevant cues that females might use, keeping in mind that there may be a multiplicity of such cues, and (2) demonstrating that the presence of such cues increases reproductive success in females. In birds, relevant male cues that appear to increase reproductive success include intensity of plumage coloration (for cardinals, Wolfenbarger 1999; but not for bluebirds, Liu et al. 2009), bower quality (in satin bower birds, Doucet and Montgomerie 2003), and song rates (in house finches, Mennill et al. 2006).

Female mate choice is less well studied in mammals (Clutton-Brock and McAuliffe 2009), and a primary focus is on how females select males with good genes. In ungulates, the relevant male factors include dominance status, prior sexual experience, and female defense. With respect to the latter, male pronghorn antelopes attempt to defend groups of females (harems) from other males during the breeding season. Nonetheless, females move independently and frequently switch harems. In one study, the males that were most attractive to females were those that could defend large groups of females for a sustained period. Furthermore, females that mated with attractive males produced offspring that were more likely to survive to weaning and beyond (Byers and Waits 2006). However, there are many factors influencing female choice that may be confounded with male quality. These include, but are not limited to, choosing a particular location in which a male is found, rather than selecting the male himself; following the choices of other females; or selecting males that benefit the female directly and not necessarily her offspring.

d. *Breeding Periods*: Although many species have a well-defined breeding season (particularly rodents and birds living in the temperate zone), others can breed opportunistically or continuously. Seasonal breeding is in part controlled by changes in day length, which in turn are also associated with various physiological changes (Lehman et al. 1997). Opportunistic breeders are those that can breed during broad periods of the year, but the actual timing within that period is related to changes in some environmental factor. Smooth-billed anis show opportunistic breeding readiness which is triffered by rainy periods in drought areas (Morais et al. 2019). Continuous breeders are those that can breed year-round, including many domesticated species (e.g., cats, dogs, cows). The picture is somewhat different for highly inbred strains of mice, like those housed at the Jackson Laboratory (Schile 2019; web site https://www.jax.org/news-and-insights/jax-blog/2013/april/how-mice-are-affected-by-the-changing-of-seasons). For most strains, maintenance under a standard light cycle (14:10) results in breeding year-round. However, some strains (e.g., A/J, SPRET/EiJ, C57BL/6J) appear to respond to unseen seasonal changes by altering their breeding activities during the winter even when maintained on the standard light cycle. These alterations include reduced litter size (A/J), reduced litter production (SPRET/EiJ), and increased gestation length (C57BL/6J).

e. *Mating Systems:* Mating system refers to the various sexual partnerships that arise during the breeding. These partnerships can be brief or last throughout the breeding period, and are generally classified as monogamous, polygynous, polyandrous, polygynandrous, or promiscuous (described above). The type of mating system is derived from an interaction of male and female mating strategies with features of the environment. Those features include both the distribution of females and the resources to which they are attracted. For example, males typically attempt to breed with as many females as possible. However, if females are widely dispersed in the environment and the cost of tracking them down is high, then monogamy (mating with a single female) may be the result. This is termed facultative monogamy, and contrasts with obligate monogamy, which may arise when there is a selective advantage for bi-parental care (and may not be related to mate density; Kvarnemo 2018).

Complicating this classification is the finding that some species can exhibit several mating systems, depending on environmental factors. Perhaps, the best example of the influence of the environment on the development of mating systems can be observed in the dunnock, a small passerine bird. Dunnocks can exhibit monogamy, polygyny, polyandry, and polygynandry within the same population. One factor that affects mating system expression in this species is the size of female territories in association with resource availability. Under resource-poor conditions, female territories are large, and the most common mating systems are polyandry and polygynandry. Under resource-rich conditions, female territory sizes are small, and monogamy is the most typical outcome (Santos and Nakagawa 2013).

One of the consequences of factors, such as sexual selection, breeding season, and mating systems, for laboratory animals is the need to consider how, when, and with whom animals should be bred. For highly inbred strains, such as rats and mice, considerable information is available at various sites to guide this process (www.jax.org, www.criver.com). For more exotic laboratory animal species (e.g., macaques, marmosets, hamsters), a careful consideration of various evolutionary selection pressures may be informative.

f. *Parental Behavior*: Just as there is a wide range of mating systems across various taxa, there are also a number of ways in which animals may care for their young. Some animals, such as many reptile species, show little to no parental behavior (see DeNardo 2021). In most bird and mammalian species, at least one parent provides care. Parental care can be performed primarily by the mother, as is the case with the majority of mammalian species, or it can be biparental, as occurs with many bird species. Other group members (e.g., siblings or others) often help in caring for the young, a behavior known as alloparenting. Interestingly, paternal-only care is relatively common in fish and amphibian species, although such care typically consists of protecting eggs, and not necessarily feeding offspring (e.g., Goldberg et al. 2020). Similar to mating strategies, parental behavior results from selection pressure, and the two are often closely tied to one another. For example, it is believed that biparental care coevolved with monogamy (Bales 2017).

Foraging and Feeding Behavior

Feeding behavior, including searching for, processing (e.g., getting the nut out of the shell), and consuming food items, comprises a significant portion of the behavioral repertoire for many species. Therefore, it is an important behavior to consider when caring for that animal. Indeed, one of the first things that people often learn about an animal's behavior is how it gets food. Animals are often categorized by their diet; carnivores survive on diets consisting primarily of live animals (i.e., prey); herbivores eat mostly plants; granivores have diets consisting of nuts, berries, and seeds; and scavengers consume diets of dead animals. There is, of course, more to feeding behavior than simply the kind of food an animal eats.

One of the obvious differences between foraging in nature and providing food in the laboratory is a substantial reduction in the costs, i.e., energy spent searching for and processing food. In many cases, laboratory animals are provisioned; thus, virtually no energy is expended in finding food, and very little energy is spent in preparing it for consumption. The net result may be relatively rapid consumption of food, and if not monitored carefully, weight gain in the consumer, along with a significant increase in "empty time". As such, providing food appropriately to laboratory animals is a focus of behavioral management for animals in captivity. Food, either the animal's main diet or additional items, such as produce, is often provided in ways that encourage species-appropriate foraging. For example, hamsters and other rodents can be provided seeds and items that encourage hoarding (see Winnicker and Prichett-Corning 2021). There are also a wide variety of available puzzle feeders and other manipulanda designed to promote foraging behavior for species, including ferrets, pigs, dogs, and nonhuman primates, among others. Because many animals prefer tasty treats to their nutritionally balanced chow, it is important that foraging strategies be implemented judiciously to avoid obesity or other dietary complications (see Schapiro 2021 for information about foraging enrichment).

Just as foraging strategies in nature are subject to constraints, any increase in foraging behavior (using a variety of devices) in the laboratory environment may pose particular constraints that need to be addressed. Two such constraints are competition for access to food resources and age-related declines in motor activity. In the former situation, using multiple devices may reduce the competition between members living in pairs or in groups. In the latter case, foraging strategies should be modified to accommodate age-related changes in motor ability to ensure that devices are used.

Communication

Communication is the act of transferring information from one individual (signaler or sender) to another (receiver), which then changes the behavior of the receiver. In order to survive, animals need to be able to exchange information about environmental factors, such as potential food sources or predators, as well as internal factors, including reproductive status and emotional states.

Communication is accomplished through the use of various signals. Animals that are active during the day often communicate with visual signals, including facial expressions or body postures. Visual signals also include stereotyped behavioral sequences, such as the intricate dance bees use to convey the location of good pollen sources, or various courtship rituals. Acoustic signals are another common way animals exchange information, particularly if they need to communicate at a distance. For example, birds chirp or sing to indicate territoriality or as part of courtship (e.g., Friedrich and Mello 2021), guinea pigs purr during courtship (Kleven 2021), and young animals vocalize to get the attention of their parents. Vocalizations are not always audible to humans; many rodent species use ultrasonic vocalizations. Chemical signals, including pheromones in scent marks, are used to communicate in many species. Scent marking, a behavior in which animals deposit these chemical cues (through secretions produced by specialized glands, or through urine or feces), is often used to signal territories or for mating, and can also be used for identification of individuals. In addition, chemical cues can be used to suppress or induce ovulation in mice and other animals (MacLellan et al. 2021). Finally, many animals transmit information via tactile communication, such as grooming. Each chapter in Part 2 discusses modes by which the species communicates.

The sensory capabilities of animals help them adapt in their natural environments. Thus, it is important to make sure that laboratory environments accommodate these sensory capabilities. For example, as stated above, scent marking is an important form of communication for many species, and those species should be provided with materials that accommodate scent marks (e.g., wood). Similarly, rodents and other species can hear sounds in ranges that we cannot. Devices, such as fluorescent lights and computer monitors, produce high frequency sounds that, while inaudible to us, can be quite loud to those species (Morgan and Tromborg 2007).

Learning

Learning, and the flexibility it provides, is an important component of species-typical behavior, and as such, it is subject to the forces of natural selection. The various types of learning range from simple associations involved in habituation and operant conditioning, to higher order mental processes that may be involved in complex tasks, such as tool use, invisible displacements (object permanence), and spatial memory.

FIGURE 2.4 Researchers observing animals in the field. (Photo by M. Novak.)

1. *Habituation and Its Relationship to Observation*: Habituation refers to a cessation of responding to repeated presentations of the same stimulus. Most novel stimuli initially produce an increase in vigilance, followed by exploration. However, repetitive presentation generally leads to a cessation in these responses. This concept is particularly important in the field of animal behavior. In many cases, observations of animals are conducted by humans out in the field, with animals moving freely, or in colony rooms of laboratory animals. Humans are typically meaningful stimuli to animals, which generally results in some sort of behavioral response to the human "stimulus", a less than ideal circumstance for "naturalistic" behavioral observations (Figure 2.4).

   Observers should habituate animals to their presence before collecting behavioral data. In the laboratory, observers should remain relatively motionless and nonreactive for a number of sessions before data are collected. In the field, observers may have to habituate animals to their movements as they follow animals through their habitat. In the absence of habituation, observers may see only a limited part of the animals' behavioral repertoires, namely, those behaviors that are directed toward strangers and potential predators.

   It should be noted that there is no easy way to prove that animals are fully and completely habituated to observers, nor is there any set number of sessions in which to achieve this habituation. The best approach is to look for a decrease in vigilance-related behavior over time when the observer is present and, where possible, compare it to the levels of vigilance behavior when the observer is absent (e.g., video data).

2. *Operant Conditioning*: Operant conditioning refers to a form of learning in which the strength of a behavior changes (increases or decreases) in response to the consequences associated with the behavior. When a behavior is reinforced (i.e., rewarded, either by providing something positive or removing something aversive), the chance that the animal will engage in that behavior again increases. This is commonly used in animal training; providing a treat (positive reinforcement) when animals engage in target behaviors increases the chance that the animal will repeat that behavior. Similarly, behaviors can be reinforced by removing something negative; for example, you might put on your seatbelt to avoid the annoying buzzer in your car. On the other hand, if the outcome of a behavior is punishment (either adding something aversive to, or removing something positive from, the animal), the chances that the animal will engage in that behavior again decrease.

   Animals must constantly learn new things to effectively exploit their environments and survive, and much of this learning is, in part, the result of operant conditioning. At its most basic level, animals must learn what is safe to eat and what is not, and this is often dependent on actual experience with the food item. Blue jays readily consume cryptically colored moths and develop a search image for these moths as a consequence of positive reinforcement (Bond and Kamil 2002). However, when naïve blue jays take a monarch butterfly, they become sick as a result of the cardiac glycosides that are present in the monarch's wings (Brower and Fink 1985). This adverse exposure, a form of punishment in the scheme above, generally causes blue jays to avoid all monarchs in the future. Thus, taste aversion learning occurs rapidly and is an important process for many species in learning what is safe to eat.

But animals must also learn that food items, whether they be plant or animal species, increase and decrease in numbers over short periods of time. This is most obvious when we consider the flowers that honey bees rely on for nectar in the temperate zone, which change from spring (e.g., lilacs and coneflowers), to summer (e.g., dandelions and sunflowers), and into the early fall (e.g., asters and goldenrods). As the relative numbers of these various flowers increase, positive reinforcement ensures return to the same area and plant. But as they fade away, extinction (a decrease in behavior that occurs with a lack of reinforcement) will take over.

Operant conditioning can play a substantial role in the well-being of laboratory animals. Procedures that initially induce fear and avoidance can be mitigated in laboratory animals by a rigorous training program that involves positive reinforcement. Animals trained using positive reinforcement techniques are typically asked to perform a behavior (e.g., "remain stationary") and are then given a reward (e.g., desired food item or praise) immediately after they engage in the behavior, which increases the likelihood that the animal will perform the behavior the next time it is asked. Many animals, including nonhuman primates, dogs, and pigs, among others, have been trained with positive reinforcement techniques to voluntarily cooperate with husbandry, clinical, and/or scientific procedures (see Schapiro 2021 for more details on positive reinforcement training). For example, rhesus macaques have been trained using positive reinforcement training to present a limb for blood sampling (e.g., Coleman et al. 2008). Such training eliminates the need for restraint and sedation, thus improving well-being for the animals and resulting in better data (Graham et al. 2013).

3. *Complex Forms of Learning*: In the past, complex forms of learning were viewed simply as a series of operant responses that had been chained together. The best example of such chaining can be seen in the squirrel mazes created by English landowners to determine exactly what squirrels will go through to obtain nuts from bird feeders (see video clip of the "mission impossible squirrel" at https://www.youtube.com/watch?v=FrPOsqzqx7k). The landowners in this case created the maze in stages, waited for the squirrels to get to the feeder in stage 1, then added stage 2 and waited for the squirrels to get to the feeder and so on.

However, there are many kinds of learning and problem solving that are not readily explained by operant conditioning alone. The term animal cognition refers to mental abilities involved in thinking and reasoning. In many cases, the assumption is that animals develop mental representations of possible solutions. Among many capabilities, animals can categorize along a number of different dimensions (e.g., abstract vs. relational), can learn basic rules by discriminating between correct and incorrect sequences of events, can use tools spontaneously without training, can infer hidden movements, and can show prodigious memory capacity.

We highlight briefly the last three capabilities listed above. Many species can spontaneously use tools to solve problems (e.g., Caledonian crows, ostriches, capuchin monkeys, macaques, and chimpanzees). The macaque in the photo (Figure 2.5) is inserting a rod into a portable tube to dislodge a carrot. When this monkey was first presented with a stationary tube task with a prune in the middle, he tried reaching in at one end, then reaching in at the other end and then biting the middle where the prune was. Immediately following those failed attempts, he walked to another part of his pen, picked up a polyvinyl chloride (PVC) rod, returned and inserted it in the tube to dislodge the treat. The entire process took 68 s. He received no training or shaping whatsoever. Rather than simply being the result of operant conditioning, it appeared as if he had some mental representation of the solution. This spontaneous solution was observed in 3 of 25 rhesus monkeys tested.

Such mental representations are also important in object permanence tasks involving invisible displacements, wherein animals must track the location of hidden objects. In this task, animals first see a treat in a container, the container is then moved behind one of three boxes and the treat transferred out of view of the animal. Finally, the container is shown empty. The animal must mentally infer the transfer of the object from the container to that particular box. Orangutans can pass invisible displacement tasks, whereas squirrel monkeys cannot. Rhesus monkeys are intermediate, in that some can

FIGURE 2.5 Rhesus macaque using a PVC rod to remove a piece of carrot from a clear tube. (Photo by M. Novak.)

solve this task, whereas others seem unable to do so (de Blois et al. 1998, 1999).

Another form of mental representation involves the development of cognitive maps. In this process, animals accumulate information about their spatial environment by identifying objects and their locations within that environment, establishing the spatial coordinates of the objects with respect to one another, and retaining a memory of these coordinates over long periods of time. Species that cache seeds and nuts over the winter include both mammals (e.g., squirrels) and birds (e.g., nutcrackers). Nutcrackers are renowned for their extensive caching of seeds, creating thousands of cache sites during the fall (Moller et al. 2001). They also remember the location of these cache sites buried beneath snow cover for as long as 7–9 months.

Studies of animal cognition can provide valuable information for enriching the environment of laboratory animals. One basic concern with laboratory animals is that in comparison to natural environments, they live in a relatively sensory-deprived environment. Providing cognitive challenges consistent with a species' existing capabilities may be an excellent way to enhance their environment and keep them active, alert, and in a state of positive well-being. New technologies are now coming on line that may facilitate this process (e.g., iPads for macaques with built-in programs).

4. *Social Learning*: Animals living in social groups often learn by observing the actions of conspecifics, a process known as social or observational learning. Most readers are likely familiar with the story of Imo, a young female Japanese macaque living on Koshima Island, Japan, in the 1950s. Imo took a sweet potato from the sand and washed it in nearby seawater, something that no other monkey in her troop had been observed to do. Soon after watching Imo, the rest of her troop began washing sweet potatoes, presumably through social learning (Kawai 1965). Of course, animals learn more than just food washing from other members of their group. By observing conspecifics, animals, particularly juveniles, learn what foods are safe, how to eat those foods, what and who to fear, and even mating strategies (for review of social learning, see Galef and Laland 2005). Not surprisingly, social learning between adults is more easily facilitated in social species, such as rats and primates, than solitary species, such as golden hamsters (Galef and Laland 2005). In an interesting study, Lupfer et al. (2003) compared social learning of food preference between golden hamsters and dwarf hamsters, a moderately social species. They found that while adult dwarf hamsters learned food preferences from adult conspecifics, golden hamsters did not. However, golden hamster pups did show preferences for food their mothers preferred (Lupfer et al. 2003), suggesting that developmental stage may be an important influence on social learning (Galef and Laland 2005). Social learning can be a useful tool in the laboratory. Animals that are new to the colony can learn how to use specific feeding equipment or enrichment devices by observing more experienced animals. They can also learn to accept handling by carestaff, particularly if the demonstrator animal is a parent. For example, young foals that were present while their mothers were gently handled by humans were more likely to tolerate handling later in life than those who did not witness the handling (Henry et al. 2005). While not widely studied in the laboratory setting, social learning can also facilitate positive reinforcement training. For example, macaques (Coleman 2017) and rats (Leidinger et al. 2018) that watch others being trained for a task are more likely than naïve animals to perform that task. Additionally, dogs can learn to perform tasks by watching a human engage in the desired behavior; this has been used as a relatively new training technique called "Do as I Do" (Fugazza and Miklósi 2014). The observer does not necessarily need to be present; budgies (and perhaps other species) can learn tasks from watching videos of conspecifics (Mottley and Heyes 2003). There can be research implications to social learning as well. Young rats that observed their littermates intoxicated, voluntarily consumed more alcohol than those that did not (Hunt et al. 2001). Finally, while less often documented, it is certainly conceivable that animals could learn 'bad" habits, as well as good. Social learning has been implicated in the transmission of abnormal behavior in chimpanzees (Hook et al. 2002) and chickens (Dixon and Lambton 2021).

Individual Variation

Even within species, the behavior of individuals can vary widely, a fact that comes as no surprise to anyone who works with or around animals. If you have spent time at a dog park, you have undoubtably seen that some dogs eagerly play with others and will approach new people without hesitation, while others hide behind their people, rarely peaking out. There are a variety of factors that can produce these sorts of behavioral differences, including the sex, age, experiences, and temperament/personality of the individuals. We briefly describe the influence of temperament on behavior below.

Temperament, or personality, can be defined as an individual's basic position toward environmental change and challenge (Lyons et al. 1988); in other words, it describes how individuals respond to, and cope with, environmental stressors. When faced with a potentially risky stimulus, such as a novel feeding patch, or, in the example above, an unfamiliar person at the dog park, some individuals (i.e., bold or exploratory) eagerly approach and inspect the novel situation, while others (i.e., shy or inhibited) actively avoid it. These sorts of differences in response to novelty have been found in a wide range of taxonomic groups, from invertebrates to fish to birds to mammals. Indeed, in every species in which differences in temperament have been investigated, they have been found, indicating the conserved nature of this trait. The study of temperament and its relationship with various behaviors has grown dramatically in the past two decades. Shy and bold individuals have been shown to differ with respect to many natural behaviors, including habitat choice and utilization, mating style, foraging behaviors, parental care, and social interactions (e.g., Réale and Montiglio 2020). For example, bold southern red-backed voles dispersed seeds at greater distances than their more reticent counterparts (Brehm et al. 2019), and shy sticklebacks had fewer interactions in social networks than bolder animals, a finding that has implications for the transfer of illness (Pike et al. 2008).

As with their wild counterparts, laboratory animals also differ with respect to temperament, which can influence various behaviors, including feeding, exploration, and socialization, among others. Further, temperament can affect the physical and behavioral health of animals. Due to their heightened stress sensitivity, shy individuals may have increased susceptibility to various illnesses, such as diarrhea (e.g., Capitanio 2017; Gottlieb et al. 2018). They also are more likely than bolder conspecifics to display behaviors indicative of negative emotional states (e.g., pigs, Asher et al. 2016; dogs, Mendl et al. 2010). Thus, temperament can influence well-being and affect many laboratory animals.

Perhaps not surprisingly, temperament can also influence how animals respond to behavioral management practices (Coleman 2020). The provision of environmental enrichment is a major part of behavioral management for most species (see Schapiro 2021). There is often an assumption made that enrichment is equally beneficial for all animals. However, individuals can differ in their response to various enrichment strategies; bold animals (e.g., pigs, rhesus macaques) are more likely than shy counterparts to utilize certain environmental enrichment items (pigs, Bolhuis et al. 2004; macaques, Coleman 2017). Further, enrichment can be potentially anxiogenic for highly inhibited individuals (e.g., orange-winged Amazon parrots, *Amazona amazonica*, Fox and Millam 2007). Temperament can also influence the outcome of positive reinforcement training. Shy, fearful macaques were less likely than bold monkeys to successfully learn tasks, including touching a target and presenting a body part (Coleman 2017; Coleman et al. 2005). Thus, when possible, individual differences in temperament should be considered when caring for laboratory animals.

Cautions: Our Limitations as Observers

As observers, it may be difficult to escape our human-centric view of the world and the animals in it. In fact, the field of animal behavior may be more vulnerable to human bias than any other field of biology. Two significant factors in this possible bias are our own limitations in sensory capabilities and in the tendency to anthropomorphize, projecting human-like traits onto nonhuman animals (e.g., my dog is jealous when I pay attention to the cat).

Sensory Motor Capabilities

As we discuss above, there are numerous examples of species having sensory capabilities that go far beyond those of humans. Bats use echolocation (similar to sonar) to detect moths; bees see ultraviolet light to locate food sources; and mice produce ultrasonic vocalizations to communicate with one another. But why are these differences important? Most biomedical animal model research is conducted on mice and rats. Although scientists can readily observe motor movements and general activities (e.g., digging, chewing, eating and the like) in mice and rats, the vocalizations they produce are inaccessible to humans without the use of sophisticated recording equipment. In some species, vocalizations can be important components of well-being assessments. It is an empirical question as to whether vocalizations in rats and mice are also significant markers for well-being.

Anthropomorphism

Anthropomorphism is especially common with pets, with mammals, and particularly, with creatures that look most like us (e.g., nonhuman primates). Scientific opinions vary on the use of anthropomorphism. Some argue that it is a form of bias, interfering with the ability to rule out simpler explanations. Others suggest that it allows us to explore higher-order emotional processes (e.g., jealousy) and higher-order cognitive processes (e.g., attribution of knowledge) in nonhuman animals. Both sides would agree that such ideas must be tested scientifically before any conclusions can be drawn.

Anthropomorphism can cause us to project our own feelings onto the animals. For example, subordinate animals that live in linear dominance hierarchies may have limited access to resources compared to the dominant individuals. This inequity is a natural part of their world; however, empathetic humans sometimes feel bad for these individuals, which can lead to decisions that are not necessarily in the affected individuals' favor, such as separating animals in a social pair because one partner receives fewer treats than the other. Lack of treats is likely to be less important to the individual than the lack of social companionship.

Summary

As you will see as you read through the chapters in this book, animals engage in some pretty amazing behaviors. Behaviors such as the way they interact with conspecifics, how they find and process food, and the manner by which they avoid predators, among others, have been shaped by natural selection to help them survive in their natural habitat. Knowing these species-appropriate behaviors is critical not only for providing the animals with optimal environmental features to promote welfare but also for being able to identify abnormal and maladaptive behaviors. Further, understanding animal behavior as an academic discipline can provide tools to help in this endeavor. This chapter is intended to give you, the reader, a very brief introduction to a very select subset of the topics and approaches that are important for the scientific understanding of animal behavior and behavioral biology.

Acknowledgments

We thank Steve Schapiro for suggestions that greatly improved this chapter, and the authors of the chapters in this book for their enthusiastic descriptions of the behavior of their species. We also thank the monkeys and other animals in our lives, for continuously surprising and amazing us. Support is acknowledged from the Oregon National Primate Research Center, 8P51OD011092 (KC), and University of Massachusetts, R24OD11180 (MAN).

REFERENCES

Altmann J. Observational study of behavior: Sampling methods. *Behaviour.* 1974;49:227–267.

Arruda MF, Araújo A, Sousa MB, Albuquerque FS, Albuquerque AC, Yamamoto ME. Two breeding females within free-living groups may not always indicate polygyny: Alternative subordinate female strategies in common marmosets (*Callithrix jacchus*). *Folia Primatol (Basel).* 2005;76(1): 10–20. doi: 10.1159/000082451.

Asher L, Friel M, Griffin K, Collins LM. Mood and personality interact to determine cognitive biases in pigs. *Biol Lett.* 2016; 12(11). doi: 10.1098/rsbl.2016.0402.

Baker KC, Bloomsmith M, Neu K, Griffis C, Maloney M, Oettinger B, Schoof VA, Martinez M. Positive reinforcement training moderates only high levels of abnormal behavior in singly housed rhesus macaques. *J Appl Anim Welf Sci.* 2009;12(3):236–252. doi: 10.1080/10888700902956011.

Bales KL. Parenting in animals. *Curr Opin Psychol.* 2017;15: 93–98. doi: 10.1016/j.copsyc.2017.02.026.

Bamshad M, Novak MA, de Vries GJ. Cohabitation alters vasopressin innervation and paternal behavior in prairie voles (*Microtus ochrogaster*). *Physiol Behav.* 1994;56(4): 751–758. doi: 10.1016/0031-9384(94)90238-0.

Barrett CE, Keebaugh AC, Ahern TH, Bass CE, Terwilliger EF, Young LJ. Variation in vasopressin receptor (Avpr1a) expression creates diversity in behaviors related to monogamy in prairie voles. *Horm Behav.* 2013;63(3):518–526. doi: 10.1016/j.yhbeh.2013.01.005.

Bechard AR, Lewis MH. Transgenerational effects of environmental enrichment on repetitive motor behavior development. *Behav Brain Res.* 2016;307:145–149. doi: 10.1016/j.bbr.2016.04.005.

Blanchard P, Lauzeral C, Chamaillé-Jammes S, Brunet C, Lec'hvien A, Péron G, Pontier D. Coping with change in predation risk across space and time through complementary behavioral responses. *BMC Ecol.* 2018;18(1):60. doi: 10.1186/s12898-018-0215-7.

Blank DA. The use of tail-flagging and white rump-patch in alarm behavior of goitered gazelles. *Behav Processes.* 2018;151:44–53. doi: 10.1016/j.beproc.2018.03.011.

Bolhuis JE, Schouten WG, de Leeuw JA, Schrama JW, Wiegant VM. Individual coping characteristics, rearing conditions and behavioural flexibility in pigs. *Behav Brain Res.* 2004;152(2):351–360. doi: 10.1016/j.bbr.2003.10.024.

Bond AB, Kamil AC. Visual predators select for crypticity and polymorphism in virtual prey. *Nature.* 2002;415(6872): 609–613. doi: 10.1038/415609a.

Borries C, Launhardt K, Epplen C, Epplen JT, Winkler P. DNA analyses support the hypothesis that infanticide is adaptive in langur monkeys. *Proc Biol Sci.* 1999;266(1422):901–904. doi: 10.1098/rspb.1999.0721.

Bosch OJ, Young LJ. Oxytocin and social relationships: From attachment to bond disruption. *Curr Top Behav Neurosci.* 2018;35:97–117. doi: 10.1007/7854_2017_10.

Brainard MS, Doupe AJ. What songbirds teach us about learning. *Nature.* 2002;417(6886):351–358. doi: 10.1038/417351a.

Brehm AM, Mortellit A., Maynard GA, Zydlewski J. Land-use change and the ecological consequences of personality in small mammals. *Ecol Lett.* 2019;22(9):1387–1395. doi: 10.1111/ele.13324.

Brower LP, Fink LS. A natural toxic defense system: cardenolides in butterflies versus birds. *Ann N Y Acad Sci.* 1985;443: 171–188. doi: 10.1111/j.1749-6632.1985.tb27072.x.

Burnett S. Perceptual worlds and sensory ecology. *Nat Ed.* 2011;3(10):75.

Burt WH. Territoriality and home range concepts as applied to mammals. *J Mammal.* 1943;24(3):346–352. doi: 10.2307/1374834.

Byers JA, Waits L. Good genes sexual selection in nature. *Proc Natl Acad Sci U S A.* 2006;103(44):16343–16345. doi: 10.1073/pnas.0608184103.

Caine NG. Antipredator behavior: Its expression and consequences in captive primates. In Schapiro SJ, ed. *The Handbook of Primate Behavioral Management.* Boca Raton, FL: CRC Press; 2017: 127–138.

Caine NG, Potter MP, Mayer KE. Sleeping site selection by captive tamarins (*Saguinus labiatus*). *Ethology.* 1992;90(1):63–71.

Caldji C, Tannenbaum B, Sharma S, Francis D, Plotsky PM, Meaney MJ. Maternal care during infancy regulates the development of neural systems mediating the expression of fearfulness in the rat. *Proc Natl Acad Sci U S A.* 1998;95(9):5335–5340. doi: 10.1073/pnas.95.9.5335.

Capitanio JP. Naturally occurring nonhuman primate models of psychosocial processes. *ILAR J.* 2017;58(2):226–234. doi: 10.1093/ilar/ilx012.

Caro T. The functions of stotting: a review of the hypotheses. *Anim Behav.* 1986;34(3):649–662. doi: 10.1016/S0003-3472(86)80051-3.

Clutton-Brock T, McAuliffe K. Female mate choice in mammals. *Q Rev Biol.* 2009;84(1):3–27. doi: 10.1086/596461. PMID: 19326786.

Coleman K. Comparative perspectives on temperament and personality in human and nonhuman animals. In Schmidt L, Poole KL, eds. *Adaptive Shyness: Multiple Perspectives.* Switzerland: Springer Publishers; 2020: 253–277.

Coleman K. Individual differences in temperament and behavioral management. In Schapiro SJ, ed. *The Handbook of Primate Behavioral Management.* Boca Raton, FL: CRC Press; 2017: 95–113.

Coleman K, Maier A. The use of positive reinforcement training to reduce stereotypic behavior in rhesus macaques. *Appl Anim Behav Sci.* 2010;124(3–4):142–148. doi: 10.1016/j.applanim.2010.02.008.

Coleman K, Pranger L, Maier A, Lambeth SP, Perlman JE, Thiele E, Schapiro SJ. Training rhesus macaques for venipuncture using positive reinforcement techniques: a comparison with chimpanzees. *J Am Assoc Lab Anim Sci.* 2008;47(1):37–41.

Coleman K, Tully LA, McMillan JL. Temperament correlates with training success in adult rhesus macaques. *Am J Primatol*. 2005;65(1):63–71. doi: 10.1002/ajp.20097.

de Blois ST, Novak MA, Bond M. Can memory requirements account for species' differences in invisible displacement tasks? *J Exp Psychol Anim Behav Process*. 1999;25(2): 168–176. doi: 10.1037/0097-7403.25.2.168.

de Blois ST, Novak MA, Bond M. Object permanence in orang-utans (*Pongo pygmaeus*) and squirrel monkeys (*Saimiri sciureus*). *J Comp Psychol*. 1998;112(2):137–52. doi: 10.1037/0735-7036.112.2.137.

DeNardo D. Behavioral biology of reptiles. In Coleman K, Schapiro SJ, eds. *Behavioral Biology of Laboratory Animals*. Boca Raton, FL: CRC Press; 2021:361–375.

Digby LJ, Ferrari SF. Multiple breeding females in free ranging groups of *Callithrix jacchus*. *Int J Primatol*. 1994;15(3): 389–397. doi: 10.1007/BF02696100.

Dixon L, Lambton S. Behavioral biology of chickens and quail. In Coleman K, Schapiro SJ, eds. *Behavioral Biology of Laboratory Animals*. Boca Raton, FL: CRC Press; 2021:299–313.

Doucet SM, Montgomerie R. Multiple sexual ornaments in satin bowerbirds: Ultraviolet plumage and bowers signal different aspects of male quality. *Behav Ecol*. 2003;14:503–509. doi: 10.1093/beheco/arg035.

Ekblom R, Farrell LL, Lank DB, Burke T. Gene expression divergence and nucleotide differentiation between males of different color morphs and mating strategies in the ruff. *Ecol Evol*. 2012;2(10):2485–505. doi: 10.1002/ece3.370.

Faulkes CG, Arruda MF, Monteiro Da Cruz MA. Matrilineal genetic structure within and among populations of the cooperatively breeding common marmoset, *Callithrix jacchus*. *Mol Ecol*. 2003;12(4):1101–1108. doi: 10.1046/j.1365-294x.2003.01809.x.

Fitzgibbon CD; Fanshawe JH. Stotting in Thomson gazelles - An honest signal of condition. *Behav Ecol Sociobiol*. 1988;23(2):69074. doi: 10.1007/BF00299889.

Fortin D, Morales JM, Boyce MS. Elk winter foraging at fine scale in Yellowstone National Park. *Oecologia*. 2005;145(2):335–343. doi: 10.1007/s00442-005-0122-4.

Fox RA, Millam JR. Novelty and individual differences influence neophobia in orange-winged Amazon parrots (*Amazona amazonica*). *Appl Anim Behav Sci*. 2007; 104(1–2):107–115. doi: 10.1016/j.applanim.2006.04.033.

Friedrich SR, Mello CV. Behavioral biology of the zebra finch. In Coleman K, Schapiro SJ, eds. *Behavioral Biology of Laboratory Animals*. Boca Raton, FL: CRC Press; 2021:315–329.

Fu P, Neff BD, Gross MR. Tactic-specific success in sperm competition. *Proc Biol Sci*. 2001;268(1472):1105–1112. doi: 10.1098/rspb.2001.1625.

Fugazza C, Miklósi Á. Should old dog trainers learn new tricks? The efficiency of the Do as I do method and shaping/clicker training method to train dogs. *Appl Anim Behav Sci*. 2014;153:53–61. doi: 10.1016/j.applanim.2014.01.009.

Galef BG, LaLand KN. Social learning in animals: Empirical studies and theoretical models. *Biosci*. 2005;55(6):489–499. doi: 10.1641/0006-3568(2005)055[0489:SLIAES]2.0.CO;2.

Getz LL, McGuire B. Demography of fluctuating vole populations: Phase homogeneity of demographic variables. *Basic Appl Ecol*. 2008;10:170–177. doi: 10.1016/j.baae.2008.03.002.

Getz LL, McGuire B, Carter CS. Social behavior, reproduction, and demography of the prairie vole, *Microtus ochrogaster*. *Ethol Ecol Evol*. 2003;15:105–118. doi: 10.2307/1381904.

Getz LL, McGuire B, Pizzuto T, Hofmann JE, Frase B. Social organization of the prairie vole (*Microtus ochrogaster*). *J Mammal*. 1993;74(1):44–58.

Gobrogge K, Wang Z. Neuropeptidergic regulation of pair-bonding and stress buffering: Lessons from voles. *Horm Behav*. 2015;76:91–105. doi: 10.1016/j.yhbeh.2015.08.010.

Gobrogge K, Wang Z. The ties that bond: neurochemistry of attachment in voles. *Curr Opin Neurobiol*. 2016;38:80–88. doi: 10.1016/j.conb.2016.04.011.

Goldberg RL, Downing PA, Griffin AS, Green JP. The costs and benefits of paternal care in fish: A meta-analysis. *Proc Biol Sci*. 2020;287(1935):20201759. doi: 10.1098/rspb.2020.1759.

Gottlieb DH, Del Rosso L, Sheikhi F, Gottlieb A, McCowan B, Capitanio JP. Personality, environmental stressors, and diarrhea in rhesus macaques: An interactionist perspective. *Am J Primatol*. 2018;80(12):e22908. doi: 10.1002/ajp.22908.

Gottlieb D, Pomerantz O. Utilizing behavior to assess welfare. In Coleman K, Schapiro SJ, eds. *Behavioral Biology of Laboratory Animals*. Boca Raton, FL: CRC Press; 2021:51–64.

Graham ML, Rieke EF, Mutch LA, Zolondek EK, Faig AW, Dufour TA…Schuurman H-J. Successful implementation of cooperative handling eliminates the need for restraint in a complex nonhuman primate disease model. *J Med Primatol*. 2013;4(2):89–106. doi: 10.1111/j.1600-0684.2011.00525.x.

Hackett SJ, Kimball RT, Reddy S, Bowie RC, Braun EL, Braun MJ…Yuri T. A phylogenomic study of birds reveals their evolutionary history. *Science*. 2008;320(5884):1763–1768. doi: 10.1126/science.1157704.

Hauber ME, Lacey EA. Bateman's principle in cooperatively breeding vertebrates: The effects of non-breeding alloparents on variability in female and male reproductive success. *Integr Comp Biol*. 2005;45(5):903–914. doi: 10.1093/icb/45.5.903.

Henry S, Richard MA, Hausberger M. Human-mare relationships and behaviour of foals toward humans. *Appl Anim Behav Sci*. 93:341–362.

Honess PE. Behavioral biology of macaques. In Coleman K, Schapiro SJ, eds. *Behavioral Biology of Laboratory Animals*. Boca Raton, FL: CRC Press; 2021:437–474.

Hook MA, Lambeth SP, Perlman JE, Stavisky R, Bloomsmith MA, Schaprio SJ. 2002. Inter-group variation in abnormal behavior in chimpanzees (*Pan troglodytes*) and rhesus macaques (*Macaca mulatta*). *Appl Anim Behav Sci*. 76(2):165–176.

Hunt PS, Hallmark RA. Increases in ethanol ingestion by young rats following interaction with intoxicated siblings: A review. *Integr Physiol Behav Sci*. 2001;36(3):239–248. doi: 10.1007/BF02734096.

Jarvis ED, Mirarab S, Aberer AJ, Li B, Houde P, Li C…Zhang G. Whole genome analyses resolve early branches in the tree of life of modern birds. *Science*. 2014;346(6215):1 320–1321. doi: 10.1126/science.1253451.

Jones AG, Ratterman NL. Mate choice and sexual selection: What have we learned since Darwin? *Proc Natl Acad Sci U S A*. 2009;106(Suppl 1):10001–10008. doi: 10.1073/pnas.0901129106.

Jorgensen MJ. Behavioral biology of vervets/African green monkeys. In Coleman K, Schapiro SJ, eds. *Behavioral Biology of Laboratory Animals*. Boca Raton, FL: CRC Press; 2021:475–494.

Kawai M. Newly-acquired pre-cultural behavior of the natural troop of Japanese monkeys on Koshima islet. *Primates*. 1965;6:1–30.

Kleven GA. Behavioral biology of guinea pigs. In Coleman K, Schapiro SJ, eds. *Behavioral Biology of Laboratory Animals*. Boca Raton, FL: CRC Press; 2021:131–146.

Kroodsma D, Hamilton D, Sanchez JE, Byers BE, Fandino-Marino H, Stemple DW... Powell GVN. Behavioral evidence for song learning in the suboscine bellbirds. *The Wilson J Ornithol*. 2013;125(1):1–14. doi: 10.1676/12-033.1.

Küpper C, Stocks M, Risse JE, Dos Remedios N, Farrell LL, McRae SB...Burke T. A supergene determines highly divergent male reproductive morphs in the ruff. *Nat Genet*. 2016;48(1):79–83. doi: 10.1038/ng.3443.

Kvarnemo C. Why do some animals mate with one partner rather than many? A review of causes and consequences of monogamy. *Biol Rev Camb Philos Soc*. 2018;93(4):1795–1812. doi: 10.1111/brv.12421.

Lank DB, Farrell LL, Burke T, Piersma T, McRae SB. A dominant allele controls development into female mimic male and diminutive female ruffs. *Biol Lett*. 2013;9(6):20130653. doi: 10.1098/rsbl.2013.0653.

Laundré JW. Behavioral response races, predator-prey shell games, ecology of fear, and patch use of pumas and their ungulate prey. *Ecology*. 2010;91(10):2995–3007. doi: 10.1890/08-2345.1.

Lehman MN, Goodman RL, Karsch FJ, Jackson GL, Berriman SJ, Jansen HT. The GnRH system of seasonal breeders: Anatomy and plasticity. *Brain Res Bull*. 1997;44(4):445–457. doi: 10.1016/s0361-9230(97)00225-6.

Leidinger CS, Kaiser N, Baumgart N, Baumgart J. Using clicker training and social observation to teach rats to voluntarily change cages. *J Vis Exp*. 2018;140:58511. doi: 10.3791/58511.

Liu M, Siefferman L, Mays H Jr, Steffen JE, Hill GE. A field test of female mate preference for male plumage coloration in eastern bluebirds. *Anim Behav*. 2009;78(4):879–885. doi: 10.1016/j.anbehav.2009.07.012.

Lupfer G, Friema, J, Coonfield D. Social transmission of flavor preferences in two species of hamsters (*Mesocricetus auratus* and *Phodopus campbelli*). *J Comp Psychol*. 2003;117(4):449–455. doi: 10.1037/0735-7036.117.4.449.

Lyons, DM, Price EO, Moberg GP. Individual differences in temperament of domestic dairy goats: constancy and change. *Anim Behav*. 1988;36:1323–1333.

MacLellan A, Adcock A, Mason G. Behavioral biology of mice. In Coleman K, Schapiro SJ, eds. *Behavioral Biology of Laboratory Animals*. Boca Raton, FL: CRC Press; 2021:89–111.

Madison DM. Space use and social structure in meadow voles, *Microtus pennsylvanicus*. *Behav Ecol Sociobiol*. 1980;7(1):65–71. doi: 10.1007/BF00302520.

Madison DM, McShea W. Seasonal changes in reproductive tolerance, spacing, and social organization in meadow voles: A microtine model. *Am Zool*. 1987;27(3):899–908.

Martin P, Bateson P. *Measuring Behaviour: An Introductory Guide*, Cambridge, UK: Cambridge University Press; 2011.

Matsumura S. The evolution of "egalitarian" and "despotic" social systems among macaques. *Primates*. 1999;40(1):23–31. doi: 10.1007/BF02557699.

McGuire B, Novak M. A comparison of maternal behavior in the meadow vole (*Microtus pennsylvanicus*), prairie vole (*Microtus ochrogaster*) and the pine vole (*Microtus pinetorum*). *Anim Behav*. 1984;32:1132–1141. doi: 10.1016/S0003-3472(84)80229-8.

Mendl M, Brooks J, Basse C, Burman O, Paul E, Blackwell E, Casey R. Dogs showing separation-related behaviour exhibit a 'pessimistic' cognitive bias. *Curr Biol*. 2010;20(19):R839–R840. doi: 10.1016/j.cub.2010.08.030.

Mennill DJ, Alexander V, Jonart LM, Hill GE. Male house finches with elaborate songs have higher reproductive performance. *Ethology*. 2006;112(2):174–180. doi: 10.1016/j.anbehav.2005.05.006.

Moller A, Pavlick B, Hile AG, Balda RP. Clark's nutcrackers *Nucifraga columbiana* remember the size of their cached seeds. *Ethology*. 2001;107(5):451–461. doi: 10.1046/j.1439-0310.2001.00679.x.

Morais MRPT, Teófilo TDS, Azevedo BKG, Cavalcanti DMLP, Fontenele-Neto JD. Drought leads to reproductive quiescence in smooth-billed anis: Phenotypic evidence for opportunistic breeding and reproductive readiness. *J Morphol*. 2019;280(7):968–981. doi: 10.1002/jmor.20995.

Morgan KN, Tromborg CT. Sources of stress in captivity. *Appl Anim Behav Sci*. 2007;102(3–4):262–302. doi: 10.1016/j.applanim.2006.05.032.

Mottley K, Heyes C. Budgerigars (*Melopsittacus undulatus*) copy virtual demonstrators in a two-action test. *J Comp Psychol*. 2003;117(4):363–370. doi: 10.1037/0735-7036.117.4.363.

National Research Council. Guide for the Care and Use of Laboratory Animals, 8th Edition. Washington, DC: National Academies Press; 2011.

Nelson DA, Marler P. Innate recognition of song in white-crowned sparrows: A role in selective vocal learning. *Anim Behav*. 1993;46(4):806–808. doi: 10.1006/anbe.1993.1258.

Nelson DA, Marler P. Selection-based learning in bird song development. *Proc Natl Acad Sci U S A*. 1994;91(22):10498–10501. doi: 10.1073/pnas.91.22.10498.

Nelson DA, Marler P, Soha JA, Fullerton AL. The timing of song memorization differs in males and females: A new assay for avian vocal learning. *Anim Behav*. 1997;54(3):587–597. doi: 10.1006/anbe.1996.0456.

Novak MA, Meyer J. Abnormal behavior in animals in research settings. In Coleman K, Schapiro SJ, eds. *Behavioral Biology of Laboratory Animals*. Boca Raton, FL: CRC Press; 2021

Nowicki S, Searcy WA, Hughes M. The territorial defense function of song in song sparrows: A test with speaker occupation design. *Behaviour*. 1998;135(5):615–628. doi: 10.1163/156853998792897888.

Pike TW, Samanta M, Lindström J, Royle NJ. Behavioural phenotype affects social interactions in an animal network. *Proc Biol Sci*. 2008;275(1650):2515–2520. doi: 10.1098/rspb.2008.0744.

Pritchett-Corning KR, Winnicker C. Behavioral biology of deer and white-footed mice, Mongolian gerbils, and prairie and meadow voles. In Coleman K, Schapiro SJ, eds. *Behavioral Biology of Laboratory Animals*. Boca Raton, FL: CRC Press; 2021:147–163.

Qiu P, Jiang J, Liu Z, Cai Y, Huang T, Wang Y...Chang HC. BMAL1 knockout macaque monkeys display reduced sleep and psychiatric disorders. *Nat Sci Rev.* 2019;6(1):87–100. doi: 10.1093/nsr/nwz002.

Ramalho WP, Guerra V, Ferraz D, Machado IF, Vieira LJS. Observations on death-feigning behaviour and colouration patterns as anti-predator mechanisms in Amazonian anurans. *Herp Notes.* 2019;12:269–272.

Réale D, Montiglio P. Evolution of adaptive individual differences in non-human animals. In Schmidt L, Poole KL, eds. *Adaptive Shyness: Multiple Perspectives.* Switzerland: Springer Publishers; 2020: 279–299.

Reamer LA, Haller RL, Thiele EJ, Freeman HD, Lambeth SP, Schapiro SJ. Factors affecting initial training success of blood glucose testing in captive chimpanzees (*Pan troglodytes*). *Zoo Biol.* 2014;33(3):212–220. doi: 10.1002/zoo.21123.

Rogers FD, Rhemtulla M, Ferrer E, Bales KL. Longitudinal trajectories and inter-parental dynamics of prairie vole biparental care. *Front Ecol Evol.* 2018;6:73. doi: 10.3389/fevo.2018.00073.

Santos ES, Nakagawa S. Breeding biology and variable mating system of a population of introduced dunnocks (*Prunella modularis*) in New Zealand. *PLoS One.* 2013;8(7):e69329. doi: 10.1371/journal.pone.0069329.

Sapolsky RM. The influence of social hierarchy on primate health. *Science.* 2005;308(5722):648–652. doi: 10.1126/science.1106477.

Schapiro SJ. An overview of behavioral management for laboratory animals. In Coleman K, Schapiro SJ, eds. *Behavioral Biology of Laboratory Animals.* Boca Raton, FL: CRC Press; 2021

Setchell JM. Sexual Selection and the differences between the sexes in Mandrills (*Mandrillus sphinx*). *Am J Phys Anthropol.* 2016;159(Suppl 61):S105–S129. doi: 10.1002/ajpa.22904.

Sibley CG, Ahlquist JE, Monroe BL. A classification of living birds of the world based on DNA-DNA hybridization studies. *Auk.* 1988;105(3):409–423. doi: 10.1093/auk/105.3.409.

Stone AI, Mathieu D, Griffin L, Bales KL. Alloparenting experience affects future parental behavior and reproductive success in prairie voles (*Microtus ochrogaster*). *Behav Processes.* 2010;83(1):8–15. doi: 10.1016/j.beproc.2009.08.008.

Sutter A, Lindholm AK. The copulatory plug delays ejaculation by rival males and affects sperm competition outcome in house mice. *J Evol Biol.* 2016;29(8):1617–1630. doi: 10.1111/jeb.12898.

Tinbergen N. On aims and methods in ethology. *Zeitschrift für Tierpsychol.* 1963;20:410–433.

Truelove MA, Martin AL, Perlman JE, Wood JS, Bloomsmith MA. Pair housing of macaques: A review of partner selection, introduction techniques, monitoring for compatibility, and methods for long-term maintenance of pairs. *Am J Primatol.* 2017;79(1):1–15. doi: 10.1002/ajp.22485.

van de Waal E, Borgeaud C, Whiten A. Potent social learning and conformity shape a wild primate's foraging decisions. *Science.* 2013;340(6131):483–485. doi: 10.1126/science.1232769.

Vehrencamp SL. A model for the evolution of despotic versus egalitarian societies. *Anim Behav.* 1983;31:667–682.

von Uexküll, J. *A Foray into the Worlds of Animals and Humans with a Theory of Meaning.* Minneapolis:University of Minnesota Press; 2010.

Waiblinger S, Boivin X, Pederse V., Tosi MV, Jancza AM, Visser EK Jones RB. Assessing the human–animal relationship in farmed species: A critical review. *Appl Anim Behav Sci.* 2006;101 (3–4):185–242.

Wild JM. Functional neuroanatomy of the sensorimotor control of singing. *Ann N Y Acad Sci.* 2004;1016:438–462. doi: 10.1196/annals.1298.016.

Winnicker C, Pritchett-Corning KR. Behavioral biology of hamsters. In Coleman K, Schapiro SJ, eds. *Behavioral Biology of Laboratory Animals.* Boca Raton, FL: CRC Press; 2021:165–171.

Winslow JT, Shapiro L, Carter CS, Insel TR. Oxytocin and complex social behavior: Species comparisons. *Psychopharmacol Bull.* 1993;29(3):409–414.

Wolfenbarger LL. Female mate choice in Northern Cardinals: Is there a preference for redder males? *Wilson Bull.* 1999;111(1):76–83. doi: 10.2307/1370195.

Wolff JO, Mech SG, Dunlap AS, Hodges KE. Multi-male mating by paired and unpaired female prairie voles (*Microtus ochrogaster*). *Behaviour.* 2002;139:1147–1160. doi: 10.1163/15685390260437308.

Yamamoto ME, Araujo A, Arruda Mde F, Lima AK, Siqueira Jde O, Hattori WT. Male and female breeding strategies in a cooperative primate. *Behav Processes.* 2014;109(Pt A): 27–33. doi: 10.1016/j.beproc.2014.06.009.

3 Abnormal Behavior of Animals in Research Settings

Melinda A. Novak and Jerrold S. Meyer
University of Massachusetts-Amherst

CONTENTS

Introduction 27
What is Abnormal Behavior? 27
    Definition 27
    Types of Abnormal Behavior 28
Prevalence of Abnormal Behavior 30
    Nonhuman Primates 30
    Other Species 31
What Causes Abnormal Behavior: Ruling out Medical Conditions as a First Step 32
    Gastrointestinal Disease 32
    Arthritic and Other Painful Disorders 32
    Neurological and Retinal Disorders 33
What Causes Abnormal Behavior: Retrospective and Correlational Approaches to Understand Underlying Mechanisms 33
    Fixed and Causal Risk Factors 33
    Variable Risk Factors (Stressors in Adolescents and Adults) 34
    Correlated Factors 35
The Persistence of Abnormal Behavior 37
    Introduction 37
    Hypotheses 37
Therapeutic Interventions 38
    Environmental Approaches 38
    Pharmacotherapy 40
Conclusions 41
Acknowledgments 42
References 42

Introduction

One of the significant challenges of maintaining animals in captivity is the development of abnormal behavior in some individuals. The presence of abnormal behavior is often construed as a marker for ill being, indicating some adverse exposure to psychosocial stress. However, the connection between abnormal behavior and health and well-being has not been fully elucidated. Addressing this issue requires a multifaceted approach that includes: defining the term "abnormal behavior" and characterizing the various forms in which it is manifested, determining the prevalence of the various conditions across species and sex, and evaluating the risk factors, namely the underlying biological/experiential mechanisms that may have contributed to the onset of behavior. Two additional approaches are needed to develop effective treatments: identifying the environmental triggers that activate abnormal behavior, once the condition has been established, and examining the persistence of such behaviors, by determining the reinforcing factors that serve to maintain the behavior. A final consideration, using the information gleaned from the approach described above, is to develop treatments that reduce or eliminate abnormal behavior in affected animals. Ideally, the ultimate goal should be to develop effective management and housing strategies that prevent the development of abnormal behavior. We discuss this approach below with an emphasis on nonhuman primates. However, while the examples are largely from nonhuman primates, concepts can be adapted to most other species used in biomedical research.

What is Abnormal Behavior?

Definition

Abnormal behavior is any pattern of activity that differs from some standard. In studying abnormal behavior of animals housed in captivity, one of three standards is often invoked implicitly. The **ethological standard** focuses on species-typical behaviors that have been shaped by natural selection. The use of this standard requires a working knowledge of all behavior patterns exhibited by a particular species (ranging from communication to cognition) in the natural environment. Behaviors that are not species-typical are then characterized as abnormal (e.g., stereotypies). An alternative is the **baseline standard**, in which an average level is established for the rates of behavior of individual animals maintained in captivity. Any significant change in the behavior of an individual, either increased or decreased from a well-established baseline, is considered abnormal (e.g., depressive behavior). In the **context standard**, behaviors that are expressed in inappropriate situations are considered abnormal (e.g., displacement activities).

None of these standards is ideal. The ethological approach can include activities that may be "normal" for the species, but are undesirable in captive animals, such as aggression. The baseline approach requires extensive observations of animals in captivity to establish base rates of various behaviors. Furthermore, the underlying assumption is that the average baseline across all animals is a reflection of normal behavior and that significantly higher or lower values in individual animals are abnormal (e.g., hyper-aggressiveness). The context standard relies not only on a detailed knowledge of species-typical behavior, but also on the determination of what constitutes an inappropriate versus an appropriate context for behavioral expression.

Types of Abnormal Behavior

Abnormal behavior patterns can be classified into four broad categories: (1) stereotypic behaviors, which include locomotor, mouth-directed, and body-directed activities, (2) dangerous self-directed behaviors that can cause tissue damage (i.e., self-injurious behavior), (3) affective behaviors indicative of abnormal mood states, and (4) displacement activities. Within these categories, the intensity of these behaviors can vary from mild to severe in individual animals.

1. *Stereotypic Behaviors*, also known as abnormal repetitive behaviors (ARBs), include a group of activities observed in some captive animals that are generally not observed in animals under more natural conditions (per the ethological standard). ARBs can be divided into: (1) **locomotor stereotypies**, e.g., pacing in macaques (Lutz et al., 2003), marmosets (Kitchen and Martin, 1996), Amazonian parrots (Cussen and Mench, 2015), and mink (Díez-León et al., 2019); back flipping in rats (Callard et al., 2000), jumping behavior (Bechard et al., 2017) and pattern running (Wolmarans et al., 2013) in deer mice; and tail chasing in dogs (Tiira et al., 2012); (2) **mouthing activities to objects**, e.g., crib biting in horses (Wickens and Heleski, 2010), bar chewing in rodents (Wurbel and Stauffacher, 1998), also including appetitive disorders, such as coprophagy in chimpanzees (Jacobson et al., 2016); and (3) **body-directed activities**, e.g., feather picking in parrots (Garner et al., 2006), flank sucking and fly biting in dogs (Frank et al., 2012), eye covering in macaques (Lutz, 2014), hair plucking in laboratory mice (Garner et al., 2004), and tongue rolling in cattle (Sato et al., 1994). Defining features of ARBs include both their repetitive nature, occurring over and over again in a form that is distinctive between individuals, and their apparent lack of functionality. In the past, ARBs were described as purposeless (Mason, 1993), as exemplified by a mouse that spends considerable time chewing back and forth across a bar in its cage. However, the more modern view is that these ARBs serve some function, and it is the task of veterinarians and behavioral scientists, working together, to determine why these behaviors develop and persist. We discuss the possible significance of these behaviors in the section on persistence.

   A major challenge to understanding and treating ARBs within particular species, such as macaques, is the incredible variability in the diversity of behaviors that exists between individuals. Some monkeys show mostly locomotor ARBs; some monkeys show mostly body-directed ARBs; and yet other monkeys exhibit several different kinds of ARBs, including both locomotor and body-directed stereotypies (Lutz et al., 2003). Given this diversity of expression, it appears unlikely that there is a common underlying etiology or function (Tynes and Sinn, 2014). Instead, it is likely that varying forms of ARBs represent different conditions (Dallaire et al., 2011) and may be associated with different affective states (Pomerantz et al., 2012a, b).

   A second challenge is concerned with the frequency and intensity of expression. In the case of locomotor ARBs in nonhuman primates, some monkeys exhibit pacing primarily in anticipation of specific events, such as feeding; whereas other monkeys may pace at high rates across the day. Should the focus be on any animal that paces, or should a severity scale be employed to determine whether intervention should be sought? Support for the development of a severity scale comes from studies of humans in the general population and from animal species. ARBs (e.g., leg swinging, face touching, fidgeting, restlessness, nail biting, etc.) occur in the general human population, at low levels in children (Smith and Van Houten, 1996) and adults (Barnes et al., 1983; Rafaeli-Mor et al., 1999). The presence of these ARBs does not appear to have any adverse consequences and is not considered abnormal. Weaving, an equine locomotor stereotypy, frequently occurs in anticipation of feeding and increased environmental activity (Cooper et al., 2000, 2005; McBride and Hemmings, 2005). Increased pacing in anticipation of feeding has also been observed in sheep (Lauber et al., 2012), polar bears (Cless and Lukas, 2017), and broiler chickens (Savory and Mann, 1997). We take the position in this review that different types of stereotypies may relate to different conditions and that some severity scale should be used with the diagnosis of ARB to determine the need for treatment.

2. *Self-Injurious Behavior (SIB)* includes activities that have the potential to cause tissue damage that may require veterinary treatment. In dogs, SIB can take the form of overgrooming (e.g., acral lick dermatitis). In rhesus macaques, SIB generally takes the form of bites directed to the body, usually the arms and/or legs (Novak, 2003). Bites can range in severity from a mild form, in which the biting does not produce tissue damage, to a more serious form capable of producing a range of outcomes from abrasions and small cuts to punctures and slashes. SIB differs from ARBs in the frequency of occurrence in individuals and in its severity. Unlike most ARBs that are iteratively expressed throughout the day in individual animals, biting behavior is considerably less frequent. Nonetheless, when it does occur, it can be harmful rather than harmless.

3. *Abnormal Affective States* include species-typical behaviors involved in motivated states that become abnormal when they are expressed at very high or very low levels relative to the baseline standard. Two such states are hyper-aggressiveness and depressed behavior.

    a. *Hyper-aggressiveness* in socially housed animals can disrupt social relations, lead to injuries that require veterinary care, and may result in the loss of subjects from research protocols. Hyper-aggressiveness has been identified as a welfare issue in rodents, poultry, and nonhuman primates.

    A recent survey of 40 rodent facilities revealed that the prevalence rate of male mouse aggression was relatively low (15 in every 1,000 males). However, when aggression did occur, the mice frequently required treatment for injuries, and staff often had to remove either the aggressor or the victim (Lidster et al., 2019). Additionally, wounds caused by fighting were the second most common clinical condition in one large mouse facility, resulting in 520 cases per year (Marx et al., 2013). Hyperaggressiveness in poultry, as manifested by pecks directed to the head of others, is also considered a substantial welfare issue. In chickens and turkeys, this behavior can lead to injury and death (Fossum et al., 2009; Marchewka et al., 2013).

    Hyperaggressiveness is a well-known, but relatively rare, phenomenon in captive macaque indoor facilities. A small percentage of monkeys, usually males, cannot be paired successfully with other monkeys. This fact is acknowledged in the United States' Animal Welfare Act, in which an exemption from social housing is allowed for extremely aggressive monkeys that appear to be incompatible with partners. Factors that contribute to extreme aggression in monkeys include genetic predisposition (Newman et al., 2005) and exposure to early life stressors (Laudenslager et al., 2013), which are common themes for human pathological aggression as well (see reviews; Provencal et al., 2015 and Siever, 2008).

    b. *Depression* can be defined as a loss of interest in the environment. In nonhuman primates, depression occurs spontaneously and is typically manifested in behaviors such as inattentiveness (failure to respond to changes in the environment), reduced activity (lethargy), and a slumped resting posture of head lower than shoulders (Shively et al., 1997). Naturally occurring depression, using the behavioral criteria above, has been identified in a small percentage of socially housed cynomolgus (Xu et al., 2015) and rhesus (Qin et al., 2015) macaques.

    Naturally occurring depression has not been studied extensively in rodents, as a result of which its prevalence is unknown. Instead, depression is typically induced using one of several paradigms including repeated social defeat (Golden et al., 2011), chronic restraint stress (Chiba et al., 2012), and chronic unpredictable stress, sometimes also called chronic mild stress (Willner, 2016). The presence of a depressed state is then evaluated using several standardized challenge tests that are discussed in a later section. There are many examples of depression induction models in the rodent literature but because our focus is on abnormal behavior as an indicator of welfare, they are outside of the scope of this chapter.

4. *Displacement Activities* consist of species-typical behaviors that, when expressed, are disconnected from the context in which they are performed relative to the context standard. Typically, the most common displacement activities in nonhuman primates involve body care (e.g., scratching or self-grooming; Maestripieri et al., 1992) and may be a form of self-calming (Troisi, 2002). In one type, the disconnected activity results from two competing arousal states, such as fight or flight. For example, scratching was elevated in baboons sitting near high-ranking males, but not when sitting in proximity to lower ranking males (Castles et al., 1999). In another type, the disconnected activity occurs when a previously learned action is thwarted or changed. In this case, scratching behavior was elevated on a reversal task, in comparison to the initial acquisition task (Judge et al., 2011). Based on these and many other findings, displacement activities appear to reflect negative emotional states involved in stress and anxiety (Aureli and Whiten, 2003).

Such displacement activities, as reflected by repetitive scratching and grooming in response to various stressors, have also been observed in rats. When exposed to the elevated plus maze test of anxiety, subordinate rats spent less time in the open arms and showed higher rates of repetitive grooming behavior than dominant rats (Pohorecky, 2008).

Despite evidence that displacement activities are elicited by stressful events, such events are generally brief, and the effects do not appear to carry over to other contexts. Thus, these behaviors will not be considered further in this review. Instead, the sections below will focus mainly on stereotypies, self-injurious behavior, and to a lesser extent, mood disorders related to abnormal affective states.

Prevalence of Abnormal Behavior

The presence of abnormal behavior is often assumed to be a welfare issue, which raises the important question of how prevalent it is in laboratory- and zoo-housed species. Answering this question requires surveying large numbers of animals across species and sex for the presence of various kinds of abnormal behavior. By far, the most comprehensive data set, both in terms of number of different behaviors assessed and the number of facilities surveyed, exists for captive nonhuman primates, especially macaques and chimpanzees. It is likely that the emphasis on these primate species has been driven, in part, by the psychological well-being provision for nonhuman primates as amended in the USDA (US Department of Agriculture), Animal Welfare Act (1991). We start first with the extensive data set on nonhuman primates and then consider some other species at the end of this section.

Nonhuman Primates

Surveys of nonhuman primates conducted over the last two decades provide a detailed view of the prevalence of abnormal behavior across species and sex in different housing environments, particularly with respect to ARBs and SIB. However, this effort has been focused almost exclusively on two groups of primates (macaques and apes). Two different kinds of information are often included in these surveys: (1) estimates of the frequency and duration of ARBs collected on individual nonhuman primates and (2) the prevalence rate, i.e., the percentage of animals exhibiting ARBs. Both of these measures are derived from behavioral assessments of individual animals. In this review, we focus only on the prevalence rates. However, prevalence data can be affected by many factors, including how the various abnormal behaviors are defined, the facility in which the animals are housed, the number of monkeys or apes in the sample, and the total number of samples obtained from each individual. For this reason, we report only on those findings that are consistent across surveys.

1. *Stereotypic Behavior:*
    a. *Species:* Surveys of various Old World species of nonhuman primates revealed different profiles of stereotypic behavior. Macaques had a high prevalence of locomotor ARBs, with a significantly lower prevalence of object-directed and body-directed ARBs (Bellanca and Crockett, 2002; Gottlieb et al., 2015; Hook et al., 2002; Lutz et al., 2003; Lutz, 2018). This pattern was stable across four different facilities containing large populations of macaques.

    In contrast, the most common abnormal behavior in chimpanzees was coprophagy (Birkett and Newton-Fisher, 2011; Brand and Marchant, 2018; Jacobson et al., 2016; Nash et al., 1999) followed by the body-directed ARB of hair pulling and rocking (Brand and Marchant, 2018; Jacobson et al., 2016). In a recent survey (Bloomsmith et al., 2019), the typical behavior profile did not change, but the prevalence of coprophagy and hair pulling was considerably lower than reported in other studies.

    These findings in chimpanzees underscore the challenges in understanding and interpreting prevalence rates. Considerable variation is evident when comparing the data from four studies on chimpanzees. The following table (Table 3.1) shows the type of facility, population size, the overall percentage of chimpanzees showing any abnormal behavior (when available), and the percentage of animals exhibiting coprophagy, hair pulling, and rocking.

    While it may be tempting to attribute this wide variation to differences in methodology, it is much more likely that it is the result of many intrinsic and extrinsic factors that interact to influence the expression of ARBs. Such factors include, but are not limited to, genetic constitution, early rearing history, past and present housing environment, past and present health status, history of stress exposure, and implementation of sophisticated enrichment programs to mitigate abnormal behavior.

TABLE 3.1

Variation in Prevalence Rates of Abnormal Behavior in Chimpanzees

Article	Facilities	Type	Animals	Overall Prevalence	Eat Feces (%)	Rocking (%)	Hair Pulling (%)
Nash et al. (1999)	Research 4 Inst	Observations	232 (140F)	Not available	38	27	19
Birkett and Newton-Fisher (2011)	Zoo 6 Inst	Observations	40 (23F)	100%	83	53	78
Jacobson et al. (2016)	Zoo 26 Inst	Questionnaire	165 (102F)	64%	41	8	32
Bloomsmith et al. (2019)	Research 6 Inst	Questionnaire	701 (378F)	37%	10	13	9

The two different profiles of abnormal behavior observed in chimpanzees and macaques are not simply related to a divide between apes and nonape primates. In two recent studies of other Old World monkeys, the stereotypic behavior profile of mangabeys was found to be somewhat similar to that of macaques (Crast et al., 2014), whereas the stereotypic profile of baboons was more similar to that of chimpanzees (Lutz, 2018). In the latter study, Lutz directly compared the stereotypic behavior of two macaque species (*Macaca mulatta* and *M. fascicularis*) with baboons (*Papio hamadryas*) using the same methodology. For both macaque species, locomotor ARBs were by far the most common form of stereotypy, whereas for baboons, appetitive disorders (e.g., coprophagy, urine drinking, hair ingestion, etc.) were most common. Additionally, macaques showed a higher prevalence of stereotypic behavior overall than baboons.

   b. *Sex:* Sex differences also varied between macaques and chimpanzees. Stereotypic behavior in macaques was much more prevalent in males than in females (Gottlieb et al., 2013; Gottlieb et al., 2015; Lutz et al., 2003; Lutz, 2018; Vandeleest et al., 2011; but see Hook et al., 2002 for a finding of no difference). In contrast, the reverse was true for chimpanzees (Bloomsmith et al., 2019; Hook et al., 2002; Jacobson et al., 2016; Nash et al., 1999). In all studies, coprophagy was higher in females than in males, and in a majority of the studies, self-directed hair pulling was also more prevalent in female chimpanzees.

2. *SIB*: Assessing the prevalence of SIB from survey data is complicated by two issues. First, some authors include SIB in a general category of body-directed stereotypies, and thus, the actual numbers of nonhuman primates exhibiting SIB cannot be determined. Second, SIB can be manifested in two forms that may not be differentiated in surveys: (1) biters that produce wounds requiring veterinary intervention, as determined from the veterinary record, and (2) biters that do not produce wounds, as determined from behavioral assessments. For example, although 25% of individually housed monkeys were identified as biters at one facility, only 11% of these monkeys actually had a veterinary record of SIB, with a median number of lifetime wounding episodes at approximately two (Lutz et al., 2003). In this data set, there was no evidence to suggest that biters were younger than monkeys that wounded themselves, or that biters progressed to wounding, as assessed over a 4-year period (Novak, unpublished data).

   a. *Species:* Survey data obtained at various facilities have focused primarily on macaques and chimpanzees. Although only a small percentage of macaques developed SIB, estimates from 1995 to 2003 were quite consistent across facilities and species (15% in rhesus macaques, Bayne et al., 1995; 15% in pigtailed macaques, Bellanca and Crockett, 2002; and 11% in rhesus macaques, Lutz et al., 2003). Since 2003, the prevalence of the wounding form of SIB has declined at many facilities due, in part, to changes in colony management practices (Baker, 2016) and pharmacotherapy (Fontenot et al., 2009; Kempf et al., 2012).

The prevalence of SIB in chimpanzees was significantly lower than observed in macaques. In three studies, characterized as having large samples, the incidence of self-wounding ranged from 0.1% to 1.4% (Bloomsmith et al., 2019; Jacobson et al., 2016; Nash et al., 1999). However, in a much smaller zoo study of 40 chimpanzees, the prevalence rate of SIB was reported as 20% (Birkett and Newton-Fisher, 2011). The reason for the high prevalence rate in this study compared to others is not known.

   b. *Sex*: It remains unclear whether males and females differ in their predilection to engage in self-injurious behavior. In macaques, males may be more vulnerable to this disorder. A strong male bias in the prevalence of SIB was detected in two studies (Lutz et al., 2003; Gottlieb et al., 2013). In a third study, a strong male bias was noted, but only in nursery-reared macaques (Rommeck et al., 2009). However, Lutz et al. (2007) failed to detect a sex difference in SIB based on assessments of both nursery-reared and mother-reared monkeys.

The low prevalence of SIB in chimpanzees makes it difficult to detect sex differences even when the sample sizes are large. In the largest study, male chimpanzees were more vulnerable to this disorder (2 females vs. 8 males), but this effect was influenced by rearing history (Bloomsmith et al., 2019).

3. *Conclusions*: Three general conclusions can be drawn from these surveys. First, profiles of stereotypic behavior differ markedly in macaques and apes. However, there is concordance on the stereotypic profile across two macaque species. Second, vulnerability to the development of ARBs varies by sex, particularly in macaques, with males being more vulnerable. Third, macaques as a group have increased vulnerability to the more severe forms of abnormal behavior (SIB) than chimpanzees.

Other Species

In contrast to the wealth of prevalence data in nonhuman primates, information on the prevalence of abnormal behavior in other species is more limited. Instead, typical studies either focus on one specific type of abnormal behavior or all abnormal behaviors are lumped together into a single stereotypy score. The measuring units are typically frequency or duration over some standard time period. A few representative examples are provided below.

1. *Mice*: Stereotypic behavior has been studied both in deer mice and C58 mice housed in standard shoebox cages. Both species develop locomotor stereotypies consisting of jumping and backward somersaulting. In deer mice, locomotor stereotypies developed prior to weaning and increased steadily across the first 2 months postnatally (Muehlmann et al., 2015). By 63 days of age, the average frequency of stereotypic behavior in deer mice, as measured over an 8-h period, exceeded 12,000 counts, corresponding to 25 instances of locomotor stereotypy per minute (Bechard et al., 2017).

    Stereotypic behaviors in laboratory mice vary significantly by strain. C58 mice showed a developmental pattern similar to deer mice with an average frequency exceeding 12,000 in an 8-h period at 2 months of age. By contrast, C57BL/6 mice showed low rates of stereotypic behavior (fewer than 1,000 counts per 8-h period) that did not vary across development (Muehlmann et al., 2012).

2. *Sheep*: Stereotypic behaviors in sheep consist mostly of pacing, and chewing or nosing the pen. Both the prevalence and incidence of these behaviors (pacing, chewing, and nosing the pen) were assessed using video recordings of 94 sheep, all of which were housed individually in indoor pens (Lauber et al., 2012). A total of 61 sheep showed stereotypic behavior that accounted for at least 10% or more of their time budgets. Stereotypic behavior varied across the day. Pacing was more common in the morning (occupying 14% of the animals' time), whereas chewing the pen was more common in the afternoon (13% of their time).

3. *Elephants*: Stereotypic behavior (using a composite score of both repetitive and self-directed ARBs) was examined in 89 elephants from 39 North American Zoos (Greco et al., 2016). Elephants were videotaped to obtain estimates of stereotypic behavior, and data include both prevalence rates and time budgets. Stereotypic behavior, usually in the form of locomotor movements, occurred in 85.4% of the elephants. Additionally, daytime stereotypic rates ranged across individuals from 0.5% to 68% of their total activity budget.

What Causes Abnormal Behavior: Ruling out Medical Conditions as a First Step

Although abnormal behavior is often thought to be psychogenic in origin, it can be a secondary consequence of some disease or medical disorder (Tynes and Sinn, 2014). Recent evidence from studies of cats, dogs, and horses, as well as case reports in human and nonhuman primates, provides examples of gastrointestinal disease, painful conditions, or neurological abnormalities giving rise to abnormal behavior in some individuals. Thus, an essential first step in understanding the etiology of abnormal behavior and selecting the proper treatment depends on ruling out medical conditions as contributing factors.

Gastrointestinal Disease

Several lines of evidence suggest a link between abnormal behavior and the gastrointestinal system. In humans, individuals exhibiting deliberate self-harm had a higher incidence of gastrointestinal disorders (gastroduodenal ulcer and ulcerative colitis) than nonharming individuals with psychiatric symptoms, as derived from emergency room data (Lidón-Moyano et al., 2019). Additionally, predictors of self-harm included gastrointestinal complications after surgery (Morgan and Ho, 2017).

Relationships between gastrointestinal disorders and abnormal behavior have also been detected in animals. Horses with crib-biting behavior were noted to have a high prevalence of colic (Escalona et al., 2014). In a study comparing foals with and without crib-biting, crib-biting foals had a higher rate of ulcerated and inflamed stomachs (Nicol et al., 2002). In dogs, at least three conditions have been linked to gastrointestinal disorders, including excessive licking of surfaces in 14 of 19 dogs (Becuwe-Bonnet et al., 2012), fly-biting behavior in 6 of 7 dogs (Frank et al., 2012), and star-gazing behavior in a case study of one dog (Poirier-Guay et al., 2014). Treatment of the underlying gastric disorder for fly-biting and star-gazing behavior led to improvements in most of the dogs.

Arthritic and Other Painful Disorders

An association between painful disorders and the development of abnormal behavior has been noted in some animals (Frank, 2014). Both repetitive and self-injurious behaviors may be secondary to somatic or neuropathic pain. Sources of neuropathic pain in animals include changes in the spinal column produced by tumors or disk herniation. Pacing and circling activity coincided with spinal meningioma in some dogs (Fingeroth et al., 1987). Anecdotally, a 17-year-old rhesus monkey, with no previous history of SIB, showed osteoarthritic changes in the spine, manifested by lameness in one limb and mild biting directed to that limb. Treatment with meloxicam and gabapentin improved locomotion and reduced the biting behavior (Novak, unpublished observations).

ARBs have also been associated with degenerative joint disease and dental disease in elderly zoo-housed animals (Bacon, 2018). This is best shown by the example of two spectacled bears living in a low-quality environment and expressing high rates of stereotypic behavior (i.e., 83% of the time). Following relocation to an enriched zoo setting in England, the stereotypic behavior was eliminated in one bear, but not the other. When the second bear was examined to determine why environmental enrichment (EE) had not produced the expected beneficial effect, it was found to be suffering from severe dental problems. Specifically, this bear had 3 broken, infected canines, and associated myelitis that required 11 extractions. After successful dental treatment, his stereotypic behavior was reduced from 55% to 16% of the time (Maslak et al., 2013).

SIB in rhesus macaques can be a secondary consequence of viral infections that are normally latent, but may become active in cases of immunodeficiency. Monkeys with cytomegalovirus (CMV) are typically asymptomatic; however, CMV-induced peripheral neuropathy can develop in monkeys infected with simian immunodeficiency virus (SIV). CMV is also known to produce neuropathic pain in HIV-infected patients that is difficult to manage with analgesic treatment (Fuller, 1992; Robert et al., 1989). In a case report, an SIV-infected macaque developed a peripheral neuropathy that led to extensive facial lesions produced by excessive scratching. Analgesics were not effective in controlling this condition. Histological examination revealed CMV infection of facial nerves, meninges, and lymph nodes (Clemmons et al., 2015). These authors suggest that CMV-induced neuropathic pain may be one of several causes of SIB in SIV-infected monkeys and should be ruled out before psychogenic or stress-related conditions are considered.

Neurological and Retinal Disorders

Abnormal behavior can be comorbid with neurological disease. In humans, increased risk of self-harm has been associated with dementia (Mitchell, et al., 2017) and epilepsy (Gorton et al., 2018). Brain tumors in 97 dogs were significantly associated with circling and head tilting behavior (Bagley et al., 1999). In one case study of a pigtailed macaque, extensive lesions of various brain regions, typical of mitochondrial disease in humans, were detected in a monkey with unremitting SIB (Bielefeldt-Ohmann et al., 2004). Recent research has also demonstrated a connection between eye poking/covering behavior in some rhesus monkeys and achromatopsia, an inherited retinal disorder resulting from a genetic mutation (Moshiri et al., 2019).

Based on the above findings, it is reasonable to assume that disease or genetic anomalies may underlie the expression of abnormal behavior in some instances. The actual percentage of cases explained by medical conditions remains unknown at this time. Additionally, in the absence of being able to determine cause and effect, it is not clear, in some cases, whether the disease produced the abnormal behavior or whether the abnormal behavior was caused by some unidentified condition that subsequently made animals more vulnerable to specific medical disorders.

What Causes Abnormal Behavior: Retrospective and Correlational Approaches to Understand Underlying Mechanisms

The focus of this chapter is on the spontaneous occurrence of abnormal behavior in some animals, in contrast to inducing the abnormal behavior experimentally, using either genetically derived strains of mice or drug induction models (e.g., amphetamine). Thus, animals are identified only after they have acquired the condition. To understand the possible underlying mechanisms, two approaches can be used: (1) examining colony management and health records *retrospectively* to identify risk factors and (2) studying animals *concurrently* with particular forms of abnormal behavior to identify other variables that may be correlated with the behavior. **Risk factors** are those "characteristics" that specifically increase the probability of developing abnormal behavior. Some risk factors are fixed (e.g., sex, genetic background), whereas others may be causal (e.g., early rearing history) or variable (e.g., age of onset of a particular type of housing). However, risk factors, by definition, must be present prior to the development of abnormal behavior and must contribute significantly to its outcome (Offord and Kraemer, 2000). **Correlated factors** are those variables that are present in animals that have previously developed abnormal behavior (e.g., anxious behavior, sleep disruption). Correlation does not imply causation; thus, one cannot conclude that a specific correlated factor led to the development of abnormal behavior.

Fixed and Causal Risk Factors

1. *Species and Sex*: Based on the survey data described above for nonhuman primates, profiles of abnormal behavior vary considerably by species group and sex. Thus, no general primate pattern is evident. In the most frequently studied group, macaques, males appear to be more vulnerable to developing abnormal behavior than females. The factors contributing to this possible sex difference and its significance are unknown at this time.

   Stereotypic behavior in laboratory mice and rats varies markedly by strain. For example, C58 mice show much higher rates of stereotypic behavior than C57BL/6 mice (Bechard et al., 2017; Muehlmann et al., 2012). Sex differences in stereotypic behavior have also been reported; however, the results are varied. In several studies, males showed higher levels of stereotypic behavior than females (e.g., in FVB/NJ mice: Benvenga et al., 2011; in ILS mice: Downing et al., 2009; and in rats: Gogos et al., 2017). However, Lewis and coworkers (2019) detected a female bias in stereotypic behavior in C58 mice; additionally, no sex difference in stereotypic behavior was detected in deer mice (Bechard et al., 2016).

2. *Genetic Factors*: Nearly all of the limited research in this area has focused on self-injurious behavior in macaques, with an emphasis on several genetic polymorphisms that have been associated with anxiety and impulse disorders in humans. These include the serotonin transporter gene (SERT or 5-HTT) and the tryptophan hydroxylase-2 gene (TPH2). The 5-HTT polymorphism consists of a long and short allele. The short (s) allele, which is less active and therefore, leads to lower uptake of serotonin (Lesch et al., 1996) appears to be associated with increased risk for (1) self-harm in humans (Hankin et al., 2015) and (2) impulsivity in monkeys (Fairbanks et al., 2004). A population of monkeys with and without SIB was genotyped for the presence of polymorphic variants in SERT and TPH2. There was no

relationship between SERT genotype or CSF concentrations of 5-HIAA and SIB in this population of monkeys (Tiefenbacher et al., 2000, 2005a). In contrast, the distribution of rhTPH2 5′-FR haplotypes differed significantly between monkeys with and without SIB, and was differentially associated with hypothalamic-pituitary-adrenocortical (HPA) axis activity (Chen et al., 2010).

Despite these interesting findings, the results should be interpreted with caution for the following reasons. The data were collected on a single, relatively small population of macaques housed at one facility. Additionally, in the human literature, findings derived from single gene studies of nonsuicidal self-injury have been inconsistent (Groschwitz and Plener, 2012), which is not surprising, considering that SIB is most likely the result of multiple genes working in concert that may be strongly influenced by exposure to challenging environments (i.e., gene by environment interactions).

3. *Early Life Stressors*: Early life stress exposure is a risk factor for the development of abnormal behavior in many species. In macaques, the most significant of these early life stressors is nursery rearing. When assessed for stress reactivity, monkeys that were nursery-reared showed a dysregulation of the HPA axis (Capitanio et al., 2005; Feng et al., 2011). Moreover, nursery-reared monkeys were much more likely to develop abnormal behavior than their mother-reared counterparts. This was particularly evident for the development of SIB (Bellanca and Crockett, 2002; Gottlieb et al., 2013; Lutz et al., 2003, 2007; Rommeck et al., 2009) and depressive behavior (Zhang et al., 2016), but not for motor stereotypy (Bellanca and Crockett, 2002; Gottlieb et al., 2013; Vandeleest et al., 2011).

Early life stress plays a significant role in the development of abnormal behavior in other species as well. Early life stressors that increase stereotypic behavior include early weaning in cats (Ahola et al., 2017), hand-rearing in giraffes (Siciliano-Martina and Martina, 2018), and artificial rearing in rats (Aguirre-Benitez et al., 2017).

Variable Risk Factors (Stressors in Adolescents and Adults)

1. *Single Cage Housing and Stereotypic Behavior*: The risk factor with the highest generality across settings and species is living without social contact. In several nonhuman primate studies, length of exposure to individual cage housing was predictive of motor stereotypy (Bellanca and Crockett, 2002; Lutz et al., 2003; Vandeleest et al., 2011), a sum of all types of stereotypic behavior (Koyama et al., 2019; Seier et al., 2011), and SIB (Lutz et al., 2003; Rommeck et al., 2009). The findings on stereotypic behavior are supported by experimental studies.

Stereotypic behavior was reduced in female, but not male, rhesus monkeys by moving the animals from individual to pair-housing (Baker et al., 2014). Additionally, pair-housed monkeys showed less abnormal behavior in response to a stressful event than individually housed monkeys, when the two conditions were directly compared (Gilbert and Baker, 2011).

Lack of a social companion is commonly held to be stressful. However, assessment of the presumed stress of single versus social housing environments has yielded mixed results. For example, a cross-sectional comparison of plasma cortisol concentrations between singly and socially housed monkeys found no significant difference (Reinhardt et al., 1991). In within-subject studies, monkeys moved from social to individual cage housing showed an increase in urinary cortisol concentrations (Koyama et al., 2019), whereas mixed results were reported when singly housed monkeys were successfully introduced to a partner. Fecal cortisol concentrations decreased (Doyle et al., 2008), but plasma cortisol concentrations did not change (Baker et al., 2012). It is possible that differences in sampling matrices may have contributed to these varied outcomes. Plasma samples represent activity of the HPA axis at the time of sample collection and were collected from monkeys using restraint, sedation, and venipuncture, whereas both fecal and urine samples did not require sedation and represent a significantly longer time frame.

The reported effects of individual housing in other species are also deleterious. Sheep housed in social isolation showed higher levels of both locomotor stereotypy and plasma cortisol concentrations than sheep maintained in groups (Guesdon et al., 2015). Dogs that were moved from group housing to individual cage housing showed increased stereotypic behavior in the form of circling and digging, and also produced more vocalizations (Beerda et al., 1999).

2. *Single-Cage Housing and Affective Disorders*: Individual cage housing has been associated not only with stereotyped behavior but also with indices of depression. Some monkeys moved from large outdoor social groups to indoor, individual cage housing developed behavioral signs of depression (Hennessy et al., 2014). In a subsequent experimental study, adult males that were removed from outdoor cages and housed indoors for two 8-day periods developed a depressive posture that was markedly exacerbated in single-housed animals compared to those housed with a partner (Hennessy et al., 2016). Rodent studies have similarly found depressive-like effects when social contacts were removed. Social isolation after weaning markedly increased vulnerability to depressive and anxious behavior in both laboratory mice (Huang et al., 2017; Ieraci et al., 2016; Martin and Brown, 2010) and rats (Fischer et al., 2012; Nelson et al., 2019).

3. *Other Factors*: Investigations conducted at multiple primate facilities have identified three other risk factors for the development of abnormal behavior: (1) number of relocations to different colony rooms (Gottlieb et al., 2013; Lutz et al., 2003), (2) enclosure position within a colony room (Gottlieb et al., 2013; Seier et al., 2011), and (3) number of medical procedures or research projects monkeys experienced (Gottlieb et al., 2013; Lutz et al., 2003; Vandeleest et al., 2011). Relocation can create stress and exacerbate abnormal behavior. In monkeys with SIB, hair cortisol concentrations and self-biting rates increased in response to relocation. Although hair cortisol concentrations eventually returned to baseline levels, rates of self-biting remained elevated for nearly a year after being moved (Davenport et al., 2008). Position of an animal's enclosure within the colony room was also identified as a factor of interest. Monkeys were more likely to show motor stereotypy and/or SIB if housed either in lower racks of an upper/lower rack cage system (Gottlieb et al., 2013; Seier et al., 2011) or near the door (Gottlieb et al., 2013). Medical procedures can be stressful, and abnormal behavior may ensue as a result of accumulating stress exposure.

Correlated Factors

Abnormal behavior in both human and nonhuman primates seldom occurs in the absence of other problems. It is often part of a more generalized condition that may include altered stress reactivity, sleep disruption, and an anxious or impulsive temperament. Determining the presence of such comorbidities is important for designing effective treatments.

The identification of comorbid features is typically dependent on studying individuals who express a particular form of abnormal behavior (usually motor stereotypy or SIB) and comparing them to individuals that do not exhibit the abnormal behavior. However, in some studies, animals are not selected for their behavioral profile; instead, variation in abnormal behavior is correlated with the variation in some other selected feature, such as cortisol concentrations. The three features most commonly associated with abnormal behavior in nonhuman primates are as follows: (1) presence of stress resulting in dysregulation of the HPA axis; (2) sleep disruption and increased nighttime activity; and (3) behavioral manifestations of anxiety, impulsivity, or perseveration. We discuss the limitations and issues surrounding these features. We then provide several examples, both in primates and nonprimate mammals, demonstrating that abnormal behavior is often part of a more generalized syndrome.

1. *Comorbidities*
   a. *Stress and HPA Axis Dysregulation*: One of the most strongly held assumptions is that abnormal behavior is a marker for stress exposure, as measured by correlated changes in the activity of the HPA axis using cortisol concentrations (see Dettmer et al., 2011, for an example). However, no consistent relationship of HPA activity with the various forms of abnormal behavior has emerged, in part because many factors can influence cortisol concentrations, such as housing, seasonal changes, reproductive condition, and dominance relationships (Mason and Latham, 2004). Additionally, it is important to consider the sample matrix used for cortisol assessment, since cortisol concentrations can be measured in blood, saliva, urine, feces, and hair (Meyer and Novak, 2012). These matrices have different measurement units and represent different time frames (see Table 3.2). Thus, blood and saliva samples are most suitable for measuring the effects of acute stressors, such as handling or venipuncture, whereas hair samples are more appropriate for assessing the effects of chronic stressors, such as relocation or daily gavage. Moreover, cortisol concentrations from these matrices are not necessarily correlated with one another. For example, exposure to chronic, predictable stress may be manifested by elevated cortisol concentrations in hair, but blunted responses to an acute stressor, as measured in blood or saliva (Herman et al., 2016).
   b. *Sleep Disruption*: Sleep disturbance is comorbid with several human mental health problems, including depression (Zhai et al., 2015), borderline personality disorder (Winsper et al., 2017), and generalized anxiety disorder (Rossler et al., 2017). Thus, it is not surprising to consider sleep disruption as a possible correlate of abnormal behavior in animals. In macaques, implanted electrodes and telemetry are required to measure the various stages of sleep (Hsieh et al., 2008). However, noninvasive procedures are frequently used, even though they do not measure sleep directly. Instead, nighttime activity is measured with actigraphy collars (Berro et al., 2016) or with surveillance software and infrared cameras (Stanwicks et al., 2017), with the assumption that periods of inactivity represent sleep. Activity monitors require the use of collars to which not all animals adjust. Surveillance

TABLE 3.2

Cortisol Sampling Matrices

Sample Matrix	Time Window	Unit
Blood	5 min	Micrograms/deciliter or ng/mL
Saliva	5 min	Micrograms/deciliter or ng/mL
Urine	Several hours; must also measure creatinine to control for urine volume	Micrograms/milligram creatinine
Feces	1–2 days; metabolites	Nanograms/gram dry fecal weight
Hair	~3 months	Picograms/milligram hair weight

software sometimes produces ghost activity, but this can be eliminated with a review of the videotapes that are produced for each action. An advantage of surveillance technology is the creation of video clips that enable one to determine the ways in which animals actually behave when they are active at night.

c. *Anxious and Impulsive Behavior:* A number of tools are available to assess temperament traits, such as anxiety and impulsivity. In nonhuman primates these include, but are not limited to, the Human Intruder Test, the Novel Object Test, and the Judgment Bias Task (see Coleman and Pierre, 2014 for a review). However, these tests may be affected by the subject's previous history. As an example, monkeys with SIB housed at large primate facilities showed a significantly reduced reaction to the Human Intruder Test compared to noninjuring controls (Peterson et al., 2017). Was this a result of blunted affect to the stranger staring at them or habituation to the increased surveillance by familiar and unfamiliar care staff? Some support for the latter hypothesis comes from our studies of 21 monkeys, both with and without SIB, all of whom received extensive human contact from many different students and care staff across each day. When tested on the Human Intruder Test, both groups of monkeys showed markedly reduced reactions to the test compared to monkeys at other facilities that lacked the same amount of contact (Peterson, Master's Thesis, 2015).

In rodents, a variety of standard tests are used to assess depressive- or anxiety-like behaviors that usually have been induced by some kind of experimental manipulation, such as those mentioned earlier. Typical tests of depressive-like behaviors are the Forced Swim Test, Tail Suspension Test (mainly used in mice), and Sucrose Preference Test (Belovicova et al., 2017). Tests of anxiety-like behaviors include the elevated plus-maze, light-dark box, open field, novelty-induced feeding, and marble burying. The validity of some of these tests has been questioned, in part because of the possible alternative interpretations attributable to the response being studied (see, for example, Anyan and Amir, 2018; Planchez et al., 2019). Therefore, assessment of the affective state produced by a particular manipulation is best accomplished using a convergent approach, in which multiple behavioral tests are performed.

d. *Perseveration:* Perseveration can be defined in many different ways and can be studied using various learning paradigms. One definition is based on resistance to extinction, i.e., an inability to inhibit responding after a reinforcing stimulus is removed. A growing literature suggests that high rates of perseverative behavior measured using various extinction procedures are linked to abnormal behavior in both humans and animals. In humans, resistance to extinction is associated with many different conditions, including addiction (Sheynin et al., 2016) and posttraumatic stress disorder (Rauch et al., 2006). In animals, extinction deficits were related to the severity of bar-biting in bank voles (Garner and Mason, 2002), whole body and self-directed stereotypies in rhesus macaques (Pomerantz et al., 2012a), and the presence of stereotypy in dogs (Protopopova et al., 2014).

2. Examples of Syndromes
   a. *Self-Injurious Behavior in Macaques:* Studies comparing monkeys with self-injurious behavior to noninjuring controls clearly support the view that abnormal behavior is part of a larger syndrome (Novak et al., 2014). For example, monkeys with SIB showed a persistent dysregulation of the HPA axis involving chronically elevated hair cortisol concentrations (Davenport et al., 2008), but a blunted plasma cortisol response to the mild stress of brief restraint, sedation, and venipuncture (Tiefenbacher et al., 2000, 2004). The same group of SIB monkeys exhibited heightened aggressiveness compared to controls (Novak, 2003) and perseverative behavior, as manifested by higher rates of lever pressing than controls during the extinction phases of an alternating reinforcement/extinction paradigm (Lutz et al., 2004). Additionally, monkeys with SIB showed increased nighttime activity indicative of sleep disruption (Davenport et al., 2008). These latter findings were subsequently replicated in another population of monkeys with SIB. Extensive analysis of nighttime activity demonstrated delayed sleep onset, longer bouts of activity, and more total activity across the night in SIB monkeys compared to controls (Stanwicks et al., 2017).

   b. *Fur Chewing in Chinchillas:* A significant abnormal behavior in some chinchillas is fur chewing that results in fur removal and bare areas of skin (Ponzio et al., 2007). Like the example above, fur chewing is part of a larger syndrome, at least in females. Fur chewing females overall showed more anxious behavior than nonchewers in the elevated plus-maze, and the females that showed the most extreme form of chewing (i.e., removing all the hair within their reach) additionally exhibited elevated urinary cortisol metabolite concentrations (Ponzio et al., 2012).

   c. *Crib-biting in Horses:* A frequent behavior problem in horses is crib-biting, a behavior that generally occurs after feeding. Crib-biting has been associated with the consumption of highly palatable cereal chow that can be consumed quickly, in contrast to the longer time spent foraging in grassy pastures (Waters et al., 2002). One of the most consistent correlates of this behavior is the presence of colic (Escalona et al., 2014) or stomach ulcers (Nicol et al., 2002). Additional correlates include both altered stress reactivity and perseverative behavior. Exposure to an ACTH (adrenocorticotropic hormone) challenge test (which stimulates cortisol production) revealed elevated salivary cortisol concentrations in crib-biters compared to controls (Briefer Freymond et al., 2015). When tested on an extinction learning paradigm, crib-biters took longer to suppress their responses during extinction than nonbiting controls (Hemmings et al., 2007).

The Persistence of Abnormal Behavior

Introduction

In the past, abnormal behaviors were viewed as either purposeless (e.g., stereotypies) or harmful (e.g., self-injurious behavior). In the current view, abnormal behaviors are thought to serve some function and to endure because of their reinforcing value. Over the last three decades, a number of hypotheses have been proposed to explain this persistence. Nearly all of these emphasize the importance of adverse environmental conditions and how abnormal behavior can be used by the animals to mitigate the effects of those conditions.

The first two hypotheses focus on the intensity and quality of environmental stimuli and are of special interest because they are mutually exclusive. In the arousal/tension reduction hypothesis, abnormal behavior is thought to arise from an overstimulating, chaotic environment and is presumed to reduce arousal in that environment (Dellinger-Ness and Handler, 2006). In contrast, the impoverishment hypothesis suggests that abnormal behavior is a form of stimulation designed to decrease the boredom associated with an unchanging, monotonous environment (Mason and Latham, 2004). Each hypothesis leads to very different interventions – either decreasing or increasing the level of stimulation available in the environment. Complicating the picture is that individual animals likely differ in their perceptions of their environment, based on their temperament and previous history. It is possible that the same environment might be too stimulating for some animals and not stimulating enough for others, thereby impeding the development of general strategies to improve the environment.

The latter two hypotheses are also environmentally driven, but are not mutually exclusive from one another or from the hypotheses presented above. In the species-typical constraints hypothesis, abnormal behavior is a variant of species-typical behavior that is altered in its form by the housing environment (Poirier and Bateson, 2017). Finally, the communication hypothesis suggests that animals engage in abnormal behavior as a form of social interaction, either to deflect animal care staff demands or to attract their attention (Iwata et al., 1994). Evidence for each of these hypotheses is presented below.

Hypotheses

1. *Arousal Reduction*: Several lines of evidence suggest that engaging in abnormal behavior may serve to reduce arousal. Much of this work is focused on SIB as a coping response to negative affect (see Dellinger-Ness and Handler, 2006, for a review of this hypothesis on both humans and nonhuman primates). Based on self-report data, some humans engage in self-harm to reduce escalating feelings of anxiety. Thus, the act of cutting or burning appears to create a cycle of behavior maintained by negative reinforcement (Chapman et al., 2006; Klonsky, 2009). SIB in monkeys may also serve to reduce arousal, although the evidence in this case is more indirect. In macaques, heart rate was elevated just prior to a biting episode and declined after the biting episode (Novak, 2003).

   There is also some evidence to suggest that engaging in stereotypic behavior may serve to modulate the effects of a physiological or psychological stressor. In one study, crib-biting and control horses were exposed to an ACTH challenge. As expected, crib-biters had higher cortisol concentrations than controls. However, when examining the crib-biting group, crib-biters that engaged in biting during this period had significantly lower salivary cortisol concentrations than crib-biters that did not show biting behavior (Briefer Freymond et al., 2015). A similar relationship was observed in high-pacing monkeys exposed to the Human Intruder Test. Monkeys that continued to pace throughout the test showed significantly lower hair cortisol concentrations (an index of chronic HPA activity) than those that suppressed their pacing in the presence of the intruder (Novak et al., 2015).

2. *Impoverishment*: Do animals in poor quality environments engage in abnormal behavior as a form of "do-it-yourself enrichment" to overcome boredom, a phrase coined by Mason and Latham (2004)? If unstimulating environments are associated with boredom, one would predict that animals in such environments would show marked interest in novel stimuli. In testing this prediction, Meagher and Mason (2012) compared the responses of mink housed in nonenriched vs. enriched environments to a variety of aversive, ambiguous, and rewarding stimuli. Nonenriched mink showed shorter latencies and spent more time with *all* stimuli compared to their enriched counterparts. One would also predict that mink in enriched environments would show lower levels of abnormal behavior than those housed in nonenriched environments, a prediction that was confirmed for stereotypic behavior in minks housed in a highly enriched environment (Díez-León et al., 2013), but only for fur chewing behavior, when animals were housed in a modestly enriched environment (Meagher et al., 2014). The above findings have some parallel with a growing literature on the effects of enrichment on abnormal behavior in captive

nonhuman primates (see Coleman and Novak, 2018, for a review). The efficacy of various forms of enrichment are discussed further under the treatment section below.

3. *Species-Typical Constraints Hypothesis*: Some forms of abnormal behavior may represent species-typical behavior that is constrained by captive conditions. This hypothesis has been applied to pacing behavior in animals housed in captivity. In their review, Poirier and Bateson (2017) posed the question: does pacing reflect a "need to walk?" If so, one would predict that the amount of pacing reported in captive animals might be correlated with the daily travel distance and/or home range size of animals in nature. In 22 zoo-maintained carnivore species, pacing frequency in captivity was correlated both with home range size and median daily travel distances (Clubb and Mason, 2003, 2007). A similar relationship was detected for daily travel distances in a study of 24 species of zoo-housed nonhuman primates. Longer daily travel times in nature were correlated with the amount of pacing in captivity (Pomerantz et al., 2013). To ensure that this effect was related to a "need to walk" and not just to a more general relationship between abnormal behavior and the activity levels of each species, Pomerantz et al. (2013) showed that there was no relationship between natural ranging behavior and hair pulling.

4. *Communication Hypothesis*: This hypothesis involves the inadvertent establishment of reinforcement contingencies between the expression of abnormal behavior and the reactions of others (usually teachers or care staff). This inadvertent reinforcement of abnormal behavior was first noted in studies of humans with developmental delays who sometimes engaged in self-injurious behavior. A review of 145 episodes of SIB revealed that 38% were associated with escape from teacher task demands (social negative reinforcement) and 26% were associated with attention or rewards from teachers (social positive reinforcement) (Iwata et al., 1994). Recent studies suggest that inadvertent reinforcement by humans may also explain some cases of abnormal behavior in animals. In some dogs, owner-directed attention can reinforce circling behavior and licking of unusual surfaces (Hall et al., 2014). In a case study of a zoo-housed baboon that exhibited SIB, Dorey and colleagues (2009) used functional analysis to identify the reinforcing stimulus, which was attention from humans. Differential reinforcement of alternative behavior (DRA), in this case lip smacking, along with extinction of the reinforcement associated with the undesired behavior, yielded a marked decrease in SIB. A similar result was reported for a chimpanzee using DRA (ring contact) with extinction to reduce the incidence of feces throwing, spitting, and screaming (Martin et al., 2011).

5. *Conclusions*: Each of the four major hypotheses regarding the function of abnormal behavior is supported by some empirical evidence, and none of the hypotheses can be considered disproven at this time. The most reasonable conclusion from the current findings is that abnormal behaviors can serve multiple functions, including possibly more than one function for a given animal. Variability in temperament, previous history of social and nonsocial living conditions, history of experimental treatments in the case of laboratory-housed animals, and the nature of the current environment likely interact in complex ways to influence what particular behaviors are elicited and what purpose(s) is(are) served by those behaviors.

Therapeutic Interventions

One of the greatest challenges confronting veterinarians and colony managers involves the development of effective treatments for animals whose abnormal behavior compromises their health and wellbeing. A number of approaches have been considered, ranging from modifications of the environment to pharmacotherapy, but none has proven to be completely effective. Additionally, the efficacy of these approaches varies as a function of the type of abnormal behavior requiring treatment (e.g., stereotypy vs. SIB) and the species being studied.

Environmental Approaches

It is often assumed that the development of abnormal behavior is related to rearing and/or housing animals in stimulus-poor environments, and there is substantial support for this view as outlined above. Thus, it is not surprising that modifications of the environment are considered the first line of defense in the treatment of abnormal behavior. These modifications include socialization, EE, and training.

1. *Social Housing*: By far, the most important mitigator of abnormal behavior in social living species is social housing. Isolation housing of laboratory rodents can lead to deleterious behavioral outcomes, including self-injury through overgrooming (Khoo et al., 2020), anxious and depressive behavior in both mice (Huang et al., 2017; Ieraci et al., 2016; Liu et al., 2019; Martin and Brown, 2010) and rats (Fischer et al., 2012; Nelson et al., 2019), and elevated plasma corticosterone levels in mice (Carnevali et al., 2020). These adverse outcomes are largely avoided with social housing.

   On the other hand, social housing of rats and mice can substantially increase the risk of aggression, particularly in males (Kappel et al., 2017). Such aggression can escalate to injury and death of some individuals and compromise scientific outcomes (Weber et al., 2017). Some victims may not be wounded but may undergo a process of repeated social defeat, which in turn has been associated with the development of a depressive-like state. These findings underscore the need to develop new housing strategies for adult rodents that reconsider their species-appropriate needs (Weber et al., 2017).

Some of the same issues are present in nonhuman primates, such as macaques. The benefits of social housing are well established; on the other hand, serious aggression, leading to significant morbidity and mortality, can occur in large groups of outdoor corral-housed macaques (Stavisky et al., 2018). In macaques housed indoors, considerable emphasis has been placed on social housing (pair or small group). Although some risks remain, careful consideration of temperament, rearing history, introduction strategies, and follow-up monitoring appears to mitigate the risk of aggression, yielding about an 80% success rate (Truelove et al., 2017). Additionally, the presence of certain behaviors early in the pairing process can be used to predict pairing success (MacAllister et al., 2020).

Pair housing is a potent means by which to reduce abnormal behavior in nonhuman primates. Social contact with other conspecifics led to reductions in stereotypy in baboons (Kessel and Brent, 2001), chimpanzees (Kalcher-Sommersguter et al., 2013), cynomolgus macaques (Koyama et al., 2019), and rhesus macaques (Baker et al., 2014; Doyle et al., 2008; Fontenot et al., 2006; Schapiro et al., 1996). Additionally, in a comparison of full vs. protected social contact, abnormal behavior was lower for rhesus macaques housed in the full contact condition (Baker et al., 2012). Self-injurious behavior in macaques was similarly reduced by group housing (Fontenot et al., 2006; Weed et al., 2003). However, abnormal behavior was not entirely eliminated in these studies.

2. EE: Enriching the environment can take many different forms, including but not limited to (1) alterations in the housing environment (e.g., nest boxes, running wheels, perches); (2) addition of items to be manipulated (e.g., toys and nest materials); and (3) provision of foraging devices designed to increase the effort to obtain food (e.g., puzzle feeders). In laboratory animals, EE has been studied most extensively in rodents and nonhuman primates.

In rodents, a majority of studies have focused on beneficial effects of EE in animal models of neurodegenerative diseases and psychiatric disorders (see review by Laviola et al., 2008). EE improved cognitive performance (He et al., 2017) and altered the cellular physiology of hippocampal neurons (Ohline and Abraham, 2019). Additionally, enrichment paradigms can reverse the effects of various prenatal and postnatal assaults on the health and well-being of laboratory rodents. For example, EE was shown to reverse hypoxia-induced cognitive deficits (Salmaso et al., 2012), protect against opioid addiction (Hofford et al., 2017), and produce marked resistance to MPTP (1-methyl-4-phenyl-1,2,3,6-tetrahydropyradine), a compound used to induce a Parkinson's disease-like syndrome in mice (Campêlo et al., 2017; Hilario et al., 2016).

Considerably less is known about the effects of enrichment on spontaneously occurring abnormal behavior in mice and rats, with the exception of the research conducted by Mark Lewis and colleagues on the deer mouse (*Peromyscus maniculatus*). As noted previously, deer mice show very high rates of stereotypic behavior, manifested by jumping and backward somersaulting (Muehlmann et al., 2015). Exposure to enriched housing environments, consisting of large, three-tiered dog kennels with multiple objects, tunnels, a hut and a running wheel, along with bird seed scattered to encourage foraging behavior, yielded substantial benefits. Enriched kennel housing produced a decrease in locomotor stereotypy in adults (Bechard et al., 2016) and also inhibited the development of locomotor stereotypy in young deer mice (Bechard et al., 2017). Furthermore, a transgenerational effect was noted, in that the benefits of kennel housing of dams were passed onto their nonenriched offspring (Bechard and Lewis, 2016). Nonenriched offspring born to kennel-housed mothers showed lower levels of stereotypic behavior than offspring born to mothers housed in standard cages. This effect was present both in first-generation and second-generation offspring (Bechard and Lewis, 2016). These beneficial effects were associated with increased activation of the indirect basal ganglia pathway, but only in males (Bechard et al., 2017).

The EE literature in captive nonhuman primates has generally not focused on symptom reversal in animal models of neuropsychiatric disease. Instead, the emphasis has been on promoting psychological well-being of captive primates by providing a physical environment that enhances species-typical behavior as legally mandated in the US by the Animal Welfare Act (1991). In some of these nonhuman primate studies, efficacy of enrichment is determined by assessing changes in abnormal behavior. However, the findings are often inconsistent, in part, because of individual, as well as, species differences in how the animals react to enrichment devices (Izzo et al., 2011; also noted in caged mink, Dallaire et al., 2011). Additional sources of outcome variability include the types of abnormal behaviors that are targeted and the ways in which enrichment is implemented.

A review of all possible forms of enrichment and their effects on abnormal behavior in nonhuman primates and other nonrodent species is beyond the scope of this chapter. However, because foraging opportunities typically promote species-typical behavior and appear to alter time budgets in many species, including macaques (Bennett et al., 2016), marmosets (Regaiolli et al., 2020), and horses (Goodwin et al., 2002), it is relevant to ask whether these opportunities also reduce abnormal behavior. We include nonprimate species in this summary. Increased foraging opportunities were associated with decreases in stereotypic behavior in brown capuchin monkeys (Boinski et al., 1999), vervet monkeys (Seier et al., 2011), chimpanzees (Baker 1997), cheetahs (Quirke and Riordan, 2011), and captive walruses (Fernandez and Timberlake, 2019), but not in squirrel monkeys (Fekete et al., 2000; Spring et al., 1997), sooty mangabeys (Crast et al., 2016), or captive coyotes (Shivik et al., 2009). In rhesus macaques, stereotypic behavior was reduced by puzzle feeders, but only during the period of usage (Novak et al., 1998), and was not affected by exposure a foraging substrate (Byrne and Suomi 1991). Similar divergence of findings was noted for SIB, with enrichment yielding lower wounding rates in cynomolgus macaques (Turner and Grantham, 2002), but not lower self-directed biting rates in rhesus monkeys provided with puzzle feeders (Novak et al., 1998). These findings reveal that the effects of foraging enrichment on abnormal behavior are highly variable. Despite the mixed findings in reducing abnormal behavior, foraging devices continue to be a very effective tool in promoting species-typical behavior.

3. *Positive Reinforcement Training (PRT)*: A primary objective of PRT is to train animals to cooperate with routine colony management practices and commonly employed veterinary/research procedures to reduce the need for sedation, restraint, and/or negative reinforcement. The benefits of training are in reducing fearful behavior in response to procedures as noted in rhesus monkeys (Clay et al., 2009) and chimpanzees (Lambeth et al., 2006), and in minimizing physiological stress responses as noted in grizzly bears (Joyce-Zuniga et al., 2016), and in bonobos and orangutans (Behringer et al., 2014). Additionally, PRT facilitates effective communication between care/research staff and the animals they work with. However, training can require an extensive investment of time and effort, and may not work for all animals (Coleman et al., 2005; Coleman, 2012).

Nonhuman primates have been successfully trained to enter boxes (Fernström et al., 2009, rhesus monkeys), shift between cages (Veeder et al., 2009, sooty mangabeys), sit at individual stations when housed in social groups (Kemp et al., 2017, rhesus macaques), and come indoors from an outdoor enclosure (Schapiro et al., 2003, chimpanzees). PRT has been extended successfully to research procedures including training rhesus monkeys for blood sampling (Coleman et al., 2008; Graham et al., 2012) and collecting saliva samples (Broche et al., 2019). With respect to veterinary care and monitoring, chimpanzees have been trained to cooperate with urine collection (Bloomsmith et al., 2015), blood glucose monitoring (Reamer et al., 2014), acupuncture therapy (Magden et al., 2013), and physical therapy (Neal Webb et al., 2020).

Given that training reduces the stressfulness of various procedures, it is relevant to ask whether the effects of PRT can extend beyond training sessions to alter time budgets and reduce abnormal behavior. There are observations and case reports suggesting that PRT can mitigate abnormal behavior in zoo-housed animals. As examples, husbandry training led to reductions in pacing behavior in three captive African wild dogs (Shyne and Block, 2010), self-slapping in a female orangutan (Raper et al., 2002), and ear covering and aggressiveness in one captive gorilla (Leeds et al., 2016).

PRT has had mixed success in reducing abnormal behavior in laboratory-housed nonhuman primates (Bloomsmith et al., 2007). The limitations of PRT in nonhuman primates are exemplified in the following three published papers. In the first study, group-housed chimpanzees at the Tel Aviv-Ramat Gan Zoological Center in Israel were trained to present various parts of their body for inspection by care staff. Using a cumulative stereotypy score, training was associated with a significant reduction in abnormal behavior, but this reduction was more pronounced for lower ranking individuals (Pomerantz and Terkel, 2009). In rhesus monkeys, training for sitting and presenting body areas was efficacious in reducing stereotypic behavior but only for those individuals that showed very high rates of stereotypy (Baker et al., 2009). In a second study of rhesus monkeys trained to touch a target and accept venipuncture, training reduced abnormal behavior in the first month, but not during the later months of training (Coleman and Maier, 2010). These findings suggest that training alone may not be the most effective way to reduce abnormal behavior.

Pharmacotherapy

Pharmacotherapy is considered a treatment of last resort for nonhuman primates and is primarily used to treat self-injurious behavior and depression. Although pharmacotherapy has efficacy in reducing abnormal behavior, it is also associated with risks. Depending on the drug, these risks include sedation effects, development of tolerance, individual differences in drug response, withdrawal symptoms on dose reduction, relapse once the drug has been removed, and increased risk of developing another disorder. In all cases, animals on these drugs must be continuously monitored and dosages adjusted when necessary. Most often, pharmacotherapy is used to treat the more serious forms of abnormal behavior that cannot be mitigated by other approaches (e.g., SIB). Over the last decade, drugs that alter the activity of the serotonergic, opioidergic, GABAergic, or adrenergic neurotransmitter systems have been employed in this effort.

1. *Serotonergic System:*
    d. *Abnormal Behavior in Nonhuman Primates:* Selective serotonin reuptake inhibitors (SSRIs) have shown some efficacy in controlling episodes of SIB in rhesus monkeys. Beneficial effects were observed for both the SSRI, fluoxetine and the 5-HT-1A receptor agonist, buspirone, but only in the early weeks of treatment (Fontenot et al., 2005). In a second study, fluoxetine was more effective in ameliorating SIB than venlafaxine (a combined SSRI and norepinephrine reuptake

inhibitor), but these monkeys were not tested for relapse (Fontenot et al., 2009). Serotonergic activity can also be manipulated by adding the precursor L-tryptophan to the diet. In rhesus monkeys, SIB decreased during L-tryptophan administration, but the monkeys relapsed once the dietary supplementation ended (Weld et al., 1998). L-tryptophan treatment of bushbabies was associated with wound healing, but again, relapse was not studied (Watson et al., 2009). Fluoxetine has also been used to treat stereotypic behavior in nonhuman primates. Fluoxetine reduced somersaulting, weaving, head tossing, and saluting in vervet monkeys (Hugo et al., 2003).

Some monkeys show depressive behavior, even when maintained in social groups (Shively et al., 2008; Qin et al., 2015; Zhang et al., 2016). The SSRI, sertraline, was examined as a possible therapy for this condition. However, treatment with sertraline neither reduced depressive behavior (Willard et al., 2015) nor improved locomotor activity (Justice et al., 2017). Instead, administration of sertraline increased the progression of coronary artery disease by 60% in treated compared to untreated depressed, female monkeys (Shively et al., 2015; Silverstein-Metzler et al., 2017). Thus, sertraline should be considered carefully prior to utilization as a treatment for depression in macaques.

e. *Other Species:* SSRIs have proven effective in reducing abnormal behaviors in several nonprimate species. For example, locomotor stereotypies were significantly reduced in deer mice using either fluoxetine (Korff et al., 2008) or another SSRI, escitalopram (Wolmarans et al., 2013). In a placebo-controlled, randomized, double-blind study, fluoxetine administration was effective in reducing acral lick dermatitis in dogs (Wynchank and Berk, 1998). This finding was extended in a second randomized, placebo-controlled, clinical trial, demonstrating reduced compulsive behaviors in dogs, including spinning, tail chasing, and flank sucking (Irimajiri et al., 2009). Importantly, medical causes were ruled out in advance of treatment.

2. *Opioidergic System:*
   a. *Abnormal Behavior in Nonhuman Primates.* Opioid compounds have shown efficacy in reducing SIB in rhesus monkeys. Administration of the long-acting opioid receptor antagonist, naltrexone, reduced SIB during treatment. More importantly, naltrexone had long-lasting benefits that were still present 110–200 days after the treatment ended (Kempf et al., 2012). In subsequent investigations, naltrexone treatment also reversed atrophy of white matter astrocytes and decreased the immune activation that had been associated with SIB (Lee et al., 2013, 2015).
   b. *Other Species:* Naltrexone was effective in reducing acral lick dermatitis in 7 of 11 dogs. However, unlike the findings for rhesus monkeys, skin lesions reoccurred once the treatment ended (White, 1990). In a case report, naltrexone reduced tail chasing in one dog, but also caused significant pruritis, resulting in termination of the treatment (Schwartz, 1993). The efficacy of naltrexone to reduce weaving behavior in one horse was compared with the SSRI paroxetine and the dopamine receptor antagonist acepromazine in an ABACAD design, with A representing a 1-month washout between periods of drug exposure. Paroxetine nearly eliminated weaving, reducing it by 95%. Both acepromazine and naltrexone showed some limited effects on weaving behavior in this horse, decreasing it by 57% and 30%, respectively (Nurnberg et al., 1997).

3. *GABAergic System:*
   a. *Abnormal Behavior in Nonhuman Primates:* In the past, the indirect GABA receptor agonist diazepam was a common drug of choice to treat SIB in monkeys, based on its anxiolytic properties and its "calming effect" (Calcaterra et al., 2014). However, when diazepam was used to treat eight monkeys with SIB, only four showed a reduction in SIB and the remaining monkeys actually got worse (Tiefenbacher et al., 2005b). Thus, there may be risk in using diazepam as a treatment for SIB. Considering that drug's long half-life, a better approach may be to use shorter acting GABA agonists, such as lorazepam or midazolam.

4. *Adrenergic System:*
   a. *Abnormal Behavior in Nonhuman Primates:* Guanfacine, an α2A-receptor agonist, has been used to treat SIB in macaques and baboons. Self-biting behavior in two rhesus monkeys and one baboon was eliminated by administration of guanfacine; however, all animals relapsed following the cessation of treatment (Macy et al., 2005). A wound scoring scale was used in a second study to evaluate the efficacy of guanfacine on SIB in rhesus macaques (Freeman et al., 2015). Treatment reduced the severity of inflicted wounds but did not completely eliminate wounding events.

Conclusions

Abnormal behavior can be a vexing problem in animal research colonies and zoo facilities. It can take many different forms, ranging from seemingly idiosyncratic, purposeless stereotypies to clinical conditions that can include depression and self-injurious behavior. In some cases, medical disorders, such as arthritis, gastritis, and neurological abnormalities, may underlie the expression of abnormal behavior. In other cases, however, there is strong evidence to suggest that stress exposure plays a significant role. Prominent risk factors for the development of abnormal behavior include nursery rearing and individual cage housing. But only some animals are vulnerable to these factors, suggesting the existence of yet unidentified genetic and epigenetic influences that may interact with environmental stressors to induce behavioral abnormalities.

Existing treatments for abnormal behavior have yielded variable results. Lack of success can be attributed to several different factors. First, some forms of abnormal behavior are syndromes with associated features. For example, SIB in macaques usually includes a dysregulation of the HPA axis and sleep disruption. SIB treatments may have to take into account these additional features. Second, it is increasingly clear that abnormal behaviors serve some function and are maintained through reinforcement contingencies associated with tension reduction, increased stimulation, or altered behavior by animal care staff. Each of these examples would lead to entirely different treatment strategies. Yet, we are not at the point where we can readily differentiate the relevant reinforcement contingencies that lead to abnormal behavior in individual animals.

Given these issues, it is not surprising that very few treatments fully eliminate abnormal behavior, especially in nonhuman primates. Strategies include enhancing the physical environment, housing animals together, training animals for potentially stressful veterinary and research procedures, and as a last resort, using pharmacotherapy. Although reductions in abnormal behavior can occur with any of these strategies, none of them result in the complete elimination of abnormal behavior. For example, social housing has led to substantial reductions in SIB and stereotypic behavior in macaques, but the behaviors continue to persist at a lower frequency. Furthermore, social housing does not appear to impact depressive behavior, inasmuch as most cases of depression have been identified in macaques housed in social groups. Therefore, a combined treatment approach, using all of the strategies described above, has the potential to produce better outcomes. And this combined approach is precisely what has been happening for a number of years now. Research facilities have developed vigorous behavioral management programs that include EE, social housing, and training (Schapiro, 2017). When necessary, pharmacotherapy may be used as a last resort. It is thought that the prevalence of abnormal behavior (e.g., SIB) is on the decline because of these programs; however, it should be noted that confirmation by statistical data analysis remains lacking at this time.

Researchers, veterinarians, and behavioral scientists have focused most of their efforts on eradicating abnormal behavior in nonhuman primates. Yet, there may be a downside to these efforts related to the potential use of nonhuman primates to model mental health disorders in humans. The most common form of abnormal behavior, locomotor stereotypy, does not appear to have any human analog at the present time. However, spontaneously occurring syndromes of SIB and depressive behavior in macaques show considerable concordance with their human analogs and are associated with a relatively extensive literature on mechanisms and treatment. In a recent survey by SAMSA, (Substance Abuse and Mental Health Services Administration published in 2019, approximately 1 in 5 adults in the US had some form of mental illness in 2018, with 1 in 25 adults in the US experiencing more serious forms of mental illness. The need for good animal models is crucial for developing effective treatments, and nonhuman primates may help fill this gap. Accordingly, we should continue to work to prevent the occurrence of abnormal behaviors, like SIB and depression, when the well-being of the animals is of paramount concern. At the same time, however, studying the etiology of these syndromes and searching for effective medications could be exploited to help improve the health, not only of the animals themselves (in cases where the syndrome has already developed), but of humans with the analogous neuropsychiatric disorders.

Acknowledgments

We thank Steve Schapiro and Kris Coleman for their insights and comments on our chapter. We are indebted to the many graduate and undergraduate students who contributed in significant ways to our research program. We are thankful for all the rhesus monkeys whose behavior challenged our thinking on animal well-being and led to many improvements in animal care. Support was provided from a grant to the University of Massachusetts and Harvard Medical School from the National Institutes of Health (R24OD11180 to MAN).

REFERENCES

Aguirre-Benítez E.L., Porras M.G., Parra L., González-Ríos J., Garduño-Torres D.F., Albores-García D., Avendaño A., Ávila-Rodríguez M.A., Melo A.I., Jiménez-Estrada I., Mendoza-Garrido M.E., Toriz C., Diaz D., Ibarra-Coronado E., Mendoza-Ángeles K., Hernández-Falcón J. Disruption of behavior and brain metabolism in artificially reared rats. *Dev Neurobiol*. 2017;77(12):1413–1429.

Ahola M.K., Vapalahti K., Lohi H. Early weaning increases aggression and stereotypic behaviour in cats. *Sci Rep*. 2017;7(1):10412. Published 2017 Sep 4. doi:10.1038/s41598-017-11173-5.

Anyan J., Amir S. Too depressed to swim or too afraid to stop? A reinterpretation of the forced swim test as a measure of anxiety-like behavior. *Neuropsychopharmacology*. 2018;43(5):931–933.

Aureli F., Whiten A. Emotions and behavioral flexibility. In D. Maestripieri (Ed.), *Primate Psychology* (pp. 289–323). 2003; Cambridge, MA: Harvard University Press.

Bacon H. Behaviour-Based Husbandry-A holistic approach to the management of abnormal repetitive behaviors. *Animals (Basel)*. 2018;8(7):103. Published 2018 Jun 27. doi:10.3390/ani8070103.

Bagley R.S., Gavin P.R., Moore M.P., Silver G.M., Harrington M.L., Connors R.L. Clinical signs associated with brain tumors in dogs: 97 cases (1992–1997). *J Am Vet Med Assoc*. 1999;9215:818–819.

Baker K.C. Straw and forage material ameliorate abnormal behaviors in adult chimpanzees. *Zoo Biol.* 1997;16:225–236.

Baker K.C. Survey of 2014 behavioral management programs for laboratory primates in the United States. *Am J Primatol.* 2016;78(7):780–796.

Baker K.C., Bloomsmith M., Neu K., Griffis C., Maloney M., Oettinger B., Schoof V.A., Martinez M. Positive reinforcement training moderates only high levels of abnormal behavior in singly housed rhesus macaques. *J Appl Anim Welf Sci.* 2009;12:236–252.

Baker K.C., Bloomsmith M.A., Oettinger B., Neu K., Griffis C., Schoof V.A. Comparing options for pair housing rhesus macaques using behavioral welfare measures. *Am J Primatol.* 2014;76:30–42.

Baker K.C., Bloomsmith M.A., Oettinger B., Neu K., Griffis C., Schoof V., Maloney M. Benefits of pair housing are consistent across a diverse population of rhesus macaques. *Appl Anim Behav Sci.* 2012;137:148–156.

Barnes T.R., Rossor M., Trauer T. A comparison of purposeless movements in psychiatric patients treated with antipsychotic drugs, and normal individuals. *J Neurol Neurosurg Psychiatry* 1983;46:540–546.

Bayne K., Haines M., Dexter S., Woodman D., Evans C. Nonhuman primate wounding prevalence: a retrospective analysis. *Lab Anim.* 1995;24:40–44.

Bechard A.R., Bliznyuk N., Lewis M.H. The development of repetitive motor behaviors in deer mice: Effects of environmental enrichment, repeated testing, and differential mediation by indirect basal ganglia pathway activation. *Dev Psychobiol.* 2017;59(3):390–399.

Bechard A.R., Cacodcar N., King M.A., Lewis M.H. How does environmental enrichment reduce repetitive motor behaviors? Neuronal activation and dendritic morphology in the indirect basal ganglia pathway of a mouse model. *Behav Brain Res.* 2016;299:122–131.

Bechard A.R., Lewis M.H. Transgenerational effects of environmental enrichment on repetitive motor behavior development. *Behav Brain Res.* 2016;307:145–149.

Becuwe-Bonnet V., Belanger M., Frank D. Gastrointestinal disorders in dogs with excessive licking of surfaces. *J Vet Behav-Clin Appl Res.* 2012;7:194–204.

Beerda B., Schilder M.B., van Hooff J.A., de Vries H.W., Mol J.A. Chronic stress in dogs subjected to social and spatial restriction. I. Behavioral responses. *Physiol Behav.* 1999;66(2):233-242.

Behringer V., Stevens J.M., Hohmann G., Möstl E., Selzer D., Deschner T. Testing the effect of medical positive reinforcement training on salivary cortisol levels in bonobos and orangutans. *PLoS One.* 2014;9(9):e108664. Published 2014 Sep 24. doi:10.1371/journal.pone.0108664.

Bellanca R.U., Crockett C.M. Factors predicting increased incidence of abnormal behavior in male pigtailed macaques. *Am J Primatol.* 2002;58:57–69.

Belovicova K., Bogi E., Csatlosova K., Dubovicky M. Animal tests for anxiety-like and depression-like behavior in rats. *Interdiscip Toxicol.* 2017;10(1):40–43.

Bennett A.J., Perkins C.M., Tenpas P.D., Reinebach A.L., Pierre P.J. Moving evidence into practice: Cost analysis and assessment of macaques' sustained behavioral engagement with videogames and foraging devices. *Am J Primatol.* 2016;78(12):1250–1264.

Benvenga S., Itri E., Hauser P., DeTolla L., Yu S.-F., Testa G., Pappalardo M.A., Trimarchi F., Amato A. Gender differences in locomotor and stereotypic behavior associated with l-carnitine treatment in mice. *Gend Med.* 2011;8(1):1–13.

Berro L.F., Andersen M.L., Tufik S., Howell L.L. Actigraphy-based sleep parameters during the reinstatement of methamphetamine self-administration in rhesus monkeys. *Exp Clin Psychopharmacol.* 2016;24:142–146.

Bielefeldt-Ohmann H., Bellanca R.U., Crockett C.M., Curnow E., Eiffert K., Gillen M., Glanister D., Hayes E., Kelley S., Minoshima S., Vogel K. Subacute necrotizing encephalopathy in a pig-tailed macaque (*Macaca nemestrina*) that resembles mitochondrial encephalopathy in humans. *Comp Med.* 2004;54:422–433.

Birkett L.P., Newton-Fisher N.E. How abnormal is the behaviour of captive, zoo-living chimpanzees? *PLoS One.* 2011;6(6):e20101. doi:10.1371/journal.pone.0020101.

Bloomsmith M.A., Clay A.W., Lambeth S.P., Lutz C.K., Breaux S.D., Lammey M.L., Franklin A.N., Neu K.A., Perlman J.E., Reamer L.A., Mareno M.C., Schapiro S.J., Vazquez M., Bourgeois S.R. Survey of Behavioral Indices of Welfare in Research Chimpanzees (*Pan troglodytes*) in t.he United States. *J Am Assoc Lab Anim Sci.* 2019;58:160–177.

Bloomsmith M.A., Jackson Marr M., Maple T.L. Addressing nonhuman primate behavioral problems through the application of operant conditioning: Is the human treatment approach a useful model? *Appl Anim Behav Sci.* 2007;102:205–222.

Bloomsmith M., Neu K., Franklin A., Griffis C., McMillan J. Positive reinforcement methods to train chimpanzees to cooperate with urine collection. *J Am Assoc Lab Anim Sci.* 2015;54:66–69.

Boinski S., Swing S.P., Gross T.S., Davis J.K. Environmental enrichment of brown capuchins (*Cebus apella*): behavioral and plasma and fecal cortisol measures of effectiveness. *Am J Primatol.* 1999;48:49–68.

Brand C.M., Marchant L.F. Prevalence and characteristics of hair plucking in captive bonobos (Pan paniscus) in North American zoos. *Am J Primatol.* 2018;80(4):e22751. doi:10.1002/ajp.22751.

Briefer Freymond S., Bardou D., Briefer E.F., Bruckmaier R., Fouché N., Fleury J., Maigrot A.L., Ramseyer A., Zuberbühler K., Bachmann I. The physiological consequences of crib-biting in horses in response to an ACTH challenge test. *Physiol Behav.* 2015;151:121–128.

Broche Jr. N., Takeshita R.S.C., Mouri K., Bercovitch F.B., Huffman M.A. Salivary alpha-amylase enzyme is a non-invasive biomarker of acute stress in Japanese macaques (*Macaca fuscata*). *Primates.* 2019;60(6):547–558.

Byrne G.D., Suomi S.J. Effects of woodchips and buried food on behavior patterns and psychological well-being of captive rhesus monkeys. *Am J Primatol.* 1991;23:141–151.

Calcaterra, N.E., Barrow, J.C. Classics in chemical neuroscience: Diazepam (valium). *ACS Chem Neurosci.* 2014;5:253–260.

Callard M.D., Bursten S.N., Price E.O. Repetitive backflipping behavior in captive roof rats (*Rattus rattus*) and the effects of cage enrichment. *Anim Welf.* 2000;9:139–152.

Campêlo C.L.C., Santos J.R., Silva A.F., Dierschnabel A.L., Pontes A., Cavalcante J.S., Ribeiro A.M., de Andrade T.G., de Oliveira Godeiro Jr. C., Silva R.H. Exposure to an enriched environment facilitates motor recovery and prevents short-term memory impairment and reduction of striatal BDNF in a progressive pharmacological model of parkinsonism in mice. *Behav Brain Res.* 2017;328:138–148.

Capitanio J.P., Mendoza S.P., Mason W.A., Maninger N. Rearing environment and hypothalamic-pituitary-adrenal regulation in young rhesus monkeys (*Macaca mulatta*). *Dev Psychobiol.* 2005;46:318–330.

Carnevali L., Statello R., Vacondio F., Ferlenghi F., Spadoni G., Rivara S., Mor M., Sgoifo A. Antidepressant-like effects of pharmacological inhibition of FAAH activity in socially isolated female rats. *Eur Neuropsychopharmacol.* 2020;32:77–87.

Castles D.L., Whiten A., Aureli F. Social anxiety, relationships and self-directed behaviour among wild female olive baboons. *Anim Behav.* 1999;58:1207–1215.

Chapman A.L., Gratz K.L., Brown M.Z. Solving the puzzle of deliberate self-harm: The experiential avoidance model. *Behav Res Ther.* 2006;44:371–394.

Chen G.-L., Novak M.A., Yang H., Vallender E.J., Miller G.M. Functional polymorphisms in the rhesus monkey *TPH2* 5'- and 3'- regulatory regions: Evidence for differential association with hypothalamic-pituitary-adrenal axis function and self-injurious behavior. *Genes Brain Behav.* 2010;9:335–347.

Chiba S., Numakawa T., Ninomiya M., Richards M.C., Wakabayashi C., Kunugi H. Chronic restraint stress causes anxiety- and depression-like behaviors, downregulates glucocorticoid receptor expression, and attenuates glutamate release induced by brain-derived neurotrophic factor in the prefrontal cortex. *Prog Neuropsychopharmacol Biol Psychiatry.* 2012;39:112–119.

Clay A.W., Bloomsmith M.A., Marr M.J., Maple T.L. Habituation and desensitization as methods for reducing fearful behavior in singly housed rhesus macaques. *Am J Primatol.* 2009;71:30–39.

Clemmons E.A., Gumber S., Strobert E., Bloomsmith M.A., Jean S.M. Self-Injurious behavior secondary to Cytomegalovirus-induced neuropathy in an SIV-infected rhesus macaque (*Macaca mulatta*). *Comp Med.* 2015;65:266–270.

Cless I.T., Lukas K.E. Variables affecting the manifestation of and intensity of pacing behavior: A preliminary case study in zoo-housed polar bears. *Zoo Biol.* 2017;36:307–315.

Clubb R., Mason G. Captivity effects on wide-ranging carnivores. *Nature* 2003;425:473–474.

Clubb R., Mason G.J. Natural behavioural biology as a risk factor in carnivore welfare: How analyzing species differences could help zoos improve enclosures. *Appl Anim Behav Sci.* 2007;102:303–328.

Coleman K. Individual differences in temperament and behavioral management practices for nonhuman primates. *Appl Anim Behav Sci.* 2012;137(3–4):106–113.

Coleman K., Maier A. The use of positive reinforcement training to reduce stereotypic behavior in rhesus macaques. *Appl Anim Behav Sci.* 2010;124:142–148.

Coleman K., Novak M.A. Environmental enrichment in the 21st Century. *ILAR J.* 2018;58:295–307.

Coleman K., Pierre P.J. Assessing anxiety in nonhuman primates. *ILAR J.* 2014;55:333–346.

Coleman K., Pranger L., Maier A., Lambeth S.P., Perlman J.E., Thiele E., Schapiro S.J. Training rhesus macaques for venipuncture using positive reinforcement techniques: A comparison with chimpanzees. *J Am Assoc Lab Anim Sci.* 2008;47:37–41.

Coleman K., Tully L.A., McMillan J.L. Temperament correlates with training success in adult rhesus macaques. *Am J Primatol.* 2005;65:63–71.

Cooper J.J., McDonald L., Mills D.S. The effect of increasing visual horizons on stereotypic weaving: implications for the social housing of stabled horses. *Appl Anim Behav Sci.* 2000;69:67–83.

Cooper J.J., McCall N., Johnson S., Davidson H.P.B. The short-term effects of increasing meal frequency on stereotypic behavior of stabled horses. *Appl Anim Behav Sci.* 2005;90:352–364.

Crast J., Bloomsmith M.A., Jonesteller T.J. Behavioral effects of an enhanced enrichment program for group-housed sooty mangabeys (Cercocebus atys). *J Am Assoc Lab Anim Sci.* 2016;55(6):756–764.

Crast J., Bloomsmith M.A., Perlman J.E., Meeker T.L., Remillard C.M. Abnormal behavior in captive sooty mangabeys. *Anim Welf.* 2014;23:167–177.

Cussen V.A., Mench J.A. The relationship between personality dimensions and resiliency to environmental stress in orange-winged amazon parrots (Amazona amazonica), as indicated by the development of abnormal behaviors. *PLoS One.* 2015;10(6):e0126170. Published 2015 Jun 26. doi:10.1371/journal.pone.0126170.

Dallaire J.A., Meagher R.K., Díez-león M., Garner J.P., Mason G.J. Recurrent perseveration correlates with abnormal repetitive locomotion in adult mink but is not reduced by environmental enrichment. *Behav Brain Res.* 2011;224:213–222.

Davenport M.D., Lutz C.K., Tiefenbacher S., Novak M.A., Meyer J.S. A rhesus monkey model of self-injury: Effects of relocation stress on behavior and neuroendocrine function. *Biol Psychiatry.* 2008;68:990–996.

Dellinger-Ness L.A., Handler L. Self-injurious behavior in human and nonhuman primates. *Clin Psychol Rev.* 2006;26:503–514.

Dettmer A.M., Novak M.A., Suomi S.J., Meyer J.S. Physiological and behavioral adaptation to relocation stress in differentially reared rhesus monkeys: Hair cortisol as a biomarker for anxiety related responses. *Psychoneuroendocrinol.* 2011; 37:191–199.

Díez-León M., Bowman J., Bursian S., Filion H., Galicia D., Kanefsky J., Napolitano A., Palme R., Schulte-Hostedde A., Scribner K., Mason G. Environmentally enriched male mink gain more copulations than stereotypic, barren-reared competitors. *PLoS One.* 2013;8(11):e80494. Published 2013 Nov 25. doi:10.1371/journal.pone.0080494.

Díez-León M., Kitchenham L., Duprey R., Bailey C.D.C., Choleris E., Lewis M., Mason G. Neurophysiological correlates of stereotypic behaviour in a model carnivore species. *Behav Brain Res.* 2019;373:112056. doi:10.1016/j.bbr.2019.112056.

Dorey N.R., Rosales-Ruiz J., Smith R., Lovelace B. Functional analysis and treatment of self-injury in a captive olive baboon. *J Appl Behav Anal.* 2009;42:785–794.

Downing C., Balderrama-Durbin C., Hayes J., Johnson T.E., Gilliam D. No effect of prenatal alcohol exposure on activity in three inbred strains of mice. *Alcohol Alcohol.* 2009;44(1):25–33.

Doyle L.A., Baker K.C., Cox L.D. Physiological and behavioral effects of social introduction on adult male rhesus macaques. *Am J Primatol.* 2008;70:542–550.

Escalona E.E., Okell C.N., Archer D.C. Prevalence of and risk factors for colic in horses that display crib-biting behaviour. *BMC Vet Res.* 2014;10(Suppl 1):S3. doi:10.1186/1746-6148-10-S1-S3.

Fairbanks L.A., Newman T.K., Bailey J.N., Jorgensen M.J., Breidenthal S.E., Ophoff R.A., Comuzzie A.G., Martin L.J., Rogers J. Genetic contributions to social impulsivity and aggressiveness in vervet monkeys. *Biol Psychiatry* 2004;55:642–647.

Fekete J.M., Norcross J.L., Newman J.D. Artificial turf foraging boards as environmental enrichment for pair-housed female squirrel monkeys. *Contemp Top Lab Anim Sci.* 2000;39(2):22–26.

Feng X., Wang L., Yang S., Qin D., Wang J., Li C., Lv L., Ma Y., Hu X. Maternal separation produces lasting changes in cortisol and behavior in rhesus monkeys. *Proc Natl Acad Sci USA.* 2011;108:14312–14317.

Fernandez E.J., Timberlake W. Foraging devices as enrichment in captive walruses (Odobenus rosmarus). *Behav Proc.* 2019;168:103943. doi:10.1016/j.beproc.2019.103943.

Fernström A.L., Fredlund H., Spångberg M., Westlund K. Positive reinforcement training in rhesus macaques-training progress as a result of training frequency. *Am J Primatol.* 2009;71(5):373–379.

Fingeroth J.M., Prata R.G., Patnaik A.K. Spinal meningiomas in dogs: 13 cases (1972–1987). *J Am Vet Med Assoc.* 1987;191:720–726.

Fischer C.W., Liebenberg N., Elfving B., Lund S., Wegener G. Isolation-induced behavioural changes in a genetic animal model of depression. *Behav Brain Res.* 2012;230(1):85–91.

Fontenot M.B., Musso M.W., McFatter R.M., Anderson G.M. Dose-finding study of fluoxetine and venlafaxine for the treatment of self-injurious and stereotypic behavior in rhesus macaques (*Macaca mulatta*). *J Am Assoc Lab Anim Sci.* 2009;48:176–184.

Fontenot M.B., Padgett 3rd E.E., Dupuy A.M., Lynch C.R., De Petrillo P.B., Higley J.D. The effects of fluoxetine and buspirone on self-injurious and stereotypic behavior in adult male rhesus macaques. *Comp Med.* 2005;55:67–74.

Fontenot M.B., Wilkes M.N., Lynch C.S. Effects of outdoor housing on self-injurious and stereotypic behavior in adult male rhesus macaques (Macaca mulatta). *J Am Assoc Lab Anim Sci.* 2006;45:35–43.

Fossum O., Jansson D.S., Etterlin P.E., Vågsholm I. Causes of mortality in laying hens in different housing systems in 2001 to 2004. *Acta Vet Scand.* 2009;51:3.

Frank D. Recognizing behavioral signs of pain and disease: A guide for practitioners. *Vet Clin North Am Small Anim Pract.* 2014;44:507–524.

Frank D., Bélanger M.C., Bécuwe-Bonnet V., Parent J. Prospective medical evaluation of 7 dogs presented with fly biting. *Can Vet J.* 2012;53:1279–1284.

Freeman Z.T., Rice K.A., Soto P.L., Pate K.A., Weed M.R., Ator N.A., DeLeon I.G., Wong D.F., Zhou Y., Mankowski J.L., Zink M.C., Adams R.J., Hutchinson E.K. Neurocognitive dysfunction and pharmacological intervention using guanfacine in a rhesus macaque model of self-injurious behavior. *Transl Psychiatry.* 2015;19(5):e567. doi:10.1038/tp.2015.61.

Fuller G.N. Cytomegalovirus and the peripheral nervous system in AIDS. *J Acquir Immune Defic Syndr.* 1992;5(Suppl 1):S33–S36.

Garner J.P., Mason G.J. Evidence for a relationship between cage stereotypies and behavior disinhibition in laboratory rodents. *Behav Brain Res.* 2002;136:83–92.

Garner J.P., Meehan C.L., Famula T.R., Mench J.A. Genetic, environmental, and neighbor effects on the severity of stereotypies and feather picking in Orange-winged Amazon parrots (*Amazona amazonica*): An epidemiological study. *Appli Anim Behav Sci.* 2006;96:153–168.

Garner J.P., Weisker S.M., Dufour B., Mench J.A. Barbering (fur and whisker trimming) by laboratory mice as a model of human trichotillomania and obsessive-compulsive spectrum disorders. *Comp Med.* 2004;54:216–224.

Gilbert M.H., Baker K.C. Social buffering in adult male rhesus macaques (Macaca mulatta): Effects of stressful events in single vs. pair housing. *J Med Primatol.* 2011;40(2):71–78.

Gogos A., Kusljic S., Thwaites S.J., van den Buuse M. Sex differences in psychotomimetic-induced behaviours in rats. *Behav Brain Res.* 2017;322(Pt A):157–166.

Golden S.A., Covington 3rd H.E., Berton O., Russo S.J. A standardized protocol for repeated social defeat stress in mice [published correction appears in Nat Protoc. 2015 Apr;10(4):643]. *Nat Protoc.* 2011;6(8):1183–1191.

Goodwin D., Davidson H.P., Harris P. Foraging enrichment for stabled horses: Effects on behaviour and selection. *Equine Vet J.* 2002;34(7):686–691.

Gorton H.C., Webb R.T., Pickrell W.O., Carr M.J., Ashcroft D.M. Risk factors for self-harm in people with epilepsy. *J Neurol.* 2018;265:3009–3016.

Gottlieb D.H., Capitanio J.P., McCowan B. Risk factors for stereotypic behavior and self-biting in rhesus macaques (*Macaca mulatta*): Animal's history, current environment, and personality. *Am J Primatol.* 2013;75:995–1008.

Gottlieb D.H., Maier A., Coleman K. Evaluation of environmental and intrinsic factors that contribute to stereotypic behavior in captive rhesus macaques (*Macaca mulatta*). *Appl Anim Behav Sci.* 2015;171:184–191.

Graham M.L., Rieke E.F., Mutch L.A., Zolondek E.K., Faig A.W., Dufour T.A., Munson J.W., Kittredge J.A., Schuurman H.J. Successful implementation of cooperative handling eliminates the need for restraint in a complex non-human primate disease model. *J Med Primatol.* 2012;41:89–106.

Greco B.J., Meehan C.L., Hogan J.N., Leighty K.A., Mellen J., Mason G.J., Mench J.A. The days and nights of zoo elephants: Using epidemiology to better understand stereotypic behavior of african elephants (*Loxodonta africana*) and asian elephants (*Elephas maximus*) in North American Zoos. *PLoS One.* 2016;11(7):e0144276. Published 2016 Jul 14. doi:10.1371/journal.pone.0144276.

Groschwitz R.C., Plener P.L. The neurobiology of non-suicidal self injury (NSSI): A Review. *Suicidol Online.* 2012;3:24–32.

Guesdon V., Meurisse M., Chesneau D., Picard S., Lévy F., Chaillou E. Behavioral and endocrine evaluation of the stressfulness of single-pen housing compared to group-housing and social isolation conditions. *Physiol Behav.* 2015;147:63–70.

Hall N.J., Protopopova A., Wynne C.D.L. The role of environmental and owner-provided consequences in canine stereotypy and compulsive behavior. *J Vet Behav.* 2014;10:24–35.

Hankin B.L., Barrocas A.L., Young J.F., Haberstick B., Smolen A. 5-HTTLPR ×interpersonal stress interaction and nonsuicidal self-injury in general community sample of youth. *Psychiatry Res.* 2015;225:609–612.

He C., Tsipis C.P., LaManna J.C., Xu K. Environmental enrichment induces increased cerebral capillary density and improved cognitive function in mice. *Adv Exp Med Biol.* 2017;977:175–181.

Hemmings A., McBride S., Hale C.E. Perseverative responding and the aetiology of equine oral stereotypy. *Appl Anim Behav Sci.* 2007;104:143–150.

Hennessy M.B., Chun K., Capitanio J.P. Depressive-like behavior, its sensitization, social buffering, and altered cytokine responses in rhesus macaques moved from outdoor social groups to indoor housing. *Soc Neurosci.* 2016;12:65–75.

Hennessy M.B., McCowan B., Jiang J., Capitanio J.P. Depressive-like behavioral response of adult male rhesus monkeys during routine animal husbandry procedure. *Front Behav Neurosci.* 2014;309:1–8.

Herman J.P., McKlveen J.M., Ghosal S., Kopp B., Wulsin A., Makinson R., Scheimann J., Myers B. Regulation of the hypothalamic-pituitary-adrenocortical stress response. *Compr Physiol.* 2016;6:603–621.

Hilario W.F., Herlinger A.L., Areal L.B., de Moraes L.S., Ferreira T.A., Andrade T.E., Martins-Silva C., Pires R.G. Cholinergic and dopaminergic alterations in nigrostriatal neurons are involved in environmental enrichment motor protection in a mouse model of Parkinson's Disease. *J Mol Neurosci.* 2016;60(4):453–464.

Hofford R.S., Chow J.J., Beckmann J.S., Bardo M.T. Effects of environmental enrichment on self-administration of the short-acting opioid remifentanil in male rats. *Psychopharmacol.* 2017;234:3499–3506.

Hook M.A., Lambeth S.P., Perlman J.E., Stavisky R., Bloomsmith M.A., Schapiro S.J. Inter-group variation in abnormal behavior in chimpanzees (*Pan Troglodytes*) and rhesus macaques (*Macaca mulatta*). *Appl Anim Behav Sci.* 2002;76:165–176.

Hsieh K.C., Robinson E.L., Fuller C.A. Sleep architecture in unrestrained rhesus monkeys (*Macaca mulatta*) synchronized to 24-hour light-dark cycles. *Sleep.* 2008;9:1239–1250.

Huang Q., Zhou Y., Liu L.Y. Effect of post-weaning isolation on anxiety- and depressive-like behaviors of C57BL/6J mice. *Exp Brain Res.* 2017;235(9):2893–2899.

Hugo C., Seier J., Mdhluli C., Daniels W., Harvey B.H., Du Toit D., Wolfe-Coote S., Nel D., Stein D.J. Fluoxetine decreases stereotypic behavior in primates. *Prog Neuropsychopharmacol Biol Psychiatry.* 2003;27(4):639–643.

Ieraci A., Mallei A., Popoli M. Social isolation stress induces anxious-depressive-like behavior and alterations of neuroplasticity-related genes in adult male mice. *Neural Plast.* 2016 doi:10.1155/2016/6212983.

Irimajiri M., Luescher A.U., Douglass G., Robertson-Plouch C., Zimmermann A., Hozak R. Randomized, controlled clinical trial of the efficacy of fluoxetine for treatment of compulsive disorders in dogs. *J Am Vet Med Assoc.* 2009;235(6):705–709.

Iwata B.A., Pace G.M., Dorsey M.F., Zarcone J.R., Vollmer T.R., Smith R.G., Rodgers T.A., Lerman D.C., Shore B.A., Mazalesk J.L., Han-Leon G., Cowdery G.E., Kalsher M.J., McCosh K.C., Willis K.D. The functions of self-injurious behavior: An experimental-epidemiological analysis. *J Appl Behav Anal.* 1994;27:215–240.

Izzo G.N., Bashaw M.J., Campbell J.B. Enrichment and individual differences affect welfare indicators in squirrel monkeys (*Saimiri sciureus*). *J Comp Psychol.* 2011;125:347–352.

Jacobson S.L., Ross S.R., Bloomsmith M.A. Characterizing abnormal behavior in a large population of zoo-housed chimpanzees: prevalence and potential influencing factors. *PeerJ.* 2016;4:e2225. doi:10.7717/peerj.2225. eCollection 2016.

Joyce-Zuniga N.M., Newberry R.C., Robbins C.T., Ware J.V., Jansen H.T., Nelson O.L. Positive reinforcement training for blood collection in grizzly bears (*Ursus arctos horribilis*) results in undetectable elevations in serum cortisol levels: A preliminary investigation. *J Appl Anim Welf Sci.* 2016;19(2):210–215.

Judge P.G., Evans D.W., Schroepfer K.K., Gross A.C. Perseveration on a reversal-learning task correlates with rates of self-directed behavior in nonhuman primates. *Behav Brain Res.* 2011;222:57–65.

Justice J.N., Silverstein-Metzler M.G., Uberseder B., Appt S.E., Clarkson T.B., Register T.C., Kritchevsky S.B., Shively C.A. Relationships of depressive behavior and sertraline treatment with walking speed and activity in older female nonhuman primates. *Gerosci.* 2017;39:585–600.

Kalcher-Sommersguter E., Franz-Schaider C., Preuschoft S., Crailsheim K. 2013. Long-term evaluation of abnormal behavior in adult ex-laboratory chimpanzees (*Pan troglodytes*) following re-socialization. *Behav. Sci.* 2013;(3):99–119.

Kappel S., Hawkins P., Mendl M.T. To group or not to group? Good practice for housing male laboratory mice. *Animals (Basel).* 2017;7(12):88–113.

Kemp C., Thatcher H., Farningham D., Witham C., MacLarnon A., Holmes A., Semple S., Bethell E.J. A protocol for training group-housed rhesus macaques (*Macaca mulatta*) to cooperate with husbandry and research procedures using positive reinforcement. *Appl Anim Behav Sci.* 2017;197:90–100.

Kempf D.J., Baker K.C., Gilbert M.H., Blanchard J.L., Dean R.L., Deaver D.R., Bohm Jr. R.P. Effects of extended-release injectable naltrexone on self-injurious behavior in rhesus macaques (*Macaca mulatta*). *Comp Med.* 2012;62:209–217.

Kessel A., Brent L. The rehabilitation of captive baboons. *J Med Primatol.* 2001;30(2):71–80.

Khoo S.Y., Correia V., Uhrig A. Nesting material enrichment reduces severity of overgrooming-related self-injury in individually housed rats [published online ahead of print, 2020 Jan 10]. *Lab Anim.* 2020.doi:10.1177/0023677219894356.

Kitchen A.M., Martin A.A. The effects of cage size and complexity on the behaviour of captive common marmosets, *Callithrix jacchus jacchus*. *Lab Anim.* 1996;30:317–326.

Klonsky E.D. The functions of self-injury in young adults who cut themselves: Clarifying the evidence for affect-regulation. *Psychiatry Res.* 2009;166:260–268.

Korff S., Stein D.J., Harvey B.H. Stereotypic behaviour in the deer mouse: pharmacological validation and relevance for obsessive compulsive disorder. *Prog Neuro-Psychopharmacol Biol Psychiatry*, 2008;32:348–355.

Koyama H., Tachibana Y., Takaura K., Takemoto S., Morii K., Wada S., Kaneko H., Kimura M., Toyoda A. Effects of housing conditions on behaviors and biochemical parameters in juvenile cynomolgus monkeys (*Macaca fascicularis*). *Exp Anim.* 2019;68:195–211.

Lambeth S.P., Hau J., Perlman J.E., Martino M., Schapiro S.J. Positive reinforcement training affects hematologic and serum chemistry values in captive chimpanzees (*Pan troglodytes*). *Am J Primatol.* 2006;68:245–256.

Lauber M., Nash J.A., Gatt A., Hemsworth P.H. Prevalence and Incidence of abnormal behaviours in individually housed sheep. *Animals (Basel).* 2012;2:27–37.

Laudenslager M.L., Natvig C., Corcoran C.A., Blevins M.W., Pierre P.J., Bennett A.J. The influences of perinatal challenge persist into the adolescent period in socially housed bonnet macaques (*Macaca radiata*). *Dev Psychobiol.* 2013;55:316–322.

Laviola G., Hannan A.J., Macrì S., Solinas M., Jaber M. Effects of enriched environment on animal models of neurodegenerative diseases and psychiatric disorders. *Neurobiol Dis.* 2008;2:159–68.

Lee K.M., Chiu K.B., Didier P.J., Baker K.C., MacLean A.G. Naltrexone treatment reverses astrocyte atrophy and immune dysfunction in self-harming macaques. *Brain Behav Immun.* 2015;50:288–297.

Lee K.M., Chiu K.B., Sansing H.A., Inglis F.M., Baker K.C., MacLean A.G. Astrocyte atrophy and immune dysfunction in self-harming macaques. *PLoS One.* 2013;8(7):e69980. doi:10.1371/journal.pone.0069980.

Leeds A., Elsner R., Lukas K.E. The effect of positive reinforcement training on an adult female western lowland gorilla's (*Gorilla gorilla gorilla*) rate of abnormal and aggressive behavior. *Anim Behav Cog.* 2016;3:78–87.

Lesch K.P., Bengel D., Heils A., Sabol S.Z., Greenberg B.D., Petri S., Benjamin J., Müller C.R., Hamer D.H., Murphy D.L.. Association of anxiety-related traits with a polymorphism in the serotonin transporter gene regulatory region. *Science.* 1996;274:1527–1531.

Lewis M.H., Rajpal H., Muehlmann A.M. Reduction of repetitive behavior by co-administration of adenosine receptor agonists in C58 mice. *Pharmacol Biochem Behav.* 2019;181:110–116.

Lidón-Moyano C., Wiebe D., Gruenewald P., Cerdá M., Brown P., Goldman-Mellor S. Associations between self-harm and chronic disease among adolescents: Cohort study using statewide emergency department data. *J Adolesc.* 2019;72:132–140.

Lidster K., Owen K., Browne W.J., Prescott M.J. Cage aggression in group-housed laboratory male mice: an international data crowdsourcing project. *Sci Rep.* 2019;9(1):15211. Published 2019 Oct 23. doi:10.1038/s41598-019-51674-z.

Liu N., Wang Y., An A.Y., Banker C., Qian Y.H., O'Donnell J.M. Single housing-induced effects on cognitive impairment and depression-like behavior in male and female mice involve neuroplasticity-related signaling [published online ahead of print, 2019 Aug 31]. *Eur J Neurosci.* 2019. doi:10.1111/ejn.14565.

Lutz C.K. Stereotypic behavior in nonhuman primates as a model for the human condition. *ILAR J.* 2014;55:284–296..

Lutz C.K. A cross-species comparison of abnormal behavior in three species of singly-housed old world monkeys. *Appl Anim Behav Sci.* 2018;199:52–58.

Lutz C.K., Davis E.B., Ruggiero A.M., Suomi S.J. Early predictors of self-biting in socially-housed rhesus macaques (*Macaca mulatta*). *Am J Primatol.* 2007;69:584–590.

Lutz C., Tiefenbacher S., Novak M.A., Meyer J.S. Extinction deficits in male rhesus macaques with a history of self-injurious behavior. *Am J Primatol.* 2004;63:41–48.

Lutz C., Well A., Novak M.A. Stereotypic and self-injurious behavior in rhesus macaques: A survey and retrospective analysis of environment and early experience. *Am J Primatol.* 2003;60:1–15.

MacAllister R.P., Heagerty A., Coleman K. Behavioral predictors of pairing success in rhesus macaques (*Macaca mulatta*). *Am J Primatol.* 2020;82(1):e23081. doi:10.1002/ajp.23081.

Macy Jr. J.D., Beattie T.A., Morgenstern S.E., Arnsten A.F. Use of guanfacine to control self-injurious behavior in two rhesus macaques (*Macaca mulatta*) and one baboon (*Papio anubis*). *Comp Med.* 2005;50:419–425.

Maestripieri D., Schino G., Aurelli F., Troisi A. A modest proposal – displacement activities as an indicator of emotions in primates. *Anim Behav.* 1992;44:967–979.

Magden E.R., Haller R.L., Thiele E.J., Buchl S.J., Lambeth S.P., Schapiro S.J. Acupuncture as an adjunct therapy for osteoarthritis in chimpanzees (*Pan troglodytes*). *J Am Assoc Lab Anim Sci.* 2013;52:475–480.

Marchewka J., Watanabe T.T., Ferrante V., Estevez I. Review of the social and environmental factors affecting the behavior and welfare of turkeys (*Meleagris gallopavo*). *Poult Sci.* 2013;92:1467–1473.

Martin A.L., Bloomsmith M.A., Kelley M.E., Marr M.J., Maple T.L. Functional analysis and treatment of human-directed undesirable behavior exhibited by a captive chimpanzee. *J Appl Behav Anal.* 2011;44:139–143.

Martin A.L., Brown R.E. The lonely mouse: Verification of a separation-induced model of depression in female mice. *Behav Brain Res.* 2010;207(1):196–207.

Marx J.O., Brice A.K., Boston R.C., Smith A.L. Incidence rates of spontaneous disease in laboratory mice used at a large biomedical research institution. *J Am Assoc Lab Anim Sci.* 2013;52(6):782–791.

Maslak R., Sergiel A., Hill S.P. Some aspects of locomotor stereotypies in spectacled bears (*Tremarctos ornatus*) and changes in behavior after relocation and dental treatment. *J Vet Behav.* 2013;8:335–341.

Mason G.J. Age and context affect the stereotypies of caged mink. *Behav.* 1993;127:191–229.

Mason G.J., Latham N.R. Can't stop, won't stop: Is stereotypy a reliable animal welfare indicator? *Anim Welf.* 2004;13:S57–S69.

McBride S.D., Hemmings A. Altered mesoaccumbens and nigrostriatal dopamine physiology is associated with stereotypy development in a non-rodent species. *Behav Brain Res.* 2005;159:113–118.

Meagher R.K., Ahloy Dallaire J., Campbell D.L.M., Ross M., Møller S.H., Hansen S.W., Díez-León M., Palme R., Mason G.J. Benefits of a ball and chain: Simple environmental enrichments improve welfare and reproductive success in farmed American mink (*Neovison vison*). *PLoS One* 2014;9(11):e110589. doi:10.1371/journal.pone.0110589.

Meagher R.K., Mason G.J. Environmental enrichment reduces signs of boredom in caged mink. *PLoS One* 2012;7(11):e49180. doi:10.1371/journal.pone.0049180.

Meyer J.S., Novak M.A. Mini-review: Hair cortisol: A novel biomarker of hypothalamic-pituitary-adrenocortical activity. *Endocrinol.* 2012;153:4120–4127.

Mitchell R., Draper B., Harvey L., Brodaty H., Close J. The survival and characteristics of older people with and without dementia who are hospitalized following intentional self-harm. *Int J Geriatr Psychiatry.* 2017;32:892–900.

Morgan D.J., Ho K.M. 2017. Incidence and risk factors for deliberate self-harm, mental Illness, and suicide following bariatric surgery: A state-wide population-based linked-data cohort study. *Ann Surg.* 2017;265:244–252.

Moshiri A., Chen R., Kim S., Harris R.A., Li Y., Raveendran M., Davis S., Liang Q., Pomerantz O., Wang J., Garzel L., Cameron A., Yiu G., Stout J.T., Huang Y., Murphy C.J., Roberts J., Gopalakrishna K.N., Boyd K., Artemyev N.O., Rogers J., Thomasy S.M. A nonhuman primate model of inherited retinal disease. *J Clin Invest.* 2019;129:863–874.

Muehlmann A.M., Bliznyuk N., Duerr I., Lewis M.H. Repetitive motor behavior: Further characterization of development and temporal dynamics. *Dev Psychobiol.* 2015;57:201–211.

Muehlmann A.M., Edington G., Mihalik A.C., Buchwald Z., Koppuzha D., Korah M., Lewis M.H. Further characterization of repetitive behavior in C58 mice: Developmental trajectory and effects of environmental enrichment. *Behav Brain Res.* 2012;235:143–149.

Nash L.T., Fritz J., Alford P.A., Brent L. Variables influencing the origins of diverse abnormal behaviors in a large sample of captive chimpanzees (*Pan troglodytes*). *Am J Primatol.* 1999;48:15–29.

Neal Webb S.J., Bridges J.P., Thiele E., Lambeth S.P., Schapiro S.J. The implementation and initial evaluation of a physical therapy program for captive chimpanzees (*Pan troglodytes*). *Am J Primatol.* 2020;82(3):e23109. doi:10.1002/ajp.23109.

Nelson S.T., Hsiao L., Turgeon S.M. Sex and housing conditions modify the effects of adolescent caffeine exposure on anxiety-like and depressive-like behavior in the rat. *Behav Pharmacol.* 2019;30(7):539–546.

Newman T.K., Syagailo Y., Barr C.S., Wendland J.R., Champoux M., Graessle M., Suomi S.J., Higley J.D., Lesch K.P. MAOA gene promoter variation and rearing experience influences aggressive behavior in rhesus monkeys. *Biol Psychiatry.* 2005;57:167–172.

Nicol C.J., Davidson H.P., Harris P.A., Waters A.J., Wilson A.D. Study of crib-biting and gastric inflammation and ulceration in young horses. *Vet Rec.* 2002;151:658–662.

Novak M.A. Self-injurious behavior in rhesus monkeys: New insights on etiology, physiology, and treatment. *Am J Primatol.* 2003;59:3–19.

Novak M.A., Kinsey J.H., Jorgensen M.J., Hazen T.J. The effects of puzzle feeders on pathological behavior in individually housed rhesus monkeys. *Am J Primatol.* 1998;46:213–227.

Novak M.A., El-Mallah S.N., Menard M.T. Use of the cross-translational model to study self-injurious behavior in human and non-human primates. *ILAR J.* 2014;55:274–283.

Novak M.A., Peterson E.J., Rosenberg K., Varner E.K., Worlein J.M., Lee G.H., Bellanca R.U., and Meyer J.S. Extreme behavioral phenotypes, hypothalamic pituitary adrenal (HPA) axis activity and anxiety in rhesus macaques (*Macaca mulatta*). *Am J Primatol.* 2015;77:49.

Nurnberg H.G., Keith S.J., Paxton D.M. Consideration of the relevance of ethological animal models for human repetitive behavioral spectrum disorders. *Biol Psychiatry.* 1997;41(2):226–229.

Offord D.R., Kraemer H.C. Risk factors and prevention. *Evidence-Based Ment Health.* 2000;3:70–71.

Ohline S.M., Abraham W.C. Environmental enrichment effects on synaptic and cellular physiology of hippocampal neurons. *Neuropharmacol.* 2019;145(Pt A):3–12.

Peterson E.J., Worlein J.M., Lee G.H., Varner E.K., Dettmer A.M., Novak, M.A. Rhesus macaques (*Macaca mulatta*) with self-injurious behavior show less behavioral anxiety during the human intruder test. *Am J Primatol.* 2017;doi:10.1002/ajp.22569.

Peterson E.J. The human intrruder test: an anxiety assessment in rhesus macaques. Masters Thesis. 2015; URL: scholarworks.umass.edu/masters_theses_2/91

Planchez B., Surget A., Belzung C. Animal models of major depression: Drawbacks and challenges. *J Neural Transm (Vienna).* 2019;126(11):1383–1408.

Pohorecky L.A. Psychosocial stress and chronic ethanol ingestion in male rats: Effects on elevated plus maze behavior and ultrasonic vocalizations. *Physiol Behav.* 2008;94:432–447.

Poirier C., Bateson M. Pacing stereotypies in laboratory rhesus macaques: Implications for animal welfare and the validity of neuroscientific findings. *Neurosci Biobehav Rev.* 2017;83:508–515.

Poirier-Guay M.P., Bélanger M.C., Frank D. Star gazing in a dog: A typical manifestation of upper gastrointestinal disease. *Can Vet J.* 2014;55:1079–1082.

Pomerantz O., Meiri S., Terkel J. Socio-ecological factors correlate with levels of stereotypic behavior in zoo-housed primates. *Behav Processes.* 2013;98:85–91.

Pomerantz O., Paukner A., Terkel J. Some stereotypic behaviors in rhesus macaques (*Macaca mulatta*) are correlated with both perseveration and the ability to cope with acute stressors. *Behav Brain Res.* 2012a;230:274–280.

Pomerantz O., Terkel J. Effects of positive reinforcement training techniques on the psychological welfare of zoo-housed chimpanzees (*Pan troglodytes*). *Am J Primatol.* 2009;71:687–695.

Pomerantz O., Terkel J., Suomi S.J., Paukner A. Stereotypic head twirls, but not pacing, are related to a pessimistic-like judgment bias among captive tufted capuchins (*Cebus paella*). *Anim Cogn.* 2012b;15:689–698.

Ponzio M.F., Busso J.M., Ruiz R.D., Fiol de Cuneo M. A survey assessment of the incidence of fur-chewing in commercial chinchilla (*Chinchilla lanigera*) farms. *Anim Welf.* 2007;16:471–479.

Ponzio M.F., Monfort S.L., Busso J.M., Carlini V.P., Ruiz R.D., Fiol de Cuneo M. Adrenal activity and anxiety-like behavior in fur-chewing chinchillas (*Chinchilla lanigera*). *Horm Behav.* 2012;61:758–762.

Protopopova A., Hall N.J., Wynne C.D. Association between increased behavioral persistence and stereotypy in the pet dog. *Behav Processes.* 2014;106:77–81.

Provencal N., Booij L., Tremblay R.E. The developmental origins of chronic physical aggression: Biological pathways triggered by early life adversity. *J Exp Biol.* 2015;218(Pt 1):123–133.

Qin D., Rizak J., Chu, X., Li Z., Yang S., Lü L., Yang L., Yang Q., Yang B., Pan L., Yin Y., Chen L., Feng X., Hu X. A spontaneous depressive pattern in adult female rhesus macaques. *Sci Rep.* 2015;5:11267. Published 2015 Jun 10. doi:10.1038/srep11267.

Quirke T., O'Riordan R.M. The effect of different types of enrichment on the behavior of cheetahs (*Acinonyx jubatus*). *Appl Anim Behav Sci.* 2011;133:87–94.

Rafaeli-Mor N., Foster L., Berkson G. Self-reported body-rocking and other habits in college students. *Am J Ment Retard.* 1999;104:1–10.

Raper J.R., Bloomsmith M.A., Stone A., Mayo L. Use of positive reinforcement training to decrease stereotypic behaviors in a pair of orangutangs (*Pongo pygmaeus*). *Am J Primatol.* 2002; 57:70–71 (abstract).

Rauch S.L., Shin L.M., Phelps E.A. Neurocircuitry models of posttraumatic stress disorder and extinction: Human neuroimaging research–past, present, and future. *Biol Psychiatry.* 2006;60:376–382.

Regaiolli B., Angelosante C., Marliani G., Accorsi P.A., Vaglio S., Spiezio C. Gum feeder as environmental enrichment for zoo marmosets and tamarins. *Zoo Biol.* 2020;39(2):73–82.

Reamer L.A., Haller R.L., Thiele E.J., Freeman H.D., Lambeth S.P., Schapiro S.J. Factors affecting initial training success of blood glucose testing in captive chimpanzees (*Pan troglodytes*). *Zoo Biol.* 2014;33:212–220.

Reinhardt V., Cowley D., Eisele S. Serum cortisol concentrations of single-housed and isosexually pair-housed adult rhesus macaques. *J Exp Anim Sci.* 1991;34:73–76.

Robert M.E., Geraghty 3rd J.J., Miles S.A., Cornford M.E., Vinters H.V. Severe neuropathy in a patient with acquired immune deficiency syndrome (AIDS). Evidence for widespread cytomegalovirus infection of peripheral nerve and human immunodeficiency virus-like immunoreactivity of anterior horn cells. *Acta Neuropathol.* 1989;79:255–261.

Rommeck I., Anderson K., Heagerty A., Cameron A., McCowan B. Risk factors and remediation of self-injurious and self-abuse behavior in rhesus macaques. *J Appl Anim Welf Sci.* 2009;12:61–72.

Rossler W., AjdacicGross V., Glozier N., Rodgers S., Haker H., Müller M. Sleep disturbances in young and middle-aged adults - Empirical patterns and related factors from an epidemiological survey. *Compr Psychiatry.* 2017;78:83–90.

Salmaso N., Silbereis J., Komitova M., Mitchell P., Chapman K., Ment L.R., Schwartz M.L., Vaccarino F.M. Environmental enrichment increases the GFAP+ stem cell pool and reverses hypoxia-induced cognitive deficits in juvenile mice. *J Neurosci.* 2012;32:8930–8939.

SAMSA (Substance Abuse and Mental Health Services Administration). Key substance use and mental health indicators in the United States: Results from the 2019 national survey on drug use and health. 2019. https://www.samhsa.gov/data/sites/default/files/reports/rpt29393/2019NSDUHFFRPDFWHTML/2019NSDUHFFR1PDFW090120.pdf

Sato S., Nagamine R., Kubo T. Tongue playing in tethered Japanese black cattle – diurnal patterns, analysis of variance and behavior sequences. *Appl Anim Behav Sci.* 1994;39:39–47.

Savory C.J., Mann J.S. Is there a role for corticosterone in expression of abnormal behaviour in restricted-fed fowls? *Physiol Behav.* 1997;62:7–13.

Schapiro S.J. (ed.). 2017 *Handbook of Primate Behavioral Management.* CRC Press, Boca Raton, FL.

Schapiro S.J., Bloomsmith M.A., Laule G.E. Positive reinforcement training as a technique to alter nonhuman primate behavior: Quantitative assessments of effectiveness. *J Appl Anim Welf Sci.* 2003;6(3):175–187.

Schapiro S.J., Bloomsmith M.A., Porter L.M., Suarez S.A. Enrichment effects on rhesus monkeys successively housed singly, in pairs, and in groups. *Appl Anim Behav Sci.* 1996;48:159–171.

Schwartz S. Naltrexone-induced pruritis in a dog with tail chasing behavior. *J Am Vet Med Assoc.* 1993; 202(2):278–280.

Seier J., de Villiers C., van Heerden J., Laubscher R. The effect of housing and environmental enrichment on stereotyped behavior of adult vervet monkeys (*Chlorocebus aethiops*). *Lab Anim* 2011;40:218–224.

Sheynin J., Moustafa A.A., Beck K.D., Servatius R.J., Casbolt P.A., Haber P., Elsayed M., Hogarth L., Myers C.E. Exaggerated acquisition and resistance to extinction of avoidance behavior in treated heroin-dependent men. *J Clin Psychiatry.* 2016;77:386–94.

Shively C.A., Laber-Laird K., Anton R.F. Behavior and physiology of social stress and depression in female cynomolgus monkeys. *Biol Psychiatry.* 1997;41:871–882.

Shively C.A., Register T.C., Adams M.R., Golden D.L., Willard S.L., Clarkson T.B. Depressive behavior and coronary artery atherogenesis in adult female cynomolgus monkeys. *Psychosom Med.* 2008;70:637–645.

Shively C.A., Register T.C., Appt S.E., Clarkson T.B. Effects of long-term sertraline treatment and depression on coronary artery atherosclerosis in premenopausal female primates. *Psychosom Med.* 2015;77:267–278.

Shivik J.A., Palmer G.L., Gese E.M., Osthaus B. Captive coyotes compared to their counterparts in the wild: Does environmental enrichment help?. *J Appl Anim Welf Sci.* 2009;12(3):223–235.

Shyne A, Block M. The effects of husbandry training on stereotypic pacing in captive African wild dogs (Lycaon pictus). *J Appl Anim Welf Sci.* 2010;13(1):56–65.

Siciliano-Martina L., Martina J.P. Stress and social behaviors of maternally deprived captive giraffes (Giraffa camelopardalis). *Zoo Biol.* 2018;37(2):80–89.

Siever L.J. Neurobiology of aggression and violence. *Am J Psychiatry.* 2008;165:429–442.

Silverstein-Metzler M.G., Justice J.N., Appt S.E., Groban L., Kitzman D.W., Carr J.J., Register T.C., Shively C.A. Long-term sertraline treatment and depression effects on carotid artery atherosclerosis in premenopausal female primates. *Menopause.* 2017;24(10):1175–1184.

Smith E.A., Van Houten R. A comparison of the characteristics of self-stimulatory behaviors in "normal" children and children with developmental delays. *Res Dev Disabil.* 1996;17:253–268.

Spring S.E., Clifford J.O., Tomko D.L. Effect of environmental enrichment devices on behaviors of single- and group-housed squirrel monkeys (*Saimiri sciureus*). *Contemp Top Lab Anim Sci.* 1997;36:72–75.

Stanwicks L.L., Hamel A.F., Novak M.A. Rhesus macaques (*Macaca mulatta*) displaying self-injurious behavior show more sleep disruption than controls. *Appl Anim Behav Sci.* 2017;197:62–67.

Stavisky R.C., Ramsey J.K., Meeker T., Stovall M., Crane M.M. Trauma rates and patterns in specific pathogen free (SPF) rhesus macaque (Macaca mulatta) groups. *Am J Primatol.* 2018;80(3):e22742. doi:10.1002/ajp.22742.

Tiefenbacher S.T., Fahey M.A., Rowlett J.K., Meyer J.S., Poliot A.L., Jones B.M., Novak M.A. The efficacy of diazepam treatment for the management of acute wounding episodes in captive rhesus macaques. *Comp Med.* 2005b;55:387–392.

Tiefenbacher S., Novak M.A., Jorgensen M.J., Meyer J.S. Physiological correlates of self-injurious behavior in captive, socially reared rhesus monkeys. *Psychoneuroendocrinol.* 2000;25:799–817.

Tiefenbacher S.T., Novak M.A., Lutz C.K., Meyer J.S. The physiology and neurochemistry of self-injurious behavior: A nonhuman primate model. *Front Biosci* 2005a;10:1–11.

Tiefenbacher S., Novak M.A., Marinus L.M., Meyer J.S. Altered hypothalamic-pituitary-adrenocortical function in rhesus monkeys (*Macaca mulatta*) with self-injurious behavior. *Psychoneuroendocrinol* 2004;29:500–514.

Tiira K., Hakosalo O., Kareinen L., Thomas A., Hielm-Björkman A., Escriou C., Arnold P., Lohi H. Environmental effects on compulsive tail chasing in dogs. *PLoS One.* 2012;7(7):e41684. doi:10.1371/journal.pone.0041684.

Troisi A. Displacement activities as a behavioral measure of stress in nonhuman primates and human subjects. *Stress.* 2002;5:47–54.

Truelove M.A., Martin A.L., Perlman J.E., Wood J.S., Bloomsmith M.A. Pair housing of Macaques: A review of partner selection, introduction techniques, monitoring for compatibility, and methods for long-term maintenance of pairs. *Am J Primatol.* 2017;79:1–15.

Tynes V.V., Sinn L. Abnormal repetitive behaviors in dogs and cats: A guide for practitioners. *Vet Clin North Am Small Anim Pract.* 2014;44:543–64.

Turner P.V., Grantham L.E. Short-term effects of an environmental enrichment program for adult cynomolgus monkeys. *Contemp Top Lab Anim Sci.* 2002;41:13–17.

USDA (U.S. Department of Agriculture). 1991. www.nal.usda.gov/awic/final-rules-animal-welfare-9-cfr-part-3-0.

Vandeleest J.J., McCowan B., Capitanio J.P. Early rearing interacts with temperament and housing to influence the risk for motor stereotypy in rhesus monkeys (*Macaca mulatta*). *Appl Anim Behav Sci.* 2011;132:81–89.

Veeder C.L., Bloomsmith M.A., McMillan J.L., Perlman J.E., Martin A.L. Positive reinforcement training to enhance the voluntary movement of group-housed sooty mangabeys (*Cercocebus atys atys*). *J Am Assoc Lab Anim Sci.* 2009;48:192–195.

Waters A.J., Nicol C.J., French N.P. Factors influencing the development of stereotypic and redirected behaviours in young horses: Findings of a four year prospective epidemiological study. *Equine Vet J.* 2002;34:572–579.

Watson S.L., McCoy J.G., Fontenot M.B., Hanbury D.B., Ward C.P. L-tryptophan and correlates of self-injurious behavior in small-eared bushbabies (*Otolemur garnettii*). *J Am Assoc Lab Anim Sci.* 2009;48:185–191.

Weber E.M., Dallaire J.A., Gaskill B.N., Pritchett-Corning K.R., Garner J.P. Aggression in group-housed laboratory mice: Why can't we solve the problem?. *Lab Anim (NY).* 2017;46(4):157–161.

Weed, J. L., Wagner, P. O., Byrum, R., Parrish, S., Knezevich, M., & Powell, D. A. (2003). Treatment of persistent self-injurious behavior in rhesus monkeys through socialization: a preliminary report. *Contemp Top Lab Anim Sci*, 42(5), 21–23.

Weld K.P., Mench J.A., Woodward R.A., Bolesta M.S., Suomi S.J., Higley J.D. Effect of tryptophan treatment on self-biting and central nervous system serotonin metabolism in rhesus monkeys (*Macaca mulatta*). *Neuropsychopharmacology* 1998;19: 314–321.

White S.D. Naltrexone for treatment of acral lick dermatitis in dogs. *J Am Vet Med Assoc.* 1990;196(7):1073–1076.

Wickens C.L., Heleski C.R. Crib-biting behavior in horses: A review. *Appl Anim Behav Sci.* 2010;128:1–9.

Willard S.L., Uberseder B., Clark A., Daunais J.B., Johnston W.D., Neely D., Massey A., Williamson J.D., Kraft R.A., Bourland J.D., Jones S.R., Shively C.A. Long term sertraline effects on neural structures in depressed and nondepressed adult female nonhuman primates. *Neuropharmacol.* 2015;99:369–378.

Willner P. The chronic mild stress (CMS) model of depression: History, evaluation and usage. *Neurobiol Stress.* 2016;6:78–93.

Winsper C., Tang N.K., Marwaha S., Lereya S.T., Gibbs M., Thompson A., Singh S.P. The sleep phenotype of Borderline Personality Disorder: A systematic review and meta-analysis. *Neurosci Biobehav Rev.* 2017;73:48–67.

Wolmarans de W., Brand L., Stein D.J., Harvey B.H. Reappraisal of spontaneous stereotypy in the deer mouse as an animal model of obsessive-compulsive disorder (OCD): response to escitalopram treatment and basal serotonin transporter (SERT) density. *Behav Brain Res.* 2013;256:545–553.

Wurbel H., Stauffacher M. Physical condition at weaning affects exploratory behaviour and stereotypy development in laboratory mice. *Behav Processes.* 1998;43:61–69.

Wynchank D., Berk M. Fluoxetine treatment of acral lick dermatitis in dogs: A placebo-controlled randomized double blind trial. *Depress Anxiety.* 1998;8(1):21–23.

Xu F., Wu Q., Xie L., Gong W., Zhang J., Zheng P., Zhou Q., Ji Y., Wang T., Li X., Fang L., Li Q., Yang D., Li J., Melgiri N.D., Shively C., Xie P. Macaques exhibit a naturally-occurring depression similar to humans. *Sci Rep.* 2015;5:9220. doi:10.1038/srep09220.

Zhai L., Zhang H., Zhang D. Sleep duration and depression among adults: A meta-analysis of prospective studies. *Depress Anxiety.* 2015;32:664–670.

Zhang Z.Y., Mao Y., Feng X.L., Zheng N., Lu L.B., Mia Y.Y., Qin D.D., Hu X.T. Early adversity contributes to chronic stress induced depression-like behavior in adolescent male rhesus monkeys. *Behav Brain Res.* 2016;306:154–159.

4
Utilizing Behavior to Assess Welfare

Daniel Gottlieb
Oregon Heath & Science University

Ori Pomerantz
University of California, Davis

CONTENTS

Introduction 51
Naturalistic Approach 52
    Species-Typical Behaviors and Welfare 52
        Species-Typical Behaviors as Indicators of an Animal's Welfare State 53
        Species-Typical Behaviors as Welfare Enhancement 53
        Species-Typical Behavioral Restriction and Frustration 53
    Limitations of Species-Typical Behaviors as Measures of Welfare 53
    Suggestions for Use of the Naturalistic Approach 54
Feelings Approach 54
    Behavioral Indicators of Affective State 55
        Negative Affective State Behaviors 55
        Abnormal Behaviors 55
        Positive Affective State Behaviors 56
        Subjective Well-Being (Qualitative Assessments) 56
        Limitations of Behavioral Indicators of Affective State and Suggested Use 57
    Behavioral Tests of Affect and Preference 57
        Measuring Relative Preference 57
        Evaluating the Degree of Preference 57
        The Cognitive Bias Paradigm 58
        Preference Test Confounds 59
Biological Functioning 59
    Behavioral Correlates of Biological Functioning 59
    Behaviors That Decrease Biological Functioning 60
Practical Applications and Conclusions 60
    Environmental Evaluations 60
    Individual Assessments 60
    Behavioral Biology and Welfare 61
Acknowledgments 61
References 61

Introduction

Animal welfare is a comprehensive scientific discipline that studies animal quality of life. Broadly, this includes an animal's physical (e.g., growth and reproduction) and psychological (e.g., affective states) health. It includes elements that contribute to the animal's quality of life, including those known as the "Five Freedoms" (Brambell 1965); freedom from hunger, thirst, and malnutrition; freedom from fear and distress; freedom from physical and thermal discomfort; freedom from pain, injury, and disease; and freedom to express normal patterns of behavior. Welfare is often used interchangeably with well-being, although there are subtle differences in the terms (see Schapiro, this volume). Welfare is often considered to be related to the *long-term* status of the animal, while well-being is often used to relate to the *current* behavioral or mental state of the animal (Morton and Hau 2010).

Why do we care about welfare assessments? At an individual level, we assess welfare in animals to ensure that they are physically and mentally healthy. Daily health checks and postprocedure monitoring are performed to guarantee that individuals are not in pain or distress. Animals that do not meet the health standards are identified for appropriate clinical and/or behavioral intervention. There are also times when we assess welfare to determine whether the environment in which an animal is being maintained meets the needs of the animal. These assessments may be performed to determine whether, for example, a new caging system improves welfare, or whether a certain procedure negatively impacts animals. In these cases, the welfare of animals in the two environments may be compared.

Measuring welfare may seem simple and straightforward at first, as many of us have natural intuitions of what it means to have good, or poor, quality of life (Duncan 2002). With both physical and psychological health being composed of multiple factors, however, there is a plethora of potential welfare measurements, each capturing a specific and unique aspect of health. As a result, there are ongoing conversations within the field regarding which measure is most pertinent when assessing quality of life. Potential measures include physiological responses (e.g., heart rate and temperature), health (e.g., presence of diarrhea), productivity (i.e., breeding), and behavior (e.g., abnormal repetitive behavior (ARB)).

In this chapter, we will focus on behavior, a common and fundamental tool used to assess animal welfare. Much like the concept of welfare as a whole, however, there are many ways in which behavior can be employed in welfare assessment. One of the most common ways in which behavior is used to assess welfare is to evaluate an animal's ability to develop and express natural behaviors, a method known as the "Naturalistic Approach". Another tactic for assessing welfare, the "Feelings Approach", focuses on the animal's feelings (e.g., distress, fear, pleasure) and/or affective states (e.g., anxious and content). Behavioral indicators of health can also serve as the basis for welfare evaluation, a concept known as the "Biological Functioning Approach" (Fraser et al. 1997). In this chapter, we will discuss how each of these approaches can be used to assess welfare of captive animals. Within each approach, we provide multiple examples of behavioral measurements and testing paradigms used to assess animal welfare. We provide descriptions of how each measurement/testing paradigm can be employed, the rationale for how they effectively measure welfare, and acknowledgements of the limitations of each approach. These approaches provide a theoretical framework for welfare assessment; they are not mutually exclusive and, as we will detail below, often overlap. However, none of these approaches can be undertaken without understanding the natural behavioral biology of the animal.

Naturalistic Approach

The naturalistic approach to assessing animal welfare is predicated on the belief that captive animals need to be able to develop and have the freedom to express behaviors naturally performed by their counterparts in the wild (Howell and Cheyne 2019). Broadly speaking, to use the naturalistic approach to evaluate welfare, an animal's behavioral repertoire and activity budget are compared against those of the same species in the wild. If animals in captivity do not exhibit key behaviors, or express those behaviors at a rate significantly different from their wild conspecifics, the captive environment is assumed to maintain suboptimal levels of welfare. While quite straightforward in concept, and an easy starting point for evaluating the welfare of a group of animals in a captive setting, this approach is limited in the scope of what it can measure. Moreover, the application of the approach is not always agreed upon (Do natural behaviors improve well-being or indicate well-being? Are they needed in an "unnatural" environment?), nor is it always clear whether all naturally occurring behaviors should be considered positive indicators of welfare. In the following section, we review the relationship between the expression of, or lack of ability to express, natural behaviors and animal welfare, discuss limitations of this approach, and provide suggestions for practical use.

Species-Typical Behaviors and Welfare

Similar to the manner by which physical traits have been shaped through the process of natural selection, behaviors have evolved in a way that increases an individual's fitness when expressed in particular environmental conditions (Manning and Dawkins 2012). Consequently, disparate selection pressures have driven development of behaviors that uniquely characterize the species (Wechsler 2007), referred to here as "species-typical behaviors" (STBs). In the literal sense, STBs comprise a characteristic set of actions that members of a particular species exhibit in their natural habitat. They include behaviors that are both similar among members of the same species and varied across different species (Haraway and Maples 1998).

It has been argued that the expression of STBs contributes to the animals' quality of life by inducing positive affect, pleasurable sensations (Špinka 2006, Bracke and Hopster 2006), and improved biological functioning (Bracke and Hopster 2006). Conversely, decreased frequency of STB expression is commonly used as a measure of poor welfare, due to its association with frustration, distress, behavioral disorders, and pain (Dantzer et al. 2008, Dawkins 2017, Kroshko et al. 2016). For example, the Institute of Medicine Committee on the Use of Chimpanzees (*Pan troglodytes*) in Biomedical and Behavioral Research stated in its report that "It is generally accepted that all species, including our own, experience a chronic stress response (comprising behavioral as well as physiological signs) when deprived of usual habitats, which for chimpanzees includes…sufficient space and environmental complexity to exhibit species-typical behavior" (National Research Council 2011a, p. 27).

Moreover, STBs are generally perceived by members of the public as more natural and therefore associated with good welfare ("if it is natural, it must be good") (Robinson 1998, Yeates 2018). Thus, the more similar the behavioral repertoire of captive animals is to that of their wild counterparts, the better the welfare of the captive population is perceived to be (Bracke and Hopster 2006). This concept was depicted in the monumental Brambell Report, which was issued by the British government in response to increasing public concerns over the welfare of farm animals (Brambell 1965). The Brambell

Committee was the first to formulate the "Five Freedoms" as a fundamental principle in animal welfare, listing certain conditions and situations that animals should be free of under captive settings in order to experience adequate welfare (Webster 2016). Indeed, one of the "Five Freedoms" mentioned in the report is the freedom to express natural behavior. Echoing this notion, the *Guide for the Care and Use of Laboratory Animals* lists the expression of STBs as one of the desired outcomes of environmental enrichment and stresses the importance of providing captive animals with appropriate conditions to exhibit such behaviors (National Research Council 2011b).

Nonetheless, there are different ways by which STBs are used in welfare assessments, and too often the distinction between different uses is not recognized. Specifically, in some cases, STBs are regarded as indicators of the animal's welfare, whereas in other cases, they are considered factors that actively influence the welfare state of the animal (e.g., coping mechanisms or intrinsically rewarding behaviors). For example, if a monkey chooses to spend more time in social grooming *because* they are relaxed, this behavior is an indicator of welfare. In contrast, if social grooming *relaxes* the monkey by causing a release in endorphins, the behavior enhances welfare. Acknowledging these differences is important when using STBs to accurately assess an animal's welfare state.

Species-Typical Behaviors as Indicators of an Animal's Welfare State

If STBs are honest indicators of animal welfare, their levels should increase in environments with adequate conditions and decrease when circumstances are suboptimal. For example, housing captive coyotes (*Canis latrans*) in larger enclosures was associated with an increase in time engaged in STBs, such as exploratory behaviors, which led the authors to conclude that providing these animals bigger enclosures enhances their welfare (Brummer et al. 2010). Along those lines, zoo-housed orangutans (*Pongo sp.*) exhibited fewer STBs, including playing and species-typical affiliative behaviors, when visitors stood close to the enclosure than when visitors were not present, suggesting that the visitors negatively impacted their welfare (Choo et al. 2011).

Species-Typical Behaviors as Welfare Enhancement

When STBs are viewed as factors that influence welfare, the act of performing the behavior is expected to improve an animal's welfare, while the inability to perform the behavior is expected to cause distress. By employing STBs, animals are able to maintain homeostasis, solve problems, and successfully cope with challenges they encounter in their natural environment (Garner 2005, Wechsler 2007). Successful coping with environmental challenges is thought to contribute to the welfare of the animal, whereas poor coping skills are linked to reduced welfare (Broom 2001, 2003, Clark 2017). It has been suggested that the expression of STBs, in particular, affiliative behaviors, in captive animals is likely to promote successful coping, thus enhancing the animals' welfare (Rose et al. 2017, Wechsler 2007). Indeed, it has been argued that the expression of STBs is highly rewarding and thus increases the likelihood of their expression (Panksepp 2010, Wechsler 2007). For instance, play behavior may actively induce positive affective states (Ahloy-Dallaire et al. 2018) in Belding's ground squirrels (*Urocitellus beldingi*); individuals that spent more time in social play became bolder, more exploratory and less fearful in comparison to individuals that were less engaged in play behavior (Vizconde et al. 2017).

Species-Typical Behavioral Restriction and Frustration

While the expression of STBs is linked with good welfare, thwarting animals from exhibiting highly motivated behaviors may induce frustration (Mason et al. 2001), even if the behaviors are no longer necessary in captive environments. For example, laboratory-housed passerines become restless during the migratory season, flapping their wings against the cage bars and trying (unsuccessfully) to leave the enclosure. These actions are associated with an increase in the physiological stress response, suggesting that the motivation to migrate does not subside, despite having minimal or no functional significance (Bateson and Feenders 2010).

From this perspective, the inability to perform STBs, even when unnecessary in captivity, can be seen as a threat to welfare. Some researchers even go so far as to claim that suffering of animals in captivity can only be avoided if animals are able to express their entire STB repertoire (e.g., Thorpe 1965). For instance, in order to survive, many wild species of animals must dedicate a significant portion of their daily activity budgets to food acquisition behaviors (hunting, foraging, manipulating, processing). In captivity, however, balanced diets are readily provided to animals, eliminating the need to "work" for food. Still, the inability to perform these highly motivated, yet functionally unnecessary, food acquisition STBs appears to cause distress in many species [e.g., zoo housed carnivores (Clubb and Mason 2003) and primates (Pomerantz et al. 2013)]. As such, providing opportunities for food acquisition behaviors, including scattering food (e.g., Lima et al. 2019), hiding food (e.g., Fischbacher and Schmid 1999), or requiring an effort to gain access to food [i.e., contrafreeloading (e.g., Bean et al. 1999, da Silva Vasconcellos et al. 2012)], has been shown to improve welfare in captive animals. These findings suggest that in addition to the functional goal of the behavior (in this case, consuming food), other motivations are likely involved. Therefore, without *a priori* knowledge of whether behaviors are either self-rewarding or will cause distress when thwarted, it is common to assume that behavioral deviations in frequency, intensity, form, or context from the typical behavioral repertoire may indicate poor welfare (Howell and Cheyne 2019).

Limitations of Species-Typical Behaviors as Measures of Welfare

Despite its appeal and wide usage, there are several limitations in using STB expression to evaluate an animal's welfare state. First, considering every STB output as desirable in captivity raises some concerns, and this concept has been heavily criticized (e.g., Howell and Cheyne 2019, Veasey et al. 1996). In particular, should behaviors that animals employ in the wild in putatively stressful situations be thought to promote the welfare of their captive counterparts, simply because they are considered typical or natural? Should we strive, for example, to enable captive animals to fight and sustain severe injuries during breeding season merely because they are part of the species' behavioral repertoire? The answer may not be as simple as one might predict. Most would agree that it is unlikely that sustaining an injury or running away from a predator is pleasurable, but what about the rewarding experience that may follow a successful escape? As pointed out by Hans Selye (1975), not all stress is inherently detrimental to the animal. Selye argued that exposure to stressors does not have to be unpleasant and subsequently conceived the term "eustress" to describe levels of stress that are beneficial and arguably pleasant to the individual (National Research Council 2008). To date, these questions remain unanswered, and further research is required in order to determine the short- and long-term values of expressing natural behaviors associated with negative affect.

Further complication with welfare assessment using STBs relates to the aspect of behavioral adaptation, namely, the interplay between the environment and the behavioral tools needed to thrive in it. Captive environments may introduce novel "problems" that animals have not evolved to solve [e.g., proximity to humans without the ability to hide or escape (Sherwen et al. 2015)]. Further, captive conditions often obviate the need to employ some goal-directed STBs, if the goals can be reached via alternate routes (e.g., animals may not need to invest similar resources in acquiring a balanced diet if nutritious food is provided by their caregivers) (Fraser et al. 1997). In such cases, does the expression of STBs, even ones that most likely induce positive affect in the wild, lead to a similar effect in captivity? For example, nest building is functionally unnecessary in hens (*Gallus gallus domesticus*) housed in conventional wire battery cages, yet hens will display increased locomotor and investigatory behavior prior to oviposition when given the opportunity (Cooper & Appleby, 2003). Does enabling this hormonally triggered, prelaying behavior improve their welfare? Without validation using alternate measures of welfare, it is difficult to confirm if a behavior actually indicates an animal's welfare state (Dawkins 2017). Further, even if a behavior is deemed to affect an animal's well-being, determining the "normal" level of a behavior (and consequently, what is "abnormal") proves to be a complicated task, as significant degrees of variability exist between groups of animals of the same species, and even among individual members of the same group (Koene 2013).

Suggestions for Use of the Naturalistic Approach

Regardless of the significant limitations described above, when used appropriately, STBs can reveal some information about the suitability of the captive environment for a particular group of animals, by serving as reference points to compare behavior in captivity against. For example, Stolba and Wood-Gush (1984) examined the behavior of Large-White pigs (*Sus scrofa*) under seminatural conditions, wherein the behavior was unrestricted by the environment, and identified features in this environment that promoted the expression of STBs. The authors then placed increasing restrictions on the environment, allowing identification of the specific features necessary to promote expression of STBs. In order to stimulate expression of the STBs, the authors incorporated the identified environmental features into more restrictive environments (e.g., paddocks and yards, family pens, and conventional fattening pens, which are similar to standard agriculture settings), and found that providing areas for nesting and rooting in regular pens was sufficient to successfully enable expression of STBs. The authors, therefore, recommended to include those areas in regular pens to stimulate and enable the expression of highly motivated STBs (Stolba and Wood-Gush 1984).

In summation, the naturalistic approach, and in particular the use of STB, is a widespread animal welfare assessment tool, and is frequently used in daily health checks. The expression of STBs is most likely affected by the interplay between the animal's current motivation and its environmental conditions, as well as the animal's experience and learning over time, making it difficult to interpret the effects of each factor directly. Still, it is a highly effective method for gaining a general understanding of the relationship between the animal and its environment, namely, how compatible the environment is with the animal's biological needs. Evaluating which behaviors are expressed under different environmental conditions and comparing them to what is expected to be found in natural settings can provide a reasonable starting point in welfare assessment. It can lead to questions that are more specific, such as: (1) Why is there is a difference in the expression of particular behaviors under different environmental conditions? (2) How motivated are animals to express them? (3) What are the welfare consequences of not being able to express them? (4) What appropriate modifications can be made to the environment in order to enable "normal" expression? Additional welfare indices, as described in the sections below, can be combined with STBs to answer these questions and provide a more complete representation of the state of the animal.

Feelings Approach

The feelings approach to assessing animal welfare is primarily focused on an animal's psychological well-being, as measured by its internal affective state. This is accomplished by evaluating behaviors [e.g., stereotypic behaviors (Gottlieb et al. 2013b)] and physiological responses [e.g., responses of the autonomous nervous system (Dum et al. 2016)] that correlate with affective states, or through testing paradigms designed to infer internal state [e.g., cognitive bias testing (Pomerantz et al. 2012b)].

For most of us, the feelings approach is an intuitive concept, and one that we employ on a daily basis; we frequently infer our pets' feelings based on observable behaviors (e.g., my dog has been panting excessively and refuses to leave my side; therefore, he is feeling anxious). While the intuitive nature of the feelings approach is part of its appeal, it is also a major weakness and a source of common pitfalls, including anthropomorphic assumptions and generalization of behaviors across species. For example, it is natural to assume that the big, toothy grin of a chimpanzee (*Pan troglodytes*) is the behavioral output of a positive emotion, such as happiness. While this might be true for a grin displayed by a Mandrill (*Mandrillus sphinx*) or a drill (*M. leucophaeus*) (Hearn et al. 1988, Setchell and Jean Wickings 2005, but see also Dixson 2016), a friendly-looking smile by a chimpanzee is actually indicative of fear, discomfort, and submission (Hammer and Marsh 2015). Further complicating the matter, visually analogous behaviors can have drastically varying implications *within a species*, depending on the context or subtle changes in behavior. A dog baring its teeth is often a sign of aggression; however, when accompanied by prosocial behaviors and body postures (e.g., a play bow), this "grin" is an invitation for positive social interactions (Lindsay 2013). Therefore, proper utilization of the feelings approach should ideally be preceded by a thorough understanding of the natural behavior of the species, and validation that a given behavior is indicative of affective state. Below, we describe observable behaviors that are commonly utilized to measure internal affective state (*Behavioral Indicators of Affective State*), as well as experimental paradigms that have been designed to elicit behaviors and choices that are indicative of affective state (*Behavioral Tests of Affect and Preference*).

Behavioral Indicators of Affective State

Negative Affective State Behaviors

Negative affective state behaviors are more likely to occur, or occur at a higher rate, when an animal is in a negative state, such as fear, conflict, frustration, uncertainty, or anxiety. Examples of such behaviors include yawning and body shaking in nonhuman primates (Schino et al. 1996), excessive drinking, scratching, and barbering in rodents (Hansen and Drake af Hagelsrum 1984, Kalueff et al. 2006), panting in dogs (Beerda et al. 1997), and freezing in pigs (Backus etal. 2017). When using a negative affective state behavior as an indicator of well-being, it is necessary to first establish the behaviors does not have an alternate explanation (for example, an increase in dog panting could also be explained by an increase in environmental temperature). On an individual level, if an animal significantly increases its rate of expressing negative affective state behaviors, this is a signal that the individual may be in a relatively compromised internal state and should prompt further evaluation of the individual. At a group level, when comparing animals in two environments, the environment in which lower overall rates of negative affective state behaviors are expressed can be assumed to promote relatively higher welfare. For example, barbering, in which an individual plucks its own, or a conspecific's fur or whiskers, is frequently seen in laboratory rodents and in some cases is a stress-evoked response to the environment (Kalueff et al. 2006). Bechard et al. (2011) compared rates of barbering (via measurable hair loss) in C57BL/6J laboratory mice that were weaned and housed in either "standard" or "enriched" conditions. The study found that mice weaned into "enriched" conditions (a relatively larger cage with a rotating enrichment item and nesting material) showed significantly less hair loss than mice weaned into standard conditions. The authors concluded that the lower incidence of barbering may indicate the enriched conditions were less stressful, more stimulating, and thus an improvement for animal welfare.

It is important to note that simply because an animal displays a behavior commonly correlated with negative affective state, this does not necessarily mean the individual is experiencing negative well-being at that moment. Fear behaviors, for example, are associated with negative affect, yet also promote actions to avoid dangerous stimuli. If an animal in a laboratory environment shows fleeing behavior in response to a caregiver or husbandry activity, and is provided opportunities to successfully flee (e.g., a burrow or visual barrier for terrestrial animals, a climbing structure for arboreal animals), this can not only counter the negative affective state but may be beneficial, even pleasurable, to the individual (see discussion of eustress in the "Naturalistic Approach" section above). In contrast, if such escape routes are not provided, and the animal shows chronic fleeing behavior without successful avoidance, this unresolved fear and stress is likely to jeopardize the animal's welfare state. In this example, fear behaviors would best be used as an indicator of well-being in cases where their expression is chronic, or an animal is unable to overcome the challenge of the negative stimuli. Overall, negative affective state behaviors are best used as indicators of well-being in situations where they are clearly nonfunctional, and all other explanations for the behaviors have been ruled out.

Abnormal Behaviors

As described in Chapter 3 of this book, abnormal behaviors can broadly be defined as any pattern of activity that differs from some standard (Novak and Meyer this volume). Abnormal behaviors are commonly used as behavioral indicators of welfare, in particular, as indicators of negative affective state, due to their frequent correlation with poor welfare conditions. These correlations include, but are not limited to, lack of stimulation (Burn 2017), barren environments (Baumans 2016), lack of opportunities for expression of key STBs (Clubb and Mason 2003, Pomerantz et al. 2013), inadequate rearing environments (Gottlieb et al. 2013a), exposure to stress (Lesse et al. 2017), and neurological malfunction (Pomerantz et al. 2012a). It is difficult, however, to generalize the utility of abnormal behaviors as an indicator of welfare and affective state, as this category of behavior is extremely broad, encompassing behaviors as benign as increased rates of vocalization, to extreme pathological behavior, such as self-mutilation.

Generally speaking, the etiology, expression, and welfare implications of abnormal behaviors vary by behavior, context, and species; when using any abnormal behavior as an indicator of welfare or affective state, it is important to understand both the behavior's causation and function (seldom an easy task). For this discussion, we will focus on one of the most commonly observed types of abnormal behaviors: abnormal repetitive behaviors (ARBs) (often referred to as stereotypic behaviors, or stereotypies).

ARBs can be defined as repetitive behaviors caused by central nervous system dysfunction, frustration, or repeated attempts to cope (Mason 2006). A wide variety of behaviors can fall under the umbrella of ARBs, including pacing [e.g., tufted capuchins (*Cebus apella*) (Pomerantz et al. 2012b), mink (*Mustela vison*) (Mason 1993)], self-hair-plucking [e.g., rhesus macaques (*Macaca mulatta*) (Lutz et al. 2013)], self-feather-plucking [e.g., parrots (Van Zeeland et al. 2009)], digging [e.g., Mongolian gerbils (*Meriones unguiculatus*) (Wiedenmayer 1997)], and sometimes, even enrichment use [e.g., wheel running in mice (*Mus musculus*) (Latham and Würbel 2006)]. Unfortunately, the relationship between ARBs and welfare is not simple; while some ARBs are indicative of a compromised welfare state, others actually benefit individuals by decreasing stress or fulfilling unmet behavioral needs [e.g., pacing as a behavioral mechanism for controlling arousal (Mason and Latham 2004)]. Thus, even though we see ARBs develop at a higher rate in suboptimal environments (due to a greater need to use ARBs to "cope"), when looking *within* a suboptimal environment, individuals that express ARBs may be better at coping than their non-ARB counterparts (Mason and Latham 2004). Furthermore, ARBs can be unrelated to an animal's current environment, and instead be a direct result of past trauma, such as early abnormal rearing, barren developmental environments, or early exposure to stressors (Novak et al. 2006). Such trauma can cause permanent neurologic damage or dysfunction, leading to the inability to inhibit inappropriate behavior and expression of ARBs (Garner 2006, Mason and Latham 2004). In this scenario, an animal can live in an optimal environment but still express ARBs due to poor past conditions.

Given the multiple and complex etiologies of ARBs, it is inappropriate to directly compare the welfare of two or more individual animals based on current expression of ARBs. Still, when used appropriately, ARBs can be effective behavioral measures of welfare. Once an individual has developed ARBs, a change in the rate of expression can be assumed to be a result of a change in environmental stimulation, control, and/or unavoidable stress (although the absence of a change in ARB cannot be used to conclude no change in welfare, as ARBs are sometimes remnants of past conditions, as noted above) (Mason 1991, Mason and Latham 2004). Therefore, on an individual level, changes in the performance of ARBs can be used to infer individual changes in welfare. Further, when comparing across environments, those correlated with high population-wide rates of ARB development can be considered relatively inferior to environments associated with low or no ARB development. For example, in a risk factor analysis of ARB expression in over 4,000 rhesus macaques, Gottlieb et al. (2013a) found ARBs to be significantly higher in animals that were indoor-reared, single-housed, housed in lower level cages, and/or housed in cages near the door. The authors further found that ARBs were positively correlated with the number of years housed indoors, social separations, experimental protocols, and relocations. Thus, the authors concluded that these environmental conditions induce decreased welfare for the monkeys and should be limited where possible.

Positive Affective State Behaviors

To date, the use of behavioral indicators of affective state has tended to focus on negative states, such as fear and anxiety. Animals in optimal conditions that would promote optimal welfare, however, should experience positive emotions, not simply an absence of negative affect. Thus, behavioral measures of positive affect are also important in adequately assessing welfare.

Behavioral indicators of positive affect are distinguished from other behaviors by both their presence in positive contexts, and their suppression in stressful or negative situations (Boissy et al. 2007). For example, some species-specific vocalizations, such as ultrasonic (~50 kHz) "chirping" in rats, are exhibited by animals only when exposed to positive stimuli, and are absent in the presence of aversive stimuli (Cloutier, this volume; Knutson et al. 2002; Panksepp and Burgdorf 2003). Similarly, play behaviors are frequently viewed as indicators of positive affective state (Fraser and Duncan 1998), since they are often suppressed in stressful and unfavorable conditions (Barrett et al. 1992, Mintline et al. 2013, Newberry et al. 1988), and typically appear only when the primary needs of the animal are met (Boissy et al. 2007, Fraser and Duncan 1998).

Play behaviors generally appear similar in form to species-specific functional behaviors (e.g., fighting, predation, predator avoidance), although they are often performed in a repetitive or exaggerated form, and in a context that provides no obvious immediate functional rewards (Boissy et al. 2007, Martin and Caro 1985). Examples include scampering in pigs (*Sus domesticus*) (Donaldson et al. 2002, Newberry et al. 1988), running and bucking in cattle (*Bos taurus*) (Mintline et al. 2013), and mock fighting in rats (*Rattus norvegicus*) (Pellis and Pellis 2017). Since play is suppressed when animals are exposed to negative stimuli, environments that promote play are considered positive from a welfare perspective. For example, hot-iron disbudding of calves is a common, painful practice in the dairy industry, and is associated with an acute decrease in play behavior (running and bucking). Mintline et al. (2013) found that disbudded calves that were treated with analgesics and anesthetics played more than nontreated individuals, and therefore concluded that both treatments were necessary to promote positive welfare during the disbudding procedure.

Subjective Well-Being (Qualitative Assessments)

Subjective well-being (SWB) assessments are radically different from the objective evaluation of animal welfare that we have discussed thus far, since they do not rely on observing or recording specific, quantifiable behaviors. Rather, observers use an established questionnaire to assess and score an individual animal's overall welfare, based on their own personal judgment. Using a Likert scale, observers rate individual animals on multiple subjective questions, often related to perceived happiness, quality of social relationships, goal achievement, and the rater's degree of empathy with the specific animal (King and Landau 2003). This method has primarily been used in nonhuman primates [common marmosets (*Callithrix jacchus*) (Inoue-Murayama et al. 2018), chimpanzees (*Pan troglodytes*) (King and Landau 2003), orangutans (*Pongo pygmaeus* and *P. abelii*) (Weiss et al. 2006), rhesus macaques (*Macaca mulatta*) (Weiss et al. 2011), brown capuchins (*Sapajus apella*) (Robinson et al. 2016)]. This technique has been used in additional species as well, including clouded leopards (*Neofelis nebulosa*), snow leopards (*Panthera uncial*), African lions (*Panthera uncia*) (Gartner et al. 2016), and Scottish wildcats (*Felis silvestris grampia*) (Gartner and Weiss 2013). Unlike traditional behavioral techniques for evaluating welfare that rely on observing and recording specific (and sometimes rare) quantifiable behaviors, SWB scores can be done quickly and in almost any environment. Despite the inherent subjectivity of the measure, multiple studies have shown high interrater reliability across observers (King and Landau 2003, Robinson et al. 2016, Weiss et al. 2011), demonstrating consistency in SWB scores. Interrater agreement alone, however, does not demonstrate validity of the measure. As with any behavioral measure of welfare, validation studies are necessary to demonstrate that SWB scores are functionally relevant for any given species, and meaningfully correlate with other established measures of well-being.

Limitations of Behavioral Indicators of Affective State and Suggested Use

A major limitation with all behavioral indicators of affective state is a lack of one-to-one correlation between behavior and individual internal state. While many behavioral indicators of negative feelings have been shown to occur at higher rates in relatively poor or stressful environments (e.g., Maestripieri et al. 1992, Mason and Latham 2004), or have been validated through alternate physiological and behavioral measures of welfare (e.g., Pomerantz et al. 2012b), it is necessary to recognize that these are correlations and general predictors at best. Animals showing an increase in negative affect behaviors are more *likely* to be in a state of anxiety, but this is not always the case. Further complicating the matter, negative affective states are not necessarily a welfare concern, if they are short-lived and promote behaviors that effectively deal with the negative stimulus. Fear-based behaviors function to help the animal avoid predation, and successfully performing such behaviors can promote eustress. Abnormal behaviors, while strongly correlated with negative experiences and poor welfare conditions, can sometimes be functional coping mechanisms, or merely indicative of past trauma. Therefore, on an individual level, behavioral indicators of affective state should be used only as *indicators*, signs that an animal may be experiencing negative welfare, prompting further evaluation. Experimentally, these behaviors are best used as population measures, comparing overall rates of behavioral indicators of positive or negative states across treatment groups, to determine which environments or stimuli generally promote positive, or negative, feelings.

Behavioral Tests of Affect and Preference

Measuring Relative Preference

Preference tests are popular in welfare assessments since they appear relatively simple to conduct and are often quite intuitive to understand. The most basic form of preference testing involves presenting the animal with several options and measuring its response. The rationale behind their use is that animals are likely to produce predictable behaviors when exposed to appetitive or aversive stimuli (e.g., approaching food and avoiding electric shock, respectively). By doing so, animals can reveal whether they find the stimulus to be pleasant or not and, consequently, provide information regarding their welfare state (Dawkins 2017). For example, using a tunnel separated into five identical chambers illuminated with varying light intensities, researchers were able to determine that laying hens (*Gallus gallus domesticus*) spent more time, and laid more eggs, in chambers with low light intensities. The authors reported that hens spent approximately 25 min/h in dark chambers, which greatly differed from the quintessential practice in commercial farms (Ma et al. 2016). These findings suggest that both the welfare of the animals and production rates are likely to improve if hens are given the preferred lighting conditions. However, while these rather simple experimental designs can provide important information regarding the *relative* value of a stimulus to an animal compared to other presented stimuli, they do not offer any insights regarding their absolute value (how important the particular resource/condition is for the animal).

Evaluating the Degree of Preference

How would one know whether the animal's preference for one option over another is strong or weak, or even meaningful? These are clearly important pieces of information with potentially significant implications for the animal. These issues can be addressed using *consumer demand, conditional place preference/aversion,* and *combined valance and arousal testing* paradigms.

Consumer demand experimental paradigms measure the degree of the animal's motivation to gain access to its preferred choice. In these paradigms, the experimenter can gradually increase the required investment by imposing increasingly "costly" tasks for the animal to perform in order to gain access to the desired resource, up to the point at which the animal is no longer willing to "pay" by performing the task (also referred to as the maximum price paid or reservation price) (Seaman et al. 2008). For instance, in order to assess the extent of preference for different types of environmental enrichment, ferrets (*Mustela putorius furo*) were placed in an apparatus consisting of a corridor connected to six chambers containing either tunnels, balls, water bowls, foraging enrichment, sleeping enrichment (structures such as buckets emulating a burrow), or social enrichment, and a seventh empty control chamber. The animals were required to push a door, the weight of which increased gradually to the point of maximum price paid, in order to enter each chamber. Ferrets were generally more motivated to enter enriched chambers over the empty one and were particularly willing to push heavier weights for sleeping enrichment. By employing this design, the researchers were able to not only detect preferences but also assess their strength (Reijgwart et al. 2016). It is important to note that studies based on the principles of consumer demand must be carefully crafted such that the task the animals are required to do is relevant to their natural capabilities. If the task is not biologically relevant to the species, the developmental stage of the individual, or its physical conditions, for example, it is very difficult to determine whether the animal's performance is due to its specific motivation to reach the options at hand, or results from its limited ability to perform the required task (Fraser and Matthews 1997, King 2003).

Conditioned place preference/aversion (also referred to as "place conditioning") is yet another version of the preference test paradigm that involves manipulating the value of previously preferred resources (by making them more or less desirable) to determine whether animals find stimuli to be rewarding or aversive (Tzschentke 2007). The first step usually consists of exposing the animals to two separate compartments connected by a doorway. Each compartment incorporates distinct environmental cues (e.g., color, temperature, texture, odor) in order to facilitate the animals' ability to discriminate between them. The animals can move freely between the chambers and baseline preference is determined according to the time spent in each of them. In step 2, the animals are exposed either to an appetitive stimulus (e.g., high-valued food, morphine, compatible social partner) in their less preferred chamber, or to an aversive stimulus (e.g., electric shock, odor of a predator) in the preferred compartment, which ultimately leads to animals reversing their baseline preference. In the third and final step, which is conducted under extinction conditions (i.e., without the stimuli added in step 2), the animals are once again allowed to explore both chambers freely, and the duration spent in each chamber is measured. The difference between the baseline preference (step 1) and the conditioned preference (step 3) is then calculated and used in the analysis. The longer the animal spends in the previously less preferred compartment, the more potent the stimulus in step 2 is considered to be, therefore, providing a measure of the animal's motivation to gain access to, or avoid a particular stimulus (Koob et al. 2014).

Combined valence and arousal testing paradigms can provide a bi-dimensional image of the welfare state by identifying both the qualitative value (i.e., valence) an animal attributes to an experience, as well as the arousal (i.e., the intensity of the experienced valence) caused by that experience (Pomerantz et al. 2012b, Yeates and Main 2008). For example, young cattle preferred gaining access to a compartment in which they were being brushed by a person over access to an empty stall. In addition, their heart rates dropped when they entered the manned compartment and were brushed, and increased when they entered the "brushing compartment", but did not get brushed (Westerath et al. 2014). These results suggest that the animals were motivated to be brushed by a human, and when they were able to achieve that goal, they were relatively relaxed and calm (presumably, since they achieved their goal and experienced low levels of arousal). Interestingly, when the cattle entered the stall expecting to be brushed, and instead were not brushed, it induced a state of frustration and anxiety (likely, since they were unable to achieve their goal and experienced high levels of arousal; Mendl et al. 2010b, Westerath et al. 2014).

The Cognitive Bias Paradigm

A promising and exciting development in the methodology of preference tests emerged with the groundbreaking publication of Harding et al. (2004). This method, known as the "cognitive bias paradigm" is based on the notion that affective states of both human and nonhuman animals create biases in their information processing, thereby manipulating decision-making and judgment, attention, and memory (Mendl et al. 2010b). Assessing the level of bias can thus inform researchers whether the animal is experiencing a positive, negative, or neutral affective state. Indeed, since the 2004 publication, there has been a sharp increase in the number of studies that have employed versions of the cognitive bias paradigm (particularly, the judgment bias paradigm) in a plethora of research species, including tufted capuchins (*Cebus apella*) (Pomerantz et al. 2012b), cattle (*Bos taurus*) (Lecorps et al. 2019), dogs (*Canis lupus familiaris*) (Duranton and Horowitz 2019), rats (*Rattus norvegicus*) (Brydges et al. 2011), rhesus macaques (*Macaca mulatta*) (Bethell et al. 2012a), and sheep (*Ovis aries*) (Verbeek et al. 2019). The judgment bias paradigm evaluates how animals interpret ambiguous stimuli. Specifically, subjects learn to differentiate between two stimuli that lie at the ends of a continuous stimulus range; one stimulus is associated with a desirable/appetitive outcome and the other with a less desirable/aversive outcome. In the next stage, the animals are presented with intermediate/ambiguous stimuli within the learned range, and their responses are recorded. If the animal responds to the intermediate stimulus similarly to the way it responded to the stimulus associated with the desirable outcome, we assume that it anticipates a positive outcome. Likewise, if the animal responds to the intermediate stimulus as if it were associated with the aversive outcome, we assume that it anticipates a negative outcome. For instance, dogs with various degrees of undesirable separation-related behavior were trained to discriminate between a "desirable" (containing food) bowl that was placed on one side of a room, and an "aversive" (empty) bowl that was placed on the opposite side (Mendl et al. 2010a). As expected, dogs were faster to approach the "desirable" bowl, compared to the empty bowl. Following that, the dogs' latency to reach a bowl that was placed intermediately between the previous "desirable" and "aversive" locations was measured. Dogs with high levels of separation anxiety were slower to get to the bowl when it was located midway between the previously learned "desirable" and "aversive" locations. These results suggest that dogs that were behaviorally categorized as anxious were less likely to anticipate finding food in the bowl when it was placed in an ambiguous location compared with less anxious individuals (Mendl et al. 2010a). The authors therefore concluded that the behavioral output of dogs categorized as anxious is indeed associated with an underlying negative affective state.

A similar variation of the cognitive bias paradigm is the attention bias test, in which the animal's tendency to divert focus to a stimulus as a derivative of its affective state is evaluated. For example, Bethell et al. (2012b) induced negative and neutral/positive affective states among male rhesus macaques by subjecting them to health examination involving restraint, routine husbandry procedures, and provision of environmental enrichment. The animals' attention bias was assessed after these events by measuring their response to two photographs of unfamiliar adult male macaques. In each trial, the study animals were simultaneously presented with two pictures: one of a frontal view of male expressing a behavior indicating a threat, and the second of a male with a neutral expression. The authors found that when monkeys experienced negative events, and presumably were in a state of negative affect, they looked away from threatening faces significantly faster than from neutral expressions. Further, after experiencing negative events, the monkeys spent less time looking at aggressive faces, compared to trials that followed a positive event. Consequently, the authors determined that changes in affective state mediate social attention, and that bias toward positively or negatively charged stimuli (in this case, threatening or neutral faces) can reveal information concerning the animal's affective state and therefore its welfare.

Finally, in mood congruent bias, subjects exhibit a tendency toward storing information that corresponds with their affective state (e.g., subjects experiencing positive mood states are more likely to recall positive events and vice-versa). While promising, this specific aspect of the cognitive bias paradigm has yet to be clearly substantiated in animals (but see Burman and Mendl 2018) and thus calls for further investigation.

Preference Test Confounds

As with any experimental paradigm, there are multiple potential confounds that should be considered when employing any of the previously described versions of preference tests. First, the determination that an animal's choice accurately reflects its preference can only be achieved once other factors that may affect the animal's performance have been ruled out. For example, an animal's preference for food is likely to be affected by how hungry it is; however, it can also be affected by other factors, such as thirst, fear, and/or illness.

Second, an individual's preferences may change in different contexts and over time, highlighting the importance of repeated trials. For instance, when testing angelfish's (*Pterophyllum scalare*) preference for proximity to dominant or subordinate individuals, it was found that the latency of subordinate fish to swim next to dominant individuals decreased over time (Gómez-Laplaza 2005). Ending the experiment after the first set of observations would have yielded very different results and conclusions. In fact, some animal welfare scientists have proposed that since choices made by individual animals are likely to change over time as the animals mature, a complete representation of an animal's preferences should only be determined by a cumulative, longitudinal set of tests (Maia and Volpato 2016). Preferences can then be assessed by analyzing the data in a manner that places more weight on the animal's most recent choices compared to earlier ones (Maia and Volpato 2016).

Intra- and interindividual differences in preference are another vital consideration in the interpretation of results. For example, in order to evaluate whether choices are consistent within and between subjects, Browne et al. (2010) familiarized hens (*Gallus gallus*) with three housing environments (wire floor, shavings floor, and peat/perch/nest-box). They then tested the hens in six T-maze trials, where each trial required a choice between two of the familiar environments. They found that at the group level, none of the environments were exclusively chosen. However, examination of individual choices revealed that the hens were quite consistent in their choices, and more birds than expected had exclusive preferences. Interestingly, hens that exhibited greater consistency also tended to make their choices more quickly than others. Therefore, interpreting the results based only on the group level could lead to operational decisions that would exclude individual preferences and decrease individual welfare. When there is intra- and interindividual variability, minority choices (i.e., trials in which the animal chose the option that ultimately was picked less) should still be considered valuable, and are still likely to promote the welfare of the individual. In fact, sometimes the mere option to choose can be advantageous to the animal (Duncan 2005).

Finally, animals do not always choose options that benefit them, both in the short and long run. Ask most dog owners, and they will corroborate that their beloved pooch will gladly consume any piece of chocolate they can put their paws on, despite the fact that it contains theobromine, which is toxic to them. For this reason, Dawkins (2017) argued that evaluating preference has to be performed in conjunction with consideration of the animal's health.

Biological Functioning

The biological functioning approach of assessing welfare is focused on the animal's health and fitness as a measure of its welfare. Commonly employed indices include disease, injury, and ability to perform normal biological functions (e.g., reproduction) (Fraser et al. 1997). Welfare is considered compromised in animals diagnosed with negative clinical conditions, even if the animal is not actively experiencing negative sensations. Through focusing purely on physiological and clinical variables, this approach does not require measurements or interpretations of subjective experiences, avoiding anthropomorphic risks inherent with both the naturalistic and feelings approaches. For example, lameness, commonly defined as gait asymmetry, can be used as an observable outcome variable to determine welfare. In a study to evaluate the relative welfare of dairy cattle housed on different substrates, it was determined that herds with access to sand had significantly lower rates of lameness than cattle housed on alternate substrates (Cook 2003). This study did not rely on animal preferences or feelings; the outcome of interest was simply which substrates resulted in the least amount of injury, and thus, from the biological functioning perspective, resulted in the highest welfare for the subjects.

Behavioral Correlates of Biological Functioning

Negative health outcomes are often preceded, or accompanied, by alterations in behavior. Pain and injury, for example, have been associated with characteristic vocalizations [e.g., castration is associated with high-pitch screaming in piglets (Leidig et al. 2009)], changes in posture [e.g., sole ulcers are associated with a pronounced back arch in dairy cattle (Flower and Weary 2006)], or changes in gait [e.g., lateral patellar ligament injury is associated with lameness in horses (Gottlieb et al. 2016)]. Nonetheless, pain and injury behaviors vary across species, can be particularly difficult to identify in stoic species, and are impacted by individual personality. For example, Ijichi et al. (2014) found that low neuroticism horses (as rated by their owner) had a higher threshold for displaying pain behaviors (lameness) in response to tissue damage, than their highly neurotic counterparts.

While behaviors associated with negative health outcomes should be considered indicators of potential compromised welfare, further evidence, such as behavior reduction in response to analgesics, is needed to categorize a behavior as a true indicator of pain. For example, broiler chickens with observed lameness are often assumed to be in pain. McGeown et al. (1999) confirmed this by administering carprofen, an analgesic, to a subset of lame chickens. Those receiving pain medication significantly improved locomotion, indicating the observed lameness was indeed caused by pain.

Behaviors That Decrease Biological Functioning

As opposed to behavioral indicators of pain and sickness, self-abusive behaviors are an extreme form of abnormal behaviors that directly *cause* harm to the animal, inherently compromising an individual's biological functioning. These behaviors are rare, but are known to develop in captivity in some species, including self-biting in nonhuman primates (Jacobson et al. 2016, Novak 2003, Polanco 2016), and feather plucking and self-mutilation in birds (Jenkins 2001, Morris and Slocum 2017). In the "Feelings Approach" section, we discussed how, through careful interpretation, abnormal behaviors can be used as indicators of affective state. Self-abusive behaviors, however, require little interpretation and, when severe, are a direct behavioral measure of compromised health and welfare.

Practical Applications and Conclusions

Assessing the welfare of captive animals is a critical part of how we care for them. Welfare assessments can be used to scientifically measure the relative quality of an animal's environment and to evaluate individual quality of life. Animal welfare is multifactorial and cannot be appropriately measured by a single behavioral outcome. The three major approaches outlined in this chapter (Naturalistic, Feelings, and Biological Functioning) represent broad categories of welfare measures. Within these categories lie numerous behavioral measures, each designed to identify a specific aspect of welfare. These behavioral measures, however, are only useful when placed in the context of an animals' natural behavior and behavioral needs.

Environmental Evaluations

The behavioral measures outlined in this chapter can be used to evaluate ways in which various environmental conditions positively or negatively influence quality of life for a group of animals. For example, suppose one was evaluating the welfare implications of different kinds of flooring substrate. One could use all three methods to assess which substrate has best promoted signs of positive welfare. Using the naturalistic approach, one would look for increases in STBs (e.g., foraging), when animals were given various substrates. Using the feelings approach, one would look for decreases in stress-related behaviors (e.g., ARBs) and increases in behaviors indicative of positive affect (e.g., play). One could also perform a preference test to determine which substrate the animals prefer, or measure cognitive bias in the animals following their experience in each environment. A biological function approach would compare illness and injury between animals given the two substrates. Welfare scientists often perform these kinds of assessments to study factors that can improve the welfare for captive animals. Such controlled studies are ideally conducted with a sample size large enough to generalize findings to a strain or species. However, these sorts of assessments can also be carried out less formally and on relatively few animals. When new enrichment items, or environmental modifications, are introduced to a group of animals, care staff typically spend some time observing the animals to examine their behavioral response (e.g., does the enrichment produce the desired behavioral outcome?) and ensure there are no untoward effects (e.g., injury). While such observations may not be part of a controlled study, they still provide valuable insights regarding the welfare implications of various environmental factors.

Individual Assessments

The behavioral tools described in this chapter can also be used to evaluate the current well-being of individuals, as happens with daily health checks performed on research animals. In these sorts of assessments, trained staff typically look for the presence of "desired" behaviors, such as species normal behaviors and those that indicate positive affect (e.g., play), and the absence of "undesired" behaviors, such as abnormal behaviors, behaviors that indicate illness or pain (e.g., inappetence, withdrawal, reluctance to move, changes in vocalizations, etc.), and even certain facial expressions (Descovich et al. 2017). "Desired" and "undesired" behaviors are species-specific, and many of these behaviors are detailed in the chapters in Part 2 of this book [for example, coat condition can indicate compromised well-being in mice (MacLellan et al., this volume)]. These daily observations are most effective when performed by those who know the behavioral characteristics of the animals, such as trained care staff or research personnel. Animals identified displaying potentially concerning behaviors can then be further evaluated using methods such as increased observations or clinical intervention, depending on the severity of the problem. If indicators of negative well-being are carefully and systematically documented, over time, these data can be used to detect patterns and identify potential causes. For example, if it is observed that an animal shows behavioral indicators of negative affect only on weekends, this could indicate an effect of different caregivers on those days.

In many cases, the use of only one or two of the approaches described in this chapter will be sufficient to assess an individual's welfare. However, there may be times, such as when the overall quality of life of the animal is in question, where it is necessary to use all three approaches. Research animals, like all animals, may become sick or injured to a point where humane euthanasia is the most appropriate option for the animal. In these scenarios, assessing whether the animal is engaging in normal or abnormal behavior, experiencing poor affect, and expressing behaviors indicative of pain are all important considerations when making end-of-life decisions. Lambeth et al. (2013) provide an excellent summary of their program using a holistic team approach to assess quality of life for chimpanzees. This multifaceted approach to assessing welfare can be adapted to other species.

Behavioral Biology and Welfare

Whether designing an experiment to determine the optimal environment for laboratory animals, or performing individual welfare checks, it is critical to have an understanding of which behaviors are "normal" (i.e., not of concern) and which behaviors are abnormal (of concern), for the specific species. While certain behaviors may be ubiquitous (for example, an animal that engages in a behavior that could cause injury to itself is always undesired), others are more ambiguous and species-specific. Neophobia with respect to food is common in many bird species, but might indicate concern in a dog. Even within a species, behaviors normal for one animal can be abnormal for another, and thus, it is also important, when possible, to account for the unique behavior patterns of the individual. The absence of behavior typical for an individual (i.e., a normally aggressive rhesus macaque is sitting in the back of the cage quietly, or a normally timid dog growls at a trusted caregiver) is a hallmark sign of stress or distress. Recognizing individual behavior patterns may be easier to accomplish in some species, such as nonhuman primates and dogs, than in others, such as mice or zebrafish. The chapters in Part 2 detail the behavioral biology of various species, as well as behaviors that are "normal" and abnormal in captive conditions.

Acknowledgments

We thank the Oregon National Primate Research Center Resources and Logistics Unit staff, the California National Primate Research Center Behavioral Health Services staff, Kristine Coleman, and Steve Schapiro for help and support of this chapter. This work was supported by NIH P51OD011092 and NIH P51OD011107.

REFERENCES

Ahloy-Dallaire, J., J. Espinosa, and G. Mason. 2018. Play and optimal welfare: Does play indicate the presence of positive affective states? *Behavioural Processes* 156:3–15.

Backus, B. L., M. A. Sutherland, and T. A. Brooks. 2017. Relationship between environmental enrichment and the response to novelty in laboratory-housed pigs. *Journal of the American Association for Laboratory Animal Science* 56 (6):735–741.

Barrett, L., R. I. Dunbar, and P. Dunbar. 1992. Environmental influences on play behaviour in immature gelada baboons. *Animal Behaviour* 44 (1):111–115.

Bateson, M., and G. Feenders. 2010. The use of passerine bird species in laboratory research: Implications of basic biology for husbandry and welfare. *ILAR Journal* 51 (4):394–408.

Baumans V. 2016. The impact of the environment on laboratory animals. In: M. Andersen, S. Tufik (eds.) *Rodent Model as Tools in Ethical Biomedical Research.* Cham: Springer. https://doi.org/10.1007/978-3-319-11578-8_3.

Bean, D., G. J. Mason, and M. Bateson. 1999. Contrafreeloading in starlings: Testing the information hypothesis. *Behaviour* 136 (10–11):1267–1282.

Bechard, A., R. Meagher, and G. Mason. 2011. Environmental enrichment reduces the likelihood of alopecia in adult c57bl/6j mice. *Journal of the American Association for Laboratory Animal Science* 50 (2):171–174.

Beerda, B., M. B. Schilder, J. A. van Hooff, and H. W. de Vries. 1997. Manifestations of chronic and acute stress in dogs. *Applied Animal Behaviour Science* 52 (3–4):307–319.

Bethell, E. J., A. Holmes, A. MacLarnon, and S. Semple. 2012a. Cognitive bias in a non-human primate: Husbandry procedures influence cognitive indicators of psychological well-being in captive rhesus macaques. *Animal Welfare* 21 (2):185–195.

Bethell, E. J., A. Holmes, A. MacLarnon, and S. Semple. 2012b. Evidence that emotion mediates social attention in rhesus macaques. *PLoS One* 7 (8):e44387.

Boissy, A., G. Manteuffel, M. B. Jensen, R. O. Moe, B. Spruijt, L. J. Keeling, C. Winckler, B. Forkman, I. Dimitrov, and J. Langbein. 2007. Assessment of positive emotions in animals to improve their welfare. *Physiology & Behavior* 92 (3):375–397.

Bracke, M. B. M., and H. Hopster. 2006. Assessing the importance of natural behavior for animal welfare. *Journal of Agricultural and Environmental Ethics* 19 (1):77–89.

Brambell, F. W. R. 1965. *Report of the technical committee to enquire into the welfare of animals kept under intensive livestock husbandry systems.* London: Her Majesty's Stationery Office.

Broom, D. 2001. Coping, stress and welfare. In Broom, D.M (ed.) *Coping with Challenge: Welfare in Animals including Humans, Proceedings of Dahlem Conference,* 1–9. Berlin: Dahlem University Press.

Broom, D. M. 2003. Causes of poor welfare in large animals during transport. *Veterinary Research Communications* 27 (1):515–518.

Browne, W. J., G. Caplen, J. Edgar, L. R. Wilson, and C. J. Nicol. 2010. Consistency, transitivity and inter-relationships between measures of choice in environmental preference tests with chickens. *Behavioural Processes* 83 (1):72–78.

Brummer, S. P., E. M. Gese, and J. A. Shivik. 2010. The effect of enclosure type on the behavior and heart rate of captive coyotes. *Applied Animal Behaviour Science* 125 (3):171–180.

Brydges, N. M., M. Leach, K. Nicol, R. Wright, and M. Bateson. 2011. Environmental enrichment induces optimistic cognitive bias in rats. *Animal Behaviour* 81 (1):169–175.

Burman, O. H. P., and M. T. Mendl. 2018. A novel task to assess mood congruent memory bias in non-human animals. *Journal of Neuroscience Methods* 308:269–275.

Burn, C. C. 2017. Bestial boredom: A biological perspective on animal boredom and suggestions for its scientific investigation. *Animal Behaviour* 130:141–151.

Choo, Y., P. A. Todd, and D. Li. 2011. Visitor effects on zoo orangutans in two novel, naturalistic enclosures. *Applied Animal Behaviour Science* 133 (1):78–86.

Clark, F. E. 2017 Cognitive enrichment and welfare: Current approaches and future directions. *Animal Behavior and Cognition* 4 (1):52–71.

Clubb, R., and G. Mason. 2003. Captivity effects on wide-ranging carnivores. *Nature* 425 (6957):473–474.

Cook, N. B. 2003. Prevalence of lameness among dairy cattle in Wisconsin as a function of housing type and stall surface. *Journal of the American Veterinary Medical Association* 223 (9):1324–1328.

Cooper, J., & Appleby, M. C. 2003. The value of environmental resources to domestic hens: A comparison of the work-rate for food and for nests as function of time. *Animal Welfare*, 12:39–52.

da Silva Vasconcellos, A., C. H. Adania, and C. Ades. 2012. Contrafreeloading in maned wolves: Implications for their management and welfare. *Applied Animal Behaviour Science* 140 (1–2):85–91.

Dantzer, R., J. C. O'Connor, G. G. Freund, R. W. Johnson, and K. W. Kelley. 2008. From inflammation to sickness and depression: When the immune system subjugates the brain. *Nature Reviews Neuroscience* 9 (1):46.

Dawkins, M. S. 2017. Animal welfare with and without consciousness. *Journal of Zoology* 301 (1):1–10.

Descovich, K. A., J. Wathan, M. C. Leach, H. M. Buchanan-Smith, P. Flecknell, D. Framingham, and S. Vick. 2017. Facial expression: An under-utilised tool for the assessment of welfare in mammals. *Altex: Alternatives to Animal Experimentation*, 34 (3):409–429.

Dixson, A. F. 2016. *The Mandrill: A Case of Extreme Sexual Selection*. Cambridge, UK: Cambridge University Press.

Donaldson, T. M., R. C. Newberry, M. Špinka, and S. Cloutier. 2002. Effects of early play experience on play behaviour of piglets after weaning. *Applied Animal Behaviour Science* 79 (3):221–231.

Dum, R. P., D. J. Levinthal, and P. L. Strick. 2016. Motor, cognitive, and affective areas of the cerebral cortex influence the adrenal medulla. *Proceedings of the National Academy of Sciences* 113 (35):9922.

Duncan, I. J. H. 2002. Poultry welfare: Science or subjectivity? *British Poultry Science* 43 (5):643–652.

Duncan, I. J. H. 2005. Science-based assessment of animal welfare: Farm animals. *Revue Scientifique et Technique-Office International des Epizooties* 24 (2):483.

Duranton, C., and A. Horowitz. 2019. Let me sniff! Nosework induces positive judgment bias in pet dogs. *Applied Animal Behaviour Science* 211:61–66.

Fischbacher, M., and H. Schmid. 1999. Feeding enrichment and stereotypic behavior in spectacled bears. *Zoo Biology* 18 (5):363–371.

Flower, F., and D. Weary. 2006. Effect of hoof pathologies on subjective assessments of dairy cow gait. *Journal of Dairy Science* 89 (1):139–146.

Fraser, D., and I. J. H. Duncan. 1998. 'Pleasures', 'pains' and animal welfare: Toward a natural history of affect. *Animal Welfare* 7 (4):383–396.

Fraser, D., and L. R. Matthews. 1997. Preference and motivation testing. In M. C. Appleby and B. O. Hughes (eds.) *Animal Welfare*, 159–173. New York: CAB International.

Fraser, D., D. M. Weary, E. A. Pajor, and B. N. Milligan. 1997. A scientific conception of animal welfare that reflects ethical concerns. *Animal Welfare* 6:187–205.

Garner, J. P. 2005. Stereotypies and other abnormal repetitive behaviors: Potential impact on validity, reliability, and replicability of scientific outcomes. *ILAR Journal* 46 (2):106–17.

Garner, J. P. 2006. Perseveration and stereotypy - Systems-level insights from clinical psychology. In G. Mason and J. Rushen (eds.), *Stereotypic Animal Behaviour: Fundamentals and Applications to Welfare*, 121–152. Wallingford: CAB International.

Gartner, M. C., D. M. Powell, and A. Weiss. 2016. Comparison of subjective well-being and personality assessments in the clouded leopard (*Neofelis nebulosa*), snow leopard (*Panthera uncia*), and African lion (*Panthera leo*). *Journal of Applied Animal Welfare Science* 19 (3):294–302.

Gartner, M. C., and A. Weiss. 2013. Scottish wildcat (*Felis silvestris grampia*) personality and subjective well-being: Implications for captive management. *Applied Animal Behaviour Science* 147 (3–4):261–267.

Gómez-Laplaza, L. M. 2005. The influence of social status on shoaling preferences in the freshwater angelfish (*Pterophyllum scalare*). *Behaviour* 142 (6):827.

Gottlieb, D. H., J. P. Capitanio, and B. McCowan. 2013a. Risk factors for stereotypic behavior and self-biting in rhesus macaques (*Macaca mulatta*): Animal's history, current environment, and personality. *American Journal of Primatology* 75 (10):995–1008.

Gottlieb, D. H., K. Coleman, and B. McCowan. 2013b. The effects of predictability in daily husbandry routines on captive rhesus macaques (*Macaca mulatta*). *Applied Animal Behaviour Science* 143 (2):117–127.

Gottlieb, R., M. B. Whitcomb, B. Vaughan, L. D. Galuppo, and M. Spriet. 2016. Ultrasonographic appearance of normal and injured lateral patellar ligaments in the equine stifle. *Equine Veterinary Journal* 48 (3):299–306.

Hammer, J. L., and A. A. Marsh. 2015. Why do fearful facial expressions elicit behavioral approach? Evidence from a combined approach-avoidance implicit association test. *Emotion* 15 (2):223–23.

Hansen, S., and L. J. Drake af Hagelsrum. 1984. Emergence of displacement activities in the male rat following thwarting of sexual behavior. *Behavioral Neuroscience* 98 (5):868.

Haraway, M. M., & Maples, E. (1998). Species-typical behavior. In G. Greenberg & M. M. Haraway (eds.), *Comparative Psychology: A Handbook* (pp. 191–197). New York: Garland.

Harding, E. J., E. S. Paul, and M. Mendl. 2004. Cognitive bias and affective state. *Nature* 427 (6972):312–312.

Hearn, G., E. Weikel, and C. Schaaf. 1988. A preliminary ethogram and study of social behavior in captive drills, *Mandrillus leucophaeus*. *Primate Report* 19:11–14.

Howell, C. P., and S. M. Cheyne. 2019. Complexities of using wild versus captive activity budget comparisons for assessing captive primate welfare. *Journal of Applied Animal Welfare Science* 22 (1):78–96.

Ijichi, C., L. M. Collins, and R. W. Elwood. 2014. Pain expression is linked to personality in horses. *Applied Animal Behaviour Science* 152:38–43.

Inoue-Murayama, M., C. Yokoyama, Y. Yamanashi, and A. Weiss. 2018. Common marmoset (*Callithrix jacchus*) personality, subjective well-being, hair cortisol level and avpr1a, oprm1, and dat genotypes. *Scientific Reports* 8 (1):1–15.

Jacobson, S. L., S. R. Ross, and M. A. Bloomsmith. 2016. Characterizing abnormal behavior in a large population of zoo-housed chimpanzees: Prevalence and potential influencing factors. *PeerJ* 4:e2225.

Jenkins, J. R. 2001. Feather picking and self-mutilation in psittacine birds. *Veterinary Clinics of North America: Exotic Animal Practice* 4 (3):651–667.

Kalueff, A., A. Minasyan, T. Keisala, Z. Shah, and P. Tuohimaa. 2006. Hair barbering in mice: Implications for neurobehavioural research. *Behavioural Processes* 71 (1):8–15.

King, J. E., and V. I. Landau. 2003. Can chimpanzee (Pan troglodytes) happiness be estimated by human raters? *Journal of Research in Personality* 37:1–15.

King, L. A. 2003. Behavioral evaluation of the psychological welfare and environmental requirements of agricultural research animals: Theory, measurement, ethics, and practical implications. *ILAR Journal* 44 (3):211–221.

Knutson, B., J. Burgdorf, and J. Panksepp. 2002. Ultrasonic vocalizations as indices of affective states in rats. *Psychological Bulletin* 128 (6):961.

Koene, P. 2013. Behavioral ecology of captive species: Using behavioral adaptations to assess and enhance welfare of nonhuman zoo animals. *Journal of Applied Animal Welfare Science* 16 (4):360–380.

Koob, G. F., M. A. Arends, and M. Le Moal. 2014. Chapter 3 - Animal models of addiction. In G. F. Koob, M. A. Arends and M. Le Moal (eds.), *Drugs, Addiction, and the Brain*, 65–91. San Diego: Academic Press.

Kroshko, J., R. Clubb, L. Harper, E. Mellor, A. Moehrenschlager, and G. Mason. 2016. Stereotypic route tracing in captive carnivora is predicted by species-typical home range sizes and hunting styles. *Animal Behaviour* 117:197–209.

Lambeth, S., S. Schapiro, B. Bernacky, and G. Wilkerson. 2013. Establishing 'quality of life' parameters using behavioural guidelines for humane euthanasia of captive non-human primates. *Animal Welfare* 22 (4):429.

Latham, N., and H. Würbel. 2006. Wheel-running: A common rodent stereotypy. In G. Mason and J Rushen (eds.), *Stereotypic Animal Behavior: Fundamentals and Applications to Welfare*, 91–92. Wallingford, England: CAB International.

Lecorps, B., B. R. Ludwig, M. A. G. von Keyserlingk, and D. M. Weary. 2019. Pain-induced pessimism and anhedonia: Evidence from a novel probability-based judgment bias test. *Frontiers in Behavioral Neuroscience* 13:54.

Leidig, M. S., B. Hertrampf, K. Failing, A. Schumann, and G. Reiner. 2009. Pain and discomfort in male piglets during surgical castration with and without local anaesthesia as determined by vocalisation and defence behaviour. *Applied Animal Behaviour Science* 116 (2–4):174–178.

Lesse, A., K. Rether, N. Gröger, K. Braun, and J. Bock. 2017. Chronic postnatal stress induces depressive-like behavior in male mice and programs second-hit stress-induced gene expression patterns of oxtr and avpr1a in adulthood. *Molecular Neurobiology* 54 (6):4813–4819.

Lima, M. F. F., C. S. de Azevedo, R. J. Young, and P. Viau. 2019. Impacts of food-based enrichment on behaviour and physiology of male greater rheas (*Rhea americana*, Rheidae, Aves). *Papéis Avulsos de Zoologia* 59:e20195911.

Lindsay, S. 2013. *Handbook of Applied Dog Behavior and Training, Etiology and Assessment of Behavior Problems*. Vol. 2: Ames, Iwoa, Blackwell Publishing Professional.

Lutz, C. K., K. Coleman, J. Worlein, and M. A. Novak. 2013. Hair loss and hair-pulling in rhesus macaques (*Macaca mulatta*). *Journal of the American Association for Laboratory Animal Science* 52 (4):454–457.

Ma, H., H. Xin, Y. Zhao, B. Li, T. Shepherd, and I. Alvarez. 2016. Assessment of lighting needs by w-36 laying hens via preference test. *Animal: An International Journal of Animal Bioscience* 10 (4):671–680.

Maestripieri, D., G. Schino, F. Aureli, and A. Troisi. 1992. A modest proposal - Displacement activities as an indicator of emotions in primates. *Animal Behaviour* 44 (5):967–979.

Maia, C. M., and G. L. Volpato. 2016. A history-based method to estimate animal preference. *Scientific Reports* 6:28328.

Manning, A., and M. S. Dawkins. 2012. *An Introduction to Animal Behaviour*. 6th ed. Cambridge, UK: Cambridge University Press.

Martin, P, and T. M. Caro. 1985. On the functions of play and its role in behavioral development. In *Advances in the Study of Behavior*, ed: Rosenblatt, J.S, Beer, C, Busnel, M and Slater, P.J.B, 59–103. Academic Press.

Mason, G. 2006. Stereotypic behaviour in captive animals: Fundamentals and implications for welfare and beyond. In G. Mason and J. Rushen (eds.), *Stereotypic Animal Behaviour: Fundamentals and Applications to Welfare*, 325–356. Wallingford: CAB International.

Mason, G. J. 1991. Stereotypies - A critical-review. *Animal Behaviour* 41:1015–1037.

Mason, G. J. 1993. Age and context affect the stereotypies of caged mink. *Behaviour* 127 (3–4):191–229.

Mason, G. J., J. Cooper, and C. Clarebrough. 2001. Frustrations of fur-farmed mink. *Nature* 410 (6824):35–36.

Mason, G. J., and N. R. Latham. 2004. Can't stop, won't stop: Is stereotypy a reliable animal welfare indicator? *Animal Welfare* 13 (S):S57–S69.

McGeown, D., T. Danbury, A. Waterman-Pearson, and S. Kestin. 1999. Effect of carprofen on lameness in broiler chickens. *Veterinary Record* 144 (24):668–671.

Mendl, M., J. Brooks, C. Basse, O. Burman, E. Paul, E. Blackwell, and R. Casey. 2010a. Dogs showing separation-related behaviour exhibit a 'pessimistic' cognitive bias. *Current Biology* 20 (19):R839–R840.

Mendl, M., O. H. P. Burman, and E. S. Paul. 2010b. An integrative and functional framework for the study of animal emotion and mood. *Proceedings of the Royal Society B: Biological Sciences* 277 (1696):2895–2904.

Mintline, E. M., M. Stewart, A. R. Rogers, N. R. Cox, G. A. Verkerk, J. M. Stookey, J. R. Webster, and C. B. Tucker. 2013. Play behavior as an indicator of animal welfare: Disbudding in dairy calves. *Applied Animal Behaviour Science* 144 (1–2):22–30.

Morris, K. L., and S. K. Slocum. 2017. Functional analysis and treatment of self-injurious feather plucking in a black vulture (*Coragyps atratus*). *Journal of Applied Behavior Analysis* 52 (4):918–927.

Morton, D. B., and J. Hau. 2010.. Welfare assessment and humane endpoints. In J. Hau and S J. Scahpiro (eds.), *Handbook of Laboratory Animal Science: Volume 1 Essential Principles and Practices*, 536. Boca Raton, FL: CRC Press.

National Research Council. 2008. *Recognition and Alleviation of Distress in Laboratory Animals*. Washington, DC: National Academies Press.

National Research Council. 2011a. *Chimpanzees in Biomedical and Behavioral Research: Assessing the Necessity*. Washington, DC: National Academies Press.

National Research Council. 2011b. *The Guide for the Care and Use of Laboratory Animals*. 8th ed. Washington, DC: The National Academies Press.

Newberry, R., D. Wood-Gush, and J. Hall. 1988. Playful behaviour of piglets. *Behavioural Processes* 17 (3):205–216.

Novak, M. A. 2003. Self-injurious behavior in rhesus monkeys: New insights into its etiology, physiology, and treatment. *American Journal of Primatology* 59 (1):3–19.

Novak, M. A., J. S. Meyer, C. Lutz, and S. Tiefenbacher. 2006. Deprived environments: Developmental insights from primatology. In G. Mason and J. Rushen (eds.), *Stereotypic Animal Behaviour: Fundamentals and Applications to Welfare*, 153–189. Wallingford: CAB International.

Panksepp, J. 2010. Affective neuroscience of the emotional brain-mind: Evolutionary perspectives and implications for understanding depression. *Dialogues in Clinical Neuroscience* 12 (4):533–545.

Panksepp, J., and J. Burgdorf. 2003. "Laughing" rats and the evolutionary antecedents of human joy? *Physiology & Behavior* 79 (3):533–547.

Pellis, S. M., and V. C. Pellis. 2017. What is play fighting and what is it good for? *Learning & Behavior* 45 (4):355–366.

Polanco, A. 2016. A Tinbergian review of self-injurious behaviors in laboratory rhesus macaques. *Applied Animal Behaviour Science* 179:1–10.

Pomerantz, O., S. Meiri, and J. Terkel. 2013. Socio-ecological factors correlate with levels of stereotypic behavior in zoo-housed primates. *Behavioural Processes* 98:85–91.

Pomerantz, O., A. Paukner, and J. Terkel. 2012a. Some stereotypic behaviors in rhesus macaques (*Macaca mulatta*) are correlated with both perseveration and the ability to cope with acute stressors. *Behavioural Brain Research* 230 (1):274–280.

Pomerantz, O., J. Terkel, S. J. Suomi, and A. Paukner. 2012b. Stereotypic head twirls, but not pacing, are related to a 'pessimistic'-like judgment bias among captive tufted capuchins (*Cebus apella*). *Animal Cognition* 15 (4):689–698.

Reijgwart, M. L., C. M. Vinke, C. F. M. Hendriksen, M. van der Meer, N. J. Schoemaker, and Y. R. A. van Zeeland. 2016. Ferrets' (*Mustela putorius furo*) enrichment priorities and preferences as determined in a seven-chamber consumer demand study. *Applied Animal Behaviour Science* 180:114–121.

Robinson, L. M., N. K. Waran, M. C. Leach, F. B. Morton, A. Paukner, E. Lonsdorf, I. Handel, V. A. D. Wilson, S. F. Brosnan, and A. Weiss. 2016. Happiness is positive welfare in brown capuchins (*Sapajus apella*). *Applied Animal Behaviour Science* 181:145–151.

Robinson, M. H. 1998. Enriching the lives of zoo animals, and their welfare: Where research can be fundamental. *Animal Welfare* 7 (2):151–175.

Rose, P. E., S. M. Nash, and L. M. Riley. 2017. To pace or not to pace? A review of what abnormal repetitive behavior tells us about zoo animal management. *Journal of Veterinary Behavior* 20:11–21.

Schino, G., G. Perretta, A. M. Taglioni, V. Monaco, and A. Troisi. 1996. Primate displacement activities as an ethopharmacological model of anxiety. *Anxiety* 2 (4):186–191.

Seaman, S. C., N. K. Waran, G. Mason, and R. B. D'Eath. 2008. Animal economics: Assessing the motivation of female laboratory rabbits to reach a platform, social contact and food. *Animal Behaviour* 75 (1):31–42.

Selye, H. 1975. Stress and distress. *Comprehensive Therapy* 1 (8):9–13.

Setchell, J. M., and E. Jean Wickings. 2005. Dominance, status signals and coloration in male mandrills (*Mandrillus sphinx*). *Ethology* 111 (1):25–50.

Sherwen, S. L., T. J. Harvey, M. J. L. Magrath, K. L. Butler, K. V. Fanson, and P. H. Hemsworth. 2015. Effects of visual contact with zoo visitors on black-capped capuchin welfare. *Applied Animal Behaviour Science* 167:65–73.

Špinka, M. 2006. How important is natural behaviour in animal farming systems? *Applied Animal Behaviour Science* 100 (1):117–128.

Stolba, A., and D. G. Wood-Gush. 1984. The identification of behavioural key features and their incorporation into a housing design for pigs. *Annales de Recherches Vétérinaires* 15 (2):287–299.

Thorpe, W. H. 1965. The assessment of pain and distress of animals in intensive livestock husbandry systems. In F. W. R. Brambell (ed.), *Report of the Technical Committee to Enquire into the Welfare of Animals Kept under Intensive Livestock Husbandry Systems*, 71–79. London, UK: Her Majesty's Stationary Office.

Tzschentke, T. M. 2007. Review on cpp: Measuring reward with the conditioned place preference (cpp) paradigm: Update of the last decade. *Addiction Biology* 12 (3–4):227–462.

Van Zeeland, Y. R., B. M. Spruit, T. B. Rodenburg, B. Riedstra, Y. M. Van Hierden, B. Buitenhuis, S. M. Korte, and J. T. Lumeij. 2009. Feather damaging behaviour in parrots: A review with consideration of comparative aspects. *Applied Animal Behaviour Science* 121 (2):75–95.

Veasey, J. S., N. K. Waran, and R. J. Young. 1996. On comparing the behaviour of zoo housed animals with wild conspecifics as a welfare indicator. *Animal Welfare* 5:13–24.

Verbeek, E., I. Colditz, D. Blache, and C. Lee. 2019. Chronic stress influences attentional and judgement bias and the activity of the HPA axis in sheep. *PLoS One* 14 (1):e0211363.

Vizconde, D. L., E. S. Gibson, J. R. Rodriguez, K. A. Marks, and S. Nunes. 2017. Play behavior and responses to novel situations in juvenile ground squirrels. *Journal of Mammalogy* 98 (4):1202–1210.

Webster, J. 2016. Animal welfare: Freedoms, dominions and "a life worth living". *Animals* 6:35.

Wechsler, B. 2007. Normal behaviour as a basis for animal welfare assessment. *Animal Welfare* 16 (2):107–110.

Weiss, A., M. J. Adams, A. Widdig, and M. S. Gerald. 2011. Rhesus macaques (*Macaca mulatta*) as living fossils of hominoid personality and subjective well-being. *Journal of Comparative Psychology,* 125 (1):72–83.

Weiss, A., J. E. King, and L. Perkins. 2006. Personality and subjective well-being in orangutans (*Pongo pygmaeus* and *Pongo abelii*). *Journal of Personality and Social Psychology* 90:501–511.

Westerath, H. S., L. Gygax, and E. Hillmann. 2014. Are special feed and being brushed judged as positive by calves? *Applied Animal Behaviour Science* 156:12–21.

Wiedenmayer, C. 1997. Causation of the ontogenetic development of stereotypic digging in gerbils. *Animal Behaviour* 53 (3):461–470.

Yeates, J. 2018. Naturalness and animal welfare. *Animals* 8 (4):53.

Yeates, J. W., and D. C. J. Main. 2008. Assessment of positive welfare: A review. *The Veterinary Journal* 175 (3):293–300.

5

An Overview of Behavioral Management for Laboratory Animals

Steven J. Schapiro
The University of Texas MD Anderson Cancer Center
University of Copenhagen

CONTENTS

Introduction 65
Definition of Behavioral Management 66
Underlying Principles – Why? 66
    Understanding the Animals and Their Natural Behavioral Biology 67
    Enhancing Welfare, Well-Being, Wellness 67
        Gentle Handling and Trust 67
        Training Humans 68
    Enhancing the Definition of the Animal Model 68
        Effects on Animal Models 68
    Functionally Simulating Natural Conditions 69
        Functionally Appropriate Captive Environments 69
    Should Behavioral Management Be Separate from Captive Management? 74
Components – How? 75
    Socialization 75
    Environmental Enrichment (What Behaviors Are You Attempting to Affect?) 75
        Feeding 76
        Physical 76
        Occupational 76
        Cognitive 76
        Sensory 76
    Positive Reinforcement Training 77
        Voluntary Participation 77
        Other Training (Learning) Approaches 77
    Empirical Assessments 78
Effects on Data and Research 78
    Costs and Benefits 78
        Validity 78
        Variability 78
        Subject Selection 78
        Expanded Capabilities 79
        Ethical Implications 79
        Exceptions 79
How to – Implementation 79
Relative to What Is in This Book 81
Future Directions 81
Conclusions 81
Acknowledgments 81
References 81

Introduction

This volume focuses on the behavioral biology of animals that are frequently found in research environments and that participate in scientific research projects. While some taxa are involved in many types of research, others are involved in a relatively small number of investigations. In all cases, the housing of animals in captivity (for research, as well as for exhibition and retirement) requires us to provide them with outstanding care; care that both allows them to (1) experience high levels of positive welfare and (2) provides high-quality data (when appropriate) that address many critical research questions. Behavioral management is an important component of captive management that enhances our ability to provide the outstanding care that our animals need and deserve. Many of the taxon-related chapters in this volume contain sections that are devoted to the general concept of behavioral management ("ways to maintain behavioral health") for the taxa of interest. This chapter will not attempt to duplicate the specific strategies discussed in each taxon's chapter, but will try to provide more of a "big picture" view of the value of implementing behavioral management programs for animals in captivity, especially those that are likely to be involved in scientific research.

This chapter will strongly emphasize and include many examples involving nonhuman primates for two reasons. First, given the substantial cognitive abilities and social complexity of many nonhuman primate species, considerable behavioral management is required in order to successfully maintain them in captivity. And second, I have been involved in the behavioral management of nonhuman primate species for most of my career and know little, if anything, about much else.

Definition of Behavioral Management

Behavioral management can be defined in a reasonably straightforward way as: management strategies that are designed to positively affect the behavior (psychology) of captive animals. As such, behavioral management is simply one component of an overall captive management approach (Bloomsmith et al. 2018). Captive management strategies are typically comprised of husbandry, veterinary, and research management programs, in addition to the behavioral management component. Behavioral management usually involves at least three interrelated tactics in various proportions (depending on taxonomic group, research goals, facility restrictions, etc.). These are socialization, environmental enrichment, and positive reinforcement training (PRT). Many behavioral management programs also include empirical assessments of both animal behavior and animal temperament (Bloomsmith et al. 2018). Successful behavioral management can add considerably to the research endeavor, and in general, behavioral management efforts are intended (1) to enhance the welfare of captive animals and, of particular importance in research settings and (2) to improve the definition of animals as investigative models. Enhanced definition of animal subjects as models has many benefits for research, including increasing experimental validity, promoting more focused testing of hypotheses, and providing opportunities to address Refinement and Reduction, two of Russell and Burch's 3Rs (Russell and Burch 1959).

In general, behavioral management strategies are designed to (1) provide captive animals with opportunities to safely engage in species-appropriate behaviors and (2) limit their opportunities to engage in abnormal behaviors. Animals that are behaving normally are considered to be appropriate models for most scientific studies, while animals that are excessively engaged in abnormal behaviors are thought to be unduly stressed and unlikely to be optimal models. As just mentioned, most behavioral management strategies are designed to *provide* the animals with *opportunities* to perform species-appropriate behaviors. It is then up to the animals to choose whether or not to take advantage of these opportunities. As one example, macaques can be housed with a compatible partner (i.e., pair-housed) in a variety of different ways (Truelove et al. 2017). Pair-housing macaques provides them with opportunities to engage in many components of their repertoire of species-typical social behaviors. The animals can choose to interact with their pairmate (or not); some interact positively with their partner, whereas others choose to ignore their enclosure mate. In both cases, behavioral management provided the animals with the opportunity to be social and the primates had the choice of whether to engage in various social behaviors (Baker et al. 2014). Obviously, behavioral managers would not allow pairmates that are incompatible (repeatedly interact aggressively) to remain together.

Another way to conceptualize the goals of behavioral management strategies is to consider behavioral management as an approach intended to minimize the amount of "empty" time experienced by captive animals. Captive animals that are bored and have too much "empty" time tend to fill that time with potentially problematic behaviors; stereotypies, aesthetically unappealing behaviors, and/or potentially damaging social or self-directed activities (Mason and Latham 2004; Novak and Meyer 2021). Thus, an important goal of most behavioral management programs is to minimize the amount of empty time that animals experience, by providing them with opportunities to perform species-typical activities. This is especially true for species-typical behaviors that are incompatible with problematic/abnormal activities (e.g., animals living with social partners tend to spend less time in self-directed behaviors, and animals provided with foraging opportunities tend to spend less time inactive and engaged in stereotypies).

In general, behavioral management strategies involve *manipulations to the environment* that are likely to alter the animals' behavior patterns, more than *direct manipulations of the target animals*. Relatively few behavioral management strategies are designed to change the animals directly (in contrast to veterinary management strategies, where treatments are given to directly change the health status of individuals). For example, socialization strategies affect the availability of conspecifics in the target animals' animate (social) environment, and most forms of environmental enrichment modify aspects of the inanimate (nonsocial) environment. PRT techniques probably do act at the level of the individual animal, as the performance of desired behaviors by the subject is affected by the trainer manipulating reinforcement contingencies. The effects of behavioral management procedures only occur when the animals "use" them, unlike veterinary management procedures, which work on the animals, regardless of their participation.

Underlying Principles – Why?

There are two extremely simple answers to the question of "why should we bother with the behavioral management of laboratory animals"? The first is simply that it is the right thing to do. In many ways, the efforts of behavioral managers should serve our animals, as an acknowledgment of their contributions to the research endeavor. It is our responsibility to make sure that our animals are living in the best possible conditions, within the constraints imposed by captivity and the research process. While guidelines exist to regulate and/or prescribe the minimum levels of behavioral management that should be provided to many captive animals, the actual care provided frequently exceeds these minimum guidelines.

The second reason to manage the behavior of laboratory animals is because such efforts enhance research. Behavioral managers not only serve the animals, they also serve the science in which the animals participate. Animals that engage in species-typical behaviors at species-appropriate levels are likely to be the most valid research models. As was mentioned already, and will be mentioned many more times within this chapter and throughout this volume, the more we know about our subjects, and the more normal and, perhaps happy (Poole 1997), they are, the better our science will be. All successful behavioral management programs/strategies/procedures are built on the foundations of our knowledge of the behavioral biology of our subjects and the way they live under natural conditions (Bloomsmith 2017; MacLellan et al. 2021 (as well as most of the taxon-related chapters in this volume)).

Understanding the Animals and Their Natural Behavioral Biology

To understand how to optimally care for animals in captivity, it is imperative to understand how the animals live in their natural situations (Bloomsmith 2017; Pruetz and McGrew 2001). Answers to questions such as:

- What do they eat (and who eats them)?
- When are they active?
- How and where do they spend their time?
- How big are their groups?
- How do they breed?
- How do they exhibit "contentment"?

among many, many others, must be known, if the goal is to study and, perhaps, breed animals in captive conditions. This should be obvious, but until relatively recently, was not always readily applied when making management decisions. One of the purposes of this volume is to provide an additional centralized resource (see Hubrecht and Kirkwood 2010 as well) to which managers, veterinarians, technicians, and investigators can refer when addressing questions related to the proper (behavioral) care of captive animals. In many captive settings, issues related to human (e.g., caregivers, researchers, veterinarians) convenience are often given higher priority than issues related to the fulfilling of animal needs. A well-designed behavioral management program, one that makes use of the available information concerning animals' behavioral biology, can help modify the balance between human convenience and animal needs (increasing emphasis on the needs of the animals), so that it is workable for caregivers, veterinarians, researchers, and the animals (Bettinger et al. 2017; Bloomsmith 2017; Bloomsmith et al. 2018; Lambeth and Schapiro 2017).

Enhancing Welfare, Well-Being, Wellness

While it may be possible to distinguish among welfare, well-being, and wellness (Broom 1991; Fraser 2009; Maple and Perdue 2013; Novak and Suomi 1988; Ross et al. 2009) for captive animals, behavioral management programs are designed to positively affect all three concepts. The use of injectable vaccines is often provided as an example of a procedure that reduces *well-being* (well-being is typically thought of as *short-term*; an animal may have to be restrained to receive a mildly painful injection of vaccine), while increasing *welfare* (welfare is typically thought of as *long-term*; the animal now has lifelong immunity to the pathogen/disease; Morton and Hau, 2021). It would seem as though there are few behavioral management strategies that enhance welfare at the expense of well-being. However, dominance hierarchy-related social interactions as a function of socially housing animals may be one example. Animals in dominance hierarchies have differential access to resources based on their dominance rank, and this can be viewed as having welfare benefits in conjunction with well-being costs. Subordinate animals may be denied access to desirable resources (diminished well-being), but they benefit from living socially (enhanced welfare). It is important to note that animals that live socially in natural circumstances have evolved social systems and coping strategies that allow them to effectively deal with inequitable access to resources. Therefore, it may not be reasonable (or appropriate) to expect equal access across animals in captivity. While humans may want to see animals "share" resources, sharing may not be a part of their natural evolutionary and behavioral biology, and enforced sharing may create issues related to status certainty for the animals (Vandeleest et al. 2016; but see studies of cooperative feeding; Bloomsmith et al. 1994; Laule et al. 2003). Wellness is a relatively new concept in terms of animal care (Maple and Perdue 2013), but as for humans, wellness for captive animals is something that we should always attempt to achieve.

Gentle Handling and Trust

Even in scenarios in which full-fledged behavioral management programs cannot be feasibly implemented, it is always a "best practice" to handle animals gently, in order to establish a trusting relationship between the animals and those with whom they interact (Coleman and Heagerty 2019; Graham 2017; Graham et al. 2012; MacLellan et al. 2021; Morton and Hau 2021; Poole 1997; Pritchett-Corning and Winnicker 2021). In many respects, gentle handling can be considered the rudimentary foundation that underlies the practice of behavioral management. While it might sound easy, simplistic, and obvious that animals should be handled gently, this has not always been the norm. Those working in many animal care programs must deal with significant time constraints, and gentle handling may not appear, at first glance, to be the most time-efficient way to work with animals. However, data now exist to support the idea that the potential short-term costs (additional time, personnel, and money) associated with instituting gentle handling and related procedures can be more than recovered with the long-term benefits of easier, faster, and safer animal handling (Bliss-Moreau et al. 2013; Coleman et al. 2008; Gouveia and Hurst 2019; Graham 2017; Graham et al. 2012; Lambeth et al. 2006; Magden 2017; Neely et al. 2018). Perhaps, more importantly, better, and additional, data can result from studies employing gentle handling and related techniques (Graham 2017; Graham et al. 2012; Lambeth et al. 2006). In fact, some breeders of laboratory primates and canines are well-known for their handling techniques that allow them to supply gentle and/or well-acclimated animals for research (Fernandez et al. 2017; Marshall Bioresources website).

Training Humans

Obviously, to properly implement behavioral management programs, some staff members will require training in addition to standard caregiver training. Initial phases of this type of training should focus on educational objectives that should result in attitudinal adjustments; education that emphasizes the natural behavior of the species, the animals' need for behavioral management, and the value to the science of the implementation of such procedures. Volumes such as the current one should be part of such an educational approach (also see Schapiro 2017 for a primate behavioral management-specific discussion). Once the importance of behavioral management has been communicated, efforts can be transitioned to training staff members to implement various socialization, enrichment, and/or training tasks. Since evaluations of the outcomes of behavioral management programs should be conducted to fine tune such programs, staff members also should be trained to systematically observe and analyze animal behavior. If additional FTEs are required to perform behavioral management duties, staff can be chosen from within the animal care crew or can be hired from other units or places to perform these tasks. When initiating a behavioral management program (at a primate facility, for example), it makes some sense to create behavioral staff position descriptions that focus on the performance of only behavioral management tasks. However, a mature behavioral management program includes behavioral management responsibilities in the position descriptions of *all members* of the animal care team (Bloomsmith et al. 2018; Lambeth and Schapiro 2017), minimizing any divisions between captive management and behavioral management personnel and responsibilities.

Enhancing the Definition of the Animal Model

One of the easiest ways to justify efforts/costs associated with enhancing the behavioral management of research animals is to focus on the ways in which behavioral management practices can yield better animal models. Socialization, enrichment, and PRT all have been shown to affect the dependent variables that are typically measured in scientific investigations (Arakawa 2018; Bailoo et al. 2018; Graham 2017; Lambeth et al. 2006; Lewejohann et al. 2006; Schapiro 2002). In most, but not all, of these studies, the data collected from animals that received behavioral management enhancements more directly addressed the hypotheses being tested.

Attempts are made to minimize as many "biological confounds" (through the use of, for example, experimental designs, control groups, inbred strains, standard diets, placebos, controlling environmental conditions) as possible in animal studies. Similarly, consistent/standardized/harmonized behavioral management practices can help minimize "behavioral/psychological confounds" (e.g., single vs. social housing, bedding vs. no bedding, voluntary vs. nonvoluntary participation) in animal studies (Arakawa 2018; Bailoo et al. 2018; Graham 2017; Graham et al. 2012; Kappel et al. 2017; Lewejohann et al. 2006; Poole 1997). In general, Refinements that minimize potential confounds can result in diminished interindividual variation in the scientific data collected, which should result in a Reduction (Russell and Burch 1959) in the number of subjects required to perform studies with appropriate statistical power. In other words, behavioral management can result in Reductions in the number of subjects required in research. While many would view a decrease in interindividual variation as a positive, and potentially necessary, product of behavioral management, there are others who subscribe to an alternative approach; one that is more interested in individual differences in responses (Capitanio 2017; Minier et al. 2012; Suomi and Novak 1991). Such investigations would be less enthusiastic about behavioral management practices that reduce variation among subjects. Standardization of conditions (including enrichment practices) across multiple laboratories to address questions related to reproducibility (Richter et al. 2010; Würbel et al. 2013) has been attempted, but the resulting data have not conclusively answered the question.

Effects on Animal Models

In addition to affecting the behavior of animals, various other biological systems are typically affected by participation in behavioral management programs. There are numerous examples of changes in physiological/immunological responses that occur as a function of socialization, enrichment, and training manipulations (Gaskill and Garner 2017; Graham 2017; Lambeth et al. 2006; Luo et al. 2020; Rommeck et al. 2011; Scarola et al. 2019; Schapiro et al. 2000). These include some rather straightforward changes in measures of stress, such as those related to glucocorticoids and the hypothalamic-pituitary-adrenal axis (Novak et al. 2013), but more importantly, also include changes in parameters that are frequently used as experimental readouts in various research projects (Lambeth et al. 2006; Schapiro 2002; Schapiro et al. 2000). Animals that are not reaping the benefits of an effective behavioral management program may be experiencing stress (actually dystress) that may interfere with the interpretation of values for certain parameters and that may overwhelm our ability to detect whether experimental treatments are effective. For example, animals of social species that are housed in socially deprived conditions (singly) have different immunological profiles than conspecifics that are housed in more socially appropriate conditions (Doherty et al. 2018; Schapiro 2002; Tuchscherer et al. 2006). This applies both early in development and during adulthood. Similarly, some physiological parameters obtained from animals living in barren/unenriched/impoverished environments differ significantly from those of animals living in enriched environments (Benaroya-Milshstein et al. 2004). Additionally, nonvoluntary participation in management and research procedures (typically involving some type of restraint) can affect several different types of data when compared to voluntary participation (usually involving PRT; Lambeth et al. 2006).

Functionally Simulating Natural Conditions

As mentioned above, most behavioral management programs are intended to fill animals' empty time by providing opportunities to safely express species-typical behavior patterns and to prevent the development and expression of abnormal behaviors. This can be accomplished through the simulation of various functional components of the animals' natural environment. Behavioral managers do not have to attempt the very difficult, potentially inconvenient, and often impossible, task of exactly duplicating the animals' natural environment in captivity, but rather should strive to simulate the way that the natural environment "works" to satisfy the needs of animals. "Trees" made of PVC, or "vines" made of hose or tubing, can be used in captivity, rather than live, natural trees, or vines. Additional examples include the provision of some arrangement of compatible social partners (for species that live socially) that allows the animals to express many, if not all, components of their repertoire of social behaviors (Baker et al. 2014; Christensen 2021; Dwyer 2021; McGrew 2017; Truelove et al. 2017; Schapiro et al. 1996). For animals that burrow, nest, or climb, suitable substrates, materials, and structures should be provided to allow the animals to engage in these always important, and oftentimes, ritualized behaviors (Gaskill et al. 2013; Gjendal et al. 2019; MacLellan et al. 2021; Manciocco et al. 2021; Pritchett-Corning and Winnicker 2021). For animals that have to search for, select, and then process their food prior to consuming it, opportunities to work for their food should be provided (Johnson et al. 2004; Reinhardt and Roberts 1997). Animals that learn and change their behavior in response to contingencies in their natural environment should be provided with opportunities to make meaningful choices and control their interactions with selected components of their captive environment (Buchanan-Smith and Badihi 2012; Sambrook and Buchanan-Smith 1997). Specific techniques to accomplish these goals will be described later in this chapter and are addressed in many of the taxa-focused chapters in this volume.

It is important to emphasize issues related to safety. All behavioral management techniques must be evaluated for safety prior to implementation. Behavioral management techniques that are hazardous to the animals will not enhance well-being or welfare. Devices that are dangerous for the animals (e.g., trapping, ingestion, or choking hazards) should not be used. In the rare event that an accident occurs with a formerly successful enrichment device, the device must either be modified to eliminate the risk, or removed from the program. Perhaps the most dangerous (but also the most potentially beneficial) behavioral management strategy involves socialization opportunities. While most behavioral managers have devised techniques for the successful housing of their animals in social settings (Baker et al. 2014; Kappel et al. 2017; Lidfors and Dahlborn 2021; McGrew 2017; Truelove et al. 2017; Vinke et al. 2021) that at least functionally simulate natural social conditions, there is always a risk that socially housed animals will be incompatible. Incompatibility can range from mild avoidance to quite severe aggression (male rabbits; Lidfors and Dahlborn 2021; female hamsters; Winnicker and Pritchett-Corning 2021), and the probability of incompatibility can be minimized by the systematic application of (1) our knowledge concerning the species and their natural environments, and (2) the implementation of well-reasoned introduction and maintenance techniques (Baker et al. 2014; McGrew 2017; Truelove et al. 2017). In some scenarios, incompatibility is only a minor inconvenience, but in other scenarios, incompatibility that results in severe aggression can ruin a study, in addition to negatively affecting the well-being of the animals. Of course, social restriction can be quite convenient for caregivers and researchers, since it does not result in incompatibility (extremely low likelihood of socially derived injuries), but it can also ruin experiments, if socially deprived animals exhibit high levels of stress and/or self-destructive abnormal behaviors (Harlow and Harlow 1962; Novak and Meyer 2021; Suomi 1997).

Functionally Appropriate Captive Environments

If captive managers can functionally simulate natural conditions in captive settings, then they are able to establish Functionally Appropriate Captive Environments (FACEs). The establishment of FACEs is another of the primary goals of behavioral management. A FACE is a captive setting that provides animals with opportunities to engage in many, if not all, of the components of their species-typical behavioral repertoire (Neal Webb et al. 2018b; 2019b; also see Bloomsmith and colleagues (2017) for a discussion of 'functionally appropriate nonhuman primate environments'). FACEs address aspects of both the animate and the inanimate environments in which animals operate, including the provision of social opportunities, foraging opportunities, and opportunities to voluntarily participate in a variety of management and research procedures (Neal Webb et al. 2018a, 2019a; see Figures 5.1–5.9, for example, of FACEs from the taxa-specific chapters in this book).

Providing Opportunities

As has been mentioned numerous times already in this chapter, behavioral managers are extremely interested in providing their animals with opportunities to express as many species-typical behaviors, at species-appropriate levels, as possible. Social behaviors and foraging behaviors are two classes of activity that are both extremely beneficial, and relatively straightforward to stimulate with behavioral management practices for many species. Physical enrichment (bedding, nesting material, structures, toys, etc.) can facilitate additional species-typical behaviors (burrowing, nesting, climbing, jumping, exploration, play, etc.; Bailoo et al. 2018; Gjendal et al. 2019; Ross et al. 2009). Again, as mentioned previously,

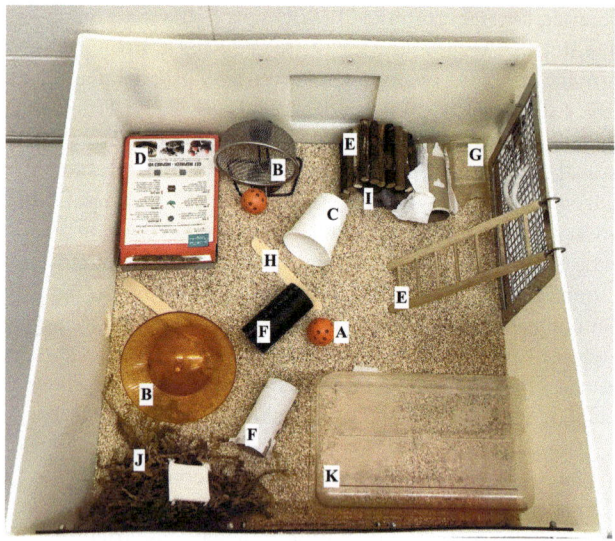

FIGURE 5.1 An example of a Functionally Appropriate Captive Environment (FACE) for mice. Identification of the devices labeled A–K can be found in Chapter 6, Table 1.

Control and Choice

Providing enrichment opportunities that animals can decide whether or not to use, gives the animals the chance to control some aspects of what happens to them and allows them to make meaningful choices. Under natural circumstances, animals have evolved to make many choices each day. Many of these choices are nontrivial, and some can have life or death consequences. When animals are brought into captivity, whether recently (e.g., last week), or 200 generations ago, they are infrequently in situations in which a choice they make can have life or death consequences. In fact, captive animals are often maintained in conditions in which they have little or no control over what happens to them. Many behavioral management strategies are designed to restore a degree of control and choice to captive animals (Buchanan-Smith and Badihi 2012; Hanson et al. 1976; Sambrook and Buchanan-Smith 1997). For example, rats can choose whether to work for, or simply receive, their food (Robertson and Anderson 1975); deer mice can control whether the light is on or off in their physical environment (Kavanau and Havenhill 1963); and chimpanzees can choose between medications in order to self-medicate for their symptoms of arthritis (Neal Webb et al. 2018a). When animals (and people) determine that their actions and choices have no consequences, they stop acting, a behavioral strategy known as learned helplessness (Seligman 1972); a clear indicator of diminished welfare and a potential indicator of depression (Klein et al. 1976).

the goal of behavioral management is to provide the animals with the opportunities; it is then up to the animals to choose whether or not to make use of these opportunities, and to exercise some degree of control over their environment.

FIGURE 5.2 An example of a Functionally Appropriate Captive Environment for rats. The dam can isolate herself from her infants by climbing onto the elevated platform.

FIGURE 5.3 An example of a Functionally Appropriate Captive Environment for dogs. An outdoor exercise yard.

FIGURE 5.5 An example of a Functionally Appropriate Captive Environment pigs. Animals are socially housed in a pen that can be divided.

Voluntary Participation Animals can be trained, using PRT techniques, to voluntarily participate in a variety of management and research behaviors (Graham 2017; Graham et al. 2012; Laule 2010; Magden 2017; Meunier 2006; Smith and Swindle 2006). Voluntary participation typically eliminates the need for physical and/or chemical restraint for routine

FIGURE 5.4 An example of a Functionally Appropriate Captive Environment for cats.

FIGURE 5.6 An example of a Functionally Appropriate Captive Environment for zebra finches.

FIGURE 5.7 An example of a Functionally Appropriate Captive Environment for zebrafish.

procedures and often allows the collection of samples/data that would be impossible to collect in the absence of PRT. Animals can be trained to voluntarily shift from one part of their enclosure to another (Bloomsmith et al. 1998), to climb into a sling for venipuncture (Edwards and Grand 2021; Grandin 1986), or to station for acupuncture treatment (Magden et al. 2013). They can also be trained to remain motionless during a functional magnetic resonance imaging (Tanji et al. 2010).

Behavioral management can also provide opportunities for animals to voluntarily participate in their own health care. They can be trained to choose to remain stationed for treatment with a medicinal laser or while a cardiac loop recorder is interrogated (Magden et al. 2016; Magden 2017). They can also be provided with opportunities to voluntarily self-medicate (Nutella and buprenorphine; Abelson et al. 2012; Kalliokoski et al. 2010) or to choose between different medications to treat their individual disease symptoms (e.g., arthritis; Neal Webb et al. 2018a).

Predictability Predictability provides a mild conundrum for behavioral managers (Bassett and Buchanan-Smith 2007). Care systems that are too predictable to the animals can result in habituation (Wirz et al. 2015) and/or issues related to anticipatory responses during the period that immediately precedes the target event (Bloomsmith and Lambeth 1995). In these types of scenarios, some degree of unpredictability may be beneficial. In fact, novelty (in a way, the opposite of predictability) is often discussed as an important component of an effective environmental enrichment program (Trickett et al. 2009).

However, in most captive care situations, behavioral management strategies designed to increase predictability can be used to make potentially problematic scenarios less problematic. If animals can predict that a particular signal reliably precedes a particular event (especially an aversive or stressful event), then they can be prepared when the event occurs, potentially reducing the stress they experience. "Doorbells" can be used to announce the entrance of caregivers into the animal room (Gottlieb et al. 2013b) or a bell can be used to signal that feeding time is imminent (Bassett 2003). It is important to note that animals are usually paying much closer attention to

FIGURE 5.8 An example of a Functionally Appropriate Captive Environment for marmosets.

FIGURE 5.9 An example of a Functionally Appropriate Captive Environment for macaques. Cynomolgus monkeys in Mauritius.

"preceding" stimuli than humans are. The squeaky wheel on the cart as it rolls down the hallway, signals to the animals that sampling is forthcoming, long before humans enter the room.

Successful PRT is based on the animals' abilities to learn and predict that the performance of the target behavior is reliably followed by the presentation of the (primary or secondary) reinforcer. It is difficult for animals to learn, if the trainer is not consistent in his/her delivery of the reward following the performance of the desired behavior. Although the relationship between the reinforcer and the target behavior has to be predictable, the "identity" of the actual positive reinforcer does not. In fact, in some situations, it is easier to train animals when they know they will receive a reinforcer for performing the target behavior, but do *not* know what the reinforcer will be (for example, apple, grape, or carrot; Schapiro, pers. obs.).

Increasing Species – Appropriate Behavior

As mentioned previously, most behavioral management strategies are designed to provide opportunities for captive animals to perform species-appropriate behaviors. Animals that are behaving in species-typical ways are thought to have adequate well-being and welfare, and to be suitable subjects for research projects. Not all species behave identically, however, and therefore, behavioral management programs must differ by species. Additionally, the behavior of animals differs as they develop, so a behavioral management strategy that is beneficial for infant and juvenile animals may not be beneficial, or even useful, for adult or geriatric members of that species (Neal Webb et al. 2019). Similarly, the behavior of male and female animals often differ (Elliott and Grunberg 2005), suggesting the need for different behavioral management strategies for the two sexes. This is most apparent in terms of socialization strategies, where it is oftentimes more difficult to house adult males with one another than it is to house adult females together (e.g., rabbits; Lidfors and Dahlborn 2021). There are clearly species differences in sociability of males and females. Adults of both sexes of species that typically live solitarily are frequently difficult to socialize, and rightly so (housing two female hamsters together would not only be difficult, but would also be against their behavioral biology, as hamsters typically live by themselves in the wild; Winnicker and Pritchett-Corning 2021). The resources available to support behavioral management programs are limited, making it quite important that investments are only made in those behavioral management enhancements that result in benefits for the animals (Bloomsmith 2017; Bloomsmith et al. 2018; Lambeth and Schapiro 2017). Systematic observations of animals "using" enhancements are critical for empirically determining whether enhancements are worth it or not (Bailoo et al. 2018). Overall, then, species, age, and sex all influence the effectiveness, and therefore, the appropriateness of the strategies that comprise a behavioral management program.

As if this were not enough, *individual animals* also differ in their responses to behavioral management procedures (Capitanio 2017). This is most apparent for certain taxa (e.g., outbred populations of a number of primate species), and for these animals, a "personalized", rather than a "one-size-fits-all" behavioral management program, is likely to be most effective (Coleman 2012; Minier et al. 2012; Suomi and Novak 1991).

Individual animals also differ in temperament (or personality), roughly defined as a relatively consistent pattern of responses to environmental stimuli (Coleman 2017). Temperament not only influences the effectiveness of behavioral management procedures, but assessment of temperament is often a component of a comprehensive behavioral management program (Capitanio 2017; Capitanio et al. 2017; Coleman 2012; 2017). Quantifying the temperament of individual animals enhances the definition of the animal as a model, facilitates the selection of subjects for studies, and allows the "personalization" of behavioral management strategies to maximize benefits (Krall et al. 2019; Michael et al. 2020; O'Malley et al. 2019). Temperament assessments can help identify animals that (1) will prosper under experimental conditions, (2) are trainable, and (3) will be compatible with conspecifics (Coleman 2012; Coleman et al. 2005; Gottlieb et al. 2013a, 2015; Michael et al. 2020).

Decreasing Abnormal Behavior

Captive animals sometimes display behavior patterns that are considered to be abnormal (Novak and Meyer 2021), and each of the taxon-related chapters in this volume contains a brief discussion of that group's abnormal behaviors in captivity. Abnormalities can either take the form of behaviors that are not expressed by wild members of the species or behaviors that are "overexpressed" in captive situations (Novak and Meyer 2021). Self-aggression is rarely, if ever, observed in wild animals, but may be performed by captive primates (Novak et al. 2017). Animals groom themselves and others in the wild but do not typically leave bald patches on either themselves (Novak and Meyer 2009 2021) or their partners (barbering in mice; Dufour and Garner 2010; Garner et al. 2004). While it is possible to view the performance of abnormal behaviors as a coping strategy to effectively deal with an abnormal environment (Mason and Rushen 2006), few would argue that animals that perform high levels of abnormal behavior (e.g., flipping, twirling, pacing, head tossing, cribbing, eye-poking) have high well-being/welfare, or that such animals are ideal models for most research projects (Mason and Latham 2004).

Overall, it is quite a bit easier to prevent abnormal behavior patterns from developing than it is to treat or cure them once they have started. Social deprivation early in life reliably produces a variety of abnormal responses (physiological, as well as behavioral) later in life that can be very, very difficult to treat and/or cure (for example, self-aggression in macaques; Novak and Meyer 2021; Novak et al. 2017; Suomi 1997). Behavioral management strategies that minimize early social deprivation also minimize later expression of abnormal behavior patterns (Schapiro 2002). Behavioral management manipulations that stimulate the performance of species-typical behaviors (e.g., foraging) that are incompatible with the performance of abnormal behaviors can be effective treatments for abnormal behavior. As mentioned previously, captive animals may behave abnormally when they are bored and/or have too much empty time. Behavioral management strategies that provide opportunities for the animals to fill this empty time with species-appropriate behaviors can be quite successful at limiting/reducing abnormal behavior, one of behavioral management's primary goals (Coleman and Maier 2010). See the chapter by Novak and Meyer (2021) and most of the chapters in Part 2 of this volume for more detailed discussions of abnormal behavior.

Individual Differences vs. Group Norms

An interesting issue in the initiation of a behavioral management program involves the establishment of the overall goals of the program: should behavioral managers attempt to design and start a program that provides a baseline of enhancements for all animals in the colony/facility or should they start a program that attempts to "help/treat" those animals that need it the most? The first perspective is one that emphasizes group norms (Bloomsmith et al. 1991), while the second perspective focuses on individual differences (Capitanio 2017). Making certain that all animals get some baseline level of behavioral management may leave little time available to "treat" animals that are experiencing behavioral problems. Conversely, treating the problem animals may leave little time for baseline behavioral management efforts for the rest of the animals. A mature, well-functioning behavioral management program is typically a combination of these two approaches (Bloomsmith 2017; Bloomsmith et al. 2018; Lambeth and Schapiro 2017).

Should Behavioral Management Be Separate from Captive Management?

Behavioral management programs have increased in prevalence and sophistication over the last three decades, and are continuing to evolve (Baker et al. 2017; Bloomsmith 2017;

Bloomsmith et al. 2018; Gottlieb et al. 2017; Schapiro 2017). The perfect behavioral management program does not yet exist, and probably never will. The evolution of behavioral management strategies parallels the evolution of animal care and research techniques, and as research becomes more and more fine-tuned, additional refinements will need to take place in care and management programs, including behavioral management. This ongoing evolution of care, management, and research practices should serve to minimize the distinction between behavioral management and captive management. At this point, and in the future, the most successful animal care and research programs will include substantial behavioral management components within the overall animal management system.

Components – How?

Most effective behavioral management programs typically comprise socialization strategies, environmental enrichment procedures, and PRT techniques (Bloomsmith et al. 1991, 2018). Especially mature programs will also allow for empirical analyses of animal behavior data, and potentially, temperament (Bloomsmith et al. 2018; Coleman 2012; Lambeth and Schapiro 2017). Multiple factors influence the actual activities that comprise the behavioral management program. These include animal species, animal numbers, characteristics of the facility, research needs, staffing levels and expertise, and funding, to name a few. While regulations and guidelines exist that focus on certain aspects of behavioral management, behavioral managers also have flexibility in the ways that they choose to implement their programs. In most instances, performance standards, rather than engineering standards, are specified. For many species, in many situations, implementation of socialization strategies is likely to yield the greatest benefits, along with the greatest risks (Bloomsmith and Baker 2001; DiVincenti and Wyatt 2011; Harlow and Harlow 1962; Kappel et al. 2017; Love and Hammond 1991).

Socialization

For animals that have evolved to live socially in the wild, there is strong evidence that housing compatible animals together is an extremely effective behavioral management procedure (Baker et al. 2014; Bloomsmith and Baker 2001; Capitanio et al. 2017; DiVincenti and Wyatt 2011; Kappel et al. 2017; Love and Hammond 1991; Schapiro 2002; Schapiro et al. 1996). Living with a compatible conspecific(s): (1) functionally simulates natural conditions (helping establish a FACE), (2) addresses the natural behavioral biology of the animals, (3) engages the sophisticated social cognition processes that are employed by animals, and (4) provides the animals with continuous opportunities to make meaningful choices. Social partners provide a constantly changing (yet, paradoxically, highly predictable) environment in which to interact. There are many different strategies that can be implemented to provide social opportunities for captive animals. While it can be difficult to mimic the social environment for animals that live in large multimale-multifemale groups, other types of social organizations (e.g., monogamous pairs, unimale-multifemale groups) can be effectively simulated in captivity. For certain species, social groupings that are atypical for that species (e.g., unisexual macaque pairs; McGrew 2017) can be employed. Many critical components of a species' behavioral repertoire can only be effectively expressed in social settings. For these reasons, the *Guide for the Care and Use of Laboratory Animals* ("the *Guide*"; NRC 2011) recommends social housing for naturally social species as the default condition; exceptions, in which social animals are housed alone, must be appropriately justified.

Not all social animals can be housed socially (in continuous full contact) in captivity. Some housing scenarios present risks of incompatibility that captive managers cannot tolerate (Lidfors and Dahlborn 2021; Winnicker and Pritchell-Corning 2021). However, there are ways to address these risks, including providing partial social contact that minimizes the potential for problems due to incompatibility. This type of solution attempts to balance well-being with welfare, as well-being is not compromised by housing incompatible animals together and welfare is achieved through partial socialization. Grooming-contact panels, used with nonhuman primates, are among the clearest examples (Crockett et al. 1997).

Environmental Enrichment (What Behaviors Are You Attempting to Affect?)

While socialization strategies can be considered enrichment of the animate, social environment, there are a number of other enrichment techniques that can be used to enhance aspects of the inanimate, non-social environment of captive animals. Many of these techniques have been mentioned in earlier sections of this chapter, so they will only be mentioned briefly here.

Behavioral managers must always have behavioral goals in mind when providing environmental enrichment for their animals; there needs to be a reason(s) why a specific enrichment opportunity is provided (Bloomsmith 2017; Bloomsmith et al. 2018; Coleman 2012; Lambeth and Schapiro 2017). Another way to look at this is to think about the behaviors that you are trying to affect by providing a chosen enrichment opportunity. Enrichment scenarios that provide opportunities for animals to search for, find, process, and ultimately, eat their food are obviously intended to increase naturalistic foraging behaviors. The provision of "diggable" substrates, nesting material, or climbable structures are intended to promote burrowing, nest-building, and climbing behavior, respectively. Participation in cognitive tests can be viewed as opportunities to control aspects of the environment (Clark and Smith 2013; Herrelko et al. 2012; Schapiro et al. 2017; Yamanashi and Hayashi 2011). While the provision of video stimulation can be interpreted as an attempt to introduce novelty into the captive environment, unfortunately, the behavior typically induced by video stimulation is inactivity (sitting and watching the video), a behavior that managers are rarely interested in increasing in captive settings. In order for video (or audio) stimulation to be an effective form of enrichment, the animals must be able to control its presentation, and especially, be able to turn the stimulus off (Bloomsmith et al. 2000; Ogura and Matsuzawa 2012).

Feeding

After socialization, feeding enrichment is usually the next most utilizable and effective behavioral management strategy. Feeding enrichment is usually intended to increase the amount of time that captive animals spend in foraging- and consumption-related behaviors. Captive managers are frequently driven by work efficiencies, so simulations of natural foraging activities, like providing mice with a seed or two every minute to cache or consume, are problematic to implement. While past feeding regimens for captive animals were driven primarily by issues related to human convenience, current feeding systems tend to put an increased priority on the needs of the animal (Schipper et al. 2008). Feeding a nutritionally complete chow/biscuit/pellet-based diet or something similar, is typical for almost all laboratory animals (Tobin and Schumacher 2021), and clearly maintains the animals in excellent physical health. However, little work is required by animals to search for, find, and process such diets; all extremely important, and time-filling, components of animals' activity patterns under natural conditions (Christensen 2021). A successful FACE will include foraging enrichment strategies that encourage animals to search for and process their food, prior to consuming it. This can be accomplished by simply distributing food items around the enclosure and/or by placing the food in devices/puzzles that must be solved in order to get access to the food (Johnson et al. 2004). Feeding enrichment can also include the provision of other food items (produce, seeds, mealworms, etc.), in addition to the nutritionally complete pellets. The best feeding enrichment procedures are those that occupy the animals for long periods, yet can be quickly provided and/or cleaned by staff (Bloomsmith et al. 2018; Lambeth and Schapiro 2017). Although animals are unlikely to habituate as quickly to feeding enrichment as they do to physical enrichment, attempts should still be made to systematically vary feeding enrichment options.

Two problems can potentially arise in association with feeding enrichment procedures. First, in social situations, dominant animals can monopolize the desirable enrichment food items, potentially preventing subordinates from getting any of this food. This may lead to obesity in dominants, and weight loss in, or aggression toward, subordinates. While behavioral managers should be concerned with weight/health consequences of monopolization, there is little need to worry about the psychological consequences to subordinates of being denied access to desirable resources. Most social species have evolved to deal with the inevitable, inequitable distribution of resources.

Second, animals may prefer to consume foraging enrichment food items to eating their nutritionally complete pellets. There is some risk of poor nutrition among animals that are not consuming enough pellets. Both of these issues can be easily addressed by carefully monitoring the utilization and results of feeding enrichment procedures.

The horse, sheep, cattle, and even guinea pig chapters in this volume (Christensen 2021; Dwyer 2021; Phillips 2021; Kleven 2021, respectively) discuss the importance of allowing these animals to live in pastures and/or to eat grass or hay. Pasture living can easily be considered a form of behavioral management, but it may be a stretch to consider pasture living as a type of feeding enrichment.

Physical

Most physical enrichment procedures are intended to increase locomotor, exploratory, and play behaviors, and to provide animals with appropriate resting/sheltering/viewing spots. Huts for rodents, climbing structures for nonhuman primates, and elevated bed boards for dogs, should all increase the performance of desired behaviors, lower the potential for abnormal behaviors, and diminish empty time. Animals tend to habituate to some types of physical enrichment (toys, in particular) relatively quickly (Gottlieb et al. 2017; Kuczaj et al. 2002; Vinke et al. 2021), so efforts must be made to systematically vary such enhancements.

Occupational

Occupational enrichment tasks are designed especially to provide animals with opportunities to engage in species-typical activities. There is a fair amount of overlap between occupational enrichment and the other types of enrichment. Occupational enrichment may or may not involve foraging opportunities as described above, or cognitive enhancements as described below. Occupational enrichment procedures are likely to be particularly valuable when attempting to minimize time spent in abnormal behaviors, as they can limit the amount of empty time available to the animals. Examples of occupational enrichment include running wheels for rodents, grooming boards for primates, and chew toys for dogs.

Cognitive

Cognitive enrichment procedures are designed to promote the use of animals' considerable cognitive abilities in species-typical ways. For animals participating in research that involves the performance of tasks (e.g., tests of learning, memory, or sensory acuity), the actual performance of the tasks can often be considered as a form of cognitive enrichment (Clark and Smith 2013; Herrelko et al. 2012; Neal Webb et al. 2019a; Yamanashi and Hayashi 2011). In some situations, animals will voluntarily perform thousands of trials a day of their "personalized testing curriculum" on touch screens (Fagot et al. 2015). While some animals may or may not be able to voluntarily participate in their tasks at the exact moment that they want to, these animals can make choices and exert control by working when other animals are not. It is probably not acceptable to make participation in experimental procedures the sole component of a behavioral management program.

Sensory

Sensory enrichment procedures are designed to expose captive animals to the wide array of sensory stimuli (sights, sounds, smells, textures, tastes) that they might experience under natural conditions, and to stimulate primarily exploratory, and in a few cases, social cohesion behaviors. Sensory enrichment is included in many behavioral management programs, but is likely to be considerably less effective than the previously mentioned forms of environmental enrichment. Sensory enrichment is also likely to be co-opted as enrichment for the care staff, rather than as enrichment for the animals (particularly in terms of audio and video enrichment). Some types of sensory stimuli can be used to mask other, disturbing stimuli that might exist in the captive environment. For example, various types of "benign" noise (e.g., music, nature sounds, waterfalls) can be used to mask the sounds of disturbing noises (e.g., animals being handled, carts being rolled, the sounds of equipment being operated). Natural odors (e.g., essential oils) can be used to counteract the effects of the highly odiferous chemicals that are traditionally used to clean animal areas, especially for species that are heavily reliant on olfactory communication (e.g., rodents, canines). Captive animals exposed to the sights, sounds, or smells of predators will often express behaviors that are less frequently seen in the relatively benign conditions of captivity (Caine and Weldon 1989; Searcy and Caine 2003). Thus, from a behavioral management perspective, it may be acceptable to mildly "scare" animals, as long as they have the opportunity to effectively respond to (huddle, etc.), and control (escape), those situations (Caine 2017).

Positive Reinforcement Training

PRT (a desired reinforcer is provided right after the animals perform the target behavior) is an extremely effective behavioral management approach for certain species in certain situations. PRT is one type of learning (there are many types of learning), and PRT techniques allow animals to voluntarily participate in a variety of husbandry, veterinary, and research management procedures (Christensen 2021; Graham et al. 2012; Laule et al. 2003; Vinke et al. 2021). In numerous circumstances, training animals can facilitate treatments and provide data that would be impossible to obtain in the absence of PRT. The reader is directed to a number of other resources for additional basic and detailed discussions of PRT with captive animals (Coleman et al. 2005, 2008; Graham 2017; Laule 2010; Magden 2017; Prescott and Buchanan-Smith 2003). A key feature that contributes to the value of PRT as a behavioral management tool is that animals have opportunities to choose whether to participate or not. The additional control provided to animals with PRT enhances well-being/welfare (Clegg et al. 2019).

Voluntary Participation

Voluntary participation in husbandry, veterinary, and research procedures helps to minimize the stress associated with such procedures. Animals that voluntarily move from one part of their enclosure to another (Bloomsmith et al. 1998; Veeder et al. 2009), present a body part for treatment (Laule 2010), and/or allow unanesthetized venipuncture (Edwards and Grand 2021; Lambeth et al. 2006) end up being "participants" in these procedures, rather than "subjects", giving them opportunities to make meaningful choices and to exercise control. Voluntary participation diminishes the "need" to use aversive procedures to obtain the desired behavior, a circumstance that was far more prevalent in the past than it is under current animal care scenarios.

Voluntary participation in health care procedures is extremely important, as animals can be trained to "station" for acupuncture and medicinal laser therapy (Magden 2017; Magden et al. 2016). While many of these procedures begin with standard PRT approaches in which food is the reinforcer, there is some evidence to suggest that the health-related benefits of the treatment (symptom relief) become reinforcing, increasing the likelihood of the occurrence of the target behavior. Animals can also be provided with opportunities to choose between medications, in essence to self-medicate (e.g., to relieve arthritis-related symptoms; Neal Webb et al. 2018a). Such procedures open lines of communication between trainers/caregivers/veterinarians and the animals that would not otherwise exist. Animals can also be trained to voluntarily provide biological samples (e.g., saliva, blood, semen) in the absence of any form of restraint (Lambeth et al. 2006; McKinley et al. 2003).

In addition to providing voluntary biological samples, animals can also voluntarily participate in a variety of research procedures, often without any form of food, water, or social restriction (Graham et al. 2012). Animals that continue to perform research tasks comprised of hundreds and hundreds of trials, especially in the absence of deprivation, are essentially responding as though the performance of the task is enriching (Fagot et al. 2015; Leidinger et al. 2017), and in certain circumstances, research tasks can be considered a component of the behavioral management program (see above and Herrelko et al. 2012; Hopper et al. 2016; Yamanashi and Hayashi 2011). Some of these tasks are performed many, many times, even when little, if any, primary reinforcement is provided (Fagot et al. 2019).

Other Training (Learning) Approaches

Although PRT fits best into the concept of maintaining and enhancing well-being/welfare/wellness, there are other techniques that can be used to train animals to perform desired behaviors. Temperament assessments have shown that there is a subset of animals that trains best using negative reinforcement training techniques (an unpleasant stimulus is removed when the animals perform the target behavior; Bliss-Moreau et al. 2013; Capitanio 2017). For this subset of animals, PRT is not the most productive training technique. Habituation and desensitization are two additional training/learning techniques that can be used as part of a comprehensive behavioral management program (Laule 2010). Habituation involves simply exposing the animals to an initially negatively charged situation in the absence of any negative consequences, until the animals no longer view the situation as negative. Desensitization on the other hand involves reinforcement of the animals as they progressively improve at tolerating the negative situation. Desensitization is often undertaken in a way that is typical of "shaping", where animals are successively reinforced for more and more "drastic" exposures to the negative situation. Animals can also learn to perform desired behaviors by observing conspecifics (and humans) perform those behaviors, either live or via video recordings (Hopper et al. 2008;

Leidinger et al. 2018). Observational learning is extremely desirable from a behavioral management perspective, as it decreases the amount of time and effort that must be invested in the initial stages of a training program (Perlman et al. 2010).

Empirical Assessments

Empirical assessments are critical components of all behavioral management programs. The implementation of behavioral management strategies must be data-based and such strategies must be regularly evaluated (Bloomsmith 2017; Bloomsmith et al. 2018; Lambeth and Schapiro 2017; Schapiro et al. 1996; 2003). Resources available for behavioral management are typically limited, and it is counterproductive to "waste" resources on strategies that do not work, or for which the costs outweigh the benefits (Bloomsmith 2017; Bloomsmith et al. 2018; Lambeth and Schapiro 2017). Systematic assessments of the effects of behavioral management techniques should be a part of the overall program, and are probably best conducted by those who possess knowledge of the animals' natural behavior and environment, and the skills to observe and record behavior. While systematic observations, using validated observational data collection techniques (Altmann 1974) are best, less rigorous observations, involving check sheets and similar tools, can also yield utilizable results. Appropriate research designs should be employed when collecting and analyzing behavioral management-related data.

Effects on Data and Research

As discussed in numerous previous sections of this chapter, behavioral management can (usually) beneficially affect the data collected and the utility of the results. Behavioral management-related refinements that enhance the definition of the animal model, increase validity, more directly address experimental hypotheses, reduce interindividual variation, and reduce the number of subjects required for projects, contribute to the ultimate goals of most research studies. Consistent, systematically applied behavioral management programs can enhance reliability, whereas nonsystematically applied behavioral management programs are likely to diminish reliability (but may enhance generalizability).

Costs and Benefits

While the initiation of behavioral management programs is likely to result in increased expenditures (in time, money, personnel, resources, etc.), an effective program is likely to "pay for itself" over time (in savings in time, money, animals, and personnel, and in the acquisition of useful measures/data that could not be collected in the absence of behavioral management practices; Graham 2017; Graham et al. 2012). Some behavioral management practices can result in safety issues. Experiments can also be affected, if data from "before" behavioral management are compared to data "during" behavioral management. One of the most common arguments from researchers for NOT implementing a behavioral management program is that their previously collected data will not be comparable to the data collected when additional behavioral management practices are in place.

Even though there can be costs associated with implementing behavioral management, overall, most socialization, enrichment, and training procedures; behavioral observations; and temperament assessments, end up benefitting the animals and research.

Validity

The overall validity of research involving animals as models is typically enhanced when behavioral management procedures are employed. Normalizing the behavioral (and physiological) responses of research animals through attempts to establish FACEs are likely to result in data that are worthwhile and that can be generalized beyond the exact experimental conditions and subjects involved.

Variability

Interindividual variability in data (sometimes referred to as error variance) can be reduced when behavioral management techniques provide refinements that help to better define animal models. Just as one attempts to control as many potential confounding factors as possible (e.g., genetics, sex, age), additional potential confounds can be controlled (e.g., housing, sampling procedures) when behavioral management techniques are employed. While many research approaches seek to minimize interindividual variation (Richter et al. 2010), there are other approaches that focus on exploring individual differences (Capitanio 2017). There are unlikely to be procedural differences for studies that focus on individual variation (subjects in these studies would still be likely to benefit from behavioral management); however, the analytical approach is likely to differ.

Subject Selection

There is the potential for subtle biases associated with behavioral management to creep into the processes that are used to select subjects for projects. The most straightforward involves training subjects for voluntary participation in experimental procedures. When training subjects for experiments, there is a tendency to only study those animals that achieve the initial training goals, allowing them to "qualify" as potential participants in the project. Those animals that cannot be trained are not eligible as subjects. It has been shown that animals with certain temperaments are more difficult to train with PRT techniques (Capitanio 2017; Coleman 2012, 2017; Coleman et al. 2005), and the possibility exists that animals with specific temperament types could be systematically excluded from research projects, diminishing the validity of the research.

When studying the behavior of animals under experimental conditions, it is quite important that the observers are "blinded" to the treatment or phase that each animal is in. This is another type of bias, which, although not necessarily directly related to behavioral management, is important for behavioral managers to consider in their programs.

Expanded Capabilities

In general, and especially in terms of PRT, behavioral management can expand the capabilities of caregivers and researchers. Animals that are (1) easily handled, using "gentle" techniques; (2) trained to voluntarily provide biological samples; and/or (3) trained to voluntarily participate in their own health care, all expand the number of minimally stressful routes by which we can interact and communicate with our animals. It would be very, very difficult to use a nebulizer multiple times each day to treat lung ailments in chimpanzees, if the animals were not trained (via PRT) to station and place their face into the nebulizer mask. Similarly, using acupuncture to treat arthritis symptoms (Magden et al. 2013) would be problematic with untrained animals. In fact, these and related techniques (Graham 2017; Graham et al. 2012) allow for the collection of valid, but not necessarily "typical" samples; samples that are considerably less likely to be affected by the stress of manual and/or chemical restraint (anesthesia; Lambeth et al. 2006).

Most behavioral management programs also result in an increase in the number of individuals that observe captive animals each day. In addition to the standard health checks performed by veterinarians and/or technicians, staff members are likely to check socially housed animals for compatibility, distribute environmental enrichment, administer cognitive tests, train animals, and/or collect behavioral data. This increase in the number of times that animals are observed, and the number of people observing, improves the healthcare of the animals, allowing for earlier detection of potential problems. Additionally, changes in use of enrichment or performance on experimental tasks can alert caregivers and researchers to issues related to declining health (Schapiro et al. 2021).

Ethical Implications

At this point in the maturation of animal care programs, it would be unethical not to provide any type of behavioral management program for laboratory animals, especially for the "higher" animal taxa (but, as you will read in the chapters that follow, even the welfare of fish can be improved with the implementation of appropriate socialization and enrichment strategies (Powell et al. 2021)). It is our responsibility to provide captive animals with the best environments possible, to refine conditions/treatments that promote their participation in our work, and to reduce the number required for specific projects. If animals are going to be maintained in laboratory settings and are going to participate in impactful research, then we owe it to them to make certain that our work has the highest probability of being of value to animals and/or humans, while "costing" the animals as little as possible (Schapiro et al. 2021). As mentioned previously, behavioral management is intended to serve the animals and the science. It is difficult to conceive of situations where the implementation of appropriate behavioral management strategies would be unethical. Perhaps, housing incompatible individuals together would be an example, but no behavioral manager is likely to do that, once incompatibility has been determined. Remember that most socially living animals have evolved ways to deal with inequitable access to resources (e.g., dominance hierarchies), so it is not unethical to have low-ranking animals in social groups.

Exceptions

There are relatively few laboratory situations in which no form of behavioral management can be provided for the animals. As mentioned at the beginning of this chapter, behavioral management programs typically involve socialization, enrichment, and training. If, for research, health, or compatibility reasons, animals must be housed individually, additional enrichment or training can be provided. It is important to note that even nonhuman primates involved in infectious disease research, or those with implants and head caps, can be successfully housed socially (DiVincenti and Wyatt 2011). Similarly, if foraging enrichment is not possible, again due to study concerns, other categories of enrichment can be emphasized.

How to – Implementation

Most of the remainder of this chapter focuses on issues related to the practical implementation of behavioral management programs. Such practical issues often have considerable influence on the quantity and quality of behavioral management enhancements provided to captive animals.

Behavioral management programs probably function best when directed by behavioral scientists, individuals who have been trained in the natural behavior of the animals, captive management techniques and principles, as well as the collection of behavioral data. In many circumstances, and at many facilities, de facto behavioral management programs are directed by veterinarians and/or researchers, which typically works quite well. Because behavioral management is simply one component of captive management, it makes sense for there to be considerable integration among the veterinary, care, behavioral, and research staffs. Behavioral managers cannot simply do whatever they want; they must work within the constraints imposed by the immutable characteristics of the captive environment and any restrictions associated with research protocols, and must be contributing members of the management team.

Behavioral management programs work best when they are cooperative endeavors involving multiple members of the care, veterinary, research, and behavior staffs. The behavioral expertise of members of the behavior staff can be optimized by making them responsible for the development of the program and the initial phases of socialization, enrichment, training, and observation. As procedures and programs mature, and the experience of other members of the team increases, responsibility for many day-to-day procedures can be transferred to members of other units on the team. For example, attempts at socialization should be identified and monitored by members of the behavioral staff, while still taking into account the familiarity of the care staff with the animals. Environmental enrichment devices can be selected, prepared, and cleaned by the behavioral staff, but can be distributed by members of the care staff during the performance of their regular duties. And finally, the initial training of new animals and/or behaviors should be performed by behavioral staff, but ultimately, the training to maintain these behaviors should be transferred to staff members from other components of the team. This is particularly important in scenarios where veterinarians and/or veterinary technicians are involved in training for the collection of biological samples, especially when venipuncture is to be performed. Veterinarians and veterinary technicians can be perceived negatively by animals, if the only times that the animals see them are when the animals are sick, restrained, or injected. It is important to desensitize the animals to the presence of the veterinary staff by providing opportunities for interactions between veterinarians and animals that are at least neutral, but hopefully, can be positive (rather than negative), for the animals. Behavioral management goals are not advanced if the animals voluntarily present for an injection in the presence of the trainer during the training process, but will not present when the veterinarian *must* give an injection.

It should be acknowledged that some behavioral management procedures can make husbandry tasks more difficult, creating extra work for the care staff. Every attempt should be made to mitigate any negative consequences of behavioral management procedures on care and veterinary staff. As an example, the provision of shelters for captive animals may make it more difficult for care staff to see the animals and to clean their enclosures. Similarly, housing animals socially increases the probability of injuries that require personnel time for observation and/or treatment.

There is no wrong time to implement behavioral management procedures; however, it is important to make certain that behavioral management only minimally disrupts the schedules of the other members of the animal care and/or research teams. This may sound as if behavioral management procedures are driven by human convenience, rather than animal needs, a situation that has already been identified as less than ideal. However, this is reality; there will be times when it is suboptimal to utilize behavioral management techniques (e.g., as animals are prepared for, or are recovering from, surgery/anesthesia; during cage changing procedures). In general, "permanent" behavioral management applications (e.g., social partner(s), physical enhancements) should be available daily, year-round, for as long as possible each day. Obviously, environmental enhancements intended to stimulate activity should be provided during the active portion of the animals' daily cycle, rather than during the inactive portion (an issue for nocturnal species that are not maintained on reversed light-dark cycles). Although virtually all facilities care for their animals every day, including weekends and holidays, many types of 'temporary/renewable' forms of behavioral management (e.g., feeding enrichment, PRT) are typically only provided on regular work/weekdays.

A way to think about the question of how much behavioral management animals need, is in terms of (1) the quantity of empty time they have, (2) the number of procedures in which they participate, and (3) the amount of stress they experience. In most cases, the more empty time, procedures, and stress the animals have, the more behavioral management they should receive, to help prevent the development of abnormal behaviors (by providing opportunities to perform species-typical behaviors at species-appropriate levels). If the typical vision of a behavioral management intervention is to promote action on the part of the animals, this may have to be modified somewhat in the case of animals that spend large portions of their day inactive under natural conditions, such as cats (Stella 2021). Of course, there are many practical constraints on the amount of behavioral management that animals experience, including those related to time, money, research participation, and personnel qualifications (among many others). It is simply not practical to provide mice with one or two seeds at a time, all night long, in order to exactly mimic their foraging behavior in the wild. Behavioral management programs that result in the animals spending most of their time in species-typical activities, and little, if any, time engaging in abnormal behaviors, have probably satisfactorily addressed the question of the proper quantity of behavioral management needed by those animals. Behavioral management actions that are not used by animals suggest that these approaches exceed what the animals need (or represent enhancements that do not address critical components of the animals' behavioral biology).

Behavioral management strategies always involve easily quantifiable costs in terms of time, money, and personnel. Such expenditures should be viewed as part of the cost of doing business (captive management), and should be included in the overall captive care budget. These costs should be re-captured as a component of per diem charges, with species, studies, or settings that require (and receive) "special" behavioral management requiring the application of supplemental charges. Over time, the costs associated with behavioral management will most likely serve as investments that yield dividends later in the animals'/program's life.

As mentioned earlier in the chapter, certain research procedures may also serve as environmental enrichment (Hopper et al. 2016). Tasks that require the animals to exercise their cognitive abilities to solve problems and/or earn rewards are likely to be reinforcing for the animals, either with or without food or liquid rewards. Even if the research tasks are not particularly cognitively demanding, they are still likely to provide beneficial opportunities for the animals to fill some of their "empty" time and to control at least that aspect of their lives (Herrelko et al. 2012; Yamanashi and Hayashi 2011). Research participation is less likely to be enriching in scenarios in which the animals are food or water "restricted" (Toth and Gardiner 2000) as a means to enhance their motivation to perform the experimental tasks. Animals are also unlikely to find situations that involve potentially stressful experimental procedures, such as the Morris water maze (D'Hooge and De Deyn 2001), particularly enriching.

Participation in behavioral research protocols has been shown to have positive effects on measures of welfare in captive chimpanzees (Neal Webb et al. 2019b). Additionally, there is a cumulative effect of PRT; animals trained for research purposes learn health care behaviors faster, and animals trained for health care behaviors learn research behaviors more quickly.

Animal and human safety are critical components of successful behavioral management programs. The potential for injury to animals and humans as a function of behavioral management procedures must be minimized. Providing enhancements to cages/enclosures is valuable, as long as the enhancements do not directly harm the animals, lead to increased aggression or injury, and/or make it difficult or dangerous to observe the animals or clean their cages/enclosures.

Relative to What Is in This Book

Each of the taxon-based chapters in this volume contain sections on the "natural history" and on "ways to maintain behavioral health" for that taxon. As was mentioned at the very beginning of this chapter, knowledge of the natural behavioral biology of your animals is an essential foundation for building an effective behavioral management program. As you read each of the taxon-based chapters, the sections on ways to maintain behavioral health are essentially describing specifics of the behavioral management programs that can, should, and are used with those animals. The current chapter is primarily intended to provide an overall framework and justification for behavioral management, as well as for attempts to create FACEs that promote welfare for captive animals and enhance the utility of the data collected.

Future Directions

Behavioral management programs are continuously evolving. As general care techniques and research techniques become ever more sophisticated, behavioral management procedures must keep pace (and even try to forge ahead). Behavioral managers should always be attempting to take captive care to the "next level". In most scenarios, social housing with compatible conspecifics is the most effective behavioral management procedure. As we better understand the behavioral biology of the animals we care for, we should become more and more successful at providing animals with social environments that maximize benefits and minimize risks; limiting, and hopefully eliminating, situations in which we must house an animal alone. We should also continue to strive to establish Functionally Appropriate Captive Environments that establish a workable balance between animal needs and human "convenience", stimulating the performance of species-typical behaviors at species-appropriate levels, while allowing humans to effectively do their work. Providing opportunities for the animals to voluntarily participate in their own care is one important step toward taking animal care to the next level. More widespread implementation of PRT techniques and the use of choice procedures that allow animals to choose their treatments (e.g., self-medication, acupuncture, laser therapy), provide additional channels of communication between human caregivers and the animals, which give the animals opportunities to control aspects of the way they live and are treated.

Conclusions

Many words in this chapter have been devoted to making some fairly straightforward statements. Behavioral management serves the animals and the science, is based on an understanding of the natural behavioral biology of the animal, is one component of overall captive care, and is essentially a combination of

- Socialization strategies
- Environmental enrichment
- PRT
- Behavioral observations
- Temperament assessments

all of which are intended to facilitate the development of Functionally Appropriate Captive Environments designed to

- Promote high levels of well-being, welfare, and wellness
- Enhance the definition of animals as valid models for research projects
- Provide opportunities for animals to express species-appropriate behaviors
- Minimize opportunities for animals to engage in abnormal behaviors by limiting the quantity of 'empty' time experienced by the animals
- Address two of the 3Rs: Refinement and Reduction

This chapter is intended to provide some overall context for subsections of later chapters that focus on the captive management of particular taxonomic groups.

Acknowledgments

Thanks are due to The University of Texas MD Anderson Cancer Center's Department of Comparative Medicine for continuing support of behavioral management initiatives at the Keeling Center. Our animals have truly benefited from the commitment of the Keeling Center's administration, veterinarians, scientists, and staff. Special thanks to the animal caregivers and behavioral management people who continue to go above and beyond for the sake of our animals.

REFERENCES

Abelson KSP, Jacobsen KR, Sundbom R, et al. Voluntary ingestion of nut paste for administration of buprenorphine in rats and mice. *Lab Animals.* 2012;46:349–351.

Altmann J. Observational study of behavior: Sampling methods. *Behaviour.* 1974;49:227–267.

Arakawa H. Ethological approach to social isolation effects in behavioral studies of laboratory rodents. *Behav Brain Res.* 2018;341:98–108.

Bailoo JD, Murphy E, Boada-Sana M, et al. Effects of cage enrichment on behavior, welfare and outcome variability in female mice. *Front Behav Neurosci.* 2018;12:232.

Baker KC, Bloomsmith MA, Coleman K, et al. The behavioral management consortium: A partnership for promoting consensus and best practices. In Schapiro SJ, ed. *Handbook of Primate Behavioral Management.* Boca Raton, FL: CRC Press, Taylor & Francis Group; 2017: 9–24.

Baker KC, Bloomsmith MA, Oettinger B, et al. Comparing options for pair housing rhesus macaques using behavioral welfare measures: Options for pair housing rhesus macaques. *Am J Primatol.* 2014;76(1):30–42.

Bassett L. Effects of predictability of feeding routines on the behaviour and welfare of captive primates. Unpublished PhD dissertation, University of Stirling, Scotland. 2003.

Bassett L, Buchanan-Smith HM. Effects of predictability on the welfare of captive animals. *Appl Anim Behav Sci.* 2007;102:223–245.

Benaroya-Milshstein N, Hollander N, Apter A, et al. Environmental enrichment in mice decreases anxiety, attenuates stress responses and enhances natural killer cell activity. *Eur J Neurosci.* 2004;20:1341–1347.

Bettinger TL, Leighty KA, Daneault RB, Richards EA, Bielitzki JT. Behavioral management: The environment and animal welfare. In Schapiro SJ, ed. *Handbook of Primate Behavioral Management.* Boca Raton, FL: CRC Press, Taylor & Francis Group; 2017:37–51.

Bliss-Moreau E, Theil JH, Moadab G. Efficient cooperative restraint training with rhesus macaques. *J Appl Anim Welf Sci.* 2013;16:98–117.

Bloomsmith MA. Behavioral management of laboratory primates: Principles and projections. In Schapiro SJ, ed. *Handbook of Primate Behavioral Management.* Boca Raton, FL: CRC Press, Taylor & Francis Group; 2017:497–513.

Bloomsmith MA, Baker KC. Social management of captive chimpanzees. In Brent L, ed. *Special Topics in Primatology (Vol 2): Care and Management of Captive Chimpanzees.* San Antonio, TX: American Society of Primatologists; 2001:16–37.

Bloomsmith MA, Baker KC, Lambeth SP, Ross SK, Schapiro SJ. Is giving chimpanzees control over environmental enrichment a good idea? In *The Apes: Challenges for the 21st Century,* conference proceedings. Brookfield, IL: Chicago Zoological Society; 2000:88–89.

Bloomsmith MA, Brent LY, Schapiro SJ. Guidelines for developing and managing an environmental enrichment program for nonhuman primates. *Lab Anim Sci.* 1991;41:372–377.

Bloomsmith MA, Hasenau J, Bohm RP. Position statement: "Functionally Appropriate Nonhuman Primate Environments" as an alternative to the term "Ethologically Appropriate Environments." *J Am Assoc Lab Anim Sci.* 2017;56(1):5.

Bloomsmith MA, Lambeth SP. Effects of predictable versus unpredictable feeding schedules on chimpanzee behavior. *Appl Anim Behav Sci.* 1995;44:65–74.

Bloomsmith MA, Laule GE, Alford PL, Thurston RH. Using training to moderate chimpanzee aggression during feeding. *Zoo Biol.* 1994;13:557–566.

Bloomsmith MA, Perlman JE, Hutchinson E, Sharpless M. Behavioral management programs to promote laboratory animal welfare. In Weichbrod RH, Thompson GA, Norton JH, eds. *Management of Animal Care and Use Programs in Research, Education, and Testing.* Boca Raton, FL: CRC Press; 2018:63–82.

Bloomsmith MA, Stone AM, Laule GE. Positive reinforcement training to enhance the voluntary movement of group-housed chimpanzees within their enclosures. *Zoo Biol.* 1998;17:333–341.

Broom DM. Animal welfare: concepts and measurement. *J Anim Sci.* 1991;69:4167–4175.

Buchanan-Smith HM, Badihi I. The psychology of control: Effects of control over supplementary light on welfare of marmosets. *Appl Anim Behav Sci.* 2012;137(3/4):166–174.

Caine NG. Antipredator behavior: Its expression and consequences in captive primates. In Schapiro SJ, ed. *Handbook of Primate Behavioral Management.* Boca Raton, FL: CRC Press, Taylor & Francis Group; 2017:127–138.

Caine NG, Weldon PJ. Responses by red-bellied tamarins (Saguinas labiatus) to fecal scents of predatory and non-predatory neotropical mammals. *Biotropica.* 1989;21:186–189.

Capitanio JP. Variation in biobehavioral organization. In Schapiro SJ, ed. *Handbook of Primate Behavioral Management.* Boca Raton, FL: CRC Press, Taylor & Francis Group; 2017:55–73.

Capitanio JP, Blozis SA, Snarr J, Steward A, McCowan BJ. Do "birds of a feather flock together" or do "opposites attract"? Behavioral responses and temperament predict success in pairings of rhesus monkeys in a laboratory setting. *Am J Primatol.* 2017;79:e22464.

Christensen JW. Behavioral biology of horses. In Coleman K, Schapiro SJ, eds. *Behavioral Biology of Laboratory Animals.* Boca Raton, FL: CRC Press, Taylor & Francis Group; 2021:285–297.

Clark FE, Smith LJ. Effect of a cognitive challenge device containing food and non-food rewards on chimpanzee well-being. *Am J Primatol.* 2013;75:807–816.

Clegg ILK, Rodel HG, Mercera B, et al. Dolphins' willingness to participate (WtP) in positive reinforcement training as a potential welfare indicator, where WtP predicts early changes in health status. *Front Psych.* 2019;10:2112.

Coleman K. Individual differences in temperament and behavioral management practices for nonhuman primates. *Appl Anim Behav Sci.* 2012;137:106–113.

Coleman K. Individual differences in temperament and behavioral management. In Schapiro SJ, ed. *Handbook of Primate Behavioral Management.* Boca Raton, FL:CRC Press, Taylor & Francis Group; 2017:95–113.

Coleman K, Heagerty A. 2019. Human-animal interactions in the research environment. In Hosey G, Melfi V, eds. *Anthrozoology.* Oxford, UK: Oxford University Press; 2019:59–80.

Coleman K, Maier A. The use of positive reinforcement training to reduce stereotypic behavior in rhesus macaques. *Appl Anim Behav Sci.* 2010;124:142–148.

Coleman K, Pranger L, Maier A, et al. Training rhesus macaques for venipuncture using positive reinforcement techniques: A comparison with chimpanzees. *J Am Assoc Lab Anim Sci.* 2008;47:37–41.

Coleman K, Tully LA, McMillan JL. Temperament correlates with training success in adult rhesus macaques. *Am J Primatol.* 2005;65:63–71.

Crockett CM, Bellanca RU, Bowers CL, Bowden DM. Grooming-contact bars provide social contact for individually caged laboratory macaques. *Contemp Topics Lab Anim Sci.* 1997;36:53–60.

DiVincenti Jr., L, Wyatt JD. Pair housing of macaques in research facilities: A science-based review of benefits and risks. *J Am Assoc Lab Anim Sci.* 2011;50:856–863.

Doherty FD, O'Mahony SM, Peterson VL, et al. Post-weaning social isolation of rats leads to long-term disruption of the gut microbiota-immune-brain axis. *Brain Behav Immun.* 2018;68:261–273.

Dufour BD, Garner JP. An ethological analysis of barbering behavior. In Kalueff AV, LaPorte, JL, Bergner CL, eds. *Neurobiology of Grooming Behavior.* Cambridge: Cambridge University Press; 2010:184–225.

Dwyer CM. The behavioral biology of sheep and implications for their use in research. In Coleman K, Schapiro SJ, eds. *Behavioral Biology of Laboratory Animals.* Boca Raton, FL: CRC Press, Taylor & Francis Group; 2021:261–271.

D'Hooge R, De Deyn PP. Applications of the Morris water maze in the study of learning and memory. *Brain Res Rev.* 2001;36:60–90.

Edwards S, Grand N. Behavioral biology of pigs and minipigs. In Coleman K, Schapiro SJ, eds. *Behavioral Biology of Laboratory Animals.* Boca Raton, FL: CRC Press, Taylor & Francis Group; 2021:243–259.

Elliott BM, Grunberg NE. Effects of social and physical enrichment on open field activity differ in male and female Sprague-Dawley rats. *Behav Brain Res.* 2005;165:187–196.

Fagot J, Boe LJ, Berthomier F, et al. The baboon: A model for the study of language evolution. *J Hum Evo.* 2019; 126:39–50.

Fagot J, Marzouki Y, Huguet P, Gullstrand J, Claidiere N. Assessment of social cognition in non-human primates using a network of computerized automated learning device (ALDM) test systems. *J Vis Exp.* 2015;99:e52798.

Fernandez L, Griffiths M-A, Honess P. Providing behaviorally manageable primates for research. In Schapiro SJ, ed. *Handbook of Primate Behavioral Management.* Boca Raton, FL: CRC Press, Taylor & Francis Group; 2017:481–494.

Fraser D. Assessing animal welfare: Different philosophies, different scientific approaches. *Zoo Biol.* 2009;28:507–518.

Garner JP, Dufour B, Gregg LE, Weisker SM, Mench JA. Social and husbandry factors affecting the prevalence and severity of barbering ('whisker trimming') by laboratory mice. *Appl Anim Behav Sci.* 2004;89:263–282.

Gaskill BN, Garner JP. Stressed out: Providing laboratory animals with behavioral control to reduce the physiological effects of stress. *Lab Anim.* 2017;46:2142–145.

Gaskill BN, Karas AZ, Garner JP, Pritchett-Corning KR. Nest building as an indicator of health and welfare in laboratory mice. *J Vis Exp.* 2013;82:e51012.

Gjendal K, Ottesen JL, Olsson IAS, et al. Burrowing and nest building activity in mice after exposure to grid floor, isoflurane or IP injections. *Phys Behav.* 2019;206:59–66.

Gottlieb DH, Capitanio JP, McCowan B. Risk factors for stereotypic behavior and self-biting in rhesus macaques (*Macaca mulatta*): Animal's history, current environment, and personality. *Am J Primatol.* 2013a;75(10):995–1008.

Gottlieb DH, Coleman K, McCowan B. The effects of predictability in daily husbandry routines in captive rhesus macaques (*Macaca mulatta*). *Appl Anim Behav Sci.* 2013b;143:117–127.

Gottlieb DH, Coleman K, Prongay K. 2017. Behavioral management of *Macaca* species (except *Macaca fascicularis*). In Schapiro SJ, ed. *Handbook of Primate Behavioral Management.* Boca Raton, FL: CRC Press, Taylor & Francis Group; 2017:279–303.

Gottlieb DH, Maier A, Coleman K. Evaluation of environmental and intrinsic factors that contribute to stereotypic behavior in captive rhesus macaques (*Macaca mulatta*). *Appl Anim Behav Sci.* 2015;171:184–191.

Gouveia K, Hurst JL. Improving the practicality of using non-aversive handling methods to reduce background stress and anxiety in laboratory mice. *Sci Rep.* 2019;9:20305.

Graham ML. Positive reinforcement training and research. In Schapiro SJ, ed. *Handbook of Primate Behavioral Management.* Boca Raton, FL: CRC Press, Taylor & Francis Group; 2017:187–200.

Graham ML, Rieke EF, Mutch LA, et al. Successful implementation of cooperative handling eliminates the need for restraint in a complex non-human primate disease model. *J Med Primatol.* 2012;41:89–106.

Grandin T. Minimizing stress in pig handling in the research lab. *Lab Anim.* 1986;15(3):15–20.

Hanson JD, Larson ME, Snowdon CT. The effects of control over high intensity noise on plasma cortisol levels in rhesus monkeys. *Behav Biol.* 1976;16:333–340.

Harlow HF, Harlow MK. Social deprivation in monkeys. *Sci Am.* 1962;207:136–150.

Herrelko ES, Vick S-J, Buchanan-Smith HM. 2012. Cognitive research in zoo-housed chimpanzees: Influence of personality and impact on welfare. *Am J Primatol.* 2012;74:828–840.

Hopper LM, Lambeth SP, Schapiro SJ, Whiten A. Observational learning in chimpanzees and children studied through 'ghost' conditions. *Proc Biol Sci.* 2008;275:835–840.

Hopper LM, Shender MA, Ross SR. Behavioral research as physical enrichment for captive chimpanzees. *Zoo Biol.* 2016;35:293–297. https://doi.org/10.1002/zoo.21297.

Hubrecht R, Kirkwood J. *The UFAW Handbook on the Care and Management of Laboratory and Other Research Animals,* 8th Edition. England, UK: Wiley-Blackwell; 2010.

Johnson SR, Patterson-Kane EG, Niel L. Foraging enrichment for laboratory rats. *Anim Welf.* 2004;13:305–312.

Kalliokoski O, Abelson KSP, Koch J, et al. The effect of voluntarily ingested buprenorphine on rats subjected to surgically induced global cerebral ischaemia. *In Vivo.* 2010;24:641–646.

Kappel S, Hawkins P, Mendl MT. To group or not to group? Good practice for housing male laboratory mice. *Animals.* 2017;7:E88. https://doi.org/10.3390/ani7120088.

Kavanau JL, Havenhill RM. Compulsory regime and control of environment in animal behaviour III. Light level preferences of small nocturnal mammals. *Behaviour.* 1963;59:203–225.

Klein DC, Fencil-Morse E, Seligman ME. Learned helplessness, depression, and the attribution of failure. *J Pers Soc Psych.* 1976;33:508–516.

Krall C, Glass S, Dancourt G, et al. Behavioural anxiety predisposes rabbits to intra-operative apnoea and cardiorespiratory instability. *Appl Anim Behav Sci.* 2019;221:104875.

Kuczaj S, Lacinak T, Fad O, et al. Keeping environmental enrichment enriching. *Int J Comp Psychol.* 2002;15:127–137.

Lambeth SP, Hau J, Perlman JE, Martino M, Schapiro SJ. Positive reinforcement training affects hematologic and serum chemistry values in captive chimpanzees (*Pan troglodytes*). *Am J Primatol.* 2006;68:245–256.

Lambeth SP, Schapiro SJ. Managing a behavioral management program. In Schapiro SJ, ed. *Handbook of Primate Behavioral Management*. Boca Raton, FL: CRC Press, Taylor & Francis Group; 2017:395–407.

Laule GE. Positive reinforcement training for laboratory animals. In Hubrecht R, Kirkwood J, eds. *The UFAW Handbook on the Care and Management of Laboratory and Other Research Animals*, 8th Edition. London, UK: Wiley-Blackwell; 2010:206–218.

Laule GE, Bloomsmith MA, Schapiro SJ. The use of positive reinforcement training techniques to enhance the care, management, and welfare of primates in the laboratory. *J Appl Anim Welf Sci*. 2003;6:163–173.

Leidinger CS, Herrmann F, Thoene-Reineke C, et al. Introducing clicker training as a cognitive enrichment for laboratory mice. *J Vis Exp*. 2017;121:e55415.

Leidinger CS, Kaiser N, Baumgart N, Baumgart J. Using clicker training and social observation to teach rats to voluntarily change cages. *J Vis Exp*. 2018;140:e58511.

Lewejohann L, Reinhard C, Schrewe A, et al. Environmental bias? Effects of housing conditions, laboratory environment and experimenter on behavioral tests. *Genes Brain Behav*. 2006;5:64–72.

Lidfors L, Dahlborn K. Behavioral biology of rabbits. In Coleman K, Schapiro SJ, eds. *Behavioral Biology of Laboratory Animals*. Boca Raton, FL: CRC Press, Taylor & Francis Group; 2021:173–190.

Love JA, Hammond K. Group-housing rabbits. *Lab Anim*. 1991;20:37–43.

Luo L, Jansen CA, Bolhuis JE, et al. Early and later life environmental enrichment affect specific antibody responses and blood leukocyte subpopulations in pigs. *Phys Behav*. 2020;217:112799.

MacLellan A, Adcock A, Mason G. Behavioral biology of laboratory mice. In Coleman K, Schapiro SJ, eds. *Behavioral Biology of Laboratory Animals*. Boca Raton, FL: CRC Press, Taylor & Francis Group; 2021:89–111.

Magden ER. 2017. Positive reinforcement training and health care. In Schapiro SJ, ed. *Handbook of Primate Behavioral Management*. Boca Raton, FL: CRC Press, Taylor & Francis Group; 2017:201–215.

Magden ER, Haller RL, Thiele EJ, et al. Acupuncture as an adjunct therapy for osteoarthritis in chimpanzees (*Pan troglodytes*). *J Am Assoc Lab Anim Sci*. 2013;52:475–480.

Magden ER, Sleeper MM, Buchl SJ, et al. Use of an implantable loop recorder in a chimpanzee (*Pan troglodytes*) to monitor cardiac arrhythmias and assess the effects of acupuncture and laser therapy. *Comp Med*. 2016;66:52–58.

Manciocco A. Behavioral biology of marmosets. In Coleman K, Schapiro SJ, eds. *Behavioral Biology of Laboratory Animals*. Boca Raton, FL: CRC Press, Taylor & Francis Group; 2021:377–394.

Maple TL, Perdue BM. Wellness as welfare. In Maple TL, Perdue BM, eds. *Zoo Animal Welfare*. New York: Springer; 2013:49–67.

Marshall Bioresources website. https://www.marshallbio.com/

Mason GJ, Latham NR. Can't stop, won't stop: Is stereotypy a reliable animal welfare indicator? *Anim Welf*. 2004;13:S57–S69.

Mason GJ, Rushen J. *Stereotypic Animal Behaviour: Fundamentals and Applications to Welfare*. 2nd Edition. Cambridge, MA: CABI; 2006.

McGrew K. Pairing strategies for cynomolgus macaques. In Schapiro SJ, ed. *Handbook of Primate Behavioral Management*. Boca Raton, FL: CRC Press, Taylor & Francis Group; 2017:255–264.

McKinley J, Buchanan-Smith HM, Bassett L, Morris K. Training common marmosets (*Callithrix jacchus*) to cooperate during laboratory procedures: Ease of training and time investment. *J Appl Anim Welf Sci*. 2003;6:209–220.

Meunier LD. Selection, acclimation, training, and preparation of dogs for the research setting. *ILAR J*. 2006;47:326–347.

Michael KC, Bonneau RH, Bourne RA, et al. Divergent immune responses in behaviorally-inhibited vs. non-inhibited male rats. *Phys Behav*. 2020;213:112693.

Minier D, Capitanio JP, Gottlieb D, McCowan B. One size may not fit all: The importance of taking an individual differences approach to behavioral management. *The Enrichment Record*. 2012;13:21–23.

Morton DB, Hau J. Welfare assessment and humane endpoints. In Hau J, Schapiro SJ, eds. *Handbook of Laboratory Animal Science*, 4th Edition. Boca Raton, FL: CRC Press; 2021:123–154.

National Research Council. *Guide for the Care and Use of Laboratory Animals*, 8th Edition. Washington, DC: National Academies Press; 2011.

Neal Webb SJ, Hau J, Schapiro SJ. Refinements to captive chimpanzee (*Pan troglodytes*) care: A self-medication paradigm. *Anim Welf*. 2018a;27(4):327–341.

Neal Webb SJ, Hau J, Schapiro SJ. Captive chimpanzee (*Pan troglodytes*) behavior as a function of space per animal and enclosure type. *Am J Primatol*. 2018b;80:e22749.

Neal Webb SJ, Hau J, Schapiro SJ. Relationships between captive chimpanzee (*Pan troglodytes*) welfare and voluntary participation in behavioral studies. *Appl Anim Behav Sci*. 2019a;214:102–109.

Neal Webb SJ, Hau J, Schapiro SJ. Does (group) size matter? Captive chimpanzee behavior as a function of group size and composition. *Am J Primatol*. 2019b;81:e22947.

Neal Webb SJ, Lambeth SP, Hau J, Schapiro SJ. Differences in behavior between elderly and non-elderly captive chimpanzees and the effects of the social environment. *J Am Assoc Lab Anim Sci*. 2019;58:783–789.

Neely C, Lane C, Torres J, Flinn J. The effect of gentle handling on depressive-like behavior in adult male mice: Considerations for human and rodent interactions in the laboratory. *Behav Neurol*. 2018;2018:2976014.

Novak MA, Hamel AF, Kelly BJ, Dettmer AM, Meyer JS. Stress, the HPA axis, and nonhuman primate well-being: A review. *Appl Anim Behav Sci*. 2013;143:135–149.

Novak MA, Hamel AF, Ryan AM, Menard MT, Meyer JS. 2017. The role of stress in abnormal behavior and other abnormal conditions such as hair loss. In Schapiro SJ, ed. *Handbook of Primate Behavioral Management*. Boca Raton:CRC Press, Taylor & Francis Group; 2017:75–94.

Novak MA, Meyer JS. Abnormal behavior in animals in research settings. In Coleman K, Schapiro SJ, eds. *Behavioral Biology of Laboratory Animals*. Boca Raton, FL: CRC Press, Taylor & Francis Group; 2021:27–50.

Novak MA, Meyer JS. Alopecia: Possible causes and treatments, particularly in captive nonhuman primates. *Comp Med*. 2009;59:18–26.

Novak MA, Suomi SJ. Psychological well-being of primates in captivity. *Am Psychol*. 1988;43(10):765–773.

Ogura T, Matsuzawa T. Video preference assessment and behavioral management of single-caged Japanese macaques (*Macaca fuscata*) by movie presentation. *J Appl Anim Welf Sci*. 2012;15:101–112.

O'Malley CI, Turner SP, D'Eath RB, et al. Animal personality in the management and welfare of pigs. *Appl Anim Behav Sci*. 2019;218:104821.

Perlman JE, Horner V, Bloomsmith MA, Lambeth SP, Schapiro SJ. Positive reinforcement training, social learning, and chimpanzee welfare. In Lonsdorf EV, Ross SR, Matsuzawa T, eds. *The Mind of the Chimpanzee: Ecological and Experimental Perspectives*. Chicago, IL: University of Chicago Press; 2010:320–331.

Phillips C. Behavioral biology of cattle. In Coleman K, Schapiro SJ, eds. *Behavioral Biology of Laboratory Animals*. Boca Raton, FL: CRC Press, Taylor & Francis Group; 2021:273–283.

Poole T. Happy animals make good science. *Lab Anim*. 1997;31:116–124.

Powell C, Fife-Cook I, Franks B. Behavioral biology of zebrafish. In Coleman K, Schapiro SJ, eds. *Behavioral Biology of Laboratory Animals*. Boca Raton, FL: CRC Press, Taylor & Francis Group; in press.

Prescott MJ, Buchanan-Smith HM. Training nonhuman primates using positive reinforcement techniques. *J Appl Anim Welf Sci*. 2003;6:157–161.

Pritchett-Corning KR, Winnicker C. Behavioral biology of deer and white-footed mice, Mongolian gerbils, and prairie and meadow voles. In Coleman K, Schapiro SJ, eds. *Behavioral Biology of Laboratory Animals*. Boca Raton, FL: CRC Press, Taylor & Francis Group; 2021:147–163.

Pruetz JDE, McGrew WC. What does a chimpanzee need? Using natural behavior to guide the care and management of captive population. In Brent L, ed. *The Care and Management of Captive Chimpanzees*. San Antonio, TX: American Society of Primatologists; 2001:16–37.

Reinhardt V, Roberts A. Effective feeding enrichment for non-human primates: A brief review. *Anim Welf*. 1997;6:265–272.

Richter SH, Garner JP, Auer C, Kunert J, Würbel H. Systematic variation improves reproducibility of animal experiments. *Nat Methods*. 2010;7:167–168.

Robertson LC, Anderson SC. The effects of differing type and magnitude of reward on the contrafreeloading phenomenon in rats. *Anim Learn Behav*. 1975;3:325–328.

Rommeck I, Capitanio JP, Strand SC, McCowan B. Early social experience affects behavioral and physiological responsiveness to stressful conditions in infant rhesus macaques (*Macaca mulatta*). *Am J Primatol*. 2011;73:692–701.

Ross SR, Schapiro SJ, Hau J, Lukas KE. Space use as an indicator of enclosure appropriateness: A novel measure of captive animal welfare. *Appl Anim Behav Sci*. 2009;121(1):42–50.

Russell WMS, Burch RL. *The Principles of Humane Experimental Technique*. London: Methuen Publishing; 1959.

Sambrook TD, Buchanan-Smith HM. Control and complexity in novel object enrichment. *Anim Welf*. 1997;6:207–216.

Scarola SJ, Trejo JRP, Granger ME, et al. Immunomodulatory effects of stress and environmental enrichment in Long-Evan rats (*Rattus norvegicus*). *Comp Med*. 2019;69:35–47.

Schapiro SJ. Effects of social manipulations and environmental enrichment on behavior and cell-mediated immune responses in rhesus macaques. *Pharmacol Biochem Behav*. 2002;73(1):271–278.

Schapiro SJ. *Handbook of Primate Behavioral Management*. Boca Raton: CRC Press, Taylor & Francis Group; 2017.

Schapiro SJ, Bloomsmith MA, Laule GE. Positive reinforcement training as a technique to alter nonhuman primate behavior: Quantitative assessments of effectiveness. *J Appl Anim Welf Sci*. 2003;6:175–187.

Schapiro SJ, Bloomsmith MA, Porter LM, Suarez SA. Enrichment effects on rhesus monkeys successively housed singly, in pairs, and in groups. *Appl Anim Behav Sci*. 1996;48(3/4):159–171.

Schapiro SJ, Brosnan SF, Whiten A, et al. Collaborative research and behavioral management. In Schapiro SJ, ed. *Handbook of Primate Behavioral Management*. Boca Raton, FL: CRC Press; 2017:243–254.

Schapiro SJ, Neal Webb SJ, Mulholland MM, Lambeth SP. Behavioral management is a key component of ethical research. *ILAR J*. 2021;59(2):ilaa023.

Schapiro SJ, Nehete PN, Perlman JE, Sastry KJ. A comparison of cell-mediated immune responses in rhesus macaques housed singly, in pairs, or in groups. *Appl Anim Behav Sci*. 2000;68:67–84.

Schipper LL, Vinke CM, Schilder MBH, Spruijt BM. The effect of feeding enrichment toys on the behaviour of kenneled dogs (*Canis familiaris*). *Appl Anim Behav Sci*. 2008;114:182–195.

Searcy YM, Caine NG. Hawk calls elicit alarm and defensive reactions in captive Geoffroy's marmosets (*Callithrix geoffroyi*). *Folia Primatol*. 2003;74:115–125.

Seligman MEP. Learned helplessness. *Ann Rev Med*. 1972;23:407–412

Smith AC, Swindle MM. Preparation of swine for the laboratory. *ILAR J*. 2006;47:358–363.

Stella J. Behavioral biology of the domestic cat. In Coleman K, Schapiro SJ, eds. *Behavioral Biology of Laboratory Animals*. Boca Raton, FL: CRC Press, Taylor & Francis Group; 2021:223–241.

Suomi SJ. Long-term effects of different earl rearing experiences on social, emotional, and physiological development in nonhuman primates. In Keshavan MS, Murray RM, eds. *Neurodevelopment & Adult Psychopathology*. Cambridge: Cambridge University Press; 1997:104–116.

Suomi SJ, Novak MA. The role of individual differences in promoting psychological well-being in rhesus monkeys. In Novak MA, Petto AJ, eds. *Through the Looking Glass: Issues of Psychological Well-Being in Nonhuman Primates*. Washington, DC: American Psychological Association; 1991:50–56.

Tanji K, Leopold DA, Ye FQ, et al. Effect of sound intensity on tonotropic fMRI maps in the unanesthetized monkey. *NeuroImage*. 2010;49:150–157.

Tobin G, Schumacher A. Laboratory animal nutrition in routine husbandry, and experimental and regulatory studies. In Hau J, Schapiro SJ eds. *Handbook of Laboratory Animal Science*, 4th Edition. Boca Raton, FL: CRC Press; 2021:269–312.

Toth LA, Gardiner TW. Food and water restriction protocols: Physiological and behavioral considerations. *J Am Assoc Lab Anim Sci*. 2000;39:9–17.

Trickett SL, Guy JH, Edwards SA. The role of novelty in environmental enrichment for the weaned pig. *Appl Anim Behav Sci*. 2009; 116:45–51.

Truelove MA, Martin AL, Perlman JE, et al. Pair housing of macaques: a review of partner selection, introduction techniques, monitoring for compatibility, and methods for long-term maintenance of pairs. *Am J Primatol*. 2017;79:e22485.

Tuchscherer M, Kanitz E, Puppe B, et al. Early social isolation alters behavioral and physiological responses to an endotoxin challenge in piglets. *Horm Behav*. 2006;50:753–761.

Vandeleest JJ, Beisner BA, Hannibal DL, et al. Decoupling social status and status certainty effects on health in macaques: A network approach. *PeerJ*. 2016;4:e2394.

Veeder CL, Bloomsmith MA, McMillan JL, Perlman JE, Martin AL. Positive reinforcement training to enhance the voluntary movement of group-housed sooty mangabeys. *J Am Assoc Lab Anim Sci*. 2009;48:192–195.

Vinke CM, Schoemaker NJ, van Zeeland YRA. The behavioral biology of the laboratory ferret (*Mustela putorius furo* T). In Coleman K, Schapiro SJ, eds. *Behavioral Biology of Laboratory Animals*. Boca Raton, FL: CRC Press, Taylor & Francis Group; 2021:191–204.

Winnicker C, Pritchett-Corning KR. Behavioral biology of hamsters. In Coleman K, Schapiro SJ, eds. *Behavioral Biology of Laboratory Animals*. Boca Raton, FL: CRC Press, Taylor & Francis Group; 2021:165–171.

Wirz A, Mandillo S, D'Amato FR, Giuliani A, Riviello MC. Response, use and habituation to a mouse house in C57BL/6J and BALB/c mice. *Exp Anim*. 2015;64:281–293.

Würbel H, Richter SH, Garner JP. Reply to: "Reanalysis of Richter *et al*. (2010) on reproducibility". *Nat Methods*. 2013;10:374.

Yamanashi Y, Hayashi M. Assessing the effects of cognitive experiments on the welfare of captive chimpanzees (*Pan troglodytes*) by direct comparison of activity budget between wild and captive chimpanzees. *Am J Primatol*. 2011;73:1231–1238.

Part 2

6

Behavioral Biology of Mice

Aileen MacLellan, Aimée Adcock, and Georgia Mason
University of Guelph

CONTENTS

Introduction	89
Typical Research Involvement	89
Behavioral Biology	90
Natural History	90
Ecology, Diets, and Threats	90
Senses and Communication	91
Social Organization and Behavior	92
Mating and Mate Choice	92
Reproduction and Maternal Behavior	93
Common Captive Behaviors	93
Normal Behaviors	93
Abnormal Behaviors	95
Ways to Maintain Behavioral Health	97
Captive Considerations to Promote Species-Typical Behavior	97
Environment	97
Social Groupings	100
Maternal Care and Optimal Weaning	101
Feeding Strategies	101
Handling and Other Ways to Reduce Fear of Humans	101
Training	101
Special Situations	102
Conclusions/Recommendations	102
Additional Resources	103
References	103

Introduction

Laboratory mice (*Mus musculus*) are the most widely used vertebrate species in research, with an estimated 85 million used per year in the USA (Schipani 2019), 8 million in Europe (European Commission 2013), and 1.4 million in Canada (CCAC 2017). Invaluable for their small size, diverse strains, and short generation times, mice have made many research triumphs possible, especially in drug discovery and the understanding of human diseases. The number of papers generated by the search term "mice" tops 1.5 million in ISI Web of Science (accessed April 2020), and Viney et al. (2015, p. 267) suggest "the laboratory mouse should be awarded an honorary Nobel Prize for its contribution to science". The first inbred strain of mice (DBA) was created about 100 years ago, but mice have been used in research for the past 300 years (Würbel et al. 2017). Since mice can breed at a rate of approximately 4 generations per year, modern day laboratory mice are therefore the product of hundreds of generations of captive breeding. Here, we (1) review the behavioral biology of their forebears, free-living wild *Mus musculus*; (2) describe the common behaviors of captive laboratory mice; (3) outline husbandry practices that aid in maintaining their behavioral health; and (4) end with some of the fascinating challenges of working with a species that is highly developmentally plastic and has a sensory world very unlike our own.

Typical Research Involvement

Mice are involved in a broad range of fields to study biological functions and processes, often in translational or applied research into the diagnosis and treatment of human diseases (e.g., cancer). Mice living beyond just 1 year of age in the laboratory can also be categorized as aged, and are useful for researchers interested in health changes during middle age and beyond (Aujard et al. 2001). Furthermore, there are approximately 55 "off-the-shelf" phenotypically different mouse strains readily available from sources like the Jackson Laboratory or Charles River Laboratories: a diverse array that provides subjects with characteristics differentially useful for specific fields of research (e.g., BALB/c mice for autism research (Brodkin 2007)). Mice also dominate studies involving genetically modified animals (accounting for 89% of such animals in the UK; Home Office 2017), due to genetic engineering technologies, including CRISPR and the establishment of mutagenesis and gene trap centers (Contet et al. 2001; Eppig et al. 2005; Singh et al. 2015). As a result, the International Mouse Strain Resource (IMSR) lists over 5,700 strains of live laboratory mice, with over 280,000 strains potentially available in various other states in repositories (e.g., as embryonic stem cells, sperm, embryos, ovaries (Eppig et al. 2015)).

Behavioral Biology

Natural History

Ecology, Diets, and Threats

The house mouse, the primary ancestor of the laboratory mouse, is one of the world's most widespread mammals (Latham and Mason 2004; Frynta et al. 2018), ubiquitous in all regions save the poles, the Sahara, and Amazonia (Phifer-Rixey and Nachman 2015). This is principally because mice have followed humans around the world. Most are "commensal", meaning dependent on humans for food and shelter, and live in residential, agricultural, and commercial structures (Phifer-Rixey and Nachman 2015), a strategy that has evolved independently at least three times in different subspecies (Singleton and Krebs 2007). Even those that are feral, living more like other wild rodents, thrive best in agricultural areas, especially those producing cereals (Morris et al. 2012). Their success stems from their great adaptability, which in turn reflects both impressive genetic potential, and phenotypic and behavioral plasticity. For example, mice can survive without water (if their food has some water content), breed in cold stores at temperatures as low as −30°C (providing they have nesting material), and even live in mines, down to depths of 600 m (Latham and Mason 2004). This immense resilience also helps explain why they have been able to become such a very common laboratory species (Mason et al. 2013).

Mice are highly omnivorous. They are well adapted to eat seeds and grains (Morris et al. 2012), but their diet can also include caterpillars, worms, and spiders (Shiels and Pitt 2014), and even vertebrates; mice on the South Atlantic's Gough Island, for example, predate seabird chicks many times their own size (Wanless et al. 2007; Cuthbert et al. 2016). Mice are typically cautious when it comes to new dietary opportunities, thanks to a long history of being poisoned by humans and an inability to vomit. Therefore, the first ingestion of a new food typically comprises small, cautious nibbles (Latham and Mason 2004; Deacon 2011). But mice are also opportunistic and exploratory, and their acceptance of novel foods or foods in novel locations can be socially facilitated. They are more likely to enter "bait stations", for instance, if these smell of other mice (Volfová et al. 2010), and are quicker to ingest novel foods if they have previously smelled these on the breath or in anogenital odors of conspecifics (Galef 1993; Valsecchi et al. 1996; Forestier et al. 2018).

Mouse lifestyles vary in more than just diet, with the differences between commensal and feral animals being among the most well studied. As reviewed by Latham and Mason (2004), feral mice are strictly crepuscular or nocturnal (to avoid predators). They typically nest in burrows (that they may dig themselves), and they breed only in spring, summer, and fall. Their home ranges can be hundreds or even thousands of square meters in size, because their food is dispersed. Feral mice thus generally only use part of their range each day, and their ranges can also vary in size and shape over time. This plasticity reflects changes in food availability and also social density, which in turn affects territoriality (reviewed by Singleton and Krebs 2007). Feral mice seem to maintain exclusive territories when social densities are low and/or during the breeding season. But they do not when social densities are relatively high and/or outside of the breeding season, instead having ranges that overlap with those of their neighbors. In contrast, commensal mice typically have much higher population densities, because they have ample food and relatively low mortality rates (see also Viney et al. 2015), and they are also consistently territorial. Commensal mice thus defend small, stable, well-marked home areas that are often three-dimensional, and may be as small as 1–2 m^3. They travel around these territories multiple times daily, sometimes even during daylight (Spoelstra et al. 2016). Their nests, structured well-insulated balls, are constructed in any suitable protected site (e.g., behind rafters or in wall cavities), and because commensal mice are buffered from seasonal changes in temperature and food availability, breeding is typically year-round.

The variation between the lifestyles of feral and commensal mice, and even within feral populations, illustrates the great flexibility of house mice. This variability reflects both genetic and experiential effects. Considering genetic adaptability first, populations of wild house mice exhibit several-fold higher genetic variation than do humans (Phifer-Rixey and Nachman 2015), and this, combined with their rapid breeding rates, means that mouse populations can adapt rapidly to new conditions. Resulting genetically mediated changes include population differences in body weight, territoriality, routine-formation, nest-building, anxiety, aggression, and burrowing (Latham and Mason 2004; Singleton and Krebs 2007; Frynta et al. 2018); both physiological and behavioral adaptations to cold temperatures (Phifer-Rixey and Nachman 2015); and the evolution of rodenticide resistance (Singleton and Krebs 2007; Song et al. 2011). Genetically mediated differences between commensal and feral populations *per se* include reduced female-female aggression (female tolerance of unfamiliar females seeming to be an adaptation for life at high densities), combined with greater tendencies to utilize vertical space (e.g., showing more climbing during behavioral tests): a useful ability for mice living in buildings (Frynta et al. 2005, 2018; Haisová-Slábováet al. 2010).

Mice also show developmental plasticity, with early, and even adult, experiences modifying their phenotypes in ways hypothesized to suit their local environments. Thus, at least in some laboratory strains (data seeming absent for wild mice), pup development is shaped by diverse pre- and postnatal events. For example, as discussed by Latham and Mason (2004) and Laviola and Terranova (1998), stress during gestation can increase pups' later response to stressors, and also masculinize female pups, while feminizing male pups. Postnatally, low food availability during gestation can then decrease weaning weight and increase later aggressiveness. The amount of milk received can affect a pup's adult weight and aggressiveness, while the amount of maternal licking and grooming they receive influences their later corticosterone responses to stress. Subsequent work has also found lasting beneficial effects of communal rearing (being raised with at least one other mother and her litter) on pups' later maternal care and social responses (Taborsky 2016). Communal rearing may also cause impressive developmental changes in pups exposed to cues signaling high future male-male competition; male pups maturing in environments rich in adult male odors become larger and develop thicker bacula (penis bones), larger testes, and faster sperm production rates than do young males maturing in environments without these odors (Ramm et al. 2015; André et al. 2018).

After weaning, several aspects of diet can also have lasting phenotypic effects. For example, experimentally raising young mice with hard diets shapes their jaw morphology in a functional way, making their mastication mechanically more effective (Anderson et al. 2014). Dietary composition also shapes gastro-intestinal development; young mice fed relatively high protein diets for a month develop large stomachs, while those fed carbohydrate-rich diets develop large intestines, caeca, and colons (Sørensen et al. 2010).

Mice also show behavioral plasticity through their lives. The omnivory, willingness to try new foods, and flexible display of territorial behavior (at least in feral populations) outlined above are three examples. Mice are also flexible in their social structures and behaviors (see below), including flexible maternal behaviors (e.g., abilities to shift between solitary and communal nesting (Ferrari et al. 2019); see below); flexible antipredator behaviors (e.g., which diminish when food availability is low (Singleton and Krebs 2007)); and in the laboratory, facultative abilities to adjust nest sizes in response to ambient temperatures (Rajendram et al. 1987; Gaskill et al. 2012). Furthermore, some of their learning abilities are impressive. Mice displaced several hundred meters can navigate back to their nest-sites, for example, and show evidence of path integration (reviewed by Latham and Mason 2004; see also Poletaeva et al. 2001; Perepelkina et al. 2014), indicative of good spatial learning abilities. Experimental work on laboratory strains also suggests that mice have episodic memory (being able to use the "what, where, and when" of past events to make decisions (Dere et al. 2005)), and even basic deductive "reasoning", in tests for intelligence (Wass et al. 2012).

The lives of wild mice are naturally short (often just a matter of months) and risk-filled. Poison threats have already been outlined above. For males, intraspecific agonism can also be a major cause of mortality, at least in commensal populations. However, predation represents the major cause of mortality for all wild mice, principally from Carnivora (such as small felids and canids), snakes, and birds of prey (Latham and Mason 2004; Singleton and Krebs 2007; Phifer-Rixey and Nachman 2015). This predation threat shapes the behavior of mice in several ways, including the manner in which they communicate (see below), and as we detail in the "Difficulty of Handling/Fear of Humans" section, the stimuli that laboratory mice find to be aversive. Thus, as mentioned, mice are typically crepuscular and nocturnal, to avoid diurnal predators (Spoelstra et al. 2016). Mice also find the odors of natural predators aversive, and avoid open spaces, especially if they are unfamiliar (Augustsson et al. 2005). Wild mice dispersing from their natal home or moving around their own territory, thus typically display "thigmotaxis", keeping their body in contact with barriers and walls (Latham and Mason 2004).

Senses and Communication

Adapted for crepuscular/nocturnal activity and communicating in a world where being predated is a major threat, the mouse's world is dominated by smell; mice have extremely sensitive olfactory systems (Wackermannová et al. 2016), made even more effective by active sniffing behaviors (Jordan et al. 2018). Odors, both volatile ones detected via the nose and nonvolatile ones detected via the vomeronasal organs on either side of the nasal septum are used to detect and assess food, conspecifics, and predators. Odors also play a crucial role in social communication (see *Social Organization and Behavior*), being released from plantar glands in the feet, lachrymal glands in the face that also secrete tears, and preputial glands in the skin surrounding the genitals, amongst others, as well as being expressed in urine and feces. These odors allow mice to both assess and convey an array of attributes, including familiarity, relatedness, dominance, sexual stage, and health status. Male mice also use urinary marking and "over-marking" (where a mouse covers another male's urine mark with his own) to indicate their dominance and delineate the edges of their territories. Investigating the urinary odors of unfamiliar males is so important to wild mice that they will persist even in the presence of predator cues (e.g., cat odor (Hughes et al. 2009)).

Mice are also very sensitive to touch. Their whiskers, actively moved or "whisked", send tactile information to specialized "whisker barrels" in the neocortex (cf. Stüttgen and Schwarz 2018), allowing mice to identify textures and avoid obstacles, even when light levels are very low, and with such fine resolution that they can distinguish different grades of sandpaper (Garner et al. 2006; Wu et al. 2013). The body's touch-sensitive guard hairs are also used to sense overhead cover and when moving along walls.

Mice have well-developed hearing, and in terms of frequency range, can hear sounds from ca. 10kHz to ultrasounds over 100kHz (Lawlor 1994). Correspondingly, they also vocalize in ultrasound, with males, females, and pups producing a variety of calls to both same and opposite sex companions (even "songs" in the case of courting males (see below and Scattoni et al. 2011; Hoier et al. 2016; Burke et al. 2018)).

Mouse vision is fairly well-developed, although their acuity is quite poor (Baker, 2013). They are very sensitive to movement and changes in light intensity, with their retinas being composed mainly of rods for sensitive vision in dim light (albeit little color perception). Furthermore, their cones lack long-wave photopigments, causing low sensitivity to red wavelengths, but they are sensitive to ultraviolet wavelengths (Sørensen 2014). Visual cues, such as postures and body language, also aid in mouse communication. Dominant mice will signal a warning by rattling their tail prior to chasing a submissive mouse (Tallent et al. 2018), and submissive mice convey subordination by raising their forepaws, thus exposing their bellies (Crawley 2000). Mice also show signs of emotional contagion, amplifying pain behavior when observing a cage mate in pain (Langford et al. 2006). Although there may be several modalities used to convey the experience of pain to conspecifics, it appears that mice demonstrate reliable changes in facial expression that reflect various emotional states (e.g., pain, malaise, fear (Langford et al. 2010; Dolensek et al. 2020)).

Finally, there is evidence that at least some laboratory strains can detect magnetic fields (Muheim et al. 2006). Further detail about murine sensory biology can be found in the following reviews: Olsson et al. (2003), Latham and Mason (2004), Shupe et al. (2006), Hurst (2009), Asaba et al. (2014), and Peirson et al. (2018).

Social Organization and Behavior

Mice are highly social (Latham and Mason 2004; Rusu and Krackow 2005; Singleton and Krebs 2007; Szenczi et al. 2012), often living in family groups called demes, which generally consist of a dominant breeding male (plus additional males occasionally), and a few related breeding females and their offspring (female offspring generally being philopatric). This social living helps them thermoregulate, find food, and raise more pups; indeed, avoiding contact with conspecifics is so atypical and maladaptive that it may indicate disease (Lopes et al. 2016). In commensal populations, demes are usually stable, and migration between demes is low, in part because the dominant male patrols and aggressively defends his territory. Here, nonterritory-holding males often adjust their behavior to be active when dominant males are asleep, and nest with other nonterritory-holders including nomad males and/or dispersing females and juveniles. In feral, typically less dense populations, territoriality tends to be more seasonal, as outlined above, and more readily replaced by a social system comprised of overlapping neighboring ranges (Singleton and Krebs 2007).

Overall, males and females exhibit very different social behavior, with males typically being more aggressive and females more affiliative. Adult females frequently raise pups communally (see *Reproduction and Maternal Behavior*), and also form huddles to conserve heat, while males are reluctant to huddle with other males unless those conspecifics are familiar and ambient temperatures are low (Groó et al. 2018). Co-nesting females will even increase the amount of allogrooming directed toward their returning nesting partner after she has been removed from the nest for an extended period (e.g., in F1–F3 generation descendants of wild-caught mice (Ferrari and König 2017)). Furthermore, adding a male to an established group of females does not incite aggression or reduce sociability between these females, showing that they likely do not compete for male partners (Weidt et al. 2018). However, aggression between female mice tends to increase with age, while sociopositive behaviors, such as grooming and sniffing, tend to decrease with age. In contrast, both aggression and sociability decrease with age for males (Szenczi et al. 2012).

Mating and Mate Choice

To reproduce, female house mice generally need to be in good physical condition and exposed to the right olfactory cues from conspecifics (typically conspecific urine), phenomena well-studied in the laboratory (though not in field conditions). For example, ovulation should be facilitated by exposure to specific odors from males (the "Whitten effect") (Latham and Mason 2004; Asaba et al. 2014). Estrous cycles last 4–6 days, and females also exhibit estrus in the 24 h following the birth of a litter (Latham and Mason 2004). Cycling females, particularly those in estrous, are attracted to male cues, such as preputial and urinary odors (e.g., the wonderfully named "darcin" protein (Roberts et al. 2010)), as well as to ultrasonic song-like vocalizations (Asaba et al. 2014; Lopes and König 2016). Males also use odor, assessing whether a female is receptive by her vaginal odors and urinary scent (Latham and Mason 2004).

Both sexes are polygamous, if opportunities permit (Montero et al. 2013; Thonhauser et al. 2013). But they are not indiscriminate; both sexes show consistent preferences for particular potential mates over others, with females being somewhat choosier than males (Montero et al. 2013; Linnenbrink and von Merten 2017). Individual-specific, idiosyncratic choices may reflect major histocompatibility complex similarity and inbreeding avoidance (Cheetham et al. 2007; Roberts et al. 2010). Females avoid males showing signs of sickness, parasitism, or food deprivation (Meikle et al. 1995; Latham and Mason 2004; Cheetham et al. 2007; Lopes and König 2016, Kavaliers and Choleris 2017), and commonly have a preference for those showing signs of dominance (Cheetham et al. 2007; Hurst 2009). In wild house mice living in seminatural conditions, the width of a male's baculum (penis bone) also predicts sexual success, indexed by the number of offspring sired (Stockley et al. 2013). By exerting mate choice, both sexes can improve their reproductive success, increasing the number and viability of their offspring (Drickamer et al. 2003; Weidt et al. 2008; Raveh et al. 2014). Females can mate multiple times in one estrous cycle, and consequently, approximately one-quarter of wild mouse litters have multiple (typically two) fathers (Thonhauser et al. 2013). This polyandry may reduce infanticide by adult males (Auclair et al. 2014).

Implantation occurs 5 days after fertilization, but can be delayed up to 12 days in lactating females, and fetal resorption may occur if an unfamiliar, but dominant, male intrudes into the territory, the so-called "Bruce effect" (reviewed in Latham and Mason 2004). This pregnancy block effect is mediated by nonvolatile scents from these males, detected by the vomeronasal organ, meaning that pregnant females can actively control their exposure to additional males by adjusting the amount of close nasal contact and sniffing they direct toward strange male odors (Becker and Hurst 2009). Scents from familiar males (such as the last sexual partner) do not have this effect, because when females mate, they learn the chemosensory identity of their partner. This prevents a male's odors from subsequently blocking his own pregnancy (Hurst 2009).

Reproduction and Maternal Behavior

Mice can reproduce at very high rates; in high-quality commensal habitats, a female can produce up to 50 pups in a year (in litters of 3–9 altricial, naked, and blind pups) and will spend much of her life in maternal care (Latham and Mason 2004). For example, when lactating, a female will spend 7–9 h per day with her litter (Ferrari et al. 2019). Until pups are 2–3 weeks old they remain in the nest, where they form a "dynamic huddle" in which individuals burrow toward the center of the pile to keep warm. They depend on their mother for thermoregulation, feeding, and to stimulate defecation. Parental care at this stage includes nursing, nest-building, licking and grooming, crouching over the pups, and retrieving them if they stray from the nest.

The parental behaviors directed to pups may predict their own behavior in adulthood (e.g., maternal behavior (Latham and Mason 2004; Würbel et al. 2017)). As mentioned, maternal behavior may also be shared between females; wild house mice often communally nest with others, especially when they are young, a behavior that generally seems to increase their lifetime reproductive success (König 1993; Heiderstadt and Blizard 2011; Weidt et al. 2014), despite higher pup mortality in communal nests than solitary nests (Ferrari et al. 2019). Females are choosy when it comes to nesting partners, being most likely to select related females and/or individuals with whom they have spent more time in the past (Weidt et al. 2018; Harrison et al. 2018). Mice of both sexes can also be infanticidal, typically to pups that are not their own (Würbel et al. 2017), which is one reason why infant mortality is very high (around 50% of litters failing to live beyond 2 weeks of age, in one well-studied commensal population (Ferrari et al. 2019)).

When pups reach 10 days of age, maternal care begins to include bringing solid food back to the nest from foraging trips, and as the pups reach 3–4 weeks of age and become ambulatory and sighted, the female then begins to accompany them on their first excursions from the nest (Latham and Mason 2004). Young do not leave their mothers until some weeks later, after they have reached sexual maturity, and females may not leave at all (reviewed by Bechard and Mason 2010). For example, wild house mice in the laboratory do not voluntarily disperse until 8+ weeks of age (Gerlach 1990). Most males do leave the natal nest in search of their own territories, however, prompted by increasingly aggressive advances from their fathers, as well as their own increasingly aggressive and exploratory behavior.

Common Captive Behaviors

The laboratory environment for a mouse is vastly different from the habitat of their wild ancestors, and the mice themselves can be quite different too, depending on strain (Poletaeva et al. 2001; Augustsson et al. 2005; Thon et al. 2010; Viney et al. 2015). Since determining which behaviors are "normal" or "abnormal" is therefore complex, we define "normal behaviors" here as those that are generally thought to be desired (or neutral) in the laboratory setting, while "abnormal" behaviors will be those that are considered undesirable from a welfare perspective. For detailed descriptions and video examples of the behaviors described below, see Stanford University's online Mouse Ethogram (mousebehavior.org), or Spangenberg and Keeling's (2016b) assessment and scoring scheme.

Normal Behaviors

Homeostasis: Eating, Drinking, and Sleeping

Laboratory mice are typically supplied with a water bottle, and hard, long-lasting food pellets in a hopper, which they are adept at gnawing. They eat and drink in short bouts multiple times a day, consuming more than their wild counterparts, partly because of larger body sizes (Viney et al. 2015 citing Bronson 1984). Deviations from this may well indicate health issues (Spangenberg and Keeling 2016a). That mice are eating adequate amounts of food can be assessed from body condition scores (Ullman-Cullere and Foltz 1999), while sufficient water intake can be assessed via tests for dehydration (e.g., skin pinch assessment on neck (Fallon 1996)). Due to their omnivory, a range of food items can also be used as enrichment and in training (see below), while their natural tendency toward novelty-suppressed feeding can be used to assay anxiety (Samuels and Hen 2011; Blasco-Serra et al. 2017). Finally, mice sleep 12–15 h in short bouts of 2–4 min throughout a 24-h day, primarily during the light period (Hawkins and Golledge 2018).

Self-maintenance: Self-grooming and Nesting

Mice spend approximately 20% of their active time self-grooming, as this is important for hygiene and insulation (Crawley 2000; Latham and Mason 2004). Self-grooming declines in mice with health problems, and therefore, scoring coat condition is useful for identifying unhealthy animals (Fallon 1996; Paster 2009). Nest-building is also essential for laboratory mice. Healthy mice are highly motivated to access nesting material and will rapidly nest-build when materials are presented (van de Weerd et al. 1997, 1998; Rock et al. 2014), with structural materials forming a dome shape and softer material lining the inside walls (Hess et al. 2008). Nest-building can be scored for nest quality and complexity, or for the time it takes mice to integrate new material into their nests (Figure 6.1; Hess et al. 2008; see Gaskill et al. 2013a for a video demonstration). Poor nest-building can reveal aggression between cage mates, sickness, or pain, while an overall increase in nest scores can indicate thermal stress (Gaskill et al. 2013a). Providing naturalistic nesting material (e.g., paper strips) allows mice to form high-quality, complex nests to better control their microenvironment (Hess et al. 2008).

Interaction with the Environment

Mice are naturally active and exploratory, and if opportunities present, they will typically engage with their environment, showing gnawing, running (e.g., on wheels), nest-building, and digging. Not doing so may indicate problems; for example, reductions in wheel running reflected the severity of restraint

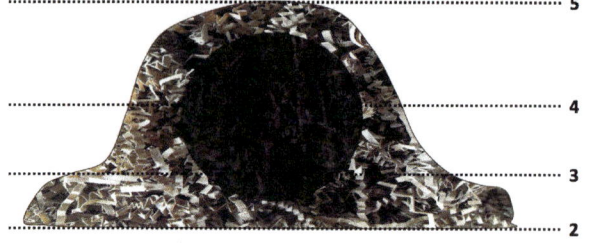

FIGURE 6.1 Nest scoring system (adapted from Hess et al. (2008) and Gaskill et al. (2013a) with kind permission from the *Journal of the American Association for Laboratory Animal Science*). Please refer to Hess et al. (2008) for detailed scoring system and additional nest images. Score 2: Flat, nesting material has been manipulated and gathered at a nest site. The nest takes the form of a flattened saucer. Score 3: Cup, nesting material has been manipulated and gathered at a nest site. The nest includes walls that form a cup or bowl. Score 4: Incomplete dome, nesting material has been manipulated and gathered at a nest site. The walls of the nest reach the widest point of the hollow interior. Score 5: Full dome, nesting material has been manipulated and gathered at a nest site. The walls of the nest completely enclose the hollow interior, except for a small entrance hole on the side or top of the structure. Not shown here are two poorer scores where nesting material remains dispersed throughout the cage; Score 0: Nesting material has not been moved or manipulated. Score 1: Nesting material has been disturbed (e.g., chewed or moved within the cage), but not collected at a nest site.

stress, as well as induced colitis, in C57BL/6 mice in one study, and this was a more sensitive indicator of distress than other clinical markers (Häger et al. 2018; see above for similar data on nest-building). Similarly, if the cage lid is removed, mice should react to the observer; however, hyper-reactivity or extreme fearfulness, such as freezing or hiding under bedding, might also indicate welfare problems (Spangenberg and Keeling 2016a).

Social Interactions

Since mice are highly social, their previous and present social environments affect their well-being, with social environments including stable, compatible groups being the most beneficial (Olsson and Westlund 2007). Social interactions are valued by mice, and they will seek the company of conspecifics (Balcombe 2006). Further, pain sensitivity in male and female mice can be influenced by the behavior of cage mates in pain. For instance, during pain assessment tests, mice will respond to noxious stimuli with more pain behavior when observing a familiar mouse exposed to the same stimuli, and pain behavior is further exacerbated when observing a familiar mouse exposed to a more severe noxious stimulus, suggesting a rudimentary form of empathy (Langford et al. 2006). Social housing can even improve recovery after surgeries for female mice (van Loo et al. 2007). Even male mice, who are commonly aggressive when housed in same-sex groups, show a preference for proximity to a same-sex conspecific over individual housing (van Loo et al. 2001a), and young male mice will exhibit helping-like behavior to other males, such as releasing cage mates and strangers from sealed tubes (Ueno et al. 2019).

Visual, auditory, and olfactory social cues are not enough for mice to experience the benefits of social interactions (van Loo et al. 2007): physical contact is also important. For instance, huddling can increase pain thresholds (i.e., decrease the pain sensitivity of mice in a beneficial way) in sibling male mice when reunited after separation (D'Amato and Pavone, 1996), and grooming of cage mates maintains social bonds (as well as facilitating the transfer of information; Brain and Benton 1983; Crawley 2000). This may also be why stroking by humans can be reinforcing (see the "Training" section below).

Mating and Parental Care

Sexual behavior generally appears similar to wild mice, despite some strain-specific differences (BALB/c males have little urinary darcin, for instance (Roberts et al. 2010), although their urine still attracts females (Liu et al. 2017)). Thus, male urine and ultrasonic vocalizations are attractive to female laboratory mice, with the modality attended to varying with experience; females with experience of males prefer olfactory signals, whereas those naïve to males prefer ultrasound (Screven and Dent 2018). Male laboratory mice even modify their ultrasonic calls according to whether females are present or just close by. Males emit complex outputs (which females find attractive) in the presence of fresh female urine, but longer, simpler calls when females are present and potentially ready to be mated (Chabout et al. 2015). Further, devocalized males attract fewer female approaches and win fewer copulations (Nomoto et al. 2018). Female laboratory mice also use odor cues to avoid both inbreeding (Yamazaki et al. 1979; Jordan and Bruford 1998) and breeding with parasitized males (Kavaliers and Choleris 2017). Finally, exposing female laboratory mice to male odors (e.g., soiled cage bedding) reliably induces estrus 2–3 days later (Roberts et al. 2010; Chabout et al. 2015, and a myriad of others), a potentially useful research technique that exploits the "Whitten effect".

The reproductive output of laboratory mice in breeding cages varies greatly. Genetic background is one factor; in the 6 months of peak reproductive output, C57BL/6 mice and other inbred strains might produce 12–30 pups, while outbred strains, like CD-1s, can produce around 80 (Gaskill et al. 2013c; Wasson 2017). Environment is another influence, as we discuss below in the "Environment" section. Finally, although the cues used are unknown, season also has predictable effects on reproduction. Even in animal facilities with unvarying lighting regimens and stable temperatures, pup output is lowest in winter-spring and highest in summer-fall (Bartiss 2014).

Like free-living house mice, maternal behaviors for laboratory mice include nest-building, pup feeding and cleaning, as well as retrieving pups that stray from the nest (reviewed in Weber and Olsson 2008; and see section "Reproduction and Maternal Behavior" above). In laboratory mice, paternal care has also been observed and studied (Alter et al. 2009). This paternal care is apparently facilitated by pheromones and ultrasonic calls from the mother (Akther et al. 2013), and in turn, maternal behavior is facilitated by male pheromones (Larsen et al. 2008).

Laboratory mice are typically weaned by animal care staff at 3 weeks of age, which is much earlier than dispersal age in the wild (see above). While laboratory mice do grow faster than their wild counterparts (Viney et al. 2015 citing Bronson 1984), they still do not appear to mature at a pace that matches these laboratory practices. For instance, when given a choice between dispersing and staying with their mothers, laboratory mouse pups spend more than half their time with their mothers, even up to 5 weeks of age (Bechard and Mason 2010).

Abnormal Behaviors

Abnormal Repetitive Behaviors

Common abnormal repetitive behaviors in laboratory mice include repeated bar-chewing, route tracing, jumping and somersaulting (sometimes collectively termed stereotypies or stereotypic behaviors), and the "barbering" of self or cage mates, where mice pluck fur or whiskers (Garner et al. 2004a). These behaviors are more prevalent and severe when mice are housed in barren than in enriched cages (Latham and Mason 2010; Tilly et al. 2010; Bechard et al. 2011; Gross et al. 2012; Nip et al. 2019), appear to be absent in wild, free-living mice, and are sometimes exacerbated by early weaning compared to more naturalistic weaning (Bechard and Mason 2010; and see the Mating and Parental Care section above). Bar-chewing appears to derive from attempts to escape from the cage (Nevison et al. 1999), and this and other stereotypic behaviors increase in cages with more aggression (Akre et al. 2011). Perseverative mice, i.e., those with greater tendencies to form routines, may be more at risk for these behaviors (Garner et al. 2011). The nucleus accumbens has been implicated in some forms of stereotypies (Phillips et al. 2016), and barbering shares many similarities with human trichotillomania, a compulsive hair-plucking disorder (Garner et al. 2004b). Whether such mice are neurologically abnormal has not yet been ascertained, however (see review by Kitchenham and Mason 2021). Nevertheless, whatever the underlying mechanisms for the expression of these abnormal repetitive behaviors, they do reflect poor welfare (Mason and Latham 2004). Barbering can also *cause* poor welfare: whisker barbering raises welfare concerns, since whiskers play a critical role in mouse perception (see above), and barbered mice demonstrate degenerative changes in the barrel cortices of the brain as a result of the lack of input (Sarna et al. 2000). Another repetitive behavior with adverse consequences is repetitive scratching, which can result in ulcerative dermatitis (George et al. 2015; Adams et al. 2016).

Inactivity

While the presence of abnormal repetitive behaviors suggests a welfare problem, their absence does not prove good welfare, because some animals instead become very inactive when their welfare is compromised. Thus, some mice in standard laboratory cages spend extensive amounts of time standing motionless with eyes open, inactive but awake. This is more marked in standard cages than in larger, environmentally enriched ones (Tilly et al. 2010; Fureix et al. 2016; Nip et al. 2019; and see Figures 6.2, 6.3, and Table 6.1 to compare cage types). Inactivity negatively covaries with stereotypic behavior, as if it is an alternative response to poor environmental conditions (Fureix et al. 2016), and is associated with time spent immobile during "forced swim tests" (a measure of learned helplessness commonly used in rodent models of depression; Fureix et al. 2016). Inactivity in the home cage could therefore be an important welfare indicator in laboratory mice, especially in strains not prone to abnormal repetitive activities.

Aggression

Aggression is not abnormal for mice, but it is undesirable in the laboratory setting, since it can cause stress and injuries (Weber et al. 2017). Male-male aggression is a common problem, especially for some strains and, paradoxically, especially if mice are given enrichment. Male aggression usually starts with a tail rattle and may escalate to a full-frontal attack by the dominant male until the subordinate flees the area, or the line of sight is broken (Tallent et al. 2018). However, in a

FIGURE 6.2 Example of standard shoebox cage. (Photo: Aileen MacLellan.)

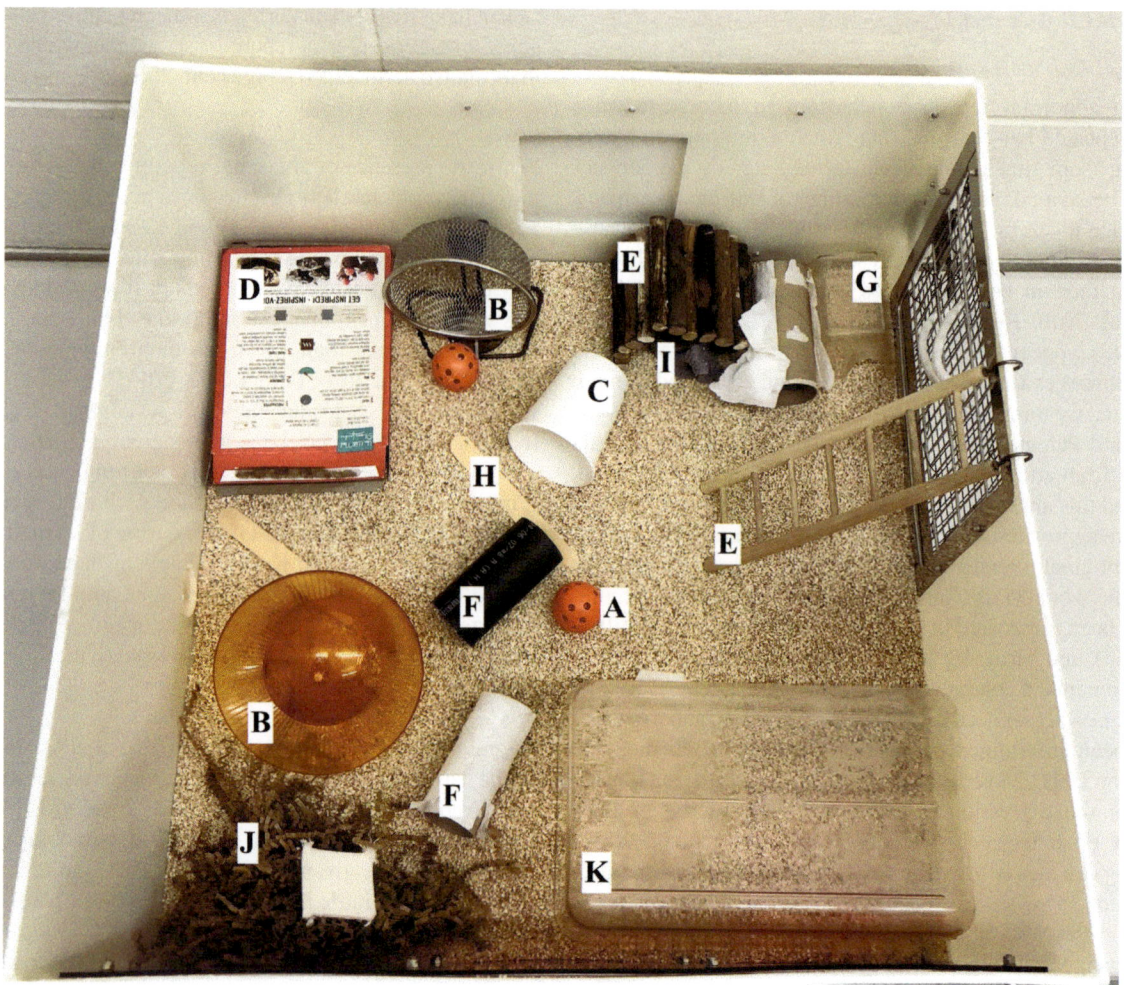

FIGURE 6.3 Example of large enriched cage showing materials added to create a complex, multi-level environment (c.f. Tilly et al. 2010; Nip et al. 2019). See Table 6.1 for item key. (Photo: Aileen MacLellan.)

TABLE 6.1

Examples of Environmental Enrichment Items That Can Be Included in Cages to Facilitate Preferred and/or Species-Typical Behaviors

Enrichment Item	Facilitated Behavior	Corresponding Figure 6.3 Label
Balls	Exploring novel objects	A
Running wheels	Exercise and locomotion	B
Paper cup	Taking cover/shelter	C
Cardboard box	Taking cover/shelter, chewing, nesting	D
Wooden climbing structure	Climbing, exploring, gnawing	E
Tubes	Taking cover, exploring (habituation for tube handling)	F
Shelter	Taking cover/shelter	G
Popsicle stick	Gnawing, exploring novel objects	H
Sock	Taking cover/shelter, nesting	I
Nesting material	Nesting	J
Nest box	Nesting	K

All items listed are disposable or can be easily cleaned or autoclaved in animal facilities, and are labeled to indicate corresponding items in Figure 6.3.

standard laboratory cage, there is extremely limited space for a subordinate mouse to flee, which can cause aggression to be dangerous. Thus, when not directly observed, bite wounds and blood stains in the cage can be used to infer in-cage aggression (Spangenberg and Keeling 2016a). In female mice, agonism is more subtle, involving mounting, rough grooming (Figure 6.4),

FIGURE 6.4 C57BL/6 mouse (right) rough grooms BALB/c mouse (left). Grimace is visible on BALB/c's face. (Photo: Emma Nip.)

chasing, and pinning, and rarely results in wounding (Nip et al. 2019). Unlike males, female aggression is decreased by environmental enrichment (see below).

Difficulty of Handling/Fear of Humans

Fearful behavior is biologically adaptive for wild mice and the natural unwillingness of mice to explore "open fields" and "elevated plus mazes", and their preference for dark chambers over brightly lit ones, are thus central to many widely used tests for rodent anxiety (Hölter et al. 2015). However, outside of such tests, fear is not desirable in laboratory settings, because it represents poor welfare. Fear responses can include hyperreactivity to humans (e.g., immediately fleeing or taking cover when a cage lid is lifted) and handling (e.g., for experiments or during cage change) (Spangenberg and Keeling 2016a), or freezing in response to perceived threats (Baumans 2010). Mitigation techniques are discussed below in the "Handling and Other Ways to Reduce Fear of Humans" section.

Sickness Behavior

Assessing mouse body condition is a rapid, noninvasive tool to detect health issues (Ullman-Culleré and Foltz 1999). During health checks and home cage observations, symptoms such as ruffled or greasy coats, weight loss, hunched postures, sunken eyes, discharge, or diarrhea are considered signs of illness. Mouse behavior can also be observed to assess health problems, with changes such as lethargy, inappetence, or labored breathing raising concern. Presentation of disease or infection symptoms may vary depending on the age, strain, and past exposure of a mouse (Baumans 2010). In addition to health checks, new technology allows for the assessment of more subtle changes. For instance, cameras that capture mouse facial expressions have been shown to detect malaise and other affective states, and such effects are corroborated by changes in neuronal activity (Dolensek et al. 2020). The presence of these symptoms should serve as signs of compromised health, necessitating close monitoring or further diagnostic testing (Foltz and Ullman-Culleré 1999; Ullman-Culleré and Foltz 1999; Baumans 2010, Burkholder et al. 2012; Spangenberg and Keeling 2016a).

Negative Parental Care

While males and nonmaternal females can contribute to parental care, female laboratory mice typically are housed with only their litters post-parturition, so their behavior will be the focus of this section. Nursing, nest-building, licking, and grooming are all parental behaviors important to pup survival (Weber and Olsson 2008). Failing to build nests or retrieve pups that stray from the nest can put pups at risk of thermal stress and potentially hypothermia (Lynch and Possidente 1978). Experimentally induced sickness (Aubert et al. 1997) or exposure to social stressors (Carini et al. 2013) can reduce pup retrieval in laboratory rodents. Such poor pup retrieval has also been associated with the depressive symptom, anhedonia, leading to the suggestion that these rodents might model postpartum depression (Carini et al. 2013).

Aggression toward offspring, avoidance, or refusal to nurse are also undesirable. In small cages, where escape from pups is difficult, female mice have been shown to demonstrate "press postures", pressing their ventrum into a corner of the cage, one function of which may be to prevent pups from accessing their nipples to nurse (Gaskill and Pritchett-Corning 2015). A more extreme example of negative parental care is infanticide, the killing of conspecific young prior to weaning (reviewed by Weber and Olsson 2008). For wild mice, such behavior might be adaptive; rearing a litter is a large investment, so if chances of pup survival are low, it may be beneficial to kill them and start over (Weber and Olsson 2008). In the laboratory, mice are protected, and resources are plentiful, so there is no real risk of starvation or predation. Yet, mice may still perceive laboratory conditions as threatening. They are housed at extremely high population densities and humans may be perceived as predators (Weber and Olsson 2008). Infanticide is considered rare in the laboratory, yet some factors can increase the risk of pup mortality. For example, restricted food (Elwood 1991), cold temperatures (Zafar and Shrivastava 2018), frequent cage changes (Reeb-Whitaker et al. 2001), and auditory stressors (Poley 1974) can decrease pup survival or increase the risk of infanticide, and some strains show a higher risk of poor maternal care (Alston-Mills et al. 1999). It is, however, worth noting that females actively killing pups is very rarely observed; instead, when pups are eaten, it is often described as infanticide. Pup mortality or maternal cannibalism may be more appropriate terms, since the pups may have died from other causes (e.g., starvation or hypothermia (Weber et al. 2013; Weber and Olsson 2008)).

Ways to Maintain Behavioral Health

Captive Considerations to Promote Species-Typical Behavior

Environment

Laboratory environments are generally designed with ergonomics, economics, hygiene, and standardization as top priorities (Olsson and Dahlborn 2002; Bailoo et al. 2018). However, not taking the motivations of the animals themselves into account comes at a cost to the behavioral health and welfare of mice. While they are undoubtedly somewhat adapted to laboratory environments, mice are still highly motivated to carry out certain species-typical behaviors, such as exploring, taking cover, nest building, and socializing. Here, aspects of the laboratory environment that impact mouse well-being will be divided into two categories: the macroenvironment, which includes parameters controlled at the room (or possibly facility) level, and the microenvironment, meaning conditions in the animal's cage.

Macroenvironment Considerations

Temperature and Humidity The National Research Council (NRC 2011) recommends housing mice at temperatures between 20°C and 26°C. However, their lower critical temperature is approximately 30°C, which means that recommended practices pose a challenge to homeostasis (Gaskill et al. 2012). It has been shown that mice from a variety of strains (CD-1, BALB/c, and C57BL/6) prefer temperatures from 26°C to 29°C and access to nesting material (Gaskill et al. 2012). Since this is a relatively wide range, and in practice, mice are typically kept between 20°C and 24°C, providing adequate nesting material is essential to allow them to control their own microclimates and avoid thermal stress (Gaskill et al. 2013b). Relative humidity should also be controlled to prevent health issues and avoid facilitating bacterial growth (Baumans 2010). For mice (and most other mammals) humidity guidelines are not stringent and recommendations range from 30% to 70% (ILAR 1996; CCAC 2003).

Visual Stimuli Lighting impacts mouse physiology and behavior (reviewed by Peirson et al. 2018), and is therefore important to control. Mice are typically kept on 12-h light: 12-h dark schedules. Since they are naturally crepuscular and nocturnal, and their active phase is during our night, these schedules can be reversed so that their active dark phase is during our daytime, thereby allowing husbandry and experimental procedures to be more humane and arguably ethologically relevant (Hawkins and Golledge 2018). Red light can be used to allow personnel to see during the dark phase, since mice are less sensitive to it than white light (Roedel et al. 2006; Peirson et al. 2018). This approach does make cage checks more challenging for human vision (Peirson et al. 2018), but in rats, it has been shown to improve welfare by reducing stress and sleep disturbances (Abou-Ismail et al. 2008).

During the light phase, mice prefer low light intensities. These should not exceed 400 Lux (NRC 2011), and for some albino strains, excessive light can cause retinal damage (LaVail et al. 1987). Placement on cage racks can strongly impact the intensity of light experienced in each cage's microenvironment (Clough 1982). Such variation can cause altered levels of "emotionality" (Ader et al. 1991) and barbering (Garner et al. 2004a). While these effects are seldom considered, it is possible that rack placement could impact some research outcomes and should be taken into account during experimental design. Providing material or an area where mice can take cover in the cage allows them to control their microenvironment and avoid aversive, intense lighting (see "Environmental Enrichment" section below).

Auditory Stimuli In mice, like other laboratory species, excessive auditory stimulation can have adverse effects on behavior and physiology (Milligan et al. 1993). In fact, auditory stress is commonly used in experimental paradigms, and such exposure can even cause convulsions (audiogenic seizures (Musumeci et al. 2000)) or elevate abortion rates (Clark et al. 1993). While efforts are generally made to limit the noise levels in laboratory environments, equipment and daily activities still generate considerable sound. Mice are most sensitive to high-pitched sounds and, as noted above, can hear ultrasounds outside the range that humans can detect (Latham and Mason 2004). Ultrasonic stimuli are rarely controlled for or assessed in laboratory environments, despite the fact that many seemingly "silent" features, like computers, light fixtures, and oscilloscopes, emit ultrasound (Sales et al. 1988). Levels of sound in the laboratory are also variable through the day, typically increasing during working hours (Milligan et al. 1993). Since mice are nocturnal, if peak noise levels occur during the light phase, this potentially disrupts their sleep patterns (Rabat 2007). This is why housing mice on a reverse light cycle may help mitigate sleep disturbances, but even then, efforts should also be made to assess ultrasonic sound and, perhaps, to reduce noise levels overall.

Olfactory Stimuli Olfactory stimuli are an important consideration for maintaining behavioral health. Cage change presents one major disturbance to the olfactory environment, since it involves the removal of familiar cues. Although it is a necessary procedure to control levels of ammonia and allergens, and to protect the health of both the mice and the staff, this process is stressful for laboratory mice (Balcombe et al. 2004; Rosenbaum et al. 2009). Cage changing frequency varies between facilities, based on cage size, housing density, bedding, and ventilation. However, it commonly occurs on a weekly or fortnightly basis. Cage change schedules should aim to minimize the frequency of changes, while still maintaining hygiene and air quality. One way to reduce the frequency of cage changes and control the levels of ammonia, humidity, and CO_2 is to use individually ventilated cages (IVC). As discussed by Baumans et al. (2002) and Burman et al. (2014), these cages also protect mice from microorganisms and pathogens, and facility staff from allergens and zoonoses. Creating a more stable microenvironment through the use of IVCs, with lower levels of infection and less frequent handling, has been suggested to reduce stress and discomfort for mice (but see "Microenvironment Considerations" section for discussion of IVC welfare costs). It has also been shown that transferring a small amount of nesting material from the used cage into the new environment can ease this transition and reduce aggression (van Loo et al. 2000).

Olfactory effects may also help explain why mice tested with different experimenters can yield different results (see "Special Situations" section); for example, mice tested by male, but not female, researchers show signs of pain inhibition as a result of stress-induced analgesia (Sorge et al. 2014). Overall, it is important to keep in mind that the olfactory worlds of mice are important to them and that only a small portion of the olfactory cues mice experience are detectable to human caregivers and researchers.

Microenvironment Considerations

Home Cage Physical Features Standard cages are typically shoebox sized, and equipped with bedding (plus food and water), but potentially little else (Figure 6.2). Space requirements according to *The Guide for the Care and Use of Laboratory Animals* (NRC 2011; see Table 6.2) are based on the available information, but the complex interactions among experimental design, strain, age, sex, housing density, etc., make it difficult to broadly apply findings, and experimental results are often conflicting, depending on the variables assessed (Whittaker et al. 2012). It is thus challenging to estimate exactly how much floor space should be provided. That said, mice in standard cages will work to gain access to additional space (Sherwin and Nicol 1996; Sherwin 2004a), and it has been suggested that restricted space might exacerbate the stressful effects of barren cage environments (Balcombe 2006). Cage space might be of particular importance for breeding mice. Pups play less in small cages, and adults of *both* sexes housed with pups in such cages demonstrate more of the "press postures" described above (see *Negative Parental Care*), not only to prevent nursing, but seemingly in an attempt to rest undisturbed (Gaskill and Pritchett-Corning 2015). In either case, Gaskill and Pritchett-Corning (2015, p. 16) interpret more press postures as "potentially indicating some sort of space oriented stress". Furthermore, in shoebox-sized cages, there is little room in which to add beneficial enrichment that might otherwise help alleviate some of the stress caused by barren environments (see below).

Some standard cages are also individually ventilated. As discussed above, IVCs help protect the health of both mice and their human caretakers, and can reduce the frequency of stressful cage changes. But such systems have welfare costs too, especially where ventilation rates range from 25 to 120 air changes hourly (Baumans et al. 2002). Drafts from high ventilation rates can create cold, dry environments. Mice find such conditions aversive, and this chronic stress can impact their behavior and welfare (Baumans et al. 2002; Burman et al. 2014). Certain factors can help limit such adverse effects. For example, mice prefer lower ventilation rates, larger cages, and air delivery closer to cage lids (as opposed to at animal level), and providing nesting material again allows mice to control their microclimates and protect themselves from thermal stress (Baumans et al. 2002; Burman et al. 2014). When IVCs are used, these factors should be considered to help the benefits of this system outweigh the costs.

Environmental Enrichment Barren cages lack the environmental features necessary for mice to carry out highly motivated behaviors, altering brain function and compromising animal welfare (Würbel 2001). Adding enrichment, a "modification in the environment of captive animals that seeks to enhance their physical and psychological well-being by providing stimuli that meet their species-specific needs" (Baumans 2010, p. 283) can help. Some have also argued that using rodents who are environmentally enriched in this way might allow for biologically relevant variation and yield results that are more robust, improving reproducibility and translatability (Sherwin 2004b; Voelkl et al. 2018). Environmental enrichment often aims to facilitate the demonstration of natural and/ or highly motivated behaviors. Some common features in enriched cages include access to more space, nesting material, shelters, hammocks, tunnels/tubes, structures for climbing, items to chew, running wheels, introduction of novel objects to explore, and scatter feeding to encourage foraging (Olsson and Dahlborn 2002; Smith and Corrow 2005; Balcombe 2006; Sztainberg and Chen 2010; Baumans and van Loo 2013; and

TABLE 6.2

Common Floor Space Recommendations for Standard Laboratory Cages

Animals	Weight (g)	ILAR/NRC: Floor Area/Animal (in², cm²)	CCAC: Floor Area/Animal (in², cm²)	ILAR/NRC: Height (in, cm)
Minimum Enclosure for 1–3 adult mice	N/A	N/A	51 (330)	5 (12.7)
Mice	Any adult mouse	N/A	15.5 (100)	
Mice	<10	6 (38.7)	Up to 9.1 (59)	5 (12.7)
	≤15	8 (51.6)	15.5 (100) if adult	5 (12.7)
	≤25	12 (77.4)	15.5 (100) if adult	5 (12.7)
	>25	≥ 15 (≥96.7)	15.5 (100) if adult	5 (12.7)
Female with any litter	N/A	51 (330)	15.5 (100) if adult	5 (12.7)
Breeding pair or trio with small litter (5 ≥ pups)	5 pups at 10g each are considered equal to one large male (CCAC 2019)	N/A	66.6 (430)	5 (12.7)
Breeding pair or trio with large litter (6 ≤ pups)	6 or more pups are considered to be equal 2 or more adults (CCAC 2019)	N/A	82.2 (530)	5 (12.7)

Data from ILAR (1996), NRC (2011), and CCAC (2019).

see Table 6.1). Figure 6.3 shows an example of a large, environmentally enriched cage. Enrichment items made out of natural materials may be particularly beneficial. Preferred natural stimuli have positive impacts on human well-being, and although this has not yet been thoroughly investigated in animals, available evidence does suggest that natural stimuli might have similar positive effects (Ross and Mason 2017). For example, rats provided with natural enrichment items (e.g., wood objects/structures, pebbles) spend more time using them than they do artificial equivalents (Lambert et al. 2016), plant odors seem to reduce signs of learned helplessness in chronically stressed mice (Nakatomi et al. 2008), and playing the sounds of tropical rainforests throughout the day has been shown to prolong the lifespan of otherwise standard-housed mice (Yamashita et al. 2018).

When effective, enrichment can enhance learning and memory (Frick and Fernandez 2003), reduce anxiety (Benaroya-Milshtein et al. 2004), increase hippocampal neurogenesis (Kempermann et al. 1997), shorten recovery and reduce self-administration of drugs postsurgery (Pham et al. 2010), improve resilience to stressors (Fox et al. 2006; Meijer et al. 2006; Crofton et al. 2015) and reduce stereotypic behaviors (Mason et al. 2007). The importance of nesting for proper thermoregulation has already been discussed (see above). Such enrichment effects can also enhance reproduction. Laboratory females have the highest reproductive outputs if they can (1) build high-quality nests that keep the young warm (Gaskill et al. 2013c) and (2) be physically active, as encouraged by running wheels (Zhang et al. 2018). These two factors, along with stress reduction, which tends to improve maternal care (Latham and Mason 2004), likely explain why environmental enrichment generally enhances pup production (Lecker and Froberg-Fejko 2016; Moreira et al. 2015).

Enrichment also reduces aggression between female cage mates (Harper et al. 2015; Nip et al. 2019). However, as outlined already, the opposite can occur with males. Thus, several studies have found that enrichment can increase aggression in male mice (Haemisch et al. 1994; Howerton et al. 2008; Weber et al. 2017), although occasionally, depending on strain and enrichment type, the opposite is observed (Pietropaolo et al. 2004; Tallent et al. 2018). The problem of male aggression is discussed further in the next section.

Social Groupings

Group housing is generally regarded as preferable because mice are social. In females, this viewpoint is well supported, since they are especially affiliative (see above), and experience many negative effects when housed individually, such as increased anxiety in the open field test (Palanza et al. 2001), anhedonia (Lamkin et al. 2012), and other depression-like effects (Martin and Brown 2010). Furthermore, group size seems to have little effect on growth, stress physiology, cognition, or behavior of female mice, even when up to 16 mice are housed together in a standard shoebox cage (Morgan et al. 2014; Paigen et al. 2016; Bailoo et al. 2018). However, at such high housing densities, ammonia levels skyrocket (Eveleigh 1993; Smith et al. 2004) and barbering is nearly doubled in cages of more than three mice (Paigen et al. 2016), factors that should be considered when selecting a social housing scheme for females. Both male and female pair-housed mice also recover faster from surgery than mice housed either individually (Jirkof et al. 2012) or separated from their cage mate via a mesh partition (Van Loo et al. 2007). However, other reports suggest that individual housing only affects performance on certain tests and in certain strains (Arndt et al. 2009).

For males, in contrast, it remains unclear whether group housing is the most appropriate system. Housing multiple males together is less natural, and can cause stress and agonism as well as physiological and behavioral changes (see Kappel et al. 2017 for review). Thus, because of the inherent aggressiveness of male mice, much of the literature on mouse social housing focuses on the best ways to house males. Individual housing may spare them from conflict (especially those from particularly aggressive strains, such as CD-1, FVB, and SJL), but comes at the cost of near complete social deprivation, which can induce anxiety and depression-related behaviors; abnormal repetitive behaviors, such as self-barbering (see above); and changes in physiology, immunocompetence, and circadian rhythm (sometimes referred to as "isolation syndrome"; reviewed by Kappel et al. 2017). Perhaps most importantly, both dominant and subordinate male mice prefer to be housed with another male, when the alternative is individual housing (Van Loo et al. 2001a; reviewed by Baumans 2010 and Kappel et al. 2017; though see Melotti et al. 2019). Mirrors have been tried as potential substitutes for social housing, but these failed to improve isolation-induced anxiety- and depression-related behaviors (Fuss et al. 2013). Despite the harms of being isolated, males of some common strains, such as BALB/c and C57BL/6, are often housing individually, even though these strains are categorized as only low to moderately aggressive (Baumans 2010; Kappel et al. 2017), and are potentially able to coexist relatively peacefully (Bisazza 1981; Gurfein et al. 2012).

Are there ways that we can minimize aggression in group-housed males? First, as we detail above, enrichment can be a trigger for aggression, likely due to resource guarding (Howerton et al. 2008; reviewed by Weber et al. 2017). Weber et al. (2017) therefore suggest that providing enrichment items "in abundance" and dispersing them throughout the cage could perhaps help reduce aggression in males, as it has been found to reduce aggression in females (Akre et al. 2011). Alternatively, enrichment could be avoided, although it is challenging to judge whether this would result in a net benefit in mouse welfare. Van Loo and colleagues (2001a) determined that smaller cage sizes (80 cm^2 per mouse) can also help reduce aggression, possibly because individuals may try to claim a territory in larger cages (Kappel et al. 2017). In at least some strains, aggression can also be mitigated by transferring nesting material to the new cage during cleaning (Van Loo et al. 2001b) or by partially dividing cages with opaque walls, creating several burrow-like areas and one bigger open space (Tallent et al. 2018). While these may seem like relatively easy ways to improve the welfare of male mice in captivity, providing an abundance of enrichment items can be costly (though it does not have to be) and often necessitates larger cages than the standard shoebox style in order to fit the extra items. This can be a deterrent to researchers when funding is scarce and colony housing space comes at a premium. Some researchers also worry that providing a more complex environment will complicate handling of the mice (e.g., when retrieving them from their cage), or even lead to unexpected research findings. As such, many researchers are hesitant to implement changes to their mouse husbandry routines.

Other tactics for reducing aggression are related to the size and nature of the social group. Forming groups with adult males that are familiar with one another from when they were juveniles (irrespective of relatedness) can help reduce aggression (Kappel et al. 2017). Thus, regrouping male mice when they are young (e.g., at weaning) can improve chances of success. Groups consisting of three individuals also exhibit less aggression than larger groups, regardless of cage size (Van Loo et al. 2001b; Poole and Morgan 1973). Weber et al. (2017) and Baumans (2010) thus recommend no more than three males per cage. Alternatively, male mice may be housed with ovariectomized females (Späni et al. 2003; Ewaldsson et al. 2016). Castrated males can also be housed together in groups as large as ten (Vaughan et al. 2014), although housing intact males with castrated males is not advisable, as it leads to severe fighting instigated by the intact males (Ewaldsson et al. 2016).

Maternal Care and Optimal Weaning

For laboratory mice in commercial or research conditions, removing pups from their mother at 3–4 weeks of age is standard practice, even though they would not voluntarily leave her this young (see above). Whether this has any harmful lasting effects, or whether mice are adaptable enough to adjust, is unclear. For example, older studies found that delaying the age of maternal separation reduced the development of stereotypic behavior, but more recent work has failed to replicate this finding (see Bechard et al. 2012), suggesting that this topic still needs further investigation.

Feeding Strategies

Under laboratory settings, mice are typically provided access to preformulated *ad libitum* hard dietary pellets (Baumans 2010), provided in a hopper built into the cage lid, where mice can gnaw at them. This practice can be refined in a number of ways. Constant access to food, combined with limited opportunities for exercise present in most cages, can render many standard-housed mice obese, and even increases the risk of developing related diseases, such as cancer, renal failure, and Type 2 diabetes (Martin et al. 2010). Caloric restriction, by reducing the quantity of pellets provided, or providing them intermittently, can help prevent or delay such conditions. The *status quo* also does not offer this naturally omnivorous animal any dietary choice, nor chances to perform diverse species-typical foraging behaviors (Baumans 2010; Baumans 2005). It has been shown that rats prefer to work for food, even in the presence of freely accessible food (Neuringer 1969; Carder and Berkowitz 1970), and scatter feedings are also widely used as enrichment for many species in the zoo world (Maple and Perdue 2013). Although this has been infrequently utilized or studied in mice, providing more diverse food items, for instance by scattering grain or seeds into their bedding, may well be beneficial (Brown 2009).

Handling and Other Ways to Reduce Fear of Humans

Providing cages with structures that allow opportunities to hide, allowing for habituation to husbandry and experimental settings, avoiding aversive stimuli like shocks (see below), and careful handling from a young age are all recommended to help mitigate fear of humans (Baumans 2010; Laule 2010). The way experimenters handle mice is important too; presumably because it resembles being predated, mice find being picked up by the tail for prolonged periods aversive (Hurst and West 2010). Using tubes or a cupped hand to handle mice reduces anxiety-related behaviors and increases their voluntary interaction with humans (Figure 6.5; Hurst and West 2010). This handling technique also makes mice easier to capture and hold, even after unpleasant procedures such as oral gavage, and leads to enhanced engagement in behavioral tests (Hurst and West 2010; Gouveia and Hurst 2013, 2017; Nakamura and Suzuki 2018). Future research should investigate whether the beneficial effects of tube handling or cupping extend to mothers housed with their litters, since early handling can negatively impact maternal care (Priestnall 1973), and frequent cage changes (which involve handling) can reduce pup survival. Indeed, as we show below (see "Training" section), human touch, if performed correctly, can even become a positive reinforcer for mice.

Training

Compared to larger, less tractable animals, such as dogs and monkeys, little has been written on training mice to improve their welfare. While there are some challenges that accompany mouse training (e.g., they satiate readily, even with small amounts of food), they can successfully learn using methods that promote behavioral health. When food rewards are to

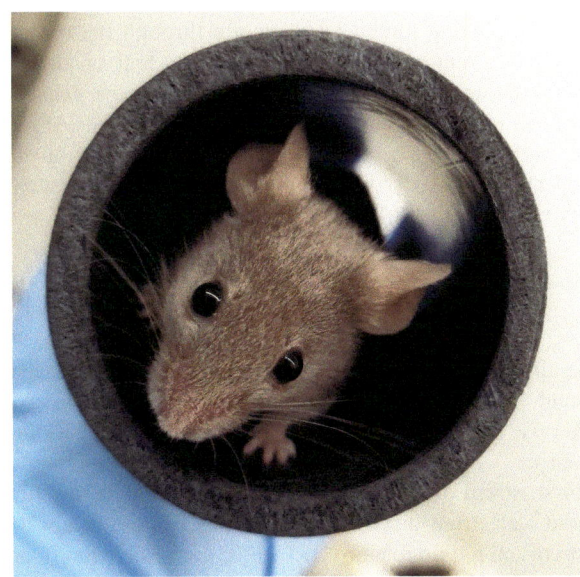

FIGURE 6.5 DBA/2 mouse during tube-handling procedure. (Photo: Emma Nip.)

be used, many researchers fast their mice long-term to motivate them to perform tasks (Van Den Bergen et al. 1997). One more welfare-friendly version of this approach is to fast mice only over the light period, when their food and water consumption are already low (Clipperton et al. 2008; Matta et al. 2017). Alternatively, tiny amounts of highly preferred food items, such as almond slivers or dried banana, could be offered as rewards for completing desired behaviors, as used successfully by pet mouse owners to train their mice in a variety of tasks, including turning on a dime and coming when called (CreekValleyCritters 2011). In laboratory practice, the use of positive reinforcement to train mice has been studied by Leidinger and colleagues (2017; and see Gaskill et al. 2011 for a more automated approach). These researchers prescreened food items for reward value via preference testing and developed a "trusting" relationship between mouse and experimenter. They then successfully used this method to first "clicker train" mice (cf. McGreevy and Boakes 2008), and once this secondary reinforcer was established, they then trained subjects to run through a tunnel, follow a target stick, and climb onto an experimenter's hand (Leidinger et al. 2017). In the University of Guelph Central Animal Facility, technicians routinely offer each mouse a small food treat (e.g., a Cheerio) during daily health checks. This makes these checks easier and faster, as mice learn to emerge from their nests or shelters when the "Cheerio cup" is rattled. Additionally, food need not be the only positive reinforcer; recent research suggests that gentle human touch or stroking is rewarding for mice (Chan 2018) and can be immune-boosting as well (Major et al. 2015). Thus, habituation and positive reinforcement are powerful tools for improving mouse welfare, for facilitating daily health checks, and even for investigating mouse cognitive abilities.

While negative reinforcers such as footshock are currently the default in research on learning and memory (Crawley 2000), the *UFAW Handbook on the Care and Management of Laboratory and Other Research Animals* (Laule 2010, p. 207) states, "there is an inherent cost to the animal's overall welfare by being forced to cooperate through the threat of a negative event or experience that elicits fear or anxiety", and recommends that all positive alternatives be exhausted before using these fear-based conditioning methods. If negative methods must be used, they should be kept to a minimum and balanced with the provision of rewards, such as tasty treats (Laule 2010).

Special Situations

The alien sensory worlds of mice, their remarkable developmental plasticity, and the incredible range of strains available raise some additional issues that may not occur for other species.

First, strain differences can manifest in almost any imaginable way, from differences in vision (Peirson et al. 2018) to nutrient requirements (NRC 1995), to behavior (Contet et al. 2001; Heinla et al. 2018), to tendencies to greater levels of anxiety (An et al. 2013). Indeed, some strains have impaired sensory systems that could affect both their welfare and how they respond during research. For example, different strains vary greatly in aspects of their vision, and some (e.g., certain albino strains, along with strains prone to retinal degeneration or aging-related blindness (Wong and Brown 2006)), have much poorer visual abilities than wild or outbred mice. Many mouse strains also show early-onset, age-related deafness; for example, DBA and C57BL/6 mice start losing audition at high frequencies by 1–2 months, and they, along with other strains, may be deaf by 4 months of age (Davis et al. 2001; O'Leary et al. 2017). Furthermore, in novel genetically modified strains, phenotypic differences cannot always be predicted, and many genetically modified strains are often not systematically characterized (Thon et al. 2010), even though they may require special housing and husbandry (Wells et al. 2006). Because of all these variables, the use of animal-based, rather than resource-based (assessing cage, environment etc.), measures during welfare assessment is crucial to protect mouse well-being (Buehr et al. 2003; Wells et al. 2006; Spangenberg and Keeling 2016b).

Another take home message from this chapter should be that the sensory worlds of mice are very different from our own. For instance, they can detect magnetic fields, rely heavily on the exceptional tactile sensitivity of their whiskers, their visual and auditory sensitivities differ greatly from our own, and their incredible olfactory sense far exceeds human capabilities. Working with and caring for a species so different from ourselves raises some interesting challenges. The sensory abilities of mice may help explain the persistence of seasonal effects, even in climate-controlled animal units, as we discuss above in the "Mating and Parental Care" section, as well as the variation in research outcomes that can exist across facilities (Crabbe et al. 1999) and individual experimenters (Chesler et al. 2002; Lewejohann et al. 2006). For example, gender aside (Sorge et al. 2014), the soap a researcher uses, whether they are vegetarian, and whether they have pet dogs or cats at home, may well all be detectable by mice, yet are not variables we think to assess or note down in our methods sections. Finally, the remarkable developmental plasticity of mice also has similar practical relevance for researchers, potentially helping to explain the presence of unwanted, unexpected supplier and site effects on mouse phenotype (cf. Crabbe et al. 1999; Latham and Mason 2004; Åhlgren and Voikar 2019).

Conclusions/Recommendations

The long history of mice in research might lead to the assumption that these animals are adapted to laboratory environments. In many ways they are, and yet, it is clear from the information described in this chapter that laboratory mice still retain many of the motivations and behaviors of their wild conspecifics. As a result, their welfare may well be compromised in standard laboratory environments, and the stress of such environments may even sometimes render them invalid as research subjects. Thus, the scientific community should aim for changes to laboratory conditions that consider the vastly different sensory world of mice and the preferences and motivations that underlie their behavior.

Additional Resources

Latham and Mason (2004), the *UFAW Handbook on the Care and Management of Laboratory and Other Research Animals* (2010), and Würbel et al. (2017) are all excellent resources for additional information on laboratory mice. There are also several useful online resources including, but not limited to:

- Stanford University's online Mouse Ethogram (mousebehavior.org) for detailed descriptions of mouse behaviors and video examples.
- NC3Rs' Resource hub (https://nc3rs.org.uk/resource-hubs) which includes guides on mouse handling, use of the Mouse Grimace Scale, blood sampling protocols, and refining the use of genetically altered mice.
- Gaskill and colleagues' (2013a) Journal of Visualized Experiments article and video (https://www.jove.com/video/51012) for implementing nest scoring as an indicator of health and welfare.
- The IMSR (http://www.findmice.org/) for a searchable database detailing mouse strains and stock available around the world. It provides researchers with the location, state (live mice, preserved embryos, etc.), and description of the strains, as well as a point of contact for acquiring mouse resources.

REFERENCES

Abou-Ismail, U. A., O. H. P. Burman, C. J. Nicol, and M. Mendl. 2008. Let sleeping rats lie: Does the timing of husbandry procedures affect laboratory rat behaviour, physiology and welfare? *Applied Animal Behaviour Science* 111, no. 3:329–41.

Adams, S. C., J. P. Garner, S. A. Felt, J. T. Geronimo, and D. K. Chu. 2016. A "Pedi" cures all: Toenail trimming and the treatment of ulcerative dermatitis in mice. *PLoS ONE* 11, no. 1 (January): e0144871. https://journals.plos.org/plosone/article?id=10.1371/journal.pone.0144871.

Ader, D. N., S. B. Johnson, S. W. Huang, and W. J. Riley. 1991. Group size, cage shelf level, and emotionality in non-obese diabetic mice: impact on onset and incidence of IDDM. *Psychosomatic Medicine* 53, no. 3:313–21.

Åhlgren, J., and V. Voikar. 2019. Experiments done in Black-6 mice: What does it mean? *Lab Animal* 48: 171–80.

Akre, A. K., M. Bakken, A. L. Hovland, R. Palme, and G. Mason. 2011. Clustered environmental enrichments induce more aggression and stereotypic behaviour than do dispersed enrichments in female mice. *Applied Animal Behaviour Science* 131, no. 3:145–52.

Akther, S., N. Korshnova, J. Zhong, M. Liang, S. M. Cherepanov, O. Lopatina, Y. K. Komleva et al. 2013. CD38 in the nucleus accumbens and oxytocin are related to paternal behavior in mice. *Molecular Brain* 6, no. 41 (September). http://www.molecularbrain.com/content/6/1/41.

Alston-Mills, B., A. C. Parker, E. J. Eisen, R. Wilson, and S. Fletcher. 1999. Factors influencing maternal behavior in the hubb/hubb mutant mouse. *Physiology & Behavior* 68, no. 1–2:3–8.

Alter, M. D., A. I. Gilani, F. A. Champagne, J. P. Curley, J. B. Turner, and R. Hen. 2009. Paternal transmission of complex phenotypes in inbred mice. *Biological Psychiatry* 66, no. 11:1061–66.

An, X. L., J. X. Zou, R. Y. Wu, Y. Yang, F. D. Tai, S. Y. Zeng, R. Jia, X. Zhang, E. Q. Liu, and H. Broders. 2013. Strain and sex differences in anxiety-like and social behaviors in C57BL/6J and BALB/cJ mice. *Experimental Animals* 60:111–23.

Anderson, P. S., S. Renaud, and E. J. Rayfield. 2014. Adaptive plasticity in the mouse mandible. *BMC Evolutionary Biology* 14:85.

André, G. I., R. C. Firman, and L. W. Simmons. 2018. Phenotypic plasticity in genitalia: Baculum shape responds to sperm competition in risk in male mice. *Proceedings of the Royal Society B* 285:20181086.

Arndt, S. S., M. C. Laarakker, H. A. van Lith, F. J. van der Staay, E. Gieling, A. R. Salomons, J. van't Klooster, and F. Ohl. 2009. Individual housing of mice – impact on behaviour and stress responses. *Physiology & Behaviour* 97, no. 3–4:385–93.

Asaba, A., T. Hattori, K. Mogi, and T. Kikusui. 2014. Sexual attractiveness of male chemicals and vocalizations in mice. *Frontiers in Neuroscience* 231, no. 8:1–13.

Aubert, A., G. Goodall, R. Dantzer, and G. Gheusi. 1997. Differential effects of lipopolysaccharide on pup retrieving and nest building in lactating mice. *Brain, Behavior, and Immunity* 11, no. 2:107–18.

Auclair, Y., B. König, and A. K. Lindholm. 2014. Socially mediated polyandry: A new benefit of communal nesting in mammals. *Behavioral Ecology* 25, no. 6:1467–73.

Augustsson, H., K. Dahlbom, and B. J. Meyerson. 2005. Exploration and risk assessment in female wild house mice (*Mus musculus musculus*) and two laboratory strains. *Physiology & Behavior* 84, no. 2:265–77.

Aujard, F., E. D. Herzog, and G. D. Block. 2001. Circadian rhythms in firing rate of individual suprachiasmatic nucleus neurons from adult and middle-aged mice. *Neuroscience* 106, no. 2:255–61.

Bailoo, J. D., E. Murphy, M. Boada-Saña, J. A. Varholick, S. Hintze, C. Baussière, K. C. Hahn et al. 2018. Effects of cage enrichment on behavior, welfare and outcome variability in female mice. *Frontiers in Behavioral Neuroscience* 12:232.

Baker, M. 2013. Neuroscience: Through the eyes of a mouse. *Nature News* 502:156–8.

Balcombe, J. P. 2006. Laboratory environments and rodents' behavioural needs: A review. *Laboratory Animals* 40, no. 3:217–35.

Balcombe, J. P., N. D. Barnard, and C. Sandusky. 2004. Laboratory routines cause animal stress. *Journal of the American Association for Laboratory Animal Science* 43, no. 6:42–51.

Bartiss, R. 2014. Predicting seasonal breeding variation in a colony of C57BL/6J mice. *Laboratory Animal Science Professional*, 46–47.

Baumans, V. 2005. Environmental enrichment for laboratory rodents and rabbits: Requirements of rodents, rabbits, and research. *ILAR Journal* 46, no. 2:162–70.

Baumans, V. 2010. The laboratory mouse. In *The UFAW Handbook on the Care and Management of Laboratory and Other Research Animals*, 8th Edition. eds. R. Hubrecht and J. Kirkwood. Oxford: Wiley-Blackwell.

Baumans, V., and P. L. P. van Loo. 2013. How to improve housing conditions of laboratory animals: The possibilities of environmental refinement. *The Veterinary Journal* 195, no. 1:24–32.

Baumans, V., F. Schlingmann, M. Vonck, and H. A. van Lith. 2002. Individually ventilated cages: Beneficial for mice and men? *Contemporary Topics in Laboratory Animal Science* 41, no. 1:13–9.

Bechard, A., A. Nicholson, and G. Mason. 2012. Litter size predicts adult stereotypic behavior in female laboratory mice. *Journal of the American Association for Laboratory Animal Science* 51, no. 4:407–11.

Bechard, A., and G. Mason. 2010. Leaving home: A study of laboratory mouse pup independence. *Applied Animal Behaviour Science* 3, no. 125:181–88.

Bechard, A., R. Meagher, and G. Mason. 2011. Environmental enrichment reduces the likelihood of alopecia in adult C57BL/6J mice. *Journal for the American Association for Laboratory Animal Science* 50:171–4.

Becker, S. D., and J. L. Hurst. 2009. Female behaviour plays a critical role in controlling murine pregnancy block. *Proceedings of the Royal Society B: Biological Sciences* 276, no. 1662:1723–9.

Benaroya-Milshtein, N., N. Hollander, A. Apter, T. Kukulansky, N. Raz, A. Wilf, I. Yaniv, and C. G. Pick. 2004. Environmental enrichment in mice decreases anxiety, attenuates stress responses and enhances natural killer cell activity. *European Journal of Neuroscience* 20, no. 5:1341–7.

Bisazza, A. 1981. Social organization and territorial behaviour in three strains of mice. *Italian Journal of Zoology* 48:157–67.

Blasco-Serra, A., E. M. González-Soler, A. Cervera-Ferri, V. Teruel-Martí, and A. A. Valverde-Navarro. 2017. A standardization of the novelty-suppressed feeding test protocol in rats. *Neuroscience Letters* 658:73–78.

Brain, P. and D. Benton. 1983. Conditions of housing, hormones and aggressive behaviour. In *Hormones and Aggressive Behavior*. ed. B. Svare, 349–72. New York: Plenum Press.

Brodkin, E. S. 2007. BALB/c mice: Low sociability and other phenotypes that may be relevant to autism. *Behavioural Brain Research* 176, no. 1:53–65.

Bronson, F. H. 1984. Energy allocation and reproductive development in wild and domestic house mice. *Biology of Reproduction* 31, no. 1:83–88.

Brown, C. 2009. Novel food items as environmental enrichment for rodents and rabbits. *Lab Animal* 38, no. 4:119–20.

Buehr, M., J. P. Hjorth, A. K. Hansen, and P. Sandøe. 2003. Genetically modified laboratory animals–what welfare problems do they face? *Journal of Applied Animal Welfare Science* 6, no. 4:319–38.

Burke, K., L. A. Screven, and M. L. Dent. 2018. CBA/CaJ mouse ultrasonic vocalizations depend on prior social experience. *PLoS One* 13, no. 6 (June): e0197774. https://journals.plos.org/plosone/article?id=10.1371/journal.pone.0197774.

Burkholder, T., C. Foltz, E. Karlsson, C. G. Linton, and J. M. Smith. 2012. Health evaluation of experimental laboratory mice. *Current Protocols in Mouse Biology* 2:145–65.

Burman, O., L. Buccarello, V. Redaelli, and L. Cervo. 2014. The effect of two different Individually Ventilated Cage systems on anxiety-related behaviour and welfare in two strains of laboratory mouse. *Physiology & Behavior* 124:92–99.

Canadian Council on Animal Care. 2003. Laboratory animal facilities- characteristics, design and development. General Guidelines. Ottawa, ON. https://www.ccac.ca/Documents/Standards/Guidelines/Facilities.pdf (accessed May 1, 2019).

Canadian Council on Animal Care. 2017. CCAC Animal Data Report. Canada. https://www.ccac.ca/Documents/AUD/2017-Animal-Data-Report.pdf (accessed May 1, 2019).

Canadian Council on Animal Care. 2019. CCAC guidelines mice. Canada. https://www.ccac.ca/Documents/Standards/Guidelines/CCAC_Guidelines_Mice.pdf (accessed April 5, 2020).

Carder, B., and K. Berkowitz. 1970. Rats' preference for earned in comparison with free food. *Science* 167, no. 3922:1273–74.

Carini, L. M., C. A. Murgatroyd, and B. C. Nephew. 2013. Using chronic social stress to model postpartum depression in lactating rodents. *Journal of Visualized Experiments* 76 (June): e50324. https://www.jove.com/t/50324/using-chronic-social-stress-to-model-postpartum-depression-lactating.

Chabout, J., A. Sarkar, D. B. Dunson, and E. D. Jarvis. 2015. Male mice song syntax depends on social contexts and influences female preferences. *Frontiers in Behavioral Neuroscience* 9, no.76 (April). https://www.frontiersin.org/articles/10.3389/fnbeh.2015.00076/full.

Chan, C. M. 2018. The rewarding value of gentle touch in mice. Master's Thesis. University of Toronto, department of cell and systems biology.

Cheetham, S. A., M. D. Thom, F. Jury, W. E. R. Ollier, R. J. Beynon, and J. L. Hurst. 2007. The genetic basis of individual-recognition signals in the mouse. *Current Biology* 17, no. 20:1771–77.

Chesler, E. J., S. G. Wilson, W. R. Lariviere, S. L. Rodriguez-Zas, and J. S. Mogil. 2002. Identification and ranking of genetic and laboratory environment factors influencing a behavioral trait, thermal nociception, via computational analysis of a large data archive. *Neuroscience and Biobehavioral Reviews* 26, no. 8:907–23.

Clark, D. A., D. Banwatt, and G. Chaouat. 1993. Stress-triggered abortion in mice prevented by alloimmunization. *American Journal of Reproductive Immunology* 29, no. 3:141–47.

Clipperton, A. E., J. M. Spinato, C. Chernets, D. W. Pfaff, and E. Choleris. 2008. Differential effects of estrogen receptor alpha and beta specific agonists on social learning of food preferences in female mice. *Neuropsychopharmacology* 33, no. 10:2362–75.

Clough, G. 1982. Environmental effects on animals used in biomedical research. *Biological Reviews* 57, no. 3:487–523.

Contet, C., J. N. Rawlins, and R. M. Deacon. 2001. A comparison of 129S2/SvHsd and C57BL/6JOlaHsd mice on a test battery assessing sensorimotor, affective and cognitive behaviours: Implications for the study of genetically modified mice. *Behavioural Brain Research* 124, no. 1:33–46.

Crabbe, J. C., D. Wahlsten, and B. C. Dudek. 1999. Genetics of mouse behavior: Interactions with laboratory environment. *Science* 284, no. 5420:1670–72.

Crawley, J. 2000. *What's Wrong with My Mouse? Behavioural Phenotyping of Transgenic and Knockout Mice*. Chichester, UK: Wiley.

CreekValleyCritters. 2011. Training time with Matilda, cute little mouse. Youtube video. https://www.youtube.com/watch?v=30Q1xEY72xI (accessed May 1, 2019).

Crofton, E. J., Y. Zhang, and T. A. Green. 2015. Inoculation stress hypothesis of environmental enrichment. *Neuroscience & Biobehavioral Reviews* 49: 19–31.

Cuthbert, R. J., R. M. Wanless, A. Angel, M. H. Burle, G. M. Hilton, H. Louw, P. Visser, J. W. Wilson, and P. G. Ryan. 2016. Drivers of predatory behavior and extreme size in house mice *Mus musculus* on Gough Island. *Journal of Mammalogy* 97, no. 2:533–44.

D'Amato, F. R., and F. Pavone. 1996. Reunion of separated sibling mice: neurobiological and behavioral aspects. *Neurobiology of Learning and Memory* 65, no. 1:9–16.

Davis, R. R., J. K. Newlander, X. Ling, G. A. Cortopassi, E. F. Krieg, and L. C. Erway. 2001. Genetic basis for susceptibility to noise-induced hearing loss in mice. *Hearing Research* 155, no. 1–2:82–90.

Deacon, R. M. J. 2011. Hyponeophagia: A measure of anxiety in the mouse. *Journal of Visualized Experiments* 51 (May): e2613. https://www.jove.com/video/2613.

Dere, E., J. P. Huston, and M. A. De Souza Silva. 2005. Integrated memory for objects, places, and temporal order: Evidence for episodic-like memory in mice. *Neurobiology of Learning and Memory* 84:214–21.

Dolensek, N., D. A. Gehrlach, A. S. Klein, and N. Gogolla. 2020. Facial expressions of emotion states and their neuronal correlates in mice. *Science* 368, no. 6486:89–94.

Drickamer, L. C., P. A. Gowaty, and D. M. Wagner. 2003. Free mutual mate preferences in house mice affect reproductive success and offspring performance. *Animal Behaviour* 65, no. 1:105–14.

Elwood, R. W. 1991. Ethical implications of studies on infanticide and maternal aggression in rodents. *Animal Behaviour* 42, no. 5:841–49.

Eppig, J. T., C. J. Bult, J. A. Kadin, J. E. Richardson, J. A. Blake, and The Mouse Genome Database Group. 2005. The Mouse Genome Database (MGD): from genes to mice—a community resource for mouse biology. *Nucleic Acids Research* 33, no. suppl_1:D471–75.

Eppig, J. T., H. Motenko, J. E. Richardson, B. Richards-Smith, C. L. Smith. 2015. The International Mouse Strain Resource (IMSR): cataloging worldwide mouse and ES cell line resources. *Mammalian Genome* 26: no. 9:448–55. http://www.findmice.org (accessed April 6, 2020).

European Commission. 2013. Seventh report on the statistics on the number of animals used for experimental and other scientific purposes in member states of the European Union. Brussels. https://eur-lex.europa.eu/legal-content/EN/TXT/PDF/?uri=CELEX:52013DC0859&from=EN (accessed May 1, 2019).

Eveleigh, J. R. 1993. Murine cage density: Cage ammonia levels during the reproductive performance of an inbred strain and two outbred stocks of monogamous breeding pairs of mice. *Laboratory Animals* 27:156–60.

Ewaldsson, B., S. F. Nunes, B. Gaskill, A. Ferm, A. Stenberg, M. Pettersson, and R. J. Kastenmayer. 2016. Who is a compatible partner for a male mouse? *Scandinavian Journal of Laboratory Animal Science* 42.

Fallon, T. 1996. Rats and mice. In *Handbook of Rodent and Rabbit Medicine*. eds. K. Laber-Laird, W. Swindle, P. Flecknell. 1–38. Charleston, SC: Elsevier Science Ltd.

Ferrari, M., A. K. Lindholm, and B. König. 2019. Fitness consequences of female alternative reproductive tactics in house mice (*Mus musculus domesticus*). *The American Naturalist* 193:106–24.

Ferrari, M., and B. König. 2017. No evidence for punishment in communally nursing female house mice (*Mus musculus domesticus*). *PloS One* 12, no. 6 (June): e0179683. https://journals.plos.org/plosone/article?id=10.1371/journal.pone.0179683.

Foltz, C. J., and M. Ullman-Cullere. 1999. Guidelines for assessing the health and condition of mice. *Lab Animal* 28, no. 4:28–32.

Forestier, T., C. Féron, and P. Gouat. 2018. Transmission of food preference between unfamiliar house mice (*Mus musculus domesticus*) is dependent on social context. *Journal of Comparative Psychology* 132, no. 3:268–79.

Fox, C., Z. Merali, and C. Harrison. 2006. Therapeutic and protective effect of environmental enrichment against psychogenic and neurogenic stress. *Behavioural brain research* 175, no. 1:1–8.

Frick, K. M., and S. M. Fernandez. 2003. Enrichment enhances spatial memory and increases synaptophysin levels in aged female mice. *Neurobiology of Aging* 24, no. 4:615–26.

Frynta, D., M. Slabova, H. Vachova, R. Volfova, and P. Munclinger. 2005. Aggression and commensalism in house mouse: A comparative study across Europe and the Near East. *Aggressive Behavior* 31:283–93.

Frynta, D., B. Kaftanová-Eliášová, H. Žampachová, P. Voráčková, J. Sádlová, and E. Landová. 2018. Behavioural strategies of three wild-derived populations of the house mouse (*Mus m. musculus* and *M. m. domesticus*) in five standard tests of exploration and boldness: Searching for differences attributable to subspecies and commensalism. *Behavioural Processes* 157: 133–41.

Fureix, C., M. Walker, L. Harper, K. Reynolds, A. Saldivia-Woo, and G. Mason. 2016. Stereotypic behaviour in standard non-enriched cages is an alternative to depression-like responses in C57BL/6 mice. *Behavioural Brain Research* 305:186–90.

Fuss, J., S. H. Richter, J. Steinle, G. Deubert, R. Hellweg, and P. Gass. 2013. Are you real? Visual simulation of social housing by mirror image stimulation in single housed mice. *Behavioural Brain Research* 243:191–98.

Galef, B. G. 1993. Functions of social learning about food: A causal analysis of effects of diet novelty on preference transmission. *Animal Behaviour* 46, no. 2:257–65.

Garner, J. P., B. Dufour, L. E. Gregg, S. M. Weisker, and J. A. Mench. 2004a. Social and husbandry factors affecting the prevalence and severity of barbering ('whisker trimming') by laboratory mice. *Applied Animal Behaviour Science* 89, no. 3:263–82.

Garner, J. P., C. M. Thogerson, B. D. Dufour, H. Würbel, J. D. Murray, and J. A. Mench. 2011. Reverse-translational biomarker validation of Abnormal Repetitive Behaviors in mice: An illustration of the 4P's modeling approach. *Behavioural Brain Research* 219, no. 2:189–96.

Garner, J. P., C. M. Thogerson, H. Würbel, J. D. Murray, and J. A. Mench. 2006. Animal neuropsychology: Validation of the intra-dimensional extra-dimensional set shifting task for mice. *Behavioural Brain Research* 173, no. 1:53–61.

Garner, J. P., S. M. Weisker, B. Dufour, and J. A. Mench. 2004b. Barbering (fur and whisker trimming) by laboratory mice as a model of human trichotillomania and obsessive-compulsive spectrum disorders. *Comparative Medicine* 54, no. 2:216–24.

Gaskill, B. N., A. Z. Karas, J. P. Garner, and K. R. Pritchett-Corning. 2013a. Nest building as an indicator of health and welfare in laboratory mice. *Journal of Visualized Experiments* 82 (December): e51012. https://www.jove.com/video/51012.

Gaskill, B. N., and K. R. Pritchett-Corning. 2015. The effect of cage space on behavior and reproduction in Crl: CD1 (Icr) and C57BL/6NCrl laboratory mice. *PLoS One* 10, no. 5 (May): e0127875. https://journals.plos.org/plosone/article?id=10.1371/journal.pone.0127875.

Gaskill, B. N., C. J. Gordon, E. A. Pajor, J. R. Lucas, J. K. Davis, and J. P. Garner. 2012. Heat or insulation: Behavioral titration of mouse preference for warmth or access to a nest. *PloS One* 7, no. 3 (March): e32799. https://journals.plos.org/plosone/article?id=10.1371/journal.pone.0032799.

Gaskill, B. N., C. Winnicker, J. P. Garner, and K. R. Pritchett-Corning. 2013c. The naked truth: Breeding performance in nude mice with and without nesting material. *Applied Animal Behaviour Science* 143:110–16.

Gaskill, B. N., J. R. Lucas, E. A. Pajor, and J. P. Garner. 2011. Little and often? Maintaining continued performance in an automated T-maze for mice. *Behavioural Processes* 86:272–78.

Gaskill, B. N., K. R. Pritchett-Corning, C. J. Gordon, E. A. Pajor, and J. R. Lucas. 2013b. Energy reallocation to breeding performance through improved nest building in laboratory mice. *PLoS One* 8, no. 9 (September): e74153. https://journals.plos.org/plosone/article?id=10.1371/journal.pone.0074153.

George, N. M., J. Whitaker, G. Vieira, J. T. Geronimo, D. A. Bellinger, and C. A. Fletcher. 2015. Antioxidant therapies for ulcerative dermatitis: A potential model for skin picking disorder. *PLoS One* 10, no. 7 (July): e0132092. https://journals.plos.org/plosone/article?id=10.1371/journal.pone.0132092.

Gerlach, G. 1990. Dispersal mechanisms in a captive wild house mouse population (*Mus domesticus* Rutty). *Biological Journal of the Linnean Society* 41:271–77.

Gouveia, K., and J. L. Hurst. 2013. Reducing mouse anxiety during handling: Effect of experience with handling tunnels. *PLoS One* 8, no. 6 (June): e66401. https://journals.plos.org/plosone/article?id=10.1371/journal.pone.0066401.

Gouveia, K., and J. L. Hurst. 2017. Optimising reliability of mouse performance in behavioural testing: The major role of non-aversive handling. *Scientific Reports* 7 (March): 44999. https://www.nature.com/articles/srep44999.

Groó, Z., P. Szenczi, O. Bánszegi, Z. Nagy, and V. Altbäcker. 2018. The Influence of familiarity and temperature on the huddling behavior of two mouse species with contrasting social systems. *Behavioural Processes* 151:67–72.

Gross, A. N., S. H. Richter, A. K. J. Engel, and H. Würbel. 2012. Cage-induced stereotypies, perseveration and the effects of environmental enrichment in laboratory mice. *Behavioural Brain Research* 234:61–68.

Gurfein, B. T., A. W. Stamm, P. Bacchetti, M. F. Dallman, N. A. Nadkarni, J. M. Milush, C. Touma, et al. 2012. The calm mouse: An animal model of stress reduction. *Molecular Medicine* 18:606–17.

Haemisch, A., T. Voss, and K. Gärtner. 1994. Effects of environmental enrichment on aggressive behavior, dominance hierarchies, and endocrine states in male DBA/2J mice. *Physiology and Behavior* 56, no. 5:1041–48.

Häger, C., L. M. Keubler, S. R. Talbot, S. Biernot, N. Weegh, and S. Buchheister. 2018. Running in the wheel: Defining individual severity levels in mice. *PLoS Biology* 16, no. 10 (October): e2006159. https://journals.plos.org/plosbiology/article?id=10.1371/journal.pbio.2006159.

Haisová-Slábová, M., P. Munclinger, and D. Frynta. 2010. Sexual size dimorphism in free-living populations of *Mus musculus*: Are male house mice bigger. *Acta Zoologica Academiae Scientiarum Hungaricae* 56, no. 2:139–51.

Harper, L., E. Choleris, K. Ervin, C. Fureix, K. Reynolds, M. Walker, and G. Mason. 2015. Stereotypic mice are aggressed by their cage-mates, and tend to be poor demonstrators in social learning tasks. *Animal Welfare* 24:463–73.

Harrison, N., A. K. Lindholm, A. Dobay, O. Halloran, A. Manser, and B. König. 2018. Female nursing partner choice in a population of wild house mice (*Mus musculus domesticus*). *Frontiers in Zoology* 15:1–13.

Hawkins, P., and H. D. R. Golledge. 2018. The 9 to 5 Rodent-Time for Change? Scientific and animal welfare implications of circadian and light effects on laboratory mice and rats. *Journal of Neuroscience Methods* 300:20–25.

Heiderstadt, K. M., and D. A. Blizard. 2011. Increased juvenile and adult body weights in BALB/cByJ mice reared in a communal nest. *Journal of the American Association for Laboratory Animal Science* 50, no. 4:484–87.

Heinla, I., J. Åhlgren, E. Vasar, and V. Voikar. 2018. Behavioural characterization of C57BL/6N and BALB/c female mice in social home cage – effect of mixed housing in complex environment. *Physiology and Behavior* 188:32–41.

Hess, S. E., S. Rohr, B. D. Dufour, B. N. Gaskill, E. A. Pajor, and J. P. Garner. 2008. Home improvement: C57BL/6J mice given more naturalistic nesting materials build better nests. *Journal of the American Association for Laboratory Animal Science: JAALAS* 47, no. 6:25–31.

Hoier, S., C. Pfeifle, and M. Linnenbrink. 2016. Communication at the garden fence–context dependent vocalization in female house mice. *PloS ONE* 11, no. 3 (March): e0152255. https://journals.plos.org/plosone/article?id=10.1371/journal.pone.0152255.

Hölter, S. M., J. Einicke, B. Sperling, A. Zimprich, L. Garrett, H. Fuchs, V. Gailus-Durner, M. de Angelis Hrabé, and W. Wurst. 2015. Tests for anxiety-related behavior in mice. *Current Protocols in Mouse Biology* 5, no. 4:291–309.

Home Office. 2017. Annual statistics of scientific procedures on living animals, Great Britain. https://assets.publishing.service.gov.uk/government/uploads/system/uploads/attachment_data/file/724611/annual-statistics-scientific-procedures-living-animals-2017.pdf. (accessed May 1, 2019).

Howerton, C. L., J. P. Garner, and J. A. Mench. 2008. Effects of a running wheel-igloo enrichment on aggression, hierarchy linearity, and stereotypy in group-housed male CD-1 (ICR) mice. *Applied Animal Behaviour Science* 1, no. 115:90–103.

Hughes, N. K., J. L. Kelley, and P. B. Banks. 2009. Receiving behaviour is sensitive to risks from eavesdropping predators. *Oecologia* 160, no. 3:609–17.

Hurst, J. L. 2009. Female recognition and assessment of males through scent. *Behavioural Brain Research* 200, no. 2:295–303.

Hurst, J. L., and R. S. West. 2010. Taming anxiety in laboratory mice. *Nature Methods* 7, no. 10:825–26.

Institute of Laboratory Animal Resources. 1996. *Guide for the care and use of laboratory animals.* Committee on Care, Use of Laboratory Animals, and National Institutes of Health (US). Division of Research Resources. Washington, D.C.: National Academies Press.

Jirkof, P., N. Cesarovic, A. Rettich, T. Fleischmann, and M. Arras. 2012. Individual housing of female mice: Influence on postsurgical behaviour and recovery. *Laboratory Animals* 46:325–34.

Jordan, R., M. Kollo, and A. T. Schaefer. 2018. Sniffing fast: Paradoxical effects on odor concentration discrimination at the levels of olfactory bulb output and behavior. *eNeuro* 5, no. 5 (September): e0148. http://www.eneuro.org/content/early/2018/09/19/ENEURO.0148-18.2018.

Jordan, W. C., and M. W. Bruford. 1998. New perspectives on mate choice and the MHC. *Heredity* 81, no. 2:127.

Kappel, S., P. Hawkins, and M. T. Mendl. 2017. To group or not to group? Good practice for housing male laboratory mice. *Animals* 7:1–25.

Kavaliers, M., and E. Choleris. 2017. Social cognition and the neurobiology of rodent mate choice. *Integrative and Comparative Biology* 57, no. 4:846–56.

Kempermann, G., H. G. Kuhn, and F. H. Gage. 1997. More hippocampal neurons in adult mice living in an enriched environment. *Nature* 386, no. 6624:493–95.

Kitchenham, L. and G. Mason. 2021. The neurobiology of environmentally induced stereotypic behaviours in captive animals: Assessing the basal ganglia pathways and corticostriatal-thalamo-cortical circuitry hypotheses. *Behaviour*, Special edition on clinical ethology: 1–52.

König, B. 1993. Maternal investment of communally nursing female house mice (*Mus musculus domesticus*). *Behavioural Processes* 30, no. 1:61–73.

Lambert, K., M. Hyer, M. Bardi, A. Rzucidlo, S. Scott, B. Terhune-Cotter, A. Hazelgrove, I. Silva, and C. Kinsley. 2016. Natural-enriched environments lead to enhanced environmental engagement and altered neurobiological resilience. *Neuroscience* 330:386–94.

Lamkin, D. M., S. K. Lutgendorf, D. Lubaroff, A. K. Sood, T. G. Beltz, and A. K. Johnson. 2012. Cancer induces inflammation and depression-like behaviour in the mouse: Modulation by social housing. *Brain, Behavior, and Immunity* 25:319–35.

Langford, D. J., A. L. Bailey, M. L. Chanda, S. E. Clarke, T. E. Drummond, S. Echols, S. Glick et al. 2010. Coding of facial expressions of pain in the laboratory mouse. *Nature Methods* 7, no. 6:447.

Langford, D. J., S. E. Crager, Z. Shehzad, S. B. Smith, S. G. Sotocinal, J. S. Levenstadt, M. L. Chanda, D. J. Levitin, and J. S. Mogil. 2006. Social modulation of pain as evidence for empathy in mice. *Science* 312, no. 5782:1967–70.

Larsen, C. M., I. C. Kokay, and D. R. Grattan. 2008. Male pheromones initiate prolactin-induced neurogenesis and advance maternal behavior in female mice. *Hormones and Behavior* 53, no. 4:509–17.

Latham, N., and G. Mason. 2004. From house mouse to mouse house: The behavioural biology of free-living *Mus musculus* and its implications in the laboratory. *Applied Animal Behaviour Science* 86, 3:261–89.

Latham, N., and G. Mason. 2010. Frustration and perseveration in stereotypic captive animals: Is a taste of enrichment worse than none at all?. *Behavioural Brain Research* 211, no. 1:96–104.

Laule, G. 2010. The laboratory mouse. In *The UFAW Handbook on the Care and Management of Laboratory and Other Research Animals*, 8th Edition. eds. R. Hubrecht and J. Kirkwood. Oxford: Wiley-Blackwell.

LaVail, M. M., G. M. Gorrin, M. A. Repaci, L. A. Thomas, and H. M. Ginsberg. 1987. Genetic regulation of light damage to photoreceptors. *Investigative Ophthalmology & Visual Science* 28, no. 7:1043–48.

Laviola, G. and M. Terranova. 1998. The developmental phychobiology of behavioural plasticity in mice: The role of social experiences in the family unit. *Neuroscience and Biobehavioral Reviews* 123:197–213.

Lawlor, M. 1994. A home for a mouse. *Humane Innovations and Alternatives* 8, 569–73.

Lecker, J., and K. Froberg-Fejko. 2016. Using environmental enrichment and nutritional supplementation to improve breeding success in rodents. *Lab Animal* 45, no. 10:406–08.

Leidinger, C., F. Herrmann, C. Thöne-Reineke, N. Baumgart, and J. Baumgart. 2017. Introducing clicker training as a cognitive enrichment for laboratory mice. *Journal of Visualized Experiments*, 121 (March):e55415. https://www.jove.com/t/55415/introducing-clicker-training-as-cognitive-enrichment-for-laboratory.

Lewejohann, L., C. Reinhard, A. Schrewe, J. Brandewiede, A. Haemisch, N. Görtz, M. Schachner, and N. Sachser. 2006. Environmental bias? Effects of housing conditions, laboratory environment and experimenter on behavioral tests. *Genes, Brain and Behavior* 5, no. 1:64–72.

Linnenbrink, M., and S. von Merten. 2017. No speed dating please! Patterns of social preference in male and female house mice. *Frontiers in Zoology* 14, no. 38 (July). https://frontiersinzoology.biomedcentral.com/articles/10.1186/s12983-017-0224-y.

Liu, Y., H. Guo, J. Zhang, and Y. Zhang. 2017. Quantitative inheritance of volatile pheromones and darcin and their interaction in olfactory preferences of female mice. *Scientific Reports* 7, no. 1 (May):2094. https://www.nature.com/articles/s41598-017-02259-1.

Lopes, P. C., and B. König. 2016. Choosing a healthy mate: Sexually attractive traits as reliable indicators of current disease status in house mice. *Animal Behaviour* 111:119–26.

Lopes, P. C., P. Block, and B. König. 2016. Infection-induced behavioural changes reduce connectivity and the potential for disease spread in wild mice contact networks. *Scientific Reports* 6 (August):31790. https://www.nature.com/articles/srep31790.

Lynch, C. B., and B. P. Possidente. 1978. Relationships of maternal nesting to thermoregulatory nesting in house mice (*Mus musculus*) at warm and cold temperatures. *Animal Behaviour* 26:1136–43.

Major, B., L. Rattazzi, S. Brod, I. Pilipović, G. Leposavić, and F. D'Acquisto. 2015. Massage-like stroking boosts the immune system in mice. *Scientific Reports* 5 (June): 10913. https://www.nature.com/articles/srep10913.

Maple T. L. and B. M. Perdue. 2013. Environmental enrichment. In *Zoo Animal Welfare: Animal Welfare*, vol. 14, 95–117. Berlin, Heidelberg: Springer.

Martin, A. L., and R. E. Brown. 2010. The lonely mouse: Verification of a separation-induced model of depression in female mice. *Behavioural Brain Research* 207:196–207.

Martin, B., S. Ji, S. Maudsley, and M. P. Mattson. 2010. "Control" laboratory rodents are metabolically morbid: Why it matters. *Proceedings of the National Academy of Sciences of the United States of America* 107, no. 14:6127–33.

Mason, G. J., and N. R. Latham. 2004. Can't stop, won't stop: Is stereotypy a reliable animal welfare indicator? *Animal Welfare* 13, no. 1:57–69.

Mason, G., C. C. Burn, J. A. Dallaire, J. Kroshko, K. McDonald, and J. M. Jeschke. 2013. Plastic animals in cages: Behavioural flexibility and responses to captivity. *Animal Behaviour* 85, no. 5:1113–26.

Mason, G., R. Clubb, N. Latham, and S. Vickery. 2007. Why and how should we use environmental enrichment to tackle stereotypic behaviour? *Applied Animal Behaviour Science* 102, no. 3:163–88.

Matta, R., A. N. Tiessen, and E. Choleris. 2017. The role of dorsal hippocampal dopamine D1-type receptors in social learning, social interactions, and food intake in male and female mice. *Neuropsychopharmacology* 42, no. 12:2344–53.

McGreevy, P., and R. Boakes. 2008. *Carrots and Sticks: Principles of Animal Training*. New York: Cambridge University Press.

Melotti, L., N. Kästner, A. K. Eick, A. L. Schnelle, P. Palme, N. Sachser, S. Kaiser, and S. H. Richter. 2019. Can live with 'em, can live without 'em: Pair housed male C57BL/6J mice show low aggression and increasing sociopositive interactions with age, but can adapt to single housing if separated. *Applied Animal Behaviour Science* 214:79–88.

Meijer, M. K., K. Kramer, R. Remie, B. M. Spruijt, L. F. M. Van Zutphen, and V. Baumans. 2006. Effect of routine procedures on physiological parameters in mice kept under different husbandry procedures. *Animal Welfare* 15, no. 1:31–38.

Meikle, D. B., J. H. Kruper, and C. R. Browning. 1995. Adult male house mice born to undernourished mothers are unattractive to oestrous females. *Animal Behaviour* 50:753–58.

Milligan, S. R., G. D. Sales, and K. Khirnykh. 1993. Sound levels in rooms housing laboratory animals: An uncontrolled daily variable. *Physiology & Behavior* 53, no. 6:1067–76.

Montero, I., M. Teschke, and D. Tautz. 2013. Paternal imprinting of mating preferences between natural populations of house mice (*Mus musculus domesticus*). *Molecular Ecology* 22, no. 9:2549–62.

Moreira, V. B., V. G. Mattaraia, and A. S. Moura. 2015. Lifetime reproductive efficiency of BALB/c mouse pairs after an environmental modification at 3 mating ages. *Journal of the American Association for Laboratory Animal Science* 54, no. 1:29–34.

Morgan, J. L., K. L. Svenson, J. P. Lake, W. Zhang, T. M. Stearns, M. A. Marion, L. L. Peters, B. Paigen, and L. R. Donahue. 2014. Effects of housing density in five inbred strains of mice. *PLoS One* 9, no. 3 (March):e90012. https://journals.plos.org/plosone/article?id=10.1371/journal.pone.0090012.

Morris, C. F., D. McLean, J. A. Engleson, E. P. Fuerst, F. Burgos, and E. Coburn. 2012. Some observations on the granivorous feeding behavior preferences of the house mouse (*Mus musculus* L.). *Mammalia* 76:209–18.

Muheim, R., N. M. Edgar, K. A. Sloan, and J. B. Phillips. 2006. Magnetic compass orientation in C57BL/6J mice. *Learning & Behavior* 34:366–73.

Musumeci, S. A., P. Bosco, G. Calabrese, C. Bakker, G. B. Sarro De, M. Elia, R. Ferri, and B. A. Oostra. 2000. Audiogenic seizures susceptibility in transgenic mice with fragile X syndrome. *Epilepsia* 41, no. 1:19–23.

Nakamura, Y., and K. Suzuki. 2018. Tunnel use facilitates handling of ICR mice and decreases experimental variation. *Journal of Veterinary Medical Science* 80:886–92.

Nakatomi, Y., C. Yokoyama, S. Kinoshita, D. Masaki, H. Tsuchida, H. Onoe, K. Yoshimoto, and K. Fukui. 2008. Serotonergic mediation of the antidepressant-like effect of the green leaves odor in mice. *Neuroscience Letters* 436, no. 2:167–70.

National Research Council. 1995. Nutrient requirements of the mouse. In *Nutrient requirements of laboratory animals*, 4th edition. https://www.nap.edu/read/4758/chapter/5. (accessed April 20, 2019).

National Research Council. 2011. *Guide for the Care and Use of Laboratory Animals*. Washington, D.C.: National Academies Press.

Neuringer, A. J. 1969. Animals respond for food in the presence of free food. *Science* 166, no. 3903:399–401.

Nevison, C. M., J. L. Hurst, and C. J. Barnard. 1999. Why do male ICR (CD-1) mice perform bar-related (stereotypic) behaviour?. *Behavioural Processes* 47, no. 2:95–111.

Nip, E., A. Adcock, B. Nazal, A. MacLellan, L. Neil, E. Choleris, L. Levison, and G. Mason. 2019. Why are enriched mice nice? Investigating how environmental enrichment reduces agonism in female C57BL/6, DBA/2, and BALB/c mice. *Applied Animal Behaviour Science* 217:73–82.

Nomoto, K., M. Ikumi, M. Otsuka, A. Asaba, M. Kato, N. Koshida, K. Mogi, and T. Kikusui. 2018. Female mice exhibit both sexual and social partner preferences for vocalizing males. *Integrative Zoology* 13, no. 6:735–44.

O'Leary, T. P., S. Shin, E. Fertan, R. N. Dingle, A. Almuklass, R. K. Gunn, Z. Yu, J. Wang, and R. E. Brown. 2017. Reduced acoustic startle response and peripheral hearing loss in the 5xFAD mouse model of Alzheimer's disease. *Genes, Brain and Behavior* 16, no. 5:554–63.

Olsson, I. A. S., C. M. Nevison, E. G. Patterson-Kane, C. M. Sherwin, H. A. Van de Weerd, and H. Würbel. 2003. Understanding behaviour: The relevance of ethological approaches in laboratory animal science. *Applied Animal Behaviour Science* 81, no. 3:245–64.

Olsson, I. A., and K. Dahlborn. 2002. Improving housing conditions for laboratory mice: A review of 'environmental enrichment'. *Laboratory Animals* 36, no. 3:243–70.

Olsson, I. A. S., and K. Westlund. 2007. More than numbers matter: The effect of social factors on behaviour and welfare of laboratory rodents and non-human primates. *Applied Animal Behaviour Science* 103, no. 3–4:229–54.

Paigen, B., J. M. Currer, and K. L. Svenson. 2016. Effects of varied housing density on a hybrid mouse strain followed for 20 months. *PLoS One* 11 (February):e0149647. https://journals.plos.org/plosone/article?id=10.1371/journal.pone.0149647.

Palanza, P., L. Gioiosa, and S. Parmigiani. 2001. Social stress in mice: Gender differences and effects of estrous cycle and social dominance. *Physiology and Behavior* 73:411–20.

Paster, E. V., K. A. Villines, and D. L. Hickman. 2009. Endpoints for mouse abdominal tumor models: Refinement of current criteria. *Comparative Medicine* 59, no. 3:234–41.

Peirson, S. N., L. A. Brown, C. A. Pothecary, L. A. Benson, and A. S. Fisk. 2018. Light and the laboratory mouse. *Journal of Neuroscience Methods* 300:26–36.

Perepelkina, O. V., V. A. Golibrodo, I. G. Lilp, and I. I. Poletaeva. 2014. Selection of laboratory mice for the high scores of logic task solutions: The correlated changes in behavior. *Advances in Bioscience and Biotechnology* 5, no. 4:294–300.

Pham, T. M., B. Hagman, A. Codita, P. L. P. Van Loo, L. Strömmer, and V. Baumans. 2010. Housing environment influences the need for pain relief during post-operative recovery in mice. *Physiology & Behavior* 99, no. 5:663–68.

Phifer-Rixey, M. and M. W. Nachman. 2015. The natural history of model organisms: insights into mammalian biology from the wild house mouse *Mus musculus*. *Elife* 4 (April): e05959. https://elifesciences.org/articles/05959.

Phillips, D., E. Choleris, K. S. Ervin, C. Fureix, L. Harper, K. Reynolds, L. Niel, and G. J. Mason. 2016. Cage-induced stereotypic behaviour in laboratory mice covaries with nucleus accumbens FosB/ΔFosB expression. *Behavioural Brain Research* 301, 238–42.

Pietropaolo, S., I. Branchi, F. Cirulli, F. Chiarotti, L. Aloe, and E. Alleva. 2004. Long-term effects of the periadolescent environment on exploratory activity and aggressive behaviour in mice: social versus physical enrichment. *Physiology & Behavior* 81, no. 3:443–53.

Poletaeva, I. I., M. G. Pleskacheva, N. V. Markina, O. V. Perepelkina, H. Shefferan, D. P. Wolfer, and H.-P. Lipp. 2001. Survival under conditions of environmental stress: Variability of brain morphology and behavior in the house mouse. *Russian Journal of Ecology* 32, no. 3:211–15.

Poley, W. 1974. Emotionality related to maternal cannibalism in BALB and C57BL mice. *Animal Learning & Behavior* 2, no. 4:241–44.

Poole, T. B., and H. D. R. Morgan. 1973. Differences in aggressive behaviour between male mice (*Mus musculus* L.) in colonies of different sizes. *Animal Behaviour* 21:788–95.

Priestnall, R. 1973. Effects of handling on maternal behaviour in the mouse (*Mus musculus*): An observational study. *Animal Behaviour* 21, no. 2:383–86.

Rabat, A. 2007. Extra-auditory effects of noise in laboratory animals: The relationship between noise and sleep. *Journal of the American Association for Laboratory Animal Science* 46, no. 1:35–41.

Rajendram, E. A., P. F. Brain, S. Parmigiani, and M. Mainardi. 1987. Effects of ambient temperature on nest construction in four species of laboratory rodents. *Italian Journal of Zoology* 54, no. 1:75–81.

Ramm, S. A., D. A. Edward, A. J. Claydon, D. E. Hammond, P. Brownridge, J. L. Hurst, R. J. Beynon, and P. Stockley. 2015. Sperm competition risk drives plasticity in seminal fluid composition. *BMC Biology* 13, no. 1 (October): 87. https://bmcbiol.biomedcentral.com/articles/10.1186/s12915-015-0197-2.

Raveh, S., S. Sutalo, K. E. Thonhauser, M. Thoß, A. Hettyey, F. Winkelser, and D. J. Penn. 2014. Female partner preferences enhance offspring ability to survive an infection. *BMC Evolutionary Biology* 14, no. 1 (January): 14. https://bmcevolbiol.biomedcentral.com/articles/10.1186/1471-2148-14-14.

Reeb-Whitaker, C. K., B. Paigen, W. G. Beamer, R. T. Bronson, G. A. Churchill, I. B. Schweitzer, and D. D. Myers. 2001. The impact of reduced frequency of cage changes on the health of mice housed in ventilated cages. *Laboratory Animals* 35, no. 1:58–73.

Roberts, S. A., D. M. Simpson, S. D. Armstrong, A. J. Davidson, D. H. Robertson, L. McLean, R. J. Beynon, and J. L. Hurst. 2010. Darcin: A male pheromone that stimulates female memory and sexual attraction to an individual male's odour. *BMC Biology* 8, no. 1 (June): 75. http://www.biomedcentral.com/1741-7007/8/75.

Rock, M. L., A. Z. Karas, K. B. G. Rodriguez, M. S. Gallo, K. Pritchett-Corning, R. H. Karas, M. Aronovitz, and B. N. Gaskill. 2014. The time-to-integrate-to-nest test as an indicator of wellbeing in laboratory mice. *Journal of the American Association for Laboratory Animal Science* 53, no. 1:24–28.

Roedel, A., C. Storch, F. Holsboer, and F. Ohl. 2006. Effects of light or dark phase testing on behavioural and cognitive performance in DBA mice. *Laboratory Animals* 40, no. 4:371–81.

Rosenbaum, M. D., S. VandeWoude, and T. E. Johnson. 2009. Effects of cage-change frequency and bedding volume on mice and their microenvironment. *Journal of the American Association for Laboratory Animal Science* 48, no. 6:763–73.

Ross, M., and Mason, G. J. 2017. The effects of preferred natural stimuli on humans' affective states, physiological stress and mental health, and the potential implications for wellbeing in captive animals. *Neuroscience and Biobehavioral Reviews* 83:46–62.

Rusu, A. S., and S. Krackow. 2005. Agonistic onset marks emotional changes and dispersal propensity in wild house mouse males (*Mus domesticus*). *Journal of Comparative Psychology* 119:58–66.

Sales, G. D., K. J. Wilson, K. E. V. Spencer, and S. R. Milligan. 1988. Environmental ultrasound in laboratories and animal houses: A possible cause for concern in the welfare and use of laboratory animals. *Laboratory Animals* 22, no. 4:369–75.

Samuels, B., and R. Hen. 2011. Novelty-suppressed feeding in the mouse. In *Mood and Anxiety Related Phenotypes in Mice*, ed. T. Gould, 107–21. Neuromethods, Vol. 63. New York: Humana Press.

Sarna, J. R., R. H. Dyck, and I. Q. Whishaw. 2000. The Dalila effect: C57BL6 mice barber whiskers by plucking. *Behavioural Brain Research* 108, no. 1:39–45.

Scattoni, M. L., L. Ricceri, and J. N. Crawley. 2011. Unusual repertoire of vocalizations in adult BTBR T+ tf/J mice during three types of social encounters. *Genes, Brain and Behavior* 10, no. 1:44–56.

Schipani, S. 2019. The history of the lab rat is full of scientific triumphs and ethical quandaries. Smithsonian Magazine. https://www.smithsonianmag.com/science-nature/history-lab-rat-scientific-triumphs-ethical-quandaries-180971533/ (accessed May 1, 2019).

Screven, L. A., and M. L. Dent. 2018. Preference in female laboratory mice is influenced by social experience. *Behavioural Processes* 157:171–79.

Sheils, A. B., W. C. Pitt. 2014. A review of invasive rodent (Rattus spp. And Mus musculus) diets on pacific islands. *USDA National Wildlife Research Center- Staff Publications*. 1781. https://digitalcommons.unl.edu/icwdm_usdanwrc/1781.

Sherwin, C. M. 2004a. The motivation of group-housed laboratory mice, *Mus musculus*, for additional space. *Animal Behaviour* 67, no. 4:711–17.

Sherwin, C. M. 2004b. The influences of standard laboratory cages on rodents and the validity of research data. *Animal Welfare* 13, no. 1:9–15.

Sherwin, C. M., and C. J. Nicol. 1996. Reorganization of behaviour in laboratory mice, *Mus musculus*, with varying cost of access to resources. *Animal Behaviour* 51, no. 5:1087–93.

Shupe, J. M., D. M. Kristan, S. N. Austad, and D. L. Stenkamp. 2006. The eye of the laboratory mouse remains anatomically adapted for natural conditions. *Brain, Behavior and Evolution* 67, no. 1:39–52.

Singh, P., J. C. Schimenti, and E. Bolcun-Filas. 2015. A mouse geneticist's practical guide to CRISPR applications. *Genetics* 199, no. 1:1–15.

Singleton, G. R., and C. J. Krebs. 2007. The secret world of wild mice. In *The Mouse in Biomedical Research*, pp. 25–51. Academic Press, Cambridge, MA.

Smith, A. L., and D. J. Corrow. 2005. Modifications to husbandry and housing conditions of laboratory rodents for improved well-being. *ILAR Journal* 46, no. 2:140–47.

Smith, A. L., S. L. Mabus, J. D. Stockwell, and C. Muir. 2004. Effects of housing density and cage floor space on C57BL/6J mice. *Comparative Medicine* 54:656–63.

Song, Y., S. Endepols, N. Klemann, D. Richter, F. R. Matuschka, C. H. Shih, M. W. Nachman, and M. H. Kohn. 2011. Adaptive introgressive hybridization with the Algerian mouse (*Mus spretus*) promoted the evolution of anticoagulant rodenticide resistance in European house mice (*M. musculus domesticus*). *Current Biology* 21, no. 15:1296–301.

Sørensen, A., D. Mayntz, S. J. Simpson, and D. Raubenheimer. 2010. Dietary ratio of protein to carbohydrate induces plastic responses in the gastrointestinal tract of mice. *Journal of Comparative Physiology B* 180, no. 2:259–66.

Sørensen, D. B. 2014. Shedding ultraviolet light on welfare in laboratory rodents: Suggestions for further research and refinement. *Animal Welfare* 23, no. 3:259–61.

Sorge, R. E., L. J. Martin, K. A. Isbester, S. G. Sotocinal, S. Rosen, A. H. Tuttle, J. S. Wieskopf et al. 2014. Olfactory exposure to males, including men, causes stress and related analgesia in rodents. *Nature Methods* 11, no. 6:629–32.

Spangenberg, E. M. F., and L. J. Keeling. 2016a. Assessing the welfare of laboratory mice in their home environment using animal-based measures–a benchmarking tool. *Laboratory Animals* 50, no. 1:Appendix 1.

Spangenberg, E. M., and L. J. Keeling. 2016b. Assessing the welfare of laboratory mice in their home environment using animal-based measures–a benchmarking tool. *Laboratory Animals* 50, no. 1:30–38.

Späni, D., M. Arras, B. König, and T. Rülicke. 2003. Higher heart rate of laboratory mice housed individually vs in pairs. *Laboratory Animals* 37:54–62.

Spoelstra, K., M. Wikelski, S. Daan, A. S. Loudon, and M. Hau. 2016. Natural selection against a circadian clock gene mutation in mice. *Proceedings of the National Academy of Sciences of the United States of America* 113, no. 3:686–91.

Stockley, P., S. A. Ramm, A. L. Sherborne, M. D. F. Thom, S. Paterson, and J. L. Hurst. 2013. Baculum morphology predicts reproductive success of male house mice under sexual selection. *BMC Biology* 11, no. 66 (May). https://bmcbiol.biomedcentral.com/articles/10.1186/1741-7007-11-66.

Stüttgen, M. C., and C. Schwarz. 2018. Barrel cortex: What is it good for? *Neuroscience* 368:3–16.

Szenczi, P., O. Bánszegi, Z. Groó, and V. Altbäcker. 2012. Development of the social behavior of two mice species with contrasting social systems. *Aggressive Behavior* 38:288–97.

Sztainberg, Y., and A. Chen. 2010. An environmental enrichment model for mice. *Nature Protocols* 5, no. 9:1535–39.

Taborsky, B. 2016. Opening the black box of developmental experiments: Behavioural mechanisms underlying long-term effects of early social experience. *Ethology* 122, no. 4:267–83.

Tallent, B. R., L. M. Law, R. K. Rowe, and J. Lifshitz. 2018. Partial cage division significantly reduces aggressive behavior in male laboratory mice. *Laboratory Animals* 52, no. 4:384–93.

Thon, R. W., M. Ritskes-Hoitinga, H. Gates, J. Prins. 2010. Phenotyping of genetically modified mice. In *The UFAW Handbook on the Care and Management of Laboratory and Other Research Animals*, 8th Edition. eds. R. Hubrecht and J. Kirkwood. Oxford: Wiley-Blackwell.

Thonhauser, K. E., M. Thoß, K. Musolf, T. Klaus, and D. J. Penn. 2013. Multiple paternity in wild house mice (*Mus musculus musculus*): Effects on offspring genetic diversity and body mass. *Ecology and Evolution* 4, no. 2:200–09.

Tilly, S. L. C., J. Dallaire, and G. J. Mason. 2010. Middle-aged mice with enrichment-resistant stereotypic behaviour show reduced motivation for enrichment. *Animal Behaviour* 80, no. 3:363–73.

Ueno, H., S. Suemitsu, S. Murakami, N. Kitamura, K. Wani, Y. Matsumoto, M. Okamoto, and T. Ishihara. 2019. Helping-like behaviour in mice towards conspecifics constrained inside tubes. *Scientific Reports* 9 (April): 5817. https://www.nature.com/articles/s41598-019-42290-y.

Ullman-Culleré, M. H., and C. J. Foltz. 1999. Body condition scoring: A rapid and accurate method for assessing health status in mice. *Laboratory Animal Science* 49, no. 3:319–23.

Valsecchi, P., G. R. Singleton, and W. J. Price. 1996. Can social behaviour influence food preference of wild mice, *Mus domesticus*, in confined field populations? *Australian Journal of Zoology* 44, no. 5:493–501.

Van de Weerd, H. A., P. L. P. Van Loo, L. F. M. Van Zutphen, J. M. Koolhaas, and V. Baumans. 1998. Strength of preference for nesting material as environmental enrichment for laboratory mice. *Applied Animal Behaviour Science* 55, no. 3:369–82.

Van de Weerd, H. A., P. L. P. Van Loo, L. F. M. Van Zutphen, J. M. Koolhaas, and V. Baumans. 1997. Nesting material as environmental enrichment has no adverse effects on behavior and physiology of laboratory mice. *Physiology & Behavior* 62, no. 5:1019–28.

Van Den Bergen, H., R. Spratt, and C. Messier. 1997. An automatic food delivery system for operant training of mice. *Physiology and Behavior* 61:879–82.

Van Loo, P. L. P., C. L. J. J. Kruitwagen, L. F. M. Van Zutphen, J. M. Koolhaas, and V. Baumans. 2000. Modulation of aggression in male mice: Influence of cage cleaning regime and scent marks. *Animal Welfare* 9, no. 3:281–95.

Van Loo, P. L. P, A. C. de Groot, B. F. M. Van Zutphen, and V. Baumans. 2001a. Do male mice prefer or avoid each other's company? Influence of hierarchy, kinship, and familiarity. *Journal of Applied Animal Welfare Science* 4:91–103.

Van Loo, P. L. P., J. A. Mol, J. M. Koolhaas, B. F. M. Van Zutphen, and V. Baumans. 2001b. Modulation of aggression in male mice. *Physiology & Behavior* 72:675–83.

Van Loo, P. L. P., N. Kuin, R. Sommer, H. Avsaroglu, T. Pham, and V. Baumans. 2007. Impact of 'living apart together' on postoperative recovery of mice compared with social and individual housing. *Laboratory Animals* 41:441–55.

Vaughan, L. M., J. S. Dawson, P. R. Porter, and A. L. Whittaker. 2014. Castration promotes welfare in group-housed male Swiss outbred mice maintained in educational institutions. *Journal of the American Association for Laboratory Animal Science* 53:38–43.

Viney, M., L. Lazarou, and S. Abolins. 2015. The laboratory mouse and wild immunology. *Parasite Immunology* 37, no. 5:267–73.

Voelkl, B., L. Vogt, E. S. Sena, and H. Würbel. 2018. Reproducibility of preclinical animal research improves with heterogeneity of study samples. *PLoS Biology* 16, no. 2 (February): e2003693. https://journals.plos.org/plosbiology/article?id=10.1371/journal.pbio.2003693.

Volfová, R., V. Stejskal, R. Aulický, and D. Frynta. 2010. Presence of conspecific odours enhances responses of commensal house mice (*Mus musculus*) to bait stations. *International Journal of Pest Management* 57, no. 1:35–40.

Wackermannová, M., L. Pinc, and L. Jebavý. 2016. Olfactory sensitivity in mammalian species. *Physiological Research* 65, no. 3:369–90.

Wanless, R. M., A. Angel, R. J. Cuthbert, G. M. Hilton, and P. G. Ryan. 2007. Can predation by invasive mice drive seabird extinctions? *Biology Letters* 3, no. 3:241–44.

Wasson, K. 2017. Retrospective analysis of reproductive performance of pair-bred compared with trio-bred mice. *Journal of the American Association for Laboratory Animal Science* 56, no. 2:190–93.

Weber, E. M., and I. A. S. Olsson. 2008. Maternal behaviour in Mus musculus sp.: An ethological review. *Applied Animal Behaviour Science* 114:1–22.

Weber, E. M., J. Ahloy Dallaire, B. N. Gaskill, K. R. Pritchett-Corning, and J. P. Garner. 2017. Aggression in group-housed laboratory mice: Why can't we solve the problem? *Lab Animal* 46:157–61.

Weber, E. M., B. Algers, J. Hultgren, and I. A. S. Olsson. 2013. Pup mortality in laboratory mice–infanticide or not? *Acta Veterinaria Scandinavica* 55, no. 1 (November): 83. https://actavetscand.biomedcentral.com/articles/10.1186/1751-0147-55-83.

Weidt, A., A. K. Lindholm, and B. König. 2014. Communal nursing in wild house mice is not a by-product of group living: Females choose. *Naturwissenschaften* 101, no. 1:73–76.

Weidt, A., L. Gygax, R. Palme, C. Touma, and B. König. 2018. Impact of male presence on female sociality and stress endocrinology in wild house mice (*Mus musculus domesticus*). *Physiology and Behavior* 189:1–9.

Weidt, A., S. E. Hofmann, and B. König. 2008. Not only mate choice matters: fitness consequences of social partner choice in female house mice. *Animal Behaviour* 75, no. 3:801–08.

Wells, D. J., L. C. Playle, W. E. J. Enser, P. A. Flecknell, M. A. Gardiner, J. Holland, B. R. Howard et al. 2006. Assessing the welfare of genetically altered mice. *Laboratory Animals* 40, no. 2:111–14.

Whittaker, A. L., G. S. Howarth, and D. L. Hickman. 2012. Effects of space allocation and housing density on measures of wellbeing in laboratory mice: A review. *Laboratory Animals* 46, no. 1:3–13.

Wong, A. A., and R. E. Brown. 2006. Visual detection, pattern discrimination and visual acuity in 14 strains of mice. *Genes, Brain and Behavior* 5, no. 5:389–403.

Wu, H. P., J. C. Ioffe, M. M. Iverson, J. M. Boon, and R. H. Dyck. 2013. Novel, whisker-dependent texture discrimination task for mice. *Behavioural Brain Research* 237:238–42.

Würbel, H. 2001. Ideal homes? Housing effects on rodent brain and behaviour. *Trends in Neurosciences* 4, no. 24:207–11.

Würbel, H., C. Burn, and N. Latham. 2017. The behaviour of laboratory mice and rats. In *The Ethology of Domestic Animals: An Introductory Text*. ed. P. Jensen, 272–86. Wallingford, UK: CAB International.

Wass, C., A. Denman-Brice, C. Rios, K. R. Light, S. Kolata, A. M. Smith, and L. D. Matzel. 2012. Covariation of learning and "reasoning" abilities in mice: Evolutionary conservation of the operations of intelligence. *Journal of Experimental Psychology: Animal Behavior Processes* 38: 109–24.

Yamazaki, K, M. Yamaguchi, L. Barnoski, J. Bard, E. A. Boyse, and L. Thomas. 1979. Recognition among mice. Evidence from the use of a Y-maze differentially scented by congenic mice of different major histocompatibility types. *The Journal of Experimental Medicine* 150, no 4:755–60.

Yamashita, Y., N. Kawai, O. Ueno, Y. Matsumoto, T. Oohashi, and M. Honda. 2018. Induction of prolonged natural lifespans in mice exposed to acoustic environmental enrichment. *Scientific Reports* 8, no. 1:1–8.

Zafar, T., and V. K. Shrivastava. 2018. Effect of cold stress on infanticide by female Swiss albino mice *Mus musculus*: A pilot study. *Journal of Animal Science and Technology* 60, no. 7 (April). https://janimscitechnol.biomedcentral.com/articles/10.1186/s40781-018-0168-6.

Zhang, Y., A. L. Brasher, N. R. Park, H. A. Taylor, A. N. Kavazis, and W. R. Hood. 2018. High activity before breeding improves reproductive performance by enhancing mitochondrial function and biogenesis. *The Journal of Experimental Biology* 221 (April): jeb-177469. http://jeb.biologists.org/content/221/7/jeb177469.abstract.

7

Behavioral Biology of Rats

Sylvie Cloutier
Animal Behavior Scientist

CONTENTS

Introduction 113
    Animal-Based Research with Rats 113
Behavioral Biology 114
    Natural History 114
        Ecology, Including Habitat Use, Predator/Prey Relations (i.e., Antipredator Behavior) 114
        Social Organization and Behavior, Including Mating and Parental Behavior 115
        Feeding Behavior 116
    Communication 117
        Olfactory 117
        Vocalizations 117
        Touch 118
        Visual 118
Common Captive Behaviors 118
    Normal 118
    Abnormal 119
Ways to Maintain Behavioral Health – Captive Considerations to Promote Species-Typical Behavior 119
    Environment–Macroenvironment 119
    Environment–Microenvironment 120
    Social Groupings 122
    Feeding Strategies 122
    Consideration for Experimental and Testing Procedures 123
    Training and Handling 123
Conclusions/Recommendations 125
References 125

Introduction

The objective of this chapter is to provide information about rats' natural behavior and ecology. It is hoped by the author that this information will promote the use of husbandry, housing, and enrichment practices adapted to their species-specific needs, thus ensuring their overall health and welfare. This knowledge will also aid in selecting from among the best available animal models, the most valid candidate to address the specific research question.

The rat is one of the most common species involved in animal-based research. In Canada, for example, rats were in the top five of all animals involved in science in 2018, and they accounted for 12% of all species selected in the European Union in 2017 (CCAC, 2018; European report, 2019).

Rats belong to the order *Rodentia*, which is by far the largest order within the class *Mammalia* (Ellenbroek and Youn, 2016). They are part of the family *Muridae*, to which mice also belong, and the subfamily *Murinae*. Although similar in many respects to mice, rats are not just "big mice"; they are different in many ways. Thus, when planning housing and enrichment for rats, one should keep in mind that what applies to mice does not necessarily work for rats, and vice versa.

Animal-Based Research with Rats

In animal models, the rat is not selected as often as the mouse, mainly because of the current availability of better molecular techniques to manipulate the mouse genome. However, recent advances in genetic tools to create knockout rat models will likely lead to their increase in a wider variety of biomedical research programs (Bryda, 2013).

Rats are known to be the first mammalian species to have been domesticated primarily for scientific purposes (Lindsey and Baker, 2006). Rats were first used in neuroanatomical, nutrition, and behavioral studies (Lindsey and Baker, 2006). Following these early studies, rats' selection as animal models expanded to many different fields: cancer and carcinogenesis, cardiovascular and stroke studies, chronic pain, nutrition, obesity, diabetes, renal diseases, immunology, pharmacology, teratology, epilepsy, dermatology, alcoholism and other addictive drugs, neonatal and gestational studies, aging, and behavior (Bryda, 2013; Hedrich, 2006; Tunstall et al., 2020). There are also numerous rat models of human nervous system disorders that include mental illnesses, such as anxiety, depression, posttraumatic stress disorder, neurodegenerative diseases, such as Alzheimer's and Parkinson's diseases, and models of acute and chronic pain (El-Ayache and Galligan, 2020). Because of the clear relationship between rat ultrasonic vocalizations (USVs) and affective states, these calls are now used as indicators of affect in affective neuroscience models (Burgdorf et al., 2020). Some rat models are applied to drug development, toxicology and toxicity, and safety testing (see for example, https://toxtutor.nlm.nih.gov/05-001.html). Because their gustatory perception is similar to humans (Burn, 2008), rats, as models, are often selected in research focused on taste. Rats' social and learning abilities also make them notable subjects in behavioral studies (Bryda, 2013; Ellenbroek and Youn, 2016). However, despite this extensive and widespread participation in numerous aspects of scientific research, deep knowledge of their natural, basic behavior is not widespread among the research community.

Behavioral Biology

Natural History

Ecology, Including Habitat Use, Predator/Prey Relations (i.e., Antipredator Behavior)

Domestication has impacted the rat in several ways, as laboratory strains are larger, show increased fecundity, and are more docile than wild rats (Pritchett-Corning, 2015). Nevertheless, laboratory rats have been shown to adapt quickly in seminaturalistic outdoor environments, expressing behaviors similar to their wild ancestors (see Berdoy, 2002). This study provides evidence that laboratory rat natural behavior is similar to that of their wild ancestor, thus cognizance of the behaviors and needs of the wild rat can help design laboratory environments associated with good welfare. Most laboratory rat strains have evolved from the Norway rat (*Rattus norvegicus*), also called the Brown rat (Winnicker et al., 2016). Thus, considering that the Norway rat is believed to be the ancestor of the laboratory rat, this chapter will focus on its natural behaviors. The Norway rat had its beginnings in Asia, but over the centuries, often in association with human migration, long-distance trading travel, and development of agriculture, spread widely across the world. Today, rats are found on every continent except Antarctica. Notably, the province of Alberta, Canada, is the largest rat-free populated area in the world, due to aggressive rat control measures started during the 1950s (https://www.alberta.ca/history-of-rat-control-in-alberta.aspx). Although Asian forests and brushy areas are known to be the native habitat of the Norway rat, currently, the species is found in a variety of habitats due to its high adaptability, often near human populations, where food can be easily found.

The Norway rat has a home range averaging 2,000 m^2, but varying widely between 10 and 8,000 m^2 (Winnicker et al., 2016). To locate resources, rats can travel up to 3–5 km, but may remain within 20 m of their nest when the food supply is abundant. The Norway rat digs and resides mainly in burrow systems (Winnicker et al., 2016). Burrows are made by both males and females, although older male rats generally were not seen burrowing in some studies (Boice, 1977; Nieder et al., 1982; Price, 1977). Burrows can be a simple chamber connected to the outside by a short tunnel, or a large complex of interconnecting tunnels, passages, and cavities (Calhoun, 1963). Cavities are used as nest chambers or food caches. Nests can be scantly lined with objects, such as leaves, or elaborate, with bowl-shaped grass structures. Burrows often include secondary entrances, called bolt holes, which may serve as escape exits in case of invasion from predators and unfamiliar conspecifics (Calhoun, 1963). Burrow entrances may be sealed with cut grass. Entrances are usually located against vertical surfaces (e.g., a wall), under flat horizontal surfaces (e.g., under a board placed on the ground), under overhead cover (e.g., shrubs, overhanging concrete, raised man-made floors), on slopes, or near sources of food and water (Winnicker et al., 2016).

Rats are primarily nocturnal or crepuscular (Antle and Mistlberger, 2004); nevertheless, they show periods of activity both during the day and at night (Cloutier and Newberry, 2010; Hurst et al., 1998). They typically show two main activity peaks, one in the early evening and one immediately before sunrise, both of which are mainly used for feeding and foraging. A third activity period may occur in the middle of the night (Koolhaas, 2010). They tend to avoid bright, open spaces (Burn, 2008), and will spend time in dark, sheltered areas, particularly during daylight (Boice, 1977; Calhoun, 1963; Cloutier and Newberry, 2010; Makowska and Weary, 2016). When outside the burrow, they typically travel on above-ground trails that are noticeable in the surrounding environment.

Rats are exploratory in nature and engage in a variety of locomotor behaviors, including walking, running, jumping (Altman and Sudarshan, 1975), climbing (Foster et al., 2011), and rearing (i.e., stretching upright on their hind legs) (Büttner, 1993; Makowska and Weary, 2016). Norway rats have been observed to ascend trees, thickets, and dry stalks to forage for berries and grain (Makowska and Weary, 2016). They are also very good swimmers (Winnicker et al., 2016), showing evidence of motor competence and coordinated swimming by 16 days of age (Schapiro et al., 1970). According to an U.S. Fish and Wildlife Service fact sheet (June 2008, https://www.fws.gov/pacificislands/publications/ratsfactsheet.pdf), rats can swim over 1,600 m in open water and can tread water for up to 3 days. However, a study on rat populations of the Falkland Islands reported that 72% of islands with a separation of 500 m or less supported populations of rats, but with a separation distance of 1,000 m or more, this proportion was reduced to just 3% (Tabak et al., 2015).

Although the Norway rat is a predator of several species of terrestrial and aquatic small animals (see "feeding" below), it is also preyed upon by several carnivorous mammals (e.g., canids and felids), birds (e.g., owls and other raptors), and reptiles (e.g., snakes) (Hamilton, 1998; Nowak and Paradiso, 1983;

Silver, 1927). Hence, rats will respond to the presence of actual, perceived, or potential threats (e.g., presence of predators, intruders, dominant individuals, threatening, novel, or aversive stimuli, such as a sudden, loud noise) in the surrounding environment by retreating to the burrow (Kitaoka, 1994). When retreat is not possible, rats may freeze and focus their attention on the threat (e.g., a predator), or they may attack (Kitaoka, 1994). Threatening situations may also trigger the expression of other defensive behaviors, such as avoidance and risk assessment (Figure 7.1), resulting in a reduction in maintenance behaviors, such as eating, drinking, exploration, sexual activity, and play (Blanchard and Blanchard, 1989). Defensive behaviors can also be elicited by painful stimuli or stimuli paired with pain (Blanchard and Blanchard, 1989). Thus, in the laboratory, expression of defensive behaviors (e.g., attack, freezing, avoidance, risk assessment) could indicate that experimental endpoints (i.e., the earliest points at which the scientific aims of the activity can be achieved, while also ensuring that pain or distress experienced by the animals is minimized) are not appropriate, and that measures should be taken to minimize stress and pain to reduce their impact on study outcomes.

Social Organization and Behavior, Including Mating and Parental Behavior

Rats are highly social animals. In the natural environment, Norway rats live in colonies that can vary widely in composition and number. Small colonies may consist of one breeding pair and their offspring, whereas larger colonies may consist of breeding demes that typically include several adult females, a few adult males, and their subadult offspring (Calhoun, 1963; Winnicker et al., 2016). Low-status animals, often juveniles, are crowded into areas not claimed by a group (Calhoun, 1963). Larger colonies have been recorded to include hundreds of individuals.

The rat has separate social systems for females and males. Females may care for their pups alone or in a shared burrow with up to six reproductive females, with each having a separate nest chamber. These females are often related through the mother's side, as they tend to remain in their home colony. Dispersal is male-biased in rats, with females typically staying closer to the natal burrow system than males (Calhoun, 1963). During adolescence, young rats begin to emerge from the natal burrow system, explore the environment, and eventually disperse from the natal area (Calhoun, 1963).

Even though females show less aggression and retaliation to attacks compared to males, their social system is associated with the defense of the litter, nest, and burrow (Hurst et al., 1996). Within social groups, only dominant females breed, and these females defend their nest area whether or not they are lactating (Calhoun 1963; Hurst et al., 1996). Subordinate or low-status females may breed, but they are usually unsuccessful in producing weaned pups, unless the female has a territory and burrow (Calhoun, 1963). As mentioned above, related dominant females may share an area, and even the same burrow, with little aggression. When females are unable to defend a nest site successfully, their adaptive response to space-related aggression is to leave and find a suitable site elsewhere (Hurst et al., 1996).

The social system of males is territorial or despotic, depending on the population density. At low population densities, male rats are territorial; each male defends a territory around a burrow housing one or more females, and mates with those

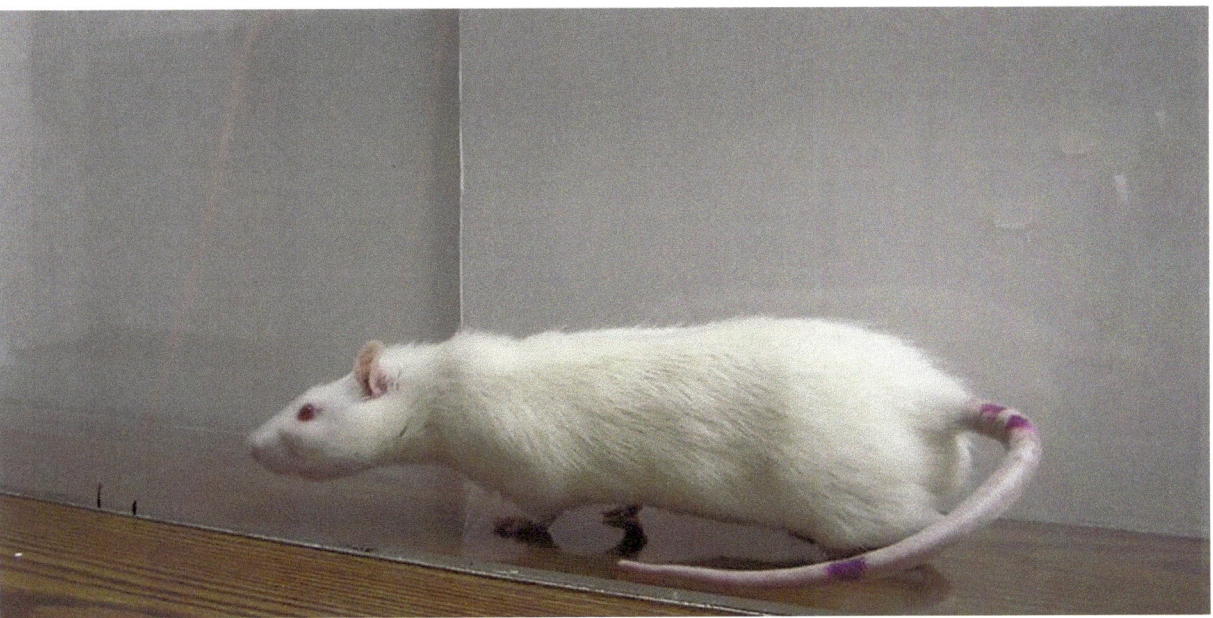

FIGURE 7.1 Example of a defensive behavior performed during a cat odor test, the stretch attend posture, where a rat stands oriented toward the stimulus with back lower than the ears, and may lean forward but without locomotion. Both stretch attend posture and flat back approach (when the rat is moving) are considered to measure risk assessment, whereby the animal acquires information about the presence of danger (Cloutier and Newberry, 2010). The cat odor test is used to evaluate treatment effects on components of natural defensive behavior exhibited in response to a partial predator stimulus, odor, as indicators of anxiety and risk assessment in rats.

females. Males will express aggressive or fear behaviors in the presence of intruder male rats in their territory. At high population densities, territories become too costly for single males to maintain, and the social system becomes a despotic dominance hierarchy. In a despotic system, one male dominates the burrow system, which includes several females, with other males existing as subordinates (Calhoun, 1963; Winnicker et al., 2016).

Aggression toward intruders seems to function to exclude them from resources available to residents, such as food, shelter, and sexual access to females, and to chase them into adjacent areas. Within the social group, males and females prioritize competition for, and defense of, resources; females focus on nest and litter defense, while males focus on protecting their access to reproductive females. Aggression between dominant and subordinate males of a group decreases over time as relationships are established (Hurst et al., 1996).

As with their social systems, rat mating systems also change with population density. At low densities, the mating system is mostly polygynous; one male mates with multiple females. As the density increases, dominant males tend to focus on an estrous female or a burrow of females. At high densities, such efforts become difficult to maintain and the mating system becomes more polygynandrous, i.e., multiple males mate with multiple females. Even though the dominant male has easier access to many females compared to subordinates males, the latter can also mate with females. In this case, large groups of males copulate sequentially with an estrous female, with the possibility of several males fathering offspring in the same litter (Winnicker et al., 2016). For a female, mating with several males minimizes the risk of infanticide, as males are unlikely to attack the litters of females with whom they have mated (Berdoy, 2002). When litter attacks do occur, an adult male kills infants to bring the mother back into estrus more quickly, so he can mate with her and produce offspring of his own. Hence, the role of infanticide of another male's progeny seems to enhance a male's reproductive success (Hrdy, 1979).

In the wild, each female in a colony has her own nesting chamber in which she will give birth and nurse her pups. Duration of gestation is 21–24 days but could be prolonged by stress (Pritchett-Corning, 2015). A few days prior to parturition, the female will build her nest and defend it against both male and female colony members. Maternal aggression will continue throughout the first week of nursing and then gradually decline (Koolhaas, 2010). Females provide all care for the pups, as no care is provided by males. Lactating females, especially those who have just given birth, may seal the entrance to their burrows using cut vegetation. When housed in dual-chamber cages that somewhat mimic a burrow system, female rats displayed a 24-h rhythm of maternal behavior, spending more time with the litter and more time nursing during daylight hours (or "the day"), and spending more time away from the pups to feed and rest at night (Grota and Ader, 1974). A progressive decline in the frequency of nursing and in time spent with the litter, which are both high immediately after birth, was also reported (Grota and Ader, 1974). Pups are weaned at approximately 21 days of age. As in most mammalian species, the transition from birth to weaning is a gradual process initiated by the mother. Weaning is characterized by decreasing amounts of contact between mother and offspring and in the frequency of nursing, and by increasing solid food consumption by the offspring.

Grooming is another important behavior performed by rats. Organized bouts of self-grooming proceed from rostral to caudal, beginning with forepaw wiping of the face and finishing with cleaning the tail (Sachs, 1988). Rats self-groom upon awakening, after eating, as a displacement activity, or when anxious (Pritchett-Corning, 2015). Grooming is also performed by mothers to their pups, from pup to pup, and between adult rats as well (Lawlor, 2002). Like self-grooming, social- or allo-grooming, which likely originated from the mother-infant grooming relationship, contributes to the maintenance of body hygiene. Allo-grooming among adult animals seems to also have a social bonding function, promoting affiliative behavior and playing a role in the regulation of relationships between members of a social group (Pritchett-Corning, 2015; Spruijt et al., 1992).

Social play, also called "rough-and-tumble play" or "play fighting", is typically a positive form of social contact. It is particularly prevalent in young animals (Pellis and Pellis, 2004; Pinelli et al., 2017), although it can be observed in adults (Panksepp, 1998; Pellis, 2002). The performance of rough-and-tumble play is first seen around 17 days of age and peaks at approximately 4–6 weeks of age (Panksepp, 1981, 1998). The two main behavioral components of rough-and-tumble play are dorsal contact and pinning. During play, rats make attempts to contact their partner's nape (dorsal contact) and recipients respond to these dorsal contacts by rolling over to a supine position, with the partner standing overtop in a pinning position (Panksepp, 1981). Participants reciprocate, taking turns being on top and beneath. The performance of dorsal contacts and pins is interspersed with jumping and chasing. Although rough-and-tumble play may look like fighting, the "target" body part is different. Rough-and-tumble play involves rats aiming for each other's nape, while serious fighting involves strikes at the rump (Pellis and Pellis, 1987). The performance of rough-and-tumble play is important for establishing normal social behavior and avoiding male–male aggression in adulthood (Pellis et al., 2010). It is also suggested to help in calibrating emotional reactions to events and frightening situations (Pellis and Pellis, 2009; Špinka et al., 2001), and to induce resilience to depression and anxiety (Burgdorf et al., 2010). In adults, the performance of rough-and-tumble play has been suggested to function as social bonding (a means of maintaining "friendly" relations with the dominant male) or as social testing (a means of assessing the weaknesses of another male) (Pellis, 2002).

Feeding Behavior

Rats are omnivorous. Hence, they can adapt their diet and foraging behavior to the habitat and food source. For example, rats from a colony established near a fish hatchery were seen catching young, small fish from hatchery ponds (Clarence, 1948). Feeding and foraging for food resources occur mainly during the night (Whishaw, 2004). Generally, rats carry food in their teeth to a safe location, before adopting a squatting posture and eating while holding the food in their forepaws

(Lawlor, 2002). Alternatively, food may be carried to the burrow and stored.

Rats are highly neophobic and as a result will show marked exploratory and sampling behavior, particularly upon discovering unfamiliar food. Sampling of unfamiliar foods will occur, especially when the novel food is found in a familiar environment (Barnett, 1956). Rats, especially young ones, also use social interactions to determine if unfamiliar food is edible. This includes (1) feeding with other conspecifics or following them to feeding sites, (2) following trails left by others to find food or sniffing odor cues left at feeding sites by conspecifics, and (3) interacting with conspecifics that have recently eaten a distinctively flavored food (Galef et al., 2006). Although it has been suggested that rats can learn food avoidance by linking feelings of sickness with the taste of recently eaten items, or by smelling a particular foodstuff on a rat that looks unwell (Berdoy, 2002), other data show that rats learn where and what to eat, rather than where and what to avoid eating, from one another (Galef et al., 2006).

Communication

Rats communicate mainly using smells and sounds. They have acute hearing, are sensitive to ultrasound, and possess a very highly developed olfactory sense. They also have keen senses of taste and touch, but poor vision. Burn (2008) provides a detailed review of the rat's sensory abilities and their implications for husbandry and research outcomes. An overview of how rats use their senses to communicate is provided below.

Olfactory

The sense of smell is highly developed in rats because they use scent marking (urine/pheromones) to communicate. Rats use marking with urine to communicate but also utilize the many scent glands on their body (Burn, 2008). Scents are used to communicate and gain information about other rats. Identity, age, sex, health, reproductive state, genetic relatedness, and social status are examples of information gathered through scent (Burn, 2008; Koolhaas, 2010). Scent also mediates mother–pup relationships. Rats use their sense of smell for finding food, and they also share information about food through odors (Burn, 2008; Galef et al., 2006). For example, the food preferences of pups are developed by sniffing their mother. Odors can also be used to obtain information about the environment, and predators, in particular. Rats, including those housed in the laboratory, are reported to have a hereditary predisposition to exhibit behavioral and physiological stress responses to the odor of certain predators, such as cats (Blanchard et al., 1990), although habituation has been reported following repeated exposure (Dielenberg et al., 2001).

Rats are known to respond differently toward different humans, mostly because of individual differences in odor among people. These odor differences occur due to both genetic and environmental factors, such as diet, smoking, perfume, soap, and deodorant, among others (Burn, 2008). Care staff or researchers may also be "marked" with odors by previously handled rodents, potentially including odors related to positive or negative situations. These odors can positively or negatively affect the response of the rat to the human and thus their relationships.

Vocalizations

Rat hearing is more sensitive than human hearing, and can detect sounds both at the human range and at higher frequencies into the ultrasonic range. The development of this sensitivity is thought to have been stimulated by their nocturnal lifestyle, for which the visual system is less helpful (Brudzynski and Fletcher, 2010). This adaptation also allows them to use ultrasonic calls to communicate, likely evolving as an adaptation to minimize the chance of being detected by predators. Evidence suggests that ultrasonic calls are also useful for short-range communication, and are effective for communicating in burrow systems (Brudzynski and Fletcher, 2010).

Young rats have been reported to produce calls in frequencies varying between 1.9 and 125 kHz, whereas adult rats produce ultrasonic calls varying in frequency between 20 and 70 kHz, although calls have been recorded as high as 100 kHz (Brudzynski and Fletcher, 2010). Rats' USVs have been divided into three main categories (40, 22, and 50 kHz). Short calls around 40 kHz, produced by pups, have been suggested as indicative of distress or isolation when young rats are isolated from the dam. Juvenile and adult rats produce calls centered around 22 and 50 kHz. The 22 kHz USVs have a frequency range of 22–29 kHz, with a narrow bandwidth of 1–4 kHz, and a duration of >100 ms (Brudzynski, 2009; Wright et al., 2010). The 50 kHz USVs have a markedly shorter average duration of 30–40 ms, with frequency ranging between 30 and 70 kHz, but primarily close to 50 kHz, with a bandwidth of 5–7 kHz (Brudzynski, 2009; Wright et al., 2010). Production of 22 kHz USVs is associated with aversive events, such as agonistic interactions and chronic pain. They are also produced during the postejaculatory period of a mating session, possibly to keep other males away, as it has been shown that rats avoid playbacks of 22 kHz USVs (Litvin, et al., 2010). The 22 kHz USVs are also used as alarm calls to alert conspecifics of a threat, such as a predator (Litvin et al., 2010; Brudzynski and Fletcher, 2010). Rats naïve to handling by humans produce these calls as well (Brudzynski and Ociepa, 1992). Overall, 22 kHz USVs are interpreted as providing information about fear and anxiety and as reflecting negative affective states (Brudzynski, 2009; Burgdorf et al., 2008). Production of 50 kHz USVs is associated with appetitive or potentially rewarding situations, such as nonaggressive social interactions, rough-and-tumble play, intake of narcotics (e.g., opioids), mating (ejaculation phase), and playful handling (tickling) by a human. Rats are attracted to playbacks of 50 kHz USVs. Thus, these calls are interpreted as indicating a positive affective state (Brudzynski, 2009; Burgdorf et al., 2005, 2008; Knutson et al., 2002; Wöhr et al., 2008). Although 14 subcategories of the 50 kHz USV have been identified (Wright et al., 2010), only the trill (also named frequency-modulated) and trills combined with one or more of the other subcategory calls (Figure 7.2) have been associated with positive events, such as play and tickling (Burgdorf et al., 2008). The less complex, "flat" category of 50 kHz USV appears to be used as a social-exploration/contact signaling mechanism, for example,

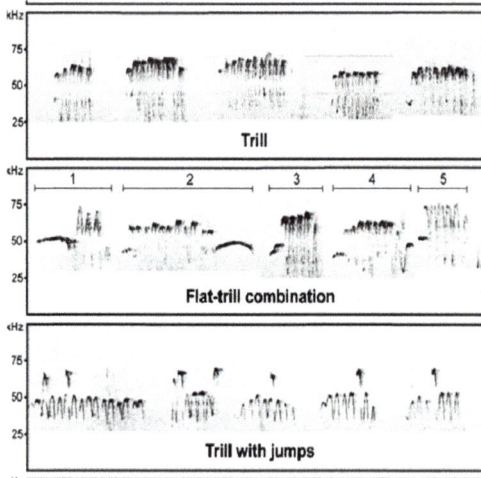

FIGURE 7.2 The 50-kHz USV trill is produced by itself or in combination with one or more of the other subcategory calls (from Wright et al., 2010).

by isolated rats attempting to make contact with conspecifics. Hence, USVs in the flat 50 kHz category are less indicative of positive affect (Burgdorf et al., 2011; Wöhr et al., 2008). The function of the other subcategories has still not been clarified.

Rats also produce communicative sounds in the human auditory range, using frequencies under 20 kHz. Vocalizations audible to humans have been associated with agonistic encounters (Kaltwasser, 1990), defensive situations (Blanchard et al., 1986), and pain and discomfort (Brudzynski, 2009).

Touch

Rats live in a world of textures! Their sense of touch is highly developed and well adapted for life in low-light conditions. They have well-developed tactile receptors on the head, around the whiskers, and on the paws and tail. They perceive their immediate environment with their whiskers or vibrissae. The whiskers are long, sensitive hairs growing from the rat's cheeks, eyebrows, and chin, known to have a sensitivity of 90 μm, which is comparable to that of primate fingertips (Koolhaas, 2010). By rapidly whisking back and forth, brushing everything nearby, whiskers allow rats to detect minute details of their immediate environment and facilitate nocturnal and underground activities. Rat whiskers are also important in social interactions, as whiskerless rats were observed to be unable to avoid bites to their faces during fighting (Blanchard et al., 1977a, b). The role of whiskers in the rat sensory system should be taken into account in the design of captive environments and enrichment programs, and in strategies implemented to avoid barbering behaviors.

Visual

Since wild rats inhabit burrows or other enclosed environments, and are mostly nocturnal or crepuscular, many of their activities occur under low-light conditions. Consequently, rats rely relatively little on vision. Compared to humans, rats have less binocular overlap and acuity, but a wider field of view and higher light intensity discrimination (Burn, 2008; Winnicker et al., 2016). Although rats are near-sighted, meaning that they view distant objects as blurry, their vision has a sensitivity to motion 2–3 times higher than that of humans (Pritchett-Corning, 2015). Thus, they are able to use their vision to detect large, moving, distant objects and to orient themselves in space. They can use body postures to communicate at close range. Rats have dichromatic color vision with one type of cone having a maximal sensitivity at 510 nm wavelength (green) and the other type being most sensitive near ultraviolet range (360 nm) (Koolhaas, 2010). Although ultraviolet receptors have been identified, the function of ultraviolet vision in rats is not well understood (Jacobs et al. 1991, 2001). It might be useful during twilight hours or to follow urine markings.

Common Captive Behaviors

Normal

Similar to their wild conspecifics, laboratory rats are nocturnal, mainly active during the dark phase of the cycle and spending most of the daylight period sleeping. They spend time hiding in shelters and burrowing, particularly during the light phase (Boice, 1977; Calhoun, 1963; Makowska and Weary, 2016). The normal sleep position is extended at full length, with the tail extended, but rats may sleep in a curled position when chilled (Lawlor, 2002). Sleeping position has also been reported to be affected by light, with the curled one being used in the light and stretched one in the dark period (van Betteray et al., 1991). Thus, the sleeping posture could be an indicator of comfort and environmental conditions.

While rats are neophobic mainly of objects and food items, they are also highly exploratory. Due to their exploratory nature, if sufficient space is provided, rats will engage in a full variety of locomotory behaviors, including walking, running, jumping, climbing, and stretching upright on their hind legs (also called rearing). The occurrence of upright stretching could be used as an indicator of the appropriateness of the cage environment (Makowska and Weary, 2016). Rats also manipulate or chew substrate, objects, and food items, which helps prevent overgrowth of the incisors and contributes to oral health.

Laboratory rats show strong motivation to maintain social contact with conspecifics and thus prefer to be housed with other rats (Patterson-Kane et al., 2002, 2004; Hurst et al., 1997, 1998). Rats housed in stable groups will even show signs of stress when some animals are removed (Burman et al., 2008c) showing their strong social trait. Male and female rats exhibit different responses to social separation and individual housing in the laboratory, which may be explained by sex differences in social roles (Hurst et al., 1997, 1998). Because of their high sociality, and with sufficient space, rats will engage in a number of social activities, including grooming (Aldridge, 2004) and play, particularly among young animals (Pellis and Pellis, 2004; Pinelli et al., 2017). There is evidence that rats show empathy, the capacity to understand or feel what another is experiencing, for other rats (Bartal et al., 2011; Sato et al., 2015). The high sociality and empathetic capability of rats could be used to promote recovery for individuals undergoing stressful procedures.

In the laboratory environment, rats show little aggression when housed in stable, single-sex groups. However, mature males will engage in agonistic behaviors when introduced to unfamiliar males, and when group housed with receptive females (Blanchard et al., 1988). Nevertheless, compared to their wild counterparts, laboratory rats show less aggression and defensive behavior, thus demonstrating less reactivity to both humans and conspecifics (Modlinska and Pisula, 2020).

Mating and other reproductive behaviors, including caring for young pups, are among the normal behaviors observed in laboratory rats. However, differences in the frequency, qualitative, or quantitative expression of behaviors compared to wild rats have been reported. Laboratory female rats still show motivation to seek out nesting material and build nests as they near parturition, but their effectiveness at performing these behaviors is affected by their exposure to nesting material as youngsters (Pritchett-Corning, 2015). Laboratory females with neonates were reported to take longer to retrieve pups, but were faster to build nests; they also show similar levels of aggression toward intruders compared to wild rats (Price and Belanger, 1977). In the laboratory, when nursing, the dam will approach the pups and gather them under her body while licking them. She will assume an upright, crouching posture, while standing over all or most of the pups with a pronounced dorsal arch (Macri and Würbel, 2006), although other postures, such as lying on the side, are also used.

Abnormal

The performance of abnormal behaviors suggests that an animal is unable to adjust behaviorally to the captive environment, increasing the likelihood of abnormal physiological responses (Garner, 2005). It also suggests that the animal's welfare is compromised. Ultimately, the performance of abnormal behaviors by experimental animals raises questions about the validity and replicability of experiments (Garner, 2005).

Stereotyped behaviors have not been extensively reported in rats, especially compared to mice, as these behaviors are performed most often at night/in the dark phase of the light cycle, when rats are most active. Hence, they may largely go unnoticed as caretakers check animals during the light phase, when rats are resting or sleeping. In studies, however, barbering, the removal of hairs and/or whiskers by conspecifics (Burn, 2008), fur chewing (Koolhaas, 2010), bar-biting, especially in females housed singly (Hurst et al., 1998), and tail manipulation, mainly in singly housed rats (Hurst et al., 1997), have been observed.

Grooming is a predominant behavior in the rat repertoire; consequently, the absence of self-grooming can indicate health or other issues. Thus, attention to the coat of the rat is important for assessing stress and other health and behavior-related issues. Rats produce porphyrin pigmented secretions, called chromodacryorrhea, from the Harderian gland near the eye. These secretions (which are often red) can be seen around the eyes and nose, and are brushed onto the fur when rats are grooming giving a white fur a pinkish tint. When produced in excess, these secretions are considered a sign of stress. A scoring system has been developed to quantify chromodacryorrhea secretion (Mason et al., 2004), although scoring is difficult on strains with dark coats, like Long-Evans.

Cannibalism (or infanticide), often used by a parturient female to dispose of weak or stillborn neonates, can be directed at healthy neonates when induced by stress or disturbances caused by cage cleaning or other activities in the room or facility (Koolhaas, 2010; Lohmiller and Swing, 2006).

A behavior termed "press posture", in which male or female rats press their ventrum into the cage floor or against a vertical surface, has been observed in adult rats and mice with litters (Gaskill and Pritchett-Corning, 2015a, b). This behavior has been associated with possible stress or crowding, as the frequency of press postures increased with litter age, and is seen most often in small cages.

Measures should be taken to minimize the expression of abnormal behaviors and mitigate their impact on rat welfare and on the validity and replicability of the experimental data. Providing a housing environment that promotes expression of species-specific behaviors and provides control to the rat can be a first step in minimizing the occurrence of abnormal behaviors.

Ways to Maintain Behavioral Health – Captive Considerations to Promote Species-Typical Behavior

Environment—Macroenvironment

The environment outside of a rat cage (i.e., macroenvironment) can have an impact on the environment within the cage (i.e., microenvironment), even in individually ventilated cage (IVC) units. Features of the macroenvironment and husbandry activities can impact rats' behaviors and communication. Hence, knowledge of behaviors, applied toward careful planning of husbandry practices and research-related activities, and organization of the room will contribute to enhancing the expression of species-specific behaviors and animal welfare.

Measures should be taken to minimize the impact of human activities related to husbandry and experimental procedures on rats' resting time and sleep, to account for their nocturnal or crepuscular active period. Thus, handling and testing rats during their active period is normally preferred.

Rats are sensitive to light in particular albino strains and can develop retinal damage due to light exposure (Koolhaas, 2010). It has been suggested to keep light intensity at the bottom of the cage [bedding surface] at 25–50 lux, based on retinal damage analysis and welfare measures (Koolhaas, 2010). Provision of a shelter could contribute to minimizing the impact of lighting, although some types of shelters, such as polyvinyl chloride (PVC) pipes, might not be as effective as others (Galef and Sorge, 2000; see below). Simulation of dawn-dusk lighting has been shown to be beneficial for other crepuscular species and, thus, could positively impact rats as well (Grover and Miller, 1985).

Vocalizations, especially those of lower frequencies, such as 22-kHz, can travel across cages, which allows rats to communicate with conspecifics in adjacent cages, although housing in

IVCs greatly reduces this ability. Thus, if one animal is upset or scared and vocalizes, the other rats in the room will hear and become upset or scared. In order to minimize unwanted effects on research results, communication through vocalization should be taken into consideration when planning rat housing (e.g., determining the type of caging to use, or the need for single housing) and experimental testing (e.g., if several test apparatuses are used concurrently in a test room, or if restraint or other procedures that can induce negative affective states and productions of distress or alarm calls are required).

In an animal facility, rats are exposed to noise from the environment (e.g., sonic and ultrasonic emissions from facility ventilation and electronics), and from human and animal activities. Noise, particularly in the ultrasonic range not audible to humans, can cause stress to rats and disrupt communication. Equipment and objects producing noise in the ultrasonic range that can potentially affect rats, or interfere with their communication system, include metal equipment, such as carts and racks, cart wheels, sinks, keys, computers and monitors, and Heating, Ventilation, and Air Conditioning (HVAC) systems. Human activities (e.g., cage changing) and behavior (e.g., heavy breathing, sniffing) can also generate ultrasounds that can potentially create a negative experience for the rat. Because handling procedures have been shown to induce the production of 22 kHz USV in naïve rats, these calls could be used to assess the effects of human interactions and activities. Recently, it has also been found that measurement of rat USVs can be used to assess chronic pain, a potential indicator that could improve our understanding of the affective component of pain (Burgdorf et al., 2019).

Odors, just like vocalizations, can travel across cages, and can be used as a source of information and communication between rats housed in a room, although IVCs reduce the effectiveness of this type of communication. Odors can affect rats' response to people and handling. For example, caretakers must be careful not to bring odors from their companion animals, particularly cats, into the rat room. Olfactory disruption following cage cleaning does not appear to strongly impact stable, nonbreeding groups of rats (Burn and Mason, 2008). Novelty and handling appears to be more disruptive than changes in odors. The time at which cage change takes place also seems to have a stronger impact than a change in odors, especially if it takes place during the day, when rats are resting (Abou-Ismail et al., 2008). When odor is used as enrichment, care should be taken to select scents with minimal adverse effects on aggressive behavior, stress, and anxiety (Burn, 2008; Winnicker et al., 2016).

Cage placement on the rack and in the room affects rat behavior. Cloutier and Newberry (2010) found that being housed in cages with higher levels of human exposure within rodent rooms negatively affected the responses of rats during behavioral testing. Providing a structure, or an area where rats can retreat, could minimize this effect. Nevertheless, cage placement should be considered in experimental design and analysis.

Environment-Microenvironment

The cage environment should allow rats to perform locomotor and exploratory behaviors, such as walking, running, jumping, climbing, and rearing. Although domestication seems to have altered the quality and thresholds needed to initiate rat species-specific behaviors, such behaviors were not eliminated by domestication and breeding in the laboratory (Makowska and Weary, 2016; Stryjek et al., 2012). Several studies have shown that seminaturalistic environments (outdoor or in-cage) promote the expression of natural behaviors in laboratory rats (Berdoy, 2002; Makowska and Weary, 2016).

Rats stand upright as they explore and socialize with other rats, and also as they stretch (Makowska and Weary, 2016). They also stretch in the lateral position while remaining parallel to the ground, especially in small cages. Evidence suggests that lateral stretching is performed as a way to compensate for the lack of vertical space, and may also be used to mitigate stiffness resulting from the low levels of mobility caused by the restricted space, especially in larger, older rats (Makowska and Weary, 2016). Considering that rats' ability to stand upright is important for their welfare (Makowska and Weary, 2016), their environment should allow freedom for this behavior.

Similar to upright standing, engagement in the full expression of social play requires the right environment and space. For example, play is decreased in locations that rats have associated with fearful events, such as exposure to a predator odor (i.e., domestic cat) (Panksepp, 1998; Siviy et al., 2006). Providing sufficient space in a safe environment for young rats to express rough-and-tumble play behavior is indispensable for their physical and emotional development. It has even been recommended to provide $0.2\,m^2$ of floor space for a group of seven or eight juvenile rats during the first 3–4 weeks after weaning (Koolhaas, 2010) to allow for expression of play behaviors and appropriate development.

The space provided by standard laboratory cages also limits the expression of maternal behaviors. The mother's ability to move away from her offspring is limited by the size and layout of the cage. Studies of laboratory rat mothers housed in dual-chamber cages showed that, when given the choice, lactating females will spend considerable amounts of time, and will consume most of their food and water, while away from their pups (Ader and Grota, 1970; Grota and Ader, 1969; Plaut, 1974). Indeed, lactating females kept in two-tiered cages, with the top tier not easily accessible to the pups, showed similar patterns of behavior; they spent more time on the upper tier as the pups aged, while other behaviors, such as nursing, appeared to be no different from females housed in single-tiered cages (Figure 7.3; Cloutier et al., unpublished data). Thus, a laboratory cage environment that includes a separate space where the mother can escape from the pups could provide a more naturalistic environment for a lactating rat and her litter than is achieved by a single compartment cage; simply adding more space will not allow the mother to fully separate herself from the pups. Additionally, this environment could positively impact aspects of offspring development (Plaut, 1974). The press posture reported by Gaskill and Pritchett-Corning (2015a, b) could be a strategy used by adults to cope with the inability to move away from the pups. Hence, the occurrence of this behavior could be an indication that the cage environment is not appropriate.

The cage environment should include bedding and nesting material, and ideally burrowing substrate, to promote

FIGURE 7.3 A lactating rat female housed in a two-tiered cage. The top tier is not easily accessible to the pups allowing her to spend time away from them.

exploration, nest building, and burrowing. Bedding should be carefully selected, as some types (e.g., aspen, corncob) have been reported to impact physiological and behavioral measures (Pritchett-Corning, 2015). Corn cob beddings is often used due to its high absorbent properties, but it is the least preferred by rats who have been reported to sleep on the cage floor rather than on corn cob bedding (Makowska and Weary, 2020). Rats have shown a preference for wood-based bedding with a larger particle size (Pritchett-Corning, 2015) and paper-based bedding (Makowska and Weary, 2020). Bedding depth should be sufficient to efficiently absorb fluids from urine and feces, and minimize ammonia level (Makowska and Weary, 2020). Alternatively, more than one type of bedding could be used as suggested by Makowska and Weary (2020). An absorbent bedding like corn cob could be placed in a litter pan, while a softer, comfortable bedding could be used in the remaining of the cage. The litter pan could be placed near or under the food hopper as there is evidence that rats prefer to eliminate near their food and water. This practice could satisfy the rat needs for defecating in a dedicated area and sleeping on a comfortable substrate. In the laboratory, both male and female rats will build nests, if given appropriate materials (e.g., long-fiber materials) and space (Jegstrup et al., 2005; Makowska and Weary, 2016; Manser et al., 1998). However, in general, this behavior only occurs if animals have been given access to nesting materials starting at a young age, and the degree of nest building varies by strain (Van Loo and Baumans, 2004; Winnicker, 2016). In addition to nest building, laboratory rats provided with appropriate substrate will express burrowing behavior. The behavioral need for burrowing appears to be increased in pregnant females, as is nest building (Winnicker et al., 2016). Consideration should be given to provide bedding material in an amount sufficient to facilitate the expression of burrowing, or to supplement it with an appropriate substrate.

As mentioned above, laboratory rats should be provided with a shelter, especially if burrowing substrate cannot be provided. They have been shown to prefer enclosed, opaque nest boxes, over clear, open-sided boxes or tubes with openings at both ends (Galef and Sorge, 2000; Patterson-Kane, 2003). They also show preferences for cages containing a structure that can be used as a shelter compared to an empty cage (Anzaldo et al., 1994; Cloutier and Newberry 2010; Patterson-Kane, 2003; Sorensen et al., 2004; Williams et al., 2008). However, shelters can make health checks difficult for caregivers and may result in increased handling or time spent checking. As an alternative to shelter provision, fully or partially colored-wall cages, allowing observation by humans, while providing rats with some sheltering, are available. The effects of such cages on rat behavior need to be further investigated, because a study found an interaction between light intensity and color tint on rat affective state, demonstrating the importance of environmental variables on behavior and welfare (LaFollette et al., 2019).

To facilitate the expression of locomotor activities in the limited space of cages running wheels could be provided. Running wheels are used by a variety of mammalian species in captivity (Dewsbury, 1980; Sherwin, 1998), and have also been reported to be used in the wild (Meier and Robbers, 2014). However, they are not extensively provided as enrichment for laboratory rats, mainly because many housing systems are not tall enough to accommodate the wheel (Sherwin, 1998). There are also concerns about the development of maladaptive (or inappropriate) behaviors, such as stereotypy and addiction to the wheel, as reported in studies that use it to assess general activity, or physical activity/exercise (de Visser et al., 2005; Kanarek et al., 2009; Richter et al., 2014; Sherwin, 1998; Stranahan et al., 2006). However, in the majority of these studies, rodents were housed individually in specially designed cages, often without visual, auditory, or olfactory contact with conspecifics. Thus, housing condition could be a contributing factor in the development of wheel-related stereotyped and addictive behaviors. Rats are highly motivated to access running wheels (de Visser et al., 2005), making it unsurprising that early housing designed by The Wistar Institute in the early 1900s, included a running wheel (Makowska and Weary, 2016). Wheel running has been reported to have beneficial effects on physical and physiological systems, learning and memory, anxiety and depression (de Visser et al., 2005; Nadel et al., 2013). It also induces a positive affective state and promotes social play and related activities (Cloutier and Newberry, unpublished data). Thus, inclusion of running wheels in the laboratory environment could satisfy some of the rats' need for locomotion. Adding climbing structures to the cage environment could also fulfill some locomotion needs, especially in smaller cages that cannot accommodate a wheel. Alternatively, providing access to a play pen or other space enriched with objects and furniture for a period of time (ranging to a few minutes to a few hours) per day (Makowska and Weary, 2020) as it is often done for primates and rabbits could fulfill these behavioral needs, and should be considered when planning rat housing and husbandry.

Like other species belonging to the order Rodentia, rats are characterized by a single pair of continuously growing incisors in each of the upper and lower jaws. Thus, they require opportunity and time for gnawing to prevent the overgrowth of these teeth. Consequently, they show a preference for material and objects that they can gnaw and chew (Manser et al., 1998; Patterson-Kane et al., 2001; Chmiel and Noonan, 1996). The provision of enrichment objects that promote the expression of gnawing behavior contributes to both the animals' psychological and dental health.

Overall, providing rats with a complex, enriched environment is crucial for the expression of species-specific behaviors. Research shows that environmental enrichment has a wide range of positive effects on laboratory rats, including behavioral, physiological, psychological, developmental, therapeutic, and recovery from neural deficits, cognitive ability and memory, physical growth, and functions and development of the brain (Abou-Ismail and Mendl, 2016). Complex, enriched environments also have a positive effect on rat welfare, including affective states (Abou-Ismail and Mendl, 2016) and various research outcomes (e.g., improved learning, decreased anxiety) (Burman et al., 2008a, b; Mendl et al., 2009).

Social Groupings

In the laboratory, rats are commonly housed socially, in all-male and all-female groups, with pair-housing the most frequent configuration. Studies suggest that laboratory rats housed in cages prefer to live in groups of three to six animals, rather than in pairs or alone (Patterson-Kane et al., 2004; Talling et al., 2002). Group sizes have been reported to affect brain development (Pellis et al., 2010) and responses in behavior tests (Botelho et al., 2007; Patterson-Kane et al., 1999). Patterson-Kane et al. (1999) reported that pair-housed rats showed intermediate activity in the Open Field test, compared to individually housed rats and those housed in groups of four. Pellis et al. (2010) reported differential development in areas of the frontal cortex between juvenile rats housed in pairs and those housed in groups of four. They suggest that some of the variation in the effects of group size, especially during development, reflects not only differing types of stimulation but also differential sensitivity of brain systems to the various experiences provided by different housing conditions (Pellis et al., 2010). These findings suggest that careful consideration should be given to social groupings and that pair housing may not reliably provide all of the benefits of social housing, although it is better than single housing. Furthermore, if possible, grouping of litter mates may reduce stress related to mixing and social interactions (Koolhaas, 2010).

Even though social housing is recommended as a standard laboratory practice for social species, such as rats (e.g., National Research Council, 2011; Canadian Council on Animal Care, 2017, 2020), individual or single housing may be used due to research constraints (e.g., food/fluid intake monitoring) or medical issues. However, individual housing impacts physiological, neurochemical, and behavioral factors, and can have permanent effects. The magnitude of the effect on those factors is influenced by the strain, sex, and age when singly housed, as well as cage type and husbandry practices (Faith and Hessler, 2006). The age or developmental stage at the time of individual housing appears to be the most critical factor. Ages at which specific social stimulation is necessary for adequate subsequent social development are mostly affected. The specific effects of individual housing differ across development and have been identified at the preweaning/neonatal, postweaning/adolescent, and adult stages (Faith and Hessler, 2006). For example, maternal separation and handling at the preweaning stage affect responses to stress and other behavioral and neurobiological processes in adulthood (Koolhaas, 2010; Plotsky et al., 2005). Single housing at the postweaning stage, particularly during postnatal weeks 5 and 6, results in deficiencies in aggressive, sexual, and other social behavior patterns, and stress reactivity in adulthood (Hol et al., 1999; Koolhaas, 2010). In particular, social deprivation in the postweaning stage prevents the expression of social (or rough-and-tumble) play. Deprivation of social play has been reported to cause abnormal patterns of social, sexual, and aggressive behavior (Einon and Potegal, 1991; Gerall, et al., 1967; Takahashi, 1986). In adulthood, individual housing increases anxiety, learning deficits, memory impairment, and locomotor activity (Koolhaas, 2010; Zorzo et al., 2019). Further, differences between individually and group-housed rats have been reported in responses to pharmacological challenges and reduction in LD50 (i.e., median lethal dose, the amount of a substance required to kill 50% of the test population) for a given drug (Faith and Hessler, 2006). Thus, all of the implications of individual housing on the animal, and the research outcomes and validity, should be carefully considered before it is implemented. When single housing is necessary, the typical practice is to add more enrichment objects to the environment. However, objects do not fulfill the social needs of the animals. Rat tickling, a handling technique that mimics aspects of rat social play, could be used as a form of social enrichment to help mitigate the negative impact of individual housing (Figure 7.4, Cloutier et al., 2013, 2018; see below for more detail).

Feeding Strategies

Rats eat during the dark period of the cycle when they are most active. This habit should be considered when planning

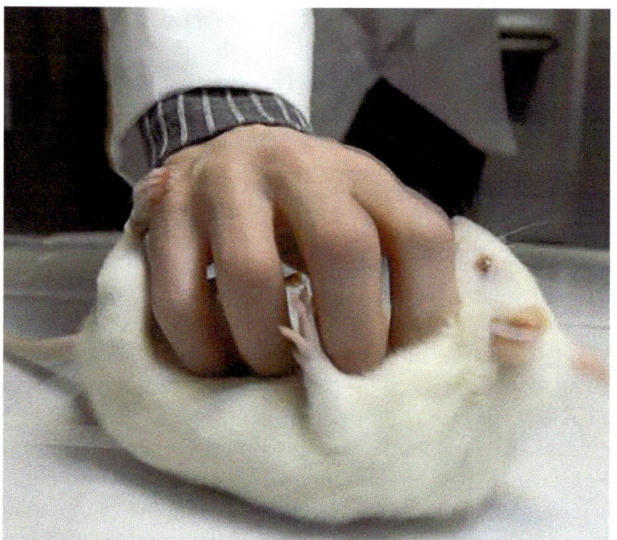

FIGURE 7.4 A rat is being given a playful pin during a tickling session. Rat tickling, a handling technique that mimics aspects of rat rough-and-tumble play, can be used as a form of social enrichment to help mitigate the negative impact of individual housing and also to habituate rats to handling and prepare them for research procedures (Photo: Henry Moore Jr. Washington State University, College of Veterinary Medicine).

studies. To promote the rat's natural squatting feeding behavior, manipulation of food items, foraging, and the appetitive phase of feeding, food enrichment provided in the cage should be dispersed in the bedding, or offered in objects, such as food puzzles. Rats can detect similar taste dimensions to humans, i.e., sweet, salty, sour, bitter, and umami (Burn, 2008). This knowledge can be used to provide rats with opportunities to sample food items that vary in flavor, nutritional content, and presentation. Additionally, food enrichment should encourage stimulation of the animals' olfactory, as well as gustatory senses, and provide feeding activities that functionally simulate natural feeding activities, including the ingestion of diverse food types (see Burn, 2008, for enrichment ideas).

Attempts to introduce novel foods will need to account for the rat's neophobia, as they have evolved to rapidly learn aversions to harmful novel foods. However, if novel foods are determined to be safe after tasting small quantities, rats will readily consume them (Burn, 2008). Social learning can also be used to facilitate the introduction of novel foods. As indicated above, rats' preference for a particular food can be enhanced by the use of visual and olfactory cues while interacting with other rats (Galef and Whiskin, 2003; Galef et al. 2006). Avoidance of a particular taste or food can be quickly gained by rats through conditioned taste aversions (Koolhaas, 2010).

Rats have a natural habit of coprophagy, ingesting about 35%–65% of their feces daily. Preventing the expression of this behavior is detrimental to rat health, as it can lead to 15%–25% reductions in weight gain (Newton, 1978), reduced efficiency in food and nutrient utilization (Cree et al., 1986), and increased susceptibility of young rats to malnutrition and abnormal eating behavior later in life (Novakova and Babicky, 1989). For these reasons, the use of wire flooring, for example, for metabolic studies, should be minimized.

Consideration for Experimental and Testing Procedures

Experimental testing and sampling should be carefully planned, taking into account rat species-specific behaviors, and sensory and communication systems. Particular attention should be given to odors, because rat behavior is strongly determined by olfactory signals or cues. The location of testing or sampling (e.g., test room or procedural space in holding room) should be carefully selected. Extraneous odors (e.g., cleaners, perfumes, personal hygiene products) and noise, especially in the ultrasonic range (e.g., computer equipment, and room lighting), that could potentially affect rat responses should be identified and minimized. Attempts should also be made at minimizing the exchange of information between rats, mainly via odor cues and vocalizations. Evidence shows that rats react negatively to conspecifics' blood or muscle tissue, but not brain odors (Pritchett-Corning, 2015). They also show signs of stress when present in a room during decapitation of conspecifics, but not to other research-related procedures (Sharp et al., 2002, 2003). Rats can learn about negative and positive experiences of conspecifics by assessing residual odors left in the test apparatus, thus identifying marked areas as aversive or attractive (Burn, 2008). These scent marks can affect subsequent rats' responses, although similar effects have not been reported for mice (Hershey et al., 2018). Thus, to minimize the impact of odor cues, cleaning equipment, such as behavioral apparatuses used by multiple rats, should be done with cleaners that are efficient at removing scent marks and pheromones without impacting rat behaviors. However, care must be taken to select detergents, disinfectants, and other cleaning products that do not affect the animals. Many of these products produce odors that can have an impact on rats' physiological and behavioral parameters and alter their responses (see Burn, 2008). Although no studies have been reported for rats specifically, there is evidence that alcohol-based products, including those used for cleaning behavioral test apparatus for example, increase avoidance in mice (Hershey et al., 2018) and reduce cortisol levels in pigs (see Burn, 2008) during testing. Similarly, if several test apparatuses are used simultaneously in a room, they should be checked for sound transmission, as rats avoid areas where alarm or distress calls, such as 22 kHz USVs, which are elicited by procedures, such as drug withdrawal and electrical shock, are produced (Wöhr and Schwarting, 2007).

Rat sensory characteristics should also be taken into account when selecting tests and measures. For example, the use of visual cues in some behavioral tests (e.g., Morris water maze, radial arm maze, visual discrimination) is questionable because of the naturally poor vision of the rat. Rat near-sightedness renders useless most markers hung on walls as navigation or discrimination cues.

Knowledge of rat feeding behavior and individual food preferences should also be used to enhance learning during the operant conditioning training phase in learning studies, rather than food deprivation. In operant conditioning experiments, two methods are commonly used to motivate laboratory rats to perform designated tasks: (1) food restriction and maintenance at 80% of body weight; (2) water restriction to 15 min/24 h. These methods are effective in motivating the animals but lead to significant behavioral changes and physiological stress (Heiderstadt et al., 2000). If food and water restrictions are necessary, then careful planning is required (Toth and Gardiner, 2000). Rats will readily work for preferred rewards, and it has been shown that using food rewards is more efficient and healthier than using food or water deprivation (Reinagel, 2018).

Consideration should be given to testing rats at night during their active period. This could improve rats' responses in studies. For example, rat motivation to earn rewards has been reported to be higher if experiments coincide with their active period—during the dark phase (Burn, 2008). In addition, because rats housed in stable groups show signs of stress when cage mates are removed (Burman et al., 2008c), and behavioral testing (e.g., Open Field test) appears to be stressful for socially housed rats compared to individually housed ones (Cloutier et al., 2013), separations for the testing of cage mates should be planned to minimize stress.

Training and Handling

The rat was the first species used in animal behavior and learning studies in the early 1900s (Lindsey and Baker, 2006). It was the species of choice for the development of behavioral research used in psychology and other fields, due to its social and adaptable characteristics. Rats contributed to the development of classical and operant conditioning tasks, frequently used as dependent measures in psychological, as well as a variety of other types of studies. They also contributed to the development of other tests (see Figure 7.5 for an example of a Y-maze) used in the assessment of learning and memory. These early studies showed that rats can be trained to press levers to obtain rewards or to avoid punishment, assess preference, and discriminate between stimuli, such as food, objects, and even humans or interaction with humans. Operant conditioning responses can also be used for a variety of studies, such as assessing the cognitive effects of a particular manipulation (e.g., lesion, stimulation, drug, etc.). Rats can be trained on the task, then given a treatment, and re-tested on the task to see if their responses (frequency, duration, and accuracy) are affected. Rats' learning abilities have also been used in cognitive bias studies (Burman et al., 2008b; Mendl et al., 2009).

Even though rats are known for their learning abilities, training is not commonly used for husbandry or habituation procedures, as is done with other laboratory species, such as dogs and nonhuman primates (Schapiro, this volume). Methods, such as Positive Reinforcement Training, are not often used with rats, because training frequently requires a large time investment, which may exceed the duration of the experiment or the experimental budget. Nevertheless, experiments could benefit from training rats, especially for habituation and handling procedures. Rats naïve to humans may find interactions frightening, as observed by their behavior (Figure 7.6) and by the production of the 22 kHz calls that are associated with negative affective states (Brudzynski and Ociepa, 1992). Rats' learning capacity and adaptability have been highlighted in a study showing that they can learn to play "hide and seek" with researchers (Reinhold et al., 2019). Thus, the learning abilities of rats could be used to develop cooperative and collaborative activities with humans that could be used in husbandry or experimental procedures as is typically done with primates. A recent example of this is the procedure developed at the Research Institute of Sweden to habituate rats to handling and procedures that involves associative learning. The procedure is implemented as soon as rats arrive at the research facility, thus minimizing the exposure to negative experiences with humans and procedures. Rats are first taught to associate a substrate (a soft fleece fabric mat) or a shelter with handling and treats. Second, handling associated with procedures (e.g., touching the tail as for injection or blood sampling, restraint) is incorporated in the handling sessions. Finally, when the rat appears comfortable, the procedure is integrated into experimental testing (e.g., blood sampling, weighing) (see https://www.ri.se/en/what-we-do/expertises/3r-refinement, accessed 18 October 2020). This example provides evidence that training activities can be used for both animals kept for long- and short-term.

Rats should be adapted to handling several days before the start of experimental testing and procedures. Initial exposure to handling for husbandry and routine medical procedures, such as injection, blood sampling, and gavage, can negatively impact the health and welfare of animals (Engelking, 2006; Gärtner et al., 1980; Hurst and West, 2010). Such handling can also result in injuries to both humans and animals, for example, when fearful animals struggle to escape, resulting in scratches or bites to the handler. Early handling of rats during the preweaning and

FIGURE 7.5 Rat at the intersection of a Y-maze during a choice test.

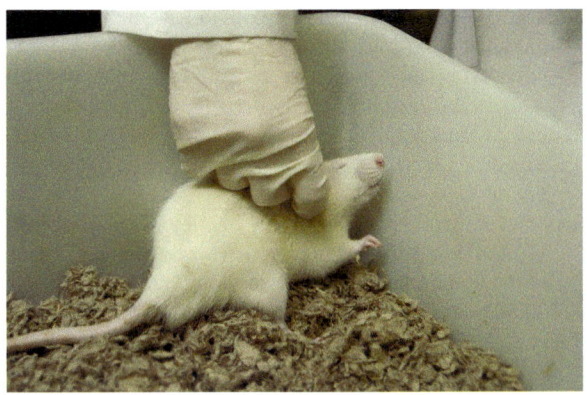

FIGURE 7.6 Rats not habituated to handling and humans can find handling aversive. When escape or avoidance is not possible, rats can remain immobile, freezing in a crouched posture with the absence of visible movement except that due to breathing. While they seem calm and easy to pick up, they are in a negative affective state.

postweaning stages has been found to have positive effects on behavior, stress, and anxiety, and results in improvements in learning and memory (Costa et al., 2012). However, considering that rats are often acquired by research facilities as adults or in the late postweaning stage, early handling is not always possible. Nevertheless, to minimize the effects of stress and injury, it is a good practice to habituate rats to human interaction. A variety of techniques are available, but many are time-intensive and costly. For example, one technique involves touching, stroking, lifting, talking to, and offering food treats to rats for a total of 20 min/day/cage for 14 days (Maurer et al., 2008), which amounts to over 40 h of work for just 10 cages. Furthermore, one of the procedures used in this technique – stroking – was found to elicit 22 kHz USVs, an indicator of negative affect, in rats naïve to handling (Brudzynski and Ociepa, 1992). Thus, habituation and handling procedures should be carefully planned and performed in a manner that minimizes the risk that a stimulus (e.g., a noise or a movement) triggers fear, and results in a negative experience associated with human interaction. Rat tickling, a technique that mimics aspects of rough-and-tumble play not human tickling, can effectively habituate rats to human interaction and induce positive affective states (Cloutier et al., 2018; LaFollette et al., 2017). It can be efficiently applied by tickling a rat for 15 s per day for 3 days (45 s total) (LaFollette et al., 2018). The technique appears to be an effective habituation protocol for use before common handling procedures, such as injection (Cloutier et al., 2012, 2013, 2014, 2015, 2018), and to provide social enrichment, especially for singly housed rats (Cloutier et al., 2013), requiring a much smaller investment of time and labor. However, the use of tickling might not be appropriate before procedures that require the rats to be calm (e.g., surgery) because it induces the expression of locomotor activities and playful behaviors. This is not surprising as tickling mimics rat social play, and the human is perceived as a play partner.

Clicker training, an operant conditioning method using a bridging stimulus (the clicker) to reward the performance of a desired behavior, has been successfully applied to cage changing procedures (Leidinger et al., 2018). Clicker training was combined with observations of experienced conspecifics (i.e., observational learning), to teach rats to voluntarily move to a clean cage through a tunnel. This combination of techniques reduced stress caused by the physical disturbance and handling associated with the cage changes, and reduced direct contact between the animals and their caregivers. This technique should be especially useful for animals that are kept for the long-term such as breeders, or in studies requiring high biosafety levels.

Conclusions/Recommendations

Understanding rat behavior is instrumental to provide housing, care, and handling practices that promote species-specific behaviors and induce positive affect in the laboratory. In the planning of housing, husbandry practices, environmental enrichment practices, and experimental design, consideration should be given to rats' natural behaviors, their sensory characteristics, and their perception of the environment, which differ from those of humans. Ensuring that the laboratory environment is adapted to satisfy both the animals' physical and psychological needs is the basis for good welfare and sound research. Thus, rats should be housed in groups, ideally in groups comprising more than two individuals. They should be provided with substrates that stimulate gnawing and burrowing. Nesting material should also be provided, as it is used by many rats, in addition to an opaque shelter that provides a retreat from human view. Rats should also be given opportunities to forage for part of their diet. They should have sufficient space to fully extend their bodies vertically, perform locomotor activities, and engage in social play. Breeding females should have access to areas of refuge from their pups. Rat tickling should be used to habituate the animals to handling and procedures, and attempts should be made to assess and control ultrasounds, and to prevent transmission of odors via caregivers (especially alarm odors). Last, but not least, whenever possible, rats' learning abilities should be used in training for husbandry, medical, and experimental procedures. Knowledge of rat behaviors will help develop environments customized to their needs. It will also allow rats to have some control over their environment, express species-specific behaviors, enhance the rat–human relationship, and improve rat welfare and the quality of research.

REFERENCES

Abou-Ismail, U.A., Burman, O.H.P., Nicol, C.J., Mendl, M. 2008. Let sleeping rats lie: Does the timing of husbandry procedures affect laboratory rat behaviour, physiology and welfare? *Appl. Anim. Behav. Sci.* 111(3–4):329–341. doi:10.1016/j.applanim.2007.06.019.

Abou-Ismail, U.A., Mendl, M.T. 2016. The effects of enrichment novelty versus complexity in cages of group-housed rats (*Rattus norvegicus*). *Appl. Anim. Behav. Sci.* 180:130–139 doi:10.1016/j.applanim.2016.04.014.

Ader, R., Grota, L.J. 1970. Rhythmicity in the maternal behavior of Rattus norvegicus. *Anim. Behav.* 18: 44–150.

Aldridge, J.W. 2004. Grooming. In *The Behaviour of the Laboratory Rat: A Handbook with Tests*. eds. I.Q. Whishaw and B. Kolb. New York: Oxford University Press.

Altman, J., Sudarshan, K. 1975. Postnatal development of locomotion in the laboratory rat. *Anim. Behav.* 23(4):896–920.

Antle, M.C., Mistlberger, R.E. 2004. Circadian rhythms. In *The Behaviour of the Laboratory Rat: A Handbook with Tests.* eds I.Q. Whishaw, and B. Kolb. New York: Oxford University Press.

Anzaldo, A.J., Harrison, P.C., Riskowski, G.L., Sebek, L.A., Maghirang, R.-G., Stricklin, W.R., Gonyou, H.W. 1994. Increasing welfare of laboratory rats with the help of spatially enhanced cages. *Anim. Welf. Inf. Center Newslett.* 5(3):5.

Barnett, S.A. 1956. Behaviour components in the feeding of wild and laboratory rats. *Behaviour* 1:24–43.

Bartal, I.B., Decety, J., Mason, P. 2011. Empathy and pro-social behavior in rats. *Science* 334:1427–1430 doi: 10.1126/science.1210789.

Berdoy, M. 2002. The laboratory rat: A natural history. Film. 27 minutes. www.ratlife.org.

Blanchard, R.J., Blanchard, D.C. 1989. Antipredator defensive behaviors in a visible burrow system. *J. Comp. Psychol.* 103(1):70–82. doi:10.1037/0735-7036.103.1.70.

Blanchard, R.J., Blanchard, D.C., Rodgers, J., Weiss, S.M. 1990. The characterization and modelling of antipredatory defensive behaviour. *Neurosci. Biobehav. Rev.* 14:463–472.

Blanchard, R.J., Flanelly, K.J., Blanchard, D.C. 1986. Defensive behaviors of laboratory and wild *Rattus norvegicus*. *J. Comp. Psychol.* 100:101–107.

Blanchard, R.J., Flannelly, K.J., Blanchard, D.C. 1988. Life-span studies of dominance and aggression in established colonies of laboratory rats. *Physiol. Behav.* 43:1–7.

Blanchard, R.J., Blanchard, D.C., Takahashi, T., Kelley, M.J. 1977a. Attack and defensive behaviour in the albino rat. *Anim. Behav.* 25:622–634.

Blanchard, R.J., Takahashi, L.K., Fukunaga, K.K., Blanchard, D.C. 1977b. Functions of the vibrissae in the defensive and aggressive behavior of the rat. *Aggressive Behav.* 3:231–240.

Boice, R. 1977. Burrows of wild and albino rats: Effects of domestication, outdoor raising, age, experience, and maternal state. *J. Comp. Physiol. Psychol.* 91(3):649–661. doi:10.1037/h0077338.

Botelho, S., Estanislau, C., Morato, S. 2007. Effects of under- and overcrowding on exploratory behavior in the elevated plus-maze. *Behav. Processes* 74:357–362.

Brudzynski, S.M. Fletcher, N.H. 2010. Rat ultrasonic vocalization: Short-range communication. Chapter 3.3. In *Handbook of Mammalian Vocalization, An Integrative Neuroscience Approach.* ed. S.M. Brudzynski, 69–76. Boston, MA: Elsevier Academic Press.

Brudzynski, S.M. 2009. Communication of adult rats by ultrasonic vocalization: Biological, sociobiological, and neuroscience approaches. *ILAR J.* 50:43–50.

Brudzynski, S.M, Ociepa, D. 1992. Ultrasonic vocalization of laboratory rats in response to handling and touch. *Physiol. Behav.* 5:655–660.

Burgdorf, J., Kroes, R.A., Beinfield, M.C., Panksepp, J., Moskal, J.R. 2010.Uncovering the molecular basis of positive affect using rough-and-tumble play in rats: A role for insulin-like growth factor I. *Neuroscience* 168:769–777.

Burgdorf, J.S., Kroes, R.A., Moskal, J.R., Pfaus, J.G., Brudzynski, S.M., Panksepp, J. 2008. Ultrasonic vocalizations of rats (*Rattus norvegicus*) during mating, play, and aggression: behavioral concomitants, relationship to reward, and self-administration of playback. *J. Comp. Psychol.* 122:357–367.

Burgdorf, J., Panksepp, J., Moskal, J.R. 2011. Frequency-modulated 50-kHz ultrasonic vocalizations: A tool for uncovering the molecular substrates of positive affect. *Neurosci. Biobehav. Rev.* 35:1831–1836.

Burgdorf, J., Panksepp, J., Brudzynski, S.M., Moskal, J.R. 2005. Breeding for 50-kHz positive affective vocalizations in rats. *Behav. Genet.* 35:67–72.

Burgdorf, J.S., Ghoreishi-Haack, N., Cearley, C.N., Kroes, R.A., Moskal, J.R. 2019. Rat ultrasonic vocalizations as a measure of the emotional component of chronic pain. *NeuroReport* 30:863–866 doi:10.1097/WNR.0000000000001282.

Burgdorf, J.S., Brudzynski, S.M., Moskal, J.R. 2020. Using rat ultrasonic vocalization to study the neurobiology of emotion: From basic science to the development of novel therapeutics for affective disorders. *Curr. Opin. Neurobiol.* 60:192–200. doi:10.1016/j.conb.2019.12.008 0959-4388.

Burn, C.C. 2008. What is it like to be a rat? Rat sensory perception and its implications for experimental design and rat welfare. *Appl. Anim. Behav. Sci.* 112:1–32. doi:10.1016/j.applanim.2008.02.007.

Burn, C.C., Mason, G.J. 2008. Effects of cage-cleaning frequency on laboratory rat reproduction, cannibalism, and welfare. *Appl. Anim. Behav. Sci.* 114:235–247.

Burman, O., Owen, D., Abou Ismail, U., Mendl, M. 2008c. Removing individual rats affects indicators of welfare in the remaining group members. *Physiol. Behav.* 93:89–96.

Burman O, Parker R, Paul E, Mendl M. 2008a. Sensitivity to reward loss as an indicator of animal affect and welfare. *Biol Lett* 4:330–3.

Burman O, Parker R, Paul E, Mendl M. 2008b. A spatial judgement task to determine background emotional state in laboratory rats (*Rattus norvegicus*). *Anim Behav* 76:801–809.

Büttner, D. 1993. Upright standing in the laboratory rat–time expenditure and its relation to locomotor activity. *J. Exp. Anim. Sci.* 36(1):19–26.

Bryda, E.C. 2013. The mighty mouse: The impact of rodents on advances in biomedical research. *Missouri Med.* 110(3):207–211.

Calhoun, J.B. 1963. The Ecology and Sociology of the Norway Rat, U.S. Department of Health, Education and Welfare, Bethesda, MD.

Canadian Council on Animal Care. 2017. CCAC guidelines: Husbandry of animals in science. Ottawa ON: Canadian Council on Animal Care.

Canadian Council on Animal Care. 2018. Annual Animal Data Report. (2019) https://www.ccac.ca/Documents/AUD/2018-Animal-Data-Report.pdf.

Canadian Council on Animal Care. 2020. CCAC guidelines: Rat. Ottawa ON: Canadian Council on Animal Care.

Chmiel Jr., D.J., Noonan, M. 1996. Preference of laboratory rats for potentially enriching stimulus objects. *Lab. Anim.* 30:97–101.

Clarence C. 1948. Aquatic habits of the Norway rat. *J. Mammal.* 29(3):299. doi:10.1093/jmammal/29.3.299.

Cloutier, S., Newberry, R.C. 2010. Physiological and behavioural responses of laboratory rats housed at different tier levels and levels of visual contact with conspecifics and humans. *Appl. Anim. Behav. Sci.* 125:69–79.

Cloutier, S., Panksepp, J., Newberry, R.C. 2012. Playful handling by care-takers reduces fear of humans in the laboratory rat. *Appl. Anim. Behav. Sci.* 140:161–171.

Cloutier, S., Baker, C., Wahl, K., Panksepp, J., Newberry, R.C. 2013. Playful handling as social enrichment for individually- and group-housed laboratory rats. *Appl. Anim. Behav. Sci.* 143:85–95.

Cloutier, S., Wahl, K., Baker, C., Newberry, R.C. 2014. The social buffering effect of playful handling on responses to repeated intra-peritoneal injections in laboratory rats. *J. Am. Assoc. Lab. Anim. Sci.* 53:161–166.

Cloutier, S., Wahl, K.L., Panksepp, J., Newberry, R.C. 2015. Playful handling of laboratory rats is more beneficial when applied before than after routine injections. *Appl. Anim. Behav. Sci.* 164:81–90. doi:10.1016/j.applanim.2014.12.012.

Cloutier, S., LaFollette, M.R., Gaskill, B.N., Panksepp, J., Newberry, R.C. 2018. Tickling, a technique for inducing positive affect when handling rats. *J. Vis. Exp.* e57190. doi:10.3791/57190.

Costa, R., Tamascia, M.L., Nogueira, M.D., Casarini, D.E., Marcondes, F.K. 2012. Handling of adolescent rats improves learning and memory and decreases anxiety. *J Am Assoc for Lab. Anim. Sci.* 51(5):548–553.

Cree, T.C., Wadley, D.M., Marlett, J.A. 1986. Effect of preventing coprophagy in the rat on neutral detergent fiber digestibility and apparent calcium absorption. *J. Nutr.* 116(7):1204–1208.

de Visser, L., van den Bos, R., Spruijt, B.M. 2005. Automated home cage observations as a tool to measure the effects of wheel running on cage floor locomotion. *Behav. Brain Res.* 160:382–388.

Dewsbury, D.A. 1980. Wheel-running behavior in 12 species of muroid rodents. *Behav. Proc.* 5:271–280.

Dielenberg, R.A., Carrive P., McGregor I.S. 2001. The cardiovascular and behavioral response to cat odor in rats: Unconditioned and conditioned effects. *Brain Res.* 897:228–237.

Einon, D., Potegal, M. 1991. Enhanced defense in adult rats deprived of play fighting experience as juveniles. *Aggressive Behav.* 17:27–40.

El-Ayache, N., Galligan, J.J. 2020. The rat in neuroscience research. Chapter 28. In *The Laboratory Rat*, Third Edition, eds. M.A. Suckow, F.C. Hankenson, R.P. Wilson, P.L. Foley, 1003–1022. Boston, MA: Elsevier Academic Press. doi:10.1016/B978-0-12-814338-4.00028-3.

Ellenbroek, B., Youn, J. 2016. Rodent models in neuroscience research: Is it a rat race? *Dis. Models Mech.* 9:1079–1087. doi:10.1242/dmm.026120.

Engelking, L.R. 2006. *QLS in Veterinary Medicine: Metabolic and Endocrine Physiology*, Second Edition. Jackson, WY: Teton New Media.

European Commission. Report from the Commission to the European Parliament and the Council. 2019 report on the statistics on the use of animals for scientific purposes in the Member States of the European Union in 2015–2017. SWD (2020) 10 final.

Faith, R.E., Hessler, J.R. 2006. Housing and environment. Chapter 10, In *The Laboratory Rat*. eds. M.A. Suckow, S.H. Weisbroth, C.L. Franklin. Boston MA: Elsevier Academic Press.

Foster, S., King, C., Patty, B., Miller, S. 2011. Tree climbing capabilities of Norway and ship rats. *N. Z. J. Zool.* 38(4):285–296. doi:10.1080/03014223.2011.599400.

Galef Jr., B.G., Sorge, R.E. 2000. Use of PVC conduits by rats of various strains and ages housed singly and in pairs. *J. Appl. Anim. Welf. Sci.* 3(4):279–292.

Galef Jr., B.G., Whiskin, E.E. 2003. Socially transmitted food preferences can be used to study long-term memory in rats. *Learn. Behav.* 31(2):160–164. doi:10.3758/BF03195978.

Galef Jr., B.G., Pretty, S., Whiskin, E.E. 2006. Failure to find aversive marking of toxic foods by Norway rats. *Anim. Behav.* 72:1427–1436. doi:10.1016/j.anbehav.2006.05.009.

Garner, J.P. 2005. Stereotypies and other abnormal repetitive behaviors: Potential impact on validity, reliability, and replicability of scientific outcomes. *ILAR*, 46(2):106–117.

Gaskill, B.N., Pritchett-Corning, K.R. 2015a. The effect of cage space on behavior and reproduction in Crl:CD1(Icr) and C57BL/6NCrl Laboratory Mice. *PLoS One* 10(5): e0127875. doi:10.1371/journal.pone.0127875.

Gaskill, B.N., Pritchett-Corning, K.R. 2015b. Effect of cage space on behavior and reproduction in Crl:CD(SD) and BN/Crl Laboratory Rats. *J. Am. Assoc. Lab. Anim. Sci.* 54: 497–506.

Gärtner, K., Buttner, D., Dohler, K., Friedel, R., Lindena, J., Trautschold, I. 1980. Stress response of rats to handling and experimental procedures. *Lab. Anim.* 14:267–274.

Gerall, H.D., Ward, I.L., Gerall, A.A. 1967. Disruption of the male rat's sexual behavior induced by social isolation. *Anim. Behav.* 15:54–58.

Grota, L.J., Ader, R. 1974. Behavior of lactating rats in a dual-chambered maternity cage. *Horm. Behav.* 5:275–282.

Grover Jr., P.B., Miller, R.J. 1985. An inexpensive microprocessor-based lighting control for simulating natural photoenvironments in the laboratory. *J. Interdisc. Cycle Res.* 16(1): 33–42.

Grota, L.J., Ader, R. 1969. Continuous recording of maternal behavior in *Rattus norvegicus*. *Anim. Behav.* 17:722–129.

Hamilton, W. 1998. *The Mammals of Eastern United States*, Third edition. Ithaca, NY: Comstock Publishing.

Hedrich, H.J. 2006. Taxonomy and stocks and strains. Chapter 3. In *The Laboratory Rat*. ed. M.A. Suckow, S.H. Weisbroth, and C.L. Franklin, 71–92. Boston MA: Elsevier Academic Press.

Heiderstadt, K.M., McLaughlin, R.M., Wrighe, D.C., Walker, S.E., Gomez-Sanchez, C.E. 2000. The effect of chronic food and water restriction on open-field behaviour and serum corticosterone levels in rats. *Lab Anim.* 34:20–28.

Hershey, J.D., Gifford, J.J., Zizza, L.J., Pavlenko, D.A., Wagner, G.C., Miller, S. 2018. Effects of various cleaning agents on the performance of mice in behavioral assays of anxiety. *J. Am. Assoc. Lab. Anim. Sci.* 57(4):335–339. doi:10.30802/AALAS-JAALAS-17-000161.

Hol, T., Van den Berg, C.L., Van Ree, J.M., Spruijt, B.M. 1999. Isolation during the play period in infancy decreases adult social interactions in rats. *Behav. Brain Res.* 100(1–2):91–97.

Hrdy, S.B. 1979. Infanticide among animals: A review, classification, and examination of the implications for the reproductive strategies of females. *Ethol. Sociobiol.* 1:13–40.

Hurst, J.L., West, R.S. 2010. Taming anxiety in mice. *Nat. Methods* 7:825–826.

Hurst, J.L., Barnard, C.J., Hare, R., Wheeldon, E.B., West, C.D. 1996. Housing and welfare in laboratory rats: Time budgeting and pathophysiology in single-sex groups. *Anim. Behav.* 52:335–360.

Hurst, J.L., Barnard, C.J., Nevison, C.M., West, C.D. 1997. Housing and welfare in laboratory rats: Welfare implications of isolation and social contact among caged males. *Anim. Welf.* 6:247–329.

Hurst, J.L., Barnard, C.J., Nevison, C.M., West, C.D. 1998. Housing and welfare in laboratory rats: The welfare implications of social isolation and social contact among females. *Anim. Welf.* 7:121–136.

Jacobs, G.H., Fenwick, J.A., Williams, G.A. 2001. Cone-based vision of rats for ultraviolet and visible lights. *J. Exp. Biol.* 204(14):2439–2446.

Jacobs, G.H., Neitz, J., Deegan, J.F. 1991. Retinal receptors in rodents maximally sensitive to ultraviolet light. *Nature* 353:655–656.

Jegstrup, I.M., Vestergaard, R., Vach, W., Ritskes-Hoitinga, M. 2005. Nest-building behaviour in male rats from three inbred strains: BN/HsdCpb, BDIX/OrIIco and LEW/Mol. *Anim. Welf.* 14(2):149–156.

Kaltwasser, M.T. 1990. Acoustic signalling in the black rat (*Rattus rattus*). *J. Comp. Psychol.* 104:227–232.

Kanarek, R.B., D'Anci, K.E., Jurdak, N., Mathes, W.F. 2009. Running and addiction: Precipitated withdrawal in a rat model of activity-based anorexia. *Behav. Neurosci.* 123:905–912. doi:10.1037/a0015896.

Kitaoka, A. 1994. Defensive aspects of burrowing behavior in rats (*Rattus norvegicus*): A descriptive and correlational study. *Behav. Processes* 31(1):13–28. doi:10.1016/0376-6357(94)90034-5

Knutson, B., Burgdorf, J., Panksepp, J. 2002. Ultrasonic vocalizations as indices of affective states in rats. *Psychol. Bull.* 128:961–977. doi:10.1037/0033-2909.128.6.961.

Koolhaas, J.M. 2010. The Laboratory Rat. Chapter 22. In *The UFAW Handbook on the Care and Management of Laboratory and Other Research Animals*, Eighth Edition, eds. R. Hubrecht, J. Kirkwood, J. Ames. Hoboken, NJ: Wiley-Blackwell.

LaFollette, M.R., O'Haire, M.E., Cloutier, S., Blankenberger, W.B., Gaskill, B.N. 2017. Rat tickling: A systematic review of applications, outcomes, and moderators. *PLoS One* 12:e0175320. doi:10.1371/journal.pone.0175320.

LaFollette, M.R., O'Haire, M.E., Cloutier, S., Gaskill, B.N. 2018. Practical rat tickling: Determining an efficient and effective dosage of heterospecific play. *Appl. Anim. Behav. Sci.* 208:82–91.

LaFollette, M.R., Swan, M.P., Smith, R.K., Hickman, D.L., Gaskill, B.N. 2019. The effects of cage color and light intensity on rat affect during heterospecific play. *Appl. Anim. Behav. Sci.* 219. doi:10.1016/j.applanim.2019.104834.

Lawlor, M.M. 2002. Comfortable quarters for rats in research institutions. In *Comfortable Quarters for Laboratory Animals*, Ninth Edition, eds. V. Reinhardt, A. Reinhardt, 26–32. Washington, DC: Animal Welfare Institute.

Leidinger, C.S., Kaiser, N., Baumgart, N., Baumgart, J. 2018. Using clicker training and social observation to teach rats to voluntarily change cages. *J. Vis. Exp.* 140:e58511, doi:10.3791/58511.

Lindsey, J. R., Baker, H. J. 2006. Historical foundations. Chapter 1. In *The Laboratory Rat*, Second Edition, ed. M.A., Suckow, S.H., Weisbroth, and C.L. Franklin, 1–52. Boston, MA: Elsevier Academic Press.

Litvin, Y., Blanchard, D.C., Blanchard, R.J. 2010. Vocalization as a social signal in defensive behavior. Chapter 5.1. In *Handbook of Mammalian Vocalization, An Integrative Neuroscience Approach*. ed. S.M. Brudzynski, Boston MA: Elsevier Academic Press. doi:10.1016/B978-0-12-374593-4.00015-2.

Lohmiller, J.J., Swing, S.P. 2006. Reproduction and breeding. Chapter 6. In *The Laboratory Rat*, Second Edition, ed. M.A., Suckow, S.H., Weisbroth, and C.L. Franklin, 147–164. Boston MA: Elsevier Academic Press.

Macrì, S., Würbel, H. 2006. Developmental plasticity of HPA and fear responses in rats: A critical review of the maternal mediation hypothesis. *Horm. Behav.* 50:667–680.

Makowska, I.J., Weary, D.M. 2016. The importance of burrowing, climbing and standing upright for laboratory rats. *R. Soc. Open Sci.* 3(6):160–136.

Makowska, I.J., Weary, D.M. 2020. A good life for laboratory rodents? *ILAR J.* 1–16. doi:10.1093/ilar/ilaa001.

Nadel, J., Huang, T., Xia, Z., Burlin, T., Zametkin, A., Smith, C.B. 2013. Voluntary exercise regionally augments rates of cerebral protein synthesis. *Brain Res.* 1537:125–131.

Manser, C.E., Broom, D.M., Overend, P., Morris, T.H. 1998. Investigations into the preferences of laboratory rats for nest-boxes and nesting materials. *Lab. Anim.* 32(1):23–35. doi: 10.1258/002367798780559365.

Mason, G., Wilson, D., Hampton, C., Würbel, H. 2004. Non-invasively assessing disturbance and stress in laboratory rats by scoring chromodacryorrhoea. *Altern. Lab. Anim.* 32 (Suppl. 1A):153–159.

Maurer, B.M., Doering, D., Scheipl, F., Kuechenhoff, H., Erhard, M.H. 2008. Effects of a gentling programme on the behaviour of laboratory rats towards humans. *Appl. Anim. Behav. Sci.* 114:554–571. doi:10.1016/j.applanim.2008.04.013.

Meier, J.H., Robbers, Y. 2014. Wheel running in the wild. *Proc. R. Soc. B* 281:20140210. doi:10.1098/rspb.2014.0210.

Mendl, M., Burman, O., Parker, R., Paul, E. 2009. Cognitive bias as an indicator of animal emotion and welfare: Emerging evidence and underlying mechanisms. *Appl. Anim. Behav. Sci.* 118:161–81.

Modlinska, K., Pisula, W. 2020. The Norway rat, from an obnoxious pest to a laboratory pet. *eLife* 9:e50651. Published online 2020 Jan 17. doi:10.7554/eLife.50651.

National Research Council 2011. Guide for the Care and Use of Laboratory Animals. Washington, DC: National Academy Press.

Newton, W.M. 1978. Environmental impact on laboratory animals. *Adv. Vet. Sci. Comp. Med.* 22:28.

Nieder, L., Cagnin, M., Parisi, V. 1982. Burrowing and feeding behaviour in the rat. *Anim. Behav.* 30(3):837–844. doi:10.1016/S0003-3472(82)80157-7

Novakova, V., Babicky, A. 1989. Coprophagy in young laboratory rat. *Physiol. Bohemoslovaca* 38(1):21–28.

Nowak, R., Paradiso, J. 1983. *Walker's Mammals of the World*, Fourth Edition. Baltimore, MD: The Johns Hopkins University Press.

Panksepp, J. 1981. The ontogeny of play in rats. *Dev. Psychobiol.* 14(4):327–332.

Panksepp, J. 1998. *Affective Neuroscience: The Foundations of Human and Animal Emotions*. New York: Oxford University Press.

Patterson-Kane, E.G. 2003. Shelter enrichment for rats. *Contemp. Top. Lab. Anim. Sci.* 42(2):46–48.

Patterson-Kane, E.G., Hunt, M., Harper, D. 1999. Behavioral indexes of poor welfare in laboratory rats. *J. Appl. Anim. Welfare Sci.* 2, 97–110.

Patterson-Kane, E.G., Harper, D., Hunt, M. 2001. The cage preferences of laboratory rats. *Lab. Anim.* 35:74–79.

Patterson-Kane, E.G., Hunt, M., Harper, D. 2002. Rats demand social contact. *Anim. Welfare* 11:327–332.

Patterson-Kane, E.G., Hunt, M., Harper, D. 2004. Short communication: Rat's demand for group size. *J. Appl. Anim. Welfare Sci.* 7:267–272.

Pellis, S.M. 2002. Keeping in touch: Play fighting and social knowledge. In *The Cognitive Animal: Empirical and Theoretical Perspectives on Animal Cognition*. eds. M. Bekoff, C. Allen, G.M. Burghardt, Cambridge: MIT Press.

Pellis, S.M., Pellis, V.C. 1987. Play-fighting differs from serious fighting in both target of attack and tactics of fighting in the laboratory rat *Rattus norvegicus*. *Aggressive Behav.* 13:227–242.

Pellis, S.M., Pellis, V.C. 2004. Play and fighting. In *The Behaviour of the Laboratory Rat: A Handbook with Tests*. eds I.Q. Whishaw and B. Kolb, Oxford: Oxford Scholarship.

Pellis, S., Pellis, V. 2009. *The Playful Brain, Venturing to the Limits of Neuroscience*. Oxford: Oneworld Publications.

Pellis, S.M., Pellis, V.C., Bell, H.C. 2010. The function of play in the development of the social brain. *Am. J. Play* 2:278–296.

Pinelli, C.J., Leri, F., Turner, P.V. 2017. Long term physiologic and behavioural effects of housing density and environmental resource provision for adult male and female Sprague-Dawley rats. *Animals* 7(6):44.

Plaut, S.M. 1974. Adult-litter relations in rats reared in single and dual-chambered cages. *Dev. Psychobiol.* 7:111–120.

Plotsky, P.M., Thrivikraman, K.V., Nemeroff, C.B., Caldji, C., Sharma, S., Meaney, M.J. 2005. Long-term consequences of neonatal rearing on central corticotropin-releasing factor systems in adult male rat offspring. *Neuropsychopharmacology* 30:2192–2204.

Price, E.O. 1977. Burrowing in wild and domestic norway rats. *J. Mammal.* 58(2):239–240. doi:10.2307/1379585.

Price, E.O., Belanger, P.L. 1977. Maternal behavior of wild and domestic stocks of Norway rats. *Behav. Biol.* 20(1):60–69. doi:10.1016/S0091-6773(77)90511-9.

Pritchett-Corning, K.R. 2015. Rats. In *Comfortable Quarters for Laboratory Animals*, Tenth Edition, eds. C. Liss, K. Litwak, D. Tilford, V. Reinhardt, 19–37. Washington, DC: Animal Welfare Institute.

Reinagel, P. 2018. Training rats using water rewards without water restriction. *Front. Behav. Neurosci.*, May 2018. doi:10.3389/fnbeh.2018.00084.

Reinhold, A.S., Sanguinetti-Scheck, J.I., Hartmann, K., Brecht, M. 2019. Behavioral and neural correlates of hide-and-seek in rats. *Science* 365:1180–1183. doi:10.1126/science.aax4705.

Richter, S.H., Gass, P., Fuss, J. 2014. Resting is rusting: A critical view on rodent wheel-running behavior. *Neuroscientist* 20(4):313–325. doi:10.1177/1073858413516798.

Sachs, B.D. 1988. The development of grooming and its expression in adult animals. *Ann. N. Y. Acad. Sci.* 525(1):1–17. doi:10.1111/j.1749-6632.1988.tb38591.x.

Sato, N., Tan, L., Tate, K. et al. 2015. Rats demonstrate helping behavior toward a soaked conspecific. *Anim. Cogn.* 18:1039–1047. doi:10.1007/s10071-015-0872-2.

Schapiro, S., Salas, M., Vukovich, K. 1970. Hormonal effects on ontogeny of swimming ability in the rat: Assessment of central nervous system development. *Science* 168:147–151.

Sharp, J.L., Zammit, T.G., Azar, T.A., Lawson, D.M. 2002. Stress-like responses to common procedures in male rats housed alone or withother rats. *Contemp. Top. Lab. Anim. Sci.* 41:8–14.

Sharp, J., Zammit, T., Azar, T., Lawson, D. 2003. Stress-like responses to common procedures in individually and group-housed female rats. *Contemp. Top. Lab. Anim. Sci.* 42:9–18.

Sherwin, C.M. 1998. Voluntary wheel running: A review and novel interpretation. *Anim. Behav.* 56:11–27.

Silver, J. 1927. The introduction and spread of house rats in the United States. *J. Mammal.* 8(1):58–60.

Siviy, S.M., Harrison, K.A., McGregor, I.S. 2006. Fear, risk assessment, and playfulness in the juvenile rat. *Behav. Neurosci.* 120:49–59.

Sorensen, D.B., Ottesen, J.L., Hansen A.K. 2004. Consequences of enhancing environmental complexity for laboratory rodents – a review with an emphasis on the rat. *Animal Welfare* 13:193–204.

Špinka, M., Newberry, R.C., Bekoff, M. 2001. Mammalian play: Training for the unexpected. *Q. Rev. Biol.* 76:141–168.

Spruijt, BM, van Hooff, J.A.R.A.M., Gispen, W.H. 1992. Ethology and neurobiology of grooming behavior. *Physiol. Rev.* 72(3):825–852.

Stranahan, A.M., Khalil, D., Gould, E. 2006. Social isolation delays the positive effects of running on adult neurogenesis. *Nat. Neurosci.* 9:526–533.

Stryjek, R., Modlińska, K., Pisula, W. 2012. Species specific behavioural patterns (Digging and swimming) and reaction to novel objects in wild type, Wistar, Sprague-Dawley and Brown Norway rats. *PLoS One* 7:e40642. doi:10.1371/journal.pone.0040642.

Tabak, M.A., Poncet, S., Passfield, K., Martinez del Rio, C. 2015. Modeling the distribution of Norway rats (*Rattus norvegicus*) on offshore islands in the Falkland Islands. *Neobiota* 24:33–48.

Takahashi, L.R. 1986. Postweaning environmental and social factors influencing onset and expression of agonistic behavior in Norway rats. *Behav. Processes* 12:237–260.

Talling, J.C., Van Driel, K.S., Inglis, I.R. 2002. Do laboratory rats choose to spend time together? In *Proceedings of the 36th International Congress of the ISAE*. Egmond aan Zee, Wageningen, The Netherlands, August 6–10, eds. P. Koene, B. Spruijt, H. Blokhuis, D. Ekkel, F. Ödberg, K. van Reenen, H. Spoolder, W. Schouten, and R. van den Bos, 168.

Toth, L.A., Gardiner, T.W. 2000. Food and water restriction protocols: Physiological and behavioral considerations. *Contemp. Top Lab. Anim. Sci.* 39(6):9–17.

Tunstall, B.J., Vendruscolo, L.F., Allen–Worthington K. 2020. Rat models of alcohol use disorder. Chapter 26. In *The Laboratory Rat*, Third Edition, eds. M.A. Suckow, F.C. Hankenson, R.P. Wilson, P.L. Foley, 967–986, Boston MA: Elsevier Academic Press. doi:10.1016/B978-0-12-814338-4.00026-X.

van Betteray, J.N.F., Vossen, J.M.H., Coenen, A.M.L. 1991. Behavioural characteristics of sleep in rats under different Light/Dark conditions. *Physiol. Behav.* 50:79–82.

van Loo, P.L., Baumans, V. 2004. The importance of learning young: The use of nesting material in laboratory rats. *Lab. Anim.* 38(1):17–24. doi:10.1258/00236770460734353.

Whishaw, I.Q. 2004. Foraging. In *The Behaviour of the Laboratory Rat: A Handbook with Tests*, eds. I.Q. Whishaw and B. Kolb, New York: Oxford University Press.

Williams, C.M., Riddell, P.M., Scott, L.A. 2008. Comparison of preferences for object properties in the rat using paired- and free-choice paradigms. *Appl. Anim. Behav. Sci.* 112:146–157.

Winnicker, C., Gaskill, B., Garner, J.P., Pritchett-Corning, K.R. 2016. A guide to the behaviour and enrichment of laboratory rodents. Charles River Laboratories.

Wöhr, M., Schwarting, R.K.W. 2007. Ultrasonic communication in rats: Can playback of 50-kHz calls induce approach behavior? *PLoS One* 2(12):e1365. doi:10.1371/journal.pone.0001365.

Wöhr, M., Houx, B., Schwarting, R.K.W., Spruijt, B. 2008. Effects of experience and context on 50-kHz vocalizations in rats. *Physiol. Behav.* 93:766–776.

Wright, J.M., Gourdon, J.C., Clarke, P.B.S. 2010. Identification of multiple call categories within the rich repertoire of adult rat 50-kHz ultrasonic vocalizations: Effects of amphetamine and social context. *Psychopharmacology* 211:1–13.

Zorzo, C., Méndez-López, M., Méndez, M., Arias, J.L. 2019 Adult social isolation leads to anxiety and spatial memory impairment: Brain activity pattern of CO_x and c-Fos. *Behav. Brain Res.* 3653:170–177.

8
Behavioral Biology of Guinea Pigs

Gale A. Kleven
Wright State University

CONTENTS

Guinea Pigs 131
    Classification 132
    History 133
Typical Research Application 133
    Infectious Disease 133
    Respiratory Disease 133
    Diabetes 133
    Otology 133
    Nutrition 133
    Reproduction/Teratology/Toxicology 133
    Behavioral Development 134
Behavioral Biology 134
    Natural History 134
        Ecology 134
        Social Organization and Behavior 134
    Common Captive Behaviors 137
        Normal 138
        Abnormal 138
Ways to Maintain Behavioral Health 139
    Captive Considerations to Promote Species-Typical Behavior 139
        Environment 139
        Social Groupings 139
        Husbandry 139
        Feeding Strategies 140
        Training 140
Special Situations 141
    Hairless Guinea Pigs 141
Conclusions/Recommendations 141
Additional Resources 141
References 142

Guinea Pigs

Neither pigs nor from Guinea, the guinea pig (*Cavia porcellus*) is rather an enigma. Even though its name is synonymous with that of the research test subject (Soanes & Stevenson, 2003), laboratory guinea pigs comprise less than 2% of all research animals (Gad, 2007). Equally perplexing, domesticated *C. porcellus* does not exist in the wild, even though closely related species (*C. aperea*, *C. tschudii*, and *C. fulgida*) are ubiquitous in South America. There was a period during the 1800–1900s, however, when guinea pigs reigned supreme as the primary research species. Some have speculated that the recent emphasis on the ease of genetic alterations in mice, and early genomic sequencing of rats and mice, has caused the guinea pig to fall from favor in the research laboratory (Wagner & Manning, 1976). Similarly, domestication of *C. porcellus* occurred over 7,000 years ago circa 5000 BCE (Morales, 1995), perhaps explaining their disappearance from the natural habitat. Since that time, guinea pigs have acquired, through selective breeding, the characteristics and variety present in modern-day guinea pig breeds (Table 8.1).

Domesticated guinea pigs are a moderate-sized (700–1,100 g) herbivorous rodent, with wild species somewhat smaller in size (500–800 g). They are quadrupedal with stocky bodies

TABLE 8.1

Common Guinea Pig Strains

Laboratory Strains[a]	
Name	Use
Hartley/Dunkin	Wide range of research
IAF hairless	Dermatology, toxicology, behavioral development
Strain 2	Immunology
Strain 13	Immunology

Fancy Strains[b]	
Name	Characteristics
American	Short silky coat
Crested	Short coat with a forehead whorl, typically white
Satin	Hollow hair shafts create a glassy sheen appearance on the coat
Silkie	Long straight coat
Texel	Long curly coat
Peruvian	Long coat with head portion growing forward creating a forelock
Coronet	Long coat with a forehead whorl
Abyssinian	Short coat covered with whorls
Teddy	Short coarse coat

[a] For a complete listing of rare strains not widely available, see Festing (1979).
[b] Adapted from ARBA (2016). Coat colorations include solid colors (white, red, brown, black), multicolored, and diluted or pastel (e.g., tan, gray).

(20–25 cm long), 4-toed forepaws, 3-toed hindpaws, and, like humans, possess a vestigial tail. Like most rodents, their teeth grow continuously and are shortened by grinding actions during mastication. They typically have short brown or red-brown hair in the wild, but albino laboratory strains and long-haired versions of domesticated guinea pigs have been achieved through selective breeding (Asher, de Oliveira, & Sachser, 2004; Eisenberg, 1989; Morales, 1995).

Classification

The modern laboratory guinea pig *Cavia porcellus* (Linnaeus 1758) is from the classification family Caviidae, and subfamily Caviinea, in the order Rodentia. The genus *Cavia* includes at least five other species in addition to *porcellus* (Figure 8.1). The ancestry of the broader category of caviomorphs is less clear. It is thought that the antecedent of both South American caviomorphs and African phiomorphs is an Asian hystricognath (Flynn, 1986; Marivaux, Vianey-Liaud, Welcomme, & Jaeger, 2002). Caviomorphs are estimated to have arrived in South America between the Middle and Late Eocene, with diversification by the Early Oligocene (Poux, Chevret, Huchon, de Jong, & Douzery, 2006). There are no fossil records to support how the migration to South America occurred, leaving speculations of migration directly from Asia through Antartica (Huchon & Douzery, 2001) or from Asia through Africa (Lavocat, 1974).

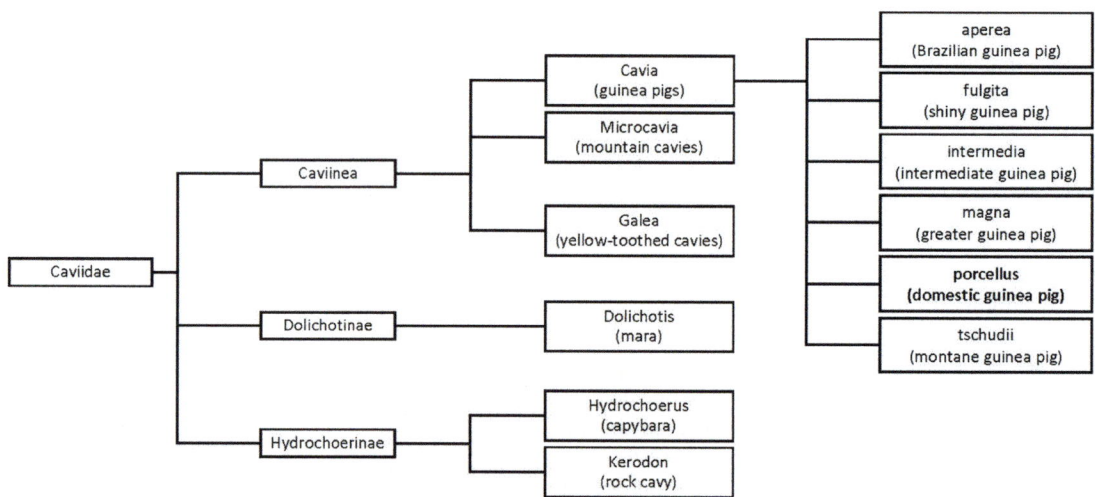

FIGURE 8.1 Classification of the domestic guinea pig *Cavia porcellus* (bold) and closely related species.

History

Spanish explorers noted that the guinea pig had already been domesticated by the Incas of Peru before their arrival in South America circa 1530 (Cumberland, 1886). Some naturalists have suggested that the guinea pig may have been domesticated as early as 5000 BCE (Wing, 1986). Initial reports on husbandry of guinea pigs by the Andean people describe strict protocols of artificial selection, from which they created a variety of breeds (Cumberland, 1886). Statues of guinea pigs have been unearthed dating back to the Moche people, circa 500 BCE (Sandweiss & Wing, 1997). To this day, the guinea pig continues to play important roles in some Andean cultures (Morales, 1994), including use as a food source, and playing a part in folklore traditions, religious ceremonies, and healing rituals.

Spanish, English, and Dutch trade delivered the guinea pig to Europe in the mid-16th century, where it was embraced as an exotic pet. Nobles and royals, including Queen Elizabeth I, were among the early fanciers (Morales, 1995). Although guinea pigs were used in the study of anatomy in the 17th century, it was not until the mid-18th century that the first scientific experiments in biology were recorded (Mason, 1939). By the end of the 19th century, scientists such as Robert Koch (Clark, Hall, & Williams, 2014), Louis Pasteur, and Emile Roux (Berche, 2012) would make the name guinea pig forever synonymous with research test subject.

Typical Research Application

Guinea pigs possess a number of attributes and similarities with humans that make them suitable research models. These models include infectious and respiratory disease, diabetes, auditory research, nutrition, reproduction, teratology, toxicology, and behavioral development.

Infectious Disease

The earliest and most extensive application of the guinea pig in research was for the study of infectious diseases. Two of the resulting discoveries, tuberculosis and diphtheria, were awarded the Nobel Prize. Other major discoveries include Yellow fever, Q fever, Rocky Mountain spotted fever, typhus, and malaria (Reid, 1959). Recent research in infectious disease utilizing the guinea pig includes Legionnaires disease, *Staphylococcus* infections, immunology, and sexually transmitted diseases, such as chlamydia and syphilis (Padilla-Carlin, McMurray, & Hickey, 2008).

Respiratory Disease

The sensitivity of the guinea pig respiratory system is well suited to research in the area of allergies (Poulsent et al., 1991), asthma, and other respiratory disease (Canning & Chou, 2008). Dermatological sensitivity and allergy have also been studied with an inbred strain of hairless guinea pig (Miyauchi & Horio, 1992).

Diabetes

Guinea pigs produce insulin at a rate much higher than other animals, and possess a rare insulin mutation which makes them suited for the study of type 2 (Brendan et al., 2017) and juvenile diabetes (Lang & Munger, 1976), as well as the production of antiinsulin antibodies (Wright, Makulu, & Posey, 1968).

Otology

Relatively easy access to the cochlea and middle ear, coupled with a docile nature, makes the guinea pig an ideal model for auditory research (Anderson & Wedenberg, 1965; Von Békésy, 1960). Additionally, guinea pigs display a Preyer reflex (pricking back of the ear in response to a Galton whistle or loud noise) that is used to detect deafness and assess hearing function (Bohmer, 1988). More recently, the development of a cochlear implant (Honeder et al., 2018), regeneration of cochlear hair cells (Kawamoto, Ishimoto, Minoda, Brough, & Raphael, 2003), as well as a device to reduce tinnitus (Marks et al., 2018) were all achieved through experimentation with guinea pigs.

Nutrition

Similar to humans, guinea pigs cannot synthesize vitamin C (ascorbic acid) and require dietary supplementation to avoid developing scurvy (Cohen & Mendel, 1918). This attribute makes guinea pigs uniquely suited to studies of vitamin C deficiency and related disease. Guinea pigs are also utilized in the development of antinausea drugs for chemotherapy treatment, and to study the brain/gut connection in the intestinal enteric nervous system (Brookes, 2001). Other nutritionally related studies involving guinea pigs include metabolic pathways and deficiencies of folic acid, arginine, potassium, and thiamine (Wagner & Manning, 1976). Guinea pigs also have lipid profiles similar to humans and, when fed high cholesterol diets, develop atherosclerosis, making the guinea pig an excellent model for studying the hypocholesterolemic effects of drugs (West & Fernandez, 2004).

Reproduction/Teratology/Toxicology

Pregnant guinea pigs share a number of commonalities with human pregnancies, such as similar endocrine control of pregnancy (Mårtensson, 1984), a discoidal hemomonochorial placenta and similar placental transport (Kaufmann & Davidoff, 1977), fetal development that can be divided into periods corresponding to human prenatal development (Bellinger, Lucas, & Kleven, 2015), the development of hypertension during pregnancy (Thompson, Pence, Pinkas, Song, & Telugu, 2016; Van Abeelen et al., 2012), and toxemia or preeclampsia (Seidl, Hughes, Bertolet, & Lang, 1979). Guinea pigs have also been employed to study fetal hypoxia (Mishra & Delivoria-Papadopoulos, 1999), interuterine growth retardation (Herrera et al., 2016), teratology (Rocca & Wehner, 2009), and preterm birth (Kelleher, Hirst, & Palliser, 2013; Shaw, Palliser, Dyson, Hirst, & Berry, 2016).

Behavioral Development

Guinea pig behavioral development has been extensively studied. Research includes fetal behavior (Bellinger et al., 2015), the effects on offspring of prenatal stress (Kaiser & Sachser, 2005), infant attachment (Hennessy, 2014), development of social behavior (Willis, Levinson, & Buchanan, 1977), and cognition (Allen-Charters, 1904; Brust & Guenther, 2017; Sylvia Kaiser, Hennessy, & Sachser, 2015; Mamczarz et al., 2016).

Behavioral Biology

Natural History

The domesticated guinea pig, *Cavia porcellus*, is not thought to exist in the wild. Related wild caviomorphs are neotropical rodents and include *C. aperea*, *C. tschudii*, *C. magna*, and *C. fulgida*. Although *C. aperea* is frequently studied and referenced as an approximation of the wild version of *porcellus*, recent molecular and anatomical investigations have also suggested *C. tschudii* is closely related to the domesticated guinea pig, and that its ancestor is the common ancestor to all *Cavia* (Dunnum & Salazar-Bravo, 2010; Spotorno et al., 2007, 2006; Weir, 1974).

Ecology

Habitat/Habitat use. Wild relatives of the domesticated guinea pig are found across most of South America, except the forested areas of the Amazon (Huckinghaus, 1961). The Brazilian guinea pig, *C. aperea*, is the most frequently studied of the wild varieties of Cavia. It ranges from Northern Argentina across Uruguay and Paraguay to Southern Brazil (Mares & Ojeda, 1983). Crepuscular herbivores, wild guinea pigs, live in the grassy plains, foraging in open areas during early morning and evening on sunny days. On overcast days, however, they may be more active throughout the day (Rood, 1972). The home range size varies greatly by species, from 500 to 1300 m² for *C. aperea* to over 11,000 m² for the greater guinea pig, *C. magna*. For most guinea pigs in the genus *Cavia*, females have an overlapping, but smaller home range than males (Adrian & Sachser, 2011). Although they are not burrowing animals, wild guinea pigs do inhabit burrows dug by other animals. Likewise, they may take advantage of natural rock crevices or vegetation shelter, such as shrubs or tall reeds (Asher et al., 2004; Castle & Wright, 1916; Rood, 1969). Wild guinea pigs can adapt to a wide range of environments. They have been found at elevations ranging from sea level to 4,000 m and can tolerate temperatures from 22°C in the daytime to −7°C at night (Cassini & Galante, 1992; Sandweiss & Wing, 1997).

Predators/Antipredator behavior. Natural predators of the guinea pig include foxes, coyotes, wolves, domestic dogs, domestic cats, ferrets, hawks, raptors, owls, snakes, and humans (Asher et al., 2004; Cassini & Galante, 1992; Morales, 1994). Predation is severe during spring and summer; one study noted a 22% loss of the adult population across this period (Asher et al., 2008). Wild guinea pigs evade predators through four methods: (1) camouflage, (2) freezing response,

FIGURE 8.2 Brazilian guinea pig *Cavia aperea*. Photo taken by Vince Smith in 2006 at Iguazú National Park, Argentina. CC BY-SA 2.0 https://creativecommons.org/licenses/by-sa/2.0/.

(3) fleeing, and (4) alarm calls. Wild guinea pigs are naturally camouflaged with a brown ticked coat of fur that allows them to blend in with their surroundings (Figure 8.2). Alternatively, hiding in tall vegetation or within burrows can also camouflage the guinea pig from predators (Asher et al., 2004; Castle & Wright, 1916; Rood, 1969). When a perceived threat is identified by individual guinea pigs, they exhibit a freezing response that is a form of hiding in the open. During this response to a loud noise or threat, guinea pigs freeze in a trance-like state, becoming vigilant, immobile, and unresponsive to external stimuli. This freezing response may last for a period of seconds, or up to several hours, and is usually exhibited when the perceived threat is quite distant (Bayard, 1957; Ratner, 1967). In contrast, individuals may react to nearby threats by fleeing. When the entire herd (group) reacts in this manner it often creates a stampede. During a stampede, individual members of the group turn in seemingly random directions and locomote at a maximum speed of 4.12 m/s (Trillmich, Bieneck, Geissler, & Bischof, 2003). This chaotic behavior may confuse predators, thus allowing escape of many, if not all, of the herd (King, 1956). Similarly, shrill alarm calls may act to confuse or even intimidate a predator (Berryman, 1976; Monticelli & Ades, 2013).

Social Organization and Behavior

The wild guinea pig, *C. aperea*, travels in pairs or small polygynous harems consisting of one male and one or more females, along with their offspring (Asher et al., 2008). Observations of other Caviomorphs, such as the *Microcavia* (mountain cavies), *C. magna* (the greater guinea pig), and some of the genus *Galea* (yellow-toothed cavies), reveal a more promiscuous social organization (Keil, Epplen, & Sachser, 1999; Kraus, Kuenkele, & Trillmich, 2002; Rood, 1969, 1972; Rood & Weir, 1970). However, it is thought that some species, perhaps *G. spixii* or *G. monasteriensis*, are monogamous (Hohoff et al., 2002; Trillmich et al., 2004). Large adult males in the wild create a linear dominance hierarchy and may display much greater aggression toward other males than domesticated *C. porcellus*. Aggression is often displayed by wild or domesticated guinea pigs through vocalizations, teeth chattering, piloerection, head thrusting, leaping attacks, hip swaying, nonsexual mounting, and biting of the ears or hair (Avery, 1925). Intermediate and smaller, nondominant offspring males roam the perimeter of the harem range (Asher et al., 2004; King, 1956). When the herd population increases to 10–15 in size, satellite groups form in the periphery of the range. These satellite harem groups consist of a single subordinate male and one or more females (Sachser, 1998). Also subordinate to the dominant male, females have a dominance hierarchy of their own (Asher et al., 2004). Their young typically follow the adults about, as the harem timidly ventures out from cover in single file formation (King, 1956; Rood, 1972).

Although they regularly self-groom, social grooming and investigation are not extensive in those wild cavies that form stable, closed groups. However, guinea pigs do seek close contact with one another when not in open areas of the habitat, and display social greetings, such as nose-to-nose contact and anogenital investigation (King, 1956; Rood, 1972).

Mating. Most wild guinea pigs of the genus *Cavia* are sexually dimorphic, with dominant males of *C. aperea* (>500 g) significantly larger than females (350–450 g). Although not different in size from females, nondominant satellite males (300–375 g) and individual males roaming on the range periphery (450 g) are also significantly smaller than the dominant males. Males are not territorial and do not defend or scent-mark the home range. However, they do scent-mark females and show aggression toward other males approaching harem females (Asher et al., 2004, 2008). Males maintain proximity to females during foraging and guard them during pregnancy. This strategy allows the dominant male to be the first to copulate with the female during postpartum estrus. Sexual behavior and courtship occurs between members of the same harem or social group, and is considered a female-defense polygynous mating system. Although nondominant males do copulate with females from the dominant male harem about 50% of the time, these matings result in only 13%–27% of the pregnancies. These pregnancy rates may reflect copulation differences, because dominant males mount females several times before ejaculation, while nondominant males ejaculate quickly during the first intromission (Adrian & Sachser, 2011).

Mating in guinea pigs requires consent of the female to complete copulation. In order to induce receptivity in the female, males demonstrate a form of courtship involving a rumble purr sound and the rhythmical swaying of the hips (rumblestrut). This behavior is exaggerated by the male when females are unreceptive. During mating, receptive females will allow investigations by the male, whereas nonreceptive females may bite, kick, or spray urine at the approaching male. Receptive females may show lordosis, rumblestrut, assume a wide stance of the hindlimbs, raise the pudendum, and show perineal dilation. Some or all of these behaviors may be observed throughout estrus in the female. The refractory period of the male, after copulation, is 1 h (Avery, 1925; Rood & Weir, 1970; Young, 1969). Mating typically takes place at night over a period of 1–2 h with 3–6 copulations (Hribal, Jewgenow, & Schumann, 2013).

Reproduction. Guinea pig females of *C. aperea* reach sexual maturity in the wild between 3 and 7 weeks of age, with earlier maturity during the longer days of summer months, and later ages apparent during the shorter days of winter (Rübensam, Hribal, Jewgenow, & Guenther, 2015; Trillmich, Laurien-Kehnen, Adrian, & Linke, 2006; Trillmich, Mueller, Kaiser, & Krause, 2009). This early onset of sexual maturity soon after weaning is still within the juvenile period of substantial growth (Rood & Weir, 1970). In the laboratory, *C. porcellus* are typically not bred until 10–12 weeks, when the females are fully grown (Richardson, 2000). Male *C. aperea* reach maturity between 55 and 80 days of age, as evidenced by descent of the testes and sperm motility (Trillmich et al., 2006).

Mature females of *C. aperea* have an estrus cycle of approximately 20.6 days, with a gestational length of 60.9 days, compared to a 16-day estrus and 70-day gestation for the domesticated *C. porcellus*. However, the gestational length of both species seems to be negatively associated with litter size, with a range of 59–64 days for *C. aperea* (Rood & Weir, 1970). Estrus is easily determined in guinea pigs by inspection of the vaginal membrane. During estrus, the vaginal membrane opens for 2–4 days. Vaginal swelling is markedly present, along with a progression of discharge that ranges from thick mucus, to thin fluid, and finally a slight bloody discharge lasting only 1–2 h (Stockard & Papanicolaou, 1917). Behavioral changes in the females during estrus are marked, with increased rates of locomotion, social sniffing, companion approach, anogenital investigation, self-grooming, and rumblestrut that are twice those typically observed in nonestrus females (Birke, 1981). Guinea pigs, like humans, are cyclical spontaneous ovulators, with ovulation occurring on the last day of estrus (Reed, Burton, & Van Diest, 1979). Ovulation can also be experimentally induced by the administration of luteinizing hormone, follicle-stimulating hormone, or guinea-pig pituitary gonadotrophins (Reed & Hounslow, 1971). However, the group synchrony of estrus that is observed in other rodents (the Lee-Boot effect) has not been achieved in laboratory breeding nor reported in wild colonies of guinea pigs (Harned & Casida, 1972; Rood & Weir, 1970).

Wild guinea pigs typically bare 3–4 L of 1–5 offspring (2.1 average) per year (Asher et al., 2004; Kraus, Trillmich, & Künkele, 2005; Trillmich et al., 2006, 2009). Although they breed throughout the year, the peak breeding seasons for wild guinea pigs are spring and summer (September through April in South America), with larger litters of offspring born in the summer months compared to winter (Rübensam et al., 2015). Depending on availability of food, smaller litters are also observed in lean seasons (Trillmich, 2000). Both wild and domesticated guinea pig females have postpartum estrus that occurs within 6–48 h of parturition (Rood & Weir, 1970).

Pregnancy complications in wild guinea pigs have not been well studied. As discussed in the section on Typical Research Application, domesticated guinea pigs share a number of pregnancy complications in common with humans. One of the most common is toxemia (hypertension), which occurs more frequently in hot climates. Signs of toxemia include anorexia, lethargy, ketone production (sweet or fruity-smelling breath), salivation, and seizure in advanced cases. Dystocia, or difficult birth, is also a potential problem during parturition in females that have not been bred before the age of 7 months. Between 6 and 10 months of age, the pubic symphysis may fuse due to calcification. If a female has not delivered a litter prior to this age, then dystocia may result, leading to death of the pregnant female during parturition (Richardson, 2000).

Parental behavior. Guinea pig offspring are born precocial, with well-developed sensory systems and adult-like locomotion within hours of birth. Although mothers will not wean the pups until 21–28 days of age, the pups are capable of consuming solid food within 3 days of birth, and nutritional independence prior to 2 weeks of age. This precocial state of the offspring is adaptive in permitting movement of the herd to new foraging locations, should food become scarce or the need to evade predators arises (Rood, 1972; Sachser, 1986).

Paternal care begins prior to birth with the guarding of pregnant females by males (Adrian & Sachser, 2011). After birth of the offspring, males tend to ignore separation distress calls from pups (Pettijohn, 1977). Similar to the domesticated *C. porcellus*, however, wild guinea pig fathers of *C. aperea* and the monogamous *G. monasteriensis* engage in frequent play behavior and grooming of offspring (Adrian, Brockmann, Hohoff, & Sachser, 2005; Hennessy, Bullinger, Neisen, Kaiser, & Sachser, 2006). These prosocial interactions are particularly important in shaping the behavior of juvenile males at the time of puberty. For example, male offspring reared in the presence of dominant herd males learn appropriate subdominant behaviors. This prosocialization of the juvenile male is necessary to avoid its use of overt aggression when introduced to other males. Under normal circumstances, adult males quickly establish dominance hierarchies without overt aggression, which might attract the attention of predators (Sachser, 1998; Sachser, Hennessy, & Kaiser, 2018).

Due to the precocial state of guinea pig offspring, minimal maternal investment is required. Maternal care in guinea pigs begins with the licking and grooming of newly delivered offspring. Other adult females often participate in this birth activity, as orphaned offspring are readily accepted and nursed by other females (Albers, Timmermans, & Vossen, 1999a). During the first few weeks, mothers also engage in licking of the anogenital region to aid pup urination and defecation. Pups typically present themselves for anogenital licking by exhibiting a filial form of lordosis that includes flattening of the back along with perineal dilation (Harper, 1972). Unlike other rodents, such as rats and mice, this maternal behavior of anogenital licking is not preferentially directed toward male offspring. Rather, male pups are attended to less frequently than females after 2 weeks of age. This reduction in maternal care may be due to identification of increased androgen and pheromone production from the male pup perineal glands after 10 days of age when testosterone levels begin to rise (Albers, Timmermans, & Vossen, 1999b; Moore, 1982; Rood & Weir, 1970).

Nursing is a fairly passive maternal endeavor in guinea pigs. Although guinea pig mothers do use the standard crouch nursing posture reported for other rodents, passive lifting of the maternal ventrum by pups is more frequent. Passive lifting describes the behavior observed as pups crawl under the mother to search for a nipple, thereby lifting the abdomen of the mother. During this process, the mothers do not accommodate the pups, but passively permit the pup to lift her ventrum and suckle (Hennessy & Jenkins, 1994). Studies in both wild and domesticated guinea pigs report little change in milk supply with increasing demand from pups. Neither varying the food supply or the litter size alters milk production from the female. It is likely that increased suckling demands result in greater social connection between the mother and pup, rather than merely providing sustenance (Fey & Trillmich, 2008; Laurien-Kehnen & Trillmich, 2002; Rehling & Trillmich, 2008).

During the preweaning period, guinea pig mothers exhibit individual differences in the pup-directed behavior they exhibit. Although consistent from litter to litter, differences in mothering styles also depend on the size of the litter, availability of the food supply, and whether the mother is also concurrently pregnant (Albers et al., 1999a). For example, females that are pregnant while lactating tend to nurse less frequently and wean the pups earlier than females that are not simultaneously pregnant and nursing (Rehling & Trillmich, 2008). These females also tend to ignore the distress separation calls of nursing offspring (Naguib, Kober, & Trillmich, 2010; Rood, 1972). In times of restricted maternal food supply, however, females nurse longer before weaning (Claudia Laurien-Kehnen & Trillmich, 2004). Together, these behaviors suggest that mothers nurse and wean according to their own nutritional needs, and that although maternal nursing behavior varies, lactation and milk production in the guinea pig do not adapt to the needs of the offspring (Claudia Laurien-Kehnen & Trillmich, 2003, 2004). Weaning is initiated by the mother after 2–3 weeks of lactation and is marked by avoidance of and aggression toward the weanling pups (Rehling & Trillmich, 2007). Offspring will remain in the harem until after maturity, at roughly 9–10 weeks of age, when males will form their own harem or be relegated to satellite roaming (Sachser, 1998; Trillmich et al., 2006).

Although guinea pig mothers are not particularly responsive to the nutritional demands of their offspring, likely due to the precocial development of the pup, they do respond to the social needs of proximity for the first few weeks after birth. When not saddled with the extra demands of concurrent lactation and pregnancy, guinea pig females readily identify and respond to the separation distress call of herd pups. Mothers preferentially respond to the calls of their own pups and can distinguish them from the calls of others. Moreover, mothers of larger litters are observed to respond strongly to separation distress, presumably to reunite the larger litter quickly, thus avoiding predator detection (Kober, Trillmich, & Naguib, 2007, 2008; Pettijohn, 1977).

Almost all maternal behaviors, except anogenital licking, appear to be initiated by the pups. From birth, guinea pig pups have a strong approach response toward the mother, despite her minimal investment in her offspring (Gaston, Stout, & Tom, 1969). This approach response is so strong that separation of the pup from the mother during the first weeks of life results in separation distress and elevated levels of cortisol in the pup (Pettijohn, 1978), even in the presence of other littermates (Ritchey & Hennessy, 1987). However, not all circumstances and environments trigger this response in pups. For example, pups who remain in familiar environments while separated for short periods from their mother do not exhibit separation distress. Likewise, pups kept in a dimly lit chamber while separated, as opposed to a brightly lit room, do not respond with separation distress. Finally, the presence of other females can also reduce the cortisol response in separated pups, but not to the degree that the mother does (Hennessy, 2003, 2014; Hennessy & Sharp, 1990; Ritchey & Hennessy, 1987).

Feeding behavior. Natural feeding behavior for guinea pigs occurs throughout the day, with increased activity near dawn and dusk (Rood, 1969, 1972). However, in the laboratory, guinea pigs have been observed feeding during the night as well (King, 1956). Unlike other rodents, guinea pigs do not manipulate food with their paws, but rather use the mouth and head exclusively during feeding (Whishaw, Sarna, & Pellis, 1998). Wild guinea pigs feed mainly on Gramineae plants (grasses) by grasping stalks in their mouths, near the roots, and pulling sharply upward (Guichon & Cassini, 1998; Rood, 1972). The grinding motion of mastication acts to not only reduce plant material for digestion but also decreases tooth length. This reduction is beneficial because, like most rodents, the teeth of guinea pigs grow continuously (Richardson, 2000).

Guinea pigs have a demonstrated preference for sweet food and water, presumably because of their need for dietary vitamin C (Jacobs & Beauchamp, 1977). However, they do not necessarily avoid bitter substances entirely but exhibit conditioned taste avoidance to natural toxicants, such as common nightshade, *Solanum dulcamora* (Jacobs & Labows, 1979; Nolte, Mason, & Lewis, 1994). Regardless of the diet, guinea pigs engage in coprophagy, an essential supplement to the diet, in a typical diurnal pattern. Special fecal pellets, called cecotropes, are consumed directly from the anus, and only rarely from the ground. These cecal pellets contain recycled B vitamins and digestive bacteria (Bjornhag & Sjoblom, 1977; King, 1956; Reid, 1959).

As would be expected in an herbivore that feeds throughout the day, food restriction is not well tolerated. In the laboratory, when guinea pigs were allowed to feed for only 2 h per day, they reduced their total food intake by 70%. Even after 3 weeks, the guinea pigs had not acclimated to reduced feeding time, suggesting the amount of food consumed is related to the amount of time allowed for consumption (Kutscher, 1969). Similarly, in studies of food deprivation, guinea pigs did not survive a 30% reduction in body weight (Campbell, Smith, Misanin, & Jaynes, 1966).

Water consumption is related to food consumption, with wild guinea pigs obtaining nearly all fluid requirements from fresh greens. Laboratory animals, however, require water supplementation in inverse proportion to the amount of green vegetation provided (Dutch & Brown, 1968; Guichon & Cassini, 1998). Drinking behavior also varies depending on the water source. When drinking from a flat surface, such as a natural water source, guinea pigs tilt back the head to swallow the water (Rood, 1969). Similar to food restriction, constraints on access to water result in weight loss from a corresponding reduction in food consumption (Dutch & Brown, 1968).

Communication. For a rodent, guinea pigs have a well-organized social structure, with a rich system of communication. Both wild and domesticated *Cavia* share similarities in communication systems, suggesting that most postures, vocalizations, and scent marking are unlikely to be the result of domestication (Monticelli & Ades, 2013).

Social postures, other than those involved in mating courtship or parental care, include both prosocial and aggressive behaviors. The most common prosocial behaviors are investigations, where the guinea pig extends the nose toward the anogenital region or the nose of a conspecific. When two guinea pigs stand nose-to-nose, they are often observed side-by-side in this posture for an extended period of time. Similarly, members of a herd may briefly touch sides as they pass by each other while foraging (Grant & Mackintosh, 1963; Rood, 1972). Aggressive postures include the head thrust, and a dominance stance, where the head is raised and the forelimbs are extended while crouching. Teeth chattering may accompany either posture, which may accelerate to circling behaviors and overt aggression, such as leaping and biting, if not resolved through the initial posturing (Avery, 1925; King, 1956).

Approximately 11 distinct vocalizations have been identified in both wild and domesticated guinea pigs. The most common vocalization is the chutt, observed mainly during exploration and foraging, but also emitted by mother and pups when locomoting together. Purring occurs during courtship or when pups approach the mother, and is often accompanied by swaying of the hips (rumba). The drrr is an alarm call that is a lower toned version similar to the purr, and is issued when a strange or abrupt sound is heard. Individuals will orient to the sound, emit the drrr, and then freeze or run. Extended freezing often is punctuated by intermittent drrrs. Tweets are emitted only by pups during anogenital stimulation and are barely audible. The remaining vocalizations (low whistle, whistle, squeal, scream, and chirrup) are voiced during threat or injury (Berryman, 1976; Monticelli & Ades, 2011, 2013). Wild and domesticated guinea pigs use calls under similar situations, with the exception of the food-anticipation whistle or wheek, which is identical to the separation whistle in young guinea pigs. This vocalization is emitted only by domesticated guinea pigs in anticipation of caregivers bringing food, and is different from the foraging call of the wild guinea pig, which occurs only after food has been located. It has recently been determined that the food-anticipation whistle is nearly identical to the distress whistle of pups during separation from the mother. It is thought that the anticipation of food represents a mild stressor in domesticated guinea pigs that must rely on caregivers, rather than foraging on their own to feed (Corat et al., 2012).

Scent marking communication of familiar territory by dragging the perineum across the ground or moving the rump side-to-side is common in guinea pigs. Both males and females engage in this behavior, although it is observed more frequently in males. Urine also is used to communicate, primarily during mating. However, the purpose of urine spraying appears to be the opposite for males and females. Males tend to mark females with urine during courtship, while females use a stream of urine to distract an unwanted male and avoid the mating ritual (Grant & Mackintosh, 1963; Rood, 1972).

Common Captive Behaviors

Domesticated guinea pigs are more docile and timid than their wild relatives, but otherwise engage in similar behaviors (Lewejohann et al., 2010; Sachser, 1998). Observed behavioral differences between *C. porcella* and wild guinea pigs are mainly the result of environmental differences, such as laboratory caging and *ad libitum* pellet diets.

Normal

Laboratory guinea pigs are crepuscular, like wild *Cavia*, being more active at dawn and dusk (Rood, 1972). Activity during the night is only slightly greater (89% of time) than that observed during the day (86% of time) and is punctuated by rest periods of 3–10 min in length. However, activity is also temperature-dependent, with activity levels dropping as ambient temperatures rise. When ambient temperatures are between 26°C and 29°C, guinea pigs reduce activity by half (Nicholls, 1922). Pregnancy and lactation will also alter activity, with mothers and pups displaying a more nocturnal pattern (Schiml & Hennessy, 1990). Domesticated guinea pigs also feed and are active throughout the day and night, with higher frequency than that seen in wild *Cavia*, perhaps due to the elimination of predation risk (King, 1956). However, when placed in a novel environment, domesticated guinea pigs are more timid and hesitant to explore than their wild relatives. *C. porcellus* engages in frequent and prolonged hiding behavior and napping, if provided a hutch or similar opportunities to sequester (Lewejohann et al., 2010). Typical movement in the home cage or testing arena will be thigmotaxic, around the exteriors and avoiding the central areas. Often mothers will move around the cage in this manner while vocalizing chutt sounds, with pups following in kind behind the mother (Berryman, 1976). Young pups also can be observed hopping, a behavior similar to the ferret war dance, and often referred to as popcorning by fanciers (King, 1956; Rood, 1972).

Social behaviors observed will mainly depend on the social housing arrangement; however, they do not differ significantly from those displayed by wild guinea pigs. *C. porcellus* is highly social and will spend a significant amount of time in contact with cage mates. Separation from conspecifics will elicit the separation cry (whistle or wheek vocalization) even in adults (Berryman, 1976). Investigation through nose-to-nose and anogenital nuzzling commonly occurs when reunited or introduced to new members of the herd.

Similar to wild guinea pigs, domesticated laboratory species create separate male and female dominance hierarchies. The creation and maintenance of these social stratifications often lead to aggression in the form of vocalizations (rumble call), chasing, aggressive stances, hind kicking, head tosses, teeth chattering, and biting (Avery, 1925; Sachser, 1998). However, dominance hierarchy in the colony is mainly determined through the same social cues (dominance postures and head thrusts) as in the wild (King, 1956; Rood, 1972).

Abnormal

Abnormal laboratory behaviors of the guinea pig include antipredation responses, symptoms of illness, distress vocalizations, aggression, and cage biting. Some of these behaviors (e.g., antipredation) would be considered normal when observing *C. aperea* in a natural habitat. However, for laboratory guinea pigs, where the environment is highly regulated, and food is provided *ad libitum*, these same behaviors become abnormal in terms of the ecology.

Antipredation. Loud sudden noises can easily startle guinea pigs, which will react by stampeding or freezing. Injury or death may result from stampeding, and events eliciting this behavior should be avoided. Freezing behavior can last for a protracted period of time, ranging from seconds to several minutes, and even longer. Tonic immobility from restraint can last over 2 h. Because both freezing responses and tonic immobility are reactions to perceived threats, increases in these behaviors suggest a high-stress environment (Bayard, 1957; King, 1956; Ratner, 1967).

Illness. Some abnormal behaviors may indicate illness. For example, head-tilting or circling behavior during activity may indicate infection of the auditory or vestibular systems (Richardson, 2000). Another common behavioral sign of illness is a hunched posture, which can occur with or without piloerection of the fur. Hunched posture is a nonspecific symptom and may indicate any one of many maladies including illness, pain, and separation sickness in preweanling pups (Hennessy, 2014). Similarly, distress vocalizations can indicate stressful situations (e.g., aggression, pain, social separation) that should be remedied quickly, in order to avoid increased stress and elevated cortisol.

Aggression. When solitary food hoppers are used for group feeding, crowding can occur. The most typical abnormal behavior observed is nudging of others to the side through lateral movement of the body. This aggressive behavior, although subtle in guinea pigs, is used to gain exclusive access to the food hopper by the aggressor. In some situations (e.g., food or water disruption or deprivation), this crowding can lead to aggressive teeth chattering and even biting (King, 1956).

Other common forms of aggression in guinea pigs usually are related to dominance hierarchies. Mild forms of aggression, such as rumble calls, chasing, kicks, and head tosses, are of little danger to the recipient and can be considered a normal part of dominance hierarchy establishment and maintenance. However, if aggression escalates to teeth chattering or biting, the animals should be separated to avoid continued aggression. Among domesticated guinea pigs, this degree of aggression is seldom seen if young guinea pigs are properly socialized by older dominant animals in the herd (Sachser et al., 2018).

Cage biting. Cage biting is a common behavior in laboratory rodents. Like other rodents, guinea pigs have teeth that are constantly growing. Consequently, they must chew in order to reduce the growing tooth length. Standard pelleted diets may not provide enough chewing action, and guinea pigs may resort to biting or chewing on caging material. When provided with alternative choices for chewing, such as wood blocks, guinea pigs typically refrain from cage biting.

Another related behavior that is often seen in domesticated guinea pigs is water bottle manipulation. During water bottle manipulation, the guinea pig grasps the tube of the water bottle and jerks its head back sharply (Guichon & Cassini, 1998). This behavior is repeated in quick succession, often causing an excess of water to be emptied into the cage. One common solution is to withdraw the water bottle until only the tip can be accessed within the cage. This method has been used successfully to teach guinea pigs to lick the water tube to obtain water

(Alvord, Cheney, & Daley, 1971). The guinea pigs might also be placed on an automatic watering system where they must extend the head through a small opening to obtain water. However, in the wild where water bottles and automatic watering systems do not exist, guinea pigs are observed bending down and opening the mouth to bring in water, then tossing the head back to swallow (Rood, 1972). This same behavioral pattern used when confronted with a water bottle suggests that it is a natural behavior that should be facilitated rather than discouraged.

Ways to Maintain Behavioral Health

Captive Considerations to Promote Species-Typical Behavior

There are five basic needs that underlie good behavioral health in the guinea pig. They are as follows: (1) a stable and predictable environment, (2) regular opportunities for social interaction with conspecifics, (3) husbandry and breeding protocols that take into account the ontogeny of social development, (4) dietary enrichment that is tailored to the foraging and neophobic characteristics of guinea pigs, and (5) mindfulness of species specifics during any behavior modification training.

Environment

The ideal environment for guinea pigs is one consistently quiet, with as little activity or intervention on the part of caregivers as possible. Any loud noise or sudden movement by humans is likely to trigger antipredator behavior in guinea pigs. These behaviors, which are considered abnormal in the laboratory environment, include stampeding and the freezing response. Both have negative consequences that are likely to affect experimental data. For example, a prolonged or frequent freezing response may cause a stress response and raise cortisol levels (Bayard, 1957). Likewise, loud noises or sudden movements on the part of caregivers may cause a herd stampede, during which animals may be injured. In situations where small pups are present, there is a risk the pup could die as a result of the trampling behavior of larger guinea pigs (King, 1956). Consequently, the best insurance against these antipredator behaviors is to educate animal care and laboratory staff about restricting loud noises and sudden movements.

Enrichment is an additional consideration in providing the proper environment for guinea pigs. The best enrichment includes aspects of the natural environment *C. porcellus* would normally experience. For example, wild guinea pigs take advantage of burrows or tall vegetation within which to hide. Consequently, a hutch or tunnel placed within the housing area can provide species-relevant enrichment for the laboratory guinea pig (Byrd, Winnicker, & Gaskill, 2016). Other species-specific considerations for enrichment are social interaction and feeding behaviors.

Social Groupings

Guinea pigs are highly social animals. As such, they require group housing, ideally replicating the natural harem arrangement of one male and several females, along with their offspring (Asher et al., 2008; Sachser, 1998). In the case where timed breeding must be controlled, a single male and two or three females can be housed together through copulation, providing not only control over the timing of the pregnancy but opportunities for social proximity similar to a harem configuration in the wild. This housing arrangement is particularly advantageous for delivering females, because guinea pigs often engage in reciprocal care of the young (Albers et al., 1999a). Males benefit from social group housing as well and do not display the aggression often observed in other rodents, unless they have not been properly socialized with other males during the juvenile period of development (Sachser et al., 2018).

In order to recreate the social dynamics of the herd environment experienced by wild guinea pigs, additional social enrichment can be provided in a testing arena or pen (Brewer, Bellinger, Joshi, & Kleven, 2014). During this form of social enrichment, a herd of 6–10 guinea pigs (3–4 females and their offspring) are placed together in a large (1 m × 2 m) arena supplied with food and water. Harem males can be placed in an adjoining area of the arena, separated by fencing that allows nose-to-nose contact during the social enrichment. Alternatively, adult and adolescent males can be placed in their own arena for enrichment. This type of enrichment is especially important for adolescent males that need to learn species-appropriate behaviors in the context of the group hierarchy (Sachser et al., 2018).

Husbandry

The highly social nature of guinea pigs also suggests species-specific husbandry requirements. For example, because guinea pig pups are born precocial and can fend for themselves nutritionally within a few days, researchers often ignore the social needs of the pups and wean them prematurely from the mother, often as early as 7 days after birth. However physically developed, the pups are not socially mature until 28 days of age, when the mother would typically wean them. Young guinea pig pups require social bonding with the mother, similar to infant attachment to the maternal caregiver in humans (Bowlby, 1953; Tinbergen, 1951). Researchers have documented the resulting lifespan changes in behavioral, immune system, and Hypothalamic-Pituitary-Adrenal axis functioning in guinea pigs separated from the mother before the normal weaning period of 28 days (Hennessy, 2003, 2014). For example, adult precocial rodents weaned at 28 days were significantly more immunocompetent than those weaned at 21 or 14 days, as measured by increased leukocyte concentrations and response to phytohemagglutinin challenge (Dlugosz et al., 2014). Reproductive changes have also been detected, with higher levels of testosterone and mating differences in males separated from the mother later (Hennessy, 2003). Consequently, guinea pig offspring should be weaned as naturally as possible, preferably at around 28 days after birth.

Likewise, researchers often take advantage of postpartum estrus for timed pregnancies, placing the male in with the female immediately after delivery of the prior pregnancy (Rowlands, 1949). Although this practice maximizes the number of pregnancies that can be produced within a specified time period (e.g., 5 per year), it does not allow time for the female to recover from the nutritional and physiological stressors of the pregnancy. A better plan, one that improves the health of the female and her offspring, is to allow a recovery period of 1–2 months after weaning of the litter (Richardson, 2000). Under this breeding strategy, females will produce 2–3 L per year.

Although beneficial to the health of the female, allowing recovery time after delivery is compounded when using timed mating procedures. With other rodents, such as rats and mice, timed mating is often achieved by placing the male and female together for a 24-h period, thus ensuring the timing of conception. However, guinea pig females require sufficient time for socialization with the male, often one or two estrus cycles, thereby adding an additional month to the time needed. If a socialization period is not provided, the female will likely refuse to mate, even though she may be in estrus (Avery, 1925; Rood & Weir, 1970). One solution is to place the male in with the female well in advance of estrus and date the pregnancy by inspecting the vaginal membrane opening daily in the female (Bellinger et al., 2015). Intromission and conception typically take place the first day of a fully open vaginal membrane. When using this system with females housed 2–3 to a cage, the male can be left in with the harem until all females have passed through estrus. Removal of the male at this time may be desired to prevent stress to the female, if the male continues to pursue her and attempt to copulate. Males give little attention to newly delivered offspring and rarely show aggression, so removal of the male is not necessary to protect offspring. Mature adult males are necessary, however, for the proper socialization of juvenile males. Consequently, social interaction between male adults and pups should be provided (Sachser, Hennessy, & Kaiser, 2018).

Feeding Strategies

Neophobia in guinea pigs is exemplified by their pattern of food choices. When presented with a novel food item, guinea pigs may sample the new item, but will not feed on the substance until several days of exposure have passed. Researchers have reported that if pups are not exposed to a range of foods, then as adults, they may refuse new items, no matter how often these novel foods are presented (Brown, Cook, Lane-Petter, Porter, & Tuffery, 1960; Quesenberry, Donnelly, & Mans, 2012). However, two foods are readily accepted by all guinea pigs, regardless of age: fresh hay (e.g., Timothy hay) and high water content leafy vegetables (e.g., Romaine lettuce). In fact, these foods are so readily accepted that they make excellent reinforcement during operant conditioning (Harder, Brock, & Rashotte, 1978).

Although the standard laboratory pelleted feed for guinea pigs is nutritionally balanced and typically contains large amounts of hay in the form of alfalfa, this food source does not provide opportunity for the same foraging behaviors that guinea pigs in the wild experience. For example, the behaviors involved in foraging for grasses in the wild include turning the head sideways, grasping the stalks near the soil, and ripping them out of the ground by jerking the head upward and back. This behavior is similar to that observed when guinea pigs drink from water bottles fitted with sipping tubes, and may satisfy an instinctual need (Guichon & Cassini, 1998; Rood, 1972).

Similarly, when consuming long grasses, guinea pigs draw them into the mouth for chewing a segment at a time, until the entire length of the grass stalk has been consumed. Enrichment with fresh hay will provide opportunity for this feeding behavior, when the bulk of the nutrition is already provided through the standard pelleted diet. One additional benefit of providing hay to supplement pellet diets is that it only minimally increases caloric intake, reducing the trend toward obesity that occurs in many laboratory animals fed *ad libitum* (Richardson, 2000). Alternatively, opportunities to engage in the types of exercise typically associated with foraging should help reduce obesity. For example, guinea pigs can be placed daily in social enrichment arenas that supply plentiful hay in two or more locations. Guinea pigs can move back and forth between the various locations to consume hay, thereby gaining much needed exercise (Brewer et al., 2014).

Similar behavioral considerations are important in maintaining proper hydration in laboratory guinea pigs. When using a water dish in the laboratory, it must be affixed or heavy enough to support the weight of the guinea pig, because they will stand on the dish while drinking. However, when confronted with a water bottle affixed with a sipping tube, guinea pigs place their mouth on the tube and jerk with a backwards toss of the head (Alvord et al., 1971). This behavior is somewhat reminiscent of the motion used to pull fresh grasses for consumption, suggesting that what is often mistaken for play behavior at the drinking station may be compensation for lack of naturalistic feeding behavior due to the standard pellet diet (Balsiger, Clauss, Liesegang, Dobenecker, & Hatt, 2017).

Training

Some researchers have incorrectly concluded that guinea pigs were not suitable for behavioral tests involving conditioning. Unfortunately, these researchers were attempting to train guinea pigs as though they were rats or mice. When presented with species-relevant reinforcement, guinea pigs perform species-appropriate behaviors and are no more difficult to condition than any other rodent (Jonson et al., 1975). For example, unlike rats or mice, guinea pigs do not use their paws to manipulate objects or food (Whishaw et al., 1998). Consequently, requiring a guinea pig to lever press to obtain reinforcement is futile, because the behavior is not appropriate for this species. Success with lever pressing for reinforcement has been achieved; however, by elevating and flipping the lever so that the guinea pig could use the natural head high behavior to raise the lever with a head toss, rather than depress it with the paws (Jonson et al., 1975). Likewise, guinea pigs are herbivores and require a constant flow of food through the digestive system. Food deprivation, which is used extensively to motivate rats and mice during conditioning, should not be attempted in guinea pigs. Rather, the natural proclivity toward constant consumption in the guinea pig should be sufficient motivation during training. Consistent success can be achieved if lettuce or hay is used as reinforcement, particularly if these foods are not provided *ad libitum* during nonconditioning situations (Richardson, 2000).

Behavioral Biology of Guinea Pigs

Special Situations

Just as there are few differences between species of guinea pigs (e.g., wild *C. aperea* versus domesticated *C. porcellus*), there are few differences between the various strains. One exception, however, is the hairless strain of domesticated guinea pig (Figure 8.3).

Hairless Guinea Pigs

Hairless guinea pigs have been derived from several sources over the years. The most common laboratory strain, however, has been the Institute Armand Frappier (IAF) hairless guinea pig Crl:HA-*Hr*^{hr} (Charles River). The IAF hairless guinea pig is an outbred strain with a recessive mutation that occurred spontaneously in a colony at the Armand Frappier Institute in Montreal. Unlike most other nude rodents, hairless guinea pigs are euthymic and have a functioning immune system (Balk, 1987). This animal model has been utilized extensively in dermatology research (Bolognia, Murray, & Pawelek, 1990; Fox, McNichols, Gowda, & Motamedi, 2004; Miyauchi & Horio, 1992; Sueki, Gammal, Kudoh, & Kligman, 2000) and, more recently, to examine behavioral development (Brewer et al., 2014; Kleven & Joshi, 2016), including that of the developing fetus (Bellinger et al., 2015).

Special considerations for this strain include a proper housing environment with sufficient humidity (40%–50%) and soft bedding, in order to avoid dryness and irritation of the sensitive skin. Some beddings, such as wood chips, are also a danger to become impacted in the vaginal canal of females during estrus, while the vaginal membrane is open. Impaction is rare in the Hartley strain, however, because of the protective properties of their fur. When humidity levels are low in the environment, nonpetroleum lotions can be used to prevent dry skin in hairless guinea pigs. Although higher ambient temperatures (e.g., 24°C) are often recommended for the hairless guinea pig, elevated temperatures often lead to dry skin and reduced activity. Activity reduction, in turn, contributes to obesity when concurrently fed an *ad libitum* diet. Finally, robustness of the offspring can be maintained by outbreeding to the Hartley strain every few generations (Balk, 1987; Banks, 1989).

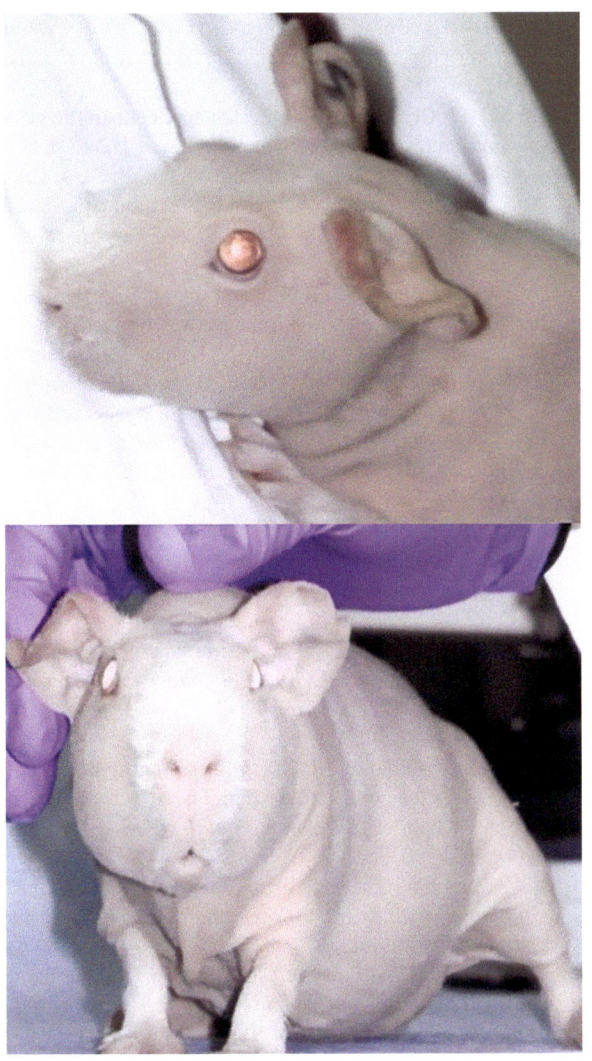

FIGURE 8.3 IAF hairless guinea pig strain Crl:HA-*Hr*^{hr} (Charles River).

Conclusions/Recommendations

Guinea pig is a versatile research model that has been widely utilized in the study of infectious disease, allergy, diabetes, otology, nutrition, pregnancy complications, teratology, and behavioral development. They are highly social animals, with a complex communication system commensurate with their social organization. Being a prey animal, they are sensitive to environmental conditions simulating predation (e.g., loud noises, sudden movement, looming figures). They are also neophobic and require special feeding considerations. Most importantly, enrichment and training/conditioning should be tailored to the species-specific needs and capabilities of the guinea pig.

ADDITIONAL RESOURCES

ARBA (2016). *Standard of Perfection*. Bloomington, IL: American Rabbit Breeders Association, Inc.

Festing, M. F. W. (1979). *Inbred Strains of Guinea-pigs Inbred Strains in Biomedical Research* (pp. 308–312). London: Macmillan Education UK.

Richardson, V. C. G. (2000). *Diseases of Domestic Guinea Pigs / V.C.G. Richardson Library of Veterinary Practice* (Vol. 2nd ed): Malden, MA: Blackwell Science.

Sachser, N., Künzl, C., & Kaiser, S. (2004). The welfare of laboratory guinea pigs. In E. Kaliste (Ed.), *The Welfare of Laboratory Animals* (pp. 181–209). Dordrecht, The Netherlands: Springer.

Terril, L. A., Clemons, D. J., & Suckow, M. A. (1998). *The Laboratory Guinea Pig*. Boca Raton, FL: CRC Press/Taylor Francis.

Wagner, J. E., & Manning, P. J. (Eds.). (1976). *The Biology of the Guinea Pig*. New York: Academic Press.

REFERENCES

Adrian, O., Brockmann, I., Hohoff, C., & Sachser, N. (2005). Paternal behaviour in wild guinea pigs: A comparative study in three closely related species with different social and mating systems. *Journal of Zoology (London), 265*(Part 1), 97–105.

Adrian, O., & Sachser, N. (2011). Diversity of social and mating systems in cavies: A review. *Journal of Mammalogy, 92*, 39–53. doi:10.1644/09-mamm-s-405.1.

Albers, P. C. H., Timmermans, P. J. A., & Vossen, J. M. H. (1999a). Evidence for the existence of mothering styles in guinea pigs (*Cavia aperea f. porcellus*). *Behaviour, 136*(4), 469–479.

Albers, P. C. H., Timmermans, P. J. A., & Vossen, J. M. H. (1999b). Maternal behaviour in the guinea pig (*Cavia aperea f. porcellus*): A comparison of multiparous, and primiparous, and hand reared primiparous mothers. *Netherlands Journal of Zoology, 49*(4), 275–287.

Allen-Charters, J. B. (1904). The associative processes of the guinea pig. *Journal of Neurology and Psychology, 14*(4), 294–361.

Alvord, J., Cheney, C., & Daley, M. (1971). Development and control of licking in the guinea pig (*Cavia porcellus*). *Behavior Research Methods & Instrumentation, 3*(1), 14–15. doi:10.3758/bf03208114.

Anderson, H., & Wedenberg, E. (1965). A new method for hearing tests in the guinea pig. *Acta Oto-Laryngologica, 60*(1–6), 375–393. doi:10.3109/00016486509127023.

Asher, M., de Oliveira, E. S., & Sachser, N. (2004). Social system and spatial organization of wild guinea pigs (*Cavia aperea*) in a natural popoulation. *Journal of Mammalogy, 85*(4), 788–796. doi:10.1644/bns-012.

Asher, M., Lippmann, T., Epplen, J. T., Kraus, C., Trillmich, F., & Sachser, N. (2008). Large males dominate: Ecology, social organization, and mating system of wild cavies, the ancestors of the guinea pig. *Behavioral Ecology and Sociobiology, 62*(9), 1509. doi:10.1007/s00265-008-0580-x.

Avery, G. T. (1925). Notes on reproduction in guinea pigs. *Journal of Comparative Psychology, 5*(5), 373–396.

Balk, M. W. (1987). Emerging models in the U.S.A.: Swine, woodchucks, and the hairless guinea pig. *Progress in Clinical and Biological Research, 229*, 311–326.

Balsiger, A., Clauss, M., Liesegang, A., Dobenecker, B., & Hatt, J. M. (2017). Guinea pig (*Cavia porcellus*) drinking preferences: Do nipple drinkers compensate for behaviourally deficient diets? *Journal of Animal Physiology and Animal Nutrition, 101*(5), 1046–1056. doi:10.1111/jpn.12549.

Banks, R. (1989). The guinea pig: Biology, care, identification, nomenclature, breeding, and genetics. *USAMRID Seminar Series*.

Bayard, J. (1957). The duration of tonic immobility in guinea pigs. *Journal of Comparative and Physiological Psychology, 50*(2), 130–133.

Bellinger, S. A., Lucas, D., & Kleven, G. A. (2015). An ecologically relevant guinea pig model of fetal behavior. *Behavioural Brain Research, 283*, 175–183. doi:10.1016/j.bbr.2015.01.047.

Berche, P. (2012). Louis Pasteur, from crystals of life to vaccination. *Clinical Microbiology and Infection, 18*, 1–6. doi:10.1111/j.1469-0691.2012.03945.x.

Berryman, J. C. (1976). Guinea-pig vocalizations their structure causation and function. *Zeitschrift Fuer Tierpsychologie, 41*(1), 80–106.

Birke, L. I. A. (1981). Some behavioral changes associated with the guinea-pig cavia-porcellus estrous cycle. *Zeitschrift Fuer Tierpsychologie, 55*(1), 79–89.

Bjornhag, G., & Sjoblom, L. (1977). Demonstration of coprophagy in some rodents. *Swedish Journal of Agricultural Research, 7*(2), 105–114.

Bohmer, A. (1988). The Preyer reflex–an easy estimate of hearing function in guinea pigs. *Acta Otolaryngol, 106*(5–6), 368–372.

Bolognia, J. L., Murray, M. S., & Pawelek, J. M. (1990). Hairless pigmented guinea pigs: A new model for the study of mammalian pigmentation. *Pigment Cell Research, 3*(3), 150–156.

Bowlby, J. (1953). *Child Care and the Growth of Love*. London: Penguin Books.

Brendan, K. P., David, F. A., Michael, A. R., James, E. D., Bruce, P., & Randall, J. B. (2017). A model of type 2 diabetes in the guinea pig using sequential diet-induced glucose intolerance and streptozotocin treatment. *Disease Models & Mechanisms, 10*(2), 151–162. doi:10.1242/dmm.025593.

Brewer, J. S., Bellinger, S. A., Joshi, P., & Kleven, G. A. (2014). Enriched open field facilitates exercise and social interaction in two strains of guinea pig, *Cavia porcellus*. *Journal of the American Association for Laboratory Animal Science, 53*(4), 344–355.

Brookes, S. J. (2001). Classes of enteric nerve cells in the guinea-pig small intestine. *The Anatomical Record, 262*(1), 58–70.

Brown, A. M., Cook, M. J., Lane-Petter, W., Porter, G., & Tuffery, A. A. (1960). Influence of nutrition on reproduction in laboratory rodents. *The Proceedings Of The Nutrition Society, 19*, 32–37.

Brust, V., & Guenther, A. (2017). Stability of the guinea pigs personality – cognition – linkage over time. *Behavioural Processes, 134*, 4–11. doi:10.1016/j.beproc.2016.06.009.

Byrd, C. P., Winnicker, C., & Gaskill, B. N. (2016). Instituting dark-colored cover to improve central space use within guinea pig enclosure. *Journal Of Applied Animal Welfare Science: JAAWS, 19*(4), 408–413. doi:10.1080/10888705.2016.1187070.

Campbell, B. A., Smith, N. F., Misanin, J. R., & Jaynes, J. (1966). Species differences in activity during hunger and thirst. *Journal of Comparative and Physiological Psychology, 61*(1), 123–127. doi:10.1037/h0022866.

Canning, B. J., & Chou, Y. (2008). Using guinea pigs in studies relevant to asthma and COPD. *Pulmonary Pharmacology & Therapeutics, 21*(5), 702–720. doi:10.1016/j.pupt.2008.01.004.

Cassini, M. H., & Galante, M. L. (1992). Foraging under predation risk in the wild guinea pig: The effect of vegetation height on habitat utilization. *Annales Zoologici Fennici, 29*(4), 285–290.

Castle, W. E., & Wright, S. (1916). *Studies of Inheritance in Guinea-Pigs and Rats*. Washington, DC: Washington Carnegie Institution.

Clark, S., Hall, Y., & Williams, A. (2014). Animal models of tuberculosis: Guinea pigs. *Cold Spring Harbor Perspectives in Medicine, 5*(5), a018572. doi:10.1101/cshperspect.a018572.

Cohen, B., & Mendel, L. B. (1918). Experimental scurvy of the guinea pig in relation to diet. *Journal of Biological Chemistry, 35*(3), 425–453. doi: https://doi.org/10.1016/S0021-9258(18)86437-7

Corat, C., Tarallo, R., Branco, C. R., Savalli, C., Tokumaru, R. S., Monticelli, P. F., & Ades, C. (2012). The whistles of the guinea pig: An evo-devo proposal. *Revista de Etologia, 11*(1), 46.

Cumberland, C. (1886). *The Guinea Pig or Domestic Cavy for Food, Fur, and Fancy*. New York: L. Upcott Gill.

Dlugosz, E. M., de Bellocq, J. G., Khokhlova, I. S., Degen, A. A., Pinshow, B., & Krasnov, B. R. (2014). Age at weaning, immunocompetence and ectoparasite performance in a precocial desert rodent. *The Journal Of Experimental Biology, 217*(Pt 17), 3078–3084. doi:10.1242/jeb.106005.

Dunnum, J. L., & Salazar-Bravo, J. (2010). Molecular systematics, taxonomy and biogeography of the genus Cavia (Rodentia: Caviidae). *Journal of Zoological Systematics and Evolutionary Research, 48*, 376–388. doi:10.1111/j.1439-0469.2009.00561.x.

Dutch, J., & Brown, L. B. (1968). Adaptation to a water-deprivation schedule in guinea pigs. *Psychological Reports, 23*(3), 737–738.

Eisenberg, J. F. (1989). *Mammals of the Neotropics*. Chicago: University of Chicago Press.

Fey, K., & Trillmich, F. (2008). Sibling competition in guinea pigs (*Cavia aperea f. porcellus*): Scrambling for mother's teats is stressful. *Behavioral Ecology & Sociobiology, 62*(3), 321–329. doi:10.1007/s00265-007-0419-x.

Flynn, L. J. (1986). Baluchimyinae: A new ctenodactyloid rodent subfamily from the Miocene of Baluchistan. In L. J. Flynn, L. L. Jacobs, & I. U. Cheema (Eds.), *American Museum Novitates* (pp. 53–58). New York: American Museum of Natural History.

Fox, M. A., McNichols, R. J., Gowda, A., & Motamedi, M. (2004). The use of the hairless guinea pig in tattoo research. *Contemporary Topics In Laboratory Animal Science, 43*(5), 35–38.

Gad, S. C. (2007). The guinea pig. In S. C. Gad (Ed.), *Animal Models in Toxicology* (2nd ed.). Boca Raton, FL: CRC/Taylor & Francis.

Gaston, M. G., Stout, R., & Tom, R. (1969). Imprinting in guinea pigs. *Psychonomic Science, 16*(1), 53–54.

Grant, E. C., & Mackintosh, J. H. (1963). A comparison of the social postures of some common laboratory rodents. *Behaviour, 21*(3–4), 246–259. doi:10.1163/156853963x00185.

Guichon, M. L., & Cassini, M. H. (1998). Role of diet selection in the use of habitat by pampas cavies *Cavia aperea* pamparum (Mammalia, Rodentia). *Mammalia, 62*(1), 23–35.

Harder, D. B., Brock, O. G., & Rashotte, M. E. (1978). Rapid operant training of guinea-pigs with lettuce as the appetitive reinforcer. *Behavior Research Methods and Instrumentation, 10*(1), 101–102.

Harned, M. A., & Casida, L. E. (1972). Failure to obtain group synchrony of estrus in the guinea-pig. *Journal of Mammalogy, 53*(1), 223–225.

Harper, L. V. (1972). The transition from filial to reproductive function of 'coitus-related' responses in young guinea pigs. *Developmental Psychobiology, 5*(1), 21–34. doi:10.1002/dev.420050104.

Hennessy, M. B. (2003). Enduring maternal influences in a precocial rodent. *Developmental Psychobiology, 42*(3), 225–236.

Hennessy, M. B. (2014). Filial attachment and its disruption: Insights from the guinea pig. *Developmental Psychobiology*. doi:10.1002/dev.21215.

Hennessy, M. B., Bullinger, K. L., Neisen, G., Kaiser, S., & Sachser, N. (2006). Social organization predicts nature of infant-adult interactions in two species of wild guinea pigs (Cavia aperea and Galea monasteriensis). *Journal of Comparative Psychology, 120*(1), 12–18.

Hennessy, M. B., & Jenkins, R. (1994). A descriptive analysis of nursing behavior in the guinea pig (*Cavia porcellus*). *Journal of Comparative Psychology, 108*(1), 23–28. doi:10.1037/0735-7036.108.1.23.

Hennessy, M. B., & Sharp, K. (1990). Voluntary and involuntary maternal separation in guinea-pig pups with mothers required to forage. *Developmental Psychobiology, 23*(8), 783–796.

Herrera, E. A., Alegria, R., Farias, M., Diaz-Lopez, F., Hernandez, C., Uauy, R., … Krause, B. J. (2016). Assessment of in vivo fetal growth and placental vascular function in a novel intrauterine growth restriction model of progressive uterine artery occlusion in guinea pigs. *The Journal of Physiology, 594*(6), 1553–1561. doi:10.1113/jp271467.

Hohoff, C., Solmsdorff, K., Lottker, P., Kemme, K., Epplen, J. T., Cooper, T. G., & Sachser, N. (2002). Monogamy in a new species of wild guinea pigs (*Galea sp.*). *The Science of Nature, 89*(10), 462.

Honeder, C., Ahmadi, N., Kramer, A.-M., Zhu, C., Saidov, N., & Arnoldner, C. (2018). Cochlear implantation in the guinea pig. *Journal of Visualized Experiments, 136*, 56829. doi:10.3791/56829.

Hribal, R., Jewgenow, K., & Schumann, K. (2013). Mating behaviour in wild guinea pigs (*Cavia aperea*) in captivity. *Reproductive Biology, 13*(1), 30.

Huchon, D., & Douzery, E. J. P. (2001). From the old world to the new world: A molecular chronicle of the phylogeny and biogeography of hystricognath rodents. *Molecular Phylogenetics and Evolution, 20*(2), 238–251. doi:10.1006/mpev.2001.0961.

Huckinghaus, F. (1961). ZurNomenklatur und abstammung des hausmeerschweinchens. *Z. Säugetierkd, 26*, 108–111.

Jacobs, W. W., & Beauchamp, G. K. (1977). Glucose preferences in wild and domestic guinea-pigs. *Physiology and Behavior, 18*(3), 491–493.

Jacobs, W. W., & Labows, J. N. (1979). Conditioned aversion bitter taste and the avoidance of natural toxicants in wild guinea-pigs. *Physiology and Behavior, 22*(1), 173–178.

Jonson, K. M., Lyle, J.G., Edwards, M.J., & Penny, R.H.C. (1975). Problems in behavioural research with the guinea pig: A selective review. *Animal Behaviour, 23*, 632–639.

Kaiser, S., Hennessy, M. B., & Sachser, N. (2015). Domestication affects the structure, development and stability of biobehavioural profiles. *Frontiers in Zoology, 12*(S1), S19. doi:10.1186/1742–9994-12-s1-s19.

Kaiser, S., & Sachser, N. (2005). The effects of prenatal social stress on behaviour: Mechanisms and function. *Neuroscience & Biobehavioral Reviews, 29*(2), 283–294. doi:10.1016/j.neubiorev.2004.09.015.

Kaufmann, P., & Davidoff, M. (1977). The guinea-pig placenta. *Advances In Anatomy, Embryology, And Cell Biology, 53*(2), 5–91.

Kawamoto, K., Ishimoto, S., Minoda, R., Brough, D. E., & Raphael, Y. (2003). Math1 gene transfer generates new cochlear hair cells in mature guinea pigs in vivo. *Journal of Neuroscience, 23*, 4395–4400.

Keil, A., Epplen, J. T., & Sachser, N. (1999). Reproductive success of males in the promiscuous-mating yellow-toothed cavy (*Galea musteloides*). *Journal of Mammalogy, 80*(4), 1257. doi:10.2307/1383176.

Kelleher, M. A., Hirst, J. J., & Palliser, H. K. (2013). Changes in neuroactive steroid concentrations after preterm delivery in the guinea pig. *Reproductive Sciences, 20*(11), 1365–1375. doi:10.1177/1933719113485295.

King, J. A. (1956). Social relations of the domestic guinea pig living under semi-natural conditions. *Ecology, 37*(2), 221–228. doi:10.2307/1933134.

Kleven, G. A., & Joshi, P. (2016). Temperature preference in IAF hairless and Hartley guinea pigs (*Cavia porcellus*). *Journal of the American Association for Laboratory Animal Science, 55*(2), 161–167.

Kober, M., Trillmich, F., & Naguib, M. (2007). Vocal mother-pup communication in guinea pigs: Effects of call familiarity and female reproductive state. *Animal Behaviour, 73*(Part 5), 917–925.

Kober, M., Trillmich, F., & Naguib, M. (2008). Vocal mother-offspring communication in guinea pigs: Females adjust maternal responsiveness to litter size. *Frontiers in Zoology, 5*, 13.

Kraus, C., Kuenkele, J., & Trillmich, F. (2002). Spatial organisation and its implications for the social system of *Cavia magna*: An almost asocial cavy? *Zoology (Jena), 105*(Supplement 5), 83.

Kraus, C., Trillmich, F., & Künkele, J. (2005). Reproduction and growth in a precocial small mammal, *Cavia magna*. *Journal of Mammalogy, 86*(4), 763.

Kutscher, C. L. (1969). Species differences in the interaction of feeding and drinking. *Annals of the New York Academy of Sciences, 157*(2), 539.

Lang, C. M., & Munger, B. L. (1976). Diabetes mellitus in the guinea pig. *Diabetes, 25*(5), 434–443.

Laurien-Kehnen, C., & Trillmich, F. (2002). Milk production in guinea pigs does not adjust to experimental manipulation of pup demands. *Zoology (Jena), 105*(Supplement 5), 6.

Laurien-Kehnen, C., & Trillmich, F. (2003). Lactation performance of guinea pigs (*Cavia porcellus*) does not respond to experimental manipulation of pup demands. *Behavioral Ecology and Sociobiology, 53*(3), 145–152.

Laurien-Kehnen, C., & Trillmich, F. (2004). Maternal food restriction delays weaning in the guinea pig, *Cavia porcellus*. *Animal Behaviour, 68*(Part 2), 303–312.

Lavocat, R. (1974). What is an histricomorph? In I. W. Rowlands & B. J. Weir (Eds.), *The Biology of Hystricomorph Rodents: The Proceedings of a Symposium Held at the Zoological Society of London* (pp. 7–20). London: Academic Press.

Lewejohann, L., Pickel, T., Sachser, N., & Kaiser, S. (2010). Wild genius - domestic fool? Spatial learning abilities of wild and domestic guinea pigs. *Frontiers in Zoology, 7*(9), 1–8.

Linneaus, C. (1758). Systema Naturae per regna tria naturae, secundum classes, ordines, genera, species, cum characteribus, differentiis, synonymis, locis. Editio Decima, Reformata. Tomus I. Holmiæ (Stockholm): Laurentii Salvii. doi:10.5962/bhl.title.542

Mamczarz, J., Pescrille, J. D., Gavrushenko, L., Burke, R. D., Fawcett, W. P., DeTolla Jr., L. J., ..., Albuquerque, E. X. (2016). Spatial learning impairment in prepubertal guinea pigs prenatally exposed to the organophosphorus pesticide chlorpyrifos: Toxicological implications. *NeuroToxicology, 56*, 17–28. doi:10.1016/j.neuro.2016.06.008.

Mares, M. A., & Ojeda, R. A. (1983). Patterns of diversity and adaptation in South American hystricognath rodents. In M. A. Mares & H. H. Genoways (Eds.), *Mammalian Biology in South America* (Vol. 32, pp. 393–432). Pittsburgh, PA: Society of Systematic Zoology.

Marivaux, L., Vianey-Liaud, M., Welcomme, J.-L., & Jaeger, J.-J. (2002). The role of Asia in the origin and diversification of hystricognathous rodents. *Zoologica Scripta, 31*(3), 225–239.

Marks, K. L., Martel, D. T., Wu, C., Basura, G. J., Roberts, L. E., Schvartz-Leyzac, K. C., & Shore, S. E. (2018). Auditory-somatosensory bimodal stimulation desynchronizes brain circuitry to reduce tinnitus in guinea pigs and humans. *Science Translational Medicine, 10*(422), eaal3175. doi:10.1126/scitranslmed.aal3175.

Mårtensson, L. (1984). The pregnant rabbit, guinea pig, sheep and rhesus monkey as models in reproductive physiology. *European Journal of Obstetrics, Gynecology, and Reproductive Biology, 18*(3), 169–182.

Mason, J. H. (1939). The date of the first use of guinea-pigs and mice in biological research. *South African Veterinary Medical Journal, 10*, 22–25.

Mishra, O. P., & Delivoria-Papadopoulos, M. (1999). Cellular mechanisms of hypoxic injury in the developing brain. *Brain Research Bulletin, 48*(3), 233–238.

Miyauchi, H., & Horio, T. (1992). A new animal model for contact dermatitis: The hairless guinea pig. *The Journal Of Dermatology, 19*(3), 140–145.

Monticelli, P. F., & Ades, C. (2011). Bioacoustics of domestication: Alarm and courtship calls of wild and domestic cavies. *Bioacoustics, 20*, 169–191.

Monticelli, P. F., & Ades, C. (2013). The rich acoustic repertoire of a precocious rodent, the wild cavy *Cavia aperea*. *Bioacoustics, 22*, 49–66. doi:10.1080/09524622.2012.711516.

Moore, C. L. (1982). Maternal behavior of rats is affected by hormonal condition of pups. *Journal of Comparative Physiology & Psychology, 96*(1), 123–129.

Morales, E. (1994). The guinea pig in the Andean economy: From household animal to market commodity. *Latin American Research Review, 29*(3), 129–142.

Morales, E. (1995). *The Guinea Pig: Healing, Food, and Ritual in the Andes*. Tucson: University of Arizona Press.

Naguib, M., Kober, M., & Trillmich, F. (2010). Mother is not like mother: Concurrent pregnancy reduces lactating guinea pigs' responsiveness to pup calls. *Behavioural Processes, 83*(1), 79–81.

Nicholls, E. E. (1922). A study of the spontaneous activity of the guinea pig. *Journal of Comparative Psychology, 2*(4), 303–330. doi:10.1037/h0074475.

Nolte, D. L., Mason, J. R., & Lewis, S. L. (1994). Tolerance of bitter compounds by an herbivore, *Cavia porcellus*. *Journal of Chemical Ecology, 20*(2), 303–308.

Padilla-Carlin, D. J., McMurray, D. N., & Hickey, A. J. (2008). The guinea pig as a model of infectious diseases. *Comparative Medicine, 58*(4), 324–340.

Pettijohn, T. F. (1977). Reaction of parents to recorded infant guinea-pig distress vocalization. *Behavioral Biology, 21*(3), 438–442.

Pettijohn, T. F. (1978). Attachment and separation distress in the infant guinea-pig. *Developmental Psychobiology, 12*(1), 73–82.

Poulsent, L. K., Lundberg, L., Sompolinsky, D., Spärck, J. V., Ploug, M., Bindslev-Jensen, C., & Skov, P. S. (1991). Inbred strains of guinea pigs in studies of allergy. In J. Ring & P. B. (Eds.), *New Trends in Allergy III* (pp. 92–99). Berlin: Springer.

Poux, C., Chevret, P., Huchon, D., de Jong, W. W., & Douzery, E. J. P. (2006). Arrival and diversification of caviomorph rodents and platyrrhine primates in South America. *Systematic Biology, 55*(2), 228–244. doi:10.1080/10635150500481390.

Quesenberry, K., Donnelly, T., & Mans, C. (2012). Biology, husbandry, and clinical techniques of guinea-pigs and chinchillas. In K. Quesenberry & J. W. Carpenter (Eds.), *Ferrets, Rabbits, and Rodents: Clinical Medicine and Surgery* (3rd ed., pp. 279–294). Saint Louis, MO: Elsevier.

Ratner, S. C. (1967). Comparative aspects of hypnosis. In J. E. Gordon (Ed.), *Handbook of Clinical and Experimental Hypnosis* (pp. 550–587). New York: Collier Macmillan Ltd.

Reed, M., Burton, F. A., & Van Diest, P. A. (1979). Ovulation in the guinea-pig. I. The ruptured follicle. *Journal Of Anatomy, 128*(Pt 1), 195–206.

Reed, M., & Hounslow, W. F. (1971). Induction of ovulation in the guinea-pig. *The Journal of Endocrinology, 49*(2), 203–211.

Rehling, A., & Trillmich, F. (2007). Weaning in the guinea pig (*Cavia aperea f. porcellus*): Who decides and by what measure? *Behavioral Ecology and Sociobiology, 62*(2), 149–157.

Rehling, A., & Trillmich, F. (2008). Changing supply and demand by cross-fostering: Effects on the behaviour of pups and mothers in guinea pigs, *Cavia aperea f. porcellus*, and cavies, *Cavia aperea*. *Animal Behaviour, 75*(Part 4), 1455–1463.

Reid, M. E. (1959). The guinea pig in research. *Academic Medicine, 34*(1), 74.

Richardson, V. C. G. (2000). *Diseases of Domestic Guinea Pigs* (Vol. 2). Malden, MA: Blackwell Science.

Ritchey, R. L., & Hennessy, M. B. (1987). Cortisol and behavioral responses to separation in mother and infant guinea pigs. *Behavioral & Neural Biology, 48*, 1–12. doi:10.1016/s0163-1047(87)90514-0.

Rocca, M. S., & Wehner, N. G. (2009). The guinea pig as an animal model for developmental and reproductive toxicology studies. *Birth Defects Research Part B: Developmental and Reproductive Toxicology, 86*(2), 92–97. doi:10.1002/bdrb.20188.

Rood, J. P. (1969). Observations on the ecology and behavior of microcavia galea and cavia. *Acta Zoologica Lilloana, 24*, 111–114.

Rood, J. P. (1972). Ecological and behavioural comparisons of three genera of Argentine cavies. *Animal Behaviour Monographs, 5*(1), 3–83.

Rood, J. P., & Weir, B. J. (1970). Reproduction in female wild guinea-pigs. *Journal of Reproduction and Fertility, 23*(3), 393–409.

Rowlands, I. W. (1949). Post-partum breeding in the guinea-pig. *The Journal of Hygiene, 47*(3), 281–287.

Rübensam, K., Hribal, R., Jewgenow, K., & Guenther, A. (2015). Seasonally different reproductive investment in a medium-sized rodent (*Cavia aperea*). *Theriogenology, 84*(4), 639–644. doi:10.1016/j.theriogenology.2015.04.023.

Sachser, N. (1986). Different forms of social organization at high and low population densities in guinea pigs. *Behaviour, 97*(3/4), 253–272. doi:10.2307/4534531.

Sachser, N. (1998). Of domestic and wild guinea pigs: Studies in sociophysiology, domestication, and social evolution. *Naturwissenschaften, 85*(7), 307–317.

Sachser, N., Hennessy, M. B., & Kaiser, S. (2018). The adaptive shaping of social behavioural phenotypes during adolescence. *Biology Letters, 14*. doi:10.1098/rsbl.2018.0536.

Sandweiss, D. H., & Wing, E. S. (1997). Ritual rodents: The guinea pigs of Chincha, Peru. *Journal of Field Archaeology, 24*(1), 47. doi:10.2307/530560.

Schiml, P. A., & Hennessy, M. B. (1990). Light-dark variation and changes across the lactational period in the behaviors of undisturbed mother and infant guinea pigs (*Cavia porcellus*). *Journal of Comparative Psychology, 104*(3), 283–288. doi:10.1037/0735-7036.104.3.283.

Seidl, D. C., Hughes, H. C., Bertolet, R., & Lang, C. M. (1979). True pregnancy toxemia (preeclampsia) in the guinea pig (*Cavia porcellus*). *Laboratory Animal Science, 29*(4), 472–478.

Shaw, J. C., Palliser, H. K., Dyson, R. M., Hirst, J. J., & Berry, M. J. (2016). Long-term effects of preterm birth on behavior and neurosteroid sensitivity in the guinea pig. *Pediatric Research, 80*(2), 275–283. doi:10.1038/pr.2016.63.

Soanes, C., & Stevenson, A. (2003). *Oxford Dictionary of English* (2nd ed.) Oxford: Oxford University Press.

Spotorno, A. E., Manríquez, G., Fernández, L. A., Marín, J. C., González, F., & Wheeler, J. (2007). Domestication of guinea pigs from a Southern Peru-Northern Chile wild species and their middle Pre-Columbian mummies. In D. A. Kelt, E. P. Lessa, & J. Salazar-Bravo (Eds.), *The Quintessential Naturalist: Honoring the Life and Legacy of Oliver P. Pearson*. Oakland, CA: University of California Press.

Spotorno, A. E., Marín, J. C., Manríquez, G., Valladares, J. P., Rico, E., & Rivas, C. (2006). Ancient and modern steps during the domestication of guinea pigs (*Cavia porcellus* L.). *Journal of Zoology, 270*(1), 57–62. doi:10.1111/j.1469-7998.2006.00117.x.

Stockard, C. R., & Papanicolaou, G. N. (1917). The existence of a typical oestrous cycle in the guinea-pig—with a study of its histological and physiological changes. *American Journal of Anatomy, 22*(2), 225–283. doi:10.1002/aja.1000220204.

Sueki, H., Gammal, C., Kudoh, K., & Kligman, A. M. (2000). Hairless guinea pig skin: Anatomical basis for studies of cutaneous biology. *European Journal Of Dermatology: EJD, 10*(5), 357–364.

Terril, L. A., Clemons, D. J., & Suckow, M. A. (1998). *The laboratory guinea pig*. Boca Raton: CRC Press/Taylor Francis.

Thompson, L. P., Pence, L., Pinkas, G., Song, H., & Telugu, B. P. (2016). Placental hypoxia during early pregnancy causes maternal hypertension and placental insufficiency in the hypoxic guinea pig model. *Biology of Reproduction, 95*(6), 128–128.

Tinbergen, N. (1951). *The Study of Instinct*. New York: Clarendon Press/Oxford University Press.

Trillmich, F. (2000). Effects of low temperature and photoperiod on reproduction in the female wild guinea pig (*Cavia aperea*). *Journal of Mammalogy, 81*(2), 586.

Trillmich, F., Bieneck, M., Geissler, E., & Bischof, H.-J. (2003). Ontogeny of running performance in the wild guinea pig (*Cavia aperea*). *Mammalian Biology, 68*(4), 214–223.

Trillmich, F., Kraus, C., Kunkele, J., Asher, M., Clara, M., Dekomien, G., ... Sachser, N. (2004). Species-level differentiation of two cryptic species pairs of wild cavies, genera *Cavia* and *Galea*, with a discussion of the relationship between social systems and phylogeny in the Caviinae. *Canadian Journal of Zoology, 82*(3), 516–524.

Trillmich, F., Laurien-Kehnen, C., Adrian, A., & Linke, S. (2006). Age at maturity in cavies and guinea-pigs (*Cavia aperea* and *Cavia aperea f. porcellus*): Influence of social factors. *Journal of Zoology (London), 268*(3), 285–294.

Trillmich, F., Mueller, B., Kaiser, S., & Krause, J. (2009). Puberty in female cavies (*Cavia aperea*) is affected by photoperiod and social conditions. *Physiology & Behavior, 96*, 476–480. doi:10.1016/j.physbeh.2008.11.014.

Van Abeelen, A. F. M., Veenendaal, M. V. E., Painter, R. C., De Rooij, S. R., Thangaratinam, S., Van Der Post, J. A. M., ... Roseboom, T. J. (2012). The fetal origins of hypertension: A systematic review and meta-analysis of the evidence from animal experiments of maternal undernutrition. *Journal of Hypertension, 30*(12), 2255–2267. doi:10.1097/HJH.0b013e3283588e0f.

Von Békésy, G. (1960). *Experiments in Hearing*. New York: McGraw-Hill.

Wagner, J. E., & Manning, P. J. (Eds.). (1976). *The Biology of the Guinea Pig*. New York: Academic Press.

Weir, B. J. (1974). Notes on the origin of the domestic guinea-pig. In I. W. Rowlands & B. J. Weir (Eds.), *The Biology of Histricomorph Rodents: Symposia of the Zoological Society of London* (pp. 437–446). London: Academic Press.

West, K. L., & Fernandez, M. L. (2004). Guinea pigs as models to study the hypocholesterolemic effects of drugs. *Cardiovascular Drug Review, 22*(1), 55–70.

Whishaw, I. Q., Sarna, J. R., & Pellis, S. M. (1998). Evidence for rodent-common and species-typical limb and digit use in eating, derived from a comparative analysis of ten rodent species. *Behavioural Brain Research, 96*(1–2), 79–91.

Willis, F. N., Levinson, D. M., & Buchanan, D. R. (1977). Development of social behavior in the guinea-pig. *Psychological Record, 27*(3), 527–536.

Wing, E. S. (1986). Domestication of Andean mammals. In F. Vuilleumier & M. Monasterio (Eds.), *High Altitude Tropical Biogeography* (pp. 246–264). New York: Oxford University Press.

Wright, P. H., Makulu, D. R., & Posey, I. J. (1968). Guinea pig anti-insulin serum. *Diabetes, 17*(8), 513.

Young, W. C. (1969). Psychobiology of sexual behavior in the guinea pig. *Advances in the Study of Behavior, 2*, 1–110. doi:10.1016/s0065-3454(08)60068-6.

9

Behavioral Biology of Deer and White-Footed Mice, Mongolian Gerbils, and Prairie and Meadow Voles

Kathleen R. Pritchett-Corning
Harvard University
University of Washington

Christina Winnicker
GlaxoSmithKline

CONTENTS

Introduction 148
    Deer Mice and White-Footed Mice (*Peromyscus*) 148
    Gerbils 148
    Voles (*Microtus*) 149
Typical Research Use 150
    Deer Mice and White-Footed Mice 150
    Gerbils 150
    Voles 150
Behavioral Biology 150
    Natural History and Ecology, Habitat Use, Predator/Prey Relations 150
        Deer Mice and White-Footed Mice 150
        Gerbils 150
        Voles 151
    Social Organization and Behavior 151
        Deer Mice and White-Footed Mice 151
        Gerbils 151
        Voles 152
    Feeding Behavior and Substrates 153
        Deer Mice and White-Footed Mice 153
        Gerbils 153
        Voles 153
    Common Captive Behaviors 153
        Normal 153
        Abnormal 154
Ways to Maintain Behavioral Health 155
    Environment 155
        Deer Mice and White-Footed Mice 155
        Gerbils 155
        Voles 155
    Social Groupings 155
        Deer Mice and White-Footed Mice 155
        Gerbils 156
        Voles 156
    Feeding Strategies 156
        Deer Mice and White-Footed Mice 156
        Gerbils 156
        Voles 156
    Training 156

Special Situations 156
    Deer Mice and White-Footed Mice 156
    Gerbils 156
    Voles 156
Conclusions/Recommendations 157
    Deer Mice and White-Footed Mice 157
    Gerbils 157
    Voles 157
Acknowledgments 157
References 158

Introduction

Although mice and rats are the most commonly used rodents in research, the contributions of other rodents as models for certain diseases, as research subjects in their own right, or for other scientific reasons should not be overlooked. Discovering information on the natural history and behavior in captivity of uncommonly used rodent species can be challenging, since there are relatively few reviews available in the literature, and aspects of behavior or biology may not be described at all. This chapter discusses several less common laboratory rodents: deer mice and white-footed mice (*Peromyscus maniculatus* and *P. leucopus*), Mongolian gerbils (*Meriones unguiculatus*), and prairie and meadow voles (*Microtus ochrogaster* and *M. pennsylvanicus*). These rodents are not closely related to one another, having differentiated several million years ago, although *Peromyscus* and *Microtus* may occupy similar ecological niches [see Figure 9.1].

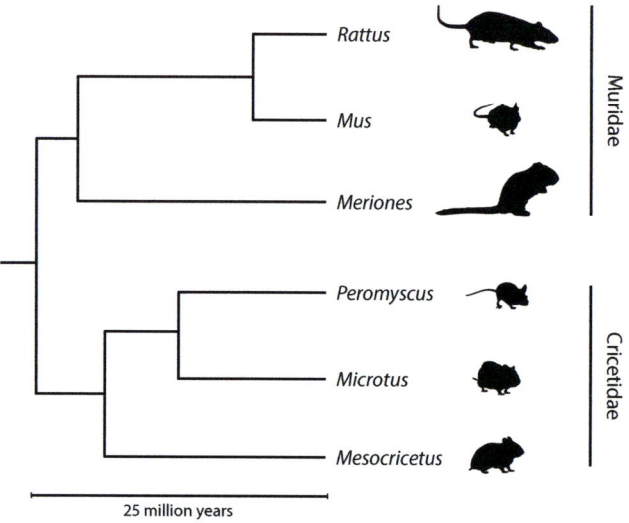

FIGURE 9.1 Evolutionary relationships between several different muroid rodents. Gerbils are more closely related to Old World mice and rats, while voles, New World mice, and hamsters are more closely related to each other than to Old World mice and rats. (The figure is adapted from information in Steppan, Adkins and Anderson (2004) presented in Bedford and Hoekstra (2015) and is further adapted for this work and used with their permission under a Creative Commons Attribution License.)

Deer Mice and White-Footed Mice (*Peromyscus*)

Peromyscus is a genus of North American rodents, currently comprise 56 recognized species that occupy many habitats throughout the continent, from the Canadian Arctic to Panama (Bedford and Hoekstra 2015). Many species of captive *Peromyscus* are used in research, and they all have different diets, body sizes, reproductive characteristics, and social structures. Considered together, they are the most common small rodents in North America. It is beyond the scope of this chapter to discuss them all, so we will focus on *P. maniculatus*, the deer mouse, and the closely related *P. leucopus*, the white-footed mouse. In this chapter, we may refer to either species of mouse by the overall genus name for ease of use (Webster and Brooks 1981). *Peromyscus* in general have large, unfurred ears and large eyes. The more arboreal species can have very long tails, which they use for balance. Deer mice have a larger home range than white-footed mice, with deer mice broadly distributed across North America and white-footed mice typically not found as far north or west as deer mice. *P. maniculatus* and *P. leucopus* are of similar size, with *P. leucopus* being 130–205 mm long and weighing 15–25 g (Lackey, Huckaby and Ormiston 1985), while *P. maniculatus* is 119–222 mm long and weighs 10–24 g (Bunker 2001). The two species are quite similar in appearance, both sporting brown to red to gray dorsums and white ventrums. In the wild, the two species are often distinguished by tail length, with *P. maniculatus*' tail being longer than *P. leucopus*' and having a more distinct demarcation between a darker dorsum and whiter ventrum. Both species are primarily nocturnal and exhibit periods of torpor. It should be noted that although *P. maniculatus* and *P leucopus* are being addressed at the species level, several subspecies are recognized, and their behavior may differ. *Peromyscus* are sympatric with *Microtus* species [see Figures 9.2 and 9.3].

Gerbils

Meriones unguiculatus (Milne-Edwards 1867), commonly known as the Mongolian gerbil or the clawed jird, is a small rodent from the steppes of Asia. Gerbils weigh 40–80 g at adulthood and have a body length of 118–138 mm, with a haired tail adding another 95–115 mm to their total length. Gerbils have been described in the literature as diurnal, nocturnal, and crepuscular, but are most likely to have a bimodal activity pattern, with the activity periods linked to specific events, such as dawn or dusk in the wild (Pietrewicz, Hoff

FIGURE 9.2 *Peromyscus maniculatus bairdii*, the deer mouse. There are currently six recognized subspecies of *P. maniculatus* that vary phenotypically based on habitat. For example, *P. maniculatus nubiterrae* (not shown) lives in forests and is an agile climber. The mouse pictured is from a laboratory colony and is ear-marked for identification.

FIGURE 9.4 *Meriones unguiculatus*, the Mongolian gerbil. Note the long, fully-haired tail, blunt muzzle, and wild-type sandy color. (Photo provided courtesy of Robert Zaccardi at Charles River Laboratories.)

FIGURE 9.3 *Peromyscus leucopus*, the white-footed mouse. Note the difference in face shape and eye size between *P. leucopus* and *P. maniculatus*. This mouse originates from a laboratory colony, hence the ear markings for identification.

FIGURE 9.5 *Microtus ochrogaster*, the prairie vole. The scientific name of this mouse is derived from Greek: the species name means "small ear", while the genus name means "yellow belly", both features readily seen in the photo. (Photo by Dr. Aubrey Kelly, courtesy of Dr. Alex Ophir. Used with their permission.)

and Higgins 1982, Roper and Polioudakis 1977, Weinert, Weinandy and Gattermann 2007). Gerbils do not hibernate. In the wild, gerbils are generally agouti, but a variety of colors are seen in domesticated animals, including nonagouti (black), pink-eyed, acromelanic albino ("Siamese"), and albino (Petrij et al. 2001) [see Figure 9.4].

Voles (*Microtus*)

Although there are several genera of voles found around the world in various habitats and of various sizes, there are two species of wild vole commonly seen in biomedical research facilities; the prairie vole, *Microtus ochrogaster* (Wagner 1842), and the meadow vole, *M. pennsylvanicus* (Ord 1815). These two species are typically contrasted in studies of bonding and parental behavior, as *M. ochrogaster* mates with one partner, remains with that partner through gestation, and displays biparental care, while *M. pennsylvanicus* does none of those things. *Microtus* are distinguished from *Peromyscus* by having darker, coarser fur, small ears (as reflected in the genus name), short tails, and beady eyes. Both species are native to North America, and both occupy similar habitats, with *M. pennsylvanicus* having the slightly larger range. Meadow voles are found roughly from North Carolina through Canada to Alaska, while prairie voles are found more in the central parts of North America, from Oklahoma and Arkansas in the south to Alberta in the north. When the two species are in the same geographical area, the meadow vole is usually found in the moister habitats. Both species are very common in their ranges and may be sympatric with *Peromyscus*. Prairie voles are 130–172 mm long and weigh 37–48 g, while meadow voles range from 140 to 195 mm in length and 34–50 g in weight. Both species are brown to chestnut, with black-tipped hairs, and have paler bellies. Both species are nocturnal in the laboratory (but are likely to have no fixed activity period in the wild (Webster and Brooks 1981)), and neither species hibernates [see Figures 9.5 and 9.6].

FIGURE 9.6 *Microtus pennsylvanicus*, the meadow vole. These small mammals have the typical microtine small ears and short tails, as well as the coarse, grayish-brown fur. Contrast the more whitish undersides of these voles with those of the prairie vole. (Photo courtesy of Dr Annelise Beery. Used with her permission.)

Typical Research Use

Deer Mice and White-Footed Mice

Deer mice are involved in research related to ecology (Wolff 1985b), hantavirus and *Borrelia* biology (Berl et al. 2018, Schwanz et al. 2011), longevity (Ungvari et al. 2008), and evolutionary questions (Bedford and Hoekstra 2015). They are also used to study torpor (Vogt and Lynch 1982, Tannenbaum and Pivorun 1987, 1988, Nestler, Dieter and Klokeid 1996) and general mouse behavior, as well as serving as a model for obsessive-compulsive behaviors in humans (Hayssen 1997, Powell et al. 2000, Wolmarans, Stein and Harvey 2017). Compared to mice and rats, few deer mice are used in research.

Gerbils

Gerbils are currently used in research as models of ischemic brain injury (Yan et al. 2014, Lee et al. 2016) and epilepsy (Heo and Kang 2012), and the study of hearing and the development of the structures of the ear (Choudhury et al. 2011, Kapuria, Steele and Puria 2017). They also play a role in the study of some infectious diseases of humans, such as *Helicobacter pylori* (Flahou et al. 2013, Beckett et al. 2018). Compared to mice and rats, few gerbils are currently used in research.

Voles

Voles are used in various research endeavors, including studies of behavior (Rice, Restrepo and Ophir 2018, Lee et al. 2019), pair bonding (Aragona and Wang 2004, Perkeybile, Griffin and Bales 2013, McGraw and Young 2010), ecology (Ostfeld 2002, Howe et al. 2006, Klatt, Getz and McGuire 2015), and as models of infectious disease (Anderson et al. 1986, Goethert and Telford 2017). As noted earlier, prairie voles and meadow voles are commonly compared in research situations (Stowe et al. 2005, McGuire, Bemis and Vermeylen 2014, Klatt, Getz and McGuire 2015). Compared to mice and rats, very few voles are being used in research.

Behavioral Biology

Natural History and Ecology, Habitat Use, Predator/Prey Relations

Deer Mice and White-Footed Mice

Peromyscus species occupy varied habitats throughout North America, including urban and suburban areas, forests, deserts, grasslands, beaches, and agricultural land. Deer mice and white-footed mice are found in forests and grasslands, since they are both terrestrial and arboreal. Deer mice and white-footed mice climb well, are strong swimmers, and easily navigate human spaces (Evans et al. 1978, Dewsbury, Lanier and Miglietta 1980). Although they are not commensal with humans, like *Mus*, they do take advantage of human dwellings and buildings, especially in the winter. Both deer mice and white-footed mice build nests near or slightly below ground; they are not extensive excavators, unlike other *Peromyscus* species (Weber and Hoekstra 2009).

Peromyscus are long-lived in the laboratory (compared to *Mus*, living 5–8 years), but in the wild have almost complete population turnover annually (Lackey, Huckaby and Ormiston 1985, Burger and Gochfeld 1992). *Peromyscus* serve as an important prey species for a wide variety of species, including raptors, reptiles, and carnivorous mammals, such as coyotes, foxes, and weasels. Antipredator strategies used by these mice include nocturnal activity, fleeing or freezing (especially in response to a looming stimulus), burrowing, and cryptic coloration. *P. leucopus* also exhibit a drumming behavior in which they thump their forefeet on logs or the ground, which may serve a warning function (Aguilar 2011).

Deer mice were first captured and bred in 1915–1916 by Dr. Francis Sumner at the Scripps Institute in California (Sumner 1917). Dr. Lee R. Dice at the University of Michigan maintained large colonies of various *Peromyscus* species between 1925 and 1975, and many lines of mice still in use can be traced back to that group (Joyner et al. 1998). In 1985, the *Peromyscus* Genetic Stock Center was formally established at the University of South Carolina (Dawson 2005). Several species of *Peromyscus* are housed at the *Peromyscus* Genetic Stock Center and this is the primary source of *Peromyscus* for researchers today. However, several large-scale research projects continue to study wild *Peromyscus* (Barrett et al. 2019), and given their ubiquity in North America, scientists may also capture their own animals to establish research colonies.

Gerbils

Gerbils are desert rodents, found in arid and semiarid regions of the world. Some species may be considered as agricultural pests. The Mongolian gerbil is an inveterate digger, constructing burrows to a depth of about 50 cm (Scheibler et al. 2006). Burrow complexity varies from a simple tube with one entrance/exit, to burrow complexes with more than ten entry/exit points (Scheibler et al. 2006). They nest within their burrows. Gerbils do not climb with dexterity, nor do they swim well. They also do not typically interact with humans or their dwellings in their home range. It is illegal to keep Mongolian gerbils as pets in California due to the availability of suitable habitat and their potential to become an agricultural pest (Cal. Food & Agric., §671.1.2.J).

In the laboratory, gerbils are moderately long-lived, with a typical lifespan of 3–4 years. In their natural habitat, gerbils are preyed upon by snakes and raptors and typical survival is only 3–4 months (Liu et al. 2009). Antipredator strategies used by gerbils include fleeing or freezing (especially in response to a looming stimulus), burrowing, and cryptic coloration. Gerbils warn conspecifics of danger by drumming the ground with their large hind feet (Randall 2001).

Gerbils kept in captivity today have their origins in a subset of 20 wild pairs captured in 1935 by D. C. Kasuga in the basin of the Amur River in eastern Mongolia. Dr. Kasuga sent these wild gerbils to the Kitasato Institute, Shirokane, Tokyo, Japan, where they were raised and studied. In 1949, offspring of the original animals were transferred to M. Nomura at the Central Laboratories for Experimental Animals in Japan. Most of the gerbils used in research today descend from a further subset of these animals, when either 4 pairs of gerbils and their offspring or 11 pairs of gerbils (accounts differ) were transferred in 1954 to Dr. Victor Schwentker at Tumblebrook Farm in New York. Only 5 females and 4 males of the transferred gerbils bred, and the closed colony of Tumblebrook Farm was descended from those nine animals. Tumblebrook Farm was acquired by Charles River in 1995. Gerbils used in research today are considered domesticated rodents, as they differ allometrically and genetically from wild gerbils (Stuermer et al. 2003, Neumann et al. 2001).

Voles

Vole species may be found around the world, although they tend to be more numerous in northern habitats. *Microtus* occupy similar ecological niches as *Peromyscus*, but show very little tendency to arboreal life (Dewsbury, Lanier and Miglietta 1980). *Microtus* seem to be comfortable swimmers (Evans et al. 1978, Sawrey, Keith and Backes 1994). In many parts of the world, voles are significant agricultural pests, girdling young fruit trees or consuming the roots of other crops (Howe et al. 2006). If prairie and meadow voles occupy the same area, prairie voles reside in shorter, drier, and more varied vegetation. Meadow voles nest above ground, while prairie voles prefer to dig underground nests (Klatt, Getz and McGuire 2015, Solomon et al. 2005).

Microtus are shorter-lived than gerbils or *Peromyscus*, with a lifespan of approximately 2.5 years in captivity. In the wild, *Microtus* typically live 2–12 months. *Microtus* are an important prey species for a wide variety of species, including raptors, reptiles, and carnivorous mammals, such as shrews, coyotes, bears, foxes, and weasels. Antipredator strategies used by *Microtus* include no fixed pattern of activity (neither nocturnal, diurnal, nor crepuscular), fleeing or freezing (especially in response to a looming stimulus), burrowing or retreating into grassy tunnels, and cryptic coloration.

There is not a commercially or publicly funded common source for voles used in research. Many investigators have captured their own colonies or have received gifts of animals from successfully maintained captive colonies.

Social Organization and Behavior
Deer Mice and White-Footed Mice

White-footed and deer mice have similar social organizations. In the wild, both sexes maintain home ranges and these home ranges overlap during breeding season. Subadult (5–8-week-old) animals disperse from the natal territory, possibly quite widely, depending on resource availability. Adults may also travel long distances (Lackey, Huckaby and Ormiston 1985, Rehmeier, Kaufman and Kaufman 2004). Territories are marked with urine, feces, and scent marks deposited by both sexes. Deer mice and white-footed mice are probably polygamous/polygynous in the wild. In the laboratory, both species are typically paired for breeding and not kept in harems, although polygamous mating may be acceptable under certain conditions (Wolff 1994).

White-footed and deer mice have approximately 6-day estrus cycles (Dewsbury, Estep and Lanier 1977, Veres et al. 2012). Both *Peromyscus* species have 23–25-day gestation periods (Svihla 1932, Layne 1968). Deer mice are reported to have no readily visible copulation plug (Veres et al. 2012), despite earlier reports (Dewsbury 1988), while white-footed mice are reported to have a copulation plug (Hartung and Dewsbury 1978, Layne 1968). Both species have postpartum estrus and may also have delayed implantation due to stress, usually related to lactation or resource scarcity (Svihla 1932, Dewsbury, Estep and Lanier 1977). *Peromyscus* have smaller litter sizes than laboratory mice, with an average of three to five pups per litter (Layne 1968). Both *Peromyscus* species can exhibit seasonal reproductive patterns under resource scarcity, and they are typically kept in the laboratory under long-day light cycles (minimum of 16L:8D) to support breeding. In both species, testes in the male move from abdominal to scrotal to indicate readiness for mating. Both species usually give birth during the day. During birth, they assume a quadrupedal or bipedal stance and aid in expulsion of fetuses by using teeth and forepaws (Clark 1937, Svihla 1932). Female parental behavior includes grooming pups, nursing, retrieving pups to the nest, and nest building (Eisenberg 1968).

Both deer mice and white-footed mice use nesting material and are motivated to make nests regardless of reproductive status (Lewarch and Hoekstra 2018). In the wild, nests may differ based on habitat or microclimate, but in the lab, these tend to be complex structures. When paired in the laboratory, both species exhibit bi-parental care, but in the wild, it is unlikely that males contribute greatly to pup care (Xia and Millar 1988). In the wild, white-footed and deer mouse females are reported to abandon litters at 20–40 days postpartum (Svihla 1932) and offspring generally disperse from the mother's home territory.

Gerbils

In the wild, gerbils live together as loosely pair-bonded (Starkey and Hendrie 1998) reproductive adults, with juveniles and subadults remaining in the group to assist with parental care. They live in a complex burrow system with multiple entrances and exits. Adjacent territories may be held by related family groups, and they may mutually maintain overlapping burrow systems. Territories are demarcated by scent for gerbils, and this scent marking is accomplished by rubbing areas in the territory with the ventral marking gland, rather than by fecal or urinary deposition. The ventral gland produces a yellow, musky substance, is larger in males than females, and males show increased marking behavior when compared to females (Whitsett and Thiessen 1972). Marking in gerbils is related to territory and territory defense, and both sexes mark and defend established territory (Ågren 1984b). Females have been observed mating with males from other territories at territorial boundaries, further supporting the loose pair bond referenced above (Ågren 1984a, Ågren, Zhou and Zhong 1989a).

After a 24–26-day gestation period, the female gives birth in either a sitting posture, resting on the rump with the hind legs splayed, or in a quadrupedal posture with the hind legs extended (Platt 1977). The postpartum estrus occurs 8–11 h after parturition in gerbils (Prates and Guerra 2005). As with other rodents, the postpartum estrus is fertile, but delayed implantation may be seen under conditions of resource scarcity or while nursing a litter (Marston and Chang 1965). Females show increased locomotor activity during estrus, and this is true during the postpartum estrus as well (Prates and Guerra 2005).

Gerbils gather nesting material, process it, and build nests, especially females close to parturition (Roper and Polioudakis 1977, Glickman, Fried and Morrison 1967, Elwood 1975). Although male gerbils do not help with parturition, they are active parents and influence the development of offspring (Clark and Galef 2000, Piovanotti and Vieira 2004, Elwood and Broom 1978). Both males and females spend a great deal of time in direct bodily contact with pups when housed in captivity, and the association with the ventral marking gland may help to identify pups as belonging to specific parents, or even function as a brood patch (Ostermeyer and Elwood 1984, Elwood 1975, Kittrell, Gregg and Thiessen 1982). Females sniff and lick pups more often than males, but neither sex shows a high frequency of pup retrieval, unless pups are deliberately separated from the nest (Elwood 1975, Waring and Perper 1979). When males are removed from the cage, females spend more time in contact with pups and build more structurally complex nests, perhaps indicating that the males serve as a crucial heat source for pups (Elwood and Broom 1978). Pup development is well-summarized by Kaplan and Hyland (Kaplan and Hyland 1972).

In captivity, gerbils are weaned at 28 days, but in the wild, subadults remain in the family burrow until, or unless, they are expelled by the parents. The subadults perform "aunting" or alloparental behaviors, such as licking, sniffing, and retrieving pups (Saltzman et al. 2009, Scheibler, Weinandy and Gattermann 2004). Since gerbils remain in family groups, the reproductive maturation of the subadult males and females is retarded by the presence of familiar adult males and females (Clark and Galef 2002, Saltzman et al. 2006, Saltzman et al. 2009). Remaining in the natal group also significantly suppresses infanticide in female gerbils, allowing them to perform alloparenting behaviors (Saltzman et al. 2009). Eventually, most offspring are expelled from the family group, usually by the dominant female (Scheibler, Weinandy and Gattermann 2006).

Voles

The prairie and meadow vole differ significantly in their social organization. *M. ochrogaster*, the prairie vole, is behaviorally (or socially) monogamous (Shuster et al. 2019), while *M. pennsylvanicus*, the meadow vole, is promiscuous (Boonstra, Xia and Pavone 1993). In the monogamous prairie vole, the primary social group is a communal nesting group formed of an original mated pair, or a singleton female breeder after her mate has died, and offspring remaining or returning to the territory (philopatry), with perhaps a few unrelated adults, based on resource availability (Getz, McGuire and Carter 2003). Social groups are formed based on season and pup survival. These social groups are territorial, and resident males defend territory vigorously (Getz, McGuire and Carter 2003, Lee et al. 2019). In the promiscuous meadow vole, both males and females maintain territories, and male territories tend to be larger than females', but territories overall are smaller than those of other rodent species (Getz, McGuire and Carter 2003). After mating, the male departs the female's territory. Meadow voles may form affiliative groups in short day length, low-resource situations (Lee et al. 2019). Unlike other species, the presence of adult female meadow or prairie voles does not suppress reproduction in younger females of either species (Wolff, Dunlap and Ritchhart 2001). Both species may breed year-round (Colvin and Colvin 1970), but success of breeding in the wild is related to resource availability, typically higher during warmer, longer days.

Estrous cycle length in *Microtus* is not quantifiable as it is in other species, since females need exposure to males to come into estrus (Taylor, Salo and Dewsbury 1992). As a group, the microtines are induced ovulators (Sawrey and Dewsbury 1985). Meadow and prairie vole males deposit copulation plugs after mating (Hamilton 1941, Fitch 1957). Both species reportedly have imperforate vulvas when not receptive to males, although this assertion is found only in historical literature on wild-caught animals and may not be relevant to laboratory populations (Hamilton 1941, Fitch 1957).

Microtus build less complex nests in the laboratory than *Peromyscus* (Hartung and Dewsbury 1979). In the wild, prairie voles construct surface nests or build underground dens with several burrows and nesting chambers (Solomon et al. 2005), while meadow voles are surface nesters in dense ground cover that do not burrow (Klatt, Getz and McGuire 2015). Both species are reported to build nests almost exclusively from grasses. After a 20–22-day gestation period (although this may be longer due to delayed implantation) (Innes and Millar 1994, Hamilton 1941, Fitch 1957), pups are born. Voles tend to give birth during the light part of the photoperiod in captivity (McGuire et al. 2003) and the postpartum estrus is fertile (McGuire et al. 2003). Prairie vole litter size is 3–5, while meadow vole litter size is slightly larger at 4–6. Parturition in both species of voles follows the typical muroid pattern of assumption of a sitting posture, expulsion of fetuses in sequence, sometimes aided by mouth and forepaws, consumption of placentae, and then gathering the newly born pups into a heap (Hamilton 1941, Fitch 1957). Both animals can exhibit seasonal reproductive patterns with low temperatures and short days inhibiting reproduction (Kriegsfeld et al.

2000), and they are typically kept in the laboratory under long-day light cycles (minimum of 14L:10D) to support breeding (Mallory and Dieterich 1985).

In the wild, meadow voles wean their young at 14–16 day (Boonstra, Xia and Pavone 1993), while prairie voles wean slightly later at 19–21 day (Innes and Millar 1994). Prairie voles may delay expulsion of young due to population density (Solomon et al. 2005), and philopatry is common. Meadow vole and prairie vole females may become reproductively active at 4 weeks of age and continue to gain mass and size during their pregnancies. Both species have exceedingly short lifespans in the wild and this early reproductive behavior allows them to replace populations quickly.

Feeding Behavior and Substrates

Deer Mice and White-Footed Mice

Peromyscus are omnivorous across all the habitats they occupy. Foods consumed vary by habitat and season. Stomach contents of wild-caught *Peromyscus* include seeds from wild and domestic plants; mast; invertebrates, including caterpillars, insects, and spiders; fruit; fungi; and miscellaneous green vegetation (Whitaker 1963, 1966, Williams 1959). Both species scatter hoard and larder hoard food in the wild (Howard and Evans 1961, Vander Wall et al. 2001, Lanier, Estep and Dewsbury 1974), and this hoarding behavior is influenced by cold temperatures and short day length (Barry 1976). In the laboratory, deer mice and white-footed mice seem to grow and reproduce adequately on standard mouse rations. Food grinding is seen frequently in colonies, however, perhaps indicating a missing nutritional or behavioral component from rations formulated for mice or rats (Pritchett-Corning et al. 2013). These rodents are typically fed from wire-bar feeders and have no issues consuming food in that fashion. Supplementing with small amounts of sunflower (*Helianthus*) seeds and mealworms may help wild animals adapt to life in the laboratory.

Gerbils

Gerbils are herbivorous root, seed, and plant eaters, with *Artemisia sieversiana* (a member of the sagebrush/wormwood/mugwort family) comprising much of their diet in one study (Ågren, Zhou and Zhong 1989b). Gerbils forage widely and hoard seeds in burrows for times of resource scarcity (Ågren, Zhou and Zhong 1989b, a). Females hoard more than males, and food deprivation will induce an increase in food hoarding (Nyby and Thiessen 1980, Nyby et al. 1973). In the laboratory, gerbils are often fed from wire bar cage lids and not given the opportunity to express the highly motivated behaviors of foraging and hoard accumulation. One study found that gerbils fed on the floor of the cage or via hanging J-feeders were significantly heavier when compared to those fed from wire cage lids, but reproductive parameters did not differ between groups (Mulder et al. 2010). Gerbils appear to be adequately nourished on a standard mouse chow diet in the laboratory, but sunflower seeds are valued by the animal if added to the diet. As a desert animal, gerbils have excellent water conservation mechanisms, do not consume large amounts of water, and produce small, dry feces, and scant amounts of urine.

Voles

Like gerbils, voles are primarily herbivorous, consuming a diet composed of mainly grasses and seeds (Brochu, Caron and Bergeron 1988, Haken and Batzli 1996). They are adapted to consuming high-fiber, low-protein, and low-mineral diets (Batzli 1985). Both species hoard food (Lanier, Estep and Dewsbury 1974) and are significant agricultural pests, consuming roots and tubers, and girdling trees (Ostfeld 2002). In the laboratory, the protein and mineral content of standard rodent diets may be too high when compared to their natural diets. At least two authors state that voles are healthier on rabbit or guinea pig chow (Jackson 1997, Batzli 1985). Given their diet in the wild, voles are likely to benefit from fresh forage, but there are significant biosecurity hazards in providing fresh forage for lab-reared animals. Irradiated hay may be of value to these animals.

Common Captive Behaviors

Normal

Deer Mice and White-Footed Mice

Captive behaviors easily observed in deer and white-footed mice are the parental behaviors described above, and other typical rodent behaviors, such as grooming and fighting. Autogrooming in deer and white-footed mice proceeds with a fixed action pattern dictating the order in which body parts are groomed (Eisenberg 1968, Spruijt, Van Hooff and Gispen 1992, Kalueff et al. 2016). Animals use forepaws, forelimbs, the tongue, and teeth to groom the pelage and remove parasites. As with other rodents, the pattern starts with the face, proceeds to the flanks and ventrum, and ends with the tail. After grooming with the forepaws or scratching with the hindlegs, the nails are cleaned by nibbling with the mouth (Eisenberg 1968). Social grooming in these mice is involved in the establishment of a dominance hierarchy and resembles autogrooming, but the action pattern is more variable. The dominant animal parts the subordinate's fur with its forepaws, licks at the fur and skin with its tongue, and nibbles at the fur with its incisors (Eisenberg 1968).

Although wild animals tend to avoid overt agonistic behavior, since it may lead to injury or death, both species of *Peromyscus* can be aggressive in defense of the nest or during territorial disputes. During agonistic encounters, animals may assume vertical postures in which they ward off attacks with the forepaws. Animals may also assume various horizontal postures, from static presentation of the white ventrum to flight and pursuit (Eisenberg 1962). Actual fighting involves direct bodily contact, with one mouse bowling the other over, each administering bites to the other, and both mice attempting to scratch at the ventrum with the hind limbs (Eisenberg 1962). A very nice ethogram covering many normal behaviors of *Peromyscus* may be found in Eisenberg (Eisenberg 1968).

Gerbils

In addition to parental behavior, as noted above, gerbils perform several typical rodent behaviors, such as grooming and fighting. The fixed action pattern of autogrooming in gerbils is like that in mice and rats: mouth/nose, face, ears, flank, ventrum, then tail (Thiessen 1988, Spruijt, Van Hooff and Gispen 1992, Kalueff et al. 2016). The grooming pattern in gerbils spreads saliva and Harderian gland secretions on the body. Differing amounts of saliva and Harderian material deposition on the pelage are seen at different temperatures (Thiessen 1988). This may serve as an aid in thermoregulation in the temperature extremes of the gerbil's natural habitat, as saliva serves to cool the gerbil and Harderian gland secretions insulate the pelage against wetness and cold (Thiessen 1988). Allogrooming is also seen in gerbils, especially between mated pairs, but rarely occurs in the wild outside of burrows (Ågren, Zhou and Zhong 1989b, Platt 1977). Allogrooming is solicited by the groomee from the groomer by a subordinate approach characterized by head lowering with ears back and eyes partially closed. The potential groomee then freezes and waits for grooming to commence. If this does not work, the groomee may nudge the groomer with the nose and wait, or present the part to be groomed (Platt 1977). Gerbils will also sand bath when given the opportunity, rolling repeatedly in a sandy area, burrowing the nose under the sand, and rolling from side to side.

Agonistic behavior in gerbils is similar to other rodents (Ginsburg and Braud 1971, Platt 1977). Aggressive interactions can be classed as mediated, in which there is aggression from one animal followed by submission, appeasement, or escape from another animal; or escalated, in which aggressive postures are returned with aggressive postures, and fighting ensues. In some cases, the subordinate animal cannot remove itself from the sight of the aggressor. In general, dominating or aggressive behaviors involve elevating postures, erect ears, staring, direct approaches, and chasing. Subordinate or submissive postures involve lowering the body, flattening the ears, closing the eyes, averting the head, freezing, and fleeing. Two strange male gerbils meeting for the first time display a sequence of possible behaviors beginning with investigation (anogenital sniffing), then progressing to agonistic postures, such as head-up, ears forward, quadrupedal threat postures; displays of the side of the body; rearing on the hind legs; and boxing with the forefeet (Eisenberg 1967, Majumdar, Schwartz and Santaspirt 1974). Some of these postures may also be defensive behaviors; for example, if one gerbil rears and begins boxing the other, the second gerbil may defend itself in a similar manner. Agonistic postures can progress to agonistic acts, for example, between two adult males or from a female defending a litter from an unfamiliar animal. In gerbils, some agonistic behaviors include rushing at another animal; biting a fleeing animal; sparring, in which both gerbils box with the forefeet; displays in which gerbils sidle toward each other; and finally, a locked fight in which both gerbils are biting and scratching (Eisenberg 1967, Platt 1977). For further information, a good review of gerbil behavior may be found in Hurtado-Parrado et al. (2015).

Voles

As with other rodents in this chapter, parental behaviors are described above. Other behaviors observed in voles are, again, highly conserved in rodents, and include grooming and agonistic behaviors (Spruijt, Van Hooff and Gispen 1992, Kalueff et al. 2016). Meadow and prairie vole autogrooming is described in Ferkin (2005) and follows the typical rodent pattern. Grooming proceeds from the head to the tail in a fixed action pattern that may be interrupted by bouts of scratching with the hind feet. Both auto- and allo-grooming are described in the literature (Wilson 1982). Agonistic behaviors in voles are well-described in Colvin (1973), in which several species of *Microtus* were paired for both inter- and intraspecific encounters. In that study, voles would approach an unfamiliar male by moving or stretching toward the unfamiliar animal and would then adopt a sideways fixed offensive stance. Lunging at another animal was described as an attack, but Colvin's ethogram did not require contact, and animals would react to the lunge by retreating (and being chased) or adopting an upright, defensive position (Colvin 1973). If one animal took an upright position, the other might as well, and then they would strike at each other with forefeet. If an animal did not retreat from an attack, physical contact might ensue, with injuries usually found on the rumps of affected animals. Animals might vocalize during any part of this agonistic encounter. *Microtus* organize agonistic encounters much as other rodents, with aggression paired with submission the usual outcome of agonistic encounters, and escalation to actual injury occurring rarely (Wilson 1982).

Abnormal

Deer Mice and White-Footed Mice

Captive deer mice have specifically been suggested as a model of human obsessive-compulsive disorder (OCD) (Wolmarans et al. 2018), although since the diagnostic criteria for OCD in humans includes intrusive thoughts (which cannot be measured in mice), the value of the deer mouse model is solely in studying the physical manifestations. Deer mice and white-footed mice are observed to perform several stereotypic behaviors similar to those of laboratory mice: flipping, bar chewing, route tracing, corner jumping, and, occasionally, stereotypic digging (Powell et al. 1999, Lacy, Alaks and Walsh 2013). In our colonies, we also observe food grinding, the non-ingestive destruction of food by chewing, which may or may not be a stereotypic behavior (Pritchett-Corning et al. 2013). Although not a stereotypy, other abnormal behaviors seen in both deer mice and white-footed mice are excessive grooming or hair and whisker plucking (also known as barbering). In our colonies, this is not as common in *Peromyscus* as in certain strains of laboratory mice, notably C57BL/6. The ontogeny of these abnormal behaviors is likely to be the same as in other rodent species, i.e., frustration caused by the inability to express highly motivated behaviors, like exploration, digging, or climbing (Würbel 2006, Würbel, Stauffacher and von Holst 1996).

Gerbils

In gerbils, stereotypic digging and bar chewing (Wiedenmayer 1997a, b) are the two most common stereotypies seen. Digging in gerbils is associated with several behaviors, such as grooming, foraging, and exploration, but its primary purpose is construction and maintenance of the burrow system (Ginsburg and Braud 1971, Eisenberg 1967). Stereotypic digging develops in gerbils as early as day 24 (Wiedenmayer 1997a), arising from bouts of normal digging that begin when the eyes open at around day 18. The goal of this digging is the excavation of a burrow leading to a nesting chamber that allows for retreat. In the wild, gerbils spend a great deal of time underground, providing safety from predators and amelioration of climatic extremes. If the animals are allowed substrate and space to dig naturalistic burrows, or given a tunnel that mimics a burrow (a tunnel with a bend so no light enters a small chamber at the end), stereotypic digging is decreased or eliminated (Wiedenmayer 1997a). In one experiment, gerbils were observed to spend approximately 92% of their time in the burrow (Wiedenmayer 1997a). The critical component of the burrow is the tube, or moving from an open area to a confined one; providing animals with darkness and confinement without a tube did not decrease stereotypical digging (Wiedenmayer 1997a). Digging behavior is not related to the substrate provided, likely due to the inability of the substrate to support a true burrow, nor it is related to cage size (Wiedenmayer 1996, 1997a).

Voles

Microtine voles are reported to spend very little of their time budget in stereotypic behavior (Baumgardner, Ward and Dewsbury 1980), but numerous stereotyped behaviors have been described in bank voles (*Clethrionomys*) (Odberg 1987). In prairie and meadow voles, jumping behavior is reportedly rare and animals do not climb (Wolff 1985a). Both prairie and meadow voles are described as rearing in cages, but this may be normal exploratory behavior and not a stereotypy. Detailed descriptions of stereotypies exhibited by microtine rodents are lacking in the literature.

Ways to Maintain Behavioral Health

Environment

Deer Mice and White-Footed Mice

At a bare minimum, both species of *Peromyscus* discussed in this chapter should be provided with solid-bottomed cages with contact bedding (wood product preferred, and shavings best), nesting material (cotton squares and crinkled paper, as they will use both), and a structure of some type. For the structure, using a tube allows for tube handling of animals, which decreases stress (Hurst and West 2010) and, in the authors' experience, improves the handling process for both the mouse and human. *Peromyscus* readily acclimate to tube handling and it decreases escape attempts during cage change, resulting in fewer escapees and a decrease in time spent changing cages. Both deer and white-footed mice have been shown to have decreased levels of stereotypic behavior when housed in enriched cages (Turner and Lewis 2003, Lacy, Alaks and Walsh 2013, Bechard, Bliznyuk and Lewis 2017). The enriched cages described in the literature are larger, multilevel, and have structures, including running wheels, tunnels, huts, and plastic toys, as well as opportunities to forage for food. As with other rodents, running wheels allow increased physical activity and their use may or may not be problematic, and should be evaluated on a case-by-case basis (Mason and Würbel 2016).

Gerbils

Minimum gerbil housing requirements include a solid-bottomed cage with enough height to allow animals to adopt normal postures, contact bedding, nesting material (preferably brown crinkled paper), and an L-shaped tube with a burrow chamber. Gerbils have a strong drive to burrow and providing a tube with a burrow chamber may significantly decrease stereotypical digging, if provided early enough (Wiedenmayer 1997b, a, Waiblinger and Koenig 2007). Gerbils actively interact with nesting material and will build nests with it, even in single-sex groups or when individually housed (Glickman, Fried and Morrison 1967). Providing sand will allow gerbils to sand bathe, but will not reduce the incidence of stereotypies (Pettijohn and Barkes 1978, Wiedenmayer 1997b). Gerbils will run on wheels (Weinert, Weinandy and Gattermann 2007, Sørensen et al. 2005), but running on a wheel may (or may not) be its own stereotypy (Mason and Würbel 2016). Although it would not harm gerbils to be handled with a tube, handling by hand is fine for these docile creatures. Using forceps or handling by the tail is not recommended.

Voles

Similar to the other species discussed in this chapter, prairie and meadow voles should be housed in solid-bottomed caging. Since they also adopt typical rodent postures such as rearing, cages should have enough height to allow for normal postural adjustments. Voles are motivated to burrow and, in the wild, create and maintain runways in grasses, and nest both above and below ground. They are likely to respond positively to nesting material (crinkled brown paper; they were reportedly less interested in cotton batting (Hartung and Dewsbury 1979)) and may both eat and nest in hay or straw if provided. As with other wild rodents, they are likely to benefit from tube handling (Hurst and West 2010), and since their tails are comparatively short, tube handling is liable to make them easier to catch and transfer between cages. Voles are reported to spend very little of their time performing stereotypic behavior, but enrichment prevents depressive responses of socially isolated prairie voles (Grippo et al. 2014). The enriched cages as described in Grippo et al. (2014) were larger than the standard shoebox cage and contained structures, including running wheels, tunnels, huts, and plastic toys, as well as opportunities to forage for food. As with other rodents, running wheels allow animals to increase their physical activity (Kenkel and Carter 2016). Their use may or may not be stereotypical and should be evaluated on a case-by-case basis (Mason and Würbel 2016).

Social Groupings

Deer Mice and White-Footed Mice

Both sexes of these mice tend to maintain discrete territories in the wild, but discrete territory does not mean an absence of social interaction. Solo housing, although at times necessary for health or research reasons, should not be the default. In the laboratory, same-sex groupings may be formed at weaning, and the usual mouse density of five adults per shoebox cage generally works well. Although females are territorial in the wild, little fighting is seen in captive colonies, and young females may be introduced to adult groups successfully. Stable groups of males formed before sexual maturity tend to remain stable, but fighting may commence related to cues imperceptible to humans.

Gerbils

Gerbils are social animals in the wild (Ågren, Zhou and Zhong 1989b, Starkey, Normington and Bridges 2007) and establishing same-sex groupings is generally acceptable. In the United States and other countries, the maximum number of gerbils per enclosure is limited by local laws. As with all rodents, introductions of adult males must be undertaken with care.

Voles

One reason prairie and meadow voles are studied is their differing social systems. In summer and with long days, meadow voles are solitary and territorial, while in winter, with short days, they live and sleep in groups. Prairie voles pair bond at mating and defend territory as a mated pair with or without related animals nearby. Prairie voles tend toward greater aggression levels with conspecifics when compared to meadow voles and this may be one way they maintain their pair bonds, although both species can tolerate adult same-sex cohabitation (Beery, Routman and Zucker 2009, Lee et al. 2019).

Feeding Strategies

Deer Mice and White-Footed Mice

Providing a variety of food and giving animals foraging opportunities would likely increase behavioral health in deer and white-footed mice. Although a commonly used enrichment food is sunflower (*Helianthus*) seeds, providing a variety of seeds might encourage mice to exhibit more naturalistic behaviors. For example, in deer mouse hoards, seeds from ragweed (*Ambrosia*), which are 1.5–3.5 mm in width, and acorns (*Quercus*), which are 1–6 cm long and 1–4 cm wide, were found in the same hoard (Howard and Evans 1961). Drickamer also found that both deer mice and white-footed mice consume sunflower, corn, multiflora rose, wheat, and elm seeds (Drickamer 1970). A seed mix could be duplicated in the lab by using wheat berries (*Triticum*), sorghum (*Sorghum*), or millet (*Pennisetum*), along with sunflower seeds and corn (*Zea*). Howard and Evans also found insect parts in the deer mouse hoards, so supplementation with mealworms or other invertebrates might be beneficial (Howard and Evans 1961).

Gerbils

Gerbils forage extensively in the wild, then hoard food for later use (Ågren, Zhou and Zhong 1989b). Although they are fine when fed from J-feeders or on the floor in captivity (Mulder et al. 2010), hiding the food in the bedding may motivate the gerbils to forage, thus eliciting a natural behavior (Forkman 1996). Sagebrush (*Artemesia*) seeds are 1 mm×2 mm (Yi et al. 2019), so providing smaller seeds for gerbils, as noted above for *Peromyscus*, may encourage foraging and hoarding behavior.

Voles

As noted earlier, voles consume a great deal of fresh, high-fiber, low-nutrient plant matter in the wild. Providing this matter in the laboratory can be challenging. There may be a way for individual patches of grasses (*Festuca*, *Vicia*) found in vole habitats to be grown and harvested specifically in the laboratory for the colony. Voles also consume monocot and dicot shoots and seeds when found in their environment (Haken and Batzli 1996), so periodic supplementation with small amounts of a seed mix, as noted above, might be welcome.

Training

No specific information could be found on training deer mice or voles. Training paradigms for gerbils tend to focus around exercise or maze training (Babcock and Graham-Goodwin 1997, Silveira et al. 2018). Positive reinforcement paradigms as used on rats and mice (Leidinger et al. 2017, Leidinger et al. 2018) may work for these animals as well (Brown 2012).

Special Situations

Deer Mice and White-Footed Mice

Deer mice are not domesticated rodents and, as such, may need extra time to acclimate to being handled. In the authors' experience, and as suggested earlier, they adapt well to tube handling and this reduces stress in both animals and handlers. They are generally agile and athletic, and will willingly run, climb, burrow, and swim.

Gerbils

As desert rodents, gerbils are hesitant swimmers and behavioral assays using water should be avoided (Platt 1977). Digging is their strong suit, and behavioral assays using digging could be valuable (Deacon 2009). Gerbils' fully-haired tails are vulnerable to degloving injuries and gerbils should never be lifted by the tail.

Voles

Voles are wild animals and may have objections to routine handling. Their shorter tails may make the use of tail-handling with forceps difficult. As suggested earlier, although not described specifically in the literature, tube handling may be of benefit to these animals. Voles are also robust animals and will run, burrow, and swim, but are not reported to be strong climbers (Evans et al. 1978, Sawrey, Keith and Backes 1994).

Conclusions/Recommendations

Deer Mice and White-Footed Mice

Deer mice and white-footed mice are available from the *Peromyscus* Stock Center at the University of South Carolina in the United States (Joyner et al. 1998). They are also found in the wild and readily captured, but may not breed well in captivity. If given rat cages, rather than mouse shoebox cages, deer and white-footed mice will readily form latrine corners, making these cages easier to clean. They should be housed in solid-bottomed cages with contact bedding, preferably a wood product. Although corncob will work as a bedding material, it is not ideal. *Peromyscus* species should be provided with nesting material, both a crinkled paper and a cotton material; they will use both to construct elaborate nests (Layne 1969, Hartung and Dewsbury 1979, Lewarch and Hoekstra 2018). These rodents adapt readily to tunnel handling and provision of a plastic tunnel also allows for self-rescue, should automatic watering be in use [see Figure 9.7]. Although they readily feed from food hoppers, they benefit from supplementation and being allowed to forage for food (Vander Wall et al. 2001). As with other captive rodents, *Peromyscus* will become overweight if supplementation is too copious or elaborate.

Gerbils

Gerbils are readily available from commercial vendors and should be considered domesticated animals. Although domesticated gerbils can exhibit a full range of wild behaviors when housed in naturalistic settings, fully or semi-naturalistic desert housing is typically unrealistic in the laboratory (Roper and Polioudakis 1977). Larger cages do not affect the development of stereotypic digging, the primary stereotypy seen in gerbils, so the provision of large cages may not be necessary (Wiedenmayer 1996), but in the United States and other countries, gerbil cage size is regulated by law. At a minimum, a gerbil's enclosure should have a solid bottom, be bedded with a wood product, allow the gerbil to stand on its hind legs with head fully erect, contain nesting material, and have a burrow chamber with an L-shaped entrance (Waiblinger 2002). Ideally, the cage would also allow the opportunity to forage for food and construct food hoards (Sørensen et al. 2005), perhaps contain a running wheel, and an area in which the gerbil could sand bathe. Gerbils are pleasant, docile animals and handling by hand is readily accepted.

Voles

There are no commercially available sources of voles in the United States. Animals are plentiful in certain areas, however, and may be captured and held in the laboratory. Depending on the location of the researcher, the species of interest may best be sourced from a colleague. In the wild, prairie and meadow voles occupy dense grasslands and build elaborate runways and burrows. They may nest both above and below ground, depending on the time of year and species, so provision of nesting material is important to these animals. As with the other rodents in this chapter, they should be housed in solid-bottomed cages bedded with a wood-product bedding. Voles are ideally fed rabbit, not mouse or rat, diets, and there is no reason not to scatter this diet in the bedding to allow for foraging and hoard construction (Jackson 1997). Cage furniture is accepted by voles and its provision may allow for self-rescue with automatic waterer malfunction. Voles are also likely to adapt well to tunnel handling and this method should be considered for stress reduction for both voles and handlers.

Acknowledgments

The authors would like to acknowledge the patience of the editors, as well as the following colleagues: Dr. Nicole Bedford for creation of Figure 9.1; Dr. Hopi Hoekstra and Chris Kirby for provision of animals and help with photography for Figures 9.2 and 9.3; Robert Zaccardi and Charles River Laboratories for Figure 9.3; Drs. Alex Ophir and Aubrey Kelly for Figure 9.5 and husbandry comments; Dr. Annelise Beery for Figure 9.6 and husbandry comments.

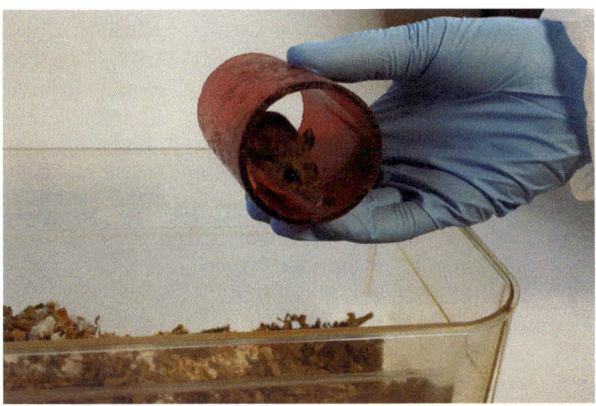 

FIGURE 9.7 Tunnel handling is readily accepted by *Peromsyscus* and anecdotally results in less stress for animals and fewer escapes (left). Also seen in the photo (right) is an example of the elaborate nests made by these mice when given enough appropriate nesting material.

REFERENCES

Ågren, G. 1984a. "Alternative mating strategies in the Mongolian gerbil." *Behaviour* 91:229–244. doi: 10.1163/156853984x00290.

Ågren, G. 1984b. "Pair formation in the Mongolian gerbil." *Animal Behavior* 32:528–535. doi: 10.1016/s0003-3472(84)80291-2.

Ågren, G., Q. Zhou, and W. Zhong. 1989a. "Ecology and social behaviour of Mongolian gerbils, *Meriones unguiculatus*, at Xilinhot, Inner Mongolia, China." *Animal Behavior* 37:11–27.

Ågren, G., Q. Zhou, and W. Zhong. 1989b. "Territoriality, cooperation and resource priority: hoarding in the Mongolian gerbil, *Meriones unguiculatus*." *Animal Behavior* 37:28–32.

Aguilar, S. 2011. "*Peromyscus leucopus*." accessed November 19th. https://animaldiversity.org/accounts/Peromyscus_leucopus/.

Anderson, J. F., R. C. Johnson, L. A. Magnarelli, F. W. Hyde, and J. E. Myers. 1986. "*Peromyscus leucopus* and *Microtus pennsylvanicus* simultaneously infected with *Borrelia burgdorferi* and *Babesia microti*." *Journal of Clinical Microbiology* 23 (1):135–137.

Aragona, B. J., and Z. Wang. 2004. "The prairie vole (*Microtus ochrogaster*): An animal model for behavioral neuroendocrine research on pair bonding." *ILAR Journal* 45 (1):35–45. doi: 10.1093/ilar.45.1.35.

Babcock, A. M., and H. Graham-Goodwin. 1997. "Importance of preoperative training and maze difficulty in task performance following hippocampal damage in the gerbil." *Brain Research Bulletin* 42 (6):415–419.

Barrett, R. D. H., S. Laurent, R. Mallarino, et al. 2019. "Linking a mutation to survival in wild mice." *Science* 363 (6426):499–504. doi: 10.1126/science.aav3824.

Barry, W. J. 1976. "Environmental effects on food hoarding in deermice (*Peromyscus*)." *Journal of Mammalogy* 57 (4):731–746. doi: 10.2307/1379443.

Batzli, G. O. 1985. "Nutrition." In *Biology of New World Microtus, Special Publication No. 8*, edited by R. H. Tamarin, 779–811. Provo, Utah: American Society of Mammalogists.

Baumgardner, D. J., S. E. Ward, and D. A. Dewsbury. 1980. "Diurnal patterning of eight activities in 14 species of muroid rodents." *Animal Learning & Behavior* 8 (2):322–330. doi: 10.3758/Bf03199612.

Bechard, A. R., N. Bliznyuk, and M. H. Lewis. 2017. "The development of repetitive motor behaviors in deer mice: Effects of environmental enrichment, repeated testing, and differential mediation by indirect basal ganglia pathway activation." *Developmental Psychobiology* 59 (3):390–399. doi: 10.1002/dev.21503.

Beckett, A. C., J. T. Loh, A. Chopra, et al. 2018. "*Helicobacter pylori* genetic diversification in the Mongolian gerbil model." *PeerJ* 6:e4803. doi: 10.7717/peerj.4803.

Bedford, N. L., and H. E. Hoekstra. 2015. "*Peromyscus* mice as a model for studying natural variation." *eLife* 4:e06813. doi: 10.7554/eLife.06813.

Beery, A. K., D. M. Routman, and I. Zucker. 2009. "Same-sex social behavior in meadow voles: Multiple and rapid formation of attachments." *Physiology & Behavior* 97 (1):52–57. doi: 10.1016/j.physbeh.2009.01.020.

Berl, J. L., A. J. Kuenzi, E. A. Flaherty, and R. K. Swihart. 2018. "Interspecific comparison of hantavirus prevalence in *Peromyscus* populations from a fragmented agro-ecosystem in Indiana, USA." *Journal of Wildlife Diseases* 54 (1):147–150. doi: 10.7589/2017-02-022.

Boonstra, R., X. H. Xia, and L. Pavone. 1993. "Mating system of the meadow vole, *Microtus pennsylvanicus*." *Behavioral Ecology* 4 (1):83–89. doi: 10.1093/beheco/4.1.83.

Brochu, L., L. Caron, and J. M. Bergeron. 1988. "Diet quality and body condition of dispersing and resident voles (*Microtus pennsylvanicus*)." *Journal of Mammalogy* 69 (4):704–710. doi: 10.2307/1381625.

Brown, S. A. 2012. "Small mammal training in the veterinary practice." *Veterinary Clinics: Exotic Animal Practice* 15 (3):469–485. doi: 10.1016/j.cvex.2012.06.007.

Bunker, A. 2001. "*Peromyscus maniculatus*." accessed November 19th. https://animaldiversity.org/accounts/Peromyscus_maniculatus/.

Burger, J., and M. Gochfeld. 1992. "Survival and reproduction in *Peromyscus leucopus* in the laboratory: viable model for aging studies." *Growth, Development and Aging* 56 (1):17–22.

Choudhury, B., O. F. Adunka, C. E. DeMason, et al. 2011. "Detection of intracochlear damage with cochlear implantation in a gerbil model of hearing loss." *Otology & Neurotology* 32 (8):1370–1378. doi: 10.1097/MAO.0b013e31822f09f2.

Clark, F. H. 1937. "Parturition in the deer mouse." *Journal of Mammalogy* 18 (1):85–87. doi: 10.2307/1374316.

Clark, M. M., and B. G. Galef, Jr. 2000. "Effects of experience on the parental responses of male Mongolian gerbils." *Developmental Psychobiology* 36 (3):177–185. doi: 10.1002/(SICI)1098-2302(200004)36:3<177::AID-DEV1>3.0.CO;2-W [pii].

Clark, M. M., and B. G. Galef, Jr. 2002. "Socially induced delayed reproduction in female Mongolian gerbils (*Meriones unguiculatus*): Is there anything special about dominant females?" *Journal of Comparative Psychology* 116 (4):363–368.

Colvin, D. V. 1973. "Agonistic behaviour in males of five species of voles *Microtus*." *Animal Behaviour* 21 (3):471–480. doi: 10.1016/S0003-3472(73)80007-7.

Colvin, M. A., and D. V. Colvin. 1970. "Breeding and fecundity of six species of voles (*Microtus*)." *Journal of Mammalogy* 51 (2):417–419. doi: 10.2307/1378508.

Dawson, W. D. 2005. "A brief history of the *Peromyscus* Genetic Stock Center." https://www.pgsc.cas.sc.edu/sites/pgsc.cas.sc.edu/files/attachments/PGSC_HISTORY_1_0.pdf

Deacon, R. M. J. 2009. "Burrowing: A sensitive behavioural assay, tested in five species of laboratory rodents." *Behavioural Brain Research* 200 (1):128–133. doi: 10.1016/j.bbr.2009.01.007.

Dewsbury, D. A. 1988. "A test of the role of copulatory plugs in sperm competition in deer mice (*Peromyscus maniculatus*)." *Journal of Mammalogy* 69 (4):854–857. doi: 10.2307/1381648.

Dewsbury, D. A., D. L. Lanier, and A. Miglietta. 1980. "A laboratory study of climbing behavior in 11 species of muroid rodents." *The American Midland Naturalist* 103 (1):66–72. doi: 10.2307/2425039.

Dewsbury, D. A., D. Q. Estep, and D. L. Lanier. 1977. "Estrous cycles of nine species of muroid rodents." *Journal of Mammalogy* 58 (1):89–92. doi: 10.2307/1379735.

Drickamer, L. C. 1970. "Seed preferences in wild caught *Peromyscus maniculatus bairdii* and *Peromyscus leucopus noveboracensis*." *Journal of Mammalogy* 51 (1):191–194. doi: 10.2307/1378564.

Eisenberg, J. F. 1962. "Studies on the behavior of *Peromyscus maniculatus gambelii* and *Peromyscus californicus parasiticus*." *Behaviour* 19 (3):177–207. doi: 10.1163/156853962X00014.

Eisenberg, J. F. 1967. "A comparative study in rodent ethology with emphasis on social behavior, I." *Proceedings of the United States National Museum* 122 (3597):1–50.

Eisenberg, J. F. 1968. "Behavior patterns." In *Biology of Peromyscus (Rodentia)*, edited by J. A. King, 451–495. The American Society of Mammalogists.

Elwood, R. W. 1975. "Paternal and maternal behavior in the Mongolian gerbil." *Animal Behaviour* 23:766–772.

Elwood, R. W., and D. M. Broom. 1978. "Influence of litter size and parental behavior on development of Mongolian gerbil pups." *Animal Behaviour* 26 (MAY):438–454. doi: 10.1016/0003-3472(78)90061-1.

Evans, R. L., E. M. Katz, N. L. Olson, and D. A. Dewsbury. 1978. "A comparative study of swimming behavior in eight species of muroid rodents." *Bulletin of the Psychonomic Society* 11 (3):168–170. doi: 10.3758/bf03336797.

Ferkin, M. H. 2005. "Self-grooming in meadow voles." *Chemical Signals in Vertebrates* 10:64–69. doi: 10.1007/0-387-25160-X_9.

Fitch, H. S. 1957. "Aspects of reproduction and development in the prairie vole (*Microtus ochrogaster*)." *University of Kansas Publications, Museum of Natural History* 10 (4):129–161.

Flahou, B., F. Haesebrouck, A. Smet, et al. 2013. "Gastric and enterohepatic non-*Helicobacter pylori* Helicobacters." *Helicobacter* 18:66–72. doi: 10.1111/Hel.12072.

Forkman, B. 1996. "The foraging behaviour of Mongolian gerbils: A behavioural need or a need to know?." *Behaviour* 133 (1/2):129–143.

Getz, L. L., B. McGuire, and C. S. Carter. 2003. "Social behavior, reproduction and demography of the prairie vole, *Microtus ochrogaster*." *Ethology Ecology & Evolution* 15 (2):105–118. doi: 10.1080/08927014.2003.9522676.

Ginsburg, H. J., and W. G. Braud. 1971. "Laboratory investigation of aggressive behavior in Mongolian gerbils (*Meriones unguiculatus*)." *Psychonomic Science* 22 (1):54–55.

Glickman, S. E., L. Fried, and B. A. Morrison. 1967. "Shredding of nesting material in the Mongolian gerbil." *Perceptual and Motor Skills* 24 (2):473–474.

Goethert, H., and S. Telford. 2017. "Retrotransposon-targeted bloodmeal remnant identification identifies meadow voles as the main host for subadult dog ticks." *American Journal of Tropical Medicine and Hygiene* 95 (5):408–409.

Grippo, A. J., E. Ihm, J. Wardwell, et al. 2014. "The effects of environmental enrichment on depressive and anxiety-relevant behaviors in socially isolated prairie voles." *Psychosomatic Medicine* 76 (4):277–284. doi: 10.1097/psy.0000000000000052.

Haken, A. E., and G. O. Batzli. 1996. "Effects of availability of food and interspecific competition on diets of prairie voles (*Microtus ochrogaster*)." *Journal of Mammalogy* 77 (2):315–324. doi: 10.2307/1382803.

Hamilton, W. J. Jr. 1941. "Reproduction of the field mouse *Microtus pennsylvanicus* (Ord)." *Cornell University Agricultural Experimental Station Memoir* 237:1–24.

Hartung, T. G., and D. A. Dewsbury. 1978. "A comparative analysis of copulatory plugs in muroid rodents and their relationship to copulatory behavior." *Journal of Mammalogy* 59 (4):717–723. doi: 10.2307/1380136.

Hartung, T. G., and D. A. Dewsbury. 1979. "Nest-building behavior in seven species of muroid rodents." *Behavioral and Neural Biology* 27 (4):532–539. doi: 10.1016/S0163-1047(79)92166-6.

Hayssen, V. 1997. "Effects of the nonagouti coat-color allele on behavior of deer mice (*Peromyscus maniculatus*): a comparison with Norway rats (*Rattus norvegicus*)." *Journal of Comparative Psychology* 111 (4):419–423.

Heo, D. H., and T. C. Kang. 2012. "The changes of ERG channel expression after administration of antiepileptic drugs in the hippocampus of epilepsy gerbil model." *Neuroscience Letters* 507 (1):27–32. doi: 10.1016/j.neulet.2011.11.043.

Howard, W. E., and F. C. Evans. 1961. "Seeds stored by prairie deer mice." *Journal of Mammalogy* 42 (2):260–263. doi: 10.2307/1376847.

Howe, H. F., B. Zorn-Arnold, A. Sullivan, and J. S. Brown. 2006. "Massive and distinctive effects of meadow voles on grassland vegetation." *Ecology* 87 (12):3007–3013. doi: 10.1890/0012-9658(2006)87[3007:Madeom]2.0.Co;2.

Hurst, J. L., and R. S. West. 2010. "Taming anxiety in laboratory mice." *Nature Methods* 7 (10):825–826. doi: 10.1038/nmeth.1500.

Hurtado-Parrado, C., C. H. González, L. M. Moreno, et al. 2015. "Catalogue of the behaviour of *Meriones unguiculatus* f. dom. (Mongolian gerbil) and wild conspecies, in captivity and under natural conditions, based on a systematic literature review." *Journal of Ethology* 33 (2):65–86. doi: 10.1007/s10164-015-0421-0.

Innes, D. G. L., and J. S. Millar. 1994. "Life histories of *Clethrionomys* and *Microtus* (Microtinae)." *Mammal Review* 24 (4):179–207. doi: 10.1111/j.1365-2907.1994.tb00142.x.

Jackson, R. K. 1997. "Unusual laboratory rodent species: Research uses, care, and associated biohazards." *ILAR Journal* 38 (1):13–21. doi: 10.1093/ilar.38.1.13.

Joyner, C. P., L. C. Myrick, J. P. Crossland, and W. D. Dawson. 1998. "Deer mice as laboratory animals." *ILAR Journal* 39 (4):322–330. doi: 10.1093/ilar.39.4.322.

Kalueff, A. V., A. M. Stewart, C. Song, et al. 2016. "Neurobiology of rodent self-grooming and its value for translational neuroscience." *Nature Reviews Neuroscience* 17 (1):45–59. doi: 10.1038/nrn.2015.8.

Kaplan, H., and S. O. Hyland. 1972. "Behavioural development in the Mongolian gerbil (*Meriones unguiculatus*)." *Animal Behaviour* 20 (1):147–154.

Kapuria, S., C. R. Steele, and S. Puria. 2017. "Unraveling the mystery of hearing in gerbil and other rodents with an arch-beam model of the basilar membrane." *Scientific Reports* 7:228. doi: 10.1038/s41598-017-00114-x.

Kenkel, W. M., and C. S. Carter. 2016. "Voluntary exercise facilitates pair-bonding in male prairie voles." *Behavioural Brain Research* 296:326–330. doi: 10.1016/j.bbr.2015.09.028.

Kittrell, E. M., B. R. Gregg, and D. D. Thiessen. 1982. "Brood patch function for the ventral scent gland of the female Mongolian gerbil, *Meriones unguiculatus*." *Developmental Psychobiology* 15 (3):197–202. doi: 10.1002/dev.420150303.

Klatt, B. J., L. L. Getz, and B. McGuire. 2015. "Interspecific interactions and habitat use by prairie voles (*Microtus ochrogaster*) and meadow voles (*M. pennsylvanicus*)." *American Midland Naturalist* 173 (2):241–252. doi: 10.1674/amid-173-02-241-252.1.

Kriegsfeld, L. J., N. J. Ranalli, M. A. Bober, and R. J. Nelson. 2000. "Photoperiod and temperature interact to affect the GnRH neuronal system of male prairie voles (*Microtus ochrogaster*)." *Journal of Biological Rhythms* 15 (4):306–316. doi: 10.1177/074873000129001413.

Lackey, J. A., D. G. Huckaby, and B. G. Ormiston. 1985. "*Peromyscus leucopus*." *Mammalian Species* (247):1–10. doi: 10.2307/3503904.

Lacy, R. C., G. Alaks, and A. Walsh. 2013. "Evolution of *Peromyscus leucopus* mice in response to a captive environment." *PLoS One* 8 (8):e72452. doi: 10.1371/journal.pone.0072452.

Lanier, D. L., D. Q. Estep, and D. A. Dewsbury. 1974. "Food hoarding in muroid rodents." *Behavioral Biology* 11 (2):177–187. doi: 10.1016/S0091-6773(74)90337-X.

Layne, J. N. 1968. "Ontogeny." In *Biology of Peromyscus (Rodentia)*, edited by J. A. King, 148–253. The American Society of Mammalogists.

Layne, J. N. 1969. "Nest-building behavior in three species of deer mice, *Peromyscus*." *Behaviour* 35:288–302. doi: 10.1163/156853969x00260.

Lee, J. C., J. H. Park, J. H. Ahn, et al. 2016. "New GABAergic neurogenesis in the hippocampal CA1 region of a gerbil model of long-term survival after transient cerebral ischemic injury." *Brain Pathology* 26 (5):581–592. doi: 10.1111/bpa.12334.

Lee, N. S., N. L. Goodwin, K. E. Freitas, and A. K. Beery. 2019. "Affiliation, aggression, and selectivity of peer relationships in meadow and prairie voles." *Frontiers in Behavioral Neuroscience* 13:52. doi: 10.3389/fnbeh.2019.00052.

Leidinger, C., F. Herrmann, C. Thoene-Reineke, N. Baumgart, and J. Baumgart. 2017. "Introducing clicker training as a cognitive enrichment for laboratory mice." *Jove-Journal of Visualized Experiments* (121):55415. doi: 10.3791/55415.

Leidinger, C. S., N. Kaiser, N. Baumgart, and J. Baumgart. 2018. "Using clicker training and social observation to teach rats to voluntarily change cages." *Jove-Journal of Visualized Experiments* (140):58511. doi: 10.3791/58511.

Lewarch, C. L., and H. E. Hoekstra. 2018. "The evolution of nesting behaviour in *Peromyscus* mice." *Animal Behaviour* 139:103–115. doi: 10.1016/j.anbehav.2018.03.008.

Liu, W., G. Wang, Y. Wang, W. Zhong, and X. Wan. 2009. "Population ecology of wild Mongolian gerbils *Meriones unguiculatus*." *Journal of Mammalogy* 90 (4):832–840. doi: 10.1644/08-mamm-a-265.1.

Majumdar, S. K., J. H. Schwartz, and J. S. Santaspirt. 1974. "Agonistic behavior in the male Mongolian gerbil." *The American Biology Teacher* 36 (8):504–506. doi: 10.2307/4444948.

Mallory, F. F., and R. A. Dieterich. 1985. "Laboratory management and pathology." In *Biology of New World Microtus, Special Publication No. 8*, edited by R. H. Tamarin, 647–684. Provo, Utah: American Society of Mammalogists.

Marston, J. H., and M. C. Chang. 1965. "The breeding, management and reproductive physiology of the Mongolian gerbil (*Meriones unguiculatus*)." *Laboratory Animal Care* 15:34–48.

Mason, G., and H. Würbel. 2016. "What can be learnt from wheel-running by wild mice, and how can we identify when wheel-running is pathological?" *Proceedings. Biological sciences* 283 (1824):B281. doi: 10.1098/rspb.2015.0738.

McGraw, L. A., and L. J. Young. 2010. "The prairie vole: An emerging model organism for understanding the social brain." *Trends Neuroscience* 33 (2):103–109. doi: 10.1016/j.tins.2009.11.006.

McGuire, B., E. Henyey, E. McCue, and W. E. Bemis. 2003. "Parental behavior at parturition in prairie voles (Microtus ochrogaster)." *Journal of Mammalogy* 84 (2):513–523. doi: 10.1644/1545-1542(2003)084<0513:Pbapip>2.0.Co;2.

McGuire, B., W. E. Bemis, and F. Vermeylen. 2014. "Parental behaviour of prairie voles (*Microtus ochrogaster*) and meadow voles (*M. pennsylvanicus*) in relation to sex of offspring." *Behaviour* 151 (4):535–553. doi: 10.1163/1568539x-00003141.

Milne-Edwards, A. 1867. "Observations sur quelques mammifères du nord de la Chine." *Annales des Sciences Naturelle* V (8):375–377.

Mulder, G. B., K. R. Pritchett-Corning, M. A. Gramlich, and A. E. Crocker. 2010. "Method of feed presentation affects the growth of mongolian gerbils (*Meriones unguiculatus*)." *Journal of the American Association for Laboratory Animal Science* 49 (1):36–39.

Nestler, J. R., G. P. Dieter, and B. G. Klokeid. 1996. "Changes in total body fat during daily torpor in deer mice (*Peromyscus maniculatus*)." *Journal of Mammalogy* 77 (1):147–154. doi: 10.2307/1382716.

Neumann, K., S. Maak, I. W. Stuermer, G. von Lengerken, and R. Gattermann. 2001. "Low microsatellite variation in laboratory gerbils." *Journal of Heredity* 92 (1):71–74.

Nyby, J., and D. D. Thiessen. 1980. "Food hoarding in the mongolian gerbil (*Meriones unguiculatus*): effects of food deprivation." *Behavioral and Neural Biology* 30 (1):39–48.

Nyby, J., P. Wallace, K. Owen, and D. D. Thiessen. 1973. "An influence of hormones on hoarding behavior in the Mongolian gerbil (*Meriones unguiculatus*)." *Hormones and Behavior* 4:283–288.

Odberg, F. O. 1987. "The influence of cage size and environmental enrichment on the development of stereotypies in bank voles (*Clethrionomys glareolus*)." *Behavioural Processes* 14 (2):155–173. doi: 10.1016/0376-6357(87)90042-8.

Ostermeyer, M. C., and R. W. Elwood. 1984. "Helpers (?) at the nest in the Mongolian gerbil, *Meriones unguiculatus*." *Behaviour* 91:61–77. doi: 10.1163/156853984x00218.

Ostfeld, R. S. 2002. "Little loggers make a big difference - Red maple seedlings don't stand a chance around meadow voles." *Natural History* 111 (4):64–71.

Perkeybile, A. M., L. L. Griffin, and K. L. Bales. 2013. "Natural variation in early parental care correlates with social behaviors in adolescent prairie voles (*Microtus ochrogaster*)." *Frontiers in Behavioral Neuroscience* 7:21. doi: 10.3389/fnbeh.2013.00021.

Petrij, F., K. van Veen, M. Mettler, and V. Brückmann. 2001. "A second acromelanistic allelomorph at the albino locus of the Mongolian gerbil (*Meriones unguiculatus*)." *Journal of Heredity* 92 (1):74–78.

Pettijohn, T. F., and B. M. Barkes. 1978. "Surface choice and behavior in adult Mongolian gerbils." *Psychological Record* 28:299–303.

Pietrewicz, A. T., M. P. Hoff, and S. A. Higgins. 1982. "Activity rhythms in the Mongolian gerbil under natural light conditions." *Physiology & Behavior* 29 (2):377–380.

Piovanotti, M. R. A., and M. L. Vieira. 2004. "Presence of the father and parental experience have differentiated effects on pup development in Mongolian gerbils (*Meriones unguiculatus*)." *Behavioural Processes* 66 (2):107–117. doi: 10.1016/j.beproc.2004.01.007.

Platt, M. M. 1977. "An ethogram of the Mongolian gerbil, *Meriones unguiculatus*." M.S., Cornell.

Powell, S. B., H. A. Newman, J. F. Pendergast, and M. H. Lewis. 1999. "A rodent model of spontaneous stereotypy: Initial characterization of developmental, environmental, and neurobiological factors." *Physiology & Behavior* 66 (2):355–363. doi: 10.1016/S0031-9384(98)00303-5.

Powell, S. B., H. A. Newman, T. A. McDonald, P. Bugenhagen, and M. H. Lewis. 2000. "Development of spontaneous stereotyped behavior in deer mice: Effects of early and late exposure to a more complex environment." *Developmental Biology* 37 (2):100–108. doi: 10.1002/1098-2302(200009)37:2<100::Aid-Dev5>3.0.Co;2-6.

Prates, E. J., and R. F. Guerra. 2005. "Parental care and sexual interactions in Mongolian gerbils (*Meriones unguiculatus*) during the postpartum estrus." *Behavioural Processes* 70 (2):104–12. doi: S0376-6357(05)00126-9 [pii]10.1016/j.beproc.2005.04.010.

Pritchett-Corning, K. R., R. Keefe, J. P. Garner, and B. N. Gaskill. 2013. "Can seeds help mice with the daily grind?" *Laboratory Animals* 47 (4):312–315. doi: 10.1177/0023677213491403.

Randall, J. A. 2001. "Evolution and function of drumming as communication in mammals." *American Zoologist* 41 (5):1143–1156. doi: 10.1093/icb/41.5.1143.

Rehmeier, R. L., G. A. Kaufman, and D. W. Kaufman. 2004. "Long-distance movements of the deer mouse in tallgrass prairie." *Journal of Mammalogy* 85 (3):562–568. doi: 10.1644/1383956.

Rice, M. A., L. F. Restrepo, and A. G. Ophir. 2018. "When to cheat: Modeling dynamics of paternity and promiscuity in socially monogamous prairie voles (*Microtus ochrogaster*)." *Frontiers in Ecology and Evolution* 6:14110. doi: 10.3389/fevo.2018.00141.

Roper, T. J., and E. Polioudakis. 1977. "The behaviour of Mongolian gerbils in a semi-natural environment, with special reference to ventral marking, dominance, and sociability." *Behaviour* 61:207–237.

Saltzman, W., S. Ahmed, A. Fahimi, D. J. Wittwer, and F. H. Wegner. 2006. "Social suppression of female reproductive maturation and infanticidal behavior in cooperatively breeding Mongolian gerbils." *Hormones and Behavior* 49 (4):527–537. doi: 10.1016/j.yhbeh.2005.11.004.

Saltzman, W., S. Thinda, A. L. Higgins, et al. 2009. "Effects of siblings on reproductive maturation and infanticidal behavior in cooperatively breeding Mongolian gerbils." *Developmental Psychobiology* 51 (1):60–72. doi: 10.1002/dev.20347.

Sawrey, D. K., and D. A. Dewsbury. 1985. "Control of ovulation, vaginal estrus, and behavioral receptivity in voles (*Microtus*)." *Neuroscience & Biobehavioral Reviews* 9 (4):563–571. doi: 10.1016/0149-7634(85)90003-X.

Sawrey, D. K., J. R. Keith, and R. C. Backes. 1994. "Place learning by three vole species (*Microtus ochrogaster, M. montanus,* and *M. pennsylvanicus*) in the Morris swim task." *Journal of Comparative Psychology* 108 (2):179–188. doi: 10.1037/0735-7036.108.2.179.

Scheibler, E., R. Weinandy, and R. Gattermann. 2004. "Social categories in families of Mongolian gerbils." *Physiology & Behavior* 81 (3):455–464. doi: 10.1016/j.physbeh.2004.02.011.

Scheibler, E., R. Weinandy, and R. Gattermann. 2006. "Male expulsion in cooperative Mongolian gerbils (*Meriones unguiculatus*)." *Physiology & Behavior* 87 (1):24–30. doi: 10.1016/j.physbeh.2005.08.037.

Scheibler, E., W. Liu, R. Weinandy, and R. Gattermann. 2006. "Burrow systems of the Mongolian gerbil (*Meriones unguiculatus* Milne Edwards, 1867)." *Mammalian Biology* 71 (3):178–182. doi: 10.1016/j.mambio.2005.11.007.

Schwanz, L. E., M. J. Voordouw, D. Brisson, and R. S. Ostfeld. 2011. "*Borrelia burgdorferi* has minimal impact on the Lyme disease reservoir host *Peromyscus leucopus*." *Vector-Borne and Zoonotic Diseases* 11 (2):117–124. doi: 10.1089/vbz.2009.0215.

Shuster, S. M., R. M. Willen, B. Keane, and N. G. Solomon. 2019. "Alternative mating tactics in socially monogamous prairie voles, *Microtus ochrogaster*." *Frontiers in Ecology and Evolution* 7:7. doi: 10.3389/fevo.2019.00007.

Silveira, A. P. C., T. T. Kitabatake, V. M. Pantaleo, et al. 2018. "Continuous and not continuous 2-week treadmill training enhances the performance in the passive avoidance test in ischemic gerbils." *Neuroscience Letters* 665:170–175. doi: 10.1016/j.neulet.2017.12.012.

Solomon, N. G., A. M. Christiansen, Y. K. Lin, and L. D. Hayes. 2005. "Factors affecting nest location of prairie voles (*Microtus ochrogaster*)." *Journal of Mammalogy* 86 (3):555–560. doi: 10.1644/1545-1542(2005)86[555:Fanlop]2.0.Co;2.

Sørensen, D. B., T. Krohn, H. N. Hansen, J. L. Ottesen, and A. K. Hansen. 2005. "An ethological approach to housing requirements of golden hamsters, Mongolian gerbils and fat sand rats in the laboratory—A review." *Applied Animal Behaviour Science* 94:181–195.

Spruijt, B. M., J. A. R. A. M. Van Hooff, and W. H. Gispen. 1992. "Ethology and neurobiology of grooming behavior." *Physiological Reviews* 72 (3):825–852.

Starkey, N. J., and C. A. Hendrie. 1998. "Disruption of pairs produces pair-bond disruption in male but not female Mongolian gerbils." *Physiology & Behavior* 65 (3):497–503. doi: 10.1016/s0031-9384(98)00190-5.

Starkey, N. J., G. Normington, and N. J. Bridges. 2007. "The effects of individual housing on 'anxious' behaviour in male and female gerbils." *Physiology & Behavior* 90 (4):545–552. doi: 10.1016/j.physbeh.2006.11.001.

Steppan, S., R. Adkins, and J. Anderson. 2004. "Phylogeny and divergence-date estimates of rapid radiations in muroid rodents based on multiple nuclear genes." *Systematic Biology* 53 (4):533–553. doi: 10.1080/10635150490468701.

Stowe, J. R., Y. Liu, J. T. Curtis, M. E. Freeman, and Z. X. Wang. 2005. "Species differences in anxiety-related responses in male prairie and meadow voles: The effects of social isolation." *Physiology & Behavior* 86 (3):369–378. doi: 10.1016/j.physbeh.2005.08.007.

Stuermer, I. W., K. Plotz, A. Leybold, et al. 2003. "Intraspecific allometric comparison of laboratory gerbils with Mongolian gerbils trapped in the wild indicates domestication in *Meriones unguiculatus* (Milne-Edwards, 1867) (Rodentia: Gerbillinae)." *Zoologischer Anzeiger* 242:249–266.

Sumner, F. B. 1917. "Modern conceptions of heredity and genetic studies at the Scripps Institution." *Bulletin of the Scripps Institution for Biological Research of the University of California* (3):1–24.

Svihla, A. 1932. "A comparative life history study of the mice of the genus *Peromyscus*." *University of Michigan. Museum of Zoology, Miscellaneous Publications* 24:1–39.

Tannenbaum, M. G., and E. B. Pivorun. 1987. "Differential effect of food restriction on the induction of daily torpor in *Peromyscus maniculatus* and *Peromyscus leucopus*." *Journal of Thermal Biology* 12 (2):159–162. doi: 10.1016/0306-4565(87)90057-X.

Tannenbaum, M. G., and E. B. Pivorun. 1988. "Seasonal study of daily torpor in southeastern *Peromyscus maniculatus* and *Peromyscus leucopus* from mountains and foothills." *Physiological and Biochemical Zoology* 61 (1):10–16. doi: 10.1086/physzool.61.1.30163731.

Taylor, S. A., A. L. Salo, and D. A. Dewsbury. 1992. "Estrus induction in four species of voles (*Microtus*)." *Journal of Comparative Psychology* 106 (4):366.

Thiessen, D. D. 1988. "Body temperature and grooming in the Mongolian gerbil." *Annals of the New York Academy of Sciences* 525:27–39.

Turner, C. A., and M. H. Lewis. 2003. "Environmental enrichment: effects on stereotyped behavior and neurotrophin levels." *Physiology & Behavior* 80 (2):259–266. doi: 10.1016/j.physbeh.2003.07.008.

Ungvari, Z., B. F. Krasnikov, A. Csiszar, et al. 2008. "Testing hypotheses of aging in long-lived mice of the genus *Peromyscus*: Association between longevity and mitochondrial stress resistance, ROS detoxification pathways, and DNA repair efficiency." *Age* 30 (2–3):121–133. doi: 10.1007/s11357-008-9059-y.

Vander Wall, S. B., T. C. Thayer, J. S. Hodge, M. J. Beck, and J. K. Roth. 2001. "Scatter-hoarding behavior of deer mice (*Peromyscus maniculatus*)." *Western North American Naturalist* 61 (1):109–113.

Veres, M., A. R. Duselis, A. Graft, et al. 2012. "The biology and methodology of assisted reproduction in deer mice (*Peromyscus maniculatus*)." *Theriogenology* 77 (2):311–319. doi: 10.1016/j.theriogenology.2011.07.044.

Vogt, F. D., and G. R. Lynch. 1982. "Influence of ambient-temperature, nest availability, huddling, and daily torpor on energy-expenditure in the white-footed mouse *Peromyscus leucopus*." *Physiological Zoology* 55 (1):56–63. doi: 10.1086/physzool.55.1.30158443.

Waiblinger, E. 2002. "Comfortable quarters for gerbils in research institutions." In *Comfortable Quarters for Laboratory Animals*, edited by V. Reinhardt and A. Reinhardt, 18–25. Washington, DC: Animal Welfare Institute.

Waiblinger, E., and B. Koenig. 2007. "Housing and husbandry conditions affect stereotypic behaviour in laboratory gerbils." *Altex-Alternativen Zu Tierexperimenten* 24:67–69.

Waring, A., and T. Perper. 1979. "Parental behavior in the Mongolian gerbil (*Meriones unguiculatus*) 1. Retrieval." *Animal Behaviour* 27 (NOV):1091–1097. doi: 10.1016/0003-3472(79)90057-5.

Weber, J. N., and H. E. Hoekstra. 2009. "The evolution of burrowing behaviour in deer mice (genus *Peromyscus*)." *Animal Behaviour* 77 (3):603–609. doi: 10.1016/j.anbehav.2008.10.031.

Webster, A. B., and R. J. Brooks. 1981. "Daily movements and short activity periods of free-ranging meadow voles *Microtus pennsylvanicus*." *Oikos* 37 (1):80–87. doi: 10.2307/3544076.

Weinert, D., R. Weinandy, and R. Gattermann. 2007. "Photic and non-photic effects on the daily activity pattern of Mongolian gerbils." *Physiology & Behavior* 90 (2–3):325–333. doi: S0031-9384(06)00423-9 [pii]10.1016/j.physbeh.2006.09.019.

Whitaker, J. O., Jr. 1963. "Food of 120 *Peromyscus leucopus* from Ithaca, New York." *Journal of Mammalogy* 44:418–419. doi: 10.2307/1377215.

Whitaker, J. O., Jr. 1966. "Food of *Mus musculus*, *Peromyscus maniculatus bairdii* and *Peromyscus leucopus* in Vigo county, Indiana." *Journal of Mammalogy* 47 (3):473–486. doi: 10.2307/1377688.

Whitsett, J. M., and D. D. Thiessen. 1972. "Sex difference in the control of scent-marking behavior in the Mongolian gerbil (*Meriones unguiculatus*)." *Journal of Comparative and Physiological Psychology* 78 (3):381–385.

Wiedenmayer, C. 1996. "Effect of cage size on the ontogeny of stereotyped behaviour in gerbils." *Applied Animal Behaviour Science* 47 (3–4):225–233. doi: 10.1016/0168-1591(95)00652-4.

Wiedenmayer, C. 1997a. "Causation of the ontogenetic development of stereotypic digging in gerbils." *Animal Behaviour* 53:461–470 doi: 10.1006/anbe.1996.0296.

Wiedenmayer, C. 1997b. "Sterotypies resulting from a deviation in the ontogenetic development of gerbils." *Behavioural Processes* 39:215–221.

Williams, O. 1959. "Food habits of the deer mouse." *Journal of Mammalogy* 40 (3):415–419. doi: 10.2307/1376568.

Wilson, S. C. 1982. "The development of social behaviour between siblings and non-siblings of the voles *Microtus ochrogaster* and *Microtus pennsylvanicus*." *Animal Behaviour* 30 (2):426–437. doi: 10.1016/S0003-3472(82)80053-5.

Wolff, J. O. 1985a. "Behavior." In *Biology of New World Microtus, Special Publication No. 8*, edited by R. H. Tamarin, 340–372. Provo, Utah: American Society of Mammalogists.

Wolff, J. O. 1985b. "Comparative population ecology of *Peromyscus leucopus* and *Peromyscus maniculatus*." *Canadian Journal of Zoology* 63 (7):1548–1555. doi: 10.1139/z85-230.

Wolff, J. O. 1994. "Reproductive success of solitarily and communally nesting white-footed mice and deer mice." *Behavioral Ecology.* 5 (2):206–209. doi: 10.1093/beheco/5.2.206.

Wolff, J. O., A. S. Dunlap, and E. Ritchhart. 2001. "Adult female prairie voles and meadow voles do not suppress reproduction in their daughters." *Behavioural Processes* 55 (3):157–162. doi: 10.1016/S0376-6357(01)00176-0.

Wolmarans, D. W., D. J. Stein, and B. H. Harvey. 2017. "Social behavior in deer mice as a novel interactive paradigm of relevance for obsessive-compulsive disorder (OCD)." *Social Neuroscience* 12 (2):135–149. doi: 10.1080/17470919.2016.1145594.

Wolmarans, D. W., I. M. Scheepers, D. J. Stein, and B. H. Harvey. 2018. "*Peromyscus maniculatus bairdii* as a naturalistic mammalian model of obsessive-compulsive disorder: current status and future challenges." *Metabolic Brain Disease* 33 (2):443–455. doi: 10.1007/s11011-017-0161-7.

Würbel, H. 2006. "The motivational basis of caged rodents' stereotypies." In *Stereotypic Animal Behavior. Fundamentals and Applications to Welfare*, edited by G. Mason and J. Rushen, 86–120. Oxfordshire, UK: CAB International.

Würbel, H., M. Stauffacher, and D. von Holst. 1996. "Stereotypies in laboratory mice — quantitative and qualitative description of the ontogeny of 'wire-gnawing' and 'jumping' in Zur:ICR and Zur:ICR nu." *Ethology* 102 (3):371–385. doi: 10.1111/j.1439-0310.1996.tb01133.x.

Xia, X., and J. S. Millar. 1988. "Paternal behavior by *Peromyscus leucopus* in enclosures." *Canadian Journal of Zoology* 66 (5):1184–1187. doi: 10.1139/z88-173.

Yan, B. C., J. H. Park, J. H. Ahn, et al. 2014. "Neuroprotection of posttreatment with risperidone, an atypical antipsychotic drug, in rat and gerbil models of ischemic stroke and the maintenance of antioxidants in a gerbil model of ischemic stroke." *Journal of Neuroscience Research* 92 (6):795–807. doi: 10.1002/jnr.23360.

Yi, F., Z. Wang, C. C. Baskin, et al. 2019. "Seed germination responses to seasonal temperature and drought stress are species-specific but not related to seed size in a desert steppe: Implications for effect of climate change on community structure." *Ecology and Evolution* 9 (4):2149–2159. doi: 10.1002/ece3.4909.

10

Behavioral Biology of Hamsters

Christina Winnicker
GlaxoSmithKline

Kathleen R. Pritchett-Corning
Harvard University
University of Washington

CONTENTS

- Introduction 165
- Typical Research Contributions 165
- Behavioral Biology 166
  - Natural History 166
    - Ecology, Habitat Use, and Predator/Prey Relations 166
    - Social Organization and Behavior 166
    - Feeding Behavior 167
    - Communication 167
  - Common Captive Behaviors 167
    - Normal 167
    - Abnormal 168
- Ways to Maintain Behavioral Health 168
  - Captive Considerations to Promote Species-Typical Behavior 168
    - Environment 168
    - Social Groupings 168
    - Feeding Strategies 169
    - Training 169
- Conclusions/Recommendations 169
- References 169

Introduction

Although they were once significant contributors to research, hamsters are now relatively uncommon research animals. According to United Kingdom sources, there were 1,400 hamsters used in research in 2018 (Home Office, 2018). Compare this to the 2.6 million mice used in research in the UK over the same time period and hamsters' rarity becomes apparent. Although statistics on the direct contributions to research of mice in the United States are not kept, statistics on hamster involvement are available, and approximately 100,000 hamsters were used in U.S. research in 2017 (USDA APHIS, 2017). In the same period, approximately double the number of guinea pigs and 1.5 times more rabbits played a part in research. Two main types of hamsters contribute to biomedical research: the Chinese (*Cricetulus griseus*) and the Syrian (*Mesocricetus auratus*). This chapter will address the Syrian exclusively, since it is by far the most numerous species in the laboratory, and is what comes to mind when hearing the word "hamster". Syrian, or golden, hamsters are thick-bodied rodents that are 15–18 cm in length and weigh between 100 and 140 g. Their short hair is reddish gold, with a pale white to gray ventrum. They have a short, blunt tail, large cheek pouches, and copious loose skin [see Figures 10.1 and 10.2]. Females are generally larger than males. Both sexes have flank glands that they use for scent marking, but males' are more prominent. Domesticated Syrian hamsters that are currently resident in laboratories are believed to have originated from a few founding animals that were littermates (Siegel, 1985). Thus, there is little genetic diversity within the laboratory hamster population.

Typical Research Contributions

Historically, large numbers of hamsters contributed to research due to the unique biology of their cheek pouches. A hamster's cheek pouch is an immunologically privileged site, so tumors could be implanted there for further study. Hamsters fell drastically in popularity with the advent of the nude and scid

FIGURE 10.1 Front view of Syrian hamster (*Mesocricetus auratus*). Note the golden coat, pale undercoat, and large cheek pouches.

FIGURE 10.2 Rear view of Syrian hamster (*Mesocricetus auratus*). Note the short tail and copious loose skin.

mouse. Mice were smaller, better-tempered, and easier to keep than hamsters, plus tumors could be implanted in multiple places on the body of immunodeficient mice.

Currently, hamsters are typically seen in chronobiology research, since as hibernating animals they are subject to jet lag (Gibson et al., 2010). Hamsters have also served as models of multiple infectious diseases: adenovirus infection (Wold et al., 2019), prion disease (Wang et al., 2019), and as an experimental model of visceral leishmaniasis (Jimenez-Anton et al., 2019). Additionally, hamsters have been useful as a model of mammalian reproductive biology, contributing to the understanding of *in vitro* fertilization and sperm microinjection (Hirose and Ogura, 2019).

Behavioral Biology

Natural History

Ecology, Habitat Use, and Predator/Prey Relations

Native to western Asia, the natural habitat of hamsters consists of either dry, rocky plains or lightly vegetated slopes where resources are relatively scarce. The summer climate is hot and dry, whereas winters can be wet and cool. Hamsters are terrestrial, and have been described in the literature as both nocturnal (in captivity) and diurnal (in nature) (Gattermann et al., 2008). For protection from both the environment and predation, hamsters dig burrows, which they may use for several years, and due to resource scarcity, these burrows are typically distant from one another (Gattermann et al., 2001). Observation of wild hamsters shows that nest burrows are generally at a depth of about 50 cm, and the smallest observed distance between inhabited burrows was 118 m (Gattermann et al., 2001). Each animal may build more than one burrow in an area; a primary burrow and several simple secondary escape burrows. Burrows are generally individually inhabited, each containing a male or a female and her litter, with two tunnels off the nest burrow: one for waste and one for food storage (Gattermann et al., 2001).

Hamsters reduce their metabolic rate and hibernate to conserve energy when temperatures drop to approximately 5°C (±2°C). Hamsters are considered true hibernators, a state in which the animal's heart rate, temperature, and metabolic rate reduce significantly. Optimal conditions for the induction of hibernation include the provision of nesting material, enough food to establish a store, and several days of cooler temperatures (Ueda and Ibuka, 1995). Hibernation usually occurs between November and February in the wild (Gattermann et al., 2001).

Social Organization and Behavior

Hamsters are territorial, and both males and females mark their territories with their flank glands. Territories are viciously defended against intruders. Adult females are larger and more aggressive than their male counterparts (Grelk et al., 1974; Kuhnen, 2002). Dominance among females appears to be stable, linear, and related to androgen levels (Drickamer et al., 1973). Agonistic encounters, similar to other rodents, generally consist of dominance behavior displayed by one animal, responded to by submissive and retreat behaviors by the subordinate animal (Lerwill and Makings, 1971).

Hamsters do play fight in early adolescence (before postnatal day (PND) 35–42) (Kyle et al., 2019) and play fighting among littermates can be distinguished from serious fighting by the location targeted on the body. In play fighting, hamsters attack and defend the cheeks and cheek pouches, whereas in serious fighting, the targeted areas tend to be the rump and lower flanks (Pellis and Pellis, 1988).

In nature, adult hamsters are solitary and only come into contact for mating. Sexual maturity is reached by 6–7 weeks of age, and optimal reproductive age is reached by 8–10 weeks in females and 10–12 weeks in males (Whittaker, 2010). Female hamsters attract mates by leaving a vaginal scent trail when they are fertile. Their interactions with the males are aggressive during late estrus and early diestrus, but as their estrous cycle proceeds, their aggressiveness changes to sexually receptive behaviors. This receptive period generally lasts only about a week surrounding mating (Marques and Valenstein, 1977). The receptive female displays lordosis, a posture of flattened back and elevated tail. Mating is brief and shortly followed by the female chasing the male from her nest site and retrieving his food supply for her own hoard (Lisk et al., 1983).

Females build nests beginning in late gestation (Bhatia et al., 1995; Kauffman et al., 2003) and give birth to altricial young that require intensive care that tapers off around PND 21, with natural weaning occurring around PND 24 (Nichita et al., 2010). Maternal behavior is similar to other rodents, in which mothers nurse and care for altricial young through approximately PND 24 (Rowell, 1961). Mother hamsters have a reputation for neglect and cannibalization that is somewhat deserved, particularly with first litters. In one study, primiparous females had a dramatic decrease in retrieval latency as their experience with pups increased over the course of raising their first litter (Swanson and Campbell, 1979a). In addition, the more experienced hamster dams retrieved pups to the nest for 18-day rather than the 15-day retrieval practiced by the primiparous dams (Swanson and Campbell, 1979b). The same study also showed a 50% rate of cannibalism and an absence of maternal care, even in multiparous females. Environmental instabilities, such as changing seasons, additional animals in the room or housing chamber, or lack of appropriate resources (space, nesting material, food, water) for the litter size, have been suggested as reasons for this (Swanson and Campbell, 1979b), and as such, disturbances to the environment of newly born litters should be avoided whenever possible.

Feeding Behavior

Hamsters are diurnal in nature, but nocturnal in captivity (Gattermann et al., 2008). Regardless of the time of day, when feeding, hamsters leave their burrows and forage for food, stuffing their cheek pouches with their finds. Food retrieved is brought back to the burrow and hoarded into a "resource hoard" [see Figure 10.3 and description below]. The diet is varied, consisting mostly of grains, plant roots and shoots, insects, and fruit (Whittaker, 2010).

Communication

Similar to other rodents, hamsters distinguish individuals by scent. Odor marks left in urine, feces, or flank gland secretions [see Figure 10.3] (Johnston et al., 1993) allow marking of territory and resources, as well as the determination of dominance rank and social order. Hamsters also communicate via ultrasonic vocalizations. During mating, males produce two distinct calls: one when in the presence of a receptive female, and one following the removal of the female after mating (Cherry, 1989). In one series of experiments, calls were influenced by odor and estrous cycle stage, and both sexes emit sounds with dominant frequencies of 32–42 kHz (Floody and Pfaff, 1977a, b, Floody, Pfaff, et al., 1977).

Common Captive Behaviors
Normal

Provided appropriate bedding substrate, laboratory hamsters will build burrows similar to their wild relatives. Regardless of bedding substrate, they will create a resource hoard, hoarding food, and nesting material alike. When provided with valuable resources, whether it be food or nesting material, distance of the material from the nest site was inversely related to the amount hoarded, suggesting that both transportability and "hoard-ability" were important qualities (Ottoni and Ades, 1991).

Like other laboratory rodents, hamsters will also make use of paper nesting material to build a nest. In one study, when given free access to straw nesting material, hamsters used from 15 to 25 g of material, building a larger nest when pregnant (Richards, 1969). When late-term pregnant hamsters were provided with tissues or crinkle paper nesting material, they used these materials to build nests of higher quality than when provided with wood shaving bedding material alone (Winnicker and Gaskill, 2014) [see Figure 10.4]. It should be noted that only crinkle paper nesting material significantly decreased pup mortality at weaning in this study.

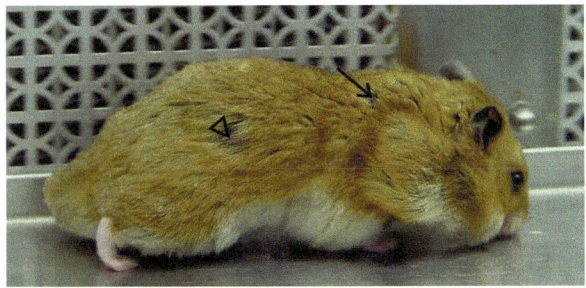

FIGURE 10.3 Side view of Syrian hamster (*Mesocricetus auratus*) showing a full cheek pouch and flank gland.

FIGURE 10.4 Hamster female using crinkle paper nesting material to build a nest for newborn pups.

Abnormal

Interindividual Aggression

Hamsters, like most domesticated rodents (Krause and Schüler, 2010), exhibit some amount of behavioral plasticity, and have a higher social tolerance threshold, allowing them to be housed in groups (Gattermann et al., 2002). While aggressiveness among hamsters is not an abnormal behavior per se, decreasing, or at least mitigating, it in the laboratory environment is desirable. The provision of structural enrichment, specifically a tunnel, has been shown to decrease aggressiveness (McClure and Thomson, 1992; Arnold and Estep, 1994). Interestingly, the more aggressive sex in hamsters appears to be the females (Gattermann et al., 2002; Krause and Schüler, 2010).

Handler-Directed Aggression

Hamsters have a reputation for being grouchy animals in laboratories. Since they are nocturnal in captivity, handling them in the day can be problematic, since they are also anecdotally sound sleepers. In one study, socially housed animals were found to attempt to bite handlers less frequently than individually housed hamsters (Arnold and Estep, 1994). In addition, a slow approach, allowing sleeping hamsters to awaken before handling them, then using a scoop technique, rather than scruffing, can decrease aggressiveness associated with handling [see Figure 10.5].

Stereotypies

Stereotypic bar-mouthing or gnawing of the wire bars or lid of the cage is an observed abnormal behavior in hamsters (Arnold and Estep, 1994; Hauzenberger et al., 2006). The provision of running wheels, at least in one study, decreased this behavior (Gebhardt-Henrich et al., 2005). As with other rodents, wheel running itself may become a stereotypic behavior and the use of running wheels must be carefully examined in relation to research goals (Mason and Würbel, 2016).

Ways to Maintain Behavioral Health

Captive Considerations to Promote Species-Typical Behavior

Environment

Solid bottom caging allows for bedding that can accommodate natural digging, burrowing and, to an extent, nest building behaviors. In one study, hamsters showed a preference for solid bottom caging with bedding versus wire bottom caging in spite of a reduction in floor space (Arnold and Estep, 1994). The provision of more biologically relevant nesting materials, such as straw and crinkle paper, allows hamsters to build a more structurally complex nest and these materials are readily utilized.

As a nocturnal animal under laboratory conditions, darkened areas, provided either by a shelter or bedding in which burrowing is possible, should be provided. If hamsters were provided with both glass Mason jars and PVC T pipe, hamsters seemed to prefer the jar, perhaps because of its spatial similarity to natural burrows (Arnold and Estep, 1994). Given the option, hamsters will choose to build a nest in a darker nest box as opposed to a lighter one, so sheltered or shaded areas should be provided. Hamsters in one study chose the dark nest box over either larger boxes or ones closer to resources (Ottoni and Ades, 1991). The ideal caging environment should take into account all of these factors: bedding substrate deep enough to burrow in, the provision of both nesting material and food in a manner in which the animals can obtain and hoard these desired resources, and areas of shadow or nest boxes, near the resources, to allow for hoarding.

Social Groupings

In spite of being a solitary species in the wild, most laboratory hamsters are group housed and do well in most group-housing conditions (Krause and Schüler, 2010). Socially housed hamsters spend more time in social contact than not, particularly if

FIGURE 10.5 Scoop technique for hamster handling (left) and restraint (right).

FIGURE 10.6 Male and female pair living harmoniously. Note they have created both a food hoard and nesting material hoard.

the animals have prior social housing experience (Arnold and Estep, 1990, 1994). Similarly, it was generally accepted that males and females only come together for breeding and should not be housed in breeding pairs. However, in at least one study, breeding hamster pairs were housed together without incident (Pritchett-Corning and Gaskill, 2015) [see Figure 10.6].

Prior experience plays a significant role in hamster behavior, and therefore significantly influences behavioral management and enrichment of the species. Socially reared male hamsters show submissive behaviors at closer distances than ones raised in isolation (Huang and Hazlett, 1974), suggesting that an increased tolerance for social situations may be a learned behavior. Early life experience appears to contribute to this social success (Huang and Hazlett, 1974). In addition, young hamsters play, and this play behavior is important in shaping adult behavior (Kyle et al., 2019). Therefore, it is likely that early housing experience affects later social preference and behavior in hamsters, and the social housing paradigms allow for a more social animal in laboratory rearing conditions.

Feeding Strategies

A typical rodent pellet works well for hamsters, as they do not seem to have special dietary needs in the laboratory. In the wild, their hoards contain fresh foodstuffs, so enrichment via diet might be a possibility for hamsters. Substrate should be provided to allow hamsters to display their natural foraging, processing, and hoarding behaviors. Placing food on the floor in their bedding should provide opportunity for these behaviors. Since hamsters have wide snouts, they cannot access pellets fed through a traditional hanging wire feeder like other laboratory rodents (Harkness et al., 1977). Alternatively, feeding in a J-feeder-style pellet delivery device will result in the hamsters removing the feed, dragging it to their hoard, and using the J-feeder as a nest site. Thus, floor feeding remains the best strategy for hamsters.

Training

Due to their nocturnal nature, hamsters woken from their daytime sleep tend to be ornery. Thus, poking or prodding them awake from sleep for handling, for procedures, or cage change, for example, can result in a bite. Gentle rousing of a cage of sleeping hamsters, and then scooping or cupping the animals with two hands, renders them least likely to bite during handling. While not a traditional behavioral model, hamsters like other small rodents can be trained via operant conditioning methods (Dibattista, 1999; Brown, 2012; Leidinger et al., 2017, 2018; Cloutier, this volume; MacLellan et al., this volume).

Conclusions/Recommendations

Hamsters, while generally considered a solitary species, will adapt to social housing, particularly if the social housing paradigm is introduced at a young age, such as weaning littermates together. Caging should be as large and deep as possible to provide ample space to have a hoard, as well as a separate toilet area. Caging space and bedding substrates that provide the opportunity to build a complex burrow, or at least the provision of a structural burrow proxy, such as bent PVC tubing or a shelter, are ideal. Food should be provided on the floor or inside the cage in some manner, in order to allow hamsters the opportunity to display the highly motivated behaviors of seeking, collecting, and hoarding valuable resources. Fibrous nesting material, particularly for solitary or paired late-term pregnant females and breeding pairs, should be provided in ways that stimulate the search for, collection, and dragging of the substrate to the hoard. Finally, hamsters display nocturnal patterns of behavior in the laboratory and can be grumpy when awoken abruptly; acclimation to handling, minimizing handling during the light portion of the cycle, gentle and gradual awakening, and scooping or other minimal contact handling techniques are highly recommended.

REFERENCES

Arnold, C. E., and Estep, D. Q. 1990. Effects of housing on social preference and behaviour in male golden hamsters (*Mesocricetus auratus*). *Applied Animal Behaviour Science.* 27 (3):253–261.

Arnold, C. E., and Estep, D. Q. 1994. Laboratory caging preferences in golden-hamsters (*Mesocricetus auratus*). *Laboratory Animals* 28 (3):232–238.

Bhatia, A. J., Schneider, J. E., and Wade, G. N. 1995. Thermoregulatory and maternal nestbuilding in Syrian hamsters: Interaction of ovarian steroids and energy demand. *Physiology & Behavior* 58 (1):141–146.

Brown, S. A. 2012. Small mammal training in the veterinary practice. *Veterinary Clinics: Exotic Animal Practice* 15 (3):469–485. doi: 10.1016/j.cvex.2012.06.007.

Cherry, J. A. 1989. Ultrasonic vocalizations by male hamsters: parameters of calling and effects of playbacks on female behaviour. *Animal Behaviour* 38 (1):138–153.

Dibattista, D. 1999. Operant responding for dietary protein in the golden hamster (*Mesocricetus auratus*). *Physiology & Behavior* 67 (1):95–98. doi: 10.1016/S0031-9384(99)00043-8.

Drickamer, L. C., Vandenbergh, J. G., and Colby, D. R. 1973. Predictors of dominance in the male golden hamster (*Mesocricetus auratus*). *Animal Behaviour* 21 (3):557–563.

Floody, O. R., and Pfaff, D. W. 1977a. Communication among hamsters by high-frequency acoustic signals: I. Physical characteristics of hamsters calls. *Journal of Comparative and Physiological Psychology* 91 (4):794–806.

Floody, O. R., and Pfaff, D. W. 1977b. Communication among hamsters by high-frequency acoustic signals: III. Response evoked by natural and synthetic ultrasounds. *Journal of Comparative and Physiological Psychology* 91 (4):820–829.

Floody, O. R., Pfaff, D. W., and Lewis, C. D. 1977. Communication among hamsters by high-frequency acoustic signals: II. Determinants of calling by females and males. *Journal of Comparative and Physiological Psychology* 91 (4):807–819.

Gattermann, R., Fritzsche, P., Neumann, K., et al. 2001. Notes on the current distribution and the ecology of wild golden hamsters (*Mesocricetus auratus*). *J. Zool.* 254 (3):359–365. doi: 10.1017/s0952836901000851.

Gattermann, R., Fritzsche, P., Weinandy, R., and Neumann, K. 2002. Comparative studies of body mass, body measurements and organ weights of wild-derived and laboratory golden hamsters (*Mesocricetus auratus*). *Laboratory Animals* 36 (4):445–454.

Gattermann, R., Johnston, R. E., Yigit, N., et al. 2008. Golden hamsters are nocturnal in captivity but diurnal in nature. *Biology Letters* 4 (3):253–255. doi: 10.1098/rsbl.2008.0066.

Gebhardt-Henrich, S. G., Vonlanthen, E. M., and Steiger, A. 2005. How does the running wheel affect the behaviour and reproduction of golden hamsters kept as pets? *Applied Animal Behaviour Science* 95:199–203.

Gibson, E. M., Wang, C., Tjho, S., Khattar, N., and Kriegsfeld, L. J. 2010. Experimental 'Jet Lag' inhibits adult neurogenesis and produces long-term cognitive deficits in female hamsters. *PLoS One* 5 (12):e15267.

Grelk, D. F., Papson, B. A., Cole, J. E., and Rowe, F. A. 1974. Influence of caging conditions and hormone treatments on fighting in male and female hamsters. *Hormones and Behavior* 5 (4):355–366.

Harkness, J. E., Wagner, J. E., Kusewitt, D. F., and Frisk, C. S. 1977. Weight loss and impaired reproduction in the hamster attributable to an unsuitable feeding apparatus. *Laboratory Animal Science* 27 (1):117–118.

Hauzenberger, A. R., Gebhardt-Henrich, S. G., and Steiger, A. 2006. The influence of bedding depth on behaviour in golden hamsters (*Mesocricetus auratus*). *Applied Animal Behaviour Science* 100 (3–4):280–294.

Hirose, M., and Ogura, A. 2019. The golden (Syrian) hamster as a model for the study of reproductive biology: Past, present, and future. *Reproductive Medicine and Biology* 18 (1):34–39. doi: 10.1002/rmb2.12241.

Home Office. 2018. Statistics of scientific prooceedures on living animals, Great Britain: 2018. Home Office, accessed 12 Dec. https://www.gov.uk/government/statistics/statistics-of-scientific-procedures-on-living-animals-great-britain-2018.

Huang, D., and Hazlett, B. A. 1974. Submissive distance in golden-hamster *Mesocricetus auratus*. *Animal Behaviour* 22 (MAY):467–472.

Jimenez-Anton, M. D., Grau, M., Olias-Molero, A. I., and Alunda, J. M. 2019. Syrian hamster as an advanced experimental model for visceral leishmaniasis. *Methods in Molecular Biology (Clifton, N.J.)* 1971:303–314. doi: 10.1007/978-1-4939-9210-2_17.

Johnston, R. E., Derzie, A., Chiang, G., Jernigan, P., and Lee, H. C. 1993. Individual scent signatures in golden hamsters: evidence for specialization of function. *Animal Behaviour* 45 (6):1061–1070.

Kauffman, A. S., Paul, M. J., Butler, M. P., and Zucker, I. 2003. Huddling, locomotor, and nest-building behaviors of furred and furless Siberian hamsters. *Physiology & Behavior* 79 (2):247–256.

Krause, S., and Schüler, L. 2010. Behavioural and endocrinological changes in Syrian hamsters (*Mesocricetus auratus*) under domestication. *Journal of Animal Breeding and Genetics* 127 (6):452–461. doi: 10.1111/j.1439-0388.2010.00861.x.

Kuhnen, G. 2002. Comfortable quarters for hamsters in research institutions. In *Comfortable Quarters for Laboratory Animals*, edited by V. Reinhardt and A. Reinhardt, 33–37. Washington, DC: Animal Welfare Institute.

Kyle, S. C., Burghardt, G. M., and Cooper, M. A. 2019. Development of social play in hamsters: Sex differences and their possible functions. *Brain Research* 1712:217–223. doi: 10.1016/j.brainres.2019.02.012.

Leidinger, C., Herrmann, F., Thoene-Reineke, C., Baumgart, N., and Baumgart, J. 2017. Introducing clicker training as a cognitive enrichment for laboratory mice. *Jove-Journal of Visualized Experiments* (121):55415. doi: 10.3791/55415.

Leidinger, C. S., Kaiser, N., Baumgart, N., and Baumgart, J. 2018. Using clicker training and social observation to teach rats to voluntarily change cages. *Jove-Journal of Visualized Experiments* (140):58511. doi: 10.3791/58511.

Lerwill, C. J., and Makings, P. 1971. Agonistic behavior of golden-hamster *Mesocricetus auratus* (Waterhouse). *Animal Behaviour* 19 (4):714–721.

Lisk, R. D., Ciaccio, L. A., and Catanzaro, C. 1983. Mating behaviour of the golden hamster under seminatural conditions. *Animal Behaviour* 31 (3):659–666.

Marques, D. M., and Valenstein, E. S. 1977. Individual differences in aggressiveness of female hamsters - response to intact and castrated males and to females. *Animal Behaviour* 25 (FEB):131–139.

Mason, G., and Würbel, H. 2016. What can be learnt from wheel-running by wild mice, and how can we identify when wheel-running is pathological? *Proceedings of the Royal Society B: Biological Sciences* 283 (1824):20150738. doi: 10.1098/rspb.2015.0738.

McClure, D. E., and Thomson, J. I. 1992. Cage enrichment for hamsters housed in suspended wire cages. *Contemporary Topics in Laboratory Animal Science* 31:33.

Nichita, I., Şereş, M., and Coman, C. 2010. Maternal behaviour on golden hamster (*Mesocricetus auratus*). *Lucrari Stiinfice Medicina Veterinara* XLIII (2):335–340.

Ottoni, E. B., and Ades, C. 1991. Resource location and structural-properties of the nestbox as determinants of nest-site selection in the golden-hamster. *Animal Learning & Behavior* 19 (3):234–240.

Pellis, S. M., and Pellis, V. C. 1988. Identification of the possible origin of the body target that differentiates play fighting from serious fighting in Syrian golden-hamsters (*Mesocricetus auratus*). *Aggressive Behavior* 14 (6):437–449.

Pritchett-Corning, K. R., and Gaskill, B. N. 2015. Lack of negative effects on Syrian hamsters and Mongolian gerbils housed in the same secondary enclosure. *Journal of the American Association for Laboratory Animal Science* 54 (3):261–266.

Richards, M. P. 1969. Effects of oestrogen and progesterone on nest building in golden hamster. *Animal Behaviour* 17: 356–361.

Rowell, T. E. 1961. The family group in golden hamsters: Its formation and break-up. *Behaviour* 17 (2–3):81–94.

Siegel, H. 1985. *The Hamster.* New York: Springer.

Swanson, L. J., and Campbell, C. S. 1979b. Maternal behavior in the primiparous and multiparous golden hamster. *Zeitschrift für Tierpsychologie* 50 (1):96–104.

Ueda, S., and Ibuka, N. 1995. An analysis of factors that induce hibernation in Syrian hamsters. *Physiology & Behavior* 58 (4):653–657. doi: 10.1016/0031-9384(95)00095-z.

USDA APHIS. 2017. Annual report animal usage by fiscal year. Accessed 12 December. https://www.aphis.usda.gov/animal_welfare/downloads/reports/Annual-Report-Animal-Usage-by-FY2017.pdf.

Wang, Z., Manca, M., Foutz, A., et al. 2019. Early preclinical detection of prions in the skin of prion-infected animals. *Nature Communications* 10. doi: 10.1038/s41467-018-08130-9.

Whittaker, D. 2010. The Syrian hamster. In: *The UFAW Handbook on The Care and Management of Laboratory and Other Research Animals* Eds: R. Hubrecht and J. Kirkwood. 8th ed. New York: Wiley-Blackwell, 348–358.

Winnicker, C., and Gaskill, B. 2014. Utilization of nesting material by laboratory hamsters. *International Society of Applied Ethology*, Vitoria-Gastiez, Spain.

Wold, W. S. M., Tollefson, A. E., Ying, B., Spencer, J. F., and Toth, K. 2019. Drug development against human adenoviruses and its advancement by Syrian hamster models. *FEMS Microbiology Reviews* 43 (4):380–388. doi: 10.1093/femsre/fuz008.

11

Behavioral Biology of Rabbits

Lena Lidfors and Kristina Dahlborn
Swedish University of Agricultural Sciences

CONTENTS

Introduction	173
Typical Research Participation	174
Behavioral Biology	174
Natural History	174
Ecology, Habitat Use, Predator/Prey Relations	175
Social Organization and Behavior Social Behavior and Reproductive Behavior	175
Reproduction	176
Feeding Behavior	177
Communication	177
Common Captive Behaviors	178
Normal	178
Abnormal	178
Ways to Maintain Behavioral Health	179
Captive Considerations to Promote Species-Typical Behavior	179
Environment	179
Social Groupings	181
Breeding	181
Feeding Strategies	182
Training	184
Special Situations	184
Welfare of Rabbits	184
Antibody Production in Rabbits	185
Conclusions/Recommendations	185
Additional Resources	186
References	186

Introduction

In the latest report on the number of animals involved in research for scientific purposes in the 27 Member States of the European Union from 2015 to 2017, around 350,000 rabbits per year were used (Report from the Commission to the European Parliament and the Council, COM, 2020), corresponding to 4.6% of the total use of mammals. This put them in third place of the most commonly used mammals, after mice and rats. Almost all rabbits were born at registered breeders in European Union countries (COM, 2020). In the United States, rabbits were the second (18.4%, 145,841 individuals) most common laboratory animal covered by the Animal Welfare Act (which does not include rats and mice), according to the United States Department of Agriculture, Animal and Plant Health Inspection Service Annual Report Animal Usage (USDA, 2017).

According to the Domestic Animal Diversity Information System at the Food and Agriculture Organization, there are over 300 different domestic breeds of rabbits (Compiled in Wikipedia, 2020). The specific characteristics of the different breeds are mainly due to selective breeding for size, fur, growth rate, etc. The standards of each breed are regulated by breeder organizations (e.g., American Rabbit Breeders' Association (ARBA) and the British Rabbit Council (BRC)). The New Zeeland White is the breed most commonly involved in research.

The rabbit is a monogastric herbivore that, together with hares and pikas, belongs to the *Lagomorpha* order and *Leporidae* family (Lumpkin and Seidensticker, 2011). This family has developed a special digestive strategy, cecotrophy, which allows them to live by consuming plant materials (Gidenne et al., 2010). This means that they produce two types of fecal pellets; some that are hard and some that are soft. The latter are referred to as cecotropes and are reingested, providing the rabbit with protein and B vitamins (Gidenne et al., 2010).

Laboratory rabbits are, by tradition, kept individually in small cages and are fed dry pellets. This has led to several physiological problems, as well as behavioral disorders, including problematically low levels of movement. Over the past 10–15 years, many laboratories have improved the housing for rabbits, both in caging and by introducing floor pens and group housing. Another improvement is the addition of hay as a food source. However, there are some aspects of rabbit housing that have yet to be addressed, as males, especially, can be very aggressive when kept in groups.

The aim of this chapter is to present the most recent knowledge about the laboratory rabbit's behavioral biology, including its natural history and common captive behaviors, ways to maintain behavioral health, and special situations in which laboratory rabbits are involved.

Typical Research Participation

Classical experimental involvements of rabbits include antibody production, development of new surgical techniques, and physiology and toxicity studies to test new drugs (Bõsze and Houdebine, 2006). The size of rabbits, compared to rodents, permits the collection of larger samples of blood, cells, and tissue from a single animal (Esteves et al., 2018). Rabbits also have a longer life span than rodents, and the immune system genes of rabbits are more similar to those of the human immune system than are rodent genes (Esteves et al., 2018). Rabbits have a more diverse genetic background than inbred and outbred mouse strains (Bõsze and Houdebine, 2006). When developing therapeutic strategies or studying complex disease models, such as atherosclerosis, this can be advantageous, since it mimics human genetic diversity (Bõsze and Houdebine, 2006).

Rabbits have participated in research studies for well over a century. Louis Pasteur (1885) is particularly renowned for his work on the vaccine against rabies, a highly contagious infection that attacks the central nervous system. By studying the tissues of infected rabbits, Pasteur was able to produce an attenuated form of the virus, which he later used to develop a vaccine against the disease. Richard Shope discovered the cottontail rabbit papillomavirus, which became the first animal model of cancer caused by a mammalian virus (Shope and Hurst, 1933). The first transgenic rabbits were obtained by pronuclear microinjection in 1985 (Brem et al., 1985). The study of rabbit immunoglobulins established much of what is known about the structure, function, and regulated expression of antibodies (Esteves et al., 2018), and the rabbit has long been the model of choice for the production of polyclonal antibodies.

Rabbits have a long history of involvement in cardiac surgery research, and in studies of hypertension, infectious diseases, virology, embryology, toxicology, experimental teratology (Hartman, 1974), arteriosclerosis (Yanni, 2004), and serological genetics (Cohen and Tissot, 1974). The Watanabe rabbit suffers from fatally high blood cholesterol levels due to a genetic defect, which mirrors the human condition, and is therefore used as a model to provide better treatments for, and general research into, high cholesterol (Watanabe et al., 1985). Young rabbits often die from a disease called mucoid enteritis, which resembles cystic fibrosis, and are therefore often used as a model for understanding these illnesses (Toofanian and Targowski, 1983).

During the last two decades, the research participation of rabbits has extended to many human diseases (Esteves et al., 2018). Examples include syphilis, tuberculosis, HIV-AIDS, acute hepatic failure, and diseases caused by noroviruses, ocular herpes, and papillomaviruses. The involvement of rabbits in vaccine development studies, which began with Louis Pasteur's rabies vaccine, continues today, with targets that include the potentially blinding HSV-1 virus infection and HIV-AIDS (Peng et al., 2015). Additionally, two viral diseases that are highly fatal to rabbits, rabbit hemorrhagic disease, and myxomatosis provide models to understand the coevolution between a vertebrate host and viral pathogens (Esteves et al., 2018).

Behavioral Biology

Natural History

Rabbits housed in laboratories descend from the European wild rabbit (*Oryctolagus cuniculus*) (Harcourt-Brown, 2002). *O. cuniculus* lived all over Europe from 1.8 million to 10,000 years ago (Flux, 1994). However, after the last Ice Age (20,000 years ago), the European wild rabbit was left only on the Iberian Peninsula, in some areas of France, and in northwest Africa (Parker, 1990; Wilson and Reeder, 1993). Humans have mainly been responsible for the later spread of the European wild rabbit (Flux, 1994), and the species currently exists in the wild all over the world, except in Asia and Antarctica (Parker, 1990; Wilson and Reeder, 1993). Climate is a major factor that prevents *O. cuniculis* from spreading; for example, in Sweden, it does not survive in the wild, north of Stockholm. Tablado et al. (2009) found that variability of the length of the breeding season, proportion of pregnant females, age of first reproduction, and number and size of litters are affected by a combination of environmental factors (i.e., temperature, resource availability, and photoperiod) and individual properties (age and body weight (BW)). A lack of predators in the areas in which rabbits were introduced has led to an explosion in the number of animals, and they now are regarded as pests (i.e., in Australia and New Zealand; see Waikato Regional Council Fact Sheet, 2015). Cooke (2012) reports that rabbits in the Australian ecosystem need to be tightly controlled, in order for the disrupted ecosystem to eventually recover.

Phoenicians brought wild rabbits from the Iberian Peninsula to Italy about 3,000 years ago, and by the 1st century BC, the Roman army held them in large enclosures to have regular access to meat (Buseth and Saunders, 2015). This spread to other inhabitants of Italy, so that by the year AD 230, large numbers of rabbits were kept in captivity (Buseth and Saunders, 2015). Monasteries in France allowed monks to eat young rabbits, even during religious fasts, which lead to an increased keeping of rabbits, and from AD 500 to 1000, selective breeding for different fur and meat quality started (Buseth and Saunders, 2015). The French monks continued breeding rabbits for different colors and patterns during the 1500s (Buseth and Saunders, 2015). In 2017, there were up to

Behavioral Biology of Rabbits

FIGURE 11.1 Male-domesticated European wild rabbit in original colors kept in an enriched outdoor enclosure at Green Chimney in USA. (Photo: Lena Lidfors.)

305 domesticated breeds of rabbits spread over 70 countries (DAD-IS, 2018). The American Rabbit Breeders' Association (2019) has breed standards for 49 different breeds. The BRC Breed Standards 2016–2020 present standards for 85 different breeds, divided into four categories (Fancy section, 20 breeds; Lop section, 9 breeds; Normal Fur section, 46 breeds; and Rex section, 10 breeds). The New Zealand White (NZW), which usually participates as a laboratory rabbit, is found within the Normal Fur section. A photo of a domesticated European wild rabbit with its original color is shown in Figure 11.1.

The behavior of domesticated rabbits has not changed much (reviewed by Bell, 1984), and NZW rabbits kept in a near-to-nature environment show behaviors similar to their wild relatives (Vastrade, 1986; Lehmann, 1991). According to Bell (1984), compared to wild rabbits, domesticated rabbits spend more time resting above the ground during the day; males chin-mark more frequently and more often in unfamiliar territory; does (females) have fewer days out of reproduction (which leads to the production of more litters with a larger mean size); and kits (newborn offspring) leave the nest about 1 week earlier (Bell, 1984; Lehman, 1991).

Ecology, Habitat Use, Predator/Prey Relations

Rabbits have been reported to be crepuscular animals that are most active at dawn and dusk (Cobb, 2011). Fraser (1992), who observed European wild rabbits in New Zealand, found that they emerged from their burrows in late afternoon, and that by sunset, 90% of the rabbits had emerged. Mykytowycz and Rowley (1958), who observed European wild rabbits during 24 h-periods in a 12-acre enclosure in Australia, found that the old bucks (males) emerged first from their burrows, about 4 h before sunset. The rabbits were visible outside their warren (a burrow system of tunnels above and below ground that rabbits dig and hide in during rest and to escape from predators) for 11–14 h/day and were found to feed all through the night, with a peak at 21:00 (Mykytowycz and Rowley, 1958). The rabbit's crepuscular behavior has probably evolved as a response to predation.

Rabbits live in habitats where they can find shelter and protection from predators. Where the soil is loose, they dig burrows, and where the soil is compact, they stay in dense vegetation (Kolb, 1994). European wild rabbits live in breeding groups that defend a territory, which has a core area with a warren within a larger home range (Bell, 1999). Breeding groups can either use multiple warrens (Daly, 1981; Parer, 1982; Cowan, 1987) or they may only occupy a single warren (Myers and Schneider, 1964; Bell, 1977). Rouco et al. (2011) offered European wild rabbits in Spain warrens of different sizes for conservation purposes, and they concluded that it is preferable to provide many small warrens, but in the event that only a few warrens can be provided, they should be large. It has been found that warrens have an average of 11.5 entrances, but this may vary from 1 to 37 entrances (Cowan, 1987).

Antipredator behaviors in rabbits include the stamping of a hind foot, which causes the other rabbits to flee underground (Mykytowycz, 1968; Black and Vanderwolf, 1969). Domesticated young rabbits have also been reported to freeze and remain motionless during handling (Jezierski and Konecka, 1996). If a predator catches a rabbit, the rabbit can emit a high distress scream (Cowan and Bell, 1986). Green and Flinders (1981) found that the alarm vocalization of the pygmy rabbit (*Brachyphalagus idahoensis*) was composed of one to seven syllables uttered in rapid succession, having complex tones with many overlying frequencies. The strongest alarm call was at 4,000 and 6,000 HZ. Green and Flinders (1981) concluded that the alarm call seemed to be elicited by a potential predator, both in laboratory and field conditions.

Social Organization and Behavior Social Behavior and Reproductive Behavior

Wild rabbits live in small, territorial breeding groups (Bell and Webb, 1991), which, in a study in England, were found to consist of 1–8 males and 1–12 females (Cowan, 1987). It was found that 89% of the males and 96% of the females lived in groups containing at least one other rabbit of the same sex (Cowan, 1987). The group defends a central warren, but can come together with other groups while grazing in an area known as their home range (Bell, 1980). In the group, there are two separate social hierarchies, one among bucks and another among does (Vastrade, 1986; Von Holst et al., 1999). The males occupy territories that they defend from one another, while the females stay in a specific area that they do not defend against other rabbits (Vastrade, 1986). Lehmann (1991) found that when free-ranging NZW rabbits were older than 70 days (not yet sexually mature, see below), aggressive and sexual behaviors were common, and a linear hierarchy among the bucks developed.

The dominant buck among European wild rabbits regularly patrols his territory and requires submission from both males and females in the territory, or else they are subject to attack (Lockley, 1961). The alpha male of free-ranging NZW rabbits also patrols his territory and can be very aggressive toward other males, but he is tolerant with females and young (Vastrade, 1986; Lehmann, 1991). Even though there are frequent aggressive encounters between males, there is always space to retreat in the wild and serious injuries are rare

(Lehmann, 1991). Free-ranging NZW bucks and does have been found to keep an average interindividual distance of 15–20 cm when active (Vastrade, 1987). Rabbits of lower rank can control their interactions by withdrawing to the periphery of their home range (Lehmann, 1991). The alpha male seeks out, and interrupts, all aggressive or sexual encounters, which may also decrease the risk for injuries (Vastrade, 1986). Females have been observed to fight as strongly as males for preferred resources, such as a nest site, and this has been found both in European wild rabbits (Myers and Poole, 1961; Lockley, 1961) and in free-ranging NZW rabbits (Lehmann, 1991). Females do not defend a territory and usually share home ranges with each other (Vastrade, 1987). Rabbits interact more amicably, and often lie in close contact with conspecifics, while resting (Lehmann, 1991). Gibb (1993) found European wild rabbits to spend about 33% of their time inactive, including resting, when they were above ground.

Reproduction

European wild rabbits are regarded as seasonal breeders, and breeding is determined by interactions among day length, climate, nutrition, population density, social status, and other factors (Bell and Webb, 1991; Bell, 1999). Tablado et al. (2009) used breeding parameters from a large number of published papers from highly separate locations around the world and analyzed seasonality in reproduction with two generalized mixed linear models. They found that both male and female European wild rabbits may reach sexual maturity at 3–4 months, gestation periods last about 28–31 days, and does have a postpartum estrus. However, the authors also found that the breeding season, proportion of pregnant females, age at first reproduction, and number and size of litters showed high spatial and temporal variation around the world (Tablado et al., 2009).

Rabbits perform courtship behaviors before mating in which they circle around each other, parade side by side, jump over each other, and sniff the genital region (Lehmann, 1991). When the doe is ready to be mated, she raises her hindquarters (Bennett, 2001). The buck mates for only a few seconds, after which he falls to one side or backward (Bennett, 2001). Mating induces the ovulation of several eggs, and conception occurs 8–10 h later (Bennett, 2001). The reproductive cycle of rabbits is shown in Figure 11.2.

Some days before the birth, the European wild rabbit doe digs a short underground tunnel or den, either within the main warren or at a separate place, and constructs her nest (Bell, 1999). High-ranking does often dig their nests as an extension to the warren, whereas some low-ranking does are chased away from the warren and have to dig nests in isolated breeding places (Mykytowycz, 1968). It was found that the young of high-ranking does had a higher survival rate than those of low-ranking does (Mykytowycz, 1968). Cowan (1987) found that females living in groups with more than one female had lower lifetime reproductive success than females living alone with a male.

Rabbit kits are born altricial, i.e., with little or no fur, and closed ears and eyes (Thoman et al., 1979). Their development in the closed nest in nature is difficult to observe; therefore our knowledge is mainly based on laboratory studies of domesticated rabbits (see "Breeding" section). The nest is typically visited by the doe only once daily for 3–5 min to feed the young (Zarrow et al., 1965; Rödel et al., 2012), after which the doe closes the entrance to the nest by moving soil over and scent marking it (Lehmann, 1991). Even though the doe spends relatively little time with her young, she is very alert to any potential threats from foxes, mustelids, and infanticide by other rabbits; playback distress calls produced by kits from nest burrows lead to the females showing increased vigilance and returning to the nest (Rödel et al., 2013). After 21 days, the European wild doe stops closing the entrance to the nest and the young emerge from it (Bell, 1999). Free-ranging NZW does stop closing the entrance to the nest when the kits are about 18 days old, so that they can move in and out from the nest as they wish, and the kits are then nursed outside (Lehmann, 1989).

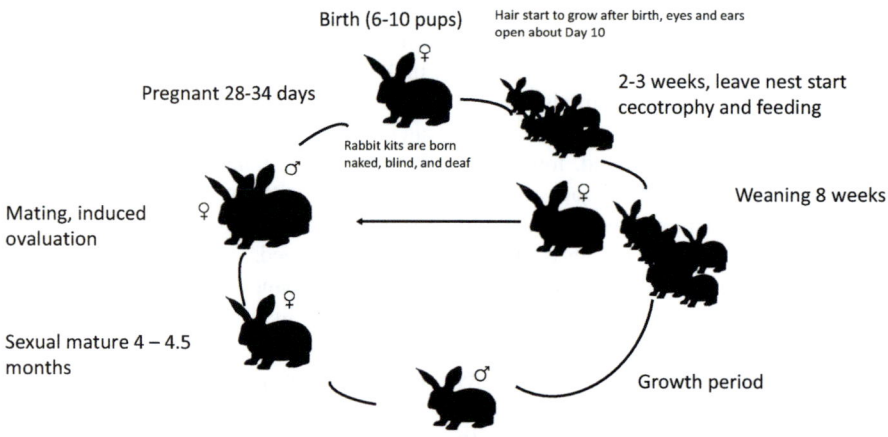

FIGURE 11.2 Reproduction cycle in rabbits based on studies on both Wild European rabbits and domesticated rabbits. (Developed by Kristina Dahlborn.)

If the female is fertilized just after parturition and concurrently sustains pregnancy and lactation, milk production decreases sharply at the end of pregnancy and stops 2–3 days before the following parturition (Fortun-Lamothe and Prunier, 1999). This frequently occurs in the spring, when the female mates again on the day of parturition. In such cases, young rabbits could be weaned at 3 weeks of age.

Feeding Behavior

Newborn rabbits have high-energy requirements and limited ability to thermoregulate. During their first 2–3 weeks of life, they are solely dependent on milk supplied by the doe. In the wild, the kits are nursed once a day (Maertens et al., 2006). Rabbit milk contains high levels of k-Casein, which is denatured in the stomach, resulting in the trapping of the milk in the stomach, and only reaching the intestine in small amounts (McClellan et al., 2008). Once the kits reach 2–3 weeks of age, feeding behavior changes in just a few days, as the young switch from a single daily meal of milk to 25–30 meals of solid food items per 24 h (Gidenne et al., 2010). The first contact with solid elements occurs during the first week of life, when the young consumes hard feces deposited by the doe in the nest during suckling (Moncomble et al., 2004). This probably introduces microbial flora to the cecum (Moncomble et al., 2004; Kovacs et al., 2004). The daily feeding behavior of the adult rabbit involves eating numerous meals (30–40/24 h; Gidenne et al., 2010).

Rabbits, like other members of the *Leporidae* family, have developed an anatomical-physiological adaptation in the gut that allows them to feed solely on plant material. All intestinal content passes the cecum, and both soft and hard fecal pellets are formed from cecal content (Björnhag, 1981). The cecotrophes consist of almost unchanged cecal contents, while the hard fecal pellets change composition drastically during their passage through the proximal (haustrated) colon (Björnhag, 1981). One example of this change is the difference in nitrogen content between discarded feces (15 mg N/g dry matter) and cecotrophes (45 mg N/g dry matter) (Björnhag, 1981). Separation and retrograde transport results in both a high average rate of transport of food and food residues through the digestive tract, and a relatively high degree of digestibility, allowing these animals to live on food rich in fiber and low in easily digested nutrients (Björnhag, 1981). Cecotrophy begins at 2–3 weeks of age, when rabbits begin to consume solid food. The intake of soft feces increases until 2 months of age and then remains constant (Gidenne et al., 2010).

European wild rabbits perform their foraging behavior within a home range, the size of which depends on food availability, age, status within the breeding group, and the size of the group (Donnelly, 1997). The home range is typically large and is not defended by the male or anyone in the breeding group. The size of the home range has been reported to be 0.4–2 ha (Cowan and Bell, 1986), 8 ha (Vastrade, 1987), or 5 ha (Myers et al., 1994). Males have been reported to have larger home ranges (0.71 ha) than females (0.44 ha) (Cowan, 1987). It has been observed that European wild rabbits can gather in large colonies of up to hundreds of animals, when feeding conditions are good or population densities are high (Myers and Poole, 1963).

Rabbits have been observed to spend 44% of their time eating, 33% inactive, 13% moving, and 10% engaged in other activities when they are above ground (Gibb, 1993). Studies on NZW rabbits in a near-to-nature environment found that young rabbits were active, on average, for 30% of daytime hours (Lehmann, 1989). Of the time spent active, one-third was spent feeding on pellets (provided by the researchers), one-third was spent grazing, and one-third was spent in exploration, gnawing, intensive locomotion, and sexual behavior (Lehmann, 1989).

Rabbits mainly eat grass and herbs, but will also consume fruit, roots, leaves, and bark (Cheeke, 1987). The food resources available to wild rabbits vary considerably and include a wide range of plants. Rabbits clearly prefer graminaceous plants (*Festuca*, *Brachypodium*, or *Digitaria* species) and graze on only a few dicotyledons, if grass are insufficient (Williams et al., 1974; Leslie et al., 2004). In the winter, rabbits are most likely to be observed eating the bark from different trees and bushes.

Rabbits need coarse fiber (roughage) in addition to rough grass for proper digestion (Brooks, 1997; Meredith, 2000). Rabbits feed in two ways: (1) plants are chewed and swallowed, while (2) cecotropes are taken from the anus and swallowed directly. Soft feces are excreted according to a circadian rhythm that is the opposite of food intake and hard feces excretion. Cecotrophy occurs mainly during the light period, whereas food intake and hard feces excretion occur during the dark period (Carabaño et al., 2010). The larger fibrous fecal pellets are deposited at specific latrines located close to territory borders (Donnelly, 1997).

Communication

Nocturnal mammals often use chemical signals to communicate, and rabbits have been studied widely in this context (Arteaga et al., 2008). They have a number of scent glands located under the chin, and in the anal and groin regions. They mark their territory with their feces or by rubbing their chins against objects (Bell, 1999). Chin-marking behavior has been studied in great detail, and is important for territorial defense and signaling social dominance (Arteaga et al., 2008). Males and dominant individuals scent mark more often than others (Lidfors and Edström, in press). Rabbits have around 100 million olfactory receptor cells in their nose, a vomeronasal organ, and large olfactory lobes in the brain (Lumpkin and Seidensticker, 2011) to help them analyze all the chemical signals during communication. Rabbits have also been found to be good at discriminating between different types of taste through their 17,000 taste buds, and they prefer sweet and salty, while disliking bitter and sour (Lumpkin and Seidensticker, 2011). They also show preference for maltose over sucrose and fructose (Lumpkin and Seidensticker, 2011).

Rabbits have good eyesight, and their maximum field of vision is almost 360° (Peiffer et al., 1994). They have dichromatic (two-color) vision, with mainly green sensitive and some blue sensitive cones (Lumpkin and Seidensticker, 2011). Their high density of rods helps them see at dawn and dusk, but they lack a tapetum lucidum, which prevents them from seeing well during periods of low light intensity (Lumpkin and Seidensticker, 2011). Rabbits have poor depth perception and poor visual acuity (Lumpkin and Seidensticker, 2011), and they cannot focus directly in front of themselves, which needs to be considered when handling them.

Rabbits have good hearing and their ears are longer in warmer climates to help them thermoregulate (e.g., lose heat) (Lumpkin and Seidensticker, 2011). The 20–25 whiskers on each side of the upper lip help the rabbits to orient in the dark and to locate where to bite vegetation (Lumpkin and Seidensticker, 2011). Rabbits can make a purring noise by "cracking jaws and quick clattering of teeth", which has been interpreted as the rabbit enjoying life (Buseth and Saunders, 2015). A low persistent hum has been interpreted as the rabbit being excited (Buseth and Saunders, 2015). When the rabbit makes a "grunt/growl", keeps the ears back, and jumps forward, it has been interpreted as a sign of irritation or fear, and that it may attack (Buseth and Saunders, 2015). When the rabbit is "grinding teeth", it indicates that the rabbit is in pain (Buseth and Saunders, 2015).

Common Captive Behaviors

Normal

Rabbits in laboratories usually rest, eat, drink, and move around in their cage or pen. Lidfors (1997) recorded the number of times un-castrated male NZW rabbits performed normal behaviors when kept individually in cages and given different types of enrichments (see Table 11.1). In all cages, including those with enrichment and without, rabbits spent the majority of observations alert, resting, eating pellets, and sniffing the cage (Table 11.1). Rabbits scent mark their cage or pen, and will spend time scent marking any new cage/pen they are placed in, especially males (Gunn and Morton, 1995), just as they scent mark their territory when living in nature.

The behavior of groups of rabbits (NZW, Lop Crosses, and Dutch breeds; both females and castrated males) housed in pens (6, 7, 15, or 20 rabbits/pen) or individually caged were recorded by Podberscek et al. (1991). Locomotor behavior (hopping) accounted for 19% of penned rabbit behavior and 23.2% of caged rabbit behavior. Maintenance behaviors (standing, resting, drinking, eating, urinating, and defecating) were more commonly observed in caged rabbits (44.5% of observed behavior) than in penned rabbits (25.3%). The reverse was true for comfort behaviors (grooming self, scratching, head shake, sneezing, stretching, yawning, and grooming another rabbit), where 23% was observed in penned rabbits and 40.6% in caged rabbits. Marking and investigatory behaviors (digging, stamping back leg, rearing up on hind legs, licking wall or urine off floor, rubbing another or object, anal-, body-, or ear sniff, and mounting) were observed more often in penned rabbits (12.6%) than in caged rabbits (3%). Agonistic behaviors (biting another and chasing) occurred only in penned rabbits (2.5% of observed behaviors), while stereotypic behaviors (clawing at cage walls or corners and biting cage bars or food hoppers) were only observed in caged rabbits (6.3%). Caged rabbits were commonly engaged in locomotor behavior early and late in the day (09–10 and 15–17 h). Resting (maintenance behavior) and grooming (comfort behavior) were observed more often in the middle of the day (10–12 h). More scratching, head shaking, sneezing, stretching, and yawning (comfort behaviors) occurred in penned rabbits than in caged rabbits. Podberscek et al. (1991) did not report if there were any differences between females and castrated males. Normal behaviors have been reported in several other studies of rabbits, when housed in floor pens (Gunn and Morton, 1995; Poggiagliolmi et al., 2011). Given appropriate environmental conditions, other laboratory rabbits should be able to engage in the same sorts of behaviors as the penned rabbits.

Abnormal

Rabbits in laboratories have traditionally been housed singly in cages, with low ceiling heights that do not allow them to stand on their hind legs to check their surroundings. One reason for keeping this naturally gregarious species socially isolated has been the problem with aggression among noncompatible animals (Morton et al., 1993). The separation from conspecifics in

TABLE 11.1

Total Number of Recorded Instances of Normal Behavior in 58 Individually Caged Male Rabbits with Five Treatments (Reprinted from Lidfors, L. 1997, *Applied Animal Behaviour Science* 52:157–169.)

Behaviors	Hay (n = 10)	Cubes (n = 12)	Stick (n = 12)	Box (n = 12)	Control (n = 12)
Being alert	1,323	2,116	2,026	1,922	2,213
Resting	638	1,374	1,360	1,260	1,235
Eating pellets	229	269	298	240	253
Drinking	184	236	254	292	240
Moving body	97	136	128	142	132
Hopping	65	47	57	69	60
Running	18	18	24	25	12
Sniffing cage	218	159	212	207	233
Sniffing bars	34	63	84	108	123
Sniffing itself	16	16	22	18	26
Coprophagy	27	7	30	52	38
Shake/scratch	12	23	27	18	24
Other		8	9	11	3
Totals	2,862	4,472	4,531	4,364	4,592

TABLE 11.2

Total Number of Recorded Instances of Abnormal Behavior in 58 Individually Caged Male Rabbits with Five Treatments (Reprinted from Lidfors, L. 1997, *Applied Animal Behaviour Science* 52:157–169.)

Behaviors	Hay ($n=10$)	Cubes ($n=12$)	Sticks ($n=12$)	Box ($n=12$)	Control ($n=12$)
Self-licking	379	442	718	766	648
Sham chewing	88	79	273	257	343
Biting bars	28	52	52	97	116
Licking cage	7	6	11	13	24
Biting cage	0	3	4	11	14
Biting itself	1		8	4	4
Digging cage	0	2	8	5	8
Other[a]	0	8		5	8
Totals	503	593	1,075	1,158	1,165

[a] Other behaviors included licking bars, sliding nose along bars, biting water nipple, and running around in same pattern.

this barren environment prevents them from performing several natural behaviors, such as digging in substrate (not abnormal digging on cage walls and plastic floor, see below), allo-grooming, and some locomotor activities. It also drastically reduces their exposure to variations in odors and diet (Gunn and Morton, 1995). This can lead to the development of abnormal behaviors, for example, excessive wall-pawing or bar-gnawing (Gunn and Morton, 1995; Held et al., 1995; Lidfors, 1997). Some abnormal behaviors recorded in individually caged male NZW rabbits, given one of four enrichments, are shown in Table 11.2 (Lidfors, 1997). The most commonly observed abnormal behaviors in that study were repeated self-licking, sham chewing, and biting bars. Stereotypic behavior seems to be most frequent at night, when rabbits are most active (Gunn and Morton, 1995). When individually kept in cages, rabbits that are more active tend to become more frustrated and show more abnormal behaviors (Gunn and Morton, 1995). Individually caged rabbits can also show signs of restlessness (Podberscek et al., 1991), and can develop intestinal disorders (Gunn and Morton, 1995). Small caging leads to limited freedom of movement, which has been shown to cause osteoporosis of the femur, osseus hypoplasia, reduced bone strength in reproducing does kept for meat production (reviewed in DiVincenti and Rehrig, 2016), and thinner tibial cortices (Buijs et al., 2014). These reproducing does kept in small and low-height cages were found to develop deformation of the vertebral column, but males in the same housing systems did not develop these problems (Drescher and Loeffler, 1996). Proks et al. (2018) examined 330 pet rabbits and found that 15.2% had congenital spinal abnormalities, mostly as a single pathology. We have not found any scientific studies that have investigated the effect of cage housing on bone development in laboratory rabbits.

Ways to Maintain Behavioral Health

Captive Considerations to Promote Species-Typical Behavior

Environment

Rabbits show more of their natural behaviors when kept in enriched floor pens that are large enough for them to show normal locomotion and avoid conflicts with conspecifics (Podberscek et al., 1991; Gunn and Morton, 1995; DiVincenti and Rehrig, 2016). However, intact, adult males have to be kept individually, due to their territorial and aggressive nature (Morton et al., 1993; DiVincenti and Rehrig, 2016). This should preferably be done in floor pens to promote normal locomotion, with the addition of environmental enrichments, as is implemented in some laboratories in the UK. By providing environmental enrichment, normal behaviors are promoted and the amount of abnormal behavior can be reduced. All refinements should have verified efficacy on the animals' welfare and be proven safe (Nevalainen et al., 2007).

Rabbits are, however, still kept individually in cages in many laboratories, probably due to tradition, lack of space, and a perceived need for individual surveillance during experimental testing. Recently, cage systems have been developed that have larger floor areas, are taller, and are equipped with a combined shelter/resting shelf, hay racks and gnawing sticks placed on the grid floor (for example, Scanbur and Tecniplast). It is also possible to connect two or more cages to offer a larger space. However, doubling the space for individually housed rabbits does not necessarily lead to increased activity, and may lead to increased abnormal behavior (Krohn et al., 1999). Larger cages do make it possible to pair house female rabbits or young males, the effects of which have been empirically tested (Chu et al., 2004; Nevalainen et al., 2007; Egedal, 2009; DiVincenti and Rehrig, 2016).

Environmental enrichments for individually housed rabbits have been tested in several different studies (Table 11.3). In summary, they show that hay, grass cubes, Bunny Stixs, cardboard rolls and rings, and a rubber ball with a ring inside lead to increased object interactions, including the chewing of the object (Table 11.3). Rabbits receiving hay spent less time lying still, showed fewer changes between behaviors, and exhibited decreases in abnormal behaviors (Table 11.3). Hay seems to be highly preferred, as rabbits chose it before grass cubes, gnawing sticks, and a box in a preference test (Lidfors, 1996). Hay results in a greater reduction in abnormal behavior than other enrichment types (Lidfors, 1997; Berthelsen and Hansen, 1999), which may be related to a lack of foraging opportunities, as the rabbits were only fed small amounts of pellets twice a day (Lidfors,

TABLE 11.3

Enrichments to Individually Caged Rabbits and their Effects on Their Normal and Abnormal Behavior in Different Studies

Enrichments	Normal Behavior	Abnormal Behavior	Reference
Hay	↑ object interaction ↓ lying still	↓ than controls	Lidfors (1997)
" - "	↓ changing between different behaviors	↓ bar gnawing, excessive grooming	Berthelsen and Hansen (1999)
Grass cubes	↑ object interaction	↓ than controls	Lidfors (1997)
Gnawing sticks	Few object interactions	= controls	Lidfors (1997)
Box	Few object interactions	= controls	Lidfors (1997)
Box/shelter	↓ restlessness, timidity	↓ excessive grooming, bar-gnawing	Hansen and Berthelsen, 2000
Mirrors	↑ exploration	↓ grooming in females	Edgar and Seaman (2010)
" - "	↑ exploration ↑ number of behaviors ↓ sitting looking out of cage	Not recorded	Jones and Phillips (2005)
" - "	↑ preference for cage half with mirror ↑ feed intake in cage half with mirror	Not recorded	Dalle Zotte et al. (2009) *(fattening rabbits 5–11 weeks, both genders)*
Cardboard rolls, cardboard rings, rubber ball with ring inside	↑ chewing object ↓ sitting	Chewing cage = controls	Poggiagliolmi et al. (2011)
Hanging toy, destructible device, dig bin	↑ active, exploration	Not recorded	Coda et al. (2020)
Food-enriched (Bunny Stix, Celery, Bunny Block)	↑ interaction with Bunny Stix ↑ activity than toys and controls	Not recorded	Harris et al. (2001)
Toy-enriched (Jingle Ball, Nylabone, Kong)	= interactions with all toys = activity as controls	Not recorded	Harris et al. (2001)

1997). Leslie et al. (2004) found that rabbits tested in a Y-maze for 3 min had a preference for grass over coarse mix if they had never been exposed to grass before, but after repeated testing, the preference diminished. In a follow-up motivation test, the rabbits were not prepared to circumnavigate more to get access to grass than for access to coarse mix (Leslie et al., 2004). One negative aspect with grass cubes was that the males receiving them increased in weight more than those given hay, gnawing sticks, or a box (Lidfors, 1997). Harris et al. (2001) also found a tendency toward increased weight in the food-enriched rabbits over the 15 days of their study.

Different types of objects or toys placed in the cages of individually housed rabbits were found to cause increased activity and exploration; abnormal behavior was not recorded (Table 11.3). A study in a safety assessment facility tested stainless-steel rabbit rattles on spring clips in individually housed rabbits and found decreased interaction with the toy over time (Johnson et al., 2003). There were no differences between rabbits with or without a toy with respect to BW, food consumption, and hematologic parameters (neutrophilia, lymphopenia, and eosinopenia) (Johnson et al., 2003). Coda et al. (2020) did not find any differences in fecal glucocorticoid levels between rabbits with (see Table 11.3) or without enrichments. Thus, there is no major risk of influencing the research results by giving rabbits different objects.

When a mirror is placed in the cage, rabbits show more exploration and variation in behaviors, less sitting and looking out of the cage, and spend more time in the area with mirrors (Table 11.3). It was also found that female rabbits performed less self-grooming with a mirror in the cage (Table 11.3). Mirrors therefore seem to offer some welfare advantages and thus may be considered an environmental enrichment (Jones and Phillips, 2005; Dalle Zotte et al., 2009; Edgar and Seaman, 2010). The responses of rabbits when tested pair-wise with four stimuli were similar among a mirror, a blank card, and a soft toy, but were different compared to a conspecific (Jones and Phillips, 2005). Placement of cages to allow rabbits to see each other can also serve as a form of environmental enrichment (Morton et al., 1993).

A box/shelter in the cage reduces abnormal behaviors and flight responses when the animals are being captured (Hansen and Berthelsen, 2000). Verga et al. (2007) propose that a rabbit that is caged alone should be given access to a refuge area with objects to play with. A dark painted rat cage placed up-side down in the rabbit cage received only few interactions from the rabbits and did not reduce abnormal behaviors (Table 11.3), but this was probably because it was too small and only had one entry (Lidfors, 1997). A larger shelter adds structure to the cage and allows the rabbits to move in a way that maintains the normal function and structure of muscles, bones, and joints (Stauffacher, 1992). Shelves provide a darker area that the animals can use when disturbed (Stauffacher, 1992) and function as important hiding places from continuous intense light, which can cause retinal damage in albino animals (Doughty, 2020). A study on female NZW rabbits' motivation to push themselves through a weighed door with increasing weights found that the motivation was lower for access to a platform (a partly closed space and a raised area) than for social contact or food (Seaman et al., 2008).

In the report on "Refinements in rabbit husbandry", Morton et al. (1993) suggested the following enrichments for caged rabbits: straw, hay, hay blocks, hydroponic grass, pieces of

FIGURE 11.3 NZW female rabbits kept in enriched floor pens at the Swedish University of Agricultural Sciences. The rabbits had more bone mass and more developed bones and higher citrate synthase in the muscles than rabbits kept in cages. (Photo and personal communication: Patricia Hedenquist.)

wood or chew sticks, hay rack, small cardboard boxes, taking them out from the cage for handling/petting or for exercise and relief of boredom, and the provision of bedding. The following enrichments for rabbits are recommended by the Council of Europe (2006) and the European Commission (2007): roughage, hay, hay blocks, and chew sticks.

Taking individually housed male rabbits out of the cage and letting them run in a floor pen with different objects has been used as an enrichment in some laboratories (evaluated by Knutsson, 2011). When rabbits are released in a floor pen, they are (1) given a change in environment, (2) exposed to odors from other male rabbits, which stimulates scent marking, (3) allowed to move over a larger area, (4) less likely to become obese, due to the exercise, and (5) handled more frequently, which may reduce fear of humans (Lidfors and Edström, in press; Knutsson, 2011).

In comparisons of cage and pen housing of rabbits, enlarging the cage size for a group (with or without variation of the density) allows more movement for the rabbits and reduces daily feed intake (Maertens and Van Herck, 2000). Additionally, being kept in floor pens can lead to stronger muscles and bones in the rabbits (Figure 11.3).

Social Groupings

Rabbits should, when possible, be kept in social groups in pens to meet their need for social behavior and exercise (Morton et al., 1993; Trocino and Xicatto, 2006; Hawkins et al., 2008). Efforts have been made to keep females in groups in pens, and breeding females in groups with a male (Stauffacher, 1992). This has proven successful, as rabbits in groups are more active and show no stereotypies (Morton et al., 1993). Rabbits in groups also express a broader behavioral repertoire, and when rabbits are kept in groups, their quality of life significantly improves (Trocino and Xicatto, 2006; Verga et al., 2007). Even though the social stress of group living may reduce welfare for some individuals, their quality of life is still greater than that of single-housed rabbits (Trocino and Xicatto, 2006; Verga et al., 2007). When group-housed female rabbits were placed individually in a preference test, where they could choose to enter enriched floor pens with or without their group mates, they showed a weak preference for the solitary pen (Held et al., 1995). However, when they were tested again, and had to choose between an enriched floor pen with their cage mates and a smaller unenriched solitary floor pen, they chose the group pen (Held et al., 1995). However, does may be aggressive and inflict wounds on each other, especially during breeding, making group housing in pens sometimes difficult to implement.

It is important to consider the compatibility of individual animals when housing rabbits in groups, as incompatible rabbits housed together will fight, especially males (Morton et al., 1993). Pair-housing may be an alternative, as presented in the review by DiVincenti and Rehrig (2016). Nevalainen et al. (2007) found that pair-housing of female rabbits led to reduced variances in growth rates and serum alkaline phosphatase levels. Chu et al. (2004) observed that pair-housed female rabbits often engaged in physical contact and that aggression between them did not increase during the 5 months together, but one pair had to be separated due to bite wounds at the end of the study.

A study of pair housing involving eight pairs of male siblings in enriched cages showed that the siblings had to be separated at a mean of 144 days of age (123–154 days), and only one pair stayed together for the entire study (until 160 days of age; Egedal, 2009). When males are going to participate in longer studies, castration is a good option, as castrated males function well in group housing (Kalagassy et al., 1999). In the review on the social nature of rabbits, DiVincenti and Rehrig (2016) concluded that social housing of young, perhaps related and castrated, rabbits in very large, enriched pens might improve their well-being. To prevent the development of abnormal behavior when group housing is not possible, extra care should be taken in order to enrich the animal's environment (Nevalainen et al., 2007), such as providing hay, shelves, exercise in larger pens, and handling (see previous section).

Breeding

The domesticated rabbit breeds of medium size are reported to be sexually mature before 5 months of age, and the smaller breeds earlier than that (Bennett, 2001). Domesticated does have been reported to be most sexually receptive immediately postpartum and after weaning (Castellini et al., 2006). Gestation lasts for 28–34 days in domesticated rabbits, but most litters are born at day 31 (Bennett, 2001). This is slightly longer than for the European wild rabbit (Bell, 1999).

There has been extensive research on maternal-offspring behavior in rabbits in laboratory environments (see review by González-Mariscal et al., 2016). González-Mariscal et al. (1994) found a correlation between blood concentrations of estradiol, progesterone, and testosterone, and the onset/offset of nest-building activities. Digging had a maximum expression in mid-gestation under high levels of estradiol and progesterone (González-Mariscal et al., 1994). Progesterone declined at the end of pregnancy and was associated with a decline in digging and a start of straw-carrying (González-Mariscal et al., 1994). Estradiol increased close to parturition, at the same time as fur-pulling (the doe pulling off her own fur from the belly and body sides and placing it in the nest), and this was followed by an abrupt secretion of prolactin (McNeilly and

Friessen, 1978). This research group has, by manipulating different hormones, further investigated the hormonal control of nest-building in rabbits (see review by González-Mariscal et al., 2016).

The domesticated rabbit uses both edible and nonedible material as a substitute for natural plant material to build a nest inside a box (González-Mariscal et al., 2016). If the doe is shaved, so that she cannot pull her own hairs, she will collect another female's, a male's, or synthetic fur from a container to line the straw nest (González-Mariscal et al., 1998). The doe appears to have a visual image of how a ready nest should look, as she stops collecting straw, if given an excessive supply and the opportunity to look at the structure. However, if the straw is removed through a back door, does continue to carry new straw into the nest box (Hoffman and Rueda-Morales, 2009). Does are recommended to be housed individually from shortly before the birth of their kits until 12 days postpartum, to reduce double litters in a nest box, crushing of kits, and infanticide by other does (Rödel et al., 2008; Braconnier et al., 2020). Regrouping does with kits at 12, 18, or 22 days did not affect the number of lesions and agonistic interactions, but these were higher during summer than winter (Braconnier et al., 2020). The doe should have access to a nest box where she can construct a nest in good time before giving birth. Canali et al. (1991) found that the better the construction of the nest, the higher the survival rate of the young.

Parturition is stimulated by a large oxytocin release into the blood and an increased sensitivity of the myometrium to oxytocin, which is associated with a decline in feeding by the doe (see González-Mariscal et al., 2016). Domesticated rabbits have litter sizes ranging from four to ten kits (Schlolaut et al., 2013), and in the laboratory breeds, six to ten kits or more have been reported (González-Mariscal et al., 2016). During parturition, the doe stands hunched over the nest, and sometimes briefly licks the kits as they are born at the rate of one kit/min (Hudson et al., 1999). The kits rapidly get free from the amniotic membranes and may start suckling while parturition is still going on (Hudson et al., 1999). The doe eats the placenta, leaves the nest, and closes it after all kits have been born (González-Mariscal et al., 2016). She does not brood or clean her kits, or retrieve them if they end up outside the nest (González-Mariscal et al., 2016). The short contact with the kits is very important for the doe to develop maternal responsiveness, because if primiparous does are separated from their kits at birth, 70% of them will neither nurse nor show interest in the offspring the next day (González-Mariscal et al., 2016).

Domesticated breeds of rabbits will, just as the European wild rabbit, nurse their kits just once per day for 3–5 min, with a 24-h periodicity (reviewed by González-Mariscal et al., 2016). Contrary to many other mammals, suckling rabbits do not stay with the same nipple, but change every 20 s during a sucking bout (Bautista et al., 2005). The kits have a circadian pattern of arousal, where they anticipate the mother's visit, and the doe emits pheromones that help the kits find the nipples (González-Mariscal et al., 2016). Rabbit kits perform a head-searching pattern and grasp a nipple (Hudson and Distel, 1983), a behavioral pattern controlled by the pheromonal signal (Schaal et al., 2003). Anosmic kits are unable to suckle and will die of starvation (Distel and Hudson, 1985).

Nursing occurs at night and the time slightly advances each day (González-Mariscal et al., 2016). If the number of kits is reduced to fewer than four, the doe will enter the nest box several times per 24 h, and manipulating the mother increases the time she stays in the nest box for nursing (González-Mariscal et al., 2016).

The development of kits has been documented in the laboratory by Thoman et al. (1979). Kits are unable to see, hear, and maintain their body temperature at birth, and they show weak and shaky walking, awkward grooming movements, and very little quiet sleep at this time. However, they show much active sleep; on day 1, rapid eye movement (REM) sleep has been recorded. At the end of the first week, their locomotion has improved, and in the beginning of the second week, they respond to sound, and can walk, groom themselves, sit, and burrow. Their eyes open between 10 and 14 days, and they then engage in more quiet sleep with eyes open than with eyes closed. After day 20 of life, the kits' states and behavior do not change much; locomotion is excellent and rapid hopping appears. Re-ingestion of feces begins during the third week and the kits are weaned and start ingesting feed pellets from the doe's food dish between 15 and 30 days of age (Thoman et al., 1979). Hudson et al. (1996) found that at day 12, the kits started to nibble at the doe's fecal pellets, and at day 14, they started eating the nest material. At day 20, the doe's milk production begins to decrease, and if pregnant again, she will stop nursing her kits on day 26 (Hudson et al., 1996; González-Mariscal et al., 2016). Does that are pregnant usually wean their young by refusing to enter the nest box and by rejecting attempts of the kits to suckle (Hudson et al., 1996; González-Mariscal et al., 2004). However, in laboratory conditions, does that were not pregnant were found to nurse their young for more than 40 days (González-Mariscal et al., 2016). Breeders of laboratory rabbits normally wean the kits from their mother by removing them from her cage between days 28 and 35, even if she is still lactating. Rabbit breeders have developed different methods to keep doe and kits separated between the daily nursing bouts in order to simulate conditions in nature, and Baumann et al. (2005) used a cat-flap at the nest entrance.

Feeding Strategies

In the feeding behavior section above, it was mentioned that wild rabbits feed on a large variety of plants, feed at dawn and dusk, use cecotrophy, and that soft pellets are primarily consumed during the daytime. The typical daily feeding behavior of the rabbit is to eat numerous meals (30–40 per 24 h; Gidenne et al., 2010). Additionally, the rabbit has developed a strategy of high feed intake (65–80 g/kg BW) and rapid transit of feed through the digestive system to meet nutritional requirements (Gidenne et al., 2010).

In breeding and research facilities, rabbits are generally fed an *ad libitum* diet of rabbit pellets. This feeding strategy can work well for rabbits with high nutritional needs, such as pregnant and lactating does, and growing young rabbits. Milk production is high in the doe (250 g/day), and the milk has the following nutrient content; dry matter content around 25%, fat 12%–13%, protein 10%–12%, with energy of 8.4 MJ/kg (Jensen, 1995; Maertens et al., 2006). The growth rate is very high, and during the lactation period, the kits can increase their bodyweight six-fold (Maertens et al., 2006). In postweaning rabbits (4 weeks old), soft feces production linearly increases with age, reaching a maximum at 63–77 days old (25 g DM/day). This period corresponds to the time of maximum growth (Gidenne et al., 2010). The intake of soft feces increases until 2 months of age and then remains steady (Gidenne et al., 2010). Expressed as fresh matter, soft feces intake increases from 10 g/day at 1 month old, to 55 g/day at 2 months of age, thus representing 15%–35% of feed intake (Gidenne and Lebas, 1987).

Between weaning (at 4–5 weeks) and 8 weeks of age, weight gain is at its highest and feed conversion is optimal. The rates of increase of feed intake and growth rate subsequently decrease, with intake stabilizing at around 12 weeks of age for current hybrid lines of domestic rabbits (Gidenne et al., 2010). Similar to other mammals, the rabbit regulates its food intake according to energy requirements. However, greater food intake was recorded for a high-, compared to a moderate-energy, diet (Debray et al., 2002; Gidenne et al., 2004). Other factors, such as the form of food, and pellet size and quality (hardness, durability), play a role in the initiation of solid food intake in young rabbits (Gidenne et al., 2010). Since the rabbit is a nonruminant herbivore, the main blood component regulating food intake is not clear, but is probably glucose (Gidenne et al., 2010).

Feeding high-quality hay and grass as the major component of the diet is a guarantee to keep the rabbit healthy. A rabbit's digestive system needs hay or grass to function properly (Gidenne et al., 2010). When it comes to laboratory rabbits, it is often important nutritionally to know exactly what they are fed; therefore, they are often fed pelleted diet. In those circumstances, it is important to provide hay as enrichment so the rabbits will benefit from the positive properties that hay offers. High-fiber food items, such as hay, help prevent diarrhea and trichobezoars (hairballs), and Patton (1994) suggests a diet with approximately 18%–25% crude fiber. Hay also provides the animals with something to manipulate and play with (Morton et al., 1993). Feeding rabbits a low-fiber diet can induce Epizootic Rabbit Enteropathy (ERE, Jin et al., 2018). To avoid ERE, feeding a restricted diet has positive effects in the postweaning period (Gidenne et al., 2012). The first outbreaks of ERE in 1997 led to up to 70% morbidity and mortality rates (Gidenne et al., 2012), and a reduction in the starch-to-fiber ratio resulted in decreased mortality. In practice, this means that, with respect to digestive health, and especially in the context of ERE, the proportion of fiber in the diet should be increased at the expense of starch content (Gidenne et al., 2010).

One of the main dietary components implicated in feed intake regulation, after weaning, is digestible energy (DE) concentration (Gidenne et al., 2010). The domestic rabbit (fed a pelleted balanced diet) is able to regulate its DE intake (and thus its growth) when the DE concentration is between 9 and 11.5 MJ/kg, or when the dietary fiber level is between 10% and 25% Acid Detergent Fiber (Gidenne et al., 2010). The intake level is thus well correlated with the dietary fiber level, compared to the dietary DE content (Gidenne et al., 2010).

According to Leslie et al. (2004), there is little evidence that rabbits prefer a grass diet to one based on compound feed. Despite this, it is often beneficial to supplement the ration of proprietary compound pellets with dietary enrichment (e.g., grass or hay), which as well as providing essential nutritional components (National Research Council, 1977), in particular fiber (Lehmann 1990), will increase time spent procuring food and reduce abnormal behaviors, such as chewing the cage (Leslie et al., 2004).

Ad libitum feeding is considered a normal practice for rodents; however, it is considered a bad management practice for rabbits, as they tend to become obese (Hart et al., 1995). When individuals of several strains of rabbits are fed *ad libitum* for a long period, they tend to become fat, especially older animals with access to limited space for locomotion (e.g., caged NZW rabbits; Ritskes-Hoitinga and Jilge, 2001). When a limited quantity of pelleted food is given to rabbits, the animals consume their daily allocation within a few hours. For example, for rabbits caged individually or in pairs, 85% of the *ad libitum* intake is ingested in a maximum of 16 h, but if the quantity available is reduced to 70% of the time required to ingest this quantity is reduced to 10 h (Bergaoui et al., 2008). In a study where rabbits had 9 h available for food intake, the total duration spent consuming feed was 1 h 20 min, compared to 1 h 45 min when rabbits of the same age had continuous access to feed (Szendrö et al., 1988).

If the nutrient intake must be reduced in order to avoid obesity, it is better to reduce the nutrition in the food then to reduce the time when food is available (Gidenne et al., 2012). Intake restriction modifies feeding behavior, as total food intake is completed within 6–10 h, instead of 24 h (Gidenne et al., 2012). In comparison, rabbits fed freely show relatively regular intake behavior, eating numerous meals (Gidenne and Lebas, 2006). Feeding restriction reduced the number of drinking bouts performed by rabbits, from 47 per day for *ad libitum* (ad lib) animals to 38 per day for restricted (R) animals (Martignon, 2010), although the quantity of water consumed was higher for R rabbits. As reported by Boisot et al. (2005), the ratio of water to feed is doubled (from 1.68 to 3.46) for rabbits restricted to 65% of *ad lib*. These authors also reported changes in feeding behavior; fecal excretion patterns, including cecotrophy, were greatly modified. For R rabbits, the fecal excretion peak occurred between 5 and 8 h after feed was given, or about 3–4 h later than the eating peak. The cecotrophy period was thus relocated to about 8–10 h after feed distribution, as previously observed by Fioramonti and Ruckebusch (1976) in adult rabbits fed once a day. However, the ratio of soft to hard feces seemed to be unaltered (Martignon, 2010). Competition to reach the feeder can be high during the first hour after feed distribution for R rabbits, potentially resulting in hostile or even aggressive behavior; however, Martignon (2010) observed no increase in aggressiveness or lesions in R rabbits.

In recent years, muesli diets have been introduced as common foods for pet rabbits. Prebble et al. (2015) compared weight gain with four diet regimens (extruded diet with ad lib hay, muesli with ad lib hay, ad lib hay only, ad lib muesli only). All groups gained weight with age, but after 9 months, relative to ad lib hay only feeding (mean 1.77 ± 0.13 kg), rabbits in the ad lib muesli only group weighed 146% (2.59 ± 0.32 kg) of the weight of hay only rabbits; rabbits fed extruded diet with ad lib hay weighed 125% (2.21 ± 0.10 kg) of hay only rabbits, and the rabbits in the muesli with ad lib hay group weighed 123% (2.18 ± 0.13 kg) of hay only rabbits (Prebble et al., 2015).

In another study (Meredith and Prebble, 2017) with similar diets, it was shown that fecal pellets were consistently smaller and lighter in rabbits fed muesli only, and the size of pellets produced by those fed muesli with hay decreased over the course of the study (Meredith and Prebble, 2017). Fecal output was greatest in rabbits with the highest hay intake and uneaten cecotrophs were found at the highest frequency in rabbits fed muesli only (Meredith and Prebble, 2017).

The effects of housing, feeding time, and diet composition on the behavior of the laboratory rabbit were studied by Krohn et al. (1999). Feeding the animals at 14:00h reduced abnormal behavior during the dark period, compared to feeding at 08:00h. No difference in behavior could be detected between feeding a high-energy and a low-energy diet at 08:00h. Animals in floor pens generally showed less abnormal behavior than caged animals. These results indicate that the welfare of caged rabbits can be improved by feeding the animals in the afternoon, rather than in the morning (Krohn et al., 1999).

Drinking water seems to be more important for rabbits, than other animal species. A 2-kg rabbit drinks as much as a 10-kg dog (Cizek, 1961; Clauss and Hatt, 2017). How the water is provided affects drinking behavior, and rabbits exhibited significantly higher water intake with open dishes compared with nipple drinkers (Tschudin et al., 2011). Water intake is also positively correlated with dry matter intake and was greatest when only hay was fed (Prebble and Meredith, 2014).

Selective feeding (only eating some of the ingredients in a "mix") occurred in pet rabbits fed muesli, whether or not hay was also available (Prebble and Meredith, 2014). The presence of selective feeding in rabbits fed muesli led to the consumption of an unbalanced diet. In addition, hay intake and water intake were lower when muesli was fed (Prebble and Meredith, 2014). Therefore, it is recommended not to feed muesli to laboratory rabbits.

Rabbits possess continuously growing teeth and growth rates were higher on a pelleted diet, compared to rabbits on a hay diet (Wyss et al., 2016). This can be of importance for rabbits kept for longer periods, e.g., breeding rabbits and rabbits used for antibody production. Tooth growth is also strongly related to tooth wear, and differs correspondingly between diets and teeth positions (Müller et al., 2014). Meredith and colleagues assessed the impact of four rabbit diets (hay only; extruded diet with hay; muesli with hay; muesli only) on length and curvature of cheek teeth, and eruption and attrition rates of incisors (Meredith et al., 2015). By month 9, a greater degree of tooth curvature was present in rabbits fed muesli only, and after 17 months, rabbits fed muesli only or muesli with hay, had longer lower first cheek teeth and larger interdental spaces between the first two molars, compared to other groups (Meredith et al., 2015). Hay seems to be the solution to avoid this problem. So far, no one has investigated the use of gnawing material (gnawing sticks, chewing blocks, wood) on tooth wear (Clauss and Hatt, 2017).

Training

Rabbits that are well acquainted with their handlers and environment are less stressed in experimental situations, which improves the outcome of the scientific work (Toth and January, 1990). Early handling of rabbits has been shown to result in more active, alert, and exploratory animals that approach novel stimuli more often (Harkness and Wagner, 1995). Dúcs et al. (2009) showed that rabbit kits approached an experimenter's hand with a shorter latency and more frequently during an approach test at 28 days of life, if they had been either handled or exposed to human smell for 5 min after nursing during the first week of life, than if they had no exposure to humans. They draw the conclusion that olfactory exposure during handling results in imprinting, even without human contact. Rabbits are rarely aggressive, if handled in a calm and steady manner, and without intimidating the rabbit by unfamiliar sounds or odors.

In Lidfors and Edström (in press), the following description, together with the photos of how to lift a rabbit out of the cage, is given:

> To pick up an animal, first speak to it and approach it quietly, then the scruff of the neck is grasped firmly with one hand (not including the ears); the animal is lifted, using the other hand to support the body, and placed on the other arm with its head in the opening between the elbow and the body of the handler."

For transport over a short distance, the animal should be allowed to rest in a normal posture on the forearm of the handler with its head hidden, while the hand of the handler holds the scruff (Stein and Walshaw, 1996). It is advisable to wear long-sleeved garments while handling rabbits, because their claws may scratch the skin of the handler's arms (Suckow and Douglas, 1997). Rabbits have a fragile skeleton and strong hind leg muscles, thus struggling may cause fracture of the spinal vertebrae (Marston et al., 1965).

Training rabbits for blood sampling and other data collection techniques in the laboratory is an effective way to reduce stress. In "Techniques Training: Rabbit 06–00040" (published by American Association of Laboratory Animal Science), there is a step-by-step guide (with pictures) that shows methods for performing basic techniques with the laboratory rabbit. Suckow et al. (2012), in their second edition of *The Laboratory Rabbit*, present basic information and common procedures in detail, as a quick reference for caretakers, technicians, and researchers in a laboratory setting.

Special Situations

Welfare of Rabbits

The welfare of laboratory rabbits is related to housing, management, and the research they are participating in. Some experiments may cause discomfort or pain, but in addition, fights between incompatible rabbits may cause nonobvious injuries, as rabbits are prey animals that try to hide if they are ill or in pain. The use of video surveillance of groups of rabbits can help find individuals that behave differently, are sitting still for long periods, or fights, which typically occur at night. A rabbit with wounds should be separated from the group as soon as possible. A rabbit with a dirty or starry coat indicates that it has not groomed itself for some time, which a healthy rabbit does regularly. Sitting or moving with a hunched posture or curled ear margins may be signs of pain.

A Rabbit Grimace Scale (RbtGS) has been developed by Keating et al. (2012), in which the rabbit's orbital tightening, cheek flattening, nostril shape, whisker shape and position, and ear shape and position are graded as not present (scale 0), moderately present (scale 1), or obviously present (scale 2) by an observer. Photos with detailed descriptions of the RbtGS and a training manual are available at the NC3Rs web page (https://www.nc3rs.org.uk/rabbit-grimace-scale). The RbtGS has been used to evaluate rabbit welfare after different laboratory procedures by Hampshire and Robertson (2015).

Other signs of poor welfare are the abnormal, stereotypic, or apathetic behaviors presented earlier. Reduced food and water intake can also be used as signs of reduced welfare, but may be difficult to observe in group-housed animals. If individual rabbits are threatened away from food or water, this may show up in reduced weight gain or no increased weight in growing animals. However, rabbits may also get obese when fed *ad libitum* with no exercise, poor diet, and lack of hay. As weighing rabbits routinely may be time consuming, using body condition scoring (BCS) regularly can help keep the rabbits within a preferred range of weights. A five-point BCS scale (very thin, thin, ideal, overweight, obese) has been developed (Rabbit Size-O-Meter, PFMA, 2020), and Prebble et al. (2015) have demonstrated a high correlation between BSC and bodyweight.

The Federation of European Laboratory Animal Science Associations (FELASA) recommendations contain detailed information on how laboratory rabbits should be housed and handled (Mähler et al., 2014).

Antibody Production in Rabbits

Today, most antibody production is carried out in Asia, with China as a major producer, or in the USA. Many of the companies involved in antibody production have their headquarters in the USA, but production is based in Asia. There are no official statistics on the number of animals used, and little information is available concerning the ways in which the animals are housed and treated during antibody production. Fuentes and Newgren (2008) found that female NZW rabbits given a subcutaneous booster of antigen without adjuvant had the same antibody production when housed in groups in floor pens as in individual steel cages. They also found that the group-housed rabbits had lower white blood cell counts, higher plasma cortisol levels and lower weight gain during the first week than the individually housed rabbits (Fuentes and Newgren, 2008). AAALAC International has given accreditation to more than 1,000 organizations, institutions, and companies in 49 countries over the world; however, it is difficult to find out whether these organizations produce antibodies. One company that has developed an animal welfare program for their goats, hens, and rabbits involved in the production of antibodies is Capra Science in Sweden. To our knowledge, several antibody-producing companies in Europe keep rabbits in group-housing in enriched floor pens, and have been doing so for many years.

Conclusions/Recommendations

Basic requirements for good rabbit management include the following:

- Housing in pens on the floor or on elevated platforms should always be considered before cage housing.
- Pens should be large enough for animals to hop, jump, and make quick changes of direction.
- Enclosures should be tall enough to allow rabbits to rear upwards for scanning, exploration, and play.
- Solid floors with substrate (2–5 cm in depth) for hygiene, comfort, and to permit foraging and digging behavior should be provided.
- Pens should contain sufficient space to permit group housing and exercise, and should contain essential enrichment that allows the animals to perform a wide range of normal behavior.
- Stable, compatible groups, established with immature animals of the same age and sex, as soon as possible after weaning, should be utilized.
- Intact male rabbits should be separated from other males at sexual maturity (12–14 weeks) and housed individually with visual and olfactory contact with other rabbits, but they cannot be placed together due to the high risk of fights and injuries.
- Hay for foraging, play and nest building, and a varied diet should be provided, where possible.
- Feeding a muesli-based diet should be avoided.
- Nest boxes with nesting material should be available for breeding females, designed such that does that have litters cannot see each other (to prevent infanticide behavior).
- A raised area to facilitate use of the vertical dimension, offer a comfortable resting place and refuge, stimulate exercise, and offer a choice of microenvironments should be provided.
- Items to gnaw (e.g., wooden blocks) for enrichment, chin-marking, and to prevent teeth from overgrowing should be provided.
- Visual barriers that allow the animals to initiate or avoid social contact should be included.
- Items that functionally simulate burrows (e.g., plastic crates, sections of 18-inch PVC pipe) should be provided to allow for retreat in fear-provoking situations and to manage social interactions.
- Pens should provide good visibility of the outside of the pen (e.g., mesh or plastic wall) so that the animals can overlook their surroundings and see approaching personnel.
- Animals should receive gentle and frequent handling from early in life.
- Animals should be habituated to procedures that will be used for repeated sampling.
- Health monitoring in the rabbit unit should be in accordance with FELASA guidelines.

ADDITIONAL RESOURCES

Agriculture Victoria State Government, Australia. Code of practice for the housing and care of laboratory mice, rats, guinea-pigs and rabbits. Page up-dated 12 July 2020. https://agriculture.vic.gov.au/livestock-and-animals/animal-welfare-victoria/pocta-act-1986/victorian-codes-of-practice-for-animal-welfare/code-of-practice-for-the-housing-and-care-of-laboratory-mice-rats-guinea-pigs-and-rabbits# (accessed August 03, 2020).

The American Rabbit Breeders Association. https://www.arba.net (accessed August 03, 2020).

Animal Research.Info. Rabbit. http://www.animalresearch.info/en/designing-research/research-animals/rabbit/ (accessed August 03, 2020).

The British Rabbit Council. https://thebritishrabbitcouncil.org/ (accessed August 03, 2020).

Capra Science. Antibody production. https://www.caprascience.com/ (accessed August 04, 2020).

DAD-IS (Domestic Animal Diversity Information System). FAO (Food and Agriculture Organization of the United Nations) at Wikipedia. List of rabbit breeds, Page up-dated 16 June 2020. https://en.wikipedia.org/wiki/List_of_rabbit_breeds (accessed August 03, 2020).

European Convention for the protection of vertebrate animals used for experimental and other scientific purposes: Appendix A: Guidelines for accommodation and care of animals (article 5 of the convention) (ETS No. 123). 2006. Approved by the Multilateral Consultation. 109 pp. http://conventions.coe.int/Treaty/EN/Treaties/PDF/123-Arev.pdf.

National Centre for the Replacement refinement and Reduction of Animals in Research (NC3Rs). Rabbit grimace scale. https://www.nc3rs.org.uk/rabbit-grimace-scale (accessed August 03, 2020).

Pet Food Manufacturers´Association (PFMA). Rabbit Size-O-Meter. http://www.pfma.org.uk/_assets/weigh-in-wednesday/pet-size-o-meter-rabbit.pdf (accessed August 03, 2020).

Report from the Commission to the European Parliament and the Council -2019 Report on the Statistics on the Use of Animals for Scientific Purposes in the Member States of the European Union in 2015–2017. Brussels, 5.2.2020. COM (2020) 16 final. Published 2020–02–05 by Directorate-General for Environment (European Commission). https://ec.europa.eu/info/sites/info/files/com-2020-16-f1-en-main-part-1.pdf (accessed August 03, 2020).

Scanbur. Conventional cages for rabbits. https://www.scanbur.com/products/housing/conventional-cages/conventional-cages (assessed August 05, 2020).

Tecniplast. R-SUITE – Rack for rabbits. https://www.tecniplast.it/en/product/r-suite-rabbit-housing-rack.html (assessed August 05, 2020).

United States Department of Agriculture Animal and Plant Health Inspection Service. Annual report animal usage, 2017. https://www.aphis.usda.gov/animal_welfare/downloads/reports/Annual-Report-Animal-Usage-by-FY2017.pdf (accessed August 03, 2020).

Waikato Regional Council Biosecurity Factsheet Series. Rabbits: *Oryctolagus cuniculus* (4283-0515). Updated June 2015. https://www.waikatoregion.govt.nz/assets/PageFiles/3508/4283%20-%20Rabbits%20factsheet%202015_web.pdf (accessed August 03, 2020).

REFERENCES

Arteaga, L., Bautista, A., Martínez-Gómez, M. et al. 2008. Scent marking, dominance and serum testosterone levels in male domestic rabbits. *Physiology and Behavior* 94:510–515.

Baumann, P., Oester, H. and Stauffacher, M. 2005. The use of a cat-flap at the nest entrance to mimic natural conditions in the breeding of fattening rabbits (*Oryctolagus cuniculus*). *Animal Welfare* 14:135–142.

Bautista, A., Mendoza-Degante, M., Coureaud, G., Martina-Gomez, M. and Hudson, R. 2005. Scramble competition in newborn domestic rabbits for an unusually limited milk supply. *Animal Behavior* 70:997–1002.

Bell, D. J. 1977. *Aspects of the Social Behaviour of Wild and Domesticated Rabbits Oryctolagus cuniculus L*. PhD diss., University of Wales.

Bell, D. J. 1980. Social olfaction in lagomorphs. *Symposium of the Zoological Society of London* 45:141–164.

Bell, D. J. 1984. The behavior of rabbits: Implications for their laboratory management. In *Proceedings of UFAW/LASA Joint Symposium. Standards in Laboratory Animal Management, Part II*, 151–162. Potters Bar: Universities Federation for Animal Welfare.

Bell, D. J. 1999. The European wild rabbit. In *The UFAW Handbook on the Care and Management of Laboratory Animals*, (7th Edition), ed. T. Poole and P. English, 389–394. Oxford: Blackwell Publishing.

Bell, D. J. and Webb, N. J. 1991. Effects of climate on reproduction in the European wild rabbit *Oryctolagus cuniculus*. *Journal of Zoology* 224:639–648.

Bennett, B. 2001. *Storey's Guide to Raising Rabbits*, ed. D. Burns and M. Salter. North Adams, MA: Storey Communications Inc.

Bergaoui, R., Kammoun, M. and Ouerdiane, K. 2008. Effects of feed restriction on the performance and carcass of growing rabbits. In *Proc. of the 9th World Rabbit Congress, Verona, Italy, 10–13 June*, ed. G. Xiccato, A. Trocino and S. D. Lukefahr, 547–550. Brescia, Italy: Fondazione Iniziative Zooprofilattiche E Zootechniche publ. http://world-rabbit-science.com/.

Berthelsen, H. and Hansen, L. T. 1999. The effect of hay on the behaviour of caged rabbits (*Oryctolagus cuniculus*). *Animal Welfare* 8:149–157.

Björnhag, G. 1981. Separation and retrograde transport in the large intestine of herbivores. *Livestock Production Science* 8:351–360.

Black, S. L. and Vanderwolf, C. H. 1969. Thumping behavior in the rabbit. *Physiology and Behavior* 4:445–449.

Boisot P, Duperray J. and Guyonvarch, A. 2005. Intérêt d'une restriction hydrique en comparaison au rationnement alimentaire en bonnes conditions sanitaires et lors d'une reproduction expérimentale de l'Entéropathie Epizootique du lapin (EEL). In *Proceedings of the 11e `mes Journe´ es deRecherches Cunicoles Franc¸aises*, ed. G. Bolet, 133–136. Paris: ITAVI Publishing.

Bõsze, Zs. and Houdebine, L. M. 2006. Application of rabbits in biomedical research: A review. *World Rabbit Science* 14:1–14.

Braconnier, M., Gómez, Y., Gebhardt-Henrich, S. G. 2020. Different regrouping schedules in semi group-housed rabbit does: Effects on agonistic behaviour, stress and lesions. *Applied Animal Behaviour Science* 228:105024.

Brem, G., Brenig, B., Godman, H. M., et al. 1985. Production of transgenic mice, rabbits and pigs by microinjection into pronuclei. *Zuchthygiene* 20:251–252.

Brooks, D. L. 1997. Nutrition and gastrointestinal physiology. In *Ferrets, Rabbits and Rodents – Clinical Medicine and Surgery*, ed. E. W. Hillyer and K. E. Quesenberry, 169–175. London: W. B. Saunders.

Buijs, S., Hermans, K., Maertens, L., Van Caelenberg, A. and Tuyttens, F. A. M. 2014. Effects of semi-group housing and floor type on pododermatitis, spinal deformation and bone quality in rabbit does. *Animal* 8:1728–1734.

Buseth, M. E. and Saunders, R. A. 2015. *Rabbit Behaviour, Health and Care*. Malta: CABI.

Canali, E., Ferrante, V., Todeschini, R. et al. 1991. Rabbit nest construction and its relationship with litter development. *Applied Animal Behaviour Sciences* 31:259–266.

Carabaño R., Piquer J., Menoyo D., Badiola I. The digestive system of the rabbit De Blas C., Wiseman J. (Eds.), Nutrition of the Rabbit (second ed.), CABI, Wallingford (2010), pp. 1–18.

Castellini, C., Dal Bosco, A. and Cardinali, R. 2006. Long term effect of post-weaning rhythm on the body fat and performance of rabbit doe. *Reproduction Nutrition Development* 46:195–204.

Cheeke, P. R. 1987. *Rabbit Feeding and Nutrition*. New York: Academic Press.

Chu, L. R., Garner, J. P. and Mench, J. A. 2004. A behavioral comparison of New Zealand White rabbits (Oryctolagus cuniculus) housed individually or in pairs in conventional laboratory cages. *Applied Animal Behaviour Science* 85:121–139.

Cizek, L. J. 1961. Relationship between food and water ingestion in the rabbit. *American Journal of Physiology* 201:557–566.

Clauss, M. and Hatt, J. M. 2017. Evidence-based rabbit housing and nutrition. *Veterinary Clinics: Exotic Animal Practice* 20:871–884.

Cobb, A. B. 2011. *Macmillan Science Library: Animal Sciences*, Vol. 1. New York: Macmillan Reference USA. 2001–2006.

Coda, K. A., Fortman, J. D. and García, K. D. 2020. Behavioral effects of cage size and environmental enrichment in New Zealand White rabbits. *Journal of the American Association for Laboratory Animal Science* 59: 356–364.

Cohen, C., and Tissot, R. G. 1974. The effect of the RL-A locus and the MLC locus on graft survival in the rabbit. *Transplantation* 18:150–154.

Cooke, B. D. 2012. Rabbits: Manageable environmental pests or participants in new Australian ecosystems. *Wildlife Research* 39:279–289.

Cowan, D. P. 1987. Group living in the European rabbit (*Oryctolagus cuniculus*): Mutual benefit or resource localization? *Journal of Animal Ecology* 56:779–795.

Cowan, D. P. and Bell, D. J. 1986. Leporid social behaviour and social organization. *Mammal Review* 16:169–179.

Dalle Zotte, E., Princz, Z., Matics, Zs. et al. 2009. Rabbit preference for cages and pens with or without mirrors. *Applied Animal Behavior Science* 116:273–278.

Daly, J. C. 1981. Effects of social organisation and environmental diversity on determining the genetic structure of a population of the wild rabbit (*Oryctolagus cuniculus*). *Evolution* 35:689–706.

Debray, L., Fortun-Lamothe, L. and Gidenne, T. 2002. Influence of low dietary starch/fibre ratio around weaning on intake behavior, performance and health status of young and rabbit does. *Animal Research* 51:63–75.

Distel H, Hudson R. 1985. The contribution of the olfactory and tactile modalities to the nipple-search behaviour of newborn rabbits. *Journal of Comparative Physiology A* 157:599–605.

DiVincenti Jr, L. and Rehrig, A. N. 2016. The social nature of European rabbits (*Oryctolagus cuniculus*). *Journal of the American Association for Laboratory Animal Science* 55:729–736.

Donnelly, T. M. 1997. Basic anatomy, physiology and husbandry. In *Ferrets, Rabbits and Rodents – Clinical Medicine and Surgery*, ed. E. W. Hillyer and K. E. Quesenberry, 147–159. London: W. B. Saunders.

Doughty, M. J. 2020. Short term effects of continuous lighting on the cornea of cage-reared laboratory rabbits. *Journal of Photochemistry and Photobiology B: Biology* 204:111764.

Drescher, B. and Loeffler, K. 1996. Scoliosis, lordosis and kyphosis in breeding rabbits. *Tierarztliche Praxis* 24:292–300.

Dúcs, A., Bilkó, Á. and Altbäcker, V. 2009. Physical contact while handling is not necessary to reduce fearfulness in the rabbit. *Applied Animal Behaviour Science* 121:51–54.

Edgar, J. L. and Seaman, S. C. 2010. The effect of mirrors on the behaviour of singly housed male and female laboratory rabbits. *Animal Welfare* 19:461.

Egedal, E. 2009. Problems when attempting to pair house un-castrated male rabbits (*Oryctolagus cuniculus*). Student report, no. 258, Swedish University of Agricultural Sciences. https://stud.epsilon.slu.se/331/.

Esteves, P. J., Abrantes, J., Baldauf, H.-M., et al. 2018. The wide utility of rabbits as models of human diseases. *Experimental and Molecular Medicine* 50:66. doi: 10.1038/s12276-018-0094-1.

Fioramonti, J. and Ruckebusch, Y. 1976. La motricite´ caecale chez le lapin. (2) Variations d'origine alimentaire. *Annales de Recherche Ve´te´rinaires* 5:201–212.

Flux, J. E. C. 1994. World distribution. In *The European Rabbit – The History and Biology of a Successful Colonizer*, ed. H. V. Thompson and C. M. King, 8–21. Oxford: Oxford University Press.

Fortun-Lamothe, L. and Prunier, A. 1999. Effects of lactation, energetic deficit and remating interval on reproductive performance of primiparous rabbit does. *Animal Reproduction Science* 55:289–298.

Fraser, K.W. 1992. Emergence behaviour of rabbits, *Oryctolagus cuniculus*, in Central Otago, New Zealand. *Journal of Zoology* 228:615–623.

Fuentes, G. C. and Newgren, J. 2008. Physiology and clinical pathology of laboratory New Zealand White rabbits housed individually and in groups. *Journal of the American Association for Laboratory Animal Science* 47:35–38.

Gibb, J. A. 1993. Sociality, time and space in a sparse population of rabbits (*Oryctolagus cuniculus*). *Journal of Zoology* 229:581–607.

Gidenne, T., Combes, S. and Fortun-Lamothe, L. 2012. Feed intake limitation strategies for the growing rabbit: Effect on feeding behaviour, welfare, performance, digestive physiology and health: A review. *Animal* 6:1407–1419.

Gidenne, T., Lapanouse, A. and Fortun-Lamothe, L. 2004. Feeding strategy for the early weaned rabbit: Interest of a high energy and protein starter diet on growth and health status. In *Proceedings of the 8th World Rabbit Congress,*

Puebla, Mexico, 7–10 September, ed. C. Becerril and A. Pro, 853–860. Colegio de Postgraduados for WRSA publ. http://world-rabbit-science.com.

Gidenne, T. and Lebas, F. 1987. Estimation quantitative de la caecotrophie chez le lapin en croissance: Variations en fonctions de l'âge. *Annales de Zootechnie* 36:225–236.

Gidenne, T. and Lebas, F. 2006. Feeding behavior in rabbits. In *Feeding in Domestic Vertebrates. From Structure to Behavior*, ed. V. Bels, 179–209. Wallingford, UK: CABI Publishing.

Gidenne, T., Lebas, F., Fortun-Lamothe, L. 2010. Feeding behaviour of rabbits. In *Nutrition of the rabbit*, ed. C. de Blas and J. Wiseman, 233–252. Wallingford, UK: CABI Publishing.

González-Mariscal, G., Caba, M., Martínez-Gómez, M., Bautista, A. and Hudson, R. 2016. Mothers and offspring: The rabbit as a model system in the study of mammalian maternal behavior and sibling interactions. *Hormones and Behavior* 77:30–41.

González-Mariscal, G., Chirino, R., Beyer, C. and Rosenblatt, J. S. 2004. Removal of the accessory olfactory bulbs facilitates maternal behavior in virgin rabbits. *Behavior and Brain Research* 152:89–95.

González-Mariscal, G., Díaz-Sánchez, V., Melo, A. I., Beyer, C. and Rosenblatt, J. S. 1994. Maternal behavior in New Zealand white rabbits: Quantification of somatic events, motor patterns and steroid plasma levels. *Physiology and Behavior* 55:1081–1089.

González-Mariscal, G., Melo, A. I., Chirino, R. Jiménez, P., Beyer, C. and Rosenblatt, J. S. 1998. Importance of mother/young contact at parturition and across lactation for the expression of maternal behavior in rabbits. *Developmental Psychobiology* 32:101–111.

Green, J. and Flinders, J. 1981. Alarm call of the Pygmy rabbit (*Brachylagus idahoensis*). *The Great Basin Naturalist* 41:158–160.

Gunn, D. and Morton, D. B. 1995. Inventory of the behavior of New Zealand White rabbits in laboratory cages. *Applied Animal Behavior Science* 45:277–292.

Hampshire, V. and Robertson, S. 2015. Using the facial grimace scale to evaluate rabbit wellness in post-procedural monitoring. *Laboratory Animals* 44:259–261.

Hansen, L. T. and Berthelsen, H. 2000. The effect of environmental enrichment on the behaviour of caged rabbits (*Oryctolagus cuniculus*). *Applied Animal Behaviour Science* 68:163–178.

Harcourt-Brown, F. 2002. Biological characteristics of the domestic rabbit. In *Textbook of Rabbit Medicine*, ed. F. Harcourt-Brown and N. H. Harcourt-Brown, 1–18. Oxford: Butterworth-Heinemann.

Harkness, J. E. and Wagner, J. E. 1995. *The Biology and Medicine of Rabbits and Rodents*, 4th edn. Baltimore: Lippincott Williams & Wilkins.

Harris, L. D., Custer, L. B., Soranaka, E. T., Burge, J. R. and Ruble, G. R. 2001. Evaluation of objects and food for environmental enrichment of NZW rabbits. *Journal of the American Association for Laboratory Animal Science* 40:27–30.

Hart, R. W., Neumann, D. A. and Robertson, R. T. 1995. *Dietary Restriction: Implications for the Design and Interpretation of Toxicity and Carcinogenicity Studies*. Washington, DC: ILSI Press.

Hartman, H. A. 1974. The foetus in experimental toxicology. In *The Biology of the Laboratory Rabbit*, ed. S. H. Weisbroth, R. E. Flatt and A. L. Kraus, 91. New York: Academic Press.

Hawkins, P., Hubrecht, R., Buckwell, A., et al. 2008. Refining rabbit care - A resource for those working with rabbits in research. Report from the UFAW/RSPCA Rabbit Behavior and Welfare Group. England. https://science.rspca.org.uk/documents/1494935/9042554/Refining+rabbit+care+-+report.pdf/a528fbaa-f11a-a5ff-f3e3-38e6e6439f4d?t=1552901950204.

Held, S. D. E., Turner, R. J. and Wootton, R. J. 1995. Choices of laboratory rabbits for individual or group-housing. *Applied Animal Behavior Science* 46:81–91.

Hoffman, K. L. and Rueda-Morales, R. I. 2009. Towards an understanding of the neurobiology of "just right" perceptions: Nest building in the female rabbit as a possible model for compulsive behavior and the perception of task completion. *Behavior and Brain Research* 204:182–191.

Hudson, R., Cruz, Y., Lucio, R. A., Ninomiya, J., and Martínez-Gómez, M. 1999. Temporal and behavioral patterning of parturition in rabbits and rats. *Physiology and Behavior* 66:599–604.

Hudson, R. and Distel, H. 1983. Nipple location by newborn rabbits: Evidence for pheromonal guidance. *Behaviour* 85:260–274.

Hudson, R., Schaal, B., Bilkó, Á. and Altbacker, V. 1996. Just three minutes a day: The behaviour of young rabbits viewed in the context of limited maternal care. In *6th World Rabbit Congress, 9-12 July 1996, Toulouse, France*. ISSN 2308-1910.

Jensen, R.G. 1995. *Handbook of Milk Composition*. Cambridge, MA: Academic Press.

Jezierski, T. A. and Konecka, A. M. 1996. Handling and rearing results in young rabbits. *Applied Animal Behaviour Science* 46:243–250.

Jin, D. X., Zou, H. W., Liu, S. Q., et al. 2018. The underlying microbial mechanism of epizootic rabbit enteropathy triggered by a low fiber diet. *Scientific Reports* 8:1–15.

Johnson, C. A., Pallozzi, W. A., Geiger, L. et al. 2003. The effect of an environmental enrichment device on individually caged rabbits in a safety assessment facility. *Journal of the American Association for Laboratory Animal Science* 42:27–30.

Jones, S. E. and Phillips, C. J. C. 2005. The effects of mirrors on the welfare of caged rabbits. *Animal Welfare* 14:195–202.

Kalagassy, E. B., Carbone, L. G. and Houpt, K. A. 1999. Effect of castration on rabbits housed in littermate pairs. *Journal of Applied Animal Welfare Science* 2:111–121.

Keating, S. C. J., Thomas, A. A., Flecknell, P. A. et al. 2012. Evaluation of EMLA cream for preventing pain during tattooing of rabbits: Changes in physiological, behavioural and facial expression responses. *PLoS One* 7(9): e44437.

Knutsson, M. 2011. Exercise pens as an environmental enrichment for laboratory rabbits. Student report 2011:48 in Veterinary program, Swedish University of Agricultural Sciences. https://stud.epsilon.slu.se/3603/.

Kolb, H. H. 1994. The use of cover and burrows by a population of rabbits (Mammalia: *Oryctolagus cuniculus*) in eastern Scotland. *Journal of Zoology* 233:9–17.

Kovacs, M., Szendrõ, Z., Csutoras, I., et al. 2004. Development of the caecal microflora of newborn rabbits during the first ten days after birth. In *Proceedings of the 8th World Rabbit Congress, Puebla, Mexico, 7–10 September*, ed. C. Becerril and A. Pro, 1091–1096. Colegio de Postgraduados for WRSA publ. http://world-rabbit-science.com.

Krohn, T. C., Ritskes-Hoitinga, J. and P. Svendsen. 1999. The effects of feeding and housing on the behavior of the laboratory rabbit. *Laboratory Animals* 33:101–107.

Lehmann, M. 1989. *Das verhalten junger hauskaninchen unter verschieden umgebungsbedingungen*. PhD Thesis, Universität Berne.

Lehmann, M. 1990. Beschäftigungsbedürfnis bei jungen hauskaninchen: Rohfaseraufnahme und tiergerechtheit. *Schweiz Archiv Tierheilkunde* 132:375–381.

Lehmann, M. 1991. Social behavior in young domestic rabbits under semi-natural conditions. *Applied Animal Behavior Science* 32:269–292.

Leslie, T. K., Dalton, L. and Phillips, C. J. C. 2004. Preference of domestic rabbits for grass or coarse mix feeds. *Animal Welfare* 13:57–62.

Lidfors, L. 1996. Behavior of male laboratory rabbits given environmental enrichment in a preference test and in an individual cage. In *Proceedings of 30th International Congress of the International Society of Applied Ethology*. Canada.

Lidfors, L. 1997. Behavioral effects of environmental enrichment for individually caged rabbits. *Applied Animal Behavior Science* 52:157–169.

Lidfors, L. and Edström, T. In press. The laboratory rabbit. In *The UFAW Handbook on the Care and Management of Laboratory and Other Animals Used in Scientific Procedures*, 9th edn, ed. R. Hubrecht and H. Golledge. Oxford: Wiley-Blackwell.

Lockley, R. M. 1961. Social structure and stress in the rabbit warren. *The Journal of Animal Ecology* 30:385–423.

Lumpkin, S. and Seidensticker, J. 2011. *Rabbits – The Animal Answer Guide*. Baltimore: The Johns Hopkins University Press.

Maertens, L., Lebas, F. and Szendrő, Zs. 2006. Rabbit milk: A review of quantity, quality and non-dietary affecting factors. *World Rabbit Science* 14: 205–230.

Maertens, L. and Van Herck, A. 2000. Performances of weaned rabbits raised in pens or in classical cages: First results. *World Rabbit Science* 8:435–440.

Mähler, M., Berard, M., Feinstein, R. et al. 2014. FELASA recommendations for the health monitoring of mouse, rat, hamster, guinea pig and rabbit colonies in breeding and experimental units. *Laboratory Animals* 48:178–192.

Marston, H.H., Rand, G. and Chang, M. C. 1965. The care, handling and anaesthesia of the snowshoe hare (*Lepus americanus*). *Laboratory Animal Care* 15:325–327.

Martignon, M. H. 2010. Consequences of a feed intake control on digestive physiopathology and on feeding behavior (In French). PhD thesis, Institut National Polytechnique de Toulouse. http://ethesis.inp-toulouse.fr/archive/00001486/.

McClellan, H., Miller, S. and Hartmann, P. 2008. Evolution of lactation: Nutrition V. Protection with special reference to five mammalian species. *Nutrition Research Reviews* 21:97–116.

McNeilly, A. S. and Friessen, H. G. 1978. Prolactin during pregnancy and lactation in the rabbit. *Endocrinology* 102:1548–1554.

Meredith, A. 2000. General biology and husbandry. In *Manual of Rabbit Medicine and Surgery*, ed. P. Flecknell, 13–23. Goucester: British Small Animal Veterinary Association.

Meredith, A.L. and Prebble, J. L. 2017. Impact of diet on faecal output and caecotroph consumption in rabbits. *Journal of Small Animal Practitioner* 58:139–145.

Meredith, A. L., Prebble, J. L. and Shaw, D. J. 2015. Impact of diet on incisor growth and attrition and the development of dental disease in pet rabbits. *Journal of Small Animal Practitioner* 56:377–382.

Moncomble, A. S., Quennedy, B., Coureaud, G., Langlois, D., Perrier, G. and Schaal, B. 2004. Newborn rabbit attraction toward maternal faecal pellets. *Developmental Psychobiology* 45:277.

Morton, D., Jennings, M., Batchelor, G. R. et al. 1993. Refinements in rabbit husbandry. Second report of the BVAAWF/FRAME/RSPCA/UFAW, joint working group on refinement. *Laboratory Animals* 27:301–329.

Müller, J., Clauss, M., Codron, D. et al. 2014. Growth and wear of incisor and cheek teeth in domestic rabbits (*Oryctolagus cuniculus*) fed diets of different abrasiveness. *Journal of Experimental Zoology Part A: Ecological Genetics and Physiology* 321:283–298.

Myers, K. and Poole, W. E. 1961. A study of the biology of the wild rabbit, *Oryctolagus cuniculus* (L.), in confined populations. II. The effects of season and population increase on behaviour. *CSIRO Wildlife Research* 6:1–41.

Myers, K. and Poole, W. E. 1963. A study of the biology of the wild rabbit, *Oryctolagus cuniculus* (L.), in confined populations. V. Population dynamics. *CSIRO Wildlife Research* 8:166–203.

Myers, K. and Schneider, E. C. 1964. Observations on reproduction, mortality and behaviour in a small, free-living population of wild rabbits. *CSIRO Wildlife Research* 9:138–143.

Myers, K., Parer, I., Wood, D. and Cooke, B. D. 1994. The rabbit in Australia. In *The European Rabbit: The History and Biology of a Successful Coloniser*, ed. H. V. Thompson and C. M. King, 108–157. Oxford: Oxford University Press.

Mykytowycz, R. 1968. Territorial marking by rabbits. *Scientific American* 218:116–126.

Mykytowycz, R. and Rowley, I. 1958. Continuous observations of the activity of the wild rabbit, *Oryctolagus cuniculus* (L.) during 24-hour periods. *CSIRO Wildlife Research* 3:26–31.

National Research Council (NRC). 1977. *Nutrient Requirements of Rabbits*. 2nd rev. edi. Washington, DC: The National Academies Press. doi: 10.17226/35/.

Nevalainen, T. O., Nevalainen, J. I., Guhad, F. A. and Lang, C. M. 2007. Pair housing of rabbits reduces variances in growth rates and serum alkaline phosphatase levels. *Laboratory Animals* 41:432–440.

Parer, I. 1982. Dispersal of the wild rabbit, *Oryctolagus cuniculus*, at Urana in New South Wales. *Australian Wildlife Research* 9:427–441.

Parker, S. 1990. *Grzimek's Encyclopedia of Mammals*. New York: McGraw-Hill Inc.

Pasteur, L. 1885. Méthode pour prévenir la rage après morsure. Comptes rendus Hebd. *des séances de l'Académie des Sciences* 101:765–774.

Patton, N. M. 1994. Colony husbandry. In *The Biology of the Laboratory Rabbit*, 2nd edn., ed. P. J. Manning, D. H. Ringler and C. E. Newcomer, 28–44. San Diego: Academic Press.

Peiffer, R. L., Pohm-Thorsen, L. and Corcoran, K. 1994. Models in ophthalmology and vision research. In *The Biology of the Laboratory Rabbit*, 2nd edn, ed. P. J. Manning, D. H. Ringler and C. E. Newcomer, 410–434. San Diego, CA: Academic Press.

Peng, X., Knouse, J. A., and Hernon, K. M. 2015. Rabbit models for studying human infectious diseases. *Comparative Medicine* 65:499–507.

Podberscek, A. L., Blackshaw, J. K. and Beattie, A. W. 1991 The behavior of group penned and individually caged laboratory rabbits. *Applied Animal Behavior Science* 28:353–363.

Poggiagliolmi, S., Crowell-Davis, S. L., Alworth, L. C. and Harvey, S. B. 2011. Environmental enrichment of New Zealand White rabbits living in laboratory cages. *Journal of Veterinary Behavior* 6:343–350.

Prebble, J. L. and Meredith, A. L. 2014. Food and water intake and selective feeding in rabbits on four feeding regimes. *Animal Physiology and Animal Nutrition* 98:991–1000.

Prebble, J. L., Shaw, D. J. and Meredith, A. L. 2015. Bodyweight and body condition score in rabbits on four different feeding regimes. *Journal of Small Animal Practitioner* 56:207–212.

Proks, P., Stehlik, L., Nyvltova, I., Necas, A., Vignoli, M. and Jekl, V. 2018. Vertebral formula and congenital abnormalities of the vertebral column in rabbits. *The Veterinary Journal* 236:80–88.

Ritskes-Hoitinga, M. and Jilge, B. 2001. Felasa – Quick reference paper on laboratory animal feeding and nutrition. https://portal.findresearcher.sdu.dk/en/publications/felasa-quick-reference-paper-on-laboratory-animal-feeding-and-nut (accessed September 11 2020).

Rödel, H. G., Dausmann, K. H., Starkloff, A., Schubert, M., von Holst, D. and Hudson, R. 2012. Diurnal nursing pattern of wild-type European rabbits under natural breeding conditions. *Mammalian Biology* 77:441–446.

Rödel, H. G., Hudson, R., and von Holst, D. 2008. Optimal litter size for individual growth of European rabbit pups depends on their thermal environment. *Oecologia* 155:677–689.

Rödel, H. G., Landmann, C., Starkloff, A., Kunc, H. and Hudson, R. 2013. Absentee mothering — not so absent? Responses of European rabbit (Oryctolagus cuniculus) mothers to pup distress calls. *Ethology* 119:1024–1033.

Rouco, C., Villafuerte, R., Castro, F. and Ferreras, P. 2011. Effect of artificial warren size on a restocked European wild rabbit population. *Animal Conservation* 14:117–123.

Schaal, B., Coureaud, G., Langlois, D., Ginies, C., Semon, E. and Perrier, G. 2003. Chemical and behavioral characterization of the rabbit mammary pheromone. *Nature* 424:68–72.

Schlolaut, W., Hudson, R. and Rödel, H. G. 2013. Impact of rearing management on health in domestic rabbits: A review. *World Rabbit Science* 21:145–159.

Seaman, S. C., Waran, N. K., Mason, G. and D'Eath, R. B. 2008. Animal economics: Assessing the motivation of female laboratory rabbits to reach a platform, social contact and food. *Animal Behaviour* 75:31–42.

Shope, R. and Hurst, W. 1933. Infectious papillomatosis of rabbits with a note on the histopathology. *Journal of Experimental Medicine* 58:607–624.

Stauffacher, M. 1992. Group housing and enrichment cages for breeding, fattening and laboratory rabbits. *Animal Welfare* 1:105–125.

Stein, S. and Walshaw, S. 1996. Rabbits. In *Rodent and Rabbit Medicine*, ed. K. Laber-Laird, M. M. Swindle and P. Flecknell, 183–211. Oxford: Elsevier.

Suckow, M. A. and Douglas, F. A. 1997. The laboratory rabbit. In *The Laboratory Animal Pocket Reference Series*, ed. M. Suckow, 71–74. Boca Raton, FL: CRC Press.

Suckow, M. A., Schroeder, V. and Douglas, F. A. 2012. *The Laboratory Rabbit*, 2nd ed. Boca Raton, FL: CRC Press.

Szendrö, Z., Szabo, S. and Hullar, I. 1988. Effect of reduction of eating time on production of growing rabbits. *Proceedings 4th World Rabbit Congress, Budapest* 3:104–114.

Tablado, Z., Revilla, E. and F. Palomares. 2009. Breeding like rabbits: Global patterns and determinants of European wild rabbit reproduction. *Ecography* 32:310–320.

Thoman, E. B., Waite, S. P., Desantis, D. T. and Denenberg, V. H. 1979. Ontogeny of sleep and wake states in the rabbit. *Animal Behaviour* 27:95–106.

Toofanian, F. and Targowski, S. 1983. Experimental production of rabbit mucoid enteritis. *American Journal of Veterinary Research* 44:705–708.

Toth, L. A. and January, B. 1990. Physiological stabilisation of rabbits after shipping. *Laboratory Animal Science* 40:384–387.

Trocino, A. and Xicatto, G. 2006. Animal welfare in reared rabbits: A review with emphasis on housing systems. *World Rabbit Science* 14:77–93.

Tschudin, A., Clauss, M., Codron, D., Liesegang, A. and Hatt, J. M. 2011. Water intake in domestic rabbits (Oryctolagus cuniculus) from open dishes and nipple drinkers under different water and feeding regimes. *Journal of Animal Physiology and Animal Nutrition* 95:499–511.

Vastrade, F. 1986. The social behavior of free ranging domestic rabbits (*Oryctolagus cuniculus* L.). *Applied Animal Behavior Science* 16:165–177.

Vastrade, M. 1987. Spacing behaviour of free-ranging domestic rabbits, *Oryctolagus cuniculus* L. *Applied Animal Behaviour Sciences* 18:185–195.

Verga, M., Luzi, F. and Carenzi, C. 2007. Effects of husbandry and management systems on physiology and behavior of farmed and laboratory rabbits. *Hormones and Behavior* 52:122–129.

von Holst, D., Hutzelmeyer, H., Kaetzke, P., Khaschei, M. and Schönheiter, R. 1999. Social rank, stress, fitness, and life expectancy in wild rabbits. *Naturwissenschaften* 86:388–393.

Watanabe, T., Hirata, M., Yoshikawa, Y., Nagafuchi, Y. and Toyoshima, H. 1985. Role of macrophages in atherosclerosis. Sequential observations of cholesterol-induced rabbit aortic lesion by the immunoperoxidase technique using monoclonal antimacrophage antibody. *Laboratory Investigation* 53:80–90.

Williams, O. B., Wells, T. C. E. and Wells, D. A. 1974. Grazing management of Woodwalton Fen: Seasonal changes in the diet of cattle and rabbits. *Journal of Applied Ecology* 11:499–516.

Wilson, D. and Reeder, D. 1993. *Mammal Species of the World: A Taxonomic and Geographic Reference*. Washington, DC: The Smithsonian Institution.

Wyss, F., Müller, J., Clauss, M., et al. 2016. Measuring rabbit (Oryctolagus cuniculus) tooth growth and eruption by fluorescence markers and bur marks. *Journal of Veterinary Dentistry* 33:39–46.

Yanni, A. E. 2004. The laboratory rabbit: An animal model of atherosclerosis research. *Laboratory Animals* 38:246–256.

Zarrow, M. X., Denenberg, V. H. and Anderson, C. O. 1965. Rabbit: Frequency of suckling in the pup. *Science* 150:1835–1836.

12
Behavioral Biology of Ferrets

Claudia M. Vinke, Nico J. Schoemaker, and Yvonne R. A. van Zeeland
University of Utrecht

CONTENTS

Introduction	191
Typical Research Participation	191
Behavioral Biology of the Ferret	192
Natural History	192
Habitat and Habitat Use	192
Home Ranges and Territoriality	192
Social Organization	192
Breeding and Parental Care	193
Feeding Behavior and Predator–Prey Relationships	193
Communication	193
Common Captive Behaviors	193
Normal Behavior	193
Abnormal Behavior and Signs of Discomfort	196
Ways to Maintain Behavioral Health	198
Captive Considerations to Promote Species-Typical Behavior	198
Early Life Considerations	198
Social Structure Preferences	198
Environmental Enrichment	198
Training	200
Special Situations	201
Conclusion and Recommendations	201
References	202

Introduction

Ferrets are often characterized as intelligent, agile, playful, lively, curious, and highly inquisitive creatures, with a natural instinct to explore (Vinke and Schoemaker, 2012). You can almost entitle them as *neophilic*, always eager for, and in search of, a new challenge. To fulfill their behavioral priorities and prevent problems, such as inter- and intraspecific aggression, fearfulness, and destructive behaviors (e.g., Applegate and Walhout, 1998; Staton and Crowell-Davis, 2003; Quesenberry and Carpenter, 2004; Bays et al., 2006; Bulloch and Tynes, 2010; Talbot et al., 2014), ferrets need various physical and mental challenges in their environment. However, providing them with appropriate and sufficient stimulation may not be easy when they are forced to spend (the majority of) their time in cages, as is often the case in a laboratory. Additionally, because of the complexity of their social organization, matching pairs (or bigger groups) of ferrets necessitates a case-by-case approach, in which the preferences of each individual should be taken into account.

Typical Research Participation

Despite relatively small numbers of ferrets participating in research, compared to mice and rats (in the Netherlands: 294 ferrets out of a total of 403,370 laboratory animals: <0.1%: https://www.centralecommissiedierproeven.nl/onderwerpen/welke-dieren), ferrets occupy a distinct niche and have some unique applications in research. Especially following the pandemic of H5N1 influenza, ferrets have become an important biomedical model to study human influenza virus, including its pathogenesis, treatment, and vaccine development (Boyce et al., 2001; Anil et al., 2002). While widely recognized for their utility in influenza virus research, ferrets are also used as a research model for a variety of other infectious and noninfectious diseases, due to their sharing many of the same anatomical, metabolic, and physiological features with humans, as well as their susceptibility to many of the same pathogens. For example, ferrets are susceptible to several other respiratory viruses recognized in humans, such as the severe acute respiratory syndrome-associated coronavirus, respiratory syncytial virus, and human metapneumovirus. In addition, they are used as models for gastrointestinal disease (most notably peptic ulcer disease associated with *Helicobacter gastritis*), cardiovascular disease (e.g., myocardial infarct models), and renal disease secondary to toxin-producing *E. coli* (e.g., Clyde, 1980; Moody et al., 1985; Fox et al., 1990; Byrd and Prince, 1997; Woods et al., 2002; Fox and Marini, 2014). In addition, ferrets have been accepted as an animal model to study neural development, visual and auditory function, reproductive physiology, nutrition (especially carotenoid metabolism in relation to cancer), toxicology, pharmacology, and endocrinology, mostly in American laboratories (Brown, 2007). All of these areas comprise a variety of applications that dictate knowledge about appropriate husbandry, nutrition, and care, in order to maintain physically healthy ferrets and to ensure the refinement of animal experiments as indicated by the 3Rs (replacement, reduction and refinement) (Russell and Burch, 1959).

Behavioral Biology of the Ferret

Natural History

Together with species such as the mink, ermine, weasel, and polecat, ferrets (*Mustela putorius furo*) belong to the carnivore family of Mustelidae. Like most other mustelids, ferrets are sexually dimorphic, with males (also referred to as hobs) generally being larger and more sturdily built than females (also referred to as jills). Respective body weight ranges vary from 1,200 to 2,100 g for male ferrets and 700–1,200 g for female ferrets (Fox et al., 2014). Ferrets' bodyweights may vary 30%–40% across the seasons, as they store extra fat during fall, which disappears the following spring (Moorman-Roest, 1993). In captivity, the average lifespan of a ferret is approximately 6–8 years (Fox and Bell, 1998), although some individuals live to be over 10 years of age (Moorman-Roest, 1993). Several color variations exist, of which the wild type (*Polecat type*, dark brown with a light colored mask), albino, and sandy color (light-brown paws and tail) are the most common (American Ferret Association, 2014). Similar to cats, a leucistic (white) coat has been linked to deafness.

Compared to other species used in laboratories, the ferret is quite an exceptional species, as it is a fully domesticated mixed-bred species for which no wild counterpart exists. There is even uncertainty about its ancestors; although based on phylogenetic studies, the European polecat (*Mustela putorius putorius*) and the Steppe polecat (*Mustela eversmanni*) are the most likely candidates (Ashton et al., 1965; Tetley, 1965; Fisher, 2006). On the geologic timescale, both these species had *M. stomeri* as a common ancestor in the Pleistocene period. Interestingly, *M. stomeri* quite likely is also the ancestor of the endangered American black footed ferret (Kurtén, 1968). However, this was long before the domestication of the ferret. The history of ferret domestication is assumed to have started approximately 350 years BC, in regions of North Africa (Fisher, 2006). It is therefore assumed that ferrets have been domesticated for at least 2,000–3,000 years (Bulloch and Tynes, 2010). Initially, they might have been kept to protect human food sources from rodents (Price, 2002), similar to the domestic cat, whereas in the Middle Ages, ferrets were mainly kept for hunting rabbits, mice, and rats, an activity also referred to as *ferreting* (Thompson, 1951). Although ferreting is still practiced, ferrets are currently mostly kept as companion animals.

As the ferret's background is not fully determined and understood, and wild ancestors are absent, a different approach is necessary to gain insight into its natural biology and behavior. In essence, two approaches can be considered. First, information can be extrapolated from other wild relatives, such as the polecat (e.g., Blandford, 1987; Lodé, 1999; Davison et al., 1999). Second, albeit scarce, information may be derived from available literature on feral ferrets and feral polecat-ferret hybrids (e.g., MacKay, 1995; Norbury et al., 1998; Caley and Morriss, 2001). The combined use of these sources may subsequently help infer knowledge on the ferret's likely habitat use, prey choices, preferences for social organization, behavioral priorities, etc.

Habitat and Habitat Use

The European polecat, the most probable ancestor of the ferret, lives in wooded and semi-wooded areas, mostly near the water (Blandford, 1987; Lodé, 1999; Davison et al., 1999). Though information is scarce, observations on feral ferrets show a preference for wooded and semi-wooded areas near water sources as well (Duda, 2003). Feral ferrets live in dens, such as old rabbit burrows and caves under stones and root systems, and spend a large proportion of their active time travelling and foraging across home ranges of up to 102 ha (Norbury et al., 1998; Fisher, 2006). The ferret is well-known for its preference for safe hiding and resting places, which matches the den choices of wild mustelids.

Home Ranges and Territoriality

Feral ferrets' mean home ranges vary across the sexes: 102±58 ha are reported for males, whereas mean home ranges of 76±48 ha are reported for females (Norbury et al., 1998). Feral ferrets' territories generally exclude territories of ferrets of the same sex (Powell, 1979), whereas territories of members of the opposite sex may overlap extensively (Moors and Lavers, 1981). According to Powell (1994), the level of territoriality of the feral ferret is likely to be related to prey abundance.

Although juvenile ferrets (male and female) can disperse over distances up to 5.0 km following weaning, only a few juveniles may completely leave their natal area (Caley and Morriss, 2001).

Social Organization

Similar to most other adult mustelids, feral ferrets are solitary and territorial animals. Except for the breeding period (see Thom et al., 2004), ferrets only incidentally meet one another at the borders of their territories and prefer to avoid conspecifics. Occasional den sharing by adult males may be observed in May and August, but otherwise, daytime resting of feral ferrets seems to be mostly a solitary activity (Norbury et al., 1998).

Breeding and Parental Care

The ferrets' breeding season lasts from March to August (in the Northern hemisphere), and is largely dependent on day length (Fox et al., 2014). A significant increase in day length, from 8 to 16 h, is necessary to induce estrus (Fox and Bell, 1998). Moreover, induction of ovulation requires the combined gripping of the jill's neck by the hob and vaginal stimulation (Bibeau et al., 1991). If not mated with, a jill will stay in estrus until the day length shortens (September) and as a result is at the risk of developing estrogen-related bone marrow suppression and pancytopenia (Bernard et al., 1983).

The gestation period in ferrets is approximately 42 days. The average litter size is 8, but there is considerable variation, and litters of up to 18 have been documented (Fox et al., 2014). Mustelids are good mothers and the mother-offspring relationship is of the utmost importance for the development of healthy ferrets. Behavioral development starts soon after birth, and the sensitive period for ferrets is assumed to be between 4 and 10 weeks, as mentioned for the polecat (Fisher, 2006). In this sensitive period, the young ferret has to learn many social skills, as well as learn and habituate to a variety of environmental stimuli. One of a ferret mother's important roles is to give her kits their first introduction to different types of environmental challenges; she teaches her kits both specific hunting skills and appropriate prey choices. The latter is particularly important as, according to Apfelbach (1986), wild polecat kits have a sensitive period for olfactory imprinting on the scent of their prey between day 60 and 90. The smell of prey species on which they have not been imprinted in that sensitive period will not be accepted as food in adulthood. Similar specific food preferences have also been observed in the ferret; imprinting of the kit on "prey" items will influence the food choices and preferences of the animals later in life.

Feeding Behavior and Predator–Prey Relationships

Under natural circumstances, mustelids spend a large proportion of their active time foraging in wide home ranges (e.g., American mink: up to 3 h/day: about 35% of their active time: Dunstone, 1993). The same is observed in feral ferrets, which can spend their active time travelling and foraging across home ranges of up to 102 ha (Norbury et al., 1998; Fisher, 2006). Foraging behaviors may include walking, running, jumping, nose pushing, digging, and overseeing the area (appetitive behaviors), and – if the ferret is successful – the finding of food or capture of prey that can be eaten (consummatory behaviors). All of these behaviors are typically carried out in a lively and agile way that closely resembles the behavioral reactivity of domestic ferrets.

Polecats feed upon a variety of vertebrate and invertebrate prey, including rodents, small birds, reptiles, amphibians, spiders, beetles, slugs, snails, and earthworms (Bulloch and Tynes, 2010). On the other hand, mustelids find themselves hunted by other predators, including wolves, wolverines, coyotes, bobcats, eagles, falcons, etc., depending on the area in which they live (American mink: Dunstone, 1993). However, mustelids are lean and fast, typically staying in their burrow until dusk, and are therefore not easily caught.

Communication

For a solitary, territorial species, olfactory information is the most important means of communication. Ferrets use anal gland secretions, feces, and urine for territorial communication, to obtain information concerning the identity, reproductive condition, and social status of conspecifics (Clapperton et al., 1988; Woodley and Baum, 2003; Berzins and Helder, 2008), including the use of olfactory cues by males to locate mating partners (Baum, 1976; Baum et al., 1983; Moors and Lavers, 1981; Kelliher and Baum, 2002).

In addition to olfactory communication, ferrets' emotional states can easily be determined from particular vocalizations, described in the "Common Captive Behaviors" section. The most prominent communicative posture that can be observed in the ferret is piloerection of the tail (bottle brush tail; Figure 12.1), which is often accompanied by an arched back posture (Fisher, 2006). Piloerection can also be observed in other species, such as rats, cats, and dogs in states of excitement. The erected hairs, which make the animal look bigger, may function as an additional communicative signal to impress an opponent or predator to avoid high risk episodes of conflict or predation.

Common Captive Behaviors

Normal Behavior

Territorial Behavior and Elimination

Many behavioral patterns of the ferret are related to their territorial nature. Territorial animals scent-mark their territories carefully, as do captive ferrets. Ferrets furthermore show other territorial behaviors, such as wiping and body rubbing (wiping across surfaces with the belly to transfer scent from the preputial sebaceous gland), anal dragging (dragging the perineum across surfaces to release scents from the peri-anal sebaceous gland), and defecating on objects (Fisher, 2006).

Domestic ferrets tend to select corners for their latrine area, which may reflect their wild ancestors' way of marking their territory perimeters. Typically, the ferret, both hobs and jills, walks backward into the corner before it defecates or urinates. An increase in defecation frequency and/or anal dragging may be seen during the breeding season or when a new ferret is introduced.

Aggressive Behavior

Just like other solitary species, ferrets can aggressively reject new conspecifics in their territory. Such aggressive interactions in Mustelidae can be rather violent, with a high risk of injuries (Poole, 1966, 1973, 1974; Dunstone, 1993) and many vocalizations (e.g., hissing, huffing, screaming loudly) being produced (Speer et al., 2018). During an aggressive confrontation, ferrets may back up with their mouth open or flip onto their back with their mouth open (Figure 12.2). Alternatively, they may actually lunge out toward the conspecific (or a person approaching them) and bite (Speer et al., 2018). When ferrets bite (people), they have a tendency not to let go. The most effective way to get the ferret to release its grip is by holding the biting ferret under a running faucet.

FIGURE 12.1 A bottle brush tail is seen in (pet) ferrets that are excited. Most frequently this behavior is seen in ferrets that are exploring new surroundings. (Photo courtesy of Birgit van der Laan.)

FIGURE 12.2 During an aggressive confrontation, ferrets may lay on their back with their mouth open in defense. It is important to closely evaluate the behavior, as this may also be seen during (rough) play and may indicate social instability between the ferrets. (Photo courtesy of Birgit van der Laan.)

Staton and Crowell-Davis (2003) reported four factors that influence the occurrence of aggressive behavior: (1) familiarity; (2) season (with higher and more serious incidents in the reproduction season); (3) sex; and (4) neutering status. Aggressive incidents are most often observed between unfamiliar adult individuals and between (intact) individuals of the same sex (Bulloch and Tynes, 2010). This is similar to the situation of the polecat, in which familiarity has also been identified as an important factor that influences the way that animals interact; animals that were raised together showed more positive, and fewer negative interactions, with overall greater tolerance toward one another (Lodé, 2008). Ferrets often show a typical meet-and-greet behavior, where they sniff the anal area and neck and shoulder region of a conspecific to obtain information on familiarity, and in the case of a male-female interaction, also to infer information about the sexual receptivity of the female.

Activity

Though mustelids spend a relatively large proportion of their daily time budget inactive, they can be very lively animals during the time they are awake and active, thereby requiring continuous physical and mental stimulation. When exploring its environment, an inquisitive ferret will periodically demonstrate scouting behavior, in the form of erect or alert posturing.

Their activity patterns include foraging times that are similar to American mink (up to 3 h/day; about 35% of their active time: Dunstone, 1993), as well as similar foraging patterns, including running, jumping, nose pushing, and digging. However, unlike polecats, ferrets are mostly active during the day, rather than at night. Under laboratory conditions, ferrets are known to sleep over 60% of the time; pet ferrets normally sleep 12–16 h (50%–66%) a day, and their activity-inactivity schedule typically reflects the schedule of their human companion (Bays et al., 2006).

Play Behavior

Like all young carnivores, juvenile ferrets play frequently and start playing as soon as their locomotor skills develop (around 4 weeks of age). Play behaviors may include dancing (*weasel war dance*), jerking, galloping, play fighting with a partner, manipulation of objects, and/or chasing artificial prey (e.g., Poole, 1978; Fisher, 2006; Bulloch and Tynes, 2010). Play patterns usually reach a maximum between 6 and 14 weeks of age (Bulloch and Tynes, 2010). Nonetheless, adult ferrets can show highly vigorous play patterns, which may escalate under certain circumstances; for example, play interactions may evolve into patterns with a more aggressive motivational background. In such circumstances, interactions typically end with the hissing and fleeing of one of the interactants (Bulloch and Tynes, 2010).

Ferrets' social play is largely comparable to that of the polecat, which is thoroughly described by Poole (1978), and may comprise chase, ambush, mounting, rolling, and wrestling behaviors. The *open mouth play face* (Figure 12.3), a ritualized pattern of inhibited biting, whereby all teeth are totally covered by the upper lips, while the mouth is open, is typically displayed during social play. Under aggressive circumstances, on the other hand, all teeth are clearly shown in the open mouth, as a true and clear signal to communicate an intention to bite.

FIGURE 12.3 During play, a ferret may show intent to bite, but will have the upper lips covering the teeth. The behavior is referred to as an *open mouth play face*. (Photo courtesy of Birgit van der Laan.)

Object manipulation, which can be seen in play-motivated ferrets as well, often starts with exploration with the mouth (Moorman-Roest, 1993), also called *mouthing* behavior; the animal carefully explores the characteristics of the object using its teeth in an inhibited way. This behavior might help the individual to test whether the object is useful as a play tool, or whether it is something useful to eat. It is possible that ferrets also explore a human's hand in this way. Despite the fact that this biting is usually gentle, such nipping and biting should not be rewarded, but rather discouraged by a temporary time out (i.e., a short time span during which no attention is paid to, or contact is made with, the animal after it has performed an undesirable behavior) or a firm "NO", to prevent the behavior from escalating. In addition, preventing nipping and biting from occurring altogether is even better; licking of the human's hand is a good indicator that a ferret may soon initiate biting, signaling a time to take action.

Abnormal Behavior and Signs of Discomfort

Most of the common behavioral problems in laboratory ferrets can be understood from the biology and specific behavioral patterns of mustelids. Almost all behavioral problems in captive ferrets are evoked by a suboptimal environment and suboptimal stimulation. In principle, these so-called behavioral problems almost all comprise functional and normal behavioral patterns that might turn out to be maladaptive in the captive environment. For example, behavioral problems, such as excessive scent-marking and aggression toward intruders (including humans), might be derived from territorial motivations, whereas problems of intraspecific aggression may be related to the mustelids' originally solitary life style and/or nonpreferred social situations.

In addition, ferrets' need for physical and mental activity and challenges, which is in line with active territory patrolling and energetic foraging patterns, may lead to problems of compulsive grooming and stress-induced alopecia, locomotor stereotypies, destructive behaviors and hyperactivity, or lethargy/inactiveness, in the event that these needs are not met in the captive environment. When such conditions become chronic, these (initially) maladaptive behaviors may turn into behavioral patterns that we classify as abnormal, such as the aforementioned compulsive grooming, self-injurious behavior, and locomotor stereotypies, which are well-known and well-described in fur-farmed American mink (e.g., Mason, 1993a, b, 1994; Vinke et al., 2008).

Aside from the previously mentioned problem behaviors, ferrets may also display distinct behavioral patterns when anxious, in pain, or experiencing some other form of discomfort.

Ferrets that are in pain are often lethargic, immobile, and anorexic; although, dependent on the individual and type of pain, some individuals may also become anxious and/or restless (Fisher, 2006). Other signs of discomfort or pain include, trembling, collapse, crying, whimpering, not assuming normal sleeping position, or absence of grooming activity (Johnson-Delaney, 2009; Brown, 2004). Spasmodic teeth grinding may also indicate pain, with ferrets holding the head down, rhythmically moving the facial muscles back and forth, and wriggling the ears (Fisher, 2006). Abdominal pain may be indicated by a hunched posture (Figure 12.4), walking with an arched back (note that healthy ferrets also commonly walk with an arched back (Figure 12.5); this is normal!), or stilted

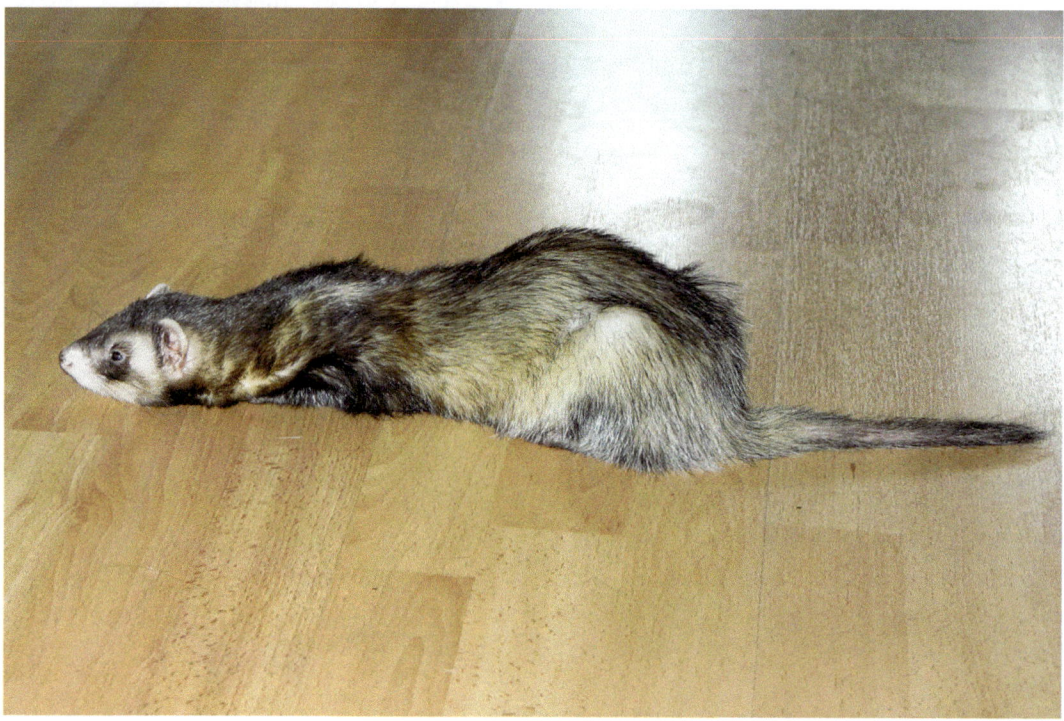

FIGURE 12.4 A hunched back, especially when a ferret is laying down, may be an indication of abdominal pain, as can be seen in the photo of this ferret. (Photo courtesy of Birgit van der Laan.)

FIGURE 12.5 Normal arched back during walking/running in a healthy ferret. It is important to be able to distinguish this normal arched back from that of a ferret with abdominal pain.

gait. Signs of discomfort or pain may furthermore include vocalizations, such as crying, whimpering, squealing, and screaming, and the expression of a facial grimace (Figure 12.6; Reijgwart et al., 2017).

Frustrated ferrets, or those in fear, may squeal and scream (Fisher, 2006). Hissing can indicate fear, but is sometimes difficult to interpret, as it can also be heard during play patterns, particularly in situations of play escalation. The same is true for barking (or barking-like sounds), which can indicate both excitement and fear. Other physical and behavioral signs indicative of stress or fear include piloerection on the tail and/ or body; standing still with an arched back; a tensed body; and orientation toward the perceived threat. Anal gland expression is seen during extreme fear in animals in which the anal gland has not been removed (anal gland removal is illegal in the Netherlands if not warranted due to a medical condition; Speer et al., 2018). Unfortunately, many of the behaviors and

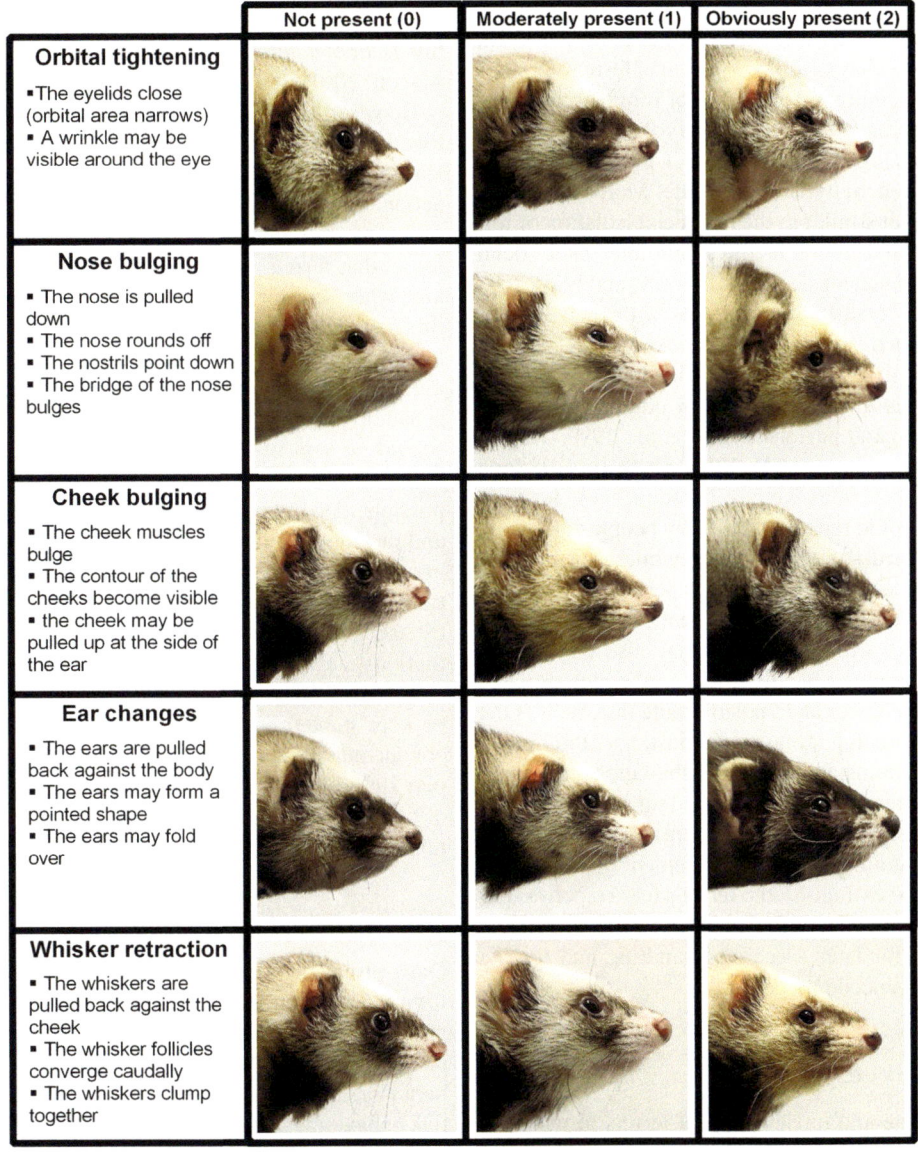

FIGURE 12.6 Photographs visualizing the normal appearance and changes (0 = not present, 1 = moderately present, 2 = obviously present) of the five Action Units that are used in the Ferret Grimace Scale (Reijgwart et al., 2017).

vocalizations just mentioned are not specific to pain or fear and may also be displayed with other forms of discomfort or disease. As a result, a thorough evaluation of the ferret's health and behavior in relation to the environmental context is needed to correctly interpret the behavior, so that appropriate action can be taken to address the problem.

Ways to Maintain Behavioral Health

Captive Considerations to Promote Species-Typical Behavior

Early Life Considerations

The first step to maintain and promote adequate behavioral health, and to prevent behavioral problems, is to ensure appropriate behavioral development early in life. This means taking care of proper socialization, allowing adequate mother-offspring relationships to develop, and providing opportunities for the expression of normal exploratory and play behaviors. In their studies of early behavioral development of ferrets, Chivers and Einon (1982) found that deprivation of rough-and-tumble social play when young led to hyperactivity that persisted into adulthood, emphasizing the importance of play behavior for normal behavioral reactivity later on in life. Moreover, Einon (1996) mentioned that similar to the rat, social isolation of ferret kits during the first month of life might alter later social interaction, sexual behavior, learning, activity, and body size, thereby arguing the case for keeping kits with their mother until they are past their weaning age (8 weeks).

Insufficient (positive) experiences with humans during the socialization phase will, just like in other animal species (e.g., dogs, cats, and parrots: Seksel et al., 1999; Turner, 2000; Wanker, 1999; Dietz et al., 2018), result in a lifelong fear of humans with serious welfare consequences. Juvenile laboratory ferrets should thus be handled by people daily from 28 days of age onward (MacKay, 2006), while ensuring that this handling is perceived as a positive experience by the ferret (i.e., handling should be calm and gentle, and preferably associated with a reward). Aggression may also be learned, for example, when previous handling was unpleasant, or when previous defensive behavior had a positive outcome, such as the removal of the threatening stimulus (for instance, the human handler). In these circumstances, gentle habituation and desensitization with counterconditioning to the human hand, while protected by a leather glove to prevent trauma and pulling back (which can lead to unintentional reinforcement of the behavior), can be helpful to extinguish the ferret's fear response. For a laboratory ferret, such socialization procedures are of the utmost importance for later successful handling and for the training of more advanced procedures.

Social Structure Preferences

The solitary life style and territoriality of ferrets provide the main background for their most frequently mentioned behavioral problem; intraspecific aggression. Frequent aggressive interactions between socially housed ferrets seriously decrease their welfare. In their study to identify factors (e.g., familiarity, sex, neutering status, time of year) associated with aggression toward conspecifics in pet ferrets, and to test a method for reducing aggression upon introduction of a new ferret to the group, Staton and Crowell-Davis (2003) advise that all introductions should be monitored carefully, with interventions taking place when fights occur. In addition, they recommend the adoption of established pairs of familiar ferrets, with male–female and male–male pairs being the most successful (without aggression and social stress) (Staton and Crowell-Davis, 2003).

Nevertheless, some individuals may suffer quietly in the presence of another ferret, a phenomenon that has also been recognized in the domestic cat, another solitary, carnivorous species. Similar to cats, housing too many ferrets in a group together, or mismatches between individuals, can result in chronic stress (Banks et al., 2010; Stella, this volume). This chronic stress may, in turn, lead to stress-induced elimination (personal communication Roest, 2011), and may increase vulnerability to certain diseases and infections, most notably *Helicobacter* infection and gastric ulcerations (Fox and Marini, 2001; Banks et al., 2010).

However, ferrets are also commonly observed to sleep together, and play with and groom each other, suggesting that at least some ferrets can appreciate certain forms of social interaction. These anecdotal findings are supported by the findings of a consumer demand study by Reijgwart et al. (2016), which found that ferrets were highly motivated to gain access to an area where conspecifics were housed. Although the reason for this motivation is not clear (i.e., a need for social interaction vs. territorial defense), current recommendations for laboratory animals still encourage social housing, particularly for young, subadult ferrets, jills, or castrated males. Preferably, ferrets should be kept in small groups (no more than four animals), although solitary housing or housing in larger groups may be possible, depending on the preference of the individual ferret and the compatibility of the individuals. The success of social grouping highly depends upon the animals' genetic predispositions, life history, familiarity, season, sex, and neutering status (Staton and Crowell-Davis, 2003), as seasonal hormonal fluctuations may increase territoriality and aggression. As ferrets may particularly reject new arrivals, pairs or groups should be kept stable once established, and introductions should be conducted with the utmost care. For individual ferrets that display signs of intraspecific aggression, arousal, or are suspected of having stress-induced disease, indicating that a companion ferret is not preferred, solitary housing should be considered.

Environmental Enrichment

Daily provision of physical and mental challenges is important for a vigorous and lively animal like the ferret. As mentioned previously, the ferrets' foraging and hunting style, and play patterns are lively. Foraging, as the appetitive phase of feeding behavior, is often regarded as a high priority behavior or essential behavioral need (i.e., an internally motivated behavioral pattern that should be performed, or otherwise may induce a state of chronic stress in the individual, which may eventually result in a physiological and/or behavioral pathology;

Dawkins, 1988; Rushen et al., 1993; Jensen and Toates, 1993). Nonetheless, laboratory (and pet) ferrets mostly (if not always) get their food for free, without foraging investment. Moreover, ferrets are often kept in relatively small enclosures, thereby limiting their ability to walk around and explore their environment. As a result, the ferrets' motivation to spend a high percentage of their active time on exploration, foraging, and play behavior may not be satisfied under routine conditions (Fisher, 2006); thus the provision of various types of food and nonfood enrichment to laboratory ferrets is necessary to fulfill these aforementioned motivations. However, these bouts of activity should alternate with regular resting opportunities in the 24 h time budget (a healthy ferret may sleep between 18 and 20 h/day; Fisher, 2006), for which a safe and comfortable resting place (e.g., hammock; Figure 12.7) is needed.

Environmental enrichments may include food enrichment (e.g., food puzzles, chewing material), which addresses the *appetitive part* of feeding (foraging) and is so often neglected in animal husbandry. Many ferret toys and food puzzles are commercially available. Nonetheless, food enrichment in socially housed animals can be challenging, as highly preferred items can lead to increased aggression if animals attempt to confiscate or monopolize a preferred toy or object (e.g., Honess and Marin, 2006; Howerton et al., 2008; Akre et al., 2011). Management procedures, such as providing multiple enrichments, dispersing valuable resources (i.e., easy to share and hard to monopolize; Akre et al., 2011), cage compartmentalization (more cage divisions where an individual can choose to eat, hide, or rest), preventing physical contact, and increased refuge opportunities may overcome this problem.

For a challenge-eager animal like the ferret, it is furthermore necessary to change enrichment objects regularly and to arrange intermittent presentation, in order to avoid habituation to the enrichment. This novelty is especially important to promote exploratory behaviors and can also be considered a form of occupational and sensory enrichment (the latter not only comprising visual stimuli but also olfactory, taste, auditory, and/or tactile opportunities).

Limited information is available on the types of enrichments that can best be provided to meet ferrets' behavioral priorities. In practice, healthy ferrets are sometimes fed live prey animals, such as mice or 1-day-old chickens. Aside from being a valuable nutrient and protein source (which is deemed essential for ferrets), feeding live prey also allows some parts of natural feeding behavior to be displayed. Nevertheless, feeding prey animals, or bone and raw food diets, also carries a risk of exposing the individual to potential pathogens and therefore should be implemented with the utmost caution. Learning from the mother and imprinting on specific food and "prey" items early in life may influence the food choices and preferences of the animal later in life.

Hiding opportunities are highly desired and the provision of tubes might be suitable for fulfilling their motivations to play and hide. Complexes of tubes can be used to construct whole pipelines that might be compared to mustelids' natural den constructions (i.e., rabbit burrows: Norbury et al., 1998). Variable bedding material might be helpful to fulfill the motivation to dig.

Recent studies on laboratory ferrets tested enrichments in a consumer-demand setup (with categories *sleeping enrichment*,

FIGURE 12.7 Ferrets love to sleep in hammocks. These hammocks are relatively cheap and can be washed easily, making them suitable to be used in laboratory situations as well.

foraging enrichment, tunnels, social enrichment, and *water bowls*). The results showed that a hammock, a water bowl, and a foraging ball were the most preferred enrichments (Reijgwart et al., 2015, 2016, 2017; Reijgwart, 2017). Though not tested in the consumer-demand setups, bedding might be preferred as well, as the author described that the provision of new bedding material often induced the weasel war dance and dooks in the ferrets, a behavior pattern and a vocalization that are both associated with positive emotional states (Reijgwart, 2017). Enriched environments are particularly important for juvenile ferrets, which should come into contact with many novel objects, environments, and situations during their early life, an important prerequisite for proper socialization.

Training

Preparatory training for handling and medical procedures is currently the most ethical way to prepare pet animals for a visit to the veterinarian and for minimizing restraint, and such training techniques are increasingly applied in laboratories as well. Training techniques can prevent mental trauma that may develop after the use of forced restraint procedures, and the process of training itself can even be perceived as enriching by the animal (Harris, 2015). Moreover, minimizing restraint procedures is better for the welfare of the animals (fewer restraint procedures) and renders research results more reliable as a stressed animal does not make a good animal model. The use of positive reinforcement training techniques results in cooperative animals not only facilitating future handling but also enhancing the safety of the person handling the animal (as defensive aggressive reactions are less likely to occur), while minimizing the risks to the animal (as forceful restraint and sedation or anesthesia are no longer required) (Brown, 2012).

Because ferrets are very intelligent animals, training using common conditioning techniques, similar to those used for training dogs, is easily accomplished. Ferrets are always eager for food (especially liquid foods such as ferret paste or Royal Canin™ convalescence support) and play; thus, reinforcers that can be used for such training procedures are readily available (Figure 12.8). Food items that can be used as reinforcers during training are summarized in Speer et al. (2018, Table 12.1). Ferrets can also be trained by the use of a clicker – the clicker training technique – which is primarily based on classical conditioning, whereby an initially neutral clicker is repeatedly paired with a primary reinforcer (e.g., food), and can be used later on as a secondary reinforcer in all kinds of operant conditioning tasks while minimizing the use of food rewards.

In principle, all procedures in a laboratory can be trained, varying from regular medical checks (e.g., weighing, inspection of mouth and ears, medication administration) to more complex procedures (e.g., used for brain or auditory research), and even simple invasive procedures. Training techniques are preferably based on positive reinforcement (rather than negative reinforcement or punishment strategies), meaning that behavioral patterns that are desired (and thus should increase in frequency) are reinforced by the use of a reward. Provision of the reward, a positive consequence, contingent upon (i.e., consistent action-consequence relationship), and immediately after (i.e., contiguous) the desired behavior pattern will help to ensure that the pattern will be shown more often in the future by the animal. Moreover, ferrets can be trained to do things

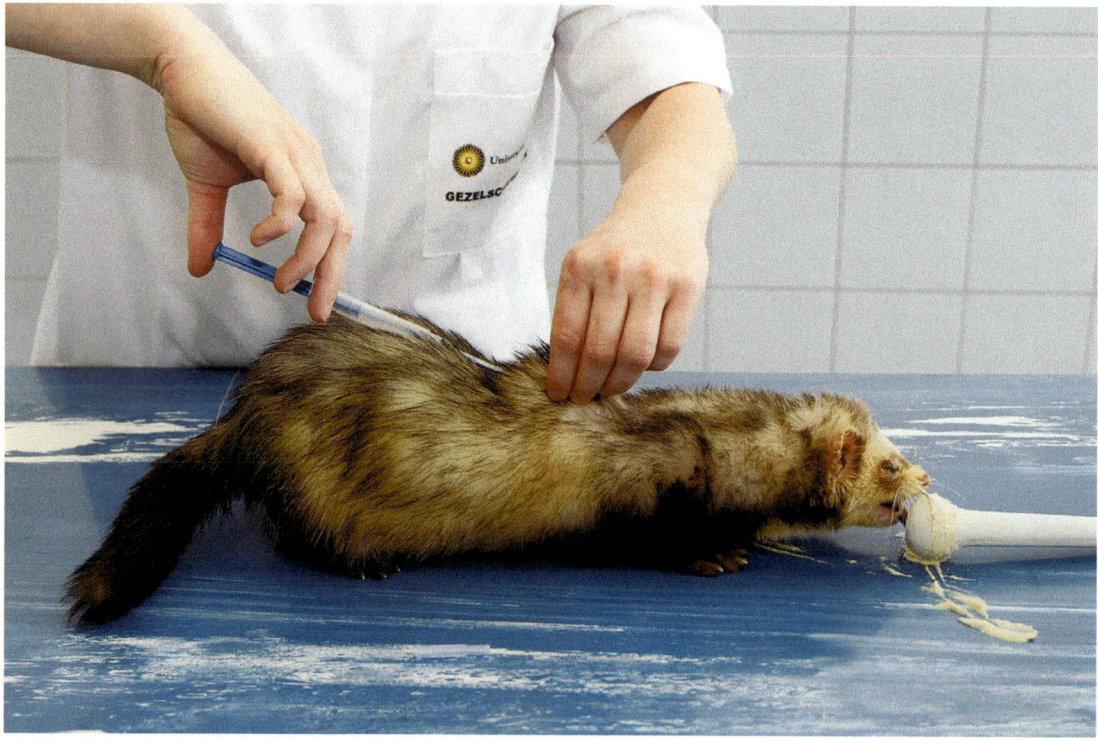

FIGURE 12.8 When a ferret is distracted with a palatable food item (e.g., Royal Canin™ convalescence support), procedures, such as placement of an implant, can be performed without sedation.

TABLE 12.1

Food Items that Can Be Used as Reinforcers during Training of Ferrets (Speer et al., 2018)

- Dried ferret food that is slightly moistened to facilitate chewing
- Cooked meat (e.g., chicken, turkey, organ meat) or eggs
- Fatty acid supplement in small squeeze bottle, syringe, or presented on a spoon (maximum ½ teaspoon a day per ferret)
- All-meat baby food on a spoon, tongue depressor, or in a syringe (diluted with water or meat broth)
- Liquid recovery diets
  - Convalescence support [Royal Canin]
  - Emeraid Carnivore [Lafeber Company]
  - Critical Care for Carnivores [Oxbow]

in response to a verbal or visual cue. As an example, target training (Speer et al., 2018), which is particularly useful for these purposes, may include the command TOUCH, whereby the ferret is taught to touch a target (e.g., target stick, hand) with its nose upon hearing this verbal cue. As soon as this command is successfully trained, it can be applied in many situations, and can be helpful to direct the ferret to assume all types of desired positions.

Special Situations

The exploratory nature of the ferret requires the careful selection of enrichments, as they readily consume and swallow (smaller) pieces of toys, particularly rubber items (Plant and Lloyd, 2010). Intake of these foreign objects is most commonly seen in younger (<2 years) animals. Though ingestion of objects may occur without symptoms, more frequently, ingestion will result in gastrointestinal obstruction and associated signs, such as anorexia, salivation, vomiting, abdominal discomfort, diarrhea, and/or melena (black feces resulting from blood loss in the gastrointestinal tract). In some cases, foreign objects will pass on their own, but more frequently, surgical removal is required (Brown, 1997; Lennox, 2005).

European guidelines for housing ferrets as animals participating in research projects state that the minimum space required is $4,500\,cm^2$, although for hobs, at least $6,000\,cm^2$ is necessary (EU Commission, 2007). However, ferrets are inquisitive, active animals, so it must be kept in mind that a laboratory setting that totally addresses their behavioral needs is difficult to achieve. Nonetheless, every effort should be made to ensure that ferrets are kept in as large an area as possible and, if and where possible, be given daily opportunities of supervised time outside of the cage. Moreover, ferrets need at least 50 cm of vertical height to stand on their hind legs and scan the surroundings, thus the cage height should exceed 50 cm to allow them to perform this natural exploratory behavior. Ferrets love to dig through substrate on the bottom of the cage. The use of hay, straw and wood shavings, however, is not recommended, as inhalation of dust may lead to chronic irritation of the upper respiratory tract (Jenkins and Brown, 1993). Instead, the use of towels, paper, or other dust-free substrates is recommended in a laboratory situation. For pet ferrets, peat moss can be used as a substrate for digging.

As ferrets are renowned escape artists, every effort should be made to ensure their enclosure is escape-proof (Schoemaker and van Zeeland, 2013). Any area through which they can squeeze their head is big enough to escape. Either mesh wire or a fully smooth surface they cannot climb or jump should prevent them from escaping from their enclosure.

Conclusion and Recommendations

Ferrets are highly agile and lively animals, which can make it challenging to meet their behavioral needs in a laboratory setting. To optimize the ferrets' housing conditions, laboratory personnel should be aware of the ferrets' strong motivations for exploration, foraging, activity, play, and resting, and of their natural social responses, especially those related to aggression and territoriality. In order to address ferrets' behavioral priorities to the fullest extent possible, variable (food) enrichments, toys and objects to play with and explore, tunnels, digging opportunities, and comfortable hiding and resting places should be provided. Personnel in laboratories furthermore need to know how to choose and variably manage valuable enriching resources; carefully introduce, match, and monitor animals kept in pairs or groups; avoid overcrowding; and respect some ferrets' preferences for a solitary life style.

Gaining fundamental knowledge on the behavior of ferrets and their motivations is advisable to adequately monitor the ferrets' welfare (e.g., behavioral patterns addressing aforementioned motivations for exploration, foraging, resting, and play, as well as fear, pain, and intraspecific aggression communication signals). Moreover, early life experiences, including socialization, are of the utmost importance to ensure that the ferret is able to live (and be handled) without stress in laboratory surroundings, including living with other ferrets. Socialization additionally increases training success for complex handling procedures in the laboratory. This means that laboratories must know exactly where their batch of laboratory ferrets came from and preferably breed the animals themselves to ensure full control over the socialization period and that necessary prerequisites are met. If laboratory ferrets are derived from companies providing laboratory animals, these companies should be encouraged to ensure that their ferrets are exposed to adequate and positive experiences early on in life and that a standard socialization protocol is in place and implemented. Only a proper start will allow the ferret to be optimally prepared for future handling procedures and social situations in a laboratory setting. Moreover, laboratories are encouraged to implement training procedures to prepare their animals for future procedures, as handling may not always be easy, given ferrets' leanness and activity level. Although such conditioning techniques may take time, they usually turn out to be a worthwhile investment to make in the laboratory, as this not only facilitates the work but also is by far the best way to ensure the laboratory ferret's welfare and well-being.

REFERENCES

Akre, A. K., Bakken, M., Hovland, A. L., Palme, R., Mason, G. 2011. Clustered environmental enrichments induce more aggression and stereotypic behaviour than do dispersed enrichments in female mice. *Applied Animal Behaviour Science* 131: 145–152.

American Ferret Association Inc (AFA). 2014 *AFA Ferret Color and Pattern Standards.* http://www.ferret.org/events/colors/colorchart.html. 27 December 2014 (not peer- reviewed publication).

Anil, S. S., Anil, L., Deen, J. 2002. Challenges of pain assessment in domestic animals. *Journal of the American Veterinary Medical Association* 220: 313–319.

Apfelbach, R. 1986. Imprinting on prey odours in ferrets (*Mustela putorius F. Furo* L.) and its neural correlates. *Behavioural Processes* 12 (4): 363–381.

Applegate, J. A., Walhout, M. F. 1998. Childhood risks from the ferret. *The Journal of Emergency Medicine* 16 (3): 425–427.

Ashton, E. H., Thomson, A. P. D., Zuckerman, F. R. S. 1965. Some characters of the skulls and skins of the European Polecat, the Asiatic Polecat and the domestic ferret. *Proceedings of the Zoological Society of London* 125 (2): 317–333 (published online 20 Aug 2009).

Banks R. E., Sharp J. M., Doss S. D., Vanderford D. A. 2010. *Exotic Small Mammal Care and Husbandry Ferrets.* New York: Wiley-Blackwell.

Baum, J. M. 1976. Effects of testosterone propionate administered perinatally on sexual behavior of female ferrets. *Journal of Comparative and Physiological Psychology* 90: 399–410.

Baum, M. J., Canick, J. A., Erskine, M. S., Gallagher, C. A., Shim, J. H. 1983. Normal differentiation of masculine sexual behaviour in male ferrets despite neonatal inhibition of brain aromatase or 5-alpha-reductase activity. *Neuroendocrinology* 36: 277–284.

Bays, T. B., Lightfoot, T., Mayer, J. 2006. *Exotic Pet Behavior. Birds, Reptiles and Small Mammals.* St. Louis, Missouri: Saunders Elsevier Inc.

Bernard, S. L., Leathers, C. W., Brobst, D. F., Gorham, J. R. 1983. Estrogen-induced bone marrow depression in ferrets. *American Journal of Veterinary Research* 44 (4): 657–661.

Berzins, R., Helder, R. 2008. Olfactory communication and the importance of different odour sources in the ferret (*Mustela putorius f. furo*). *Mammalian Biology* 73 (5): 379–387.

Bibeau, C. E., Tobet, S. A., Anthony, E. L., Carroll, R. S., Baum, M. J., King, J. C. 1991. Vaginocervical stimulation of ferrets induces release of luteinizing hormone-releasing hormone. *Journal of Neuroendocrinology* 3 (1): 29–36.

Blandford, P. R. S. 1987. Biology of the polecat *Mustela putorius*: A literature review. *Mammal Review* 17: 155–198.

Boyce, S., Zingg, B., Lightfoot, T. 2001. Behavior of *Mustela putorius furo* (the domestic ferret). *Veterinary Clinics of North America: Exotic Animal Practice* 4: 697–712.

Brown, S. A. 2004. Basic anatomy, physiology and husbandry. In *Ferrets, Rabbits and Rodents, Clinical Medicine and Surgery,* ed. E. V. Hillyer and K. E. Quesenberry, 3–14. Philadelphia, PA: W.B. Saunders company.

Brown, S. A. 2007. History of the Ferret. Small Animal Health Series. *Veterinary Partner.* http://www.veterinarypartner.com/Content.plx?P=AandA=496 (not peer-reviewed publication).

Brown, S. A. 2012. Small mammal training in the veterinary practice. *Veterinary Clinics of North America - Exotic Animal Practice* 15: 469–485.

Bulloch, M. J., Tynes, V. V. 2010. Ferrets. In *Behaviour of Exotic Pets*, ed. V. V. Tynes. Hoboken, NJ: Wiley-Blackwell Publishing Ltd.

Byrd, L. G., Prince, G. A. 1997. Animal models of respiratory syncytial virus infection. *Clinical Infectious Diseases* 25 (6): 1363–1368.

Caley, P., Morriss, G. 2001. Summer/autumn movements, mortality rates and density of feral ferrets (*Mustela furo*) at a farmland site in North Canterbury, New Zealand. *New Zealand Journal of Ecology* 25 (1): 53–60.

Chivers, S. M., Einon, D. F. 1982. Effects of social early experience on activity and object investigation in the ferret, *Mustela furo*. *Developmental Psychobiology* 15: 75–80.

Clapperton, B. K., Minot E. O., Crump D. R. 1988. An olfactory recognition system in the ferret *Mustela furo* L. (Carnivora: Mustelidae). *Animal Behaviour* 36 (2): 541–553.

Clyde Jr, W. A. 1980. Experimental models for study of common respiratory viruses. *Environmental Health Perspectives* 35: 107–112.

Davison, A., Birks, J., Griffiths, H., Kitchener, A., Biggens, D. 1999. Hybridization and the phylogenetic relationship between polecats and the domestic ferret in Britain. *Biological Conservation* 87: 155–161.

Dawkins, M. S. 1988. Behavioural deprivation: A central problem in animal welfare. *Applied Animal Behaviour Science* 20: 209–225.

Dietz, L., Arnold, A., Goerlich-Jansson, V., Vinke, C. M. 2018. The importance of early life experiences for the development of behavioural disorders in domestic dogs. *Behaviour* 155: 83–114.

Duda, J. 2003. *Mustela putorius furo* domestic ferret. *Animal Diversity Web.* Link (d.d. 16-2-2020): http://animaldiversity.org/accounts/Mustela_putorius_furo/.

Dunstone, N. 1993. *The Mink*. London: T and AD Poyser.

Einon, D. 1996. The effects of environmental enrichment in ferrets. In *Environmental enrichment information resources for laboratory animals. 1965–1995: birds, cats, dogs, farm animals, ferrets, rabbits and rodents*, 113–126. AWIC Resource Series no. 2. Beltsville, MD: U.S. Department of Agriculture and Universities' federation for animal welfare (UFAW).

EU Commission recommendation of 18 June 2007 on Guidelines for the accommodation and care of animals used for experimental and other scientific purposes, 2007/526/EC. Link (d.d. 16-2-2020): https://eur-lex.europa.eu/legal-content/EN/TXT/?qid=1581859224708&uri=CELEX:32007H0526.

Fisher, P. G. 2006. Ferret behavior. In *Exotics Pet Behavior. Birds, Reptiles, and small Mammals*, ed. T. B. Bays, T. Lightfoot and J. Mayer, 163–205. St. Louis, MO: Saunders, Elsevier Inc.

Fox, J. G., Bell, J. A. 1998. Growth, reproduction, and breeding. In *Biology and Diseases of the Ferret*, 211–227. 2nd ed. London: Lippincott Williams and Wilkins.

Fox, J. G., Bell, J. A., Broome, R. 2014. Growth and reproduction. In *Biology and Diseases of the Ferret*, ed. J. G. Fox and R. P. Marini, 187–210. 3rd ed. Ames, Iowa: Wiley & Sons.

Fox, J. G., Correa, P., Taylor, N. S., Lee, A., Otto, G., Murphy, J. C., Rose, R. 1990. Helicobacter mustelae-associated gastritis in ferrets: An animal model of Helicobacter pylori gastritis in humans. *Gastroenterology* 99 (2): 352–361.

Fox, J. G., Marini, R. P. 2001. Helicobacter mustelae infection in ferrets: pathogenesis, epizootiology, diagnosis and treatment. *Seminars in Avian and Exotic Pet Medicine* 10: 36–44.

Fox, J. G., & Marini, R. P. (Eds.). (2014). *Biology and Diseases of the Ferret*. New York: John Wiley & Sons.

Harris, L. M. 2015. Ferret wellness management and environmental enrichment. *Veterinary Clinics of North America: Exotic Animal Practice* 18: 233–244.

Honess, P. E., Marin, C. M. 2006. Enrichment and aggression in primates. *Neuroscience and Biobehavioral Reviews* 30: 413–436.

Howerton, C. L., Garner, J. P., Mench, J. A. 2008. Effects of a running wheel-igloo enrichment on aggression, hierarchy linearity, and stereotypy in group-housed male CD-1 (ICR) mice. *Applied Animal Behaviour Science* 115: 90–103.

Jenkins J. R., Brown, S. A. 1993. *A practitioner's guide to rabbits and ferrets*. USA: American Animal Hospital Association.

Jensen, P., Toates, F.93. Who needs "behavioural needs"? Motivational aspects of the needs of animals. *Applied Animal Behaviour Science* 37: 161–181.

Johnson-Delaney, C. A. 2009. Ferrets: Anesthesia and analgesia. In *BSAVA Manual of Rodents and Ferrets*, ed. E. Keeble and A. Meredith, 245–253. England: BSAVA.

Kelliher, K. R., Baum, M. J. 2002. Effect of sex steroids and coital experience on ferrets' preference for the smell, sight and sound of conspecifics. *Physiology and Behavior* 76: 1–7.

Kurtén, B. 1968. *Pleistocene Mammals of Europe*. First publication in 1968 by Transaction Publishers. Republishing in 2017. New York: Routledge, Taylor & Francis Group.

Lodé, T. 1999. Comparative measurements of terrestrial and aquatic locomotion in *Mustela lutreola* and *Mustela putorius*. *Zeit. Saugertier* 64: 110–115.

Lodé, T. 2008. Kin recognition versus familiarity in a solitary mustelid, the European polecat *Mustela putorius*. *Comptes Rendus - Biologies* 331: 248–254.

Lennox, A. M. 2005. Gastrointestinal diseases of the ferret. *Veterinary Clinics of North America: Exotic Animal Practice* 8: 213–225.

MacKay, J. 1995. *Complete Guide to Ferrets*. Shrewsbury: Swan Hill Press.

MacKay, J. 2006. *Ferret breeding*. Shrewsbury: Swan Hill Press.

Mason, G. J. 1993a. Age and context affect the stereotypical behaviours of caged mink. *Behaviour* 127: 191–229.

Mason, G. J. 1993b. Forms of stereotypic behaviour. In *Stereotypic Animal Behaviour: Fundamentals and Applications to Welfare*, ed. A. B. Lawrence and J. Rushen, 7–40. Oxon: CAB International.

Mason G. J. 1994. Tail-biting in mink (*Mustela vison*) is influenced by age at removal from the mother. *Animal Welfare* 3: 305–311.

Moody, K. D., Bowman, T. A., Lang, C. M. 1985. Laboratory management of the ferret for biomedical research. *Laboratory Animal Science* 35 (3): 272–279.

Moorman-Roest, J. 1993. De fret. In *Diergeneeskundig memorandum. Handleiding voor bijzondere dieren*, 82–88. The Netherlands: Gezamelijke uitgave van Mycofarm, Janssen Pharmaceutica en Solvay Duphar.

Moors, P. J., Lavers, R. B. 1981. Movements and home range of ferrets (*Mustela furo*) at Puke puke lagoon. *New Zealand Journal of Zoology* 8: 413–423.

Norbury, G. L., Norbury, D. C., Heyward, R. P. 1998. Space use and denning behaviour of wild ferrets (*Mustela furo*) and cats (*Felis catus*). *New Zealand Journal of Ecology* 22: 149–159.

Plant, M., Lloyd, M. 2010. The ferret. In *The UFAW Handbook on the Care and Management of Laboratory and Other Research Animals*, ed. R. Hubrecht and J. Kirkwood, 418–432. 8th ed. London: Wiley-Blackwell.

Poole, T. B. 1966. Aggressive play in polecats. *Proceedings of the Zoological Society of London* 18: 23–44.

Poole, T. B. 1973. The aggressive behaviour of individual male polecats (*Mustela putorius, M furo* and hybrids) towards familiar and unfamiliar opponents. *Journal of Zoology* 170: 395–414.

Poole T. B. 1974. Detailed analysis of fighting in polecats (Mustelidae) using ciné-film. *Journal of Zoology* 173: 369–393.

Poole, T. B. 1978. An analysis of social play in polecats (Mustelidae) with comments on the form and evolutionary history of the open mouth play face. *Animal Behaviour* 26: 36–49.

Powell, R. A. 1979. Mustelid spacing patterns: Variations on a theme by Mustela. *Zeitschrift fur Tierpsychologie* 90: 153–165.

Powell, R. A. 1994. Structure and spacing of Martes populations. In *Biology and Conservation of Martens, Sables and Fishers*, ed. S. W. Buskirk, A. Harestad, M. Raphael, and R. Powell, 101–121. Ithaca, NY: Cornell University Press.

Price, E. O. 2002. *Animal Domestication and Behaviour*. Wallington, Oxon: CABI Publishing, CAB International.

Quesenberry, K. E., Carpenter, J. W. 2004. *Ferrets, rabbits and rodents: clinical medicine and surgery*. 2nd edition. Philadelphia, USA: WB Saunders Co.

Reijgwart, M. L. 2017. *Refinement of the Care and Use of Laboratory Ferrets*. PhD Thesis. Utrecht, The Netherlands: Utrecht University.

Reijgwart, M. L., Schoemaker, N. J., Pascuzzo, R., Leach, M. C., Stodel, M., de Nies, L., Hendriksen, C. F. M., van der Meer, M., Vinke, C. M., van Zeeland, Y. R. A. 2017. The composition and initial evaluation of a grimace scale in ferrets after surgical implantation of a telemetry probe. *PLoS One* 12(11): e0187986.

Reijgwart, M. L., Vinke, C. M., Hendriksen, C. F. M., van der Meer, M., Schoemaker, N. J., van Zeeland, Y. R. A. 2015. Workaholic ferrets: does a two-chamber consumer demand study give insight in the preferences of laboratory ferrets (Mustela putorius furo)? *Applied Animal Behaviour Science* 171: 161–169.

Reijgwart, M. L., Vinke, C. M., Hendriksen, C. F. M., van der Meer, M., Schoemaker, N. J., van Zeeland, Y. R. A. 2016. Ferrets' (Mustela putorius furo) enrichment priorities and preferences as determined in a seven-chamber consumer demand study. *Applied Animal Behaviour Science* 180: 114–121.

Rushen, J., Lawrence, A. B., Terlouw, C. E. M. 1993. The motivational basis of stereotypies. In *Stereotypic Animal Behaviour: Fundamentals and Applications to Welfare*, ed. A. B. Lawrence and J. Rushen, 41–64. Wallingford, Oxon: CAB International.

Russell, W. M. S., Burch, R. L. 1959. *The Principles of Humane Experimental Technique*. London: Methuen & Co, Ltd.

Schoemaker, N. J., van Zeeland, Y. R. A. 2013. Fret. In *Diergeneeskundig memorandum Bijzondere gezelschapsdieren*, ed. F. Pasmans, 189–222. Oosterhout, The Netherlands: Leonard Strategische Communicatie BV.

Seksel, K., Mazurski, E. J., Taylor, A. 1999. Puppy socialisation programs: short and long term behavioural effects. *Applied Animal Behaviour Science* 62: 335–349.

Speer, B. L., Hennigh, M., Muntz, B., van Zeeland, Y. R. A. 2018. Low stress handling techniques in birds and small mammals. *Veterinary Clinics of North America: Exotic Animal Practice* 21: 261–285.

Staton, V. W., Crowell-Davis, S. L. 2003. Factors associated with aggression between pairs of domestic ferrets. *Journal of the American Veterinary Medical Association* 222: 1709–1712.

Talbot S., Freire, R., Wassens, S. 2014. Effect of captivity and management on behaviour of the domestic ferret (*Mustela putorius furo*). *Applied Animal Behaviour Science* 151: 94–101.

Tetley, H. 1965. Notes on British Polecats and ferrets. *Proceedings of the Zoological Society of London* 155 (1–2): 212–217 (published online 21 Aug 2009).

Thom, M. D., MacDonald, D. W., Mason, G. J., Pedersen, V., Johnson, P. J. 2004. Female American mink, *Mustela vison*, mate multiply in a free-choice environment. *Animal Behaviour* 67: 975–984.

Thompson, A. D. 1951. A history of the ferret. *Journal of the History of Medicine and Allied Sciences* 6: 471–480.

Turner, D. C. 2000. The human-cat relationship. In *The Domestic Cat. The Biology of Its Behaviour*, ed. D.C. Turner and P. Bateson. 2nd ed. Cambridge: Cambridge University Press.

Vinke, C. M., Hansen, S. W., Mononen, J., Korhonen, H., Cooper, J. J., Mohaibes, M., Bakken, M., Spruijt, B. M. 2008. An interpretation of farmed mink's motivation for a water bath: to swim or not to swim? *Applied Animal Behaviour Science* 111: 1–27.

Vinke C. M., Schoemaker, N. J. 2012. The welfare of pet ferrets (*Mustela putorius furo* T). A review on the housing and management of pet ferrets. *Applied Animal Behaviour Science* 139: 155–168.

Wanker, R. 1999. Socialization in spectacled parrotlets (*Forpus conspicillatus*): how juveniles compensate for the lack of siblings. *Acta Ethologica* 2: 23–28.

Woodley, S. K., Baum, M. J. 2003. Effects of sex hormones and gender on attraction thresholds for volatile anal scent gland odors in ferrets. *Hormones and Behavior* 44: 110–118.

Woods, J. B., Schmitt, C. K., Darnell, S. C., Meysick, K. C., O'Brien, A. D. 2002. Ferrets as a model system for renal disease secondary to intestinal infection with Escherichia coli O157: H7 and other Shiga toxin-producing E. coli. *The Journal of Infectious Diseases* 185 (4): 550–554.

13

Behavioral Biology of Dogs

Laura Scullion Hall
University of Stirling

Mark J. Prescott
National Centre for the Replacement, Refinement and Reduction of Animals in Research (NC3Rs)

CONTENTS

Introduction ... 205
Typical Research Use ... 206
Behavioral Biology ... 207
    Natural History ... 207
        Ecology ... 207
        Social Organization and Behavior ... 208
        Feeding Behavior ... 208
        Senses and Communication ... 208
    Common Captive Behaviors ... 209
        Normal Behavior ... 209
        Abnormal Behavior ... 209
Ways to Maintain Behavioral Health ... 210
    Captive Considerations to Promote Species-Typical Behavior ... 210
        Environment ... 210
        Social Groupings ... 211
        Feeding Strategies ... 213
        Training ... 214
Special Situations ... 216
    Single Housing ... 216
    The Use of Nonbeagle Dogs ... 216
Conclusions/Recommendations ... 216
Additional Resources ... 217
Appendix: Ethogram ... 217
References ... 219

Introduction

Dogs are extensively used in scientific research aimed at understanding their biology and behavior. The largest use of dogs worldwide is for the development and testing of pharmaceuticals for veterinary and human use. In all cases, a good understanding of the behavioral biology of the dog will help to ensure their humane use and care, which will ultimately benefit the science the animals are used for. In this chapter, we review the natural history and domestication of dogs and summarize their sensory and communication capabilities. We present information to assist with the identification of normal and abnormal behavior within the laboratory, and ways to maintain behavioral health. Provision of appropriate housing, with adequate space, visibility, complexity and choice, environmental enrichment, and opportunities for exercise, will permit performance of species-typical behavior, reduce stereotypies and other problems, and improve welfare. Housing in compatible social groups is also fundamental for dog welfare and should be the cornerstone of a dedicated enrichment program. Increasingly, requirements for single housing during regulatory studies are being challenged. Food is a highly significant resource for dogs and an effective reinforcer during training for cooperation with husbandry, handling, restraint, and scientific procedures. A dog training program will not only minimize stress associated with these events but can also improve the efficiency of staff activities and the quality of scientific data collected. Refinement of the lifetime experiences of dogs used in research will help to meet the 3Rs (Russell &

Burch, 1959) and animal welfare expectations of the public, legislators, and research funders, and is an ongoing endeavor, dependent on access to the latest scientific knowledge and practical information (see www.nc3rs.org.uk).

Typical Research Use

An estimated 140,000 dogs worldwide are used in research and testing every year. An accurate figure is not available because some user countries do not have mandatory reporting requirements on the use of animals in scientific procedures. The most current statistics from the US Department of Agriculture show that 64,707 dogs were used for research, testing, teaching, or experimentation in 2017 (USDA, APHIS, 2018). The Canadian Council on Animal Care reports that 18,216 dogs were used in science in 2018 (Canadian Council on Animal Care, 2019). 13,688 dogs were used for experimental purposes in European Union (EU) Member States in 2017, the latest date for which combined figures are available (European Commission, 2020). China, Japan, and India are also likely to use dogs in large numbers, given their growing investment in pharmaceutical research and development.

Most dogs used in research and testing are purpose-bred beagles. There are sound scientific and welfare reasons for using purpose-bred dogs (suppliers of which are referred to as class A dealers in the USA) (National Research Council, 2009; Prescott et al., 2004). The beagle was probably originally chosen for use in the laboratory because of its relatively small size and friendly temperament (and hence ease of dosing and of measuring body responses), short hair, ease of access to the cephalic vein for blood sampling, and because it can easily be housed in kennels (Prescott et al., 2004). Today there is a significant body of background data for the beagle, which increases its value as a defined research animal. Other breeds and cross-bred dogs are also used in research, mainly for reasons related to genetic diversity or for research into specific breed-related diseases, such as mast cell tumors in Labrador and golden retrievers, or histiocytic ulcerative colitis in boxers (Hasiwa et al., 2011). For more information, see *"The use of nonbeagle dogs"* section of this chapter.

Dogs are used to study conditions to which they are predisposed, such as hip dysplasia, and also as models of human inherited disorders, such as Duchenne muscular dystrophy and many cancers (Shearin & Ostrander, 2010; Zurlo et al., 2011; McGreevy et al., 2015). The various dog breeds carry unique breed-specific variations in morphological and behavioral traits, and more than 600 genetic disorders have been described, with approximately 70% of the diseases having corresponding human conditions (Starkey et al., 2005; Hasiwa et al., 2011). However, the vast majority of scientific procedures on dogs are conducted during the research, development, and testing of new pharmaceuticals, including veterinary medicines and vaccines. Dogs are used to test for both the intended beneficial effects (efficacy) and any adverse effects (safety). The majority of these tests are for human medications and are required by regulatory authorities before permission can be granted to test the medication in human volunteers or patients (ICH, 2020). Nonpharmaceutical products, such as agricultural products (mainly pesticides and biocides), food additives, and industrial chemicals, are also tested extensively for safety according to legal requirements (OECD, 2019). Dogs are occasionally used for these tests, although the numbers used are small in comparison to the numbers used to test pharmaceutical products, especially since the requirement for a 1-year chronic toxicity study in the safety assessment of pesticides has been challenged (Spielmann, 2019).

In regulatory toxicity testing of pharmaceuticals, dogs are one of the nonrodent species commonly used in conjunction with rodents to provide data to assess the potential for adverse drug effects. Groups of dogs are administered a test drug and clinical parameters are measured at various time-points throughout the study, after which the dogs are euthanized and their organs removed for detailed examination. Use of two species (a rodent and a nonrodent) is recommended in international guidelines (ICH, 2020), and has been adopted and broadly followed by individual countries worldwide. Although the study requirements and designs are harmonized, different perceptions and interpretation of requirements for individual test drugs can lead to variations in the number of animals used for similar studies. Consequently, there are many opportunities to reduce the number of dogs or to refine the procedures that are performed on a case-by-case basis (Sewell et al., 2017).

Dogs have experienced convergent evolution with humans, which has resulted in the environment having similar effects on health and disease processes of both species. An International Life Sciences Institute study found that approximately 70% of the human toxicities observed in clinical studies for pharmaceuticals were predictable from one or more preclinical animal toxicology studies, with rodents alone (primarily the rat) predicting 43% of observed adverse effects in humans, and nonrodents alone (primarily the dog) predicting 63% (Olson et al., 2000). More recently, an IQ-DruSafe study found that the proportion of positive animal toxicity findings that translated to positive clinical findings increased when the same target organ was identified in both the rodent model (primarily the rat) and nonrodent model (primarily the dog) (Monticello et al., 2017). The dog is considered to be especially predictive of human gastrointestinal and cardiovascular events, such as QT prolongation and arrhythmias (Ewart et al., 2014; Clark & Steger-Hartmann, 2018). Dogs are preferred over alternative second species, such as macaques, for scientific, ethical, and practical reasons. Dogs are better adapted to live in kennel situations in close contact with humans, and thus may experience less stress as a result of captivity.

Dogs have also long been used in psychological research. From Pavlov's (1928) studies of learning to Fox and Stelzner's (1966) exploration of early experiences, much psychological research has aimed to uncover processes that explain human psychology. Today's modern cognitive research in dogs is a far cry from older research, such as Seligman's (1972) learned helplessness experiments, in which dogs were subjected to extreme distress in order to study their responses. There has been a great deal of refinement in cognitive research over the years, and dogs are now studied while awake in MRI scanners, observing facial expressions from screens, and following gestures (Reid, 2009; Huber & Lamm, 2017). In much of this research, they are trained to cooperate with the noninvasive procedures.

The husbandry and care of laboratory dogs may be influenced by the particular research areas. Most clinical veterinary research takes place with pet dogs attending veterinary clinics, but basic research that aims to better understand the processes underlying diseases may take place in research facilities with purpose-bred populations. Whatever the purpose of the investigation, the needs of the dog, including opportunities to engage in normal behavior and social interactions, should be paramount. These can be accommodated in most circumstances, provided they are recognized and understood by animal care staff and investigators (Prescott et al., 2004).

Behavioral Biology

Natural History

Ecology

Evidence suggests that dogs were domesticated between 14,000 and 33,000 years ago (Sablin & Khlopachev, 2002; Druzhkova et al., 2013). Dogs share a common wolf ancestor with modern grey wolves (*Canis lupus*), although there is some debate about the identity of that now extinct species. Domestication does not appear to have been a single event, with DNA evidence suggesting multiple domestications, and perhaps interbreeding, between dogs and wolves over a period of time (Axelsson et al., 2013).

It was thought that dogs were purposefully domesticated by humans, who selected for tameness and kept the ancestors of the modern dog for hunting or livestock guarding purposes. Indeed, evidence to support this theory was found in the long-running experiments of Dmitri Belyaev (Dugatkin et al., 2017), who began a study of arctic foxes in the 1950s that led to greater understanding of the role of artificial selection for the criteria of "tameness". Foxes that displayed higher levels of "tameness" (willingness to approach and interact with humans) displayed differential physical characteristics across subsequent generations, including floppy ears, curly tails, pedomorphic facial characteristics, and physiological differences, such as reduced glucocorticoid levels and increased serotonin levels (Dugatkin et al., 2017). These changes appear to have mimicked the differences seen in domestic dogs when compared to contemporary wolf populations.

It is now thought that natural selection also played a role in the domestication of dogs (Coppinger & Coppinger, 2001; Freedman et al., 2014). It is difficult to find truly wild counterparts to the dog upon which to draw comparisons, with even the closest genetic relative of the dog, the grey wolf, having many differences in social behavior and cognition (e.g., Miklósi, 2016). Few dogs live apart from humans, with many "free-roaming" dogs released from human homes during the day returning to their owners at night. Dogs in feral or unowned free-roaming states are often found in proximity to human settlements and are in some way influenced by human activity (e.g., availability of food or shelter). It is hypothesized that natural selection has influenced the domestication of dogs by selecting for those that lived in proximity to human settlements, benefiting from the availability of food upon which to scavenge, for example, from landfill sites. It is notable that dogs tend to value human company more greatly than that of other dogs, and that even dogs that have not been raised around humans are quick to lose their fear of humans.

Selective breeding by humans to produce, preserve, or remove specific characteristics has resulted in a great diversity of dog breeds, ranging from toy breeds that weigh less than 1 kg to giant breeds that weigh over 140 kg (Young, 1994). This diversity of breeds also leads to diversity in behavior; dogs bred for livestock guarding may show greater levels of vigilance and territorial behavior, while those bred for assisting with hunting (e.g., hounds) may show a greater propensity for scenting and chasing of prey.

The clearest differences between dogs and wolves can be found in physical characteristics, such as size, physical power, and dentition. Dogs are, on the whole, much smaller than wolves, although size varies considerably among breeds. In free-ranging dogs not purposefully bred by humans, size may tend toward a medium to large build, although this, in turn, may be influenced by the environment (e.g., with larger dogs dominating in cold climates) (Bonanni & Cafazzo, 2014).

Dogs are also less powerful than wolves (Coppinger & Coppinger, 2001), perhaps a cause of, or resulting from, the differences in food sources. While wolves may hunt to bring down large prey, such as ungulates, dogs are more frequently found in proximity to human settlements, scavenging for scraps, rather than hunting. While there is evidence of dogs hunting smaller ungulates (e.g., sheep), this tends to be less coordinated than the pack hunts of wolves and does not form the mainstay of their diet (Miklósi, 2016). Where free-roaming dogs do hunt, they tend to take medium prey in packs (Avis, 1999). The diet of free-ranging dogs tends to consist of human food waste, various carrion, and small prey animals (e.g., rodents, reptiles), although they may predate a range of local species (e.g., see Young et al., 2011).

Dogs obtain food in a different manner from wolves, primarily scavenging in small groups. This difference may also have influenced group structure, with dogs more often forming small, fluid groups, rather than the larger, stable, family groups found in wolf populations. Lower levels of cooperative behavior are also found in dogs, perhaps due to the reduced pressure for group communication during a hunt, or the need to live in close proximity to a tight-knit group. Stable groups are unlikely to form, and where groups are formed, dogs may still forage alone (Pal, 2001; Bonanni & Cafazzo, 2014). Free-roaming dogs tend to form groups where food is more available, resulting in a higher population density (Bonanni & Cafazzo, 2014). Dogs are less likely to form groups with other dogs when they are owned by humans; dogs that are unowned and free-roaming are more likely to form associations with other dogs. This indicates that human relationships may be particularly important and mediate intraspecific relationships. Dog group composition and behavior are likely to be influenced by the proximity of human settlements, and so dogs have developed enhanced behavioral flexibility to facilitate success in a variety of environments. This is reflected in the presence of dogs as companion animals and working animals across the world.

Social Organization and Behavior

While wolves form clear hierarchies within a family unit of the breeding pair (parents) and their offspring (Mech, 2000), the same is not true of dogs. The differences in access to food mean that dogs are more likely to be found in small groups, and the fluidity of group compositions may make it difficult for stable hierarchies to form. However, it does appear to be true that where stable groups do form, older dogs appear to be higher ranked than younger dogs, and male dogs are higher ranked than females, unless they are younger. Dogs that are higher ranking tend to be more likely to act as the leaders in a situation, determining group movements. They also receive more affiliative and agonistic interactions than other dogs (Bonanni & Cafazzo, 2014).

Unlike in wolf family groups (Mech, 1999), all members of a dog group may breed, and although extended monogamy is not apparent (Pal et al., 1999), a range of mating styles may be observed in dogs (Pal, 2011). Breeding occurs regularly, with female dogs (bitches) coming into estrus approximately twice per year. The regularity of food availability from human settlements may be the cause of this regular breeding, as puppies are able to start feeding early, by scavenging for scraps with their parents, and are not reliant on parental hunting for food. Another difference in the rearing of young is found in cooperative behavior, with dog puppies only being fed by the mother, rather than by the family unit, as is the case with wolves.

Bitches reach sexual maturity between 6 and 14 months of age, while male dogs reach sexual maturity from 10 months of age, with peak fertility occurring at approximately 2 years of age (England & von Heimendahl, 2010). A bitch will allow a dog to mate for several days both before and after ovulation, which in bitches is spontaneous. A successful mating can be observed through "tying", in which the dog and bitch are briefly attached following mating. Breeders of laboratory dogs will typically use one of three mating systems: observed mating, in which a bitch is housed with a stud dog for a period of time while fertile; harem mating, in which the male lives continuously with a group of females; and artificial insemination, commonly used where the impracticalities of transporting dogs for breeding prohibit other methods.

The duration of pregnancy is typically around 63 days and can be 58–72 days when measured from the date of breeding. At 2 weeks prior to parturition, moving the bitch to a whelping area is recommended (Prescott et al., 2004). One to two days before parturition, bitches will seek a quiet area in which to whelp. At this stage, they should be individually housed in a pen containing a nesting box and nesting material, with a platform to allow the bitch respite from the pups, within sight and smell of other bitches (Prescott et al., 2004).

Smith (2012) describes in detail the stages of normal parturition. Briefly, Stage I (dilation of the cervix) typically lasts 6–12 (but as many as 36) h; Stage II (delivery of pups) lasts from 3 to 12 (and no more than 42) h; while Stage III (delivery of the placenta) should be complete within 15 min of delivery of the final pup.

At birth, dogs have limited sensory and motor abilities (Fox & Stelzner, 1966). They are born with eyes and ears closed. However, they have olfactory abilities as well as tactile and thermal sensory capabilities (Jones, 2007). Desensitization to unfamiliar stimuli during the period in which the startle response emerges may also be beneficial to the pups' future responses to unfamiliar stimuli.

Exposure to age-appropriate, gentle stimulation can encourage maturation of the brain, heart, and motor skills (Fox & Stelzner, 1966). Animals with more developed nervous systems are more likely to be able to adapt to the environment or stressors – a quality which is necessary in laboratory animals. Pups which experience daily handling from birth are more emotionally stable, less likely to yelp or vocalize in a novel environment, and more likely to explore the environment (Gazzano et al. 2008).

Dogs have home ranges between 0.001 and 70 km^2 (Bonanni & Cafazzo, 2014), considerably smaller than the typical home ranges of wolves, which are between 30 and 6,000 km^2 and are influenced by the density of prey animals (Mech & Boitani, 2003). Territorial behavior among dogs is dependent on the density of the dog population; where many groups live within a densely populated area, territorial behaviors, such as boundary marking with urine, are seen more frequently, in both male and female dogs. However, territories may overlap, and several groups may scavenge for food in the same area.

Feeding Behavior

Dogs are opportunistic feeders, relying on scavenging. Canids have broad feeding habits and are known to eat plants, including berries and other fruit, as well as small and large animal prey. Although dogs scavenge for food, they can also consume their metabolic requirements for the day in a single meal. When given the choice, dogs will eat over 4–8 meals (Mugford & Thorne, 1978), predominantly during daylight hours.

Senses and Communication

Dogs have a complex repertoire of social behaviors that are used to communicate with conspecifics. The laboratory environment may impede these behaviors. Dogs rely heavily on olfaction to assess their surroundings and to communicate. They can detect odors in the range of parts per trillion to parts per quadrillion (in contrast, humans can detect odors only in the range of parts per million; Ong et al., 2017). Therefore, as well as sight and sound, care must be taken to ensure that dogs are able to make olfactory contact with one another, and that smells associated with adverse events are minimized. Dogs use scent to communicate information about their identity and social status. They examine the feces, urine, genitals, ears, and mouths of others for information (Fox, 1971). Dogs also use olfaction to identify humans; staff should take care to avoid strong scents that may interfere with the dogs' olfactory communication.

Dogs also communicate via a range of vocalizations. Barking may indicate arousal, fear, or the appearance of a stranger (Miklósi, 2016). Other vocalizations, such as whining and yowling, may indicate high arousal negative states. Staff should be able to distinguish between vocalizations associated with play behavior, affiliative behavior, and adverse social interactions, in order to monitor colony harmony.

Dogs also have sensitive hearing, particularly at high frequencies, making them more sensitive to sounds in the laboratory environment. Dogs' hearing is most sensitive between 1 and 20 kHz, in comparison to 1–5 kHz in humans. The upper limit of dog hearing is approximately 50 kHz, in comparison to 20 kHz in humans (Sales et al., 1997). The dog bark peaks in the 500 Hz–16 kHz range, where their hearing is 24 dB more sensitive than that of humans (Sales et al., 1997). Vision is dichromatic, with the greatest sensitivity between 429 and 555 nm; objects are primarily seen in blues, yellows, and greens (Neitz et al., 1989). Dogs have a high sensitivity for moving objects, allowing them to discriminate between moving objects at distances of up to 900m (Miklósi 2016). Visual discrimination is greatest when objects are moving, and dogs can detect a change in movement that exists in a single diopter of space within their eye. The field of vision is wide, allowing dogs to be aware of movement around them. Canid species prefer high vantage points, allowing them to monitor movement around them. At close distances, vision is less sensitive, so in this situation, dogs rely on touch and smell.

Behaviors, including postural changes, are also used to communicate with both conspecifics and humans. For example, dogs may use the "play bow" to instigate play interactions, while a rapid change from high posture (body drawn tall, ears up) to low posture (body low to the ground, ears back), may be seen during play (Horowitz, 2002). Other behaviors, such as turning the head away, may be used to avert conflict.

Touch, for example, resting in contact with conspecifics, or manipulating objects with the mouth or paws, is used to explore the environment. Touch can have an impact on both behavior and physiology, which can be pleasant or unpleasant depending on the circumstances (e.g., physical restraint, see: Hall, 2014; Hall et al., 2015). Positive interactions between humans and dogs (e.g., petting) can result in decreased blood pressure and salivary cortisol levels, increased release of beta-endorphins and increased dopamine levels in dogs (Odendaal & Meintjes, 2003; Bergamasco et al., 2010). Juvenile dogs will typically spend approximately 10% of their day resting in physical contact with pen mates (Hubrecht et al., 1992).

Shared evolutionary history between humans and dogs has led to dogs developing abilities to communicate with, and understand, humans. Dogs demonstrate a strong ability to understand human eye gaze, attention, and pointing gestures, and are greatly influenced by human-given social cues (Hare & Tomasello, 1999). The ability to use social referencing (the use of social cues from other individuals to regulate behavior toward the environment) means that interactions with dogs, particularly when referencing potentially aversive stimuli, can be strongly influenced by the handler. For example, a handler who is experiencing stress during the conduct of a procedure is likely to be communicating to the dog that the situation is fearful, while in contrast, a relaxed handler is likely to communicate that the situation is not one that requires the dog to be anxious. In the context of dog use in scientific research, it is important to exploit this social referencing ability to ensure that positive associations are formed with staff, equipment, and procedures.

Common Captive Behaviors

Normal Behavior

Dogs are macrosmatic, neophyllic, active, and cursorial animals. Captive dogs should exhibit a range of normal, species-specific behaviors. Ideally, they should display a relaxed demeanor in the home pen, including loose, relaxed, body posture; calm locomotion; and interaction with pen mates and the surroundings (Hall, 2014). Dogs should be willing to approach familiar staff and should not display anxious behavior in the presence of humans (e.g., jumping, barking, pacing). Dogs should make use of a range of effective enrichment items in the home pen. A lack of interaction with enrichment indicates that it is either unsuitable, uninteresting, or there is a welfare concern for the dog (Hubrecht & Kirkwood, 2010). A range of refinements, including improvements to the design of the animal room and pens, training, and provision of appropriate enrichment items, encourages the display of normal behaviors.

The importance of staff familiarity with the behavior of individual dogs is critical here; any deviations from normal behavior may be indicative of illness, adverse reactions to scientific procedures, poor welfare, or conflicts with pen mates. Staff should be able to enter and exit the animal room without causing increased arousal in the dogs, thus enabling them to observe normal behavior. Beagles are an excitable breed (Fogel, 1990) that tend to display "agreeable" personality dimensions and are not prone to displaying aggressive behavior (Kraeuter, 2001). These traits make them well-adapted for an environment that can involve intensive handling and group living. Beagles have a strong scenting drive, being bred for scenting of prey animals, making their motivation for training strong when food is used as the reinforcer. Their reliance on scent also means that beagles are easily distracted by competing scents, and therefore, handlers and trainers should take this into account when interacting with or training dogs.

Abnormal Behavior

Abnormal behaviors that are seen in dogs with poor welfare are not defined by their presence or absence in many cases, but by the frequency of their presentation. This is because many of these behaviors are seen in the normal behavioral repertoire, but prolonged exhibition of these behaviors is likely to indicate poor welfare. For example, prolonged vigilance (alert, oriented, high posture) is a welfare concern (Hall, 2014), but this vigilance can be seen as part of a normal play bout. Similarly, it is normal for dogs to use olfaction to explore their pen, but displaying this behavior regularly may indicate that the dog's environment is disturbed, that its olfactory territorial markings have been cleaned away, or even that it is using scenting as a displacement behavior due to stress (Hall, 2014).

For this reason, use of a quantitative behavioral scoring tool is recommended. Examples can be found in Hall et al. (2015). Measurement of behavior at multiple time points, rather than at a single time point, allows the differentiation between transient behaviors and prolonged abnormal behaviors. Staff should also be encouraged to become familiar with their dogs' normal behavior, so that they can notice increases in the incidence of undesirable behaviors and address the cause.

Behavioral indicators of poor welfare are most likely to become evident when dogs do not have their social or environmental needs met. The behavioral response may be adaptive (e.g., seeking contact with pen mates or staff) or maladaptive (e.g., stereotypic behavior that becomes divorced from the original cause). The most common behavior associated with reduced welfare status includes vigilant behavior, which manifests as a perpetually alert state in the absence of a stimulus (Scullion Hall et al., 2017). This behavior is particularly evident in dogs that are singly housed or that are showing anticipation of regulated procedures or other unpleasant events. Other behaviors that indicate increased arousal include pacing, jumping, and standing with the forepaws against the walls of the enclosure (Scullion Hall et al., 2016).

Indicators of acute stress or anticipation can be most commonly seen when dogs are removed from the home pen or handled by staff. These include the sudden raising of a forepaw, lip smacking or licking, and panting in the absence of excessive heat or exercise (Hall, 2014). These behaviors are also commonly seen during times of disturbance in the animal room, for example, during cleaning, if the dogs are not moved to another room.

Ways to Maintain Behavioral Health

Captive Considerations to Promote Species-Typical Behavior

Environment

Laboratory-housed dogs spend the majority of their day in their home enclosures (the exceptions being removal for scientific procedures, pen cleaning, and exercise), so the design of the enclosure and the animal room has a major impact on their welfare. The aim should be to provide an environment that allows the dogs to perform the widest possible range of normal behavior and to exercise a degree of choice and control (e.g., to be seen by neighboring dogs or not seen) (Prescott et al., 2004). Ease of husbandry by caregiving staff and maintenance of the animals' physical health have traditionally been given higher priority compared to features that would suit what is known about the natural history and behavior of dogs, and compared to their behavioral and welfare needs (Hubrecht, 1995, 2002). However, there is a gradual move away from traditional pen designs toward modern designs featuring greater space, increased visibility and choice, and more enrichment for their occupants (see Figure 13.1). Scullion Hall et al. (2017) showed that beagles that were (1) housed in a purpose-built facility with modern home pen design, (2) exposed to regular staff contact, and (3) trained, demonstrated more behavioral signs of positive welfare (such as resting), and fewer signs of negative welfare (such as vigilance and stereotypies), as well as higher mechanical pressure thresholds.

The amount of space provided for dogs is of paramount importance because it dictates not only their ability to perform species-typical behaviors but also the size of the social group and opportunities to provide environmental enrichment. EU legislation mandates minimum pen sizes of 2.25 m^2 floor area per dog (10–20 kg) for group-housed dogs, 4.5 m^2 for singly-housed dogs, and 1.5 m^2 for postweaned breeding stock (European Union, 2010). EU-style pen housing is considered the "gold standard"; however, even under these conditions, some dogs develop behavioral abnormalities (Hubrecht, 1995). In other regions, mandatory minimum enclosure sizes can be significantly smaller; for example, the ILAR *Guide for the Care and Use of Laboratory Animals*, which is used in the USA and elsewhere, recommends 0.74 m^2 for dogs of a similar size (National Research Council, 2011). In some countries, laboratory dogs remain singly housed in small, metal cages that negatively impact their welfare and are not consistent with the 3Rs (see Beerda et al., 1999a, b, for the effects of spatial restriction on dog behavior and physiology). The impact of housing systems and home pen design should be considered part of the experimental protocol and be factored into the harm-benefit analysis (Sherwin, 2007).

A well-designed dog pen will have good visibility for dogs and staff (e.g., through use of toughened glass; horizontal, rather than vertical, bars; raised platforms), a choice of resting places, ease of entry for staff, ease of partitioning dogs, and use of noise-reducing materials (e.g., plastic, resin coatings) (Hubrecht et al., 1992; Sales et al., 1997; Prescott et al., 2004). Poor visibility can lead to socially facilitated (allelomimetic) barking, resulting in considerable amounts of noise, especially in units with many hard, smooth surfaces that reflect sound (Prescott et al., 2004). Sound levels in dog units can be so high (85–122 dB), as to cause stress for both dogs and humans, and risk damage to hearing, requiring staff to wear protective ear plugs or defenders (Sales et al., 1997). The amount of barking can be an immediate and striking indicator of the degree to which the housing design meets the behavioral needs of the dogs within it.

Modern home pens are typically constructed with opaque or transparent plastics and stainless steel. Plastics are warmer to the touch and noise-absorbing. Stainless steel is easy to clean and preferable to galvanized metal, as it is less likely to flake or be ingested by dogs, an important consideration as dogs spend a great deal of time foraging or interacting with their surroundings. Wood is not a suitable material for constructing dog pens because it is difficult to sanitize and can be damaged by chewing. Flooring should be solid, with inbuilt drainage for ease of cleaning. Concrete is less preferable than epoxy flooring, as it is not easily adapted, and the porous surface is not amenable to thorough cleaning. A suitable substrate (e.g., clean sawdust, shredded paper) in sufficient amounts should be provided to absorb urine and excreta, as well as to provide warmth and comfort. Open flooring systems (e.g., slatted or mesh floors) are not recommended because dogs prefer solid flooring and because there is a greater risk of pain, injury, or disease (e.g., from pressure sores or entrapped toes, dew claws, or collars) (Gärtner et al. 1994; Prescott et al. 2004). Dogs kept on open flooring should always be provided with a comfortable solid surface for resting and sleeping. Canids typically defecate and urinate away from the sleeping area, and the design of the home pen should be such that this is possible

FIGURE 13.1 The design and layout of a modern dog pen. Note the horizontal bars, clear plastic walls, and ledges of varying heights.

(Fox, 1971). Small pens, or those without designated sleeping areas, are unlikely to provide this ability.

The design of home pens should be flexible, allowing dogs a choice of location within the pen and the ability to withdraw from pen mates. Hatches allow adjoining pens to be connected to provide dogs with more space, and can also be used to temporarily separate dogs when required for the science (e.g., for feeding or postdose observation in toxicology studies). Shelves or platforms at varying heights allow dogs further choice within the pen and increase the visibility of the animal room (Figure 13.1). A dry, warm, draught-free bedding area should be provided. Solid plastic beds are suitable and should not be overly chewed unless insufficient enrichment is provided. Bedding materials, such as Vetbed®, shredded paper, or corrugated cardboard, encourage nesting and exploratory behavior and are typically well-utilized (e.g., see Figure 13.2).

Since dogs are macrosmatic, neophyllic, active, and cursorial, the denial of exercise and sufficient space to run and explore may have detrimental welfare implications. A separate exercise area should be provided to all dogs, which can be indoors or outdoors, depending on the design of the facility. Outdoor exercise areas have the additional benefits of variable weather and temperature, and increased olfactory stimuli. The exercise area should contain a range of enrichment items (e.g., steps, ramps, tunnels, toys) that are not typically available in the home pen, to offer the dogs novel stimuli and allow expression of a wider range of normal behavior. This is considerably more beneficial than simply turning the dogs out into the corridor for exercise when cleaning pens. Dogs should have daily access to the exercise area, and staff responsibilities and work levels should be regularly reviewed to maximize the amount of time that the dogs have access. Twenty minutes per day is a minimum beneficial time period (Prescott et al., 2004; Meunier & Beaver, 2014). Regardless of the type of exercise areas provided, staff contact should be available during exercise times. Dogs are more active in the presence of staff, and this provides an excellent opportunity to create positive associations with staff members. Examples of the contents of indoor and outdoor exercise areas can be seen in Figures 13.2 and 13.3.

There is a large body of literature on the benefits of environmental enrichment for the welfare of kenneled dogs (see Wells, 2004b, for review). Increased social contact (e.g., Mertens & Unshelm, 1996) and toys are most frequently recommended (e.g., Wells, 2004a). Laboratory dogs show considerable interest in toys, particularly novel toys, those that can be chewed (e.g., rawhide, Nylabone ™, cardboard boxes), generate noise, or are scented (Hubrecht 1993; 1995; Wells 2009). Positive changes in behavior are seen following the introduction of feeding toy enrichment (Hall, 2014); for example, Schipper et al. (2008) found that provision of a Kong™ toy stuffed with food treats stimulated appetitive behaviors, increased activity (exercise), and decreased barking frequency. It is recommended to have a selection of preferred toys and to alternate their use (Meunier, 2006; Tarou & Bashaw, 2007).

Social Groupings

Dogs are a highly social species and there is strong evidence to support improved welfare when they are housed socially (Hetts et al., 1992; Hubrecht et al., 1992; Hubrecht, 1995; Mertens & Unshelm, 1996). Social housing should therefore

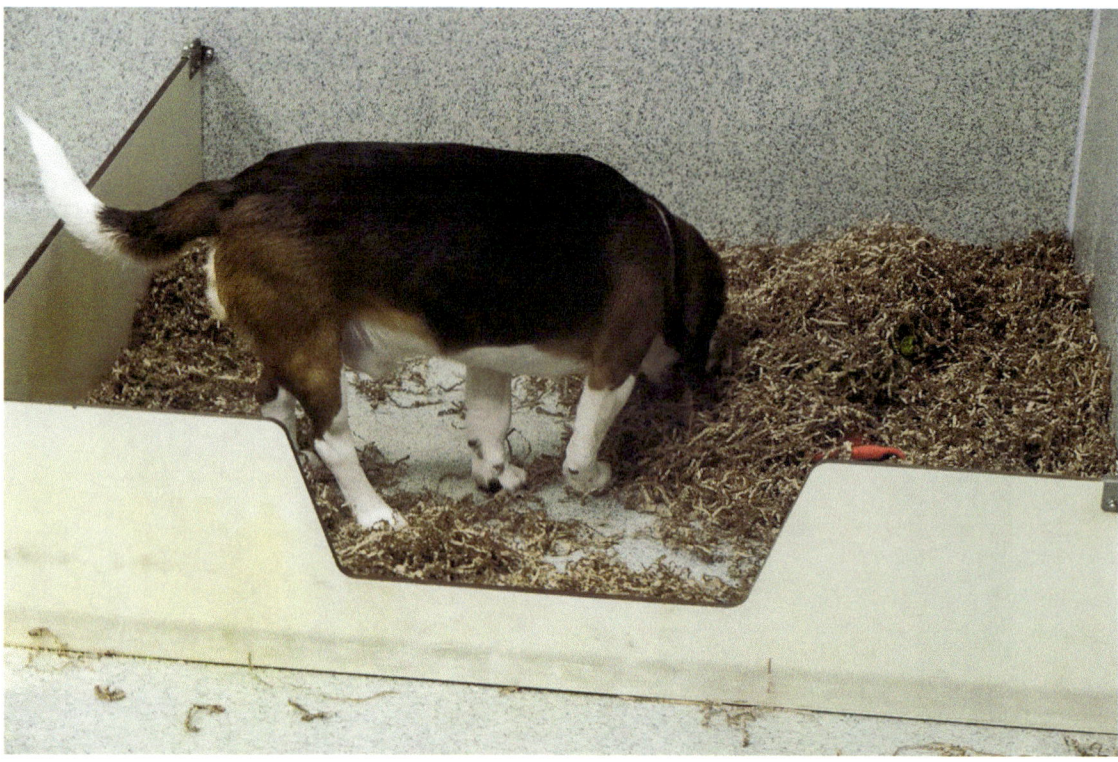

FIGURE 13.2 A dog forages for toys in a shredded paper bed (Enviro-Dri, LBS) located in an internal exercise room.

FIGURE 13.3 Dogs interact with enrichment items in an external exercise area.

be the default housing configuration for dogs in research settings, and social enrichment should be the cornerstone of a dedicated enrichment program. Social housing also provides olfactory enrichment, which may encourage increased interaction with the environment. Single housing and social isolation are detrimental to the welfare of laboratory dogs and should be avoided whenever possible. For more information, see section "Single housing".

The available space in dog facilities may limit the size of social groups; guidance on space requirements should be followed to ensure adequate space for all animals (Home Office, 1995). Housing in very large groups may lead to conflict, especially where resources, such as resting areas or toys, are in limited supply. Free-ranging dogs would naturally form a small, fluid group, and this composition should be taken into account when deciding upon stocking density. It is not advisable for large numbers of dogs to be forced to share an area where they cannot avoid group members.

Care should be taken to ensure that dogs are matched based on compatibility. Pairing or grouping of incompatible individuals can lead to aggression or competition over resources, which may affect research outcomes. Ideally, compatible groups should be established at the breeding facility and information regarding compatibility transferred with the dogs. Dogs on regulatory studies are typically randomized into study groups using physical data, such as age, sex, weight, and sibling status. Given the impact of social interactions on welfare, compatibility with other colony members should be considered during the randomization process. This will prevent difficulties in reassigning incompatible dogs once a study is underway. Staff should become familiar with individual dogs

FIGURE 13.4 Group-housed dogs interacting, depicting a number of normal behaviors, including affiliative and exploratory behaviors.

and their social relationships, and records should be kept of harmonious groupings so that these can be maintained during the dogs' lifetimes. It is common for animal rooms to be fitted with video recording equipment that permits the monitoring of social groups in the absence of human activity. See Figure 13.4 for depictions of a range of positive, affiliative behaviors in group-housed dogs.

Social housing is particularly important during puppyhood, with considerable deficits in social and physical development seen in puppies exposed to varying degrees of isolation (Fox & Stelzner, 1966). Behavioral problems involving social conflict seen in puppyhood (e.g., interdog aggression, resource guarding) may persist into adulthood, so staff should be appropriately trained to deal with these issues and/or seek advice from an expert dog behaviorist. Providing dogs with sufficient pen enrichment, choice of location within the pen, and the ability to withdraw from others will help ameliorate compatibility issues later in life. For further general advice on preventing and managing aggression, see Prescott et al. (2004).

Feeding Strategies

Food is a highly significant resource for dogs and should be considered as part of their environment, as well as for nutrition and energy. Most laboratory dogs are likely to be fed a commercially prepared, standard, dry diet, which should be nutritionally adequate for the stage of life, uncontaminated, and palatable. However, dogs prefer moist or fresh diets over dehydrated diets (Bhadra et al., 2016), and this should be taken into consideration when selecting a diet. Dogs are also likely to crave novelty, and additions should be made to the standard diet, e.g., food treats that can be used in training, or biscuits that can be given to dogs in the home pen. Variations in size, textures, and odors provide novelty, and are likely to stimulate interest. Changing the presentation of food is also likely to be enriching. Staff should be aware that aversion to the diet may occur if it becomes associated with adverse study outcomes (e.g., from administration of test compounds). In such cases, alternative diets should be used to prevent weight loss.

The two most common methods of feeding laboratory dogs are *ad libitum* and restricted. *Ad libitum* feeding provides dogs with a more naturalistic eating pattern, as per Mugford & Thorne (1978). A stainless-steel food hopper mounted within the pen allows multiple dogs to access unrestricted food. However, approximately 30%–40% of dogs will become obese on unrestricted feeding (Mugford & Thorne, 1978), so it is important to monitor the weight of dogs fed *ad libitum* to ensure obesity does not occur. If weight increases are seen, smaller meals should be provided throughout the day. Weight should always be regularly monitored to ensure that all dogs are able to access sufficient food and therefore maintain body weight. Free-ranging dogs spend a considerable amount of time foraging; providing free access to food in the laboratory may therefore remove a significant source of enrichment in the environment. Consider providing some food in feeding toys (e.g., Kong™) as an alternative means of presentation.

Restricted feeding, usually a single meal provided for 2–4 h, is typically used when study requirements necessitate monitoring of food intake, or food restriction for other reasons, such as fasting before sampling. Restricted feeding typically takes place while dogs are temporarily separated, in order to achieve accurate measurements of individuals' food consumption. Food should be provided in stainless steel bowls, which can be mounted on brackets within the pen. Modern pens have hatches that allow staff to remove these bowls without opening the pen door. This design feature minimizes disturbance to the dogs, reduces the risks associated with staff entering pens (e.g., escape), and prevents food spillage or waste (see Figure 13.5).

FIGURE 13.5 A food bowl attached to a feeding hatch, allowing food to be placed in the home pen from outside.

Training

Early life experiences can have a profound impact on dog welfare later in life. Due to the timing of sensitive developmental periods (within the first 12 weeks of life), there is a particular responsibility for breeding facilities to ensure that appropriate rearing programs are used to provide dogs with the skills necessary to adapt to later challenges in laboratory settings. Typical use of dogs in safety assessment studies occurs at 5–9 months of age. Stressors associated with transport and acclimatization to a new facility mean that early opportunities for desensitization in the breeding facility are particularly important to future welfare.

In a review, Meunier (2006) suggested that the most important factor is to develop a program that succeeds in reducing distress through the implementation of training, desensitization, and socialization tailored to the individual future experiences of the dogs. Early training programs should introduce the examination table and health checks. Where dogs are likely to experience jacketing and instrumentation during studies, early acclimatization to the jackets is likely to prevent later issues. The challenges are greater when dogs are transported from commercial breeders to experimental facilities.

Interactions with staff should be calm and not rely on the use of force. Dogs should freely exit the home pen for the purposes of husbandry, activity, and procedures. Training to exit the pen removes the need for manual handling, reducing stress for both handler and dog (Hall & Robinson, 2016). A dog that must be physically removed from the home pen is fearful, requires additional socialization/training, and is not suitable for use in experimental procedures.

Healthy, unstressed dogs are highly food motivated and food should be used as the primary reinforcer when introducing training. Dogs that are not motivated to take food from a handler are often either ill or experiencing stress, which inhibits appetite, and these factors should be resolved before training can take place. Like many hounds, beagles can be energetic, scent-driven, and there may be a perception that they are more difficult to train than other breeds. The use of a clicker as a secondary reinforcer allows for rapid and precise training of many behaviors in beagles, as well as other animals (Laule, 2010).

A frequent concern, particularly in regulatory environments, is that using food treats as positive reinforcement will adversely impact study data. However, the benefits gained by improving welfare should be considered to be greater than any unwanted variability introduced by additional food treats (Prescott et al., 2004), and as such, they should be adopted into study design.

In a survey of UK dog facilities, it was found that the most commonly cited reasons for implementing a training program were the needs of the sponsor or the study design (unpublished data, Refining Dog Care survey, 2015), although improving animal welfare, increasing cooperation with husbandry and procedures, reducing stress, and increasing the efficiency of routine tasks were also important considerations. Barriers to training that may prevent an effective training program from being implemented include, limited staff knowledge and confidence, time pressures, and concerns about the impact of training on scientific outcomes. Managers should account for the need for, and benefits of, a training program in staff workload allocation.

Training is most commonly used for handling, husbandry, and procedures. Poor handling techniques lead to problems for the handler, caused by carrying an unbalanced weight or the risk of the dog jumping away, while also inducing fear and anxiety in the dog. Good handling techniques should incorporate securing the dog so that its weight is supported. Scruffing or other rough handling techniques should never be used, and the dog's body should be kept close to the handler's body to prevent strain on the handler's back. Staff should be given training in correct handling methods and should be observed conducting handling until competency is achieved. The correct methods for handling dogs are depicted in Figure 13.6.

Laboratory dogs experience frequent interactions with staff, during husbandry, restraint, and regulated procedures, all of which have the potential to cause stress. Training should minimize the stress associated with these events, not only to promote positive welfare, but also to avoid unwanted effects of stress on the data obtained from the animals' use. The incorporation of training activities into routine husbandry (e.g., pen removal, health checks) will increase the efficiency of all activities.

Restraint is used for a number of reasons, with the most common being to reduce the ability of dogs to move during procedures and to protect staff from injury. Despite the positive effects of touch on dogs (e.g., Coppola et al., 2006), physical restraint is frequently stressful and can induce a state of distress in dogs (Hall, 2014; Hall et al., 2015). In particular, excessive stress can result when animals are restrained for prolonged periods (Wolfe, 1990; Stokes & Marsman, 2014). Restraint can be replaced with a number of other techniques that afford the same level of safety for dogs and staff, without inducing stress. Dogs can easily be trained to hold a position during procedures, using techniques such as stationing and targeting. Such techniques are trained using positive reinforcement in the prestudy period and can be trained rapidly using shaping and clicker training. The resulting behavior requires reduced physical force from staff, meaning that only minimal restraint is needed, and dogs actively comply, rather than being forced into positions that often result in struggling or escape behavior (Hall et al., 2015).

Physical restraint for longer periods of time may require the use of restraint equipment, for example, slings or tethers for aerosol inhalation delivery (Authier et al., 2009). Dogs can be trained to accept harnesses or slings, using positive reinforcement training and habituation, with voluntary restraint gradually increasing in duration. The resulting trained procedure requires fewer staff to supervise dogs, and also results in reduced stress, offsetting the initial training period needed prior to the beginning of the study.

Dogs can also be trained to comply with more invasive procedures, such as dosing by oral gavage. It is common in industry for dogs to be trained with one or two sham doses, including a "mock" dose in which the dog is gavaged with either no compound or an inert substance. Hall et al. (2015) found that simple sham dosing sensitized, rather than habituated, the dogs to the gavage procedure. Dogs that received four training sessions of

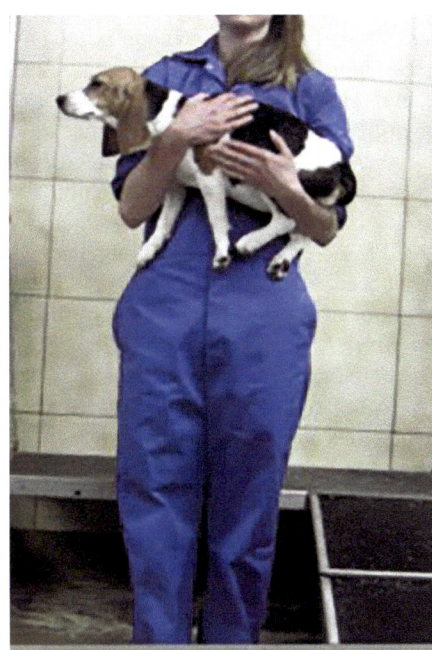 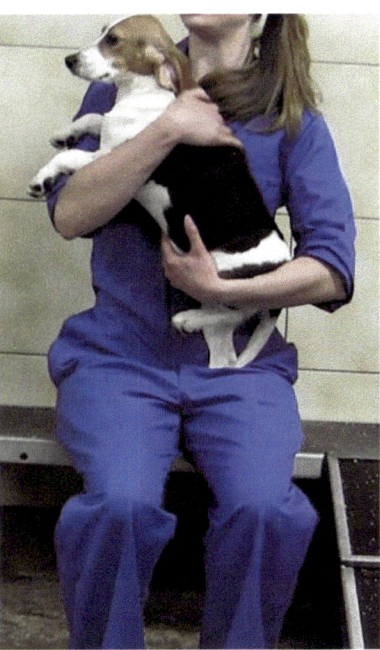 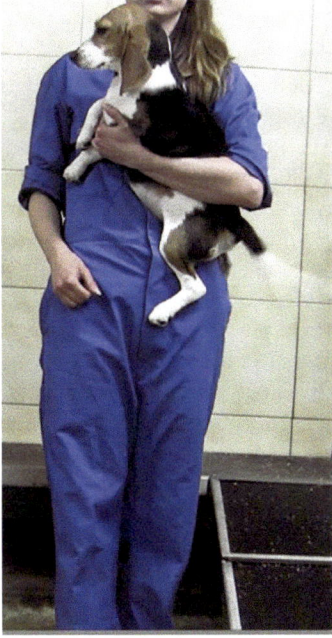

FIGURE 13.6 Figures illustrating the correct handling methods for dogs. Note that the forelimbs and hindquarters of the dog should be gently but securely restrained, and the dog's weight should be held against the handler's body to prevent unnecessary movement. Left: the dog's body is supported by two hands and held securely. Middle: the dog is held across the handler's body and the hindquarters supported. Right: the dog is supported against the handler's body using a single arm.

less than 5 min each plus two modified sham doses (desensitization plus flavored paste) displayed improved welfare in the pen and during procedures, as well as faster dosing, than dogs that received only a standard sham dose.

Staff working with dogs, or responsible for study design, are encouraged to develop evidence-based, positive reinforcement training programs for the range of husbandry and procedure activities involved in their work. More detailed protocols for training can be found at www.refiningdogcare.com.

Special Situations

Single Housing

Single housing, the separation of a dog from its conspecifics for a period of time, occurs for a variety of reasons in the laboratory environment. Temporary separations may occur when dogs are fed individually (e.g., to avoid food competition and ensure adequate intake), for veterinary treatment, or following administration of test substances. Longer periods of separation may be required for specific studies, such as those performed to assess the absorption, distribution, metabolism, and excretion (ADME) properties of new drugs, which require the collection of all urine and feces from dosed animals. Pair or group housing of dogs is the standard in the UK pharmaceutical and contract research industries, and is used within GLP studies, recognizing that barriers to social housing are not sufficient to prevent its use in the regulatory environment. Short periods of temporary separation may be necessary for feeding, monitoring of food consumption, and postdose observations. Directive 2010/63/EU (European Union, 2010) limits these periods to 4 h at a time, though regulatory bodies in EU member states may grant exemptions where justified (e.g., for metabolism studies).

Dogs are highly social, group-living animals that actively seek out the company of conspecifics or humans. Single housing, except for brief periods, is detrimental to their welfare, and the impact of this on study outcomes should be recognized. Singly housed dogs demonstrate a number of abnormal behaviors, including contact seeking, repetitive pacing, circling, jumping, and other stereotypic behaviors (Hetts et al., 1992; Hubrecht et al., 1992; Beerda et al., 1999b; Hall, 2014). The stress associated with single housing is likely to cause abnormal physiology as well (Beerda et al., 1999a), including increases in salivary and plasma cortisol.

Increasingly, requirements for single housing are being challenged, with a number of recent studies demonstrating that there is no detriment to the quality of scientific data from pair-housed dogs in situations that have traditionally involved single housing. For example, modern telemetry equipment transmits on multiple frequencies, enabling dogs to be socially housed during data collection (recording and non-recording days) for safety pharmacology and toxicology studies. In a survey of 39 facilities worldwide, Prior et al. (2016) found that 84% (safety pharmacology) and 94% (toxicology) of respondents pair housed dogs during telemetric recordings. A majority of respondents observed reduced stress and abnormal behaviors with pair housing and noted that the data were comparable, or of higher quality, than data from singly housed dogs. Kendrick et al. (2020) modified metabolism cages to allow pair housing of dogs on ADME studies, and noted the animals appeared calmer and barked less when housed in pairs. Mean recovery rate of radioactivity in collected excreta was comparable for pair- and single-housed dogs, with the data collected from pair-housed dogs within the variability range of single-housed dogs. The authors concluded that this supports pair housing in future ADME studies, as the data produced are suitable for regulatory submission.

The justification for single housing should therefore be assessed on a case-by-case/study-by-study basis, and only permitted when there are compelling scientific, welfare, or veterinary justifications. When single housing is unavoidable, measures must be taken to ameliorate its negative effects (Prescott et al., 2004). Singly housed dogs should be provided with additional human contact (e.g., extra exercise periods, with staff present for interaction and play), as well as a variety of additional enrichment items.

The Use of Nonbeagle Dogs

This chapter focuses on the beagle; however, other breeds of dog are used in research for a variety of reasons, including anatomical and physiological features. Hounds (e.g., greyhounds, mongrel hounds) have deep chests and larger hearts and lungs than beagles, making them popular for cardiovascular and pulmonary research. For the same reasons of size, large breed dogs are also popular in orthopedic research, particularly in the development of prostheses. There may also be a need for aged animals, animals with particular diseases (see Hasiwa et al., 2011), or genetic diversity. Legislation governing the sourcing of these dogs will vary from country to country.

There is a considerable volume of historical data on the purpose-bred beagle, and this, combined with other features, such as specific pathogen free status, results in enhanced research value, as well as higher costs, associated with breeding these dogs. In countries where regulations do not prohibit the use of random-source dogs (which may include surrendered pets, strays, and dogs bred by so-called "backyard breeders"), they are sometimes used because they can be sourced at lower cost. In general, the risks of acquiring and using such dogs outweigh the benefits. In particular, the lack of preparation for adaptation to the laboratory environment is likely to lead to poorer welfare for these dogs. There are regulated breeders of purpose-bred mongrels and hounds, and where scientific justification for the use of these types of dogs can be made, they should be sourced from these breeders, rather than using random-source dogs.

Conclusions/Recommendations

The dog is a gregarious, social carnivore, which in its free-roaming state, occupies a wide variety of environments and forms complex social relationships with conspecifics. The domestic dog has become embedded in human societies around the globe, performing not only a range of working roles but also living with many families as a pet. Despite having complex social and environmental needs, dogs are readily trainable and can adapt to perform a range of behaviors needed in the laboratory, when appropriate rearing and training are provided.

The continuing, extensive use of dogs in scientific research worldwide requires that evidence-based refinements are implemented to improve and maintain dog welfare. We provide the following recommendations:

- The environment should include refinements that reflect the natural history of the dog, including opportunities to forage or scavenge, and to make use of a variety of spaces and environments.
- Dogs should be provided with access to outdoor exercise areas where possible, and exercise outside of the animal room should be provided regularly.
- Dogs should receive basic positive reinforcement training from an early age, increasing in complexity in the prestudy period when procedures are known.
- Regular human contact should be provided, given dogs' affinity for human contact, in addition to, or in preference to, conspecific contact.
- Social housing should be the default, and dogs should only be singly housed when scientific or veterinary justifications can be made. Staff should pay attention to social groupings and ensure dogs are compatible.
- Staff should be trained in welfare monitoring, using a validated tool, and should be familiar with the behaviors that are indicative of positive or negative welfare changes.
- Behavioral management and the 3Rs literature should be continually reviewed. We provide below a table of resources that provide relevant information.

Additional Resources

Table 13.1 provides a list of books and websites that provide relevant and comprehensive information on the care of dogs in laboratories.

TABLE 13.1

Table of Additional Resources

Resource	Description
Refining Dog Care (https://www.refiningdogcare.com)	A website providing resources for those working with dogs in research.
NC3Rs 3Rs resources (https://nc3rs.org.uk/3rs-resources)	A portal providing a range of resources on the application of the 3Rs.
Prescott et al. (2004) *Joint Working Group on Refinement paper on Refining Dog Husbandry and Care*	An expert working group report including recommendations on refinements to practice.
Bradshaw (2011) *In Defense of Dogs: Why Dogs Need Our Understanding*	An insight into dog behavior from a leading expert.
Miklósi (2016) *Dog Behaviour, Evolution and Cognition*	A collation of primary research literature on dog behavior, evolution and cognition.
Pryor (2006) *Don't Shoot the Dog: The New Art of Teaching and Training*	A clear introduction to the theories of positive reinforcement training for dogs.

Appendix: Ethogram

TABLE A1

Behavioral Measures of Positive Welfare in the Home Pen

Behavior	Description	Source
Resting head up	Sitting or lying, not apparently asleep but not orientated towards any stimulus	Beerda et al. (1998); Haverbeke, et al. (2008)
Resting head down	Lying, may be apparently asleep, not orientated towards any stimulus	Beerda et al. (1998); Hubrecht et al. (1992); Spangenberg et al. (2006)
Interact with environment	Sniffing or investigating pen or objects	Haverbeke et al. (2008) Beerda et al. (1998)
Amicable	Lick, play, allogroom dog, often with tail wag	Hubrecht et al. (1992); Spangenberg et al. (2006)
Solicit play	Bow, metaplay	Hubrecht et al. (1992); Hubrecht (1993); Spangenberg et al. (2006)
Play (self)	Usually involving toys or other objects Bouncing gait, play face, wrestle, play chase	
Play (social)		Hubrecht et al. (1992); Hubrecht (1993); Spangenberg et al. (2006)
Calm locomotion	Walk, 4 beat gait and 3 feet on the ground at any one time	Horowitz (2002) Overall (2014)

TABLE A2

Behavioral Measures of Negative Welfare in the Home Pen

Behavior	Description	Source
Stand against walls	Stands on hind legs with forelegs against wall	Beerda et al. (1998); Hubrecht et al. (1992); Hubrecht (1993); Beerda et al. (1999b); Haverbeke et al. (2008); Spangenberg et al. (2006)
Circling	Repetitive movement around pen	Beerda et al. (1998); Hubrecht et al. (1992); Hubrecht (1993); Beerda et al. (1999b); Haverbeke et al. (2008); Spangenberg et al. (2006)
Pace	Repetitive pacing, usually along a boundary	Hubrecht et al. (1992); Hubrecht (1993); Haverbeke et al. (2008)
Social pace	Repetitive pacing, in parallel with a dog on other side of boundary	Hubrecht et al. (1992); Hubrecht (1993)
Sit alert	Dog orientated toward stimulus while in a sitting position	Ley, Bennett, and Coleman (2008)
Stand alert	Dog orientated toward stimulus while in a standing position, usually accompanied by high posture	Ley et al. (2008)
Rapid locomotion	Trot, 2 beat gait, diagonally opposite legs move together	Overall (2014)

TABLE A3

Postural Measures of Welfare in the Home Pen

Behavior	Description	Source
High	Breed specific posture as shown under neutral conditions, with the addition of high tail, head, and ear position	Beerda et al. (1998); Beerda et al. (1999b)
Neutral	Breed specific posture as shown under neutral conditions	Beerda et al. (1998); Overall (2014)
Half-low	Two features from: low position of tail, backward bending of ears, bent legs	Beerda et al. (1998); Beerda et al. (1999b); Haverbeke et al. (2008)
Low	As above, all three features present	Beerda et al. (1998); Beerda et al. (1999b); Haverbeke et al. (2008)
Very low	As above, with body close to ground	Beerda et al. (1998); Haverbeke et al. (2008); Normando et al. (2009)
Tail wag high	Repetitive movements with the tail held high	Beerda et al. (1998); Beerda et al. (1999b); Haverbeke et al. (2008);
Tail wag low	Repetitive movements with the tail held low	Beerda et al. (1998); Beerda et al. (1999b); Haverbeke et al. (2008); Normando et al. (2009)

TABLE A4

Behavioral Events Indicating Negative Welfare in the Home Pen

Behavior	Description	Source
Oral behaviors	Includes tongue out, snout licking, swallowing, lip smacking	Beerda et al. (1998); Beerda et al. (1999b); Haverbeke et al. (2008)
Paw lift	Sudden raising of one limb, usually foreleg, and usually in response to stimulus	Beerda et al. (1998); Beerda et al. (1999b); Haverbeke et al. (2008); Spangenberg et al. (2006); Stephen and Ledger (2005)

TABLE A5

Other Behavioral Measures in the Home Pen

Behavior	Description	Source
Crouch	Bent legs, body lowered towards ground	Beerda et al. (1998)
Tremble	Clear shivering of the body	Beerda et al. (1998)

TABLE A6

Additional Behavioral Measures for Challenges

Behavior	Description	Source
Struggle	Dog attempts to avoid restraint and/or human	McGreevy et al. (2012)
Escape attempts	All occurrences of attempts to leave table/procedure station	

REFERENCES

Authier, S., Legaspi, M., Gauvin, D., & Troncy, E. (2009). Respiratory safety pharmacology: Positive control drug responses in Sprague–Dawley rats, beagle dogs and cynomolgus monkeys. *Regulatory Toxicology and Pharmacology*, 55, 229–235.

Avis, S. P. (1999). Dog pack attack: Hunting humans. *The American Journal of Forensic Medicine and Pathology*, 20, 243–246.

Axelsson, E., Ratnakumar, A., Arendt, M.-L., Maqbool, K., Webster, M. T., Perloski, M., Liberg, O., Arnemo, J. M., Hedhammar, Å., & Lindblad-Toh, K. (2013). The genomic signature of dog domestication reveals adaptation to a starch-rich diet. *Nature*, 495, 360–364.

Bhadra, A., Bhattacharjee, D., Paul, M., Singh, A., Gade, P. R., Shrestha, P., & Bhadra, A. (2016). The meat of the matter: A rule of thumb for scavenging dogs?. *Ethology Ecology & Evolution*, 28(4), 427–440.

Beerda, B., Schilder, M. B., Van Hooff, J. A., De Vries, H. W., & Mol, J. A. (1998). Behavioural, saliva cortisol and heart rate responses to different types of stimuli in dogs. *Applied Animal Behaviour Science*, 58(3–4), 365–381.

Beerda, B., Schilder, M., Bernadina, W., Van Hooff, J., De Vries, H., & Mol, J. (1999a). Chronic stress in dogs subjected to social and spatial restriction. II. Hormonal and immunological responses. *Physiology and Behavior*, 66, 243–254.

Beerda, B., Schilder, M., Van Hooff, J., De Vries, H., & Mol, J. (1999b). Chronic stress in dogs subjected to social and spatial restriction. I. Behavioral responses. *Physiology and Behavior*, 66, 233–242.

Bergamasco, L., Osella, M. C., Savarino, P., Larosa, G., Ozella, L., Manassero, M., Badino, P., Odore, R., Barbero, R., & Re, G. (2010). Heart rate variability and saliva cortisol assessment in shelter dog: Human-animal interaction effects. *Applied Animal Behaviour Science*, 125, 56–68.

Bonanni, R., & Cafazzo, S. (2014). The social organisation of a population of free-ranging dogs in a suburban area of rome: A reassessment of the effects of domestication on dogs' behaviour. In *The Social Dog* (pp. 65–104). Amsterdam: Elsevier.

Bradshaw, J. (2011). *In Defence of Dogs*. UK: Penguin.

Canadian Council on Animal Care (2019). *Animal data report 2018*. URL: https://www.ccac.ca/Documents/AUD/2018-Animal-Data-Report.pdf.

Clark, M., & Steger-Hartmann, T. (2018). A big data approach to the concordance of the toxicity of pharmaceuticals in animals and humans. *Regulatory Toxicology and Pharmacology*, 96, 94–105.

Coppinger, R., & Coppinger, L. (2001). *Dogs: A Startling New Understanding of Canine Origin, Behavior & Evolution*. New York: Simon and Schuster.

Coppola, C. L., Grandin, T., & Enns, R. M. (2006). Human interaction and cortisol: Can human contact reduce stress for shelter dogs? *Physiology & Behavior*, 87, 537–541.

Druzhkova, A. S., Thalmann, O., Trifonov, V. A., Leonard, J. A., Vorobieva, N. V., Ovodov, N. D., Graphodatsky, A. S., & Wayne, R. K. (2013). Ancient DNA analysis affirms the canid from Altai as a primitive dog. *PloS One*, 8, e57754.

Dugatkin, L. A., Trut L., & Trut. L. N. (2017). *How to Tame a Fox (and Build a Dog): Visionary Scientists and a Siberian Tale of Jump-Started Evolution*. Chicago, IL: University of Chicago Press.

England, G.C., & von Heimendahl, A. (2010). *BSAVA Manual of Canine and Feline Reproduction and Neonatology*. Ed. 2. Quedgeley: British Small Animal Veterinary Association.

European Union. (2010). *Directive 2010/63/EU of the European Parliament and of the Council of 22 September 2010 on the protection of animals used for scientific purposes*. Brussels, Luxembourg: Official Journal of the European Union, L 276/33.

European Commission (2020). *Report from the Commission to the European Parliament and the Council. 2019 report on the statistics on the use of animals for scientific purposes in the member states of the European Union in 2015–2017*, URL: https://eur-lex.europa.eu/legal-content/EN/TXT/?uri=CELEX:52013SC0497.

Ewart, L., Aylott, M., Deurinck, M., Engwall, M., Gallacher, D. J., Geys, H., Jarvis, P., Ju, H., Leishman, D., Leong, L., et al. (2014). The concordance between nonclinical and Phase I clinical cardiovascular assessment from a cross-company data sharing initiative. *Toxicological Sciences*, 142, 427–435.

Fogel, B. (1990). *The Dog's Mind: Understanding Your Dog's Behavior*. New York: Macmillan.

Fox, M. (1971). *Behaviour of Wolves Dogs and Related Canids*. Wenatchee, WA: Dogwise Publishing.

Fox, M., & Stelzner, D. (1966). Behavioural effects of differential early experience in the dog. *Animal Behaviour*, 14, 273–281.

Freedman, A. H., Gronau, I., Schweizer, R. M., Ortega-Del Vecchyo, D., Han, E., Silva, P. M., Galaverni, M., Fan, Z., Marx, P., Lorente-Galdos, B., Beale, H., Ramirez, O., Hormozdiari, F., Alka, C., Vilá, C., Squire, K., Geffen, E., Kusak, J., Boyko, A. R., Parker, H. G., Lee, C., Tadigotla, V., Siepel, A., Bustamante, C. D., Harkins, T. T., Nelson, S. F., Ostrander, E. A., Marques-Bonet, T., Wayne, R. K., & Novembre, J. (2014). Genome sequencing highlights the dynamic early history of dogs. *PLoS Genetics*, 10, e1004016.

Gärtner, K., Baumans, V., Brian, P., Hackbarth, H., Militzer, K., Morton, D., Nebendahl, K., Netto, J., Poole, T., & Whittaker, D. (1994). Dogs. In *The Accommodation of Laboratory Animals in Accordance with Animal Welfare Requirements: Proceedings of an International Workshop held at Bundesgesundheitsamt, Berlin*. Bonn: Bundesministerium für Ernährung, Landwirtschaft und Forsten.

Gazzano, A., Mariti, C., Notari, L., Sighieri, C., & McBride, E.A. (2008). Effects of early gentling and early environment on emotional development of puppies. *Applied Animal Behaviour Science*, 110, 294–304.

Hall, L. E. (2014). *A practical framework for harmonising welfare and quality of data output in the laboratory-housed dog*. Unpublished PhD thesis: University of Stirling.

Hall, L. E., Robinson, S., & Buchanan-Smith, H. M. (2015). Refining dosing by oral gavage in the dog: A protocol to harmonise welfare. *Journal of Pharmacological and Toxicological Methods*, 72, 35–46.

Hare, B., & Tomasello, M. (1999). Domestic dogs (*Canis familiaris*) use human and conspecific social cues to locate hidden food. *Journal of Comparative Psychology*, 113(2), 173.

Hasiwa, N., Bailey, J., Clausing, P., Daneshian, M., Eileraas, M., Farkas, S., Gyertyan´, I., Hubrecht, R., Kobel, W., Krummenacher, G., Leist, M., Lohi, H., Miklósi, A., Ohl, F., Olejniczak, K., Schmitt, G., Sinnett-Smith, P., Smith, D., Wagner, K., Yager, J., Zurlo, J., & Hartung, T. (2011). Critical evaluation of the use of dogs in biomedical research and testing in europe. *ALTEX*, 28, 326–340.

Haverbeke, A., Laporte, B., Depiereux, E., Giffroy, J. M., & Diederich, C. (2008). Training methods of military dog handlers and their effects on the team's performances. *Applied Animal Behaviour Science*, 113(1–3), 110–122.

Hetts, S., Derrell Clark, J., Calpin, J. P., Arnold, C. E., & Mateo, J. M. (1992). Influence of housing conditions on beagle behaviour. *Applied Animal Behaviour Science*, 34, 137–155.

Home Office (1995). *Code of practice for the housing and care of animals used in scientific procedures*. Her Majesty's Stationary Office.

Horowitz, A. C. (2002). *The behaviors of theories of mind, and a case study of dogs at play*. Ph.D. thesis University of California, San Diego.

Huber, L., & Lamm, C. (2017). Understanding dog cognition by functional magnetic resonance imaging. *Learning & Behavior*, 45, 101–102.

Hubrecht, R. (1993). A comparison of social and environmental enrichment methods for laboratory housed dogs. *Applied Animal Behaviour Science*, 37, 345–361.

Hubrecht, R. (1995). Enrichment in puppyhood and its effects on later behaviour of dogs. *Laboratory Animal Science*, 45, 70–75.

Hubrecht, R. (2002). Comfortable quarters for dogs in research institutions. *Comfortable quarters for laboratory animals*, (pp. 56–64).

Hubrecht, R., & Kirkwood, J. (2010). Introduction. In R. Hubrecht, & J. Kirkwood (Eds.), *The UFAW Handbook on the Care and Management of Laboratory and Other Research Animals: Eighth Edition* (pp. 206–218). Wheathampstead, UK: Universities Federation for Animal Welfare.

Hubrecht, R., Serpell, J., & Poole, T. (1992). Correlates of pen size and housing conditions on the behaviour of kennelled dogs. *Applied Animal Behaviour Science*, 34, 365–383.

ICH (2020). Safety guidelines. International Council for Harmonisation of Technical Requirements for Pharmaceuticals for Human Use. Available at: www.ich.org/page/safety-guidelines [accessed 22 March 2020].

Jones, A. (2007). Sensory development in puppies (*Canis lupus f. familiaris*): Implications for improving canine welfare. *Animal Welfare*, 16 (3), 319–329.

Kendrick, J., Stow, R., Ibbotson, N., Adjin-Tettey, G., Murphy, B., Bailey, G., Miller, J., Helleberg, H., Finderup Grove, M., Øvlisen, K. et al. (2020). A novel welfare and scientific approach to conducting dog metabolism studies allowing dogs to be pair housed. *Laboratory Animals*, (p. 0023677220905330). doi:10.1177/0023677220905330.

Kraeuter, K. (2001). *Training your beagle*. U.S.A.: Barron's Educational Series.

Laule, G. (2010). Positive reinforcement training for laboratory animals. In R. Hubrecht, & J. Kirkwood (Eds.), *The UFAW Handbook on the Care and Management of Laboratory and Other Research Animals: Eighth Edition* (pp. 206–218). Wheathampstead, UK: Universities Federation for Animal Welfare.

Ley, J., Bennett, P., & Coleman, G. (2008). Personality dimensions that emerge in companion canines. *Applied Animal Behaviour Science*, 110, 305–317.

McGreevy, J., Hakim, C., McIntosh, M., & Duan, D. (2015). Animal models of Duchenne muscular dystrophy: From basic mechanisms to gene therapy. *Disease Models & Mechanisms*, 8, 195–213.

McGreevy, P. D., Starling, M., Branson, N. J., Cobb, M. L., & Calnon, D. (2012). An overview of the dog–human dyad and ethograms within it. *Journal of Veterinary Behavior*, 7(2), 103–117.

Mech, L. D. (1999). Alpha status, dominance, and division of labor in wolf packs. *Canadian Journal of Zoology*, 77, 1196–1203.

Mech, L. D. (2000). Leadership in wolf, *canis lupus*, packs. *Canadian Field Naturalist*, 114, 259–263.

Mech, L. D., & Boitani, L. (2003) Wolf social ecology. In: Mech L. D., Boitani L. (eds) *Wolves: Behaviour, Ecology, and Conservation*. Chicago: University of Chicago Press, pp. 1–34.

Mertens, P. A., & Unshelm, J. (1996). Effects of group and individual housing on the behavior of kennelled dogs in animal shelters. *Anthrozoos*, 9, 40–51.

Meunier, L. D. (2006). Selection, acclimation, training, and preparation of dogs for the research setting. *ILAR Journal*, 47, 326–347.

Meunier, L. D., & Beaver, B. V. (2014). Dog and cat welfare in a research environment. In K. Bayne, & P. V. Turner (Eds.) *Laboratory Animal Welfare* (pp. 213–231). Amsterdam: Elsevier.

Miklósi, Á. (2016). *Dog Behaviour, Evolution, and Cognition*. Oxford: OUP Oxford.

Monticello, T.M., Jones, T.W., Dambach, D.M., Potter, D.M., Bolt, M.W., Liu, M., Keller, D.A., Hart, T.K., & Kadambi, V.J. (2017). Current nonclinical testing paradigm enables safe entry to First-In-Human clinical trials: the IQ consortium nonclinical to clinical translational database. *Toxicology and Applied Pharmacology*, 334, 100–109.

Mugford, R., & Thorne, C. (1978). Comparative studies of meal patterns in pet and laboratory housed dogs and cats. In *Nutrition of the Dog and Cat: Proceedings of the International Symposium on the Nutrition of the Dog and Cat, Arranged by the Institute of Animal Nutrition in Conjunction with the 200-year Anniversary of the Veterinary School, Hannover, 26 June 1978/editor, RS Anderson*. Oxford; New York: Pergamon Press.

National Research Council (2009). *Scientific and Humane Issues in the Use of Random Source Dogs and Cats in Research*. Washington, DC: National Academies Press (US).

National Research Council (2011). *Guide for the Care and Use of Laboratory Animals*. Washington DC, National Academies Press.

Neitz, J., Geist, T., & Jacobs, G. H. (1989). Color vision in the dog. *Visual Neuroscience*, 3, 119–125.

Normando, S., Corain, L., Salvadoretti, M., Meers, L., & Valsecchi, P. (2009). Effects of an enhanced human interaction program on shelter dogs' behaviour analysed using a novel nonparametric test. *Applied Animal Behaviour Science*, 116(2–4), 211–219.

Odendaal, J., & Meintjes, R. (2003). Neurophysiological correlates of affiliative behaviour between humans and dogs. *The Veterinary Journal*, 165, 296–301.

OECD (2019). *OECD test guidelines programme. Organisation for Economic Cooperation and Development*, https://www.oecd.org/chemicalsafety/testing/oecd-guidelines-testing-chemicals-related-documents.htm.

Olson, H., Betton, G., Robinson, D., Thomas, K., Monro, A., Kolaja, G., Lilly, P., Sanders, J., Sipes, G., Bracken, W., et al. (2000). Concordance of the toxicity of pharmaceuticals in humans and in animals. *Regulatory Toxicology and Pharmacology, 32*, 56–67.

Ong, T.-H., Mendum, T., Geurtsen, G., Kelley, J., Ostrinskaya, A., & Kunz, R. (2017). Use of mass spectrometric vapor analysis to improve canine explosive detection efficiency. *Analytical Chemistry, 89*, 6482–6490.

Overall, K. L. (2014). The ethogram project. *Journal of Veterinary Behavior: Clinical Applications and Research*, 1(9), 1–5.

Pal, S., Ghosh, B., & Roy, S. (1999). Inter-and intra-sexual behaviour of free-ranging dogs (*Canis familiaris*). *Applied Animal Behaviour Science, 62*, 267–278.

Pal, S. K. (2001). Population ecology of free-ranging urban dogs in West Bengal, India. *Acta Theriologica, 46*, 69–78.

Pal, S. K. (2011). Mating system of free-ranging dogs (*Canis familiaris*). *International Journal of Zoology, 2011*, 314216.

Pavlov, I. P. (1928). *Lectures on conditioned reflexes. (Translated by W.H. Gantt)*. London, U.K: Allen and Unwin.

Prescott, M., Morton, D. B., Anderson, D., Buckwell, A., Heath, S., Hubrecht, R., Jennings, M., Robb, D., Ruane, B., Swallow, J., & Thompson, P. (2004). Refining dog husbandry and care: Eighth report of the BVAAWF/FRAME/RSPCA/UFAW joint working group on refinement. *Laboratory Animals, 38*, S1:1–S1:94.

Prior, H., Bottomley, A., Champe´roux, P., Cordes, J., Delpy, E., Dybdal, N., Edmunds, N., Engwall, M., Foley, M., & Hoffmann, M. (2016). Social housing of non-rodents during cardiovascular recordings in safety pharmacology and toxicology studies. *Journal of Pharmacological and Toxicological Methods, 81*, 75–87.

Pryor, K. (2006). *Don't Shoot the Dog: The New Art of Teaching and Training*. New York: Simon & Schuster.

Reid, P. J. (2009). Adapting to the human world: Dogs' responsiveness to our social cues. *Behavioural Processes, 80*, 325–333.

Russell, W.M.S., & Burch, R.L. (1959). *The Principles of Humane Experimental Technique*. Wheathampstead, U.K.: Universities Federation for Animal Welfare.

Sablin, M., & Khlopachev, G. (2002). The earliest ice age dogs: Evidence from eliseevichi. *Current Anthropology, 43*, 795–799.

Sales, G., Hubrecht, R., Peyvandi, A., Milligan, S., & Shield, B. (1997). Noise in dog kennelling: Is barking a welfare problem for dogs? *Applied Animal Behaviour Science, 52*, 321–329.

Schipper, L. L., Vinke, C. M., Schilder, M. B., & Spruijt, B. M. (2008). The effect of feeding enrichment toys on the behaviour of kennelled dogs (*Canis familiaris*). *Applied Animal Behaviour Science, 114*, 182–195.

Scullion Hall, L., & Robinson, S. (2016). Implementing a successful positive reinforcement training protocol in laboratory-housed dogs. *Animal Technology and Welfare, 15(2)*, 83–88.

Scullion Hall, L. E., Robinson, S., Finch, J., & Buchanan-Smith, H. M. (2017). The influence of facility and home pen design on the welfare of the laboratory-housed dog. *Journal of Pharmacological and Toxicological Methods, 83*, 21–29.

Seligman, M.E.P. (1972). Learned helplessness. *Annual Review of Medicine*, 23, 407–412.

Sewell, F., Edwards, J., Prior, H., & Robinson, S. (2017). Opportunities to apply the 3rs in safety assessment programs. *ILAR Journal, 57*, 234–245.

Shearin, A. L., & Ostrander, E. A. (2010). Leading the way: Canine models of genomics and disease. *Disease Models & Mechanisms, 3*, 27–34.

Sherwin, C. (2007). Animal welfare: reporting details is good science. *Nature, 448*, 251.

Smith, F.O. (2012). Guide to emergency interception during parturition in the dog and cat. *Veterinary Clinics: Small Animal Practice, 42*, 489–499.

Spangenberg, E. M. F., Björklund, L., & Dahlborn, K. (2006). Outdoor housing of laboratory dogs: Effects on activity, behaviour and physiology. *Applied Animal Behaviour Science*, 98(3–4), 260–276.

Spielmann, H. (2019). Progress in eliminating one-year dog studies for the safety assessment of pesticides. In H. Kojima, T. Seidle, & H. Spielmann (Eds.) *Alternatives to Animal Testing* (pp. 50–56). Singapore: Springer.

Starkey, M. P., Scase, T. J., Mellersh, C. S., & Murphy, S. (2005). Dogs really are man's best friend—canine genomics has applications in veterinary and human medicine! *Briefings in Functional Genomics, 4*, 112–128.

Stephen, J. & Ledger, R. (2005). An audit of behavioral indicators of poor welfare in kennelled dogs in the United Kingdom. *Journal of Applied Animal Welfare Science, 8(2)*, 79–96.

Stokes, W. S., & Marsman, D. S. (2014). Animal welfare considerations in biomedical research and testing. In K. Bayne, & P. V. Turner (Eds.) *Laboratory Animal Welfare* (pp. 115–140). Amsterdam: Elsevier.

Tarou, L. R., & Bashaw, M. J. (2007). Maximizing the effectiveness of environmental enrichment: Suggestions from the experimental analysis of behavior. *Applied Animal Behaviour Science, 102*, 189–204.

USDA, APHIS (2018). *Annual Report Animal Usage by Fiscal Year -2017. United States Department of Agriculture*.

Wells, D. (2004a). The influence of toys on the behaviour and welfare of kennelled dogs. *Animal Welfare, 13*, 367–373.

Wells, D. (2004b). A review of environmental enrichment for kennelled dogs, canis familiaris. *Applied Animal Behaviour Science, 85*, 307–317.

Wells, D. (2009). Sensory stimulation as environmental enrichment for captive animals: a review. *Applied Animal Behaviour Science, 118*, 1–11.

Wolfe, T. (1990). *Policy, Program and People: The three P's to Well-Being*. Washington, DC: National Academy of Sciences.

Young, M.C. (1994). *The Guinness Book of Records 1995*. New York: Facts on File.

Young, J. K., Olson, K. A., Reading, R. P., Amgalanbaatar, S., & Berger, J. (2011). Is wildlife going to the dogs? Impacts of feral and free-roaming dogs on wildlife populations. *BioScience, 61*, 125–132. doi:10.1525/bio.2011.61.2.7.

Zurlo, J., Bayne, K., Cimino Brown, D., Burkholder, T., Dellarco, V., Ellis, A., Garrett, L., Hubrecht, R., Janus, E., Kinter, R., Luddy, E., Meunier, L., Scorpio, D., Serpell, J., & Todhunter, R. (2011). Critical evaluation of the use of dogs in biomedical research and testing. *ALTEX, 4*, 355.

14

Behavioral Biology of the Domestic Cat

Judith Stella
Good Dog, Inc.

CONTENTS

Introduction 223
Typical Research Involvement 223
Behavioral Biology 224
    Natural History 224
    Social Organization and Behavior 225
    Mating Behavior 226
    Feeding Behavior 227
    Communication 228
Common Captive Behaviors 229
    Normal Behavior 229
    Stress-Related and Sickness Behavior 229
    Ways to Maintain Behavioral Health 230
    Environment 231
    Social Groupings 233
    Feeding Strategies 235
    Training 235
    Human–Cat Interactions 235
Conclusions 235
References 236

Introduction

The process of domesticating the cat likely began 12,000 years ago in the Fertile Crescent with the onset of human agriculture. Once revered in ancient Egypt, cats were persecuted during the European Middle Ages, and not until the 19th century in Europe were cats admired again by artisans and intellectuals for their independence, presaging their widespread adoption into society as fashionable middle-class pets. In the United States, there are an estimated 94 million cats living in homes (APPA 2017). Yet, despite their popularity, attitudes toward cats are often ambivalent. They are mysterious and appear to be half wild-half domestic, independent wanderers. This attitude is reflected in the estimated 30–40 million free-roaming cats in the United States (HSUS 2018). Although cats are the most popular companion animal in the US and in the world, they do not meet all the criteria for complete domestication. Yet, the cat has been a successful animal model for many decades. In 2018, approximately 18,000 cats participated in biomedical research (USDA 2018). Due to their relatively small size, longevity, and naturally occurring diseases/conditions that have correlates to human diseases/conditions, they have been an ideal model. Cats have been instrumental to our understanding of the stress response system (SRS), development of the visual system, the aging process, and various diseases, including interstitial cystitis (IC) and HIV/AIDS.

However, cats face many challenges to their welfare when confined in laboratories. Environmental factors, such as ambient temperature and lighting, noise, quality and quantity of space provided, and aspects of the social environment, can affect cat behavior and physiology, and ultimately research outcomes. This chapter will describe cat behavior, including the natural ecology of free-roaming populations, known or presumed environmental factors that cause distress, and methods and resources for alleviating distress in the laboratory housing environment.

Typical Research Involvement

The domestic cat has been involved in biomedical research to promote both human and feline health for decades. They have been an important animal model for the study of neurological, cardiovascular, and respiratory diseases; the immune system; and for understanding the function of the neuron, the visual system, and the organization of the brain. The eminent physiologist Walter B. Cannon's description of the "fight or flight" response resulted from studies of cats conducted during the first two decades of the 20th century (Cannon 1914, 1927, 1942). In 1981, David Hubel and Thorsten Wiesel were awarded the Nobel Prize in Physiology or Medicine "for their discoveries concerning information processing in the visual system". Using kittens, they discovered that visual neurons need to be stimulated by light for the eyes, optic nerve, and visual centers of the brain to properly develop (The Nobel Prize 1981).

The genetic diversity of cats and the ability to do whole genome sequencing have led to the identification of mutations resulting in diseases in cats (Lyons 2010, 2015), as well as over 200 hereditary diseases with closely correlated conditions in humans (O'Brien et al. 2008). This has led to the participation of cats in comparative research aimed at advancing human and veterinary medicine (Gurda et al. 2017). Cats have been genetically based models in the study of diseases, including hypertrophic cardiomyopathy (e.g., Fries et al. 2008; Godiksen et al. 2011; Meurs et al. 2005) and the metabolic disorder, mucopolysaccharidosis (Sleeper et al. 2008; Yogalingam et al. 1996). Like humans, cats are prone to obesity and accompanying type 2 diabetes mellitus (DM), which are increasing in prevalence in both species (Chandler et al. 2017; Hoenig 2006; Tarkosova et al. 2016). Many clinical, physiological, and pathological similarities between human type 2 diabetes and feline DM have been identified, including age of onset, association with obesity, declining insulin secretion, development of islet amyloid deposits, loss of β-cell mass, and nerve pathology (Henson and O'Brien 2006).

Cats are also a good model for research into human aging, having comparatively long lives of up to 20 years, compared to the 2–3 year lifespans for rodents. Some elderly cats exhibit symptoms like those seen in people with Alzheimer's disease, including confusion and difficulty recognizing familiar people. Research has found that cats and humans develop similar pathologies, including amyloid β (Aβ) accumulation, neurofibrillary tangle formation, and neuronal loss, making them a naturally occurring model for the disease (Chambers et al. 2015).

Clinical signs of lower urinary tract (LUT) disease or dysfunction are common in cats, reportedly occurring in approximately 1.5% of owned cats (Lund et al. 1999). IC, a similar syndrome in humans, is characterized by pain, inflammation, and increased urgency and frequency to urinate, and affects up to 1 in 4.5 women and some men (Hanno et al. 2010). Feline idiopathic cystitis (FIC) is the term that has been used to describe the naturally occurring condition in cats (Izgi and Daneshgari 2013), characterized by chronic or recurrent LUTS, along with characteristic cystoscopic findings of the bladder (Buffington et al. 1999). However, evidence suggests that IC/FIC is more complex, with the LUT symptoms presenting as a result of interactions among the bladder, nervous system, adrenal glands, husbandry practices, and the environment (Buffington 2011; Buffington et al. 2014). This discovery has led to innovative treatments, focused on minimizing stress, to alleviate symptoms and promote well-being. These include environmental management and decreasing the perception of threat from environmental factors in cats (Stella et al. 2011), and cognitive behavioral therapy or psychotherapy in humans (Atchley et al. 2015).

Lastly, cats have been critical in the study of HIV and AIDS. Feline immunodeficiency virus (FIV) and HIV are both lentiviruses that cause suppression of the immune system, leading to similar disease progression, including neuropathology (Fletcher et al. 2011), and ultimately death in infected individuals (Elder et al. 2010). Cats are the smallest animal model for study of naturally occurring lentivirus, and since FIV does not infect humans, FIV research is safer than research with the SIV-infected NHP model (e.g., Peeters et al. 2002). Research involving FIV-infected cats has focused on vaccine development (e.g., Yamamoto et al. 2010) and the testing of antiviral drugs (e.g., Elder et al. 2008), with the goal of preventing both FIV and HIV.

Despite the long history of the cat model in biomedical research, in the United States, there is currently a trend toward decreasing use of cats. From 2013 to 2017, there was a 30% decrease in the number of cats involved in biomedical research, from 24,221 to 18,146 (Grimm 2018; USDA 2018). One likely factor for this decline is public perceptions of animal participation in biomedical research in general, and specifically, because like dogs, cats are increasingly viewed and valued as companion animals (Ormandy and Schuppli 2014).

Behavioral Biology

Natural History

The modern domestic cat (*Felis silvestris catus*) is the product of at least 12,000 years of natural selection (Driscoll et al. 2009). The oldest known remains of a cat found buried in association with a human are from the site of Shillourokambos in Cyprus, dated approximately 7500 BCE (before the common era) (Vigne et al. 2004). Since that time, cats have successfully "colonized" every continent except Antarctica, with an estimated global population of approximately 272 million, of which 58% are free-roaming or feral (Hiby et al. 2014).

Although cats are increasingly kept as pets and confined to the indoors in many parts of the world, they have been described as "exploited captives" (Clutton-Brock 1999) that are not yet truly domesticated. The process of domestication begins when animals are incorporated into human social groups or communities, and become objects of ownership, inheritance, and purchase, and ends when the population's breeding, organization of territory, and food supply come under permanent human control, isolating them from the wild progenitor species (Clutton-Brock 1999). Wildcats (*Felis silvestris libyca*) are improbable candidates for domestication; they have specialized diets (obligate carnivores), a solitary social system, and defend exclusive territories, making them more attached to places than people (Driscoll et al. 2007, 2009). There is some evidence that early civilizations sought out wildcats to tame as pets, but the more likely explanation for domestication is that the cats exploited the anthropogenic environment and were tolerated by people, which eventually led to divergence from the wild form. Rodents likely were the first commensal species and became a reliable food source for native wildcats. Wildcats that adapted to this "urban" environment then became a human commensal themselves. This explanation is supported by phylogenetic and phylogeographic evidence that the speciation of the domestic cat from the wildcat occurred simultaneously alongside human civilizations (Driscoll et al. 2007).

So, is the cat truly domesticated? The criteria for complete domestication, a dependence on humans for food, shelter, and control of breeding, are not satisfied in all cats (Driscoll et al. 2009). Over 97% of the nearly 272 million cats are self-bred, thus the overwhelming majority choose their own mate. Most feral and free-roaming cats obtain their food without human assistance and the domestic cat varies morphologically very little from the wildcat (Driscoll et al. 2011). Moreover, domestic cats are not permanently isolated from their wild ancestral species (Cameron-Beaumont et al. 2002; Driscoll et al. 2007). In areas inhabited by both wild and domestic cat species, there is considerable interbreeding. But domestic cats are polyestrus, having multiple litters each year, whereas wildcats, although polyestrus in some climates, usually have only one litter per year (Kitchener 1991). The coat colors of domestic cats are often very different than the mackerel tabby of the wildcat. And most notably, domestic cats are tolerant of humans, a key attribute of domestication. Perhaps a more accurate description is symbiotic carnivores, being "domestic" in the sense of being a "friendly symbiont" of humans, as proposed by Downs (1960).

Understanding domestication is essential to the study of behavior and welfare of animals in captivity. An animal's behavioral organization is shaped by evolution to use information obtained from the environment to react to an event or to interact with an environmental feature, in order to form rules of response for similar events or stimuli. The extent to which these "decision rules" of the ancestral species can become altered by domestication may influence the negative subjective experiences (suffering) of an animal, especially when there is a mismatch between the current environment and the environment in which the decision rules evolved (Cameron-Beaumont et al. 2002; Inglis 2000). When confined, the limited space, proximity of conspecifics and other predator (e.g., humans and dogs) and prey (e.g., rodents, rabbits) animals, combined with limited resources and opportunities to express species-typical behavior, potentially influence perceptions of control and threat, and ultimately, the animals' welfare.

Social Organization and Behavior

The social behavior of domestic cats exhibits great plasticity, both individually and as a species. It appears to be influenced by ontogeny, such that kittens socialized to other cats, humans, dogs, etc., during the sensitive period of socialization, are likely to adapt to life in social groups more readily than kittens raised by their mother alone (Mendl and Harcourt 2000). Cats are unique in that the wild progenitor species is solitary, while the domesticate is relatively social. In free-roaming populations, territory size and density are influenced by the availability of resources, with food and shelter most often the limiting resources for females, and with females the limiting resource for males. The Resource Dispersion Hypothesis proposes that the size of the territory is related to the concentration of resources, so that the smallest territory able to maintain the primary social unit (mother and offspring) may be able to maintain additional cats if resources are abundant (MacDonald et al. 2000). Therefore, when cats have access to a clumped food resource, they tend to live in groups, whereas those living on wild prey, such as rodents, will tend to be solitary (MacDonald et al. 2000). Population density can range from 100 cats/km^2 in areas where food is clumped, such as garbage dumps or where provisioned by humans, to less than 5 cats/km^2 in areas where prey is scarce or dispersed (Liberg et al. 2000; Turner 2014). Territories for free-roaming females can range from 0.27 ha in a city to 170 ha in the bush of Australia, while the home range of males average between 0.72 and 990 ha, roughly three times larger than that of females (Liberg et al. 2000). Domestic cats should be considered a relatively solitary species, often choosing population densities of less than 50 cats/km^2 (Liberg et al. 2000). Even lions, arguably the most social cat species, have a fission-fusion society, where females without young are more often found alone than with the pride (Mosser and Packer 2009).

The backbone of cat society consists of groups of related adult females and several generations of their offspring, called lineages. Each lineage will maintain a primary or home area, and within that primary area, individuals will maintain overlapping home ranges. Members of the lineage will defend their territory and the resources within from unrelated females. A large colony near a rich food source typically incorporates several lineages, with the largest occupying the central or core area closest to the resources, and the smaller lineages living on the periphery (Kerby 1987). The resources are time-shared to avoid confrontations. Females will exhibit social preferences, choosing to spend more time near, and engage in more affiliative interactions with, favored individuals. These associations are influenced by age, sex, and relatedness, so that closer relationships are developed with members of their own lineage (MacDonald et al. 2000). Adult males are more solitary, not attached to any lineage, and tend to live on the margins of a group or several groups of females. Adolescent males, and some females, disperse from the natal group when they reach 2–3 years of age (Kerby and MacDonald 1988; Turner 2014). Females that disperse likely establish new lineages and territories with their offspring, provided adequate resources are available.

One of the cat's unique characteristics is that almost all owned populations are sympatric with and interbreed freely with feral populations, with mate choice decided by the cat. The cats' relationships with humans exist on a continuum, from a fully dependent pet, to commensal, to complete independence, with movement from one end of the spectrum to the other occurring within a few generations. The fluidity of the cat population is well recognized by cat guardians. In studies asking people how they acquired their cats, 34%–44% of respondents reported that they had found their cat as a stray or as an orphaned kitten (Chu et al. 2009; Stella and Croney 2016a). Explanations for the maintenance of feral populations have historically been that the small number of generations

FIGURE 14.1 Normal mating behavior of domestic cats. The tom is atop the queen who is exhibiting lordosis behavior.

since domestication has been too few for complete domestication. This would have favored and maintained the wild characteristics of feral cats. Alternatively, cats have, until recently, been maintained as both a pet and for rodent control, obviating the need for selection pressure to produce two distinct populations. Recently, a third possible explanation has been proposed. The cat is an obligate carnivore and this nutritional characteristic may have allowed for selection of both the ability to hunt and to scavenge to meet nutritional needs. Until recently, humans did not have the knowledge or ability to meet the nutritional requirements of cats, so they maintained the ability to feed themselves (Bradshaw et al. 1999).

Mating Behavior

Domestic cats will reach puberty and be sexually mature between 6 and 10 months of age. Female estrous behavior is easily recognizable; she will perform body rolls, creep, belly-crawl, and/or display lordosis behavior, characterized by elevating the hind end, pressing the chest to the ground, arching the back, and raising or side placement of the tail, with accompanying vocalizations that can be described as howling or yelling. Cats have induced ovulation, and estrus will typically end within 24 h of mating. If she is not bred, estrus can last up to 10 days and will repeat every 18–24 days. In the presence of a male, the female will display estrous behavior in front of the tom while he approaches her, until she allows him to mount (see Figure 14.1). After ejaculation, withdrawal of his barbed penis causes the female to experience pain, so she will turn to bite and scratch at his face and head. The female will repeatedly mate (15–20 times per 24 hrs), and several males may be involved, such that a single litter of kittens can be of mixed paternity (Leyhausen 1979; Liberg 1983). In time, she loses her sex drive and ovulation will occur 24 h later, allowing the sperm to mature and become fertile in her oviduct (Fraser 2012).

Gestation in domestic cats is about 63 days (Hemmer 1979), approximately 3–7 days longer than in the wildcat (*Felis silvestris libyca*) (Haltenorth and Diller 1980). Related females will often crèche their litters and alloparent the kittens (MacDonald et al. 1987). In the first 2 weeks of life, the neonatal kitten's sensorium is predominated by tactile, thermal, and olfactory stimuli. Behavior consists mainly of suckling, sleeping, and soliciting maternal care, with learning limited to simple associations with touch, taste, or smell (Bateson 2014). Kittens that are separated at a very young age (<2 weeks) from their mother and littermates, and hand-reared by humans, exhibit developmental, behavioral, emotional, and physical abnormalities. As adults, they are fearful and aggressive toward unfamiliar cats and people, exhibit large amounts of abnormal locomotor activity, and have cognitive deficits (Bateson 2014; Seitz 1959). Table 14.1 summarizes milestones in early kitten development.

TABLE 14.1

Milestones in Kitten Development (as Reviewed in Bateson 2014, Original References Provided)

Age	Milestone	Comment	Reference
0–2 weeks	Suckling, sleeping, solicitation of maternal care	Sensorium predominated by tactile, thermal and olfactory stimuli	Bateson (2014)
0–3 weeks	Entirely dependent on milk for nutrition	Queen returns to nest frequently to feed kittens; nursing bouts initiated by queen	Deag et al. (2000); Martin (1986)
	Can orient to the nest	Mainly use olfactory cues but thermal too	Luschekin and Shuleikina (1989)
5 days	Respond to sound		Olmstead and Villablanca 1980
7–10 days	Eyes open	Large individual variability in age of eye-opening ranging from 2 to 16 days	Olmstead and Villablanca (1980); Braastad and Heggelund (1984)
2 weeks	Teeth begin to erupt	Continues until 5 weeks	Hemmer (1979)
3 weeks	Walking begins		Moelk (1979)
	Thermoregulation begins		Jenson et al. (1980)
4 weeks	Able to use visual cues	Will locate and follow mother, visually orient to the nest	Rosenblatt (1976); Thorn et al. (1976)
	Adult-like orienting response to sounds		Olmstead and Villablanca (1980)
	Weaning begins	Kittens start to consume solid food; Kittens increasingly initiate nursing bouts; Queen begins to provide live prey	Baerends-van Roon and Baerends (1979); Martin and Bateson (1985)
	Social play involving chasing begins	Continues at high levels until 12–14 weeks of age	West (1974); Barrett and Bateson (1978); Caro (1981)
5 weeks	Hunting	Some kittens can already kill mice	Baerends-van Roon and Baerends (1979)
	Running begins		Moelk (1979)
5–6 weeks	Voluntary elimination developed	Kittens no longer need maternal stimulation to urinate or defecate. Will begin to use litter pan.	Fox (1970)
	Crouch when moving toward another kitten		Dumas and Dore (1991)
	Begin to search for an object that has moved out of sight		Dumas and Dore (1991)
6 weeks	Air-righting reaction fully developed	Air-righting is the ability to right the body mid-air while falling	Martin (1982)
7 weeks	Weaning typically completed	Intermittent suckling may continue for several months, especially if the kitten is a singleton.	Martin (1986); Bateson (1994); Leyhausen (1979)
	All adult gaits exhibited		Moelk (1979)
	Full thermoregulation attained		Olmstead et al. (1979)
6–8 weeks	Begin to show adult-like responses to threatening social stimuli	Visual and olfactory stimuli	Kolb and Nonneman (1975)
7–8 weeks	Adult-like sleep patterns have developed		McGinty et al. (1977)
	Object play markedly increases	Enabled by concurrent development of eye-paw coordination	Barrett and Bateson (1978)
	Locomotor play develops		Martin and Bateson (1985)

Males do not provision the female or young with food or provide any other parental care. Infanticide has been documented in domestic cats, especially when a new (unrelated) male has taken over a territory near a female with kittens (Hart and Hart 2016; MacDonald et al. 1987). It has been reported that males will normally avoid females with kittens and are attacked if they get too close (Natoli 1985a). Therefore, caution must be exercised, and it is not recommended for males to be housed with a female and her offspring.

Feeding Behavior

The domestic cat, as a subspecies of the wildcat, evolved from, and still is, a solitary hunter of small prey, and the prey of larger carnivores. Cats are opportunistic hunters of small prey, primarily rodents, but also birds, amphibians, reptiles, and insects, eating frequent small meals per day. While cats may live in groups and share overlapping hunting territories, they retain their solitary hunting style, unlike lions who cooperatively hunt (Turner and Meister 1988; Turner and Mertens 1986). Domestic cats are quite efficient hunters, with a successful kill occurring every 40–180 min (Turner 2014). Where wildcats are primarily nocturnal or crepuscular hunters, domestic cats are more flexible, hunting throughout the day, but with approximately one-third of their prey captured at night (Turner 2014). Dietary studies assessing gut contents or scat have reported that remains of small mammals make up 33%–90%, whereas birds are a smaller (10%–36%), yet consistent, part of the diet (Fitzgerald and Turner 2000; Leyhausen 1979). It has been reported that cats are responsible for 1 billion bird deaths per year, but there is no strong evidence to support the viewpoint that cats are a serious threat to wildlife. In fact, the population level effects of cat predation on bird populations in continental environments are poorly documented and have not been demonstrated (Fitzgerald and Turner 2000). One reason why the risk to song birds may be perceived to be so high is because bird hunting takes place during daylight hours, whereas rodent predation typically is a nocturnal activity (Fitzgerald and Turner 2000). It is likely that the songbird population on continents (e.g., North America, Europe) can withstand predation by cats, since they have coexisted for hundreds of generations. If they could not, songbirds would have disappeared long ago. The effect of free-roaming and feral cats on island wildlife is a special case, and should be considered separately in any discussion of predation, since the fauna on these islands evolved in the absence of predators and are thus more vulnerable (Fitzgerald and Turner 2000).

Kittens begin to develop predatory behavior at about 4 weeks of age, when their mother begins to provision live prey and encourage her kittens to learn the skills needed to be efficient hunters. By 5 weeks of age, some kittens can already kill mice (Baerends-van Roon and Baerends 1979). Increased predatory behavior is exhibited by kittens in the presence of their mother, with the mother acting in a manner that leads the kittens to interact with the prey (Caro 1980; Leyhausen 1979). Prey and food preferences also develop at this time, with significant maternal influence. Kittens tend to prefer the prey they have seen their mother kill (Kuo 1930), have been shown to be more likely to eat a novel food in the presence of their mother than without her (Wyrwicka and Long 1980), and are also more likely to eat a novel food (mashed bananas) than a preferred food (meat) if their mother eats the novel food in their presence (Wyrwicka 1978). This type of social learning has been shown to extend to other behavior as well; kittens who watched their mother perform an operant task (pressing a lever for food) learned the behavior faster than those observing an unfamiliar female, whereas kittens exposed to a trial and error condition never learned the task (Chesler 1969). Additionally, social learning through observation extends to other species, including humans, and continues into adulthood (Bateson 2014).

Communication

The process of domestication has influenced communication in cats, enabling them to successfully live in social groups. The solitary wildcat rarely encounters another wildcat, so the main form of communication is olfactory, via scent marks that can remain in the environment for several days. This type of communication, also important to domestic populations, allows cats to avoid potentially injurious aggressive encounters. There are several sources of odors used for marking, including urine, feces, and skin glands. Urine may be the most important scent marker, typically sprayed on a vertical surface, usually accompanied by a quivering of the tail. While most frequently observed in adult males, females will also spray (Bradshaw and Cameron-Beaumont 2000). This may be suppressed in juveniles and females in high-density colonies, leaving only "dominant" males to mark (Natoli 1985b). All cats, but male cats especially, spend considerable time investigating spray marks by sniffing, often followed by flehmen, a curling of the mouth that transmits pheromones to the vomeronasal organ located on the roof of the mouth. While the territorial function of urine-spraying is unknown, it seems to contain social information pertinent to communication (Bradshaw and Cameron-Beaumont 2000). The use of feces to scent-mark appears to be less important. Cats typically bury feces in the core area of their territory, while leaving it exposed on the periphery (MacDonald et al. 1987).

Cats also have skin glands in several locations, including interdigital, at the base of the tail, and on the head in the submandibular, perioral, and temporal regions (Bradshaw and Cameron-Beaumont 2000). Scratching, which typically occurs in well-travelled areas of the territory on vertical and horizontal surfaces, will deposit scent from the interdigital glands, as well as leave a visual signal (Feldman 1994). Rubbing the face on objects or individuals, known as bunting behavior, will leave scent marks from the glands located on the head (Houpt and Wolski 1982).

Vocalizations, another important form of communication, occur in four types of circumstances; agonistic (aggressive and defensive), sexual, mother-offspring, and cat-human (Bradshaw and Cameron-Beaumont 2000). Vocalizations used to communicate aggression include growl, howl, yowl, and snarl, while defensive vocalizations include hiss and spit. The purr, meow, and trill, or chirrup are used when greeting or making contact, and the tom (an intact male) and queen (an intact female) each have a sex-specific sexual call. The function of purring in communication is not well understood. It is often considered to be an indication of a positive affective state or contentment, but this is not always the case. Cats will also purr when they are afraid or in pain (Beaver 1992). Kittens can purr soon after birth, and do so while nursing, possibly to solicit maternal care. Meowing is rarely observed in cat–cat communication, but is very common in cat–human communication, and is thought to be a learned attention-soliciting behavior (Turner 2017; Yeon et al. 2011).

Visual communication between cats is mainly used to modulate aggressive or agonistic interactions. To show aggression, cats will make themselves seem as big as possible, by standing tall with arched back and piloerection; the classic "Halloween cat" posture. A cat that wants to avoid a confrontation will crouch to the ground and flatten its ears. Affiliative behaviors include rolling; stretching; rubbing of the head or body on another cat, human, or object; head-butting; allogrooming; and lying in proximity, with or without physical contact. The tail is very effective in signaling. Cats will tuck the tail under them when adopting a submissive or defensive posture, lash it side to

Behavioral Biology of the Domestic Cat

FIGURE 14.2 Ethogram of facial and body postures of the domestic cat.

side as a display of aggression, and hold it straight up (tail-up) when approaching another cat or human to signal affiliative or friendly intent (Bradshaw and Cameron-Beaumont 2000). Figure 14.2 (adapted from UFAW 1995) illustrates cat facial expressions and body postures as they relate to affective states.

Common Captive Behaviors

Normal Behavior

The wildcat is a strictly nocturnal hunter. However, domestic cats appear to organize their activity and feeding behavior around human activity. It has been hypothesized that both genetic and ecological factors influence behavior in some species (Kavanau 1969); cats may be a nocturnal species genetically, but when living with humans adopt a diurnal lifestyle (Randall et al. 1985). This is supported by studies of feral cats; when not provisioned by humans, roaming and hunting tend to be nocturnal activities (MacDonald and Apps 1978), but when ample food is supplied by humans, diurnal feeding and activity patterns are adopted (Dards 1979). Further, cats in a laboratory environment have also been reported to organize activity and feeding around human presence, but when isolated, 80% of the subjects exhibited nocturnal behavior (Randall et al. 1985).

Few studies have aimed to assess daily time budgets for cats. Podberscek et al. (1991) observed group-living laboratory cats for 8 h a day from 8:00 to 16:00 h. Cats engaged in maintenance behaviors (e.g., resting, drinking, eating, eliminating) 36% of the time; 30% was spent in comfort behaviors (e.g., grooming); and 24.5% in locomotory behaviors. A study assessing shelter cats in enriched or control housing environments reported cats spent 6% of their time active. But differences between the groups were found for time spent alert resting (65% vs 70%), sitting (9% vs 14%), and in restful sleep (20% vs 11%) for the enriched and control conditions, respectively (Kry and Casey 2007). A study of singly housed adult male laboratory cats found a fivefold increase in active behaviors after addition of a ball for enrichment (de Monte and Le Pape 1997). Finally, Eckstein and Hart (2000) reported grooming behavior occupied about 4% of the time budget. In addition to resources and enrichment items, personality, sex, age, and breed likely affect daily time budgets.

Stress-Related and Sickness Behavior

Ethological studies in research laboratories (Carlstead et al. 1993; Stella et al. 2011, 2012, 2014, 2017) and quarantine boarding facilities (Rochlitz et al. 1998) report that cats subjected to impoverished or unpredictable environments decrease their activity and increase hiding behaviors. For example, Carlstead et al. (1993) imposed a 21-day psychological stressor on singly housed cats that included unpredictable caretaking and mildly aversive handling. This protocol proved to be a potent psychological stressor. Compared to controls, stressed cats exhibited decreased activity, increased attempts to hide, and had an increase in adrenal-cortical output (increased urine cortisol concentrations), enhanced adrenal sensitivity to adrenocorticotropic hormone (ACTH), and reduced pituitary sensitivity to luteinizing hormone-releasing hormone. The researchers concluded that hiding was an important behavior for regulating hypothalamic-pituitary-adrenal (HPA) axis activation caused by an unpredictable environment, and a good caretaker-cat relationship may be essential for cats to adapt and achieve good welfare.

Activation of the SRS affects immune function, one consequence of which is the development of sickness behavior (SB). SB is a behavioral switch from activities, such as feeding, social contact, or grooming, to processes that conserve energy to boost immune function to fight pathogens (Dantzer et al. 2008; Raison and Miller 2003). This likely occurs with activation of the SRS leading to release of corticotrophin-releasing factor, activating the sympathetic nervous system and the immune system, and causing the release of proinflammatory cytokines, while increasing vigilance and suppressing maintenance behaviors (Marques-Deak et al. 2005; Sapolsky 2004) (see Figure 14.3). This is a well-documented response that has been reported in many species, including rodents (Broom 2006) and dairy cattle (Fogsgaard et al. 2012). Additionally, psychological stress has been associated with immune activation and proinflammatory cytokine release (Marques-Deak et al. 2005), and research has linked SB, cytokine activation, mood symptoms, and pathologic pain (Raison and Miller 2003).

Cats appear to exhibit SB in response to environmental disturbances. Stella et al. (2011) reported that colony-housed cats exhibited increased SB in response to environmental stressors that occurred during routine management of the colony. These events included transient (one-week) discontinuation of contact or interactions with the cats' primary caretaker, changes in time of day of routine husbandry, unfamiliar caretakers, and a delay of 3 h in feeding time. These events resulted in a 3.2-fold increased relative risk (RR) for SB compared to control weeks, and larger increases in risk for decreased food intake (RR=9.3) and eliminations (RR=6.4), and an increased risk for elimination of feces (RR=9.8) and urine (RR=1.6) outside the litter pan. In a follow-up study (Stella et al. 2012), changes in immune system function associated with stressors were

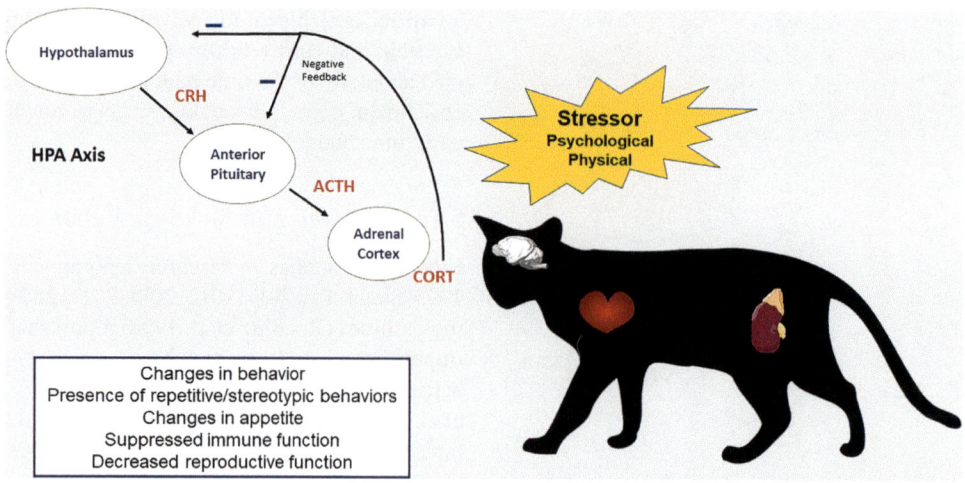

FIGURE 14.3 Representation of the interaction between external events (stressors), internal physiological response (HPA axis activation), and the physical and behavioral outcomes expressed by the cat.

found, including a decrease in circulating lymphocytes and an increase in the neutrophil to lymphocyte ratio from baseline. Additionally, gene expression for the cytokines IL-6 and TNF-α was altered.

In domestic cats, common responses to stress include decreased appetite; vomiting of hair, food, or bile; eliminating out of the litter pan; decreased social interactions; decreased grooming behavior; and an increase in the frequency and intensity of attempts to hide (Stella et al. 2011, 2012, 2014, 2017 see Figure 14.4). Grooming behavior is a normal response to stressors in cats, possibly a self-soothing behavior resulting in release of endorphins. However, psychogenic alopecia is reported to be uncommon in cats and difficult to distinguish from other causes of hypersensitivity (Waisglass et al. 2006). Optimizing the housing environment, social interactions, and allowing cats to maintain some degree of predictability and control will decrease their perception of threat and dramatically decrease the incidence of SB (Stella et al. 2011, 2012, 2014, 2017). A more in-depth discussion is outlined below.

The motivational state of the SB response has a physiologic basis and should be considered in welfare assessments, in addition to other motivational states, such as fear, hunger, and thirst. Seeking of rest, withdrawal from the environment, and caring for one's self are adaptive responses to infection that are as normal as arousal and escape are in response to a threat (Dantzer and Kelley 2007). However, when this motivational state is caused by chronic environmental disturbances with which the individual is unable to cope, it is a sign of impaired welfare and should be addressed. Thus, daily monitoring of cats for SB may offer a practical, noninvasive method to assess stress responses and thus gauge overall welfare.

Ways to Maintain Behavioral Health

In the lives of captive animals, the perception or actual lack of ability to control their surroundings is perhaps the greatest stressor they experience. When confined, they have little or no control over who their social partners are; how much space they can put between themselves and others; the type, amount, or availability of food; or the quality or quantity of environmental stimuli, including lights, noise, odors, and temperature that they experience (Morgan and Tromborg 2007). Predictability, or the lack thereof, is another aspect of the captive environment that may be stressful. Studies have shown that, given a choice, animals will choose predictability over unpredictability, especially in relation to aversive events (Morgan and Tromborg 2007; Weiss 1971, 1972). Predictability refers both to temporal aspects, as well as familiarity of caretakers and the environment. A consistent, predictable daily routine is essential, particularly when an animal is confined. Daily cleaning and feeding procedures conducted at the same time of day and performed by a familiar person allow cats to predict potentially aversive events. Studies have shown a variety of animals, from macaques to rats, show physiological and behavioral stress responses to cage-cleaning (Morgan and Tromborg 2007). In a study of laboratory-housed cats, changing the time of daily husbandry and feeding resulted in increased SB, also indicative of a stress response (Stella et al. 2011, 2012). Lack of control can lead to symptoms of chronic stress, including anorexia, weight loss, inhibition of exploratory behaviors, learned helplessness, stereotypies, and aberrant immune responses (Bassett and Buchanan-Smith 2007). But absolute control is not necessary, and, in fact, predictable, but not invariant, routines may lead to optimal welfare (Broom 1991). Impaired welfare can be considered a chronic

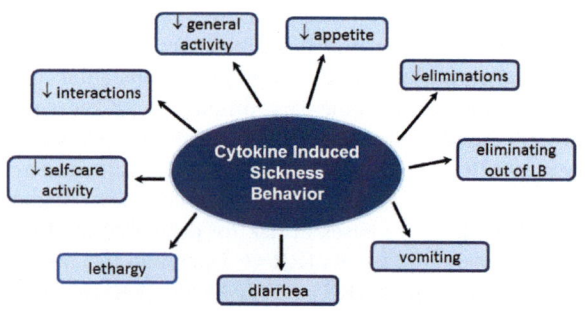

FIGURE 14.4 Common SBs of the domestic cat.

imbalance between positive and negative experiences resulting in chronic stress and failure to cope. It is now assumed that, similar to humans, chronic stress may induce mental suffering in animals, with or without physical health problems.

Environment

The housing space of confined cats is generally reduced in both quantity and quality in comparison to options available to their free-roaming counterparts. There are many aspects of the captive environment that may impact the welfare of cats in biomedical research. Individuals must adapt to the physical environment, and often the captive environment does not match the physical environment in which the species evolved. Captive housing environments are typically built and maintained for human comfort and convenience. Cats perceive the environment very differently than humans do, so many environmental factors may be aversive or stressful to them. Factors pertaining to the macroenvironment (the room), the microenvironment (the individual cage or restricted area available to the cat), predictability and control of the environment, and the quality of the social environment (including the human–cat relationship, as well as interactions with conspecifics and other animals in the environment, such as dogs housed in auditory or olfactory proximity), all influence cats' perception of threat (Amat et al. 2016; Stella et al. 2014; Stella and Croney 2016b; Vinke et al. 2014). The impact of stress caused by human intervention(s) should be minimized for maintenance of both the psychological and physiological (immunologic) health of the individual. Understanding the cats' sensorium and how they are likely to perceive stimuli will aid caretakers in monitoring and improving the housing environment. Finally, the captive environment should be behaviorally relevant, with the quantity and quality of space provided allowing for the development and normal expression of species-typical behavior patterns.

Macroenvironmental considerations: The room environment appears to be at least as salient to confined cats as the cage environment (Stella et al. 2014, 2017), so attention to factors including temperature, lighting, odors, and noise is essential to protect cat welfare. The thermoregulatory environment is an important aspect of animal welfare, with temperature extremes being well-known stressors, often used to provoke a physiologic and behavioral stress response. Studies suggest that ideal temperature ranges are unique for each species; are impacted by characteristics of the individual, including weight and age; and can alter social behavior (reviewed in Morgan and Tromborg 2007). The thermoneutral zone for cats is 30°C–36°C (NRC 2006), whereas housing rooms are typically maintained at 18°C–29°C (NRC 2011). It is likely that this difference in the preferred versus the experienced temperature results in many cats commonly experiencing cold stress or thermal discomfort. In mice, this difference in preferred versus experienced temperature has been shown to increase metabolism, decrease growth and organ weight, and impair immune function (Gaskill et al. 2013). Providing cats access to adequate resources that allow them to behaviorally thermoregulate, such as blankets, towels, shredded paper, heating pads, lamps, or SnuggleSafe®, will enable them to more easily cope with the thermal environment.

Several aspects of lighting in the vivarium are of importance, including photoperiod, intensity, and spectrum. Cats should never be exposed to continuous light or darkness, as this disrupts circadian rhythms and sleep patterns (Ikeda et al. 2000; Vanbetteray et al. 1991), and is of particular importance for reproduction. Additionally, turning lights on and off with human activity in the housing area is also disruptive and potentially stressful. Wherever possible, natural lighting through a sky light or window should be provided. Light intensity is also important, as cats have a lower threshold for detecting light than humans. It is recommended to maintain light intensity between 130 and 325 lux (NRC 2011), while cats can see in 0.125 lux. Therefore, they may be more sensitive to bright light and prefer lower intensity light levels in their housing environment. Potentially compounding this factor is that caging is most often stainless steel, a very reflective surface, so attention to the light level *within* the cage, where the cat is experiencing it, becomes important. Using less reflective caging, avoiding light shining directly into cages, and providing the ability to adjust behaviorally through access to a "dark" and a "light" area within the enclosure will aid in minimizing associated stress. Wavelength may also negatively impact welfare. Several species have been reported to show activation of the SRS and altered behavior when deprived of critical wavelengths (reviewed in Morgan and Tromborg 2007). Finally, fluorescent lights may be aversive to cats, as they emit light discontinuously and therefore have a flicker frequency. Species differ at what rate they can no longer detect the flicker, with some species showing a preference for higher frequency fluorescent lights. To my knowledge, the "optimal" wavelength and flicker frequency are not known for domestic cats.

Most animals, apart from primates, have a very keen sense of smell. Therefore, many of the odors that cats are commonly and continuously exposed to in the housing room may be aversive and a chronic source of stress. These include alcohol hand sanitizer, cleaning chemicals, unfamiliar conspecifics, natural predators (most commonly dogs), and citrus scents. Use of unscented cleaning products and replacing use of hand sanitizer with hand washing are some ways to minimize aversive odors. Wherever possible, care staff should be devoted to a single species. When that is not possible, then the order in which species are cared for should avoid movements from predator to prey, as in dog to cat, or cat to rodent. Use of scents, such as vanilla, lavender, catnip, mint, valerian, honeysuckle, and the pheromone, Feliway®, can be added to the environment as olfactory enrichment to encourage exploratory behavior and minimize stress (Ellis and Wells 2010).

The final macroenvironmental factor of importance is sound. The auditory frequency range of cats exceeds that of humans, yet the welfare consequences of chronic exposure to high-frequency noise is not well documented and difficult to assess. Cats evolved on the savanna, where background sound pressure level is 20–40 dB, whereas sound levels in laboratory housing have been reported to exceed 110 dB (Sales et al. 1999). The SRS of *rats* has been shown to be activated with 30 min of exposure to sound levels >73 dB, resulting in a 100%–200% increase in circulating corticosterone levels (Prabhakaran et al. 1988). *Cats* have a similar auditory frequency range, as they utilize the ultrasonic vocalizations emitted by rodents to hunt. Therefore, it is likely they will respond to this level of noise with a corresponding activation of the SRS. Housing areas should aim for maximum noise levels of <60 dB, i.e., a quiet conversational level. Speaking in quiet voices when in the presence of cats, conducting conversations outside of the cat housing areas whenever possible, housing barking dogs as far from cats as possible, closing cage and room doors quietly, and avoiding sudden and unexpected noises (which are especially disturbing to cats), will help minimize the stress associated with noise. Auditory enrichment can be provided in the form of cat-specific music developed to match the frequency range of cat vocalizations. Cats have been shown to respond to this type of music more than to human music, with an increase in activity and prosocial behavior (Snowden et al. 2015). Cat-specific music should be played at a low level (<60 dB) and intermittently.

Micro-environmental factors of importance to cats include hiding and perching opportunities, elimination facilities, food (type and presentation), and outlets for the expression of other species-typical behaviors. The type and presentation, as well as the availability, of these features of the environment can be either a source of stress or enrichment (see Figure 14.5).

It is important to remember that domestic cats are both predator and prey, so the opportunities to hide and to perch are important species-typical behaviors. Hiding is an adaptive response to avoid predation and this behavior has undergone almost no change in the course of domestication. In fact, it has been suggested that thwarting attempts to hide can contribute disproportionately to any overall measure of stress (Overall and Dyer 2005), and the ability to hide is essential to successful coping when cats are exposed to stressors (McCune 1994; Roclitz 2000; Smith et al. 1994). Hiding behavior has been correlated with enhanced ACTH response and increased urinary cortisol levels, and has been identified as a reliable indicator of stress. Cats hide to cope with and alleviate stress, so the welfare of a stressed and fearful cat that can hide is likely better than that of a stressed and fearful cat that is unable to hide. Studies have found that the provision of a hide box resulted in cats that approached humans more often, retreated less often, engaged in more restful sleep, and exhibited less vigilant behavior (Kry and Casey 2007). Several studies have demonstrated that when given the opportunity, cats will spend considerable amounts of time (up to 77%) in their hide box (Gourkow and Fraser 2006; Kry and Casey 2007; Rochlitz 2000; Stella et al. 2014, 2017). Additionally, housing environments, with disturbances from stimuli including loud music from radios, barking dogs, or unpredictable husbandry activities, increased hiding behavior, likely due to an increase in perceived threat (Stella et al. 2014, 2017). Another factor relevant to hiding behavior may be individual coping style. Stella and Croney (2019) identified individual preferences for hiding and perching behavior in response to confinement in a cage, with some cats spending significantly more time hiding than others. Together, these results suggest that hiding behavior is a complex, multifactorial response to stress, as well as a coping mechanism. Cats appear to have a need to partially isolate themselves from conspecifics and humans, but also differ in the strength of the motivation and the amount of time they will

FIGURE 14.5 Enriched cage for a singly housed cat that includes hiding and perching opportunities, toys, and soft bedding.

FIGURE 14.6 Cat resting in hide box.

spend hiding, which will also be impacted by the quality of the environment (see Figure 14.6). All cats should be provided a hiding area when confined, and caretakers should observe individuals to better understand each individual's "normal" hiding behavior.

As a prey species, perching is also an important behavior for cats. They prefer to observe their environment from an elevated position. Therefore, well-designed cat housing will include elements, such as climbing opportunities, hammocks, platforms, raised walkways, shelves, and/or window seats. Observations of cats acutely confined in cages have found that as cats acclimate to a new housing environment, they tend to switch from primarily hiding to primarily perching behavior (Stella 2013). Additionally, individual differences have been identified, indicating that personality traits may play a role in cat preferences for hiding or perching (Stella and Croney 2019). This highlights the importance of addressing individual differences, by providing an enriched cage with both hiding and perching opportunities to confined cats (Gourkow and Fraser 2006; Kry and Casey 2007; Roclitz 2000; Stella et al. 2014, 2017).

Finally, appealing objects (e.g., boxes, scratching posts, mats) and substrates (e.g., cardboard, sisal, carpet, wood) must be provided to permit expression of scratching and marking behavior. This will enable cats to maintain claw health and to leave visual and pheromonal markings throughout their "territory" (Roclitz 2000; Stella and Buffington 2016). Offering different types, textures, and orientations (horizontal and vertical) of appealing objects will allow cats to express their individual preferences.

Elimination areas are another microenvironmental factor of importance to cats. Several points should be kept in mind when assessing the elimination area. It should be a quiet area to permit use without being disturbed by the approach of humans or machinery that could come on unexpectedly and disrupt the normal elimination sequence. Litter pans should be scooped at least daily, although twice daily is recommended, and cleaned with mild dish soap at least weekly so that some smell familiar to the cat remains. The general recommendation is to provide a litter pan 1.5 times the length of the cat, where space allows, to ensure it is large enough for them to engage in normal eliminative behavior (Stella and Buffington 2016). The type of litter is important as well. Most cats prefer unscented and finely particulate litter material, such as clumping litter (Horwitz 1997), although individual preferences have been documented (Borchelt 1991). Be sure that the litter is deep enough for the cats to dig (without hitting the bottom of the litter pan) and then bury their eliminations. Place elimination areas as far from resting and feeding areas as possible. One option for caged cats is to have compartmented cages so that the litter pan is in an area separate from the feeding and resting area. Resting in, or eliminating out of, the litter pan are both aberrant behaviors and good indicators that something is amiss. If all the above criteria are met, this behavior is likely a sign of distress, illness, or conflict between cats.

Optimal space allocation (cage or enclosure size) for cats is not known, and is likely impacted by factors such as age, diet, health status, social group, individual differences in activity levels, and the quality of the housing environment. When provided a housing room with minimal disturbances; enriched cages; and daily free play time out of the cages to socialize with conspecifics, exercise, and explore; cats have been successfully maintained in stainless steel cages with $0.56\,m^2$ of floor space for many years with few adverse behavioral or physiological effects (Stella et al. 2011, 2013). Therefore, these forms of enrichment may compensate for or be more salient to the cat than increased cage space (see Figure 14.7).

Social Groupings

Whether social or single housing is better for cats is an area of debate. One of the greatest differences between free-roaming and confined environments is the reduction in choices available. When free-roaming, the selection of social partners, for mating or other reasons, is largely determined by the individual, and escape or avoidance of aggressive conspecifics is limited only by natural barriers. This is drastically altered in captivity. Increased population density, unnatural social groups (both sex and age groups are more uniform than would occur in nature), and frequent regrouping can all potentially cause social stress. In a laboratory environment, cats are typically housed with unrelated conspecifics. While cats are less gregarious than other domesticated species as discussed above,

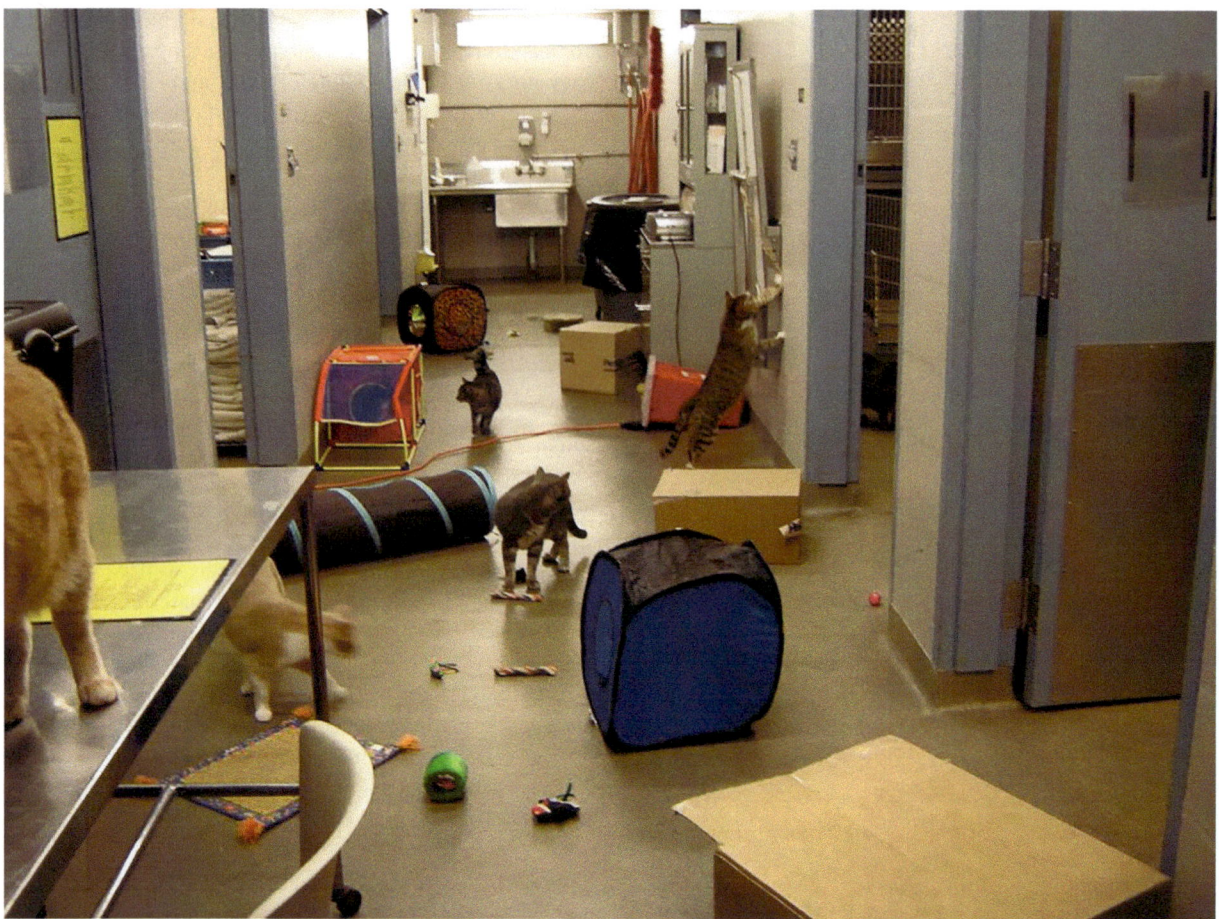

FIGURE 14.7 Free playtime for singly housed colony cats with opportunities for exercise, social interactions, scratching, climbing, and toys for play.

they do exhibit great plasticity in social behavior. So, while group living in itself may not pose welfare issues, for some individuals, being confined in close proximity with unrelated cats may be stressful. The demographic and behavioral composition of the group, and the quality and quantity of the housing space will greatly impact the welfare implications for the cats.

Studies have reported no effect of group size (single or multi) on urinary cortisol:creatinine (Lichtsteiner and Turner 2008), moderately increased stress in group-housed compared to singly housed cats (Ottway and Hawkins 2003), and high numbers of agonistic interactions among group-housed cats (Dantas-Divers et al. 2011). Rates of allogrooming and proximity to other cats have been shown to be significantly associated with familiarity and relatedness (Curtis 2003). Barry and Crowell-Davis (1999) reported that cats living in two-cat households spent 50% of their time within 1–3 m, but visually obstructed, from one another. Further, a study of 14 cats living in a home reported that the cats maintained individual, overlapping home ranges, with favored resources (e.g., resting areas) time shared by small groups (Bernstein and Strack 1996). de Monte and le Pape (1997, page 63) concluded that for adult cats "social isolation may not be considered a totally unfortunate housing situation" if they have positive human interaction and can see and hear other cats. A systematic review of the literature comparing single and multicat housing revealed mixed results, with recommendations for housing as follows: (1) if social history is unknown, the cat should be housed singly in an enriched cage; (2) minimize environmental changes and maintain consistent husbandry; (3) if cats are to be group housed, all cats in the group should be well socialized to conspecifics; and (4) if housed in groups of initially unfamiliar conspecifics, groups should be stable (i.e., no addition or removal of individuals) (Finka et al. 2014). Based on the available literature to date, maintaining good welfare for group-housed cats will likely take some additional management effort. Suggestions for success include: (1) start with young cats, ideally littermates, and keep them together in a stable group of 4–6 individuals; (2) provide abundant resources throughout the enclosure and utilize the three-dimensional space to increase complexity; (3) provide a litter pan for each cat, plus one additional pan, all out of sight of each other (Neilson 2004); and (4) provide separate feeding areas, one per cat, distributed throughout the enclosure on multiple levels and out of sight of other cats' feeding areas, to facilitate "solitary" feeding. Additionally, each cat should have its own perch and hide box. Ample, well-distributed resources will reduce the risk of conflict, and if the cats are maintaining individual home ranges within the enclosure, all cats will have their own set of resources. Finally, observe the group closely, especially when first established, using video monitoring, if possible, to minimize observer effects on the cats' behavior. This will help to identify individuals who may not be coping well in the group environment. As in other species, individual differences in temperament or personality (Adamec 1991; Feaver et al. 1986) and stress susceptibility (De Boer et al. 2016; Koolhaas et al. 2010), as well as individual variation in experience (Adamec et al. 1998; Boissy 1995; Gottlieb and Halpern 2002), will influence how individual cats respond to group housing. Finally, some cats may be better suited to single housing, so this option should be available for these individuals.

Feeding Strategies

When feeding confined cats, it is important to consider both what and how food is presented. Until recently, cats have been maintained as both a pet and for rodent control, and humans did not have the knowledge or ability to meet their nutritional requirements. Therefore, cats may have maintained both hunting and scavenging abilities in order to feed themselves (Bradshaw et al. 1999; Bradshaw 2006). Evidence suggests that diet and nutrition are environmental factors that may lead to chronic health problems in domestic cats (Zoran and Buffington 2011). Studies have demonstrated that cats have a "target" intake of total energy comprised of 52% protein, 36% fat, and 12% carbohydrates, and this "target" could only be met by provision of wet foods in addition to dry (Hewson-Hughes et al. 2011, 2013). Additionally, it has been reported that there is a limit to the amount of carbohydrates cats will ingest, termed the "carbohydrate ceiling", so that cats fed only high-carbohydrate dry diets (50% energy from carbohydrates), cannot meet their target intake of protein and fat (Hewson-Hughes et al. 2011, 2013). Further, the macronutrient profile of the diets that cats chose to consume in these experiments was similar to the profile reported for free-ranging feral cats (Plantinga et al. 2011). This suggests that domestic cats have retained the capacity to regulate macronutrient intake, despite being fed dry commercial diets that have little resemblance to the "natural" diets of feral and wildcat populations (Hewson-Hughes et al. 2013). Offering both dry and wet food, in separate feeding containers, that are nutritionally complimentary is recommended to allow cats to consume a diet that meets their target macronutrient profile. Additionally, cats seem to be less sensitive to thirst and dehydration than other species, such as dogs, so when eating commercial dry foods, water intake will be roughly half the amount compared to when fed wet diets. Increased water intake will increase urine volume, resulting in decreased concentration of urolith-forming minerals in the urine (Zoran 2002), a major factor in LUTS (Cannon et al. 2007). Taken together, these findings suggest that cats may prefer wet food over dry, and that providing wet food as part of the diet may have health benefits as well.

Food presentation is also important to cats. Free-roaming cats spend a considerable portion of their day engaged in food acquisition, eating several small meals throughout the day. Cats with free access to food usually prefer to eat several small meals, as opposed to one or two large meals, and most will hunt for prey when given the option (Bradshaw and Thorne 1992). Although free access to food may allow for frequent feeding sessions, this feeding strategy does not allow the cat the opportunity to express natural predatory instincts (Morris 2002), leading to boredom and overeating, and may contribute to the development of obesity and other health problems (Kienzle and Bergler 2006). One strategy for meeting nutrient needs in a species-typical way is to mimic their natural feeding preferences. Puzzle toys, such as balls or other devices designed specifically for cats, which release dry food or treats when physically manipulated, provide outlets for natural predatory behavior and increase daily activity, supplying both mental and physical enrichment (Ellis 2009).

Training

Despite their reputation, cats are relatively easy to train. In fact, cats have participated in psychological studies of operant conditioning for decades, with one of the first published references being Thorndike's *Animal Intelligence* (1911). Using positive reinforcement techniques and avoiding punishment will be most effective. Positive reinforcement techniques have been utilized to clicker train cats in shelters (Kogan et al. 2017), and cats used in research have been trained to accept blood collection from the jugular vein, while lying in a dorsally recumbent position (Lockhart et al. 2013).

Human–Cat Interactions

The quality of human–cat interactions is extremely important to confined cats. Animals often perceive contact with humans as predatory encounters, resulting in fear (Waiblinger et al. 2006). Positive interactions, such as gentle handling or feeding of a preferred food treat, can decrease fear (Claxton 2011). In effective cat management, a familiar, consistent caretaker appears to be essential. Positive human–cat interactions have been shown to decrease abnormal behaviors and increase reproductive success in wild felids in zoos (Mellen 1991). Additionally, in clouded leopards, a decrease in fecal cortisol metabolites (FGM) was found to be associated with increasing amounts of time spent with their primary keeper, and an increase in FGM was associated with increasing numbers of keepers (Wielebnowski et al. 2002). This suggests that inconsistency in keepers may have prevented the cats from forming and maintaining a predictable relationship with any single keeper, thus increasing the stress of captivity. If cats can recognize and discriminate among the humans they regularly encounter, then these same persons can become predictors of a salient event (e.g., food) and be used to improve their welfare.

Conclusions

Domestic cats have made invaluable contributions to medical and veterinary science over the years. With adequate understanding of how cats have evolved, how they sense the world, and the behaviors they are highly motivated to express, caretakers will be equipped to provide a resource-rich, high-quality environment that will meet their needs. Importantly, this will improve their welfare, as well as the quality of the research. Additionally, with the changing attitude of the public toward animal participation in biomedical research, and the associated trend toward rehoming cats and dogs, it is important to set cats up for a successful retirement. Several states, including Minnesota, California, Connecticut, Delaware, Illinois,

Maryland, Nevada, New York, and Rhode Island (NAVS, 2019), now require rehoming of dogs and cats leaving biomedical research in universities, and this trend is likely to continue in the future. Therefore, providing housing that minimizes stress from environmental factors (e.g., noise, odors, lighting, temperature) and social factors (e.g., incompatible or unstable groups), and allows cats to engage in species-typical behaviors through environmental enrichment, will maintain their physical and behavioral health, improving their chances of successfully transitioning into a home.

REFERENCES

Amat, M., Camps, T., and X. Manteca. 2016. Stress in owned cats: Behavioural changes and welfare implications. *Journal of Feline Medicine and Surgery*. 18:577–586.

Adamec, R.E. 1991. Anxious personality in the cat: Its ontogeny and physiology. In: *Psychopathology and the Brain,* eds. B. J. Carroll and J. E. Barrett, 153–168. New York: Raven Press.

Adamec, R., Kent, P., Anisman, H., Shallow, T., and Z. Merali. 1998. Neural plasticity, neuropeptides and anxiety in animals-Implications for understanding and treating affective disorder following traumatic stress in humans. *Neuroscience and Biobehavioural Reviews*. 23:301–318.

APPA, American Pet Products Association. The 2017–2018 National Pet Owners Survey Debut. https://americanpetproducts.org/Uploads/MemServices/GPE2017_NPOS_Seminar.pdf. (accessed April 8, 2019).

Atchley, M. D., Shah, N. M., and K. E. Whitmore. 2015. Complementary and alternative medical therapies for interstitial cystitis: An update from the United States. *Translational Andrology and Urology*. 4:662–667.

Baerends-van Roon, J. M., and G. P. Baerends 1979. *The Morphogenesis of the Behavior of the Domestic Cat*. Amsterdam: North-Holland.

Barrett, P., and P. Bateson. 1978. The development of play in cats. *Behaviour*. 66:106–120.

Barry, K. J., and S. L. Crowell-Davis. 1999. Gender differences in the social behavior of the neutered indoor-only domestic cat. *Applied Animal Behaviour Science*. 64:193–211.

Bassett, L., and H. M. Buchanan-Smith. 2007. Effects of predictability on the welfare of captive animals. *Applied Animal Behaviour Science*. 102:223–245.

Bateson, P. 1994. The dynamics of parent-offspring relationships in mammals. *Trends in Ecology and Evolution*. 9:399–403.

Bateson, P. 2014. Behavioural development in the cat. In *The Domestic Cat: The Biology of Its Behavior* (3rd ed.), ed. D. C. Turner and P. Bateson, 12–26. Cambridge: Cambridge University Press.

Beaver, B. V. 1992. *Feline Behaviour: A Guide for Veterinarians*. St. Louis: C.V. Mosby.

Bernstein, P. L., and M. Strack. 1996. A game of cat and house: Spatial patterns and behavior of 14 domestic cats (*Felis catus*) in the home. *Anthrozoos*. 9:25–39.

Boissy, A. 1995. Fear and fearfulness in animals. *Quarterly Reviews in Biology*. 70:165–191.

Borchelt, P. L. 1991. Cat elimination behavior problems. *Veterinary Clinics of North America Small Animal Practice*. 21:257–264.

Bradshaw, J. W. S., and C. J. Thorne. 1992. Feeding behaviour. In *Waltham Book of Dog and Cat Behavior*, ed. C. J. Thorne, 115–129. Oxford: Pergamon.

Bradshaw, J. W. S., Horsfield, G. F., Allen, J. A., and I. H. Robinson. 1999. Feral cats: Their role in the population dynamics of *Felis catus*. *Applied Animal Behaviour Science*. 65:273–283.

Bradshaw, J., and C. Cameron-Beaumont. 2000. The signaling repertoire of the domestic cat and its undomesticated relatives. In *The Domestic Cat: The Biology of Its Behavior* (2nd ed.), eds. D. C. Turner and P. Bateson, 67–93. Cambridge: Cambridge University Press.

Bradshaw, J. W. S. 2006. The evolutionary basis for the feeding behavior of domestic dogs (*Canus familiaris*) and cats (*Felis catus*). *Journal of Nutrition*. 136:1927S–1931S.

Braastad, B.O., and P. Heggelund. 1984. Eye-opening in kittens: Effects of light and some biological factors. *Developmental Psychobiology*. 17:675–681.

Broom, D. M. 1991. Animal welfare: Concepts and measurements. *Journal of Animal Science*. 69:4167–4175.

Broom, D. M. 2006. Behaviour and welfare in relation to pathology. *Applied Animal Behaviour Science*. 97:73–83.

Buffington, C. A. 2011. Idiopathic cystitis in domestic cats—beyond the lower urinary tract. *Journal of Veterinary Internal Medicine*. 25:784–796.

Buffington, C. A. T., Chew, D. J., and B. E. Woodworth. 1999. Feline interstitial cystitis. *Journal of the American Veterinary Medical Association*. 215:682–687.

Buffington, C. A. T., Westropp, J. L., and D. J. Chew. 2014. From FUS to Pandora syndrome: Where are we, how did we get here, and where to now? *Journal of Feline Medicine and Surgery*. 16:385–394.

Cameron-Beaumont, C., Lowe, S. E., and J. W. S. Bradshaw. 2002. Evidence suggesting preadaptation to domestication throughout the small Felidae. *Biological Journal of the Linnean Society*. 75:361–366.

Cannon, W. B. 1914. The interrelations of emotions as suggested by recent physiological researches. *The American Journal of Psychology*. 25:256–282.

Cannon, W. B. 1927. The James-Lange theory of emotions: A critical examination and an alternative theory. *The American Journal of Psychology*. 39:106–124.

Cannon, W. B. 1942. "Voodoo" Death. *American Anthropologist*. 44:169–181.

Cannon, A. B., Westropp, J. L., Ruby, A. L., and P. H. Kass. 2007. Evaluation of trends in urolith composition in cats: 5,230 cases (1985–2004). *Journal of the American Veterinary Medical Association*. 231:570–576.

Carlstead, K., Brown, J. L., and W. Strawn. 1993. Behavioral and physiological correlates of stress in laboratory cats. *Applied Animal Behaviour Science*. 38:143–158.

Caro, T. M. 1980. Predatory behavior in domestic cat mothers. *Behaviour*. 74:128–148.

Caro, T.M. 1981. Predatory behavior and social play in kittens. *Behaviour*. 76:1–24.

Chambers, J. K., Tokuda, T., Uchida, K., et al. 2015. The domestic cat as a natural model of Alzheimer's disease. *Acta Neuropathologica Communications*. 3:78.

Chandler, M., Cunningham, S., Lund, E. M., et al. 2017. Obesity and associated comorbidities in people and companion animals: A one health perspective. *Journal of Comparative Pathology*. 156:296–309.

Chesler, P. 1969. Maternal influence in learning by observation in kittens. *Science*. 166:901–903.

Chu, K., Anderson, W. M., and M. Y. Rieser. 2009. Population characteristics and neuter status of cats living in households in the United States. *Journal of the American Veterinary Medical Association*. 234:1023–1030.

Claxton, A. M. 2011. The potential of the human-animal relationship as an environmental enrichment for the welfare of zoo-oused animals. *Applied Animal Behaviour Science*. 133:1–10.

Clutton-Brock, J. 1999. *A Natural History of Domesticated Mammals*. New York: Cambridge University Press.

Curtis, S. 2003. What constitutes animal well-being. In *Animal Stress*, ed. G. Moberg, 1–11. Bethesda, MD: American Physiological Society.

Dards, J. L. 1979. The population ecology of feral cats (*F. catus* L.) in portsmouth dockyard. PhD thesis, University of Southampton.

Dantas-Divers, L., Crowell-Davis, S. L., Alford, K., Genaro, G., D'Almeida, J. M., and R. L. Paixao. 2011. Agonistic behavior and environmental enrichemnt of cats communally housed in a shelter. *Journal of the American Veterinary Medical Association*. 239:796–802.

Dantzer, R., and K. W. Kelley. 2007. Twenty years of research on cytokine-induced sickness behavior. *Brain Behavior and Immunity*. 21:153–160.

Dantzer, R., O'Connor, J. C., Freund, G. G., Johnson, R. W., and K. W. Kelley. 2008. From inflammation to sickness and depression: When the immune system subjugates the brain. *Nature Reviews Neuroscience*. 9:46–56.

Deag, J. M., Manning, A., and C. E. Lawrence. 2000. Factors influencing the mother-kitten relationship. In *The Domestic Cat: The Biology of Its Behavior* (2rd ed.), eds. D. C. Turner and P. Bateson, 23–45. Cambridge: Cambridge University Press.

De Boer, S. F., Buwalda, B., and J. M. Koolhaas. 2016. Untangling the neurobiology of coping styles in rodents: Towards neural mechanisms underlying individual differences in disease susceptibility. *Neuroscience and Biobehavioral Reviews*. 74:401–422.

de Monte, M., and G. Le Pape. 1997. Behavioural effects of cage enrichment in single-housed adult cats. *Animal Welfare*. 6:53–66.

Downs, J. 1960. Domestication: An examination of the changing social relationship between man and animals. *Kroeber Anthropological Society Papers*. 22:08–67.

Driscoll, C. A., Menotti-Raymond, M., Roca, A. L., et al. 2007. The near Eastern origin of cat domestication. *Science*. 317:519–523.

Driscoll, C. A., MacDonald, D. W., and S. J. O'Brien. 2009. From wild animals to domestic pets, an evolutionary view of domestication. *Proceedings of the National Academy of Sciences of the United States of America*. 106:9971–9978.

Driscoll, C. A., Yamaguchi, N., O'Brien, S. J., and D. W. MacDonald. 2011. A suite of genetic markers useful in assessing wildcat (*Felis silvestris* ssp). – domestic cat (*Felis silvestris catus*) admixture. *Journal of Heredity*. 102:S87–S90.

Dumas, C., and F. Y. Dore. 1991. Cognitive-development in kittens (*Felis catus*)- an observational study of object permanence and sensorimotor intelligence. *Journal of Comparative Psychology*. 105:357–365.

Eckstein, R. A., and B. L. Hart. 2000. The organization and control of grooming in cats. *Applied Animal Behaviour Science*. 68:131–140.

Elder, J. H., Sundstrom, M., de Rozieres, S., de Parseval, A., Grant, C. K., and Y. C. Lin. 2008. Molecular mechanisms of FIV infection. *Veterinary Immunology and Immunopathology*. 123:3–13.

Elder, J. H., Lin, Y. C., Fink, E., and C. K. Grant. 2010. Feline immunodeficiency virus (FIV) as a model for study of lentivirus infections: Parallels with HIV. *Current HIV Research*. 8:73–80.

Ellis, S. L. 2009. Environmental enrichment: Practical strategies for improving feline welfare. *Journal of Feline Medicine and Surgery*. 11:901–912.

Ellis, S. L. H., and D. L. Wells. 2010. The influence of olfactory stimulation on the behaviour of cats housed in a rescue shelter. *Applied animal Behaviour Science*. 123:56–62

Feaver, J., Mendl, M., and P. Bateson. 1986. A method for rating the individual distinctiveness of domestic cats. *Animal Behaviour*. 34:1016–1025.

Feldman, H. 1994. Methods of scent marking in the domestic cat. *Canadian Journal of Zoology*. 72:1093–1099.

Finka, L. R., Ellis, S. L. H., and J. Stavinsky. 2014. A critically appraised topic (CAT) to compare the effects of single and multi-cat housing on physiological and behavioral measures of stress in domestic cats in confined environments. *BMC Veterinary Research*. 10:73.

Fitzgerald, B. M., and Turner, D. C. 2000. Hunting behaviour of domestic cats and their impact on prey populations. In *The Domestic Cat: The Biology of Its Behavior*. (2nd ed.), eds. D. C. Turner and P. Bateson, 151–175. Cambridge: Cambridge University Press.

Fletcher, N. F., Meeker, R. B., Hudson, L. C., and J. J. Callanan. 2011. The neuropathogenesis of feline immunodeficiency virus infection: Barriers to overcome. *Veterinary Journal*. 188:260–269.

Fogsgaard, K. K., Røntved, C. M., Sørensen, P., and S. M. Herskin. 2012. Sickness behavior in dairy cows during *Escherichia coli* mastitis. *Journal of Dairy Science*. 95:630–638.

Fraser, A. 2012. *Feline Behavior and Welfare*. Cambridge, MA: CABI.

Fries, R., Heaney, A. M., and K. M. Meurs. 2008. Prevalence of the myosin-binding protein C in Maine Coon cats. *Journal of Veterinary Internal Medicine*. 22:8933–8936.

Fox, M. W. 1970. Reflex development and behavioral organization. In *Developmental Neurobiology*, ed. W.A. Himwich, Springfield, IL: Thomas.

Gaskill, B. N., Gordon, C. J., Pajor, E. A., Lucas, J. R., Davis, J. K., and J. P. Garner. 2013. Impact of nesting material on mouse body temperature and physiology. *Physiology & Behavior*. 110:87–95.

Godiksen, M., Granstrom, S., Koch, J., and M. Christiansen. 2011. Hypertrophic cardiomyopathy in young Maine Coon cats caused by the p.A31P cMyBP-C mutation- the clinical significance of having the mutation. *Acta Veterinaria Scandinavica*. 53:7.

Gourkow, N., and D. Fraser 2006. The effect of housing and handling practices on the welfare, behaviour and selection of domestic cats (*Felis sylvestris catus*) by adopters in an animal shelter. *Animal Welfare*. 15:371–377.

Gottlieb, G., and C. T. Halpern. 2002. A relational view of causality in normal and abnormal development. *Devevelopmental Psychopathology.* 14:421–435.

Grimm, D. 2018. U.S. Labs using a record number of monkeys. *Science* 362:630.

Gurda, B. L., Bradbury, A. M., and C. H. Vite. 2017. Canine and feline models of human genetic diseases and their contributions to advancing clinical therapies. *Yale Journal of Biology and Medicine.* 90:417–431.

Haltenorth, T., and H. Diller. 1980. *A Field Guide to the Mammals of Africa including Madagascar.* London: Collins.

Hanno, P., Lin, A., Nordling, J., et al. 2010. Bladder pain syndrome committee of the international consultation on incontinence. *Neurourology and Urodynamics.* 29:191–198.

Hart, B. L., and L. A. Hart. 2016. Normal and problematic reproductive behaviour in the domestic cat. In *The Domestic Cat: The Biology of its Behavior.* (3rd ed.), eds. D. C. Turner and P. Bateson, 27–36. Cambridge: Cambridge University Press.

Hemmer, H. 1979. Gestation period and postnatal development in felids. *Carnivore.* 2:90–100.

Henson, M. S., and T. D. O'Brien. 2006. Feline models of type 2 diabetes mellitus. *ILAR Journal.* 47:234–42.

Hewson-Hughes, A. K., Hewson-Hughes, V. L., Miller, A.T., Hall, S. R., Simpson, S. J., and D. Raubenheimer. 2011. Geometric analysis of macronutrient selection in the adult domestic cat, *Felis catus. The Journal of Experimental Biology.* 214:1039–1051.

Hewson-Hughes, A., Hewson-Hughes, V., Colyer, A., et al. 2013. Consistent proportional macronutrient intake selected by adult domestic cats (*Felis catus*) despite variations in macronutrient and moisture content of foods offered. *Jouranl of Comparative Physiology B.* 183:525–536.

Hiby, E., Eckman, H., and I. MacFarlaine. 2014. Cat population management. In *The Domestic Cat: The Biology of Its Behavior* (3rd ed.), eds. D. C. Turner and P. Bateson, 215–230. Cambridge: Cambridge University Press.

Hoenig, M. 2006. The cat as a model from human nutrition and disease. *Current Opinion in Clinical Nutrition and Metabolic Care.* 9:584–8.

Horwitz, D. F. 1997. Behavioral and environmental factors associated with elimination behavior problems in cats: A retrospective study. *Applied Animal Behaviour Sciemce.* 52:129–137.

Houpt, K. J., and T. R. Wolski. 1982. *Domestic Animal Behaviour for Veterinarians and Animal Scientists.* Ames: Iowa State University Press.

HSUS, Humane Society of the United States. Pets by the numbers. https://www.humanesociety.org/resources/pets-numbers (accessed April 8, 2019).

Ikeda, M., Dagara, M., and S. Inoue. 2000. Continuous exposure to dim illumination uncouples temporal patterns of sleep, body temperature, locomotion and drinking behavior in the rat. *Neuroscience Letters.* 279:185–189

Inglis, I. R. 2000. Review: The central role of uncertainty reduction in determining behaviour. *Behaviour.* 137:1567–1599

Izgi, K., and Daneshgari, F. 2013 Animal models for interstitial cystitis/painful bladder syndrome (IC/PBS). *Erciyes Medical Journal.* doi:10.5152/etd.2013.41.

Jenson, R. A., Davis, J. L., and A. Snerson. 1980. Early experience facilitates the development of temperature regulation in the cat. *Developmental Psychobiology.* 13:1–6.

Kavanau, J. 1969. Influences of light on activity of small mammals. *Ecology.* 50:548–557.

Kerby, G. 1987. The social organization of farm cats (Felis catus L.) D. Phil. Thesis, University of Oxford.

Kerby, G., and D. W. MacDonald. 1988. Cat society and the consequences of colony size. In *The Domestic Cat: The Biology of Its Behavior* (1st ed.), ed. D. C. Turner and P. Bateson, 67–82. Cambridge: Cambridge University Press.

Kienzle, E., and R. Bergler. 2006. Human-animal relationship of owners of normal and overweight cats. *Journal of Nutrition.* 136:1947S–1950S.

Kitchener, A. 1991. *The Natural History of Wild Cats.* Ithaca, NY: Comstock Publishing Associates.

Kogan, L., Kolus, C., and R. Schoenfeld-Tacher. 2017. Assessment of clicker training for shelter cats. *Animals.* 10:73.

Kolb, B., and A. J. Nonneman. 1975. The development of social responsiveness in kittens. *Animal Behaviour.* 23:368–374.

Koolhaas, J. M., de Boer, S. F., Coppens, C. M., and B. Bewalda. 2010. Neuroendocrinology of coping styles: Towards understanding the biology of individual variation. *Frontiers in Neuroendocrinology.* 31:307–321.

Kry, K., and R. Casey. 2007. The effect of hiding enrichment on stress levels and behaviour of domestic cats (*Felis sylvestris catus*) in a shelter setting and the implication for adoption potential. *Animal Welfare.* 16:375–383.

Kuo, Z.Y. 1930. The genesis of the cat's response to the rat. *Journal of Comparative Psychology.* 11:1–35.

Leyhausen, P. 1979. *Cat Behavior: The Predatory and Social Behavior of Domestic and Wild Cats.* New York, NY: Garland STPM Press.

Liberg, O. 1983. Courtship behaviour and sexual selection in the domestic cat. *Applied Animal Ethology.* 10:117–132.

Liberg, O., Sandell, M., Pontier, D., et al. 2000. Density, spatial organization and reproductive tactics of the domestic cat and other felids. In *The Domestic Cat: The Biology of Its Behaviour* (2nd ed.), ed. D. C. Turner and P. Bateson, 119–147. Cambridge: Cambridge University Press.

Lichtsteiner, M., and D. C. Turner. 2008. Influence of indoor-cat group size and dominance rank on urinary cortisol levels. *Animal Welfare.* 17:215–237.

Lockhart, J., Wilson, K., and C. Lanman. 2013. The effects of operant training on blood collection for domestic cats. *Applied Animal Behaviour Science.* 143:128–134.

Lund, E. M., Armstrong, P. J., Kirk, C. A., Kolar, L. M., and J. S. Klausner. 1999. Health status and population characteristics of dogs and cats examined at private practices in the United States. *Journal of the American Veterinary Medical Association.* 214:1336–1341.

Luschekin, V. S., and K. V. Shuleikina. 1989. Some sensory determinates of home orientation in kittens. *Developmental Psychobiology.* 22:601–616.

Lyons, L. 2010. Feline genetics: Clinical applications and genetic testing. *Topics in Companion Animal Medicine.* 24:203–212.

Lyons, L. 2015. DNA mutations of the cat: The good, the bad and the ugly. *Journal of Feline Medicine and Surgery.* 17:203–219.

MacDonald, E., and P. Apps. 1978. The social behavior of a group of semi-dependent farm cats, *Felis catus*: A progress report. *Carnivore Genetics Newsletter.* 3:256–268.

MacDonald, D. W. Apps, P. J., Carr, G. M., and G. Kerby. 1987. Social dynamics, nursing coalitions and infanticide among farm cats, *Felis catus. Ethology.* 28(Suppl), 66.

MacDonald, D. W., Yamaguchi, N., and G. Kerby. 2000. Group-living in the domestic cat: Its sociobiology and epidemiology. In *The Domestic Cat: The Biology of Its Behaviour* (2nd ed), ed. D. C. Turner and P. Bateson, 95–118. Cambridge: Cambridge University Press.

Marques-Deak A., Cizza G., and E. Sternberg. 2005. Brain-immune interactions and disease susceptibility. *Molecular Psychiatry.* 10:239–250.

Martin, P. 1982. Weaning and behavioural development in the domestic cat. *Behaviour.* 99:221–249.

Martin, P. 1986. An experimental study of weaning in the domestic cat. *Behaviour.* 99:221–249.

Martin, P., and P. Bateson. 1985. The ontogeny of locomotor play behavior in the domestic cat. *Animal Behavior.* 33:502–510.

McCune, S. 1994. Caged cats: Avoiding problems and providing solutions. *Companion Animal Behaviour Therapy Study Group Newsletter.* 7:33–40.

McGinty, D. J., Stevenson, M., Hoppenbrouwers, T., et al. 1977. Polygraphic studies of kitten development: Sleep state patterns. *Developmental Psychobiology.* 10:455–469.

Mellen, J. D. 1991. Factors influencing reproductive success in small captive exotic felids (*Felis spp*): A multiple regression analysis. *Zoo Biology.* 10:95–110.

Mendl, M., and R. Harcourt. 2000. Individuality in the domestic cat: Origins, development and stability. In *The Domestic Cat: The Biology of its Behaviour* (2nd ed.), ed. D. C. Turner and P. Bateson, 47–64. Cambridge: Cambridge University Press.

Meurs, K. M., Sanchez, X., David, R.M., et al. 2005. A cardiac myosin binding protein C mutation in Maine Coon cat with familial hypertrophic cardiomyopathy. *Human Molecular Genetics.* 14:3587–93.

Moelk, M. 1979. The development of friendly approach behavior in the cat: A study of kitten-mother relations and the cognitive development of the kitten from birth to eight weeks. In *Advances in the Study of Behavior* (Vol. 10), ed. J. S. Rosenblatt, R. A. Hinde, C. Beer, et al. New York, NY: Academic Press.

Morgan, K. N., and C. T. Tromborg. 2007. Sources of stress in captivity. *Applied Animal Behaviour Science.* 102:262–302.

Morris, J. G. 2002. Idiosyncratic nutrient requirements of cats appear to be diet-induced evolutionary adaptations. *Nutrition Research Reviews.* 15:153–168.

Mosser, A., and C. Packer. 2009. Group territoriality and the benefits of sociality o the African lion, *Panthera leo. Animal Behaviour.* 78:359–370.

Natoli, E. 1985a. Spacing patterns in a colony of urban stray cats (*Felis catus*, L) in the historic centre of Rome. *Applied Animal Ethology.* 14:289–304.

Natoli, E. 1985b. Behavioural responses of urban feral cats to different types of urine marks. *Behaviour.* 94:234–243.

NAVS, National Anti-vivisection Society. 2019. Adopting cats and dogs form research facilities. https://www.navs.org/what-we-do/keep-you-informed/legal-arena/research/adopting-cats-and-dogs-from-research-facilities/#.XH2HMIhKiUk (accessed April 9, 2019).

Neilson, J. 2004. Thinking outside the box: Feline elimination. *Journal of Feline Medicine and Surgery.* 6:5–11.

NRC, National Research Council. 2006. Thermoregulation in cats. In *Nutrient requirements of dogs and cats.* Washington, D.C: National Academy Press.

NRC, National Research Council. 2011. *Guide for the Care and Use of Laboratory Animals.* Washington, D.C: National Academy Press.

O'Brien, S. J., Johnson, W., Driscoll, C., Pontius, J., Pecon-Slattery, J., and M. Menotti-Raymond. 2008. State of cat genomics. *Trends in Genetics.* 24:268–79.

Olmstead, C. E., Villablanca, J. R., Torbiner, M., et al. 1979. Development of thermoregulation in the kitten. *Physiology & Behavior.* 23:489–495.

Olmstead, C. E., and J. R. Villablanca. 1980. Development of behavioral audition in the kitten. *Physiology & Behavior.* 24:705–712.

Ormandy, E. H., and C. A. Schuppli. 2014. Public attitudes toward animal research: A review. *Animals* 4:391–408.

Ottway, D. S., and D. M. Hawkins. 2003. Cat housing in rescue shelters: A welfare comparison between communal and discrete-unit housing. *Animal Welfare.* 12:173–189.

Overall, K. L., and D. Dyer. 2005. Enrichment strategies for laboratory animals from the viewpoint of clinical veterinary behavioral medicine: Emphasis on cats and dogs. *National Research Council Journal, Institute of Laboratory Animal Resources.* 46:202–215.

Peeters, M., Courgnaud, V., Abela, B., et al. 2002. Risk to human health from a plethora of simian immunodeficiency viruses in primate bushmeat. *Emerging Infectious Diseases.* 8:451–457.

Plantinga, E. A., Bosch, G., and W. H. Hendriks., 2011. Estimation of the dietary nutrient profile of free-roaming feral cats: Possible implications for nutrition of domestic cats. *British Journal of Nutrition.* 106:S35–S48.

Podberscek, A. L., Blackshaw, J. K., and A. W. Beattie. 1991. The behaviour of laboratory colony cats and their reactions to a familiar and unfamiliar person. *Applied Animal Behaviour Science.* 31:119–130.

Prabhakaran, K., Suthanthirarajan, N., and A. Namasivayam. 1988. Biochmical changes in acute noise stress in rats. *Indian Journal of Physiology and Pharmacology.* 32:100–104.

Randall, W., Johnson, R. F., Randall, S., and J. T. Cunnigham. 1985. Circadian rhythms in food intake and activity in domestic cats. *Behavioral Neuroscience.* 99:1162–1175.

Raison, C. L., and A. H. Miller. 2003. When not enough is too much: The role of insufficient glucocorticoid signaling in the pathophysiology of stress-related disorders. *American Journal of Psychiatry.* 160:1554–1565.

Rochlitz, I, Podberscek, A.L., and Broom. 1998. Welfare of cats in a quarantine cattery. *The Veterinary Record.* 143:36–39.

Rochlitz, I., 2000. Feline welfare issues. In *The Domestic Cat: The Biology of Its Behaviour* (2nd ed.), ed. D. C. Turner and P. Bateson, 207–226. Cambridge: Cambridge University Press.

Rosenblatt, J. C. 1976. Stages in the early behavioural development of altricial young of selected species of non-priamte animals. In *Growing Points in Ethology*, ed. P. Bateson and R.A. Hinde, 345–383. Cambridge: Cambridge University Press.

Sales, G., Milligan, S. R., and K. Khirnykh. 1999. Sources of sound in the laboratory animal environment: A survey of the sounds produced by procedures and equipment. *Animal Welfare.* 8:97–115.

Sapolsky, R., 2004. *Why Zebras Don't Get Ulcers*. 3rd ed. New York: Henry Holt and Company, LLC.

Seitz, P. F. D. 1959. Infantile experience and adult behavior in animal subjects. II. Age of seperation from the mother and adult behavior in the cat. *Psychosomatic Medicine.* 21:353–378.

Sleeper, M. M., Kusiak, C. M., Shofer, F. S., et al. 2008. Clinical characterization of cardiovascular abnormalities associated with feline mucopolysaccharidosis I and IV. *Journal of Inherited Metabolic Disease.* 31:424–431.

Smith, D. F. E., Durman, K. J., Roy, D. B., et al. 1994. Behavioral aspects of the welfare of rescued cats. *Journal of the Feline Advisory Bureau.* 31:25–28.

Snowden, C. T., Teie, D., and M. Savage. 2015. Cats prefer species-appropriate music. *Applied Animal Behaviour Science.* 166:106–111.

Stella, J. L. Lord, L., and C. A. T. Buffington. 2011. Sickness behaviors in response to unusual external events in healthy cats and cats with feline interstitial cystitis. *Journal of the American Veterinary Medical Association.* 238:67–73.

Stella, J. L., Croney, C. C., and C. A. T. Buffington. 2012. Effects of stressors on the physiology and behavior of domestic cats. *Applied Animal Behaviour Science.* 143:157–163.

Stella, J. L. 2013. An investigation of environmental factors that affect the behavior and welfare of domestic cats (*Felis silvestrics catus*). PhD thesis, The Ohio State University.

Stella, J. L., Croney, C. C., and T. B. Buffington. 2014. Environmental factors that affect the behavior and welfare of domestic cats (*Felis silvestris catus*) housed in cages. *Applied Animal Behaviour Science.* 160:94–105.

Stella, J. L., and C. C. Croney. 2016a. Management practices of cats owned by faculty, staff, and students at two Midwest veterinary schools. *Science World Journal.* Article ID 7108374, 13 p.

Stella, J. L., and C. C. Croney. 2016b. Environmental aspects of domestic cat care and management: Implications for cat welfare. *Science World Journal.* Article ID 6296315, 7 p.

Stella J. L., and T. Buffington. 2016. Environmental strategies to promote health and wellness. In *August's Consultations in Feline Internal Medicine* (1st ed.), ed. S. Little, 718–736. St. Louis, MO: Elsevier.

Stella, J. L., Croney, C. C., and T. B. Buffington. 2017. Behavior and welfare of domestic cats housed in cages larger than U.S. norm. *Journal of Applied Animal Welfare Science.* 3:296–312.

Stella, J. L., and C. C. Croney. 2019. Coping styles in the domestic cat (*Felis silvestris catus*) and implications for cat welfare. *Animals.* 9:370 doi:10.3390/ani9060370.

Tarkosova, D., Story, M. M., Rand, J. S., and M. Svoboda. 2016. Feline obesity- prevalence, risk factors, pathogenesis, associated conditions and assessment: A review. *Veterinarni Medicina.* 61:295–307.

The Nobel Prize. The Nobel Prize in Physiology or Medicine. 1981. https://www.nobelprize.org/prizes/medicine/1981/press-release/ (accessed March 2019).

Thorn, F., Gollender, M., and P. Erikson. 1976. The development of the kitten's visual optics. *Vision Research.* 16:1145–1149.

Thorndike, E. L. 1911. *Animal Intelligence*. New York: MacMillan.

Turner, D.C., and Mertens, C. 1986. Home range size, overlap and exploitation in domestic farm cats (*Felis catus*). *Behavior.* 99:22–45.

Turner, D.C., and Meister, O. 1988. Hunting behavior of the domestic cat. In *The Domestic Cat: The Biology of Its Behaviour* (1st ed.), ed. D. C. Turner and P. Bateson, 111–121. Cambridge: Cambridge University Press.

Turner, D. 2014. Social organization and behavioral ecology of the free-ranging domestic cat. In *The Domestic Cat: The Biology of Its Behaviour* (3rd ed.), eds. D. C. Turner and P. Bateson, 63–70. Cambridge: Cambridge University Press.

Turner, D. 2017. A review of over three decades of research on cat-human and human-cat relationships. *Behavioural Processes.* 141:297–304

UFAW, Universities Federation for Animal Welfare. 1995. An ethogram for behavioural studies of the domestic cat (*Felis silvestris catus* L.) UK. https://www.ufaw.org.uk/shop/publications/product/an-ethogram-for-behavioural-studies-of-the-domestic-cat (accessed April 9, 2019).

USDA, United States Department of Agriculture. Annual report of animal usage by fiscal year. 2018. http://www.aphis.usda.gov/aphis/ourfocus/animalwelfare/SA_Obtain_Research_Facility_Annual_Report (accessed March 21, 2019).

Vanbetteray, J. N. F., Vossen, J. M. H., and A. M. L. Coenen. 1991. Behavioural characteristics of sleep in rats under different light dark conditions. *Physiology & Behavior.* 50:79–82.

Vigne, J. D., Briois, F., Zazzo, A., et al. 2004. Early taming of the cat in Cyprus at least 10,600 y ago. *Proceedings of the National Academy of Sciences of the United States of America.* 109:8445–8449.

Vinke, C. M., Godijn, L. M., and W. J. R. van der Leij. 2014. Will a hiding box provide stress reduction for shelter cats? *Applied Animal Behaviour Science.* 160, 86–93.

Waiblinger, S., Boivin, X., Pederson, V., et al. 2006. Assessing the human-animal relationship in farmed species: A critical review. *Applied Animal Behaviour Science.* 101:185–242.

Waisglass, S. E., Landsberg, G. M., Yager, J. A., and J. A. Hall. 2006. Underlying medical conditions in cats with presumptive psychogenic alopecia. *Journal of the American Veterinary Medical Association.* 228:1705–1709.

Weiss, J. M. 1971. Effects of coping behavior in different warning signal conditions on stress pathology in rats. *Journal of Comparative Physiology and Psychology.* 77:1–13.

Weiss, J. M. 1972. Psychological factors in stress and disease. *Scientific American.* 226:104–113.

West, M. J. 1974. Social play in the domestic cat. *American Zoologist.* 14:2427–436.

Wielebnowski, N. C., Fletchall, N., Carlstead, K., Busso, J. M., and J. L. Brown. 2002. Noninvasive assessment of adrenal activity associated with husbandry and behavioral factors in the North American clouded leopard population. *Zoo Biology.* 21:77–98.

Wyrwicka, W. 1978. Imitation of mother's inappropriate food preferences in weanling kittens. *Pavlovian Journal of Biological Science.* 13:55–72.

Wyrwicka, W., and A. M. Long. 1980. Observations on the initiation of eating of new food by weanling kittens. *Pavlovian Journal of Biological Science.* 15:115–122.

Yogalingam, G., Litjens, T., Bielicki, J., et al. 1996. Feline mucopolysaccharidosis type VI. Characterization of recombinant *N-acetylgalactosamine 4-sulfatase* and identification of a mutation causing the disease. *Journal of Biological Chemistry.* 271:27259–27265.

Yamamoto, J. K., Sanou, M. P., Abbott, J. R., and J. K. Coleman. 2010. Feline immunodeficiency virus model for designing HIV/AIDS vaccines. *Current HIV Research.* 8:14–25.

Yeon, S. C., Kim, Y. K., Park Se, J., et al. 2011. Differences between vocalization evoked by social stimuli in feral cats and house cats. *Behavioural Processes.* 87: 183–189.

Zoran, D. L. 2002. The carnivore connection to nutrition in cats. *Journal of the American Veterinary Medical Association.* 220:1559–156.

Zoran, D. L., and C. T. Buffington. 2011. Effects of nutrition choices and lifestyle changes on the well-being of cats, a carnivore that has moved indoors. *Journal of the American Veterinary Medical Association.* 239:596–606.

15 Behavioral Biology of Pigs and Minipigs

Sandra Edwards
Newcastle University

Nanna Grand
Scantox A/S

CONTENTS

Introduction 243
Involvement in Research 244
    Pharmaceutical Research 244
    Disease Models 244
Behavioral Biology 245
    Natural History 245
        Ecology and Feeding Behavior 245
        Social Organization and Behavior 245
        Reproductive Behaviors 246
    Common Captive Behaviors 247
        Social Behavior 247
        Foraging and Exploratory Behavior 247
        Reproductive Behavior 248
        Farrowing and Maternal Behavior 248
        Maintenance Behaviors 249
Ways to Maintain Behavioral Health in Pigs and Minipigs Involved in Research 249
    Housing 250
    Enclosure Environment and Enrichment 250
    Feeding 251
    Human–Animal Interaction and Training 251
    Commonly Applied Study Procedures and Welfare Issues 252
        Dosing 252
        Other Study-Related Procedures 253
Special Situations 254
Conclusions and Recommendations 255
Additional Resources 255
References 256

Introduction

Domestic pigs are descended from the wild boar in Europe, *Sus scrofa*, and Asia, *Sus scrofa cristatus*. Domestication first began 9,000 years ago, when foraging wild pigs were encouraged to associate with human habitation and were then enclosed (Porter 1993). Selective breeding initially focused on characteristics desirable for human food production, particularly large size, fatness, and greater prolificacy. Local selection meant that many different regional breeds developed, varying in size and color, with each adapted to their local environment. As production intensity increased in response to demand for plentiful and cheap food, breeding in many countries became the preserve of specialist companies, which have subsequently grown to multinational scale. Using more sophisticated genetic approaches and large population sizes, a small number of breeds came to predominate, with different genetic lines within these breeds selected for specific combinations of desirable traits. Amongst these, fast growth, efficient feed conversion, carcass leanness, and prolificacy, predominated.

While the use of the pig as a meat animal drove selection for many centuries, similarities to humans in body size and many biological characteristics made it an ideal model for many biomedical studies, and specialist lines were developed for this purpose. These lines prioritized smaller size and tractability for more efficient housing and easy handling. Some strains of minipigs are of natural origin (e.g., the Yucatan minipig), but many of the existing minipig breeds (e.g., the Göttingen minipig) have been purpose bred for use in different areas of research. A range of minipigs now exists, varying from 30 to 90 kg in mature body size, compared with 300–400 kg in their agricultural counterparts. However, despite the diversity in physical appearance of domesticated pigs, research has repeatedly shown that the behavioral biology of their wild ancestors has been genetically conserved. Even behaviors whose functions have been rendered unnecessary in domestic circumstances are still present in the ethogram of modern pigs, and are readily exhibited when they are returned to a seminatural environment (Stolba and Wood-Gush 1989). The strong motivation to perform behaviors that had a high survival value during the evolution of the species can give rise to behavioral problems in the domestic environment, if these motivational and behavioral needs are not met.

Involvement in Research

Both large pig breeds and minipigs are used as experimental subjects in biomedical research today. Larger pigs are often involved in experimental settings, where the size of the pig is not a disadvantage, for example when serving as surgical models, whereas minipigs are often the preferred choice in preclinical studies that are necessary for the development of potential new pharmaceuticals.

Pharmaceutical Research

Due to (1) the many similarities between minipigs and humans in terms of anatomy and physiology (summarized by Bode et al. (2010)), (2) their smaller size and lower body weight, and (3) the ever-increasing quantity of background data, the minipig is often used in regulatory studies, including toxicity and safety pharmacology studies. Regulatory acceptance of the minipig as an alternative model in preclinical studies has increased their involvement in experimental settings over the last decades.

The minipig breed that is most frequently used in pharmaceutical research today is the Göttingen minipig (Heining and Ruysschaert 2016). Ellegaard Göttingen minipigs have an adult weight of 35–45 kg, although breeding sows may be larger, reaching a bodyweight of approximately 60 kg. Other minipig breeds that have been frequently involved in biomedical research are the Sinclair, Hanford, and Yucatan minipigs. More rarely, the Clawn minipig and the Ossabaw minipig are used, but the involvement of these minipig breeds is limited, locally oriented, or for very specific scientific purposes (Köhn 2012).

Pigs and minipigs are the preferred species for dermal research and in wound-healing studies (Mortensen et al. 1998; Glerup 2012; Willard-Mack et al. 2016). Skin penetration of several drugs has been reported to be similar between pig and human skin, with low intra- and intervariation in vivo between Göttingen minipig, domestic pig, and human skin (Lin et al. 1992; Meyer and Neurand 1976; Reifenrath et al. 1984; Qvist et al. 2000). The wound healing process in porcine skin is considered similar to that seen in humans (Sullivan et al. 2001; Seaton et al. 2015). Wounds, including split-thickness wounds (removal of epidermis/superficial part of the dermis) and full-thickness wounds (removal of epidermis, dermis, and subcutis), can be established on fully grown minipigs.

The minipig is also considered to be an acceptable alternative nonrodent species for testing of immunomodulatory drug candidates (Heining and Ruysschaert 2016), and the predictive value of immunogenicity when testing biopharmaceuticals (e.g., antibodies) has been reported as being comparable to the predictive value of nonhuman primates (van Mierlo et al. 2013, 2014).

More specialized preclinical studies comprise reproductive and embryofetal toxicity studies, as well as juvenile studies. Reproductive studies and embryofetal studies are designed to identify potential risks to reproductive function/fertility and fetal development (EMA 2020). Göttingen minipigs are susceptible to known human teratogens (Jørgensen 1998) and are considered an appropriate alternative nonrodent species in reproductive toxicity studies for small molecules. Published data on the background incidence of congenital malformations and variations also emphasize their possible use (Ellemann-Laursen et al. 2016). However, pigs and minipigs are often considered unsuitable for investigations of biopharmaceutical proteins due to the protection of the fetus by the placenta, hindering fetal uptake of these compounds (van der Laan et al. 2010). Juvenile studies in animals support the development of drugs intended for the human pediatric population. They are included in drug development when the safety profile of the potential new pharmaceutical is expected to be different from that seen in adults, and to investigate if there is a risk of adverse effects on postnatal growth and development (EMA 2008; Baldrick 2013; Roth et al. 2013), including development of cognitive functions. Pregnant sows are often purchased from the supplier and piglets are born at the research facility. Mating and farrowing can be synchronized, and cross-fostering of piglets, which ensures comparable study groups and limits experimental bias, is easily performed. Minipigs are often preferable to the larger pig in reproductive, embryofetal, and juvenile studies, due to the smaller size of sows and piglets.

Disease Models

A number of disease models have been developed in the minipig, including the following examples.

Diabetes models and metabolic syndrome models: Spontaneous type 1-like diabetes is rare in pigs, but diabetes type I models have been developed surgically, by pancreatomy, and chemically, via streptozotocin- or alloxan-induced damage of the pancreatic beta cells (Larsen and Rolin 2004; Rolandsson et al. 2002; Stump et al. 1988). Models of milder forms of diabetes, characterized by fasting and postprandial hyperglycemia, as well as reduced insulin secretion and pancreatic β-cell mass, have also been described (Larsen et al. 2002). The metabolic syndrome is a cluster of conditions including, but not limited to, obesity and decreased insulin sensitivity. This syndrome is an increasing health concern in many countries and is considered to increase the risk of people developing type II diabetes and cardiovascular disease (Reaven 1995; Wilson et al. 2005). A model for the metabolic syndrome has been developed in minipigs (Johansen et al. 2001).

Atherosclerosis models: Atherosclerosis is the narrowing and hardening of arteries due to formation of plaques composed of fat, cholesterol, and other substances. It is one of the leading causes of stroke, heart attack, and peripheral vascular disease in humans. Atherosclerosis can be induced in porcine models with predilection sites similar to those in humans (Gerrity et al. 2001; Ludvigsen et al. 2015).

Melanoma models: Melanoma models have been established in Sinclair, MeLiM and MMS-Troll minipigs. There are notable differences between the swine melanoma models and the human disease, but as the swine melanoma models also show many similarities to malignant melanoma in humans, these models are useful in the continued study of melanoma development and host-tumor interaction (Bourneuf 2017; Hook et al. 1982; Millikan et al. 1974; Oxenhandler et al. 1982).

Behavioral Biology

The behavioral biology of the domestic pig can best be understood by studying the modern relatives of its wild ancestors in their natural environment. It was under these conditions that the porcine functional behavioral repertoire evolved to facilitate survival and reproductive success, and thus became genetically ingrained.

Natural History

Ecology and Feeding Behavior

Wild boar live in scrubland and forest margins in diverse coastal, woodland, and mountain regions around the world (Fradrich 1974). They forage in forests and surrounding grasslands, with incursions into farmland that cause significant damage. A group will have a typical home range of 5–20 km^2, depending on the density of food resources (Leaper et al. 1999). Different groups may have overlapping home ranges, but rarely interact. At night, group members sleep in close proximity in nesting areas protected by cover in thickets or copses. They emerge at dawn and dusk to spread out for synchronized foraging, punctuated by a rest period in the middle of the day (Gundlach 1968). Wild boar have relatively few natural predators currently, as large carnivores have been marginalized by human influence in many countries, but are hunted by humans for both sport and pest control. In areas where hunting intensity is high, the group may show a more nocturnal activity pattern to avoid contact with humans (Lemel et al. 2003).

The wild boar is a true omnivore and their diet changes according to the seasonal availability of different foodstuffs (Schley and Roper 2003). In the summer, they will graze on grasses and shrubs, while in the autumn they will feed on fallen fruit and nuts. Throughout the year, much of their nutrition is obtained by digging for roots and invertebrates. Although typically 90% of the diet comes from plant material (Pinna et al. 2007), they will also eat small amphibians, reptiles, and rodents, and will scavenge on dead carcasses (Gundlach 1968; Dardaillon 1987; Massei et al. 1996). Finding sufficient food requires them to forage for the majority of their active time each day (Fradrich 1974).

Social Organization and Behavior

Wild boar live in matriarchal family groups, typically consisting of 2–6 related adult females and their subadult offspring (Mauget 1981; Dardallion 1988). Group sizes vary, depending on habitat and available food supply, but can contain up to 30–40 individuals. Males leave the group at 6–10 months of age, as they reach puberty, and form small bachelor groups of typically 2–3 individuals, or roam as solitary adult males (Fradrich 1974). Their home ranges may overlap with a number of female groups, but they only interact with the females during the breeding season. The matriarchal group is dynamic in structure, with breeding females isolating themselves for farrowing in spring, and re-entering the group when their new litter of piglets is 2–3 weeks of age. As the piglets are progressively weaned, they begin to operate increasingly as a separate group made up of the juveniles from all litters of that year. This grouping associates loosely and shares a home range with the maternal group. When females reach puberty at 8–10 months of age, they may rejoin the core maternal group or split off to form a new group (Kaminski et al. 2005).

Within the family group of the adult females, juvenile offspring, and, for periods of the year, an adult male, a stable dominance hierarchy exists. This is based on age and body size, giving the higher ranking animals priority access to resources (Mauget 1981). There is usually little overt aggression, with dominance being enforced by more subtle signaling, although aggression can occur during competition for food or breeding opportunities. Aggression between unfamiliar animals takes the form of pushing and biting, as the pigs stand in inverse parallel position, giving rise to wounds on the shoulder, head, and ears (Jensen, 1980), which can be severe if inflicted by the sharp tusks and lower canines of male pigs. Conflicts end when the defeated animal flees and manages to elude further pursuit if, as is sometimes the case, it is pursued by the victor.

Wild boar use all of their senses in communication, but as they evolved to spend much of their time amongst trees and undergrowth, the olfactory and acoustic channels predominate. Other individuals are recognized by their characteristic odor (Fradrich 1974; Mendl et al. 2002), and information about a pig's emotional state can also be transmitted by chemical signals (McGlone 1985a; Vieuille-Thomas and Signoret 1992). Pheromones play an important role in the behavioral biology of the species. They are deposited by rubbing on trees and bushes to mark trails within the home range (Spitz 1986). The male pheromones produced by the preputial and salivary glands are important in stimulating estrus and receptivity in the female (Booth and Signoret 1992), while a maternal pheromone is also produced by glands on the udder of the sow, which has a calming effect on the offspring (McGlone and Anderson 2002). Pigs also exhibit a variety of vocal signals used in different contexts (Tallet et al. 2013). These range from contact grunts within the group, to barks signaling alarm, and squeals and screams signaling distress. Specific grunting patterns are employed by the boar during courtship and by the sow during nursing. Although vision is the least important of their senses, pigs do use visual signals, such as threats and submissive gestures, in daily social interactions (Jensen 1982), and respond to sudden movements in their environment that might indicate the presence of a threat.

Reproductive Behaviors

Wild boar are seasonal breeders, with reproductive activity regulated by photoperiod (Fradrich 1974; Tast et al. 2001). During the long days of summer, the females enter an anestrus phase. As daylength begins to shorten, the females come into estrus, with the precise timing of this dependent on food supply (Spitz 1986). If food is plentiful, breeding begins in early autumn, giving the possibility of a second annual litter, conceived after an early spring farrowing, and before the summer anestrus period. If food is scarce, breeding is delayed until late autumn and only a single litter is reared in that year (Mauget 1981). Estrus is synchronized within the group by pheromonal cues, and during this time, the group is approached by adult males who compete aggressively to become the breeding male for that group with access to the females. Estrus lasts for approximately 48 h and recurs at intervals of approximately 21 days until the female is pregnant. A female in estrus will seek out the boar to solicit mating. Courtship behavior of the boar includes, champing to spread pheromone-laden saliva; a rhythmic grunting, known as chanting; and vigorous nudging of the body of the female (Fradrich 1974). This induces the female, when at the peak of estrus, to show a "standing response", in which she will stand rigidly and allow mounting by the boar. Once intromission occurs, the penis of the boar locks in the cervix of the female and ejaculation occurs. Mating can take 3–20 min before the male disengages and reoccurs at intervals over the receptive period (Signoret 1970). Once all females in the group are pregnant, the successful breeding male will leave the group to resume a solitary existence until the next breeding season.

After a gestation period of approximately 115 days, the female gives birth in the spring (Mauget 1981). A few days before the time of parturition, the female will start to isolate herself from the group and wander long distances in search of a suitable nest site. Nest site selection, governed by concealment from predators and protection from the weather, is typically among bushes and enclosed by slopes, rocks, or branches, with a clear outlook for danger (Gundlach 1968; Spitz 1986). The sow shows intense nest building behavior in the last 24 h before farrowing, initially rooting and pawing to hollow out the site, and then gathering grass, leaves, and twigs to bring back to the nest site to arrange into a large circular bowl, often covered with longer branches (Gundlach 1968; Fradrich 1974).

At the onset of parturition, the female remains in the nest, periodically standing and turning around as contractions begin. When the first offspring is born, the mother will often stand and inspect it, before lying down again for further deliveries that occur at intervals averaging 20–30 min. Unlike the behavior shown by most ungulates, the porcine young are never licked, but are occasionally nosed if they approach the head of the sow or if she briefly stands between deliveries (Gundlach 1968; Fradrich 1974). As more piglets are born, the female typically becomes very inactive, which reduces the risk of crushing the newborns and facilitates their quick access to the udders to suckle. Litter size in the wild boar is typically 3–7 piglets, depending on food availability and age of the dam (Mauget 1981; Leaper et al. 1999), far below the litter size of 10–20 piglets now commonly seen in domestic pigs. First-time wild boar mothers in a farmed environment may sometimes show fear of the offspring and behave aggressively, even biting and killing them (Harris et al. 2001), although it is not known how common this is under natural conditions.

The newborn piglets are active within a few seconds, standing and nosing the body of the mother as they seek the udder (Gundlach 1968; Fradrich 1974). They are guided in their search by the sound of maternal grunting, by the hair pattern on the mother's body, and by the warmth of the naked udder in contrast to other hair-covered body regions. Once at the udder, they will locate an unoccupied teat and suckle, often moving from teat to teat as they seek those with the best yield. After initial competition between the siblings, the best teats are appropriated by the strongest individuals and, over the course of a few hours, a stable teat order begins to develop, with each piglet showing fidelity to its own teat (Lewis and Hurnik 1985). For the first few hours, the flow of milk is continuous, but then starts to become episodic, with short periods of milk ejection interspersed by longer periods when no milk is available, setting up the cyclical suckling that characterizes the highly specific nursing pattern for this species (Fraser 1980). The nursing bout begins when the sow summons the piglets with nursing grunts, although as the piglets age, they may try to initiate a bout by massaging the udder. The piglets then line up at their respective teats and start vigorous massaging activity. This stimulates oxytocin production and milk ejection, which is signaled by an increasing grunt rate from the sow, and the piglets change their behavior to quiet intense suckling. The ejection lasts only for 30–40 s, after which the piglets again show vigorous massaging behaviors, believed to stimulate blood flow to the gland and enhance future milk yield. The nursing bout terminates when the piglets fall asleep at the udder, or when the sow rolls over or stands up to remove udder contact. Nursing initially occurs at approximately 20 min intervals, with the interval increasing as the piglets become older. This characteristic pattern of nursing behavior, seen in wild boar and domestic pigs, has evolved to maximize piglet survival in natural conditions. The strong teat fidelity, with unused teats drying up within a few days, ensures that piglets in small litters exploit the most productive teats, and minimizes subsequent needless energy expenditure in teat disputes. The lack of unused productive teats, and the short and well-synchronized nursing bouts, reduces the risk of competition at the udder from piglets of other sows once the litter rejoins the family group.

For the first 2 days after giving birth, the sow remains in the nest with the piglets. She then begins to make short trips away from the nest to feed, drink, and defecate, returning to nurse at regular intervals. Toward the end of the first week, the piglets start to follow their mother away from the nest for increasing periods of time, until finally, at approximately 2 weeks of age, the sow and litter quit the nest and rejoin the other sows in the family group (Gundlach 1968; Fradrich 1974). Nursing continues for up to 4 months, with piglets mainly suckling from their own mother, who they recognize by her odor and call. The milk supply of the sow decreases over time and the interval between sucklings increases, as the mother initiates fewer sucklings and terminates them more frequently. This encourages the piglets to forage for solid food, often roaming together with the other juveniles in the family group, and their nutritional dependence on the mother gradually decreases (Gundlach 1968; Newberry and Wood-Gush 1988). While suckling is frequent and intense, the sow is prevented from coming back into estrus. As the suckling intensity diminishes, the decreasing autumn photoperiod and stimuli provided by the return to the group of a mature boar induce a resumption of estrus and the mother may become pregnant again in the later stages of lactation.

Development of species-typical behavior patterns begins in the young from an early age. Foraging behavior, including searching (approach and investigation of novel areas), rooting (using the nose to dig in the ground or move objects), and food sampling (ingestion of substrates), can be seen as soon as the piglets leave the nest (Gundlach 1968). As they become older, while still interacting with their mother, the piglets spend increasing time with other juveniles carrying out independent exploration around the home range of the group. Social interactions occur with piglets from other litters, although there are long-lasting preferences for association with littermates. Play behaviors include play fighting, which may play a role in determining dominance relationships, as well as locomotor play and object play. Brown et al. (2005) give a fuller ethogram of play behaviors. While play can be solitary, social facilitation within the juvenile group is often seen, by which many individuals are stimulated to perform the same activity at the same time (Fradrich 1974). Play is most intense in the first 2 months of life, before decreasing as weaning progresses, and is rarely seen in adults.

The pigs maintain a clear distinction between resting, activity, and elimination areas. As they become older, rather than resting with their mother, the juveniles nest together adjacent to the adult group. All individuals move away from their nest sites to eliminate in corridors amongst nearby bushes. Maintenance behaviors include skincare through rubbing on bushes and wallowing in mud (Gundlach 1968; Fradrich 1974). This latter behavior also serves a thermoregulatory function (Bracke 2011) and increases in hot weather, while in cold conditions, the pigs will huddle together for warmth and increase the amount of vegetation insulating their sleeping nests.

Common Captive Behaviors

It has been repeatedly demonstrated that, even after many generations of domestication, modern pigs and minipigs still retain the same behavioral predispositions as their wild ancestors and relatives. If placed in a natural or seminatural environment, they will show the full spectrum of behaviors described in section *Natural History*, with an ethogram reported to comprise 103 different elements (Stolba and Wood-Gush 1989; D'Eath and Turner 2009). However, opportunities to express many of these behaviors are constrained by the housing and management procedures typically employed at modern pig farms or research facilities. This can give rise to the development of abnormal behavior patterns indicative of impaired welfare. Abnormal behaviors can also sometimes be damaging toward other individuals in the group.

Social Behavior

Domestic pigs, with the exception of the lactating sow and litter, are usually kept in pens where all members of the group are of uniform size and age. They also commonly experience regrouping with unfamiliar pigs. These situations, differing markedly from the natural social environment, increase the frequency of injurious fighting, as individuals seek to establish new dominance relationships against well-matched opponents (Rushen 1990). A quantitative ethogram of aggressive and submissive behaviors in recently regrouped pigs is given for growing pigs by McGlone (1985b), and for sows by Jensen (1980). Aggression can be further increased by a lack of space, which impairs avoidance behavior and the exhibition of species-typical submissive behaviors to avert attacks (Jensen 1982). Resources, such as feed, may be limited in quantity or accessibility in captivity, giving rise to competition and aggression.

Foraging and Exploratory Behavior

Pigs in the wild live in large complex environments and spend the majority of their active time foraging and exploring. The scope for exhibiting these behaviors is more limited for domestic pigs. Morrison et al. (2003) present a quantitative ethogram for pigs housed in large, deep-bedded pens or smaller unenriched pens. When straw is available, pigs show an extensive ethogram of more than 26 different straw-related behavioral elements (Day et al. 2001). When housing is restrictive in area and relatively barren in structure, the strong exploratory drive of pigs can be redirected toward groupmates as massaging and chewing behaviors, which can develop into injurious tail, ear, and flank biting in extreme situations (Taylor et al. 2010; see Table 15.1). Foraging and exploratory motivation is increased in situations where the animal experiences hunger. This can be the case if growing animals are food-restricted for commercial or experimental reasons. However, it occurs most commonly in breeding animals that receive a nutrient-dense, compound diet once or twice daily, in quantities that can be consumed within 5–10 min. This dietary restriction is necessary to prevent obesity but, whilst nutritionally adequate for health and productive performance, it leaves the animal unsatisfied and highly motivated to obtain more food. The consequent foraging behavior has no appropriate means of expression in a barren environment and can become channeled into a variety of abnormal oral behaviors, such as bar biting, chain chewing, or vacuum chewing (Lawrence and Terlouw 1993; see Table 15.1). Over time, such behaviors can develop into stereotypies – repetitive behaviors, fixed in form and performed for large periods of the day (Mason 1991). These stereotypies are widely recognized to be indicative of impaired welfare resulting from environmental inadequacy.

TABLE 15.1

An Ethogram of Abnormal Behaviors in Domestic Pigs

Behavior	Description
Apathy	The pig stands or sits inactive for long periods of time and is unresponsive to external events
Stereotyped Oral Behaviors:	
Bar biting/chain chewing	The pig takes a bar of the pen, or other solid object such as a hanging chain, into its mouth and performs a prolonged repetitive sequence of mouthing, chewing or licking elements
Vacuum chewing	The pig performs a prolonged repetitive sequence of chewing or sucking mouth movements without any material in the mouth
Injurious Oral Behaviors:	
Tail biting	The pig takes the tail of another animal into its mouth and chews or bites it, eventually causing bleeding lesions and partial or total amputation
Ear biting	The pig chews or bites the ear of another animal, eventually causing bleeding lesions
Flank biting	The pig massages and chews repeatedly at a specific site on the flank of another animal until a raw and bleeding circular lesion develops
Vulva biting	The pig bites the vulva of another animal, causing a bleeding lesion – seen in pregnant sows, most commonly in late gestation
Other Redirected Exploratory/Foraging Behaviors:	
Belly nosing	The pig massages vigorously with its nose on the belly of another animal for a prolonged period of time – most commonly seen in early weaned piglets, but can persist even into adulthood
Anal massage	The pig massages vigorously with its nose on the anus of another animal for a prolonged period
Tail in mouth behavior	The pig takes the tail of another animal into its mouth and sucks or chews it repeatedly – this may develop further into tail biting behavior
Other Abnormal Behaviors:	
Excessive aggression	One or more pigs repeatedly attack and bite another individual, causing lesions, exhaustion, and sometimes death
Excessive mounting	The pig repeatedly places both front legs over the front or back of another animal, thrusts with its hips and sometimes achieves anal intromission and ejaculation – most commonly seen in uncastrated male growing pigs and may cause lameness and spinal injury in repeated recipients
Maternal aggression	A sow attacks her newborn piglets and may injure or kill them, sometimes accompanied by cannibalism – most commonly seen in first-time mothers

Reproductive Behavior

While domestic pigs retain some influences from their ancestry as seasonal breeders, they now breed all year round. However, their breeding patterns are usually strictly controlled and it is common to use exogenous hormones to induce or synchronize estrus. Natural mating may still occur, by brief introduction of a boar when females are in estrus; however, it is much more common to use artificial insemination. Semen is collected from boars, which are usually housed in individual pens with limited social contact with any other pigs. The sows are then artificially inseminated while restrained in stalls and receive minimal experience of the natural cues of courtship behavior. This lack of mating stimulation can be detrimental to reproductive function (Hemsworth et al. 1978).

Farrowing and Maternal Behavior

While the wild boar has relatively few piglets, domestic pigs have been selected for greater prolificacy. However, the increase in piglet number has come at the expense of individual birthweight and vigor, and results in greater competition for teat access and milk supply. Consequently, mortality from chilling, starvation, and crushing by the dam is increased, and frequently exceeds 10% of all piglets born (Baxter and Edwards 2018). To minimize these losses, specialized housing systems have been developed in which the sow is closely confined within a crate to reduce the risk of piglet crushing and to make it easier to provide supplementary heat and human intervention to help weak piglets. This close confinement of the sow, from a few days before farrowing, prevents the normal behaviors, hormonally induced at this time, of increased ambulation and nest site searching. Furthermore, together with the typical absence of plentiful substrate for building a nest, it precludes the appropriate expression of nest-building behavior as farrowing approaches (Baxter et al. 2018). At this time, sows in crates become very restless, pawing at the ground and biting the bars of the crate. This elevated activity can continue into the delivery period and, together with the elevated levels of stress hormones observed, can prolong the farrowing process and increase the risk of stillbirth and crushing. As farrowing progresses, the sow normally becomes calmer and less active but, in some cases, and especially with first-time mothers, fearfulness of the newborn piglets and the inability to move away from them can lead to fatal savaging events (Chen et al. 2008; see Table 15.1).

With the larger litter sizes in domestic pigs, the number of piglets born can sometimes exceed the number of functional teats on the sow, and there is high variability in size at birth among littermates. As a result, the smaller piglets may be unable to compete and obtain adequate milk. To address this problem, it is common to cross-foster piglets between litters to increase survival chances. Cross-suckling has been reported in wild boar (Mauget 1981), but it is not a common occurrence.

Unless carried out immediately after farrowing, cross-fostering can lead to short-term distress of both piglets and sows, who know from odor cues that alien individuals are present. The disruption of the stable teat order also leads to increased fighting between piglets at the udder, causing facial lacerations and interrupted suckling bouts (Robert and Martineau 2001).

In order to maximize the reproductive output of the sow, her offspring are weaned after a short lactation period to remove the suckling-induced suppression of estrus and initiate the next breeding cycle. Typically, the piglets are weaned abruptly at 3–5 weeks of age and moved to nursery accommodation, where they may be mixed with unfamiliar piglets from other litters for the first time. This situation contrasts markedly with the gradual weaning and social integration processes that occur in nature. As a result, elevated aggression is seen, and piglets that have little previous experience of finding and consuming solid food may take several days before achieving energy balance and establishing a stable eating pattern. Immaturity of the digestive system results in high susceptibility to enteric disease during this period (Pluske et al. 2018). The abrupt cessation of maternal contact and suckling also has consequences for the behavioral development of the piglets (Fraser et al. 1998). The massaging and sucking behaviors normally directed toward the udder during nursing bouts are redirected toward other piglets and can develop into stereotyped belly nosing, occurring for long periods of time and disturbing resting in the group (Table 15.1). Once developed, these behaviors can persist for many weeks after weaning (Gonyou et al. 1998).

Maintenance Behaviors

Domestic pigs are usually kept in very controlled housing conditions, and the limited space and environmental uniformity can make it more difficult to correctly establish functional areas within the pen. All pigs normally maintain a clear distinction between resting, activity, and elimination areas (Petherick 1983; Andersen et al. 2020), but this can become blurred in overcrowded situations, resulting in poor hygienic conditions. Because of the uniformity of the environment, they are less able to make appropriate behavioral choices when faced with environmental challenges. This is most apparent in the case of thermal challenge, since they often have no, or inadequate, bedding to make insulating nests when cold, and no mud baths for wallowing and cooling off at high temperatures. As a result, they can only lose heat by wallowing in excreta, which can compromise hygiene (Huynh et al. 2005).

Ways to Maintain Behavioral Health in Pigs and Minipigs Involved in Research

Generally, more information and research are available on large pig welfare within farm environments, but this knowledge is useful for pig and minipig caregivers in experimental facilities, as basic behavior patterns are again rooted in the behavior of the wild boar (Ellegaard et al. 2010). Experimental protocols used in research facilities will, however, most likely expose pigs to stressors different from those seen in farm situations, and a direct comparison may not always be possible. Stressors in the research environment include handling and dosing with test items, and minipigs may also be affected by the test item in ways that compromise welfare.

As no single definition exists for defining animal welfare, it can be challenging to decide which parameters must be fulfilled to optimize animal welfare in research facilities. Often three main elements are included when good animal welfare is discussed: the normal biological functioning of the animal, the emotional state of the animal, and the animal's ability to express certain normal behaviors (Fraser 2008). Animal welfare is often discussed using different concepts ("the concept of natural living", "the perspective of biological functioning", and "the concept of affective state"), each focusing on different aspects considered to be important for well-being of the animal. The "concept of natural living" (Learmonth 2019), implies that animals should be kept under conditions that mimic the natural environment the animals would live in, if not captive. This is very rarely possible in experimental settings. However, the ability to perform natural behaviors, if it is not possible to recreate a natural environment, is also embraced within this concept (Stolba and Wood-Gush 1984; Fraser 2008; Marchant-Forde 2015), and it is important that inherent behavioral and/or physiological needs are considered in research facilities to avoid compromising welfare. In the "perspective of biological functioning", measurable parameters are used to define animal welfare (Marchant-Forde 2015) and, thus, animals in good condition, with good health and growth and, if needed in reproductive toxicology or juvenile studies, good reproductive performance, are considered to have their needs met. This is a convenient approach to take in experimental facilities, but is regarded as an oversimplified approach to define animal welfare. The "concept of affective state" is more subjective and refers to whether the animals are experiencing pleasant or unpleasant emotions (Marchant-Forde 2015). Absence of negative emotions, such as chronic pain or fear, is an important feature when discussing good animal welfare (Fraser et al. 1997), but is not always easy to monitor objectively. Knowledge and experience of pig and minipig behavior are necessary to detect any changes that may imply a welfare issue. It must also be remembered that any induced stress is also considered likely to affect the outcome of studies (Bailey 2018; Everds et al. 2013), and it is therefore very important to limit the stress induced on the pigs/minipigs as much as possible in regulatory studies.

Use of animals in research, including welfare aspects, is governed by both national and international legislation. In Europe, Council Directive 2010/63/EU on the protection of animals used for scientific purposes is based firmly on the 3Rs principle (Replacement, Reduction and Refinement), which were introduced by Russell and Burch in "The Principles of Humane Experimental Technique" in 1959, and later further discussed and evolved. Today "Replacement" refers to methods that avoid or replace the use of animals. "Reduction" refers to methods that minimize the number of animals used in experiments, but that still provide data that are robust and reproducible. "Refinement" refers to methods that minimize animal suffering and improve welfare (http://altweb.jhsph.edu/pubs/books/humane_exp/het-toc; Tannenbaum and Bennett 2015; https://www.nc3rs.org.uk/the-3rs). It is recommended to consider and include elements from all three welfare concepts when establishing an animal welfare program in research facilities, and procedures should be in place to ensure that action is taken if welfare is compromised due to the test item or test procedure.

Housing

Pigs and minipigs are social animals. According to guidelines, group housing is preferable when housing pigs and minipigs involved in research (Council of Europe, ETS123 Appendix A, 2006; National Research Council 2011); however, the natural social grouping should always be considered. Young minipigs can be kept together in gender-separated groups. As minipigs used for research are often obtained from a breeder, and not bred in-house, information (if available) on which animals have been housed together before arrival should be taken into account. If the minipigs already know their penmates, mixing with unknown pigs may be avoided, limiting the need for them to establish a new hierarchy. Often females can be kept in groups of 2–3 (or even more, if appropriate) during the complete length of the study. Young males can also be kept in smaller groups, and for studies of shorter duration, they may not need to be separated. It is not possible to give an exact age range where males can be housed together, as they may start to fight suddenly, even when together with well-known penmates. It has been reported likely that males less than 1 year of age can be housed together; however, when performing studies of longer duration, it must be expected that separation of the males will be necessary (Council of Europe, ETS123 Appendix A, 2006; Ellegaard et al. 2010). This should already be considered when allocating pens and space for the study. Even though a stable hierarchy appears to form rather quickly in minipig groups (Søndergaard et al. 2007), it should be kept in mind that the lowest ranking minipig may be chronically stressed due to dominance behavior shown by the other pigs in the group (Ruis et al. 2001a). If a subordinate minipig appears to show hierarchy-related stress, consideration should be given to splitting the group or housing the subordinate minipig alone for the remaining part of the study. Individual housing may also be necessary in some studies applying routine experimental procedures. In dermal studies, the test item is often applied on a large area on the back of the minipig, which is then covered for a variable period of time. If minipigs are group housed, this may lead to removal of the cover dressing from individual pigs and unintended exposure to penmates. Isolation of pigs has been associated with chronic stress (Ruis et al. 2001b) and, when single housing is necessary during a study, the minipigs should always be able to see, smell and, if possible, even touch other minipigs. Pens partly divided by bars make contact between pigs possible (Figure 15.1). When minipigs are able to see, smell, and even touch other minipigs, they do not generally appear to be adversely affected by single housing during experimental studies. However, it is most important that caregivers pay attention to any change in the behavior of single housed minipigs and to discover if single housing is affecting individual animals negatively.

It may also be desirable to separate minipigs during certain procedures or at feeding. Pens can be designed so they can be used for both single and group housing, and this allows for easy separation of the animals (Figure 15.2).

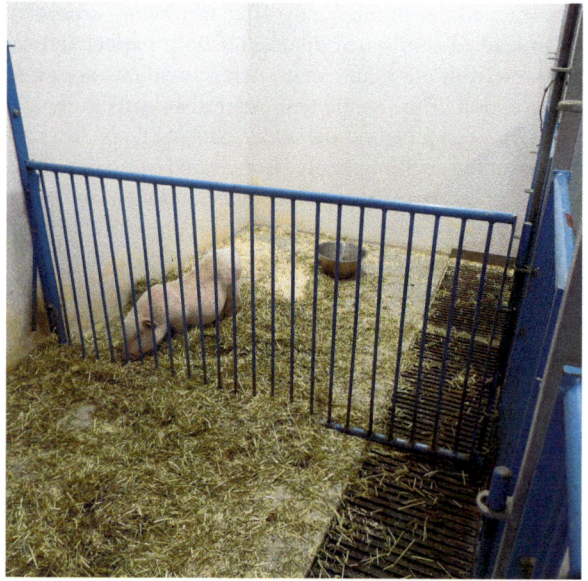

FIGURE 15.1 Minipig pens with barred partition allowing visual, auditory, olfactory and limited tactile contact between pigs.

FIGURE 15.2 Minipig pens with plastic divider/slide allowing temporary separation of animals.

Enclosure Environment and Enrichment

When housing minipigs, it is important to consider the thermoneutral zone (also described as the metabolic neutral temperature), as they are sensitive to environmental temperature and temperature changes. The thermoneutral zone varies between age groups and also slightly between small and large swine breeds. Minipig newborn piglets should have access to a resting place with an environmental temperature of approximately 32°C and adolescent or adult minipigs should be kept at a temperature between 15°C and 25°C (Bollen and Ritskes-Hoitinga 2004, Ellegaard et al. 2010). For larger pig breeds, the thermal comfort zone has been reported to be between 32°C and 38°C for newborn piglets and between 10°C and

21°C for adolescent and adult pigs (Stewart and Cebezón, 2016). It has been proposed to either keep the whole room at a temperature within the thermoneutral zone, or to allow different microclimates within the pen that provide a choice of climate. Different microclimates can be introduced by various methods, for example, using bedding in one part of the pen or using a heat lamp. When temperature is controlled, the relative humidity is believed to be relatively unimportant, as long as it is kept within a range of 45%–75% (Bollen et al. 1998; Ellegaard et al. 2010).

Minipigs are exploratory by nature, and rooting, chewing, and checking scents are considered exploratory behaviors (Ellegaard et al. 2010). According to Studnitz et al. (2007), materials that are complex, changeable, destructible, and that contain some edible parts are best at stimulating exploratory behavior in pigs. Thus, even if it is not possible to create an environment that mimics the natural environment in research facilities, it is possible to introduce elements aimed at satisfying exploratory motivation. The use of bedding is one way to stimulate and satisfy rooting and foraging behavior, and to provide material for nest building and comfortable resting. Various types of bedding exist, from long straw bedding to chopped straw, wood shavings, and shredded paper. Of these, long straw bedding is often regarded as optimal (Lahrmann et al. 2015). If possible, floor feeding can also be used to stimulate foraging. Barriers that subdivide the pen, or ramps that the minipigs can stand on, are also examples of enrichment that can be applied in experimental facilities. However, such enhancements may make it more difficult to clean the pen.

Additional enrichment materials or "toys" have been investigated for use in research facilities, as bedding material has occasionally not been considered feasible to use in experimental settings. Generally, minipigs are not easily kept active using toys, as interest drops when objects are familiar (Moustgaard et al. 2002). "Toys" that may be used include rubber sticks or metal chains that are attached to the pen walls to stimulate chewing. Additional materials can be attached to the chains to make them more interesting. If metal chains without attachments are used, cheek teeth may wear down over time, which must be considered if the minipigs are used for dental research. Treat dispensing chew balls, other plastic balls, plastic cones, or even scented plastic items, if the study design allows it, can also be used for enrichment (Huntsberry et al. 2008; Smith et al. 2009) (Figure 15.3). However, such "toys" should preferably only be supplied as an addition to bedding material and are not considered to be suitable as the only kind of enrichment offered to pigs (EFSA 2007; Bracke and Koene 2019).

Feeding

Special diets have been developed for minipigs used in research. For regular preclinical toxicity and pharmaceutical studies, high-fiber diets can be used to satisfy hunger and prevent obesity. *Ad libitum* feeding is often not possible when housing minipigs for research purposes, and some of the most widely available breeds will become obese when fed *ad libitum*. Female Göttingen minipigs are more prone to obesity when receiving excess diet than males. Göttingen minipig males may develop serous fat atrophy in the marrow cavity of bones, if they are kept on a restricted diet, and therefore, it is recommended to feed males a greater quantity of diet than females (Bollen and Skydsgaard 2006). Floor feeding can increase activity and time spent searching for and eating food. In research facilities, and especially in preclinical studies, floor feeding is often not possible, as food consumption may be a parameter that is measured and included in the study evaluation. However, if possible, it should be considered and, at the least, used for any stock animals.

Human–Animal Interaction and Training

Human–animal interaction and the socialization of pigs or minipigs at the breeding facility is important. It is recommended that procedures for human-animal interaction,

FIGURE 15.3 Pen enrichment for minipigs using (left) hanging toys and (right) plastic rooting ball.

socialization, and even training are described in standard operating procedures to ensure that they are performed uniformly. Transport of pigs and minipigs from the breeding facility to the research institution may be of variable duration, and they need an acclimatization period before being included in a study. It is advisable to make good use of the acclimatization period to socialize the pigs or minipigs, so that they develop a comfortable relationship with their caregivers. For minipigs, social interactions can be initiated on the day after arrival, but caregivers must be aware of the correct ways to approach them, as abrupt movements and loud voices/noises may frighten them. Food pellets can also be offered. A training program can be initiated as soon as the animals are no longer frightened of humans and will accept treats from them. Ellegaard et al. (2010) have described a workable acclimatization and training program. Minipigs are curious animals and are often very responsive to human contact and are easy to train for many tasks (Tsutsumi et al. 2001; Gieling et al. 2011). They can be trained in procedures unrelated to dosing, such as being (1) walked to another room, (2) weighed, and (3) placed in a sling for ophthalmoscopy or electrocardiogram measurement. Often, it is advisable to train all experimental procedures, but preclinical studies may involve a large number of minipigs, and training may not always be feasible for all of the procedures in the protocol. Training should then occur for those few procedures that will be performed many times during the course of the study. For example, animal subjects may be trained to walk onto a scale for weighing or into a crate for dermal application of a test compound (Figure 15.4), procedures that they will experience on a daily or weekly basis. They are less likely to be trained for blood sampling, or other procedures, that may only be performed a few times during the entire study. Clicker training can be used effectively in minipigs (Blye et al. 2006), and can be used both for dosing-related procedures (e.g., intranasal dosing) and procedures not directly related to dosing (e.g., walking onto a scale or into a crate for dermal dosing). Normal diet can often be used for rewards during training, and minipigs most often learn quickly if the training is performed correctly.

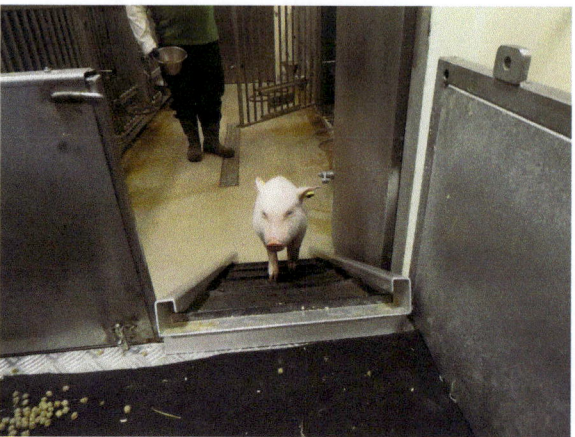

FIGURE 15.4 Minipig trained to walk into a dermal dosing crate.

Commonly Applied Study Procedures and Welfare Issues

Dosing

Dosing of minipigs in a preclinical study is performed in various ways. Commonly used administration methods are oral dosing, dermal dosing, and subcutaneous/intramuscular dosing. Less common dosing routes are buccal administration, nasal dosing, or dosing via implanted catheters, placed either in blood vessels or abdominal cavities/organs. The handling and restraining of the minipig will impose some stress on the animals during the experimental studies, and when training can be applied, it creates a calmer environment, both for the minipigs and the technicians.

Oral Dosing

Oral dosing by gavage is generally the preferred scientific method when it is necessary to ensure that the pig or minipig receives an accurate amount of test substance, which is often the situation in experimental research. When administering capsules, tablets or pills to minipigs, capsule/tablet/pill size must be considered, as it is generally believed that minipigs are not able to swallow the largest available capsules/tablets/pills. Depending on the size of the minipig, the largest capsule that can most likely be administered is a size "0" or "00". When oral dosing is performed, a mouth gag/bite bar and a gavage tube (liquid substances) or dosing gun (tablet/pills) are used. Oral dosing by any administration method is considered to be a stressful procedure, and the animals may vocalize and struggle. In some cases, it may be possible to offer the test substance mixed with the diet and thus avoid the more stressful procedure of oral gavage or tablet/pill administration. However, if the test formulation reduces palatability of the diet and affects the uptake of the formulation, or if the test substance formulation has to mimic the formulation intended for humans (e.g., pills or tablet), or when the exact amount of test substance received by each animal subject during a preclinical study has to be documented and evaluated, mixing of the test formulation in the diet is often not possible. Experienced caretakers limit the stress experienced by the animal subjects when dosing by oral gavage or by tablet/pill administration, and they also minimize the time required for these procedures.

Injections and Sprays/Drops

Injections may be performed by various routes: intradermal, subcutaneous, intramuscular, intravenous, and intraperitoneal. Intranasal and intratracheal application by spray or droplets are also possible. Restraint of the minipig is necessary for most parenteral injections/applications and often training is not performed, as dosing and restraint will only be performed once per day or less. However, a clicker training program has been developed for intranasal application, and in the experience of the author, minipigs were readily trained to accept the intranasal administration without any apparent stress. Subcutaneous or intramuscular injections are often performed with restrained animals. However, when volume and study design allow, these injections can also be

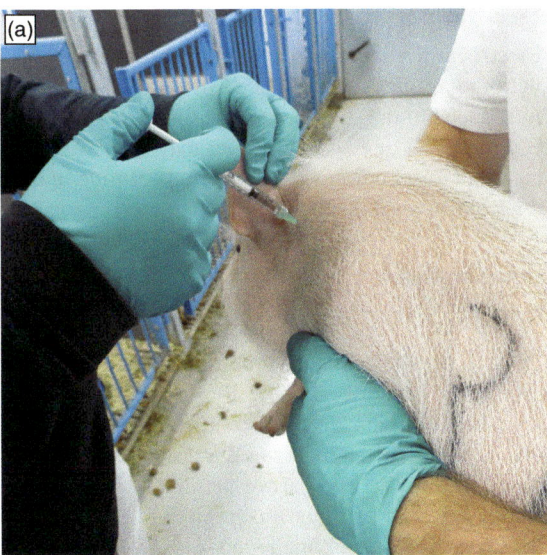
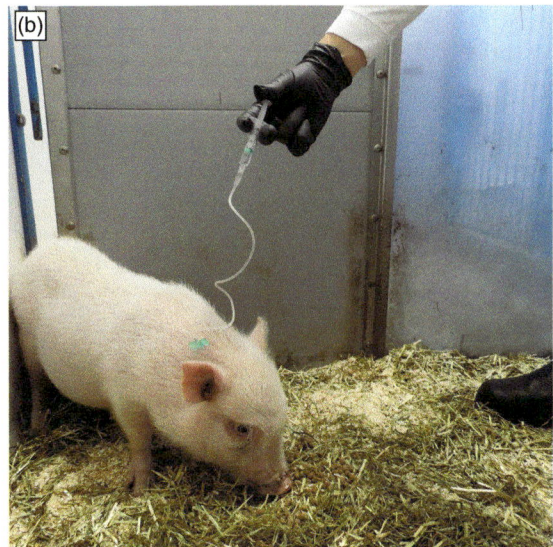

FIGURE 15.5 Minipig dosing protocols showing (left) restraint and intramuscular injection, and (right) subcutaneous injection of a free moving pig via a butterfly needle.

performed using an infusion set, allowing the minipig to freely walk around during dosing, which is considered to be less stressful (Figure 15.5).

Dermal Application

Minipigs are often the model of choice in dermal research. When a topical dermal test item is applied, the minipigs are often trained to run from their pen into the crate used for dosing. During dosing (or other procedures that can be performed in the crate, such as washing off the test item), the animal subject is offered some of its diet. Minipigs generally readily accept this procedure and appear unstressed during test item administration. Upon completion of the procedure, the subject is trained to run back into its pen, where it receives the remainder of its diet.

Special Dosing Procedures

Sometimes, very specific procedures are necessary when conducting preclinical studies. This may include complicated surgical protocols, or animal subjects may have to be dosed using specific devices that may affect how they are housed and/or handled. Specific protocols may include intravenous administration of a test item using an indwelling catheter (Figure 15.6) or vascular access port, continuous subcutaneous administration via implanted catheters or pumps, or intraperitoneal or gastrointestinal administration using permanent catheters placed in various locations in the abdominal cavity/intestines. When catheters are implanted surgically, minipigs need to be housed singly, possibly for a long period, depending on the protocol. It is important for the animals to have visual, auditory and/or tactile contact with other minipigs during such a study. Minipigs need to be kept warm during and after surgical procedures, until they are fully recovered, as they are prone to hypothermia. This is a particular issue if surgery is performed on very young pigs or on piglets.

Wound healing studies, which are widely performed with minipigs, may also be regarded as a special dosing procedure, as animal subjects need to be anesthetized on multiple occasions for evaluation of the wounds.

Other Study-Related Procedures

Blood Sampling

For blood sampling, V-shaped restraint benches have been developed especially for minipig blood sampling (Figure 15.7), a and b), but can possibly also be used for larger pig breeds when the animals are young, depending on body size of the animal.

Placement of animals in slings when blood sampling is performed is also possible for minipigs and young farm pigs, depending on the body size. Both the V-shaped bench and the slings for minipigs are commercially available. A pig snare should never be used when blood sampling minipigs. Blood samples are most often obtained from the jugular vein in both minipigs and larger pigs; however in larger pig breeds, blood sampling from ear veins may also be possible. In the Göttingen minipig, the ear veins are often small and fragile, and taking blood samples from these veins is often difficult, if not impossible. To blood sample pigs and minipigs from the jugular veins can be challenging, as the veins are located deep in the neck and may be difficult to access. When the V-shaped restraint bench is used, the blood sample is obtained with the animal in dorsal recumbency. As placement of the minipig in dorsal recumbency is considered to be stressful in itself, it may seem more appealing to place the animal in a sling for blood sampling. However, blood sampling of the animal subjects in the sling requires very skilled personnel, due to the location of the jugular veins, as superficial landmarks used to insert the needle are not as easy to see as when the pig is in dorsal recumbency. The actual blood sampling is also in itself considered stressful, but as blood sampling is often only performed a few times during a study, it may be more feasible to have the procedure performed quickly (but calmly) by experienced caregivers, rather than to apply a training program. The correct

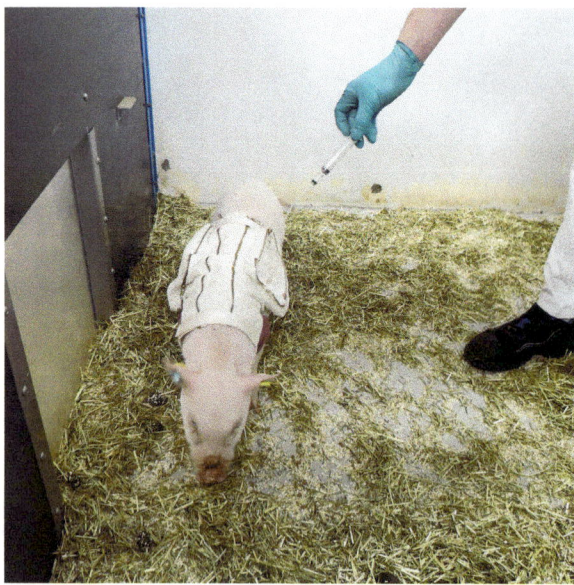

FIGURE 15.6 Dosing of a free moving minipig via a surgically implanted intravenous catheter.

Electrocardiograms and Ophthalmoscopy

Electrocardiograms and ophthalmoscopy can be performed with the animal subject placed in a sling. Animals should be introduced to the sling before the first experimental procedure is carried out and some training may be necessary to habituate them to spending variable periods of time in the sling (see www.minipigs.dk).

Special Situations

Special situations include reproductive toxicity studies and, especially, juvenile studies. Reproductive/embryofetal studies using minipigs are often terminated at gestation day 70 or 110. Sows are often mated at the test facility and handled and dosed during the first part of pregnancy. Mating may involve use of boars or artificial insemination. Both approaches call for experienced personnel, who may not necessarily be available in-house. The sows that are typically used in these studies are adolescents or adults, often weighing more than 20 kg, making handling more difficult. Handling and dosing will often be transiently stressful, especially if the sows are orally dosed or injected, without the use of a catheter/infusion set. In addition, blood sampling of sows without catheters imposes stress which may adversely affect their pregnancy state, so all procedures should be carried out calmly and with these factors in mind.

In juvenile studies, pregnant sows are often purchased from the breeder and piglets are born at the research facility. Piglets are especially sensitive to cold temperatures as, unlike most other newborn animals, they do not have any brown fat (Mellor and Cockburn 1986; Trayhurn et al. 1989). They should have access to a separate area where they can lie together in a warmer microclimate, provided for example, by a heat lamp. The separate area should be connected to the main pen, providing easy access for the piglets, but preventing access by the sow. When a separate area is provided, it is also easier to start feeding the piglets a solid diet, without their food being eaten by the sow.

procedure for placing the minipig in the restraint bench should be followed carefully. To minimize stress, careful planning of studies to allocate enough time for blood sampling is advised. This is especially important when toxicokinetic samples are to be taken, as these are often scheduled to be obtained at very specific time points before and after dosing. If many samples are to be taken in a large study with many pigs, time may be an important stress factor for personnel, which will quickly stress the minipigs, and could end up compromising the outcome of the blood sampling (inability to obtain the samples, samples of poor quality, etc.).

If multiple samples need to be obtained during a study, jugular or ear vein catheters may be implanted. However, catheters have to be regularly maintained and they pose a risk for infection. Minipigs usually have to be single housed after catheter implantation.

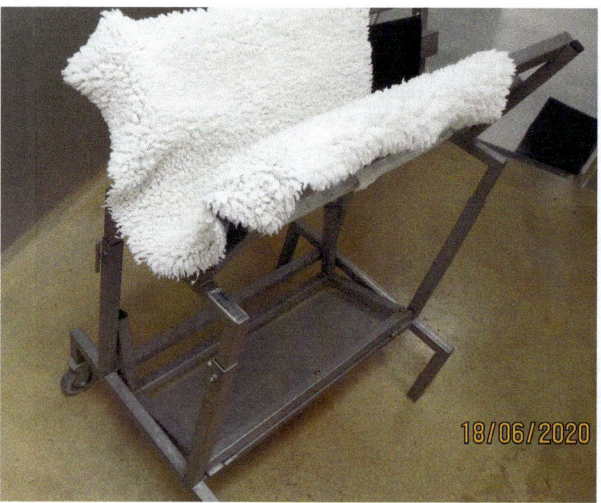
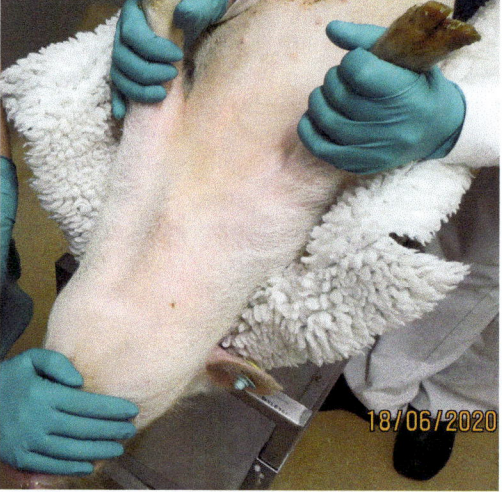

FIGURE 15.7 V-shaped restraint bench for blood sampling.

Different strategies can be applied to a juvenile study, where piglets are dosed before weaning (Kim et al. 2017). If the litter is used as a "study group", there is minimal disturbance of the family unit, but litter size and gender distribution are unpredictable and all piglets in the group are related. If the piglets in a litter are assigned to different treatments, the family unit is again minimally disturbed; however, there are risks of cross-contamination and other secondary factors influencing the outcome of the study (e.g., competition with stronger siblings). Cross-fostering is considered the best approach, both scientifically and for practical reasons. It allows for the constitution of optimal groups (equal litter sizes and gender distributions) and all piglets in the group receive the same treatment (reducing a number of potential confounds). Cross-fostering is easily performed when using Göttingen minipigs, as the sows usually accept the introduction of new piglets. However, there is some risk that the sow may be aggressive toward newly introduced piglets, and to reduce this risk, all the piglets can be put into one crate at reallocation to mix their odors and the sow may be lightly sedated before returning any piglets.

In a juvenile study, dosing of piglets may start at a very early age, sometimes when they are only a few days old, and individuals may be handled multiple times daily from the beginning. This may easily induce stress, both on the piglets and on the sow, and there is a risk that the sow may become so stressed that milk production is affected. To reduce handling stress, pen design should allow for easy separation of the piglets from the sow. Furthermore, procedures should be carried out calmly, and the piglets should be returned to the sow as quickly as possible. If surgery is performed on piglets, attention must be paid when reintroducing them to the sow to avoid aggression. The sow may be lightly sedated to facilitate the reintroduction of the piglets. Weaning age should also be considered carefully in juvenile studies. A weaning age of 4 weeks follows the earliest recommended and accepted weaning age of piglets in farm settings (Council Directive 2008/120/EEC), which is early compared to the natural weaning process of piglets (Jensen and Recén 1989). Weaning time is especially stressful for the piglets, as they are separated from the sow and are fed only solid diet. Piglets can be introduced to a solid diet from 14 days of age and should be eating well and able to grow and thrive without being dependent on the sow's milk when they are weaned. In juvenile studies, weaning age should be considered on a case-by-case basis in research facilities, and if at all possible, it is recommended to wean piglets at a later time point. If various procedures (e.g., blood sampling) for data collection are performed around 4 weeks of age in the juvenile studies, it is actually often feasible to postpone weaning of the piglets in these studies. Thus, all information and necessary data collection should be considered carefully when deciding on the weaning age of the juvenile minipigs.

Conclusions and Recommendations

It is very clear, from both observational studies of modern pigs placed in seminatural conditions and from experimental studies of preference and motivation, that domestic pigs and minipigs retain many behavior patterns inherited from their wild ancestors. The high survival value which these behavior patterns confer has caused them to be strongly embedded into the genetic makeup of the species, and subsequent domestication and selective breeding for production characteristics has resulted in little attenuation of their influence. This gives rise to a number of important behavioral needs that must be taken into account in the housing and management of domestic pigs, if welfare is not to be compromised. Foremost among these needs are a stable social environment, a supported weaning transition, and a stimulating physical environment that provides opportunities for establishing separate functional areas and appropriate expression of foraging, exploratory, and nest-building behavior. When these needs are not adequately met, abnormal behaviors can develop. Some of these, such as oral stereotypies, are indicative of welfare impairment in those individuals exhibiting them. Others, such as elevated aggression and tail, ear, and flank biting, may additionally be injurious to other individuals in the group.

Generally, information on welfare needs in larger breeds of pigs used for farming is also applicable to minipig strains that participate in biomedical research, as basic behavior patterns for these smaller pigs are also rooted in the behavior of the wild pig. Housing and enrichment guidelines for large pigs can generally be adapted by research facilities to accommodate minipigs, but experimental protocols often expose minipigs to stressors (handling, dosing with test item, possible reactions to the test item) that are different from those seen in farm situations. Therefore, specific considerations must be applied to pen design and handling protocols, including appropriate training for procedures, as compromised welfare may not only affect the individual minipig, but may also compromise the outcome of the study.

Additional Resources

A detailed description of the evolutionary history of the pig can be found in:

Porter, V. 1993. *Pigs: a handbook to the breeds of the world.* Ithaca: Cornell University Press.

The normal biology of the pig and its physiological systems are described in:

Pond, W.G., and H.J. Mersmann. Ed. 2001. *Biology of the domestic pig.* Ithaca: Cornell University Press.

Further detail on the behavior and welfare of the domestic pig can be found in:

Marchant-Forde, J.N. Ed. 2009. *The welfare of pigs.* Springer.

Spinka, M. Ed. 2018. *Advances in pig welfare.* Duxford: Woodhead Publishing.

Information on selected minipig strains is available at:

Ellegaaard Göttingen Minipigs: https://minipigs.dk/

Sinclair minipigs: http://www.sinclairbioresources.com/miniature-swine/sinclair/

Yucatan minipigs: http://www.sinclairbioresources.com/miniature-swine/yucatan/

Techniques Training: Minipig – A visual guide to Research Techniques, American Association for Laboratory Animal Science

Information on the participation of minipigs in biomedical research can be found in:

McAnulty, P.A., A.D. Dayan, N-C. Ganderup and K.L. Hastings. Eds. 2012. *The Minipig in Biomedical Research*, Boca Raton: CRC Press, Taylor and Francis Group.

A basic training guide for minipigs used in biomedical research can be acquired from the website of the American Association for Laboratory Animal Science (https://www.aalas.org):

REFERENCES

Andersen, H.M., A.G. Kongsted, and M. Jakobsen. 2020. Pig elimination behavior - a review. *Applied Animal Behaviour Science* 222: 104888.

Bailey, J. 2018. Does the stress of laboratory life and experimentation on animals adversely affect research data? A critical review, *ATLA* 46: 291–305.

Baldrick, P. 2013. The evolution of juvenile animals testing for small and large molecules. *Regulatory Toxicology and Pharmacology* 67: 125–135.

Baxter, E.M., and S.A. Edwards. 2018. Piglet mortality and morbidity: Inevitable or unacceptable? In *Advances in Pig Welfare*, ed. M. Spinka, 73–100. Duxford: Woodhead Publishing.

Baxter, E.M., S.A. Edwards, and I.L. Andersen. 2018. Sow welfare in the farrowing crate and alternatives. In *Advances in Pig Welfare*, ed. M. Spinka, 27–72. Duxford: Woodhead Publishing.

Blye, R., R. Burke, C. James, R. Vorce, A. Fitzgerald, and M. Cox. 2006. The use of operant conditioning (clicker training) of Göttingen minipigs for topical safety studies. *Poster*, doi:10.13140/RG.2.2.11934.43849.

Bode, G., P. Clausing, F. Gervais, J. Loegsted, J. Luft, V. Nogues, and J. Sims. 2010. The utility of the minipig as an animal model in regulatory toxicology. *Journal of Pharmacological and Toxicological Methods* 62: 196–220.

Bollen, P., A. Andersen, and L. Ellegaard. 1998. The behaviour and housing requirements of minipigs. *Scandinavian Journal of Laboratory Animal Science* 25: 23–26.

Bollen, P. and M. Ritskes-Hoitinga. 2004. The welfare of pigs and minipigs. In *The Welfare of Laboratory Animals*, ed. E. Kaliste, 275–290. The Netherlands: Kluwer Academic Publishers

Bollen, P., and M. Skydsgaard. 2006. Restricted feeding may induce serous fat atrophy in male Göttingen minipigs. *Experimental Toxicology and Pathology* 57: 347–349.

Booth, W.D., and J.P. Signoret. 1992. Olfaction and reproduction in ungulates. *Oxford Reviews of Reproductive Biology* 14: 263–301.

Bourneuf, E. 2017. The MeLiM minipig: An original spontaneous model to explore cutaneous melanoma genetic basis. *Frontiers in Genetics* 8: 1–12.

Bracke, M.B.M. 2011. Review of wallowing in pigs: Description of the behaviour and its motivational basis *Applied Animal Behaviour Science* 132: 1–13.

Bracke, M.B.M., and P. Koene. 2019. Expert opinion on metal chains and other indestructible objects as proper enrichment for intensively farmed pigs. *PLoS One* 14: e0212610.

Brown S.M., M. Klaffenböck, I.M. Nevison, and A.B. Lawrence. 2015. Evidence for litter differences in play behaviour in pre-weaned pigs. *Applied Animal Behaviour Science* 172: 17–25.

Chen, C., C.L. Gilbert, G. Yang, Y. Guo, A. Segonds-Pichon, J. Ma, G. Evans, B. Brenig, C. Sargent, N. Affara, L. Huang. 2008. Maternal infanticide in sows: Incidence and behavioural comparisons between savaging and non-savaging sows at parturition. *Applied Animal Behaviour Science* 109: 238–248.

Council Directive 2008/120/EEC. Current consolidated version 14.12.2019. Laying down minimum standards for the protection of pigs.

Council of Europe. 2006. *European Convention for the protection of Vertebrate Animals used for Experimental and Other Scientific Purposes (ETS123)*, Appendix A.

Dardaillon, M, 1987. Seasonal feeding habits of the wild boar in a Mediterranean wetland, the Camargue (Southern France). *Acta Theriologica* 32: 389–401.

Dardallion, M. 1988. Wild boar social groupings and their seasonal changes in the Camargue, southern France. *Zeitschrift fur Säugetierkunde* 53: 22–30.

Day, J.E.L., X. Whittaker, H.A.M. Spoolder, and S.A. Edwards. 2001. A note on the development of a complex ethogram and the reliability of behavioural measurement using multiple observers. *Pig News and Information* 22: 49N–54N.

D'Eath, R.B., and S.P. Turner. 2009. The natural behaviour of the pig. In *The Welfare of Pigs*, ed. J.N. Marchant-Forde, 13–45. New York: Springer.

EFSA. 2007. Scientific opinion of the panel on animal health and welfare on a request from the commission on animal health and welfare in fattening pigs in relation to housing and husbandry. *The EFSA Journal* 564: 1–14.

Ellegaard, L., A. Cunningham, S. Edwards, N. Grand, T. Nevalainen, M. Prescott, and T. Schuurman. 2010. Welfare of the minipig with special reference to use in regulatory toxicology studies. *Journal of Pharmacological and Toxicological Methods* 62: 167–183.

Ellemann-Laursen, S., E. Marsden, B. Peter, N. Downes, D. Coulby, and A.B. Grossi. 2016. The incidence of congenital malformations and variations in Göttingen minipigs. *Reproductive Toxicology* 64: 162–168.

EMA. 2008. Guideline on the need for non-clinical testing in juvenile animals of pharmaceuticals for pediatric indications: 1–9.

EMA. 2020. ICH S5 (R3) guideline on reproductive toxicology: Detection of Toxicity to Reproduction for Human Pharmaceuticals: 1–127.

Everds, N.E., P.W. Snyder, K.L. Bailey, B. Bolon, D.M. Creasy, G.L. Foley, T.J. Rosol, and T. Sellers. 2013. Interpreting Stress Responses during Routine Toxicity Studies: A Review of the Biology, Impact, and Assessment. *Toxicologic Pathology*, 41: 560–614

Fradrich, H. 1974. A comparison of behaviour in the Suidae. In *The Behaviour of Ungulates and Its Relation to Management*, eds. V. Geist, and F. Walther, 133–143. Rome: IUCN Publishers.

Fraser, D. 1980. A review of the behavioural mechanisms of milk ejection in the domestic pig. *Applied Animal Ethology* 6: 247–255.

Fraser, D. 2008. Understanding animal welfare. *Acta Veterinaria Scandinavica* 50 (Suppl 1): S1.

Fraser, D., B.N. Milligan, E.A. Pajor, P. Phillips, A.A. Taylor, and D.M. Weary. 1998. Behavioural perspectives on weaning in domestic pigs. In *Progress in Pig Science*, ed. J. Wiseman, M.A. Varley and J. Chadwick, 121–140. Nottingham: Nottingham University Press.

Fraser, D., D.M. Weary, E.A. Pajor, and B.N. Milligan. 1997. A scientific conception of animal welfare that reflects ethical concerns. *Animal Welfare* 6: 187–205.

Gerrity, R.G., R. Natarajan, J.L. Nadler, and T. Kimsey. 2001. Diabetes-induced accelerated atherosclerosis in swine. *Diabetes* 50: 1654–1665.

Gieling, E.T., R.E. Nordquist, and F.J. van der Staay. 2011. Assessing learning and memory in pigs. *Animal Cognition* 14: 151–173.

Glerup, P. 2012. Wound healing models. In *The Minipig in Biomedical Research*. eds. P.A. McAnulty, A.D. Dayan, N.-C. Ganderup, and K.L. Hastings, 525–531. Boca Raton, FL: CRC Press, Taylor and Francis Group.

Gonyou, H.W., E. Beltranena, D. L. Whittington, and J.F. Patience. 1998. The behaviour of pigs weaned at 12 and 21 days of age from weaning to market. *Canadian Journal of Animal Science* 78: 517–523.

Gundlach, H. 1968. Brutfursorge, Brutpflege, Verhaltensontogenese und Tagesperiodik beim Europaischen Wildschwein (*Sus scrofa* L.). *Zeitschrift fur Tierpsychologie* 25: 955–995.

Harris, M.J., R. Bergeron, and H.W. Gonyou. 2001. Parturient behaviour and offspring-directed aggression in farmed wild boar of three genetic lines. *Applied Animal Behaviour Science* 74: 153–163.

Heining, P., and T. Ruysschaert. 2016. The use of minipig in drug discovery and development: Pros and cons of minipig selection and strategies to use as a preferred nonrodent species. *Toxicologic Pathology* 44: 467–473.

Hemsworth, P.H., R.G. Beilharz, and W.J. Brown. 1978. The importance of the courting behaviour of the boar on the success of natural and artificial mating. *Applied Animal Ethology* 4: 341–347.

Hook, R.R., J. Berkelhammer, and R.W. Oxenhandler. 1982. Melanoma: Sinclair swine melanoma. *American Journal of Pathology* 108: 130–133.

Huntsberry, M.E., D. Charles, K.M. Adams, and J.L. Weed. 2008. The foraging ball as a quick and easy enrichment device for pigs (*Sus scrofa*). *Laboratory Animal* 37: 411–414.

Huynh, T.T.T., A.J.A. Aarnink, W.J.J. Gerrits, M.J.H. Heetkamp, T.T. Canh, H.A.M. Spoolder, B. Kemp, and M.W.A. Verstegen. 2005. Thermal behaviour of growing pigs in response to high temperature and humidity. *Applied Animal Behaviour Science* 91: 1–16.

Jensen, P. 1980. An ethogram of social interaction patterns in group-housed dry sows. *Applied Animal Ethology* 6: 341–350.

Jensen, P. 1982. An analysis of agonistic interaction patterns in group-housed dry sows - aggression regulation through an "avoidance order". *Applied Animal Ethology* 9: 47–61.

Jensen, P., and B. Recén. 1989. When to wean. *Applied Animal Behaviour Science* 23: 49–60.

Johansen, T., H.S. Hansen, B. Richelsen, and K. Malmlöf. 2001. The obese Göttingen minipig as a model of the metabolic syndrome: Dietary effects on obesity, insulin sensitivity, and growth hormone profile. *Comparative Medicine* 51: 150–155.

Jørgensen, K.D. 1998. Teratogenic activity of tretinoin in the Göttingen minipig. *Scandinavian Journal of Laboratory Animal Science* 26 (Suppl. 1): 235–243.

Kaminski, G., S. Brandt, E. Baubet, and D.C. Baudoin. 2005. Life-history patterns in female wild boars (Sus scrofa): Mother–daughter postweaning associations. *Canadian Journal of Zoology* 83: 474–480.

Kim, N.N., R.M. Parker, G.F. Weinbauer, A.K. Remick, and T. Steinbach. 2017. Points to consider in designing and conducting juvenile toxicology studies. *International Journal of Toxicology* 36: 325–339.

Köhn, F. 2012. History and development of miniature, micro - and minipigs. In *The Minipig in Biomedical Research*, eds. P.A. McAnulty, A.D. Dayan, N.-C. Ganderup, and K.L. Hastings, 3–15. Boca Raton, FL: CRC Press, Taylor and Francis Group.

Lahrman, H.P., L.C. Oxholm, H. Steinmetz, M.B. Nielsen, and R.B. D'Eath. 2015. The effect of long or chopped straw on pig behaviour. *Animal* 9: 862–870.

Larsen, M.O., M. Wilken, C.F. Gotfredsen, R.D. Carr, O. Svendsen, and B. Rolin. 2002. Mild streptozotocin diabetes in the Göttingen minipig: A novel model of moderate insulin deficiency and diabetes. *American Journal of Physiology Endocrinology and Metabolism* 282: E1342–E1351.

Larsen, M., and Rolin, B. 2004. Use of the Göttingen minipig as a model of diabetes, with special focus on type 1 diabetes research. *ILAR Journal* 45: 303–313.

Lawrence, A.B., and E.M.C. Terlouw. 1993. A review of the behavioural factors involved in the development and continued performance of stereotypic behaviours in pigs. *Journal of Animal Science* 71: 2815–2825.

Leaper, R., G. Massei, M.L. Gorman, and R. Aspinall. 1999. The feasibility of reintroducing wild boar (*Sus scrofa*) to Scotland. *Mammal Review* 29: 239–259.

Learmonth, J.A. 2019. Commentary: Dilemmas for natural living concepts of zoo animal welfare. *Animals* 9: 318.

Lemel, J., J. Truve, and B. Soderberg. 2003. Variation in ranging and activity behaviour of European wild boar *Sus scrofa* in Sweden. *Wildlife Biology* 9 (Suppl 1): 29–36.

Lewis, N.J., and J.F. Hurnik. 1985. The development of nursing behaviour in swine. *Applied Animal Behaviour Science* 14: 225–232.

Lin, S.Y., S.J. Hou, T.H. Hsu, and F.L. Yeh. 1992. Comparisons of different animal skins with human skin in drug percutaneous penetration studies. *Methods and Findings in Experimental Clinical Pharmacology* 14: 645–654.

Ludvigsen, T.P., R.K. Kirk, B.Ø. Christoffersen, H.D. Pedersen, T. Martinussen, J. Kildegaard, P.M.H. Heegaard, J. Lykkesfeldt, and L.H. Olsen. 2015. Göttingen minipig model of diet induced atherosclerosis: influence of mild streptozotocin induced diabetes on lesion severity and markers of inflammation evaluated in obese, obese and diabetic, and lean control animals. *Journal of Translational Medicine* 13: 1–12.

Marchant-Forde, J.N. 2015. The science of animal behaviour and welfare: Challenges, opportunities, and global perspective. *Frontiers in Veterinary Science* 2: 16.

Mason, G.J. 1991. Stereotypies: A critical review. *Animal Behaviour* 41: 1015–1037.

Massei, G., P.V. Genov, and B.W. Staines. 1996. Diet, food availability and reproduction of wild boar in a Mediterranean coastal area. *Journal of Zoology* 242: 411–423.

Mauget, R. 1981. Behavioural and reproductive strategies in wild forms of *Sus scrofa* (European wild boar and feral pigs). In *The Welfare of Pigs*, ed. W. Sybesma, 3–13. The Hague: Martinus Nijhoff.

McGlone, J.J. 1985a. Olfactory cues and pig agonistic behaviour: Evidence for a submissive pheromone. *Physiology and Behaviour* 34: 195–198.

McGlone, J.J. 1985b. A quantitative ethogram of aggressive and submissive behaviors in recently regrouped pigs. *Journal of Animal Science* 61: 556–566.

McGlone, J.J., and D.L. Anderson. 2002. Synthetic maternal pheromone stimulates feeding behaviour and weight gain in weaned pigs. *Journal of Animal Science* 80: 3179–3183.

Mellor, D.J., and F. Cockburn. 1986. A comparison of energy metabolism in the new-born infant, piglet and lamb. *Quarterly Journal of Experimental Physiology* 71: 361–379.

Mendl, M., K. Randle, and S. Pope. 2002. Young female pigs can discriminate individual differences in odours from conspecific urine. *Animal Behaviour* 64: 97–101.

Meyer, W., and K. Neuran. 1976. The distribution of enzymes in the skin of the domestic pig. *Laboratory Animals* 10: 237–247.

Millikan, L.E., J.L. Boylon, R.R. Hook, and P.J. Manning. 1974. Melanoma in Sinclair swine: A new animal model. *The Journal of Investigative Dermatology* 62: 20–30.

Morrison, R.S., P.H. Hemsworth, G.M. Cronin, and R.G. Campbell. 2003. The social and feeding behaviour of growing pigs in deep-litter, large group housing systems. *Applied Animal Behaviour Science* 82: 173–188.

Mortensen, J.T., P. Brinck, and J. Lichtenberg. 1998. The minipig in dermal toxicology: A literature review. *Scandinavian Journal of Laboratory Animal Science* 25 (Suppl. 1): 77–83.

Moustgaard, A., N.M. Lind, R. Hemmingsen, and A.K. Hansen. 2002. Spontaneous object recognition in the Göttingen minipig. *Neural Plasticity* 9: 255–259.

National Research Council. 2011. *Guide for the Care and Use of Laboratory Animals.* 8th edition. Washington, DC: National Academies Press.

Newberry, R., and D.G.M. Wood-Gush. 1988. Development of some behaviour patterns in piglets under semi-natural conditions. *Animal Production* 46: 103–109.

Oxenhandler, R.W., J. Berkelhammer, G.D. Smith, and R.R. Hook. 1982. Growth and regression of cutaneous melanomas on Sinclair miniature swine. *American Journal of Pathology* 109: 259–269.

Petherick, J.C. 1983. A biological basis for the design of space in livestock housing. In *Farm Animal Housing and Welfare*, ed. S.H. Baxter, M.R. Baxter, and J.A.D. MacCormack, 103–120. The Hague: Martinus Nijhoff.

Pinna, W., G. Nieddu, G. Moniello, and M.G. Cappai. 2007. Vegetable and animal food sorts found in the gastric content of Sardinian Wild Boar (*Sus scrofa meridionalis*). *Journal of Animal Physiology and Animal Nutrition* 91: 252–255.

Pluske, J.R., D.L. Turpin, and J.C. Kim. 2018. Gastrointestinal tract (gut) health in the young pig. *Animal Nutrition* 4: 187–196.

Porter, V. 1993. *Pigs: A Handbook to the Breeds of the World.* Ithaca: Cornell University Press.

Qvist, M.H., U. Hoeck, B. Kreilgaard, F. Madsen, and S. Frokjaer. 2000. Evaluation of Göttingen minipig skin for transdermal in vitro permeation studies. *European Journal of Pharmaceutical Sciences* 11: 59–68.

Reaven, G.M. 1995. Pathophysiology of insulin resistance in human medicine. *Physiological Review* 75: 473–486.

Reifenrath, W.G., E.M. Chellquist, E.A. Shipwash, and W.W. Jederberg. 1984. Evaluation of animal models for predicting skin penetration in man. *Fundamental and Applied Toxicology* 4: S224–S230.

Robert, S., and G.P. Martineau. 2001. Effects of repeated cross-fosterings on preweaning behaviour and growth performance of piglets and on maternal behaviour of sows. *Journal of Animal Science* 79: 88–93.

Rolandsson, O., M.F. Haney, E. Hägg, B. Biber, and Å. Lernmark. 2002. Streptozotocin induced diabetes in minipig: A case report of a possible model for type 1 diabetes. *Autoimmunity* 35: 261–264.

Roth, W.J., C. B. Kissinger, R.R. McCain, B.R. Cooper, J.N. Marchant-Forde, R.-C. Vreeman, S. Hannou, and G.T. Knipp. 2013. Assessment of juvenile pigs to serve as human pediatric surrogates for preclinical formulation pharmacokinetic testing. *The AAPS Journal* 15: 763–774.

Ruis, M.A.W., J. de Grootte, J.H.A. Brake, E.D. Ekkel, J.A. van de Burgwal, J.H.F. Erkens, B. Engel, W.G. Buist, H.J. Blokhuis, and J.M. Koolhaas. 2001a. Behavioural and physiological consequences of acute social defeat in growing gilts: Effects of the social environment. *Applied Animal Behaviour Science* 70: 201–225.

Ruis, M.A.W., J.H. Brake, B. Engel, W.G. Buist, H. Blokhuis, and J.M. Koolhaas. 2001b. Adaption to social isolation. Acute and long-term stress responses of growing gilts with different coping characteristics. *Physiology and Behaviour* 73: 541–551.

Rushen, J. 1990. Social recognition, social dominance and the motivation of fighting by pigs. In *Social Stress in Domestic Animals*, eds. R. Zayan and R. Danzer, 135–142. Dordrecht: Kluwer Academic Press.

Schley, L., and T.J. Roper. 2003. Diet of wild boar *Sus scrofa* in Western Europe, with particular reference to consumption of agricultural crops. *Mammal Review* 33: 43–56.

Seaton, M., A. Hockin, and N.S. Gibran. 2015. Porcine models of cutaneous wound healing. *ILAR Journal* 56: 127–138.

Signoret, J.P. 1970. Reproductive behaviour of pigs. *Journal of Reproduction and Fertility* 11: 105–117.

Smith, M.E., N.V. Gopee, and S.A. Ferguson. 2009. Preferences of minipigs for environmental enrichment objects. *Journal of the American Association for Laboratory Animal Science* 48: 391–394.

Spitz, F. 1986. Current state of knowledge on wild boar biology. *Pig News and Information* 7: 171–175.

Stewart, K., and F. Cebezón. 2016. Fact sheet: Heat Stress Physiology in Swine. https://www.extension.purdue.edu/extmedia/AS/AS-362-W.pdf: 1–2.

Stolba, A., and D.G.M. Wood-Gush. 1989. The behaviour of pigs in a semi-natural environment. *Animal Production* 48: 419–425.

Stolba, A., and D.G.M. Wood-Gush. 1984. The identification of behavioural key features and their incorporation into a housing design for pigs. *Annales de Recherches Vétérinaires* 15: 287–302.

Studnitz, M., M.B. Jensen, and L.J. Pedersen. 2007. Why do pigs root and in what will they root: A review on the exploratory behaviour of pigs in relation to environmental enrichment. *Applied Animal Behaviour Science* 107: 183–197.

Stump, K.C., M.M. Swindle, C.D. Saudek, and J.D. Strandberg. 1988. Pancreatomized swine as a model of diabetes mellitus. *Laboratory Animal Science* 38: 439–443.

Sullivan, T.P., W.H. Eaglstein, S.C. Davis, and P. Mertz. 2001. The pig as a model for human wound healing. *Wound Repair and Regeneration* 9: 66–76.

Søndergaard, L.V., K.H. Jensen, R. Hemmingsen, A.K. Hansen, and N.M. Lind. 2007. Characterization of spontaneous behaviour in Göttingen minipigs in the home pen. *Scandinavian Journal of Laboratory Animal Science* 34: 91–103.

Tallet, C., P. Linhart, R. Policht, K. Hammerschmidt, P. Simecek, P. Kratinova, and M. Spinka. 2013. Encoding of situations in the vocal repertoire of piglets (*Sus scrofa*): A comparison of discrete and graded classifications. *PLoS One* 8: e71841.

Tannenbaum, J., and B.T. Bennett. 2015. Russell and Burch's 3Rs then and now: the need for clarity in definition and purpose. *Journal of the American Association for Laboratory Animal Science* 54: 120–132.

Tast, A., O. Halli, S. Ahlstrom, H. Andersson, R.J. Love, and O.A.T. Peltoniemi. 2001. Seasonal alterations in circadian melatonin rhythms of the European wild boar and domestic gilt. *Journal of Pineal Research* 30: 43–49.

Taylor, N., D.C.J. Main, M. Mendl, and S.A. Edwards. 2010. Tail biting: A new perspective. *The Veterinary Journal* 186: 137–147.

Trayhurn, P., N.J. Temple, and J. van Aerde. 1989. Evidence from immunoblotting studies on uncoupling protein that brown adipose tissue is not present in the domestic pig. *Canadian Journal of Physiology and Pharmacology* 67: 1480–1485.

Tsutsumi, H., N. Morikawa, R. Niki, and M. Tanigawa. 2001. Acclimatization and response of minipigs toward humans. *Laboratory Animals* 35: 236–242.

van der Laan, J.W., J. Brightwell, P. McAnulty, J. Ratky, and C. Stark. 2010. Regulatory acceptability of the minipig in the development of pharmaceuticals, chemicals and other products. *Journal of Pharmacological and Toxicological Methods* 62: 184–195.

van Mierlo, G.J.D., N.H.P. Cnubben, C.F. Kuper, J. Wolthoorn, A.P. van Meeteren-Kreikamp, M.M. Nagtegaal, R. Doornbos, N.-C. Ganderup, and A. Penninks. 2013. The Göttingen minipig® as an alternative non-rodent species for immunogenicity testing: A demonstrator study using the IL-1 receptor antagonist anakinra. *Journal of Immunotoxicology* 10: 96–105.

van Mierlo, G.J.D., N.H.P. Cnubben, D. Wouters, G.J. Wolbink, M.H.L. Hart, T. Rispens, N.-C. Ganderup, C.F. Kuper, L. Aarden, and A. Penninks. 2014. The minipig as an alternative non-rodent model for immunogenicity testing using the TNFα blockers adalimumab and infliximab. *Journal of Immunotoxicology* 11: 62–71.

Vieuille-Thomas, C., and J.P. Signoret. 1992. Pheromonal transmission of an aversive experience in domestic pigs. *Journal of Chemical Ecology* 18: 1551–1557.

Willard-Mack, C., T. Ramani, and C. Auletta. 2016. Dermatotoxicology: Safety evaluation of topical products in minipigs: Study designs and practical considerations. *Toxicologic Pathology* 44: 382–390.

Wilson, P.W.F., R.B.D. D'Agostino, H. Parise, L. Sullivan and J.B. Meigs. 2005. Metabolic syndrome as a precurser of cardiovascular disease and type 2 diabetes mellitus. *Circulation* 112: 3066–3072.

16

Behavioral Biology of Sheep

Cathy M. Dwyer
Scotland's Rural College (SRUC)

CONTENTS

Introduction 261
Typical Research Contributions 261
Behavioral Biology 262
    Natural History 262
        Ecology and Habitat Use 262
        Social Organization and Behavior 263
        Feeding Behavior 264
        Communication 265
    Common Captive Behaviors 265
        Normal 266
        Abnormal 266
Ways to Maintain Behavioral Health 266
    Captive Considerations to Promote Species-Typical Behavior 266
        Environment 266
        Social Groupings 267
        Feeding Strategies 268
        Training 268
Special Situations 269
Conclusions/Recommendations 269
Additional Resources 269
References 269

Introduction

Sheep were one of the first species to be domesticated, over 10,000 years ago, and are farmed for meat, milk, wool, and their skins. There are more than a billion sheep worldwide, and more than 1,000 different breeds, ranging from temperate, woolly sheep breeds to tropically adapted "hair" sheep, which may also be classified as "fat-tailed" or "fat-rumped", as part of their adaptation to conserving water. Sheep are very common in Africa, Asia, and the Middle East, with the most sheep per country found in China, Australia, and India (161, 72, and 63 million respectively; FAOSTAT, 2017). Due to their hardiness and adaptability, and the different products that can be harvested, sheep play an important role in sustaining poor, rural communities in some of the harshest climates on the globe. However, sheep are also involved extensively in research, where their small size, relative docility, ease of handling, and precocial offspring make them suitable to many areas of research.

Typical Research Contributions

Sheep have a number of features that make them attractive as research animals. They are smaller than other large animal models (pigs or cattle), docile and nonaggressive, and with good handling, have a relatively short flight distance from humans that can be reduced further with training. Despite their ill-deserved reputation for stupidity, sheep are actually very capable of learning, and can be readily trained with positive rewards to tolerate close human presence and to cooperate with experimental setups. As a domesticated species, many aspects of their behavior and general husbandry are well known, which can facilitate good experimental management. They are also relatively long-lived (typically reaching 7–8 years of age on a farm and can live up to 15 years), which is useful for long-term studies.

Sheep research has covered areas relevant to sheep production and basic biology, such as seasonal control of reproductive activity, parturition, neonatal care, lactation, growth, ruminant metabolism, and physiology (e.g., Manikkam et al., 2004; Chillard et al., 2005; De Nicolo et al., 2008). Sheep have also been involved in research as models for human development, particularly fetal and neonatal development, and growth and disease (e.g., Hassenbusch et al., 1999; Johansen et al., 2004; Chang et al., 2009), including neurodegenerative diseases (Huntington's disease and Batten's disease, for example; Morton et al., 2014; Perentos et al., 2015), where their relatively large brains, similar neural axis structure to humans, and longevity make them better models than rodents. Sheep have participated in the development of transplants and artificial organs, large-scale production of antibodies, and improvements in vaccine development (e.g., Ramirez et al., 2008; Kortekaas, 2014). However, perhaps the best-known example of sheep research is in genetics, and "Dolly the sheep" was famously the first mammal to be successfully cloned from an adult cell at the Roslin Institute in 1996 (Wilmut et al., 1997).

Behavioral Biology

Sheep are a highly adaptable and social ruminant ungulate (hoofed) species. Although they have been domesticated for many years, and were among the first species to be domesticated, there are still a number of wild sheep species and feral animals that can provide good evidence for the natural history and behavior of sheep. These include Mountain Bighorn and Dall's sheep in North America, Mouflon in Europe and Asia, and Urial and Argali in Asia. In addition, the feral and primitive breed, the Soay sheep, has been well studied on the Scottish islands of St. Kilda.

Natural History

Ecology and Habitat Use

Predation risk and food availability have shaped habitat use, and social and foraging behavior in sheep (Dwyer, 2004). Sheep are largely defenceless against large predators, so their main antipredator behavior is flight toward cover or other escape-friendly terrain, where they can more successfully evade predators than in open areas. Thus, their preferred habitat is elevated rocky and steep slopes close to hilly areas (wild sheep are rarely more than 200m from escape terrain), and they are adapted to live in upland regions, often in areas where the grazing is too poor to sustain other species. Wild sheep species are generally found above 1,000m elevation, but below the snowline, preferring meadows, broom moorland, shrub areas, and open woodlands, with closed forests being actively avoided (Dwyer, 2008a). Feral or island sheep are generally found at much lower elevations, but still with access to escape terrain, such as rocky and secluded areas.

Other important features of preferred habitats are suitable forage and a water source. As ruminants, they are efficient users of grass and forage, which means they can survive in harsh climates, with variable growing seasons and low-quality herbage. Sheep are very adaptable, and wild and feral species can be found in a wide variety of habitats: in desert areas in the Middle East; in high mountain regions in the Himalayas and Rocky mountains, where temperature extremes between summer and winter can exceed 40°C; and in the Arctic and sub-Antarctic. They are able to utilize a wide variety of forage types (Dwyer, 2008a), including cacti, fruit, trees, shrubs, and even seaweed (eaten by the sheep on the island of North Ronaldsay in Scotland). Although sheep have colonized desert habitats, they must cluster around watering points, as even desert-adapted sheep need to drink every day.

Sheep show distinct seasonal and diurnal patterns of grazing, ruminating, and resting, which follow the patterns of forage growth, and the presence of shelter. They usually rest on higher ground at night, often in more inaccessible parts of their range with good visibility, and move down to lower ground to feed in the morning. There are distinct periods of synchronized grazing and ruminating during the day, before they move back up to resting grounds at dusk. Although most grazing is done during the day, sheep may also show short periods of grazing during the night (Dwyer, 2008a). Seasonal patterns of movement follow the growing seasons for different plants, and the availability of shade or opportunities to shelter from wind. During the summer, sheep will graze higher up in their habitat than during the winter, and use natural features, such as rocks, shrubs, and trees, for shade and shelter.

Sheep are a predated species and are hunted by a wide variety of large predators including wolves, bears, lynx, and lions. The lambs may be predated by larger avian predators, such as eagles, and smaller predators, such as foxes, jackals, and wolverines. Avoidance of predation and antipredator responses have shaped much of the behavior of sheep, coupled to the need to find sufficient forage (Dwyer, 2004). As part of their antipredator defenses, sheep are highly social, and their synchronized behavior and following responses, or allomimetic tendencies, are part of a suite of behaviors to reduce predator detection. In open ground, their main response to the presence of a predator is to flock tightly together, and run to higher and more inaccessible parts of their range, although ewes have been seen to protect their lambs and drive off small predators with their horns. In escape terrain, sheep may be more likely to stand their ground when confronted by a predator, and spend less time alert in vigilance postures and more time feeding and engaging in social interactions, because in these areas the threat of predation is reduced (Jansen et al., 2006). The propensity to flee is influenced by the assessment of risk; e.g., sheep are more likely to run from human walkers with dogs than walkers alone, and they are more likely to flee when further from escape terrain. Rams are less vulnerable to predation, and are therefore bolder and less likely to flee than ewes.

Behavioral synchronization and following responses are also an important part of sheep antipredator behavior. This reduces the chance that any single sheep might be detected, since predators will often select the weakest or most obvious member of the flock, and maintains group cohesion in movement about the environment. In addition, when foraging in an environment where resources are dispersed and may be of poor quality, following can help younger animals learn (1) the location of food, water, and shelter, and (2) appropriate social behaviors from older, more successful adults.

Following and flocking responses have been useful in sheep management during domestication, and these behaviors have been retained in domestic sheep, and selected for, perhaps unconsciously. Thus, domesticated sheep continue to show a high degree of sociality and, regardless of whether a predator is present, will become very fearful and distressed when separated from the flock (Dwyer, 2008a).

Social Organization and Behavior

Sheep have a matrilineal social structure, with daughters remaining in the social group of their mothers, while male offspring separate from the ewe flock at weaning, and form bachelor flocks. Sheep have a home range (an area of land where they live and are familiar with the resources), but are nonterritorial, as they do not defend their range. Ewe home ranges consist of ewes and their daughters and lambs, which overlap with the home ranges of several male flocks (Dwyer, 2008a). The environment, physiological state, and breed influence the preferred size of the social groups. In open areas, where forage is abundant, the social group may be large (30–50 individuals), whereas in hill, desert, or wooded areas, social groups may be considerably smaller (Dwyer, 2008a). Domestication seems to have had an impact on preferred social group size, as well as the closeness of social relationships, as more intensively reared or lowland sheep, aggregate into large flocks with shorter interindividual distances than less selected, hill-adapted sheep breeds (Dwyer and Lawrence, 2000). However, lactating ewes of all breeds are less gregarious when accompanied by their lambs, and may move further from the social group until the lambs are weaned.

Dominance and demonstration of social status mainly rely on visual displays, especially in the male flock. The size and degree of curvature of horns can denote age and status in rams, and in wild and feral species, males will grow a beard or ruff when entering the breeding season. The social dynamics in the ewe flock may be less overt, and are maintained by eye contact, head threats (intentions to butt or actual butting events), and displacements (moving other animals from preferred resting or feeding places by actual aggression, threatening behaviors, or movements toward subordinates). When resources, particularly food items, are less abundant, there are increased displacements, with younger and older ewes reported to be less aggressive than intermediate-aged ewes (Arnold and Maller, 1974).

Another important component of sheep social behavior is leadership. These animals, typically one or two per group, are those that are most likely to start the movement from one part of the terrain to another, but are generally not the most dominant sheep in the flock. Older and more confident animals are most likely to act as leaders, which may reflect a greater familiarity with the environment than other members of the flock. Although leader animals are those that habitually start the movement, a consistent movement order has not been demonstrated in sheep flocks. However, dominant animals are usually in the middle of the movement order, the safest place in the movement, and subordinates are usually at the back, which is the riskiest location (Lynch et al., 1992).

Sheep have a promiscuous breeding strategy, and rams only associate with the ewe flock during the rut. Ewes are seasonally anestrus and, in temperate regions, enter estrus when day length shortens, dictating the onset of the mating period (Nowak et al., 2008). Male sheep will fight for access to females, usually by butting or clashing heads. Many species of sheep are horned, but horns are largely used as symbols of male fitness and dominance, rather than as weapons in fights. Rams court estrous females through a well-defined sequence of courtship, comprised of reciprocal proceptive and receptive behaviors, and will stay with a receptive female until mating has taken place (Nowak et al., 2008). Typically, ewes will mate for the first time in their second year, although in faster growing domestic sheep breeds, mating may be possible in ewe lambs (within their first year; Nowak et al., 2008).

Sheep have a 145-day pregnancy in domestic breeds, and typically give birth to one, two, or three lambs, although four or even five lambs are possible in very prolific breeds. Twinning is most common in domestic breeds, whereas singleton pregnancies are more likely in wild sheep or more primitive breeds. As ewes prepare to lamb, they move away from the social group, when they are able to do so, and withdraw to more inaccessible parts of their home range where predator detection is more difficult. Labor lasts between 15 min and 2 h normally, and lambs are usually presented in the birth canal lying head first with the forelegs extended (Dwyer, 2017). Prior to birth, ewes are largely indifferent to the presence of lambs, and sheep do not normally show maternal behavior spontaneously when presented with a lamb. The onset of maternal behavior is governed by a precise series of hormonal cues and changes in relative hormone concentrations in late pregnancy, peripheral stimulation during labor, and cues from the lamb and amniotic fluids (Dwyer, 2008b). These serve initially to stimulate interest in amniotic fluid where it is spilled on the ground during labor, with this interest then transferred to the coat of the lamb, which is soaked in amniotic fluid at birth. Early maternal behavior is expressed as a period of intense interest and licking or grooming of the newborn lamb, accompanied by frequent low-pitched bleats or "rumbles" for the first 4–6 h after birth (Figure 16.1). Licking helps to clean, dry, and stimulate the lamb, which facilitates effective thermoregulation by the lamb. Licking also serves another important function, as ewes form an exclusive attachment to their own lamb, based initially on learning their offspring's unique olfactory "signature", by licking the newborn lamb's coat (Poindron et al., 2007). This initial olfactory recognition can develop within 30 min of birth. Thereafter, the ewe will restrict her maternal care only to her own offspring, and the lambs of other ewes will be rejected if they attempt to suck. This means that in a mobile and synchronously breeding social group, where many young lambs are present at the same time, the ewe is able to ensure that only her own offspring have access to her resources.

Lambs are precocious at birth, and will stand within a few minutes of birth and move toward the udder, guided by the ewe's licking behavior, as well as a set of reflex upward movements of the head and nose to detect the change from warm and woolly body to the cooler, bare skin of the udder. Typically, lambs will suck within the first hour of birth and ingest the colostrum, which provides fuel for thermoregulation, immunological protection from the immunoglobulins that pass over the immature gut, and that plays a role in the lamb learning

FIGURE 16.1 The early development of the ewe-lamb bond is characterized by licking and grooming behavior, and movements by the lamb to stand and then to find the udder, as shown by the sequence here. (Photos: Ann McLaren.)

to recognize its own mother (Nowak et al., 2007). Once ewes and their lambs have learned to recognize each other by smell, which can take 12–24 h for the lamb, the lamb will follow the ewe closely throughout lactation. This is an important learning period for the lamb, particularly for the ewe lambs that will remain on the home range of their mothers, as they learn the location of water, shelter, and forage from following their mother. Over time, the initial olfactory recognition becomes multimodal, with ewes and lambs recognizing each other from visual and auditory cues when at a distance, and olfactory cues when at close quarters. Lambs may form play bands with other similarly aged lambs, and spend short periods away from the ewes in play, which may involve running, jumping, and chasing, as well as play fighting and mounting.

Initially, ewes will approach the lambs, and suckling frequency is high in the first few weeks after birth. However, the ewe-lamb dynamic changes at approximately 4 weeks of age, when responsibility for approaching shifts to the lamb, who is called to the ewe by vocal and postural signals (Pickup and Dwyer, 2011). From this age onward, lambs are only allowed to suck when the ewe has signaled her readiness to feed the lamb. As lambs get older, they become less dependent on milk for nutrition, but will still approach the ewe and suck for comfort, remaining psychologically attached to the ewe until weaning (Dwyer, 2017). Ewes will wean the lamb at around 6 months of age, when her milk supply has declined, and before entering the rut for the next breeding season. At this point, male lambs will have to leave the social group of their mother and form new relationships with other males. Ewe lambs, on the other hand, remain in the ewe group, although it is not clear whether long-term bonds with their mother are maintained. These are survival strategies for sheep, as the behavioral needs and priorities of young sheep and older sheep, or of males and females, differ, and allomimetic responses would become more difficult.

Feeding Behavior

Sheep are predominantly grazing animals, although they will sometimes browse, feeding on leaves from shrubs and trees, for example. However, they are remarkably adaptable, as shown by the seaweed diet of feral North Ronaldsay sheep in Scotland. These animals have been excluded from accessing inland pastures by fencing, and survive entirely on seaweed (Dwyer, 2008a). This has influenced their rumen microbiota and their diurnal patterns of behavior, as the seaweed is only exposed at low tide, and illustrates the ability of sheep to adapt their feeding behavior to consume what forage is available.

Sheep have no upper incisors, which have been replaced by a hard dental pad, allowing them to graze closer to the ground than other grazing species. Feeding behavior is characterized by cycles of grazing and foraging, interspersed with rumination. Although there is seasonal variation in food abundance, and animals living in the desert may need to forage in a nutrient-poor environment where food patches are scarce, sheep living in temperate environments have access to relatively abundant food. However, sheep still forage selectively, and choose between different plants, and even parts of the plant (e.g., leaf or stem), when feeding (Rutter, 2006). The choice of which forage to select, diet choice, is influenced by the preingestive characteristics of the food (such as taste, smell, or color) and by the postingestive consequences experienced after eating (such as the nutrient value of the food or the toxicity of the plant). Studies in domestic sheep suggest that postingestive consequences are the main factors influencing the daily food intake of sheep, but preingestive factors, such as palatability, influence eating patterns, and short-term diet selection (Favreau et al., 2013). Sheep also show a diurnal pattern of diet selection, selecting different food items in the morning to those consumed in the afternoon. Dietary preferences seem to be influenced by novelty (for example, sampling of foods that they have not eaten recently), and sheep prefer a mixed diet to a single plant species (Rutter, 2006). The strong flocking and social behavior of sheep also influence feeding; lambs learn about the environment and food sources from observing their mothers feeding on particular plants (Saint-Dizier et al., 2007), and exposure to specific plant species during early development has an important influence on what is consumed as an adult (Provenza et al., 1992). This allows the lamb to learn about the location and relative seasonal abundance of food in the home range, and to avoid ingestion of poisonous plants.

Sheep will spend about 8 h a day grazing, which can increase to 12 h if nutritional resources are limiting. However, as ruminants, sheep need to devote a significant proportion of their day to ruminating, when food is regurgitated and rechewed to allow the cellulose cell walls of plant cells to be broken open, so that the ruminal microbes can perform fermentation. Rumination usually takes place while the sheep is lying down and, as with many behaviors, is socially facilitated, such that this usually occurs synchronously across the flock.

Communication

Sheep, as a social species, use a variety of signals to maintain group cohesion. They appear to rely predominantly on visual cues to recognize one another, although auditory and olfactory information is also used (Geist, 1971). Sheep are considered to use graded communication signals, mainly based on postural and behavioral signaling. In wild sheep populations, a number of social behavior patterns used for signaling have been identified, including 17 different signals in Mountain Bighorn sheep (Geist, 1971). These are signals, postures, and actions (1) performed in a stereotyped, distinct manner to conspecifics; (2) occasionally accompanied by sounds or odors;

FIGURE 16.2 Example of the "head-up" posture by a lactating ewe, where the head is held alert, rigid, and above the level of the back oriented toward the threat (in this case the photographer). (Photo: Marianne Farish.)

(3) characterized by unusual body conformations; and (4) that may be faster, slower, jerkier, or stiffer than normal and are oriented toward conspecifics. The best-described postural signal is the vigilance or alarm posture, sometimes described as a "head-up" posture (Figure 16.2), where animals stand rigidly with the head and neck alert and elevated, staring in the direction of a threat. This posture communicates threat and alarm to other members of the flock and may be followed by flight. Ewes also use this posture to bring their lambs to them to suckle when they are young (Pickup and Dwyer, 2011).

As a predated species, sheep do not use vocal communication extensively and have only three defined call types. These consist of a high-pitched protest or distress *bleat*, made with the mouth open; a low-pitched maternal *rumble*, made by the ewe with the mouth closed when communicating to the lamb at close quarters; and a similar, but harsher *low-pitched call*, made by rams when courting and mating with ewes. The high-pitched bleat is used in a variety of contexts; to express alarm, to communicate between social group members when separated, and during extreme fear or pain. However, vocal communication is often inhibited in the presence of a predator, when sheep may remain silent to avoid detection. Sheep do show individual variation in the structure of their bleats in sound spectrograms, which suggests that individual recognition would be possible from vocal cues, although only ewe-lamb recognition from vocal cues has been convincingly demonstrated so far (Sèbe et al., 2007).

Common Captive Behaviors

As a species that has been domesticated for a long time, sheep have adapted to captive life. However, they are usually husbanded in extensive grazing situations or at pasture, and some sheep breeds (generally hill and more primitive breeds) are less tolerant of close confinement and crowded housing than others. On large and extensive pastures, sheep are able to show home range behavior, in which ewes of the same social group restrict themselves to particular areas in which they become familiar with the location of resources, such as food, water, and shelter. In more confined and fenced pastures, home range behavior is not possible, but sheep will form subgroups, choosing to associate predominantly with preferred social companions (Lynch et al., 1992). The size and closeness of subgrouping is influenced by (1) breed, as more highly selected breeds form larger subgroups and flock closer together (Dwyer and Lawrence, 2000), and (2) environment, with some sheep more likely to subgroup if the environment is more complex (Stolba et al., 1990). With commercial farming, many sheep are never "housed", spending their entire lives at pasture, thus their responses to close confinement have been less studied than their behavior at grass.

Normal

Given sufficient space, social companions, and adequate resources, sheep are generally able to adapt their natural behavior very well to a captive environment. Years of domestication have led to reduced alertness and attentiveness to the environment, and increases in social tolerance and affiliative responses. For example, domestic sheep are more vocal than wild sheep (Lynch et al., 1992), which may have arisen from the need to have more complex social signals in larger groups, and a relaxation of selection pressure on vocal behavior from the reduced risk of predation.

Sheep show less pronounced circadian rhythms of feeding and resting behavior in a housed environment, although behavioral synchrony can still be seen in feeding, ruminating, and resting responses (Lynch et al., 1992). Sheep that have not been well accustomed to close contact with humans will express fear responses and potentially, flight to the back of the pen, when people approach. As described below, sheep should be habituated to the presence of handlers to reduce the stress of frequent contact and can be readily trained to tolerate human presence and handling.

Abnormal

Abnormal or stereotypical behavior is rarely observed in sheep, perhaps because they are not commonly housed in the sorts of restrictive environments where these behaviors typically arise. Alternative explanations suggest that rumination may also act as a buffer in stressful situations, contributing to the low level of stereotypy in group-housed sheep (Nowak et al., 2008). However, individually housed sheep have been shown to demonstrate stereotypical oral behaviors, such as mouthing bars, chewing slats or chains, rattling or chewing buckets, biting and chewing pen fixtures, mandibulation (licking lips and mouthing air), and repetitive licking (Lauber et al., 2012). Locomotor stereotypies have also been reported, including rearing against the pen, repetitive butting, star-gazing (arching the head and neck over the back), leaping vertically up and down, weaving, and route-tracing (Done-Currie et al., 1984; Marsden and Wood-Gush 1986). Feed restriction, particularly the provision of limited energy, or a diet lacking in fiber, increases the frequency of abnormal oral behaviors (Vasseur et al., 2006).

In housed sheep at high stocking density, the most commonly described or observed aberrant behavior is wool-biting (sometimes called wool-pulling, or wool-chewing), often carried out by the more dominant animal on subordinates. Typically, the wool is pulled or plucked from one animal's back by another, using the mouth to pull strands of wool free, which may then be ingested. As a consequence, the bitten animal becomes progressively denuded on the back area. The causes of wool-biting are not completely understood, but appear to be related to overcrowding and social dominance, although nutritional deficiencies may also contribute. One study suggests that wool-biting can be decreased by increasing the amount of fiber provided in the diet (Vasseur et al., 2006), suggesting that this may be primarily a redirected oral response of housed sheep deprived of activity or oral stimulation.

Aggression is not commonly seen in unconfined and established social groups, except between males during the breeding season. However, even ewes will show aggression, such as butting and chasing, if unfamiliar animals are introduced into the group and resources are limited, or animals are unable to get away (Lynch et al., 1992). In females, this rarely leads to serious injury, but males, and especially those with horns, can cause significant damage to other rams, resulting in injury or death.

Ways to Maintain Behavioral Health

Captive Considerations to Promote Species-Typical Behavior

Domestication of sheep has largely allowed natural behavior patterns to continue, or has selected for animals that express high levels of some behaviors, such as flocking, as this makes management easier. However, successful management does rely on understanding the behavior of the species and providing opportunities for natural behaviors to be expressed.

Environment

Sheep should be kept in bedded pens with sufficient space and deep straw to allow all animals to lie down at the same time on a dry surface. Sheep spend less time lying down and more time engaged in social behaviors, both positive and negative, with an increase in displacement activity when the space per sheep is reduced (Bøe et al., 2006; Averós et al., 2014). In the European Union (EU), the minimum recommendation for space per animal is 1.5–1.8 m^2 for adult sheep (depending on body weight; Council of Europe, 2010), and housing densities of approximately 1 m^2 per animal have been shown to cause a significant increase in displacements and aggression (Averós et al., 2014). Sheep also require a minimum of 0.4 m (for sheep weighing between 35 and 60 kg) or 0.5 m (for sheep over 60 kg) of trough space, and 0.12 m per animal of access to *ad libitum* feeding, for forage. Although sheep often lie against the walls of the pen and against structures in outdoor environments, adding additional wall space or cubicles does not appear to increase resting time, reduce aggression, or promote better behavioral synchrony (Jørgensen et al., 2009). The opportunity for all individuals to lie synchronously and the ability to see other pen mates seem to be more important than access to walls.

Sheep generally show a preference for a bedded lying surface over a solid floor, particularly if they are shorn (Færevik et al., 2005), and spend more time lying down on bedded surfaces.

Studies comparing different types of bedding suggest that wood-chips, sawdust, straw, or rice-husks can all provide suitable bedding material for sheep (Lynch et al., 1992), although the use of these substrates may be affected by local availability, manure management, and hygiene requirements. Sheep will avoid unbedded surfaces, and metal, wooden, or plastic slats are not preferred when bedded surfaces are available. Slatted surfaces can cause damage to feet, lesions, or calluses on the joints, and may affect thermal comfort at cold temperatures. In the EU, the recommendations (Council of Europe, 2010 ETS 123 revised Appendix A) are that the entire housing enclosure for experimental sheep should have a solid floor, with appropriate bedding.

Sheep normally rely on walking on hard or abrasive surfaces to wear down their hooves and prevent long and overgrown toes. Prolonged housing in smaller areas, which reduces activity, with only soft-bedded floors, can lead to deformed or overgrown hooves and lameness. Hoof trimming is considered painful for sheep. The restraint and inversion required to trim hooves is stressful, and hoof trimming can be a route to pass bacterial infections (such as *Dichelabacter nodosus*, which causes foot rot) from one animal to another (Green and Clifton, 2018). Therefore, allowing sheep to walk on hard surfaces would allow the normal process of hoof wear to occur and prevent the need for painful or stressful foot treatments.

Specific consideration should be given to whether individual penning is required for experimental purposes, and should be avoided if possible, as sheep are highly social and find separation very stressful. If individual penning is required, pens should be arranged such that animals are able to see and hear, and preferably touch one another. Sheep require a period of acclimatization to a new environment, and reducing the degree of novelty of any changes will improve their ability to cope. There have been very few studies of enrichment in sheep, and animals with access to forage, space, and social companions may not require the provision of "toys" or other enrichment devices. The use of a mirror to substitute for social companions in isolated sheep is only partially successful (Parrott et al., 1988) and is unlikely to completely mitigate the absence of conspecifics.

Social Groupings

Sheep should be kept in groups of at least 3–4 animals, with the sexes segregated. The introduction of new animals should be kept to a minimum and the formation of stable social groups should be encouraged to reduce aggression and stress. If there are sufficient resources (feeder space, lying space, etc.), sheep will be able to show synchronous behaviors, and even subordinate animals will be able to feed and rest at the same time as other animals. When space allowances decrease, increases in displacements from feeders and preferred lying locations, and frequent affiliative and aggressive social contacts, should be expected (Averós et al., 2014). Subordinate animals will be less able to access food and rest, and may experience impaired growth and higher disease susceptibility.

Social isolation and separation from the group are extremely stressful for sheep, and can impair their ability to cope with the other demands made upon them. This is potentially problematic for experimental studies, as sheep may be anxious and fearful when initially isolated, and subsequently become withdrawn and apathetic (Done-Currie et al., 1984). Wherever experimental protocols allow, sheep should be allowed to see and hear other conspecifics at all times, and to have tactile contact, if possible. For example, if animals are involved in individual testing, the use of "buddy" animals or visual contact with other sheep should be employed to reduce stress.

Sheep often regard a human presence in a similar way to a predator, which can make moving and using experimental sheep challenging, particularly when only one animal might need to be moved. The strong flocking and following responses of sheep, which facilitate movement of a group of animals, means that sheep are very reluctant to move away from the group alone, and the primary response of an isolated sheep is to attempt to return to the social group. This can lead the animals to panic and potentially injure themselves or their human handlers. Whenever possible, sheep should be moved as a group, as this will help the animals remain calm, and greatly ease the work involved in moving them. Lactating ewes, however, become significantly less gregarious during lactation, and are mostly concerned about the presence of, and following, their lambs. These animals can be moved most easily by carrying the lambs and encouraging the ewe to follow, and can be kept away from the flock as a family.

Hand-reared animals, or lambs reared with close, positive contact with humans (such as stroking and hand-feeding), can come to rely on their human care-givers as social companions in an isolation situation (Boivin et al., 2000), which may be required in some experimental setups. Sheep are very capable of learning to distinguish between familiar and unfamiliar handlers, and will react more positively, and with fewer distress vocalizations, in the presence of a familiar person (Boivin et al., 1997). To achieve these familiarization-type responses, the lamb needs to be exposed to frequent positive human contact at a young age, and to be most effective, the lamb should be removed from the ewe and raised artificially, because handling is much less effective if the mother is present (Boivin et al., 2002). Gentling or taming of older animals is also possible and reduces the stress responses (such as elevated heart rates and flight) shown by sheep when approached by humans; Boivin et al., 2000). Although older animals that have been gentled are less reactive to human presence, they may still show similar struggling behaviors to those seen in unhandled animals when experiencing restraint, inversion, and/or confinement.

The period of time that the lamb spends with its mother is important for development and for it to learn social behaviors and responses; essentially how to behave like a sheep. In the management of farmed meat sheep, lambs spend 8–12 weeks with their mothers before they are weaned, although dairy lambs may be weaned at earlier ages. Young lambs can be successfully reared in mixed-sex peer groups and readily learn to suckle milk from a bucket or automatic feeder. Compared to dairy calves, fewer studies have been conducted on the behavioral responses of lambs reared without their mothers. However, in cattle, the separation of the calf from the mother causes an alteration in behavior and judgment bias that suggests that the calves are pessimistic (Daros et al., 2014), a response similar to that shown by calves after de-horning. Calves that were reared in isolation had poorer social skills and an impaired ability to learn cognitive tasks, particularly those involving reversal learning (Meagher et al., 2015). It is quite likely that lambs will show similar responses, and recent studies using magnetic resonance imaging to assess lamb brain development suggest that early separation from the dam may be associated with alterations in neurological development (Chaillou et al., personal communication). Thus, whenever possible, lambs should be reared by their mothers, preferably for at least 8 weeks, or within a peer group, if experimental requirements dictate.

Feeding Strategies

Sheep, as ruminants, have a specialized digestive system that is designed to extract and metabolize nutrients from plants, and the maintenance of ruminal health is important for sheep well-being. The rumen is a delicate ecosystem of micro-organisms, bacteria, and other microbes that can break down cell walls and ferment the contents to produce energy. Sudden changes in diet, particularly a large increase in nutrient-rich foods, such as grains, can upset this balance and cause potentially fatal digestive consequences. In the wild, sheep normally cover a considerable distance each day, walking up to 12 km to water sources and to select a diet from a mixed array of plant species available (Dwyer, 2008a). In captive situations, they are able to move less and are often given a monotonous diet, or a diet that is denser in nutrients than they would normally be able to find by grazing. This causes an increase in time spent lying in indoor-housed sheep. To mimic grazing, animals should be given *ad libitum* access to a forage-based diet, such as hay, freshly cut grass, or silage, such that the amount of time spent feeding can be prolonged and functionally simulate their evolved feeding responses. The allomimetic tendencies of sheep mean that they also require sufficient space to allow all animals in the group to feed at the same time.

Sheep sometimes require additional protein or energy than can be achieved from forage alone (for example, in late pregnancy; AHDB Better Returns Programme, 2018). This is often provided by feeding a concentrated feed, in addition to the forage diet, in small quantities, once or twice a day. The concentrated feed is often highly palatable and thus, if offered to a group with insufficient feeder space, can lead to aggression and displacements at the feeder, as everyone attempts to eat the maximum amount. Subordinate animals may be displaced from the feeder entirely and unable to eat their ration, so special attention will be needed to ensure that they can eat their nutritional requirements.

Training

Sheep have an undeserved reputation for lack of intelligence, perhaps influenced by their strong flight responses and following behaviors. However, behavioral and neurobiological research has shown that sheep have significant perceptual and cognitive abilities, are able to show sophisticated use of visual and olfactory cues in recognition and learning, and have distinct individual personality traits (Lynch et al., 1992; Kendrick et al., 2001; Nowak et al., 2008). Sheep have excellent visual recognition abilities for both social and other cues, and can learn to identify and remember the facial characteristics of at least 50 different sheep faces and 10 different human faces, and retain that memory for at least 2 years (Peirce et al., 2000; Kendrick et al., 2001). They are also capable of discriminating the emotional content in sheep or human faces, showing preferences for relaxed and smiling human faces over anxious or angry faces (Tate et al., 2006). Further, sheep have been shown to be able to recognize a familiar human from a 2D representation of the person (Knolle et al., 2017), suggesting they are able to form a 3D mental representation from a 2D image.

Sheep can be trained to perform operant tasks, usually by pressing a panel with their nose to obtain a food reward. During the 1970s and 1980s in particular, a range of operant studies were conducted with sheep, which demonstrated their ability to press panels to obtain radiant heat in a cold environment, to determine their preferences for different levels of illumination, and to discriminate between different shapes or feed types (Baldwin, 1981; Hutson and van Mourik, 1981). The ability of sheep to learn these discrimination tasks depends to some extent on the objects they are being asked to distinguish. Sheep show much quicker learning for familiar types of objects (such as food or sheep faces) than when required to learn to discriminate between abstract, less salient shapes (Baldwin, 1981; Kendrick et al., 1996).

Vertical transmission of information from mother to young, a form of observational learning, is an important source of learning in sheep, as has been discussed above. In addition, lambs appear to be able to learn to drink milk from a bucket by observing lambs that already know how to do so (Veissier and Stefanova, 1993). There is less evidence for whether sheep can learn from human gaze or pointing gestures (Nawroth et al., 2019). However, farm sheep that have not been specifically habituated to humans show behaviors that are indicative of a negative responsiveness to the gaze of humans (Beausoleil et al., 2006), suggesting that they regard human direct gaze as a warning cue. Whether sheep that have been reared by humans, and thus have a more affiliative and positive view of humans, can follow human gaze or learn from a human demonstrator is not yet known.

As described above, sheep can be fearful of humans, unless they have had specific positive contact with humans at an early age, or a prolonged period of habituation to humans. They also show fear or aversion to novel environments. These fears, in addition to their strong social tendencies, can mean that training for paradigms that require the animal to be alone in an apparatus can be difficult, as fear and panic may overwhelm any learning. In these situations, animals may need to be desensitized to the components of the trial, such as novelty of a testing area and being alone while in the presence of humans. Forcing or rushing animals to enter an unknown space can be counter-productive, but sheep can quickly habituate to new environments when allowed the time to explore *in the company of conspecifics*. Sheep are receptive to food rewards and, in animal handling situations, their willingness to move into an arena improved after pairing even aversive procedures with preferred foods, such as barley (Hutson, 1985), although this was less effective with more unpleasant treatments. Sheep have excellent memories and were able to retain the memory of their training for at least 1 year (Hutson, 1985).

Special Situations

Sheep are adaptable and versatile animals that can cope well with a captive environment. Their strong social tendencies mean that experimental paradigms that require prolonged periods of social isolation, or physical restraint, should be carefully considered, as animals will find these procedures extremely stressful. This can impact experimental outcomes, as well as adversely affect animal welfare. Should these procedures be necessary, then hand-rearing animals, such that human caregivers can be substitutes for the presence of other sheep, would reduce the impact of fear in these situations. However, it is important to remember that hand-rearing, especially if this is in isolation from other animals, has other negative consequences for welfare (as outlined above). Therefore, the need to conduct these types of studies, and whether the experimental design can be altered to achieve the outcomes without social isolation, should be carefully considered.

In general, sheep are relatively docile and, when working within their evolved desires to follow and stay within the social group, can be easy to handle. Rams, who are frequently horned, can be the exception, particularly when near ewes in estrus. Rams are generally less fearful of humans, more aggressive, and more dangerous than ewes, and therefore should be handled with additional care.

Conclusions/Recommendations

Sheep have been domesticated for a long time, their behavior has adapted to the captive environment, and there is a wealth of knowledge about their behavior, healthcare needs, and feeding requirements, making them relatively easy to manage. A number of aspects of their behavior, such as promiscuous breeding, maternal care, and relative docility, have been characteristics exploited by humans for production purposes, and these can also be favorable for management in other situations. However, part of the adaptation of sheep during domestication for farming purposes has been the strengthening of sociopositive responses, and reinforcement of flocking and following behavior. These behaviors make it straightforward to move and manage animals as a group, but very difficult to maintain them individually. Because some of the social responses of sheep must be learned from associations acquired by the lamb from following and mimicking the mother, positive handling and early life contact with humans can be beneficial for the use of sheep in research. This is particularly true for studies in which animals may need to be alone or human care-givers can provide social support.

Sheep have a good ability to learn, especially when relevant cues are used, and have excellent memories, retaining information about training for at least a year and social information for at least 2 years. These traits can be very positive, as trained animals are then able to participate in many studies over a number of years, due to their long life span. However, animals are also very capable of learning about unpleasant associations, with these memories also being retained for some time. Sheep can form conditioned place aversions very readily and have also been shown to form associations between specific people and unpleasant experiences (Nowak et al., 2008), so movements to new locations and exposure to novel environments should always be paired with positive rewards, such as food. If routine aversive procedures, such as inversion for foot trimming, need to be carried out, then ensuring these are done by people different from those responsible for experimental handling may be beneficial.

Additional Resources

The Welfare of Sheep (2008), ed. C.M. Dwyer, AWNS 6, Springer

Advances in the Welfare of Sheep (2017), ed. D. Ferguson, C. Lee, A. Fisher, Woodhead Publishing.

AHDB Better Returns Programme: Feeding the Ewe. (2018) http://beefandlamb.ahdb.org.uk/wp-content/uploads/2018/03/Feeding-the-ewe.pdf (accessed 30 April 2020).

REFERENCES

Arnold, G.W., Maller, R.A. (1974) Some aspects of competition between sheep for supplementary feed. *Animal Production* 19, 309–319.

Averós, X., Lorea, A., Beltran de Heredia, I., Ruiz, R., Marchewka, J., Arranz, J., Estevez, I. (2014) The behaviour of gestating dairy ewes under different space allowances. *Applied Animal Behaviour Science*. 150, 17–26.

Baldwin, B.A. (1981) Shape discrimination in sheep and calves. *Animal Behaviour*, 29, 830–834

Beausoleil, N.J., Stafford, K.J., Mellor, D.J. (2006) Does direct human eye contact function as a warning cue for domestic sheep (*Ovis aries*)? *Journal of Comparative Psychology* 120, 269–279.

Bøe, K.E., Berg, S., Andersen, I.L. (2006) Resting behaviour and displacements in ewes – effects of reduced lying space and pen shape. *Applied Animal Behaviour Science*, 98, 249–259.

Boivin, X., Nowak, R., Desprès, G., Tournadre, H., Le Neindre, P. (1997) Discrimination between shepherds by lambs reared under artificial conditions. *Journal of Animal Science* 75, 2892–2898.

Boivin, B., Tournadre, H., Le Neindre, P. (2000) Hand-feeding and gentling influence early-weaned lambs' attachment responses to their stockperson. *Journal of Animal Science*, 78, 879–884.

Boivin, X., Boissy, A., Nowak, R., Henry, C., Tournadre, H., Le Neindre, P. (2002) Maternal presence limits the effects of early bottle feeding and petting on lambs' socialisation to the stockperson. *Applied Animal Behaviour Science*, 77, 311–328.

Chang, D.W., Satterfield, W.C., Son, D., et al. (2009) Use of vascularized periosteum or bone to improve healing of segmental allografts after tumor resection: An ovine rib model. *Plastic and Reconstructive Surgery* 123, 71–78.

Chillard, Y., Delavaud, C., Bonnet, M. (2005) Leptin expression in ruminants: Nutritional and physiological regulations in relation to energy metabolism. *Domestic Animal Endocrinology* 29, 3–22.

Council of Europe (2010). Directive 2010/63/EU of the European Parliament and of the Council of 22 September 2010 on the protection of animals used for scientific purposes.

Daros, R.R., Costa, J.H.C., Von Keyserlingk, M.A.G., Hotzel, M.J., Weary, D.M. (2014) Separation from the dam causes negative judgement bias in calves. *PLoS One* 9, e98429.

De Nicolo, G., Morris, S.T., Kenyon, P.R., Morel, P.C.H.H., Parkinson, T.J. (2008) Melatonin-improved reproductive performance in sheep bred out of season. *Animal Reproduction Science* 109, 124–133.

Done-Currie, J.R., Hecker, J.F., Wodzicka-Tomaszewska, M. (1984) The behaviour of sheep transferred from pasture to an animal house. *Applied Animal Behaviour Science* 12, 121–130.

Dwyer, C.M. (2004) How has the risk of predation shaped the behavioural responses of sheep to fear and distress? *Animal Welfare* 13, 269–281.

Dwyer, C.M. (2008a) Environment and the sheep. In: *The Welfare of Sheep* (ed. C.M. Dwyer), Springer, Berlin.

Dwyer, C.M. (2008b) Individual variation in the expression of maternal behaviour: a review of the neuroendocrine mechanisms in the sheep (*Ovis aries*). *Journal of Neuroendocrinology* 20, 526–535.

Dwyer, C.M. (2017) Reproductive management (including impacts of prenatal stress on offspring development). In: *Advances in the Welfare of Sheep* (ed. D. Ferguson, C. Lee, A. Fisher), Woodhead Publishing, Sawston.

Dwyer, C.M., Lawrence, A.B. (2000) Effects of maternal genotype and behaviour on the behavioural development of their offspring in sheep. *Behaviour* 137, 1629–1654.

Færevik, G., Andersen, I.L., Bøe, K.E. (2005) Preferences of sheep for different types of pen flooring. *Applied Animal Behaviour Science* 90, 265–276.

FAOSTAT. (2017) Food and Agriculture Organization of the United Nations. http://www.fao.org/faostat/en/#data/QA/visualize

Favreau-Peigné, A., Baumont, R., Ginane, C. (2013) Food sensory characteristics: their unconsidered role in the feeding behaviour of domestic ruminants. *Animal* 7, 806–813.

Geist, V. (1971) *Mountain Sheep: A Study in Behaviour and Evolution*. University of Chicago Press, Chicago; London

Green, L., Clifton, R. (2018) Diagnosing and managing footrot in sheep: An update. *In Practice* 40, 17–26.

Hassenbusch, S.J., Satterfield, W.C., Gradert, T.L. (1999) A sheep model for continuous intrathecal infusion of test substances. *Human and Experimental Toxicology* 18, 82–87.

Hutson, G.D. (1985) The influence of barley food rewards on sheep movement through a handling system. *Applied Animal Behaviour Science* 14, 263–273.

Hutson, G.D., van Mourik, S.C. (1981) Food preferences of sheep. *Australian Journal of Experimental Agriculture and Animal Husbandry* 21, 575–582.

Jansen, B.D., Krausman, P.R., Heffelfinger, J.R., Devos, J.C. (2006) Bighorn sheep selection of landscape features in an active copper mine. *Wildlife Society Bulletin*, 34, 1121–1126.

Johansen, M.J., Gradert, T.L., Satterfield, W.C., et al. (2004) Safety of continuous intrathecal midazolam infusion in the sheep model. *Anesthesia and Analgesia* 98, 1528–1535.

Jørgensen, G.H.M., Andersen, I.L., Bøe, K.E. (2009) The effect of different pen configurations on the behaviour of sheep. *Applied Animal Behaviour Science* 119, 66–70.

Kendrick, K.M., Atkins, K., Hinton, M.R., Heavens, P., Keverne, E.B. (1996) Are faces special for sheep? Evidence from facial and object discrimination learning tests showing effects of inversion and social familiarity. *Behavioural Processes* 38: 19–35.

Kendrick, K.M., da Costa, A.P., Leigh, A.E., Hinton, M.R., Peirce, J.W. (2001) Sheep don't forget a face. *Nature*, 414, 165–166.

Kortekaas, J. (2014) One Health approach to Rift Valley fever vaccine development. *Antiviral Research* 106, 24–32.

Knolle, F., Goncalves, R.P., Morton, A.J. (2017) Sheep recognise familiar and unfamiliar human faces from two-dimensional images. *Royal Society Open Science* 4, 171228.

Lauber, M., Nash, J.A., Gatt, A., Hemsworth, P.H. (2012) Prevalence and incidence of abnormal behaviours in individually housed sheep. *Animals* 2, 27–37.

Lynch, J.J., Hinch, G.N., Adams, D.B. (1992) *The Behaviour of Sheep: Biological Principles and Implications for Production*. CSIRO Publishing, Australia.

Manikkam, M., Crespi, E.J., Doop, D.D., Herkimer, C., Lee, J.S., Yu, S., Brown, M.B., Foster, D.L., Padmanabhan, V. (2004) Fetal programming: Prenatal testosterone excess leads to fetal growth retardation and postnatal catch up growth in sheep. *Endocrinology* 145, 790–798.

Marsden, M., Wood-Gush, D. (1986) The use of space by group-housed sheep. *Applied Animal Behaviour Science* 15, 178–182.

Morton, A.J., Rudiger, S.R., Wood, N.I., Sawaik, S.J., Brown, G.C., McLaughlan, C.J., Kuchel, T.R., Snell, R.G., Faull, R.L.M., Bawden, C.S. (2014) Early and progressive circadian abnormalities in Huntington's disease sheep are unmasked by social environment. *Human Molecular Genetics* 23, 3375–3383.

Meagher, R.K., Daros, R.R., Costa, J.H.C., von Keyserlingk, M.A.G., Hötzel, M., Weary, D.M. (2015) Individual housing impairs reversal learning and increases fear of novel objects in dairy calves. *PLoS ONE* 10, e0132828.

Nawroth, C., Langbein, J., Coulon, M., Gabor, V., Oesterwind, S., Benz-Schwarzburg, J., von Borell, E. (2019) Farm animal cognition – Linking behavior, welfare and ethics. *Frontiers in Veterinary Science*, 6, 24.

Nowak, R., Keller, M., Val-Laillet, D., Lévy, F. (2007) Perinatal visceral events and brain mechanisms involved in the development of mother-young bonding in sheep. *Hormones & Behavior*, 52, 92–98.

Nowak, R., Porter, R.H., Blache, D., Dwyer, C.M. (2008) Behaviour and the welfare of sheep. In: *The Welfare of Sheep* (ed. C.M. Dwyer), Springer, Berlin.

Parrott, R.F., Houpt, K.A., Mission, B.H. (1988) Modification of the responses of sheep to isolation stress by the use of mirror panels. *Applied Animal Behaviour Science* 19, 331–338.

Pickup, H.E., Dwyer, C.M. (2011) Breed differences in the expression of maternal care at parturition persist throughout the lactation period in sheep. *Applied Animal Behaviour Science* 132, 33–41.

Peirce, J.W., Leigh, A.E., Kendrick, K.M. (2000) Configurational coding, familiarity and the right hemisphere advantage for face recognition in sheep. *Neuropsychologia*, 38, 475–483.

Perentos, N., Martins, A.Q., Watson, T.C., Bartsch, U., Mitchell, N.L., Palmer, D.N., Jones, M.W., Morton, A.J. (2015) Translational neurophysiology in sheep: Measuring sleep and neurological dysfunction in CLN5 Batten disease affected sheep. *Brain* 138, 862–874.

Poindron, P., Lévy, F., Keller, M. (2007) Maternal responsiveness and maternal selectivity in domestic sheep and goats: The two facets of maternal attachment. *Developmental Psychobiology* 49, 54–70.

Provenza, F.D., Pfister, J., Cheney, C.D. (1992) Mechanisms of learning in diet selection with reference to phytotoxicosis in herbivores. *Journal of Rangeland Management* 45, 36–45.

Ramirez, E.R., Ramirez, D.K., Pillari, V.T., Humberto Vasquez, D.M.V., Ramirez, H.A. (2008) Modified uterine transplant procedure in the sheep model. *Journal of Minimially Invasive Gynecology* 15, 311–314.

Rutter, S.M. (2006) Diet preference for grass and legumes in free-ranging domestic sheep and cattle: Current theories and future application. *Applied Animal Behaviour Science* 97, 17–35.

Saint-Dizier, H., Lévy, F., Ferreira, G. (2007) Influence of the mother in the development of flavoured-food preference in lambs. *Developmental Psychobiololgy* 49, 98–106.

Sèbe, F., Nowak, R., Poindron, P., Aubin, T. (2007) Establishment of vocal communication and discrimination between ewes and their lamb in the first two days after parturition. *Developmental Psychobiololgy* 49, 375–386.

Stolba, A., Hinch, G.N., Lynch, J.J., Adams, D.B., Munro, R.K., Davies, H. I. (1990) Social organisation of Merino sheep of different ages, sex and family structure. *Applied Animal Behaviour Science,* 27, 337–349.

Tate, A.J., Fischer, H., Leigh, A.E., Kendrick, K.M. (2006) Behavioural and neurophysiological evidence for face identity and face emotion processing in animals. *Philosophical Transactions of the Royal Society London B-Biological Sciences*, 361, 2155–2172.

Vasseur, S., Paull, D.R., Atkinson, S.J., Colditz, I.G., Fisher, A.D. (2006) Effects of dietary fibre and feeding frequency on wool biting and aggressive behaviour in Merino sheep. *Australian Journal of Experimental Agriculture*, 46, 777–782.

Veissier, I., Stefanova, I. (1993) Learning to suckle from an artificial teat within groups of lambs: Influence of a knowledgeable partner. *Behavioural Processes* 30, 75–82.

Wilmut, I., Schnieke, A.E., McWhir, J., Kind, A.J. Campbell, K.H.S. (1997) Viable offspring derived from fetal and adult mammalian cells. *Nature* 385, 310–313.

17

Behavioral Biology of Cattle

Clive Phillips
Curtin University

CONTENTS

Introduction	273
Typical Research Participation	274
Behavioral Biology	275
Ecology and Habitat Use	275
Social Organization and Behavior	276
Feeding Behavior	276
Communication	277
Common Captive Behavior	277
Ways to Maintain Behavioral Health	279
Captive Considerations to Promote Species-Typical Behavior	279
Environment	279
Social Groupings	280
Feeding Strategies	280
Training	280
Cattle	280
Handlers	280
Special Situations	281
Conclusions	281
Additional Resources	281
References	281

Introduction

Cattle are used infrequently as laboratory animals, compared with smaller animals, such as rodents and zebrafish, which are used extensively for medical and genetic research, respectively (Lambert et al., 2019). When cattle are used, it is usually to develop systems of production that increase profitability for the farmers that manage them. This may include benefit to the animals themselves, such as testing new methods of disease prevention, but it may be to increase productivity levels, which, unless associated with means of providing for the additional nutrient demand, may have adverse effects on the animals' welfare (Nalon and Stevenson, 2019). Dairy cows have to support their high milk yields in early lactation by loss of body reserves, as intake is not usually sufficient to provide adequate nutrients at this stage in their lactation cycle (Phillips, 2018).

The most defendable use of cattle in research is where they themselves benefit, such as in testing remedies for a disease from which an animal is suffering; also defendable in many people's view would be research that benefits the species as a whole, such as finding a cure for a disease that many other cattle suffer from, such as mastitis or lameness. Recently, a new type of cattle research has emerged, which seeks to address the damaging effect that cattle have on the local and global environment, including global climate change. However, many poor people, mainly subsistence farmers in developing countries, are dependent on cattle production systems for their livelihood, and reducing their contribution to the environmental problems is urgent and can only be done by investigating alternative systems experimentally, such as agroforestry systems. Numerically, these by far outnumber cattle in intensive farms; and per quantity of product in particular, their pollution potential is very significant (Phillips, 2018, p. 7). Less defendable cattle research would be instances in which the disease is primarily of zoonotic concern, such as bovine tuberculosis, or where cattle are being used for testing methods of increasing their productivity in a farming system. Breeding and managing cattle with enhanced growth rates or milk yields is a common use of cattle in research.

Cattle parts are also used in research, obtained usually at abattoirs. Hooves have been used to determine slipperiness of floors (Phillips et al., 2000), and rumen liquor may be used to test the digestibility of different feeds (Phillips et al., 2001).

Another alternative system attracting much attention is the growth of beef *in vitro*, using biopsied samples of cattle muscle tissue (Wilks and Phillips, 2017). One of the contentious aspects of this system of producing beef meat is the extent to which donor cows have to supply muscle samples.

Typical Research Participation

Early research in the 20th century often focused on the digestive function of cattle, since feed represents about 70% of the economic input into a cattle farm. From simple feeding trials, recording the growth rate, or milk yield of cattle on different feeding regimes more sophisticated studies were developed, in which the workings of the rumen and intestines were investigated in detail. A key advance was determining individual feed intake. Techniques to measure individual intakes in a group of cattle were devised, using individual access gates to the feed for each animal (Broadbent et al., 1970), even though the resulting lack of competition between animals reduces intake by about 5%, compared with a group feeding situation. Later, placing load cells under the corners of feed bins, coupled with individual cow recognition, allowed individual intake to be determined in a group feeding situation.

Comparison of the feeding value of individual feeds when feeding a mixed diet, a technique whereby small amounts of each feed were placed in the rumen in nylon bags and the residues weighed after sequential extraction of the bags at set times was used to determine the rate of utilization of nutrients by rumen micro-organisms (Ørskov and McDonald, 1979). Bags were initially inserted and removed *per fistulam*, accessed via a cannula permanently sewn into the rumen wall and attached to the animal's epidermal tissue. Extreme care had to be taken to avoid contamination of the site of attachment and invasion of the site by ectoparasites, in particular flies. However, such cattle were frequently kept for research for several years, usually indoors, because the sight of cattle with cannulae in their sides was considered too disturbing for members of the public to view, if the cattle were outside grazing. Recently, a more humane technique of oral insertion of the bags, followed by *postmortem* collection, has been developed (Pagella et al., 2018), avoiding the need for surgery.

Much laboratory research with cattle has been conducted by restraining them individually in metabolism crates. The floor of the crate is raised off the ground so that a tray can be placed underneath to collect excreta. It is the most accurate way to measure intake and output of faeces and urine, which are separated by allowing them to land on a sloping tray. The method also allows the operator to easily restrain the cattle for sample collection. By monitoring feed intake for about 1 week, or more if major changes in diet are being tested (in which case the rumen micro-organisms need longer to adapt), and multiplying it by nutrient concentrations, nutrient intakes can be estimated. Then by subtracting nutrients in excreta, whole tract feed and nutrient digestibilities can be determined (e.g., Batista et al., 2016). Further detail of small intestine function can be gained by cannulating this section of the digestive tract, but this is more invasive and less commonly used than rumen fistulation.

By combining restraint of cattle in metabolism crates with rumen fistulation and insertion of the feed into nylon bags, digestion of nutrients in the rumen can also be determined; further measurements of nutrient contents in the rumen can provide information on solubilization of nutrients on entry into the rumen (Chiy and Phillips, 1993). Detailed information on digestion of feed particles can also be obtained by photographing herbage particles with an electron microscope at various stages in the digestion process. Hence, much information can be obtained from metabolism studies; however, they impose significant movement restrictions on the animals, which can only stand up and lie down, in order to keep the excreta to one end of the crate. Turning may be prevented by tethering the cattle. Cattle should be removed from the crates for exercise at regular intervals, preferably weekly, otherwise their joints become stiff. Rubber mats should be used when moving cattle to stop them slipping. The major restrictions on cattle behavior and welfare, including that associated with isolation of individual animals, mean that such research should only be allowed if there are major benefits envisaged from the research. Given that nutrition research has been conducted with cattle in metabolism crates for at least 50 years, further advances in feeding systems that originate from metabolism crate studies are likely to be limited. The development of plant wax markers (n-alkanes) to determine the intake of grazing cattle (Jurado et al., 2019) has made it unnecessary to estimate the intake of grazing cattle by feeding them cut grass, usually while tethered. One area that has gained recent attention from cattle researchers is the quest to find plant-based strategies to limit the release of methane, a potent greenhouse gas (Martin et al., 2010; Kingston-Smith et al., 2010).

A related area of research is the measurement of gaseous exchange in cattle, which involves placing them in open circuit respiration (isolation) chambers, where all entering and exiting gases can be measured (Li et al., 2019). Originally used in attempts to improve the efficiency of cattle feeding, by reducing losses of energy in methane emissions, for example, these systems can now be used to attempt to reduce gaseous emissions to the atmosphere that support the "greenhouse effect", especially methane. This field of research is described in more detail later under Special Situations.

Grazing cattle, known to be selective in their grazing habits, have also been fistulated in their esophagus to determine the digestibility of herbage actually consumed by the animal, rather than that present in the field (Alder, 1969; Van Dyne and Torrell, 1964). Consumed herbage tends to be of higher quality than the average in the pasture because cattle are selective in their choice of herbage to eat. However, the cannula, in this case, can irritate the esophagus, causing an accumulation of fibrous tissue, eventually leading to nerve damage, loss of peristaltic function, and even death. At the other end, excreta collection is possible in grazing cattle, by attaching a bag that covers the vulva and/or rectum (Magner et al., 1988). Again, the discomfort induced in the animal must be considered, as well as the potential benefits of the research.

It is important that an adequate number of cattle are used in any research project. Hence a power calculation, based on previously observed variation in responses and the size of the

Behavioral Biology of Cattle

FIGURE 17.1 *Bos gaurus* cattle in the highlands of Malaysia.

effect it is desired to detect, is highly recommended before commencing any cattle research.

More justifiable research projects involve the observation of wild-type cattle to learn about the needs of the species. Although the forebears of modern *Bos taurus* and *B. indicus* cattle, *B. primigenius* cattle, became extinct in Europe about 500 years ago, there remain close relatives that can be observed in parks and wildlife sanctuaries around the world, such as *B. gaurus gaurus* in Asia (Figure 17.1). Observations of these cattle show that cattle are not exclusively grazing animals, but also browse bushes and shrubs in the forest fringes (Ahrestani et al., 2012).

Behavioral Biology

A brief outline of the characteristics of cattle is presented here, with more detail added later. Cattle are gregarious, polygynous, large ungulates, characteristics which dictate their behavior to a considerable extent. Their nutritional ecology (discussed below) determines that they spend considerable time feeding and ruminating. They are sexually dimorphic, with males being larger and possessing strong forequarters to enable them to fight for access to females (Phillips, 2002). Their polygynous lifestyle means that the largest and oldest males serve the majority of cows, and in semi-wild situations, the males will be evicted from the matriarchal group at puberty, only to return to serve cows if they are able to prevail against competition from other males. Cows attract males by mounting each other, often forming into sexually active groups, with the mounted cow/cows being at the right stage of the estrous cycle for service.

The sensory apparatus of cattle makes extensive use of chemoreception, particularly in sexual behavior, with a well-developed vomeronasal organ and olfactory epithelium that have evidence of sexually dimorphic genetic expression (Kubo et al., 2016). Their sight offers a good field of vision, which enables them to see behind them, and their retinas have a tapetum to help them see in the dark (Phillips, 2002).

In the wild, cattle protect themselves against predators by using their horns, which are present in both males and females, since their relatively large, heavy rumen makes them not particularly fleet of foot. Extensive herds of cattle operate a crèche system for young calves, in which one or more older cows will keep constant watch. If attacked, adults will form a corral around vulnerable calves to protect them (see below). Ectoparasites are discouraged from landing on the skin of cattle by vigorous use of the tail, and commensal birds (such as egrets) will guard cattle against such predation.

Ecology and Habitat Use

Cattle evolved from *Bos primigenius*, the auroch cattle that emerged in the Pliocene period, as inhabitants of the grasslands that were spreading during the cooler climate at that time (van Vuure, 2005). These horned, agile rovers of forests and grasslands had few predators and once ranged widely in Eurasia (Park et al., 2015). One of their unique features, along with other bovids, included an enlarged forestomach, or rumen, which stored herbaceous material and allowed it to be digested slowly by micro-organisms, bacteria, fungi, and protozoans, in conjunction with addition of large amounts of alkaline saliva that facilitated mastication and the swallowing of food boluses, while buffering the acidity of the end-products of digestion. Efficient sorting of long and short particles by a flotation-sedimentation system in the reticulum allows the latter to exit the reticulorumen with the rumen fluid (Clauss and Hummel, 2017). This mechanism and a rapid fluid throughput, associated with high saliva production, enables modern day cattle to eat large quantities of herbaceous material, but still digest them efficiently with the assistance of mercyism (regurgitation and remastication), more effectively than the camilids, which generally have a slower metabolism and eat less (Clauss and Hummel, 2017). Regurgitation assists in the comminution of the feed and adds the necessary saliva, with food boluses regurgitated into the buccal cavity at regular intervals, through reverse peristalsis in the esophagus. The addition of saliva during this process also aids in lubricating the prolonged chewing that ensues following regurgitation of each bolus of food. This process is best accomplished whilst the animal is lying down, and disruption of normal rumination activity can be taken as an indication that the animal's welfare is poor (e.g., Whay and Shearer, 2017).

From this rugged, hardy fighter, humans tamed and adapted modern domesticated cattle, mainly of two types, *Bos taurus* in the Near East and *B. indicus* in Southwest Asia (Park et al., 2015). The rumen was enlarged, agility naturally reduced, and some breeds were developed without horns, for ease of management. Later, *B. taurus* cattle were further differentiated into high milk-producing breeds, with enlarged udders, and cattle for meat production and tillage of the land.

Social Organization and Behavior

Cattle evolved as highly social animals, aggregating for safety because their calves were vulnerable to predation by wolves and bears. By communing in groups of 15–20 animals, i.e., herds, they could defend the most vulnerable in the population by corralling themselves into a circle with the strongest on the outside.

Nowadays, strong sexual dimorphism, with the bulls having bigger horns and strengthened shoulders and backbones, allows the herd to be defended against attack, rather than having to rely on flight, as some other herbivores do. Synchronizing the production of offspring to one season, spring to early summer, reduces the availability of vulnerable members of the herd. Calves are highly precocious, standing within a few hours of birth. Rather than following their mothers, calves tend to hide, by lying in long grass in groups, referred to as crèches, a characteristic that can still be observed in some extensive herds today (Sato et al., 1987).

Cattle today usually have their home range determined by human managers; even under the most extensive of conditions, their movements, resting places, and pasture access are determined for them (Araujo et al., 2018). Cattle cannot be considered truly territorial, i.e., they do not defend their home territory against incursion of other cattle. However, home ranges can be detected and are stable across years, averaging 20–40 km^2 in beef cattle grazing in boreal forest in Norway (Tofastrud et al., 2020).

Cattle groups were essentially matriarchal in nature over the course of their evolution, with males leaving the herd on attainment of puberty at about 300 days of age and forming bachelor herds of their own, or ranging in isolation. The matriarchal groups therefore consisted of a group of cows with their offspring, and whichever bulls were consorts for these cows at the time. Cows developed a habit of mounting each other when in estrus, a signal to bulls nearby that there were cows in the right stage of the estrus cycle to be impregnated. Bulls then attach themselves to the cows for a few days and have a legendary capacity for breeding with, and impregnating them (de Araujo et al., 2003), a fertility factor recognized by ancient Egyptians in their bull worship (Robins, 2019).

In this polygynous society, both males and females today maintain a strict hierarchy. Only the strongest and most experienced males, i.e., the dominant males, breed with the cows, and both cows and bulls will fight to gain access to resources, primarily opportunities to mate with the cows in the case of the bulls. However, since overt aggression can be costly, much aggression is ritualized – a deferent lowering of the head in the case of the subordinate animal when approached by a more dominant animal. Social grooming helps to cement relationships and gives subordinate cows an opportunity to curry favor with dominant ones, by extensive licking of their forequarters. The withers are very sensitive in this regard and cattle sometimes still have whiskers there and around the mouth, an adaptation that was probably to help them find their way through forest tracks (Phillips, 2002).

Traditionally, there was an annual production of (usually) a single calf from each cow in spring to early summer. Cattle naturally separate themselves from the rest of the herd when giving birth. In extensive systems of grazing cattle, a calf crèche is formed, guarded by one or two cows. The calves suckle the cows for about 6–9 months. The mother–young bond used to be thought to be relatively weak, because there was little evidence of stress in the mothers when separated from their calf (Phillips, 2002). However, more recent psychological research, using novel cognitive bias tests developed from research with humans, has determined that separation of the calf from the dam does cause the dam to become unduly pessimistic (Daros et al., 2014), potentially indicative of depression. Calves naturally stay with their mothers until puberty and engage in play behavior regularly in order to learn various behaviors, and to develop motosensory responses in particular (Phillips, 2002). Play can be categorized as mock fleeing, mock aggression, mock copulation, and environmental exploration.

Feeding Behavior

Consumption of coarse grass is a slow process, due to the time required for selection and ingestion of the grass, and comminution of the grass particles in the buccal cavity. The latter occurs not only during rumination; but during the ingestion, there is also extensive chewing and mixing of grass with saliva into a bolus that is suitable for swallowing. Cattle bite the grass by wrapping their tongue around a bundle of leaves, compressing it against the upper palate, which is toothless, and jerking their head backwards and upwards to tear the grass from the sward. They do this 40,000–60,000 times a day. If the grass is short, they increase their biting rate, which averages about once per second. In total, cattle spend about 8–10 h a day in this grazing activity, and another 6–8 h a day ruminating or chewing the cud, mostly while lying down. Thus, even though cattle do not usually flee in response to danger, they are quite active throughout the day, often walking 8–20 km/d when grazing in rangeland conditions (McGavin et al., 2018).

Although cattle have the capacity to digest coarse grasses, these have insufficient nutrients for cattle bred to produce large volumes of milk daily, and "milk" cows need supplementation with feeds of higher energy and protein content. Cattle also have a strong appetite for sodium, an element often deficient in their diet, especially if they graze far from the sea. Modern husbandry systems aim to provide mainly feed that has been processed and stored (conserved by either drying or pickling in acid) indoors, or if grazing systems are used, short, nutritious herbage is provided, to encourage the animals to grow rapidly or produce milk. Enhancement of protein content in the herbage may be achieved by including legumes in the pasture, especially clovers, but excessive protein content can stimulate bloating, if digested too rapidly. Under laboratory conditions, cattle are rarely fed fresh grass; rather it is usually preserved as hay or silage, or replaced altogether with pelleted feed. This can lead to oral stereotypies, as such food items are rapidly consumed and the cattle do not have the opportunity to spend their usual time in the grazing process. One such stereotypy occurs when the cattle hang their tongues out of their mouth and roll them repeatedly (tongue rolling or tonguing), also evidence of a deficient diet.

Communication

B. primigenius cattle were potentially prey animals, particularly the calves, and had the sensory apparatus typical of herbivores that needed to protect themselves from predation, as well as horns in the adult animals. The eyesight of cattle today therefore differs from ours in many respects. They have large eyes on the sides of their head, ideal for wrap-around vision. Visual acuity at dawn and dusk, when these crepuscular animals have their most active grazing periods, is further enhanced by the presence of a reflective tapetum on the retina (Bortolam et al., 1974), allowing light to pass through the receptor cells twice. Their cones are oriented into a "visual streak" in the retina, which probably facilitates detection of objects on the horizon (Rehkamper et al., 2000), in contrast to our own fovea, which gives us excellent acuity at a point source. Cattle, therefore, may not perceive features in a built environment as accurately as we can, with the result that people may consider them clumsy. Their dichromatic retina are most sensitive to long wavelengths (red-colored), which they can readily distinguish from medium (green/yellow) or short (blue) wavelengths (Phillips and Lomas, 2001). They have limited ability to distinguish medium from short wavelengths. An ability to detect red may be adaptive by facilitating the recognition of the presence of blood or the reddened vulva of a cow in estrus.

Auditory communication is important to cattle, who vocalize regularly, at frequencies ranging from a low-pitched "moo" to a high-pitched bellow. The higher the pitch and louder the call, the more aroused the animal is. For example, a cow in distress, such as following calf removal; or one that is generally excited and anxious, such as during transport or in a novel location, such as an abattoir, will produce more calls and higher pitched calls (Bristow and Holmes, 2007; Kim et al., 2019). The ears of cattle have erect pinnae and are designed for detection of sounds emanating from a long way off. Accurate localization of the sound is not as good as in humans, but this is generally more important for predators than prey animals. Cattle are, however, much better at detecting high-pitched sounds than humans, which means we should be careful that they are not distressed by such noise, in milking parlors, for example.

Olfactory communication is very important to cattle, because they are prey animals needing covert methods of informing conspecifics of their mood and sexual state. As well as processing odors inhaled via the nasal apertures, cattle have a well-developed vomeronasal organ at the base of their nasal septum, accessed through nasopalatine canals (Salazar et al., 2008), particularly when the animal bares its lips and opens its mouth in a grimace known as flehmen (Figure 17.2). This is particularly used for sexual communication between cattle.

Cattle use visual displays to communicate sexual receptivity. Because wild cattle herds evicted their male members after they reached puberty, cows in estrus needed a signal to provide for males grazing at a distance, either in bachelor groups or alone, that they were receptive to breed. Hence, they developed a unique homosexual mounting behavior between females, in which the mounted cow is able to signal to males that she is ready for insemination (Phillips, 2002). This may have been accentuated during the process of domestication, because cow owners needed to detect when their cows were ready for a village bull to be brought to breed with the cow(s) in estrus. Cattle also signal their mood with their tail and ears. An erect tail indicates that the animal is excited, a relaxed hanging tail indicates that she is calm, and a tail tucked between the hind legs indicates that she is anxious. The position of the ears of cattle also provides information on their state of mind; ears held low and backwards indicate that cattle are relaxed, compared with erect and forward ears, which indicate the animal is alert. Tail and ear positions can be used when assessing the welfare of cattle under laboratory conditions (de Oliveira and Keeling, 2018).

Common Captive Behavior

Normal behaviors include feeding, lying, standing, walking, affiliating, excreting, respiring, and ruminating. These are essential for the survival of the animal, however poor the captive conditions are. In captivity, of these essential behaviors, there are most often limits placed on walking behavior, since cattle are usually either brought into close confinement, sometimes even tethered, or under close supervision by their managers. They also lie down for just short bouts when indoors, indicating that they are not comfortable (Atkins et al., 2020) and are more aggressive to other cows (Arnott et al., 2017). Disease prevalence often increases when the cows are housed

FIGURE 17.2 Flehmen display in a cow.

indoors, particularly lameness and mastitis, but also hock lesions and uterine disease (Arnott et al., 2017, Charlton and Rutter, 2017). Abnormal behaviors mainly derive from inadequate feeding, space, thermal environment, and companionship (Phillips, 2002). They include stereotyped oral behaviors, such as tongue rolling, in which the tongue is extruded and rolled on a repeated basis (Redbo, 1992); feed tossing, in which the feed is repeatedly thrown out in front of them; and intersucking, the sucking of the teat of another cow, which may be performed when natural suckling behavior is thwarted. Of these, tongue rolling is particularly common in cattle tethered for long periods (Redbo, 1992). Aggressive intent is often indicated by stamping behavior, a sure sign that the animal is irritated, and pawing behavior, performed by bulls as an intention charge, often accompanied by an orientation of the head to the ground. Mounting between steers is a form of redirected aggression and is not usually of a sexual nature. In stressful situations, cattle stamp more with their left feet than their right feet, indicating right brain hemisphere processing (the side that deals with flight-or-fight responses) (Idrus, 2020). Cattle that have a history of being abused by humans rarely allow themselves to be approached closely without fleeing. Thus, their powers of remembering people's faces and other characteristics, such as clothing color, are excellent (Rybarczyk et al., 2001).

If cattle are too hot, they reduce the time they spend lying and increase the time they spend standing in order to increase airflow to their lower body and hence, heat loss. Their respiration rate increases and the open-mouthed panting that ensues will be accompanied by drooling and hanging their tongue out of their mouth at extreme temperatures. At this point, further respiration rate increases may be counterproductive, as they generate more heat than is lost during the extra air exchange, so the animal may revert to deep respiration, attempting to get air exchange into the lower regions of the lungs (Phillips, 2016). Feed intake declines rapidly at high temperatures, an excellent survival strategy, as feed digestion produces much heat. This heat enables cattle to cope well at cold temperatures, but in extremely high temperatures, or if the animal is sick or preruminant, they may suffer. Depending on the floor covering, they may lie down less when it is cold, especially if the floor is wet, because they lose too much heat through their torso. When cows are cold, they stand and orientate their body to the sun, capturing its radiant heat. In the long term, over a few weeks, cattle in very hot temperatures reduce skin thickness and the pelage becomes shiny, while in very cold temperatures, they increase the hairiness of their coat (Phillips, 2018). Experimental feedlots may not be subject to regulated provision of shade if they only contain a small number of cattle, but shade should be provided to all cattle if temperatures exceed their thermoneutral zone, i.e., for an adult cow, about 26°C–27°C, dry bulb temperature.

Cattle are naturally stoic animals, and it can be quite difficult to determine when a cow is in pain. However, a pain scale for cattle has been developed that uses many of the more subtle measures, including their attention toward their surroundings, head position, ear position, facial expression, response to approach, and back position (Gleerup et al., 2015, Table 17.1).

As well as recognizing negative affect in cattle, it is important to understand what might constitute positive affect, but for this, there is less scientific evidence in the literature. In general, it is agreed that animal autonomy, play behavior, benign human-animal relationships, and social interaction with

TABLE 17.1

Summarized Behaviors Used in Assessment of Pain in Cows (from Gleerup et al., 2015; See this Paper for Pain Assessment Scheme)

Category	Definition of Behaviors
Attention	The cow is not attentive towards its surroundings. Cow is not active, performing normal activities, such as eating, ruminating or sleeping. Cow is facing wall, away from conspecifics.
Head bearing	Below withers, at withers or above withers
Ear position	Forward or frequently moving (relaxed) or low ears or both ears consistently backwards (in pain)
Facial expression	Changes in muscle tension along the sides of the head and above the eyes manifested as oblique lines or above the nostrils manifested as wrinkles
Eye white	Proportion visible in eyes
Nostril cleanliness	Presence of nasal discharge and cleaning with tongue
Chewing	When without feed in mouth
Tooth grinding (bruxism)	Pressing teeth hard together
Vocalization	Moaning or grunting
Shivering	Muscle tremors
Tenesmus	Abdominal straining with little production of feces of urine
Piloerection	On back of neck and on back
Response to approach	Human approach, measuring the distance that cow can be approached before fleeing
Back position	Contour of back line
Weight shifting	Frequent unprovoked stepping and kicking with hind limbs

conspecifics are likely to be important (Vigors and Lawrence, 2019), but it is also possible that exploration, excitement, and novelty are valued by cattle. Social interaction is often evidenced by grooming between animals close in the dominance hierarchy. Cattle also self-groom regularly and this can be facilitated by providing a rotating brush at approximately head height (Ninomiya, 2019).

Ways to Maintain Behavioral Health

Cattle must be given the chance to exercise regularly. At pasture they normally walk several kilometers per day, even when highly stocked. They should not be restrained for more than a few days, for example in a metabolism crate, without being released for exercise. They should be fed a diet with adequate fiber, energy, protein, minerals, and vitamins, according to their physiological status. High-yielding cows, in particular, need regular, high-energy feed to be provided, as well as milking twice or more per day, if excessive build-up of pressure in the udder is to be avoided. Cattle on short pasture may be unable to ingest sufficient herbage to support rapid growth or milk production, and therefore, require supplementary forage and concentrates.

Captive Considerations to Promote Species-Typical Behavior

Environment

Environmental enrichment is usually in the form of provision of the physical facilities, e.g., space and pasture, that cattle need, rather than for any cognitive enrichment purposes. In relation to the latter, provision of mirrors, for example, for enrichment of isolated cattle, does not reduce stress (Mandel et al., 2019). However, cattle evolved as grazing and browsing animals and should have access to pasture if they are to engage in normal feeding behavior. If they do have to be fed inside, they should be provided with adequate fiber to support normal rumination activity, demonstrating adequate functioning of the rumen digestion processes. As indicated above, cattle are prone to heat stress and should not be kept indoors without good ventilation. In cooler climates, this can be natural ventilation, but in warm conditions, where temperatures are likely to reach or exceed their upper critical temperature of 26°C–27°C dry bulb, there should be artificial ventilation. This can be in the form of fans, but in very hot conditions, the animals must receive air conditioning. Cattle only suffer from cold stress if they are sick, small, nonruminant calves, not growing or producing milk, or if temperatures fall below −15°C to −20°C, depending on the type of cattle involved.

Cattle behavior and physiology are affected by photoperiod (Phillips, 1992), and they should have opportunities for crepuscular activity, especially feeding, as this allows them to maintain their rumen in a satiated state where microbial fermentation can continue without additional feed at night. Naturally, they graze until just after dark, lie down overnight (unless they are high-yielding cows, in which case they often rise for additional grazing in the middle of the night), and then get up to start feeding again at dawn. However, in intensive housing of cattle, the hierarchies that emerge mean that some subordinate cattle will be most comfortable feeding at times when the dominant cows are resting, such as in the middle of the night. In such circumstances, it is prudent to light the feeding area and passageways continuously, so that animals can move around with ease at any time of the day.

Social Groupings

Cattle are gregarious animals, seeking protection from predators by aggregating into herds. Hence, isolating cattle for detailed scientific measurements can stress them and make the measurements unreliable. When working with groups of cattle, they are usually kept in single sex groups, because of the risk of unwanted pregnancies and the risk of fighting between males for access to females. The exceptions are prepubescent cattle, rangeland cattle, and some small dairy and beef herds, which often have some bulls with cows for natural mating.

Just as in any scientific investigation, researchers working with cattle need to use statistical analyses to determine the likelihood that their measurements are influenced by random variation. They do this by comparing the variation due to fixed effects, such as different types of treatments, with random variation. Group size is often constrained to the numbers of cattle required for this comparison, since keeping cattle under experimental conditions is an expensive operation. Depending on the parameters being measured, this is usually between 5 and 20 cattle in each treatment group. Fewer cattle are required if each animal is exposed to several different treatments sequentially, but although residual effects of treatment can be estimated if a balanced design is used, determining long-term effects is difficult. If animals are kept in a group, as opposed to isolated, they may influence each other, especially in terms of behavior. This includes feeding behavior. Hence, there may be "social facilitation" to reduce the range of values observed, compared to the range that would occur in the same number of individually isolated cattle. Some behaviors, such as urination and breathing, may be considered independent of this allelomimetic effect and can be reliably measured in groups. Other behaviors, such as aggression between animals, are clearly not independent of the group effect (Phillips, 1998). If cattle are kept in isolation, they should have sight of other cattle, and it should be noted that calves are prohibited from being kept in individual pens after 8 weeks of age in the European Union.

Replicating cattle groups in different facilities, or within a facility, to gain an appreciation of the true variation in responses, may be practically hard to achieve or prohibitively expensive. It can be done in feedlots without too much difficulty, but in a grazing dairy herd, for example, it may be difficult. Some leniency, with recognition of a potential group effect, is advocated for these difficult situations, in order that, over time, a complete understanding is gained through different trials. Combining the results of several studies in a meta-analysis can usefully determine effects more accurately than individual experiments. If this is not possible, systematic reviews of the literature, which include assessment of sources of bias and reliability of observed effects, are increasingly used to determine whether effects observed in one experiment can be generalized, or are peculiar to the circumstances of that experiment.

Feeding Strategies

Cattle should be fed at least daily, with feed that contains sufficient nutrients for maintenance and any production requirements, such as for milk, growth, or pregnancy. Competition for feed should be minimized in groups of cattle by ensuring that all members of the group are of similar size, and their feed should be offered in suitable troughs or bunks.

Training

Training is required for both animal and research workers to produce a successful research outcome.

Cattle

Cattle may be required to wear equipment or encounter unfamiliar environments, diets, or people during the course of the experimentation. They habituate readily to novel stimuli, given the right training (Rorvang et al., 2017). For example, when measuring activity by attaching a leg-mounted monitor, it is recommended that a 2-day adjustment period is included (MacKay et al., 2012). In other situations, appropriate training to ensure that this does not influence the results of the experiment can take as long as the experiment itself. It also reduces between animal variations, resulting in fewer animals being required for the experiment (Raundal et al., 2015). It is usually achieved by gradually introducing the animals to the novel stimuli, for initially a short period each day, as well as allowing them to explore the stimuli without human assistance in their own time. For example, cattle may be required to press a lever to indicate how much they value a resource (e.g., Rybarczyk et al., 2001; Cooper et al., 2010). It is particularly important that research cattle are familiar with their handlers, a process which should begin well before the actual experiment commences.

Handlers

Personnel managing experimental cattle should be trained to handle them quietly and safely. The research environment is often more stressful for cattle than the farm environment, as animals may be isolated and/or have procedures done to them that are painful. Hence, more careful handling and better facilities are needed. Handlers should be familiar with the signs of pain and frustration in cattle, for example, stamping on the ground or withholding a limb during walking (Table 17.1). Technical staff looking after the cattle must have a regular management schedule for the work: cleaning, feeding, hosing down, etc. Scientists' ambitions to conduct research that is ground-breaking should not be allowed to drive experiments so that they pose major risks to the animals, with little chance of a successful outcome (Frasch, 2016). The scientists should be trained in ethical appraisal, and the research should be reviewed and approved by an ethics committee before being commenced. This may be an institutional committee, which ideally will have scientists, veterinarians, and also lay members on it, or it may be centrally managed by a government agency. Either way, there should be sufficient resources for the committee to spend enough time considering each application in full. A utilitarian approach is usually advocated, so that the harm suffered by the cattle is less than the benefits to human society. This is hard to evaluate and a rights-based approach may be preferable and more humane.

Special Situations

Some aspects of the use of cattle in research require special situations that involve particular methods of restraint. Indirect calorimetry, in which cattle are individually confined in respiration chambers measuring gaseous exchange (Machado et al., 2016), is increasingly used to research the animals' production of methane, a potent greenhouse gas, and thermal responses to hot climates. Adaptation to the facility can take as long as 2 months (Posada-Ochoa et al., 2017). Field calorimetric studies can be conducted with cattle wearing a face mask, which obviously requires training. Respiratory rate can be measured to assess their habituation to the device (Camerro et al., 2016).

In other research, samples are taken from tissues or organs by biopsy, or muscle tissue from cattle may be used to grow *in vitro* meat (Ghosh, 2013). In this case, an ethical review is essential before allowing samples to be taken from cattle. Pregnant cattle are also used for the collection of fetal calf serum (or fetal bovine serum, FBS) in abattoirs, which is then used for cell and tissue culture, e.g., for monoclonal antibodies. Worldwide, about 0.5 million liters are collected annually from 1 to 2 million fetuses (van der Valk, 2019). The serum is obtained directly from the heart of the living fetus, about 20–60 min after the mother has been euthanized. The ethics of such special situations must be questioned, and a value placed on the life of an unborn fetus. Cows themselves can be used as factories for polyclonal antibodies, which would recognize several parts of a virus, not just one as in monoclonal antibodies. In cases such as SARS-Co-V-2 virus (the virus that causes COVID-19), where large quantities are required in a short period of time, cows provide an opportunity to produce large quantities of antibodies most expeditiously. Potentially every month cows could produce enough vaccine to treat several hundred patients.

Conclusions

Cattle are not used nearly as commonly in research as some rodent and fish species, but where they are used, the implications for the welfare of the animals can be quite serious. It is suggested that research is more justifiable when cattle themselves, and especially, the individual animal(s) being used for the research, are the beneficiaries of any advances from the work. The natural behavior of cattle, as grazing and browsing animals, means that they are not easily kept in a research facility that provides few opportunities for natural behavior. Facilities used today should be of the highest quality, and management practices should be particularly focused on minimizing pain, discomfort, and distress in the animals, by allowing them the capacity to perform natural behavior.

Additional Resources

Phillips, C.J.C. 2018. *Principles of Cattle Production.* Third Edition. CAB International, Wallingford.

Phillips, C.J.C. 2002. *Cattle Behaviour and Welfare.* Blackwell's Scientific, Oxford.

REFERENCES

Ahrestani, F.S., Heitkonig, I.M.A., and Prins, H.H.T. 2012. Diet and habitat-niche relationships within an assemblage of large herbivores in a seasonal tropical forest. *Journal of Tropical Ecology* 28: 385–394.

Alder, F.E. 1969. The use of cattle with oesophageal fistulae in grassland experiments. *Grass and Forage Science* 24: 6–12.

Araujo, A.G.J., Monteiro, A.M.V., Oliveira, G.S., et al. 2018. Beef cattle production systems in South Pantanal: considerations on territories and integration scales. *Land* 7(4), Article Number: 156.

Arnott, G., Ferris, C.P., and O'Connell, N.E. 2017. Review: Welfare of dairy cows in continuously housed and pasture-based production systems. *Animal* 11: 261–273.

Atkins, N.E., Cianchi, C., Rutter, S.M., Williams, S.J., Gauld, C., Charlton, G.L., and Sinclair, L.A. 2020. Performance, milk fatty acid composition and behaviour of high-yielding Holstein dairy cows given a limited grazing period. *Grass and Forage Science.* doi:10.1111/gfs.12471 (early access).

Batista, E.D., Hussein, A.H., Detmann, E., Miesner, M.D., and Titgemeyer, E.C. 2016. Efficiency of lysine utilization by growing steers. *Journal of Animal Science* 94: 648–655.

Bortolam, R., Callegar E., and Lucchi, M.L. 1974. Some ultrastructural features of tapetum lucidum of cat, lamb, horse and cattle. *Bolletino della Societa Italiana di Bioloogia Sperimentale* 50: 272–274.

Broadbent, P.J., McIntosh, J.A.R., and Spence, A. 1970. The evaluation of a device for feeding group-housed animals individually. *Animal Production* 12: 245–252.

Bristow, D.J., and Holmes, D.S. 2007. Cortisol levels and anxiety-related behaviours in cattle. *Physiology and Behaviour* 90: 626–628

Camerro, L.Z., Campos Maia, A.S., Chiquitelli, N.M., et al. 2016. Thermal equilibrium responses in Guzerat cattle raised under tropical conditions. *Journal of Thermal Biology* 60: 213–221.

Charlton, G.L., and Rutter, S.M. 2017. The behaviour of housed dairy cattle with and without pasture access: A review. *Applied Animal Behaviour Science* 192: 2–9.

Chiy, P.C., and Phillips, C.J.C. 1993. Sodium fertilizer application to pasture. 4. Effects on mineral uptake and the sodium and potassium status of steers. *Grass and Forage Science* 48: 260–270.

Clauss, M., and Hummel, J. 2017. Physiological adaptations of ruminants and their potential relevance for production systems. *Revista Brasileira de Zootecnia* 46: 606–613.

Cooper, M.A., Arney, D.R., and Phillips, C.J.C. 2010. The motivation of high and low yielding dairy cows to obtain supplementary concentrate feed. *Journal of Veterinary Behaviour: Clinical Applications and Research* 5: 75–81.

Daros, R.R., Costa, J.H.C., von Keyserlingk, M.A.G., Hötzel, M.J., and Weary, D.M. 2014. Separation from the dam causes negative judgement bias in dairy calves. *PLoS One* 9: 5 Article Number: e98429.

de Araujo, J.W., Borgwardt, R.E., Sween, M.L., Yelich, J.V., and Price, E.O. 2003. Incidence of repeat-breeding among Angus bulls (*Bos taurus*) differing in sexual performance. *Applied Animal Behaviour Science* 81: 89–98.

de Oliveira, D., and Keeling, L.J. 2018. Routine activities and emotion in the life of dairy cows: Integrating body language into an affective state framework. *PLoS One* 13, Article Number: e0195674.

Frasch, P.D. 2016. Gaps in US animal welfare law for laboratory animals: Perspectives from an animal law attorney. *ILAR Journal* 57: 285–292.

Ghosh, P. 2013. *"World's first lab-grown burger is eaten in London"*. BBC News. 2013-08-05. https://www.bbc.com/news/science-environment-23576143. Accessed 27 September, 2019.

Gleerup, K.B., Anderson, P.H., Munksgaard, L., and Fjorkman, B., 2015. Pain evaluation in cattle. *Applied Animal Behaviour Science* 171: 25–32.

Idrus, M. 2020. *Behavioural and physiological responses of beef cattle to hot environmental conditions.* PhD thesis, University of Queensland, School of Veterinary Sciences.

Jurado, N., Tanner, A.E., Blevins, S., Fiske, D., Swecker, W.S., McNair, H.M., and Lewis, R.M. 2019. Choices between red clover and fescue in the diet can be reliably estimated in heifers post-weaning using n-alkanes. *Animal* 13: 1907–1916.

Kim, N.Y., Kim, S.J., Jang, S.Y., Seong, H.J., Yun, Y.S., and Moon, S.H. 2019. Characteristics of vocalisation in Hanwoo cattle (*Bos taurus coreanae*) under different call-causing conditions. *Animal Production Science* 59: 2169–2174.

Kingston-Smith, A.H., Edwards, J.E., Huws, S.A., Kim, E.J., and Abberton, M. 2010. Plant-based strategies towards minimising 'livestock's long shadow'. *Proceedings of the Nutrition Society* 69: 613–620.

Kubo, H., Otsuka, M., and Kadokawa, H. 2016. Sexual polymorphisms of vomeronasal 1 receptor family gene expression in bulls, steers, and estrous and early luteal-phase heifers. *Journal of Veterinary Medical Science* 78: 271–279.

Lambert, K., Kent, M., and Vavra, D. 2019. Avoiding Beach's Boojum Effect: Enhancing bench to bedside translation with field to laboratory considerations in optimal animal models. *Neuroscience and Biobehavioural Reviews* 104: 191–196.

Li, J., Green-Miller, A.R., and Shike, D.W. 2019. Integrity assessment of open-circuit respiration chambers for ruminant animal indirect calorimetry. *Transactions of the American Society of Agricultural and Biological Engineers* 62: 1185–1193.

Machado, F.S., Tomich, T.R., Ferreira, A.L., et al. 2016. Technical note: A facility for respiration measurements in cattle. *Journal of Dairy Science* 99: 4899–4906.

MacKay, J.R.D., Deag, J.M., and Haskell, M.J. 2012. Establishing the extent of behavioural reactions in dairy cattle to a leg mounted activity monitor. *Applied Animal Behaviour Science* 139: 35–41.

Magner, T., Sim, W.D., and Bardsley, D.H. 1988. Apparatus for urine collection from female cattle in metabolism crates. *Australian Journal of Experimental Agriculture* 28: 725–727.

Mandel, R., Wenker, M.L., van Reenen, K., et al. 2019. Can access to an automated grooming brush and/or a mirror reduce stress of dairy cows kept in social isolation? *Applied Animal Behaviour Science* 211: 1–8.

Martin, C., Morgavi, D.P., and Doreau M. 2010. Methane mitigation in ruminants: From microbe to the farm scale. *Animal* 4: 351–365.

McGavin, S.L., Bishop-Hurley, G.J., Charmley, Ed., et al. 2018. Effect of GPS sample interval and paddock size on estimates of distance travelled by grazing cattle in rangeland, Australia. *Rangeland Journal* 40: 55–64.

Nalon, E., and Stevenson, P. 2019. Protection of dairy cattle in the EU: State of play and directions for policymaking from a legal and animal advocacy perspective. *Animals* 9, Article Number: 1066. doi:10.3390/ani9121066.

Ninomiya, S. 2019. Grooming device effects on behaviour and welfare of Japanese black fattening cattle. *Animals* 9: Article Number: 186.

Ørskov, E.R., and McDonald, I. 1979. The estimation of protein degradability in the rumen from incubation measurements weighted according to rate of passage. *Journal of Agricultural Science* 92: 499–503.

Pagella, J.H., Mayes, R.W., Perez-Barberia, F.J., and Ørskov, E.R. 2018. The development of an intraruminal nylon bag technique using non-fistulated animals to assess the rumen degradability of dietary plant materials. *Animal* 12: 54–65.

Park, S.D.E., Magee, D.A., McGettigan, P.A. et al. 2015. Genome sequencing of the extinct Eurasian wild aurochs, *Bos primigenius*, illuminates the phylogeography and evolution of cattle. *Genome Biology* 16: 234. doi:10.1186/s13059-015-0790-2.

Phillips, C.J.C. 2016. The welfare risks and impacts of heat stress on sheep shipped from Australia to the Middle East. *The Veterinary Journal* 218: 78–85.

Phillips, C.J.C. 2002. *Cattle Behaviour and Welfare*. Blackwell's Scientific, Oxford. 264 p.

Phillips, C.J.C. 1998. The use of individual dairy cows as replicates in the statistical analysis of their behaviour at pasture (letter to the editor). *Applied Animal Behaviour Science* 60: 365–369.

Phillips, C.J.C. 1992. Environmental factors influencing the production and welfare of farm animals: Photoperiod. Ch. 3 in *Farm Animals and the Environment*, ed. C.J.C. Phillips and D. Piggins, 49–65. Wallingford: CABI.

Phillips, C.J.C., and C.A. Lomas. 2001. The perception of color by cattle and its influence on behaviour. *Journal of Dairy Science* 84: 801–813.

Phillips, C.J.C., Chiy, P.C., Bucktrout, M.J., Collins, S.M., Gasson, C.J., Jenkins, A.C., and Paranhos da Costa, M.J.R. 2000. Frictional properties of cattle hooves and their conformation after trimming. *Veterinary Record* 147: 607–609.

Phillips, C.J.C., Tenlep, S.Y.N., Pennell, K., Omed, H., and Chiy, P.C. 2001. The effect of applying sodium fertilizer on the rate of digestion of perennial ryegrass and white clover incubated in rumen and faecal liquors, with implications for ruminal tympany in cattle. *The Veterinary Journal* 161: 63–70.

Posada-Ochoa, S.L., Noguera, R.R., Rodriguez, N.M., et al. 2017. Indirect calorimetry to estimate energy requirements for growing and finishing Nellore bulls. *Journal of Integrative Agriculture* 16: 151–161.

Raundal, P.M., Andersen, P.H., Toft, N., et al. 2015. Pre-test habituation improves the reliability of a handheld test of mechanical nociceptive threshold in dairy cows. *Research in Veterinary Science* 102: 189–195.

Redbo, I. 1992. The influence of restraint on the occurrence of oral stereotypies in dairy cows. *Applied Animal Behaviour Science* 35: 115–123.

Rehkamper, G., Perrey, A., Werner, C.W., Opfermann-Rungeler, C., and Gorlach, A. 2000. Visual perception and stimulus orientation in cattle. *Vision Research* 40: 2489–2497.

Robins, A., 2019. The Alpha Hypothesis: Did lateralized cattle-human interactions change the script for Western culture? *Animals* 9: Article Number: 638.

Rorvang, M.V., Jensen, M.B., and Nielsen, B.L. 2017. Development of test for determining olfactory investigation of complex odours in cattle. *Applied Animal Behaviour Science* 196: 84–90.

Rybarczyk, P., Koba, Y., Rushen, J., et al. 2001. Can cows discriminate people by their faces? *Applied Animal Behaviour Science* 74: 175–189. doi:10.1016/S0168-1591(01)00162-9.

Salazar, I., Sánchez-Quinteiro, P., Alemañ, N., and Prieto, D. 2008. Anatomical, immnunohistochemical and physiological characteristics of the vomeronasal vessels in cows and their possible role in vomeronasal reception. *Journal of Anatomy* 212: 686–696.

Sato, S., Woodgush, D.G.M., and Wetherill, G. 1987. Observations on crèche behaviour in suckler calves. *Behavioural Processes* 15: 333–343.

Tofastrud, M., Hessle, A., Rekdal, Y., and Zimmermann, B. 2020. Weight gain of free-ranging beef cattle grazing in the boreal forest of south-eastern Norway. *Livestock Science* 233: Article Number: UNSP 103955.

Van der Valk 2019. Foetal Bovine Serum – background. www.piscltd.org.uk. Accessed 28 May, 2020.

Van Dyne, G.M., and Torell, D.T. 1964. Development and use of an esophageal fistula – a review. *Journal of Range Management* 17: 7–19.

Van Vuure, T. (Cis) 2005. *Retracing the Aurochs – History, Morphology and Ecology of an extinct wild Ox*. Sofia-Moscow: Pensoft Publishers. ISBN 954-642-235-5.

Vigors, B., and Lawrence, A. 2019. What are the positives? Exploring positive welfare indicators in a qualitative interview study with livestock farmers. *Animals* 9, Article Number: 694. doi:10.3390/ani9090694.

Whay, H.R., and Shearer, J.K. 2017. The impact of lameness on welfare of the dairy cow. *Veterinary Clinics of North America Food Animal Practice* 33: 153–164.

Wilks, M., and Phillips, C.J.C. 2017. Attitudes to *in vitro* meat: A survey of potential consumers in the United States. *PLoS One*, 12 (2): 14. doi:10.1371/journal.pone.0171904.

18

Behavioral Biology of Horses

Janne Winther Christensen
Aarhus University

CONTENTS

- Introduction 285
- Typical Research Participation 286
- Behavioral Biology 287
  - Ecology, Habitat Use, and Antipredator Behavior 287
  - Social Organization and Behavior 288
    - Sexual Behavior of Stallions 289
    - Sexual Behavior of Mares 289
    - Mare–Foal Interactions 289
    - Behavior of the Foal 290
    - Separation and Weaning 290
    - Feeding Behavior 290
    - Communication 291
  - Common Captive Behaviors 291
- Ways to Maintain Behavioral Health 292
  - Captive Considerations to Promote Species-Typical Behavior 292
  - Environment 292
    - Social Groupings 293
    - Feeding Strategies 293
    - Training 294
- Special Situations 294
- Conclusions and Recommendations 294
- References 294

Introduction

Horses have a special status in many human societies. Valued for their power, trainability, reliability, speed, and grace, horses have played a significant role in human history. Originally used as a source of food, horses came to play a main role for human migration when domesticated approximately 5,000 years ago. New documentation suggests that horses were first tamed and ridden by the Botai people, a hunter–gatherer population east of the Ural mountains, but several different human populations likely domesticated horses independently of each other (de Barros Damgaard et al., 2018). Domestication resulted in decreased reactivity and flight distance, increased adaptation to human husbandry, and a gradual enhancement of traits that yield human advantage, such as speed and strength. Until the Middle Ages, horses were mainly used for the purposes of transport, war, and agriculture; however, in the Middle Ages, horse riding became a noble pursuit, practiced in the Royal courts as entertainment (McGreevy et al., 2018a). Nowadays, horses in Western countries are mainly used for sports and leisure, although they are still used as working animals in many areas around the world.

A close relative of domestic horses is the Przewalski horse (*Equus ferus przewalskii*), a species that was at the brink of extinction, but was preserved through an international breeding network. Przewalski horses survive today in zoos and reserves, and have also been successfully reintroduced to the wild in Mongolia and China (Boyd and Bandi, 2002; Xia et al., 2014; Sarkissian et al., 2015). We are fortunate to be able to study these animals in various contexts. In addition to Przewalski horses, other feral horse groups that receive minimal management provide important information on natural horse behavior. The behavioral repertoire and behavioral needs of horses appear relatively unchanged by domestication (Figure 18.1; Christensen et al., 2002b; Van Dierendonck, 2006). Consequently, some of the constraints imposed on horses in captivity conflict with their naturally evolved behavior and can lead to welfare problems (Minero and Canali, 2009).

FIGURE 18.1 Social behavior and social needs in horses remain relatively unchanged by domestication. Grooming Przewalski stallions (a); plains zebra stallions (b); domestic stallions (c); and foals (d).

Typical Research Participation

Most behavioral research on domestic horses focuses on horse–human interactions, social behavior, effects of housing and management restrictions, and their cognitive abilities. Due to the long and close relationship between foal and mare, horses appear to be ideal model animals for studies into maternal transmission of behavior, a topic that has been mainly studied in rodents and primates until now. Fear reactions are a common cause of horse-human accidents, and the possible modulation of fearfulness through maternal mediation is particularly promising (Figure 18.2; Henry et al., 2005; Christensen, 2016).

Until recently, the majority of studies on horse cognition have used positively reinforced cognitive tasks, e.g., the animal receives a food reward when making a correct choice. This reflects the tasks used for cognitive research on other studied species. However, it was suggested that horse performance in positively reinforced tasks may be influenced by food motivation (Olczak et al., 2018). In addition, practical horse training is primarily based on negative reinforcement, through the application of rein tension and pressure from the rider's legs, which is removed when the horse shows the desired response. Horses could indeed be an important animal model for studies into negative reinforcement learning (Ahrendt et al., 2015; McLean and Christensen, 2017). In combination with their close relationship with humans and our dependency on horse learning abilities, the factors that affect this relationship are an interesting area for future research.

An increasing body of research focuses on the use of different types of equipment and their effects on horse welfare, for example, the use of whips (during horse racing; McGreevy et al., 2018b), spurs, bits (e.g., Manfredi et al., 2005), or tight nosebands (e.g., Doherty et al., 2017). These studies question the traditional methods and equipment used in horse sports and give rise to ethical debates on the use of animals for human entertainment. Similarly, inappropriate use of negative reinforcement (e.g., resulting from relentless and strong rein tension) could lead to some degree of learned helplessness, due to the horse's inability to control the aversive stimulus.

Therapeutic involvement of animals to help humans with physical or psychological disorders has also received increased interest in recent years. Horses frequently function as therapy

FIGURE 18.2 The mother is an important source of information for the young foal. The habituated mare can help decrease fearfulness in the foal if they are exposed to usually fear-eliciting situations together and the mare remains calm.

animals, especially for interventions targeting children and young people. However, there are still large gaps in our knowledge of how animals affect vulnerable persons and how to safeguard the welfare of therapy animals (Thodberg et al., 2014).

Research on horses should take behavioral needs and prior experiences into account. For example, it is vital to allow ample time to habituate naïve horses to the test conditions, especially if this includes social separation and the use of previously unknown equipment. If one fails to do so, reactions to social separation and novelty are likely to interact with treatment effects. Prior experience also needs to be considered, because horses are likely to learn from all prior interactions with humans; thus, especially in research on training, but also in all other experiments that involve human handling or presence, treatment effects are likely affected by expectations based on positive or negative prior experiences (Hausberger et al., 2008). It is further necessary to consider the environmental conditions of animals used for research and to identify individuals with behavioral disorders. For example, it was reported that stereotypic (crib-biting) horses are more resistant to behavioral extinction (Roberts et al., 2015), which may skew data in cognitive studies if not taken into account. Since many horse studies are based on small sample sizes, it is particularly important to carefully consider inclusion/exclusion criteria, and to attempt to balance horses across treatments. Development of valid and reliable ethograms, with high degrees of intra- and interobserver reliability, is fundamental to behavioral research. McDonnell's (2003) ethogram of horse behavior has been a valuable source used by many researchers, and more recently, ethograms of ridden horse behavior are being developed (Pierard et al., 2019).

Finally, it should be noted that horse behavior is best assessed in combination with physiological responses, because prior experience can alter behavioral responses. For example, a well-trained horse can show a blunted behavioral fear response, despite being comparably frightened as an untrained horse. Noninvasive measurements, such as salivary cortisol, fecal cortisol metabolites, and heart rate have been validated as physiological measures of stress, and new methods, such as infrared thermography, are currently being investigated (McGreevy et al., 2018a).

Behavioral Biology

Ecology, Habitat Use, and Antipredator Behavior

The process of development that transformed the dog-sized, forest-dwelling Eohippus into the modern horse is a textbook example of evolution. The early ancestors of the modern horse were adapted to the moist and soft grounds of the primeval forests and walked on several spread-out toes. As the environment changed and steppes began to appear, the early horses needed to be able to outrun predators. Increased speed was attained through the lengthening of limbs and the lifting of some toes from the ground, so the weight of the body was gradually placed on one toe, which developed into a hoof. The amazing diversity of the family *Equidae*, comprising asses, zebras, and horses, reflects isolation of populations that have independently adapted to their environment.

Equids have coevolved with their predators for millions of years in the wild, and like many other prey species, they have evolved antipredator responses both to actual encounters with predators and to generalized threatening stimuli, such as loud noises and sudden events. Equids demonstrate two main predator defense strategies, which depend on environmental conditions and the type of predator (Goodwin, 2007). Predation by Pantherine predators (large cats), which commonly kill by leaping onto the back, appear to be associated with the evolution of the rapid flight response that keeps the vulnerable head area as far away from the predator as possible. The alternative predator defense strategy appears to be directed toward canid predators or when environmental conditions prevent rapid flight responses. In these conditions, horses defend themselves with foreleg strikes, which can smash the skull of a dog or wolf. Like other prey species, equids readily learn to distinguish hunting behavior from other behavior in constantly present predators, and can be observed grazing close to resting predators. Equids prefer the company of their own species, but will accept other species as companions as well, e.g., zebra and wildebeest are commonly observed in large herds on the African savannah. As prey species, group living is an important survival strategy, as it increases the probability of detecting approaching predators. Domestic equids are, therefore, preadapted to forming associations with other species, and to respond to warning signals in the body language of other species, including humans (Goodwin, 2007).

In the wild, equids typically live in open grasslands with a good view of the surrounding environment, and adaptations of the visual system provide them with an effective early warning system for the detection of approaching predators. The lateral position of the equine eye provides an extensive visual field, the majority of which is monocular. Equids are particularly sensitive to subtle changes in illumination and stimulus motion (Hall, 2007). The resultant fast response of the horse to sudden movement in the peripheral visual field is a useful adaptation to escape predators, but it is also an unwelcome response in domestic horses, as it tends to persist regardless of the level of training (McGreevy et al., 2018a). Equids are also sensitive to auditory signals of danger, such as sounds of predators, and they have a good sense of hearing (Heffner and Heffner, 1983). In addition, equids use olfaction as another modality through which predators can be detected and possibly identified. Chemical signals have been suggested to be involved in several equine processes, such as individual identification, coordination and spacing of individuals (both within and between social groups), mare-foal communication, navigation and orientation, sexual arousal, and alarm signaling (Hothersall et al., 2010; Jezierski et al., 2018; Deshpande et al., 2018). Christensen and Rundgren (2008) found that predator odor *per se* did not frighten domestic horses, but if presented in combination with an additional fear-eliciting stimulus, odor-exposed horses reacted with a more pronounced flight reaction than did horses exposed only to the fear-eliciting stimulus without predator odor.

Social Organization and Behavior

The family *Equidae*, including horses, zebras, and asses, represents two general types of social organization (Klingel, 1975). One type appears to be an adaptation to semi-desert conditions, where males are territorial and adults do not form lasting bonds. This type of social system is typically shown by Grevy's zebra (*Equus grevyi*) and some asses (*E. africanus* and *E. asinus*). The other type of social organization appears to be an adaptation to more unpredictable environmental conditions with a changing food supply, where the animals typically migrate (Goodwin, 2007). This type of social organization is shown by feral (*E. caballus*) and wild horses (*E. przewalski*), and by some zebras (*E. burchelli* and *E. zebra*), which live in year-round social groups and show a female defense polygynous mating system, where one male forms bonds, and breeds, with several females. Groups are relatively stable, as horses tend to live in the same group for many years (Klingel, 1975; Boyd and Keiper, 2005). Stallions (i.e., adult males) actively defend all members of their group against predators, and the permanent bond between stallions and mares (i.e., adult females) may have evolved as a response to large and cooperatively hunting predators (Feh, 2005). Young horses typically disperse from their natal group at the age of 1–2 years, and surplus stallions aggregate into bachelor groups. Group members maintain close proximity to each other and show low levels of overt aggression (Christensen et al., 2002b; Boyd and Keiper, 2005). Several studies suggest that domestic horses express the same movement and social behaviors as wild horses, if provided with an appropriate physical and social environment in which to show their full behavioral repertoire (Christensen et al., 2002b; Boyd and Keiper, 2005; Feh, 2005; Figure 18.1).

Horse groups usually stabilize with a social order, which is challenged when new members arrive. A new social order is typically formed within a few days to weeks. Within the group, horses usually have one or a few preferred social partners, and they spend more time in close proximity and in friendly social interactions with these partners (Van Dierendonck, 2006). Living in groups has a number of advantages, mainly in relation to social transmission of behavior, locating food and water, and as a defense strategy. As an example, all horses of a group rarely lie down together; one will remain standing and guard the group. Horses will generally become anxious and insecure when isolated from other horses. In domestic horses, lack of social contact both early and later in life may cause development of abnormal or stereotypic behavior. Social relationships between horses in a group are not as straightforward as they were once thought to be (with a clear linear hierarchy). Indeed, there are great difficulties in determining social hierarchies within groups of horses. We can measure the relative ease with which one horse can displace another from resources, but the outcomes of such interactions usually depend on the resource in question, as well as the context and, hence, reflect the current motivation of the individual to access the resource (Weeks et al., 2000). Furthermore, hierarchies are often not simply linear, because coalitions among horses within an established social group mean that the presence of key affiliates affects the ability of individuals to retain and access resources (McGreevy et al., 2018a).

Sexual Behavior of Stallions

In nature, stallions mature sexually when they are 2.5–3 years old (Monfort et al., 1994), which is later than domestic stallions, where some may be sexually mature at only 1 year of age. Under natural conditions, colts leave the family group voluntarily, or are chased away by the harem stallion, when they reach sexual maturity. Hereafter, the young stallion joins a bachelor group, and it is usually the highest-ranking stallion in the bachelor group that will challenge existing harem stallions when groups meet. Wild/free-ranging stallions therefore rarely get the opportunity to reproduce before the age of 5–7 years, when they are fully mature and experienced. Weak stallions rarely get the chance to reproduce, ensuring that only genes from the most fit individuals are passed on to the next generation.

The characteristic flehmen response (i.e., curling of the upper lip, facilitating transfer of pheromones to the vomeronasal organ, Figure 18.3) is a normal part of sexual olfactory investigation, and it is frequently shown when stallions investigate urine from mares in estrus. In nature, stallions typically urinate on top of urine from a mare, probably to hide the smell from other stallions, and possibly, to mark that the mare is his. The same behavior can be observed in bachelor groups, where the highest-ranking stallion covers the urine of others. This behavior is likely related to social status, since the highest-ranking stallion is always the last to urinate. In the same way, dominant stallions may try to cover manure from group members, by defecating or urinating on top. Masturbation naturally occurs in feral and domestic stallions (McDonnell, 1992). Likewise, covering other stallions, with or without erection, is a common occurrence, and the covering stallion is typically the highest ranking of the two.

Sexual Behavior of Mares

Sexual behavior in mares is limited to their estrous periods, which occur as early as 2 years of age in nature, depending on the physical condition of the mare. However, most mares are approximately 4 years of age at their first gestation (Monfort et al., 1994). A mare is regularly in estrus approximately every 3 weeks throughout the breeding season (during the spring and summer), unless she becomes pregnant. Most mares do not show regular estrus during winter, but when spring approaches, estrous periods become regular again. With a gestation period of 11 months, the majority of foals are born in the spring, when food availability and survival chances in nature are the highest.

The physical activity of the mare increases during estrus, and she may be more agitated or aggressive, influencing social relationships within the group. Other behavior patterns can also be influenced by estrus; lying time may be shortened and food intake may decrease, for example. A mare in estrus typically shows increased interest in unknown horses and interindividual distances decrease. In nature, an adult mare typically will not accept mating attempts by a young stallion (under 5 years of age), nor by small or weak stallions. Mares that are tightly bonded (e.g., mare and adult daughter) may try to prevent the stallion from covering the other, and mares of high social status may interfere with the covering of lower ranking mares (Rutberg and Greenberg, 1990).

Mare–Foal Interactions

The behavior of the mare is only weakly affected by pregnancy during the 11-month gestation period (Estep et al., 1993). Mares appear to have a certain control over the onset of foaling, which is largely under parasympathetic control (Nagel et al., 2014). In that way, the mare is able to give birth during an undisturbed period. Shaw et al. (1988) found that 86% of all foalings in domestic horses took place during a quiet period, between 18:00 and 06:00 hours. Normally, foaling happens very quickly (approximately 20 min), and both mare and foal are soon on their feet. The mare typically stands 15–20 min after foaling, and most foals are on their feet within the first hour after birth (Rosales et al., 2017). Usually, only a single foal is born, and in the rare case of twins, only one normally survives. In nature, mare and foal quickly leave the place of birth, because the smell of blood and fetal membranes may attract predators, and if the birth takes place at night, by dawn, the foal and mare rejoin the group. Directly after birth an

FIGURE 18.3 The flehmen response facilitates transfer of pheromones and other scents into the vomeronasal organ and is typically shown by stallions when investigating urine from mares in estrus. Horses may also show the flehmen response in other situations when they perceive an unknown smell.

important process starts, during which mare and foal learn each other's smells, and the foal learns to follow the mare. The mare learns the smell of her foal and gets bonded to it immediately after foaling, whereas it takes a few days before the foal is able to recognize its mother. Mares therefore tend to be very protective toward their foal, especially during the first days of the foal's life (Carson and Wood-Gush, 1983). A mare may attack a higher-ranking individual in order to protect her foal, creating instability in the social hierarchy, and she may prefer to stay close to other mares with foals, instead of her usual social partners. Mares with their first foal are especially likely to become less socially active in the group, which probably helps to ensure the bonding between mare and foal, and prevents other horses from disturbing the process (Estep et al., 1993).

Behavior of the Foal

Foals are well-developed when born (precocial); all their senses function and they are able to follow the mare shortly after birth. The foal stays in close contact with the dam and suckles hourly in the first weeks after birth, and unless weaned artificially, it continues to suckle at a gradually decreasing rate until the mare rejects its attempts a few months prior to the birth of her next foal. If the mare does not give birth to another foal, her offspring may continue to suckle until the age of 2 years or older. Foals start to graze as early as their first week of life, and it is assumed that foals learn to graze and discriminate between different plants partly by social facilitation, i.e., by observing foraging behavior in their mother and other horses, and partly by tasting different species of grass (Houpt, 1990). It is common for foals to exhibit coprophagy (eating feces) during the first several months. Normally, fresh feces from the mare are consumed, and it is assumed that the foal is getting supplements of vitamins and manure bacteria, which are essential to the intestinal system (Carson and Wood-Gush, 1983).

Foals spend a considerable amount of their time playing, and play behavior is important for the social and physical development of the young horse (Cameron et al., 2008). The presence of other foals or horses facilitates normal development of the young horse. Play behavior is primarily solitary in young foals, but around 4 weeks of age, foals start to play with each other. Male foals typically play more than females and they show more play fighting (Carson and Wood-Gush, 1983; Rho et al., 2007). The social status of a foal depends on its mother; the foal will typically attain similar status as its mother. If there are many foals in a group, a foal most frequently plays with the foal of the mother's preferred partner, i.e., foals of mares that stay close together, and then become each other's preferred playing partners (Araba and Crowell-Davis, 1994; Weeks et al., 2000).

Separation and Weaning

In nature, the weaning process occurs gradually, as the foal moves further and further away from the dam and suckling ceases. At approximately 2 years of age, the young stallions (called colts) either leave the family group voluntarily, or are chased away by the harem stallion. If there are more than one colt in the group, they typically leave the family group together and form a bachelor group, with or without other colts and stallions. Young females (called fillies) often stay in their natal group, and if so, mother and daughter show strong bonds their entire lives. A harem stallion may chase fillies away from the group if he is their father, potentially to avoid inbreeding, or a filly may sometimes leave with a group of bachelors.

Feeding Behavior

Horses are herbivores with a digestive system adapted to a diet of low nutrient grasses and other plants, consumed steadily throughout the day. Therefore, they have a relatively small stomach, but very long intestines to facilitate a steady flow of nutrients. Horses are adapted to move slowly forward while eating, and they frequently move between patches of grass, even when grazing on a field with plenty of grass. Thus, restriction of movement and periods of fasting cause welfare problems in captive horses.

As is true for many social species, horses prefer to eat together (Sweeting et al., 1985; Rifá, 1990). Their natural feeding behavior is to slowly move forward with the head lowered, tearing off grass with the teeth. Preferred grass areas are typically grazed to less than 20 mm (Ödberg and Francis-Smith, 1976), and plants are carefully selected, probably using the visual, olfactory, and gustatory senses. Under natural conditions, the summer diet consists mainly of different species of grass (80%–95%), whereas branches, twigs, cortex, and roots may constitute a significant part of the winter diet (Houpt, 1990). Chewing wood, which is considered an inconvenient behavior in domestic horses, probably originates from this natural behavior pattern, which increases survival chances during food scarcity in the wild. The horse's digestive system has evolved to process food items of low nutritive value and feral horses spend a large part of their day feeding. Depending on food availability, horses in the wild may spend up to 60% of their time (i.e., up to 15 h/day) searching for and ingesting food (Figure 18.4).

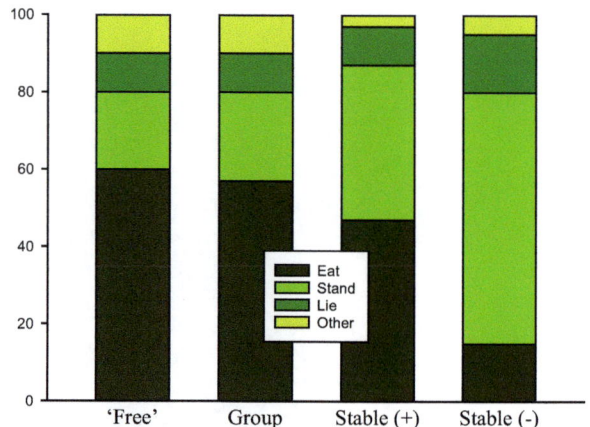

FIGURE 18.4 Average time budget (% of 24 h) for horses in different environments. "Free", feral Camargue horses; Group, group stabled horses with hay ad libitum; Stable (+), singly stabled horses with *ad lib* hay, which are able to see and touch each other; Stable (-), singly stabled horses, which are unable to see other horses and receive a diet based primarily on concentrates. (Modified from Kiley-Worthington, 1997.)

Food and water intake are interdependent, with foodstuffs with high water content reducing the need for water intake, and insufficient water intake reducing food intake. The average daily water intake of Przewalski horses (approx. 300 kg body weight) was 2.4–8.3 l in a reserve, where grasses with a high water content constituted the main part of their diet (Scheibe et al., 1998). An estimate of a horse's daily water intake is 2–3 l/ kg dry matter consumed (Hintz, 1994), but water uptake depends on the individual horse and is influenced by environmental factors, such as temperature and humidity. It is assumed that an adult domestic horse drinks 20–50 l water/24 h. Horses with free access to water, drink more frequently in warm weather (approx. every 1.8 h at 30°C–35°C), and the drinking frequency is lowest in the early morning and highest in late afternoon. Foals may drink water from the age of 3 weeks, but some foals do not drink water until after weaning (Crowell-Davis et al., 1985). Natural drinking behavior is to stretch the head forward and downward, and take in water by sucking. Horses prefer to drink from an open water surface (Nyman and Dahlborn, 2001). Water quality is important for horses, and they tend to be skeptical toward new water sources, where the water smells or tastes different from what they are used to.

Communication

Horses communicate via visual, auditory, olfactory, and tactile signals. Some types of communicative behavior are innate, but other types require learning, which may be linked to certain life (sensitive) periods. If the right cues are not present at the right time, communication problems can occur. When young horses were deprived of social contact for a period of time (single housing), they were subsequently more aggressive and showed less "mouth clapping", a submissive signal, compared to previously group-housed horses (Christensen et al., 2002a).

Some aspects of communication between horses are obvious to humans, including whinnying or aggressive behavior. However, a large part of horse communication is based upon body language, including small changes in body posture, which can be difficult for humans to detect. Horses are sensitive to human body language, such as nervousness (Keeling et al., 2009), and also respond to warning signals from other prey species.

Common Captive Behaviors

If provided with appropriate conditions, domestic horses show similar behavior as wild and feral horses. For example, social, feeding, and comfort behaviors (e.g., rolling, self-grooming, stretching) tend to be expressed in the same way and with the same frequency as in free-ranging horses. However, domestic horses are often housed and managed in ways that differ significantly from the natural conditions to which they have adapted through millions of years. Thus, domestic horses can experience a wide range of external and internal stressors, including lack of fulfillment of biological needs, as well as psychological stressors, such as loss of control and predictability. One of the most common stressors in domestic horses is social stress.

Horses are sometimes stabled with limited social contact with other horses, and research has shown that singly stabled horses have a strong motivation for social contact (Christensen et al., 2002a; Søndergaard et al., 2011). Chronically isolated horses even show a depressed cortisol response in a corticotropin-releasing hormone challenge test, suggesting desensitization of the hypothalamic-pituitary-adrenal axis (Visser et al., 2008). Thus, keeping horses in social isolation is a severe stressor that will lead to physical and psychological damage. Similarly, social instability may act as a stressor that increases aggression (Christensen et al., 2011).

Another common class of stressors in domestic horses involves feeding. Sports horses are often fed starch-rich, high-energy diets with reduced fibrous components that are very different from the diet available in the wild (McGreevy, 2004). Concentrated meals are usually provided twice a day and are rapidly consumed while the animal is confined within a stable. Under these circumstances, salivation is reduced, leading to a compromised buffering effect on gastric acidity, and the horse may still be motivated to forage, thus potentially leading to behavior problems, including crib-biting, sham-chewing, or wood chewing (Waters, 2002; Hothersall and Nicol, 2009; Wickens and Heleski, 2010). In addition, feed with a high starch content results in a higher glycemic response (increase in blood glucose concentrations), which has been hypothesized to cause high reactivity in horses (Bulmer et al., 2015). Nicol et al. (2005) found that, compared with foals fed on starch and sugar diets, foals fed after weaning on fat and fiber diets were more investigative and were more likely to approach unfamiliar people. They also galloped about less, and were calmer and less distressed immediately after weaning. Bulmer et al. (2015) found that a high-starch diet led to increased heart rate responses in novel object and handling tests, compared to when the same horses were fed a low-starch and high-fiber diet. Thus, fiber-based diets may result in calmer patterns of behavior, and due to the increased chewing time, likely better fulfill the need for prolonged feeding.

Another significant class of stressors for domestic horses is related to training, including when the training intensity is not adapted to the individual horse, the rider is too heavy, equipment is ill-fitting, goals are unrealistic (leading to harsh training methods), and trainers and riders are ignorant of the correct application of learning principles (McGreevy et al., 2018a). Furthermore, studies have shown that certain training methods are more stressful than others (e.g., riding with hyperflexed head and neck positions; Christensen et al., 2014; Zebisch et al., 2014). Transport can also be a significant stressor for domestic horses (Schmidt et al., 2010a, b), and even when the horse has been habituated to transport, frequent transport and environmental change when taking part in competitions are additional stressors for sports horses.

Domestic horses may be subject to stressful experiences for various lengths of time. Active behavioral responses are characteristic of *acute* (short-term) stress. The animal becomes more vigilant and hyper-reactive when the stressor is escapable. Acute stress responses may appear disguised when they manifest as redirected aggression, displacement activities, or as "freezing" or quiescence. However, stress typically manifests hyper-reactively as conflict behaviors that range from increased muscle tonus and body tension to bolting, rearing, bucking, or shying (McGreevy et al., 2018a). When stressful situations are regular, conflict behaviors can become ritualized, as the effects of prolonged stress accumulate (Ladewig, 2000). Passive coping frequently characterizes *chronic* stress, resulting in disengagement, decreased vigilance, and quiescence. Heart rate and blood pressure may be lowered and the horse may appear dull. In such situations, trainers may mistakenly believe that the horse is now more accepting of current events, but the quiescence is really a result of inescapable stress and lack of control (McGreevy et al., 2018a). Prolonged exposure to stressors results in physiological degradation (such as immunological disturbances and gastric disorders) and behavioral disturbances (such as development of stereotypies and injurious behaviors, including self-mutilation and increased aggression). McGreevy (2004) provides an exhaustive list of stereotypies, including crib-biting, wind-sucking, box-walking, weaving, and head-shaking. A stabled horse may perform more than one stereotypy, while no free-living feral horse has ever been observed to exhibit stereotypies; thus, these behaviors are considered to be caused by housing and management restrictions. The prevalence of crib-biting and weaving has been found to differ between breeds; however, it is difficult to separate the effects of breed from the effects of management, as they frequently overlap (Wickens and Heleski, 2010). Frustrated motivation is considered a possible underlying cause common to all stereotypies. Management factors that might frustrate behavioral needs of horses include, the amount and type of forage, quantity and quality of social contacts, and the amount of time spent in the stable (Minero and Canali, 2009). Unfortunately, owners often "treat" stereotypic horses ineffectively, by isolating them or by physically preventing them from performing the behaviors. These practices further impair the welfare of stereotypic horses and should be discouraged (McGreevy, 2004). On the contrary, increased time on pasture, more fiber in the diet, and increased social contact can all improve the welfare of stereotypic horses.

Horses may, however, continue to perform abnormal and stereotypic behavior, even if the environment is improved. Consequently, the occurrence of stereotypic behavior cannot easily be used to evaluate housing and management systems, because the behavior can result from previous conditions. In addition, some horses are more sensitive than others, and all horses do not respond in the same way to stressful conditions. A foal does not "inherit" a stereotypy from its mother, but there is a risk that it inherits her stress sensitivity, and if placed in the same environment, the foal is likely to develop the same stereotypy (Gillham et al., 1994; Wickens and Heleski, 2010). Likewise, horses do not pass stereotypies on to each other, but a stereotyping horse may cause disturbance in the stable, because other horses react to its unusual behavior, for example, with aggression. Thus, stereotypic horses should preferably be kept on pasture together with other horses, with *ad lib* access to grass or other roughage, which can eventually lead to a decrease in the stereotypic behavior.

Ways to Maintain Behavioral Health

Captive Considerations to Promote Species-Typical Behavior

Environment

In order to fulfill the behavioral needs of horses, the environment should reflect natural conditions in terms of access to full physical contact with other horses, access to free movement on pasture, and free access to roughage. Stall confinement should be avoided as much as possible. Some countries and organizations have implemented specific regulations for housing of horses, e.g., for box sizes. It is important that horses can lie down in a natural position with legs stretched, and head and neck extended (Figure 18.5). Since the length of a horse is approximately 1.5 times its height at its withers, when standing in a natural position, it is clear that it is longer

FIGURE 18.5 The natural lying position of horses with extended legs, neck, and head. Indoor housing facilities, such as single boxes, should provide sufficient space for horses to rest in their natural position.

when lying with the head and neck extended. Thus, to give the horse sufficient space, it has been suggested that the floor area in the box should be at least (2×horse height at withers in m²), i.e., a 1.7 m warmblood horse should have a box size of at least 11.56 m² (Søndergaard et al., 2004). Horses are able to sleep both standing up and lying down; however, the deep sleep, known as Rapid Eye Movement sleep, requires that the horse lies down. Failure to lie down, either due to physical pain or housing restrictions (e.g., inadequate box sizes, suboptimal bedding material, poor social management), can cause horses to collapse. Such collapses can cause severe injuries, and disturbance in sleep patterns are a sign of reduced welfare in horses (Fuchs et al., 2018). Some countries have national regulations concerning access to pasture or exercise, e.g., 2h of exercise at least 5 days/week (The Danish Horse Law, LBK no. 304 of 30/03/2017), and the EU Guide to Good Animal Welfare Practice for the Keeping, Care, Training, and Use of Horses recommends that horses have daily access to free exercise with other horses on paddock or pasture.

Social Groupings

Keeping horses in groups is recognized to best fulfill their behavioral and physical needs (Hartmann et al., 2012). It is essential that the groups have sufficient space, so that lower ranking individuals can move away without getting trapped. The groups should preferably be relatively stable and of mixed age and sex, wherever possible (Sigurjonsdottir and Haraldsson, 2019). Stallions should preferably be kept in harem or bachelor groups. Most behavior problems in domestic stallions are caused by inappropriate housing or management. Some individually stabled stallions develop an abnormal self-directed aggressive behavior (self-mutilation). This behavior is probably caused by lack of fulfillment of biological needs and can often be reduced by turn-out, social companions, and appropriate feeding (grass/roughage instead of concentrates), and sometimes by castration (Houpt, 1983; McDonnell, 2008). Christensen et al. (2002a) found that previously individually stabled stallions were more aggressive toward unknown stallions, compared to previously group-stabled stallions. If stallions are used to interacting with other stallions (group housing), aggression-related problems are rare, and a stallion that is temporarily removed from a bachelor group in order to work as a breeding stallion, can be reintroduced into the group without problems (pers. com., Jette Laursen, Denmark, Welsh breeder). However, mares should not be present within visual or olfactory range of the stallions, because the presence of mares in estrus causes an increase in aggressive and frustration behavior. In nature, older stallions that have lost their harem to another stallion, also join bachelor groups (Christensen et al., 2002b).

Domestic stallions that are castrated prior to puberty (at 1–2 years of age) do not develop normal sexual behavior, whereas stallions that have reached sexual maturity before castration may retain stallion sexual behavior for a long time after the operation (McDonnell, 1992). It is therefore not uncommon to see a gelding covering a mare. Castration is generally recommended for stallions that are not intended for breeding, because castration makes them easier to handle and keep together with other horses.

Parental behavior in domestic horses is very similar to their wild relatives. Mare and foal should ideally be left undisturbed, unless birth problems require human assistance, to allow sufficient time for bonding between the two (Henry et al., 2009). It is essential that mares and foals are allowed free exercise on pasture, preferably with other mares and foals, to allow for normal physical and behavioral development. In contrast to natural conditions, domestic foals are typically weaned early, at approximately 6 months of age. Both mare and foal respond vigorously to abrupt separation, which is common in captivity, indicating that the situation is very stressful for them both (Merkies et al., 2016). Waters et al. (2002) reported that weaning was associated with the onset of abnormal and stereotypic behavior, and that individual box weaning was more likely to induce these problems compared to group weaning. It is assumed that weaning is least stressful if natural conditions are imitated (i.e., by leaving the foal with the dam for a longer period, e.g., until shortly before the birth of her next foal, and applying a gradual weaning process, preferably with other foals or horses from the same group). Under all circumstances, it is essential that the foal is also kept with conspecifics after weaning, since young horses develop their skills and gain social advantages by growing up in a group with both young and adult horses (Bourjade et al., 2008).

Feeding Strategies

It is essential to provide domestic horses with sufficient roughage, ideally so that they can eat for at least 60% of the 24h period (Figure 18.4). This can be challenging with horse breeds that are prone to obesity and requires large pastures with low-nutrient grass, the feeding of special low-nutrient roughage, and/or feeding in ways such that the horse has to spend time getting access to roughage, for example, from slow feeders (Rochais et al., 2018). Optimal feeding of captive horses serves two purposes; partly to supply the horse with the appropriate amount of energy, protein, minerals, vitamins, and roughage to ensure its health; and partly to satisfy the natural behavioral need for prolonged feed intake (Minero and Canali, 2009; Harris et al., 2017). It is therefore important that a large part of what is fed consists of roughage, instead of concentrates. When provided with the opportunity, horses choose specific plants according to their energy needs and individual preferences; however, the diet of stabled horses is usually monotonous, with only one kind of roughage available (Goodwin et al., 2005). Thus, well-managed pastures that are not overgrazed, where horses can move freely and interact with other horses, best satisfy their behavioral needs.

Wherever possible, horses avoid eating manure-contaminated feed and do not graze close to dung piles. In that way, they keep endoparasite pressure at a minimum, because endoparasites are excreted via manure, and the risk of contamination is higher in defecation areas. A pattern of defecation areas and areas used for grazing therefore soon appears in domestic horse enclosures. In small enclosures and/or if kept at high densities, horses are "forced" to graze the field more intensively, thereby increasing endoparasite pressure (Medica et al., 1996).

The best way to avoid endoparasites is to make sure that enclosures are sufficiently large to allow separation of defecation and grazing areas, to remove manure from enclosures, and/or to change enclosures on a regular basis. However, it is normal for domestic foals to show coprophagy (manure-eating) during the first months of life. It is usually fresh feces from the dam that is consumed, and this behavior should not be prevented, as it is assumed to play a role in priming the gastrointestinal tract, and possibly, to assist with food selection (Marinier and Alexander, 1995).

Training

Training of horses should always align with learning theory, regardless of whether they are trained for research purposes, for sports, leisure, agriculture, or transport. The application of learning theory in horse training is described in detail elsewhere (e.g., McLean and Christensen, 2017; McGreevy et al., 2018a). There are a number of recognized desensitization techniques that can be applied to habituate horses to the various situations and stimuli in the training/research environment. Flooding techniques, where horses are exposed to an overwhelming amount of a fear-eliciting stimulus, should be avoided (McGreevy et al., 2018a). Fear responses should be avoided during training, because of their rapid acquisition and their resistance to extinction, and because they can be a significant safety risk for both trainer and horse. Horse-riding differs from training of most other species, due to the reliance on pressure and release (negative reinforcement), and animal welfare considerations put the onus on trainers to use minimal pressure and release it immediately when the horse shows the desired response. First principles in horse training are that the pressure motivates the horse to respond, and its removal trains the response. Thus, the timing of release of pressure is critical. Poor timing accounts for many behavioral problems in ridden and led horses, which can manifest as conflict behaviors and descend into learned helplessness. Through classical conditioning, the horse can learn to respond to light versions of the motivating pressure or other cues, e.g., auditory or visual signals. Negative reinforcement can be augmented by primary or secondary positive reinforcement and this combined reinforcement can enhance the success of training. It has been shown that the use of positive reinforcement (food) can enhance the horse-human relationship (Sankey et al., 2010a, b). Training methods based solely on positive reinforcement, e.g., clicker training, are also gaining ground in practical horse training, although the training often includes a certain amount of negative reinforcement, for example, if the horse is on a lead rope, or if human body posture or movement is used to apply a certain amount of pressure on the animal.

Special Situations

As mentioned above, horses are unique in terms of (1) their relationship with humans, and (2) the fact that they are used for agriculture and transport in some areas, for sports and recreation in other areas, as well as for animal-assisted therapy. Horse breeds are highly diverse, both in terms of physical appearance and physical abilities (e.g., jumping abilities, gait traits), as well as in temperament. Regardless, all horses share the same basic behavioral needs and failure to meet these needs can lead to stress, behavioral disorders, and health issues.

Conclusions and Recommendations

The behavioral repertoire and behavioral needs of horses remain relatively unchanged by domestication. Thus, domestic horses are likely to develop behavioral disorders when captive conditions fail to meet their needs. It is recommended (1) to keep horses in stable social groups, or at least ensure that they have access to full physical contact with other horses during part of the day; (2) to ensure that they receive sufficient roughage; and (3) to ensure that they have access to free movement, at least during part of the day. When behavioral needs are met, horses provide unique opportunities for research into social relationships (horse-horse and horse-human), communication, maternal transmission of behavior, learning, and training.

REFERENCES

Ahrendt, L.P., Labouriau, R., Malmkvist, J., Nicol, C., Christensen, J.W., 2015. Development of a standard test to assess negative reinforcement learning in horses. *Applied Animal Behaviour Science* 169, 38–42.

Araba, B.D., Crowell-Davis, S.L., 1994. Dominance relationships and aggression of foals (*Equus caballus*). *Applied Animal Behaviour Science* 41, 1–25.

Bourjade, M., Moulinot, M., Richard-Yris, M.-A., Hausberger, M., 2008. Could adults be used to improve skills of young horses, *Equus caballus*. *Developmental Psychobiology* 50, 408–417.

Boyd, L., Bandi, N., 2002. Reintroduction of takhi, *Equus ferus przewalskii*, to Hustai National Park, Mongolia: time budget and synchrony of activity pre- and post-release. *Applied Animal Behaviour Science* 78, 87–102.

Boyd, L., Keiper, R., 2005. Behavioural ecology of feral horses. In: *The Domestic Horse. The Evolution, Development and Management of its Behaviour*. Ed. by D. Mills & S. McDonnell, Cambridge University Press, Cambridge, 55–82.

Bulmer, L., McBride, S., Williams, K., Murray, J.-A., 2015. The effects of a high-starch or high-fibre diet on equine reactivity and handling behaviour. *Applied Animal Behaviour Science* 165, 95–102.

Cameron, E.Z., Linklater, W.L., Stafford, K.J., Minot, E.O., 2008. Maternal investment results in better foal condition through increased play behaviour in horses. *Animal Behaviour* 76, 1511–1518.

Carson, K., Wood-Gush, D.G.M., 1983. Equine behaviour: I. A review of the literature on social and dam-foal behaviour. *Applied Animal Ethology* 10, 165–178.

Christensen, J.W., 2016. Early-life exposure with a habituated mother reduces fear reactions in foals. *Animal Cognition* 19, 171–179.

Christensen, J.W., Ladewig, J., Søndergaard, E., Malmkvist, J., 2002a. Effects of individual versus group stabling on social behaviour in domestic stallions. *Applied Animal Behaviour Science* 75, 233–248.

Christensen, J.W., Zharkikh, T., Ladewig, J., Yasinetskaya, N., 2002b. Social behaviour in stallion groups (*Equus przewalskii* and *Equus caballus*) kept under natural and domestic conditions. *Applied Animal Behaviour Science* 76, 11–20.

Christensen, J.W., Rundgren, M., 2008. Predator odour per se does not frighten domestic horses. *Applied Animal Behaviour Science* 112, 136–145.

Christensen, J.W., Søndergaard, E., Thodberg, K., Halekoh, U., 2011. Effects of repeated regrouping on horse behaviour and injuries. *Applied Animal Behaviour Science* 133, 199–206.

Christensen, J.W., Beekmans, M., van Dalum, M., Van Dierendonck, M., 2014. Effects of hyperflexion on acute stress responses in ridden dressage horses. *Physiology & Behavior* 128, 39–45.

Crowell-Davis, S.L., Houpt, K.A., Carnevale, J., 1985. Feeding and drinking behavior of mares and foals with free access to pasture and water. *Journal of Animal Science* 60, 883–886.

de Barros Damgaard, P., Martiniano, R., Kamm, J., et al., 2018. The first horse herders and the impact of early Bronze Age steppe expansions into Asia. *Science* 360, eaar7711.

Deshpande, K., Furton, K.G., Mills, D.E.K., 2018. The equine volatilome: Volatile organic compounds as discriminatory markers. *Journal of Equine Veterinary Science* 62, 47–53.

Doherty, O., Casey, V., McGreevy, P. Arkins, S., 2017. Noseband use in equestrian sports – An international study. *PLoS One* 12, e0169060.

Estep, D.Q., Crowell-Davis, S.L., Earl-Costello, S.A., Beatey, S.A., 1993. Changes in the social behaviour of drafthorse (*Equus caballus*) mares coincident with foaling. *Applied Animal Behaviour Science* 35, 199–213.

Feh, C., 2005. Relationships and communication in socially natural horse herds. In: *The Domestic Horse. The Evolution, Development and Management of its Behaviour*. Ed. by D. Mills & S. McDonnell, Cambridge University Press, Cambridge, 83–93.

Fuchs, C., Kiefner, L.C., Reese, S., Erhard, M., Wöhr, A.C., 2018. Equine recumbent sleep deprivation: effects on mental and physical health. *Proceedings of the 14th Equitation Science Conference*, p. 44.

Gillham, S.B., Dodman, N.H., Shuster, L., Kream, R., Rand, W., 1994. The effect of diet on cribbing behavior and plasma β-endorphin in horses. *Applied Animal Behaviour Science* 41, 147–153.

Goodwin, D., 2007. Horse behaviour: Evolution, domestication and feralisation. In: *The Welfare of Horses*. Ed. by N. Waran, Springer, Dordrecht, The Netherlands, 1–18.

Goodwin, D., Davidson, H.P.B., Harris, P., 2005. Sensory varieties in concentrate diets for stabled horses: Effects on behaviour and selection. *Applied Animal Behaviour Science* 90, 337–349.

Hall, C.A., 2007. The impact of visual perception on equine learning. *Behavioural Processes* 76, 29–33.

Harris, P., Ellis, A.D., Fradinho, M.J., Jansson, A., Julliand, V., Luthersson, N., Santos, A.S., Vervuert, I., 2017. Review: Feeding conserved forage to horses: Recent advances and recommendations, *Animal* 11, 958–967.

Hartmann, E., Søndergaard, E., Keeling, L.J., 2012. Keeping horses in groups: A review. *Applied Animal Behaviour Science* 136, 77–87.

Hausberger, M., Roche, H., Henry, S., Visser, E.K., 2008. A review of the human–horse relationship. *Applied Animal Behaviour Science* 109, 1–24.

Heffner, R.S., Heffner, H.E., 1983. Hearing in large mammals; horses (*Equus caballus*) and cattle (*Bos taurus*). *Behavioral Neuroscience* 97, 299–309.

Henry, S., Hemery, D., Richard, M-A., Hausberger, M., 2005. Human–mare relationships and behaviour of foals toward humans. *Applied Animal Behaviour Science* 93, 341–362.

Henry, S., Richard-Yris, M-A., Tordjman, S., Hausberger, M., 2009. Neonatal handling affects durable bonding and social development. *PLoS One* 4, e5216.

Hintz, H.F., 1994. Nutrition and feeding. In: *Przewalski's Horse: The History and Biology of and Endangered Species*. Ed. by L.E. Boyd & K.A. Houpt, State University of New York Press, Albany, 115–129.

Hothersall, B., Nicol, C., 2009. Role of diet and feeding in normal and stereotypic behaviors in horses. *Veterinary Clinics of North America: Equine Practice* 25, 167–181.

Hothersall, B., Harris, P., Sörtoft, L., Nicol, C.J., 2010. Discrimination between conspecific odour samples in the horse (*Equus caballus*). *Applied Animal Behaviour Science* 126, 37–44.

Houpt, K.A., 1983. Self-directed aggression: A stallion behaviour problem. *Equine Practice* 5, 6–8.

Houpt, K.A., 1990. Ingestive behavior. *Veterinary Clinic of North America: Equine Practice* 6, 319–337.

Jezierski, T., Jaworski, Z., Sobczynska, M., Ensminger, J., Górecka-Bruzda, A., 2018. Do olfactory behaviour and marking responses of Konik polski stallions to faeces from conspecifics of either sex differ? *Behavioural Processes* 155, 38–42.

Keeling, L.J., Jonare, L., Lanneborn, L., 2009. Investigating horse-human interactions: The effect of a nervous human. *Veterinary Journal* 181, 70–71.

Kiley-Worthington, M., 1997. *The Behaviour of Horses in Relation to Management and Training*. J.A. Allen, London.

Klingel, H., 1975. Social organization and reproduction in equids. *Journal of Reproduction and Fertility, Supplement* 23, 7–11.

Ladewig, J., 2000. Chronic intermittent stress: a model for the study of long-term stressors. In: *The Biology of Animal Stress. Basic Principles and Implications for Animal Welfare*. Eds. by G.P. Moberg, J.A. Mench, CABI Publishing, Wallingford, England, 159–169.

Manfredi, J., Clayton, H.M., Derksen, F.J., 2005. Effects of different bits and bridles on frequency of induced swallowing in cantering horses. *Equine and Comparative Exercise Physiology* 2, 241–244.

Marinier, S.L., Alexander, A.J., 1995. Coprophagy as an avenue for foals of the domestic horse to learn food preferences from their dams. *Journal of Theoretical Biology* 173, 121–124.

McDonnell, S.M., 1992. Normal and abnormal sexual behavior. *Veterinary Clinics of North America: Equine Practice* 8, 71–89.

McDonnell, S.M., 2003. *The Equid Ethogram. A Practical Field Guide to Horse Behavior.* The Blood Horse Inc, Lexington, Kentucky.

McDonnell, S.M., 2008. Practical review of self-mutilation in horses. *Animal Reproduction Science* 107, 219–228.

McGreevy, P.D., 2004. *Equine Behaviour: A Guide for Veterinarians and Equine Scientists.* W.B. Saunders, Edinburgh.

McGreevy, P., Christensen, J.W., König von Borstel, U., McLean, A., 2018a. *Equitation Science.* Wiley Blackwell, Oxford.

McGreevy, P., Griffiths, M.D., Ascione, F.R., Wilson, B., 2018b. Flogging tired horses: Who wants whipping and who would walk away if whipping horses were withheld? *PLoS One* 13, e0192843.

McLean, A.N., Christensen, J.W., 2017. The application of learning theory in horse training. *Applied Animal Behaviour Science* 190, 18–27.

Medica, D.L., Hanaway, M.J., Ralston, S.L., Sukhdeo, M.V.K., 1996. Grazing behaviour of horses on pasture: predisposition to strongylid infection. *Journal of Equine Veterinary Science* 16, 421–427.

Merkies, K., DuBois, C., Marshall, K., Parois, S., Graham, L., Haley, D., 2016. A two-stage method to approach weaning stress in horses using a physical barrier to prevent nursing. *Applied Animal Behaviour Science* 183, 68–76.

Minero, M., Canali, E., 2009. Welfare issues of horses: an overview and practical recommendations. *Italian Journal of Animal Science* 8, 219–230.

Monfort, S.L., Arthur, N.P., Wildt, D.E., 1994. Reproduction in the Przewalski's horse. In: *Przewalski's Horse: the History and Biology of an Endangered Species.* Ed. by L.E. Boyd, K.A. Houpt, State University of New York Press, Albany, 173–193.

Nagel, C., Erber, R., Ille, N., von Lewinski, M., Aurich, J., Möstl, E., Aurich, C., 2014. Parturition in horses is dominated by parasympathetic activity of the autonomous nervous system. *Theriogenology* 82, 160–168.

Nicol, C.J., Badnell-Waters, A.J., Bice, R., Kelland, A., Wilson, A.D., Harris, P.A., 2005. The effects of diet and weaning method on the behaviour of young horses. *Applied Animal Behaviour Science* 95, 205–221.

Nyman, S., Dahlborn, K., 2001. Effect of water supply method and flow rate on drinking behavior and fluid balance in horses. *Physiology & Behavior* 73, 1–8.

Ödberg, F.O., Francis-Smith, K., 1976. A study on eliminative and grazing behaviour - the use of the field by captive horses. *Equine Veterinary Journal* 8, 147–149.

Olczak, K., Christensen, J.W., Klocek, C., 2018. Food motivation in horses appears stable across different test situations. *Applied Animal Behaviour Science* 204, 60–65.

Pierard, M., McGreevy, P., Geers, R., 2019. Reliability of a descriptive reference ethogram for equitation science. *Journal of Veterinary Behavior: Clinical Applications and Research* 29, 118–127.

Rho, J.R., Srygley, R.B., Choe, J.C., 2007. Sex preferences in Jeju pony foals (*Equus caballus*) for mutual grooming and play-fighting behaviors. *Zoological Science* 24, 769–773.

Rifá, H., 1990. Social facilitation in the horse (*Equus caballus*). *Applied Animal Behaviour Science* 25, 167–176.

Roberts, K., Hemmings, A., Moore-Colyer, M., Hale, C., 2015. Cognitive differences in horses performing locomotor versus oral stereotypic behaviour. *Applied Animal Behaviour Science* 168, 37–44.

Rochais, C., Henry, S., Hausberger, S., 2018. "Hay-bags" and "Slow feeders": Testing their impact on horse behaviour and welfare. *Applied Animal Behaviour Science* 198, 52–59.

Rosales, C., Krekeler, N., Tennent-Brown, B., Stevenson, M.A., Hanlon, D., 2017. Periparturient characteristics of mares and their foals on a New Zealand Thoroughbred stud farm. *New Zealand Veterinary Journal* 65, 24–29.

Rutberg, A.T., Greenberg, S.A., 1990. Dominance, aggression frequencies and modes of aggressive competition in feral pony mares. *Animal Behaviour* 40, 322–331.

Sankey, C., Richard-Yris, M-A., Henry, S., Fureix, C., Nassur, F., Hausberger, M., 2010a. Reinforcement as a mediator of the perception of humans by horses (*Equus caballus*). *Animal Cognition* 13, 753–764.

Sankey, C., Richard-Yris, M-A., Leroy, H., Henry, S., Hausberger, M., 2010b. Positive interactions lead to lasting positive memories in horses, *Equus caballus. Animal Behaviour* 79, 869–875.

Sarkissian, C.D., Ermini, L., Schubert, M., et al., 2015. Evolutionary genomics and conservation of the endangered Przewalski's horse. *Current Biology* 25, 2577–2583.

Scheibe, K.M., Eichhorn, K., Kalz, B., Streich, W.J., Scheibe, A., 1998. Water consumption and watering behavior of Przewalski Horses (*Equus ferus przewalskii*) in a semireserve. *Zoo Biology* 17, 181–192.

Schmidt, A., Aurich, J., Möstl, E., Müller, J., Aurich, C., 2010a. Changes in cortisol release and heart rate and heart rate variability during the initial training of 3-year-old sport horses. *Hormones and Behavior* 58, 628–636.

Schmidt, A., Biau, S., Möstl, E., Becker-Birck, M., Morillon, B., Aurich, J., Faure, J.-M., Aurich, C., 2010b. Changes in cortisol release and heart rate variability in sport horses during long-distance road transport. *Domestic Animal Endocrinology* 38, 179–189.

Shaw, E.B., Houpt, K.A., Holmes, D.F., 1988. Body temperature and behaviour of mares during the last two weeks of pregnancy. *Equine Veterinary Journal* 20, 199–202.

Sigurjonsdottir, H., Haraldsson, H., 2019. Significance of group composition for the welfare of pastured horses. *Animals* 9, 14.

Søndergaard, E., Clausen, E., Christensen, J.W., Schougaard, H., 2004. Housing of horses. Danish Institute of Agricultural Sciences, Tjele, Denmark.

Søndergaard, E., Jensen, M.B., Nicol, C.J., 2011. Motivation for social contact in horses measured by operant conditioning. *Applied Animal Behaviour Science* 132, 131–137.

Sweeting, M.P., Houpt, C.E., Houpt, K.A., 1985. Social facilitation of feeding and time budgets in stabled ponies. *Journal of Animal Science* 60, 369–374.

Thodberg, K., Berget, B., Lidfors, L., 2014. Research in the use of animals as a treatment for humans. *Animal Frontiers* 4, 43–48.

Van Dierendonck, M.C., 2006. The importance of social relationships in horses. Degree Diss., Utrecht University, The Netherlands.

Visser, E.K., Ellis, A.D., van Reenen, C.G., 2008. The effect of two different housing conditions on the welfare of young horses stabled for the first time. *Applied Animal Behaviour Science* 114, 521–533.

Waters, A.J., Nicol, C.J., French, N.P., 2002. Factors influencing the development of stereotypic and redirected behaviours in young horses: Findings of a four year prospective epidemiological study. *Equine Veterinary Journal* 34, 572–579.

Weeks, J.W., Crowell-Davis, S.L., Caudle, A.B., Heusner, G.L., 2000. Aggression and social spacing in light horse (*Equus caballus*) mares and foals. *Applied Animal Behaviour Science* 68, 319–337.

Wickens, C.L., Heleski, C.R., 2010. Crib-biting behaviour in horses: a review. *Applied Animal Behaviour Science* 128, 1–9.

Xia, C., Jie, C., Zhang, H., et al., 2014. Reintroduction of Przewalski's horse (*Equus ferus przewalskii*) in Xinjiang, China: The status and experience. *Biological Conservation* 177, 142–147.

Zebisch, A., May, A., Reese, S., Gehlen, H., 2014. Effect of different head-neck positions on physical and psychological stress parameters in the ridden horse. *Journal of Animal Physiology and Animal Nutrition* 98, 901–907.

19

Behavioral Biology of Chickens and Quail

Laura Dixon
Scotland's Rural College (SRUC)

Sarah Lambton
University of Bristol

CONTENTS

Introduction 299
Typical Research Involvement 300
Behavioral Biology 300
    Natural History of Fowl 300
        Ecology 300
        Social Organization and Behavior 301
        Feeding Behavior 302
        Communication 302
    Natural History of Quail 302
        Ecology 302
        Social Organization 302
        Feeding Behavior 303
        Communication 303
Common Captive Behaviors 303
    Normal Behaviors 303
    Abnormal Behaviors 304
Ways to Maintain Behavioral Health 305
    Environment 305
    Social Groupings 306
Conclusions 308
Recommendations 308
References 308

Introduction

Poultry are housed under laboratory conditions, although not always the kind of laboratories one usually imagines, because they are useful in a variety of fields of scientific research. Their housing conditions are hugely variable, and the term poultry itself encompasses a wide variety of different species, referring to any domesticated birds farmed for meat, eggs, or feathers. However, here we will concentrate on the Galliformes; an order of ground-dwelling birds that encompasses chickens, quail, partridges, pheasants, and turkeys. The species most commonly used in research are members of the genera *Gallus* (domestic fowl and junglefowl) and *Coturnix* (quail); therefore, these will be the focus of this chapter. We will first describe their use in research, before discussing the behavioral biology of these species in the wild. We will use this information to put in context the behaviors of these species in captivity, both normal and abnormal, before considering the ways in which these species may be managed in captivity in order to maintain their behavioral health.

The domestic fowl (*Gallus gallus domesticus*) is a subspecies of red junglefowl (*Gallus gallus gallus*). Archaeological evidence shows chickens were domesticated at least 8,000 years ago (West and Zhou 1988), while phylogenetic analysis suggests that domestic fowl diverged from red jungle fowl 58,000 ± 16,000 years ago (Sawai et al. 2010). The two species can successfully interbreed to produce fertile offspring, leading to populations of feral fowl (Callaway 2016). Domestic fowl, both captive and feral, perform a similar range of behaviors to junglefowl, although they may differ in frequency (Duncan et al. 1978, Kruijt 1964, McBride et al. 1969, Andersson et al. 2001).

FIGURE 19.1 Japanese quail in the laboratory. (Photo: Michael Domjan).

Similarly, domestic quail (Figure 19.1) have been derived from Japanese quail (*Coturnix japonica*). Domestication of quail took place around 900–1,000 years ago (Mills et al. 1997, Kovach 1974). The two are not distinct species; they interbreed successfully, performing many of the same behaviors, albeit at different frequencies (Chang et al. 2009, Wang et al. 2003, Mills et al. 1997). Therefore, studies of red junglefowl and of Japanese quail can be used to make inferences about the "natural" behavior of the corresponding poultry species, and inform our understanding of their welfare in captivity.

Typical Research Involvement

Birds make up a small percentage of the total animals involved in laboratory studies. In 2019 in Great Britain, the Home Office found that 3.6% of recorded experimental work used chickens and 0.01% used quail, equating to more than 124,000 animals (Home Office 2019). These figures only account for studies involving regulated procedures, for which a project license is required, including studies exploring fundamental research, which look to improve scientific understanding and theories of various phenomena, or in applied research, where these theories are used to develop techniques to alter the phenomena. The actual number of poultry housed under laboratory conditions in the UK is likely to be higher than the figures quoted above, as not all studies require a project license. A search of the literature reveals many studies conducted using poultry across Europe, the USA, Japan, China, and Australia, which we surmise equates to a large number of individuals kept in this way.

Chickens and quail are found to be useful laboratory models because of their relatively small size compared to most mammals, their short reproduction cycle and fast maturation, the low cost of housing and maintaining them, and the accessibility of the embryo from the laid egg. Research on chickens and quail is conducted for a variety of different purposes. For example, studies of behavior, development, nutrition, pharmacology and toxicology, endocrinology, and immunology have all used chicken or quail models at times. This research may be on the young or adult animal, or it may be done using the unhatched embryo in the egg. The removal of a part of the eggshell allows access to the embryo for developmental observations or manipulations (e.g., Huss et al. 2008).

Both chickens and quail have similar enough development and physiology to humans to make them a good model for studying aging and disease. Both species are considered an excellent model for toxicological studies, since effects of toxins can be shown as abnormalities in egg development or cessation of laying (Baher 2008). They are also used in antibody production, as the yolk of the egg contains antibodies, and embryos can be genetically modified to produce human antibodies in the eggs that will later hatch (Davis 2003). Quail are often used to study sexual development and reproductive behavior, because of the clear sex differentiation of female and male birds. They are also used for developmental biology questions, such as how early social experiences can affect mate preferences, or for physiological questions, such as how photoperiod or gonadal hormones affect sexual behavior (e.g., Ball and Balthazart 2010, Mills et al. 1997).

Chickens and quail are the main avian species used to study embryology (Huss et al. 2008) due to the similarity of the chick and human embryo at molecular, cellular, and anatomical levels (Vergara and Canto-Soler 2012). Chick embryos also develop quickly, and the ability to remove part of the shell allows for easy inspection and manipulation of the embryo (Vergara and Canto-Soler 2012). Crucial information on how the cells that form the nervous system develop and differentiate, and on the molecular development of limbs, has been gained from such studies (Understanding Animal Research 2015). Another major laboratory use of a fertilized chicken egg is in the production of vaccines (Chen et al. 2010). Quail embryos are also used in research, and are considered an excellent model for human fetal alcohol syndrome (Smith 2008) and to study the effects of microgravity on development during short-duration space flights (Barrett et al. 2000).

Both chicken and quail can be used to test the effects of different feed formulations or feed additives on growth and production parameters for applied use in the agriculture sector (Elangovan et al. 2003). Finally, they can also be studied to learn more about their behavior in different environments and the ways in which different housing and management systems affect their welfare, with the aims of minimizing suffering and promoting positive welfare for captive birds (Odeh et al. 2003).

The wide variety of study objectives listed above inevitably means chickens and quail are housed in a wide variety of conditions, ranging from single birds or small groups in cages, to floor pens, to larger flocks in houses or barns more akin to those found in commercial farming systems. Consequently, their ability to behave "naturally", and thus, their welfare, will be affected differently, depending on the type of study and housing.

Behavioral Biology

Natural History of Fowl

Ecology

Four species of junglefowl (red, grey, Ceylon, and green) have different (in some cases, overlapping) home ranges in Asia, with five subspecies of red junglefowl occupying the greatest area in north and northeast India, and southeast Asia (Desta 2019). Red junglefowl inhabit Sal (moist deciduous) forests, scrub jungle, bamboo, and grassland (Javed and Rahmani

2000); areas with an upper story of trees and an understory of tall herbs, shrubs, and grass (Collias and Collias 1967). Range sizes are variable, likely depending on resource availability and energy requirements, and change across seasons. In areas with abundant populations of red jungle fowl, Collias and Collias (1967) found that the average range size was around 1 ha per adult bird, while Subhani et al. (2010) found much larger average range sizes, from 6.4 to 16.5 ha per bird. In a small study of three males and one female, Arshad and Zakaria (1999) found that daily and monthly home range size was highly variable, ranging from 0.3 to 17.3 ha daily and 2.6–230 ha monthly for males, and 0.3–5.5 ha daily and 1.3–9.1 ha monthly for the female. Within their ranges, junglefowl occupy stable roosting areas (Collias and Collias 1967), although roosts moved after a disturbance. Flocks roost overnight on the branches of trees 6–9 m above the ground; however, hens with chicks roosted slightly lower, around 4 m above ground (Arshad and Zakaria 2009), often returning to the same branch in the same tree (McBride et al. 1969). Birds leave the roost before sunrise and return before sunset, moving and foraging during the early and later parts of the day (Arshad and Zakaria 2009, Collias and Collias 1967). Collias and Collias (1967) observed that flocks moved in predictable patterns through their territory, incorporating watering holes into their daily route, and returning to their roost to rest during the hottest part of the day.

The habitat structure described allows red junglefowl to move through the undergrowth with relative ease, preferring to walk rather than fly (Collias and Collias 1967). At the same time, this habitat provides shelter from predators, including dogs, cats, civets, jackals, birds of prey, large lizards, and snakes (Arshad and Zakaria 1999, Collias and Collias 1967). Fowl are cautious animals; dominant males engage in vigilance behavior (McBride et al. 1969, Pizzari 2003), emitting alarm calls that are modified according to whether the threat detected is from a ground-based or aerial predator (Collias 1987), which cause the flock to fly up into the trees or retreat to the undergrowth, respectively. Additionally, hens with chicks emit alarm calls and perform noisy displays designed to ward off and/or divert the attention of potential predators while chicks crouch and hide (McBride et al. 1969, Collias and Collias 1967).

Social Organization and Behavior

Red junglefowl live in small flocks, the size of which varies with habitat. Javed and Rahmani (2000) reported that Sal forests, mixed forest, and forest grassland were the only habitats that had flock sizes that exceeded ten birds (mean flock sizes were 2.4, 2.5, and 2.5, respectively, in the summer, and 5.7, 4.0, and 3.2, respectively, in the winter). The smallest flock sizes were found in grasslands and riparian forest (means 1.6 and 1.8 birds per flock, respectively, in the summer). Birds were often solitary, with 80% of observations being of a single bird, although group sizes increased pre- and postbreeding. Red jungle fowl are polygynous and a flock frequently comprises a single male and a harem of females, with further subordinate males nearby (Collias and Collias 1967, Anwar et al. 2016). Javed and Rahmani (2000) observed male:female ratios of 0.75:1.0 in winter and 0.72:1.0 in summer, while in a larger study, Anwar et al. (2016) found an average harem size of three females with each male. Domestic fowl establish social hierarchies from as early as 5 or 6 weeks of age by means of displays of aggression, including pecking and fighting, most commonly performed by males, and directed toward both males and females (Guhl 1968). Birds in a flock with a stable hierarchy showed fewer agonistic behaviors, and hierarchies have been found in flocks of up to 100 birds (Guhl 1958). The ability to form stable hierarchies depends on the ability of birds to recognize their flockmates (Smith et al. 2016). A study of large flocks of 300 and 700 domestic hens found that hens do not appear to recognize individuals from their own flock, or attempt to form social hierarchies within those flocks (Hughes et al. 1997), suggesting that this explains an apparent lack of aggression in large flocks, compared with small flocks, where birds may attempt to establish a hierarchy (Nicol et al. 1999).

In populations of wild junglefowl, dominant males maintain territories, engaging in predawn crowing, patrolling boundaries daily during the breeding season, chasing off subordinate males, and monopolizing access to females during the breeding season (Anwar et al. 2016, McBride et al. 1969, Collias and Collias 1967). Dominant males typically have longer tail feathers that are held more erect than subordinate males (Collias and Collias 1967). Although Collias and Collias (1967) rarely observed fights between males, in a flock of red junglefowl kept under seminatural conditions, more aggressive males mated with more different females, and more frequently with each female (McDonald et al. 2017). Males initiate courtship by "waltzing"; sidling toward the female with the feathers of the outer wing spread out and down toward the ground (Kruijt 1966). Males were observed to attract hens with a food call, emitted while holding a food item in their beak or while scratching and pecking at the food on the ground (known as tid-bitting; Collias and Collias 1967, Pizzari 2003, Kruijt 1966). Pizzari (2003) observed a free-ranging flock of feral domestic fowl and found that dominant males were more likely to engage in courtship feeding and vigilance behaviors than subordinate males, and the likelihood of copulation increased with male social status. Females choose which males to mate with, soliciting copulation by crouching in front of the chosen male and raising their tail (Pizzari 2003). In response to unwanted copulations, female feral domestic fowl emit distress calls, attracting the attention of dominant males (Pizzari 2001). Females also utilize cryptic choice; they are able to eject the sperm of subordinate males (Pizzari and Birkhead 2000) or to store sperm for several weeks, resulting in competition between the sperm of different males (reviewed by Garnham and Løvlie 2018, Løvlie and Pizzari 2007).

Red junglefowl lay their eggs in nests located under cover of dense vegetation (Anwar et al. 2016). Nests comprise a small depression of 2.5–4 inches in the soil, lined with leaves, feathers, and twigs (Anwar et al. 2016, Collias and Collias 1967). Red junglefowl hens have been observed to lay between one and nine eggs per clutch (Anwar et al. 2016, Arshad and Zakaria 1999). In captivity, red Burmese junglefowl laid 5–7 eggs in their first clutch, and 6–11 in their second (Meijer and Siemers 1993), over the course of 2–5 days. Once the last egg was laid, hens incubated (sat on) the eggs for 19–20 days, only leaving them for periods of 30–60 min every day or two, until the chicks were hatched. In the wild, 99% of red junglefowl eggs were observed to hatch and the hen stayed with the nest until the last chick had dried (Arshad and Zakaria 1999).

After hatching, red junglefowl hens spent most of their time with their chicks (Collias and Collias 1967) in thick vegetation cover and patches of grass (Anwar et al. 2016), which provide cover and food in the form of insects. Red junglefowl hens brood their chicks (chicks are warmed by sheltering them under the wings and next to the body of the hen), for at least the first 9 days after hatching, until the chicks are able to effectively thermoregulate (Sherry 1981, McBride et al. 1969). Brooding was initiated by chicks pecking at the breast feathers of the hen and nuzzling under her for up to 75% of the observed time, depending on chick age and air temperature (Sherry 1981). The presence of a hen influences the behavioral development of the chicks (Shimmura et al. 2010). Domestic chicks imprint on the first moving object they encounter (Nakamori et al. 2013), a process that is reinforced by maternal vocalizations (Sluckin and Salzen 1961, Nicol 2015), and they emit distress calls when separated from the hen (McBride et al. 1969).

Feeding Behavior

Junglefowl and their domestic cousins are omnivores, consuming a variety of nuts, seeds, fruits, insects, crustaceans, molluscs, and small vertebrates (Arshad et al. 2000), and chicks adjust their foraging behavior according to the most nutritionally appropriate food for their age (Murphy et al. 2014). Adult junglefowl allocate a large proportion of their time to foraging activities; hens were observed ground pecking in 60% of minutes during the active part of the day, and ground scratching in 34% of minutes (Dawkins 1989). These behaviors are also performed as part of the dustbathing behavior sequence (Kruijt 1964, Collias and Collias 1967), which hens perform for less than 30 min every other day, to maintain their plumage condition (Vestergaard 1982). Chicks are also guided toward appropriate food by junglefowl and domestic hens that use vocalizations and pecking movements directed toward food items to attract their chicks to food (Sherry 1977, Stokes 1971). They modify their behavior/display, depending on the quality of the food and the behavior of the chicks (Moffatt and Hogan 1992, Nicol and Pope 1996). In this way, chicks learn appropriate foraging behaviors, although this learning is modulated by the chicks' own experiences (Moffatt and Hogan 1992, Nicol 2004, Hogan 1973).

Communication

Flock living is central to many aspects of the lives of junglefowl and their counterparts. The behavior of the flock is mediated by complex communication and social cognition (Marino 2017). We have described several forms of communication, defined as the transfer of information from a signaler to a receiver, causing a change in behavior of the receiver (alarm calling, feed calling, courtship displays, crowing, and maternal vocalizations). Jungle fowl communication consists of at least 24 different vocal signals, as well as visual displays (Collias 1987, Collias and Collias 1967). Fowl employ complex referential communication, i.e., meaning is attached to each signal, which is flexible and informed by cognitive and social awareness (Marino 2017). Domestic fowl are able to recognize and discriminate between conspecifics (Bradshaw 1992, Guhl 1958), and modify their behavior according to their perceptions of others (Marino 2017). For example, subordinate males modified their courtship display according to the perceived attention of a dominant male (Smith et al. 2011) and hens increased their feeding displays in response to feeding errors by their chicks (Nicol and Pope 1996). Domestic fowl engage in social learning; learning a behavior through observation of the behavior and its consequences (Marino 2017). Naïve hens that watched a trained hen perform a task learned that task more reliably than hens that observed a naïve demonstrator (Nicol and Pope 1992, 1994). In particular, foraging behaviors may spread through a flock by a process of social learning (Nicol 2004), and in captive groups of domestic fowl, social learning may play a role in the transmission of abnormal behaviors, such as feather pecking (pecking at the feathers of a conspecific) and cannibalism (Cloutier et al. 2002, McAdie and Keeling 2002, Nicol 1995). Red junglefowl and domestic fowl both exhibit behavioral synchronization; individuals performing the same behaviors at the same time (Eklund and Jensen 2011). This acts as an antipredator strategy, but may also facilitate the complex social processes described above. Behavioral synchronization in domestic fowl chicks is disrupted if they are not brooded by a hen (Riber, Nielsen, et al. 2007), and the lack of behavioral synchronization in domestic fowl has been associated with the development of feather pecking (Jensen et al. 2006).

Natural History of Quail

Ecology

Research on quail behavior has been considerably less extensive. Furthermore, much of what we do know comes from studies of quail housed in cages (e.g., Chang et al. 2009). Quail, like junglefowl, originate from southeast Asia (Kovach 1974) and their preferred habitat is grasslands and shrubby areas. Quail prefer to stay in dense vegetation, and respond to an approaching threat first by freezing, and then by flying up vertically for a short distance, and then taking cover (Buchwalder and Wechsler 1997). Perhaps reflecting this habitat preference, quail housed in seminatural outdoor aviaries rarely used elevated structures and did not roost at night (Schmid and Wechsler 1997). Quail are migratory, moving south to lower elevations in the winter months, although remaining within Asia (Kovach 1974).

Social Organization

Male quail are territorial and mark their territory by crowing from an elevated spot within that area (Kovach 1974). When housed together in seminatural outdoor aviaries, aggressive interactions were observed between males (Schmid and Wechsler 1997). There is some evidence that quail are monogamous, forming stable pair-bonds during the breeding season (Kovach 1974, Schmid and Wechsler 1997, Stevens 1961). Males initiate mating with a courtship display, puffing up their feathers and strutting with a particular posture (Farris 1967),

FIGURE 19.2 Quail chick. (Photo: Mathieu Giraudeau, Centre de Recherches Ecologiques et Evolutives sur le Cancer, France.)

and females crouch in response to allow mating to take place. Female quail nest in the undergrowth, laying their eggs in shallow depressions scratched in the ground (Stevens 1961, Schmid and Wechsler 1997). Quail hens living in a large semi-natural enclosure were able to brood two clutches during the breeding season. Clutch size ranged from 4 to 14 eggs, with an average of 67% of eggs hatching. Like fowl, quail chicks (Figure 19.2) show filial imprinting (Mills et al. 1997).

Feeding Behavior

Like fowl, quail are ground-feeding omnivores; they eat seeds, berries, peas, grains, tender shoots and leaves, insects, and grubs (Taka-Tsukasa 1967, cited by Schmid and Wechsler 1997). When housed under seminatural conditions in aviaries, quail spent 8.3% of their time engaged in exploratory and foraging behaviors (Schmid and Wechsler 1997), despite having *ad lib* access to food. Like junglefowl, most foraging activity took place shortly after dawn and in the 3 h preceding darkness (Woodard and Wilson 1970). Quail, like fowl, also perform ground scratching and pecking during dustbathing. Quail housed indoors spent between 2.6% and 4.5% of their observed time dustbathing, and all of the birds observed, dust-bathed each day (Statkiewicz and Schein 1980).

Communication

Studies have documented between 15 and 28 distinct vocalizations in quail, but unlike in laying hens, their role in communication is less well-defined (Mills et al. 1997). As noted above, males crow to mark their territory and attract females. Males and females in mating pairs perform separation calls to locate each other (Lawrence 1975), and are able to recognize each others' calls (reviewed by Mills et al. 1997). Hens perform a variety of calls to their chicks, which serve variously to attract the chicks to the hen, direct the chicks to food, and cause the chicks to hide (reviewed by Mills et al. 1997). Evidence also suggests that quail, like fowl, can engage in social learning (Akins and Zentall 1996, Sanavio and Savardi 1980), performing a task more readily, if they have previously observed a demonstrator perform the same task. Hens brood their chicks after hatching and present food items to them (often live prey) (Orcutt and Orcutt 1976), suggesting that quail chicks and fowl chicks learn foraging behavior similarly.

Common Captive Behaviors

Normal Behaviors

Many of the behaviors described above for junglefowl are reflected in the behavioral needs (instinctive behaviors performed even when an appropriate environment or resource is not available) listed by Weeks and Nicol (2007) for laying hens. These include perching, nesting, foraging, and dustbathing, and the ability of captive fowl to perform these behaviors affects their welfare. Animals in captivity should have the freedom to express normal behavior (FAWC 1993); therefore, the goal of a captive environment should be to promote these behaviors, in particular, those considered to be behavioral needs. Domestic chickens and quail show similar behavioral repertoires to their wild or feral counterparts when housed in appropriate conditions (Andersson et al. 2001). Clearly, some normal behaviors observed in wild animals occur in stressful situations, such as avoiding predators, and situations that promote these behaviors may not need to be replicated in husbandry systems (Hawkins 2010). Having said this, both fowl and quail are nervous species, prone to alarm at minor disturbances. Domestic fowl, if startled, will take flight or run for shelter, while quail exhibit vertical flight behavior (Buchwalder and Wechsler 1997), and the inability to perform these behaviors may itself cause stress (Jones 2007).

Foraging or food searching behavior takes up a large part of the daily time budget for chickens and quail, and the birds are motivated to perform this behavior, even when food is freely available (Duncan and Hughes 1972). In the wild, foraging behavior is performed as a precursor to feeding behavior; the birds must forage to find food. In captivity, food is frequently easily accessible and provided *ad libitum* in a concentrated form, which allows daily nutritional requirements to be met in a relatively short time, with little foraging necessary (Keeling 2002). However, foraging activities are internally motivated (changes in internal states, such as hormones, that direct changes in behavior (Colgan 1989)), meaning that even if food is provided in a way that does not require work to access it, the animal will still be motivated to perform the behavior (Hughes and Duncan 1988).

Both chickens and quail perform dustbathing behavior in captivity. Dustbathing helps to remove debris and ectoparasites from the feathers, and is also internally motivated, meaning that birds will still be motivated to perform the behavior, even in the absence of a suitable substrate (Olsson and Keeling 2005). Chickens housed in cages with wire floors perform bouts of sham dustbathing (Hughes and Duncan 1988), and even genetically featherless birds will dustbathe, demonstrating the motivation to dustbathe exists even in the absence of plumage condition to stimulate the behavior (Vestergaard et al. 1999). In captive environments, there is evidence that early experience influences the perception of substrates as suitable for both foraging and dustbathing. It is clear that maternal behavior, where present, influences learning about foraging substrates (Edgar et al. 2016). Some have suggested chicks imprint on substrates to which they are exposed in the first 10 days of life, regardless of maternal influence (Braastad 1990, Hess 1964, Johnsen et al. 1998, Nicol et al. 2001, Vestergaard and Baranyiova 1996, Vestergaard and Lisborg 1993).

Both chickens and quail are social species and behaviors, such as foraging and dustbathing, are often synchronized or can be initiated through social facilitation and social learning (Nicol 1995, Olsson, Duncan, et al. 2002). This is often seen during feeding time, with most birds wanting to feed simultaneously, where facilities and social groupings allow. In small groups of ten, captive chickens will form dominance hierarchies (Rushen 1982), although this behavior may be suspended in larger groups (Hughes et al. 1997, Widowski and Duncan 1994).

Sexually mature chickens and quail also perform nesting behaviors, if the captive environment has the resources required. This involves searching for an appropriate nesting area in the environment – one study found hens walked 2.5 km/day while exploring the environment (Keppler and Folsch 2000) – and then building the nest. In captivity, domestic fowl in particular, will readily "nest" in nest boxes, if they are suitably secluded and lined with an appropriate substrate (Guinebretière et al. 2012, Struelens et al. 2005, 2008, Wall and Tauson 2013, Zupan et al. 2008). Quail, on the other hand, are notoriously reluctant to nest in captivity, but can be persuaded to do so with the right stimulation and environmental conditions (Michel 1989, Mills et al. 1997, Orcutt and Orcutt 1976). Unlike wild species, eggs are usually removed daily from captive chicken and quail enclosures, so the females do not brood the eggs. While some breeds would incubate their eggs if given the opportunity, broody behavior has been bred out of, or reduced in, other breeds through genetic selection. For example, White Leghorn chickens have been bred in such a way that genes related to brooding behavior were inhibited or selected against, and broodiness is now rarely shown in this breed (Romanov 2001).

Laying hens are highly motivated to perch and will push through weighted doors in order to access a perch to roost on overnight (Olsson and Keeling 2002). Elevated perches can also allow birds to avoid or escape from each other, as well as providing exercise for hens (Newberry et al. 2001). Proper perch shape is important, as keel bone deformations can occur if the perch does not have a flattened top surface (Duncan et al. 1992, Schmid and Wechsler 1997).

In this section, we have discussed a number of normal or "natural" behaviors that domestic fowl and quail remain motivated to perform in captivity. If conditions in captivity affect the performance of natural or internally motivated behaviors, such as foraging, dustbathing, nesting, and perching, this can lead to stress, frustration, and the development of abnormal behavior patterns (Mason et al. 2007). For example, birds may show "time-filling" or displacement behaviors. Alternatively, normal behaviors may become damaging, if performed at abnormal frequencies or redirected toward inappropriate objects (Cooper and Albentosa 2003). The performance of abnormal behaviors is often used as an indicator of poor welfare and a sign that the environment is not meeting the animal's needs (Hawkins 2010).

Abnormal Behaviors

Although chickens and quail are social animals, if social groups are not formed and managed properly, abnormally high levels of aggression and fighting can result, with birds becoming injured or killed (Duncan 1999). In the wild, feral domestic fowl have the space to maintain appropriate social distance from each other, which can be over 2 m (McBride et al. 1969). In captive environments, birds are unable to choose their social groups and have limited opportunities to escape from each other, both of which can lead to increased stress and injury (Channing et al. 2001). When birds are housed in mixed-sex groups in captive environments, male chickens can harass the females and decrease their feeding and use of space (Pizzari 2016). Further, in mixed-sex groups, particularly if males do not have the space or opportunity to perform appropriate courtship behaviors, repeated, forced, mating attempts can cause damage to the females and can result in head wounds, eye damage, or body damage (Cheng et al. 2010). Additionally, female quail are larger than males, and males may get injured during aggressive encounters with females. This can be exacerbated in cages where the males are unable to get away from the females (Gerkin and Mills 1993). Male quail should not be housed in single-sex groups, as they can be aggressive toward, and try to mount, one another. Also, as mentioned above, under natural conditions mother hens incubate, hatch, and brood their chicks, yet chicks are often reared without a broody hen in captivity. Chicks not brooded by a hen show lower levels of behavioral synchronization, may not learn foraging behaviors as efficiently, and are more likely to perform harmful abnormal behaviors, like feather pecking (discussed below) (Edgar et al. 2016, Riber, Nielsen, et al. 2007, Riber, Wichman, et al. 2007).

One of the most common abnormal behaviors found in domestic fowl and quail is feather pecking, although quail seem to express lower levels of this behavior than chickens (Gerkin and Mills 1993). Feather pecking is distinct from aggressive pecking, and involves the pecking at, and possible removal of, feathers of one bird by another. Feather removal is painful for the recipient bird and blood from the removed feather can stimulate cannibalism (Dixon 2008). There are many factors that can contribute to the development of feather pecking, but lack of appropriate foraging substrates and stress are considered key risk factors (University of Bristol 2013, Rodenburg et al. 2013). Diet is also considered to be a major risk factor for feather pecking (reviewed by Rodenburg et al. 2013).

Cannibalism can be stimulated by blood from removed feathers, as mentioned above, and it can also be directed to the vent of the bird after egg laying, when mucous membranes of the internal structures are exposed after egg expulsion (Savory 1995). The risk factors for cannibalism are similar to those of feather pecking, such as lack of foraging opportunities. Due to the correlation between feather pecking and cannibalism, it can be difficult to untangle their different causes (Lambton et al. 2015). Domestic fowl can also show stereotyped spot pecking – repetitive pecking at elements in the environment – if there is a deficiency in the nutritional elements of food or the quantity of food is insufficient to meet requirements (D'Eath et al. 2009).

Disruptions to the dustbathing behavioral system are related to the absence of adequate litter (substrate). Although sham or vacuum dustbathing is performed in the absence of adequate litter, there is some debate over whether performance of the behavior is adequate in itself, or if it is more satisfactory if performed on an appropriate substrate (Lindberg and Nicol 1997, Weeks and Nicol 2007). Hens that engaged in sham or vacuum dustbathing when deprived of a substrate did not subsequently perform less dustbathing when a substrate was provided (Olsson, Keeling, et al. 2002).

Both chickens and quail show prelaying restlessness or perform stereotypic pacing when housed in environments without appropriate nesting areas (Cooper and Appleby 1997). If appropriate nesting material is not present, nest searching may be prolonged and nest building absent (Wood-Gush and Gilbert 1969). Prelaying restlessness, pacing, and prolonged nest searching are thought to be indicators of stress and frustration. Since hens are willing to work for access to a nest area (Cooper and Appleby 1996, 2003), lack of an appropriate nesting area is considered one of the most important welfare issues for laying birds (Appleby and Hughes 1991).

Domestic fowl housed in barren conditions often show a higher fear response to novelty than birds in more enriched enclosures (Young 2003), and this can also affect learning. Birds from enriched conditions were less fearful and made fewer mistakes when learning a preference test task than birds from a less enriched environment. Animals repeatedly exposed to aversive environments or stimuli can become stressed, and increased activation of the hypothalamic-pituitary-adrenal axis can affect physiological measures, such as body weight and blood chemistry, which may adversely affect the reliability of experimental findings, and impact animal welfare (Laurence et al. 2015). Additionally, the vertical flight response of startled quail, which in the wild allows them to escape predators, causes them to bang their head on cage ceilings, potentially causing head and neck injuries (Gerkin and Mills 1993).

Ways to Maintain Behavioral Health

Environment

There are minimum housing requirements for laboratory animals, and in general, items that increase the complexity of the housing environment should be provided, as they can reduce fearfulness, increase activity, and give domestic fowl an opportunity to adjust their social interactions (Bizeray et al. 2002, Newberry 1995). However, depending on the goal of the experiment, animals may be kept in relatively barren or sterile conditions, with little to no enrichment provided, no social companions, and limited opportunities to perform motivated behaviors (Hawkins 2010). Early life exposure to natural or bright colored objects has been found to reduce fear of humans and novel objects in quail, and may also help to reduce aggression in adult birds (Jones et al. 1991). For example, balls, tubing, cylinders, and cubes reduced fear of humans and novel stimuli (Jones et al. 1991), and the provision of stones and pinecones reduced aggression (Ottinger and Rattner 1999). Quail chicks prefer colors in the short end and middle range of the color spectrum. For example, they prefer green and yellow over blue and red (Kovach 1978).

As chickens and quail are prey animals that mainly live on the ground, providing cover and opportunities to escape from perceived threats can help reduce stress. Even an environment that uses gradients of light intensity to mimic light gradients in the undergrowth can reduce stress in laying hens compared to environments that use uniform light intensity (Newberry and Shackleton 1997). Laying hens are motivated to perch, and in the wild, perching is used to escape predators, so the absence of an appropriate perch can cause stress and frustration (Newberry et al. 2001). Laying hens will roost on perches overnight, and perches provide a means for birds to escape aggressive conspecifics and provide opportunities for exercise (See Figure 19.3). However, the benefits of perch provision are not observed in quail (Schmid and Wechsler 1997).

FIGURE 19.3 Adult laying hens housed in a commercial scale research facility. (Photo: Michael Toscano, University of Bern, Switzerland.)

Ideally, quail should be housed in aviary-type environments with plenty of ground cover – this will decrease the chances of a vertical flight response, reduce the chance of injuries, and help reduce stress (Buchwalder and Wechsler 1997). If the quail must be kept in a cage environment, a maximum cage ceiling height of 20–30 cm has been recommended, as has the use of softer material for the cage ceiling (Gerkin and Mills 1993, Hawkins et al. 2001). Quail can achieve high momentum during the flight response, so cages with ceilings higher than 20–30 cm result in increased damage to the quail, when they do strike the ceiling, as their momentum increases with vertical distance.

Appropriate cover may also help reduce prelaying restlessness shown by quail in captive environments. Chickens and quail are motivated to nest in an appropriate area; chickens in particular, choosing to lay in secluded areas, with floors covered by peat or artificial turf (Guinebretière et al. 2012, Struelens et al. 2005, 2008, Wall and Tauson 2013, Zupan et al. 2008). Quail prefer nesting material, such as hay or chaff, to nest boxes containing artificial turf (Schmid and Wechsler 1998). Therefore, the provision of an appropriate nesting location will increase normal behavior and reduce stress.

To minimize damage from feather pecking, birds are often beak trimmed (removal of up to one-third of the sharp tip of the beak). However, beaks are highly innervated and beak trimming can cause acute and chronic pain (Gentle et al. 1990). There are also antipecking "bits" available, which fit through the nostrils and go between the jaws, preventing the birds from fully closing their beaks. These bits have a negative impact on the ability of the birds to perform natural behaviors and should be avoided (Gerkin and Mills 1993). Beak trimming and antipecking bits are "band-aid" solutions, as they do not address the reason(s) feather pecking occurs in the first place. They are simply a palliative treatment, making it more difficult for the birds to damage each other, rather than curing the problem. In order to effectively treat the problem, it is vital to encourage normal pecking behavior. The goal should always be to house birds in as stimulating an environment as possible, thereby promoting natural behavior patterns, taking into account the constraints of research projects.

Modifications to the environment to promote normal behaviors and reduce abnormal behaviors are often referred to as environmental enrichment. These can be grouped into categories of social stimulation, structural complexity, foraging, and novelty or sensory stimulation (Newberry 1995). There is good evidence that animals in enriched environments have larger behavioral repertoires, more complex brain structure, improved learning ability, and reduced abnormal behaviors than animals in unenriched environments (de Jong et al. 2000, Laurence et al. 2015, Praag et al. 2000). For example, hens from enriched environments had a shorter latency to start a discrimination task (i.e., perform in a test maze), explored the test maze more quickly, and made fewer mistakes in discrimination tasks (Krause et al. 2006). The benefits of enrichment are greater when introduced during the rearing phase, and it is recommended that enrichment be added as early as possible (Krause et al. 2006). Stress can also impact reproductive function, leading to lower egg production, fertility, and hatchability of the chicks (Marin et al. 2002, Leone and Estévez 2008).

FIGURE 19.4 Young domestic chickens housed in a research facility. (Photo: Lucy Asher, Newcastle University, UK.)

If cages are necessary to achieve the experimental goals, then the time the animals spend housed in these environments should be minimized.

Fowl prefer to be housed on solid flooring as opposed to wire flooring (FAWC 1997). Rearing on a wire floor is associated with the development of feather pecking (Johnsen et al. 1998). Birds housed on wire floors can also develop painful swelling in the feet (often called bumblefoot in chickens), and this limits their ability to move around and interact with other birds normally (Hawkins et al. 2001). Whenever possible, birds should be housed on solid flooring with litter (Figure 19.4). But if wire floor cages are needed for sample collections, a small solid-floored area should still be included to provide the birds with a break from standing on the wire (Hawkins et al. 2001), to facilitate foraging opportunities that encourage natural behavior, reduce pacing, and reduce the occurrence of feather pecking and cannibalism (Miller and Mench 2005). If possible, feed should be scattered on the floor, hidden around the environment, and put into foraging toys. Hay can be provided on the floor or in a hay rack, alfalfa cubes can be provided, or vegetation can be hung from the roof of the enclosure to encourage pecking (Hawkins et al. 2001). Straw bales can also be used to promote foraging, and as a way for subordinate birds to get away from dominant ones. Providing additional opportunities for pecking, such as hanging bundles of string (McAdie et al. 2005) can also help re-direct pecking onto a more suitable outlet than the feathers of other birds.

Dustbath provision can also decrease harmful pecking (Vestergaard and Lisborg 1993). Materials that have a fine particulate size, such as sand or peat, are preferred to materials with more coarse particles, like wood shavings (Sanotra et al. 1995). Dustbathing is socially facilitated, and the sight of one bird dustbathing can stimulate others to join in (Vestergaard 1981). This means dustbaths need to be big enough for all animals in the environment to use simultaneously (Olsson and Keeling 2005). If a dustbath cannot be provided at all times, then it could be offered for a period every day to allow the birds to fulfil their dustbathing motivation.

Social Groupings

As mentioned above, chickens and quail are social species and should be housed in appropriate social groups, with as much space as possible. Careful consideration should be given when housing groups of males and females together, to minimize aggression and excessive mating. If intending to breed birds, then normal courtship displays and mating behaviors should be taken into account to facilitate successful breeding, and to minimize distress to the hen. Housing chicks with a broody hen may safeguard the development of normal behaviors and reduce the risk of, for example, feather pecking. However, as this is often impossible, consideration should be given to changes that can be made to mimic maternal behavior, such as playing recordings of maternal vocalizations and providing dark brooders (Edgar et al. 2016). Decreased stocking levels reduce aggressive encounters and lower agitation and excitement of the birds (Onbasilar and Aksoy 2005). If this is not possible, birds should always be housed within sight of a companion, and the time the bird is socially isolated should be kept to a minimum.

Housing social animals with a companion can reduce fearfulness and abnormal behaviors, while increasing activity and social behavior (Jones and Merry 1988). However, as aggression can increase with unnatural social groupings, housing should be designed such that birds are able to escape from and avoid each other. Placing vertical barriers in pens can help birds get away from each other and this has been shown to reduce aggression and feather pecking, and to improve immune parameters (Laurence et al. 2015, Young 2003).

The enrichment approaches that we use to maintain the behavioral health of poultry in captivity are summarized, with their benefits, in Table 19.1. Some experiments may preclude the use of all types of enrichment; however as many as possible should be implemented to encourage normal behavior and reduce abnormal behavior patterns.

TABLE 19.1

Enrichments That May Be Employed for the Housing of Fowl and Their Potential Benefits in Terms of Maintaining Behavioral Health

Item	Benefit
Maternal care[a–d] (mother or foster mother rears chicks)	Encourages feeding behavior
	Decreases aggression
	Decreases fearfulness
	Decreases feather pecking and cannibalism
Panels or partitions[e,f]	Improves feeding behavior
	Allows birds to avoid each other
	Decreases male harassment of females
	Decreases aggression and feather pecking
Litter[g,h]	Fulfills foraging motivation
• Wood shavings or chopped straw	Improves mobility
	Decreases incident of foot swelling
	Decreases pacing
	Decreases feather pecking and cannibalism
	Decreases pre-lay restlessness
Feeding enrichments[h]	Increases feeding time
• Scatter feeding	Increases foraging behavior
• Hiding food in the environment	Decreases feather pecking and cannibalism
• Using food toys	
• Operant feeders (singly housed birds)	
Foraging enrichments[i,j]	Fulfills foraging motivation
• Baskets, hay racks or hanging items from ceiling	Can also be used as nesting material
• Hay, chopped straw, grass, fruit	Decreases feather pecking and cannibalism
	Decreases stress and frustration
	Decreases pre-lay restlessness
Straw bale[h]	Fulfills foraging motivation
	Allows birds to avoid each other
Bundles of string[k]	Redirects pecking onto string
	Decreases feather pecking and cannibalism
Dustbath[l]	Fulfills dustbathing motivation
• Fine particulate material, such as sand or peat	Increases foraging behavior
	Improves feather condition
	Decreases sham dustbathing
Appropriate social housing[m]	Fulfills social motivation
	Increases activity
	Decreases fearfulness
	Decreases performance of abnormal behavior

(Continued)

TABLE 19.1 (*Continued*)
Enrichments That May Be Employed for the Housing of Fowl and Their Potential Benefits in Terms of Maintaining Behavioral Health

Item	Benefit
Reduced stocking density[n]	Improves social behaviors
	Increased space to exercise and stretch wings
	Improves bone strength
	Decreases aggression
Perches[o]	Fulfills perching motivation
(chickens only)	Provides exercise
	Allows birds to get away from each other
Novel objects[p]	Decreases fearfulness
• Balls, tubing, cylinders, stones, pine cones, etc.	Decreases aggression
Handling habituation[q]	Decreases fear and stress
Nest box[h,r]	Fulfills nesting motivation
	Decreases pre-lay restlessness
	Decreases pacing
	Decreases stress

[a] Falt (1978).
[b] Roden and Wechsler (1998).
[c] Campo (2014).
[d] Riber, Wichman, et al. 2007 (2007).
[e] Leone and Estévez (2008).
[f] Laurence et al. (2015).
[g] Miller and Mench (2005).
[h] Hawkins et al. (2001).
[i] Dixon (2008).
[j] Huber-Eicher and Wechsler (1998).
[k] McAdie et al (2005).
[l] Olsson and Keeling (2005).
[m] Jones and Merry (1988).
[n] Onbasilar and Aksoy (2005).
[o] Newberry et al. (2001).
[p] Ottinger and Rattner (1999).
[q] Jones (1993).
[r] Buchwalder and Wechsler (1997).

Conclusions

Domestic fowl are studied under laboratory conditions for a variety of purposes, including advancing human medical science and improvements in commercial fowl housing and management. They are descended from wild fowl that live in mixed-sex groups in forests (chickens), jungles (chickens) and grasslands (chickens and quail), where the undergrowth and tree cover provide protection from predators and allow for a range of behavior patterns, including foraging, nesting, dust-bathing, perching, and mating. Domestic fowl show similar behavioral patterns to wild fowl when housed in appropriate environments that allow: for secluded nesting, the ability to separate from and develop proper social structures with flock mates, foraging and dustbathing opportunities, and perching (for chickens). Many of these behaviors are highly motivated and the inability to perform them will negatively impact domestic fowl welfare. This can result in increased frustration and stress, and the development of abnormal behavior patterns, such as feather pecking, restlessness, and pacing; high levels of aggression and injury; increased fear responses; and a decrease in cognitive abilities. Providing a complex, enriched environment, with cover, foraging and dustbathing opportunities, objects to peck, suitable social groups, and sufficient space with areas that allow a bird to avoid its flock mates, will improve the behavioral health of captive fowl. The following list of recommendations comprises a wish list for experimenters to strive toward whenever possible, within the constraints of their experiment.

Recommendations

- Provide overhead cover or a secluded nesting area with nesting material
- Scatter feed, hide food in the environment, hang vegetation
- Provide forage materials, like hay or straw bales
- Provide other items that encourage pecking and interaction, like bundles of string, balls, stones, or pinecones
- Provide litter and dustbaths
- House birds in appropriate social groups with plenty of space and vertical barriers
- For fowl, provide perches
- For quail, house birds in aviaries wherever possible, otherwise ensure a maximum ceiling height of 20–30 cm

REFERENCES

Akins, C., and T. Zentall. 1996. True imitative learning in male Japanese quail (*Coturnix japonica*). *Journal of Comparative Psychology* 110:316–320.

Andersson, M., E. Nordin, and P. Jensen. 2001. Domestication effects on foraging strategies in fowl. *Applied Animal Behaviour Science* 72 (1):51–62. https://doi.org/10.1016/S0168-1591(00)00195-7.

Anwar, M., S. Ali, M. Rais, and T. Mahmood. 2016. Breeding ecology of red jungle fowl (*Gallus gallus*) in Deva Vatala National Park, Azad Jammu and Kashmir, Pakistan. *Journal of Applied Agriculture and Biotechnology* 1 (1):59–65.

Appleby, M. C., and B. O. Hughes. 1991. Welfare of laying hens in cages and alternative systems: Environmental, physical and behavioural aspects. *World's Poultry Science Journal* 47 (2):109–128. https://doi.org/10.1079/WPS19910013.

Arshad, M., M. Zakaria, A. S. Sajap, and A. Ismail. 2000. Food and feeding habits of red junglefowl. *Pakistan Journal of Biological Sciences* 3:1024–1026. https://doi.org/10.3923/pjbs.2000.1024.1026.

Arshad, M. I., and M. Zakaria. 2009. Roosting habits of red junglefowl in orchard area. *Pakistan Journal of Life and Social Sciences* 7 (1):86–89.

Arshad, Z., and M. Zakaria. 1999. Breeding ecology of red junglefowl (*Gallus gallus spadiceus*) in Malaysia. *Malayan Nature Journal* 53:355–365.

Baher, J. M. 2008. The chicken as a model organism. In *Sourcebook of Models for Biomedical Research*, edited by P. M. Conn, 161–167. Totowa, NJ: Humana Press.

Ball, G. F., and J. Balthazart. 2010. Japanese quail as a model system for studying the neuroendocrine control of reproductive and social behaviors. *ILAR Journal* 51 (4):310–325. https://doi.org/10.1093/ilar.51.4.310.

Barrett, J. E., D. C. Well, A. Q. Paulsen, and G. W. Conrad. 2000. Embryonic quail eye development in microgravity. *Journal of Applied Physiology* 88:1614–1622.

Bizeray, D., I. Estevez, C. Leterrier, and J. M. Faure. 2002. Effects of increasing environmental complexity on the physical activity of broiler chickens. *Applied Animal Behaviour Science* 79 (1):27–41. https://doi.org/10.1016/S0168-1591(02)00083-7.

Braastad, B. O. 1990. Effects on behaviour and plumage of a key-stimuli floor and a perch in triple cages for laying hens. *Applied Animal Behaviour Science* 27 (1–2):127–139.

Bradshaw, R. 1992. Conspecific discrimination and social preference in the laying hen. *Applied Animal Behaviour Science* 33 (1):69–75.

Buchwalder, T., and B. Wechsler. 1997. The effect of cover on the behaviour of Japanese quail (*Coturnix japonica*). *Applied Animal Behaviour Science* 54 (4):335–343. https://doi.org/10.1016/S0168-1591(97)00031-2.

Callaway, E. 2016. When chickens go wild. *Nature* 529:270–273.

Campo, J.L., Garcia Davila, S., and M. Garcia Gil. 2014. Comparison of the tonic immobility duration, heterophil to lymphocyte ratio, and fluctuating asymmetry of chicks reared with or without a broody hen, and of broody and non-broody hens. *Applied Animal Behaviour Science* 151:61–66.

Chang, G. B., X. P. Liu, H. Chang, et al. 2009. Behavior differentiation between wild Japanese quail, domestic quail, and their first filial generation. *Poultry Science* 88 (6):1137–1142. https://doi.org/10.3382/ps.2008-00320.

Channing, C. E., B. O. Hughes, and A. W. Walker. 2001. Spatial distribution and behaviour of laying hens housed in an alternative system. *Applied Animal Behaviour Science* 72 (4):335–345.

Chen, Z., W. Wang, H. Zhou, et al. 2010. Generation of live attenuated novel influenza virus A/California/7/09 (H1N1) vaccines with high yield in embryonated chicken eggs. *Journal of Virology* 84 (1):44–51. https://doi.org/10.1128/jvi.02106-09.

Cheng, K. M., D. C. Bennett, and A. D. Mills. 2010. The Japanese quail. In *The UFAW Handbook on the Care and Management of Laboratory and Other Research Animals*, edited by R. Hubrecht and J. Kirkwood, 655–673. West Sussex: Wiley-Blackwell.

Cloutier, S., R. C. Newberry, K. Honda, and J. R. Alldredge. 2002. Cannibalistic behaviour spread by social learning. *Animal Behaviour* 63 (6):1153–1162. https://doi.org/10.1006/anbe.2002.3017.

Colgan, P. 1989. *Animal Motivation*. London: Chapman and Hall.

Collias, N. 1987. The vocal repertoire of the red junglefowl: A spectrographic classification and the code of communication. *The Condor* 89:510–524.

Collias, N. E., and E. C. Collias. 1967. A field study of the red jungle fowl in north-central India. *The Condor* 69 (4):360–386. https://doi.org/10.2307/1366199.

Cooper, J. J., and M. J. Albentosa. 2003. Behavioural priorities of laying hens. *Avian and Poultry Biology Reviews* 14 (3):127–149.

Cooper, J. J., and M. C. Appleby. 1996. Demand for nest boxes in laying hens. *Behavioural Processes* 36 (2):171–182. https://doi.org/10.1016/0376-6357(95)00027-5.

Cooper, J. J., and M. C. Appleby. 1997. Motivational aspects of individual variation in response to nestboxes by laying hens. *Animal behaviour* 54 (5):1245–1253.

Cooper, J. J., and M. C. Appleby. 2003. The value of environmental resources to domestic hens: A comparison of the work-rate for food and for nests as function of time. *Animal Welfare* 12:39–52.

D'Eath, R. B., B. J. Tolkamp, I. Kyriazakis, and A. B. Lawrence. 2009. 'Freedom from hunger' and preventing obesity: The animal welfare implications of reducing food quantity or quality. *Animal Behaviour* 77 (2):275–288. https://doi.org/10.1016/j.anbehav.2008.10.028.

Davis, K. 2003. The experimental use of chickens and other birds in biomedical and agricultural research. Accessed 3rd November 2020. http://www.upc-online.org/experimentation/experimentalConclusion.htm.

Dawkins, M. S. 1989. Time budgets in red junglefowl as a baseline for the assessment of welfare in domestic fowl. *Applied Animal Behaviour Science* 24 (1):77–80. https://doi.org/10.1016/0168-1591(89)90126-3.

de Jong, I. C., T. Prelle, J. A. van de Burgwal, E. Lambooij, H. J. Blokhuis, and J. M. Koolhaas. 2000. Effects of environmental enrichment on behavioral responses to novelty, learning and memory, and the circadian rhythm in cortisol in growing pigs. *Physiology and Behavior* 68 (4):571–578. https://doi.org/10.1016/S0031-9384(99)00212-7.

Desta, T. T. 2019. Phenotypic characteristic of junglefowl and chicken. *World's Poultry Science Journal* 75 (1):69–82. https://doi.org/10.1017/S0043933918000752.

Dixon, L. M. 2008. Feather pecking behaviour and associated welfare issues in laying hens. *Avian Biology Research* 1 (2). https://doi.org/10.3184/175815508X363251.

Duncan, I. J. H. 1999. The domestic fowl. In *The UFAW Handbook on the Care and Management of Laboratory Animals*, edited by T. Poole and P. English, 677–696. Oxford: Blackwell Science.

Duncan, I. J. H., M. C. Appleby, and B. O. Hughes. 1992. Effect of perches in laying cages on welfare and production of hens. *British Poultry Science* 33:25–35.

Duncan, I. J. H., and B. O. Hughes. 1972. Free and operant feeding in domestic fowls. *Animal Behaviour* 20 (4):775–777. https://doi.org/10.1016/S0003-3472(72)80150-7.

Duncan, I. J. H., C. J. Savory, and D. G. M. Wood-Gush. 1978. Observations on the reproductive behaviour of domestic fowl in the wild. *Applied Animal Ethology* 4 (1):29–42. https://doi.org/10.1016/0304-3762(78)90091-3.

Edgar, J., S. Held, C. Jones, and C. Troisi. 2016. Influences of maternal care on chicken welfare. *Animals* 6 (1):2. https://doi.org/10.3390/ani6010002.

Eklund, B., and P. Jensen. 2011. Domestication effects on behavioural synchronization and individual distances in chickens (*Gallus gallus*). *Behavioural Processes* 86 (2):250–256. https://doi.org/10.1016/j.beproc.2010.12.010.

Elangovan, A. V., S. V. S. Verma, V. R. B. Sastry, and E. Al. 2003. Laying performance of Japanese quail fed on different seed meals in diet. *Indian Journal of Animal Nutrition* 19:244–250.

Farris, H. E. 1967. Classical conditioning of courting behavior in the Japanese quail, *Coturnix coturnix japonica*. *Journal of the Experimental Analysis of Behavior* 10 (2):213–217. https://doi.org/10.1901/jeab.1967.10-213.

FAWC. 1993. Second report on priorities for research and development in farm animal welfare. Tolworth, UK: FAWC.

FAWC. 1997. Report on the welfare of laying hens. Tolworth, UK: FAWC.

Falt B 1978. Differences in aggressiveness between brooded and non-brooded domestic chicks. *Applied Animal Ethology* 4:211–221.

Garnham, L., and H. Lovlie. 2018. Sophisticated fowl: The complex behaviour and cognitive skills of chickens and red junglefowl *Behavioral Sciences* 8 (1):13. https://doi.org/10.3390/bs8010013.

Gentle, M. J., D. Waddington, L. N. Hunter, and R. B. Jones. 1990. Behavioural evidence for persistent pain following partial beak amputation in chickens. *Applied Animal Behaviour Science* 27 (1–2):149–157. https://doi.org/10.1016/0168-1591(90)90014-5.

Gerkin, M., and A. D. Mills. 1993. Welfare of domestic quail. *Fourth European Symposium on Poultry Welfare*, Potters Bar, UK.

Guhl, A. M. 1958. The development of social organisation in the domestic chick. *Animal Behaviour* 6 (1):92–111. https://doi.org/10.1016/0003-3472(58)90016-2.

Guhl, A. M. 1968. Social behavior of the domestic fowl. *Transactions of the Kansas Academy of Science (1903-)* 71 (3):379–384. https://doi.org/10.2307/3627156.

Guinebretière, M., A. Huneau-Salaün, D. Huonnic, and V. Michel. 2012. Cage hygiene, laying location, and egg quality: The effects of linings and litter provision in furnished cages for laying hens. *Poultry Science* 91 (4):808–816.

Hawkins, P. 2010. The welfare implications of housing captive wild and domesticated birds. In *The Welfare of Domestic Fowl and Other Captive Birds*, edited by I. J. H. Duncan and P. Hawkins, 53–102. London: Springer.

Hawkins, P., D. B. Morton, D. Cameron, et al. 2001. Laboratory birds: Refinements in husbandry and procedures. *Laboratory Animals* 35 (S1):1–163.

Hess, E. H. 1964. Imprinting in birds. *Science* 146 (3648):1128–1139.

Hogan, J. A. 1973. Development of food recognition in young chicks: I. Maturation and nutrition. *Journal of Comparative and Physiological Psychology* 83 (3):355.

Home Office. 2019. Annual statistics of scientific procedures on living animals: Great Britain 2019. https://www.gov.uk/government/statistics/statistics-of-scientific-procedures-on-living-animals-great-britain-2019.

Huber-Eicher, B., and B. Wechsler. 1998. The effect and availability of foraging materials on feather pecking in laying hen chicks. *Animal Behaviour* 55:861–873.

Hughes, B. O, and I. Duncan. 1988. The notion of ethological 'need', models of motivation and animal welfare. *Animal Behaviour* 36 (6):1696–1707.

Hughes, B. O., N. L. Carmichael, A. W. Walker, and P. N. Grigor. 1997. Low incidence of aggression in large flocks of laying hens. *Applied Animal Behaviour Science* 54 (2):215–234. https://doi.org/10.1016/S0168-1591(96)01177-X.

Huss, D., G. Poynter, and R. Lansford. 2008. Japanese quail (Coturnix japonica) as a laboratory animal model. *Lab Animal* 37 (11):513–519. https://doi.org/10.1038/laban1108-513.

Javed, S., and A. Rahmani. 2000. Flocking and habitat use pattern of the red junglefowl *Gallus gallus* in Dudwa National Park, India. *Tropical Ecology* 41:11–16.

Jensen, A., B. Forkman, and C. Ritz. 2006. The effect of broody hens on chicks' activity cycles and synchrony. *Proceedings of the 40th International Congress of the ISAE*, Bristol, UK.

Johnsen, P. F., K. S. Vestergaard, and G. Nørgaard-Nielsen. 1998. Influence of early rearing conditions on the development of feather pecking and cannibalism in domestic fowl. *Applied Animal Behaviour Science* 60 (1):25–41. https://doi.org/10.1016/S0168-1591(98)00149-X.

Jones, R.B. 1993. Reduction of the domestic chick's fear of human beings by regular handling and related treatments. *Animal Behaviour* 46:991–998.

Jones, R. B. 2007. Fear and adaptability in poultry: Insights, implications and imperatives. *World's Poultry Science Journal* 52 (2):131–174. https://doi.org/10.1079/WPS19960013.

Jones, R. B., A. D. Mills, and J. M. Faure. 1991. Genetic and experimental manipulation of fear-related behaviour in Japanese quail chicks (*Coturnix coturnix japonica*). *Journal of Comparative Psychology* 105 (1):15–24. https://doi.org/10.1037/0735-7036.105.1.15.

Jones, R. B., and B. J. Merry. 1988. Individual or paired exposure of domestic chicks to an open field: Some behavioural and adrenocortical consequences. *Behavioural Processes* 16 (1–2):75–86.

Keeling, L. 2002. Behaviour of fowl and other domesticated birds. In *The Ethology of Domestic Animals. An Introductory Text*, edited by P. Jensen, 101–117, Wallingford, UK: CABI.

Keppler, C., and D. W. Folsch. 2000. Locomotive behaviour of hens and cocks (*Gallus gallus f. Domesticus*): Implications for housing systems. *Archiv für Tierschutz* 43:184–188.

Kovach, J. K. 1974. The behaviour of Japanese quail: Review of literature from a bioethological perspective. *Applied Animal Ethology* 1 (1):77–102.

Kovach, J. K. 1978. Sources of individual variation in the naive color preferences of quail chicks: Age, stimulus information, and genotypes. *Behaviour* 64 (3–4):173–183.

Krause, T. E., M. Naguib, F. Trillmich, and L. Schrader. 2006. The effects of short term enrichment on learning in chickens from a laying strain (Gallus gallus domesticus). *Applied Animal Behaviour Science* 101 (3–4):318–327. https://doi.org/10.1016/j.applanim.2006.02.005.

Kruijt, J. P. 1964. Ontogeny of social behaviour in Burmese red junglefowl (*Gallus gallus spadiceus*) bonnaterre. *Behaviour Supplement* I: 1–201.

Kruijt, J. P. 1966. The development of ritualized displays in junglefowl. *Philosophical Transactions of the Royal Society of London. Series B, Biological Sciences* 251 (772):479–484.

Lambton, S. L., T. G. Knowles, C. Yorke, and C. J. Nicol. 2015. The risk factors affecting the development of vent pecking and cannibalism in free-range and organic laying hens. *Animal Welfare* 24:101–111.

Laurence, A., C. Houdelier, L. Calandreau, et al. 2015. Environmental enrichment reduces behavioural alterations induced by chronic stress in Japanese quail. *Animal* 9 (2):331–338. https://doi.org/10.1017/S1751731114002523.

Lawrence, M. P. 1975. An experimental analysis of the use of location calls by Japanese quail, *Coturnix coturnix japonica*. *Behaviour* 54 (3–4):153–179. https://doi.org/10.1163/156853975X00245.

Leone, E. H., and I. Estévez. 2008. Economic and welfare benefits of environmental enrichment for broiler breeders. *Poultry Science* 87 (1):14–21. https://doi.org/10.3382/ps.2007-00154.

Lindberg, A. C., and C. J. Nicol. 1997. Dustbathing in modified battery cages: Is sham dustbathing an adequate substitute? *Applied Animal Behaviour Science* 55 (1):113–128. https://doi.org/10.1016/S0168-1591(97)00030-0.

Løvlie, H., and T. Pizzari. 2007. Sex in the morning or in the evening? Females adjust daily mating patterns to the intensity of sexual harassment. *The American Naturalist* 170 (1):E1–E13. https://doi.org/10.1086/518180.

Marin, R. H., D. C. Satterlee, G. G. Cadd, and R. B. Jones. 2002. T-maze behavior and early egg production in Japanese quail selected for contrasting adrenocortical responsiveness. *Poultry Science* 81 (7):981–986.

Marino, L. 2017. Thinking chickens: A review of cognition, emotion, and behavior in the domestic chicken. *Animal Cognition* 20 (2):127–147. https://doi.org/10.1007/s10071-016-1064-4.

Mason, G., R. Clubb, N. Latham, and S. Vickery. 2007. Why and how should we use environmental enrichment to tackle stereotypic behaviour? *Applied Animal Ethology* 102 (3–4):163–188. https://doi.org/10.1016/j.applanim.2006.05.041.

McAdie, T. M., and L. J. Keeling. 2002. The social transmission of feather pecking in laying hens: Effects of environment and age. *Applied Animal Behaviour Science* 75 (2):147–159. https://doi.org/10.1016/S0168-1591(01)00182-4.

McAdie, T. M., L. J. Keeling, H. J. Blokhuis, and R. B. Jones. 2005. Reduction in feather pecking and improvement of feather condition with the presentation of a string device to chickens. *Applied Animal Behaviour Science* 93 (1–2):67–80. https://doi.org/10.1016/j.applanim.2004.09.004.

McBride, G., I. P. Parer, and F. Foenander. 1969. The social organization and behaviour of the feral domestic fowl. *Animal Behaviour Monographs* 2:125–181 https://doi.org/10.1016/S0066-1856(69)80003-8.

McDonald, G. C., L. G. Spurgin, E. A. Fairfield, D. S. Richardson, and T. Pizzari. 2017. Pre- and postcopulatory sexual selection favor aggressive, young males in polyandrous groups of red junglefowl. *Evolution* 71 (6):1653–1669. https://doi.org/10.1111/evo.13242.

Meijer, T., and I. Siemers. 1993. Incubation development and asynchronous hatching in junglefowl. *Behaviour* 127 (3–4):309–322.

Michel, R. 1989. Influence of experience and visual stimulation in Japanese quail laying site selection under experimental condition. *Behavioural Processes* 18 (1):155–171. https://doi.org/10.1016/S0376-6357(89)80013-0.

Miller, K. A., and J. A. Mench. 2005. The differential effects of four types of environmental enrichment on the activity budgets, fearfulness, and social proximity preference of Japanese quail. *Applied Animal Behaviour Science* 95 (3–4):169–187. https://doi.org/10.1016/j.applanim.2005.04.012.

Mills, A. D., L. L. Crawford, M. Domjan, and J. M. Faure. 1997. The behavior of the Japanese or domestic quail Coturnix japonica. *Neuroscience & Biobehavioral Reviews* 21 (3):261–81. https://doi.org/10.1016/s0149-7634(96)00028-0.

Moffatt, C. A., and J. A. Hogan. 1992. Ontogeny of chick responses to maternal food calls in the Burmese red junglefowl (*Gallus gallus spadiceus*). *Journal of Comparative Psychology* 106 (1):92–96. https://doi.org/10.1037/0735-7036.106.1.92.

Murphy, K. J., T. J. Hayden, and J. P. Kent. 2014. Chicks change their pecking behaviour towards stationary and mobile food sources over the first 12 weeks of life: Improvement and discontinuities. *PeerJ* 2:e626–e626. https://doi.org/10.7717/peerj.626.

Nakamori, T., F. Maekawa, K. Sato, K. Tanaka, and H. Ohki-Hamazaki. 2013. Neural basis of imprinting behavior in chicks. *Development, Growth & Differentiation* 55 (1):198–206. https://doi.org/10.1111/dgd.12028.

Newberry, R. C. 1995. Environmental enrichment: Increasing the biological relevance of captive environments. *Applied Animal Behaviour Science* 44 (2–4):229–243. https://doi.org/10.1016/0168-1591(95)00616-Z.

Newberry, R. C., and D. M. Shackleton. 1997. Use of visual cover by domestic fowl: A Venetian blind effect? *Animal Behaviour* 54 (2):387–395. https://doi.org/10.1006/anbe.1996.0421.

Newberry, R. C., I. Estevez, and L. J. Keeling. 2001. Group size and perching behaviour in young domestic fowl. *Applied Animal Behaviour Science* 73 (2):117–129. https://doi.org/10.1016/S0168-1591(01)00135-6.

Nicol, C. J. 1995. The social transmission of information and behaviour. *Applied Animal Behaviour Science* 44 (2–4):79–98.

Nicol, C. J. 2004. Development, direction, and damage limitation: Social learning in domestic fowl. *Animal Learning & Behavior* 32 (1):72–81. https://doi.org/10.3758/BF03196008.

Nicol, C. J. 2015. *The Behavioural Biology of Chickens*. 2nd ed. Oxford, UK: CABI.

Nicol, C. J., A. C. Lindberg, A. J. Phillips, S. J. Pope, L. J. Wilkins, and L. E. Green. 2001. Influence of prior exposure to wood shavings on feather pecking, dustbathing and foraging in adult laying hens. *Applied Animal Behaviour Science* 73 (2):141–155. https://doi.org/10.1016/S0168-1591(01)00126-5.

Nicol, C. J., N. G. Gregory, T. G. Knowles, I. D. Parkman, and L. J. Wilkins. 1999. Differential effects of increased stocking density, mediated by increased flock size, on feather pecking and aggression in laying hens. *Applied Animal Behaviour Science* 65 (2):137–152. https://doi.org/10.1016/S0168-1591(99)00057-X.

Nicol, C. J., and S. J. Pope. 1992. Effects of social learning on the acquisition of discriminatory keypecking in hens. *Bulletin of the Psychonomic Society* 30 (4):293–296.

Nicol, C. J., and S. J. Pope. 1994. Social learning in small flocks of laying hens. *Animal Behaviour* 47 (6):1289–1296.

Nicol, C. J., and S. J. Pope. 1996. The maternal feeding display of domestic hens is sensitive to perceived chick error. *Animal Behaviour* 52 (4):767–774. https://doi.org/10.1006/anbe.1996.0221.

Odeh, F. M., G. G. Cadd, and D. G. Satterlee. 2003. Genetic characterisation of stress responsiveness in Japanese quail 1. Analyses of line effects and combining abilities by diallel crosses. *Poultry Science* 82:25–30.

Olsson, I. A. S., I. J. H. Duncan, L. J. Keeling, and T. M. Widowski. 2002. How important is social facilitation for dustbathing in laying hens? *Applied Animal Behaviour Science* 79 (4):285–297. https://doi.org/10.1016/S0168-1591(02)00117-X.

Olsson, I. A. S., and L. J. Keeling. 2002. The push-door for measuring motivation in hens: Laying hens are motivated to perch at night. *Animal welfare* 11 (1):11–19.

Olsson, I. A. S., and L. J. Keeling. 2005. Why in earth? Dustbathing behaviour in jungle and domestic fowl reviewed from a tinbergian and animal welfare perspective. *Applied Animal Behaviour Science* 93 (3–4):259–282. https://doi.org/10.1016/j.applanim.2004.11.018.

Olsson, I. A. S., L. J. Keeling, and I. J. Duncan. 2002. Why do hens sham dustbathe when they have litter? *Applied Animal Behaviour Science* 76 (1):53–64.

Onbasilar, E. E., and F. T. Aksoy. 2005. Stress parameters and immune response of layers under different cage floor and density conditions. *Livestock Production Science* 95 (3):255–263. https://doi.org/10.1016/j.livprodsci.2005.01.006.

Orcutt, F. S., and A. B. Orcutt. 1976. Nesting and parental behavior in domestic common quail. *The Auk* 93 (1):135–141.

Ottinger, M. A., and B. A. Rattner. 1999. Husbandry and care of quail. *Poultry and Avian Biology Reviews* 10 (2):117–120.

Pizzari, T. 2001. Indirect partner choice through manipulation of male behaviour by female fowl, Gallus *gallus domesticus*. *Proceedings. Biological sciences* 268 (1463):181–186. https://doi.org/10.1098/rspb.2000.1348.

Pizzari, T. 2003. Food, vigilance, and sperm: The role of male direct benefits in the evolution of female preference in a polygamous bird. *Behavioral Ecology* 14 (5):593–601. https://doi.org/10.1093/beheco/arg048.

Pizzari, T. 2016. The wood-gush legacy: A sociobiology perspective to fertility and welfare in chickens. *Applied Animal Behaviour Science* 181:12–18.

Pizzari, T., and T. R. Birkhead. 2000. Female feral fowl eject sperm of subdominant males. *Nature* 405 (6788):787–789. https://doi.org/10.1038/35015558.

Praag, H., G. Kempermann, and F. H. Gage. 2000. Neural consequences of environmental enrichment. *Nature Reviews Neuroscience* 1:191–198.

Riber, A. B., A. Wichman, B. O. Braastad, and B. Forkman. 2007. Effects of broody hens on perch use, ground pecking, feather pecking and cannibalism in domestic fowl (*Gallus gallus domesticus*). *Applied Animal Behaviour Science* 106 (1–3):39–51.

Riber, A. B., B. L. Nielsen, C. Ritz, and B. Forkman. 2007. Diurnal activity cycles and synchrony in layer hen chicks (*Gallus gallus domesticus*). *Applied Animal Behaviour Science* 108 (3–4):276–287. https://doi.org/10.1016/j.applanim.2007.01.001.

Roden C, and B. Wechsler. 1998. A comparison of the behaviour of domestic chicks reared with or without a hen in enrichment pens. *Applied Animal Behaviour Science* 55:317–326.

Rodenburg, T. B., M. M. Van Krimpen, I. C. De Jong, et al. 2013. The prevention and control of feather pecking in laying hens: Identifying the underlying principles. *World's Poultry Science Journal* 69 (2):361–374. https://doi.org/10.1017/S0043933913000354.

Romanov, M. N. 2001. Genetics of broodiness in poultry - A review. *Asian-Australasian Journal of Animal Sciences* 14 (11):1647–1654. https://doi.org/10.5713/ajas.2001.1647.

Rushen, J. 1982. The peck orders of chickens: How do they develop and why are they linear? *Animal Behaviour* 30 (4):1129–1137. https://doi.org/10.1016/S0003-3472(82)80203-0.

Sanavio, E., and U. Savardi. 1980. Observational learning in Japanese quail. *Behavioural Processes* 5 (4):355–361.

Sanotra, G. S., K. S. Vestergaard, J. F. Agger, and L. G. Lawson. 1995. The relative preferences for feathers, straw, wood-shavings and sand for dustbathing, pecking and scratching in domestic chicks. *Applied Animal Behaviour Science* 43 (4):263–277. https://doi.org/10.1016/0168-1591(95)00562-7.

Savory, C. J. 1995. Feather pecking and cannibalism. *World's Poultry Science Journal* 51 (2):215–219. https://doi.org/10.1079/WPS19950016.

Sawai, H., H. Kim, K. Kuno, et al. 2010. The origin and genetic variation of domestic chickens with special reference to junglefowls Gallus g. gallus and G. varius. *PLoS One* 5 (5). https://doi.org/10.1371/journal.pone.0010639.

Schmid, I., and B. Wechsler. 1997. Behaviour of Japanese quail (*Coturnix japonica*) kept in semi-natural aviaries. *Applied Animal Behaviour Science* 55 (1):103–112. https://doi.org/10.1016/S0168-1591(97)00039-7.

Schmid, I., and B. Wechsler. 1998. Identification of key nest site stimuli for Japanese quail (*Coturnix japonica*). *Applied Animal Behaviour Science* 57 (1–2):145–156. https://doi.org/10.1016/S0168-1591(97)00107-X.

Sherry, D. F. 1977. Parental food-calling and the role of the young in the Burmese red junglefowl (*Gallus gallus spadiceus*). *Animal Behaviour* 25:594–601 https://doi.org/10.1016/0003-3472(77)90109-9.

Sherry, D. F. 1981. Parental care and the development of thermoregulation in red junglefowl. *Behaviour* 76 (3–4):250–279.

Shimmura, T., E. Kamimura, T. Azuma, N. Kansaku, K. Uetake, and T. Tanaka. 2010. Effect of broody hens on behaviour of chicks. *Applied Animal Behaviour Science* 126 (3):125–133. https://doi.org/10.1016/j.applanim.2010.06.011.

Sluckin, W., and E. A. Salzen. 1961. Imprinting and perceptual learning. *Quarterly Journal of Experimental Psychology* 13 (2):65–77. https://doi.org/10.1080/17470216108416476.

Smith, C. L., A. Taylor, and C. S. Evans. 2011. Tactical multimodal signalling in birds: Facultative variation in signal modality reveals sensitivity to social costs. *Animal Behaviour* 82 (3):521–527. https://doi.org/10.1016/j.anbehav.2011.06.002.

Smith, C. L., J. Taubert, K. Weldon, and C. S. Evans. 2016. Individual recognition based on communication behaviour of male fowl. *Behavioural Processes* 125:101–105. https://doi.org/10.1016/j.beproc.2016.02.012.

Smith, S. M. 2008. The avian embryo in fetal alcohol research. In *Alcohol: Methods and Protocols*, edited by L. E. Nagy, 75–84. Totowa, NJ: Humana Press.

Statkiewicz, W. R., and M. W. Schein. 1980. Variability and periodicity of dustbathing behaviour in Japanese quail (*Coturnix coturnix japonica*). *Animal Behaviour* 28 (2):462–467. https://doi.org/10.1016/S0003-3472(80)80053-4.

Stevens, V. C. 1961. Experimental study of nesting by coturnix quail. *The Journal of Wildlife Management* 25 (1):99–101. https://doi.org/10.2307/3797004.

Stokes, A. W. 1971. Parental and courtship feeding in red jungle fowl. *The Auk* 88 (1):21–29. https://doi.org/10.2307/4083958.

Struelens, E., A. Van Nuffel, F. A. Tuyttens, et al. 2008. Influence of nest seclusion and nesting material on pre-laying behaviour of laying hens. *Applied Animal Behaviour Science* 112 (1–2):106–119.

Struelens, E., F. Tuyttens, A. Janssen, et al. 2005. Design of laying nests in furnished cages: Influence of nesting material, nest box position and seclusion. *British Poultry Science* 46 (1):9–15.

Subhani, A., M. S. Awan, M. Anwar, U. Ali, and N. I. Dar. 2010. Population status and distribution pattern of red jungle fowl (*Gallus gallus murghi*) in deva Vatala National Park, Azad Jammu & Kashmir, Pakistan: A pioneer study. *Pakistan Journal of Zoology* 42:701–706.

Taka-Tsukasa, N. 1967. *The Birds of Nippon*. Tokyo: Maruzen.

Understanding Animal Research. 2015. Chicken. Accessed 3rd November 2020. http://www.animalresearch.info/en/designing-research/research-animals/chicken/.

University of Bristol. 2013. Featherwel: Promoting bird welfare. Accessed 3rd November 2020. www.featherwel.org.

Vergara, M. N., and M. V. Canto-Soler. 2012. Rediscovering the chick embryo as a model to study retinal development. *Neural Development* 7 (1):1–19. https://doi.org/10.1186/1749-8104-7-22.

Vestergaard, K. S. 1981. The well-being of the caged hen — an analysis based on the normal behaviour of fowls. In *The Behaviour of Fowl*, edited by D. W. Fölsch and K. S. Vestergaard 145–165, New York, USA: Springer.

Vestergaard, K. S. 1982. Dust-bathing in the domestic fowl — Diurnal rhythm and dust deprivation. *Applied Animal Ethology* 8 (5):487–495. https://doi.org/10.1016/0304-3762(82)90061-X.

Vestergaard, K. S., B. I. Damm, U. K. Abbott, and M. Bildsøe. 1999. Regulation of dustbathing in feathered and featherless domestic chicks: The Lorenzian model revisited. *Animal Behaviour* 58 (5):1017–1025. https://doi.org/10.1006/anbe.1999.1233.

Vestergaard, K. S., and E. Baranyiova. 1996. Pecking and scratching in the development of dust perception in young chicks. *Acta Veterinaria Brno* 65 (2):133–142.

Vestergaard, K. S., and L. Lisborg. 1993. A model of feather pecking development which relates to dustbathing in the fowl. *Behaviour* 126 (3–4):291–308.

Wall, H., and R. Tauson. 2013. Nest lining in small-group furnished cages for laying hens. *Journal of Applied Poultry Research* 22 (3):474–484.

Wang, H. Y., H. Y. Chang, W. Xu, et al. 2003. Preliminary study on the level of evolutionary differentiation between domestic quails and wild Japanese quails. *Asian-Australasian Journal of Animal Sciences* 16 (2):266–268. https://doi.org/10.5713/ajas.2003.266.

Weeks, C. A., and C. J. Nicol. 2007. Behavioural needs, priorities and preferences of laying hens. *World's Poultry Science Journal* 62 (2):296–307. https://doi.org/10.1079/WPS200598.

West, B., and B.-X. Zhou. 1988. Did chickens go north? New evidence for domestication. *Journal of Archaeological Science* 15 (5):515–533. https://doi.org/10.1016/0305-4403(88)90080-5.

Widowski, T. M., and I. J. H. Duncan. 1994. Do domestic fowls form groups when resources are unlimited? *28th Congress of the International Society for Animal Ethology, ISAE*, Foulum, Denmark.

Wood-Gush, D. G. M., and A. B. Gilbert. 1969. Observations on the laying behaviour of domestic hens in battery cages. *British Poultry Science* 10 (1):29–36. https://doi.org/10.1080/00071666908415739.

Woodard, A. E., and W. O. Wilson. 1970. Behavioral patterns associated with oviposition in Japanese quail and chickens. *Journal of Interdisciplinary Cycle Research* 1 (2):173–180. https://doi.org/10.1080/09291017009359215.

Young, R. J. 2003. *Environmental Enrichment for Captive Animals*. England, UK: UFAW/WIley Blackwell.

Zupan, M., A. Kruschwitz, T. Buchwalder, B. Huber-Eicher, and I. Štuhec. 2008. Comparison of the prelaying behavior of nest layers and litter layers. *Poultry Science* 87 (3):399–404.

20
Behavioral Biology of the Zebra Finch

Samantha R. Friedrich and Claudio V. Mello
Oregon Health & Science University

CONTENTS

Introduction 315
Typical Research Contributions 316
　Vocal Learning 316
　Adult Neurogenesis 316
　Sex Dimorphism and Steroid Action 316
　Neuronal Gene Expression 317
Behavioral Biology 317
　Natural History 317
　　Ecology 317
　　Social Organization and Behavior 317
　　Mating and Reproduction 318
　　Feeding Behavior 320
　　Vocal Communication 320
　Common Captive Behaviors 321
　　Normal Behaviors 321
　　Abnormal Behaviors 322
Ways to Maintain Behavioral Health in Captivity 323
　Environment 323
　Social Groupings 324
　Feeding Strategies 324
　Training 324
　Breeding 325
Conclusions 325
Additional Resources 325
References 326

Introduction

Zebra finches belong to the clade Poephilinae of the family Estrildidae (waxbills) within the order of perching songbirds, Passeriformes (Olsson and Alström, 2020). There are two distinct subspecies of zebra finch: *Taeniopygia guttata guttata* and *Taeniopygia guttata castanotis*. Known as the Timor zebra finch, *T. g. guttata* are native to the island of Timor and other islands of the Lesser Sunda Islands. Compared to *T. g. castanotis*, *T. g. guttata* finches are smaller, have a distinct timbre (less nasal) due to fewer harmonics in their vocalizations, and have less conspicuous male plumage, especially the stripes that adorn the throat and chest. This chapter will focus on *T. g. castanotis*, the subspecies native to the mainland of Australia and that gave rise to domesticated zebra finches (Figure 20.1). *T. g. castanotis* zebra finches were transported from Australia to Europe in the mid-1800s, and were being bred widely in captivity by the late 1800s (Zann, 1996). The ease with which zebra finches were bred in captivity, in conjunction with Australia's 1960 export ban on native species, severely diminished the influx of wild zebra finches into Europe and North America. Around this time, it was estimated that captive zebra finches in Europe had not outbred with wild-caught birds for 50–80 generations (Zann, 1996). A more recent study demonstrates that domesticated zebra finch populations likely have not undergone severe bottlenecks, since at this point in their domestication, they retain substantial genetic variability (Forstmeier et al., 2007).

Nonetheless, domestication has altered several biological characteristics of zebra finches. Relative to wild birds, domesticated females are larger and take longer to reach full body size and sexual maturity (Zann, 1996). When properly cared for, domesticated zebra finches live on average 5–7 years

FIGURE 20.1 Sexually dimorphic plumage and common plumage color variants. (a) Adult zebra finches are sexually dimorphic in plumage and bill color. The male (top) is characterized by his red bill, orange-brown cheek patches, striped throat and chest, and spotted flanks. The female (left) has an orange bill. Both sexes have black "tear drops" under their eyes, a white "face" area, and barred tail feathers. Juveniles (right, sitting on feeder) have black bills and monomorphic plumage until ~40 dph when male-specific plumage begins to show. (b) A chestnut-flanked white juvenile next to a wild-type juvenile; note the white body, wings, and cheek patch in the former. (c) A pied juvenile. Note the white patches on the throat, head, and wing, and lack of pigmentation in parts of the beak. (Photo credit: Samantha Friedrich.)

(Burley, 1985), with some reports of even longer lived birds (Walton et al., 2012), while the maximum lifespan of wild zebra finches is approximately 5 years (Zann and Runciman, 1994). One of the most noticeable effects of domestication is the change in eye color, which is a striking red in wild birds, and red-brown to dark brown in domesticated birds (Figure 20.1). Domestication has also revealed over 30 plumage color variants, including "fawn", "chestnut flanked white", and "pied" (Figure 20.1b and c). The behavioral alterations that have resulted from domestication will be discussed later under the section titled "Common Captive Behaviors".

Typical Research Contributions

Research involving zebra finches has led to advances across a wide array of biomedical fields and continues to expand since its humble beginnings in the 1950s (Griffith and Buchanan, 2010a). Because zebra finches take so well to captivity, they support experimental ingenuity and controlled manipulations that are not feasible with wild avian models. Due to their imitative vocal abilities, zebra finches are an indispensable model for studying the speech and language deficits associated with several human conditions, including developmental verbal dyspraxia (Haesler et al., 2007), fragile X syndrome (Winograd et al., 2008), Huntington's disease (Tanaka et al., 2016), and Parkinson's disease (George et al., 1995; Miller et al., 2015). Their unique reproductive behaviors and biology have shed light on mate choice (Swaddle and Cuthill, 1994), maternal effects (reviewed in Griffith and Buchanan, 2010b), and sperm evolution (Birkhead et al., 2005). Their bright red bills have prompted studies on the genetics and immune function of carotenoids (Blount et al., 2003; Mundy et al., 2016). In addition to those mentioned above, there are numerous other research tracks that have benefited from zebra finches, but perhaps the most studied aspect of the zebra finch is their neurobiology (Mello, 2014; reviewed in Zeigler and Marler, 2008). The following paragraphs highlight some of the ways zebra finches have informed our understanding of vocal learning, neurogenesis, sex dimorphism and steroid action, and neuronal gene expression.

Vocal Learning

While it may seem counterintuitive to study a complex human trait using avian models, songbirds have illuminated the biological mechanisms of human speech more than any other animal model (Brainard and Doupe, 2002). Vocal learning, or the ability to learn vocalizations through imitation, is a trait found in a select few animal groups, including humans, dolphins, whales, possibly bats, elephants, and pinnipeds, as well as three groups of birds; songbirds, parrots, and hummingbirds (Petkov and Jarvis, 2012). The behavioral and neurobiological processes underlying avian vocal learning share remarkable similarity with human speech acquisition. Much like human infants learning to speak, young zebra finches produce unstructured vocalizations ("subsong") analogous to babbling, require a functioning auditory system and feedback to acquire and modify vocal sounds, learn optimally during critical periods of development, and exhibit considerable vocal variability from one individual to the next (reviewed in Doupe and Kuhl, 1999). Despite major differences in gross brain organization between birds and mammals, the neural circuitry driving human speech and zebra finch song is wired with parallel blueprints, and homologous cortical and striatal vocal circuit elements express similar groups of genes (Karten, 2013; Pfenning et al., 2014).

Adult Neurogenesis

Songbirds have played a pivotal role in our understanding of adult neurogenesis. Despite evidence of adult neurogenesis in higher vertebrates surfacing in the 1960s and 1970s, the phenomenon was broadly neglected due to widespread skepticism. Research on adult neurogenesis was reinvigorated in the 1980s with the discovery of adult-born neurons that integrate into the song system (Alvarez-Buylla et al., 1988; Goldman and Nottebohm, 1983; Paton and Nottebohm, 1984). These findings established that neurogenesis is not limited to early development in vertebrate brains and occurs in many regions of the avian forebrain (reviewed in Brenowitz and Larson, 2015). Studies of zebra finches have demonstrated, among other contributions, that neuronal death (Scharff et al., 2000), as well as forebrain auditory and sensorimotor circuit activity (Pytte et al., 2012; Wang et al., 1999), play essential roles in the survival of newborn neurons. Furthermore, studies of long-lived (11-year-old) zebra finches showed that neurogenesis is a lifelong process that continues well into the late stages of life (Walton et al., 2012).

Sex Dimorphism and Steroid Action

Zebra finches are highly dimorphic in their plumage, behavior, and brain structure. Most notably, males produce learned songs, in addition to learned and unlearned calls, while females produce only unlearned calls (Zann, 1990). This dimorphism in vocal behavior is reflected in the morphology of the underlying brain circuitry, called the song system. Compared to males, most nuclei of the song system are much smaller, if not absent, in adult female zebra finches (Bottjer et al., 1985; Nottebohm and Arnold, 1976). Intriguingly, much of the song system of young (10 days posthatch) males and females is very similar in size and morphology, meaning this sex dimorphism develops some weeks after hatching, and manifests over the course of sensorimotor learning (Konishi and Akutagawa, 1985; Nixdorf-Bergweiler, 1996). These sex differences in neuroanatomical structure and singing behavior are under the control of steroid signaling. Among studies addressing the role of steroids in singing, one surprising finding was that zebra finches synthesize estrogen *de novo* within the brain (Holloway and Clayton, 2001; Schlinger and Arnold, 1992). Experiments using various steroid treatments and gonadectomies have produced singing female zebra finches with enlarged song nuclei and connectivity, while similar manipulations in males have had little to no effect on song nuclei or singing (reviewed in Arnold, 1997; Wade and Arnold, 2004). To date, no steroid manipulation has completely masculinized the song system of a genetic female nor feminized the song system of a genetic male, implying that genomic factors are also at play.

Neuronal Gene Expression

As the second avian species to have its entire genome sequenced (Warren et al., 2010), the zebra finch is well positioned to address questions surrounding genomic and transcriptomic influences on brain function. These endeavors encompass high-throughput studies using microarrays and RNA-seq, as well as more focused approaches using *in situ* hybridization. The activation of auditory and song system regions in response to song was revealed by immediate early gene expression, providing direct links between a specific behavior and concurrent gene regulation in the circuit elements driving that behavior (Jarvis et al., 1998; Mello et al., 1992; Mello and Clayton, 1994). This exploration of neurogenomic dynamics underlying a learned behavior was further expanded through the identification of additional genes and gene networks activated by singing and hearing song (Dong et al., 2009; Hilliard et al., 2012; Wada et al., 2006; Whitney et al., 2014). Furthermore, comparative transcriptomic analyses of song nuclei in birds and homologous speech processing regions in humans revealed shared molecular specializations, adding support to the theory that human and avian vocal learning systems are built on convergent neural mechanisms (Pfenning et al., 2014).

Behavioral Biology

Natural History

Ecology

Zebra finches belong to the Estrildidae, a family of small, predominantly seed-eating passerine birds native to Australasia and the Old World Tropics. They are one of two species that make up the genus *Taeniopygia*, the other being the double-barred finch (*Taeniopygia bichenovii*). Zebra finches are the most abundant and widespread estrildid finch in Australia, and they are classified as a Least Concern species (BirdLife International, 2016). Two subspecies are recognized: *T. g. guttata* is found on the islands of the Lesser Sundas archipelago, and *T. g. castanotis* is found throughout the interior of mainland Australia. Zebra finch habitats span the vast majority of the interior of the Australian continent, excluding only parts of the coastal margins in the north, east, and south. They live in grasslands, scrublands, grassy woodlands, cultivated lands, and edges of forests; anywhere that provides grass seeds, bush or tree cover, and surface water (Goodwin and Woodcock, 1982). They are hardy birds adapted to survive in arid climates, low overnight temperatures, and unpredictable bouts of rainfall or drought. The maximum lifespan of a wild zebra finch is approximately 5 years. Only 9% of eggs survive to breeding age, and the estimated annual mortality rate for adults is between 72% and 96%; however, this estimate of adult mortality may be inflated due to colony emigration (Zann, 1996; Zann and Runciman, 1994). Zebra finches quickly reach sexual maturity around 90 days post hatch (dph). Their nests are susceptible to predation by lizards, snakes, rodents, marsupials, corvids, honeyeaters, and birds of prey (Zann, 1996). Adult zebra finches are hunted by snakes and carnivorous birds, including raptors, butcherbirds, and kookaburras, particularly while drinking and bathing at watering holes (Zann, 1996). The highest rate of mortality occurs in the window after the birds fledge, but are still dependent on their parents for feeding (Zann and Runciman, 1994).

Generally, zebra finches are most active in the morning and evening, and take to the shade during the higher temperatures midday. They move about by hopping with both feet and flying and do not display bipedal walking or running gaits like other birds, such as crows or chickens. Among several shared features common to estrildid finches, they have stout, conical beaks and short legs. They have a long molting cycle, continuously replacing feathers, and are never flightless (Zann, 1985). There are no significant size differences between males and females, although they are highly sexually dimorphic in plumage and bill color (Figure 20.1a). These conspicuous sex-specific traits have been proposed to expedite sex identification, and subsequent pair bond formation and breeding, which is adaptive, considering the zebra finch's relatively short lifespan, unpredictable breeding opportunities, and high turnover in colony membership (Zann, 1996).

Social Organization and Behavior

Given that they nest in colonies and forage in noisy flocks that can number well into the hundreds, it is not surprising that the word "gregarious" is commonly used to describe zebra finches. Although large flocks have been sighted at precious water sources during periods of drought, zebra finches typically perform their daily activities in pairs or small groups of three to ten birds (McCowan et al., 2015; Zann, 1996). The fundamental unit in zebra finch social organization is the pair; adult zebra finches form lifelong pair bonds and only re-pair with new individuals after prolonged separation from, or death of, their partner. Breeding colonies range in size from a few to 40–50 pairs, and in non-arid regions of Australia, colonies are occupied year-round (Zann, 1996). Nesting in colonies is thought to offer its occupants a "selfish herd" advantage, but only when predation levels are moderate (Zann, 1996). Population sizes fluctuate drastically from year to year, but correlate positively with rainfall in arid regions. The sex ratio of adult zebra finches shows slight male bias (52%), most likely due to higher mortality rates for young females (Burley et al., 1989).

Each zebra finch population occupies an extensive home range that encompasses multiple nesting colonies that all depend on shared water sources. Zebra finches are highly mobile across their home range and frequently move from colony to colony. Emigration from natal colonies is the norm, as less than one-quarter of a colony's adult population consists of birds that hatched there (Zann, 1996). Most arriving birds are transient visitors, but a small proportion form pairs, establish nests, and become residents. As a result of rainfall fluctuation and high individual mobility, colony size and membership are constantly in flux.

Zebra finches live virtually every part of their life in close proximity to conspecifics. This high degree of sociality is apparent in the notable synchronicity of their daily routines, as flocks will forage, sleep, preen, and bathe together. Both adults and juveniles exhibit "clumping" which involves perching shoulder to shoulder in a squat posture and fluffing the feathers, especially those around the feet (Figure 20.2a). Siblings often clump together, and sick birds will exaggerate their clumping posture (sitting lower and fluffing feathers wider) to invite other birds to clump with them. Partners and offspring will also preen one another, a behavior called "allopreening" (Figure 20.2b). One bird may prompt a conspecific to allopreen by perching close, fluffing up the feathers of the head or neck, and orienting their head or neck toward the other bird.

The first scientific journal article on zebra finches included detailed descriptions of their aggressive behaviors (Morris, 1954). Although wild zebra finches do not form persistent dominance hierarchies, aggressive behaviors surface in the defense of food, mates, nesting material, nests, and perches (Evans, 1970; Zann, 1996). Breeding and roosting pairs defend their nest from intruders, and this is the only marked display of territoriality for the species. An extremely common sign of aggression is the sleeking of the feathers against the body combined with either a tall, imposing posture to tower over a nearby opponent, or a long, low posture oriented toward a more distant opponent, sometimes threatening with a gaping bill. Both males and females will quarrel by bill fencing, in which a closed bill is used to prod and swipe at the opponent's head and bill. An aggressive bird will aim to displace opponents from their perch using a "supplanting attack", flying straight at the opponent and landing on their back if the opponent does not move in time. This supplanting attack is commonly accompanied by a Wsst call from the aggressor (the Wsst and other calls mentioned in *Natural History* are described below under *Vocal Communication*). Males and females use supplanting attacks (which looks like chasing when sequential) to defend their mate from same-sex rivals and to defend their nest. Females sometimes use supplanting attacks to ward off unwanted solicitations from courting males. Scuffles can also occur in midair, where birds tangle together and flap to the ground where they may continue to fight. While less common and more extreme, aggressive birds may clamp an opponent's foot or wing in the bill and dangle the opponent from a perch.

Mating and Reproduction

Zebra finches breed opportunistically. Unpredictable rainfall and resource availability, combined with a short lifespan, create urgency to pair and produce offspring whenever conditions are favorable. Breeding periods are long (8–15 months) and dictated by the availability of ripening grass seed that is crucial to the diet of newborns (Zann et al., 1995). The production of fresh grass seed follows a seasonal cycle in coastal regions, but is sporadic and unpredictable in central Australia. Due to these rainfall patterns, zebra finches located closer to the coasts show some seasonality, with breeding peaks in spring and sometimes fall, while those that live in the arid interior do not show seasonal breeding peaks (Zann, 1996).

Zebra finches are socially monogamous and form highly stable pair bonds that are maintained both during and outside of the breeding period (McCowan et al., 2015; Zann, 1996). All parental care activities are shared between the male and female. The lifelong duration of the pair bond and intricacies of parental cooperation highlight the importance of choosing a compatible mate. Females prefer males that sing frequently and have deep red bills; these attributes are directly correlated with testosterone level and may signal good physical condition (Blount et al., 2003; Burley and Coopersmith, 1987). While the choice to engage with a potential mate is dominated by outward physical characteristics, pair bond formation is dependent upon the interactions that follow this initial assessment. Compared to other closely related estrildids, zebra finches form pair bonds more rapidly, less selectively, and can do so without engaging in a nest site display (Zann, 1996, 1977).

Pair bond formation is generally initiated by a single male who approaches a female and performs a courtship waltz, sings to her, or proceeds directly to an attempted mount. An interested female will respond by twisting her head and tail toward the male ("head-tail twist"), while a disinterested female will ignore, flee, or supplant the courting male. If there is mutual interest, courtship continues over the next day or two and usually does not involve copulation (Zann, 1996). Instead, the pair bond is developed through affiliative contact behaviors, including allopreening, clumping, perching near one another, and behavioral synchrony (Caryl, 1976; Silcox and Evans, 1982). Once formed, the pair bond is maintained for life, in part through continuation of these affiliative behaviors.

FIGURE 20.2 Common affiliative behaviors. (a) Juvenile siblings clumping together on a millet spray stem. (b) A father allopreening his offspring. (Photo credit: Samantha Friedrich.)

Partners keep very close to one another and use calls to remain in auditory contact when they are not in visual contact (Silcox and Evans, 1982; Zann, 1996).

Nest building and copulation begin soon after pair bond formation. Nest building drive is very strong in zebra finches year-round; even after a roosting or breeding nest is complete, they will continue to renovate and expand the nest, and sometimes build nests that go unused (Zann, 1996). The search for a nesting site is an important first step. The male of a pair initiates by leading the female to potential nest sites for her appraisal, while indicating them with Ark and Kackle calls. A male may incorporate several different elements into his nesting display, including undirected song, nodding toward the site, mandibulating (clicking the beak open and closed), and a bowing posture, in which he fans his tail while rotating his body side to side. Females are extremely choosy about nesting sites and express disinterest by moving away from the site. Once she finds a site suitable, a female may display a head-tail twist, mandibulate, and/or hop around the site. In response, the male enhances his display and produces Whine calls.

Next, the pair will begin a nest ceremony in which they sit shoulder to shoulder at the exact spot where they will build their nest while mandibulating toward one another and producing Whine calls. This behavior is often interspersed with bouts of hopping around the site while making Kackle calls. Then, the male will set out to find and bring back the first item of nesting material, the incorporation of which is a momentous affair met with more Whine calls and much fussing over its exact placement. These nest ceremony behaviors wane with each subsequent deposit of nesting material. From there, both sexes contribute to nest building; the male forages for materials and delivers them to the nest, and the female arranges the collected materials to shape the nest. Zebra finches exhibit very stereotyped building movements that are conserved with other estrildids (Zann, 1996). While foraging for nesting material nearby, the male sings undirected song and produces Distance calls to remain in auditory contact with the female. Grasses and twigs are used to construct nests that have an entrance tunnel attached to a spherical egg chamber, resembling the shape of a water skin. Zebra finch pairs build nests in trees and bushes, preferring thorny vegetation to discourage predation, and they are the only Australian estrildid to build nests for roosting as well as breeding (Zann, 1996).

Copulation usually occurs a couple of days after a pair meets, once nest building is underway. As is common with many songbirds, the mating behaviors of the zebra finch involve elaborate songs and coordinated body movements (Morris, 1954; Williams, 2001; Zann, 1996). Dead branches are preferred for performance sites, as they minimize structural interference with the choreography. Generally, males initiate the sequence by flying to and greeting the female with a tall posture and Distance or Kackle calls. The male and female then begin a waltz, in which they hop back and forth between nearby perches, while producing progressively louder and more frequent calls. Each time they land, they bend their head and tail in a head-tail twist and bow, sometimes wiping their bill on the perch. Once the female stops waltzing, the male approaches and solicits her with song and posturing. He stands erect, fluffs his feathers, and orients himself to sing toward the female. He often performs a series of hop-pivots between perches, while singing and keeping his tail bent toward the female. If performing on the ground, he hops in a semicircle around her, pivoting at each end. If the female accepts the male, she will adopt a low crouching posture and quiver her tail while eliciting Copulation calls, inviting the male to mount. The male mounts atop her back, flapping his wings to keep balance, while producing Copulation calls. Copulation happens in 1 or 2 s, through brief cloacal contact, and males sometimes quiver their tails after dismounting. For the production of a single clutch, each pair typically copulates about 15 times, starting as early as 11 days prior to the first egg being laid (Birkhead et al., 1988a). Though less common than male-initiated copulation, females sometimes use tail quivers to initiate copulation (Birkhead et al., 1989).

Females are fertile for a period of 14–15 days with the second-to-last egg marking her final day of fertility. Both males and females may solicit extra-pair copulation with the opposite sex, but only 2.4% of offspring, and 8% of broods, were found to derive from extra-pair paternity in the wild (Birkhead et al., 1990). More common than extra-pair paternity is conspecific brood parasitism, where a female lays an egg in the nest of another female; 36% of all broods contain at least one parasitic egg, and 10.9% of all offspring hatch from parasitic eggs (Birkhead et al., 1990), although a more recent study reported 17.5% of all broods and 5.4% of all offspring (Griffith et al., 2010). The final copulation fertilizes the majority of the clutch, termed "last male sperm precedence" (Birkhead et al., 1988b; Birkhead and Hunter, 1990). Females can store sperm in storage tubules for up to 13 days (Birkhead et al., 1989). Thus, paired males will secure their paternity by frequently copulating and vigorously defending their mate from intruding males while she is fertile (Birkhead et al., 1988a). Eggs are fertilized about 24 h prior to being laid and are laid one a day, usually in the morning (Birkhead et al., 1989). After an egg is laid, there is an hour fertilization window for sperm to fertilize the next day's egg, which might explain why most copulation occurs in the morning (Birkhead and Møller, 1993). Clutch size ranges from two to eight eggs, with four to six eggs being typical.

Both males and females partake in all parental care behaviors, though the balance is not exactly even; females provide slightly more care than males during the day and are responsible for all nocturnal incubation (Burley, 1988; Zann and Rossetto, 1991). After a clutch is laid, the incubation period begins where eggs must be kept warm by sitting atop them, and females develop brood patches to increase heat transfer. The sitting bird positions itself so that it can peer out of the nest entrance to keep watch, and often occupies itself by repositioning stems in the nest. During incubation, parents alternate between sitting in the nest and joining the flock. This intermittent separation requires a temporary loosening of the pair bond that strengthens again once the nestlings are about a week old (Zann, 1996).

Hatching begins after ~12 days of incubation, and the parents usually consume the eggshells. The amount of time spent brooding remains constant for ~5 days after the first egg hatches, then gradually decreases, and ends altogether at approximately 11 or 12 dph (Zann, 1996). Zebra finches do exhibit some degree of nest hygiene, swallowing nestling droppings for the first 7 days after hatching, then transitioning to transporting dried droppings away from the nest (Zann, 1996). In the days leading up to fledging, chicks will take turns peering out from the nest entrance (Figure 20.3d). Young birds fledge the nest around 17–20 dph and remain nutritionally dependent on their parents until about 35 dph. Sexually dimorphic male plumage begins to show around 40 dph, although dark striping of male breast feathers can be seen as early as 20 dph.

Feeding Behavior

Zebra finches feed in tight flocks that range in size from dozens to hundreds of birds, depending on their environment and breeding status (Zann, 1996). They readily relocate in response to changes in rainfall to find more favorable foraging grounds. They consume the most seed within two diurnal windows, in the morning and just before roosting in the evening. Given their widespread distribution, zebra finches, as a species, consume a wide variety of grass seeds; however, any given individual in a constrained region feeds primarily on a single type of seed (Zann, 1996). Their beaks are strong and dexterous, allowing them to manipulate hard grass seeds. They collect ripened seeds from the ground or pluck them from grass stalks. To uncover buried seeds, they move their bills in a sweeping sideways motion to dig. Seeds are dehusked, swallowed, and collected in the crop, an esophageal pouch that stores and softens food, before entering the gizzard for further chemical and mechanical digestion. Small rocks and grit are found in the gizzard and thought to aid in breaking down swallowed food. While grass seeds make up the bulk of the zebra finch diet, they also consume green food, such as sprouted seed, and occasionally insects.

Drought-resistant birds, zebra finches can survive long periods without water, but drink often throughout the day when water is readily available (Goodwin and Woodcock, 1982). They prefer to drink from, and bathe in, small shallow pools unobstructed by vegetation, presumably to avoid being ambushed by lurking predators. When water is scarce, as in arid zones, zebra finches will drink wherever they can find water. They will fly dozens of kilometers to reach water sources, and their wings evolved to maximize distance over speed (Zann, 1996). Unlike many other avian species, zebra finches can drink without tipping their head back to swallow, instead using the larynx to move water down the esophagus (Heidweiller and Zweers, 1990). This adaptation allows them to drink faster and from harder to reach sources, such as rock cervices. Amazingly, laboratory birds with access to unlimited seed can survive without water for hundreds of days, though not without a deterioration in health (Cade et al., 1965; Lee and Schmidt-Nielsen, 1971).

Until around 35 dph, young zebra finches are dependent on their parents to feed them, and signal hunger using Begging calls and movements of the tongue and head (Muller and Smith, 1978). Parents begin feeding nestlings a day after hatching by first regurgitating food stored in their crop, which manifests behaviorally as head shaking, throat pumping, and bill gaping. They then enter the nest and feed the young from behind, keeping an eye on the nest entrance. The parent places its bill inside that of the nestling and uses its tongue to push food into the begging mouth. Nestlings and fledglings beg for food with Begging calls, bill gaping, tongue movements, and head rotations, and some features of these begging behaviors change over development (Zann, 1996).

Vocal Communication

Zebra finches are very vocal birds and their vocalizations include both learned songs and unlearned calls. The tonal quality of the zebra finch voice is nasal and scratchy, because the syllables of both songs and calls are characterized by harmonic stacks, as opposed to pure tone whistles, such as those sung by the canary. They produce calls regularly throughout their daily activities to establish contact, alert conspecifics to danger, defend mates, beg for food, and coordinate nest activities. Zebra finches also use calls to determine an individual's identity (D'Amelio et al., 2017; Elie and Theunissen, 2018).

Zebra finch calls have been classified into 12 different call types (see Table 20.1) that were originally described by Zann (1996) and recently corroborated using a more data-driven approach (Elie and Theunissen, 2016). **Tet calls** are the most common call type and are emitted almost reflexively as zebra finches move around, particularly while hopping. Tet calls are not directed at any one individual, do not provoke any specific response from conspecifics, and are produced even in social isolation. **Distance calls** are the loudest of the call types and used quite flexibly while the bird is flying, greeting, courting, between songs, about to take off, raising alarm, or separated from its familiars. It is thought to carry identity information, signal when a bird is lost, and coordinate flight. Distance calls are sexually dimorphic, being shorter and more frequency modulated in males, which allows zebra finches to identify the sex of a conspecific without visual cues (Zann, 1990). The third most common call is the **Stack call**, which is louder and higher in pitch than the Tet call. Zebra finches emit Stack calls as they take off, as well as while in flight, and males give Stack calls when leading females to potential nesting sites. **Wsst calls** sound more like a hiss than a tone and denote aggression, as they are often emitted just prior to a supplanting attack. **Thuk calls** are named phonetically for their shorter, deeper sound. They are alert calls produced by parents to warn their offspring of potential danger. Parents produce an abundance of Thuk calls in the days surrounding fledging. **Kackle**, **Ark**, and **Whine calls** are the signature sounds produced by males and females during nest site searching, nest ceremony, and nest building. The Kackle and Ark calls are short and staccato, while Whine calls are longer, higher pitched, and more tonal. **Copulation calls** are a series of whines emitted by both sexes during copulation. The female produces them while quivering her tail, and the male produces them as he mounts her. **Distress calls** are sharp, high-pitched squeals made when a bird is frightened or in pain. Nestlings and fledglings sometimes emit Distress calls when handled, and adults most commonly produce Distress calls during particularly ferocious fights. **Begging calls** begin as soft, high-pitched peeps around the third day posthatch and

TABLE 20.1

Zebra Finch Calls

Call Type	Behavioral Context
Tet	Hopping, foraging, moving about
Distance	Flying, greeting, courting, singing, alerting, when isolated
Stack	Taking off, flying, searching for nesting site
Wsst	Threatening, attacking, fighting
Thuk	Alerting to danger
Kackle	Searching for nesting site, nest ceremony, nest building
Ark	Searching for nesting site, nest ceremony, nest building
Whine	Searching for nesting site, nest ceremony, nest building
Copulation	Tail quivering, mounting
Distress	Frightened, in pain, fighting
Begging (juveniles)	Hunger
Long tonal (juveniles)	Hunger, contacting parents, when isolated
Song Type	*Behavioral Context*
Directed	Courtship display, mating
Undirected	Attracting mates, practicing song, maintaining auditory contact with mate
Subsong	Juveniles "babbling" early in song development

gradually grow louder, longer, and harsher with age. They increase in frequency and noisiness as a function of hunger. **Long tonal calls** are produced only by young zebra finches and are distinguishable from noisier Begging calls. They are emitted between Begging call bouts, in response to Distance calls from parents, and when a juvenile is isolated from its parents.

Zebra finch song is a learned vocalization used to attract mates, identify individual conspecifics, establish pair bonds, and maintain auditory contact (Zann, 1996). Unlike many other songbirds, zebra finches do not use song for territorial defense. Singing is a sexually dimorphic behavior in zebra finches, as only the males sing. Each male's song is made up of several complex syllables that occur in a faithfully repeated sequence, called a motif. Multiple motifs sung in succession that are often preceded by a set of repeated introductory notes constitute a song bout (Price, 1979). Interestingly, a motif sometimes contains a component syllable that is indistinguishable from a Distance call, suggesting the latter are sometimes incorporated into the song structure (Zann, 1990). Juveniles learn their song through imitation, a complex trait called vocal learning (reviewed in Zeigler and Marler (2008)). The song learning process takes place over ~20–90 dph, beginning with a sensory phase, in which juveniles listen to and acquire an auditory memory of the tutor's song, which is usually the father's song. Around 35 dph, young males begin to produce highly variable and noisy syllables called "subsong", which is considered analogous to the babbling phase in human speech development. This subsong progressively evolves into multiple distinct syllables sung in succession as recognizable song motifs, around the time that juveniles enter the plastic song phase at ~50 dph. At this point, shared song elements between the tutor and juvenile also become recognizable. Juveniles continue to practice and polish their song, and by 90 dph, the song is "crystalized" into its final, stable form. Every male's song is unique, but similar to the tutor's song, thus song serves as an acoustic signature of individual identity, as well as signaling kinship (Geberzahn and Derégnaucourt, 2020).

Singing occurs in directed and undirected contexts. In both singing contexts, the acoustic nature of the song is largely preserved, with detectable, but small, changes in song tempo, variability across renditions, and number of introductory notes. The associated behaviors and motivations are what mostly differentiate directed and undirected song. Directed song is sung specifically to a mate as part of a courtship display. It is always accompanied by circumscribed posturing and movements, and its ultimate goal is to elicit copulation solicitation from a female. Undirected song is sung without the corresponding courtship dance, to no conspecific in particular, and in a wider array of contexts. Paired males sing undirected song just outside the nest to signal the female to stay put, and unpaired males sing undirected song to advertise their availability. Undirected singing in adults is also thought to reflect vocal practice (Jarvis et al., 1998). Visually and acoustically isolated zebra finches will sing undirected song, and there is some evidence to suggest that singing is inherently rewarding (Riters and Stevenson, 2012).

Common Captive Behaviors

Normal Behaviors

The majority of behaviors described above for wild zebra finches are also observed in captive birds. However, domestication has undoubtedly shifted some elements of zebra finch behavior from their free-living state. This section will highlight behaviors that differ between wild and domesticated zebra finches in terms of quality, intensity, timing, or frequency. While they deviate from wild zebra finch behavior, these behaviors are not indicators of poor welfare or cause for concern. Nevertheless, it is important to consider domestication-induced behavioral differences when designing or drawing conclusions from studies aimed at exploring aspects of natural, species-typical behavior.

Several aspects of vocal, social, and sexual behavior are significantly altered in domesticated zebra finches. Vocalizations differ between wild Australian finches and captive ones, including changes in call structure (Okanoya et al., 1993) and the speed at which syllables are sung (Zann, 1993). While females readily clump and allopreen in the wild, it is rare to see unpaired males allopreening or clumping together. In contrast, allopreening and clumping between adult males are quite common in an aviary setting. Domesticated zebra finches have shed considerable light on the central importance of the pair bond. While same-sex pair bonds have not been observed in wild colonies, they are common in aviaries, where they form at rates proportional to skewed sex ratios, forming most often among individuals that are housed in single-sex cages (Elie et al., 2011). Same-sex pair bonds are stable and characterized by the same affiliative and sexual behaviors as male-female pair bonds (Elie et al., 2011). In fact, even when females are presented, 25%–50% of males from same-sex pairs show preference for their male partner (Elie et al., 2011; Tomaszycki and Zatirka, 2014). These findings suggest that in zebra finches, pair bonds can be facultative or a choice, sometimes taking precedence over reproductive outcomes. Courtship and copulation are also altered in domesticated zebra finches; a pair may shorten or skip the waltz phase of courtship (Morris, 1954), and partners copulate 11–12 times per clutch compared to 15 or more with wild birds (Zann, 1996). Extra-pair paternity occurs at double the rate in aviary settings (5.6% of young) compared to the wild, likely due to nests being in closer proximity (Birkhead et al., 1989; Zann, 1996). Conspecific brood parasitism is common in captivity, with 21% of nests containing at least one parasitic egg (Schielzeth and Bolund, 2010), and 3.6% of offspring arising from parasitized eggs (Burley et al., 1996). These rates of conspecific brood parasitism are comparable to, or somewhat less than, those reported for wild birds (Birkhead et al., 1990; Griffith et al., 2010).

Domestication has also impacted nesting behaviors, particularly in males. The nest ceremony is typically abbreviated or omitted, although this does not prevent nest building or egg laying. The nest building interval is shorter, with domesticated zebra finches investing 1–5 days in nest construction, compared to 5–13 days in wild birds (Zann, 1996). When building nests inside nesting boxes, zebra finches do not build the entrance tunnel or roof but this omission is innocuous with respect to breeding. Breeders may build flimsy, poorly constructed nests, attempt to build a nest outside the provided nesting box, or fail to build a nest at all. One strategy to encourage egg laying, despite these shortcomings, is to place matted nesting material in the nesting box to provide a preliminary foundation. Additional nesting material can still be provided in the cage to encourage nest building behavior. While nest quality may be reduced in captivity, a strong nest building drive is generally maintained. Even when supplied with nesting material, such as coconut husk, zebra finches typically incorporate a variety of materials into their nest, including feathers, cage liner, stems and pieces of millet spray, and leafy greens (Figure 20.3a and c). Finally, domesticated zebra finches are more flexible when it comes to incubation timing. Unlike wild zebra finches, domesticated birds are less keen to delay incubation until the majority of the clutch is laid. As a result, clutches hatch far less synchronously over longer incubation periods, with more time between individual egg hatchings (Gilby et al., 2013).

Abnormal Behaviors

Compared to other avian species, there is very little research or detailed description of abnormal behavior in captive finches. What seems clear is that behavioral problems stem from overcrowding, social instability, insufficient environmental stimulation, unmet dietary needs, or some combination of these factors. In contrast to the behaviors described above, the behaviors outlined in this section may pose a threat to animal safety and/or indicate poor welfare, and preventative or ameliorative measures are proposed where applicable.

Captive enclosures may bring out higher levels of aggression in domesticated zebra finches. Sometimes, one or two particularly aggressive individuals emerge in a group; however, the dominant status of these individuals is constantly challenged and thus usually temporary (Evans, 1970; Ratcliffe and Boag, 1987). In our lab, we have found that aggressive behavior sometimes persists in certain individuals, and these aggressors can severely injure other birds. Placing the aggressive bird into a different group cage can disrupt this pattern of behavior. While many avian species respond to major stress by plucking their own feathers, this coping behavior is not characteristic of zebra finches. Rather, plucked feathers in a group cage, especially from the tail and neck, are indicative of aggressive conflict, and may require reassigning birds to other group cages. Breeding pairs sometimes pluck feathers from each other or their offspring to obtain nest lining material. This minor plucking behavior in breeding cages is normal and only becomes of concern if plucking is overzealous and leads to bald patches, or if plucking is accompanied by Wsst calls and supplanting attacks. Breeder aggression toward offspring is most likely to appear once the female begins laying a second clutch. If fledglings must remain housed with their parents beyond nutritional independence for experimental reasons, this aggression is usually abated by removing the nest box to prevent further breeding.

Zebra finches are incredibly active birds that spend a significant portion of their day flying or hopping between perches, and vocalizing frequently with Tet and Stack calls. Though they rest periodically throughout the day, a bird that is consistently quiet and stagnant is a sign that something is amiss. Hyper-regular movements in a zebra finch are a sign of distress or neurological failure. This includes route-tracing, when a bird repeatedly follows the same exact route through their enclosure, and spot-picking, when a bird repeatedly touches the bill to a particular spot on its body or in the enclosure (Law et al., 2018). Sometimes breeders get carried away with nest building and continue layering nesting material atop a clutch of eggs, leading to the loss of that clutch. To discourage this tendency, no further nesting material should be provided once the first egg is laid.

Behavioral Biology of the Zebra Finch

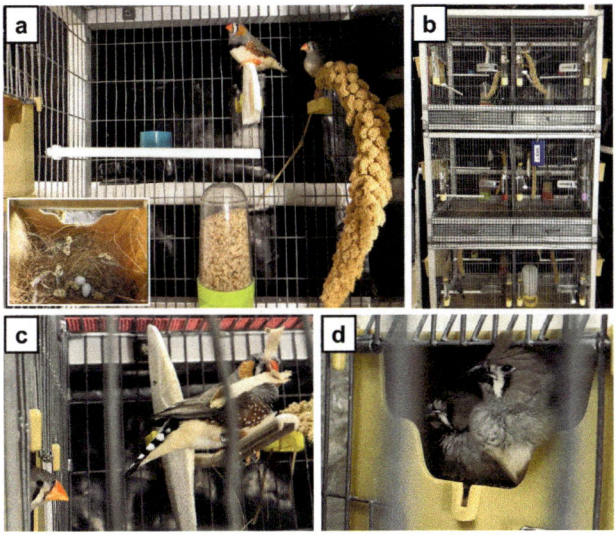

FIGURE 20.3 Breeding cages and nesting behaviors. (a) A male and female pair in a breeding cage. The nesting box on the left is mounted from the outside of the cage. The inset shows a top-down view of the inside of the nesting box which contains two eggs and various nesting materials including coconut husk, pieces of millet spray, and torn bits of paper cage liner. (b) A rack of six individual breeding cages. Solid dividers between adjacent cages and black tarps between racks provide visual isolation of each breeding cage to encourage juveniles to learn their father's song, and not the songs of neighboring males. This ability to control tutor song is useful in song learning studies, for example, when comparing how similar a juvenile's song is to the tutor's song. (c) A male collecting nesting material and a female poking her head out of the nesting box opening. (d) Two juveniles peeking out from the nesting box prior to fledging. (Photo credit for panels (a) and (b): Monica Masse; (c) and (d): Samantha Friedrich.)

Ways to Maintain Behavioral Health in Captivity

Generally speaking, zebra finches are quite robust and easily bred in captivity. Despite the popularity of the zebra finch as a model organism and pet, very few studies have been conducted that specifically address the impacts of conventional husbandry practices on zebra finch health and behavior. The recommendations outlined here are a summation of best practices from a collection of laboratories and organizations that work with zebra finches. Additional studies designed to assess and optimize husbandry methodologies are needed in order to evaluate and improve the current guidelines.

Environment

Zebra finches thrive in thoughtfully designed captive environments that accommodate their needs for ample space, light, and structural enrichment. Because zebra finches spend most of their time in close proximity to each other, they can live well in higher densities than other bird species. However, enclosures that provide sufficient space are vital to encourage natural foraging and flocking behaviors, as well as to keep aggression at a minimum.

During the day, zebra finches spend much of their active time flying between perches. Maximizing horizontal flight is the highest priority when designing a flight cage or aviary, thus enclosure length is more important than height or width. In addition, perches, food containers, and other items placed around the perimeter keep the interior free of obstructions to allow for more and longer flightpaths. A solid enclosure floor encourages naturalistic foraging behavior as zebra finches prefer to forage for seed on the ground. The floor may be covered with a paper liner or loose substrate, such as wood shavings. Loose substrate creates an enriched foraging environment, but care should be taken in choosing and refreshing the substrate.

Although zebra finches are small, they are extremely active birds that require spacious enclosures to fly and forage. The Joint Working Group Report and European Commission make specific recommendations for cage size requirements (Nager and Law, 2010). For a solo or breeding cage, the recommended minimums are $0.5\,m^2$ of floor space and 0.3–0.4 m in height. Up to six birds can be kept in a cage with a minimum of $1\,m^2$ floor space and 1 m height. Enclosures housing more than six birds should meet a minimum height requirement of 2 m and floor spaces of 1.5 or $2\,m^2$ for 7–12 or 13–20 birds, respectively. A good rule of thumb for cohousing more than 20 zebra finches is that each additional bird requires another $0.05\,m^2$ of space. When designing zebra finch enclosures and deciding on occupant densities, it is important to consider the tradeoff between the size of the enclosure and how accessible each bird is to a caretaker or experimenter (i.e., birds are more easily retrieved from smaller cages with fewer individuals).

Zebra finches' health and fertility is directly impacted by the quality and amount of light they receive. Bright natural sunlight is the ideal light source, but zebra finches can be maintained and successfully bred with properly tuned full spectrum bulbs that have some ultraviolet component. Energy-efficient compact fluorescent light bulbs work well for this application, and flicker at much higher rates (10–40 kHz) than avian flicker fusion thresholds (80–100 Hz), thus avoiding flicker-induced stress (Nager and Law, 2010). Lights should be faded on and off at the beginning and end of the light cycle to simulate dawn and dusk. The dusk transition period is important to allow birds to settle into their roosts before dark. Zebra finches may be particularly sensitive to sleep disruption, as they show complex sleep patterns similar to those of mammals, including rapid eye movement sleep that increases in cycle length as the night progresses (Low et al., 2008). Birds sometimes startle in the night and can fall off their perch, so a small nightlight or dim light source should be left on during the dark cycle to prevent nocturnal injuries.

Perches are essential features of zebra finch enclosures. To mimic the trees and bushes of their natural environment, it is best to provide perches of various diameters and at different levels of the cage. In order to preserve open space for flight, it is important to avoid cluttering the enclosure with an excess of perches. Natural branch perches and swinging perches challenge birds to balance and exercise their stability muscles. Captive zebra finches tend to favor the highest perches, but lower perches help them get to and from the ground, making them indispensable for newly fledged birds. Perches should be of sufficient length to accommodate several birds, as zebra finches often clump together while resting.

Bathing is a joyous activity for the zebra finch. Placing shallow containers of water on the cage floor is an easy way to provide enrichment, as well as support feather and skin health by increasing humidity. Baths should be offered in a covered bathing container or just before floor cover is changed, as zebra finches bathe ecstatically and in groups, spraying substantial amounts of water in the process. While fledglings are keen to bathe, baths should be withheld from breeding cages until all fledglings are flying and perching without losing balance or falling to avoid the possibility of drowning.

Social Groupings

Zebra finches are highly social creatures that live in colonies and move about in pairs or small groups in the wild. Taking measures to accommodate their social nature in captive housing design will lead to better health, less aggression, and more natural behaviors. A large, mixed-sex aviary will most closely approximate natural conditions, although this option may not be feasible for any number of reasons. The exact arrangement of birds into social housing is flexible and various strategies may be used to accommodate space limitations, experimental needs, and accessibility. Unwanted breeding is usually avoided by housing males and females separately in single-sex enclosures; however, same-sex groups are not observed in wild zebra finches, and further studies are needed to determine the effects of this widely adopted, yet unnatural, housing strategy (McCowan et al., 2015). Fledglings of both sexes can be housed together after they reach nutritional independence. Their social environment during adolescence has lasting impacts on social and sexual behavior (Ruploh et al., 2013, 2014), including the extent to which each juvenile models his song after that of his father or his peers (Derégnaucourt and Gahr, 2013; Mann and Slater, 1995; Tchernichovski and Nottebohm, 1998).

Zebra finches in the wild are virtually always in the company of conspecifics and show signs of stress including increased corticosterone and altered vocal behavior when separated from other birds (Perez et al., 2012). Social isolation may also increase susceptibility to illness (Martin, 2000). The lack of a social environment has lasting effects on the development and health of captive birds. For example, song development is highly abnormal in isolated birds (Price, 1979), and there is some evidence that isolation alters brain gene expression (George et al., 2019). The impact of social environment on brain function also bears out in rates of adult neurogenesis in zebra finch forebrain, as neuronal recruitment is greater in group-housed birds compared to birds that are housed solo or paired (Adar et al., 2008; Barnea et al., 2006; Lipkind et al., 2002). Thus, solo housing should be avoided, except for very specific purposes, such as surgery recovery, injury healing, or studies specifically aimed at addressing aspects of social interaction.

Feeding Strategies

Zebra finches in captivity can be kept and bred in good health on a diet of seeds, fresh foods, egg food, and mineral supplements. On average, a captive zebra finch consumes 3g of seed per day, with millet being the preferred seed (Zann, 1996). Feeders placed on the floor of the cage are recommended, as foraging on the ground is more naturalistic than taking seed from wall-mounted containers. Stocking multiple feeders in larger enclosures will help to minimize aggression during foraging. Seed can be supplied in large quantities, as zebra finches are not prone to overeating. Hanging dried millet sprays from the cage ceiling away from perches provides both food and enrichment, encouraging birds to forage more naturally and gymnastically, and the stems are often repurposed as nesting materials.

Various dietary supplements are necessary to maintain zebra finch health, because seeds alone do not provide sufficient vitamins, minerals, and protein. Fresh fruits and vegetables, particularly dark leafy greens, should be offered at least weekly. Live insects, such as mealworms, can be offered sparingly as a concentrated source of protein and foraging enrichment. Calcium in conjunction with vitamin D are extremely important nutrients to both breeding and non-breeding zebra finches, and deficiencies in either can lead to life-threatening problems with bone health (e.g., fractures and abnormal growth) and egg laying (e.g., thin shells and egg binding). Dried cuttlefish bone and shells from hard-boiled chicken eggs are readily consumed calcium supplements. It is a common misconception that zebra finches require insoluble granules called grit to digest seed (Taylor, 1996). Grit is not essential, because zebra finches use their beak to remove the husk from seeds before swallowing. However, offering soluble commercial grit (e.g., Kaytee High-Calcium Grit Supplement) in a small container is a great way to ensure zebra finches ingest sufficient calcium.

Because of the different dietary needs and higher energetic demands of breeding and rearing, soft, protein- and calcium-rich food should be offered at least once or twice a week to breeders before eggs are laid and until chicks are nutritionally independent. Female breeders must significantly increase their intake of calcium during egg production, so soluble grit, cuttlebone, and other calcium supplements should be closely monitored and replenished regularly. Many labs and aviaries also provide breeders "egg food", a combination of dry nestling food (e.g., 92A Nestling Food from ABBA Products Corp.) mixed with mashed hard-boiled egg. Egg food is voraciously consumed by breeding pairs and their offspring, and can be supplied to group cages on a weekly basis to enrich the diets of all captive birds. Dry seed can also be soaked and sprouted to create a live, nutrient-rich food that approximates the half-ripe seeds that zebra finches feed their young in the wild.

Every enclosure should offer multiple sources of clean drinking water that are checked daily for water level and sanitation. Drinking bottles mounted to the sides of the cage allow for easy access during water changes with minimal disruption to the birds. Water may also be provided in dishes on the cage floor, but this requires more frequent water changes per day as zebra finches will bathe in and foul the water.

Training

Zebra finch behaviors are typically studied from an ethological standpoint under naturalistic conditions that do not require training. Unlike some avian species, domesticated zebra finches do not become tame or accustomed to handling, unless handled consistently from a very young age. However, zebra finches can be trained with minimal handling through operant conditioning procedures to test auditory perceptual and discrimination abilities, as well as song preference. Hopping to specific perches or feeders and key pecking are behaviors that can serve as operant responses in birds and may be trained using go/no-go or choice paradigms. In auditory discrimination tasks, an operant response is prompted by the presentation of auditory stimuli played through a speaker in the operant chamber (Cynx and Nottebohm, 1992; Sturdy et al., 2001; Vernaleo and Dooling, 2011). Song preference can be evaluated using a choice paradigm in which landing on one perch triggers playback of one song and landing on another perch triggers playback of a different song (Coleman et al., 2019). Song preference has also been tested using nonoperant procedures that measure untrained responses, such as time spent in the chamber closest to the speaker playing the preferred song (Clayton, 1988). Other operant conditioning methods have been designed to assess social aspects of song learning (Derégnaucourt et al., 2013; Houx and Ten cate, 1999), guide song development (Lipkind et al., 2013), and test visual discrimination abilities (Melgar et al., 2015). While many of these training procedures rely on food reward, visual social contact is an effective reinforcer that does not require depriving the bird of food, water, or social interaction (Macedo-Lima and Remage-Healey, 2020).

Breeding

The key to successfully breeding zebra finches is strong pair bonds. Because parental responsibilities are shared between zebra finch parents, the compatibility of two birds is essential to produce healthy offspring. Some pair bonds form on the scale of minutes or weeks, depending on age, reproductive history, and partner compatibility, but on average pair bonds form by the second day after introduction (Silcox and Evans, 1982). Tactile contact is extremely important to pair bond development, as birds deprived of tactile contact and only allowed auditory and visual contact are far less likely to form pair bonds (Clayton, 1990; Silcox and Evans, 1982).

One method of creating pairs is to simply place one male and one female in a breeding cage together (Figure 20.3a and b). This method allows the experimenter to strategically pair individual birds but removes all mate choice and thus runs a greater risk of incompatibility. An incompatible pair will express rejection through Wsst calls, supplanting attacks, chasing, pecking, bill fencing, and keeping their distance. If these behaviors are observed after the second day postpairing, the dyad should be separated and re-paired with different individuals. A more efficient method to pair zebra finches is to place at least four adult finches of each sex in a flight cage and allow them to choose mates over several days, before moving each established pair (evident from clumping, allopreening, perching adjacent, etc.) to a breeding cage. One consideration with this method is that females may be fertilized by more than one male in the pairing process. To avoid extra-pair paternity, partners of newly established pairs can be separated and placed in same-sex group housing for the duration of sperm storage, which is 13 days in zebra finches (Birkhead et al., 1989). Due to the highly stable nature of established pair bonds, partners will breed readily when reunited in breeding cages following this separation period. If unconstrained parentage and breeding is an option, breeding can proceed in larger enclosures with distributed nesting boxes. Installing more nesting boxes than there are pairs will reduce aggressive competition for nesting sites.

Pairs that have produced a clutch and subsequently been separated will always re-pair with each other (Zann, 1996). Furthermore, pair bonds persist despite auditory and visual separation, even if additional pair bonds are formed with other individuals in the interim (Silcox and Evans, 1982). Because of the resilience of these pair bonds, breeding will be most efficient with established pairs. Though a newly formed pair may successfully produce a clutch in the nest, inexperienced parents often fail to raise chicks, due to any number of missteps, including insufficient brooding, failure to feed nestlings, injuring fragile nestlings, and "chick tossing", which is expulsion of chicks from the nest. Finding dead chicks in or around the nest does not make it necessary to separate the pair or declare the pair unfit breeders; most pairs that struggle during the first attempt succeed the next time around. Also, nestling mortality is 13% in domesticated zebra finches, so some loss of nestlings is to be expected (Zann, 1996).

Conclusions

The zebra finch is a favorite bird among aviculturists and researchers for their vocal abilities, social behaviors, and readiness to breed. Understanding much of zebra finch biology and behavior follows from appreciating their unique ecological niche and pronounced social nature. They are resilient and resourceful birds that have adapted to live under harsh and unpredictable environmental conditions. Though they are nonmigratory, they are distinctly mobile birds that breed opportunistically. Their high degree of sociality is evident from their breeding colonies and foraging flocks, and the pair bond between mates is the strongest conspecific relationship. Aggressive behavior is mostly limited to defending a nest or mate, and zebra finches do not form dominance hierarchies in the wild. Although they remain genetically diverse, domestication has altered some behaviors and physical traits of zebra finches. Designing enclosures that maintain behavioral health requires careful attention to the needs of zebra finches for space to fly and forage, a nutritious diet, and opportunities for unconstrained social interaction.

ADDITIONAL RESOURCES

Fee, M.S., Scharff, C., 2010. The Songbird as a Model for the Generation and Learning of Complex Sequential Behaviors. *ILAR J* 51, 362–377. https://doi.org/10.1093/ilar.51.4.362.

Olson, C.R., Wirthlin, M., Lovell, P.V., Mello, C.V., 2014. Proper Care, Husbandry, and Breeding Guidelines for the Zebra Finch, *Taeniopygia guttata*. *Cold Spring Harb Protoc* 2014, pdb.prot084780. https://doi.org/10.1101/pdb.prot084780.

Patterson, M.M., Fee, M.S., 2015. Zebra Finches in Biomedical Research, in: *Laboratory Animal Medicine*. Elsevier, pp. 1109–1134. https://doi.org/10.1016/B978-0-12-409527-4.00023-7.

Schmidt, M.F., 2010. An IACUC Perspective on Songbirds and Their Use in Neurobiological Research. *ILAR J* 51, 424–430.

REFERENCES

Adar, E., Nottebohm, F., Barnea, A., 2008. The relationship between nature of social change, age, and position of new neurons and their survival in adult zebra finch brain. *J. Neurosci.* 28, 5394–5400. https://doi.org/10.1523/JNEUROSCI.5706-07.2008.

Alvarez-Buylla, A., Theelen, M., Nottebohm, F., 1988. Birth of projection neurons in the higher vocal center of the canary forebrain before, during, and after song learning. *Proc. Natl. Acad. Sci.* 85, 8722–8726. https://doi.org/10.1073/pnas.85.22.8722.

Arnold, A.P., 1997. Sexual differentiation of the zebra finch song system: positive evidence, negative evidence, null hypotheses, and a paradigm shift. *J. Neurobiol.* 33, 572–584.

Barnea, A., Mishal, A., Nottebohm, F., 2006. Social and spatial changes induce multiple survival regimes for new neurons in two regions of the adult brain: an anatomical representation of time? *Behav. Brain Res.* 167, 63–74. https://doi.org/10.1016/j.bbr.2005.08.018.

BirdLife International, 2016. *Taeniopygia castanotis. IUCN Red List Threat. Species 2016.* e.T103818044A104212010. https://doi.org/10.2305/IUCN.UK.2016-3.RLTS.T103818044A104212010.en.

Birkhead, T.R., Burke, T., Zann, R., Hunter, F.M., Krupa, A.P., 1990. Extra-pair paternity and intraspecific brood parasitism in wild zebra finches *Taeniopygia guttata*, revealed by DNA fingerprinting. *Behav. Ecol. Sociobiol.* 27, 315–324.

Birkhead, T.R., Clarkson, K., Zann, R., 1988a. Extra-pair courtship, copulation and mate guarding in wild zebra finches *Taeniopygia guttata*. *Anim. Behav.* 36, 1853–1855. https://doi.org/10.1016/S0003-3472(88)80133-7.

Birkhead, T.R., Hunter, F.M., 1990. Mechanisms of sperm competition. *Trends Ecol. Evol.* 5, 48–52. https://doi.org/10.1016/0169-5347(90)90047-H.

Birkhead, T.R., Hunter, F.M., Pellatt, J.E., 1989. Sperm competition in the zebra finch, *Taeniopygia guttata*. *Anim. Behav.* 38, 935–950. https://doi.org/10.1016/S0003-3472(89)80135-6.

Birkhead, T.R., Møller, A.P., 1993. Sexual selection and the temporal separation of reproductive events: sperm storage data from reptiles, birds and mammals. *Biol. J. Linn. Soc.* 50, 295–311. https://doi.org/10.1111/j.1095-8312.1993.tb00933.x.

Birkhead, T.R., Pellatt, E.J., Brekke, P., Yeates, R., Castillo-Juarez, H., 2005. Genetic effects on sperm design in the zebra finch. *Nature* 434, 383–387. https://doi.org/10.1038/nature03374.

Birkhead, T.R., Pellatt, J., Hunter, F.M., 1988b. Extra-pair copulation and sperm competition in the zebra finch. *Nature* 334, 60–62. https://doi.org/10.1038/334060a0.

Blount, J.D., Metcalfe, N.B., Birkhead, T.R., Surai, P.F., 2003. Carotenoid modulation of immune function and sexual attractiveness in zebra finches. *Science* 300, 125–127. https://doi.org/10.1126/science.1082142.

Bottjer, S.W., Glaessner, S.L., Arnold, A.P., 1985. Ontogeny of brain nuclei controlling song learning and behavior in zebra finches. *J. Neurosci.* 5, 1556–1562.

Brainard, M.S., Doupe, A.J., 2002. What songbirds teach us about learning. *Nature* 417, 351–358. https://doi.org/10.1038/417351a.

Brenowitz, E.A., Larson, T.A., 2015. Neurogenesis in the adult avian song-control system. *Cold Spring Harb. Perspect. Biol.* 7, a019000. https://doi.org/10.1101/cshperspect.a019000.

Burley, N., 1985. Leg-band color and mortality patterns in captive breeding populations of zebra finches. *The Auk* 102, 647–651.

Burley, N., 1988. The differential-allocation hypothesis: an experimental test. *Am. Nat.* 132, 611–628.

Burley, N., Coopersmith, C.B., 1987. Bill color preferences of zebra finches. *Ethology* 76, 133–151. https://doi.org/10.1111/j.1439-0310.1987.tb00679.x.

Burley, N., Zann, R.A., Tidemann, S.C., Male, E.B., 1989. Sex ratios of zebra finches. *Emu - Austral Ornithol.* 89, 83–92. https://doi.org/10.1071/MU9890083.

Burley, N.T., Parker, P.G., Lundy, K., 1996. Sexual selection and extrapair fertilization in a socially monogamous passerine, the zebra finch (*Taeniopygia gullata*). *Behav. Ecol.* 7, 218–226. https://doi.org/10.1093/beheco/7.2.218.

Cade, T.J., Tobin, C.A., Gold, A., 1965. Water economy and metabolism of two estrildine finches. *Physiol. Zool.* 38, 9–33. https://doi.org/10.1086/physzool.38.1.30152342.

Caryl, P.G., 1976. Sexual behaviour in the zebra finch *Taeniopygia guttata*: Response to familiar and novel partners. *Anim. Behav.* 24, 93–107. https://doi.org/10.1016/S0003-3472(76)80103-0.

Clayton, N.S., 1988. Song discrimination learning in zebra finches. *Anim. Behav.* 36, 1016–1024. https://doi.org/10.1016/S0003-3472(88)80061-7.

Clayton, N.S., 1990. Mate choice and pair formation in Timor and Australian Mainland zebra finches. *Anim. Behav.* 39, 474–480. https://doi.org/10.1016/S0003-3472(05)80411-7.

Coleman, M.J., Saxon, D., Robbins, A., Lillie, N., Day, N.F., 2019. Operant conditioning task to measure song preference in zebra finches. *J. Vis. Exp. JoVE*. https://doi.org/10.3791/60590.

Cynx, J., Nottebohm, F., 1992. Role of gender, season, and familiarity in discrimination of conspecific song by zebra finches (*Taeniopygia guttata*). *Proc. Natl. Acad. Sci.* 89, 1368–1371. https://doi.org/10.1073/pnas.89.4.1368.

D'Amelio, P.B., Klumb, M., Adreani, M.N., Gahr, M.L., ter Maat, A., 2017. Individual recognition of opposite sex vocalizations in the zebra finch. *Sci. Rep.* 7, 1–10. https://doi.org/10.1038/s41598-017-05982-x.

Derégnaucourt, S., Gahr, M., 2013. Horizontal transmission of the father's song in the zebra finch (Taeniopygia guttata). *Biol. Lett.* 9. https://doi.org/10.1098/rsbl.2013.0247.

Derégnaucourt, S., Poirier, C., van der Kant, A., van der Linden, A., Gahr, M., 2013. Comparisons of different methods to train a young zebra finch (*Taeniopygia guttata*) to learn a song. *J. Physiol. Paris* 107, 210–218. https://doi.org/10.1016/j.jphysparis.2012.08.003.

Dong, S., Replogle, K.L., Hasadsri, L., Imai, B.S., Yau, P.M., Rodriguez-Zas, S., Southey, B.R., Sweedler, J.V., Clayton, D.F., 2009. Discrete molecular states in the brain accompany changing responses to a vocal signal. *Proc. Natl. Acad. Sci.* 106, 11364–11369. https://doi.org/10.1073/pnas.0812998106.

Doupe, A.J., Kuhl, P.K., 1999. Birdong and human speech: common themes and mechanisms. *Annu. Rev. Neurosci.* 22, 567–631. https://doi.org/10.1146/annurev.neuro.22.1.567.

Elie, J.E., Mathevon, N., Vignal, C., 2011. Same-sex pair-bonds are equivalent to male–female bonds in a life-long socially monogamous songbird. *Behav. Ecol. Sociobiol.* 65, 2197–2208. https://doi.org/10.1007/s00265-011-1228-9.

Elie, J.E., Theunissen, F.E., 2016. The vocal repertoire of the domesticated zebra finch: a data-driven approach to decipher the information-bearing acoustic features of communication signals. *Anim Cogn* 19, 285–315. https://doi.org/10.1007/s10071-015-0933-6.

Elie, J.E., Theunissen, F.E., 2018. Zebra finches identify individuals using vocal signatures unique to each call type. *Nat. Commun.* 9, 4026. https://doi.org/10.1038/s41467-018-06394-9.

Evans, S.M., 1970. Aggressive and territorial behaviour in captive Zebra Finches. *Bird Study* 17, 28–35. https://doi.org/10.1080/00063657009476252.

Forstmeier, W., Segelbacher, G., Mueller, J.C., Kempenaers, B., 2007. Genetic variation and differentiation in captive and wild zebra finches (*Taeniopygia guttata*). *Mol. Ecol.* 16, 4039–4050. https://doi.org/10.1111/j.1365-294X.2007.03444.x.

Geberzahn, N., Derégnaucourt, S., 2020. Individual vocal recognition in zebra finches relies on song syllable structure rather than song syllable order. *J. Exp. Biol.* 223, jeb220087. https://doi.org/10.1242/jeb.220087.

George, J.M., Bell, Z.W., Condliffe, D., Dohrer, K., Abaurrea, T., Spencer, K., Leitão, A., Gahr, M., Hurd, P.J., Clayton, D.F., 2019. Acute social isolation alters neurogenomic state in songbird forebrain. *Proc. Natl. Acad. Sci.* 117, 23311–23316. https://doi.org/10.1073/pnas.1820841116.

George, J.M., Jin, H., Woods, W.S., Clayton, D.F., 1995. Characterization of a novel protein regulated during the critical period for song learning in the zebra finch. *Neuron* 15, 361–372. https://doi.org/10.1016/0896-6273(95)90040-3.

Gilby, A.J., Mainwaring, M.C., Griffith, S.C., 2013. Incubation behaviour and hatching synchrony differ in wild and captive populations of the zebra finch. *Anim. Behav.* 85, 1329–1334. https://doi.org/10.1016/j.anbehav.2013.03.023.

Goldman, S.A., Nottebohm, F., 1983. Neuronal production, migration, and differentiation in a vocal control nucleus of the adult female canary brain. *Proc. Natl. Acad. Sci.* 80, 2390–2394. https://doi.org/10.1073/pnas.80.8.2390.

Goodwin, D., Woodcock, M., 1982. *Estrildid finches of the world*. British Museum (Natural History), London.

Griffith, S.C., Buchanan, K.L., 2010a. The zebra finch: the ultimate Australian supermodel. *Emu - Austral Ornithol.* 110, v–xii. https://doi.org/10.1071/MUv110n3_ED.

Griffith, S.C., Buchanan, K.L., 2010b. Maternal effects in the zebra finch: a model mother reviewed. *Emu - Austral Ornithol.* 110, 251–267. https://doi.org/10.1071/MU10006.

Griffith, S.C., Holleley, C.E., Mariette, M.M., Pryke, S.R., Svedin, N., 2010. Low level of extrapair parentage in wild zebra finches. *Anim. Behav.* 79, 261–264. https://doi.org/10.1016/j.anbehav.2009.11.031.

Haesler, S., Rochefort, C., Georgi, B., Licznerski, P., Osten, P., Scharff, C., 2007. Incomplete and inaccurate vocal imitation after knockdown of FoxP2 in songbird basal ganglia nucleus Area X. *PLoS Biol.* 5, e321. https://doi.org/10.1371/journal.pbio.0050321.

Heidweiller, J., Zweers, G.A., 1990. Drinking mechanisms in the zebra finch and the Bengalese finch. *The Condor* 92, 1–28. https://doi.org/10.2307/1368379.

Hilliard, A.T., Miller, J.E., Horvath, S., White, S.A., 2012. Distinct neurogenomic states in basal ganglia subregions relate differently to singing behavior in songbirds. *PLoS Comput. Biol.* 8, e1002773. https://doi.org/10.1371/journal.pcbi.1002773.

Holloway, C.C., Clayton, D.F., 2001. Estrogen synthesis in the male brain triggers development of the avian song control pathway in vitro. *Nat. Neurosci.* 4, 170–175. https://doi.org/10.1038/84001.

Houx, A.B., Ten cate, C., 1999. Song learning from playback in zebra finches: is there an effect of operant contingency? *Anim. Behav.* 57, 837–845. https://doi.org/10.1006/anbe.1998.1046.

Jarvis, E.D., Scharff, C., Grossman, M.R., Ramos, J.A., Nottebohm, F., 1998. For whom the bird sings: Context-dependent gene expression. *Neuron* 21, 775–788. https://doi.org/10.1016/S0896-6273(00)80594-2.

Karten, H.J., 2013. Neocortical evolution: neuronal circuits arise independently of lamination. *Curr. Biol.* 23, R12–R15. https://doi.org/10.1016/j.cub.2012.11.013.

Konishi, M., Akutagawa, E., 1985. Neuronal growth, atrophy and death in a sexually dimorphic song nucleus in the zebra finch brain. *Nature* 315, 145–147. https://doi.org/10.1038/315145a0.

Law, G., Nager, R., Wilkinson, M., 2018. Zebra finches (*Taeniopygia guttata*), in: Yeates, J. (Ed.), *Companion animal care and welfare*. John Wiley & Sons, Ltd, Chichester, UK, pp. 318–337. https://doi.org/10.1002/9781119333708.ch15.

Lee, P., Schmidt-Nielsen, K., 1971. Respiratory and cutaneous evaporation in the zebra finch: Effect on water balance. *Am. J. Physiol.* 220, 1598–1605. https://doi.org/10.1152/ajplegacy.1971.220.6.1598.

Lipkind, D., Marcus, G.F., Bemis, D.K., Sasahara, K., Jacoby, N., Takahasi, M., Suzuki, K., Feher, O., Ravbar, P., Okanoya, K., Tchernichovski, O., 2013. Stepwise acquisition of vocal combinatorial capacity in songbirds and human infants. *Nature* 498, 104–108. https://doi.org/10.1038/nature12173.

Lipkind, D., Nottebohm, F., Rado, R., Barnea, A., 2002. Social change affects the survival of new neurons in the forebrain of adult songbirds. *Behav. Brain Res.* 133, 31–43. https://doi.org/10.1016/S0166-4328(01)00416-8.

Low, P.S., Shank, S.S., Sejnowski, T.J., Margoliash, D., 2008. Mammalian-like features of sleep structure in zebra finches. *Proc. Natl. Acad. Sci.* 105, 9081–9086. https://doi.org/10.1073/pnas.0703452105.

Macedo-Lima, M., Remage-Healey, L., 2020. Auditory learning in an operant task with social reinforcement is dependent on neuroestrogen synthesis in the male songbird auditory cortex. *Horm. Behav.* 121, 104713. https://doi.org/10.1016/j.yhbeh.2020.104713.

Mann, N.I., Slater, P.J.B., 1995. Song tutor choice by zebra finches in aviaries. *Anim. Behav.* 49, 811–820. https://doi.org/10.1016/0003-3472(95)80212-6.

Martin, H.J., 2000. *Zebra finches*. Barron's, New York.

McCowan, L.S.C., Mariette, M.M., Griffith, S.C., 2015. The size and composition of social groups in the wild zebra finch. *Emu - Austral Ornithol.* 115, 191–198. https://doi.org/10.1071/MU14059.

Melgar, J., Lind, O., Muheim, R., 2015. No response to linear polarization cues in operant conditioning experiments with zebra finches. *J. Exp. Biol.* 218, 2049–2054. https://doi.org/10.1242/jeb.122309.

Mello, C.V., 2014. The zebra finch, *Taeniopygia guttata*: an avian model for investigating the neurobiological basis of vocal learning. *Cold Spring Harb. Protoc.* 2014, pdb.emo084574. https://doi.org/10.1101/pdb.emo084574.

Mello, C.V., Clayton, D.F., 1994. Song-induced ZENK gene expression in auditory pathways of songbird brain and its relation to the song control system. *J. Neurosci.* 14, 6652–6666. https://doi.org/10.1523/JNEUROSCI.14-11-06652.1994.

Mello, C.V., Vicario, D.S., Clayton, D.F., 1992. Song presentation induces gene expression in the songbird forebrain. *Proc. Natl. Acad. Sci.* 89, 6818–6822. https://doi.org/10.1073/pnas.89.15.6818.

Miller, J.E., Hafzalla, G.W., Burkett, Z.D., Fox, C.M., White, S.A., 2015. Reduced vocal variability in a zebra finch model of dopamine depletion: implications for Parkinson disease. *Physiol. Rep.* 3, e12599. https://doi.org/10.14814/phy2.12599.

Morris, D., 1954. The reproductive behaviour of the zebra finch (*Poephila guttata*), with special reference to pseudofemale behaviour and displacement activities. *Behaviour* 6, 271–322.

Muller, R.E., Smith, D.G., 1978. Parent-offspring interactions in zebra finches. *The Auk* 95, 485–495. https://doi.org/10.1093/auk/95.3.485.

Mundy, N.I., Stapley, J., Bennison, C., Tucker, R., Twyman, H., Kim, K.-W., Burke, T., Birkhead, T.R., Andersson, S., Slate, J., 2016. Red carotenoid coloration in the zebra finch is controlled by a cytochrome p450 gene cluster. *Curr. Biol.* 26, 1435–1440. https://doi.org/10.1016/j.cub.2016.04.047.

Nager, R.G., Law, G., 2010. The zebra finch, in: Hubrecht, R., Kirkwood, J. (Eds.), *The UFAW Handbook on the Care and Management of Laboratory and Other Research Animals*. Wiley-Blackwell, Oxford, UK, pp. 674–685. https://doi.org/10.1002/9781444318777.ch43.

Nixdorf-Bergweiler, B.E., 1996. Divergent and parallel development in volume sizes of telencephalic song nuclei in and female zebra finches. *J. Comp. Neurol.* 375, 445–456. https://doi.org/10.1002/(SICI)1096-9861(19961118)375:3<445::AID-CNE7>3.0.CO;2-2.

Nottebohm, F., Arnold, A.P., 1976. Sexual dimorphism in vocal control areas of the songbird brain. *Science* 194, 211–213.

Okanoya, K., Yoneda, T., Kimura, T., 1993. Acoustical variations in sexually dimorphic features of distance calls in domesticated zebra finches (*Taeniopygia guttata castanotis*). *J. Ethol.* 11, 29–36. https://doi.org/10.1007/BF02350003.

Olsson, U., Alström, P., 2020. A comprehensive phylogeny and taxonomic evaluation of the waxbills (Aves: Estrildidae). *Mol. Phylogenet. Evol.* 146, 106757. https://doi.org/10.1016/j.ympev.2020.106757.

Paton, J.A., Nottebohm, F.N., 1984. Neurons generated in the adult brain are recruited into functional circuits. *Science* 225, 1046–1048. https://doi.org/10.1126/science.6474166.

Perez, E.C., Elie, J.E., Soulage, C.O., Soula, H.A., Mathevon, N., Vignal, C., 2012. The acoustic expression of stress in a songbird: does corticosterone drive isolation-induced modifications of zebra finch calls? *Horm. Behav.* 61, 573–581. https://doi.org/10.1016/j.yhbeh.2012.02.004.

Petkov, C.I., Jarvis, E., 2012. Birds, primates, and spoken language origins: behavioral phenotypes and neurobiological substrates. *Front. Evol. Neurosci.* 4, 12. https://doi.org/10.3389/fnevo.2012.00012.

Pfenning, A.R., Hara, E., Whitney, O., Rivas, M.V., Wang, R., Roulhac, P.L., Howard, J.T., Wirthlin, M., Lovell, P.V., Ganapathy, G., Mountcastle, J., Moseley, M.A., Thompson, J.W., Soderblom, E.J., Iriki, A., Kato, M., Gilbert, M.T.P., Zhang, G., Bakken, T., Bongaarts, A., Bernard, A., Lein, E., Mello, C.V., Hartemink, A.J., Jarvis, E.D., 2014. Convergent transcriptional specializations in the brains of humans and song-learning birds. *Science* 346, 1256846. https://doi.org/10.1126/science.1256846.

Price, P.H., 1979. Developmental determinants of structure in zebra finch song. *J. Comp. Physiol. Psychol.* 93, 260–277. https://doi.org/10.1037/h0077553.

Pytte, C.L., George, S., Korman, S., David, E., Bogdan, D., Kirn, J.R., 2012. Adult neurogenesis Is associated with the maintenance of a stereotyped, learned motor behavior. *J. Neurosci.* 32, 7052–7057. https://doi.org/10.1523/JNEUROSCI.5385-11.2012.

Ratcliffe, L.M., Boag, P.T., 1987. Effects of colour bands on male competition and sexual attractiveness in zebra finches (*Poephila guttata*). *Can. J. Zool.* 65, 333–338. https://doi.org/10.1139/z87-052.

Riters, L.V., Stevenson, S.A., 2012. Reward and vocal production: song-associated place preference in songbirds. *Physiol. Behav.* 106, 87–94. https://doi.org/10.1016/j.physbeh.2012.01.010.

Ruploh, T., Bischof, H.-J., von Engelhardt, N., 2013. Adolescent social environment shapes sexual and aggressive behaviour of adult male zebra finches (*Taeniopygia guttata*). *Behav. Ecol. Sociobiol.* 67, 175–184. https://doi.org/10.1007/s00265-012-1436-y.

Ruploh, T., Bischof, H.-J., von Engelhardt, N., 2014. Social experience during adolescence influences how male zebra finches (*Taeniopygia guttata*) group with conspecifics. *Behav. Ecol. Sociobiol.* 68, 537–549. https://doi.org/10.1007/s00265-013-1668-5.

Scharff, C., Kirn, J.R., Grossman, M., Macklis, J.D., Nottebohm, F., 2000. Targeted neuronal death affects neuronal replacement and vocal behavior in adult songbirds. *Neuron* 25, 481–492. https://doi.org/10.1016/S0896-6273(00)80910-1.

Schielzeth, H., Bolund, E., 2010. Patterns of conspecific brood parasitism in zebra finches. *Anim. Behav.* 79, 1329–1337. https://doi.org/10.1016/j.anbehav.2010.03.006.

Schlinger, B.A., Arnold, A.P., 1992. Plasma sex steroids and tissue aromatization in hatchling zebra finches: Implications for the sexual differentiation of singing behavior. *Endocrinology* 130, 289–299. https://doi.org/10.1210/endo.130.1.1727704.

Silcox, A.P., Evans, S.M., 1982. Factors affecting the formation and maintenance of pair bonds in the zebra finch, *Taeniopygia guttata*. *Anim. Behav.* 30, 1237–1243. https://doi.org/10.1016/S0003-3472(82)80216-9.

Sturdy, C.B., Phillmore, L.S., Sartor, J.J., Weisman, R.G., 2001. Reduced social contact causes auditory perceptual deficits in zebra finches, *Taeniopygia guttata*. *Anim. Behav.* 62, 1207–1218. https://doi.org/10.1006/anbe.2001.1864.

Swaddle, J.P., Cuthill, I.C., 1994. Preference for symmetric males by female zebra finches. *Nature* 367, 165–166. https://doi.org/10.1038/367165a0.

Tanaka, M., Alvarado, J.S., Murugan, M., Mooney, R., 2016. Focal expression of mutant huntingtin in the songbird basal ganglia disrupts cortico-basal ganglia networks and vocal sequences. *Proc. Natl. Acad. Sci.* 113, E1720–E1727. https://doi.org/10.1073/pnas.1523754113.

Taylor, E.J., 1996. An evaluation of the importance of insoluble versus soluble grit in the diet of canaries. *J. Avian Med. Surg.* 10, 248–251.

Tchernichovski, O., Nottebohm, F., 1998. Social inhibition of song imitation among sibling male zebra finches. *Proc. Natl. Acad. Sci.* 95, 8951–8956. https://doi.org/10.1073/pnas.95.15.8951.

Tomaszycki, M.L., Zatirka, B.P., 2014. Same-sex partner preference in zebra finches: Pairing flexibility and choice. *Arch. Sex. Behav.* 43, 1469–1475. https://doi.org/10.1007/s10508-014-0377-0.

Vernaleo, B.A., Dooling, R.J., 2011. Relative salience of envelope and fine structure cues in zebra finch song. *J. Acoust. Soc. Am.* 129, 3373–3383. https://doi.org/10.1121/1.3560121.

Wada, K., Howard, J.T., McConnell, P., Whitney, O., Lints, T., Rivas, M.V., Horita, H., Patterson, M.A., White, S.A., Scharff, C., Haesler, S., Zhao, S., Sakaguchi, H., Hagiwara, M., Shiraki, T., Hirozane-Kishikawa, T., Skene, P., Hayashizaki, Y., Carninci, P., Jarvis, E.D., 2006. A molecular neuroethological approach for identifying and characterizing a cascade of behaviorally regulated genes. *Proc. Natl. Acad. Sci.* 103, 15212–15217. https://doi.org/10.1073/pnas.0607098103.

Wade, J., Arnold, A.P., 2004. Sexual differentiation of the zebra finch song system. *Ann. N. Y. Acad. Sci.* 1016, 540–559. https://doi.org/10.1196/annals.1298.015.

Walton, C., Pariser, E., Nottebohm, F., 2012. The zebra finch paradox: song is little changed, but number of neurons doubles. *J. Neurosci.* 32, 761–774. https://doi.org/10.1523/JNEUROSCI.3434-11.2012.

Wang, N., Aviram, R., Kirn, J.R., 1999. Deafening alters neuron turnover within the telencephalic motor pathway for song control in adult zebra Finches. *J. Neurosci.* 19, 10554–10561. https://doi.org/10.1523/JNEUROSCI.19-23-10554.1999.

Warren, W.C., Clayton, D.F., Ellegren, H., Arnold, A.P., Hillier, L.W., Künstner, A., Searle, S., White, S., Vilella, A.J., Fairley, S., Heger, A., Kong, L., Ponting, C.P., Jarvis, E.D., Mello, C.V., Minx, P., Lovell, P., Velho, T.A.F., Ferris, M., Balakrishnan, C.N., Sinha, S., Blatti, C., London, S.E., Li, Y., Lin, Y.-C., George, J., Sweedler, J., Southey, B., Gunaratne, P., Watson, M., Nam, K., Backström, N., Smeds, L., Nabholz, B., Itoh, Y., Whitney, O., Pfenning, A.R., Howard, J., Völker, M., Skinner, B.M., Griffin, D.K., Ye, L., McLaren, W.M., Flicek, P., Quesada, V., Velasco, G., Lopez-Otin, C., Puente, X.S., Olender, T., Lancet, D., Smit, A.F.A., Hubley, R., Konkel, M.K., Walker, J.A., Batzer, M.A., Gu, W., Pollock, D.D., Chen, L., Cheng, Z., Eichler, E.E., Stapley, J., Slate, J., Ekblom, R., Birkhead, T., Burke, T., Burt, D., Scharff, C., Adam, I., Richard, H., Sultan, M., Soldatov, A., Lehrach, H., Edwards, S.V., Yang, S.-P., Li, X., Graves, T., Fulton, L., Nelson, J., Chinwalla, A., Hou, S., Mardis, E.R., Wilson, R.K., 2010. The genome of a songbird. *Nature* 464, 757–762. https://doi.org/10.1038/nature08819.

Whitney, O., Pfenning, A.R., Howard, J.T., Blatti, C.A., Liu, F., Ward, J.M., Wang, R., Audet, J.-N., Kellis, M., Mukherjee, S., Sinha, S., Hartemink, A.J., West, A.E., Jarvis, E.D., 2014. Core and region-enriched networks of behaviorally regulated genes and the singing genome. *Science* 346. https://doi.org/10.1126/science.1256780.

Williams, H., 2001. Choreography of song, dance and beak movements in the zebra finch (*Taeniopygia guttata*). *J. Exp. Biol.* 204, 3497–3506.

Winograd, C., Clayton, D., Ceman, S., 2008. Expression of fragile X mental retardation protein within the vocal control system of developing and adult male zebra finches. *Neuroscience* 157, 132–142. https://doi.org/10.1016/j.neuroscience.2008.09.005.

Zann, R., 1977. Pair-bond and bonding behaviour in three species of grassfinches of the genus *Poephila* (Gould). *Emu - Austral Ornithol.* 77, 97–106. https://doi.org/10.1071/MU9770097.

Zann, R., 1985. Slow continuous wing-moult of Zebra Finches *Poephila guttata* from southeast Australia. *IBIS* 127, 184–196. https://doi.org/10.1111/j.1474-919X.1985.tb05054.x.

Zann, R., 1990. Song and call learning in wild zebra finches in south-east Australia. *Anim. Behav.* 40, 811–828. https://doi.org/10.1016/S0003-3472(05)80982-0.

Zann, R., 1993. Structure, sequence and evolution of song elements in wild Australian zebra finches. *The Auk* 110, 702–715. https://doi.org/10.2307/4088626.

Zann, R., 1996. *The zebra finch: a synthesis of field and laboratory studies*. Oxford University Press, Cambridge.

Zann, R., Morton, S.R., Jones, K.R., Burley, N.T., 1995. The timing of breeding by zebra finches in relation to rainfall in central Australia. *Emu* 95, 208–222. https://doi.org/10.1071/mu9950208.

Zann, R., Rossetto, M., 1991. Zebra finch incubation: brood patch, egg temperature and thermal properties of the nest. *Emu* 91, 107–120. https://doi.org/10.1071/mu9910107.

Zann, R., Runciman, D., 1994. Survivorship, dispersal and sex ratios of zebra finches *Taeniopygia guttata* in southeast Australia. *IBIS* 136, 136–143. https://doi.org/10.1111/j.1474-919X.1994.tb01077.x.

Zeigler, H.P., Marler, P., 2008. *Neuroscience of birdsong*. Cambridge University Press, New York.

21

Behavioral Biology of Zebrafish

Christine Powell, Isabel Fife-Cook, and Becca Franks
New York University

CONTENTS

Introduction 331
   Typical Research 332
Natural History 332
   Ecology 332
      Geography 332
      Diet 333
      Growth and Mortality 333
      Predation 334
   Social Organization and Behavior 334
      Reproductive Behaviors 334
      Shoaling 334
      Agonistic Behaviors 335
   Feeding Behaviors 335
   Communication 335
Captive Behaviors 336
   Normal Behaviors 336
   Abnormal Behaviors in Captivity 336
Ways to Maintain Behavioral Health 337
   Environment to Promote Species-Typical Behaviors 337
   Social Environment to Promote Species-Typical Behaviors 337
   Feeding Strategies to Promote Species-Typical Behaviors 338
   Training to Promote Species-Typical Behaviors 339
Special Situations 339
Conclusion 339
References 340

Introduction

The more we study zebrafish, the more we learn about (1) their specific behavioral biology and associated welfare requirements and (2) how little we know about their specific behavioral biology and associated welfare requirements. Zebrafish welfare deserves careful consideration for a number of reasons (Graham et al., 2018b), including the growing consensus that zebrafish feel pain (Maximino, 2011; Sneddon, 2012), show evidence of consciousness through emotional fever (Rey et al., 2015), and have high cognitive capacities (Al-imari & Gerlai, 2008; Colwill et al., 2005; Lindeyer & Reader, 2010). Further, the "refinement" principle of the 3Rs ("refinement", "replacement", and "reduction") of animal research compels caretakers to promote good welfare through husbandry and housing (Russell & Burch, 1959; Sneddon et al., 2017). However, even as zebrafish are involved in scientific research in larger numbers than ever before (Lawrence, 2016), our knowledge of their behavioral biology may be limited by the fact that most of the information we collect about zebrafish occurs in restrictive laboratory environments (see Figure 21.1).

Good welfare extends beyond physiological needs to include the freedom to exercise highly motivated behaviors (Mellor, 2016); an understanding of natural behavior and motivation is, therefore, a prerequisite for ensuring acceptable welfare and for generating sound scientific data (Bracke & Hopster, 2006). The current standard in laboratory housing is designed to meet the minimal physiological needs of zebrafish, while minimizing the cost of research; resulting in spaces that are small, static, and barren in comparison to natural home ranges. Behavior in such environments may differ from behavior in the wild. As such, after briefly detailing typical laboratory husbandry, the first half of this chapter describes the behavior of zebrafish in natural environments to shift baseline

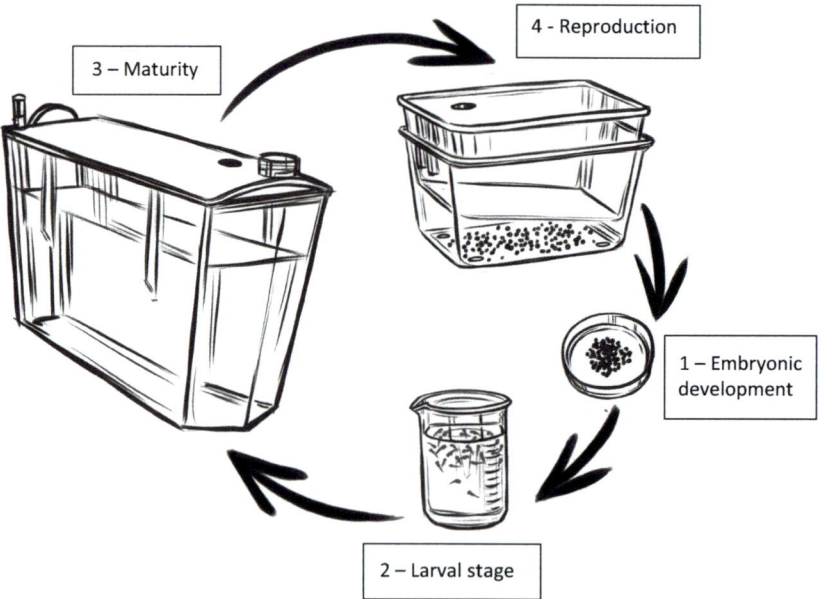

FIGURE 21.1 Standard laboratory housing conditions for zebrafish throughout their lifecycle. (1) Zebrafish embryos are moved from the spawning tank to a petri dish, where they are either used immediately or incubated at 28°C for 72 h until hatched (Reed & Jennings, 2011). (2) Larval zebrafish may remain in the petri dish for up to 5 days postfertilization before being moved to a larger container (such as the graduated cylinder depicted above). Alternately, larvae may be transferred directly into a holding tank (3), provided the water flow can be restricted (Reed & Jennings, 2011). (3) Juvenile and adult zebrafish are typically housed in racked tanks, ranging in size from 3 to 10 L, at stocking densities between 1 and 5 individuals/L (Lawrence, 2016). Fish who are permitted to reach adulthood will reach sexual maturity at approximately 3 months and achieve peak fecundity between 7 and 18 months of age (Singleman & Holtzman, 2014). (4) Between two and ten adult zebrafish (recommended sex ratio is fewer males than females) are placed in a spawning tank, where the eggs are deposited, fertilized, and removed (Tsang et al., 2017). (Illustrations by I. Fife-Cook.)

standards back to the ecological context in which they were formed. The second half of the chapter applies our understanding of these natural behaviors to inform the way we care, and improve care, for zebrafish in captivity.

Typical Research

Fish have become one of the most popular taxa in scientific research, with zebrafish the most widely involved species ("CCAC Animal Data Report," 2014). Including breeding fish, the number of zebrafish kept in laboratories worldwide can be conservatively estimated to be 3.5 million, but the exact number is likely higher and expected to rise (Lidster et al., 2017). Compared to the other most common research model, mice, zebrafish are 1,000 times less expensive to maintain (Goldsmith, 2004). This price difference is due to the species' short life span, high fecundity, and minimal labor costs associated with current housing practices.

As scientific models, zebrafish offer certain advantages. Transparent embryos allow for detailed anatomical characterization (Fetcho & Liu, 1998). The zebrafish genome is somewhat homogenous to humans and is now the most fully mapped of all vertebrates (Howe et al., 2013). Because of these technological developments, zebrafish are typically involved in research for molecular biology, developmental biology, neurobiology, and genetics (Reed & Jennings, 2011), but also increasingly behavioral research investigating psychological disorders and pharmaceutical effects (Gerlai, 2003).

Enthusiasm for zebrafish as an ideal scientific model, however, should be tempered by our relatively limited knowledge regarding how environmental parameters and variation in housing conditions affect their biology, behavior, and welfare. The majority of published papers do not report reliable or reproducible descriptions of the conditions in which the zebrafish live (Lawrence, 2016), and a recent survey by Lidster et al. of laboratories across 20 different countries suggests that variability in husbandry procedures is great. For example, some laboratories used flow-through systems, but the majority had recirculation systems. Stocking densities also ranged greatly, from one to more than five individuals per liter of water. Finally, 20% of laboratories reported that their tanks contain plants and gravel, but the majority of survey respondents answered that these items were either "preferable" to current conditions or "not considered an option", indicating that most laboratory zebrafish live in barren tanks (Lidster et al., 2017). Importantly, while some husbandry parameters are quite variable, the general picture of typical zebrafish housing within a laboratory is one of monotony and low-complexity (see Figure 21.1). These characteristics contrast greatly with what we know about their lives in the wild (see Figure 21.2).

Natural History

Ecology

Geography

Danio rerio are native to the Ganges and Brahmaputra river basins across northeastern India, Bangladesh, and Nepal. They typically live in slow-moving or stagnant pools that connect

Behavioral Biology of Zebrafish

FIGURE 21.2 The community and habitat in which wild zebrafish live. (1) Indian pond heron (*Ardeola grayii*), a known predator of zebrafish (*Danio rerio*); (2) Common kingfisher (*Alcedo atthis*), another predator of zebrafish; (3) Zebrafish feeding on mosquito larvae; (4) Flying barb (*Esomus danricus*), a species commonly found with zebrafish; (5) Dwarf snakehead (*Channa andrao*), an aquatic predator of zebrafish. (Illustrations by I. Fife-Cook.)

to main rivers during monsoon season, as well as man-made lakes, ponds, and irrigation channels constructed for agricultural use (Spence et al., 2006). Suriyampola et al. (2015) found that group sizes are larger in faster-moving bodies of water (where fish form groups averaging 80 individuals) than in stagnant environments (where average group size is ten fish). They can be found in a wide range of elevations (8–1,576 m) and water conditions (12°C–39°C, pH from 5.9 to 9.8, and salinities of 0.01–0.8; Engeszer et al., 2007; Spence et al., 2006). As a nonmigratory species, zebrafish have evolved to withstand seasonal environmental fluctuations within their home ranges.

A location analysis of 28 sites along the Ganges and Brahmaputra floodplains found that zebrafish live in shallow bodies of water with overhanging and aquatic vegetation, silt substrate, and visibility of approximately 30 cm. On the rare occasion that zebrafish are identified in rocky-bottomed riverbeds, they are found in small groups and remain near the shoreline, underneath overhanging vegetation (Engeszer et al., 2007). In the wild, zebrafish occupy the entire water column and spend time in open water, as well as among aquatic vegetation (Spence et al., 2006).

Diet

Zebrafish are omnivorous and opportunistic feeders. Gut-content analysis has reported that zooplankton and insects form the foundation of their diets, supplemented by a variety of organic and inorganic materials, including collections of phytoplankton, algae, plant material, spores, invertebrate eggs, fish scales, arachnids, detritus, sand, and mud (McClure et al., 2006; Spence et al., 2008). These substances suggest that zebrafish feed throughout the entire water column and scavenge for food. However, a large proportion of terrestrial insects found in gut-content analyses indicate that although zebrafish may forage within the entire water column, they primarily feed at the surface (McClure et al., 2006; Spence et al., 2007). Food items appear to change with seasonal availability.

Growth and Mortality

Zebrafish grow at a rapid rate for their first 3 months, but do not reach full maturity until around 18 months of age (Figure 21.3). In the wild, this initial growth spurt typically coincides with the onset of monsoon season (June to September), when high temperatures and ample rainfall create an abundance of food

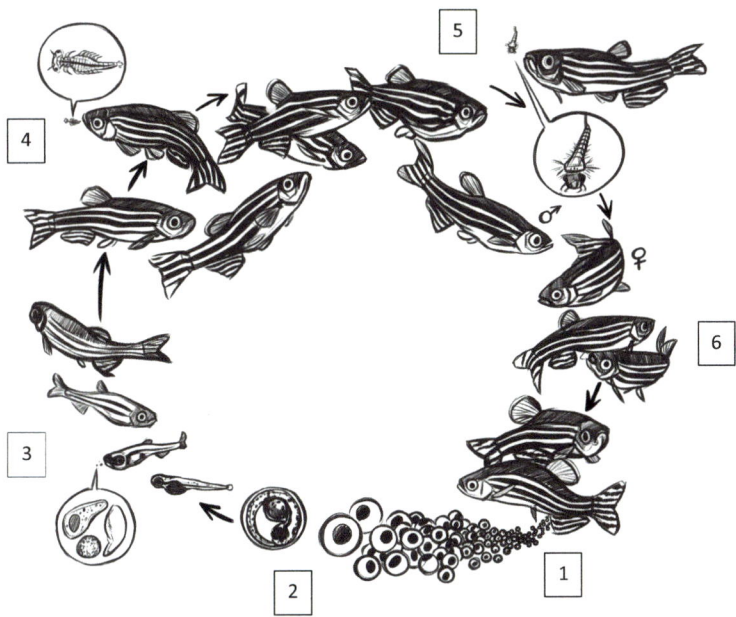

FIGURE 21.3 The life cycle of the zebrafish, including sexual dimorphism and diet. (1) *D. rerio* spawning; (2) Embryonic development; (3) *D. rerio* fry feeding on phytoplankton and paramecia; (4) Juvenile *D. rerio* feeding on brine shrimp; (5) Adult *D. rerio* feeding on *Culicidae* larvae; (6) A sexually mature male *D. rerio* courting a female. (Illustrations by I. Fife-Cook.)

(Spence et al., 2007). The mean length of zebrafish from a lake population in Bangladesh is 25mm, and the maximum length is 35mm, the latter of which is comparable to the typical length observed in laboratory strains. While there are few reliable data on lifespan in the wild, Gerhard et al. (2002) reported a mean life span of 3.5 years in domesticated zebrafish, with the oldest individual living to be more than 5.5 years old.

Predation

Many predatory fishes live in areas with zebrafish, including snakeheads, *Channa sp*; the freshwater garfish, *Xenentodon cancila*; and nocturnal catfish, *Mystus bleekeri*. Avian predators may include the Indian pond heron, *Ardeola grayii*, and the common kingfisher, *Alcedo atthis*, among many others (see Figure 21.2b). Engeszer et al. (2007) provides an extensive list of potential predators, but states that further field research (including gut-content analysis) is required to identify these species as confirmed predators.

Laboratory studies have shown that zebrafish display distinct reactions to both visual and olfactory predatory cues. Zebrafish show fear reactions to even trace quantities of the pheromone, schreckstoff, that is released as a result of injury to the epidermal cells (Wisenden et al., 2004). Visual cues associated with specific predatory animals also elicit fear responses in zebrafish; when shown a video recording of a sympatric predator (the Indian Leaf Fish, *Nandus nandus*), zebrafish, without previous exposure to similar threats, attempted to flee from the stimulus (Bass & Gerlai, 2008). Specific escape behaviors include increased shoaling, erratic movement, zig-zagging, and freezing (Kalueff et al., 2013). Alarm behaviors include an increase in shoal cohesion and either agitated swimming or freezing on the substratum, a decrease in feeding rate, and increased aggression. Social learning also plays a role in minimizing the risk of predation. Zebrafish have been shown to observe others to assess risk (Zala et al., 2012) and identify novel predators (Brown & Laland, 2003; Kelley & Magurran, 2003). Lindeyer and Reader found that zebrafish can learn escape routes from conspecifics and then transmit this information across generations (2010).

Social Organization and Behavior

As zebrafish are an especially gregarious species, understanding their social behavior is essential to accessing their umwelt (Graham et al., 2018b). The broad features of their social life in nature are fairly well-understood (Engeszer et al., 2007; Spence et al., 2008); for instance, we know that zebrafish in the wild spend their time in groups (Suriyampola et al., 2015), that they breed in the summer months (Spence et al., 2006), and that they are a shoaling species that displays high coordination for purposes of finding food and evading predators (Miller & Gerlai, 2012). However, less is known about their fine-grained, daily social interactions under natural conditions. Furthermore, much of our current understanding of the details of zebrafish social behavior comes from laboratory studies involving fish housed in barren tanks with high stocking densities. These artificial environments may alter zebrafish social behavior and thereby inhibit our ability to understand their normal social behavior (Graham et al., 2018b). Nevertheless, some general patterns of their social interactions appear to be constant regardless of environment, providing a broad outline of aspects of their social lives. In addition, recent work in natural or seminatural settings has extended our knowledge of the range of their social dynamics.

Reproductive Behaviors

Wild zebrafish breed in the summer months during monsoon season, when the rains create shallow, still pools tangential to streams, with abundant vegetation and minimal predation risk (Parichy, 2015; Spence et al., 2006). Spawning occurs in the morning and lasts about an hour (Hutter et al., 2010; Spence et al., 2007, 2008). It is a common practice for laboratories to induce year-round breeding by adding more daylight hours into the photoperiod. Pheromones appear to be another proximate cause of breeding (van den Hurk et al., 1987). Zebrafish prefer to release their eggs into vegetation (Hutter et al., 2010) and over gravel (Spence et al., 2007).

Courtship begins with an increase in activity as the fish swim low in the water column. A male will chase a female, touching her belly and sides with his nose, attempting to lead her to a spawning site (Spence et al., 2008) (Figure 21.2). Females can assert their choice of mate by chasing these males away, but receptive females will engage with males, and the two will circle each other. Once over the spawning site, the two swim in parallel, with the male in contact, but slightly behind her. To end the process, the male quivers close to the female over the spawning site, egg and sperm are released, and the two part abruptly (Spence & Smith, 2005). This process may be repeated several times in one spawning period (Hutter et al., 2010). Both sexes change color during spawning, with more conspicuously colored males engaging in courtship more often than less colorful males (Hutter et al., 2012). Additionally, social status, olfactory cues, and body size also appear to influence reproductive success (Filby et al., 2010; Spence et al., 2008). After spawning, zebrafish provide no care to their offspring (Spence et al., 2008), and at least in laboratory conditions, filial cannibalism is common in both sexes (Hutter et al., 2010; Spence et al., 2008).

Shoaling

Wild zebrafish are rarely observed alone, instead preferring to swim in a shoal (Engeszer et al., 2007; Suriyampola et al., 2015). In the wild, zebrafish have been observed shoaling in groups as small as 4 individuals or as large as 300 individuals, with the largest groups found in fast-moving bodies of water (Suriyampola et al., 2015). Shoaling allows fish to forage while decreasing the attention needed for predator detection (Pitcher & Parrish, 1993), thus aiding predator evasion; the attack success of predators is inversely related to shoal size (Wright et al., 2006). Shoaling provides concrete benefits to the individual; compared to solitary zebrafish, shoaling zebrafish show lower indicators of stress in the presence of chemical alarm cues (Faustino et al., 2017), and individuals in a group are more likely to explore a novel object or food (Kareklas et al., 2018). As a result, the drive to shoal is strong and rewarding to zebrafish (Al-imari & Gerlai, 2008).

Laboratory studies have found that zebrafish imprint on the visual phenotype they are exposed to when young, regardless of their own phenotype, and prefer to shoal with these fish throughout their lives (Engeszer et al., 2004). In the wild, zebrafish often shoal with their kin, but have also been observed with other species of a similar appearance, such as *Rasbora daniconius*, and the flying barb, *Esomus danricus* (Engeszer et al., 2007; Mann et al., 2011; Suriyampola et al., 2015). Females were once thought not to shoal together (Delaney et al., 2002), but more recent studies found that sex ratios after spawning were even (Hutter et al., 2010; Spence et al., 2007).

In certain locations, shoals of zebrafish were observed displaying highly synchronized movements, a phenomenon known as schooling (Suriyampola et al., 2015). Schooling provides group members with hydrodynamic benefits that improve foraging and predator avoidance (Miller & Gerlai, 2012; Wiwchar et al., 2018). Often studied in response to threatening stimuli (Soares et al., 2018), evidence is growing that schooling may occur for more complex reasons. For example, zebrafish housed in a seminaturalistic laboratory setting exhibited increased coordination when given the opportunity to explore a previously inaccessible area of their environment (Graham et al., 2018a). In naturalistic tanks, zebrafish have also shown episodes of spontaneous heightened-shoaling, a behavior that appears to be self-reinforcing and driven by internal group dynamics (Franks et al., 2018b).

Agonistic Behaviors

Zebrafish engage in agonistic behavior in the form of chasing, fin-raises, mouth-opening, body color changes (with the aggressor becoming darker in color), biting, charging, and circling conspecifics (Kalueff et al., 2013). In escalated agonistic behavior, a pair may progress to physical contact, including biting, flicking with the tail, and lateral displays (Spence et al., 2008). Dominance hierarchies have yet to be studied in the wild, but are present in both sexes in laboratory studies (Dahlbom et al., 2012; Paull et al., 2010; Spence et al., 2008). In both sexes, more dominant individuals are larger and darker in color than subordinates (Filby et al., 2010); whether the relationship between size and social status is causal or correlational has yet to be studied. Notwithstanding, once formed, social status appears to be stable over time (Filby et al., 2010). Importantly, observations of wild zebrafish found low overall levels of aggression (Suriyampola et al., 2015). When moved to the lab, however, wild zebrafish have been found to become increasingly agonistic with increased time spent in captivity (Martins & Bhat, 2014).

Captive environments can increase levels of aggression in animals for at least two reasons: (1) confinement increases the number of interactions between conspecifics and (2) the limited resources in captivity (e.g., preferred locations) increase competition (Grant, 1993; Schoener, 1987). Indeed, the quantity of resources has been found to be a mitigating factor in zebrafish aggression (Basquill et al., 1998), and is especially important during spawning, when fish require access to spawning sites. As a result, territorial behavior around plants and substrate has been observed by male zebrafish during spawning in laboratory settings (Spence et al., 2007; Spence & Smith, 2005). Many studies investigating the relationship between territoriality and reproduction provide just one spawning site (Paull et al., 2010; Pyron, 2003; Spence & Smith, 2005), but little defensive territoriality was observed in Hutter and colleague's 1,100 L aquaria, with a stocking density of one fish per 137.5 L (Hutter et al., 2010). Captive environments are inherently less complex than natural habitats and complex environments have also been shown to lower agonistic behavior (Carfagnini et al., 2009). For example, in cichlids, both larger tanks and increased structural complexity decreased the frequency and intensity of agonistic interactions (Barley & Coleman, 2010; Oldfield, 2011). It is possible, therefore, that barren laboratory environments may provide a skewed picture of zebrafish agonistic behavior.

In sum, zebrafish are a highly social species that display a wide range of environmentally sensitive social behaviors. Some of the social behaviors observed under laboratory conditions (e.g., aggression levels) differ from those observed under naturalistic conditions, which raises questions about standard laboratory conditions as a reference for normal social behavior in this species. Studying zebrafish in more natural environments (including naturalistic laboratory tanks), therefore, will contribute critical knowledge to the scientific literature and improve our ability to understand their social needs and repertoire.

Feeding Behaviors

As a gregarious species, zebrafish typically forage in groups. Laboratory research has shown that, as with predator detection, zebrafish can learn details about the location and quality of food from their conspecifics. For example, Zala and Määttänen (2013) found that fish used the behavior of a knowledgeable demonstrator to anticipate the location of food delivery. Aspects of feeding strategies may also depend on water flow and predation threat; populations from high-flow and high-predation habitats take more risks to feed than populations from low-predation and stagnant habitats (Bhat et al., 2015). In general, males are bolder than females and smaller individuals take greater risks to feed than larger individuals. When given a choice among open, open and vegetated, open and unvegetated, and covered feeding areas, zebrafish prefer feeding where there is overhead coverage (Hamilton & Dill, 2002), possibly because the perceived risk of predation is lower.

Communication

Zebrafish communicate with one another through an impressive variety of modalities. Behavioral research has shown that visual exposure to conspecifics engaged in alarm behavior produces alarm responses in the observing fish (Suboski et al., 1990). Similarly, olfactory cues from a threatened fish can also elicit alarm responses in fish not directly exposed to the threat (Bailey et al., 2013), and handling stress can be communicated between fish through chemical signaling (Barcellos et al., 2011). Olfactory communication through pheromone release can also instigate reproductive behaviors (van den Hurk et al., 1987). For spatial navigation and group-coordination, zebrafish receive information from each other through the lateral line, an organ that runs from their head to their tail (Partridge & Pitcher, 1980). The movements of hair cells, structurally similar to those in the inner ear of human beings, allow fish to sense changes in water movement (Suli et al., 2012), and disturbance to the lateral line alters schooling behavior (Partridge & Pitcher, 1980).

Captive Behaviors

Behavioral observations are useful indicators of welfare in fish captivity, because they can be quickly administered and performed on living individuals. Moreover, as they are noninvasive and nonintrusive, the risk of exacerbating the animal's condition through the testing process is reduced (Huntingford et al., 2006).

Normal Behaviors

Typically, a behavior is considered "normal" if it is consistent with the full behavioral repertoire for the species observed in nature or in natural conditions (Hill & Broom, 2009). Indeed, the drive to perform certain natural behaviors is strong enough to be considered necessary and must be included in definitions of well-being (Bracke & Hopster, 2006). That being said, normal behavior is not always synonymous with good welfare in captive environments; some behaviors considered to be "natural" may harm the welfare of animals living in captivity (Špinka, 2006). For example, in zebrafish, individuals may act aggressively toward, and isolate, conspecifics that show signs of illness (personal observation), a behavior that would be adaptive for the groups' health in the wild, but may harm the isolated individual. Ultimately, we as caretakers are responsible for avoidable suffering that may result from such "normal behaviors" in laboratories.

Healthy zebrafish engage in food-seeking behaviors. As scavengers, zebrafish continuously forage throughout the day, which may appear as investigating substrate or nibbling items in the water column (Kalueff et al., 2013). If offered food, zebrafish show interest by approaching, increased activity levels, and ingesting with jaw movements. Zebrafish also sometimes taste their food by ingesting pieces and then spitting them out (Reed & Jennings, 2011).

Movements and activity levels can be indicative of welfare state in fish (Huntingford et al., 2006). In the daytime, zebrafish are constantly moving and active. Zebrafish should display similar activity levels within groups; steady, high levels of activity are considered normal (Suriyampola et al., 2017). At night, zebrafish sleep, which is characterized by reduced movement, reduced opercular rate, reduced mouth opening frequency, and alternating bouts of swimming and inactivity with a "droopy" caudal fin (Kalueff et al., 2013). Sleep behavior should cease quickly in response to external stimuli. Unstressed zebrafish are also known to occupy the entire water column (Cachat et al., 2011; Spence et al., 2006), but will spend more time near the bottom in the mornings during spawning (Hutter et al., 2010).

Social behaviors are sensitive indicators of welfare in zebrafish. When given opportunities to explore, zebrafish engage in agonism less frequently (Graham et al., 2018a). This is consistent with observations in the wild, where levels of aggression have been observed to be low compared to captivity (Suriyampola et al., 2015). Thus, low levels of agonistic behavior should be considered the norm. Affiliative behaviors, such as shoaling and schooling, are also observed in the wild (Suriyampola et al., 2015), may be indicative of positive affective states, and should be displayed in good welfare conditions. When given exploration opportunities, zebrafish showed an increase in affiliative behaviors, coordination, and cohesion (Graham et al., 2018a); providing zebrafish with opportunities for new experiences may promote engagement in such behaviors and improve well-being. Finally, we found that in naturalistic tanks, zebrafish regularly engage in heightened shoaling, an affiliative behavior that is marked by tight group cohesion and low aggression (Franks et al., 2018b). It appears to be self-reinforcing and driven by internal group dynamics, and may therefore be indicative of acceptable welfare. More research is needed to add to the toolkit caretakers have to assess positive welfare in zebrafish, and affiliative behaviors provide a promising area for future research (Fife-Cook & Franks, 2019; Franks et al., 2018c).

Abnormal Behaviors in Captivity

In general, a healthy appetite and food-seeking behavior is consistent with welfare, but on its own, feeding behavior may not always be a sensitive indicator of well-being in fish; as ectothermic organisms, zebrafish save energy by allowing their environments to heat their bodies for them, and so there are many reasons why a fish may not eat (Huntingford et al., 2006). However, acute and chronic stress does affect the feeding behavior. Zebrafish show an increased latency to feed after exposure to an aversive stimuli, such as exposure to an alarm chemical, being netted and placed in a novel environment (Moretz et al., 2007), or simply closing the tank lid to disturb the water's surface (Oswald & Robison, 2008).

Pain is known to have several effects on zebrafish behavior, including changes in activity levels (e.g., lethargy, especially with dorsal fins held close to the body), inhibited exploration, changes in feeding behavior, and pale body color (Kalueff et al., 2013). Increased ventilation (Reilly et al., 2008), irregular movements (Maximino, 2011), and rubbing (Reilly et al., 2008) have all been found to be associated with pain in other species of fish. However, zebrafish are a prey species and may attempt to hide their expressions of pain in threatening situations (Maximino, 2011), increasing the chances of caretakers missing indicators of pain in stressed fish.

Erratic swimming and freezing are two common behavioral indicators of anxiety in zebrafish (Blaser et al., 2010; Cachat et al., 2011). Freezing behavior is characterized by "a complete cessation of movement except for the gills and eyes" at the bottom of the tank (Kalueff et al., 2013, p. 75). Erratic movement is a pattern of darting, zig-zagging movements that frequently occurs directly before or after the freezing behavior. Zebrafish are also known to change their body color, and will become pale when displaying other signs of fear (Gerlai et al., 2001), and after handling (Hutter et al., 2012). Increased ventilation appears as gill movement in fish, as measured by the opercular rate. Increased opercular rates are well studied in association with stress and are commonly used to assess welfare

(Huntingford et al., 2006). In states of stress, fish increase their respiration as the body prepares for activity in a fight-or-flight response (Reilly et al., 2008).

When threatened, zebrafish are hesitant to enter open areas, staying close to the walls, a behavioral pattern known as "thigmotaxis" (Cachat et al., 2011; Stewart et al., 2010). Diving to the tank bottom is also often seen upon exposure to a novel, threatening environment (Cachat et al., 2011). Interestingly, diving appears to be not so much a preference to be close to the substrate, but a motivation to escape from the surface of the water, potentially to minimize exposure to overhead threats (Blaser & Goldsteinholm, 2012). In this way, the position of the fish in the water column is frequently used as a measure of stress in zebrafish, and termed "bottom-dwelling" (Blaser et al., 2010). Bottom-dwelling also increases with exposure to stimuli thought to induce pain, and decreases with the administration of analgesics (Schroeder & Sneddon, 2017).

Ataxia, or a disruption of equilibrium, can be displayed by fish swimming on their side or spiraling, and is associated with exposure to toxins. These behaviors, along with flashing, or rubbing of the body against the sides of the tank or structures, may be signs of pathogenic conditions (Huntingford et al., 2006; Martins et al., 2012). Other indicators of illness are similar to those indicative of pain, including lethargy, inhibited food-seeking or food intake, and inhibited exploration (Kalueff et al., 2013).

Stereotypic behavior is defined as any behavior that is "repetitive, invariant, and has no obvious goal or function" (Mason, 1991, p. 103). Well studied in terrestrial animals, repetitive behaviors arise in suboptimal environments and are often used as negative welfare indices. However, the interpretation of these behaviors can be more ambiguous; stereotypic behaviors may be coping strategies of individuals to such environments (Cronin et al., 1985). More recent studies in fish reveal a link between stereotypic behaviors and known stressors in fish (e.g., high-density conditions and environments without substrate) (Martins et al., 2012). In zebrafish, stereotypic behavior has been observed in the form of jaw movements and repetitive swimming patterns. Jaw movements are described as "non-foraging, mouth opening behavior" (Kalueff et al., 2013, p. 75). Stereotypic swimming behavior has recently become a topic of study as researchers attempt to model human neurobehavioral disorders (D'Amico et al., 2015; Liu et al., 2018). Included in these behaviors were movements termed "figure 8", "circling", and "walling", in which individuals swim along a wall repeatedly. But again, the welfare implications of such behaviors have yet to be studied in this species.

Ways to Maintain Behavioral Health

Environment to Promote Species-Typical Behaviors

A growing body of research suggests that housing captive animals in seminaturalistic environments is not only beneficial but also essential to promote psychological and physiological health. Animals living in restrictive, barren conditions often exhibit behaviors indicative of suffering, including chronic stress, heightened sensitivity to pain, abnormal behavioral patterns, and compromised immunity (Dawkins, 1988; Kistler

FIGURE 21.4 Enriched zebrafish tank. (Illustrations by I. Fife-Cook.)

et al., 2011; Lawrence, 2011). Conversely, enriched environments stimulate natural behaviors (Fraser et al., 1997; Newberry, 1995) and mitigate negative behaviors associated with poor physical and mental health (Williams et al., 2009). Such benefits have been observed in zebrafish (Franks et al., 2018b; Graham et al., 2018a, b). For these reasons, laboratory managers have reason to make every effort to provide zebrafish with housing that resembles the species' natural habitat as closely as possible (Williams et al., 2009; see Figure 21.4).

Some enrichment strategies designed to imitate the native habitat and promote natural behavior in captive zebrafish include housing animals in appropriately sized social groups with sufficient space to engage in natural behaviors; furnishing tanks with gravel substrate as well as live and/or artificial plants; and providing shelters (Kistler et al., 2011; Schroeder et al., 2014). Table 21.1 details features of the natural habitat (Engeszer et al., 2007; Spence et al., 2006, 2008) and the methods to replicate those features when designing captive environments (Schroeder et al., 2014; Williams et al., 2009).

Social Environment to Promote Species-Typical Behaviors

Zebrafish consistently prefer to shoal with a larger group (Mansur et al., 2014; Pritchard et al., 2001; Seguin & Gerlai, 2017). Larger groups consisting of eight individuals showed less thigmotaxis than groups comprised of six or four (Mansur et al., 2014), indicating that group size may be inversely related to levels of anxiety-like behaviors. Additionally, fewer instances of aggression have been observed in larger groups of zebrafish in the wild (Suriyampola et al., 2015).

Ideal group size likely depends on the amount of space available. The size of the environment is an essential factor in determining the size of a shoal; overcrowding can result in chronic stress, poor health outcomes (Piato et al., 2011; Ramsay et al., 2006), elevated territorial behaviors, and aggression (Gillis & Kramer, 1987; Spence & Smith, 2005). Zebrafish have also been seen to spend time spread out from one another, if given more space (Shelton et al., 2015). At the same time, episodes of heightened shoaling and increased affiliative behaviors were observed in large, 110 L aquaria with groups of ten (Franks et al., 2018b; Graham et al., 2018a). Environments with more space may decrease competition for resources, easing agonism and facilitating social cohesion. Thus, while zebrafish will most often select a larger shoal,

TABLE 21.1

Environmental Features to Promote Species-Typical Behaviors

Natural Feature	Laboratory Substitute
Aquatic vegetation (submerged)	Real or artificial (plastic/silk) vegetation affixed to weighted base and placed strategically to provide animals with a variety of vegetative densities.
Overhanging vegetation	Floating real or artificial (plastic/silk) vegetation.
Pebble, sand, or silt substrate	Pebble substrate with underlying structure designed to simulate a natural slope.
Protection provided by rocks, vegetation, undercut roots, etc.	Shelter materials appropriate for the experimental design, such as aquarium wood (bogwood, mopani, redmoor, etc.) for breeder tanks, and PVC piping or ceramic pots for toxicology studies (see "Special Situations" section).
Somewhat turbid environment with average visibility of approximately 30 cm.	Reduce light penetration in transparent habitats by placing the habitat on a dark surface and covering outside of tank sides with an opaque material.
Depths up to 30 cm.	Tank design should maximize surface area while also providing animals with a variety of depths.

physical space is another limited resource in captivity that can lead to competition. The need for an appropriately sized social group and adequate space are competing interests that must be balanced when designing zebrafish housing.

Zebrafish kept in isolation display elevated levels of blood cortisol, a chemical associated with both physical and psychological stress (Ziv et al., 2013), and abnormal brain function (Shams et al., 2015). Isolation induces behavioral changes, such as decreased shoal cohesion, thigmotaxis, and erratic movements (Collymore et al., 2015; Shams et al., 2018). Isolated zebrafish display a higher stress response and present more variable behaviors than zebrafish kept in groups of three (Pagnussat et al., 2013). Research on similar social fish species suggests that social isolation stunts cognitive development; for example, cichlids perform more poorly in cognitive challenges when housed in isolation (Brandão et al., 2015).

Housing zebrafish in pairs should also be avoided. Fish are known to form dominance hierarchies in both sexes, as characterized by agonistic behaviors (Larson et al., 2006), which can be dangerous for subordinate individuals (Filby et al., 2010), especially for extended periods of time and without the provision of hiding spots (Keck et al., 2015). After establishment of a hierarchy, the amount of agonism decreases over time (Ricci et al., 2013), but subordinate fish show altered brain serotonergic and dopaminergic neurotransmitter systems, indicative of chronic stress (Dahlbom et al., 2012). Larger groups may better allow the recipient of agonism to escape the aggressor and avoid repeated interactions.

How important is the make-up of the social group? Females prefer shoals of at least three, rather than with a single conspecific, regardless of the group's sex ratio (Ruhl & McRobert, 2005; Ruhl et al., 2009). However, females housed in same-sex groups can become "eggbound", a condition in which the oviduct becomes clogged with eggs that have not been released. Exposure to male pheromones causes females to spawn, but without a male present this condition can become fatal (Spence et al., 2008); mixed-sex groups are recommended. Active shoals are preferred over lethargic shoals (Pritchard et al., 2001), perhaps because activity signals that the group is healthy, or because active shoals are more effective at finding food. Further, zebrafish prefer to shoal with a group that is well-fed, rather than hungry (Krause et al., 1999). Familiar shoals of fish display more predator aversion tactics and greater cohesion than unfamiliar shoals (Ward & Hart, 2003); maintaining consistent groups may allow for more affiliative behaviors, indicative of positive welfare (Fife-Cook & Franks, 2019; Franks et al., 2018b). Familiar shoals should also be preserved to avoid the increases in aggression that arise from the establishment of dominance hierarchies; aggressive interactions are most frequent when fish are first introduced, and decrease in frequency and duration with time (Larson et al., 2006; Ricci et al., 2013). More research is needed to further study how to introduce fish to a shoal while minimizing subsequent agonism; the optimal number of fish to be introduced and timing of these changes are factors that may influence results. Until then, we recommend relying on known mitigators of agonism, such as plants or shelters (Keck et al., 2015).

Researchers should also be cognizant of group size when designing and implementing experimental methods. Fish tested in groups, pairs, or isolation consistently differ in their response to the novel tank test (Parker et al., 2012b; White et al., 2017). Additionally, fish tested in groups of three took at most 10 trials to learn a maze, whereas isolated fish have been reported to need up to 70 trials (Spence et al., 2011). Testing fish in dyads may also influence results; Schroeder et al. (2014) found that dominant individuals guarded enriched areas of experimental tanks, relegating subordinates to barren environments. Baseline behavior for zebrafish is firmly situated in the context of their social dynamics.

Feeding Strategies to Promote Species-Typical Behaviors

In most laboratories, juvenile and adult fish are given a combination of live or frozen food, and a commercial dry feed two to three times per day (Williams et al., 2009). While commercial fish feeds are a financially attractive alternative to dried, frozen, or live food, they are not species-specific. As stated by Lawrence (Lawrence, 2007, p. 8), "the specific nutritional requirements of zebrafish are still unstudied, and may be fundamentally different from even closely related species", making it unwise to rely solely on these artificial foods. Instead, live, whole animal foods, such as paramecium and rotifers (for fry), and bloodworms (*Artemia nauplii*), *Drosophilia* larvae, and krill (for adults) are preferred, due to their positive effect on survivability and fecundity, and should thus, comprise the backbone of the zebrafish's diet in captivity (Lawrence, 2016, 2007). Additionally, live food may also be used to stimulate natural hunting and foraging behaviors. Ideal feeding frequency has also yet to be systematically studied in zebrafish, but because of their small body size and high activity levels, multiple, small meals per day may be best practice (Lawrence, 2007).

Training to Promote Species-Typical Behaviors

Zebrafish have the ability to be trained in classical and operant learning paradigms, and can even learn reversal discrimination tasks (Colwill et al., 2005; Parker et al., 2012a), which are considered to be challenging for many mammalian species (Rajalakshmi & Jeeves, 1965). Training zebrafish through, for example, associative learning tests and conditioned place preference (Darland & Dowling, 2001), is becoming commonplace. Researchers also use operant conditioning paradigms, including tests that use both positive punishment, such as active avoidance tasks (Aoki et al., 2015; Xu et al, 2007), and positive reinforcement tasks that use rewards of food or access to conspecifics (Al-imari & Gerlai, 2008; Williams et al., 2002).

While no research has examined its effects directly, these studies suggest that zebrafish are trainable and that zebrafish living in captivity may benefit from positive reinforcement training (PRT) as a form of enrichment. In other species, PRT has been used to teach captive animals to voluntarily participate in necessary husbandry events. PRT increases the predictability and therefore the cognitive control animals have, mitigating the stress of these procedures (Bassett & Buchanan-Smith, 2007). In addition to reducing stress, PRT may also enhance positive welfare. Many mammals voluntarily interact with learning devices, and cognitive activities, such as learning, have been associated with signs of positive affect (Franks, 2018a). Zebrafish are not likely to be an exception; they too have been found to seek out cognitive stimulation (Graham et al., 2018a). Because of their cognitive abilities and willingness to engage in learning activities, PRT may be rewarding in and of itself.

Special Situations

Standard housing systems for zebrafish in laboratories lack complexity, may provide inadequate social environments, and use husbandry techniques that can harm fish. As reported by Lidster et al. (2017), the majority of laboratory zebrafish housing lacks physical structures (i.e., plants, gravel) and substrate in recirculating or flow-through systems. In fish, barren environments have been found to negatively impact health, fecundity, brain size, behavior, stress response, and aggression (Carfagnini et al., 2009; Giacomini et al., 2016; Näslund & Johnsson, 2014; Wafer et al., 2016). Simplistic environments are also likely to restrict the range of social dynamics (Graham et al., 2018b). Indeed, the effects of barren environments can be so extreme as to change the body and head shape of fish (Garduño-Paz et al., 2010). The social environments in which zebrafish live are also key to their well-being. At more than five individuals per liter of water, standard housing practice runs the risk of overcrowding (Lidster et al., 2017), which can result in chronic stress (Ramsay et al., 2006). In addition to housing fish at high densities, common laboratory practices that isolate or maintain fish in pairs also have dangerous welfare implications (see "Social Environment to Promote Species-Typical Behaviors" section).

Certain research paradigms require that test subjects live in specific, restrictive conditions. Toxicology studies require that the container that the fish lives in remains inert and efforts must be made to avoid microbial growth (e.g., minimized surface area) and to maintain concentrations of the test chemical (Williams et al., 2009). However the existing evidence suggests that improvements can be made, even under these restrictive test conditions. For example, small external changes, such as dark, opaque backgrounds and photos of gravel lining the tank bottoms, in addition to live feed, have been found to improve welfare indicators efficiently without compromising health (Watts et al., 2016). Researchers are encouraged to consider these modifications as practical ways to improve the health and welfare of zebrafish and thus, improve the generalizability and validity of the scientific results.

Unnecessarily confining the experimental environment to restricted, simplistic housing risks reducing external validity by narrowing the range of phenotypes studied (Richter et al., 2009). The external validity of an experiment is likely to benefit from a more heterogeneous test population (Richter et al., 2009). Indeed, the simplicity of standard laboratory housing may be so extreme as to create an "abnormal" model that should not be treated as a sample from a normal population (Sherwin, 2004). Broadening the range of phenotypes studied under laboratory conditions may ultimately reduce the cost of research by reducing the need to repeat studies, while also reducing the number of lives negatively impacted by the research.

Conclusion

When we consider the features of the zebrafish's natural habitat, it is difficult to ignore the remarkable variation in each aspect of their lives. Every year, their environments undergo fluctuations due to the monsoon season, including changing water depths and the creation of new waterways to explore (Spence et al., 2008). They find food by scavenging, ingesting aquatic plants, and hunting insects; each method requiring a unique skillset that must be learned (McClure et al., 2006; Spence et al., 2008). While feeding, they must also protect themselves from predation, which can come from a variety of fishes, amphibians, and birds (Engeszer et al., 2007). Meanwhile, they have the social flexibility to treat some nonconspecifics as allies and are readily found in mixed-species shoals. In sum, zebrafish are generalists who can thrive in complex and challenging environments (see Figure 21.2).

Throughout this diversity, the shoal maintains a consistently important role; recent research suggests that the internal dynamics of the shoal may be more complex than we originally imagined. For example, research has shown that zebrafish can recognize each other as individuals and can form specific relationships (Paull et al., 2010). Moreover, zebrafish are known to discriminate between their peers' color, pheromones, stress levels, and hunger levels, and have preferences surrounding each of these factors (Hutter et al., 2012; Krause et al., 1999; Saszik & Smith, 2018; Spence et al., 2007). These relationships create the basis of complex dominance hierarchies (Dahlbom et al., 2012; Filby et al., 2010), which can determine an individual's success in a number of arenas, including reproduction and access to food (Paull et al., 2010). Some of the patterns of zebrafish social dynamics have been found to be driven by factors internal to the group alone, suggesting modes of communication that we are only beginning to understand (Franks et al., 2018b). Thus, not only can zebrafish thrive in a wide variety of conditions they also have the potential to lead social lives that are complex in and of themselves.

It is our hope that the complexity of zebrafish behaviors will be more fully understood by caretakers going forward. A better understanding of zebrafish behavior in the wild will benefit the science by widening the behaviors that can be studied. Because zebrafish studies serve as models for human diseases, these data would improve comparisons between species. The promotion of natural behaviors is worthwhile, because in addition to promoting biological function, some natural behaviors are pleasurable for their own sake (Bracke & Hopster, 2006). Natural behaviors also provide insight into the animals' needs in captivity. In animal welfare science, the consensus about what makes up a "good life" has been extended beyond mere survival (Mellor, 2016); this definition now includes adequate cognitive stimulation, agency, and social affiliation (Fife-Cook & Franks, 2019; Franks, 2017; Špinka & Wemelsfelder, 2018). Attending to their full potential fosters a deeper appreciation of the animals in our care, improves science, and extends knowledge about the world.

REFERENCES

Al-imari, L., & Gerlai, R. (2008). Sight of conspecifics as reward in associative learning in zebrafish (*Danio rerio*). *Behavioral Brain Research*, *189*(1), 216–219. https://doi.org/10.1016/j.bbr.2007.12.007.

Aoki, R., Tsuboi, T., & Okamoto, H. (2015). Y-maze avoidance: An automated and rapid associative learning paradigm in zebrafish. *Neuroscience Research*, *91*, 69–72. https://doi.org/10.1016/j.neures.2014.10.012.

Bailey, J., Oliveri, A., & Levin, E. D. (2013). Zebrafish model systems for developmental neurobehavioral toxicology. *Birth Defects Research Part C - Embryo Today: Reviews*, *99*(1), 14–23. https://doi.org/10.1002/bdrc.21027.

Barcellos, L. J. G., Volpato, G. L., Barreto, R. E., Coldebella, I., & Ferreira, D. (2011). Chemical communication of handling stress in fish. *Physiology and Behavior*, *103*(3–4), 372–375. https://doi.org/10.1016/j.physbeh.2011.03.009.

Barley, A. J., & Coleman, R. M. (2010). Habitat structure directly affects aggression in convict cichlids Archocentrus nigrofasciatus. *Current Zoology*, *56*(1), 52–56.

Basquill, S. P., Grant, J. W. A., Basquill, Sean, P., & Grant, J. W. A. (1998). An increase in habitat complexity reduces aggression and monopolization of food by zebrafish (*Danio rerio*). *Canadian Journal of Zoology*, *76*(4), 770–772. https://doi.org/10.1139/cjz-76-4-770.

Bass, S. L. S., & Gerlai, R. (2008). Zebrafish (*Danio rerio*) responds differentially to stimulus fish: The effects of sympatric and allopatric predators and harmless fish. *Behavioural Brain Research*, *186*(1), 107–117. https://doi.org/10.1016/j.bbr.2007.07.037.

Bassett, L., & Buchanan-Smith, H. M. (2007). Effects of predictability on the welfare of captive animals. *Applied Animal Behaviour Science*, *102*(3–4), 223–245. https://doi.org/10.1016/j.applanim.2006.05.029.

Bhat, A., Greulich, M. M., & Martins, E. P. (2015). Behavioral plasticity in response to environmental manipulation among zebrafish (*Danio rerio*) populations. *PLoS One*, *10*(4), 1–13. https://doi.org/10.1371/journal.pone.0125097.

Blaser, R. E., Chadwick, L., & McGinnis, G. C. (2010). Behavioral measures of anxiety in zebrafish (*Danio rerio*). *Behavioural Brain Research*, *208*(1), 56–62. https://doi.org/10.1016/j.bbr.2009.11.009.

Blaser, R. E., & Goldsteinholm, K. (2012). Depth preference in zebrafish, *Danio rerio*: Control by surface and substrate cues. *Animal Behaviour*, *83*(4), 953–959. https://doi.org/10.1016/j.anbehav.2012.01.014.

Bracke, M. B. M., & Hopster, H. (2006). Assessing the importance of natural behavior for animal welfare. *Journal of Agricultural and Environmental Ethics*, *19*(1), 77–89. https://doi.org/10.1007/s10806-005-4493-7.

Brandão, M. L., Braithwaite, V. A., & Gonçalves-de-Freitas, E. (2015). Isolation impairs cognition in a social fish. *Applied Animal Behaviour Science*, *171*, 204–210. https://doi.org/10.1016/j.applanim.2015.08.026.

Brown, C., & Laland, K. N. (2003). Social learning in fishes: A review. *Fish and Fisheries*, *4*, 280–288. https://doi.org/10.1046/j.1467-2979.2003.00122.x.

Cachat, J. M., Canavello, P. R., Elegante, M. F., Bartels, B. K., Elkhayat, S., Hart, P., … Kalueff, A. V. (2011). Modeling Stress and Anxiety in Zebrafish. In *Zebrafish Models in Neurobehavioral Research* (Neuromethods, pp. 73–88). Totowa, NJ: Humana Press. https://doi.org/10.1007/978-1-60761-922-2_3.

Carfagnini, A. G., Rodd, F. H., Jeffers, K. B., & Bruce, A. E. E. (2009). The effects of habitat complexity on aggression and fecundity in zebrafish (*Danio rerio*). *Environmental Biology of Fishes*, *86*(3), 403–409. https://doi.org/10.1007/s10641-009-9539-7.

CCAC Animal Data Report. (2014). Canadian Council on Animal Care.

Collymore, C., Tolwani, R. J., & Rasmussen, S. (2015). The behavioral effects of single housing and environmental enrichment on adult zebrafish (*Danio rerio*). *Journal of the American Association for Laboratory Animal Science*, *54*(3), 280–285. https://www.ncbi.nlm.nih.gov/pmc/articles/PMC4460940/?report=abstract.

Colwill, R. M., Raymond, M. P., Ferreira, L., & Escudero, H. (2005). Visual discrimination learning in zebrafish (*Danio rerio*). *Behavioural Processes*, *70*(1), 19–31. https://doi.org/10.1016/j.beproc.2005.03.001.

Cronin, G. M., Wiepkema, P. R., & van Ree, J. M. (1985). Endogenous opioids are involved in abnormal stereotyped behaviours of tethered sows. *Neuropeptides*, *6*(6), 527–530. https://doi.org/10.1016/0143-4179(85)90114-3.

D'Amico, D., Estivill, X., & Terriente, J. (2015). Switching to zebrafish neurobehavioral models: The obsessive-compulsive disorder paradigm. *European Journal of Pharmacology, 759*, 142–150. https://doi.org/10.1016/j.ejphar.2015.03.027.

Dahlbom, S. J., Backström, T., Lundstedt-Enkel, K., & Winberg, S. (2012). Aggression and monoamines: Effects of sex and social rank in zebrafish (*Danio rerio*). *Behavioural Brain Research, 228*(2), 333–338. https://doi.org/10.1016/j.bbr.2011.12.011.

Darland, T., & Dowling, J. E. (2001). Behavioral screening for cocaine sensitivity in mutagenized zebrafish. *Proceedings of the National Academy of Sciences, 98*(20), 11691–11696. https://doi.org/10.1073/pnas.191380698.

Dawkins, M. S. (1988). Behavioural deprivation: A central problem in animal welfare. *Applied Animal Behaviour Science, 20*(3–4), 209–225.

Delaney, M., Follet, C., Ryan, N., Hanney, N., Lusk-yablick, J., & Gerlach, G. (2002). Social interaction and distribution of female zebrafish (*Danio rerio*) in a large aquarium. *Biological Bulletin, 203*(2), 240–241.

Engeszer, R. E., Patterson, L. B., Rao, A. A., & Parichy, D. M. (2007). Zebrafish in the wild: A review of natural history and new notes from the field. *Zebrafish, 4*(1), 21–40. https://doi.org/10.1089/zeb.2006.9997.

Engeszer, R. E., Ryan, M. J., & Patrichy, D. M. (2004). Learned social preference in zebrafish. *Current Biology, 14*(10), 881–884. https://doi.org/10.1016/j.cub.2004.04.042.

Faustino, A. I., Tacão-Monteiro, A., & Oliveira, R. F. (2017). Mechanisms of social buffering of fear in zebrafish. *Scientific Reports, 7*(March), 44329. https://doi.org/10.1038/srep44329.

Fetcho, J. R., & Liu, K. S. (1998). Zebrafish as a model system for studying neuronal circuits and behavior. *Annals of the New York Academy of Sciences, 860*, 333–345. https://doi.org/10.1111/j.1749-6632.1998.tb09060.x.

Fife-Cook, I., & Franks, B. (2019). Positive welfare for fishes: Rationale and areas for future study. *Fishes, 4*(31), 1–14. https://doi.org/doi:10.3390/fishes402003.

Filby, A. L., Paull, G. C., Bartlett, E. J., Van Look, K. J. W., & Tyler, C. R. (2010). Physiological and health consequences of social status in zebrafish (*Danio rerio*). *Physiology and Behavior, 101*(5), 576–587. https://doi.org/10.1016/j.physbeh.2010.09.004.

Franks, B. (2018a). Cognition as a cause, consequence, and component of welfare. In J. A. Mench (Ed.), *Advances in Agricultural Animal Welfare* (pp. 3–24). Woodhead Publishing.

Franks, B., Graham, C., & Von Keyserlingk, M. A. G. (2018b). Is heightened-shoaling a good candidate for positive emotional behavior in zebrafish? *Animals, 8*(9), 152. https://doi.org/10.3390/ani8090152.

Franks, B., Sebo, J., & Horowitz, A. (2018c). Fish are smart and feel pain: What about joy? *Animal Sentience, 3*(21), 16.

Fraser, D., Weary, D. M., Pajor, E. A., & Milligan, B. N. (1997). A scientific conception of animal welfare that reflects ethical concerns. *Animal Welfare, 6*(3), 187–205.

Garduño-Paz, M. V., Couderc, S., & Adams, C. (2010). Habitat complexity modulates phenotype expression through developmental plasticity in the threespine stickleback. *Biological Journal of the Linnean Society, 100*(2), 407–413. https://doi.org/10.1111/j.1095-8312.2010.01423.x.

Gerhard, G. S., Kauffman, E. J., Wang, X., Stewart, R., Moore, J. L., Kasales, C. J., … Cheng, K. C. (2002). Life spans and senescent phenotypes in two strains of Zebrafish (*Danio rerio*). *Experimental Gerontology, 37*(8–9), 1055–1068. https://doi.org/10.1016/s0531-5565(02)00088-8.

Gerlai, R. (2003). Zebrafish: An uncharted behavior genetic model. *Behavior Genetics, 33*(5), 461–468.

Gerlai, R., Lahav, M., Guo, S., & Rosenthal, A. (2001). Drinks like a fish: Zebrafish (*Danio rerio*) as a behavior genetic model to study alcohol effects. *Pharmacology, Biochemistry and Behavior, 67*(4), 773–782.

Giacomini, A. C. V. V., Abreu, M. S., Zanandrea, R., Saibt, N., Friedrich, M. T., Koakoski, G., … Barcellos, L. J. G. (2016). Environmental and pharmacological manipulations blunt the stress response of zebrafish in a similar manner. *Scientific Reports, 6*(January), 1–6. https://doi.org/10.1038/srep28986.

Gillis, D. M., & Kramer, D. L. (1987). Ideal interference distributions: Population density and patch use by zebrafish. *Animal Behaviour, 35*(6), 1875–1882. https://doi.org/10.1016/S0003-3472(87)80080-5.

Goldsmith, P. (2004). Zebrafish as a pharmacological tool: The how, why and when. *Current Opinion in Pharmacology, 4*(5), 504–512. https://doi.org/10.1016/j.coph.2004.04.005.

Graham, C., Von Keyserlingk, M. A. G., & Franks, B. (2018a). Free-choice exploration increases affiliative behaviour in zebrafish. *Applied Animal Behaviour Science, 203*(September 2017), 103–110. https://doi.org/10.1016/j.applanim.2018.02.005.

Graham, C., Von Keyserlingk, M. A. G., & Franks, B. (2018b). Zebrafish welfare: Natural history, social motivation and behaviour. *Applied Animal Behaviour Science, 200*(April 2017), 13–22. https://doi.org/10.1016/j.applanim.2017.11.005.

Grant, J. W. A. (1993). Whether or not to defend? The influence of resource distribution. *Marine Behaviour and Physiology, 23*(1–4), 137–153. Retrieved from http://books.google.com/books?id=L_Hedxcy8O4C&pg=PA258&dq=poeciliid+fish&hl=en&ei=KwlrTaHJBcp8AaxyaCKCw&sa=X&oi=book_result&ct=result&resnum=9&ved=0CFAQ6AEwCA#v=onepage&q=poeciliid fish&f=false.

Hamilton, I. M., & Dill, L. M. (2002). Monopolization of food by zebrafish (*Danio rerio*) increases in risky habitats. *Canadian Journal of Zoology-Revue Canadienne De Zoologie, 80*(12), 2164–2169. https://doi.org/10.1139/z02-199.

Hill, S. P., & Broom, D. M. (2009). Measuring zoo animal welfare: Theory and practice. *Zoo Biology, 28*(6), 531–544. https://doi.org/10.1002/zoo.20276.

Howe, K., Clark, M. D., Torroja, C. F., Torrance, J., Berthelot, C., Muffato, M., … Stemple, D. L. (2013). The zebrafish reference genome sequence and its relationship to the human genome. *Nature, 496*(7446), 498–503. https://doi.org/10.1038/nature12111.

Huntingford, F., Adams, C., Braithwaite, V. A., Kadri, S., Pottinger, T. G., Sandoe, P., & Turnbull, J. F. (2006). Current issues in fish welfare. *Journal of Fish Biology, 68*(1), 332–372. https://doi.org/10.1111/j.1095-8649.2005.01046.x.

Hutter, S., Hettyey, A., Penn, D. J., & Zala, S. M. (2012). Ephemeral sexual dichromatism in zebrafish (*Danio rerio*). *Ethology, 118*(12), 1208–1218. https://doi.org/10.1111/eth.12027.

Hutter, S., Penn, D. J., Magee, S., & Zala, S. M. (2010). Reproductive behaviour of wild zebrafish (*Danio rerio*) in large tanks. *Behaviour*, *147*(5–6), 641–660. https://doi.org/10.1163/000579510X12632972473944.

Kalueff, A. V., Gebhardt, M., Stewart, A., Cachat, J. M., Brimmer, M., Chawla, J. S., … Neuhauss, S. C. F. (2013). Towards a comprehensive catalog of zebrafish behavior 1.0 and beyond. *Zebrafish*, *10*(1), 70–86. https://doi.org/10.1089/zeb.2012.0861.

Kareklas, K., Elwood, R. W., & Holland, R. A. (2018). Grouping promotes risk-taking in unfamiliar settings. *Behavioural Processes*, *148*(January), 41–45. https://doi.org/10.1016/j.beproc.2018.01.003.

Keck, V. A., Edgerton, D. S., Hajizadeh, S., Swift, L. L., Dupont, W. D., Lawrence, C., & Boyd, K. L. (2015). Effects of habitat complexity on pair-housed zebrafish. *Journal of the American Association for Lab Animal Science*, *54*(4), 378–383.

Kelley, J. L., & Magurran, A. E. (2003). Learned predator recognition and antipredator responses in fishes. *Fish and Fisheries*, *4*(3), 216–226. https://doi.org/10.1046/j.1467-2979.2003.00126.x.

Kistler, C., Hegglin, D., Würbel, H., & Konig, B. (2011). Preference for structured environment in zebrafish (*Danio rerio*) and checker barbs (*Puntius oligolepis*). *Applied Animal Behaviour Science*, *135*(4), 318–327. https://doi.org/10.1016/j.applanim.2011.10.014.

Krause, J., Hartmann, N., & Pritchard, V. (1999). The influence of nutritional state on shoal choice in zebrafish, *Danio rerio*. *Animal Behaviour*, *57*(4), 771–775. https://doi.org/10.1006/anbe.1998.1010.

Larson, E. T., O'Malley, D., & Melloni, R. H. (2006). Aggression and vasotocin are associated with dominant-subordinate relationships in zebrafish. *Behavioural Brain Research*, *167*(1), 94–102. https://doi.org/10.1016/j.bbr.2005.08.020.

Lawrence, C. (2007). The husbandry of zebrafish (*Danio rerio*): A review. *Aquaculture*, *269*(1–4), 1–20. https://doi.org/10.1016/j.aquaculture.2007.04.077.

Lawrence, C. (2011). Advances in zebrafish husbandry and management. *Methods in Cell Biology*, *104*(1), 429–451. https://doi.org/10.1016/B978-0-12-374814-0.00023-9.

Lawrence, C. (2016). New frontiers for zebrafish management. *Methods in Cell Biology*, 135, 483–508. https://doi.org/10.1016/bs.mcb.2016.04.015.

Lidster, K., Readman, G. D., Prescott, M., & Owen, S. F. (2017). International survey on the use and welfare of zebrafish *Danio rerio* in research. *Journal of Fish Biology*, *90*(5), 1891–1905. https://doi.org/10.1111/jfb.13278.

Lindeyer, C. M., & Reader, S. M. (2010). Social learning of escape routes in zebrafish and the stability of behavioural traditions. *Animal Behaviour*, *79*(4), 827–834. https://doi.org/10.1016/j.anbehav.2009.12.024.

Liu, C. X., Li, C. Y., Hu, C. C., Wang, Y., Lin, J., Jiang, Y. H., … Xu, X. (2018). CRISPR/Cas9-induced shank3b mutant zebrafish display autism-like behaviors. *Molecular Autism*, *9*(1), 1–13. https://doi.org/10.1186/s13229-018-0204-x.

Mann, K. D., Turnell, E. R., Atema, J., & Gerlach, G. (2011). Kin recognition in juvenile zebrafish (*Danio rerio*) based on olfactory cues. *Biology Bulletin*, *205*(October 2003), 224–225. Retrieved from http://www.ncbi.nlm.nih.gov/entrez/query.fcgi?cmd=Retrieve&db=PubMed&dopt=Citation&list_uids=14583541.

Mansur, B. de M., Dos Santos, B. R., Dias, C. A., Pinheiro, M., & Gouveia, A., Jr. (2014). Effects of the number of subjects on the dark/light preference of zebrafish (*Danio rerio*). *Zebrafish*, *11*(6), 560–566. https://doi.org/10.1089/zeb.2014.0977.

Martins, C. I. M., Galhardo, L., Noble, C., Damsgård, B., Spedicato, M. T., Zupa, W., … Kristiansen, T. (2012). Behavioural indicators of welfare in farmed fish. *Fish Physiology and Biochemistry*, *38*(1), 17–41. https://doi.org/10.1007/s10695-011-9518-8.

Martins, E. P., & Bhat, A. (2014). Population-level personalities in zebrafish: Aggression-boldness across but not within populations. *Behavioral Ecology*, *25*(2), 368–373. https://doi.org/10.1093/beheco/aru007.

Mason, G. J. (1991). Stereotypies and suffering. *Behavioural Processes*, *25*(2–3), 103–115. https://doi.org/10.1016/0376-6357(91)90013-P.

Maximino, C. (2011). Modulation of nociceptive-like behavior in zebrafish (*Danio rerio*) by environmental stressors. *Psychology and Neuroscience*, *4*(1), 149–155. https://doi.org/10.3922/j.psns.2011.1.017.

McClure, M. M., McIntyre, P. B., & McCune, A. R. (2006). Notes on the natural diet and habitat of eight danionin fishes, including the zebrafish *Danio rerio*. *Journal of Fish Biology*, *69*(2), 553–570. https://doi.org/10.1111/j.1095-8649.2006.01125.x.

Mellor, D. J. (2016). Updating animal welfare thinking: Moving beyond the "five freedoms" towards "A life-worth living." *Animals*, *6*(3), 21. https://doi.org/10.3390/ani6030021.

Miller, N., & Gerlai, R. (2012). From schooling to shoaling: Patterns of collective motion in zebrafish (*Danio rerio*). *PLoS One*, *7*(11), 8–13. https://doi.org/10.1371/journal.pone.0048865.

Moretz, J. A., Martins, E. P., & Robison, B. D. (2007). Behavioral syndromes and the evolution of correlated behavior in zebrafish. *Behavioral Ecology*, *18*(3), 556–562. https://doi.org/10.1093/beheco/arm011.

Näslund, J., & Johnsson, J. I. (2014). Environmental enrichment for fish in captive environments: Effects of physical structures and substrates. *Fish and Fisheries*, *17*(1), 1–30. https://doi.org/10.1111/faf.12088.

Newberry, R. C. (1995). Environmental enrichment: Increasing the biological relevance of captive environments. *Applied Animal Behaviour Science*, *44*(2–4), 229–243.

Oldfield, R. G. (2011). Aggression and welfare in a common aquarium Fish, the Midas Cichlid. *Journal of Applied Animal Welfare Science*, *14*(4), 340–360. https://doi.org/10.1080/10888705.2011.600664.

Oswald, M., & Robison, B. D. (2008). Strain-specific alteration of zebrafish feeding behavior in response to aversive stimuli. *Canadian Journal of Zoology*, *86*(10), 1085–1094. https://doi.org/10.1139/Z08-085.

Pagnussat, N., Piato, Â. L., Schaefer, I. C., Blank, M., Tamborski, A., Guerim, L. D., … Lara, D. (2013). One for all and all for one: The importance of shoaling on behavioral and stress responses in zebrafish. *Zebrafish*, *10*, 338–342. https://doi.org/10.1089/zeb.2013.0867.

Parichy, D. M. (2015). Advancing biology through a deeper understanding of zebrafish ecology and evolution. *ELife*, *4*, 1–11. https://doi.org/10.7554/eLife.05635.

Parker, M. O., Gaviria, J., Haigh, A., Millington, M. E., Brown, V. J., Combe, F. J., & Brennan, C. H. (2012a). Discrimination reversal and attentional sets in zebrafish (*Danio rerio*). *Behavioural Brain Research*, 232(1), 264–268. https://doi.org/10.1016/j.bbr.2012.04.035.

Parker, M. O., Millington, M. E., Combe, F. J., & Brennan, C. H. (2012b). Housing conditions differentially affect physiological and behavioural stress responses of zebrafish, as well as the response to anxiolytics. *PLoS One*, 7(4). https://doi.org/10.1371/journal.pone.0034992.

Partridge, B. L., & Pitcher, T. J. (1980). The sensory basis of fish schools: Relative roles of lateral line and vision. *Journal of Comparative Physiology A*, 135(4), 315–325. https://doi.org/10.1007/BF00657647.

Paull, G. C., Filby, A. L., Giddins, H. G., Coe, T. S., Hamilton, P. B., & Tyler, C. R. (2010). Dominance Hierarchies in zebrafish (*Danio rerio*) and their relationship with reproductive success. *Zebrafish*, 7(1), 109–117. https://doi.org/10.1089/zeb.2009.0618.

Piato, Â. L., Capiotti, K. M., Tamborski, A., Oses, J. P., Barcellos, L. J. G., Bogo, M. R., ... Bonan, C. D. (2011). Unpredictable chronic stress model in zebrafish (*Danio rerio*): Behavioral and physiological responses. *Progress in Neuro-Psychopharmacology and Biological Psychiatry*, 35(2), 561–567. https://doi.org/10.1016/j.pnpbp.2010.12.018.

Pitcher, T. J., & Parrish, J. K. (1993). Functions of shoaling behaviour in teleosts. In T.J. Pitcher (Ed.), *The Behavior of Teleost Fishes* (pp. 363–439). Springer US. https://doi.org/10.1007/978-1-4684-8261-4_12.

Pritchard, V. L., Lawrence, J., Butlin, R. K., & Krause, J. (2001). Shoal choice in zebrafish, *Danio rerio*: The influence of shoal size and activity. *Animal Behaviour*, 62(6), 1085–1088. https://doi.org/10.1006/anbe.2001.1858.

Pyron, M. (2003). Female preferences and male–male interactions in zebrafish (*Danio rerio*). *Canadian Journal of Zoology*, 81(1), 122–125. https://doi.org/10.1139/z02-229.

Rajalakshmi, R., & Jeeves, M. A. (1965). The relative difficulty of reversal learning (reversal index) as a basis of behavioural comparisons. *Animal Behaviour*, 13(2–3), 203–211. https://doi.org/10.1016/0003-3472(65)90035-7.

Ramsay, J. M., Feist, G. W., Varga, Z. M., Westerfield, M., Kent, M. L., & Schreck, C. B. (2006). Whole-body cortisol is an indicator of crowding stress in adult zebrafish, *Danio rerio*. *Aquaculture*, 258(1–4), 565–574. https://doi.org/10.1016/j.aquaculture.2006.04.020.

Reed, B., & Jennings, M. (2011). *Guidance on the Housing and Care of Zebrafish Danio rerio*. RSPCA.

Reilly, S. C., Quinn, J. P., Cossins, A. R., & Sneddon, L. U. (2008). Behavioural analysis of a nociceptive event in fish: Comparisons between three species demonstrate specific responses. *Applied Animal Behaviour Science*, 114(1–2), 248–259. https://doi.org/10.1016/j.applanim.2008.01.016.

Rey, S., Huntingford, F. A., Boltaña, S., Vargas, R., Knowles, T. G., & Mackenzie, S. (2015). Fish can show emotional fever: Stress-induced hyperthermia in zebrafish. *Proceedings of the Royal Society B: Biological Sciences*, 282(1819), 0–6. https://doi.org/10.1098/rspb.2015.2266.

Ricci, L., Summers, C. H., Larson, E. T., O'Malley, D., & Melloni, R. H. (2013). Development of aggressive phenotypes in zebrafish: Interactions of age, experience and social status. *Animal Behaviour*, 86(2), 245–252. https://doi.org/10.1016/j.anbehav.2013.04.011.

Richter, S. H., Garner, J. P., & Würbel, H. (2009). Environmental standardization: Cure or cause of poor reproducibility in animal experiments? *Nature Methods*, 6(4), 257–261. https://doi.org/10.1038/nmeth.1312.

Ruhl, N., & McRobert, S. P. (2005). The effect of sex and shoal size on shoaling behaviour in *Danio rerio*. *Journal of Fish Biology*, 67(5), 1318–1326. https://doi.org/10.1111/j.0022-1112.2005.00826.x.

Ruhl, N., McRobert, S. P., & Currie, W. J. S. (2009). Shoaling preferences and the effects of sex ratio on spawning and aggression in small laboratory populations of zebrafish (*Danio rerio*). *Lab Animal*, 38(8), 264–269. https://doi.org/10.1038/laban0809-264.

Russell, W. M. S., & Burch, R. L. (1959). *The Principles of Humane Experimental Technique*. Methuen Publishing London.

Saszik, S. M., & Smith, C. M. (2018). The impact of stress on social behavior in adult zebrafish (*Danio rerio*). *Behavioural Pharmacology*, 29(1), 53–59. https://doi.org/10.1097/FBP.0000000000000338.

Schoener, T. W. (1987). Time budgets and territory size : Some simultaneous optimization models for energy maximizers. *American Zoologist*, 27(2), 259–291.

Schroeder, P., Jones, S., Young, I. S., & Sneddon, L. U. (2014). What do zebrafish want? Impact of social grouping, dominance and gender on preference for enrichment. *Laboratory Animals*, 48(4), 328–337. https://doi.org/10.1177/0023677214538239.

Schroeder, P., & Sneddon, L. U. (2017). Exploring the efficacy of immersion analgesics in zebrafish using an integrative approach. *Applied Animal Behaviour Science*, 187, 93–102. https://doi.org/10.1016/j.applanim.2016.12.003.

Seguin, D., & Gerlai, R. (2017). Zebrafish prefer larger to smaller shoals: Analysis of quantity estimation in a genetically tractable model organism. *Animal Cognition*, 20(5), 813–821. https://doi.org/10.1007/s10071-017-1102-x.

Shams, S., Amlani, S., Buske, C., Chatterjee, D., & Gerlai, R. (2018). Developmental social isolation affects adult behavior, social interaction, and dopamine metabolite levels in zebrafish. *Developmental Psychobiology*, 60(1), 43–56. https://doi.org/10.1002/dev.21581.

Shams, S., Chatterjee, D., & Gerlai, R. (2015). Chronic social isolation affects thigmotaxis and whole-brain serotonin levels in adult zebrafish. *Behavioural Brain Research*, 292, 283–287. https://doi.org/10.1016/j.bbr.2015.05.061.

Shelton, D. S., Price, B. C., Ocasio, K. M., Martins, E. P., Shelton, D. S., Price, B. C., ... Martins, E. P. (2015). Density and group size influence shoal cohesion, but not coordination in zebrafish (*Danio rerio*). *Journal of Comparative Psychology*, 129, 72–77. https://doi.org/10.5061/dryad.90n5f.

Sherwin, C. M. (2004). The influences of standard laboratory cages on rodents and the validity of research data. *Animal Welfare*, 13(Suppl.), 9–15.

Singleman, C., & Holtzman, N. G. (2014). Growth and maturation in the zebrafish, *Danio rerio*: A staging tool for teaching and research. *Zebrafish*, 11(4), 396–406. https://doi.org/10.1089/zeb.2014.0976.

Sneddon, L. U. (2012). Clinical anesthesia and analgesia in fish. *JEPM*, 21(1), 32–43. https://doi.org/10.1053/j.jepm.2011.11.009.

Sneddon, L. U., Halsey, L. G., & Bury, N. R. (2017). Considering aspects of the 3Rs principles within experimental animal biology. *The Journal of Experimental Biology*, *220*(17), 3007–3016. https://doi.org/10.1242/jeb.147058.

Soares, M. C., Cardoso, S. C., Carvalho, T. dos S., & Maximino, C. (2018). Using model fish to study the mechanisms of cooperative behaviors. *Progress in Neuro-Psychopharmacology and Biological Psychiatry*, *82*, 205–215.

Spence, R., Ashton, R., & Smith, C. (2007). Oviposition decisions are mediated by spawning site quality in wild and domesticated zebrafish, *Danio rerio*. *Behaviour*, *144*(8), 953–966. https://doi.org/10.1163/156853907781492726.

Spence, R., Fatema, M. K., Reichard, M., Huq, K. A., Wahab, M. A., Ahmed, Z. F., & Smith, C. (2006). The distribution and habitat preferences of the zebrafish in Bangladesh. *Journal of Fish Biology*, *69*(5), 1435–1448. https://doi.org/10.1111/j.1095-8649.2006.01206.x.

Spence, R., Gerlach, G., Lawrence, C., & Smith, C. (2008). The behaviour and ecology of the zebrafish, *Danio rerio*. *Biological Review*, *83*(1), 13–34. https://doi.org/10.1111/j.1469-185X.2007.00030.x.

Spence, R., Magurran, A. E., & Smith, C. (2011). Spatial cognition in zebrafish: The role of strain and rearing environment. *Animal Cognition*, *14*(4), 607–612. https://doi.org/10.1007/s10071-011-0391-8.

Spence, R., & Smith, C. (2005). Male territoriality mediates density and sex ratio effects on oviposition in the zebrafish, *Danio rerio*. *Animal Behaviour*, *69*(6), 1317–1323. https://doi.org/10.1016/j.anbehav.2004.10.010.

Špinka, M. (2006). How important is natural behaviour in animal farming systems? *Applied Animal Behaviour Science*, *100*(1–2), 117–128. https://doi.org/10.1016/j.applanim.2006.04.006.

Špinka, M. & Wemelsfelder, F. (2018). Environmental Challenge and Animal Agency. In M. C. Appleby (Ed.), *Animal Welfare* (3rd ed., pp. 39–55). CAB International Oxford, UK.

Stewart, A., Cachat, J., Wong, K., Gaikwad, S., Gilder, T., DiLeo, J., … Kalueff, A. V. (2010). Homebase behavior of zebrafish in novelty-based paradigms. *Behavioural Processes*, *85*(2), 198–203. https://doi.org/10.1016/j.beproc.2010.07.009.

Suboski, M. D., Bain, S., Carty, A. E., McQuoid, L. M., Seelan, M. I., & Seifert, M. (1990). Alarm reaction in acquisition and social transmission of simulated- predator recognition by zebra danio fish (Brachydanio rerio). *Journal of Comparative Psychology*, *104*(1), 101–112. https://doi.org/10.1037/0735-7036.104.1.101.

Suli, A., Watson, G. M., Rubel, E. W., & Raible, D. W. (2012). Rheotaxis in larval zebrafish is mediated by lateral line mechanosensory hair cells. *PLoS One*, *7*(2). https://doi.org/10.1371/journal.pone.0029727.

Suriyampola, P. S., Shelton, D. S., Shukla, R., Roy, T., Bhat, A., & Marins, E. P. (2015). Zebrafish Social Behavior in the Wild. *Zebrafish*, *28*(1), 49–100. https://doi.org/10.1089/zeb.2015.1159.

Suriyampola, P. S., Sykes, D. J., Khemka, A., & Shelton, D. S. (2017). Water flow impacts group behavior in zebrafish (*Danio rerio*). *Behavioral Ecology*, *28*(1), 94–100. https://doi.org/10.1093/beheco/arw138.

Tsang, B., Zahid, H., Ansari, R., Lee, R. C.-Y., Partap, A., & Gerlai, R. (2017). Breeding zebrafish: A review of different methods and a discussion on standardization. *Zebrafish*, *14*(6), 561–573. https://doi.org/10.1089/zeb.2017.1477.

van den Hurk, R., Schoonen, W. G. E. J., van Zoelen, G. A., & Lambert, J. G. D. (1987). The biosynthesis of steroid glucuronides in the testis of the zebrafish, Brachydanio rerio, and their pheromonal function as ovulation inducers. *General and Comparative Endocrinology*, *68*(2), 179–188. https://doi.org/10.1016/0016-6480(87)90027-X.

Wafer, L. N., Jensen, V. B., Whitney, J. C., Gomez, T. H., Flores, R., & Goodwin, B. S. (2016). Effects of environmental enrichment on the fertility and fecundity of zebrafish (*Danio rerio*). *Journal of the American Association for Laboratory Animal Science*, *55*(3), 291–294.

Ward, A. J. W., & Hart, P. J. B. (2003). The effects of kin and familiarity on interactions between fish. *Fish and Fisheries*, *4*(4), 348–358. https://doi.org/10.1046/j.1467-2979.2003.00135.x.

Watts, S. A., Lawrence, C., Powell, M., & D'Abramo, L. R. (2016). The vital relationship between nutrition and health in zebrafish. *Zebrafish*, *13*(S1), S72–S76. https://doi.org/10.1089/zeb.2016.1299.

White, L. J., Thomson, J. S., Pounder, K. C., Coleman, R. C., & Sneddon, L. U. (2017). The impact of social context on behaviour and the recovery from welfare challenges in zebrafish, *Danio rerio*. *Animal Behaviour*, *132*, 189–199. https://doi.org/10.1016/j.anbehav.2017.08.017.

Williams, F. E., White, D., & Messer, W. S. (2002). A simple spatial alternation task for assessing memory function in zebrafish. *Behavioral Processes*, *58*, 125–132. https://doi.org/10.1016/s0376-6357(02)00025-6.

Williams, T. D., Readman, G. D., & Owen, S. F. (2009). Key issues concerning environmental enrichment for laboratory-held fish species. *Laboratory Animals*, *43*(2), 107–120. https://doi.org/10.1258/la.2007.007023.

Wisenden, B., Vollbrecht, K. A., & Brown, J. L. (2004). Is there a fish alarm cue? Affirming evidence from a wild study. *Animal Behaviour*, *67*(1), 59–67. https://doi.org/10.1016/j.anbehav.2003.02.010.

Wiwchar, L. D., Gilbert, M. J. H., Kasurak, A. V., & Tierney, K. (2018). Schooling improves critical swimming performance in zebrafish (*Danio rerio*). *Canadian Journal of Fisheries and Aquatic Sciences*, *75*(4), 653–661. https://doi.org/10.1139/cjfas-2017-0141.

Wright, D., Ward, A. J. W., Croft, D. P., & Krause, J. (2006). Social organization, grouping, and domestication in fish. *Zebrafish*, *3*(2), 141–155. https://doi.org/10.1089/zeb.2006.3.141.

Xu, X., Scott-Scheiern, T., Kempker, L., & Simons, K. (2007). Active avoidance conditioning in zebrafish (*Danio rerio*). *Neurobiology of Learning and Memory*, *87*(1), 72–77. https://doi.org/10.1016/j.nlm.2006.06.002.

Zala, S. M., & Määttänen, I. (2013). Social learning of an associative foraging task in zebrafish. *Naturwissenschaften*, *100*(5), 469–472. https://doi.org/10.1007/s00114-013-1017-6.

Zala, S. M., Määttänen, I., & Penn, D. J. (2012). Different social-learning strategies in wild and domesticated zebrafish, *Danio rerio*. *Animal Behaviour*, *83*(6), 1519–1525. https://doi.org/10.1016/j.anbehav.2012.03.029.

Ziv, L., Muto, A., Schoonheim, P. J., Meijsing, S. H., Strasser, D., Ingraham, H. A., … Baier, H. (2013). An affective disorder in zebrafish with mutation of the glucocorticoid receptor. *Molecular Psychiatry*, *18*(6), 681–691. https://doi.org/10.1038/mp.2012.64.

22

Behavioral Biology of Amphibians

Charlotte A. Hosie and Tessa E. Smith
University of Chester

CONTENTS

Introduction 345
Typical Research Participation 346
Behavioral Biology 347
    Natural History 347
    Ecology 348
    Feeding Behavior 349
    Social Organization and Behavior 349
    Courtship and Mating Behavior 350
    Communication 351
Common Captive Behaviors 351
    Normal and Abnormal 351
    Ways to Maintain Behavioral Health 352
        Captive Considerations to Promote Species-Typical Behavior 352
Conclusions/Recommendations 354
References 354

Introduction

Amphibians make up a wonderfully diverse group of rather small ectothermic vertebrates, generally characterized by having soft, permeable skin and a requirement for water or moist environments. The group is made up of three distinctly different orders. The Anura (frogs and toads), of which there are more than 5,000 extant species, have a squat body shape and long, powerful hind limbs (Duellman & Trueb 1994; Vitt & Caldwell 2014). The Urodela (salamanders) have long bodies with tails, limbs of roughly equal sizes, and number around 655 extant species. The limbless Apoda (caecilians) are relatively poorly known, but highly adapted for a burrowing life and are thought to number around 200 extant species.

Across the group, amphibians occupy habitats from deserts through damp forest, temperate and tropical, to fast flowing streams, small ponds, and large lakes. They are relatively long-lived and exhibit a great range with respect to life history strategy, morphology, physiology, behavior, and ecology. They lead "amphi bios"; two lives, which can have two meanings. One refers to the common strategy of terrestrial living except for reproduction, which may require water for mating and egg development. The other meaning refers to the fact that most amphibians lead "two lives": (1) a (generally) aquatic larval form, which then undergoes metamorphosis to (2) an adult form, often with a different morphology, ecology, and lifestyle. This adult form may be terrestrial, semiaquatic, or rather uncommonly, fully aquatic (Duellman & Trueb 1994; Vitt & Caldwell 2014). Interestingly, the fully aquatic species are heavily represented in the group used as laboratory models, for reasons discussed below.

As a group, amphibian basic physiology is well understood (Burggren & Warburton 2007; Wells 2010), but its behavior is less well-studied, compared to mammals and most other vertebrates. The behavior of only a handful of species has been studied extensively. More than 50 years of elegant field and laboratory manipulations with the Eastern Red-backed salamander, *Plethodon cinereus*, have elucidated the fine detail of much of its behavioral ecology (e.g., Jaeger et al., 2016). Thirty plus years of focus on the tungara frog (*Engystomops pustulosus*), in particular mate choice and sexual selection, has yielded fascinating insight into the complexities of male advertisement calling behavior, female choice, and the trade-offs with predation risk (Ryan 1985; Lea & Ryan 2015). Extensive work on the European "*Triturus*" group of newts (now *Triturus, Lissotriton, and Ichthyosaurus*) over four decades has yielded deep insight into courtship behavior (sexual selection and mate choice, Halliday 1990) and feeding behavior/niche partitioning in this group (Beebee & Griffiths 2000). These species have been studied for their considerable intrinsic interest and have made highly valuable contributions to our understanding of general vertebrate behavioral and ecological principles. Although some of this work has been conducted in a laboratory environment, to enable careful control of experimental and observational conditions, these are not generally considered laboratory animals. However, this research, and similar work on a broad range of amphibians, provides valuable insights for researchers working in the laboratory; some of the characteristics that make amphibians so interesting to biologists are key to their versatility and robust reliability as laboratory models.

Four model species will provide the main focus for this chapter, discussed here in varying degrees of depth, with reference to a wider range of species for context: (1) *Xenopus laevis* (the African clawed frog) is best known by far, and the species upon which many thousands of papers have depended in the last 40+ years; (2) the long-studied Axolotl *Ambystoma mexicanum*, like *Xenopus*, is rather unusual in being fully aquatic (but also pedomorphic, and so distinctly atypical); (3) *Lithobates catesbeianus* (American bullfrog) is a more typical Anuran; and (4) *Pleurodeles waltl* (Iberian ribbed newt), a rather more typical Urodele. These species share certain characteristics as biological research models, but differ greatly in their evolutionary position, life history characteristics, morphology, physiology, and behavior (Burggren & Warburton 2007). Little unites this diverse group (e.g., Feder (1992) discusses "the myth of the typical amphibian"), but together they provide a reasonably comprehensive picture of how amphibian models are represented in research laboratories worldwide.

Typical Research Participation

There is no consensual, definitive, list of amphibian models in research, but *Xenopus laevis* tops every citation as the most commonly participating species (O'Rourke 2007; Burggren & Warburton 2007; Brod et al., 2019; see Figure 22.1). A model biological organism since the 1940s, it has contributed to biomedical research across a wide range of fields, including cell and molecular biology, developmental (and regenerative) biology, physiology, neurobiology (Pearl et al., 2012; Babošová et al., 2018), and human genetic disease (Blum & Ott 2018). It has also contributed to immunological and toxicological work (Robert & Ohta 2009; Babošová et al., 2018), in particular, endocrine disruption and bioaccumulation research (e.g., Yost et al., 2016).

Females are large (around 100 mm long) and, when injected with hormones, can produce a good number of uniform, large (and thus easily manipulated) oocytes and fertilized eggs, as well as embryonic material, highly suitable for embryological and biochemical research (Xenbase; Pearl et al., 2012). The species is capable of breeding in large numbers, as is often required, but it can take a year to reach adult sexual maturity (Xenbase; Jafkins et al., 2012).

Xenopus species are also models in genetics studies. The whole genome has been sequenced for two *Xenopus* species, *X. laevis* and *X. tropicalis*. Both display remarkable similarity with the human genome, hence their great value to research. *X. laevis* is allotetraploid (four sets of chromosomes; whole genome evolutionarily reduplicated; e.g., see Session et al. (2016)), so there is an "excess" of genetic material, which presents certain challenges for genomic work (Pearl et al., 2012; Session et al., 2016). As an "easier" diploid (two sets of chromosomes) species, *X. tropicalis* is fast becoming a valuable alternative model or complement to *X. laevis* (Xenbase 2020; Harland & Grainger 2011; Jafkins et al., 2012). Being diploid allows a range of genomic and genetics work to be completed in *X. tropicalis* that would not be possible with *X. laevis*. For example, creating specific genetic lines that are robust and consistent. *X. tropicalis* also grows to maturity more rapidly (4 months), is smaller (females around 40–50 mm), and can be easier to keep (Harland & Grainger 2011; Grainger 2012) than *X. laevis*.

For decades, the bullfrog *Lithobates* (erstwhile *Rana*) *catesbeianus* (see Figure 22.2) has played a wide-ranging role as a research organism, particularly in physiological studies, including ecotoxicology (e.g., water quality; Ossana & Salibian 2013) and effective anesthesia for amphibians (Williams et al., 2017). It is also a focus of study as a highly invasive pest, particularly where it has been released from ranacultural systems (i.e., where it is bred as food for humans). While its genome is yet to be fully elucidated, Hammond et al. (2017) point out that as a member of the largest frog family, Ranidae (true frogs), with a global distribution, bullfrogs are far more representative of anuran

FIGURE 22.1 The African clawed frog, *Xenopus laevis*. (https://upload.wikimedia.org/wikipedia/commons/7/73/Gemeiner_Krallenfrosch_-_Xenopus_laevis_-_aus_Afrika.jpg.)

FIGURE 22.2 The American bullfrog, *Lithobates catesbeianus*.

amphibians than the evolutionarily distant, and rather specialized *Xenopus* (Family Pipidae). Recently, a draft genome of this species was reported (Hammond et al., 2017), which will not only enable a wider range of research in developmental biology but will also foster more research on amphibian biology *per se*, e.g., evolutionary relationships and disease resistance (see below). This will be vital in conservation work, as approximately two-thirds of ~7,000 extant species are currently in decline.

Axolotls are aquatic salamanders with an intriguing pedomorphic lifestyle (neoteny), including retention of larval features, such as large, feathery external gills, into sexual maturity (e.g., see Denoël (2017) and Figure 22.3). The developmental stages of the axolotl have been carefully characterized (Bordzilovskaya et al., 1989; Laudet 2011), and therefore, these animals have been extensively used in developmental research, including elucidating the role of thyroid hormones in metamorphosis (Laudet 2011). A lack of thyroxine underpins their neoteny. Much work since the early 20th century has focused on their ability to regenerate limbs, research that continues to thrive (Elewa et al., 2017; Vance 2017). Despite its widespread occurrence in laboratories, the wild species now exists only in a single lake near Mexico City (see below), and research has turned urgently toward its conservation (Vance 2017).

The Iberian ribbed newt, *Pleurodeles waltl*, has been used to a smaller extent historically, but is re-emerging as an excellent model for regenerative research (e.g., see Elewa et al (2017); see Figure 22.4) and brain development work. Slower developmental times compared with anuran (e.g., *Xenopus*) systems mean stage-specific interactions at the molecular level can be more precisely isolated, enabling work on regionalization and specification of the vertebrate brain (Busse & Seguin 1993; Joven et al., 2013). This species has also helped understand the immunosuppressive impacts of space flight, accompanying astronauts on the Mir space station (Bascove et al., 2009). Following years of primarily anecdotal study into its reproductive behavior, courtship has recently been characterized and proposed as a behavioral assay for pheromonal research (Janssenswillen & Bossuyt 2016). The wider value of amphibians as behavioral models, in this and other areas, has yet to be fully explored.

Other amphibian species have also participated widely as research models, and more have been recently highlighted: *Rhinella* (was *Bufo*) *marina*, the huge Cane toad, has participated in sperm cryopreservation studies (Fitzsimmons et al.,

FIGURE 22.3 The axoltl, *Ambystoma mexicanum*. (https://commons.wikimedia.org/wiki/File:Axolotl_ganz.jpg.)

FIGURE 22.4 The Iberian sharp-ribbed newt, *Pleurodeles waltl.*

2007) and endocrine stress assessment (Narayan et al., 2012a; Narayan et al., 2012b); *Hyla chrysoscelis* (the southern gray tree frog) has participated in a range of work on auditory communication (e.g., Baugh et al., 2019); and *Bombina* spp. (fire-bellied toads) are involved in studies investigating learning and memory (Lewis et al., 2020). Taken together, these showcase the participation by amphibians across a widely diverse range of research fields, and illustrate the enormous contribution they have made, and continue to make, to biology. Crucially, they share a tolerance of a wide range of laboratory conditions (Burggren & Warburton 2007; Lieggi et al., 2020), but while many species are collected from the wild (e.g., *Rana* sp., *Hyla* sp.) when required, it is a long-lived robustness in captive conditions that unites the four focal species in this chapter.

Behavioral Biology

Natural History

Amphibians generally begin life as large, water-dependent eggs, laid in either water bodies or damp environments. Eggs hatch and spend days or weeks as fully-functioning, free-living, aquatic animals in their larval stage. They grow rapidly, feeding voraciously until metamorphosis is triggered by a combination of external (temperature, season, pond drying up, etc.) and internal hormonal (thyroxine particularly) factors. Urodeles (salamanders) lose their external gills and undergo other phenotypic changes, metamorphosing into the adult form, with a long body and legs of similar length. The adult form may be similar to that of the larva, but the physiology and anatomy are now those of lung-breathing terrestrial animals, fully equipped to live, feed, and breed on land (with a return to a water source for oviposition). Behavior also changes after metamorphosis, especially as sexual maturity is reached (Wells 2010).

Anurans (frogs and toads) undergo an even more dramatic transformation from (usually) herbivorous, gilled aquatic larvae with no limbs, to developing four limbs, to emerging from the water in the adult form. During this process, the larval gut and mouth are resorbed, and a large jaw and a tongue are formed, which are vital for efficient prey capture. Various neural networks also change, and the lateral line system of the aquatic larva is lost and stereoscopic vision develops.

As they usually share occupancy of water bodies, larvae often compete for access to food and also have to contend with a variety of predation risks (e.g., Ferrari et al., 2008). Recent work suggests behavior at the individual level ("personality") may be important in mediating these competitive relationships (Kelleher et al., 2018, and see below). As adults, outside of the breeding season, terrestrial amphibians are generally rather solitary, are active diurnally or nocturnally, and face a range of challenges quite different from those experienced as a young animal. Metamorphosis is a particularly vulnerable life stage for amphibians, and many perish, though both larvae and adults do provide an enormously important food source in many terrestrial and aquatic ecosystems, as well as serving as detritivores, helping to maintain water quality.

In a few groups, such as tropical forest-dwelling Eleutherodactylid frogs and Plethodontid (lungless) salamanders (which inhabit moist temperate forests), direct development has evolved. In these species, offspring pass through the tadpole stage within the egg, and emerge fully metamorphosed, freeing them from dependence on water bodies and reducing vulnerability to predation and competition.

Once adulthood is reached, amphibians can live for many years. Many aspects of behavior depend on amphibians' "cold-bloodedness" (ectothermic poikilothermy), including habitat choice, activity levels diurnally and seasonally, and life expectancy (Vitt & Caldwell 2014). Long-term skeletochronological studies (in which cross sections of phalanges are "ring-counted") on the European alpine newt (*Ichthyosaura alpestris*) in high alpine ponds over the last 20 years have given valuable and generalizable insights into amphibian lifespans. Described as "medium-sized" for a newt (around 110 mm max length, including 30–40 mm of tail, and body weight of a few grams), individuals from cold alpine ponds were estimated in a 1994 study to be 10 years old or more (Wagner et al., 2011). Recent work, resampling the same ponds and individual animals (with important refinements to the technique), suggests these newts can live beyond 30 years (Wagner et al., 2011). Translated to a laboratory animal under human care (e.g., *X. laevis* can live for 15–20 years in captivity), this longevity has implications for the best possible husbandry over a potentially long period.

Ecology

Most amphibians have a fundamental requirement for damp or wet conditions/waterbodies for breeding, courtship, mating, oviposition (discussed below), and larval development (as outlined above), although some can live in extreme xeric and/or cold conditions (Wells 2010). The broad ecology of *Xenopus* is adaptable, and they occupy all kinds of, often muddy, water bodies in sub-Saharan Africa, where they are generally active at night (Tinsley et al., 1996). Bullfrogs are similarly adaptive, widespread, and able to tolerate a wide range of conditions. In contrast to the wide distributions of these two anurans, axolotls have a very restricted range, and, along with other Mexican *Ambystoma*, they are an EDGE (Evolutionarily Distinct and Globally Endangered) species. They are, as identified above, now restricted to a single "lake", which has been reduced to a series of canals close to Mexico City, much affected by pollution and habitat loss (Vance 2017). *Pleurodeles waltl* is restricted to Spain, Portugal, and Morocco, and inhabits small indistinctive ponds or pools, remaining aquatic unless water dries up (Joven et al., 2015). Numbers of this species are also declining in its natural range, mainly through habitat loss, and it is now classed as Endangered.

Water is important for maintaining the healthy cutaneous layers, as amphibians generally have soft, permeable skins across which gas exchange may occur, although many also have lungs (Duellman & Trueb 1994; Vitt & Caldwell 2013). Maintenance of the skin is an amphibian's first line of defense against pathogens, and it (or associated glands) may have important chemical or microbiotic/antifungal properties. For example, bullfrog skin has been found to contain gastrointestinal hormones that vary seasonally, with active and torpid phases (Wang et al., 2016), and may help regulate skin physiology. Lactic acid bacteria, which provide protection from infections, such as *Bd* (*Batrachochytrium dendrobatidis*), has also been found in bullfrog skin (Niederle et al., 2019), suggesting positive lines of inquiry in the fight against at least one group of bacterial pathogens that has devastated amphibian populations globally.

As prey species, particularly in the larval stages, amphibians exhibit many antipredator strategies that may strongly influence their distributions (Morrison & Hero 2003; Koprivnikar & Penalva 2015). These may be anatomical features, such as, for *Xenopus*, eyes on the dorsal surface of the head to detect predators (and food) on the surface of the water and nostrils at the end of snout (Chum et al., 2013). An important adaptation is warning coloration, seen in many species globally (Santos et al., 2016; Saporito & Grant 2018). Vivid coloration may be a form of mimicry, or may be associated with various physiological adaptations, including slippery skin mucus or toxic secretions, known as the "peptide arsenal" (Roelants et al., 2013; Santos et al., 2016). These secretions can be toxic or mildly aversive substances. For example, *Xenopus* have been shown to be distasteful to snakes, making them yawn and allowing the frog to escape (Barthalmus & Zielinski 1988). North American rough-skinned newts (*Taricha* spp.), on the other hand, produce a neuropeptide almost identical to that produced by pufferfish, which is deadly to all vertebrate groups of predator. Interestingly however, in some areas in which the species co-exists with the common garter snake (*Thamnophis sirtalis*), the snakes have developed immunity and feed on them widely. Many toad species are chemically defended by bufadienolides, lethal to most predators, but, again immunity has co-evolved, this time in the Western garter snake (*T. elegans*) (Mohammadi et al., 2017). Rapid evolutionary responses are evident in amphibian prey–predator relations and may persist through the life stages (Bucciarelli et al., 2017). In many amphibians, camouflage coloration is a key adaptation, as broadly seen in bullfrogs, *Xenopus*, *Pleurodeles*, and the axolotl. *Xenopus* and *Pleurodeles* also have physical structures to assist with predator defense; claws on the hind feet in *Xenopus* and sharp ribs protruding through the abdomen in *Pleurodeles waltl*, giving the species its common name, the Iberian sharp-ribbed newt (Joven et al., 2015).

Behavioral defense strategies mainly involve detection of the predator, followed by flight or immobility (e.g., Garcia et al., 2009; Hossie et al., 2017). Nuanced responses may also have evolved alongside diel activity patterns (Santos & Grant 2011), or, indeed, moon phases (Grant et al., 2013). While predator defense is a key evolutionary driver, avoidance of (excess) UV-B light may also be a key factor in shaping behavior (Marco et al., 2001; Blaustein & Belden 2003; Garcia et al., 2009). Of course, combinations of threats are important, and research has revealed how amphibians may avoid both predators and the danger of UV light (Garcia et al., 2009), or trade off the risks of predation and parasitism (Koprivnikar & Penalva 2015).

Feeding Behavior

Feeding encompasses the behavior of foraging, searching for and consuming food, and diet selection. Generally, amphibians are carnivorous and will attempt to eat anything that moves past them, using a "sit and wait" strategy. They generally have a wide-ranging, opportunistic diet, whether they are terrestrial or aquatic. Some will move to track down a source of prey if it is detected, by smell or visually. How amphibians locate and then capture prey has long been of interest (Martinez-Conde & Macknik 2008) in anurans (e.g., Monroy & Nishikawa 2009) and salamanders (e.g., Borghuis & Leonardo 2015). The vertebrate midbrain, which has a role in many oriented responses, including prey-capture, escape, and eye movement, may have topographically ordered maps of an animal's surroundings, based on visual, acoustic, somatosensory, electrosensory, and lateral line inputs (Claas & Dean 2006). Some (mainly aquatic) amphibians, including *Xenopus* spp. (Elepfandt 1996), have a lateral line (somatosensory) system that also helps detect prey items (Chum et al., 2013). This is likely used in combination with visual (and olfactory, Lieggi et al., 2020) cues in *Xenopus*, as their visual and lateral line maps correspond (Claas & Dean 2006). The acute sense of smell and powerful hind limbs of *Xenopus* species enable them to chase and capture prey in their aquatic habitat, catching a wide range of whatever can be caught (Chum et al., 2013). They lack extendable tongues and, therefore, feed with their mouths, sometimes using their forelimbs to pull food apart and push it into the mouth, assisted by a hyobranchial pump that helps suck the food in. This species has achieved invasive species status, like the bullfrog (see below), largely due to rapid reproduction rates, robust tolerance of a wide range of conditions, and the ability to eat or scavenge a huge diversity of prey.

Over 4000 species of anuran use a soft, strong, and very sticky tongue to capture a located prey item in a terrestrial environment. Recent work with *Rana pipiens* shows the tongue has special adaptations to make it super-soft and stretchy, but the saliva is also highly adapted to ensure maximum stickiness on a variety of surfaces, such as hard insect exoskeletons, furry mammals, or feathery chicks (Noel et al., 2017). Noel and Hu (2018) review this adaptation and the intriguing human-world applications it presents for adhesives research. The bullfrog is known to be a major pest (alien invasive) species in many locations outside of its already extensive range (Snow & Witmer 2010). Their large size and generalized diet means they exploit a widely diverse range of prey (e.g., Jancowski & Orchard 2013), demonstrating highly effective predatory, and invasive, behavior.

Salamanders have a range of broadly similar feeding strategies (Reilly & Lauder 1992), generally locating prey by smell and sight. A range of invertebrates and small fish are captured by axolotls, which snap at possible prey with their jaws, while also employing a sucking mechanism. A widely diverse diet was also found for *Pleurodeles waltl* (Escoriza et al., 2020), where it was also noted that, unsurprisingly, larger individuals/later life stages maintained a wider diet.

Larval amphibian stages usually eat a different diet from that of adults, reflecting not only their different sizes but also their different physiological requirements. Where adults and larvae of the same or different species share the same natural habitat, some form of niche partitioning occurs; spatially, temporally, and behaviorally. This is a vibrant area of current research with anurans (e.g., Székely et al., 2020) and urodeles (e.g., Salvidio et al., 2019), or both (Arribas et al., 2015). More recently, diet variation has been examined both ontogenetically and between individuals (e.g., Schriever & Dudley Williams 2012), exploring how animals may exploit seasonal abundances or changing environments (Arribas et al., 2015; Székely et al., 2020).

Social Organization and Behavior

While sociality, as understood for mammals and birds, may not apply to amphibians, some clearly have the capacity to recognize kin and respond accordingly; individuals of many species may associate and some may form hierarchies. It is also important to explore kin recognition separately in larval and fully adult animals, as the behavior may have very different origins and functions. Waldman and Adler (1979) showed many years ago that spadefoot toad tadpoles appeared to associate with kin, but whether this was due to real "recognition" or simultaneous habitat selection is hard to discern (and see Blaustein & Waldman 1992). Key to this is the question of whether amphibian larvae have the capacity to recognize their siblings, or whether they are simply sharing resources. Work by Pfennig et al. (1993) demonstrated cannibalism by kin and suggests larvae are most likely to recognize kin where it really matters, e.g., in crowded environments where resources may be/become limited, as may often be the case for larval stages in a communal water body.

Pizzatto et al. (2016) examined the influence of a sympatric species on kin recognition in Australian green and golden bell frog tadpoles (*Litoria aurea*), as part of a captive breeding study for conservation. In a series of laboratory choice tests, tadpoles aggregated preferentially with conspecifics, rather than sympatric heterospecifics. In the absence of conspecifics, however, they preferred to aggregate with the heterospecific tadpoles than remain alone, and preferred unfamiliar kin over unfamiliar conspecific nonkin. In addition, they showed no preferences for familiar, rather than unfamiliar, siblings. These results suggest siblings (whether familiar or unfamiliar) are preferred over conspecific kin, but these are preferred over conspecific nonkin. If no conspecifics are present, then they will prefer to mix with tadpoles of another species rather than remain alone, presumably to aid in predator defense.

Kin recognition has been long established in adult amphibians (Waldman & Adler 1979), in both anurans (Blaustein & Waldman 1992) and urodeles (Pfennig et al., 1994; Jaeger et al., 2016). As the extensive work on *Plethodon cinereus* shows (Jaeger et al., 2016), there are myriad social relationships between and within sexes, and across life stages. Groupings in lab/captive conditions do not often take into account what little is known, although much more baseline work, for larval and adult stages, is required to inform these decisions.

Individual personality or behavioral syndromes (see Sih et al., 2004 for discussion of terminology/definitions) are fast gaining currency as a valuable measure in both welfare and conservation research. Relatively new as a focus for amphibians (for review, see Kelleher et al. (2018)), most work has focused on the larval stage in anurans (e.g., Carlson & Langkilde 2013 for *Bufo sp*). However, recent studies (Wilson & Krause 2012 for an anuran, *Rana ridibunba*; Koenig & Ousterhout 2018 for the salamander, *Ambystoma maculatum*) have measured behavioral parameters, including activity, boldness, and exploration, for both larval and juvenile stages, to see if individual measures persisted through metamorphosis; i.e., are retained through ontogeny. They found within-life-stage effects, i.e., more active individual larvae were consistently bolder and showed more "exploratory" behavior. They also found a correlation between activity, boldness, and exploration in larvae, and boldness as juvenile metamorphs, suggesting that animals may develop particular traits and keep those through life. This could have valuable implications for individuals in laboratory populations (see Coleman and Novak, this volume; Schapiro, this volume).

Courtship and Mating Behavior

Fundamental sexual selection and mate choice theory underpin much of what we understand about amphibians' mating and parenting behavior (Halliday 1990; Houck & Arnold 2003; Wells 2010; Bee et al., 2013). Anurans generally use an "explosive breeding" strategy (Wells 2010; Ulloa et al., 2019), where males congregate at a pond/water body (often their natal pond) at the start of a typically brief breeding season. Here they employ advertisement calls to compete with other males for access to females, and/or to attract good quality females (e.g., Ryan's work on the tungara frog discussed above). Vocal competition may escalate, involving advertisement and aggressive calls, or conversely involve suppression of the calls of certain males (usually satellites/subordinates) (Tobias et al., 2010). Once in contact with a female, amplexus ensues, in which the male clasps the female with his forelimbs to try and assure paternity of the offspring. If the female is receptive, she releases eggs into the water and the male fertilizes them externally. Such behavior, with male aggregations at a pond to call and intercept females, is typical for the long-studied bullfrog (Emlen 1976; Howard 1978; Howard 1984).

For fully aquatic *Xenopus*, the mating strategy is rather different. Both sexes live together in the same pond all year, and sexual receptivity in males continues throughout the breeding season, which lasts approximately 6 months (Tobias et al., 2004, 2010). During this time, females become sexually receptive only briefly, and asynchronously (unlike explosive strategies). This results in a highly skewed operational sex ratio, in which males are competing for rare, sexually receptive females (Tobias et al., 1998). Male–male social interactions occur via clasping behavior (akin to amplexus observed in mating pairs) and vocal behavior. Such vocalization involves a variety of calls, including chirping, growling, answer, and advertisement calling. As this occurs under water, with no visible vocal sac movements, identifying callers is very difficult. Vocal suppression occurs between males (Tobias et al., 2010) according to status. As adult *X. laevis* retain their lateral line system, they may also use information from this source, in addition to olfactory and visual cues, to locate and detect competing males. Females locate and swim toward advertising males, so males that can vocally out-signal competitors are likely to have a reproductive advantage (Tobias et al., 2004, 2010).

Salamander reproduction, by contrast, can involve lengthy and complex behavior, with stimulation of unreceptive females by pheromones. For terrestrial species, this may be via tactile methods (e.g., see Houck 2009; Propper & Moore 1991; Wack et al., 2013; Woodley 2015). This may happen similarly in some aquatic species, such as *Pleurodeles waltl*, that show lengthy "pinwheeling" behavior (Janssenswillen & Bossuyt 2016), and the axolotl (Maex et al., 2016), where males nudge females repeatedly to initiate mating. Other species use mechanical transmission of vibrations and pheromones through the water by waving/fanning their tail (Halliday 1990; Houck & Arnold 2003). Most aquatic breeding salamanders have external sperm transfer; the male deposits a packet of sperm (via a spermatophore) on the substrate where the female may pick it up in her cloaca. Here it is stored (perhaps mixing with those of other males, Halliday 1990), and the eggs are then fertilized internally as they are laid.

The importance of female choice in amphibians has been well studied, though mainly as a mechanism in sexual selection (e.g., Halliday 1990; Jaquiery et al., 2010). The features preferred by females (e.g., high-ranking males in *Xenopus*, perhaps) may provide valuable information for laboratory breeding, and help facilitate the generation of large numbers of animals. This could result in a reduction in the overall numbers of females required for breeding programs/laboratory work (Holmes et al., 2018; Brod et al., 2019).

Amphibian parental care has been studied intensively over the last few decades (e.g., Crump 1996), with recent comprehensive reviews of developments, future perspectives (Ringler et al., 2013; Shulte et al., 2020), and the diversity of its evolution (Furness & Capellini 2019). Parental care varies widely from none/very little to a great deal. Most anurans, the focal species in this chapter, show little parental care. Some amphibians may deposit the large clutch(es) of eggs in a shady part of a pond (e.g., some anurans), or individually wrap eggs in a leaf (e.g., *Lissotriton, Triturus*; Norris & Hosie 2005; Tóth et al., 2011), or protect a small clutch of eggs with the body (many terrestrial salamanders; Crump 1996; Jaeger et al., 2016).

A small number of species have highly specialized strategies involving care at a level that outshines even some mammals. Neotropical Dendrobatid frogs, for example, exhibit tadpole transport, which is obligatory in almost all species (Ringler et al., 2013). In tadpole transport, the male (usually, but sometimes both sexes of a species) carries the tadpoles to a pool, a safe water-filled leaf axil, or bromeliad, or indeed keeps them with him (Ringler et al., 2013). Provisioning may also take place in these frogs, where the adult provides food to the developing offspring. The cognitive capacities for such strategic planning, flexibility, learning, and memory required for this complex behavior are little understood in this, supposedly unsophisticated, animal (Ringler et al., 2013). Male parental care has also been reported in an unusual aquatic salamander (*Siren intermedia*) (Reinhard et al., 2013) and the Japanese giant salamander *Andrias japonicus* (Okada et al., 2015).

Communication

Most research on communication in amphibians has focused on reproductive behavior; male advertisement calling and elevation of receptivity in females through chemical or tactile modes, all of which are discussed in the above section.

An intriguing possibility for visual detection and/or communication has been opened up by new work on bioluminescence in amphibians (Lamb & Davis 2020). This work has revealed that fluorescent green coloration in response to blue excitation light is strikingly widespread across the salamander amphibian radiation, and some with bold patterns and colors fluoresce brightly (e.g., *Ambystoma tigrinum*). Biofluorescence in amphibians could potentially contribute to perception by conspecifics and heterospecifics (competitors and predators) in low-light environments. Amphibians occupy a variety of habitats with complex ambient light environments that vary with vegetation structure, weather, and time of day (Endler 1993; Lamb & Davis 2020). During twilight (when crepuscular and nocturnal amphibians are more active), the ambient spectra in terrestrial habitats shifts to predominantly blue light, which results in the green fluorescence detected by Lamb and Davis (2020). Perhaps some of the intra- and intersexual territorial and sexual dynamics known to exist in salamanders may be shown, in the future, to be (at least partly) mediated by this "new" mode of communication. This work illustrates that there is still much to learn about amphibian perception and detection.

Common Captive Behaviors

Normal and Abnormal

When not feeding or reproducing, most amphibians are generally inactive and appear to show limited behavioral repertoires, in both the wild and the laboratory. There is also a perception that amphibians tend to respond to a variety of stressors with inactivity (Wack et al., 2013). What constitutes "normal" or "abnormal" captive behavior can therefore be hard to pin down. It is now reasonably clear that amphibians have the capacity to feel pain (Sneddon et al., 2014; Sneddon 2015; Sneddon 2017), but whether/how they experience poor welfare remains difficult to distinguish (Arena et al., 2012; Sneddon et al., 2014).

Behavior is, of course, fundamentally important, as it tells us about the precise, day-to-day interactions between an animal and its environment, enabling us to identify and measure what drives change over time. Using behavioral measures alongside concomitant physiological measures would clearly provide the best possible picture (Dawkins 1998, 2003), but behavior is much easier to quickly assess. Use of behavior to monitor welfare issues in captive environments has been repeatedly called for in amphibian research (e.g., Burghardt 2013; Tapley et al., 2015; Harding et al., 2015), but a real lack of baseline descriptions of captive behavior for most species (Michaels et al., 2014; Holmes et al., 2016, 2018) impedes progress here. What is known about abnormal behavior for the focal species of this chapter, including levels of expression of otherwise "normal" behavior, will be discussed below in the "Ways to Maintain Behavioral Health" section. Shulte et al. (2020) argue strongly (also echoed in the recently updated ARRIVE guidelines; Percie du Sert et al. 2020) that the development of comprehensive, fully validated ethograms would enable much better focus on vital baseline amphibian behavior research, and help ensure comparability of behavioral measures across experimental conditions (e.g., Norris & Hosie 2005; Shulte et al., 2020; Lieggi et al., 2020); this should enable behavior to be used as a reliable tool for monitoring welfare in laboratories.

In the context of limited captive amphibian behavioral knowledge overall, it is not surprising that sex differences are rarely studied or identified in published work (Percie du Sert et al., 2020 Ogurtsov et al., 2018). In two separate studies, sex differences were found for *X. laevis* in physiological (corticosterone or CORT, see below), but not behavioral measures of welfare. Holmes et al. (2018) found that males released significantly higher levels of CORT than females after transportation, though both sexes showed an increase. The impact of different colored tank backgrounds (Holmes et al., 2016) is discussed below.

In the Lake Zacapu salamander, *Ambystoma andersoni* (close relative of the axolotl), Emmans (2015) established a baseline ethogram for males and females separately, and measured behavioral variation over different times of day. Males were found to show significantly more "searching" behavior compared to females, but only in the morning. However, this is but one study. Clearly, there is a need for more studies establishing baseline behaviors for amphibian species, and how they may vary, for example, over development (as seen above for personality), by sex, and across different times of day and seasonally, in order to appropriately attend to individual species' and animals' husbandry requirements under laboratory conditions.

As noted above, different levels of "normal" behaviors exhibited may suggest more or less stress – e.g., general swimming behavior in *X. laevis* (see below). However, some specific behaviors do seem to be indicative of stress and could be usefully explored further, once baselines have been established. "Walling" (repeated rapid, erratic swimming up and down, bumping into the tank sides, which can lead to lesions on the snout), a term originally used for similar fish behavior, has been shown to be elevated, at least for a time, under stressful conditions for *X. laevis*. This measure could be useful for *X. tropicalis*, or indeed, other fully aquatic species. More "burst breathing" (ascent to the water surface to quickly take a breath) may occur in *X. laevis* under stressful conditions (Chum et al., 2013), which could be monitored. Similarly, gill beat rates in larval amphibians and adults that possess them (*Ambystoma* species, see below) are likely to rise under stressful conditions to increase oxygen uptake. The behavior "resting bent", an alert, unrelaxed body posture found to be elevated for the Dwarf African frog *Hymenochirus boettgeri* under more stressful conditions (see below), could also be used for this, or other species, as a putative stress measure.

Ways to Maintain Behavioral Health

Captive Considerations to Promote Species-Typical Behavior

Studies on captive welfare, and behavioral needs in particular, for amphibians have lagged far behind other laboratory vertebrates (Reed 2005; Burghardt 2013; Holmes et al., 2016). While it is relatively novel to consider amphibian welfare in the same way as mammalian welfare, it is fast growing in importance, particularly in laboratory environments (e.g., Brod et al., 2019), but also for conservation (Harding et al., 2015). Some laboratory amphibian species may live for many years (5–10+) in barren captive conditions, whereas laboratory mice may live only 1.5 years in conditions that promote high levels of welfare. A level of "domestication" is sometimes assumed after generations in captivity (e.g., suggested for *Ambystoma mexicanum*, Voss et al., 2009), but this has not been properly addressed in amphibians, nor do we know whether this might positively or negatively affect their long-term health and wellbeing. Providing husbandry conditions that promote natural behavior is vitally important to help ensure that robust, minimally stressed animals can perform consistently and reliably in experimental protocols (Lieggi et al., 2020; Brod et al., 2019; Holmes et al., 2016). This generates potential cost savings for research labs but, perhaps more importantly, means that these individuals may experience welfare standards similar to those accorded to other laboratory vertebrates.

Assessing Welfare

Determining best possible laboratory conditions for any species requires sensitive and reliable welfare assessment tools. Development of appropriate techniques for amphibians has been limited (Reed 2005) until very recently. Corticosterone (CORT; similar to cortisol in many mammals and fish) is now considered the main amphibian "stress hormone". The last few years have seen a surge of studies utilizing field-based corticosterone measuring in amphibians (for reviews, see Narayan 2013, Walls & Gabor 2019) to address critical conservation issues. However, extensive discussion in the endocrine literature urges caution in interpretation of CORT measures, particularly in isolation from other measures, and urges the validation of these with behavioral, and additional physiological, measures (and see above). This may be particularly important for amphibians, given that at particular life stages, or when certain behavior is being expressed (e.g., mating), high levels of CORT are normal, as they promote developmental processes (e.g., trigger larval stages) or coordinate other hormones and behavior, and are not indicative of "stress". Knowledge of these "baseline" patterns of CORT release and associated behaviors are limited for many vertebrates, but almost completely lacking for amphibians (e.g., see Falso et al., 2015). This knowledge is vitally important, however, particularly for understanding the influence of current laboratory environments, and properly assessing potential improvements. Better understanding of variation in behavioral expression in laboratory/captive settings is important in other contexts as well. For example, changes in behavior may indicate infection, disease, and/or a compromised immune system, so behavioral assays could be usefully developed to assess captive environments or as health and welfare indicators.

One example of a study in which behavioral and physiological markers have been used to assess stress in amphibians involves the fully aquatic Lake Zacapu salamander, *Ambystoma andersoni* (a close relative of the axoltl). In this study, "gill beat" rates (similar to opercular beat rate often measured in fish as a "stress indicator") were measured after handling (for a routine health check). Importantly, CORT was measured concomitantly. Gill beat rates were found to be significantly elevated following handling, and CORT was also elevated (Emmans 2015). Thus, elevated gill beat rates appear to be a behavioral indicator of stress, validated by physiological CORT measures, which could be used to monitor health and welfare in this species, as well as the very closely related axolotl. Similar work is needed on other species.

Environment

Elements of the captive environment for the four focal species have been investigated to very different degrees; e.g., little or no evidence exists for the "best practice" lab environments for bullfrogs, axolotls, or *Pleurodeles waltl* (but see Joven et al., 2015). Provision of tank environments that promote species-typical behavior is strongly encouraged, although what this might be is open to interpretation, without validating data. Substrates should be considered carefully. As amphibians can be unfussy about what they eat, small stones, gravel, and other "furniture" that could be ingested should be avoided. For fully aquatic species, sufficient water depth to allow a full range of behaviors is important, as is an ecologically appropriate temperature range. Standard cleaning practices and other husbandry protocols may conflict with ideal/preferred environments, for example adding enrichment objects may encourage undesirable algal growth (McNamara et al., 2018). Clear evidence for the beneficial impact of interventions is needed to garner support for their wide implementation.

A study with the small (~20–25 mm) "pet" species, the Dwarf African frog, *Hymenochirus boettgeri* (Hosie et al., unpubl data), in our laboratory attempted to determine positive tank environments by testing three different enrichments: external shading (with no contact), and two sizes (small and large) of upturned pots, located inside the tank, that the animals could use for physical shelter. Behavior was measured in the three conditions but, crucially, CORT measures were collected concomitantly. The data showed highly significant reduction of one behavior, "resting bent", for each of three different types of enrichments, when compared to having no enrichment (see Figure 22.5). In addition, "resting bent" behavior was strongly associated with elevated levels of CORT (Figure 22.5), and so could prove to be a useful "stress-indicator" behavior. This study demonstrates the importance of providing refuge, either external cover or structures, to amphibians.

Much work has been done to establish effective husbandry and suitable laboratory environments for *X. laevis* (and more recently, *X. tropicalis*) (e.g., see protocols such as McNamara et al., 2018). However, while these may meet minimal needs, Reed (2005) identified that many elements of standard housing and care are based on what has worked "well enough", rather than any underpinning science. Lieggi et al. (2020) identify

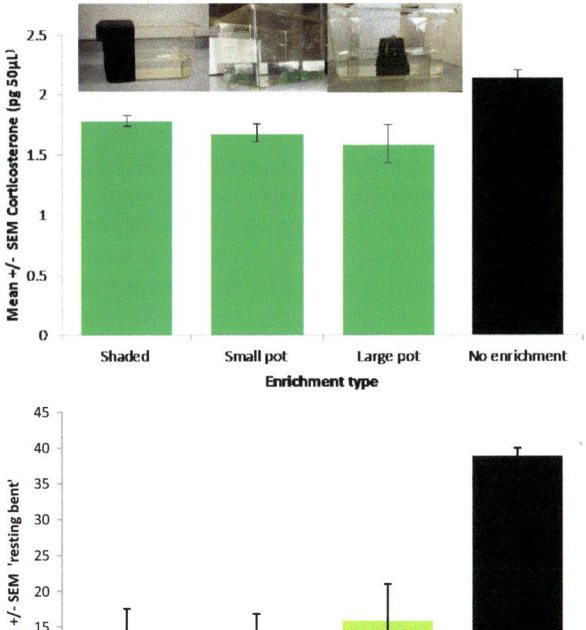

FIGURE 22.5 Testing three different enrichments in the Dwarf African frog *Hymenochirus boettgeri*. CORT release was significantly lower with enrichments, compared to none (ANOVA: $F(3, 45) = 6.85$; $p < 0.001$); "resting bent' behavior, a putative "stress indicator" was also significantly lower with enrichments (ANOVA: $F(3, 45) = 11.75$; $p < 0.001$). (See text for details.)

that what might be optimal for many aspects of tank design (shape, size, color, material), water flow, and structures within the tank still needs to be established.

Enrichment is often suggested for amphibians (e.g., Burghardt 2013), but rather little evidence has emerged to identify what has positive impacts. Standard protocols for *X. laevis* suggest enrichment provision (e.g., McNamara et al., 2018), but these are often relatively simple or minimal, to enable easy cleaning. Brown and Nixon (2004) found *X. laevis* preferred tanks with shelter over no shelter, and this environment produced rather more, and better quality, eggs. Torreilles and Green (2007) found provision of pieces of drainpipe as enrichment reduced the number of harmful bites between *X. laevis* individuals. It should be noted that these animals were housed at densities that were probably far higher than they would experience in the wild, so this work also suggests that appropriate densities should be maintained (see also Chum et al., 2013). Archard (2013) observed female *X. laevis* and found they preferred to use a refuge if present, used them more during the day, and would share with another female. Presence of a refuge modified daily activity compared with no refuge; refuge females were less active.

Our laboratory has investigated the impact of tank background coloration for *X. laevis*, by measuring behavior validated by physiological measures. Dark (ecologically relevant) shading placed outside of the tanks was compared with opaque white (nonecologically relevant) backgrounds. Adult *X. laevis* were placed into tanks for 48 h, each experiencing both the white and black conditions (counterbalanced and with rest between). Female frogs released significantly more CORT when they were placed in tanks with a white background when compared to placement in tanks with a black background; however, males did not show this effect. Another physiological indicator, body mass loss, was significantly greater for both sexes on white backgrounds compared with black. Significantly more "walling" behavior was measured for both sexes immediately after placement in a tank with a white background compared to black, but this declined after 2 days in both black and white conditions (Holmes et al., 2016). In contrast "swimming" (i.e., general activity) was found to be exhibited to a similar extent following immediate placement in tanks with white and black backgrounds. Unlike walling behavior, however, this declined significantly after 48 h on black backgrounds, but not white. This indicates how two different behaviors (swimming and walling) can be used to determine nuanced responses to the laboratory environment, perhaps enabling the impact of different timeframes to be elucidated. On a practical level, this work suggests that shading, or light reduction, may reduce stress for *Xenopus* and could be implemented relatively economically.

As indicated above, providing refuge is an important aspect of the environment. In a study evaluating refuge preferences in *X. tropicalis*, animals used a refuge with a transparent red cover in a similar way to the same refuge with a black cover (Cooke 2018). This shows that color and opacity may both be important in determining best environments (see points above about fluorescence), and highlights that fundamental understanding of the perceptual capabilities of amphibians is still lacking (Arena et al., 2012).

Overall, there is evidence that provision of "enrichment" or at least an ecologically appropriate environment is likely to benefit the welfare of laboratory amphibians. Enrichment suggestions include shelter/cover objects, dim lighting, dark paper enclosed tanks (three sides), areas of land and water where appropriate, and/or adequate water depth. Measuring behavior and/or physiological parameters to experimentally test these provisions would be a great step forward.

Social Groupings

No evidence has been established for best practice stocking densities or how to accommodate the little that is known about social behavior (see earlier section) in the four species discussed here. Though stocking densities are suggested/recommended by standard protocols (e.g., McNamara et al., 2018), *optimal* stocking densities to maximize good welfare for *X. laevis* (or *tropicalis*) have not yet been determined. There is limited mention of social behavior in *Xenopus* in early literature (e.g., Tinsley & Kobel 1996). Field observations and limited laboratory studies suggest that they are territorial (see above section) and may develop a hierarchical system in the laboratory (Chum et al., 2013). Adult *X. laevis* in our laboratory seem to show distinct hierarchies, where certain individuals are invariably first to feed (Hosie, pers. obs.). Establishing whether/how this behavior should be accounted for in most species is urgently needed to inform best practice. In the interim, sensible stocking densities (ensuring minimal aggression/injury), and sex ratios, should be established experimentally as far as possible. These should be evaluated over time, to accommodate seasonal variations in behavior, and also acknowledge length of time in captivity, as long captive lives should be good lives. Keeping more than one species together should be avoided, as interspecific competition is likely to result (see earlier section).

Feeding Strategies

In keeping with their active feeding behavior and wide-ranging diet, most laboratory amphibians can be easily maintained on standard commercially produced food, perhaps supplemented with live invertebrates. Competition for food is likely to be normal and may be "used" by *X. laevis* keepers to promote/stimulate feeding activity/frenzies in a group (Chum et al., 2013). However, care should be taken over feeding to avoid aggression and injury. As explained by Chum et al. (2013), animals that eat too much or are stressed too quickly after eating may regurgitate their stomach contents, although what might be stressful has not been established. Clearly more work is needed to provide optimal feeding recommendations. Searching for live invertebrates may well provide a source of interest or enrichment (Chum et al., 2013) as it stimulates more natural behavior.

Training

Learning and memory has been little studied in amphibians (Lewis et al., 2020), but they are recognized to show different forms of learning. *Bufo* species have been shown to discriminate between food types (Schmajuk et al., 1980) and show learned exploratory behavior (Miller et al., 2018). Recently, Lewis et al. (2020) used prey capture conditioning in *Bombina orientalis* to investigate comparative aspects of learning and memory. They suggest that this species could be a novel model for learning and memory work in amphibians. Animal training *per se* has been little explored, however, one study examined predator recognition training in larval hellbenders (the endangered *Cryptobranchus alleganiensis*). Crane and Mathis (2011) exposed larvae to adult hellbender plus trout secretions and trout secretions alone in a training regime. In a subsequent test, larvae froze significantly less when exposed to the hellbender secretions than to a nonrelevant predator (trout), suggesting the training enabled learning of the relevant potential predator (Crane & Mathis 2011). Why more movement, rather than freezing, might indicate an appropriate predator response is unclear. This does, at least, open up the possibility of using training in a laboratory environment. Exploring personality types may also enable wider application of novel approaches in laboratory amphibian care.

Conclusions/Recommendations

Behavior in wild amphibians has been relatively poorly studied though, as discussed above, with some notable exceptions with regard to key species and areas of focus. This patchy coverage of a very diverse group has resulted in rather little overall evidence to underpin best ways to maintain good welfare for laboratory amphibians. Growing evidence suggests that provision of shelter is important to reduce stress, either externally, with tank shading or with internal objects/refugia; this reflects natural environments and predator avoidance behavior. Few studies have investigated feeding behavior directly, but evidence suggests enrichment may be possible with live food. Where food/provision is standardized (for experimental reasons), care must be taken to avoid overfeeding and aggressive competition. In line with this, optimum stocking densities, for best welfare, need to be established and aggressive/injurious interactions avoided. Erring on the side of caution would be desirable while the evidence accrues. The tools needed to investigate best practice are now being developed, and measuring behavior is an important part of this. Establishing baseline behaviors for laboratory amphibians is crucial, taking account of sex, developmental stage, and diurnal and seasonal factors. Alongside sensitive use of physiological measures, this should enable rapid development of improved husbandry for laboratory amphibians. The long captive life of many laboratory amphibians is worth reemphasizing. Future work should focus particularly on improving provision for a wide range of natural behaviors (Mellor 2016) in this group of fascinating animals that has quietly served biological science for many decades.

REFERENCES

Arena, P. C., Steed, C., & Warwick, C. (2012). Amphibian and reptile pet markets in the EU; an investigation and assessment. APA.org.

Archard, G. A. (2013). Refuge use affects daily activity patterns in female *Xenopus laevis*. *Appl Anim Behav Sci* 2013; 145(3): 123–128. https://doi.org/10.1016/j.applanim.2013.02.006.

Arribas, R., Díaz-Paniagua, C., Caut, S., & Gomez-Mestre, I. (2015). Stable isotopes reveal trophic partitioning and trophic plasticity of a larval amphibian guild. *PLoS One*, 10(6): e0130897. https//doi.org/10.1371/journal.pone.0130897.

Babošová, M., Vašeková, P., Porhajašová, J. I., & Noskovič, J. (2018). Influence of temperature on reproduction and length of metamorphosis in Xenopus laevis (Amphibia: Anura). *The European Zoological Journal*, 85(1): 150–157. https://doi.org/10.1080/24750263.2018.1450456.

Barthalmus, G. T., & Zielinski, W. J. (1988). *Xenopus* skin mucus induces oral dyskinesias that promote escape from snakes. *Pharmacology, Biochemistry & Behavior*, 30: 957–959. https://doi.org/10.1016/0091-3057(88)90126-8.

Bascove, M., Huin-Schohn, C., Gueguiou, N., Tschirhart, E., & Frippiat, J. (2009). Spaceflight-associated changes in immunoglobulin VH gene expression in the amphibian Pleurodeles waltl. *The FASEB Journal*, 23: 1607–1615. https://doi.org/10.1096/fj.08-121327.

Baugh, A. T., Bee, M. A., & Gall, M. D (2019). The paradox of hearing at the lek: auditory sensitivity increases after breeding in female gray treefrogs (*Hyla chrysoscelis*). *Journal of Comparative Physiology A*, 205: 629–639. https://doi.org/10.1007/s00359-019-01354-0.

Bee, M. A., Schwartz, J. J., & Summers, K. (2013). All's well that begins wells: celebrating 60 years of animal behaviour and 36 years of research on anuran social behaviour. *Animal Behaviour*, 85: 5–18. https://doi.org/10.1016/j.anbehav.2012.10.031.

Beebee, T., & Griffiths, R. (2000). *Amphibians and Reptiles*. New Naturalist Library 87. HarperCollins, London.

Blaustein, A. R., & Belden, L. K. (2003). Amphibian defenses against ultraviolet-B radiation. *Evolution & Development*, 5: 89–97. https://doi.org/10.1046/j.1525-142X.2003.03014.x.

Blaustein, A. R & Waldman, B. (1992). Kin recognition in anuran amphibians. *Animal Behaviour*, 44: 207–221. https://doi.org/10.1016/0003-3472(92)90027-7.

Blum, M., & Ott, T. (2018). Xenopus: an undervalued model organism to study and model human genetic disease. *Cells Tissues Organs*, 205: 303–313. https://doi.org/10.1159/000490898.

Bordzilovskaya, N. P., Dettlaff, T. A., Duhon, S. T., Malacinski, & G. M. (1989). Developmental-stage series of axolotl embryos. In: Armstrong, J. B., & Malacinski, G. M. (eds.), *Developmental Biology of the Axolotl*. Oxford University Press, New York, pp. 201–219.

Borghuis, B. G., & Leonardo, A. (2015). The role of motion extrapolation in amphibian prey capture. *Journal of Neuroscience*, 35(46): 15430–15441. https://doi.org/10.1523/JNEUROSCI.3189-15.2015.

Brod, S., Brookes, L., & Garner, T. W. J. (2019). Discussing the future of amphibians in research. *Lab Animal*, 48: 16–18. https://doi.org/10.1038/s41684-018-0193-6.

Brown, M. J., & Nixon, R. M. (2004). Enrichment for a captive environment –the Xenopus laevis. *Animal Technology & Welfare* 3: 87–95.

Bucciarelli, G. M., Shaffer, H. B., Green, D. B. et al. (2017). An amphibian chemical defense phenotype is inducible across life history stages. *Scientific Reports*, 7: 8185. https://doi.org/10.1038/s41598-017-08154-z.

Burggren, W. W., & Warburton, S. (2007). Amphibians as animal models for laboratory research in physiology. *ILAR* 48(3): 260–269. https://doi.org/10.1093/ilar.48.3.260.

Burghardt, G. M. (2013). Environmental enrichment and cognitive complexity in reptiles and amphibians: concepts, review, and implications for captive populations. *Applied Animal Behaviour Science*, 147: 286–298. https://doi.org/10.1016/j.applanim.2013.04.013.

Busse, U., & Seguin, C. (1993). Molecular analysis of the Wnt-1 proto-oncogene in *Ambystoma mexicanum* (axolotl) embryos. *Differentiation*, 53: 7–15. https://doi.org/10.1111/j.1432-0436.1993.tb00640.x.

Carlson, B. E., & Langkilde, T. (2013). Personality traits are expressed in bullfrog tadpoles during open-field trials. *Journal of Herpetology*, 47(2): 378–383. https://doi.org/10.1670/12-061.

Chum, H., Felt, S., Garner, J., & Green, S. (2013). Biology, behavior, and environmental enrichment for the captive African clawed frog (*Xenopus spp*). *Applied Animal Behaviour Science*, 143: 150–156 https://doi.org/10.1016/j.applanim.2012.10.013.

Claas, B., & Dean, J. (2006). Prey-capture in the African clawed toad (*Xenopus laevis*): comparison of turning to visual and lateral line stimuli. *Journal of Comparative Physiology A*, 192: 1021–1036 https//doi.org/10.1007/s00359-006-0137-2.

Cooke, G. M. (2018). Use of a translucent refuge for *Xenopus tropicalis* with the aim of improving welfare. *Laboratory Animals*, 52(3): 304–307. https://doi.org/10.1177/0023677217737798.

Crane, A., & Mathis, A. (2011). Predator-recognition training: a conservation strategy to increase post-release survival of hellbenders in head-starting programs. *Zoo Biology*, 30: 611–622. https://doi.org/10.1002/zoo.20358.

Crump, M. L. (1996). Parental care among the amphibia. *Advances in the Study of Behavior*, 25: 109–144. https://doi.org/10.1016/S0065-3454(08)60331-9.

Dawkins, M. S. (1998). Evolution and animal welfare. *Quarterly Review of Biology*, 73: 305–328.

Dawkins, M. S. (2003). Behaviour as a tool in the assessment of welfare. *Zoology*, 106: 383–387. https://doi.org/10.1078/0944-2006-00122.

Denoël, M. (2017). On the identification of paedomorphic and overwintering larval newts based on cloacal shape: review and guidelines, *Current Zoology*, 63: 165–173. https://doi.org/10.1093/cz/zow054.

Duellman, W. E., & Trueb, L. (1994). *Biology of Amphibians*. John Hopkins University Press, Baltimore, MD.

Elepfandt, A. (1996). Sensory perception and the lateral line system in the clawed frog *Xenopus*. In: Tinsley, R. C., & Kobel, H. R. (eds.), *The Biology of Xenopus*. Oxford University Press, New York, pp. 97–120.

Elewa, A., Wang, H., Talavera-López, C. et al. (2017). Reading and editing the *Pleurodeles waltl* genome reveals novel features of tetrapod regeneration. *Nature Communications*, 8: 2286. https://doi.org/10.1038/s41467-017-01964-9.

Emlen, S. T. (1976). Lek organization and mating strategies in the bullfrog. *Behavioral Ecology and Sociobiology*, 1: 283–313 https://doi.org/10.1007/BF00300069.

Emmans, C. J. (2015). Using behaviour and glucocorticoids as non-invasive measures of stress and welfare in captive *Ambystoma andersoni*, a critically endangered species of salamander found in Lake Zacapu, Mexico. Unpublished MSc thesis, University of Chester.

Endler, J. A. (1993). The color of light in forests and its implications. *Ecological Monographs*, 63: 1–27. https://doi.org/10.2307/2937121.

Escoriza, D., Hassine, J. B., Boix, D., & Sala, J. (2020). Diet of larval Pleurodeles waltl (Urodela: Salamandridae) throughout its distributional range. *Limnetica*, 39(2): 667–676. https://doi.org/10.23818/limn.39.43.

Falso, P. G., Noble, C. A., Diaz, J. M., & Hayes, T. B. (2015). The effect of long-term corticosterone treatment on blood cell differentials and function in laboratory and wild-caught amphibian models. *General and Comparative Endocrinology*, 212: 73–83. https://doi.org/10.1016/j.ygcen.2015.01.003.

Feder, M. E. (1992). A perspective on environmental physiology of the amphibians. In: Feder, M. E., & Burggren, W. W. (eds.), *Environmental Physiology of the Amphibia*. University of Chicago Press, Chicago, IL, pp. 1–8.

Ferrari, M. C. O., Messier, F., & Chivers, D. P. (2008). Larval amphibians learn to match antipredator response intensity to temporal patterns of risk, *Behavioral Ecology*, 19: 980–983. https://doi.org/10.1093/beheco/arn056.

Fitzsimmons, C., McLaughlin, E. A., Mahony, M. J., & Clulow, J. (2007). Optimisation of handling, activation and assessment procedures for *Bufo marinus* spermatozoa. *Reproduction, Fertility and Development*, 19: 594–601. https://doi.org/10.1071/RD06124.

Furness, A. I., & Capellini, I. (2019). The evolution of parental care diversity in amphibians. *Nature Communications*, 10: 4709. https://doi.org/10.1038/s41467-019-12608-5.

Garcia, T., Paoletti, D. J., & Blaustein, A. R. (2009). Correlated trait responses to multiple selection pressures in larval amphibians reveal conflict avoidance strategies. *Freshwater Biology*, 54: 1066–1077. https://doi.org/10.1111/j.1365-2427.2008.02154.x.

Grainger, R. M. (2012). *Xenopus tropicalis* as a model organism for genetics and genomics: past, present, and future. In: Hoppler, S., & Vize, P. (eds.), *Xenopus Protocols. Methods in Molecular Biology (Methods and Protocols)*, vol 917. Humana Press, Totowa, NJ, pp. 3–15. https://doi.org/10.1007/978-1-61779-992-1_1.

Grant, R. A., Halliday, T. R., & Chadwick, E. A. (2013). Amphibians' response to the lunar synodic cycle—a review of current knowledge, recommendations, and implications for conservation. *Behavioral Ecology*, 24: 53–62. https://doi.org/10.1093/beheco/ars135.

Halliday, T. R. (1990). The evolution of courtship behavior in newts and salamanders. *Advances in the Study of Behavior*, 19: 137–169. https://doi.org/10.1016/S0065-3454(08)60202-8.

Hammond, S., Warren, R. L., Vandervalk, B. P. et al. (2017). The North American bullfrog draft genome provides insight into hormonal regulation of long noncoding RNA. *Nature Communications*, 8: 1433. https://doi.org/10.1038/s41467-017-01316-7.

Harding, G., Griffiths, R. A., & Pavajeau, L. (2015). Developments in captive breeding and reintroduction programs. *Conservation Biology*, 30: 340–349. https://doi.org/10.1111/cobi.12612.

Harland, R. M., & Grainger, R. M. (2011). *Xenopus* research: metamorphosed by genetics and genomics. *Trends Genet*, 27: 507–515. https://doi.org/10.1016/j.tig.2011.08.003.

Holmes, A. M., Emmans, C. J., Jones, N., Coleman, R., Smith, T. E., & Hosie, C. A. (2016). Impact of tank background on the welfare of the African clawed frog, *Xenopus laevis* (Daudin). *Applied Animal Behaviour Science*, 185: 131–136. https://doi.org/10.1016/j.applanim.2016.09.005.

Holmes, A. M., Emmans, C. J., Coleman, R., Smith, T. E., & Hosie, C. A. (2018). Effects of transportation, transport medium and re-housing on *Xenopus laevis* (Daudin). *General and Comparative Endocrinology*, 266: 21–28. https://doi.org/10.1016/j.ygcen.2018.03.015.

Hossie, T., Landolt, K., & Murray, D. L. (2017). Determinants and coexpression of anti-predator responses in amphibian tadpoles: a metaanalysis. *Oikos*, 126: 173–184. https://doi.org/10.1111/oik.03305.

Houck, L. D. (2009). Pheromone communication in amphibians and reptiles. *Annual Review of Physiology*, 71: 161–176. https://doi.org/10.1146/annurev.physiol.010908.163134.

Houck, L. D., & Arnold, S. J. (2003). Courtship and mating behavior. In: Jamieson, B. G. M. (ed.), *Reproductive Biology and Phylogeny of Urodela*. CRC Press, Boca Raton, FL, pp. 383–424.

Howard, R. D. (1978). The evolution of mating strategies in bullfrogs, *Rana catesbeiana*. *Evolution*, 32: 850–871. https://doi.org/10.2307/2407499.

Howard, R. D. (1984). Alternative mating behaviors of young male bullfrogs, *American Zoologist*, 24: 397–406. https://doi.org/10.1093/icb/24.2.397.

Jaeger, R. G., Gollmann, B., Anthony, C. D., Gabor, C. R., & Kohn, N. R. (2016). *Behavioral Ecology of the Eastern Red-backed Salamander: 50 Years of Research*. Oxford University Press, New York.

Jafkins, A., Abu-Daya, A., Noble, A., Zimmerman, L. B., & Guille, M. (2012). Husbandry of *Xenopus tropicalis*. In: Hoppler, S., & Vize, P. (eds.), *Xenopus Protocols. Methods in Molecular Biology (Methods and Protocols)*, vol 917. Humana Press, Totowa, NJ, pp. 17–31. https://doi.org/10.1007/978-1-61779-992-1_2.

Jancowski, K., & Orchard, S. A. (2013). Stomach contents from invasive American bullfrogs Rana catesbeiana (= Lithobates catesbeianus) on southern Vancouver Island, British Columbia, Canada. *NeoBiota* 16: 17–37 https//doi.org/10.3897/neobiota.16.3806.

Janssenswillen, S., & Bossuyt, F. (2016). Male courtship pheromones induce cloacal gaping in female newts (Salamandridae). *PLoS One*, 11(1): e0144985. https://doi.org/10.1371/journal.pone.0144985.

Jaquiery, J., Broquet, T., Aguilar, C., Evanno, G., & Perrin, N. (2010). Good genes drive female choice for mating partners in the lek-breeding European treefrog. *Evolution*, 64: 108–115. https://doi.org/10.1111/j.1558-5646.2009.00816.x.

Joven, A., Morona, R., Gonzalez, A., & Moreno, N. (2013). Spatiotemporal patterns of Pax3, Pax6, and Pax7 expression in the developing brain of a Urodele amphibian, *Pleurodeles waltl*. *Journal of Comparative Neurology*, 521: 3913–3953. https://doi.org/10.1002/cne.23385.

Joven, A., Kirkham, M., & Simon, A. (2015). Husbandry of Spanish ribbed newts (*Pleurodeles waltl*). In: Kumar, A., & Simon, A. (eds.), *Salamanders in Regeneration Research. Methods in Molecular Biology*, vol 1290. Humana Press, New York, pp. 47–70. https://doi.org/10.1007/978-1-4939-2495-0_4.

Kelleher, S. R., Silla, A. J., & Byrne, P. G. (2018). Animal personality and behavioral syndromes in amphibians: a review of the evidence, experimental approaches, and implications for conservation. *Behavioral Ecology and Sociobiology*, 72: 79. https://doi.org/10.1007/s00265-018-2493-7.

Koenig, A. M., & Ousterhout, B. H. (2018). Behavioral syndrome persists over metamorphosis in a pond-breeding amphibian. *Behavioral Ecology and Sociobiology*, 72: 184. https://doi.org/10.1007/s00265-018-2595-2.

Koprivnikar, J., & Penalva, L. (2015). Lesser of two evils? Foraging choices in response to threats of predation and parasitism. *PLoS One*, 10(1): e0116569. https//doi.org/10.1371/journal.pone.0116569.

Lamb, J. Y., & Davis, M. P. (2020). Salamanders and other amphibians are aglow with biofluorescence. *Scientific Reports*, 10: 2821. https://doi.org/10.1038/s41598-020-59528-9.

Laudet, V. (2011). The origins and evolution of vertebrate metamorphosis. *Current Biology*, 21: R726–R737. https://doi.org/10.1016/j.cub.2011.07.030.

Lea, A. M., & Ryan, M. J. (2015). Irrationality in mate choice revealed by túngara frogs. *Science*, 349: 964–966. https://doi.org/10.1126/science.aab2012.

Lewis, V., Laberge, F., & Heyland, A. (2020). Temporal profile of brain gene expression after prey catching conditioning in an anuran amphibian. *Frontiers in Neuroscience*, 13: article 1407. https://doi.org/10.3389/fnins.2019.01407.

Lieggi, C., Kalueff, A. V., Lawrence, C., & Collymore, C. (2020). The influence of behavioral, social, and environmental factors on reproducibility and replicability in aquatic animal models. *ILAR Journal*, 60: 270–288. https://doi.org/10.1093/ilar/ilz019.

Maex, M., Van Bocxlaer, I., Mortier, A. et al. (2016). Courtship pheromone use in a model Urodele, the Mexican axolotl (*Ambystoma mexicanum*). *Scientific Reports*, 6: 20184. https://doi.org/10.1038/srep20184.

Marco, A., Lizana, M., Alvarez, A., & Blaustein, A. R. (2001). Egg-wrapping behaviour protects newt embryos from UV radiation. *Animal Behaviour*, 61: 639–644. https://doi.org/10.1006/anbe.2000.1632.

Martinez-Conde, S., & Macknik, S. L. (2008). Fixational eye movements across vertebrates: comparative dynamics, physiology, and perception. *Journal of Vision*, 8(14): 28. https://doi.org/10.1167/8.14.28.

Mellor, D. J. (2016). Updating animal welfare thinking: moving beyond the "five freedoms"towards "A lifeworth living." *Animals*, 6(3), 21. https://doi.org/10.3390/ani6030021.

McNamara, S., Wlizla, M., & Horb, M. E. (2018). Husbandry, general care, and transportation of *Xenopus laevis* and *Xenopus tropicalis*. In: Vleminckx, K. (eds.), *Xenopus. Methods in Molecular Biology*, vol 1865. Humana Press, New York, pp. 1–17. https://doi.org/10.1007/978-1-4939-8784-9_1.

Michaels, C. J., Antwis, R. E., & Preziosi, R. F. (2014). Impact of plant cover on fitness and behavioural traits of captive red-eyed tree frogs (*Agalychnis callidryas*). *PLoS One*, 9(4): e95207. https://doi.org/10.1371/journal.pone.0095207.

Miller, A. J., Page, R. A., & Bernal, X. E. (2018). Exploratory behavior of a native anuran species with high invasive potential. *Animal Cognition*, 21: 55–65. https//doi.org/10.1007/s10071-017-1138-y.

Mohammadi, S., Savitzky, A. H., Lohr, J., & Dobler, S. (2017). Toad toxin-resistant snake (*Thamnophis elegans*) expresses high levels of mutant Na+/K+-ATPase mRNA in cardiac muscle. *Gene*, 614: 21–25. https://doi.org/10.1016/j.gene.2017.02.028.

Monroy, J. A., & Nishikawa, K. C. (2009). Prey location, biomechanical constraints, and motor program choice during prey capture in the tomato frog, *Dyscophus guineti*. *Journal of Comparative Physiology A*, 195: 843–852. https://doi.org/10.1007/s00359-009-0463-2.

Morrison, C., & Hero, J. (2003). Geographical variation in life-history characteristics: a review. *Journal of Animal Ecology*, 72: 270–279. https://doi.org/10.1046/j.1365-2656.2003.00696.x.

Narayan, E. J. (2013). Non-invasive reproductive and stress endocrinology in amphibian conservation physiology. *Conservation Physiology*, 1, cot011. https://doi.org/10.1093/conphys/cot011.

Narayan, E. J., Hero, J., & Cockrem, J. F. (2012a). Inverse urinary corticosterone and testosterone metabolite responses to different durations of restraint in the cane toad (*Rhinella marina*). *General and Comparative Endocrinology*, 179: 345–349. https://doi.org/10.1016/j.ygcen.2012.09.017.

Narayan, E. J., Molinia, F. C., Cockrem, J. F., & Hero, J. (2012b). Individual variation and repeatability in urinary corticosterone metabolite responses to capture in the cane toad (*Rhinella marina*). *General and Comparative Endocrinology*, 175: 345–349. https://doi.org/10.1016/j.ygcen.2011.11.023.

Niederle, M. V., Bosch, J., Ale, C. E., et al. (2019). Skin-associated lactic acid bacteria from North American bullfrogs as potential control agents of *Batrachochytrium dendrobatidis*. *PLoS One* 14(9): e0223020. https://doi.org/10.1371/journal.pone.0223020.

Noel, A., Hao-Yuan, G., Mandica, M., & Hu, D. L. (2017). Frogs use a viscoelastic tongue and non-Newtonian saliva to catch prey. *Journal of the Royal Society Interface*, 14: 20160764. https://doi.org/10.1098/rsif.2016.0764.

Noel, A., & Hu, D. L. (2018). The tongue as a gripper. *Journal of Experimental Biology*, 221: jeb176289. https//doi.org/10.1242/jeb.176289.

Norris, K. M., & Hosie, C. A. (2005). A quantified ethogram for oviposition in *Triturus* newts: description and comparison of *T. helveticus* and *T. vulgaris*. *Ethology*, 111: 357–366. https://doi.org/10.1111/j.1439-0310.2005.01071.x.

Ogurtsov, S. V., Antipov, V. A., & Permyakov, M. G. (2018). Sex differences in exploratory behaviour of the common toad, *Bufo bufo*. *Ethology Ecology & Evolution*, 30: 543–568. https//doi.org/10.1080/03949370.2018.1459864.

Okada, S., Fukuda, Y., & Takahashi, M. K. (2015). Parental care behaviors of Japanese giant salamander *Andrias japonicus* in natural populations. *Journal of Ethology*, 33: 1–7. https://doi.org/10.1007/s10164-014-0413-5.

O'Rourke, D. P. (2007). Amphibians used in research and teaching. *ILAR*, 48: 183–187. https://doi.org/10.1093/ilar.48.3.183.

Ossana, N. A., & Salibian, A. (2013). Micronucleus test for monitoring the genotoxic potential of the surface water of Luján River (Argentina) using erythrocytes of *Lithobates catesbeianus* tadpoles. *Ecotoxicology and Environmental Contamination*, 8: 67–74. https://doi.org/10.5132/eec.2013.01.010.

Pearl, E. J., Grainger, R. M., Guille, M., & Horb, M. E. (2012). Development of *Xenopus* resource centers: the national *Xenopus* resource and the European *Xenopus* resource center. *Genesis*, 50: 155–163. https://doi.org/10.1002/dvg.22013.

Percie du Sert, N., Hurst, V., Ahluwalia, A., et al. (2020). The ARRIVE guidelines 2.0: updated guidelines for reporting animal research. *PLoS Biology*, 18(7): e3000410. https://doi.org/10.1371/journal. pbio.3000410.

Pfennig, D. W., Reeve, H. K., & Sherman, P. W. (1993). Kin recognition and cannibalism in spadefoot toad tadpoles. *Animal Behaviour*, 46: 87–94. https://doi.org/10.1006/anbe.1993.1164.

Pfennig, D. W., Sherman, P. W., & Collins, J. P. (1994). Kin recognition and cannibalism in polyphenic salamanders. *Behavioral Ecology*, 5: 225–232. https://doi.org/10.1093/beheco/5.2.225.

Pizzatto L, Stockwell, M., Clulow, S., Clulow, J., & Mahony, M (2016). How to form a group: effects of heterospecifics, kinship and familiarity in the grouping preference of green and golden bell frog tadpoles. *The Herpetological Journal*, 26(2) 157–164.

Propper, C. R., & Moore, F. L. (1991). Effects of courtship on brain gonadotropin hormone-releasing hormone and plasma steroid concentrations in a female amphibian (*Taricha granulosa*). *General and Comparative Endocrinology*, 81: 304–313. https://doi.org/10.1016/0016-6480(91)90015-X.

Reed, B. (2005). Guidance for the housing and care of the African clawed frog *Xenopus laevis*. RSPCA Report. Available from: www.rspca.org.

Reilly, S. M., & Lauder, G. V. (1992). Morphology, behavior, and evolution: comparative kinematics of aquatic feeding in salamanders. *Brain, Behavior and Evolution*, 40: 182–196. https://doi.org/10.1159/000113911.

Reinhard, S., Voitel, S., & Kupfer, A. (2013). External fertilisation and paternal care in the paedomorphic salamander *Siren intermedia* Barnes, 1826 (Urodela: Sirenidae). *Zoologischer Anzeiger*, 253: 1–5. https://doi.org/10.1016/j.jcz.2013.06.002.

Ringler, E., Pašukonis, A., Hödl, W. et al. (2013). Tadpole transport logistics in a Neotropical poison frog: indications for strategic planning and adaptive plasticity in anuran parental care. *Frontiers in Zoology*, 10: 67. https://doi.org/10.1186/1742-9994-10-67.

Robert, J., & Ohta, Y. (2009). Comparative and developmental study of the immune system in *Xenopus*. *Developmental Dynamics*, 238: 1249–1270. https://doi.org/10.1002/dvdy.21891.

Roelants, K., Fry, B. G., Ye, L., et al. (2013). Origin and functional diversification of an amphibian defense peptide arsenal. *PLoS Genetics*, 9(8): e1003662. https//doi.org/10.1371/journal.pgen.1003662.

Ryan, M. J. (1985). *The Tungara Frog: A Study in Sexual Selection and Communication*. University of Chicago Press. Chicago, IL.

Salvidio, S., Costa, A., & Crovetto, F. (2019). Individual trophic specialisation in the alpine newt increases with increasing resource diversity. *Annales Zoologici Fennici*, 56: 17–24. https://doi.org/10.5735/086.056.0102.

Santos, J. C., Tarvin, R. D., & O'Connell, L. A. (2016). A review of chemical defense in poison frogs (dendrobatidae): ecology, pharmacokinetics, and autoresistance. In: Schulte, B. A., Goodwin, T. E., Ferkin, M. H. (eds.), *Chemical Signals in Vertebrates 13*. Springer International Publishing, New York, pp. 305–337.

Santos, R. R., & Grant, T. (2011). Diel pattern of migration in a poisonous toad from Brazil and the evolution of chemical defenses in diurnal amphibians. *Evolutionary Ecology*, 25: 249–258. https://doi.org/10.1007/s10682-010-9407-0.

Saporito, R. A., & Grant, T. (2018). Comment on Amézquita et al. (2017) "Conspicuousness, color resemblance, and toxicity in geographically diverging mimicry: the pan-Amazonian frog *Allobates femoralis*". *Evolution*, 72: 1009–1014. https://doi.org/10.1111/evo.13468.

Schmajuk, N. A., Segura, E. T., & Reboreda, J. C. (1980). Aetitive conditioning and discriminatory learning in toads. *Behavioral and Neural Biology*, 28: 392–397. https//doi.org/10.1016/s0163-1047(80)91698-2.

Schriever, T., & Dudley Williams, D. (2012). Ontogenetic and individual diet variation in amphibian larvae across an environmental gradient. *Freshwater Biology*, 58: 223–236. https://doi.org/10.1111/fwb.12044.

Session, A., Uno, Y., Kwon, T. et al. (2016). Genome evolution in the allotetraploid frog *Xenopus laevis*. *Nature*, 538: 336–343. https://doi.org/10.1038/nature19840.

Shulte, L. M., Ringler, E., Rojas, B., & Stynoski, J. (2020). Developments in amphibian parental care research: history, present advances, and future perspectives. *Herpetological Monographs*, 34: 71–97. https://doi.org/10.1655/HERPMONOGRAPHS-D-19-00002.1.

Sih, A., Bell, A. M., Chadwick Johnson, J., & Ziemba, R. E. (2004). Behavioral syndromes: an integrated overview. *The Quarterly Review of Biology*, 79: 241–277. https://doi.org/0033-5770/2004/7903-0001.

Sneddon, L. U., Elwood, R. W., Adamo, S. A. & Leach, M. C. (2014). Defining and assessing animal pain. *Animal Behaviour*, 97: 201–212. https://doi.org/10.1016/j.anbehav.2014.09.007.

Sneddon, L. U. (2015). Pain in aquatic animals. *Journal of Experimental Biology*, 218: 967–976. https//doi.org/10.1242/jeb.088823.

Sneddon, L. U. (2017). Comparative Physiology of Nociception and Pain. *Physiology* 33: 63–73. https://doi.org/10.1152/physiol.00022.2017

Snow, N. P., & Witmer, G. (2010). American bullfrogs as invasive species: a review of the introduction, subsequent problems, management options, and future directions. *Proc. 24th Vertebr. Pest Conf.*, Timm, R. M., & Fagerstone, K. A. (eds.), University of California, Davis, 86–89. https://doi.org/10.5070/V424110490.

Székely, D., Cogălniceanu, D., Székely, P., & Denoel, M. (2020). Adult—juvenile interactions and temporal niche partitioning between life-stages in a tropical amphibian. *PLoS One*, 15(9): e0238949. https://doi.org/10.1371/journal.pone.0238949.

Tapley, B., Bradfield, K. S., Michaels, C. et al. (2015). Amphibians and conservation breeding programmes: do all threatened amphibians belong on the ark? *Biodiversity and Conservation*, 24: 2625–2646. https://doi.org/10.1007/s10531-015-0966-9.

Tinsley, R. C., & Kobel, H. R. (eds.). (1996). *The Biology of Xenopus*. Oxford University Press, New York.

Tinsley, R. C., Loumont, C., & Kobel, H. R. (1996). Geographical distribution and ecology. In: Tinsley, R. C., Kobel, H. R. (eds.), *The Biology of Xenopus*. Oxford University Press, New York, pp. 35–59.

Tobias, M. L., Viswanathan, S., & Kelley, D. B. (1998). Rapping, a female receptive call, initiates male–female duets in the South African clawed frog. *Proceedings of the National Academy of Sciences*, 95: 1870–1875. https://doi.org/10.1073/pnas.95.4.1870.

Tobias, M. L., Barnard, C., O'Hagan, R., Horng, S. H., Rand, M., Kelley, D. B. (2004). Vocal communication between male *Xenopus laevis*. *Animal Behaviour*, 67: 353–365. https://doi.org/10.1016/j.anbehav.2003.03.016.

Tobias, M. L., Corke, A., Korsh, J. et al. (2010). Vocal competition in male *Xenopus laevis* frogs. *Behavioral Ecology and Sociobiology*, 64: 1791–1803. https://doi.org/10.1007/s00265-010-0991-3.

Torreilles, S. L., & Green, S. L., (2007). Refuge cover decreases the incidence of bite wounds in laboratory South African clawed frogs (*Xenopus laevis*). *Journal of the American Association for Laboratory Animal Science*, 46: 33–36.

Tóth, Z., Hoi, H., & Hettyey, A. (2011). Intraspecific variation in the egg-wrapping behaviour of female smooth newts, *Lissotriton vulgaris*. *Amphibia-Reptilia*, 32: 77–82. https://doi.org/10.1163/017353710X543001.

Ulloa, J. S., Aubin, T., Llusia, D. et al (2019). Explosive breeding in tropical anurans: environmental triggers, community composition and acoustic structure. *BMC Ecology*, 19: 28. https://doi.org/10.1186/s12898-019-0243-y.

Vance, E. (2017). Biology's beloved amphibian — the axolotl — is racing towards extinction. *Nature*, 551(7680): 286–289. https://doi.org/10.1038/d41586-017-05921-w.

Vitt, L., & Caldwell, J. P. (2013). *Herpetology: An Introductory Biology of Amphibians and Reptiles.* 4th edn. Academic Press, San Diego, CA.

Voss, S. R., Epperlein, H. H., &Tanaka, E. M. (2009). *Ambystoma mexicanum*, the Axolotl: a versatile amphibian model for regeneration, development, and evolution studies. *Spring Harbor Protocols*, 2009(8): pdb.emo128. https//doi.org/10.1101/pdb.emo128Cold.

Wack, C. L., Lovern, M. B., & Woodley, S. K. (2013). Transdermal delivery of corticosterone in terrestrial amphibians. *General and Comparative Endocrinology*, 169: 269–275. https://doi.org/10.1016/j.ygcen.2010.09.004.

Wagner, A. Schabetsberger, R. Sztatecsny, M., & Kaiser, R. (2011). Skeletochronology of phalanges underestimates the true age of long-lived Alpine newts (*Ichthyosaura alpestris*). *The Herpetological Journal*, 21: 145–148.

Waldman, B., & Adler, K. (1979). Toad tadpoles associate preferentially with siblings. *Nature*, 282: 611–613. https://doi.org/10.1038/282611a0.

Walls, S. C., & Gabor, C. R. (2019). Integrating behavior and physiology into strategies for amphibian conservation. *Frontiers in Ecology and Evolution*, 7: 234. https//doi.org/10.3389/fevo.2019.00234.

Wang, Y., Zhang, Y., Lee, W., & Zhang, Y. (2016). Novel peptides from skins of amphibians showed broad-spectrum antimicrobial activities. *Chemical Biology & Drug Design*, 87: 419–424. https://doi.org/10.1111/cbdd.12672.

Wells, K. D (2010). *The Ecology and Behavior of Amphibians.* University of Chicago Press, Chicago, IL.

Williams, C. J. A., Alstrup, A. K. O., Bertelsen, M. F., Jensen, H. M., Leite, C. A. C., & Wang, T. (2017). When local anesthesia becomes universal: Pronounced systemic effects of subcutaneous lidocaine in bullfrogs (*Lithobates catesbeianus*). *Comparative Biochemistry and Physiology Part A: Molecular & Integrative Physiology*, 209: 41–46. https://doi.org/10.1016/j.cbpa.2017.03.019.

Wilson, A. D. M., &Krause, J. (2012). Personality and metamorphosis: is behavioral variation consistent across ontogenetic niche shifts? *Behavioral Ecology*, 23: 1316–1323. https://doi.org/10.1093/beheco/ars123.

Woodley, S. K. (2015). Chemosignals, hormones, and amphibian reproduction. *Hormones & Behavior*, 68: 3–13. https://doi.org/10.1016/j.yhbeh.2014.06.008.

Xenbase. (2020). *Xenopus* frog model organism database. Available from: http://www.xenbase.org.

Yost, A. T., Thornton, L. M., Venables, B. J., & Sellin Jeffries, M. K. (2016). Dietary exposure to polybrominated diphenyl ether 47 (BDE-47) inhibits development and alters thyroid hormone-related gene expression in the brain of *Xenopus laevis* tadpoles. *Environmental Toxicology and Pharmacology*, 48: 237–244. https://doi.org/10.1016/j.etap.2016.11.002.

23

Behavioral Biology of Reptiles

Dale F. DeNardo
Arizona State University

CONTENTS

Introduction	361
Typical Participation in Research	362
Behavioral Biology	362
Natural History	362
Ecology	362
Sensing Their Environment	363
Thermoregulation	364
Social Organization and Behavior	364
Parental Care	365
Feeding Behavior	367
Drinking Behavior	367
Communication	368
Common Captive Behaviors	368
Normal	368
Thermoregulation	368
Feeding and Drinking Behavior	370
Reproductive Behavior	370
Abnormal	370
Ways to Maintain Behavioral Health	371
Captive Considerations to Promote Species-Typical Behavior	371
Environment	371
Social Groupings	371
Feeding and Watering Strategies	371
Training	372
Special Situations	372
Conclusions/Recommendations	372
References	372

Introduction

Reptiles represent an immensely diverse group of organisms, with the major groups commonly referred to as the crocodilians, turtles, lizards, and snakes. As of April 2020, there were 11,242 species, and this number will surely be much higher by the time of publication, since this is 106 species more than were described as of December 2019 (Uetz et al., 2020). Phylogenetically, reptiles are not a monophyletic group (i.e., a group that consists of all the descendants of a common ancestor). In fact, crocodilians, and even turtles, are more closely related to birds than they are to lizards and snakes (Crawford et al., 2014; Figure 23.1). Consequently, there is no identifying characteristic for reptiles. Instead, they are best identified by a combination of what they have and do not have. In the simplest of terms, reptiles are the amniotes that have an integument made of scales, but no feathers or hair.

Given the phylogenetic diversity, reptiles are also extremely diverse morphologically, ecologically, and behaviorally. Reptiles are distributed over most of the globe, with the exception of extremely high latitudes and elevations. They also occupy a diverse array of habitats, including oceans, deserts, tropics, temperate regions, and mountains. Within a given habitat, they fill a wide variety of niches, with microhabitats that are arboreal, terrestrial, aquatic, and fossorial. Their activity periods vary between diurnal, crepuscular, and nocturnal, sometimes within the same individual over the various seasons. Reptiles can be carnivorous, insectivorous, herbivorous, or omnivorous.

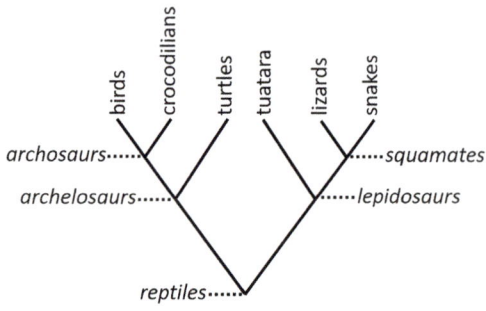

FIGURE 23.1 Relationship of the living major taxa of reptiles and birds. Higher order groups are labeled in italics.

Given the great diversity within the group we commonly refer to as reptiles, it is not surprising that there are no standard reptile behaviors. Herein lies the challenge of writing a chapter on "reptile behavior". As a result, this chapter will point out some of the behavioral diversity among reptiles, link behavioral tendencies to ecological preferences, and emphasize the importance of various reptile behaviors.

One commonality among all reptiles is that none are domesticated. Each is adapted to its natural habitat, and its behaviors relate to meeting its needs. Behaviors are the means by which an organism enables itself to use the biotic and abiotic resources available to it or to cope with resource limitations. Providing a reptile with each of its resource needs does not ensure that a reptile will use them, if it cannot use its typical behaviors to do so. In this chapter, I focus on the primary resource needs (i.e., heat, energy, water, and mates) and the behaviors associated with them.

Typical Participation in Research

While not considered standard species that participate in research, reptiles serve as valuable study systems in a variety of research efforts. The most direct goal of reptiles participating in research is in ecological studies aimed at learning about an individual species of interest, or examining variation among species based on phylogenetic or environmental differences. While the majority of such ecological work is conducted in the field, studies may include housing in the lab, either short-term for processing (e.g., collection of morphometric data or implantation of a radiotransmitter) or for lengthier periods to conduct a laboratory component of the study, evaluating the species' behavior or physiology under a more controlled and observation-friendly environment. In such cases, for animal well-being and data quality, it is important to enable the captive individuals to perform their natural behaviors.

Reptiles are also commonly used as model systems for understanding the dynamics of physiological processes. Reptiles tolerate more varied internal conditions than do most mammals and birds, and this enables researchers to examine deviations from optimal homeostatic conditions, both in terms of altered set points as a result of environmental or physiological conditions, and effects of undesirable deviations in internal conditions. Physiological processes in which reptiles are commonly used as study species include thermoregulation and digestion. As ectotherms, reptiles possess a desired body temperature, but achieving that temperature is heavily dependent on external heat sources (for review, see Angilletta, 2009). As a result, reptile body temperature can vary considerably over time. Because most physiological processes are highly temperature sensitive, such variation in body temperature can have substantial effects on performances. Due to this relationship, reptiles are the predominant group of animals used to better understand thermoregulation and the effects of changing environmental temperatures. This has become especially important given the rapid changes in environmental temperatures forecasted by most climate models (Stocker et al., 2013).

Reptiles are also used in studies of digestion, particularly activation and deactivation of digestive processes (Ott and Secor, 2007; Perry et al., 2019). Snakes, in particular, serve as excellent models for such studies, because they can go extended periods (e.g., months) without feeding, leading to morphological and biochemical downregulation of the digestive system. This period of extended fasting is then followed by the ingestion of a meal that can weigh as much as 100% of the snake's body weight. Digesting such a meal requires immediate and extensive upregulation of digestive processes, as well as supporting processes, including cardiovascular functioning (Enok et al., 2016). Studying such dramatic change can provide insight into less discrete digestive and cardiovascular changes in other species that may be associated with varied food intake and disease.

Perhaps counterintuitively, reptiles have also contributed to research on the evolution of endothermy (Brashears et al., 2017). In general, reptiles are ectothermic, but facultative endothermy has been demonstrated in some species of python while egg brooding (Van Mierop and Barnard, 1978; Brashears and DeNardo, 2013) and, more recently, in tegu lizards (Tattersall et al., 2016). However, endothermy by these reptiles is less effective than what is seen in birds and mammals, and it is this limitation that provides researchers with the opportunity to study potential benefits of what would be considered equivalent to critical early steps in the evolution of homeothermic endothermy.

Many lizards show characteristic behavioral displays for intraspecific communication (for review, see Fox et al. (2003)). Such behaviors are easily studied and have provided insight into our broad understanding of intraspecific organization and communication.

More recently, reptiles are oftentimes used in biomimicry studies, particularly related to locomotion (Abdel-Aal and Hisham, 2013; Tang et al., 2017); but also to other properties, including adhesion (Autumn et al., 2014) and surface texture (Greiner and Schäfer, 2015). By studying natural processes, the aim is to discover revelations that can be used in engineering design, including robotics.

Behavioral Biology

Natural History

Ecology

Most reptiles, especially those commonly used in research, tend to be intermediate on the food chain, preying on some species, while being the prey of others. Therefore, reptiles tend to carefully select microenvironments that enable them to acquire food while avoiding being eaten.

However, such microhabitats can vary considerably, with reptiles occupying aquatic, semi-aquatic, arboreal, terrestrial, saxicolous, and fossorial microhabitats. The type of microhabitat that a species inhabits not only influences its behavior, but also its morphology and physiology, and this has been most thoroughly examined in the Greater Antillean *Anolis* lizards, where several microhabitat-related ecomorphs have been identified (Langerhans et al., 2006).

Most lizards are diurnal and relatively active; therefore, most species tend to be somewhat conspicuous. In contrast, snakes tend to be quite secretive and spend much of their time in refugia. Even when active, snakes still tend to be inconspicuous by remaining close to available structures such as rocks and plants, as well as often being nocturnal.

Numerous reptile species, especially crocodilians and turtles, have a highly aquatic lifestyle. However, they also exit the water for various reasons, including to thermoregulate and to periodically dry off, which can reduce the risk of skin disease. In fact, some turtle species, in addition to the purely terrestrial tortoises, spend a large proportion of their time on land (Blob et al., 2016).

Sensing Their Environment

While this book specifically focuses on animal behavior, understanding reptile behaviors requires some level of understanding of how reptiles sense their environment. It is important to understand that reptiles, and in fact, most animals, have sensing capabilities that are quite different from those of humans, and the relative use of the various senses varies greatly among the different reptiles. Here, I will discuss the primary senses of the different groups of reptiles, but, for brevity, will minimally discuss the anatomy and physiology of the reptile senses. When discussing behavior, it is important to know what senses the various reptiles use, more so than the details of the mechanisms behind these senses.

Lizards have the best sight among the reptiles, with most species seeing color better than humans. Not surprisingly, sight is therefore the major sense through which lizards find their prey and communicate among themselves. Nocturnal lizards show adaptations to fit their activity period; they see colors poorly, but have good vision in low light. Many lizards also possess a parietal eye on top of their head. While not an organ of sight, per se, the parietal eye senses light levels and serves in helping to regulate daily activity rhythms, including thermoregulation (Hutchison and Kosh, 1974). More recently, it has been demonstrated that the lizard parietal eye can detect green and blue, possibly helping them identify dawn and dusk (Su et al., 2006).

While sight plays the primary role in lizard communication, many lizards also have significant chemosensory abilities that help them locate prey and sense conspecifics. Lizards possess a vomeronasal (i.e., Jacobson's) organ in the roof of their mouth that can differentiate chemicals. The extended tongue can therefore collect chemicals from both surfaces and from the air, and transport these chemicals to the vomeronasal organ for chemodetection (Halpern et al., 1989). While lizards do hear, having a tympanum (i.e., ear drum) but lacking a pinna, for most species it is not a predominant sense for navigating their environment.

Having evolved from within the lizards, snakes have sensing organs similar to those of lizards, but details of their sense organ structures and their abilities vary among snakes with activity period being a key influencer. Snakes tend to see three visual pigments – blue, green, and ultraviolet (UV). Sensitivity to UV light enables nocturnal snakes to see well in low light conditions. In diurnal snakes, the eye lens blocks UV light, sharpening their vision (Simões et al., 2016). Despite the ability to see in low light levels, most snakes are highly reliant on scent for evaluating their environment. Like lizards, snakes have a vomeronasal organ, and it is highly developed, as is their tongue, which is slimmer, longer, and more forked than that of lizards. Snakes do not hear well, as they lack a tympanum, but they do have an inner ear. In order to transmit sound to the inner ear, rather than detecting sounds pressures as a tympanum would do, snakes detect sound-induced vibrations transmitted directly from the air to the skull (Christensen et al., 2012).

Pit vipers, boas, and pythons possess infrared-sensing pit organs that enable them to sense subtle differences in heat from several meters away, and the nerve fibers are estimated to have a temperature sensitivity as low as 0.001°C (Bakken and Krochmal, 2007). This heat-sensing ability is thought to benefit prey detection, thermoregulation, and predator detection (Krochmal et al., 2004).

Crocodilians and turtles face an addition challenge compared to most lizards and snakes in that they need to sense in both air and underwater environments. Because of the density differences between air and water, a sensing system that does well in air typically does poorly in water and vice versa.

The primary senses of turtles are sight and smell. Turtles have good color vison, including the ability to see UV (Grotzner et al., 2020). Because they occupy both aquatic and terrestrial environments, turtles have the potential to detect smells via three organs – the olfactory system in the nostrils, the vomeronasal organ in the roof of the mouth, and barbels, which are thick whisker-like projections from the chin of many turtles (Shoji and Kurihara, 1991; Legler, 1993). The presence and relative importance of each of these mechanisms depends on the species, the type of chemical being detected, and the environment that the turtle is in (i.e., under or out of water).

Crocodilians have a highly developed nasal olfactory system, but, unlike the other reptile groups, lack a vomeronasal organ (Schwenk, 2008). While crocodilians are active by day and night, their vision is adapted to low-light situations. Their retinas are dominated by rods, and they have well-developed tapeta lucida, which are responsible for the classic bright eye shines seen when crocodilians are illuminated with a light at night. The dim light adaptations of the eye enable crocodilians to forage underwater where light levels are low both day and night. Despite the rod domination in the retina, crocodilians have trichromatic vison. When out of water during the day, crocodilians contract their pupils to slits to limit the amount of light reaching their highly light-sensitive retinas. Compared to other reptiles, crocodilians have a keen sense of hearing, as they use it for prey detection, and socially, for territoriality, courtship, and parental care (Grigg and Gans, 1993).

Thermoregulation

As with all vertebrates, the physical and physiological performances of reptiles are optimized over a relatively narrow range of temperatures (for review, see Angilletta, 2009). As ectotherms, reptiles rely on their environment for heat (i.e., reptiles possess a finely tuned thermostat, but lack a furnace to elevate their body temperature). A reptile uses the variable heat sources and the resulting complex thermal heterogeneity of its environment to adjust its body temperature as needed. Even when seeking a refuge, reptiles often make their choice based on thermal conditions (Huey et al., 1989; Stahlschmidt et al., 2011). Because of the importance of careful regulation of temperature, thermoregulatory behavior is a vital component of a reptile's normal milieu of behaviors.

Behavioral thermoregulation comprises both spatial and postural adjustments. Spatial adjustments are the movements that the reptile makes among various components of its environment, selecting warmer locations when needing to increase its body temperature, and cooler locations when needing to reduce its temperature (Sears et al., 2016). Depending on the species, reptiles favor different thermal opportunities over others. For example, most diurnal species typically bask in full or dappled sunlight to obtain heat. In contrast, nocturnal (e.g., most geckos) or secretive (e.g., many small snakes) species tend to utilize heat sources that limit their vulnerability. For aquatic species, thermoregulation typically requires them to leave the water, which can increase predation risk (Ibanez et al., 2015). Because of the risk associated with basking, whether for aquatic or terrestrial species, basking reptiles tend to be highly vigilant and will forego their basking event to seek shelter when disturbed (Mukherjee et al., 2018; Heppard and Buchholz, 2019).

Regardless of the location where a reptile chooses to thermoregulate, they can alter the rate of heat gain through postural adjustments and, thus, expand thermal opportunities beyond those provided by spatial heterogeneity alone (Brewster and Beaupre, 2019). Reptiles can increase heat gain by exposing more of their body surface area to either a radiant or conductive heat source. This can be achieved by positioning their bodies perpendicular to a radiant heat source or pressing their abdomens and limbs against a conductive heat source (Figure 23.2).

While reptile temperature control is commonly, and rightfully, referred to as behavioral thermoregulation, it is important to realize that the utilized behaviors are often accompanied by physiological changes that enhance thermoregulatory effectiveness (Seebacher and Franklin, 2005). For example, radiant heat gain is accelerated by increasing peripheral blood flow to areas receiving radiant heat. By doing this, the reptile transports heat gained at the skin rapidly to deeper tissues. As a result, reptiles can heat faster than they cool (Bakken, 1976). Some reptiles can also alter their skin color, darkening when they want to elevate their temperature, and lightening when they want to gain heat more slowly (Smith et al., 2016).

Similarly, when trying to eliminate heat from their body, reptiles combine behavior and physiology. When their body temperature exceeds their optimal performance temperature, reptiles commonly seek cooler locations, such as a burrow or crevice. However, to guard their territory, lizards often prefer

FIGURE 23.2 Lesser earless lizard (*Holbrookia maculata*) basking on a rock. Note the high extention from the hot rock as well as the lifting of the toes off of the rock surface in order to limit heat gain. (Photo by M. Felder.)

to remain active, despite temperatures exceeding those which are desired, and this sometimes requires them to continue to select sunny locations. To combat heat gain, lizards may gape or pant to facilitate evaporative cooling from the mucous membranes of the mouth and respiratory tract. Interestingly, because of the proximity of the mouth to the brain, this form of cooling can emphasize cooling of the brain over that of the rest of the body (Tattersall et al., 2006).

Social Organization and Behavior

In a strict sense, social organization is extremely rare in reptiles (Bull et al., 2017). However, many reptiles, especially small lizards, have frequent intraspecific interactions (Gardner et al., 2015). Most commonly, these relate to mate acquisition. Accordingly, tolerance of conspecifics often strongly depends on whether the individuals are mature and whether it is the reproductive season. Intraspecific intolerance is typically much more prevalent during the mating season (Ruby, 1978). Many male lizards are territorial, and, during the breeding season, this enables them to isolate mates and other resources (e.g., feeding and basking sites). During this season, male lizards often increase interindividual communication through behavioral displays (e.g., push-ups and headbobs) and enhanced secondary sex characteristics, including changes in the color of particular areas of skin (e.g., throat, ventral torso; Figure 23.3) and the production of secretory materials from femoral or perianal pores used for pheromonal marking of their environment (Ferguson, 1977).

Snakes are typically solitary with the exception of reproductive behavior and, for some species living in colder climates, overwintering. During overwintering, individuals will often aggregate at locations that provide protection from freezing temperatures, as well as opportunities to bask at the time of emergence (Gienger and Beck, 2011). During emergence, individuals may spend one to several weeks intermittently basking not far from their refuge, so they may retreat into the refuge at night to avoid freezing. During overwintering aggregation, both males and females tolerate the presence of other

FIGURE 23.3 Male brown anole (*Anolis sagrei*) displaying its dewlap to signal other males of its territory. (Photo by author.)

individuals, usually with minimal interactions. However, in some species (e.g., red-sided garter snakes, *Thamnophis sirtalis parietalis*, and some temperate rattlesnakes, *Crotalus* spp.), the emergence period is used for reproductive interactions both between males, and between males and females (Aleksiuk and Gregory, 1974).

While snakes do not defend territories, during the reproductive season, males may become aggressive to other males that they encounter. Some snake species use ritualistic combat, such as that seen in rattlesnakes, which wrestle by intertwining in an upright position (Senter et al., 2014). However, male–male combat in other snakes (e.g., some pythons) can inflict serious injury through the repeated use of the teeth to slice the opponent's skin.

Reptiles are mostly polygynous, with individual males mating with as many females as they can. When males encounter a reproductive female, they often use ritualized behaviors to encourage receptivity by the female.

In crocodilians, dominant males maintain territories that provide them exclusive access to multiple females. Territories are maintained through the use of complex communication that involves visual and auditory signals that transmit through air and water (Garrick and Lang, 1977). The details of the communication vary among species, but can include a combination of bellowing/roaring, snapping the jaws shut, splashing, creating subaudible vibrations at the surface, and blowing bubbles just under the surface.

Courtship and copulation in crocodilians typically occur in the water. In addition to the visual and auditory signals that are also used in territory defense, males court females by swimming around the female and rubbing against her. If she is receptive, the male will mount the female, grasp her with his hindlimbs, and then wrap his tail around her so that his cloaca opposes hers, enabling internal copulation.

Turtle courtship can entail two distinct behaviors that are not mutually exclusive; aggressive biting/head butting and a nonaggressive fluttering of the foreclaws (Liu et al., 2013). In both aquatic and terrestrial chelonians, males will often bite at the head and limbs of the female to induce her to withdraw into her shell and thus prohibit her from escaping from him. Because of the strength of turtle jaws and the sharpness of their beaks, this behavior can inflict significant injury on the female. Alternatively, or additionally, male aquatic turtles may court females while underwater, by positioning themselves in front of and facing the female. They will then extend their forelimbs forward, turn them so that the dorsal surfaces oppose each other, and then rapidly vibrate their digits directly in front of the female's head.

In lizards, males often use the bright colors and postural movements that are used in territorial defense to court females (Fleishman, 1992). If the female permits the male's approach, he will mount her and ensure his position, oftentimes by biting and holding on to the nape of her neck. If the female struggles, or is mounted repeatedly, the nape can be cut and abraded as a result of the biting (Figure 23.4). Once in position, like crocodilians, the male lizard will wrap his tail around the female's pelvic area and oppose his cloaca against the female's cloaca to enable insertion of one of his hemipenes (i.e., the paired copulatory organs of lizards and snakes).

In snakes, males court females by rubbing the chin against the dorsum of the female and sliding his tail up and down the female at the area of her cloaca (Senter et al., 2014). If the female is receptive, she will raise her tail and allow the male to oppose cloacae and insert one of his hemipenes. The duration of intromission can vary considerably, ranging from a few minutes to more than 24 h.

Parental Care

Parental care is defined as any nongenetic contribution by a parent that appears likely to increase the fitness of its offspring (modified from Clutton-Brock, 1991). Thus, in the broad sense, parental behaviors can occur while the embryos/fetuses still reside within either the mother or an egg. All turtles and crocodilians are oviparous, and there is limited embryonic development while the egg is still within the female (Rafferty and Reina, 2012). However, in lizards and snakes, 25%–30% of embryonic development can occur while the eggs are still within the female (Andrews, 2004). This, combined with the fact that viviparity has independently evolved in squamates over 100 times (Shine, 1985), makes preparitive (used to define the time when the developing embryos are with the females, regardless of whether the species is oviparous or viviparous;

FIGURE 23.4 Marine iguanas (*Amblyrhynchus cristatus*) mating. Note the male gripping the female's nape and the injury that it causes to the female (inset). (Photo by author.)

Smith, 1975) parental behaviors worthy of mention. Females that are gravid with eggs or pregnant with embryos are often aphagic, and they select body temperatures that exceed those of nonreproductive females and maintain that body temperature more precisely (Lourdais et al., 2008). Such efforts require increased basking time and, thus, can subject the female to greater risk of predation (Lorioux et al., 2013).

Most reptile eggs, especially those of all snakes, most lizards, and many turtles, are parchment-shelled (Hallmann and Griebeler, 2015). Rather than having a rigid shell, their shell is leathery and more permeable. Therefore, the hydric conditions of the nest are vital to the successful incubation of most reptile eggs (Deeming, 2018). Accordingly, females of oviparous species carefully select a microhabitat that best accommodates the thermal and hydric needs of the developing offspring. Failure to locate a suitable oviposition site can lead to dystocia.

While preparitive parental behavior is virtually ubiquitous among reptiles, postparitive parental care, the more traditionally considered form, is rare among reptiles (Gans, 1996). That is, eggs or newly born offspring are typically on their own once the female deposits them at the selected partition site. However, there are exceptions (for reviews, see Somma, 2003; Stahlschmidt and DeNardo, 2010). In no species of reptile is postparitive parental care required for the successful incubation of the eggs or survival of the offspring. However, parental behaviors can enhance egg/offspring survival and quality. In most cases, postparitive parental behavior in reptiles is limited to egg or nest guarding, and while rare, it is seen in representative species of all the higher taxa of reptiles, including the tuatara, *Sphenodon punctatus*. In some species, such as the Burmese mountain tortoise, *Manouria emys*, and the tuatara, the defense of the nest is short-term, lasting only a week or two. In other species, such as some skinks, the king cobra, pythons, and crocodilians, the female may guard the eggs or nest for the duration of incubation.

Parental efforts beyond the guarding of the eggs or nest are very rare among reptiles, but there are two noteworthy examples. All species of python coil tightly around their eggs (Stahlschmidt and DeNardo, 2010; Figure 23.5). This brooding behavior not only provides the eggs with protection from predators but aids the eggs in maintaining water balance and temperature regulation (Stahlschmidt et al., 2008). Brooding behavior insulates the eggs during nighttime cooling (Stahlschmidt and DeNardo, 2009). Additionally, females may periodically leave their secluded clutch to bask (Shine, 2004; Alexander, 2018). Basking elevates the female's body temperature, and she then returns to her clutch to transfer some of this gained heat to the eggs. Additionally, a few species of pythons use facultative endothermy during brooding to warm their clutches by as much as 8.3°C (14.9°F; Van Mierop and Barnard, 1978). While it is widespread dogma that facultative endothermy is characteristic of all pythons, the majority of species that have been examined are not facultatively endothermic during brooding (Brashears and DeNardo, 2015). Facultative endothermy appears to be limited to those species of pythons that live at the highest latitudes (i.e., the Burmese

FIGURE 23.5 Female water python (*Liasis fuscus*) brooding its eggs. (Photo by Z. Stahlschmidt.)

python, *Python molurus*, in the northern hemisphere and the diamond python, *Morelia spilota spilota*, and southwest carpet python, *M. s. imbricata*, in the southern hemisphere). Recently, it has been shown, for at least the South African rock python, *P. natalensis*, that the female will continue to stay with her newly hatched offspring for the first several weeks, until they have undergone their first ecdysis (shedding of their skin; Alexander, 2018). As observing python nests in the wild is a difficult challenge that has been rarely undertaken, it is uncertain whether, but it is perhaps likely that, other python species also attend to their newly hatched offspring.

Offspring attendance is widely prevalent across numerous species of rattlesnakes (for review, see Greene et al., 2002). Female rattlesnakes give birth to live young and then stay with them while they intermittently bask, until, like the rock pythons, the offspring undergo their first ecdysis (Figure 23.6). During the period of guarding, female rattlesnakes have been documented to move closer to their offspring when a threat is perceived, and will even regather a neonate that ventures too far from the nest.

FIGURE 23.6 Female Arizona black rattlesnake (*Crotalus cerberus*) with her recently born offspring. Note the dullness of the neonates' spectacles indicating that they have not yet shed their skin for the first time. (Photo by M. Felder.)

By far, the most extensive parental care by reptiles is seen in the crocodilians. Female crocodilians lay their eggs in a well-constructed nest mound and then will guard them throughout incubation (Kushlan and Kushlan, 1980). As the embryos reach full development, they will begin to chirp, encouraging the mother to excavate the nest, retrieve her hatchlings, and carefully bring them to the water in her jaws, traveling back and forth between the nest and water until all hatchlings have been safely transported (Hunt, 1987). Once at the water, the hatchlings will remain close to their mother for protection, and even use her back as a place to bask. Offspring will oftentimes remain with their mother for up to 2 years, mainly using her for protection.

Feeding Behavior

Little can be summarized about reptile feeding behavior, because there is great species and even ontogenetic variability in diet and feeding strategies (but see Knowlton, 1946). While all crocodilians and snakes are carnivorous, turtles and lizards show great dietary variation, with individual species being carnivorous, herbivorous, or omnivorous. Similarly, there is great variation among reptiles in terms of the specificity of their diet, with some species being highly specialized for certain prey (e.g., snail-eating snakes, algae-eating marine iguanas), while others rely on a diverse diet, especially many of the herbivorous species.

Not surprisingly, feeding behavior is equally as diverse as diet. Among the carnivorous species, regardless of higher taxon, some feed predominantly through active foraging, while others use a sit-and-wait foraging approach to ambush their prey. Additionally, among the snakes, multiple families (Viperidae, Elapidae) have evolved venoms to help subdue their prey (Fry et al., 2009).

Drinking Behavior

Water balance relies on water intake matching water expenditures for processes such as waste elimination, respiration, and evaporation. Water acquisition can be via three sources; free-standing water, dietary water, and metabolic water. While reptiles tend to have very low water expenditures, water balance can be challenging for species living in xeric environments, where free-standing water is rarely available. The infrequency of rainfall, combined with the porous, sandy soils that are typical of deserts, may limit standing water to as little as a few days per year. In such environments, reptiles have adapted specialized drinking behaviors with accompanying morphological adaptations.

When free-standing water is available to desert species, it is oftentimes only in the form of dew or scattered rain droplets on impermeable or low permeability surfaces, such as vegetation and rocks. Some snakes emerge from their refugia during rains, even in cold temperatures, in order to collect water droplets (Bonnet and Brischoux, 2008; Repp and Schuett, 2008; Figure 23.7). In fact, dozens of reptile species have evolved chemical and structural properties of the integument that facilitate drinking of water collected on their skin (for review, see Comanns, 2018). Even in the tropics, where pools and puddles of water are frequently available, reptiles may forego drinking from such sources, favoring fresh droplets from recent rains.

FIGURE 23.7 Two male western diamond-backed rattlesnakes (*Crotalus atrox*) that have emerged from their overwintering den to drink off the rock during a rain event. (Photo by M. Feldner.)

Dietary water, that is the water present in ingested food, plays an important role in water balance for some lizards, including insectivorous (Znari and Nagy, 1997) and herbivorous (Minnich and Shoemaker, 1970) species. However, reptiles that are carnivorous on larger prey species (e.g., rodents) must consume free-standing water to meet hydric needs (Wright et al., 2013; Murphy and DeNardo, 2019). In fact, in species with these diets, meal consumption increases the requirement for drinking (Lillywhite, 2017). Because of the low metabolic rates of reptiles, metabolic water only provides a very limited source of water (Minnich and Shoemaker, 1970).

Some reptiles can cope with limited water resources by storing water (e.g., in their urinary bladder) and tolerating levels of dehydration that would be detrimental to most other vertebrates (Peterson, 1996; Davis and DeNardo, 2007). However, even when devoid of clinical signs, elevated osmolality can alter normal activity and behavior (Davis and DeNardo, 2009).

Communication

As predominantly asocial species, reptiles use minimal intraspecific communication beyond that associated with courtship. The most extensive examples of communication within the reptiles are those of territorial lizards. Lizard displays are predominantly visual, often combining movements and colors to communicate with, and fend off, competing males or to attract mates (Cooper and Burns, 1987). As most small lizards serve as prey for many other animals, color signals are often only pronounced during the breeding season and located so that they can be flaunted or hidden depending on the lizard's preference. Color signals are often limited to the throat or ventral surface of the torso and hidden most of the time, but they are exposed through species-specific choreographic push-ups or head-bobs that are used for territory defense and courtship. Anolis lizards enhance the color signaling from their throats through the use of an expandable dewlap that is colored in a species-specific manner (Nicholson et al., 2007). Unlike the static seasonal color signals typical of most lizards, chameleons use rapid, physiological color change for complex intraspecific communications (Ligon and McGraw, 2013).

Reptiles may also use chemical communication (for review, see Mason and Parker (2010)). Lizards often use pheromones present in waxy secretions that are produced by glands located in the underside of the hindlimbs or the pericloacal region (Martín and López, 2015). These glands recrudesce during the breeding season and the secretions from the glands are rubbed on prevalent sites to inform conspecifics of the individual's presence.

During the reproductive season, female snakes produce pheromones from their skin that are shed onto the environment. These pheromone trails are then used by males to seek out reproductive females (Ford, 1986). Given that snakes are solitary and often in relatively low population densities, the existence of this pheromonal communication helps to ensure the presence of a mate when the female is receptive.

Sounds are a rare form of communication among reptiles. However, some geckos (e.g., *Gekko gecko*, *Ptenopus* spp.) vocalize to ward off competing males and attract females (Rohtla et al., 2019). Similarly, as discussed above, many crocodilians will grunt and cause sonic vibrations at the water's surface to communicate with conspecifics during the reproductive season (Vergne et al., 2009).

Interspecific communication is rare among reptiles. Most notably, venomous reptiles will often use specific behaviors to communicate their inherent danger by rattling their tail (rattlesnakes), spreading their neck into a hood (e.g., cobras), or opening their mouth (Leon and Manjarrez, 2017; Figure 23.8). Interestingly, nonvenomous species mimic these signals to falsely suggest a level of danger (Allf et al., 2016; Davis Rabosky et al., 2016; Figure 23.9).

Common Captive Behaviors

Normal

One shortcoming of maintaining nondomesticated species in captivity is that the captive environment, regardless of enclosure size or design, can never duplicate the complexity of the natural environment in which the species lives. Critically, it is the complexity of the environment that enables an animal to acquire its needed resources, using species-specific behaviors that have taken millennia to evolve. Despite this limitation, it is possible to humanely and effectively house reptiles long-term. The focus of captive husbandry should be to provide an environment, whether complex or simple, that enables the reptile to utilize natural behaviors to obtain its nutritional, hydric, and thermal resources, yet avoid situations that induce abnormal behavior or pathologic conditions. While providing captive reptiles with large and complex enclosures is often considered the gold standard, doing so also introduces a critical challenge. That is, large, complex enclosures can lead to difficulty in monitoring the condition of the animal as the animals can seek secluded, inaccessible shelters. Therefore, such enclosures should be restricted to occupants that have been acclimated to the captive setting and are feeding regularly.

Thermoregulation

Because of its influence on nearly all aspects of a reptile's behavioral and physiological performances, the most critical aspect of a reptile's captive environment is ensuring that

FIGURE 23.8 Behaviors used by some reptiles to warn potential predators of their danger. (a) Mojave rattlesnake (*Crotalus scutulatus*) raised, neck cocked, and rattling its tail (b) coral cobra (*Aspidelaps lubricus*) with its head raised and neck flattened, (c) cottonmouth (*Agkistrodon piscivorus*) opening its mouth exposing its highly contrasting white buccal cavity, and (d) Gila monster (*Heloderma suspectum*) opening its mouth and sticking out its black tongue. (Photos a and b by M. Feldner; photos c and d by author.)

the reptile can perform at its optimal temperature. This can be achieved by providing captive reptiles with a homogenous thermal environment that is set to what has been determined to be their preferred temperature. However, this approach is undesirable, unless research objectives mandate that the reptile be at a constant temperature. Providing a reptile with a homogeneous thermal environment is problematic in two ways, one is physiological and the other behavioral.

Physiologically, a reptile's optimal temperature often varies over time, fluctuating with changes in its internal state. While performances tend to have similar optimal temperatures, they are not identical. For example, state-based differences in thermal preference have been demonstrated with reproductive state (Lourdais et al., 2008), digestive state (Beck, 1996), and immune activity (Merchant et al., 2007). A homogenous environment does not enable a reptile to alter its body temperature based on its physiological needs at the moment.

Behaviorally, thermoregulatory behavior is a dominant component of a reptile's activity budget (Blazquez, 1996). Failure to perform such behaviors can be detrimental to the reptile's long-term well-being, regardless of whether the reptile is being maintained at a favorable temperature. Hence, providing thermal heterogeneity in a reptile enclosure is important. Equally important is that heat resources provided in the captive setting match those that the species utilizes in its natural environment. Species that typically use radiant heat sources are best provided lights and, for most species, an elevated perch on which to bask under the light (e.g., branches for arboreal species or

FIGURE 23.9 Sonoran gopher snake (*Pituophis catenifer affinis*) mimicking a rattlesnake by raising its body into a strike pose, flattening its head in a somewhat triangular shape, and shaking its tail, which makes noise by striking the substrate around it. (Photo by M. Comroe.)

rocks for saxicolous species). In contrast, nocturnal or secretive species should be provided a thermogradient created by conductive sources, such as subsurface heating elements.

Feeding and Drinking Behavior

A reptile's decision to eat is determined not only by what is offered to it but by how the meal is offered. Recently captured or newborn reptiles oftentimes can be difficult to feed in captivity. While this may be a result of poor environmental conditions (i.e., improper thermal environment), it may also be a result of the animal not recognizing what is being offered as a suitable meal. The meal may be the wrong prey item or may be the right meal being offered in an unnatural way. For many reptiles, the behavior of the subdued prey can be an important stimulus for components of the feeding response (Bealor et al., 2013). It is often preferred to feed reptiles dead prey items because of ethical concerns for the prey species or health reasons for the reptile (i.e., preventing the prey item from inflicting an injury). However, for reluctant feeders, providing live food to promote natural foraging behaviors can be fruitful in getting these individuals to acclimate to the captive environment.

The feeding response, and the associated behavior, can be quite fixed once initiated. Once a reptile identifies that a meal is present, it may initiate a focused feeding response with little specificity of what it acts upon. As a result, reptiles that have cued on feeding can present risks to both cagemates and caregivers. Reptiles will often bite cagemates or caregivers that are nearby, mistaking them for their food, resulting in bite wounds. Feeding-associated trauma is especially concerning in constricting and venomous snakes, as their typical feeding responses can result in the death, and even consumption, of a cagemate. To avoid deleterious outcomes, snakes should typically be separated from cagemates prior to triggering the focused feeding response.

Reptiles whose natural behavior includes frequent searching for food (i.e., active foragers) require more enclosure space than those that ambush their prey, and even with a relatively large space, active foragers are susceptible to rostral trauma from rubbing against the enclosure walls. This is especially true for walls that are transparent since reptiles often try persistently to get past these very unnatural barriers. Providing active foragers with relatively large and complex captive environments, as well as mostly opaque walls, can reduce the occurrence of rostral trauma.

While many reptiles will drink from a water bowl, some will not because of their natural behavior to drink fresh rain droplets, rather than from puddles or pools. Therefore, water must be provided fresh and in a manner that is consistent with the means by which the species naturally obtains its water.

Reproductive Behavior

Observing reproductive behavior in captive reptiles is good evidence that the animals are in good body condition, unstressed, and being kept in an environment that provides for their thermal and hydric needs. While indicating good husbandry practices, reproductive behavior can alter what might be considered stable and functional husbandry practices; individuals that have been highly compatible can become agonistic to each other during reproductive periods.

Onset of cohabitant aggression can result from individuals reaching sexual maturity or seasonal stimulation related to reproductive activity. Cohoused adult males will usually become aggressive toward each other, and in all taxonomic groups, male–male aggression can cause serious morbidity, and even mortality, in captive conditions, since the subordinate individual cannot escape from the dominant one. Even if no physical injury is inflicted, dominance may lead to subordinate animals ceasing normal and critical feeding and thermoregulatory behaviors, which can be fatal.

Similarly, male–female courtship and mating behaviors can lead to detrimental outcomes in captivity, because confinement prevents the female from escaping from the male's advances. This is especially true for those species that have aggressive components of their courtship and copulatory behaviors, including biting of the head and limbs in chelonians, and biting of the nape in lizards. Even in species that use few or no aggressive components, the persistence of the male can stress the female and lead to her failing to perform critical normal behaviors.

To avoid negative consequences of agonistic behaviors among cagemates that cannot flee from each other, in most cases, it is best to house reptiles singly. Even when reproduction is desired, it is safest, and often most productive (because it is more natural), to only periodically pair potential mates and, when paired, carefully monitor the female for avoidance behavior and injury. If seen, the animals should be separated and reintroduced at a future time or to an alternate male, if available.

When reproductively active, female reptiles, especially squamates, have altered behavior. They are often aphagic and tend to behaviorally focus on thermoregulation, usually resulting in more precise regulation at a body temperature that is usually slightly higher than when they are nonreproductive (Lourdais et al., 2008).

Abnormal

When provided the proper environment, diet, and enclosure design, most reptiles adapt well to captivity and perform species-typical behaviors to the extent permitted by their captive environment. Abnormal behaviors include repetitive escape behavior and chronic inattentiveness, during what should be their activity period.

Repetitive escape behavior is chronic pacing or scratching at the side of the enclosure. This can be most severe when the enclosure is made of transparent material (e.g., glass, Plexiglas, wire mesh) as transparent barriers are rare in nature. Such behavior needs to be corrected through the modification of the animal's husbandry, since it is abnormal and can often cause injury, most commonly rostral abrasions. Many active lizards, such as basilisks (*Basiliscus* spp.) and monitor lizards (*Varanus* spp.), may perform repetitive escape behavior to the extent of wearing their rostrum down to bone and causing jaw deformities that are much easier to prevent than to correct (Scheeliings and Hellebuyck, 2019). Such behavior can be reduced or eliminated by making sure that the animal has an enclosure that provides the proper thermal environment (excessive temperatures can lead to escape behavior); is of sufficient size and complexity, so as to provide ample basking and refuge sites; and has minimal transparent barriers.

When awake, reptiles normally remain alert to their surroundings, even if relatively sedentary. A captive reptile typically responds to the approach of a person by either moving or at least turning its head to get a better look. Such responses may be less obvious in some species (e.g., snakes), and the inattentiveness of a particular animal is best assessed by comparisons to conspecifics. An inattentive reptile tends to move very little and ignores, or does not even detect, movement around it. This abnormal behavior can be caused by a primary disease, but more commonly, it is due to poor husbandry. Inadequate thermal conditions, insufficient enclosure complexity (i.e., lack of sufficient refuge sites), or the presence of incompatible cagemates can all lead to inattentiveness. Even if no aggression is seen among cagemates, an individual may assume a submissive role and, since it cannot escape from the threat, it will choose to be as inconspicuous as possible. Regardless of the cause, chronic inattentive behavior leads to poor health, as the individual will forego sufficient thermoregulatory and feeding behaviors. When an animal shows chronic inattentive behavior, it should not be ignored, as it typically indicates an environmental or physiological problem that requires action to correct. It is essential to perform a thorough examination for an underlying cause and to isolate the animal from others. Isolation removes any negative social cues and serves as a quarantine, should an infectious agent be responsible. One common mistake made when isolating a poor-doing reptile, for whatever reason, is that the new environment is overly simplistic. Poor-doing individuals are especially vulnerable in suboptimal housing conditions, so enclosures for isolated animals should be carefully designed to make sure that the new/isolation enclosure provides for all the needs of the animal.

Ways to Maintain Behavioral Health

Captive Considerations to Promote Species-Typical Behavior

Environment

Reptiles represent an especially diverse group of animals, and therefore, caution must be taken to not make assumptions based on generalizations, even among closely related species. Fortunately, in the last few decades, there has been amazing progress in herpetoculture (i.e., the care of reptiles in captivity), and numerous scientific articles, books, and web sites exist to provide the appropriate captive environment for a vast array of species. Scientific papers provide the most reliable accuracy, but few cover general behaviors, especially under captive conditions. Books and the internet provide almost endless resources, but the reader must use caution as most of these resources are not peer reviewed.

One often overlooked resource for developing an enclosure with the appropriate environment is the animal itself. For example, a reptile that spends the majority of its time at a basking location is indicating the enclosure is not providing sufficient heat for adequate thermoregulation. In contrast, one that spends most of its time far from the heat source, or spends a lot of time trying to escape, may be doing so because the enclosure is too hot. A proper captive environment is one where the reptile remains alert at all times, except when sleeping, and uses its entire cage at various times. Realize, in order to understand what the captive reptile is "telling" you, it is critical to be familiar with the normal behaviors of the species. What may be an abnormal behavior for one species might be part of the normal repertoire of another.

In addition to the primary environment of the captive reptile, one must also consider the surroundings. Unless a reptile has acclimated to high traffic areas, it is best to house them in relatively quiet areas. Frequent disturbances can lead to reduced basking and frequent escape behavior, both of which can have negative consequences on the reptile's health and well-being.

Social Groupings

As most reptiles are not social, it is usually best to house reptiles individually. If there is a need to cohouse reptiles, whether for reproductive reasons or efficiency, it is critical to frequently and carefully monitor cohabitants for injuries, weight loss, and abnormal behaviors. This is most critical in the initial days when reptiles are first put together, but some extent of monitoring needs to continue for the long term, as compatibility can change with age and seasons.

Feeding and Watering Strategies

An appropriate feeding strategy provides the captive reptile with the correct food, presented in the correct amount, and in an effective way. Failure to eat may reflect inappropriate food items, poor animal health, poor environmental conditions, social stress, and/or reproductive activity (the latter being the only healthy reason for aphagia). Vast resources are available that can help in understanding each of these, but maintaining a healthy reptile may require some persistence in identifying foods that the reptile will eat, while still providing a balanced diet. It is critical to know the proper body proportions of the species (reptile body form can vary considerably). In general, a lizard or snake in good body condition will have sufficient soft tissue around its pelvis or spine, respectively, so that the outlines of the bones are not visible. Visual assessments of the body condition of turtles can be the most difficult, because the shell prevents the observation of most of the animal's soft tissue.

It is important to realize that, even in reptiles, obesity, not just malnourishment, can cause altered behavior and well-being. Therefore, in addition to qualitative observation of body proportions, it is valuable to maintain weight logs for captive reptiles. However, given the low metabolic rates of reptiles, frequency of weighing can be lower than what is typical of mammals (i.e., weeks for lizards, months for snakes and turtles).

Reptiles consume limited amounts of water, so it would take a reptile an extended period of time to empty a water bowl. Consistent with reptiles' preference for fresh water, it is important to change a reptile's water bowl periodically, rather than refilling it as needed. Many reptiles, including tortoises, snakes, and large lizards, seem to prefer to defecate in their water bowl, creating a sanitation concern. Given the low water intake of many reptiles, one strategy is to only offer water periodically, rather than ad libitum. This approach provides a more natural situation for most nonaquatic reptiles and maximizes the likelihood that water is fresh when the animal drinks.

Training

Regardless of how acclimated a captive reptile is to its enclosure or to handling, none are domesticated. Similarly, while recent studies have demonstrated that reptiles can learn relatively simple tasks (Bridgeman and Tattersall, 2019; Cooper et al., 2019), the ability to train them is very limited. One thing that many reptile keepers have discovered is that reptiles can learn to associate keeper actions with feeding. This is especially true for reptiles that are minimally handled, such as those in a research setting. Many reptiles learn to associate the opening of its cage with feeding, if the cage is rarely opened for reasons other than feeding. The linking of cage opening with feeding can create a dangerous situation if it pertains to larger or venomous species. Reptiles that are deemed as aggressive, because they lunge or strike when their cage is opened, oftentimes are merely prematurely initiating a feeding response. This feeding behavior may be heightened if the animal detects the scent of a meal, as may occur when feeding a large number of reptiles in a room.

Special Situations

The housing of reptiles for research purposes often requires special husbandry considerations. Most commonly, research aims may require environmental consistencies among individuals and over time. For some studies, the enclosures for a study group of reptiles need to be identical to avoid unnecessary, uncontrolled variables. Also, because most body processes are sensitive to temperature, it is sometimes necessary to house reptiles at a constant temperature, both spatially and temporally. Doing so alters the normal thermoregulatory behavior of reptiles, and thus, use of this approach needs to be scientifically justified. When constant thermal environments are used, extra effort should be made to monitor the well-being and behavior of the animals.

Also, if the research requires frequent handling, especially if acquisition must be quick in order to obtain timely samples, reptile enclosures may need to be smaller and less complex than what is optimal for best preserving normal behaviors. In such situations, it is imperative to distinguish complexities that make the enclosure aesthetically pleasing from those that serve vital roles for animal well-being, including normal behavioral function. Designing simplistic enclosures is typically more challenging than designing complex ones, if the enclosure is to effectively provide for the needs of both the animal and the research. When a simplistic approach is taken, it is crucial to incorporate frequent qualitative (e.g., close observation of behavior) and quantitative (e.g., weighing) monitoring of the animals.

Conclusions/Recommendations

Reptiles, in general, especially those that commonly participate in research, usually adapt well to captive environments and typically require less attention and husbandry provisioning (e.g., feeding, cage cleaning) than do more traditional laboratory animal species. However, this convenience of low maintenance can also be a downfall. Reptile care, since it is often not a daily task beyond observation, can be overlooked without an established schedule and documentation of activities. Additionally, poor husbandry practices for reptiles tend to lead to a relatively slow demise of the animal, which can be missed. Therefore, it is vital to possess a solid understanding of a species' normal behavior, since behavioral abnormalities tend to precede physical ones.

REFERENCES

Abdel-Aal, H. A., and Hisham, A. 2013. On surface structure and friction regulation in reptilian limbless locomotion. *J. Mech. Behav. Biomed. Mater.* 22: 115–135.

Aleksiuk, M., and Gregory, P. T. 1974. Regulation of seasonal mating behavior in *Thamnophis sirtalis parietalis*. *Copeia* 1974: 681–689.

Alexander, G. J. 2018. Reproductive biology and maternal care of neonates in southern African python (*Python natalensis*). *J. Zool.* 305: 141–148.

Allf, B. C., Durst, P. A. P., and Pfennig, D. W. 2016. Behavioral plasticity and the origins of novelty: The evolution of the rattlesnake rattle. *Am. Nat.* 188: 475–483.

Andrews, R. M. 2004. Embryonic development. In: D. C. Deeming, ed. *Reptilian Incubation: Environment, Evolution, and Behaviour.* Nottingham: Nottingham University Press, 75–102.

Angilletta, M. 2009. *Thermal Adaptation: A Theoretical and Empirical Synthesis.* Oxford: Oxford University Press.

Autumn, K., Niewiarowski, P. H., and Puthoff, J. B. 2014. Gecko adhesion as a model system for integrative biology, interdisciplinary science, and bioinspired engineering. *Annu. Rev. Ecol. Evol. Syst.* 45: 445–470.

Bakken, G. S. 1976. An improved method for determining thermal conductance and equilibrium body temperature with cooling curve experiments. *J. Therm. Biol.* 1:169–175.

Bakken, G. S., and Krochmal, A. R. 2007. The imaging properties and sensitivity of the facial pits of pitvipers as determined by optical and heat-transfer analysis, *J. Exp. Biol.* 210: 2801–2810.

Bealor, M. T., Miller, J. L., de Queiroz, A., and Chiszar, D. A. 2013. The evolution of the stimulus control of constricting behaviour: Inferences from North American gartersnakes (*Thamnophis*). *Behaviour* 150: 225–253.

Beck, D. D. 1996. Effects of feeding on body temperatures of rattlesnakes: A field experiment. *Physiol. Zool.* 69: 1442–1455.

Blazquez, M. C. 1996. Activity and habitat use in a population of *Ameiva ameiva* in Southeastern Colombia. *Biotropica* 28: 714–719.

Blob, R. W., Mayerl, C. J., Rivera, A. R. V., Rivera, G., and Young, V. K. H. 2016. "On the fence" versus "all in": insights from turtles for the evolution of aquatic locomotor specialization and habitat transitions in tetrapod vertebrates. *Integr. Comp. Biol.* 56: 1310–1322.

Bonnet, X., and Brischoux, F. 2008. Thirsty sea snakes forsake refuge during rainfall. *Austral Ecol.* 33: 911–921.

Brashears, J. A., and DeNardo, D. F. 2013. Revisiting python thermogenesis: brooding Burmese pythons (*Python bivittatus*) cue on body, not clutch, temperature. *J. Herpetol.* 47: 440–444.

Brashears, J., and DeNardo, D. F. 2015. Facultative thermogenesis during brooding is not the norm among pythons. *J. Comp. Physiol. A* 201: 817–825.

Brashears, J. A., Hoffman, T. C. M., and DeNardo, D. F. 2017. Modeling the costs and benefits associated with the evolution of endothermy using a robotic python. *J. Exp. Biol.* 220: 2409–2417.

Brewster, C. L., and Beaupre, S. J. 2019. The effect of body posture on available habitat and activity-time in a lizard: Implications for thermal ecology studies. *J. Therm. Biol.* 82: 10–17.

Bridgeman, J. M., and Tattersall, G. J. 2019. Tortoises develop and overcome position biases in a reversal learning task. *Anim. Cogn.* 22: 265–275.

Bull, C. M., Gardner, M. G., Sih, A., Spiegel, O., Godfrey, S. S., and Leu, S. T. 2017. Why is social behavior rare in reptiles? lessons from sleepy lizards. *Adv. Stud. Behav.* 49: 1–26.

Christensen, C. B., Christensen-Dalsgaard, J., Brandt, C., and Madsen, P. T. 2012. Hearing with an atympanic ear: Good vibration and poor sound-pressure detection in the royal python, *Python regius*. *J. Exp. Biol.* 215: 331–342.

Clutton-Brock, T. H. 1991. *The Evolution of Parental Care*. Princeton, NJ: Princeton University Press.

Comanns, P. 2018. Passive water collection with the integument: Mechanisms and their biomimetic potential. *J. Exp. Biol.* 221. doi:10.1242/jeb.153130.

Cooper, T., Liew, A., Andrle, G., Cafritz, E., Dallas, H., Niesen, T., Slaters, E., Stockert, J., Vold, T., Young, M., and Mendelson III, J. 2019. Latency in problem solving as evidence for learning in varanid and helodermatid lizards, with comments on foraging techniques. *Copeia* 107: 78–84.

Cooper Jr., W. E., and Burns, N. 1987. Social significance of ventrolateral coloration in the fence lizard, *Sceloporus undulatus*. *Anim. Behav.* 35: 526–532.

Crawford, N. G., Parham J. F., Sellas, A. B., Faircloth, B. C., Glenn, T. C., Papenfuss, T. J., Henderson, J. B., Hansen, M. H., and Simison, W. B. 2014. A phylogenomic analysis of turtles. *Mol. Phylogenetics Evol.* 83: 250–257.

Davis, J. R., and DeNardo, D. F. 2007. The urinary bladder as a physiological reservoir that moderates dehydration in a large desert lizard, the Gila monster (*Heloderma suspectum*). *J. Exp. Biol.* 210: 1472–1480.

Davis, J. R., and DeNardo, D. F. 2009. Water supplementation affects the behavioral and physiological ecology of Gila monsters (*Heloderma suspectum*) in the Sonoran Desert. *Physiol. Biochem. Zool.* 82: 739–748.

Davis Rabosky, A. R., Cox, C. L., Rabosky, D. L., Title, P. O., Holmes, I. A., Feldman, A., and McGuire, J. A. 2016. Coral snakes predict the evolution of mimicry across New World snakes. *Nat. Commun.* 7. doi:10.1038/ncomms11484.

Deeming, D. C. 2018. Nesting environment may drive variation in eggshell structure and egg characteristics in the Testudinata. *J. Exp. Zool.* 329: 331–342.

Enok, S., Leite, G. S. P. C., Leite, C. A. C., Gesser, H., Hedrick, M. S., and Wang, T. 2016. Improved cardiac filling facilitates the postprandial elevation of stroke volume in *Python regius*. *J. Exp. Biol.* 219: 3009–3018.

Ferguson, G. W. 1977. Display and communications in reptiles: An historical perspective. *Amer. Zool.* 17: 167–176.

Fleishman, L. J. 1992. The influence of the sensory system and the environment on motion patterns in the visual displays of anoline lizards and other vertebrates. *Am. Nat.* 139: S36–S61.

Ford, N. 1986. The role of pheromone trails in the sociobiology of snakes. In: D. Duvall, D. Müller-Schwarze, and R.M. Silverstein, eds. *Chemical Signals in Vertebrates*, vol. 4. Boston, MA: Springer, 261–78.

Fox, S. F., McCoy, J. K., and Baird, T.A. eds. 2003. *Lizard Social Behavior*. Baltimore, MD: Johns Hopkins University Press.

Fry, B. G., Vidal, N., van der Weerd, L., Kochva, E., and Renjifo, C. 2009. Evolution and diversification of the Toxicofera reptile venom system. *J. Proteom.* 72: 127–136.

Gans, C. 1996. An overview of parental care among the Reptilia. *Adv. Stud. Behav.* 25: 145–157.

Gardner, M. G., Pearson, S. K., Johnston, G. R., and Schwarz, M. P. 2015. Group living in squamate reptiles: a review of evidence for stable aggregations. *Biol. Rev. Camb. Philos. Soc.* doi:10.1111/brv.12201.

Garrick, L. D., and Lang, J. W. 1977. Social signals and behaviors of adult alligators and crocodiles. *Amer. Zool.* 17: 225–239.

Gienger, C. M., and Beck, D. D. 2011. Northern pacific rattlesnakes (*Crotalus oreganus*) use thermal and structural cues to choose overwintering hibernacula. *Can. J. Zool.* 89: 1084–1090.

Greene, H. W., May, P. G., Hardy, D. L., Sr., Sciturro, J. M., and Farrell, T. M. 2002. Parental behavior by vipers. In: G. W. Schuett, M. Höggren, M. E. Douglas, and H. W. Greene, eds. *Biology of the Vipers*. Eagle Mountain, UT: Eagle Mountain Publishing, 179–205.

Greiner, C., and Schäfer, M. 2015. Bio-inspired scale-like surface textures and their tribological properties. *Bioinspiration Biomim.* 10: 044001.

Grigg, G., and Gans, C. 1993. Morphology and physiology of the crocodylia. In: C. G. Glasby, G. J. B. Ross, and P. L. Beesley, eds. *Fauna of Australia - Volume 2A Amphibia and Reptilia*. Canberra: Australian Govt. Pub. Service, 40, 1–26.

Grotzner, S. R., Rocha, F. A. D., Corredor, V. H., Liber, A. M. P., Hamassaki, D. E., Bonci, D. M. O. and Ventura, D. F. 2020. Distribution of rods and cones in the red-eared turtle retina (*Trachemys scripta elegans*). *J. Comp. Neurol.* 528: 1548–156.

Hallmann, K., and Griebeler, E. M. 2015. Eggshell types and their evolutionary correlation with life-history strategies in squamates. *PLoS One* 10: doi:10.1371/journal.pone.0138785.

Halpern, M., Graves, B. M., and Halpern, M. 1989. Chemical access to the vomeronasal organs of the lizard *Chalcides ocellatus*. *J. Exp. Zool.* 249: 150–157.

Heppard, J. M., and Buchholz, R. 2019. Impact of human disturbance on the thermoregulatory behaviour of the endangered ringed sawback turtle (*Graptemys oculifera*). *Aquat. Conserv.: Mar. Freshw. Ecosyst.* 29: 990–1001.

Huey, R., B., Peterson, C. R., Arnold, S. J., and Porter, W. P. 1989. Hot rocks and not-so-hot rocks: Retreat-site selection by garter snakes and its thermal consequences. *Ecology* 70: 931–944.

Hunt, R. H. 1987. Nest excavation and neonate transport in wild *Alligator mississippiensis*. *J. Herpetol.* 21: 348–350.

Hutchison, V. H., and Kosh, R. J. 1974. Thermoregulatory function of the parietal eye in the lizard *Anolis carolinensis*. *Oecologia* 16: 173–177.

Ibanez, A., Marzal, A., Gonzalez-Blazquez, M., Lopez, P., and Martin, J. 2015. Basking activity is modulated by health state but is constrained by conspicuousness to predators in male spanish terrapins. *Ethology* 121: 335–344.

Knowlton, G. F. 1946. Feeding habits of some reptiles. *Herpetologica* 3: 77–80.

Krochmal, A. R., Bakken, G. S., and LaDuc, T. J. 2004. Heat in evolution's kitchen: Evolutionary perspectives on the functions and origin of the facial pit of pitvipers (Viperidae: Crotalinae). *J. Exp. Biol.* 207: 4231–4238.

Kushlan, J. A., and Kushlan, M. S. 1980. Function of nest attendance in the American alligator. *Herpetologica* 36: 27–32.

Langerhans, R. B., Knouft, J. H., and Losos, J. B. 2006. Shared and unique features of diversification in Greater Antillean *Anolis* ectomorphs. *Evolution* 60: 362–369.

Legler, J. M. 1993. Morphology and physiology of the chelonia. In: C. G. Glasby, G. J. B. Ross, and P. L. Beesley, eds. *Fauna of Australia - Volume 2A Amphibia and Reptilia*. Canberra: Australian Govt. Pub. Service, 16, 1–23.

Leon, O. I. M. V., and Manjarrez, J. 2017. Forewarned is forearmed the use of aposematic signals in snakes. *Ciencia Ergo-Sum* 24: 267–272.

Ligon, R. A., and McGraw, K. J. 2013. Chameleons communicate with complex colour changes during contests: Different body regions convey different information. *Biol Lett.* 9: doi:10.1098/rsbl.2013.0892.

Lillywhite, H. B. 2017. Feeding begets drinking: Insights from intermittent feeding in snakes. *J. Exp. Biol.* 220: 3565–3570.

Liu, Y.-X., Davy, C. M., Shi, H.-T., Murphy, R. W. 2013. Sex in the half-shell: A review of the functions and evolution of courtship behavior in freshwater turtles. *Chelonian Conserv. Bi.* 12: 84–100.

Lorioux, S., Lisse, H., and Lourdais, O. 2013. Dedicated mothers: Predation risk and physical burden do not alter thermoregulatory behaviour of pregnant vipers. *Anim. Behav.* 86: 401–408.

Lourdais, O., Heulin, B., and DeNardo, D. F. 2008. Thermoregulation during gravidity in the Children's python (*Antaresia childreni*): A test of the pre-adaptation hypothesis for maternal thermophily in snakes. *Biol. J. Linnean Soc.* 93: 499–508.

Martín, J., and López, P. 2015. Condition-dependent chemosignals in reproductive behavior of lizards. *Horm. Behav.* 68: 14–24.

Mason, R. T., and Parker, M. R. 2010. Social behavior and pheromonal communication in reptiles. *J. Comp. Physiol. A* 196: 729–749.

Merchant, M., Williams, S., Trosclair, P. L., Elsey, R. M., and Mills, K. 2007. Febrile response to infection in the American alligator (*Alligator mississippiensis*). *Comp. Biochem. Physiol. A* 148: 921–925.

Minnich, J. E., and Shoemaker, V. H. 1970. Diet, behavior and water turnover in the desert iguana, *Dipsosaurus dorsalis*. *Am. Midl. Nat.* 84: 496–509.

Mukherjee, A., Kumara, H. N., and Bhupathy, S. 2018. Sun-basking, a necessity not a leisure: Anthropogenic driven disturbance, changing the basking pattern of the vulnerable Indian rock python in Keoladeo National Park, India. *Glob. Ecol. Conserv.* 13. doi:10.1016/j.gecco.2017.e00368.

Murphy, M. S., and DeNardo, D. F. 2019. Rattlesnakes must drink: Meal consumption does not improve hydration state. *Physiol. Biochem. Zool.* 92: 381–385.

Nicholson, K. E., Harmon, L. J., and Losos, J. B. 2007. Evolution of *Anolis* lizard dewlap diversity. *PLoS One* 2: e274.

Ott, B. D., and Secor, S. M. 2007. Adaptive regulation of digestive performance in the genus *Python*. *J. Exp. Biol.* 210: 340–356.

Perry, B. W., Andrews, A. L., Kamal, A. M., Card, D. C., Schield, D. R., Pasquesi, G. I. M., Pellegrino, M. W., Mackessy, S. P., Chowdhury, S. M., Secor, S. M., and Castoe, T. A. 2019. Multi-species comparisons of snakes identify coordinated signalling networks underlying post-feeding intestinal regeneration. *Proc. Royal Soc. B* 286. https://doi-org.ezproxy1.lib.asu.edu/10.1098/rspb.2019.0910.

Peterson, C. C. 1996. Anhomeostasis: Seasonal water and solute relations in two populations of the desert tortoise (*Gopherus agassizii*) during chronic drought. *Physiol. Zool.* 69: 1324–1358.

Rafferty, A. R., and Reina, R. D. 2012. Arrested embryonic development: A review of strategies to delay hatching in egg-laying reptiles. *Proc Biol Sci.* 279: 2299–2308.

Repp, R. A., and Schuett, G. W. 2008. Western diamond-backed rattlesnakes, *Crotalus atrox* (Serpentes: Viperidae), gain water by harvesting and drinking rain, sleet, and snow. *Southwest. Nat.* 53: 108–114.

Rohtla, E. A., Russell, A. P., and Bauer, A. M. 2019. Sounding off: Relationships between call properties, body size, phylogeny, and laryngotracheal form of geckos. *Herpetologica* 75: 175–197.

Ruby, D. E. 1978. Seasonal changes in the territorial behavior of the iguanid lizard *Sceloporus jarrovi*. *Copeia* 1978: 430–438.

Sears, M. W.; Angilletta, M. J., Schuler, M. S., Borchert, J., Dilliplane, K. F., Stegman, M., Rusch, T. W., and Mitchell, W. A. 2016. Configuration of the thermal landscape determines thermoregulatory performance of ectotherms. *Proc. Natl. Acad. Sci. U. S. A.* 113: 10595–10600.

Scheelings, T. F., and Hellebuyck, T. 2019. Dermatology – skin. In: Divers, S. J., Stahls, S. J., eds. *Mader's Reptile and Amphibian Medicine and Surgery*, 3rd edition. St. Louis: Elsevier, 709.

Schwenk, K. 2008. Comparative anatomy and physiology of chemical senses in nonavain reptiles. In: J. G. M. Thewissen and S. Nummela, eds. *Sensory Evolution on the Threshold Adaptations in Secondarily Aquatic Vertebrates*. Berkeley, CA: University of California Press, 65–81.

Seebacher, F., and Franklin, C. E. 2005. Physiological mechanisms of thermoregulation in reptiles: A review. *J. Comp. Physiol. B* 175: 533–541.

Senter, P., Harris, S. M. and Kent, D. L. 2014. Phylogeny of courtship and male-male combat behavior in snakes. *PLoS One* 9. doi:10.1371/journal.pone.0107528.

Shine, R. 1985. The evolution of viviparity in reptiles: An ecological analysis. In: C. Gans and F. Billet, eds. *Biology of the Reptilia*, Vol. 15. New York, NY: John Wiley and Sons, 605–694.

Shine, R. 2004. Incubation regimes of cold-climate reptiles: The thermal consequences of nest-site choice, viviparity and maternal basking. *Biol. J. Linn. Soc.* 83: 145–155.

Shoji, T., and Kurihara, K. 1991. Sensitivity and transduction mechanisms of responses to general odorants in turtle vomeronasal system. *J. Gen. Physiol.* 98: 909–919.

Simões, B. F., Sampaio, F. L., Douglas, R. H., Kodandaramaiah, U., Casewell, N. R., Harrison, R. A., Hart, N. S., Partridge, J. C., Hunt, D. M., Gower, D. J. 2016. Visual pigments, ocular filters and the evolution of snake vision. *Mol. Biol. Evol.* 33: 2483–2495.

Smith, H. M. 1975. Grist for the mills of herpetophiles in Mexico. *Bull. Maryland Herp. Soc.* 11: 40–44.

Smith, K. R., Cadena, V., Endler, J. A., Kearney, M. R., Porter, W. P., and Stuart-Fox, D. 2016. Color change for thermoregulation versus camouflage in free-ranging lizards. *Am. Nat.* 188: 668–678.

Somma, L. A. 2003. *Parental Behavior in Lepidosaurian and Testudinian Reptiles: A Literature Survey*. Malabar, FL: Krieger Publishing Company.

Stahlschmidt, Z. R., and DeNardo, D. F. 2009. Effect of nest temperature on egg-brooding behavior, metabolism, and clutch-nest thermal relations in Children's pythons (*Antaresia childreni*). *Physiol. Behav.* 98: 302–306.

Stahlschmidt, Z. R., and DeNardo, D. F. 2010. Parental care in snakes. In: R. D. Aldridge and D.M. Sever, eds. *Reproductive Biology and Phylogeny of Snakes*. Enfield, NH: Science Publishers Inc., 673–702.

Stahlschmidt, Z. R., Brashears, J. A., and DeNardo, D. F. 2011. The role of temperature and humidity in python nest-site selection. *Anim. Behav.* 81: 1077–1081.

Stahlschmidt, Z. R., Hoffman, T. C. M., and DeNardo, D. F. 2008. Postural shifts during egg-brooding and their impact on egg water balance in Children's pythons (*Antaresia childreni*). *Ethology* 114: 1113–1121.

Stocker, T. F., Qin, D., Plattner, G.-K., Alexander, L. V., Allen, S.K., Bindoff, N.L., Bréon, F.-M., Church, J.A., Cubasch, U., Emori, S., Forster, P., Friedlingstein, P., Gillett, N., Gregory, J. M., Hartmann, D. L., Jansen, E., Kirtman, B., Knutti, R., Krishna Kumar, K., Lemke, P., Marotzke, J., Masson-Delmotte, V., Meehl, G. A., Mokhov, I. I., Piao, S., Ramaswamy, V., Randall, D., Rhein, M., Rojas, M., Sabine, C., Shindell, D., Talley, L. D., Vaughan, D. G., and Xie, S.-P. 2013. Technical summary. In: T. F. Stocker, D. Qin, G.-K. Plattner, M. Tignor, S. K. Allen, J. Boschung, A. Nauels, Y. Xia, V. Bex, and P. M. Midgley, eds. *Climate Change 2013: The Physical Science Basis. Contribution of Working Group I to the Fifth Assessment Report of the Intergovernmental Panel on Climate Change*. Cambridge, UK, and New York: Cambridge University Press.

Su, C.-Y., Luo, D.-G., Terakita, A., Shichida, Y., Liao, H.-W., Kazmi, M. A., Sakmar, T. P., and Yau, K.-W. 2006. Parietal-eye phototransduction components and their potential evolutionary implications. *Science* 311: 1617–1621.

Tang, Z., Qi, P., and Dai, J. 2017. Mechanism design of a biomimetic quadruped robot. *Ind. Robot* 44: 512–520.

Tattersall, G. J., Cadena, V., and Skinner, M. C. 2006. Respiratory cooling and thermoregulatory coupling in reptile. *Resp. Physiol. Neurobi.* 154: 302–318.

Tattersall, G. J., Leite, C. A., Sanders, C. E., Cadena, V., Andrade, D. V., Abe, A. S., and Milsom, W. K. 2016. Seasonal reproductive endothermy in tegu lizards. *Sci. Adv.* 2: e1500951.

Uetz, P., Freed, P., and Hosek, J., eds. 2020. The Reptile Database, http://www.reptile-database.org, accessed July 11, 2020.

Van Mierop, L. H. S., and Barnard, S. M. 1978. Further observations on thermoregulation in the brooding female *Python molurus bivittatus* (Serpentes: Boidae). *Copeia* 1978: 615–621.

Vergne, A. L., Mathevon, N., Pritz, M. B. 2009. Acoustic communication in crocodilians: From behaviour to brain. *Biol. Rev.* 84: 391–411.

Wright, C. W., Jackson, M. L., and DeNardo, D. F. 2013. Meal consumption is ineffective at maintaining or correcting water balance in a desert lizard *Heloderma suspectum*. *J. Exp. Biol.* 216: 1439–1447.

Znari, M., and Nagy, K. A. 1997. Field metabolic rate and water flux in free-living Bibron's agama (*Agama impalearis*, Boettger, 1874) in Morocco. *Herpetologica* 53: 81–88.

24

Behavioral Biology of Marmosets

Arianna Manciocco
Istituto di Scienze e Tecnologie della Cognizione

Sarah J. Neal Webb and Michele M. Mulholland
The University of Texas MD Anderson Cancer Center

CONTENTS

Introduction 377
Research Contributions 377
Behavioral Biology 379
    Ecology 379
    Social Organization and Behavior 380
    Feeding Behavior 380
    Mating and Reproduction 381
Common Captive Behaviors 383
    Normal Behaviors 383
    Abnormal and Fear-Related Behaviors 383
Ways to Maintain Behavioral Health in Captivity 384
    Environmental Complexity 384
    Social Groupings 385
    Feeding 387
    Training 388
Special Situations 388
    Care of Aged Animals 388
    Social Problems 388
    Wasting Syndrome 388
Conclusions 389
References 389

Introduction

Marmosets are small Neotropical primates that inhabit South America; most are found in Brazil, but some species inhabit Bolivia, Paraguay, Colombia, Ecuador, and Peru. Marmoset distribution in South America has been altered due to habitat destruction and, more importantly, anthropogenic introductions of marmoset species outside of their natural areas, causing sympatry across marmoset species. For example, black-tufted-ear marmosets (*Callithrix penicillata*) and common marmosets (*C. jacchus*) have been introduced to, and/or invaded, areas that were historically outside of their natural ranges, resulting in interbreeding of species and hybrid marmoset forms (Malukiewicz, 2019). Marmosets are currently categorized into four genera (*Callithrix, Cebuella, Callibella,* and *Mico*) consisting of 21 species, though this organization is still being debated (Rosenberger, 2011; Schneider and Simpaio, 2015; Garbino, 2015). As of 2020, the buffy-headed marmoset (*Callithrix flaviceps*) and buffy-tufted-ear marmoset (*C. aurita*) are the only species classified as critically endangered or endangered (respectively) on the IUCN Red List, but populations of all marmoset species are decreasing and data are insufficient to evaluate the conservation status for six of the *Mico* spp. (IUCN, 2020). Together with squirrel monkeys (*Saimiri sciureus*.), common marmosets (*Callithrix jacchus*) are the New World Monkeys (NWMs) most commonly involved in biomedical research. *C. jacchus* are listed as "Least Concern" on the IUCN Red List (2020), but it is important to note that captive research subjects are not taken from the wild, but rather from established captive breeding colonies (Abbott et al., 2003).

Research Contributions

Research interest in callitrichids (marmosets and tamarins) began in the 1960s, with the first behavioral study on captive common marmosets (*C. jacchus*) (Epple, 1967). Preliminary biological information became available in the 1970s, when Bridgwater (1972) and Hershkovitz (1977) published complete descriptions of the biology and systematics of marmosets. The first descriptions of natural marmoset behavior and ecology came later, when Stevenson (1977) and Rylands (1981) published their field research on common marmosets and black tassel-eared marmosets (*C. humeralifer*, now *Mico humeralifer*), respectively, in Brazil. As common marmosets are the marmoset species that is most frequently involved in biomedical research, they will be discussed in the greatest detail in this chapter.

Marmosets are involved in a wide range of laboratory research, including immunology, neuroscience, reproductive biology, behavioral sciences, regenerative medicine, pharmacology, toxicology, aging, and genetic animal modeling (Mansfield, 2003; NIH, 2018). Although marmosets are more phylogenetically removed from humans than Old World Monkeys, they exhibit a number of unique characteristics that make them attractive models for research (Okano et al., 2012; Bert et al., 2012; Kishi et al., 2014; Saito, 2015; Hashikawa et al., 2015; Eliades and Miller, 2017). Their small size and relatively rapid breeding in captivity (where twins are common) facilitate housing relatively large numbers of subjects in relatively small amounts of space, reductions in feeding-related costs, and the likelihood of developing a breeding colony in a short time. Moreover, their low body mass results in decreased costs associated with the synthesis of compounds for pharmacological research (Ward and Vallender, 2012). In addition, husbandry, veterinary care, and enrichment provision are relatively less complicated compared to Old World Monkeys, yet marmosets still have a lifespan of up to 16 years in captivity, allowing them to serve as a viable animal model for longitudinal and aging studies. Further, marmosets, like all NWMs, do not carry herpes B virus (*Macacine alphaherpesvirus 1*), reducing the degree of some of the biosafety practices that are required when working with macaques (Tokuno et al., 2015).

In addition, some progressive autoimmune and neurodegenerative diseases (i.e., multiple sclerosis, Parkinson's disease) can be experimentally induced in marmosets, and better resemble the human pathology than similar manipulations in rodents and rhesus monkeys (Mansfield 2003). The brain of the marmoset is compact, making it relatively easy to analyze (Tokuno et al., 2015), with frontal lobes (the brain areas involved in many human psychiatric diseases) that are more developed than those of other animals with similarly sized brains (Cyranoski, 2014). Indeed, neuroscience research with marmosets is becoming more popular, with the United States National Institutes of Health (NIH) specifically providing funding to expand existing marmoset colonies in order to meet the increasing demand for these animals in this line of research (NIH, 2019). The widespread participation of marmosets in biomedical research around the world has led to the creation of the International Marmoset Research Group (IMRG; 't Hart et al., 2012), which aims to integrate and harmonize the existing activities of "local" marmoset research groups in the Americas, Europe, and Asia.

The genetic and physiological differences between Old and NWMs, and even those within NWM species, make the validity of marmoset experimental models dependent on the specifics of the particular study. This is especially evident in relation to some of the unique physiological characteristics of callithrichids. For example, marmosets and tamarins exhibit hematopoietic chimerism, resulting in important genetic constraints and immunological challenges for studies related to susceptibility to pathogens (Carrion and Patterson, 2012; Ward and Vallender, 2012). The fact that marmosets typically give birth to twins (or even triplets) is advantageous in preclinical pharmacological and toxicological studies, where bone marrow chimeric twins/triplets can serve as subjects in therapy trials to assess drug efficacy (Orsi et al., 2011). Common marmosets were also the first nonhuman primates from which embryonic stem cells were derived, leading to the first derivation of human embryonic cells that express specific phenotypic characteristics and can be used in studies to replace organs and tissues (Thomson et al., 1996; 't Hart et al., 2012). The development of research tools, including anatomical and MRI atlases (Hikishima et al., 2013), genome sequencing (Sato et al., 2015), and other genetics-related technologies (Sasaki, 2015, Sato et al., 2016) have made marmosets an extremely valuable model for studies that explore strategies for treating and eliminating major brain disorders. In 2014, the Japanese project, Brain/MINDS, was initiated to build a multiscale marmoset brain map and to create transgenic marmosets to model human brain diseases (Okano et al., 2016). Furthermore, marmosets are the world's first transgenic primate with germline transmission to have been developed (Sasaki et al., 2009). This research approach is being used to generate genetically modified marmosets that model neurological disorders, including Alzheimer's Disease, Rett syndrome, and tuberous sclerosis complex (e.g., Chahrour and Zoghbi, 2007; Ess, 2010; Fischer and Austad, 2011). Indeed, researchers have shown success in using CRISPR/Cas9 technology to knock in or out specific genes in marmoset embryos, including the introduction of genetic mutations associated with Parkinson's disease (Kumita et al., 2019; Vermilyea et al., 2020). However, there are numerous ethical issues that must be addressed prior to the widespread participation of transgenic marmosets (and of primates in general) in research ('t Hart et al., 2012).

Characteristics of marmoset social structure, such as strong family bonds, shared parental care, and well-structured visual and vocal communication, make these monkeys valuable experimental models for human diseases characterized by the breakdown of social behaviors (e.g., autism), as well as investigations into the evolution of human language (Miller et al., 2016). Further, the behavioral and physiological mechanisms by which the dominant breeding female suppresses the reproduction of the other females in her family group has facilitated interdisciplinary studies incorporating neuroimaging, behavioral, and endocrinological methods ('t Hart et al., 2012). In the last decade, behavior and cognition studies involving common marmosets have increased, including investigations on topics such as hand preference, cooperation, and personality (Adriani et al., 2013; Zoratto et al., 2014; Tomassetti et al., 2019). Given the sociality of marmosets, they serve as an effective model for the evolution and neurobiology of social signaling, active communication, and gaze following (Miller et al., 2016). Recently, touchscreen apparatus that easily attach to marmoset home cages and have flexible, adaptable software have allowed the study of visual discrimination, reversal learning, delayed match-to-sample, reaction time, and other learning tasks, as well as age-related changes in such cognitive measures (Takemoto et al., 2011; Kangas et al., 2016).

Behavioral Biology

Ecology

In general, marmosets weigh between 236 and 256 g, with a body length of about 185–188 mm (Mittermeier et al., 1988; IUCN, 2020). Pygmy marmosets (*Cebuella pygmaea*), as one might expect, are smaller (136 mm) and weigh less (119 g; Soini, 1988) than other marmoset species. In captivity, a captive diet and limited physical activity may result in obesity problems for zoo- and laboratory-housed marmosets. For example, captive and wild *Callithrix jacchus* show consistent weight differences, with the captive animals typically heavier (230–360 g) than their wild counterparts (160–320 g; Araújo et al., 2000). Due to the difficulties in obtaining longitudinal data in the field, less is known about the life span of marmosets in the wild (Ash and Buchanan-Smith, 2014); however, the average lifespan in captivity is about 6 years, with a maximum of 16 years (*C. jacchus*; Tardif et al., 2003; Abbott et al., 2003).

Marmosets are arboreal, diurnal primates, with nonprehensile tails that are as long as, or longer than, their bodies. The tail of an adult common marmoset can reach 280 mm and provides balance and stability when jumping or moving along tree branches (Stevenson and Rylands, 1988). Marmosets are adapted to a variety of different environments, living in habitats that range from rainy to dry, and from dense to sparse vegetation. For example, pygmy marmosets (*Cebuella pygmaea*) inhabit tropical evergreen forests of the western Amazon basin, while the "Southern" marmosets (*Callithrix aurita* and *C. flaviceps*) live in montane subtropical forest (Ferrari et al., 1996). Common marmosets can be found inhabiting contrasting environments, such as the humid Atlantic forest and the semiarid Caatinga, and are therefore a highly adaptive species with a wide geographical radiation (Schiel and Suoto, 2017). Schiel and Souto (2017) suggest that the ecological success of common marmosets is the result of several specific characteristics, including efficient gummivory, a high rate of reproduction, a cooperative breeding system, an omnivorous diet, and excellent manual dexterity, as well as highly developed cognitive capabilities.

Seasonal and longitudinal variations in resource abundance differentially affect the activity budgets, daily ranging patterns, and foraging behaviors of most marmoset species. In the semiarid conditions of the Caatinga, common marmosets have been observed to spend most of their time engaged in resting behaviors and obtaining water through consumption of cacti (Abreu et al., 2016). In *C. jacchus*, tree-gouging activity (to obtain exudate, or gum) remains consistent across seasons and habitats, while in *C. aurita*, exudate feeding depends on fruiting patterns, with gum consumption increasing sharply when fruit availability is limited (Martins and Setz, 2000). *C. geoffroyi* has been reported to engage in predatory behavior more frequently during the dry than the wet season, which corresponds with the greater presence of insects during the dry season (Passamani, 1998). Furthermore, *C. geoffroyi* significantly increased time spent resting and decreased time spent foraging during the rainy season (Passamani, 1998). In addition to these environmental factors, anthropomorphic impact, tourism, and human presence have all been reported to cause behavioral changes in marmosets, including altering their presence in the lower strata of the forest, reducing social activity, and changing vocalization rates (de la Torre et al., 2000).

Marmosets' home ranges vary from 0.5 to 30 ha, depending on factors, such as habitat type, distribution and abundance of trophic resources, and fragmentation of forest (which adversely affects home range suitability and quality). The size of the home range can impact daily time budgets, sleeping site choices, and antipredator behaviors. Home range size also seems to affect the number of sleeping sites used; in general, large home ranges have been associated with frequent changes to sleeping-sites whereas fewer sleeping sites are usually maintained in a small home range (Ferrari and Ferrari, 1990). For example, *C. jacchus* groups with home ranges between 0.5 and 1 ha often use the same sleeping site repeatedly, whereas other marmoset species with much larger home ranges (often over 12 ha; e.g., *C. kuhlii*, *C. humeralifer*) tend to change sleep location almost every night (Ferrari and Ferrari, 1990).

The small size of marmosets makes them particularly vulnerable to both aerial and terrestrial predators, and as such, predator interactions are an important part of marmoset ecology (Caine, 2017). These mostly include raptors (e.g., hawks, falcons), snakes (e.g., boids, viperids), and arboreal carnivores (e.g., ocelot, raccoon, tayra). Moreover, capuchin monkeys and birds, such as toucans, are also potential predators that may stimulate antipredator responses in marmosets (Ferrari, 2009). Predation pressure is likely to have influenced many of the physical, ecological, and behavioral characteristics of marmoset species. These include cryptic coats, preference for dense vegetation, vocal repertoire, and social living, among others (Ferrari and Lopes Ferrari, 1990). Small, diurnal, arboreal species, such as marmosets, have the highest predation rates of all primates (Hart, 2007); though predation rates are often based on observations that are anecdotal in nature and may be underestimated (Caine, 1990). Predation rates might appear artificially low, as historically, it has been difficult to observe such unpredictable events under natural circumstances (particularly nocturnal predation). Although the published literature rarely reports successful and fatal attacks on marmosets by aerial and terrestrial predators, evidence of successful attacks by snakes, monkeys, hawks, and tayras have been documented (e.g., Corrêa and Coutinho, 1997; Bezerra et al., 2009a; Ferrari and Beltrao-Mendes, 2011; Albuquerque et al., 2014; Barnett et al. 2015; Teixeira et al., 2016).

Marmosets are uniquely adapted to predator confrontations, and such predation has widespread effects on behavior, including selection of sleeping sites, vigilance behavior, and specialized alarm calls. Marmosets sleep snugly, very close to one another on specific preferred plants (e.g., vines, tree forks, epiphytic bromeliads) that are often concealed and in high locations. The choice of such sleeping sites is often determined by previous encounters with predators, many of which are nocturnal hunters (e.g., felids) (Ferrari and Lopes Ferrari, 1990). Additionally, the performance of "vigilance" behavior prior to retirement has been observed in some species (e.g., *C. flaviceps, C. humeralifer*; Ferrari and Lopes Ferrari, 1990). The presence of a predator likely influences marmoset behavior for extended periods of time, even after the predator is no longer present. Indeed, Hankerson and Caine (2004) found that *C. geoffroyi* altered their morning routine (i.e., they performed more vigilance checks and exhibited a delay in foraging on the ground) following presentation of a snake model outside of their enclosure the evening before. Marmosets adopt different antipredator strategies depending on the type of predator. Generally, sighting aerial predators stimulate specific alarm calls (i.e., one short, high-pitched call that likely makes it difficult for aerial predators to localize the caller) and rapid withdrawal to lower and/or denser vegetation, followed by freezing behavior (Ferrari and Lopes Ferrari, 1990). The approach of a terrestrial predator (e.g., snake), on the other hand, elicits vocalizations that serve to recruit group members, and mobbing behavior, where multiple individuals vocalize and move (in a stop-and-go manner) toward the predator together, to drive it away (Ferrari and Lopes Ferrari 1990; Bezerra and Souto 2008; Clara et al., 2008; Petracca and Caine 2013). Clearly, predation plays a large role in marmoset behavioral ecology.

Social Organization and Behavior

Marmosets live in cohesive family groups consisting of approximately 3–15 adults (including the breeding pair), subadults, juveniles, and infants (Stevenson and Rylands, 1988; Ferrari and Lopes Ferrari, 1989). These individuals are often related to each other, as the most typical marmoset group composition consists of the breeding pair and their adult, juvenile, and infant offspring. Adult offspring of the breeding pair often remains with their natal group past maturity and aid in the care of dependent offspring (i.e., their siblings) and territory defense at the cost of their own reproduction (de Sousa et al., 2009). Encounters with neighboring groups (often as a result of territorial confrontations) may offer the opportunity for adult offspring within the group to assess breeding and dispersal opportunities with nonnatal group members (Lazaro-Perea, 2001; Arruda et al., 2005). Generally, female common marmosets tend to emigrate more often than males, and this has been hypothesized to be due to social tension within the group (de Sousa et al., 2009). Subordinate females may begin to experience increased social competition (and therefore, increased stress) with the breeding female, primarily over limited food resources. However, Ferrari (2009b) reported that emigration in *C. flaviceps* was unrelated to resource competition and intragroup aggression, and suggests that emigration seems to be driven primarily by the availability of potential mates.

In addition to immigration, groups may include nonkin due to the fusion of two unrelated groups following the death or disappearance of a breeder (Lazaro-Perea et al., 2000). Additionally, groups may fission along sexual lines, wherein members from a natal group form two new groups, one composed of males and one composed of the females from the original group (Lazaro-Perea et al., 2000). Those new groups may then migrate to a new territory, immigrate into a different, existing group, or another male or female may immigrate into their group (Lazaro-Perea et al., 2000). Sociality among marmosets is characterized by high interindividual tolerance of one another, with strong bonds existing among members of the family. Because of these traits, phenomena, such as active and consistent involvement of all individuals in infant care, efficient and widespread transfer of information within the group, and considerable food sharing with immatures, are common in marmoset groups (Caldwell and Whiten, 2003).

Feeding Behavior

Marmoset species are omnivorous, frugivorous-insectivorous, or gummivorous-insectivorous, with diets consisting of a wide range of food items (e.g., fruits, exudates, seeds, flowers, insects, small frogs and lizards, and bird eggs and nestlings; Stevenson and Rylands, 1988). Common marmosets have also been observed to consume leaves, bromeliads, and cactus fruits (Amora et al., 2013; Abreu et al., 2016). Wild common marmosets spend most of their daily activity budget engaged in foraging activities, regardless of weather conditions (Menárd et al., 2013; Abreu et al., 2016).

Foraging patterns vary across marmoset species, since seasonal changes and territorial characteristics influence diet composition with respect to the consumption of fruit, insects, and flowers. Reports on marmoset nutrition in the wild found that for *C. aurita*, the animal part of the diet is almost entirely invertebrate, while for *C. flaviceps*, a species that is predominantly mycophagous-insectivorus, fungi are the preferred type of food (Corrêa et al., 2000, Hilário and Ferrari, 2010). *C. jacchus* and *C. penicillata* spend a great deal of time-consuming exudates. They actively gouge holes in tree trunks to elicit gum flow, a foraging activity that does not seem to be affected by season or habitat differences. Exudates represent an important source of carbohydrates, proteins, minerals, and water (Francisco et al., 2014). In common marmosets, exudate extraction represents about 40% of daily food intake and is made possible through specific adaptations, including the shape of their teeth and jaw, and their ability to cling vertically on tree trunks using their claw-shaped nails (Stevenson and Rylands, 1988). *C. aurita* has been observed to feed on *Acacia* branches, but this species fails to gouge holes that cause gum exudation, foraging opportunistically on gum exuded after wind breakage or insect activity, instead (Martins and Setz, 2000). Consumption of gum and live prey is generally inversely correlated with fruit availability; when fruits are available, they are typically the preferred resource, given the energetic costs associated with gummivory and prey capture. The wide variety of food items consumed by marmosets, and their capacity to modify their diet on a seasonal basis, suggest that they are a generalist species, with considerable plasticity and the ability to cope with potentially difficult situations, such as fragmented and human-influenced environments.

In addition to foraging, marmosets are predators that search for and eat insects (e.g., crickets and grasshoppers), spiders, snails, small amphibians, lizards, and even birds (Stevenson and Rylands, 1988). Predatory behavior appears to be similar in wild and captive marmosets, consisting mostly of both wide and narrow scanning of the ground and the surrounding environment. These scans are regularly punctuated by short bursts of locomotion followed by a pounce to catch small, quick, and often cryptic, prey (Rylands and de Faria, 1993; Caine, 1996; Souto et al., 2007). Visual searching requires considerable concentration, resulting in substantial vulnerability to predation. Schiel and colleagues (2010) report four distinct age-dependent and prey-specific hunting strategies for common marmosets, suggesting that the hunting methodology applied depends on the prey and the hunter's age-related ability to choose the correct strategy.

Similar to many other NWMs, marmosets possess visual polymorphism, with dichromats and trichromats present within the species; all males are dichromats and females can be either dichromats or trichromats (Jacobs et al., 1987; Moreira et al., 2015). Marmosets (and tamarins) present six separate color vision phenotypes including three types of trichromacy and three types of dichromacy, each appearing to have different color-discriminative abilities (Pessoa et al., 2005). The assumption is commonly made that trichromats have an advantage in foraging for ripe fruits (since they can see red), whereas dichromats are more efficient at detecting camouflaged prey (e.g., insects) and foraging in low-light conditions (e.g., in the shade) (Morgan et al., 1992; Caine and Mundy, 2000; Caine et al., 2010).

Mating and Reproduction

For a long time, it was thought that the mating system of these primates was most correctly described as monogamy. More current evidence suggests that while there is usually an identifiable breeding pair consisting of one adult male and one adult female, sexually and/or reproductively active group members other than the breeding pair may be observed (Saltzman, 2004). Dominant/breeder marmosets are able to behaviorally and chemically suppress the reproduction of both subordinate females and males, although the different species vary in the degree to which this suppression is manifested (Abbott, 1984; Abbott et al., 2003). In *C. jacchus*, the breeding female inhibits the reproduction of subordinate females primarily by urinating and rubbing her perianal glands on surrounding environmental features (e.g., trees, sleeping sites, fruit) (Abbot et al., 1990; Ziegler, 2013). Interestingly, despite reproductive suppression via this method, the dominant female does not necessarily engage in scent-marking more than subordinate females (Lazaro-Perea et al., 1999). However, *C. jacchus* ovulation suppression does not seem to be entirely effective, as conceptions and births by subordinate females may occur (Yamamoto et al., 2009). Long-term field studies suggest that polygyny may not be advantageous for subordinate females, due to high infant mortality (mostly for unknown causes, but an infanticide case by a dominant female has been reported; Bezerra et al., 2007). Nevertheless, it is not uncommon to observe copulations between subordinate *C. jacchus* and a dominant individual, when the other partner is absent (Manciocco, pers. obs.), and free-ranging subordinate females have also been observed to copulate with extra-group males (Arruda et al., 2005). Furthermore, although ovulation suppression of subordinate females is almost universal in captive *peer* groups (Abbott, 1984), eldest daughters have been reported to ovulate in captive *family* groups (Saltzman et al., 1997). It is thought that this is mainly to avoid incest, rather than due to rank-related suppression imposed by the dominant male (typically the father). Nondominant males appear behaviorally and physiologically reproductively competent and do copulate with unrelated females when the possibility occurs (e.g., during encounters with nonnatal groups/extra-group females; Baker et al., 1999; Lazaro-Perea, 2001). However, there is some evidence of physiological suppression of their reproductive activity by the dominant male, given the decrease in production of sperm cells by subordinates (Abbott, 1984). Thus, the breeding behavior observed between related individuals in laboratories and zoos would seem to be a function of captivity, due to the lack of unrelated females in the social group and the inability of males to emigrate. As a consequence, mother–son or father–daughter matings may be observed, following the "disappearance" of the other breeding adult (Anzenberger and Simmen, 1987).

Marmosets give birth most frequently to twins or triplets, with triplets more frequent in captivity than in the wild. Females display postpartum estrus, which may result in conception and the production of 2 litters per year. Twin offspring increase the overall costs of infant care, potentially resulting in increased reproductive competition among females (Digby, 1995) and large roles for other group members in caring for the young. The responsibilities for infant care are distributed via a cooperative breeding system, in which all group members (helpers) assist with the care of neonates, primarily by carrying them (Ferrari, 1992). Neonates are transferred between two monkeys when they come into close proximity to one another, and the infant(s) safely moves from the back of one to the back of the other. In common marmosets, the breeding male is the individual most involved in the initial care of the offspring (other than nursing) in the days immediately after birth. In fact, interactions between the mother and her infant(s) are almost exclusively limited to nursing and stimulation of neonate defecation (Rothe et al., 1993). These behavioral patterns are consistent in both captive and wild populations. Father involvement does appear to be more conspicuous in captivity than in the wild, but this small difference could be due to larger group sizes (and more helpers) in wild populations, or as a result of paternity certainty in captive populations with no male–male competition (Yamamoto et al., 1996).

Approximately 1 month after birth, the other adult and immature individuals (older siblings) in the group share in the carrying of neonates during daily activities, as well as in the introduction of solid food into the diet (Rothe et al., 1993). When infants start to eat solid food, they have been observed to actively pull food items out of the mouth of an adult, or solicit food transfer by vocalizations. This phenomenon, known as food sharing, highlights the heightened social cooperation and tolerance among marmosets (Rothe et al., 1993). Food sharing is hypothesized to supplement offspring nutrition, provides information to infants about novel foods, and increases infant survival rates, since the development of foraging skills in marmoset infants is relatively slow compared to other nonhuman primates (Snowdon and Ziegler, 2007). Marked tolerance for one another during feeding makes social learning an important factor in the life of these animals, primarily in the early stages of development. During this time, infants are carried on the back of adults and occupy an advantageous position to observe, maintain social interactions, and learn. The close proximity of "demonstrator" and "learner" during interesting activities allows knowledge to be transferred both passively and actively (Heyes, 1994). By observing the foraging behaviors of adults, infant and juvenile marmosets may reduce the risks associated with eating unpalatable or toxic foods. Infant *C. jacchus* attend to the feeding behaviors of older individuals, especially dominant females, although this attentiveness diminishes as they themselves age (Schiel and Huber, 2006). For helpers, successful alloparental care seems to be a learned process that improves with practice, while simultaneously enhancing (1) the helpers' probability of succeeding as a future breeder and (2) the infants' chances of survival (Tardif et al., 1984).

Communication. Marmosets communicate using acoustic, visual, olfactory (chemical), and tactile signals (Stevenson and Poole, 1976). Vocalizations comprise an important communication modality for arboreal nonhuman primates, such as marmosets, that inhabit forests where visibility is often limited, and that live in social systems in which coordination among individuals is fundamental. Marmosets have a complex repertoire of vocalizations that are context- and age-specific (Pistorio et al., 2006), with consistent interspecific variability in the acoustic morphology of calls (Snowdon, 1993). At least 13 distinct vocalizations have been identified in wild common marmosets (Bezerra and Souto, 2008). For example, contact calls include both low- and high-frequency vocalizations that are emitted depending on whether the signal needs to cover a long or a short distance. These calls (e.g., long or *phee* call, trill call) are used in a variety of social contexts (Bezerra and Souto, 2008), including maintaining contact among group members as they move across the forest. *Phee* calls are also used by *C. jacchus* for territorial defense and mate attraction (Snowdon, 1993). The degradation of the acoustic signal permits the receiver to evaluate the distance of the caller, and characteristics, such as call frequency and temporal parameters (e.g., duration of call), allow the receiver to recognize the identity of the caller (de la Torre and Snowdon, 2002). Several studies have shown that marmoset *phee* or long calls contain an individual vocal signature that is relatively stable over time, yet still includes natural and context-specific variation (Jones et al., 1993; Rukstalis et al., 2003; Miller and Thomas, 2012). Individual specificity and variability of calls may provide ecological advantages. In the wild, the ability to recognize the identity of the caller provides information on neighboring groups and opportunities for emigration, and may allow evaluations of familiar and unfamiliar animals, costs of potential fights and injuries, and mating opportunities. Additionally, alterations to the acoustic structure of *phee* vocalizations, specifically, may be associated with stress-related stimuli (e.g., social isolation) and may serve as a shared signal to facilitate new social groupings. Across marmoset species, alarm (*tsik*) calls, and associated mobbing behavior, are used to alert groupmates to the presence of danger, and the calls have been found to remain stable over time, permitting the identification of the caller, even if changes in the "voice" of individuals have occurred (Mulholland and Caine, 2019). Call structures and rates progressively change as marmosets age, with adults vocalizing more than juveniles and infants (Bezerra et al., 2009b).

Changes of body posture, gestures, and facial expression represent another modality of communication for marmosets. Unlike vocal communication, communication on the visual channel requires two or more actors to be in proximity to one another, so that individuals can observe changes in the face and/or body of conspecifics, and adjust their behavior accordingly. Even though their small body size and poor visibility in their arboreal environments limit the effectiveness of visual communication, they still possess a diverse repertoire of body and facial expressions (see www.marmosetcare.com for detailed ethograms). They possess specialized facial muscles and well-developed visual pathways in the nervous system, making them capable of performing and perceiving a variety of visual displays (Burrows, 2008). These include movements of colored ear tufts, pilo-erection of the fur and/or tail, tongue-flicking, arched back displays, and side-to-side movements (Stevenson and Poole, 1976). Social interactions, including aggression, play, fear, and mating, are commonly associated with visual displays in marmosets (de Boer et al., 2013).

Marmosets have a well-developed epidermal scent gland system (circumgenital glands, anogenital glands, sternal glands) and elaborate scent-marking behaviors, suggesting that chemical signals play a vital role in marmoset communication. Scent-marking is an important activity for both wild and captive marmosets, as it functions to mark food sources, territorial boundaries, and serves as intra- and intergroup communication (Lazaro-Perea et al., 1999; Lucio Nogueira et al., 2001). For example, females of *C. penicillata* scent mark more along buffer zones of their home ranges than in their core areas (Oliveira and Macedo, 2010). In addition, male common marmosets distinguish between the scents of ovulatory and nonovulatory females, showing different behavioral and physiological responses to each and, in fact, appear to be able to identify females individually (Smith et al., 2001; Ziegler et al., 2005; Smith, 2006). Furthermore, in *C. jacchus*, scent-marking has been correlated with feeding activity; there are two peaks, one observed at the beginning and one at the end of the daily light period, coinciding with increased foraging activities in the early morning and shortly before sleep (De Sousa et al., 2006).

Finally, tactile communication is used in a variety of contexts. Some have postulated that in *C. jacchus*, allogrooming may be a strategy used by the dominant female to encourage other females to remain in the group and assist with offspring care, rather than a strategy aimed at increasing social tolerance or reducing aggressive behavior (Lazaro-Perea et al., 2004, Campennì et al., 2015). Marmosets also engage in face nuzzling, a social greeting behavior that consists of rubbing the muzzle against a groupmate (Stevenson and Poole, 1976). Licking is also a social greeting behavior between juveniles and adults, distinguished from licking of infants, which all groupmates tend to do for hygienic purposes (Stevenson and Poole, 1976). Biting also serves as tactile communication. The intensity and context of the biting communicate different messages, with inhibited biting on various parts of the body often occurring during play, and snap biting occurring when an adult scolds a younger individual (Stevenson and Poole, 1976).

Common Captive Behaviors

Normal Behaviors

With respect to the number and variety of behaviors, captive and wild marmosets generally exhibit similar behavioral repertoires, especially when captive animals are provided with outdoor enclosures and appropriate environmental enrichment. The complexities associated with seasonal changes in the natural environment influence the amount of time marmosets spend in behaviors, such as searching for, gathering, and consuming food resources, avoiding predators, interacting with neighboring groups, and finding nesting sites. Seasonal variations in the availability of edible fruit affect time spent traveling, resting, and socializing, as well as time spent feeding on specific plants. For a detailed ethogram of captive common marmoset behavior, see Stevenson and Poole (1976 and www.marmosetcare.com).

Feeding. Foraging includes not only consumption of the food item, but activities such as locomotion to reach the resource and processing the food items as well. Given that wild common marmosets spend most of their daily activity budget engaged in foraging activities, regardless of weather conditions (Abreu et al., 2016), it is important that captive marmosets are given the opportunity to forage for food in order to stimulate species-typical behavior and time budgets.

Resting. In the wild, time spent resting may be influenced by several factors, including distribution of food items, antipredator strategies, and ambient temperature (Stevenson and Rylands, 1988; Ferrari and Rylands, 1994; Abreu et al., 2016). High levels of inactivity may be observed in marmosets housed in barren captive environments, likely due to boredom and/or low motivation to display species-typical behaviors (Kitchen and Martin, 1996). Manciocco and colleagues (2009) have shown that the provision of environmental stimulation reduces time spent inactive and increases time spent in positive social activities, such as play and grooming behavior. At the same time, improved husbandry practices may result in increased resting behavior, perhaps indicating increased relaxation (Badihi, 2006). These disparate findings related to time spent resting demonstrate that data on a variety of behaviors must be collected to accurately assess marmoset behavior in captivity.

Play. Play, consisting of chasing, wrestling, play biting, and particular vocalizations, most often occurs between siblings, particularly between twins (Stevenson and Poole, 1982). Play is a common expression of positive well-being, and play is often decreased when levels of stress are high (Biben and Champoux, 1999; Mellou, 2006). Therefore, play in captive settings should be regarded as indicative of positive psychological well-being.

Antipredator Behavior. Although marmosets may not experience predators in the captive setting, some, but perhaps not all, antipredator behaviors are maintained. Fear responses that normally occur during confrontation with a predator, including alarm calls, piloerection, freezing, and withdrawal, may occur in research contexts, such as during experimental procedures, capture, restraint, and in response to particular researchers or caregivers (Caine, 2017). Additionally, captive marmosets, although they are not confronted with predators on a regular basis (or at all), still prefer to sleep in elevated, concealed areas.

Grooming. Social grooming (or allogrooming) comprises approximately 15% of the activity budget of common marmosets (Lazaro-Perea et al., 2004). However, social grooming is most common between breeding individuals, with the dominant male usually the most active groomer in the group (Lazaro-Perea et al., 2004). Autogrooming (or self-grooming) is a natural self-care activity. However, as described below, self-grooming at a higher frequency than that which is found in the wild can indicate welfare issues (Olsson et al., 2018). Similarly, a lack of self-grooming, resulting in a disheveled appearance, may indicate detriments in health or welfare (Neal Webb, pers. obs.).

Exploration. Exploration is generally considered a positive behavior for many nonhuman primates, including marmosets (Chamove and Anderson, 1989). Indeed, head cocking in marmosets is an expression of exploration (allowing them to inspect objects from multiple angles), and is considered an indicator of positive welfare (Kaplan and Rogers, 2006). Marmosets that were administered an antianxiety drug increased exploration of a novel environment, suggesting that increased exploration is likely indicative of lower levels of stress (Barros et al., 2001). Similarly, marmoset infants that received less anogenital licking from their mothers (a behavior that lowers infant stress levels) tended to perform less novel object exploration later in life (Kaplan and Rogers, 1999). In the wild, exploration is likely an important activity, as marmosets explore potential foods, objects, and animals within their home range. Increased environmental complexity and enrichment (discussed below) is important in stimulating such exploratory behaviors in captive marmosets (Badihi, 2006).

Abnormal and Fear-Related Behaviors

The presence of a single specific behavior is not sufficient to conclude that a marmoset is experiencing stressful conditions. The context in which the behaviors are observed, their frequency, and the performance of additional undesirable behaviors are all important factors that are necessary to determine whether animals are behaving abnormally or not. Indeed, many "abnormal" behaviors are simply normal behaviors that are performed at different frequencies than those typically observed in the wild (Olsson et al., 2018; Novak and Meyer, this volume). As such, many of the normal behaviors listed above may become abnormal if they are performed at "abnormal" frequencies or in an "abnormal" form.

Behaviors, such as scratching and self-grooming, are normal ("natural") activities. However, if expressed at unnatural frequencies/durations, or if they appear out of context or to function as displacement activities, they may be indicative of reduced welfare. Observation of a combination of these behaviors would be considered a reliable indicator of reduced welfare. Scratching, specifically, has often been used as an indicator of anxiety in marmosets, as studies have found increases in scratching rates under anxiety-provoking circumstances (Barros et al., 2000). However, recent evidence suggests that scratching does not always increase with arousal, and scratching may increase under positively arousing circumstances, such as play (Cagni et al., 2009; Kato et al., 2014; Neal and Caine, 2016). As such, the contexts in which scratching (and other "abnormal" behaviors) occur must be taken into consideration to determine an association with negative welfare.

Slit-stare, anogenital presentation, pilo-erection, and *tsik* and geckering vocalizations are considered to be normal aggressive or alert displays, and freezing behavior (described as the animal remaining motionless, with ear-tufts down and eyes semi-closed, for long periods) is unequivocally an indicator of fear and anxiety in marmosets (Barros et al., 2000). Each of these can be considered anxiety/fear-related activities as well. Specifically, if these behaviors are exhibited in response to certain situations in captivity (e.g., freezing or alarm calling in response to a particular caregiver or husbandry routine), the situation should be addressed to reduce fear and anxiety for the individual. Scent-marking, a normal behavior with multiple functions, increases following stressful husbandry events, and shows dose-dependent increases and decreases following administration of anxiogenic and anxiolytic drugs, respectively (Cilia and Piper, 1997; Barros et al., 2001; Bassett et al., 2003). Similarly, scent-marking is absent in captive marmosets with no visual contact with neighboring marmoset groups (Sutcliffe and Poole, 1978). As such, scent-marking seems to be linked to stress, anxiety, and social tension (Badihi, 2006).

Stereotyped motor behaviors, such as repetitive and unvarying locomotion, and repetitive jumping, have long been considered indicators of diminished well-being in captive marmosets (Kitchen and Martin, 1996). These are behaviors that differ from the behavior of their wild counterparts in form and/or intensity (Olsson et al., 2018). Such stereotypic motion is hypothesized to derive from aversive living conditions, including limited physical space, limited social or physical stimulation, or frustration (Olsson et al., 2018). However, recent evidence in other nonhuman primates (e.g., rhesus macaques) suggests that individuals use stereotypic behaviors as a coping mechanism, and individuals that engage in stereotypic behavior in the most aversive environments have higher welfare than those who do not engage in such behaviors (Mason and Mench, 1997; Mason and Latham, 2004; Pomerantz et al., 2012; Olsson et al., 2018).

Ways to Maintain Behavioral Health in Captivity

It is important that marmoset groups are monitored daily by qualified caregivers to identify potential health and behavioral problems. Standardized check sheets that can be easily and rapidly filled in, and that include information on some of the main welfare indicators (e.g., coat condition, locomotion, eating, signs of nasal discharge, diarrhea), should be utilized.

Environmental Complexity

One indicator of good behavioral health is the ability for animals to display many/most/all behaviors in their natural repertoire (Suomi and Novak, 1991; Tasker and Buchanan-Smith, 2016; Bettinger et al., 2017). Through proper management (veterinary, husbandry, and behavioral) programs, it is possible to provide environmental complexity and cognitive stimulation that promote the expression of species-typical behaviors (Buchanan-Smith, 2011; Williams and Ross, 2017). Currently, it is well-known that the total **quantity** of space afforded to nonhuman primates is less important than the **quality** of that space (Reinhardt et al., 1996; Young, 2003; Hosey, 2005). Functionally appropriate captive environments (see Schapiro, this volume) for marmosets must be furnished with at least horizontal, vertical, and diagonal "branches", platforms, ropes, nest boxes, movable structures, and related apparatus to optimize the quality of the usable space in the environment and stimulate the expression of multiple components of the species-typical behavioral repertoire (see Figure 24.1a and b). Furthermore, at least some of these structures should be made of a material that allows for and encourages scent-marking and gouging (such as actual branches or wood) and should not go through the cage wash as frequently as other items.

All attempts at environmental enrichment should be based on an understanding of the biology and behavior of the target taxonomic group; in this case marmosets (Williams and Ross, 2017). As an example, understanding the vocal and body gesture communication signals of captive marmosets will improve our management of them, as knowledge concerning these signals allows us to promptly assess whether any welfare problems exist. Additionally, knowledge of the personality traits and the life experiences of each individual animal by care staff will permit the optimization of all management strategies, including environmental enrichment. Therefore, it is strongly recommended that, when possible, caregivers be given the opportunity to "learn about" their marmosets.

As mentioned above, the provision of physical furnishings in the areas inhabited by marmosets functionally simulates their natural environments, allowing them to exhibit locomotion patterns similar to those in the wild. Appropriate environmental enrichment plans include foraging devices that provide marmosets with opportunities to exhibit natural behavior patterns, such as tree gouging, predatory strategies, searching for and gathering fruits, and additional cognitive challenges to obtain food (see Figure 24.2). Furthermore, the presence of plants (carefully chosen to include only those that are not toxic, e.g., bamboo) in outdoor areas, both stimulate marmosets to exhibit multiple aspects of their locomotor repertoire and give them the opportunity to choose to be out of view of conspecifics or people when desired (see Figure 24.3).

For social animals such as monkeys, opportunities to exhibit social behaviors are important components of functionally appropriate captive environments, and the performance of social behaviors is an indicator of well-being, both at the individual and the group level. In captivity, aspects of the surrounding environment influence the frequency and duration of social behaviors (e.g., play, allo-grooming, affiliative behavior, agonistic displays). Maintaining multiple groups of marmosets in adjacent cages or enclosures, so that visual (but not physical) contact among groups is possible, permits animals to display additional components of their behavioral repertoire, including agonistic displays (e.g., pilo-erection, mobbing movements and calls, ear-tuft retraction, ano-genital glands display). However, this may lead to additional stress. As such,

FIGURE 24.1 An adult male (a) and young (b) common marmoset in an enriched laboratory cage. (Photos: Arianna Manciocco.)

FIGURE 24.2 Adult common marmosets engaged with a challenging food device. (Photo: Arianna Manciocco.)

if groups are housed in visual contact with one another, then the housing should include a partial visual barrier (e.g., hedge, curtain) that allows them to choose whether or not they are in visual contact with their neighbors (Williams and Ross, 2017). The capacity to control and modify aspects of their environment by their own behavior represents an important requirement for captive marmosets (Basset and Buchanan-Smith, 2007; Buchanan-Smith and Badihi, 2012). Increased complexity within the environment inherently provides individuals with increased choice, which enhances their welfare (Badihi, 2006). For example, marmosets that were moved from a cage to a larger, more complex room (thus representing an increase in available choices) exhibited increased exploration and play behavior (Badihi, 2006).

There is one particularly important item that should be included in all captive marmoset housing: a nest box (Layne and Power, 2003). Nest boxes allow individuals to hide from conspecifics or human view and are used as sleeping sites (e.g., the whole family group will snuggle together and sleep inside the box). Recall that wild marmosets often choose elevated sleeping sites that are hidden from predator view. Indeed, Caine et al. (1992) found that tamarins (a species group that is closely-related to marmosets) preferred nest boxes that provided the most concealment. That is, they preferred the nest box with just one opening in the front, compared to boxes with an opening at the top and those that were not fully enclosed. Additionally, they consistently preferred the nest box at the higher height. Depending on the size of the marmoset group, these nest boxes can be made from gallon plastic jugs (for a singly housed marmoset) or 5-gallon plastic tubs (for larger family groups), both of which allow for easy cleaning.

Social Groupings

A compatible social partner is the single most important factor for promoting the welfare of captive primates, and this is especially true for monogamous marmoset species (Williams and Ross, 2017). Social isolation during early development adversely affects communication skills in marmosets, with important repercussions for the individual's quality of life. In these species, learning plays a central role in vocal development, and the lack of adult feedback during development causes later disability in "conversational" skills (Chow et al., 2015). As mentioned previously, marmoset groups are typically composed of related individuals. Captive common

FIGURE 24.3 Examples of dense vegetation in outdoor enclosures: (a) the Falconara Zoo (Italy) (Photo: Arianna Manciocco) and (b) research facility affiliated with California State University San Marcos (Valley Center, California).

marmosets spend a considerable amount of time in contact with one another in stationary and relaxed positions, an important intragroup affiliative behavior (Finkenwirth et al., 2015). Wild *C. jacchus* have been observed embracing a long-term partner, demonstrating the strong bond that can exist between individuals (Bezerra et al., 2014). Family member dyads with strong bonds also show synchronized changes in urinary oxytocin levels, again demonstrating the strong bonds that exist among group members (Finkenwirth et al., 2015).

It would not be surprising if embracing and other physical contact resulted in an increase in well-being for the receiver through the release of beta-endorphins, as has been reported for other species of monkeys when receiving social grooming (Keverne et al., 1989).

As indicated above, marmosets should be housed socially, at least in pairs, but preferably in breeding pairs or family groups (Williams and Ross, 2017). If this is not possible, nonbreeding pairings may be a short-term alternative. While housing nonbreeding individuals together can be challenging, it is not impossible to pair same-sex individuals or siblings together (Williams and Ross, 2017). For example, two adult males (both ex-breeding males) were successfully housed together for several months, after all of their former groupmates were lost due to a number of different, and unlucky, events (Manciocco, pers. obs.). However, male–male pairs may begin to exhibit aggression toward one another, if a cycling female is present in the room (Williams and Ross, 2017). Regardless of sex, it should be noted that these nonbreeding pairings are likely to be successful for only a limited time and, as such, are not good long-term housing option (Majolo et al., 2003; Williams and Ross, 2017). Single-housing of marmosets must be limited to experimental, veterinary, or specific husbandry situations, and only for the minimum required time. There are two situations in which it might be appropriate to house a marmoset alone; (1) to allow safe recovery following substance administration, surgical procedure, or injury; and (2) to prevent repeated attacks toward a specific individual by the group (see below for details). The negative effects of single housing can be somewhat mitigated by housing the individual within visual and auditory contact of conspecifics (Layne and Power, 2003); however, the lack of physical contact can still lead to compromised welfare for that animal (Smith et al., 2011).

In addition to social housing, social enrichment can also be achieved by promoting human–marmoset interaction (Bassett et al., 2003; Manciocco et al., 2009). Habituation to the presence of humans improves the overall quality of life for laboratory marmosets (Rennie and Buchanan-Smith, 2006; Manciocco et al., 2009). When the presence of familiar people does not cause fear in marmosets, they frequent every zone of their enclosures, including those in proximity to people and those located on the ground (Rennie and Buchanan-smith, 2006). Furthermore, in cases where it is difficult to introduce a formerly singly housed subject to a new group, caregiver interactions (e.g., offering preferred food, softly talking with the individual, gentle grooming) may be a useful strategy to minimize the negative effects of isolation on animal health (Manciocco, pers. obs., see Figure 24.4a and b).

Feeding

Captive marmoset diets must address all of the nutritional needs of these species and encourage normal feeding behaviors. Commercial marmoset diets are available and are typically supplemented with fresh fruits. Additionally, foods rich in protein, such as live prey (mealworms, crickets), boiled eggs, and/or yogurt, should be periodically offered. Furthermore, as marmosets in captivity may suffer from vitamin D3 deficiency, which is believed to be related to some diseases (e.g., wasting syndrome, metabolic bone disease; Layne and Power 2003), it is critical that captive diets consistently supply the proper amount of vitamin D3 (contained in most commercial marmoset diets).

In captivity, the time spent in foraging behaviors is typically reduced, at least in part, due to the decreased dimensions of laboratory or zoo enclosures (compared to natural home ranges), even if animals are provided with outdoor access (Kitchen and Martin, 1996). Well-designed environmental enhancement plans (EEP) can help increase travel time between food sources (e.g., by heterogeneous distribution of food across the enclosure, inclusion of foraging apparatus), but are unlikely to exactly match natural conditions. Consumption of food in the wild requires marmosets to process fruits, hunt prey, gouge tree trunks, etc. Many of these activities can be functionally simulated in captivity as part of the EEP. For example, the provision

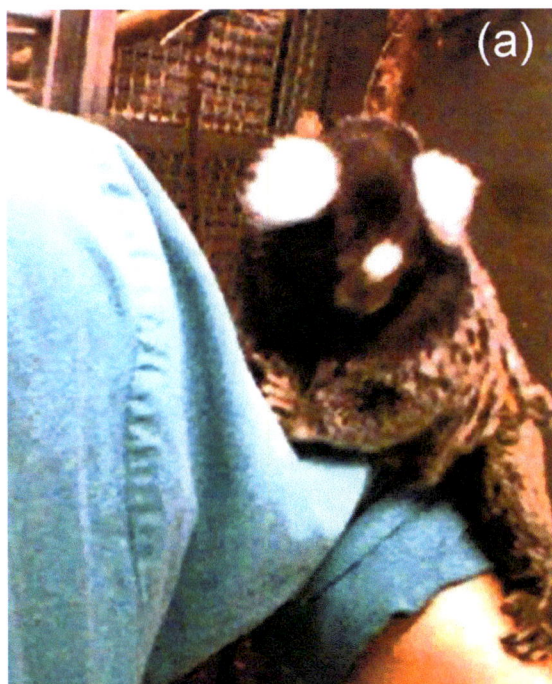 

FIGURE 24.4 Two examples of positive interactions between marmosets and caretakers. Such interactions can be rewarding for both parties. (Photos: Arianna Manciocco.)

of live prey can stimulate the animals to scan their environment, while slowly locomoting/searching for prey, resulting in rapid startle/attack behaviors when prey has been located. Given that marmosets are also gummivores (recall that approximately 40% of wild marmoset daily food intake is through exudate extraction), an important component of environmental enrichment is providing the opportunity for individuals to engage in tree-gouging behavior. This can be achieved by offering fresh-cut branches, or by spreading gum arabic on existing branches or enrichment devices (Williams and Ross, 2017).

Training

A positive, collaborative, and fear-free relationship between humans and monkeys can be established and maintained through positive reinforcement training (PRT). PRT allows one to perform routine husbandry (e.g., moving animals for cage cleaning) and experimental (e.g., blood, urinary, salivary collection, capturing) procedures while minimizing the restraint, handling, and stress experienced by the marmosets (McKinley et al., 2003; Williams and Ross, 2017). It has been reported that training may take longer for marmosets than other primate species (e.g., chimpanzees, macaques), but that daily training sessions are effective (Spinelli et al., 2004; Stevens et al., 2005). Nevertheless, once marmosets have been successfully trained to perform a behavior, the learning of additional behaviors typically proceeds quickly (Spinelli et al., 2004; Stevens et al., 2005; Williams and Ross, 2017). Aggressive, dominant, or aged marmosets may be more difficult to train, and therefore may require additional time or alternate training strategies. However, marmosets present unique advantages when it comes to training, as the close nature of the social group decreases chances of aggression during training, allows caregivers to train in the home cage within the social group, and facilitates social learning (Williams and Ross, 2017). As is the case with most PRT, a positive relationship between caregiver and animal is fundamental for the success of PRT. Establishing this relationship requires that caregivers have an adequate amount of time to spend with animals outside of training sessions, during training sessions, and during maintenance phases of training (Rennie and Buchanan-Smith, 2003). To date, marmosets have been successfully trained to move between different cages to be weighed or transported, for urine and saliva collection (Cross et al., 2004; McKinley et al., 2003), and for hand-feeding while being palpated (Savastano et al., 2003). There are practical limitations to what can be trained in these species; invasive procedures, such as venipuncture for blood samples, can be difficult given the animals' small body size.

Special Situations

Care of Aged Animals

The presence of elderly individuals (older than 10 years) in marmoset social groups requires special care and monitoring by all staff (veterinarians, caregivers, researchers, etc.). Enclosures should be furnished in a way that accounts for the impaired mobility of elderly individuals; platforms, branches, and especially nest boxes and foraging sites should be located so that older animals have easy access to them. Food should be distributed to minimize interanimal competition, and particular attention should be paid to maintaining proper temperature and humidity conditions, as thermoregulatory abilities decrease as animals approach old age (see Bridges et al., 2015, 2016, for example, in chimpanzees).

Social Problems

It is not rare in captive conditions to witness repeated and increasingly aggressive attacks by one or more individuals against a specific family member. This phenomenon is distinguishable from "normal" aggression by several characteristics: (1) there appears to be no specific reason for the attacks; and (2) once an initial aggressive episode has occurred, additional increasingly severe attacks follow (Kitchen and Martin, 1996; De Filippis et al., 2009). Furthermore, there is a negative correlation between cage size and complexity, and aggression levels (Kitchen and Martin, 1996). Both males and females of any age can be the aggressors, but aggression most often involves siblings or twins (De Filippis et al., 2009). The prompt removal of the victim from the group is the best solution, and it is essential that such situations are quickly recognized and interventions occur rapidly, as in a short time frame, the severity of attacks against the victim can increase. Additional group mates may become aggressors and the death of the rejected individual may result. Removing the aggressor, rather than the victim, is not a good solution, as the victim may soon be attacked by other members of the group (Manciocco, pers. obs.; Williams and Ross, 2017). To reduce the impacts of single housing on the victim while working toward a long-term solution, we have had success rotating the aggressor and victim in and out of the family group on a bi-weekly schedule (Neal Webb and Mulholland, pers. obs).

Wasting Syndrome

Marmoset Wasting Syndrome (MWS) is a disease that causes morbidity and mortality in captive colonies (Ludlage and Mansfield, 2003). Currently, there is disagreement on the etiology of, and treatment for, MWS (Ludlage and Mansfield, 2003). Gluten sensitivity has been proposed as a potential cause and gluten-free diets have been introduced (Kuehnel et al., 2013). MWS has also been linked to infestation with *Trichospirura leptostoma* and with stress (Ludlage and Mansfield, 2003; Beglinger et al., 1988; Cabana et al., 2018). Weight loss, alopecia, diarrhea, hindlimb paresis, and paralysis are the clinical signs typically associated with MWS (Ludlage and Mansfield, 2003). Histologically, postmortem examinations report a lymphoplasmacytic infiltration of the intestines and numerous gastroenteric lesions (Chalifoux et al., 1982). Additionally, low serum albumin has been reported. Therapy with glucocorticoids, such as budesonide, has been effective in ameliorating symptoms in marmosets affected by chronic (i.e., long-developing) MWS, but no positive effects were observed in marmosets with acute onset of the disease (i.e., indicated by sudden and severe symptoms) (Otovic et al., 2015).

Conclusions

To optimize the welfare of captive marmosets, one must pay attention to the quality of the housing they live in, the management practices that influence their daily lives, the refinements to those procedures, the EEP, and the monitoring practices used to assess their behavioral health. Such an approach addresses ethical and scientific concerns, as it is our moral obligation to strive to minimize the stress experienced by captive animals, thereby enhancing the validity and reliability of the scientific data collected.

As reported throughout this chapter, marmoset monkeys show many interspecific differences, ranging from ecological (e.g., foraging strategies) to cognitive (e.g., learning capacity) factors. While common marmosets represent the species that is most widespread in research laboratories, other marmoset species are broadly represented in zoos. Therefore, it is essential for the improvement of the care and management of captive populations that additional data on the natural behavior, ecology, and sociobiology of different species is collected in the wild.

REFERENCES

Abbott, D. H., Barnett, D. K., Colman, R. J., Yamamoto, M. E., Schultz-Darken, N. J. 2003. Aspects of common marmoset basic biology and life history important for biomedical research. *Comparative Medicine* 53: 339–350.

Abbott, D. H. 1984. Behavioral and physiological suppression of fertility in subordinate marmoset monkeys. *American Journal of Primatology* 6: 169–186.

Abbott, D. H., George, L. M., Barrett, J., Hodges, J. K., O'byrne, K. T., Sheffield, J. W.,... & Lunn, S. F. 1990. Social control of ovulation in marmoset monkeys: a neuroendocrine basis for the study of infertility. In T. E. Ziegler & F. B. Bercovitch (Ed.), *Monographs in Primatology, Vol. 13. Socioendocrinology of Primate Reproduction* (pp. 135–158). Wiley-Liss: New York.

Abreu, F., De La Fuente, M.F.C., Schiel, N., Souto, A. 2016. Feeding ecology and behavioral adjustments: Flexibility of a small neotropical primate (*Callithrix jacchus*) to survive in a semiarid environment. *Mammal Research* 1: 1–9.

Adriani, W., Romani, C., Manciocco, A., Vitale, A., Laviola, G. 2013. Individual differences in choice (in) flexibility but not impulsivity in the common marmoset: An automated, operant-behavior choice task. *Behavioural Brain Research* 256: 554–563.

Albuquerque, N. M., Silvestre, S. M., Cardoso, T. S., et al. 2014. Capture of a common marmoset (*Callithrix jacchus*) by a capuchin monkey (*Sapajus* sp.) in the Ibura National Forest, Sergipe (Brazil). *Neotropical Primates* 21: 218–221.

Amora, T.D., Beltrao-Mendes, R., Ferrari, S.F. 2013. Use of alternative plant resources by common marmosets (*Callithrix jacchus*) in the semi-arid Caatinga scrub forests of Northeastern Brazil. *American Journal of Primatology* 75: 333–341.

Anzenberger, G., Simmen, C. 1987. Father-daughter incest in a family of common marmosets (*Callithrix jacchus*). *International Journal of Primatology* 8:524.

Araújo, A., Arruda, M. F., Alencar, A. I., Albuquerque, F., Nascimento, M. C., Yamamoto, M. E. 2000. Body weight of wild and captive common marmosets (*Callithrix jacchus*). *International Journal of Primatology* 21: 317–324.

Arruda, M. F., Araújo, A., Sousa, M. B. C., Albuquerque, F. S., Albuquerque, A. C. S. R., Yamamoto, M. E. 2005. Two breeding females within free-living groups may not always indicate polygyny: Alternative subordinate female strategies in common marmosets (*Callithrix jacchus*). *Folia Primatologica* 76: 10–20.

Ash, H., Buchanan-Smith, H. M. 2014. Long-term data on reproductive output and longevity in captive female common marmosets (*Callithrix jacchus*). *American Journal of Primatology* 76: 1062–1073.

Badihi, I. 2006. The effects of complexity, choice and control on the behaviour and the welfare of captive common marmosets (*Callithrix jacchus*). PhD diss., University of Stirling, Stirling, UK.

Baker, J. V., Abbott, D. H., Saltzman, W. 1999. Social determinants of reproductive failure in male common marmosets housed with their natal family. *Animal Behaviour* 58: 501–513.

Barnett, A. A., Andrade, E. S., Ferreira, M. C., Soares, J. B. G., da Silva, V. F., & de Oliveira, T. G. 2015. Primate predation by black hawk-eagle (*Spizaetus tyrannus*) in Brazilian Amazonia. *Journal of Raptor Research*, 49: 105–107.

Barros, M., Boere, V., Huston, J. P., Tomaz, C. 2000. Measuring fear and anxiety in the marmoset (*Callithrix penicillata*) with a novel predator confrontation model: Effects of diazepam. *Behavioural Brain Research*, 108: 205–211.

Barros, M., Mello, E.L., Huston, J.P. and Tomaz, C. 2001 Behavioral effects of buspirone in the marmoset employing a predator confrontation test of fear and anxiety. *Pharmacology, Biochemistry and Behavior* 68: 255–262.

Bassett, L., Buchanan-Smith, H. M. 2007. Effects of predictability on the welfare of captive animals. *Applied Animal Behaviour Science* 102: 223–245.

Bassett, L., Buchanan-Smith, H. M., McKinley, J., Smith, T. E. 2003. Effects of training on stress-related behavior of the common marmoset (*Callithrix jacchus*) in relation to coping with routine husbandry procedures. *Journal of Applied Animal Welfare Science* 6: 221–233.

Beglinger, R., Illgen, B., Pfister, R., Heider, K. 1988. The parasite *Trichospirura leptostoma* associated with wasting disease in a colony of common marmosets, *Callithrix jacchus*. *Folia Primatologica* 51: 45–51.

Bert, A., Abbott, D. H., Nakamura, K., Fuchs, E. 2012. The marmoset monkey: A multi-purpose preclinical and translational model of human biology and disease. *Drug Discovery Today* 17: 1160–1165.

Bettinger, T. L., Leighty, K. A., Daneault, R. B., Richards, E. A., Bielitzki, J. T. 2017. Behavioral management: The environment and animal welfare. In S.J. Schapiro (Ed.), *Handbook of Primate Behavioral Management* (pp. 37–51). CRC Press, Taylor Francis Group: Boca Raton, FL.

Bezerra, B. M., Barnett, A. A., Souto, A., Jones, G. 2009a. Predation by the tayra on the common marmoset and the pale-throated three-toed sloth. *Journal of Ethology* 27: 91.

Bezerra, B. M., da Silva Souto, A., de Oliveira, M. A. B., Halsey, L. G. 2009b. Vocalisations of wild common marmosets are influenced by diurnal and ontogenetic factors. *Primates* 50: 231–237.

Bezerra, B. M., Keasey, M. P., Schiel, N., da Silva Souto, A. 2014. Responses towards a dying adult group member in a wild New World Monkey. *Primates* 55: 185–188.

Bezerra, B. M., Souto, A. 2008. Structure and usage of the vocal repertoire of *Callithrix jacchus*. *International Journal of Primatology* 29: 671.

Bezerra, B.M., Souto, A.S., Schiel, N. 2007. Infanticide and cannibalism in a free-living polygynous group of common marmosets (*Callithrix jacchus*). *American Journal of Primatology* 69: 1–8.

Biben, M., Champoux, M. 1999. Play and stress: Cortisol as a negative correlate of play in *Saimiri*. In S. Reifel (Ed.), *Play and Culture Studies* (pp. 191–208). Ablex Publishing: Stamford, CT.

Bridges J. P., Haller R. L., Buchl S. J., Magden E. R., Lambeth S. P., Schapiro S. J. 2015. Establishing a behavioral management program for geriatric chimpanzees. *American Journal of Primatology* 77: 111.

Bridges J. P., Powers K. M., Lambeth S. P., Schapiro S. J. 2016. Assessment of structure use and structural modifications for geriatric chimpanzees (*Pan troglodytes*). Presented at the Joint meeting of the International Primatological Society and the American Society of Primatologists: August 21–27, 2016, Chicago, IL.

Bridgwater, D. D. 1972. Saving the lion marmoset. In *The Wild Animal Propagation Trust Conference on the Golden Marmoset, National Zoological Park, Washington, DC*, 1972.

Buchanan-Smith, H. M. 2011. Environmental enrichment for primates in laboratories. *Advances in Science and Research* 5: 41–56.

Buchanan-Smith, H. M., Badihi, I. 2012. The psychology of control: Effects of control over supplementary light on welfare of marmosets. *Applied Animal Behaviour Science* 137: 166–174.

Burrows, A. M. 2008. The facial expression musculature in primates and its evolutionary significance. *BioEssays* 30: 212–225.

Cabana, F., Maguire, R., Hsu, C. D., Plowman, A. 2018. Identification of possible nutritional and stress risk factors in the development of marmoset wasting syndrome. *Zoo Biology* 37: 98–106.

Caine, N. G. 1996. Foraging for animal prey by outdoor groups of Geoffroy's marmosets (*Callithrix geoffroyi*). *International Journal of Primatology* 17: 933–945.

Caine, N. G., Mundy, N. I. 2000. Demonstration of a foraging advantage for trichromatic marmosets (*Callithrix geoffroyi*) dependent on food colour. *Proceedings of the Royal Society of London. Series B: Biological Sciences* 267: 439–444.

Caine, N. G., Osorio, D., Mundy, N. I. 2010. A foraging advantage for dichromatic marmosets (*Callithrix geoffroyi*) at low light intensity. *Biology Letters* 6: 36–38.

Caine, N. G. (2017). Antipredator Behavior. In S. J. Schapiro (Ed.) *Handbook of Primate Behavioral Management* (pp. 127–138). CRC Press: Boca Raton, FL.

Caine, N. G., Potter, M. P., Mayer, K. E. 1992. Sleeping site selection by captive tamarins (*Saguinus labiatus*). *Ethology* 90: 63–71.

Caine, N. G. 1990. Unrecognized anti-predator behaviour can bias observational data. *Animal Behaviour* 39: 195–197.

Caldwell, C. A., Whiten, A. 2003. Scrounging facilitates social learning in common marmosets, *Callithrix jacchus*. *Animal Behaviour* 65: 1085–1092.

Campennì, M., Manciocco, A., Vitale, A., Schino, G. 2015. Exchanging grooming, but not tolerance and aggression in common marmosets (*Callithrix jacchus*). *American Journal of Primatology* 77: 222–228.

Carrion Jr., R., Patterson, J. L. 2012. An animal model that reflects human disease: The common marmoset (*Callithrix jacchus*). *Current Opinion in Virology* 2: 357–362.

Chahrour, M., Zoghbi, H.Y. 2007. The story of Rett syndrome: From clinic to neurobiology. *Neuron* 56: 422–437.

Chalifoux, L. V., Bronson, R. T., Escajadillo, A., McKenna, S. 1982. An analysis of the association of gastroenteric lesions with chronic wasting syndrome of marmosets. *Veterinary Pathology* 19: 141–162.

A.S. Chamove, J. Anderson. 1989. Examining environmental enrichment. In E.F. Segal (Ed.), *Housing, Care and Psychological Well-being of Captive and Laboratory Primates* (pp. 183–202) Noyes Publication: New Jersey.

Chow, C. P., Mitchell, J. F., Miller, C. T. 2015. Vocal turn-taking in a non-human primate is learned during ontogeny. *Proceedings of the Royal Society B: Biological Sciences* 282: 20150069.

Cilia, J. and Piper, D.C. 1997. Marmosets conspecific confrontation: An ethologically-based model of anxiety. *Pharmacology Biochemistry and Behavior* 58: 85–91.

Clara, E., Tommasi, L., Rogers, L. J. 2008. Social mobbing calls in common marmosets (*Callithrix jacchus*): Effects of experience and associated cortisol levels. *Animal Cognition* 11: 349–358.

Corrêa, H. K. M., Coutinho, P. E. 1997. Fatal attack of a pit viper, *Bothrops jararaca*, on an infant buffy-tufted ear marmoset (*Callithrix aurita*). *Primates* 38: 215–217.

Corrêa, H. K. M., Coutinho, P. E., Ferrari, S. F. 2000. Between-year differences in the feeding ecology of highland marmosets (*Callithrix aurita* and *Callithrix flaviceps*) in south-eastern Brazil. *Journal of Zoology* 252: 421–427.

Cross, N., Pines, M. K., Rogers, L. J. 2004. Saliva sampling to assess cortisol levels in unrestrained common marmosets and the effect of behavioral stress. *American Journal of Primatology* 62: 107–114.

Cyranoski, D. 2014. Marmosets are stars of Japan's ambitious brain project. *Nature News* 514: 151.

de Boer, R. A., Overduin-de Vries, A. M., Louwerse, A. L., Sterck, E. H. 2013. The behavioral context of visual displays in common marmosets (*Callithrix jacchus*). *American Journal of Primatology* 75: 1084–1095.

De Filippis, B., Chiarotti, F., Vitale, A. 2009. Severe intragroup aggressions in captive common marmosets (*Callithrix jacchus*). *Journal of Applied Animal Welfare Science* 12: 214–222.

de la Torre, S., Snowdon, C. T. 2002. Environmental correlates of vocal communication of wild pygmy marmosets, *Cebuella pygmaea*. *Animal Behaviour* 63: 847–856.

de la Torre, S., Snowdon, C. T., Bejarano, M. 2000. Effects of human activities on wild pygmy marmosets in Ecuadorian Amazonia. *Biological Conservation* 94: 153–163.

de Sousa, M. B. C., da Rocha Albuquerque, A. C. S., Yamamoto, M. E., Araújo, A., de Fátima Arruda, M. 2009. Emigration as a reproductive strategy of the common marmoset (*Callithrix jacchus*). In *The Smallest Anthropoids* (pp. 167–182). Springer: Boston, MA.

de Sousa, M. B. C., Moura, S. L. N., de Lara Menezes, A. A. 2006. Circadian variation with a diurnal bimodal profile on scent-marking behavior in captive common marmosets (*Callitrhix jacchus*). *International Journal of Primatology* 27: 263–272.

Digby, L. 1995. Infant care, infanticide, and female reproductive strategies in polygynous groups of common marmosets (*Callithrix jacchus*). *Behavioral Ecology and Sociobiology* 37: 51–61.

Eliades, S. J., Miller, C. T. 2017. Marmoset vocal communication: behavior and neurobiology. *Developmental Neurobiology* 77: 286–299.

Epple, G. 1967. Vergleichende untersuchungen über sexual- und sozialverhalten der krallenaffen (Hapalidae). *Folia Primatologica* 7: 37–65.

Ess, K. C. 2010. Tuberous sclerosis complex: A brave new world? *Current Opinion in Neurology* 23: 189–193.

Ferrari, S. F. 1992. The care of infants in a wild marmoset (*Callithrix flaviceps*) group. *American Journal of Primatology* 26: 109–118.

Ferrari, S. F. 2009. Predation risk and antipredator strategies. In P.A. Garber, A. Estrada, J.C. Bicca-Marques, E.W. Heymann and K.B. Strier (Eds.), *South American Primates* (pp. 251–277). Springer: New York.

Ferrari, S. F. 2009b. Social hierarchy and dispersal in free-ranging buffy-headed marmosets. In *The Smallest Anthropoids* (pp. 155–165). Springer: Boston, MA.

Ferrari, S. F., Beltrao-Mendes, R. 2011. Do snakes represent the principal predatory threat to callitrichids? Fatal attack of a viper (*Bothrops leucurus*) on a common marmoset (*Callithrix jacchus*) in the Atlantic Forest of the Brazilian Northeast. *Primates* 52: 207.

Ferrari, S. F., Lopes Ferrari, M. A. 1989. A re-evaluation of the social organisation of the Callitrichidae, with reference to the ecological differences between genera. *Folia Primatologica* 52: 132–147.

Ferrari, S. F., Ferrari, M. A. L. 1990. Predator avoidance behaviour in the buffy-headed marmoset, *Callithrix flaviceps*. *Primates* 31: 323–338.

Ferrari, S. F., Kátia, H., Corrêa, M., Coutinho, P. E. 1996. Ecology of the "southern" marmosets (*Callithrix aurita* and *Callithrix flaviceps*). In M. A. Norconk, A. L. Rosenberger and P. A. Garber (Eds.) *Adaptive Radiations of Neotropical Primates* (pp. 157–171). Springer: Boston, MA.

Ferrari, S. F., Rylands, A. B. 1994. Activity budgets and differential visibility in field studies of three marmosets (*Callithrix spp.*). *Folia Primatologica* 63: 78–83.

Finkenwirth, C., van Schaik, C., Ziegler, T. E., Burkart, J. M. 2015. Strongly bonded family members in common marmosets show synchronized fluctuations in oxytocin. *Physiology and Behavior* 151: 246–251.

Fischer, K. E., Austad, S. N. 2011. The development of small primate models for aging research. *ILAR Journal* 52: 78–88.

Francisco, T. M., Couto, D. R., Zanuncio, J. C., Serrão, J. E., de Oliveira Silva, I., Boere, V. 2014. Vegetable exudates as food for *Callithrix spp.* (Callitrichidae): Exploratory patterns. *PloS One*, 9: e112321.

Garbino G. S. 2015. How many marmoset (Primates: Cebidae: Callitrichinae) genera are there? A phylogenetic analysis based on multiple morphological systems. *Cladistics* 31: 652–678.

Hankerson, S.J, Caine, N.G. 2004. Pre-retirement predator encounters alter the morning behavior of captive marmosets (*Callithrix geoffroyi*). *American Journal of Primatology* 63: 75–85.

Hart, D. 2007. Predation on primates: A biogeographical analysis. In *Primate Anti-predator Strategies* (pp. 27–59). Springer, Boston, MA.

Hashikawa, T., Nakatomi, R., Iriki, A. 2015. Current models of the marmoset brain. *Neuroscience Research* 93: 116–127.

Hershkovitz, P. 1977. *Living New World Monkeys (Platyrrhini)*. University of Chicago Press, Chicago, IL.

Heyes, C. M. 1994. Social learning in animals: Categories and mechanisms. *Biological Reviews* 69: 207–231.

Hikishima, K., Sawada, K., Murayama, A. Y., et al. 2013. Atlas of the developing brain of the marmoset monkey constructed using magnetic resonance histology. *Neuroscience* 230: 102–113.

Hilário, R. R., Ferrari, S. F. 2010. Feeding ecology of a group of buffy-headed marmosets (*Callithrix flaviceps*): Fungi as a preferred resource. *American Journal of Primatology* 72: 515–521.

Hosey, G. R. 2005. How does the zoo environment affect the behaviour of captive primates? *Applied Animal Behaviour Science* 90: 107–129.

IUCN, International Union for Conservation of Nature 2020. *The IUCN Red List of Threatened Species.* Accessed at http://www.iucnredlist.org on September 1, 2020.

Jacobs, G. H., Neitz, J., Crognale, M. 1987. Color vision polymorphism and its photopigment basis in a callitrichid monkey (*Saguinus fuscicollis*). *Vision Research* 27: 2089–2100.

Jones, B. S., Harris, D. H., Catchpole, C. K. 1993. The stability of the vocal signature in phee calls of the common marmoset, *Callithrix jacchus*. *American Journal of Primatology* 31: 67–75.

Kangas, B. D., Bergman, J., Coyle, J. T. 2016. Touchscreen assays of learning, response inhibition, and motivation in the marmoset (*Callithrix jacchus*). *Animal Cognition* 19: 673–677.

Kaplan, G., Rogers, L. J. 1999. Parental care in marmosets (*Callithrix jacchus jacchus*): Development and effect of anogenital licking on exploration. *Journal of Comparative Psychology*, 113: 269–276.

Kaplan G., Rogers L. J. 2006. Head-cocking as a form of exploration in the common marmoset and its development. *Developmental Psychobiology* 48: 551–560.

Kato, Y., Gokan, H., Oh-Nishi, A., Suhara, T., Watanabe, S., Minamimoto, T. 2014. Vocalizations associated with anxiety and fear in the common marmoset (*Callithrix jacchus*). *Behavioral Brain Research* 275: 43–52.

Keverne, E. B., Martensz, N. D., Tuite, B. 1989. Beta-endorphin concentrations in cerebrospinal fluid of monkeys are influenced by grooming relationships. *Psychoneuroendocrinology* 14: 155–161.

Kishi, N., Sato, K., Sasaki, E., Okano, H. 2014. Common marmoset as a new model animal for neuroscience research and genome editing technology. *Development, Growth Differentiation* 56: 53–62.

Kitchen, A. M., Martin, A. A. 1996. The effects of cage size and complexity on the behaviour of captive common marmosets, *Callithrix jacchus jacchus*. *Laboratory Animals* 30: 317–326.

Kuehnel, F., Mietsch, M., Buettner, T., Vervuert, I., Ababneh, R., Einspanier, A. 2013. The influence of gluten on clinical and immunological status of common marmosets (*Callithrix jacchus*). *Journal of Medical Primatology* 42: 300–309.

Kumita, W., Sato, K., Suzuki, Y., Kurotaki, Y., Harada, T., Zhou, Y., … Sakakibara, Y. 2019. Efficient generation of Knock-in/Knock-out marmoset embryo via CRISPR/Cas9 gene editing. *Scientific Reports* 9: 1–13.

Layne, D. G., Power, R. A. 2003. Husbandry, handling, and nutrition for marmosets. *Comparative Medicine* 53: 351–359.

Lazaro-Perea, C. 2001. Intergroup interactions in wild common marmosets, *Callithrix jacchus*: Territorial defence and assessment of neighbours. *Animal Behaviour* 62: 11–21.

Lazaro-Perea, C., Castro, C. S., Harrison, R., Araujo, A., Arruda, M. F., Snowdon, C. T. 2000. Behavioral and demographic changes following the loss of the breeding female in cooperatively breeding marmosets. *Behavioral Ecology and Sociobiology* 48: 137–146.

Lazaro-Perea, C., De Fatima Arruda, M., Snowdon, C.T. 2004. Grooming as a reward? Social function of grooming between females in cooperatively breeding marmosets. *Animal Behaviour* 67:627–636.

Lazaro-Perea, C., Snowdon, C. T., de Fátima Arruda, M. 1999. Scent-marking behavior in wild groups of common marmosets (*Callithrix jacchus*). *Behavioral Ecology and Sociobiology* 46: 313–324.

Lúcio Nogueira, S., Cordeiro de Sousa, M. B., de Medeiros Neto, C. F., & da Paz de Oliveira Costa, M. 2001. Diurnal variation in scent marking behavior in captive male and female common marmosets, *Callithrix jacchus*. *Biological Rhythm Research* 32: 169–177.

Ludlage, E., Mansfield, K. 2003. Clinical care and diseases of the common marmoset (*Callithrix jacchus*). *Comparative Medicine* 53: 369–382.

Majolo, B., Buchanan-Smith, H. M., Morris, K. 2003. Factors affecting the successful pairing of unfamiliar common marmoset (*Callithrix jacchus*) females: Preliminary results. *Animal Welfare* 12: 327–338.

Malukiewicz, J. 2019. A review of experimental, natural, and anthropogenic hybridization in Callithrix marmosets. *International Journal of Primatology* 40: 72–98.

Manciocco, A., Chiarotti, F., Vitale, A. 2009. Effects of positive interaction with caretakers on the behaviour of socially housed common marmosets (*Callithrix jacchus*). *Applied Animal Behaviour Science* 120: 100–107.

Mansfield, K. 2003. Marmoset models commonly used in biomedical research. *Comparative Medicine* 53: 383–392.

Martins, M. M., Setz, E. Z. 2000. Diet of buffy tufted-eared marmosets (*Callithrix aurita*) in a forest fragment in southeastern Brazil. *International Journal of Primatology* 21: 467–476.

Mason, G. J., Latham, N. R. 2004. Can't stop, won't stop: is stereotypy a reliable animal welfare indicator? *Animal Welfare* 13: S57–S70.

Mason, G. J., & Mench J. 1997. Using behavior to assess animal welfare. In M. C. Appleby & B. O. Hughs (Eds.) *Animal Welfare* (pp. 127–141). CABI International: Wallingford, Oxfordshire.

McKinley, J., Buchanan-Smith, H. M., Bassett, L., Morris, K. 2003. Training common marmosets (*Callithrix jacchus*) to cooperate during routine laboratory procedures: Ease of training and time investment. *Journal of Applied Animal Welfare Science* 6: 209–220.

Mellou, E. 2006. Play theories: A contemporary review. *Early Child Development and Care* 102: 91–100.

Menárd, N., Motsch, P., Delahaye, A., et al. 2013. Effect of habitat quality on the ecological behaviour of a temperate-living primate: time-budget adjustments. *Primates* 54: 217–228.

Miller, C. T., Freiwald, W. A., Leopold, D. A., Mitchell, J. F., Silva, A. C., Wang, X. 2016. Marmosets: A neuroscientific model of human social behavior. *Neuron* 90: 219–233.

Miller, C. T., Thomas, A. W. 2012. Individual recognition during bouts of antiphonal calling in common marmosets. *Journal of Comparative Physiology A* 198: 337–346.

Mittermeier, R., Rylands, A., Coimbra-Filho, A., Bouchardet, G., Fonseca, O. 1988. *Ecology and Behavior of Neotropical Primates*, Vol. 2, World Wildlife Fund: Washington, D.C.

Moreira, L.A.A., Oliveira, D.G.R., Sousa, M.B.C., Pessoa, D.M.A. 2015. Parturition signaling by visual cues in female marmosets (*Callithrix jacchus*). *PLoS One* 10:e0129319

Morgan, M.J., Adam, A., Mollon, J.D. 1992. Dichromats detect colour-camouflaged objects that are not detected by trichromats. *Proceedings of Royal Society of London B Biological Sciences* 248: 291–295.

Mulholland, M. M., Caine, N. G. 2019. Stability and change in the vocal signatures of common marmoset mobbing calls. *Bioacoustics* 28: 210–223.

National Institutes of Health. 2019. BRAIN Initiative: Marmoset Colonies for Neuroscience Research. Retrieved from https://grants.nih.gov/grants/guide/rfa-files/rfa-mh-20-145.html.

National Institutes of Health, Office of Research Infrastructure Programs. 2018, October 12. Nonhuman primate evaluation and analysis. Part 1: Analysis of future demand and supply. Retrieved from: https://orip.nih.gov/about-orip/research-highlights/nonhuman-primate-evaluation-and-analysis-part-2-report-expert-panel

Neal, S. J., Caine, N. G. 2016. Scratching under positive and negative arousal in common marmosets (*Callithrix jacchus*). *American Journal of Primatology* 78: 216–226.

Okano, H., Hikishima, K., Iriki, A., Sasaki, E. 2012. The common marmoset as a novel animal model system for biomedical and neuroscience research applications. In *Seminars in Fetal and Neonatal Medicine*, Vol. 17, No. 6 (pp. 336–340). WB Saunders, Philadelphia, PA.

Okano, H., Sasaki, E., Yamamori, T., et al. 2016. Brain/MINDS: A Japanese national brain project for marmoset neuroscience. *Neuron* 92: 582–590.

Oliveira, D. G., Macedo, R. H. 2010. Functional context of scent-marking in *Callithrix penicillata*. *Folia Primatologica* 81: 73–85.

Olsson, I. A. S., Wurbel, H., Mench, J. A. 2018. Behaviour. In M. C. Appleby, I. A. S. Olsson, F. Galindo (Eds.) *Animal Welfare* (3rd edition) (pp. 160–180). CABI: Boston, MA.

Orsi, A., Rees, D., Andreini, I., Venturella, S., Cinelli, S., Oberto, G. 2011. Overview of the marmoset as a model in nonclinical development of pharmaceutical products. *Regulatory Toxicology and Pharmacology* 59: 19–27.

Otovic, P., Smith, S., Hutchinson, E. 2015. The use of glucocorticoids in marmoset wasting syndrome. *Journal of Medical Primatology* 44: 53–59.

Passamani, M. 1998. Activity budget of Geoffroy's marmoset (*Callithrix geoffroyi*) in an Atlantic forest in southeastern Brazil. *American Journal of Primatology* 46: 333–340.

Pessoa, D. M., Tomaz, C., Pessoa, V. F. 2005. Color vision in marmosets and tamarins: Behavioral evidence. *American Journal of Primatology* 67: 487–495.

Petracca, M. M., Caine, N. G. 2013. Alarm calls of marmosets (*Callithrix geoffroyi*) to snakes and perched raptors. *International Journal of Primatology* 34(2): 337–348.

Pistorio, A. L., Vintch, B., Wang, X. 2006. Acoustic analysis of vocal development in a New World primate, the common marmoset (*Callithrix jacchus*). *The Journal of the Acoustical Society of America* 120: 1655–1670.

Pomerantz, O., Paukner, A., Terkel, J. 2012. Some stereotypic behaviors in rhesus macaques (*Macaca mulatta*) are correlated with both perseveration and the ability to cope with acute stressors. *Behavioural Brain Research* 230: 274–280.

Reinhardt, V., Liss, C., Stevens, C. 1996. Comparing cage space requirements for nonhuman primates in the United States and in Europe. *Animal Welfare Information Center Newsletter (USA)*.

Rennie, A. E., Buchanan-Smith, H. M. 2006. Refinement of the use of non-human primates in scientific research. Part I: The influence of humans. *Animal Welfare* 15: 203.

Rosenberger A. L. 2011. Evolutionary morphology, platyrrhine evolution, and systematics. *The Anatomical Record: Advances in Integrative Anatomy and Evolutionary Biology* 294: 1955–1974.

Rothe, H., Darms, K., Koenig, A., Radespiel, U., Juenemann, B. 1993. Long-term study of infant-carrying behavior in captive common marmosets (*Callithrix jacchus*): Effect of nonreproductive helpers on the parents' carrying performance. *International Journal of Primatology* 14: 79–93.

Rukstalis, M., Fite, J. E., French, J. A. 2003. Social change affects vocal structure in a callitrichid primate (*Callithrix kuhlii*). *Ethology* 109: 327–340.

Rylands, A. B. 1981. Preliminary field observations on the marmoset, *Callithrix humeralifer intermedius* (Hershkovitz, 1977) at Dardanelos, Rio Aripuana, Mato Grosso. *Primates* 22: 46–59.

Rylands, A. B., de Faria, D. S. 1993. Habitats, feeding ecology, and home range size in the genus *Callithrix*. In A.B. Rylands (Ed.), *Marmosets and Tamarins. Systematics, Behaviour, and Ecology* (pp. 262–272), Oxford University Press: Oxford.

Saito, A. 2015. The marmoset as a model for the study of primate parental behavior. *Neuroscience Research* 93: 99–109.

Saltzman, W., Schultz-Darken, N. J., Abbott, D. H. 1997. Familial influences on ovulatory function in common marmosets (*Callithrix jacchus*). *American Journal of Primatology* 41: 159–177.

Saltzman, W., Pick, R. R., Salper, O. J., Liedl, K. J., & Abbott, D. H. 2004. Onset of plural cooperative breeding in common marmoset families following replacement of the breeding male. *Animal Behaviour*: 68: 59–73.

Sasaki, E. 2015. Prospects for genetically modified non-human primate models, including the common marmoset. *Neuroscience Research* 93: 110–115.

Sasaki, E., Suemizu, H., Shimada, A., et al. 2009. Generation of transgenic non-human primates with germline transmission. *Nature* 459: 523–527.

Sato, K., Kuroki, Y., Kumita, W., et al. 2015. Resequencing of the common marmoset genome improves genome assemblies and gene-coding sequence analysis. *Scientific Reports* 5: 16894.

Sato, K., Oiwa, R., Kumita, W., et al. 2016. Generation of a non-human primate model of severe combined immunodeficiency using highly efficient genome editing. *Cell Stem Cell* 19: 127–138.

Savastano, G., Hanson, A., McCann, C. 2003. The development of an operant conditioning training program for New World primates at the Bronx Zoo. *Journal of Applied Animal Welfare Science* 6: 247–261.

Schiel, N., Huber, L. 2006. Social influences on the development of foraging behavior in free-living common marmosets (*Callithrix jacchus*). *American Journal of Primatology* 68: 1150–1160.

Schiel, N., Souto, A. 2017. The common marmoset: An overview of its natural history, ecology and behavior. *Developmental Neurobiology* 77: 244–262.

Schiel, N., Souto, A., Huber, L., Bezerra, B. M. 2010. Hunting strategies in wild common marmosets are prey and age dependent. *American Journal of Primatology* 72: 1039–1046.

Schneider, H., Sampaio, I. 2015. The systematics and evolution of New World primates–A review. *Molecular Phylogenetics and Evolution* 82: 348–357.

Smith, T. 2006. Individual olfactory signatures in common marmosets (*Callithrix jacchus*). *American Journal of Primatology* 68:585–604.

Smith, A. S., Birnie, A. K., French, J. A. 2011. Social isolation affects partner-directed social behavior and cortisol during pair formation in marmosets, *Callithrix geoffroyi*. *Physiology & Behavior* 104: 955–961.

Smith, T. E., Tomlinson, A. J., Mlotkiewicz, J. A., Abbott, D. H. 2001. Female marmoset monkeys (*Callithrix jacchus*) can be identified from the chemical composition of their scent marks. *Chemical Senses* 26: 449–458.

Snowdon, C. T. 1993. A vocal taxonomy of Callitrichid primates. In A.B. Rylands (Ed.), *Marmosets and Tamarins: Systematics, Behaviour, and Ecology.* Oxford University Press, Oxford.

Snowdon, C. T., Ziegler, T. E. 2007. Growing up cooperatively: Family processes and infant care in marmosets and tamarins. *Journal of Developmental Processes* 2: 40–66.

Soini, P. 1988. The pygmy marmoset, *Cebuella*. In R. A. Mittermeier, A. B. Rylands, A. Coimbra-Filho, G.A.B. Fonseca (Eds.) *Ecology and Behavior of Neotropical Primates* (pp. 79–129), World Wildlife Fund: Washington, D.C.

Souto, A., Bezerra, B.M., Schiel, N., Huber, L. 2007. The salutatory search in free-living *Callithrix jacchus*: Environmental and age influences. *International Journal of Primatology* 28: 881–893.

Spinelli, S., Pennanen, L., Dettling, A. C., Feldon, J., Higgins, G. A., Pryce, C. R. 2004. Performance of the marmoset monkey on computerized tasks of attention and working memory. *Cognitive Brain Research* 19:123–137.

Stevens, D. J., Hornby, R. J., Cook, D. L., Griffiths, G. D., Scott, E. A. M., Pearce, P. C. 2005. A simple method for assessing muscle function in common marmosets. *Laboratory Animals* 39: 162–168.

Stevenson, M. F. 1977. The behaviour and ecology of the common marmoset (*Callithrix jacchus jacchus*) in its natural environment. *Primate Eye* 9:5–6.

Stevenson, M. F., Poole, T. B. 1976. An ethogram of the common marmoset (*Callithrix jacchus jacchus*): General behavioural repertoire. *Animal Behavior* 24: 428–451.

Stevenson, M. F., Poole, T. B. 1982. Playful interactions in family groups of the common marmoset (*Callithrix jacchus jacchus*). *Animal Behaviour* 30: 886–900.

Stevenson, M. F., Rylands, A. B. 1988. The marmosets, genus *Callithrix*. In A.B. Rylands, A.F. Coimbra-Filho and G.A.B. da Fonseca, R.A. Mittermeier (Eds.), *Ecology and Behavior of Neotropical Primates* (pp. 131–222). World Wildlife Fund: Washington, DC.

Suomi, S. J., Novak, M. A. 1991. The role of individual differences in promoting psychological well-being in rhesus monkeys. In M.A. Novak, A.J. Petto (Eds.), *Through the Looking Glass: Issues of Psychological Well-Being in Nonhuman Primates* (pp. 50–56). American Psychological Association: Washington, DC.

Sutcliffe, A. G. and Poole, T. B. 1978 Scent marking and associated behaviour in captive common marmosets (*Callithrix jacchus*) with a description of the histology of scent glands. *Journal of Zoology (London)* 185: 41–56.

't Hart, B. A., Abbott, D. H., Nakamura, K., Fuchs, E. 2012. The marmoset monkey: a multi-purpose preclinical and translational model of human biology and disease. *Drug Discovery Today* 17: 1160–1165.

Takemoto, A., Izumi, A., Miwa, M., Nakamura, K. 2011. Development of a compact and general-purpose experimental apparatus with a touch-sensitive screen for use in evaluating cognitive functions in common marmosets. *Journal of Neuroscience Methods* 199: 82–86.

Tardif, S. D., Richter, C. B., Carson, R. L. 1984. Effects of sibling-rearing experience on future reproductive success in two species of Callitrichidae. *American Journal of Primatology* 6: 377–380.

Tardif, S. D., Smucny, D. A., Abbott, D. H., Mansfield, K., Schultz-Darken, N., Yamamoto, M. E. 2003. Reproduction in captive common marmosets (*Callithrix jacchus*). *Comparative Medicine* 53: 364–368.

Tasker, L., Buchanan-Smith, H. M. 2016. Linking welfare and quality of scientific output through refinement of enhanced socialization with carestaff in cynomolgus macaques (*Macaca fascicularis*) used for regulatory toxicology. Presented at the *Joint Meeting of the International Primatological Society and the American Society of Primatologists*: August 23, 2016, Chicago, IL, USA.

Teixeira, D. S., Dos Santos, E., Leal, S. G., et al. 2016. Fatal attack on black-tufted-ear marmosets (*Callithrix penicillata*) by a Boa constrictor: A simultaneous assault on two juvenile monkeys. *Primates* 57: 123–127.

Thomson, J. A., Kalishman, J., Golos, T. G., Durning, M., Harris, C. P., Hearn, J. P. 1996. Pluripotent cell lines derived from common marmoset (*Callithrix jacchus*) blastocysts. *Biology of Reproduction* 55: 254–259.

Tokuno, H., Watson, C., Roberts, A., Sasaki, E., Okano, H. 2015. Marmoset neuroscience. *Neuroscience Research*, 93: 1.

Tomassetti, D., Caracciolo, S., Manciocco, A., Chiarotti, F., Vitale, A., De Filippis, B. 2019. Personality and lateralization in common marmosets (*Callithrix jacchus*). *Behavioural Processes* 167: 103899.

Vermilyea, S. C., Babinski, A., Tran, N., To, S., Guthrie, S., Kluss, J. H., ... Murphy, M. E. 2020. In vitro CRISPR/Cas9-Directed gene editing to model LRRK2 G2019S Parkinson's Disease in common marmosets. *Scientific Reports* 10: 1–15.

Ward, J. M., Vallender, E. J. 2012. The resurgence and genetic implications of New World primates in biomedical research. *Trends in Genetics* 28: 586–591.

Williams, L. E., Ross, C. N. 2017. Behavioral management of neotropical primates: Aotus, Callithrix, and Saimiri. In S. J. Schapiro (Ed.) *Handbook of Primate Behavioral Management* (pp. 409–434). CRC Press: Boca Raton, FL.

Yamamoto, M. E., Box, H. O., Albuquerque, F. S., de Fátima Arruda, M. 1996. Carrying behaviour in captive and wild marmosets (*Callithrix jacchus*): A comparison between two colonies and a field site. *Primates* 37: 297–304.

Yamamoto, M. E., de Fátima Arruda, M., Alencar, A. I., de Sousa, M. B. C., Araújo, A. 2009. Mating systems and female–female competition in the common marmoset, *Callithrix jacchus*. In S.M. Ford, L.M. Porter and L.C. Davis (Eds.), *The Smallest Anthropoids* (pp. 119–133). Springer: Boston, MA.

Young, R. J. 2003. *Environmental Enrichment for Captive Animals*. Blackwell Science Ltd.: Malden, MA.

Ziegler, T. E., Schultz-Darken, N. J., Scott, J. J., Snowdon, C. T., Ferris, C. F. 2005. Neuroendocrine response to female ovulatory odors depends upon social condition in male common marmosets, *Callithrix jacchus*. *Hormones and Behavior* 47: 56–64.

Ziegler, T. E. 2013. Social effects via olfactory sensory stimuli on reproductive function and dysfunction in cooperative breeding marmosets and tamarins. *American Journal of Primatology* 75: 202–211.

Zoratto, F., Sinclair, E., Manciocco, A., Vitale, A., Laviola, G., Adriani, W. 2014. Individual differences in gambling proneness among rats and common marmosets: An automated choice task. *BioMed Research International* 2014: 927685.

25

Behavioral Biology of Squirrel Monkeys

Anita I. Stone
California Lutheran University

Lawrence Williams
The University of Texas MD Anderson Cancer Center

CONTENTS

Introduction 395
Typical Research Participation 395
Behavioral Biology 396
    General Ecology 396
    Social Organization and Social Behaviors 397
    Life History and Reproduction 398
    Feeding Behavior 399
    Communication 399
Common Captive Behavior 399
    Normal Behavior 399
    Behaviors Affected by Captivity 399
Ways to Maintain Behavioral Health 400
    Environment 400
    Social Groupings 400
    Enrichment and Feeding Strategies 401
    Communication 401
    Training 402
    Special Situations 402
Conclusions and Recommendations 402
References 402

Introduction

Squirrel monkeys (*Saimiri* spp.) are one of the top three most commonly involved Neotropical primates in biomedical research (Abee 2000; Tardif et al. 2012). Their physical characteristics, including small size and ease of handling, contribute to their desirability as research subjects. Squirrel monkeys easily adapt to laboratory housing and can be maintained in smaller spaces and less expensive cages compared to larger primates, such as macaques and baboons. This characteristic is especially important in facilities with space limitations.

Animals maintained in laboratory environments often live in restrictive and limiting conditions compared with their natural habitats. Although such environments meet the physical needs of the animals, they are usually not sufficient to allow the animals to express the full range of their species-typical behavior. The environments of most laboratory animals are extremely complex, involving unpredictable elements, as well as routine, predictable tasks, such as feeding and cleaning. Many laboratory environments lack some or all of the stimuli that will promote desirable natural behavior. Therefore, knowledge of the natural behavior of squirrel monkeys is critical for anyone who plans to work with these animals in a laboratory environment. Understanding the natural history of a species, as well as how members of the species react to stressors, helps researchers to adjust their animals to a captive environment, providing a better model for research. Animals stressed by inadequate housing or inappropriate environmental enhancement may exhibit a host of physiological and psychological manifestations that will influence the way they react to experimental stimuli.

Typical Research Participation

Although there are 8 species of squirrel monkeys currently recognized (Chiou et al. 2011; Lynch Alfaro et al. 2015), the most common species involved in laboratory research are *Saimiri sciureus* and *S. boliviensis*, both South American species.

Squirrel monkeys offer several advantages over Old World monkeys for laboratory research (Abee 2000). Physically they are smaller and easier to handle. Doses of experimental compounds that need to be administered by body weight can be lower. There are no known viral transmission issues with squirrel monkeys. They also adapt to laboratory conditions very well and most procedures can be carried out on unsedated animals relatively easily.

The squirrel monkey is an important animal model for malaria vaccine development studies (Anderson et al. 2017; de Souza et al. 2017; Gunalan et al. 2019; Peterson et al. 2019). Because *Plasmodium* spp., the parasite that causes malaria, are host-specific, animal models used to study human malaria must be susceptible to the same strains that cause the disease in humans. The Bolivian squirrel monkey (*Saimiri boliviensis*) has been shown to be a superior model for studies of the pathogenesis of *P. falciparum* Indochina I (Whiteley et al. 1987), developing lesions and clinical signs similar to those reported in human disease. Squirrel monkeys are used to screen *P. falciparum* vaccine candidates for humans (Alves et al. 2015; Cunha et al. 2017; Tougan et al. 2018; Yagi et al. 2014). Squirrel monkeys (in particular *S. boliviensis* and *S. sciureus*) are susceptible to infection with different strains of *P. vivax*, but each squirrel monkey species/subspecies responds differently depending on the strain of the parasite used (Anderson et al. 2015; Anderson et al. 2017; Gunalan et al. 2019; Peterson et al. 2019). Obaldia et al. (2018) reported positive findings using *S. boliviensis* as a model for *P. vivax* vaccine testing. Differences in susceptibility and response to experimental malaria infections underscore the importance of species identification when using squirrel monkeys. Collins et al. (2009) reported that *S. boliviensis* is a suitable model for testing sporozoite and liver stage vaccines against the Salvador I strain of *P. vivax*. Malaria remains one of the top three diseases causing human mortality worldwide.

Squirrel monkeys are also involved extensively in neuroscience, vision, and hearing research. The squirrel monkey model offers advantages over rodent models in that cortical organization of the squirrel monkey brain more closely resembles that of humans (Gao et al. 2016; Schilling et al. 2018). Studies involving squirrel monkeys have also contributed to our understanding of injury-induced reorganization of the sensory cortex and somatotopic reorganization of the brain stem and thalamus (Cheung et al. 2017; Mowery et al. 2015; Yang et al. 2014). The field of behavioral pharmacology, particularly drug addiction research, has involved squirrel monkeys for many years (Achat-Mendes et al. 2012; Rowlett, et al. 2005; Valdez et al. 2007). Squirrel monkeys have made contributions to studies on drug addiction and its behavioral and physiological sequelae (Desai et al. 2016a; Kangas et al. 2016; Kangas and Bergman 2014).

Contributions to vision research include gene therapy in color vision (Mancuso et al. 2009; Neitz and Neitz, 2014; Shapley et al. 2009), adaptation of the vestibule-ocular reflex (Migliaccio et al. 2010), and topography of cones, rods, and optic nerve axons. Auditory research has included auditory frequency discrimination (Malone et al. 2015, 2017; Marsh et al. 2015) and mapping of the auditory cortex and projections throughout the brain (Cheung et al. 2017).

Recent research has asked the question of whether squirrel monkeys could be considered an important sylvatic reservoir of the Zika virus in South America. Vanchiere et al. (2018) confirmed that *S. boliviensis* were susceptible to the Zika virus, documenting seroconversion in five out of six animals. There is now a validated fecal assay for Zika virus to be used in South America for rapid assessment of hosts in the forest (Milich et al. 2018).

Behavioral Biology

General Ecology

Squirrel monkeys are distributed widely in the Amazon basin, with disjunct populations occurring in Central America. Eight species are currently recognized: *S. oerstedii* in Central America, and seven South American forms: *S. boliviensis*, *S. sciureus*, *S. collinsi*, *S. ustus*, *S. vanzolinii*, *S. cassiquiarensis*, and *S. macrodon* (Chiou et al. 2011; Lynch Alfaro et al. 2015). Long-term studies in the wild have been conducted on *S. oerstedii* in Costa Rica (Boinski 1986), *S. boliviensis* in Peru (Mitchell 1990), *S. sciureus* in Suriname (Boinski et al. 2002), and more recently on *S. collinsi* in Brazil (Stone 2007a, b) and *S. vanzolinii* in Brazil (Paim et al. 2017). Until recently, *S. collinsi* was grouped with *S. sciureus* as one species (Merces et al. 2015).

Squirrel monkeys live in various habitats, including primary or secondary rain forests, riverine areas, flooded forests, and mosaic habitats with anthropogenic disturbances (Boinski 1989; Carretero-Pinzón et al. 2017; Lima and Ferrari 2003; Paim et al. 2013; Peres 1994; Pontes 1997; Terborgh 1983). They also do well in areas of secondary growth, perhaps due to their high consumption of insects (Stone 2007a). Compared to other Neotropical primates, lower strata of the forest are often preferred (Fleagle and Mittermeier 1980; Mittermeier and van Roosmalen 1981). Squirrel monkeys also show terrestrial behavior, coming to the forest floor, primarily when foraging for larger insects in the leaf litter, fallen ripe fruit, and flowers (*S. collinsi*: Stone 2007a; *S. cassiquiarensis*: M. Araujo, pers. communication). Squirrel monkey home ranges are large, varying from 150 to 250 ha (Carretero-Pinzón et al. 2017; Mitchell 1990; Stone 2007a), but these are not defended, and there can be overlap in the ranges of neighboring troops (Stone 2007a). In addition, these primates can travel 2–5 km in a day (Mitchell 1990).

Where they are sympatric, *Saimiri* and capuchin monkeys (genus *Sapajus*) can form mixed-species associations, traveling and foraging together. The most common association occurs between *Saimiri* and *Sapajus apella* (Levi et al. 2013; Pinheiro, et al. 2011; Podolsky 1990; Terborgh 1983), although squirrel monkeys can also form polyspecific associations with *Sapajus olivaceous* (Pontes 1997) and cuxiús (genus *Chiropotes*; Pinheiro, et al. 2011). These associations may occur due to enhanced access to fruit resources (Podolsky 1990) or antipredator benefits for both species involved (Levi et al. 2013; Terborgh 1983).

Because of their small size, squirrel monkeys are at high predation risk, particularly infants traveling on their mother's back (Boinski 1986). In fact, it has been argued that the high birth synchrony within wild squirrel monkey groups is an antipredator adaptation (Boinski 1987; Stone and Ruivo, 2020). The main predators of squirrel monkeys are raptors (Boinski 1987; Stone 2004) and constricting snakes (Ribeiro-Junior et al. 2016). Fear of snakes appears to be learned, as laboratory-raised squirrel monkeys do not react when presented with live snakes (Murray and King 1973). In the wild, seasonality has a strong effect on predator-sensitive foraging in squirrel monkeys. Experimental evidence shows that during the wet season (high fruit availability), squirrel monkeys avoid taking risks when foraging, with juveniles not foraging in open areas (Stone 2007b). Frechette et al. (2014) showed that *S. sciureus* in Suriname use both environmental and social information to assess predation risk, and are more reactive to potential predators at lower heights. Furthermore, field data on *S. oerstedii*, *S. boliviensis*, and *S. sciureus* living in distinct forest types indicates that avian attacks on squirrel monkeys primarily target individuals unprotected by understory vegetation (Boinski et al. 2003).

Social Organization and Social Behaviors

Squirrel monkey groups are generally large, ranging from 25 to 75 individuals (Boinski, 1999). However, unlike patterns of basic ecology, such as diet and habitat use, the social organization in the genus *Saimiri* is extremely diverse. For example, in some species, males are dominant over females (e.g., *S. sciureus*), while others are female-dominant (e.g., *S. boliviensis*), and yet in others, females and males are codominant (*S. oerstedii* and possibly *S. collinsi*; Boinski 1999; Pinheiro and Lopes 2018; Zimbler-Delorenzo and Stone 2011). Such interspecific differences in social structure have been documented both in wild and captive settings, and extend to characteristics, such as group size, intersexual dominance, within-sex dominance hierarchies, and dispersal patterns (Table 25.1). However, it should be noted that, due to the difficulty of recognizing individual monkeys in the wild, much remains unknown about social structure and interactions in the genus.

Costa Rican squirrel monkeys (*S. oerstedii*), in addition to showing egalitarian interactions between sexes, also show weak female bonds and hierarchies (Boinski 1999); both males and females disperse from the group, though males disperse over longer distances (Blair and Melnick 2012).

Squirrel monkey groups studied in Peru (*S. boliviensis*) showed dominance hierarchies within each sex, along with strong female-bonding and female philopatry (Mitchell et al. 1991). In contrast, *S. sciureus* are characterized by high levels of intragroup aggression; male dominance over females; rigid, linear dominance hierarchies within each sex; weak female social bonds; and both sexes may disperse from their natal group (Boinski et al. 2005). Boinski et al. (2002) propose that fruit patch size and density in each species' habitat determine within-group competition regimes, which in turn shapes social structure (the socioecological model; Wrangham 1980). Another species, *S. macrodon* (studied in Ecuador) is characterized by a high level of kinship among females in a group, with individual females maintaining spatial proximity to closer kin. Like *S. boliviensis*, males disperse and females are philopatric (Montague et al. 2014).

Male–male interaction patterns are also variable among species. Costa Rican monkeys rarely display male–male within-group aggression, even during mating season, and males cooperate in aggressive interactions with neighboring groups and in the defense of infants from predators (Boinski 1994). Conversely, in *S. sciureus*, *S. boliviensis*, and *S. collinsi*, male agonism is high, particularly over access to females (Boinski et al. 2002; Mitchell 1990; Stone 2014). Male *S. boliviensis* and *S. sciureus* often form bachelor groups with conspecifics (Carretero-Pinzón et al. 2016; Mitchell 1990), although *S. collinsi* do not (Stone 2004). However, both *S. boliviensis* and *S. collinsi* show marked sexual segregation within groups, with males remaining at the periphery of the troop, especially during birthing season (Leger et al. 1981; Mendoza et al. 1978; Ruivo et al. 2017; Stone 2014).

Most social interactions occur within age-sex classes, such as play among infants and juveniles, and agonism among adult males (Zimbler-DeLorenzo and Stone 2011). Squirrel monkeys rarely engage in social grooming, and social bonds appear to be reinforced mostly by spatial proximity (Montague et al. 2014; Stone, pers. observation) and vocalizations (Biben and Suomi 1993; Biben and Symmes 1991; Boinski and Mitchell 1997). Play behavior in infants and juveniles is strongly tied to seasonality and resource availability, with less play occurring during the dry season or low food conditions (Stone 2008; Baldwin and Baldwin 1976).

TABLE 25.1

Comparison of Socioecological Parameters across Four Squirrel Monkey Species (*Saimiri* sp).

Trait	*S. oerstedii*	*S. boliviensis*	*S. sciureus*	*S. collinsi*	References
Group size	35–65	45–75	15–25	35–50	Boinski (1999)
Fruit patches fed on	Small, low density	Large, moderate density	Small, extremely dense	Medium, extremely dense	Boinski et al. (2002), Stone (2007a,b)
Dominance pattern	Egalitarian	Female dominant	Males dominant	Female dominant or co-dominant	Boinski et al. (2002), Stone (2004)
Female bonding	Weak	Strong	Weak	Unknown	Boinski (1999), Boinski et al. (2002)
Male–male aggression	Rare	Common during mating season	Common during mating season	Common during mating season	Baldwin (1969), Boinski and Mitchell (1994), Stone (2014)
Dispersal pattern	Both sexes disperse	Males disperse	Both sexes disperse	Unknown	Blair and Melnick (2012), Boinski et al. (2002), Boinski and Mitchell (1994)

Life History and Reproduction

Squirrel monkeys are characterized by a suite of reproductive and life history traits that distinguish them from other similarly sized New World primates (Garber and Leigh 1997; Hartwig 1996). For example, the genus is characterized by extremely high pre- and postnatal maternal investment, even though singletons are produced. Squirrel monkeys rank highest among Neotropical primates under 2 kg in infant care costs, including the cost of transporting heavy infants over long day ranges (Tardif 1994). The gestation period is long in squirrel monkeys (5 months; Stone 2004), similar to that of the larger capuchins (Hartwig 1996). Even when compared to the twin-producing callitrichines, estimated prenatal growth rates are high (Leighty et al. 2004; Ross 1991).

Not surprisingly, neonates are large, weighing 16%–20% of the mother's body weight (Elias 1977; Kaack et al. 1979; Milligan et al. 2008; Stone, unpublished trapping data; Williams et al. 1994) and have 60% of adult brain mass at birth (Elias 1977). In the laboratory squirrel monkeys produce twins at a rate of approximately 1 in 150 pregnancies (Williams, personal observation). When this occurs, one neonate is usually smaller than the other and does not survive. The first 3 months of life are characterized by rapid somatic and brain growth (Kaplan 1979; Leigh et al. 2004; Manocha 1979). Somatic growth rates are reduced at approximately 6 months of age, which corresponds to weaning in wild *S. collinsi* (Stone 2006). Interestingly, weaning age is highly variable among squirrel monkey species, ranging from 4 months in wild *S. oerstedii* (Boinski and Fragaszy 1989) to up to 18 months in *S. boliviensis* (Mitchell 1990). Interbirth intervals vary from 1 to 2 years, depending on the species (Zimbler-DeLorenzo and Stone 2011). Despite having expensive infants, females generally receive little extra-maternal support. In some species, other adult females and juveniles can provide some degree of allocare. For example, in *S. collinsi* and *S. boliviensis*, juvenile females often carry infants for short periods (< 5 min), although never during rapid group travel (Baldwin 1969; Mitchell 1990; Rosenblum 1972; Stone 2004). Mothers often "exchange" infants for short periods and females who lose their infants may carry the infants of other mothers (Ruivo et al. 2017). Allonursing has not been reported for wild populations, but has been observed in captivity for *S. collinsi* (Pinheiro 2015) and *S. boliviensis* (Williams et al. 1994). In contrast to rapid development during infancy, the juvenile stage (approximately 6 months to 3–5 years) is characterized by an extended period of relatively slow growth (Scollay 1980; Stone 2004). Females become sexually mature between 2.5 and 3.5 years of age, and males breed later, at 4.5–6 years (with variability among species; Zimbler-DeLorenzo and Stone 2011).

Squirrel monkeys show a polygamous mating system (Boinski 1987; Mitchell 1990, Stone 2014), and show the most pronounced pattern of seasonal breeding of any Neotropical primate (Di Bitetti and Janson 2000; Table 25.2). The pronounced birth seasonality seen in the wild may or may not persist under captive conditions (P. Castro, personal communication; Trevino 2007; Williams et al. 1986; Williams and Abee 1986; Williams et al., 2010). Within-group birth synchronicity is also common in wild populations. In Costa Rica, females in a group give birth within a 1-week period (Boinski 1987), and in Eastern Amazonia, 80% of the females in a group can give birth in 1–2 nights (Stone 2004). Squirrel monkey males are approximately 30% larger than females (Boinski 1999; Mitchell 1990; Stone, unpublished trapping data), though sexual dimorphism increases in the mating season. This occurs because males gain water weight and adipose tissue on the shoulders, arms, and torso, acquiring a "fattened appearance" (Du Mond and Hutchinson 1967; Mendoza et al. 1978). Male fattening is associated with seasonal changes in hormone levels, rather than with increased food consumption, and may have evolved via sexual selection (Boinski 1992a; Stone 2014). An interesting phenomenon consistently observed during the mating season of *S. collinsi* is infant aggression toward adult males that approach the infant's mother. Infants often chase adult males that are attempting to mate, and males show a submissive response (Ruivo and Stone 2014).

TABLE 25.2

Comparison of Life History and Reproductive Traits across Four Squirrel Monkey Species (*Saimiri* sp).

Trait	*S. oerstedii*	*S. boliviensis*	*S. sciureus*	*S. collinsi*	References
Gestation length (months)	5	5	5	5	Mitchell (1990), Stone (2004)
Weaning age (months)	4	18	12–18	8	Boinski and Fragaszy (1989), L. Kaufmann (pers. comm.), Mitchell (1990), Stone (2006)
Age at first reproduction for female (in years)	2.5	3.5	3.5	3.5	Boinski (1987), Boinski et al. (2005), Mitchell (1990), Stone (2004)
Age at first reproduction for males (in years)	4	6	4.5	4.5	Boinski (1987), Boinski et al. (2005), Mitchell (1990), Stone (2004)
Interbirth interval (yrs)	1	2	1	1-2	Boinski (1987), Boinski et al. (2005), Mitchell (1990), Stone (2004)
Duration of mating season (weeks)	8	8	10	9	Boinski (1987), Boinski et al. (2005), Mitchell (1990), Stone (2014)
Duration of birth season (weeks)	2	8	8	8	Boinski (1987), L. Kauffman (pers. comm.), Mitchell (1990), Stone (2004)
Infant care patterns	Allocare uncommon	Allocare and allonursing	Allocare and allonursing	Allocare common	Boinski (1986), L. Kauffman (pers. comm.), Mitchell (1990), Pinheiro (2015), Stone (2004)

Feeding Behavior

Squirrel monkeys are very active and spend most of their day foraging (Mitchell 1990; Paim et al. 2017; Pinheiro et al. 2013; Stone 2007a). They consume fruits and insects, and show opportunistic consumption of items, such as bird eggs and lizards (Janson and Boinski 1992). Squirrel monkeys are the most insectivorous of Neotropical primates, with up to 75% of foraging/feeding time spent on arthropods (Zimbler-DeLorenzo and Stone 2011). Diet varies seasonally, with prey consumption increasing during periods of low fruit availability (Boinski 1986; Lima and Ferrari 2003; Paim et al. 2017, Stone 2007a). Foraging substrates used while searching for insects include live foliage, dead foliage, and leaf litter on the ground (Stone 2007a, Terborgh 1983). Direct intragroup feeding competition is generally low in *Saimiri*, perhaps due to the high proportion of insect prey in their diet, a resource that is difficult to monopolize. However, even when foraging on clumped fruit resources, such as *Scheelea* or *Attalea* palm fruits, individuals are generally tolerant of each other, eating in close proximity or plucking a fruit and leaving the infructescence to eat on a nearby tree branch (Baldwin and Baldwin 1972; Stone 2007b).

Sex differences in diet and foraging behavior have been observed in squirrel monkeys, with females spending more time feeding and foraging, and less time resting, compared to males (Boinski 1988; Montague et al. 2014). In addition, *S. collinsi* females consume more protein in the form of insects than do males (Ruivo et al. 2017), as well as a greater variety of fruit species (Stone 2018). Acquisition of foraging skills, including for large insects, occurs rapidly; in one field study, 1-year old juveniles were as efficient as adults at capturing prey (Stone 2006).

Communication

Few field studies have been conducted on communication (vocal or olfactory) in squirrel monkeys. As pointed out by Zimbler-DeLorenzo and Stone (2011), studying communication in *Saimiri* is more feasible in captivity because of the large social groups in the wild. Five main groups of calls have been identified: peeps; twitters; chuck calls; cackles; and noisy calls (Zimbler-DeLorenzo and Stone 2011). Studies of wild populations have shown that back-and-forth peeps and chuck calls are used in affiliative contexts, particularly by adult females that are not in close proximity (Boinski 1991; Boinski and Mitchell 1992). Chuck calls also may contain information on the caller's identity (Boinski and Mitchell 1997). In particular, these calls increase auditory contact in densely forested areas where visual contact is difficult, and such vocalizations also affect group cohesion and coordination of group movement (Boinski 1991; Boinski and Mitchell 1992). A specialized maternal vocalization of *S. boliviensis* is the "caregiver call", a call from adult females directed to their infants that is used to coordinate nursing bouts until infants are fully weaned (Boinski and Mitchell 1995). Alarm calls have also been described for wild squirrel monkeys (Stone 2004). One type of high-pitched alarm call is elicited when aerial predators approach the group. A second call is the "nasal-honk" call, which is given in response to terrestrial predators, such as large constricting snakes, and even humans (Stone 2004).

Olfactory communication accounts for less than 1% of daily behavior time in field populations of *S. oerstedii* (Boinski 1992b). In *S. oerstedii* and *S. collinsi*, males use olfaction when evaluating the estrous condition of females, either by genital sniffing or by sniffing substrates recently used by females (Boinski 1992a; Stone 2014; Williams personal observation). Squirrel monkeys can discriminate urine odors of conspecifics (Laska and Hudson 1995), which has led to the suggestion that urine washing behavior (urinating on palms and rubbing the urine on the soles) functions as a mechanism of social communication, particularly sexual communication (Miller et al. 2008). However, recent fieldwork on *S. collinsi* by Jasper et al. (2020) has shown that urine washing is predominantly performed by juveniles, and may serve as an anxiety displacement mechanism (Miller et al. 2008;, Schino et al. 2011).

Common Captive Behavior

Normal Behavior

Squirrel monkeys continue to show species-typical behavior in most captive settings and can adapt to many social settings; pair housing, as well as small and large social groups (Tardif et al. 2012; Williams et al. 2010). The patterns of male–female and female–female interactions described above can clearly be seen in captive colonies.

Behaviors Affected by Captivity

While squirrel monkeys are not prone to the abnormal behaviors seen in Old World monkeys, there are several behavioral and reproductive traits that need to be considered. The large neonatal sizes mentioned elsewhere in the chapter may necessitate c-sections and nursery rearing of the infants. Although nursery-reared infants can function in a breeding colony, producing and caring for infants at rates similar to dam-reared individuals (Williams, personal observation), they are behaviorally different, particularly when interacting with care staff. In a study that compared Primate Neonatal Neurobehavioral Assessment scores, Mulholland et al. (2020) reported that nursery-reared infants differed from dam-reared infants in motor maturity and control states. Nursery-reared infants had worse balance, slower responses, and were generally quieter than were dam-reared infants.

Squirrel monkeys in captivity live to their mid- to upper-twenties (Akkoç and Williams 2005), much longer than their wild counterparts. Older female squirrel monkeys, particularly those with many pregnancies over the course of their life, can develop an alopecia that resembles hair loss seen as a function of dominance and stressful interactions in other primate species. However, we have demonstrated that this alopecia is due to a chronic telogen effluvium, related to high levels of estrogen hormones, not to physical hair plucking (Horenstein et al. 2005).

Ways to Maintain Behavioral Health

Environment

Several factors related to Neotropical primate sociality and natural living conditions play a role in determining functionally appropriate captive environments (FACEs, see Schapiro, this volume) and related management procedures. The social structure of squirrel monkeys, while superficially similar, is significantly different from the typical social structures of Old World monkeys, and housing conditions need to reflect this. Decisions related to enclosure size need to reflect the fact that squirrel monkeys are primarily arboreal animals and the enclosure should include vertical space for climbing, quadrupedal running, and jumping. The *Guide for the Care and Use of Laboratory Animals* (National Research Council 2011) classifies squirrel monkeys as type 2 primates, requiring 3 ft^2 (0.28 m^2) for each animal. Figure 25.1 illustrates squirrel monkey housing with multiple layers of travel paths and perches. All squirrel monkeys should be housed with a temperature range of approximately 75°F–82°F (24°C–27°C), with relative humidity between 40% and 60%. Lower temperatures or humidity levels can lead to upper respiratory problems (Abee 1985).

Since squirrel monkeys are arboreal, the provision of three-dimensional space and differing perch arrangements can vary the cage environment and increase the activity of the animals. Runs or large pens can be used to allow the squirrel monkeys to move over longer distances. Pens measuring 4′ wide by 6′ tall by 14′ deep can house up to 15 animals. Periodically rearranging the perches and providing new materials is an easy way to provide novelty.

Social Groupings

Saimiri should be housed in species-typical social groups in captivity whenever possible. In the wild, squirrel monkeys live in large social groups containing 20–50 animals (Baldwin and Baldwin 1971; Boinski, 1999), with multiple males and females of different ages. Our experience in captivity is that it is difficult for more than one male to live in a social group when space is limited. *S. boliviensis* females typically stay within the natal social group and males emigrate after they reach 1.5–2 years of age. *S. sciureus* females are more flexible and may emigrate after their first breeding season (Boinski et al. 2002). Removing the males from their natal group when they reach approximately 1.5–2 years old can help reduce intragroup aggression and prevent inbreeding, as the males continue to mature sexually. At this age, the males will usually form an all-male group without any aggression.

Adult females generally maintain social stability in squirrel monkey groups. Saltzman et al. (1989) reported that all age-sex classes in *S. sciureus* spend more time in proximity to adult

FIGURE 25.1 Squirrel monkey (*Saimiri sciureus*) housing with multiple layers of perches and resting points that give them different travel paths. Off the back, there are feeding enrichment devices; a black grooming mat (just below the mid-point) and a feeding trough (just above the upper perch). There is also a piece of destructible enrichment on the ceiling; a paper bag that was filled with food enrichment.

females than any other age-sex class. During the nonbreeding season, South American squirrel monkey social groups will segregate by sex, with males remaining near the periphery of the social group (Baldwin and Baldwin 1972; Hopf 1978; Lyons et al 1992; Stone 2014; Vaitl 1977). Seasonal sexual segregation in captivity is generally expressed in vertical separation of the males and females; with females occupying the upper levels of the housing, relegating the males to perches and structures closer to the ground.

S. sciureus and *S. boliviensis* will fight unfamiliar conspecifics (Williams and Abee 1988), suggesting that frequent movements of animals between social groups should be avoided. Squirrel monkeys that must live separately from their social group for experimental or clinical purposes should be pair housed. If pair housing is not possible, animals should be provided with visual and/or auditory access to conspecifics (Schuler and Abee 2005). Forming new pairs or small groups of females is rarely a problem with squirrel monkeys. Introducing all the females to one another simultaneously will lessen the probability of an individual animal becoming the target of aggression. Male pairs or triplets can be housed together if care is taken when introducing the animals. Juvenile males will bond together quickly, and adult males will tolerate others, if there are no females housed in visual contact. When introducing new animals to an established social group, it is important to give members of the social group contact with the new animal in small subgroup situations, prior to introducing the entire group. Placing the new animal in a neutral cage with a small subset of the social group, and swapping out social group members periodically, will give the new animal an opportunity to establish relationships, prior to living in the whole group. This process will work for females during the entire year and for males during the breeding season only. Males can be difficult to introduce to a new group of females during the nonbreeding season, when they are normally segregated from the female social core. Increased levels of aggression should be expected at the beginning and end of the breeding season, when hormonal changes lead to increases in social interactions between males and females. The squirrel monkey breeding season occurs during short days; in the northern hemisphere, it typically runs from January through April for outdoor-housed colonies, and into May and June in indoor colonies (Schiml et al. 1996; Trevino 2007; Williams et al. 2010).

Enrichment and Feeding Strategies

Squirrel monkeys are omnivores in the wild and may spend up to 75% of their time eating as they move through the forest. Squirrel monkeys' captive diets can be supplemented with many fruits and vegetables, depending on seasonal availability. Additional items such as Gatorade®, multivitamins, and food pellet rewards may be added for experimental or clinical reasons.

Squirrel monkeys are naturally active and readily manipulate objects in their enclosure. Destructible enrichment and devices that require time to open or manipulate can be provided to occupy the animals' time. Small objects, such as plastic golf balls and polyvinyl chloride (PVC) plumbing joints, can be employed, with or without food enrichment, to increase

FIGURE 25.2 *Saimiri boliviensis* juvenile manipulating food enrichment from a device made from an old syringe cover. The device has holes cut into the side just bigger than the food item to be retrieved.

activity rates. Feeding or foraging boards can also be used, as the presence of artificial turf forage boards has been shown to increase activity (Fekete et al. 2000). Because squirrel monkeys are naturally inquisitive and active, the provision of multiple perches, swings, and other hanging devices will increase activity levels. Figure 25.2 shows a squirrel monkey puzzling over a mealworm feeder. In all caging conditions, food items and perch arrangements can be used to vary the environment of the animals and, consequently, increase activity levels and sensory stimulation. Perches and manipulanda should be rotated and replaced routinely to prevent habituation.

Communication

Communication between animals within a given social group includes olfactory, visual, and vocal signals. Olfaction is most important during the breeding and birthing seasons. Vocal signals are very important for squirrel monkeys, with infants possessing a wide repertoire of sounds, including grumbles, tucks, purrs, chucks, and peeps (Biben and Symmes 1986) to communicate their wellbeing to the mother. In adults, several calls are used to signal alarm, display general disturbance or excitement, or establish contact when group members are visually separated. Visual cues are of primary importance in a number of displays intended to assert dominance or reduce tension between individuals. Visual cues involving facial expressions can be associated with fear or aggression. In a captive enclosure, visual barriers that a submissive monkey can use to break the stare of a more aggressive monkey should be provided. These "hide-boxes" can be made from a piece of a large diameter PVC pipe or another opaque material. However, remember that caregivers and researchers must be able to see the animals for health checks and other observations. Males use scent to determine the receptivity of females, and females use scent to identify individual infants. When cleaning squirrel monkey enclosures, it is preferable to avoid heavily scented products.

Training

Squirrel monkeys have a long history in cognitive research (Gazes et al. 2018; Thomas and Frost 1983; Woodburne and Rieke 1966) and can readily be trained using positive reinforcement training techniques (Gillis et al. 2012; Rogge et al. 2013). Squirrel monkeys can be trained to voluntarily participate in many aspects of their care, including the presentation of body parts, sitting on a scale, or moving from one cage to another on command. Rogge et al. (2013) found that it only took a few sessions to train the animals to present their hands for inspection on command, even when they were trained in their social group. They can also be trained to participate in more complex tasks, such as using a joystick and computer screen (Judge et al. 2005; Washburn et al. 1989), or engaging in token exchange procedures (Hopper et al. 2013; Talbot et al. 2011,). When training squirrel monkeys, it is important to remember to use very small portions of food as reinforcers, because the animals can quickly become satiated and stop working if given too much food.

Special Situations

Investigators not accustomed to using small nonhuman primates in their research should also be aware of squirrel monkey characteristics that may require alteration of experimental procedures. Squirrel monkeys are predisposed to hypoglycemia (Abee 1985; Brady 1990, Brady et al. 1991). The normal blood glucose concentration for adult squirrel monkeys is 80 ± 28 mg/dL (Loeb and Quimby 1989). Prolonged research procedures, fasting, or debilitating conditions that result in anorexia may place animals at risk for a hypoglycemic crisis. Their small size, low body fat reserves, and long extremities all contribute to rapid heat loss when they are anaesthetized or debilitated. The body temperature of at-risk animals should be monitored for body temperature on a regular basis. Awake, restrained squirrel monkeys normally have a body temperature of 38°C–39.5°C, while healthy, active, struggling animals may rapidly increase their body temperatures to as high as 41°C.

Although social housing of squirrel monkeys is desirable, fight-related injuries do occur in such housing, especially when social groups are first established. Close attention to group compatibility, careful supervision of newly established groups, and provision of hiding places in animal enclosures can help reduce such injuries or prevent them altogether (Williams and Abee 1988). When wounds do occur, they can usually be divided into crushing-type injuries (more commonly inflicted by female squirrel monkeys) or lacerations (commonly inflicted by males) (Ruiz et al. 2005).

Conclusions and Recommendations

Squirrel monkeys have many qualities that make them an important biomedical research model. Their small size, ease of handling, and lack of viruses that are pathogenic in humans are particularly desirable for studies requiring frequent handling without sedation. Our aim in this chapter is to provide a background through which you can better appreciate the behavior and biology of squirrel monkeys. Each species has a unique set of environmental and social needs that must be met for the animals to thrive in captivity. Even within a single taxonomic order, such as primates, there are major differences among genera in requirements for functionally appropriate captive environments. When designing appropriate housing for squirrel monkeys it is imperative to consider their ecology, social structure, and social organization. Squirrel monkeys are small arboreal insectivores, that are constantly in motion, and that have complex social needs. By designing their homes to match their biological needs, you will have a healthier, less stressed research animal.

REFERENCES

Abee, C. R. 1985. "Medical care and management of the squirrel monkey." In *Handbook of Squirrel Monkey Research*, edited by L. Roseblum and C. Coe, 447–488. New York: Plenum Press.

Abee, C. R. 2000. "Squirrel monkey (Saimiri spp.) research and resources." *ILAR Journal* 41 (1):2–9. doi:10.1093/ilar.41.1.2.

Achat-Mendes, C., Platt, D.M. and Spealman, R.D. 2012. "Antagonism of metabotropic glutamate 1 receptors attenuates behavioral effects of cocaine and methamphetamine in squirrel monkeys. Journal of Pharmacology and Experimental." *Therapeutics*, 343 (1):214–224.

Akkoç, C. C., and L. E. Williams. 2005. "Population modeling for a captive squirrel monkey colony." *American Journal of Primatology* 65 (3):239–254.

Alves, F. A., M. Pelajo-Machado, P. R. Totino, M. T. Souza, E. C. Goncalves, M. P. Schneider, J. A. Muniz, M. A. Krieger, M. C. Andrade, C. T. Daniel-Ribeiro, and L. J. Carvalho. 2015. "Splenic architecture disruption and parasite-induced splenocyte activation and anergy in Plasmodium falciparum-infected Saimiri sciureus monkeys." *Malar Journal* 14:128. doi:10.1186/s12936-015-0641-3.

Anderson, D. C., S. A. Lapp, S. Akinyi, E. V. Meyer, J. W. Barnwell, C. Korir-Morrison, and M. R. Galinski. 2015. "Plasmodium vivax trophozoite-stage proteomes." *Journal of Proteomics* 115:157–76. doi:10.1016/j.jprot.2014.12.010.

Anderson, D. C., S. A. Lapp, J. W. Barnwell, and M. R. Galinski. 2017. "A large scale Plasmodium vivax-Saimiri boliviensis trophozoite-schizont transition proteome." *PLoS One* 12 (8):e0182561. doi:10.1371/journal.pone.0182561.

Baldwin, J. D. 1969. "The ontogeny of social behavior of squirrel monkeys (Saimiri sciureus) in a seminatural environment." *Folia Primatologica* 11:35–79.

Baldwin, J. D., and J. Baldwin. 1972. "The ecology and behavior of squirrel monkeys (Saimiri oerstedi) in a natural forest in western Panama." *Folia Primatologica* 18 (3):161–184. doi:10.1159/000155478.

Baldwin, J. D., and J. I. Baldwin. 1976. "Effects of food ecology on social play: A laboratory simulation." *Z Tierpsychologie* 40 (1):1–14. doi:10.1111/j.1439-0310.1976.tb00922.x.

Baldwin, J. D., and J. I. Baldwin. 1971. "Squirrel monkeys (*Saimiri*) in natural habitats in Panama, Colombia, Brazil, and Peru." *Primates* 12 (1):45–61.

Biben, M., and S. J. Suomi. 1993. "Lessons from primate play." In *Parent-Child Play: Descriptions and Implications*, edited by K. MacDonald, 185–196. New York: SUNY Press.

Biben, M., and D. Symmes. 1986. "Play vocalizations of squirrel monkeys (*Saimiri sciureus*)." *Folia Primatologica* 46 (3):173–182. doi:10.1159/000156250.

Biben, M., and D. Symmes. 1991. "Playback studies of affiliative vocalizing in captive squirrel-monkeys - familiarity as a cue to response." *Behaviour* 117 (1/2):1–19. doi:10.1163/156853991x00094.

Blair, M. E., and D. J. Melnick. 2012. "Genetic evidence for dispersal by both sexes in the Central American Squirrel Monkey, Saimiri oerstedii citrinellus." *American Journal of Primatology* 74 (1):37–47. doi:10.1002/ajp.21007.

Boinski, S. 1987. "Birth synchrony in squirrel-monkeys (Saimiri-Oerstedi) - a strategy to reduce neonatal predation." *Behavioral Ecology and Sociobiology* 21 (6):393–400. doi:10.1007/Bf00299934.

Boinski, S. 1988. "Sex-differences in the foraging behavior of squirrel-monkeys in a seasonal habitat." *Behavioral Ecology and Sociobiology* 23 (3):177–186. doi:10.1007/Bf00300352.

Boinski, S. 1989. "The effects of body size, morphological specializations, and food distribution on foraging behavior in the cebinae." *American Journal of Physical Anthropology* 78 (2):195–195.

Boinski, S. 1991. "The coordination of spatial position - a field-study of the vocal behavior of adult female squirrel-monkeys." *Animal Behaviour* 41:89–102. doi:10.1016/S0003-3472(05)80505-6.

Boinski, S. 1992a. "Monkeys with inflated sex appeal." *Natural History* 101(7):42–49.

Boinski, S. 1992b. "Olfactory communication among costa-rican squirrel-monkeys - a field-study." *Folia Primatologica* 59 (3):127–136. doi:10.1159/000156650.

Boinski, S. 1994. "Affiliation patterns among male costa-rican squirrel-monkeys." *Behaviour* 130:191–209. doi:10.1163/156853994x00523.

Boinski, S. 1999. "The social organizations of squirrel monkeys: Implications for ecological models of social evolution." *Evolutionary Anthropology* 8 (3):101–112. doi:10.1002/(Sici)1520-6505(1999)8:3<101::Aid-Evan5>3.0.Co;2-O.

Boinski, S., E. Ehmke, L. Kauffman, S. Schet, and A. Vreedzaam. 2005. "Dispersal patterns among three species of squirrel monkeys (Saimiri oerstedii, S-boliviensis and S-sciureus): II. Within-species and local variation." *Behaviour* 142:633–677. doi:10.1163/1568539054352860.

Boinski, S., and D. M. Fragaszy. 1989. "The ontogeny of foraging in squirrel-monkeys, saimiri-oerstedi." *Animal Behaviour* 37:415–428. doi:10.1016/0003-3472(89)90089-4.

Boinski, S. H. 1986. "The ecology of squirrel monkeys in Costa Rica." *Dissertation Abstracts International* B47 (5):1893.

Boinski, S., and C. L. Mitchell. 1992. "Ecological and social-factors affecting the vocal behavior of adult female squirrel-monkeys." *Ethology* 92 (4):316–330. doi:10.1111/j.1439-0310.1992.tb00969.x.

Boinski, S., and C. L. Mitchell. 1994. "Male residence and association patterns in Costa Rican squirrel monkeys (Saimiri oerstedi)." *American Journal of Primatology* 34 (2):157–169. doi:10.1002/ajp.1350340207.

Boinski, S., and C. L. Mitchell. 1995. "Wild squirrel monkey (Saimiri sciureus) "caregiver" calls: Contexts and acoustic structure." *American Journal of Primatology* 35 (2):129–137. doi:10.1002/ajp.1350350205.

Boinski, S., and C. L. Mitchell. 1997. "Chuck vocalizations of wild female squirrel monkeys (*Saimiri sciureus*) contain information on caller identity and foraging activity." *International Journal of Primatology* 18 (6):975–993. doi:10.1023/A:1026300314739.

Boinski, S., K. Sughrue, L. Selvaggi, R. Quatrone, M. Henry, and S. Cropp. 2002. "An expanded test of the ecological model of primate social evolution: Competitive regimes and female bonding in three species of squirrel monkeys (*Saimiri oerstedii, S. boliviensis, and S. sciureus*)." *Behaviour* 139:227–261. doi:10.1163/156853902760102663.

Brady, A. G., G. E. Hutto, L. E. Williams, and C. R. Abee. 1991. "Comparison of two tests for identifying squirrel monkey infants at risk for hypoglycemia." *AALAS Bulletin* 30(4):28–29.

Brady, A. G., L. E. Williams, and C. R. Abee. 1990. "Hypoglycemia of squirrel monkey neonates: implications for infant survival." *Laboratory Animal Science* 40 (3):262–5.

Carretero-Pinzón, X., T. R. Defler, and M. Ruiz-Garcia. 2016. "How does the Colombian squirrel monkey cope with habitat fragmentation? Strategies to survive in small fragments." In: *Phylogeny, Molecular Population Genetics, Evolutionary Biology and Conservation of the Neotropical Primates*, edited by M. Ruiz-Garcia and J.M. Shostell, 2–16. Nova Science Publisher: New York.

Carretero-Pinzón, X., T.R. Defler, C.A. McAlpine, and J.R. Rhodes. 2017. "The influence of landscape relative to site and patch variables on primate distributions in the Colombian Llanos." *Landscape Ecology* 32 (4):883–896. doi:10.1007/s10980-017-0493-z.

Cheung, S. W., C. A. Atencio, E. R. J. Levy, R. C. Froemke, and C. E. Schreiner. 2017. "Anisomorphic cortical reorganization in asymmetric sensorineural hearing loss." *Journal of Neurophysiology* 118 (2):932–948. doi:10.1152/jn.00119.2017.

Chiou, K. L., L. Pozzi, J. W. Lynch Alfaro, and A. Di Fiore. 2011. "Pleistocene diversification of living squirrel monkeys (*Saimiri* spp.) inferred from complete mitochondrial genome sequences." *Molecular Phylogenetics and Evolution* 59 (3):736–745. doi:10.1016/j.ympev.2011.03.025.

Collins, W. E., J. S. Sullivan, E. Strobert, G. G. Galland, A. Williams, D. Nace, T. Williams, and J. W. Barnwell. 2009. "Studies on the Salvador I strain of Plasmodium vivax in non-human primates and anopheline mosquitoes." *American Journal of Tropical Medicine Hygiene* 80 (2):228–235.

Cunha, J. A., L. J. M. Carvalho, C. Bianco-Junior, M. C. R. Andrade, L. R. Pratt-Riccio, E. K. P. Riccio, M. Pelajo-Machado, I. J. da Silva, P. Druilhe, and C. T. Daniel-Ribeiro. 2017. "Increased plasmodium falciparum parasitemia in non-splenectomized saimiri sciureus monkeys treated with clodronate liposomes." *Front Cell Infect Microbiol* 7:408. doi:10.3389/fcimb.2017.00408.

de Souza, H., E. H. Costa-Correa, C. Bianco-Junior, M. C. R. Andrade, J. D. C. Lima-Junior, L. R. Pratt-Riccio, C. T. Daniel-Ribeiro, and P. R. R. Totino. 2017. "Detection of signal regulatory protein alpha in *Saimiri sciureus* (squirrel monkey) by anti-human monoclonal antibody." *Frontiers in Immunology* 8:1814. doi:10.3389/fimmu.2017.01814.

Desai, R. I., M. R. Doyle, S. L. Withey, and J. Bergman. 2016a. "Nicotinic effects of tobacco smoke constituents in nonhuman primates." *Psychopharmacology* 233 (10):1779–1789. doi:10.1007/s00213-016-4238-5.

Desai, R. I., K. A. Sullivan, S. J. Kohut, and J. Bergman. 2016b. "Influence of experimental history on nicotine self-administration in squirrel monkeys." *Psychopharmacology* 233 (12):2253–2263. doi:10.1007/s00213-016-4274-1.

Di Bitetti, M. S., and C. H. Janson. 2000. "When will the stork arrive? Patterns of birth seasonality in neotropical primates." *American Journal of Primatology* 50 (2):109–130. doi:10.1002/(sici)1098-2345(200002)50:2<109::Aid-ajp2>3.0.Co;2-w.

Du Mond, F. V., and T. C. Hutchinson. 1967. "Squirrel monkey reproduction: The "fatted" male phenomenon and seasonal spermatogenesis." *Science* 158:1067–1070 doi:10.1126/science.158.3804.1067.

Elias, M. F. 1977. "Relative maturity of cebus and squirrel monkeys at birth and during infancy." *Developmental Psychobiologyl* 10 (6):519–528. doi:10.1002/dev.420100605.

Fekete, J. M., J. L. Norcross, and J. D. Newman. 2000. "Artificial turf foraging boards as environmental enrichment for pair-housed female squirrel monkeys." *Contemporary Topics in Laboratory Animal Science* 39 (2):22–26.

Fleagle, J. G., and R. A. Mittermeier. 1980. "Locomotor behavior, body size, and comparative ecology of seven Surinam monkeys." *American Journal of Physical Anthropology* 52 (3):301–314. doi:10.1002/ajpa.1330520302.

Frechette, J. L., K. E. Sieving, and S. Boinski. 2014. "Social and personal information use by squirrel monkeys in assessing predation risk." *American Journal of Primatology* 76 (10):956–66. doi:10.1002/ajp.22283.

Gao, Y., P. Parvathaneni, K. G. Schilling, F. Wang, I. Stepniewska, Z. Xu, A. S. Choe, Z. Ding, J. C. Gore, L. M. Chen, B. A. Landman, and A. W. Anderson. 2016. "A 3D high resolution ex vivo white matter atlas of the common squirrel monkey (*Saimiri sciureus*) based on diffusion tensor imaging." *Proceedings SPIE – The International Society for Optical Engineering* 9784. doi:10.1117/12.2217325.

Garber, P. A., and S. R. Leigh. 1997. "Ontogenetic variation in small-bodied New World primates: Implications for patterns of reproduction and infant care." *Folia Primatolica* 68 (1):1–22. doi:10.1159/000157226.

Gazes, R. P., A. R. Billas, and V. Schmitt. 2018. "Impact of stimulus format and reward value on quantity discrimination in capuchin and squirrel monkeys." *Learning and Behavior* 46 (1):89–100. doi:10.3758/s13420-017-0295-9.

Gillis, T. E., A. C. Janes, and M. J. Kaufman. 2012. "Positive reinforcement training in squirrel monkeys using clicker training." *American Journal of Primatology* 74 (8):712–720. doi:10.1002/ajp.22015.

Gunalan, K., J. M. Sa, R. R. Moraes Barros, S. L. Anzick, R. L. Caleon, J. P. Mershon, K. Kanakabandi, M. Paneru, K. Virtaneva, C. Martens, J. W. Barnwell, J. M. Ribeiro, and L. H. Miller. 2019. "Transcriptome profiling of Plasmodium vivax in Saimiri monkeys identifies potential ligands for invasion." *Proceedings of the National Academy of Sciences of the United States of America* 116 (14):7053–7061. doi:10.1073/pnas.1818485116.

Hartwig, W. C. 1996. "Perinatal life history traits in New World monkeys." *American Journal of Primatology* 40 (2):99–130. doi:10.1002/(Sici)1098-2345(1996)40:2<99::Aid-Ajp1>3.0.Co;2-V.

Hopf, S. 1978. "Huddling subgroups in captive squirrel monkeys and their changes in relation to ontogeney." *Biology of Behavior* 3:147–162.

Hopper, L., A. Holmes, L. Williams, and S. Brosnan. 2013. "Dissecting the mechanisms of squirrel monkey (*Saimiri boliviensis*) social learning." *PeerJ* 1:e13. doi:10.7717/peerj.13.

Horenstein, V.D., L. E. Williams, A.R. Brady, C.R. Abee, and M.G. Horenstein. 2005. "Age-related diffuse chronic telogen effluvium-type alopecia in female squirrel monkeys (*Saimiri boliviensis boliviensis*)." *Comparative Medicine* 55 (2):169–174.

Janson, C. H., and S. Boinski. 1992. "Morphological and behavioral adaptations for foraging in generalist primates: The case of the cebines." *American Journal of Physical Anthropology* 88 (4):483–498. doi:10.1002/ajpa.1330880405.

Jasper, J.M, Delacth, N.A., Aghababian,, S.C., Brown, C.E. and Stone, A.I. 2020. Why do squirrel monkeys urine wash? A field study of Saimiri collinsi in Eastern Amazonia, Brazil. *American Journal of Primatology* 82 (S1): 58.

Judge, P. G., T. A. Evans, and D. K. Vyas. 2005. "Ordinal representation of numeric quantities by brown capuchin monkeys (*Cebus apella*)." *Journal of Experimental Psychology: Animal Behavior Process* 31 (1):79–94. doi:10.1037/0097-7403.31.1.79.

Kaack, B., L. Walker, K. R. Brizzee, and R. H. Wolf. 1979. "Comparative normal levels of serum triiodothyronine and throxine in nonhuman primates." *Laboratory Animal Science* 29:191–194.

Kangas, B. D., and J. Bergman. 2014. "Operant nociception in nonhuman primates." *Pain* 155 (9):1821–1828. doi:10.1016/j.pain.2014.06.010.

Kangas, B. D., M. Z. Leonard, V. G. Shukla, S. O. Alapafuja, S. P. Nikas, A. Makriyannis, and J. Bergman. 2016. "Comparisons of Delta9-tetrahydrocannabinol and anandamide on a battery of cognition-related behavior in nonhuman primates." *Journal of Pharmacology Experimental Therapeutics* 357 (1):125–33. doi:10.1124/jpet.115.228189.

Kaplan, J. N. 1979. "Growth and development of infant squirrel monkeys during the first six months of life." In *Nursery Care of Nonhuman Primates*, edited by G.C. Ruppenthal and D.J. Reese, 153–164. Springer US: Boston, MA.

Laska, M., and R. Hudson. 1995. "Ability of female squirrel-monkeys (saimiri-sciureus) to discriminate between conspecific urine odors." *Ethology* 99 (1):39–52.

Leger, D. W., W. A. Mason, and D. M. Fragaszy. 1981. "Sexual segregation, cliques, and social power in squirrel-monkey (*Saimiri*) groups." *Behaviour* 76 (3–4):163–181. doi:10.1163/156853981x00068.

Leighty, K. A., G. Byrne, D. M. Fragaszy, E. Visalberghi, C. Welker, and I. Lussier. 2004. "Twinning in tufted capuchins (*Cebus apella*): Rate, survivorship, and weight gain." *Folia Primatologica* 75 (1):14–8. doi:10.1159/000073425.

Levi, T., K. M. Silvius, L. F. B. Oliveira, A. R. Cummings, and J. M. V. Fragoso. 2013. "Competition and facilitation in the capuchin-squirrel monkey relationship." *Biotropica* 45 (5):636–643. doi:10.1111/btp.12046.

Lima, E. M., and S. F. Ferrari. 2003. "Diet of a free-ranging group of squirrel monkeys (*Saimiri sciureus*) in eastern brazilian amazonia." *Folia Primatologica* 74 (3):150–158. doi:10.1159/000070648.

Loeb, W. F., and F. W. Quimby, eds. 1989. *The Clinical Chemistry of Laboratory Animals*. Pergamon Press: New York.

Lynch Alfaro, J. W., J. P. Boubli, F. P. Paim, C. C. Ribas, M. N. Silva, M. R. Messias, F. Rohe, M. P. Merces, J. S. Silva Junior, C. R. Silva, G. M. Pinho, G. Koshkarian, M. T. Nguyen, M. L. Harada, R. M. Rabelo, H. L. Queiroz, M. E. Alfaro, and I. P. Farias. 2015. "Biogeography of squirrel monkeys (genus *Saimiri*): South-central Amazon origin and rapid pan-Amazonian diversification of a lowland primate." *Molecular Phylogenetics and Evolution* 82 Pt B:436–454. doi:10.1016/j.ympev.2014.09.004.

Lyons, D. M., S. P. Mendoza, and W. A. Mason. 1992. "Sexual segregation in squirrel monkeys (*Saimiri sciureus*): A transactional analysis of adult social dynamics." *Journal of Comparative Psycholy* 106 (4):323–30. doi:10.1037/0735-7036.106.4.323.

Malone, B. J., R. E. Beitel, M..Vollmer, M. A. Heiser, and C. E. Schreiner. 2015. "Modulation-frequency-specific adaptation in awake auditory cortex." *Journal of Neuroscience* 35 (15):5904–5916. doi:10.1523/JNEUROSCI.4833-14.2015.

Malone, B. J., M. A. Heiser, R. E. Beitel, and C. E. Schreiner. 2017. "Background noise exerts diverse effects on the cortical encoding of foreground sounds." *Journal of Neurophysiology* 118 (2):1034–1054. doi:10.1152/jn.00152.2017.

Mancuso, K., W. W. Hauswirth, Q. Li, T. B. Connor, J. A. Kuchenbecker, M. C. Mauck, J. Neitz, and M. Neitz. 2009. "Gene therapy for red-green colour blindness in adult primates." *Nature* 461 (7265):784–787. doi:10.1038/nature08401.

Manocha, S. L. 1979. "Physical growth and brain development of captive-bred male and female squirrel monkeys, *Saimiri sciureus*." *Experientia* 35 (1):96–98. doi:10.1007/BF01917901.

Marsh, H. L., A. Q. Vining, E. K. Levendoski, and P. G. Judge. 2015. "Inference by exclusion in lion-tailed macaques (*Macaca silenus*), a hamadryas baboon (*Papio hamadryas*), capuchins (*Sapajus apella*), and squirrel monkeys (*Saimiri sciureus*)." *Journal of Comparative Psychology* 129 (3):256–267. doi:10.1037/a0039316.

Mendoza, S. P., E. L. Lowe, J. M. Davidson, and S. Levine. 1978. "Annual cyclicity in the squirrel-monkey (*Saimiri sciureus*) - relationship between testosterone, fatting, and sexual-behavior." *Hormones and Behavior* 11 (3):295–303. doi:10.1016/0018-506x(78)90033-8.

Merces, M. P., J. W. Lynch Alfaro, W. A. Ferreira, M. L. Harada, and J. S. Silva Junior. 2015. "Morphology and mitochondrial phylogenetics reveal that the Amazon River separates two eastern squirrel monkey species: *Saimiri sciureus* and *S. collinsi*." *Molecular Phylogeneics and Evolution* 82 (Pt B):426–435. doi:10.1016/j.ympev.2014.09.020.

Migliaccio, A. A., L. B. Minor, and C. C. Della Santina. 2010. "Adaptation of the vestibulo-ocular reflex for forward-eyed foveate vision." *Journal of Physiology* 588 (Pt 20):3855–3867. doi:10.1113/jphysiol.2010.196287.

Milich, K. M., B. J. Koestler, J. H. Simmons, P. N. Nehete, A Di Fiore, L. E. Williams, J. P. Dudley, J. Vanchiere, and S. M. Payne. 2018. "Methods for detecting Zika virus in feces: A case study in captive squirrel monkeys (*Saimiri boliviensis boliviensis*)." *PLoS One* 13 (12):e0209391. doi:10.1371/journal.pone.0209391.

Miller, K.R., K. Laszlo, and S.J. Suomi. 2008. "Why do captive tufted capuchins (*Cebus apella*) urine wash?" *American Journal of Primatology* 70 (2):119–126. doi:10.1002/ajp.20462.

Milligan, L. A., S. V. Gibson, L. E. Williams, and M. L. Power. 2008. "The composition of milk from Bolivian squirrel monkeys (*Saimiri boliviensis boliviensis*)." *American Journal of Primatology* 70 (1):35–43. doi:10.1002/ajp.20453.

Mitchell, C. L. 1990. "The ecological basis for female social dominance: A behavioral study of the squirrel monkey (*Saimiri sciureus*) in the wild." *Dissertation Abstracts International* B51(4):1614.

Mitchell, C.L., S. Boinski, and C. P. van Schaik. 1991. "Competitve regimes and female bonding in two species of squirrel monkeys (*Saimrir oerstedi* and *S. sciureus*)." *Behavioral Ecology and Sociobiology* 28:55–60.

Mittermeier, R. A., and M. G. van Roosmalen. 1981. "Preliminary observations on habitat utilization and diet in eight Surinam Monkeys." *Folia Primatologica* 36 (1–2):1–39. doi:10.1159/000156007.

Montague, M.J., T.R. Disotell, and A. Di Fiore. 2014. "Population genetics, dispersal, and kinship among wild squirrel monkeys (*Saimiri sciureus macrodon*): Preferential association between closely related females and its implications for insect prey capture success." *International Journal of Primatology* 35 (1):169–187. doi:10.1007/s10764-013-9723-7.

Mowery, T. M., R. M. Sarin, P. V. Kostylev, and P. E. Garraghty. 2015. "Differences in AMPA and GABAA/B receptor subunit expression between the chronically reorganized cortex and brainstem of adult squirrel monkeys." *Brain Research* 1611:44–55. doi:10.1016/j.brainres.2015.03.010.

Mulholland, M.M., L. E. Williams, and C. R. Abee. 2020. "Rearing condition may alter neonatal development of captive Bolivian squirrel monkeys (*Saimiri boliviensis boliviensis*)." *Developmental Psychobiology*. doi:10.1002/dev.21960.

Murray, S. G., and J. E. King. 1973. "Snake avoidance in feral and laboratory reared squirrel-monkeys." *Behaviour* 47 (3–4):281. doi:10.1163/156853973x00120.

National Research Council. 2011. *Guide for the Care and Use of Laboratory Animals*. 8th ed. National Academy Press: Washington, DC.

Neitz, M., and J. Neitz. 2014. "Curing color blindness--mice and nonhuman primates." *Cold Spring Harbor Perspectives in Medicin* 4 (11):a017418. doi:10.1101/cshperspect.a017418.

Obaldia, N., E. Meibalan, J. M. Sa, S. Ma, M. A. Clark, P. Mejia, R. R. Moraes Barros, W. Otero, M. U. Ferreira, J. R. Mitchell, D. A. Milner, C. Huttenhower, D. F. Wirth, M. T. Duraisingh, T. E. Wellems, and M. Marti. 2018. "Bone marrow is a major parasite reservoir in plasmodium vivax infection." *mBio* 9 (3):e00625-18. doi:10.1128/mBio.00625-18.

Paim, F. P., J. D. E. Silva, J. Valsecchi, M. L. Harada, and H. L. de Queiroz. 2013. "Diversity, geographic distribution and conservation of squirrel monkeys, saimiri (primates, cebidae), in the floodplain forests of central amazon." *International Journal of Primatology* 34 (5):1055–1076. doi:10.1007/s10764-013-9714-8.

Paim, F. P., C. A. Chapman, H. L. de Queiroz, and A. P. Paglia. 2017. "Does resource availability affect the diet and behavior of the vulnerable squirrel monkey, *Saimiri vanzolinii*?" *International Journal of Primatology* 38 (3):572–587. doi:10.1007/s10764-017-9968-7.

Peres, C. A. 1994. "Primate responses to phenological changes in an amazonian terra-firme forest." *Biotropica* 26 (1):98–112. doi:10.2307/2389114.

Peterson, M. S., C. J. Joyner, R. J. Cordy, J. L. Salinas, D. Machiah, S. A. Lapp, H. C. Ma, E. V. S. Meyer, S. Gumber, and M. R. Galinski. 2019. "*Plasmodium vivax* parasite load is associated with histopathology in *Saimiri boliviensis* with findings comparable to *P vivax* pathogenesis in humans." *Open Forum Infectious Diseases* 6 (3):ofz021. doi:10.1093/ofid/ofz021.

Pinheiro, T., S. F. Ferrari, and M. A. Lopes. 2011. "Polyspecific associations between squirrel monkeys (*Saimiri sciureus*) and other primates in eastern Amazonia." *American Journal of Primatology* 73 (11):1145–1151. doi:10.1002/ajp.20981.

Pinheiro, T., S. F. Ferrari, and M. A. Lopes. 2013. "Activity budget, diet, and use of space by two groups of squirrel monkeys (*Saimiri sciureus*) in eastern Amazonia." *Primates* 54 (3):301–308. doi:10.1007/s10329-013-0351-9.

Pinheiro, T., and M. A. Lopes. 2018. "Hierarchical structure and the influence of individual attributes in the captive squirrel monkey (*Saimiri collinsi*)." *Primates* 59 (5):475–482. doi:10.1007/s10329-018-0668-5.

Pinheiro, T.M. 2015. "Sistema social do macaco-de-cheiro (*Saimiri collinsi*) em cativeiro." Ph.D. thesis, Universidade Federal do Pará, Belem, Brazil.

Podolsky, R. D. 1990. "Effects of mixed-species association on resource use by *Saimiri sciureus* and *Cebus apella*." *American Journal of Primatology* 21 (2):147–158. doi:10.1002/ajp.1350210207.

Pontes, A. R. M. 1997. "Habitat partitioning among primates in Maraca island, Roraima, northern Brazilian Amazonia." *International Journal of Primatology* 18 (2):131–157. doi:10.1023/A:1026364417523.

Ribeiro-Junior, M. A., S. F. Ferrari, J. R. Lima, C. R. da Silva, and J. D. Lima. 2016. "Predation of a squirrel monkey (*Saimiri sciureus*) by an Amazon tree boa (*Corallus hortulanus*): even small boids may be a potential threat to small-bodied platyrrhines." *Primates* 57 (3):317–322. doi:10.1007/s10329-016-0545-z.

Rogge, J., K. Sherenco, R. Malling, E. Thiele, S. Lambeth, S. Schapiro, and L. Williams. 2013. "A comparison of positive reinforcement training techniques in owl and squirrel monkeys: time required to train to reliability." *Journal of Applied Animal Welfare Science* 16 (3): 211–220. doi:10.1080/10888705.2013.798223.

Rosenblum, L. A. 1972. "Sex and age differences in response to infant squirrel monkeys." *Brain and Behavior Evolution* 5 (1):30–40. doi:10.1159/000123735.

Ross, C. 1991. "Life-history patterns of new-world monkeys." *International Journal of Primatology* 12 (5): 481–502. doi:10.1007/Bf02547635.

Rowlett, J.K., Platt, D.M., Lelas, S., Atack, J.R. and Dawson, G.R. 2005. "Different GABAA receptor subtypes mediate the anxiolytic, abuse-related, and motor effects of benzodiazepine-like drugs in primates." *Proceedings of the National Academy of Sciences* 102 (3):915–920.

Ruivo, L. V. P., A. I. Stone, and M. Fienup. 2017. "Reproductive status affects the feeding ecology and social association patterns of female squirrel monkeys (*Saimiri collinsi*) in an Amazonian rainforest." *American Journal of Primatology* 79 (6):e22657. doi:10.1002/ajp.22657.

Ruivo, L. V. P., and A.I. Stone. 2014. "Jealous of mom? Interactions between infants and adult males during the mating season in wild squirrel monkeys." *Neotropical Primates* 21:165–170.

Ruiz, J. C., A. G. Brady, S. L. Gibson, L. E. Williams, and C. R. Abee. 2005. "Morbidity and mortality of adult female Bolivian squirrel monkeys in a 10-year-period." *Contemporary Topics in Laboratory Animal Science* 44(4):74.

Saltzman, W., S. P. Mendoza, and W. A. Mason. 1989. "Social dynamics of sequential group formation among female squirrel monkeys." *American Journal of Primatology* 18(2):164.

Schilling, K. G., V. Janve, Y. Gao, I. Stepniewska, B. A. Landman, and A. W. Anderson. 2018. "Histological validation of diffusion MRI fiber orientation distributions and dispersion." *Neuroimage* 165:200–221. doi:10.1016/j.neuroimage.2017.10.046.

Schiml, P. A., S. P. Mendoza, W. Saltzman, D. M. Lyons, and W. A. Mason. 1996. "Seasonality in squirrel monkeys (*Saimiri sciureus*): Social facilitation by females." *Physiology and Behavior* 60 (4):1105–13. doi:10.1016/0031-9384(96)00134-5.

Schino, G., M. Palumbo, and E. Visalberghi. 2011. "Factors affecting urine washing behavior in tufted capuchins (*Cebus apella*)." *International Journal of Primatology* 32 (4):801–810. doi:10.1007/s10764-011-9502-2.

Schuler, A. M., and C. R. Abee. 2005. "Squirrel monkeys (Saimiri)." In *Enrichment of Nonhuman Primates*, 73–88. NIH Office of Laboratory Animal Welfare: Bethesda.

Scollay, P. A. 1980. "Cross-sectional morphometric data on a population of semifree-ranging squirrel-monkeys, saimiri-sciureus (iquitos)." *American Journal of Physical Anthropology* 53 (2):309–316. doi:10.1002/ajpa.1330530216.

Shapley, R. 2009. "Vision: Gene therapy in colour." *Nature* 461 (7265):737–739. doi:10.1038/461737a.

Stone, A.I. 2004. "Juvenile feeding ecology and life history in a neotropical primate, the squirrel monkey (*Saimiri sciureus*)." Ph.D. Dissertation, Program in Ecology and Evolutionary Biology, University of Illinois at Urbana-Champaign.

Stone, A. I. 2006. "Foraging ontogeny is not linked to delayed maturation in squirrel monkeys (*Saimiri sciureus*)." *Ethology* 112 (2):105–115. doi:10.1111/j.1439-0310.2005.01121.x.

Stone, A. I. 2007a "Responses of squirrel monkeys to seasonal changes in food availability in an eastern Amazonian forest." *American Journal of Primatology* 69 (2):142–57. doi:10.1002/ajp.20335.

Stone, A. I. 2007b. "Ecological risk aversion and foraging behaviors of juvenile squirrel monkeys (*Saimiri sciureus*)." *Ethology* 113 (8):782–792. doi:10.1111/j.1439-0310.2007.01377.x.

Stone, A. I. 2008. "Seasonal effects on play behavior in immature *Saimiri sciureus* in Eastern Amazonia." *International Journal of Primatology* 29:195–205.

Stone, A. I. 2014. "Is fatter sexier? Reproductive strategies of male squirrel monkeys (*Saimiri sciureus*)." *International Journal of Primatology* 35 (3–4):628–642. doi:10.1007/s10764-014-9755-7.

Stone, A.I. 2018. "The foraging ecology of male and female squirrel monkeys (*Saimiri collinsi*) in Eastern Amazonia, Brazil." In *La primatología en Latinoamérica 2*, edited by B. Urbani, M. Kowalewski, R.G.T. Cunha, S. de la Torre and L. Cortés-Ortiz, 229–237. Instituto Venezolano de Investigaciones Científicas: Caracas, Venezuela.

Stone, A.I. and L.V.P. Ruivo. 2020. "Synchronization of weaning time with peak fruit availability in squirrel monkeys (*Saimiri collinsi*) living in Amazonian Brazil." *American Journal of Primatology* 82 (7): e23139. doi:10.1002/ajp.23139.

Talbot, C. F., H. D. Freeman, L. E. Williams, and S. F. Brosnan. 2011. "Squirrel monkeys' response to inequitable outcomes indicates a behavioural convergence within the primates." *Biology Letters* 7 (5):680–682. doi:10.1098/rsbl.2011.0211.

Tardif, S. D. 1994. "Relative energetic cost of infant care in small-bodied neotropical primates and its relation to infant-care patterns." *American Journal of Primatology* 34 (2):133–143. doi:10.1002/ajp.1350340205.

Tardif, S., A. Carville, D. Elmore, L.E. Williams, and K. Rice. 2012. "Reproduction and breeding of nonhuman primates." *Nonhuman Primates in Biomedical Research* 2012: 197

Terborgh, J. 1983. *Five New World primates*. Princeton University Press: Princeton, NJ.

Thomas, R. K., and T. Frost. 1983. "Oddity and dimension-abstracted oddity (Dao) in squirrel-monkeys." *American Journal of Psychology* 96 (1):51–64. doi:10.2307/1422208.

Tougan, T., N. Arisue, S. Itagaki, Y. Katakai, Y. Yasutomi, and T. Horii. 2018. "Adaptation of the Plasmodium falciparum FCB strain for in vitro and in vivo analysis in squirrel monkeys (*Saimiri sciureus*)." *Parasitology International* 67 (5):601–604. doi:10.1016/j.parint.2018.05.012.

Trevino, H. S. 2007. "Seasonality of reproduction in captive squirrel monkeys (*Saimiri Sciureus*)." *American Journal of Primatology* 69 (9):1001–1012. doi:10.1002/ajp.20409.

Vaitl, E. A. 1977. "Social context as a structuring mechanism in captive groups of squirrel monkeys (*Saimiri sciureus*)." *Primates* 18:861–874.

Valdez, G.R., Platt, D.M., Rowlett, J.K., Rüedi-Bettschen, D. and Spealman, R.D. 2007. "κ Agonist-induced reinstatement of cocaine seeking in squirrel monkeys: a role for opioid and stress-related mechanisms." *Journal of Pharmacology and Experimental Therapeutics* 323 (2):525–533.

Vanchiere, J. A., J. C. Ruiz, A. G. Brady, T. J. Kuehl, L. E. Williams, W. B. Baze, G. K. Wilkerson, P. N. Nehete, G. B. McClure, D. L. Rogers, S. L. Rossi, S. R. Azar, C. M. Roundy, S. C. Weaver, N. Vasilakis, J. H. Simmons, and C. R. Abee. 2018. "Experimental zika virus infection of neotropical primates." *American Journal of Tropical Medicine and Hygiene* 98 (1):173–177. doi:10.4269/ajtmh.17–0322.

Washburn, D. A., W. D. Hopkins, and D. M. Rumbaugh. 1989. "Automation of learning-set testing: the video-task paradigm." *Behavior Research Methods Instruments and Computers* 21 (2):281–284. doi:10.3758/bf03205596.

Whiteley, H. E., J. I. Everitt, I. Kakoma, M. A. James, and M. Ristic. 1987. "Pathologic changes associated with fatal Plasmodium falciparum infection in the Bolivian squirrel monkey (*Saimiri sciureus boliviensis*)." *American Journal of Tropical Medicine and Hygiene* 37 (1):1–8. doi:10.4269/ajtmh.1987.37.1.

Williams, L.E., A.G. Brady, and C.R. Abee. 2010. "Squirrel monkeys." In *The UFAW Handbook on the Care and Management of Laboratory and Other Research Animals*, Eighth Edition, edited by R. Hubrecht and J. Kirkwood, 564–578. Wiley-Blackwell: Oxford, UK.

Williams, L., Vitulli, W., McElhinney, T., Wiebe, R.H. and Abee, C.R. 1986. "Male behavior through the breeding season in *Saimiri boliviensis boliviensis*." *American journal of primatology* 11 (1):27–35.

Williams, L. E., and C. R. Abee. 1988. "Aggression with mixed age-sex groups of bolivian squirrel-monkeys following single animal introductions and new group formations." *Zoo Biology* 7 (2):139–145. doi:10.1002/zoo.1430070207.

Williams, L., S. Gibson, M. McDaniel, J. Bazzel, S. Barnes, and C. Abee. 1994. "Allomaternal interactions in the Bolivian squirrel monkey (*Saimiri boliviensis boliviensis*)." *American Journal of Primatology* 34 (2):145–156. doi:10.1002/ajp.1350340206.

Williams, L., W. Vitulli, T. Mcelhinney, R. H. Wiebe, and C. R. Abee. 1986. "Male-behavior through the breeding-season in *Saimiri boliviensis boliviensis*." *American Journal of Primatology* 11 (1):27–35. doi:10.1002/ajp.1350110104.

Woodburne, L.S., and G.K. Rieke. 1966. "Response to symbols by squirrel monkeys." *Psychonomic Science* 5 (11):429–430.

Wrangham, R. W. 1980. "An ecological model of female-bonded primate groups." *Behaviour* 75 (3/4):262–300. doi:10.1163/156853980x00447.

Yagi, M., G. Bang, T. Tougan, N. M. Palacpac, N. Arisue, T. Aoshi, Y. Matsumoto, K. J. Ishii, T. G. Egwang, P. Druilhe, and T. Horii. 2014. "Protective epitopes of the *Plasmodium falciparum* SERA5 malaria vaccine reside in intrinsically unstructured N-terminal repetitive sequences." *PLoS One* 9 (6):e98460. doi:10.1371/journal.pone.0098460.

Yang, P. F., H. X. Qi, J. H. Kaas, and L. M. Chen. 2014. "Parallel functional reorganizations of somatosensory areas 3b and 1, and S2 following spinal cord injury in squirrel monkeys." *Journal of Neuroscience* 34 (28):9351–9363. doi:10.1523/JNEUROSCI.0537-14.2014.

Zimbler-Delorenzo, H. S., and A. I. Stone. 2011. "Integration of field and captive studies for understanding the behavioral ecology of the squirrel monkey (*Saimiri* sp.)." *American Journal of Primatology* 73 (7):607–622. doi:10.1002/ajp.20946.

26
Behavioral Biology of Owl Monkeys

Alba García de la Chica
Universidad de Barcelona

Eduardo Fernandez-Duque
Yale University

Lawrence Williams
The University of Texas MD Anderson Cancer Center

CONTENTS

Introduction 409
Typical Research Involvement 410
Behavioral Biology 410
    General Ecology 410
    Social Organization and Behavior 411
    Feeding Behavior 412
    Communication 413
Common Captive Behavior 413
    Normal Behavior 413
    Abnormal Behavior 413
Ways to Maintain Behavioral Health 413
    Environment/Housing 413
    Social Groupings 414
    Enrichment 415
    Feeding Strategies 415
    Training 415
    Special Considerations - Lighting 415
Conclusions and Recommendations 415
References 416

Introduction

Owl monkeys (genus *Aotus*) consist of 11 species that range from Panama in Central America to the northern portions of Argentina. Our understanding of this Neotropical genus' behavior, ecology, and evolution has increased substantially in the last two decades, even though much remains to be learned. Although a picture is emerging about the social organization, behavior, and ecology of this, mostly nocturnal, genus, many aspects of these primates need to be examined before any broad generalizations can be made. Owl monkeys are the only anthropoids with nocturnal habits, and the study of their remarkable activity patterns has benefited enormously from an integrated approach that combined field research with research in seminatural conditions and the laboratory. In fact, given their nocturnality, small size, and arboreality, we will never fully understand owl monkey biology unless we complement field research with detailed observations and measurements of individuals housed in laboratory or captive settings.

Until a few decades ago, our understanding of pair-living and sexual monogamy, and the involvement of the male in infant care (Fernandez-Duque et al. 2020, Huck and Fernandez-Duque 2017), two defining characteristics of the genus, had been primarily informed by studies of captive individuals (Cicmanec and Campbell 1977, Dixson and Fleming 1981, Jantschke et al. 1995, Meritt and Dennis 1980). In some cases, this may be problematic if captive animals are housed in ways that do not reflect the size and composition of wild groups (Fernandez-Duque et al. 2020). Although owl monkeys are pair-living and sexually monogamous in the wild, the fast rate of adult replacement in an *A. a. azarae* population in Argentina suggests that serial monogamy may be the norm (Huck and Fernandez-Duque 2013). Thus, the long-held assumption of stable, lasting pair bonds in pair-living, sexually monogamous primates continues to be challenged and may need further revision (Fernandez-Duque et al. 2020, Huck and Fernandez-Duque 2017). Consequently, it may also become necessary to reevaluate many of our assumptions about the evolutionary forces leading to pair-living, sexual monogamy, and paternal care. Advancing our understanding of the evolutionary forces favoring pair-living and monogamy in owl monkeys will require a comparative approach that considers some of the other more tropical, strictly nocturnal owl monkey species, as well as some of the other strictly pair-living, sexually monogamous taxa.

Owl monkeys have the potential to be an excellent model to accomplish a thorough integration of zoo, laboratory, and field research (Erkert 2008, Fernandez-Duque 2012). In the future, a truly integrated zoo-laboratory–field approach that focuses on certain areas that cannot be examined in only one or the other setting (e.g., reproductive biology, communication, energetics) will offer unique opportunities for synergistic interactions that will have both basic science and applied benefits.

Typical Research Involvement

The owl monkey offers similar research advantages as the squirrel monkey. Its small size allows for ease in handling and reduces the size of cages and special equipment needed for housing, permitting more animals to be housed in facilities with space limitations. The personnel protective equipment required for handling owl monkeys is less extensive than for most nonhuman primates, due to the reduced biohazard risks associated with working with a Neotropical primate. In addition, much smaller doses of synthesized compounds or reagents are needed for research projects.

Owl monkeys are the third most common Neotropical primate to participate in biomedical research, after squirrel monkeys and marmosets. Owl monkeys have been extensively involved in the study of malaria, including maintenance of various malarial strains, antigen production, studies of host–vector relationships, and parasite life cycles (Aikawa et al. 1988, Moreno-Perez et al. 2017). Owl monkey species valuable in malaria research typically include *Aotus lemurinus grisiemembra*, *A. nancymaae*, and *A. vociferans*, since they are highly susceptible to human malaria, *Plasmodium vivax* and *P. falciparum* in particular, and can transmit these infections to mosquitoes, which can then transfer infections back to naïve monkeys. Owl monkeys are also valuable models for testing vaccines (Douglas et al. 2015), screening antimalarial compounds (Suarez et al. 2017), studying host–parasite interactions and mechanisms of pathogenesis (Paul et al. 2015), and, most recently, as a model for HIV infection (Meyerson et al. 2015). They are particularly valuable models in malaria vaccine testing because they respond to vaccines similarly to humans, and experimental vaccine trials in owl monkeys have been predictive of vaccine efficacy in human trials (Moreno-Perez et al. 2017, Obaldia et al. 2018).

The unique characteristics of the owl monkey eye have made it a valuable model in vision research as well (Jacobs et al. 1993, Koga et al. 2017, Levenson et al. 2007, Mundy et al. 2016). The genus is primarily nocturnal, yet it appears to have evolved from a diurnal ancestor, because the eyes retain vestigial features characteristic of diurnal vision, such as a retinal fovea. Unlike other nocturnal mammals, owl monkeys lack a tapetum lucidum. The retina contains both rods and cones, with a markedly decreased number of cones compared to other primates. Rod convergence is higher in the owl monkey retina than in diurnal primates (Nishihara et al. 2018, Silveira et al. 2001). Owl monkeys perform better than diurnal New World monkeys at seeing and following moving objects and at spatial resolution at low light levels (Berezovskii and Born 2000, Wright 1989). The owl monkey has been extensively involved in studies of the architecture and function of both normal and lesioned visual cortex (Sereno et al. 2015, Shostak et al. 2003).

Research involving the owl monkey has led to important contributions to the field of neuroscience. Current areas of investigation include studies on sensory representation and motor control (Shostak et al. 2003), spinal cord atrophy and reorganization following limb loss (Cerkevich et al. 2014), and pioneering work on remote control of robotic devices fundamental to the development of technology that will allow people to control prosthetic devices with their mind (Yin et al. 2016).

Behavioral Biology

General Ecology

A solid understanding of the distribution of owl monkeys has been obtained through more than 30 years of field research. Owl monkeys range from Panama to northern Argentina, and from the foothills of the Andes to the Atlantic Ocean (Aquino et al. 2013, Cornejo et al. 2008, Fernandez-Duque 2011, 2012, Shanee et al. 2013, Svensson et al. 2010). Over this vast range, they inhabit a variety of forests of both primary and secondary growth, sometimes up to 3,200 m above sea level (Defler 2004, Fernandez-Duque 2012).

Owl monkeys most commonly live in lowland forest that is flooded seasonally, or terra firma forest that does not flood. In these forests, they can be found in every strata of the trees, from 7 to 35 m above the ground. At the southern end of their range, owl monkeys are distributed across the South American Gran Chaco of Argentina, Bolivia, and Paraguay, where they can be found in dry forests that receive only 500 mm of annual rainfall (Brooks 1996, Campos et al. 2004, Cervera et al. 2018, Stallings 1989).

Owl monkeys are the only nocturnal primates in America and the only nocturnal Anthropoids. They concentrate their activities during the dark portion of the 24-h cycle, with peaks of activity at dawn and dusk. This has been confirmed by observational studies of free-ranging *Aotus nigriceps*, *A. a. boliviensis*, *A. a. azarae*, *A. vociferans*, and *A. miconax* (Cornejo et al. 2008, Fernandez-Duque 2016, Fernandez-Duque et al. 2008, Khimji and Donati 2014, Shanee et al. 2013). While the species in the tropics are primarily nocturnal, at least one shows a remarkable temporal plasticity in its activity patterns. Azara's owl monkeys (*A. a. azarae*) of the Argentinean and

Paraguayan Chaco are cathemeral, showing bouts of activity during the day, as well as during the night (Fernandez-Duque et al. 2010).

An understanding of the mechanisms regulating cathemerality and nocturnality in *Aotus* is emerging, thanks to the integration of studies from the wild, from seminatural conditions, and from captivity. We know that nocturnal activity of owl monkeys (also called night monkeys) is strongly influenced by the amount of available moonlight; activity is maximal during nights with a full moon and minimal during nights without moonlight. In all six species studied, the amount of available moonlight is strongly and positively related to movement through their home range. Animals travel greater distances during a full moon night than during a new moon one (Fernandez-Duque 2011).

Evidence also shows that owl monkeys adjust their periods of activity in association with changes in the ambient temperature, although the effects of temperature are dependent on ambient luminance. They also appear to have a narrow thermoneutral zone, ranging between 28°C and 30°C, and relatively low resting metabolic rates (Le Maho et al. 1981, Morrison and Simoes 1962). Studies of captive *A. lemurinus* showed they were most active when ambient temperature was 20°C and least active when temperature was 30°C (Erkert 1991), while the activity of free-ranging *A. a. azarae* was maximal between 15°C and 25°C. In *A. a. azarae*, activity was reduced when temperatures were lower than 15°C or higher than 25°C, and there was almost no activity when the temperature was 5°C or lower, even under optimal luminance conditions (Erkert 1991, Fernandez-Duque et al. 2010).

Owl monkey groups are territorial and maintain home ranges of 4.0–17.5 ha. Each group typically occupies a home range that overlaps by 48% (±15%) with the outer parts of the neighboring group's home range and 11% (±15%) with the core areas used by neighboring groups (Fernandez-Duque 2011, Wartmann et al. 2014). Groups regularly encounter other groups at range boundaries (Garcia and Braza 1987, Garcia-Yuste 1989, Wright 1978, 1994), vocalizing and chasing each other.

In the Argentinean population of *A. a. azarae*, intergroup encounters are more likely to occur when the moon is full or directly overhead and can also occur during daylight. The function that territories serve is unknown, as are the behavioral mechanisms that maintain them. It has been suggested that territories function to defend feeding sources, but some encounters seem to be related to reproductive interactions with other groups or with solitary individuals (Fernandez-Duque 2016, Huck and Fernandez-Duque 2017, Mundy et al. 2016).

Owl monkeys are arboreal, and most rest together in sleeping trees during daylight. *A. vociferans*, *A. nigriceps*, and *A. a. boliviensis*, like many other nocturnal primates, have all been found to use a relatively small number of sleeping trees (Aquino and Encarnacion 1986, Fernandez-Duque et al 2008, Wright 1978). The sleeping habits of the small, nocturnal owl monkey species in the tropics contrast sharply with the use of sleeping trees by *A. a. azarae* in the Chaco of Argentina and Paraguay. While tropical owl monkeys use a few specific trees and sleep inside tree holes, owl monkeys in the Chaco regularly use numerous arboreal sleeping sites, and always sleep in vine tangles or open branches, rather than in tree holes (Savagian and Fernandez-Duque 2017). The sleeping habits of Azara's owl monkey seem compatible with minimizing predation risk and maximizing thermoregulatory advantages during the relatively cold winter days of the Chaco.

Social Organization and Behavior

Owl monkeys are consistently described in the literature as pair-living and sexually monogamous. From Panama to Argentina, from evergreen rain forests to semi-deciduous dry forests, whether nocturnal or cathemeral, owl monkeys are always found in small groups that are generally composed of an adult heterosexual pair, one infant, and one or two non-infant individuals that are smaller than the adults (Cornejo et al. 2008, Aquino et al. 2013, Fernandez-Duque 2011, Fernandez-Duque et al. 2020, Huck and Fernandez-Duque 2012, Shanee et al. 2013, Svensson et al. 2010). However, studies of a cathemeral Azara's owl monkey population in northern Argentina have shown that a substantial number of dispersing adults (25–30%) do not belong to a pair-living social group, but range solitarily (Huck and Fernandez-Duque 2013 2017). These solitary "floater" individuals are either young adults that have recently emigrated from their natal groups or relatively old adults that have been evicted from their groups by incoming adults. Interactions between members of social groups and floaters can be aggressive, sometimes leading to the death of individuals (Fernandez-Duque 2009, Huck and Fernandez-Duque 2017). Thus, while pair-living and sexual monogamy is the rule in all species of *Aotus*, evidence from an Argentinean population indicates that there is frequent turnover of partners. This suggests their mating and social system might best be described as serial sexual monogamy. A survey of *Aotus* spp. in northern Colombia also found an important number of solitary individuals (Villaseñor 2003) and fights were reported in free-ranging *A. nigriceps* (Wright 1985) and *A. nancymaae* (Aquino and Encarnación 1989). It seems likely that less conspicuous solitary individuals will be detected in other populations and species as more studies of identified individuals are conducted.

Depending on the species, owl monkeys achieve sexual maturity when they are between 2 and 5 years old (Corley et al. 2017, Dixson 1994, Dixson et al. 1980, Gonzalo and Montoya 1990, Huck et al. 2011). However, in free-ranging populations, no individuals disperse before they are 26 months old. Data on dispersal obtained from free-ranging, radio-collared *A. a. azarae* suggest that dispersal may occur before, after, or around the time of sexual maturation in both sexes (Corley et al. 2017). Most animals disperse at about 3 years of age, but some remain with their natal group until they are four or five. The dispersal of young individuals appears to be associated with changes in group composition, such as the eviction of the resident adult male or female. The evidence suggests that the timing of dispersal may be flexible and dependent on the social context. In the free-ranging *A. a. azarae* population, the peripheralization of subadults is observed prior to dispersal. This behavior may begin as early as 2 or 3 years of age, when juveniles begin lagging behind the group, sleeping in different trees, and occasionally leaving the group for a few days at a time. With regard to the housing of animals in captivity, it is important to realize that very frequently, dispersing young adults spend substantial time (i.e., from days to 15 months) as solitary individuals, ranging among the home ranges of pair-living adults.

Mating is infrequently observed in the wild, but the relatively low frequency of mating that has been reported from free-ranging populations needs to account for the fact that the frequency of mating during the night cannot be adequately estimated. In other words, it is quite likely that the frequency estimates obtained from observations of mating during the daylight hours underestimate the true frequency. Mating takes place as soon as a new adult enters a group. The observation of mating following pair formation agrees well with results from pair-testing experiments in captivity. *A. lemurinus* and *A. nancymaae* mated more frequently in the presence of a new partner than when paired with their regular mate (Dixson 1994, Williams et al. 2017, Wolovich and Evans 2007). Females in new pairs of free-ranging *A. a. azarae* do not produce offspring for at least 1 year, even if the pair is formed during the mating season (Fernandez-Duque, pers. obs.).

Gestation length ranges from 117 to 159 days, depending on the species (Fernandez-Duque 2011, Fernandez-Duque et al. 2020, Hunter 1981, Wolovich et al. 2008,). Pregnancy length in wild Azara's owl monkeys has been estimated to last 132–140 days (Fernandez-Duque et al. 2020). In the *A. azarae* population of Argentina, the majority (80%–90%) of births occur between October and December. Females almost always produce one infant per year; twin births have been rarely observed in captivity, and only three times in a free-ranging population (Fernandez-Duque et al. 2020). Infant survival tends to be higher than juvenile survival in both captivity and the wild.

The intensive involvement of the male in infant care is one of the most fascinating aspects of the social organization of the genus (Fernandez-Duque 2009, Fernandez-Duque et al. 2020, Huck and Fernandez-Duque 2013, Huck et al. 2019). Comparable only to the pattern described in titi monkeys (*Callicebus* spp.), in owl monkeys, there is rarely any involvement of the siblings in the care of infants (Fernandez-Duque et al. 2009, Huck and Fernandez-Duque 2013). Sibling care has been recorded infrequently in captive groups of *A. lemurinus* and *A. a. boliviensis* (Dixson and Fleming 1981, Jantschke et al. 1998, Wright 1984), and only once in free-ranging *A. a. azarae*, following the eviction from the group of the putative father of the infant (Fernandez-Duque et al. 2008).

In wild *A. a. azarae*, males carry the infant 84% of the time after the first week of life (Rotundo et al. 2005). The male not only carries the infant most of the time, but also plays, grooms, and shares food with it. Food sharing, prevalent among captive *A. nancymaae* and *A. lemurinus*, was also observed in free-ranging *A. a. azarae*, and sharing is more frequently observed from the male to the infant than from the female to the infant (Wolovich et al. 2006, 2007). The closer relationship of the male with the infant continues as the infant matures and becomes a juvenile (Huck and Fernandez-Duque 2013, Hunter and Dixson 1983).

The evolution and function of male infant care remains largely unexplained. Adaptive hypotheses have been proposed, including that it may increase offspring survival, or provide increased opportunities for lactating females to forage (Huck et al. 2019). An alternative hypothesis, that male care functions as mating effort, is somewhat supported by observations in the field. For example, intruding *A. a. azarae* males care for the putative offspring of the evicted male. A study found no effect when comparing the survival of infants and juveniles in stable groups to those in groups where one putative parent was replaced (Huck and Fernandez-Duque 2012). Additionally, the replacement of a putative parent had no effect on the age of dispersal (Huck and Fernandez-Duque 2017).

Feeding Behavior

Owl monkeys are primarily frugivorous. Fruits were the most consumed item by *A. nigriceps* (Wright 1994, 1986, 1985), *A. a. azarae* (Arditi 1992, Fernandez-Dunque et al. 2002, Wright 1985), and *A. vociferans* (Puertas et al. 1992). *Ficus spp.* fruits are a highly preferred item in all examined species (Fernandez-Duque 2011). Still, there are no satisfactory quantitative estimates of diet composition and foraging for any of the strictly nocturnal species. Wright's work (Wright 1986, 1994) on *A. nigriceps* in Manú National Park, Perú, was the only attempt at quantifying diet in one of the nocturnal species, but the problems of obtaining quantitative estimates in a nocturnal primate were numerous. Cathemeral owl monkeys have provided opportunities to examine the diet in some detail during daylight hours (Huck and Fernandez-Duque 2013) but determining their foraging habits during the night remains a challenge.

The absence of leaf veins or leaf refuse in 36 fecal samples is the only evidence suggesting that leaf eating may not be important in *A. nigriceps* (Wright 1985). Still, it remains possible that leaf eating in strictly nocturnal owl monkeys occurs but remains undetected. However, leaf consumption has been regularly observed in the cathemeral *A. a. azarae*, especially during the harsh dry season (Fernandez-Duque and van der Heide 2013).

Flowers may be an important food item for *A. a. azarae* and *A. lemurinus* during certain times of the year (Ganzhorn and Wright 1994, Giménez and Fernandez-Duque 2003, Marín-Gómez 2008, Wright 1985). Azara's owl monkeys of Argentina and Paraguay regularly feed on *Tabebuia heptaphylla* flowers during the winter season, when the range of available food items is narrower (Fernandez-Dunque et al. 2002, Van der Heide et al. 2009, Wright 1989).

While insects are undoubtedly part of their diet, quantitative assessment of insect consumption has proven difficult to accomplish in the wild. Most authors have observed owl monkeys eating insects (Arditi 1992, Giménez and Fernandez-Duque 2003, Moynihan 1964, Puertas et al. 1992, Wright 1985, Wolovich et al. 2010), but no one has been able to obtain quantitative estimates of its prevalence. Wright observed them eating lepidopterans (moths 2–8 cm), coleopterans (beetles, 2–4 cm), and spiders; but she "could not collect quantitative data on insect-foraging" (Wright 1985, p. 60). Fernandez-Duque (personal communication) observed free-ranging *A. a. azarae* eating cicadas a few times during the daylight hours, and it is frequently reported that pet owl monkeys keep houses clean of spiders. Captive owl monkeys housed outdoors attempt to capture insect prey approximately ten times per hour and eat small vertebrate prey. There have been observations of an unsuccessful attempt to capture a tree frog, and successful capture and consumption of lizards, geckos and, on one occasion, a small bird (Wolovich et al. 2010).

Communication

Vocal, olfactory, and visual communication are undoubtedly important for owl monkeys (Herrera et al. 2011, Macdonald et al. 2008, Wright 1989, Wolovich et al. 2007, Zito et al. 2003). Still, the challenges of studying certain aspects of communication (e.g., olfactory) in the field make this aspect of owl monkey biology fertile for a captivity-field approach. In fact, most detailed descriptions of communication have been derived from captive animals, although the field work has helped us to understand its functional value.

Like all primates, owl monkeys are visual animals. Behaviors, such as piloerection (hairs of the body and tail are erect) and swaying (lateral movements of their bodies while staying static on the same branch), are visual signals often associated with arousal, excitement, or aggression, and may be accompanied by vocalizations (Moynihan 1964, Wolovich et al. 2007). They make several distinct vocalizations, including owl-like hoots, soft metallic clicks, roars (said to be reminiscent of jaguars), gong-like noises, and low-pitched, resonating tonal grunts. Unlike many other monogamous primates, owl monkeys do not perform vocal duets. The vocal repertoire of *Aotus* can be divided into eight different categories. The two most salient calls are resonant whoops and hoots (García de la Chica et al. 2020, Moynihan 1964).

Resonant whoops consist of a series of 10–17 low notes that build to a climax, are usually produced by both sexes during intergroup encounters, and occur together with visual displays, such as swaying or arching. Hoots are low-frequency calls, given by one individual in the social group or by a solitary individual, that convey information over long distances (García de la Chica et al. 2020, Wright 1989). Playback of these calls to free-ranging *A. a. azarae* elicits responses from animals in the area, both groups and solitaries (Depeine et al. 2008), suggesting that the function of the hoot may not necessarily be to identify and locate a mate, but it may be a more generalized contact call.

Owl monkeys, when communicating, make heavy use of olfactory cues, by using urine and cutaneous secretions to relay information between mates and other conspecifics. In order to do so, they perform scent marking behaviors, i.e., they rub their anogenital region with a substrate, while sliding the rear part of their bodies forward or laterally. In addition to the scent marking that has been well observed in the wild, other behaviors that are likely to be associated with olfactory communication have been observed in captive male owl monkeys. Captive individuals have been observed rubbing their subcaudal glands against their mate's fur, drinking their urine, and self-anointing with items, such as plants and millipedes (Wolovich et al. 2007, Zito et al. 2003). The secretions from their sternal and subcaudal glands most likely play a role in mediating social interactions, as suggested by differences across sexes, ages, and families in the chemical constitution of the secretions (MacDonald et al. 2008, Spence-Aizenberg et al. 2018). Olfaction also plays a prominent role in sexual recognition and aggression (Hunter and Dixson 1983, Hunter et al. 1984). During confrontations involving *A. lemurinus*, some form of olfactory communication always preceded contact aggression between same-sex individuals. Blocking olfactory input led to a reduction in inter-male aggression (Hunter and Dixson 1983).

Common Captive Behavior

Normal Behavior

Owl monkeys in captivity do much the same things as owl monkeys in the field. Family groups sit together and explore their environment as a group. They are not prone to manipulate objects in their cage, as are squirrel monkeys and macaques; if it is not food, then they are not interested. Owl monkeys are neophobic and will not readily accept new enrichment food or devices, new technicians, or even new clothing on the technicians. Normal behavior for an established pair includes sitting together while being vigilant, eating, and sleeping. If one of the pair begins to move independently from the other and, most particularly, begins to sleep in a different site, this is a sign that the pair bond is breaking. Leaving the animals together can result in agonistic interactions, and one of the pair developing stress responses, such as sitting alone and not eating.

Abnormal Behavior

Unlike other nonhuman primate species, there are very few reports of abnormal behavior in *Aotus* spp. In 20 years of managing a large, reproductive colony, we have observed very few instances of stereotypic or abnormal behavior. What we have seen is almost always associated with a split between the two adults in the social group. Female-male pairs of *Aotus nancymaae* have a median longevity of approximately 3 years in our environment (Williams et al. 2017).

Ways to Maintain Behavioral Health

When designing an environment for a captive animal, it is important to understand the animal's natural history and to consider species ecology and social structure. In the beginning of this chapter, we presented a concise description of owl monkeys and their interactions in the field. What follows is how we have used that information to design a captive environment that encourages species-typical behavior.

Environment/Housing

Given the arboreal habits of wild owl monkeys, captive enclosures should include perches, swings, and ladders to provide them as much three-dimensional space as possible (Figure 26.1). Perches and feeding structures should be positioned at multiple places throughout their environment, so that the monkeys are encouraged to move about their housing to get from food sources to water sources and resting platforms.

FIGURE 26.1 Owl monkey housing with multiple layers of perches and nest boxes that provide three-dimensional travel paths.

FIGURE 26.2 Owl monkey housing. Cages (right) are designed so that the owl monkeys cannot see another family group. The water (left) contains a waterfall that produces white noise to mask auditory stimuli from other groups. The airflow is designed to move the air from the registers in the ceiling, down and out the return air vents at the wall base. This reduces the spread of olfactory cues across the room.

During their inactive time, owl monkeys will seek to retreat to enclosed or protected spaces. They will also retreat to an enclosed space when threatened or even when presented with a novel toy or food item (Hooff and Wolovich 2008, Tardif et al. 2006). Therefore, hiding places should be provided within their environment. Enclosed spaces, such as tubes, boxes, or "tangles" of polyvinyl chloride (PVC) chain can provide sleeping sites and places of safety.

Given the reliance of owl monkeys on olfactory communication, and especially their reliance on olfactory cues to locate food, it is important to use care when cleaning and sanitizing their housing (Bolen and Green 1997, Nevo and Heymann 2015, Spence-Aizenberg et al. 2018). One way to maintain some degree of olfactory consistency is to use a modified cleaning schedule. Alternating the weeks that nest boxes are cleaned with weeks that the rest of the housing is sanitized is an effective way to make sure that some part of the owl monkey's environment always smells like "home." If wooden or other porous materials are used to capture olfactory cues, they should be replaced when they become excessively soiled or deteriorated, using the same alternating strategy mentioned above.

An important feature of owl monkey housing is their exposure to visual and olfactory contact with other social groups and/or solitary individuals (Huck and Fernandez-Duque 2013, 2017). Owl monkey groups are territorial (Fernandez-Duque 2012, Wartmann et al. 2014), rarely interacting with nongroup members, except during an aggressive attempt to oust an adult from a stable social group. A typical housing arrangement for captive owl monkeys has family groups housed in direct visual contact with other owl monkeys living in the same room, a situation that is very different from the way in which they interact in the wild, and therefore, possibly a source of stress. Owl monkeys have high incidences of hypertension and cardiomyopathy in captivity (Brady et al. 2005, Rajendra et al. 2010, Smith and Astley 2007), which has been hypothesized to result from constant interactions with extra-group conspecifics. Enclosures should be constructed to minimize visual and olfactory contact between groups, so that it is nearly impossible for one owl monkey to see or smell an owl monkey in another group (Figure 26.2).

Social Groupings

Owl monkeys are generally housed as monogamous pairs, with young, for breeding (Fernandez-Duque 2012, Malaga et al. 1997, Moynihan 1964, Wright et al. 1989). The most successful pairing strategy for adult owl monkeys is to pair a male with a female; however, if there are no compatible opposite sex partners, owl monkeys will sometimes choose to live in iso-sexual pairs (Williams et al. 2017). Eighty-two percent of mixed-sex pairs were living together and sleeping in the same nest box within a week, while iso-sexual pairs, resulted in lower pairing success rates (62% for FF pairs and 40%

for MM pairs). Given the strong social bond between male and female adult owl monkeys, mixed-sex pairs are easier to form and, hence, preferred. Studies have shown that captive pairs increase their reproductive success when they have the same species. However, even with lower success, mated pairs of different karyotypes are also productive (Malaga et al 1997, Williams personal communication). In the wild, adult owl monkeys may be expelled from a social group by intruding, relatively young adults (Fernandez-Duque and Huck 2013; Huck and Fernandez-Duque 2017), events which generally do not occur in captivity. However, captive pairs do seem to have a finite compatibility "life span"; in *A. nancymaae* male–female pairs, the mean compatibility is close to 3 years (Williams et al. 2017). Because pairing owl monkeys is not a "once and done" process, facilities should be prepared to (1) separate animals that become overly aggressive to one another as their time together approaches 3 years, and (2) form new pairs as needed. Given that owl monkeys in their natural habitat do experience a period of solitary life as young adults (Corley et al. 2017, Fernandez-Duque et al. 2009, Huck and Fernandez-Duque 2017), the housing of animals individually may be acceptable for short periods, particularly for relatively young adults.

Enrichment

Enrichment should focus on those facets of the genus that contribute most to its uniqueness. Owl monkeys are known to be slow to approach novel objects (Ehrlich 1970, Williams, personal obs.). When introducing new enrichment devices or food, or even new housing, it is important to do so slowly and incrementally. When shifting light cycles to allow for more active dark time during our workday, we noticed a drop in body weights across the affected animals (Williams, personal communication). Only by slowly moving the light on/off times by 1 hour per month were we able to accomplish the desired change without negatively impacting the animals' health.

Feeding Strategies

When considering foraging enrichment, it should be remembered that owl monkeys are primarily frugivores that supplement their diets with flowers, insects, and leaves (Fernandez-Duque 2011, 2012, Fernandez-Duque and van der Heide 2013). Beyond the standard laboratory diets, owl monkeys will respond positively to a wide variety of food items. Fresh fruits and vegetables, rotated on a seasonal basis, provide novelty and nutritional value. Owl monkeys will use a foraging board to glean seeds and soft food items, like vegetable purees, peanut butter, or yogurt, and will transfer food items both between members of the breeding pair, and to their young (Wolovich et al. 2008). However, food that is difficult for the adult to obtain is not transferred to the young, so providing small, easily accessible food items will encourage positive social interactions between the infants and adults. Unlike squirrel monkeys, owl monkeys are not destructive foragers and will not work particularly hard on devices that require large amounts of manipulation to obtain the reward.

Training

Owl monkeys can be trained, using positive reinforcement training techniques, to participate in several behavioral tasks. Because they are generally neophobic, training can take longer than is typical for Old World primates; on one task, owl monkeys took twice as long to learn to reliably touch a target than did squirrel monkeys (Rogge et al. 2013). However, once targeting was learned, training for additional behaviors, such as to present a hand or foot, occurred just as quickly as for squirrel monkeys. Since they live in small family groups or pairs, it is easy to train them in their home cage. Clickers and verbal cues both work as a bridge between the behavioral response and the reward. Primary reinforcement should consist of sweet food items, including raisins, grapes, and marshmallows. Additionally, owl monkeys can be trained to jump from old to new housing by placing a jackpot (i.e., large) reward in the new housing. Given the cardiomyopathy and hypertension seen in owl monkeys (Smith and Astley 2007), using training to either acclimate the animals to new procedures, or to reward them for actively participating in routine procedures, should reduce stress-related lesions.

Special Considerations - Lighting

Since owl monkeys are nocturnal, wellness checks should be made during the dark or lights-out period. In most facilities, this means an altered light/dark schedule from the normal workday. Without a reversed cycle, not only will all the cleaning, feeding, and clinical observations be done during the animals' natural resting (sleep) period, but there will also be few opportunities for observation of social interactions or movement, both of which are crucial for identifying sick or injured animals. Shifting lights on/off in captivity to a schedule similar to lights on at midnight and off at noon, provides the monkeys with an undisturbed period of rest from midnight until the staff arrives, while still providing staff with time after lights out for direct observations during the animals' active period. At least one species, *A. azarae*, shows a high degree of cathemerality in the field (Fernandez-Duque 2003, Fernandez-Duque et al. 2010) suggesting that an appropriate light cycle for this species would include times of transition from complete light to complete darkness (dusk), and vice versa (dawn).

Conclusions and Recommendations

Our understanding of owl monkey behavior, ecology, and evolution is limited, but is rapidly growing as the number of research field sites increases. A comprehensive picture is emerging about the social organization, behavior, and ecology of the genus; however, other aspects need to be examined before any broad generalizations can be made. The nocturnal owl monkeys are slow to accept new conditions, so any changes in their environment or husbandry routines should be introduced gradually. Pairing of individuals should be a slow process, giving several days before deciding whether a pair has been successfully made.

Although owl monkeys are undoubtedly pair-living and sexually monogamous, the unexpectedly rapid rate of adult replacement in the *A. a. azarae* population suggests that serial monogamy may be the norm. Thus, the long-held assumption of stable, lasting pair bonds in pair-living primates will need to be revised, both for managing captive populations and for advancing our understanding of the evolution and maintenance of social monogamy. We know in the captive environment that pairs will break up, so you must remain vigilant for signs of a split. These can include very visible signs like increased aggression to more subtle changes in feeding and sleeping patterns.

The social system of owl monkeys has facilitated their management and breeding in captivity (Cicmanec and Campbell 1977, Erkert 1999, Gozalo and Montoya 1990, Malaga et al. 1997, Williams et al. 2017), allowing for the housing of a single pair of reproducing adults and offspring. Pair housing of males and females is the recommended housing situation. Still, in captivity, high levels of aggression have been reported (Evans et al. 2009, Hunter and Dixson 1983) that may have been initially difficult to reconcile with the traditional, and incorrect, picture of "low-intensity, low-frequency competition" described in the literature (Plavcan and van Schaik 1992). Intense aggression has been observed during experimental testing, but also in well-established family groups. Management of captive individuals and groups will benefit from (1) consideration of the period of life when subadults or young adults disperse and spend time alone, and (2) knowledge that the social system of owl monkeys may be better described as serial monogamy, resulting in social groups that consistently include step-parents and step-siblings.

The function of territoriality in owl monkeys will also need careful examination. To identify some of the relevant factors driving and maintaining territoriality, it will be necessary to develop a semi-experimental approach to examine some of the unresolved issues. For example, in the wild, playback experiments to simulate intruders, or food-provisioning experiments to manipulate available food resources, will need to be considered and implemented following similar experimental studies with captive animals. Finally, advancing our understanding of the evolutionary forces favoring pair-living, sexual monogamy, biparental behavior, and nocturnality in owl monkeys will benefit from a comparative approach that considers some of the other more tropical, strictly nocturnal owl monkey or prosimian species, as well as some of the other socially monogamous primates. Such a comparative approach should include research in both captive and field settings.

REFERENCES

Aikawa, M., G. Jacobs, H. E. Whiteley, I. Igarashi, and M. Ristic. 1988. "Glomerulopathy in squirrel monkeys with acute Plasmodium falciparum infection." *Am J Trop Med Hyg* 38 (1):7–14. doi:10.4269/ajtmh.1988.38.7.

Aquino, R., F. M. Cornejo, E. Pezo, and E. W. Heymann. 2013. "Distribution and abundance of white-fronted spider monkeys, Ateles belzebuth (Atelidae), and threats to their survival in Peruvian Amazonia." *Folia Primatol (Basel)* 84 (1):1–10. doi:10.1159/000345549.

Aquino, R., and F. Encarnacion. 1986. "Characteristics and use of sleeping sites in Aotus (Cebidae: Primates) in the Amazon lowlands of Peru." *Am J Primatol* 11 (4):319–331. doi:10.1002/ajp.1350110403.

Aquino, R., and F. Encarnación. 1989. "Aspectos de la dinámica poblacional de *Aotus nancymai* (Cebidae: Primates)." *La primatología en Latinoamérica*, 187–192.

Arditi, S.I. 1992. "Variaciones estacionales en la actividad y dieta de Aotus azarae y Alouatta caraya en Formosa, Argentina." *B Primat Latinoam* 3 (1):11–30.

Berezovskii, V. K., and R. T. Born. 2000. "Specificity of projections from wide-field and local motion-processing regions within the middle temporal visual area of the owl monkey." *J Neurosci* 20 (3):1157–1169. doi:10.1523/Jneurosci.20-03-01157.2000.

Bolen, R. H., and S. M. Green. 1997. "Use of olfactory cues in foraging by owl monkeys (*Aotus nancymai*) and capuchin monkeys (*Cebus apella*)." *J Comp Psych* 111 (2):152–158.

Brady, A. G., V. L. Parks, S. V. Gibson, and C. R. Abee. 2005. "Cardiomyopathy of owl monkeys: A study of seven cases." *Contemp Top Lab Anim Sci* 44(4):57.

Brooks, D. M. 1996. "Some observations on primates in Paraguay." *Neotropical Primates* 4 (1):15–19.

Campos, J. M., I. Benítez, and D. A. Meritt Jr. 2004. "On the occurrence of the owl monkey (*Aotus azarae*) in Cerro León, Chaco, Paraguay." *Neotrop. Primates* 12 (2):55–56.

Cerkevich, C. M., C. E. Collins, and J. H. Kaas. 2014. "Cortical inputs to the middle temporal visual area in New World owl monkeys." *Eye Brain* 2015 (7):1–15. doi:10.2147/EB.S69713.

Cervera, L., S. de la Torre, G. Zapata Ríos, F. A. Cortés, S. Álvarez Solas, O. Crowe, R. Cueva, A. de la Torre, I. Duch Latorre, and M. F. Solórzano. 2018. "working together towards one goal: results of the first primate census in western ecuador." *Primate Conserv* 32: 49–56.

Cicmanec, J. C., and A. K. Campbell. 1977. "Breeding the owl monkey (*Aotus trivirgatus*) in a laboratory environment." *Lab Anim Sci* 27 (4):512–517.

Corley, M. K., S. Xia, and E. Fernandez-Duque. 2017. "The role of intragroup agonism in parent-offspring relationships and natal dispersal in monogamous owl monkeys (*Aotus azarae*) of Argentina." *Am J Primatol* 79 (11):e22712. doi:10.1002/ajp.22712.

Cornejo, F. M., R. Aquino, and C. Jimenez. 2008. "Notes on the natural history, distribution and conservation status of the Andean night monkey, Aotus miconax Thomas, 1927." *Primate Conserv.* 23 (1):1–4.

García de la Chica, A., M. Huck, C. Depeine, M. Rotundo, P. Adret, and E. Fernandez-Duque. 2020. "Sexual dimorphism in the loud calls of Azara's owl monkeys (*Aotus azarae*): Evidence of sexual selection?" *Primates* 61 (2):309–319.

Defler, T. R. 2004. "Primates of Colombia." *Conserv Intern Trop Field Guide Series* 5:1–550.

Depeine, C.D., M. Rotundo, C.P. Juárez, and E. Fernandez-Duque. 2008. "Hoot calling in owl monkeys (*Aotus azarai*) of Argentina: Sex differences and function." *Am J Primtol* 70(S1): 69.

Dixson, A. F., and D. Fleming. 1981. "Parental behavior and infant development in owl monkeys (*Aotus-Trivirgatus-Griseimembra*)." *J Zool* 194:25–39. doi:10.1111/j.1469-7998.1981.tb04576.x.

Dixson, A.F. 1994. "Reproductive biology of the owl monkey." In *Aotus: The Owl Monkey*, edited by J.F. Baer, R.E. Weller and I. Kakoma, 113–132. San Diego, CA: Academic Press.

Dixson, A.F., J.S. Gardner, and R.C. Bonney. 1980. "Puberty in the male owl monkey (*Aotus trivirgatus griseimembra*): A study of physical and hormonal development." *Inter J Primato* 1 (2):129–139.

Douglas, A. D., G. C. Baldeviano, C. M. Lucas, L. A. Lugo-Roman, C. Crosnier, S. J. Bartholdson, A. Diouf, K. Miura, L. E. Lambert, J. A. Ventocilla, K. P. Leiva, K. H. Milne, J. J. Illingworth, A. J. Spencer, K. A. Hjerrild, D. G. Alanine, A. V. Turner, J. T. Moorhead, K. A. Edgel, Y. Wu, C. A. Long, G. J. Wright, A. G. Lescano, and S. J. Draper. 2015. "A PfRH5-based vaccine is efficacious against heterologous strain blood-stage Plasmodium falciparum infection in aotus monkeys." *Cell Host Microbe* 17 (1):130–139. doi:10.1016/j.chom.2014.11.017.

Ehrlich, A. 1970. "Response to novel objects in three lower primates: Greater galago, slow loris, and owl monkey." *Behaviour* 37 (1–2):55–63.

Erkert, H. G. 1991. "Influence of ambient-temperature on circadian-rhythms in colombian owl monkeys, aotus-lemurinus-griseimembra." *Primatol Today*, 435–438.

Erkert, H. G. 2008. "Diurnality and nocturnality in nonhuman primates: Comparative chronobiological studies in laboratory and nature." *Bio Rhythm Res* 39 (3):229–267. doi:10.1080/09291010701683391.

Erkert, H. G. 1999. "Owl monkeys." In *The UFAW Handbook on the Care and Management of Laboratory Animals*, vol. 1, edited by T. Poole, 574–590. Hoboken, NJ: Wiley-Blackwell.

Evans, S., C. Wolovich, and J. Herrera. 2009. "Aggression in captive owl monkeys." *Am J Primatol* 71:97.

Fernandez-Dunque, E., M. A. Rotundo, and P. Ramirez-Llorens. 2002. "Environmental determinants of birth seasonality in night monkeys (*Aotus azarai*) of the Argentinean Chaco." *Int J Primatol* 23 (3):639–656.

Fernandez-Duque, E. 2003. "Influences of moonlight, ambient temperature, and food availability on the diurnal and nocturnal activity of owl monkeys (*Aotus azarai*)." *Behav Ecol Sociobiol* 54 (5):431–440. doi:10.1007/s00265-003-0637-9.

Fernandez-Duque, E. 2009. "Natal dispersal in monogamous owl monkeys (*Aotus azarai*) of the Argentinean Chaco." *Behaviour* 146:583–606 doi:10.1163/156853908x397925.

Fernandez-Duque, E. 2012. "Owl monkeys *Aotus* spp. in the wild and in captivity." *Int Zoo Yearb* 46 (1):80–94. doi:10.1111/j.1748-1090.2011.00156.x.

Fernandez-Duque, E. 2016. "Social monogamy in wild owl monkeys (*Aotus azarae*) of Argentina: the potential influences of resource distribution and ranging patterns." *Am J Primatol* 78 (3):355–371. doi:10.1002/ajp.22397.

Fernandez-Duque, E., H. de la Iglesia, and H. G. Erkert. 2010. "Moonstruck primates: owl monkeys (*Aotus*) need moonlight for nocturnal activity in their natural environment." *PLoS One* 5 (9):e12572. doi:10.1371/journal.pone.0012572.

Fernandez-Duque, E., & Huck, M. 2013. "Till death (or an intruder) do us part: intra-sexual competition in a monogamous primate." *PLoS One* 8 (1):e53724.

Fernandez-Duque, E., M. Huck, S. Van Belle, and A. Di Fiore. 2020. "The evolution of pair-living, sexual monogamy, and cooperative infant care: Insights from research on wild owl monkeys, titis, sakis, and tamarins." *Am J Phys Anthropol* 171 (Suppl 70):118–173. doi:10.1002/ajpa.24017.

Fernandez-Duque, E., C. P. Juarez, and A. Di Fiore. 2008. "Adult male replacement and subsequent infant care by male and siblings in socially monogamous owl monkeys (*Aotus azarai*)." *Primates* 49 (1):81–4. doi:10.1007/s10329-007-0056-z.

Fernandez-Duque, E., C. R. Valeggia, and S. P. Mendoza. 2009. "The biology of paternal care in human and nonhuman primates." *Annu Rev Anthropol* 38 (1):115–130. doi:10.1146/annurev-anthro-091908-164334.

Fernandez-Duque, E., & van der Heide, G. 2013. "Dry season resources and their relationship with owl monkey (*Aotus azarae*) feeding behavior, demography, and life history." *Int J Primatol* 34 (4):752–769. doi:10.1007/s10764-013-9689-5.

Fernandez-Duque, E. 2011. "The Aotinae: Social monogamy in the only nocturnal anthropoid." In *Primates in Perspective*, edited by C.J. Campbell, A. Fuentes, K.C. MacKinnon, S.K. Bearder and R.M. Stumpf, 140–154. Oxford: Oxford University Press.

Ganzhorn, J. U., and P. C. Wright. 1994. "Temporal patterns in primate leaf eating: The possible role of leaf chemistry." *Folia Primatol (Basel)* 63 (4):203–208. doi:10.1159/000156820.

Garcia, J. E., and F. Braza. 1987. "Activity rhythms and use of space of a group of aotus-azarae in bolivia during the rainy season." *Primates* 28 (3):337–342. doi:10.1007/Bf02381016.

Garcia Yuste, J.E. 1989. "Patrones etológicos y ecológicos del mono nocturno, *Aotus azarai boliviensis*." *Diss Abstr Int* 50:473.

Giménez, M., and E. Fernandez-Duque. 2003. "Summer and winter diet of night monkeys in the gallery and thorn forests of the Argentinean Chaco." *Rev. Etol.* 5(suppl.):164.

Gozalo, A., and E. Montoya. 1990. "Reproduction of the owl monkey (*Aotus nancymai*) (primates:Cebidae) in captivity." *Am J Primatol* 21 (1):61–68. doi:10.1002/ajp.1350210107.

Herrera, J.P., L.L. Taylor, and S. Evans. 2011. "Use of auditory and olfactory signals in night monkeys (*Aotus nancymaae*)." *Am J of Phys Anthropol* 144:161.

Hooff, S.B., and C.K. Wolovich. 2008. "Captive owl monkeys respond to novel flavors with neophobia, discrimination, and food sharing." *Am J of Primatol* 70:48.

Huck, M., and E. Fernandez-Duque. 2012. "Children of divorce: Effects of adult replacements on previous offspring in Argentinean owl monkeys." *Behav Ecol Sociobiol* 66 (3):505–517. doi:10.1007/s00265-011-1297-9.

Huck, M., and E. Fernandez-Duque. 2017. "The floater's dilemma: Use of space by wild solitary Azara's owl monkeys, *Aotus azarae*, in relation to group ranges." *Anim Behav* 127:33–41. doi:10.1016/j.anbehav.2017.02.025.

Huck, M., A. Di Fiore, and E. Fernandez-Duque. 2019. "Of apples and oranges? The evolution of "Monogamy" in non-human primates." *Front Ecol* 7:472.

Huck, M., Rotundo, M., & Fernandez-Duque, E. 2011. "Growth, development and age categories in male and female wild monogamous owl monkeys (*Aotus azarai*) of Argentina." *Int J Primatol* 32:1133–1152. doi:10.1007/s10764-011-9530-y.

Huck, M., and E. Fernandez-Duque. 2013. "When dads help: Male behavioral care during primate infant development." In *Building Babies*, edited by K. B. H. Clancy, K. Hinde and J. N. Rutherford, 361–385. New York: Springer.

Hunter, A. J., and A. F. Dixson. 1983. "Anosmia and aggression in male owl monkeys (*Aotus trivirgatus*)." *Physiol Behav* 30 (6):875–879. doi:10.1016/0031-9384(83)90251-2.

Hunter, A. J., D. Fleming, and A. F. Dixson. 1984. "The structure of the vomeronasal organ and nasopalatine ducts in aotus-trivirgatus and some other primate species." *J Anat* 138 (Mar):217–225.

Hunter, A. J. 1981. "Chemical communication, aggression and sexual behaviour in the owl monkey (*Aotus trivirgatus griseimembra*)." *Unpublished PhD Thesis, University of London*.

Jacobs, G. H., J. F. Deegan 2nd, J., Neitz, M. A. Crognale, and M. Neitz. 1993. "Photopigments and color vision in the nocturnal monkey, *Aotus*." *Vision Res* 33 (13):1773–1783. doi:10.1016/0042-6989(93)90168-v.

Jantschke, B., C. Welker, and A. Klaiber-Schuh. 1998. "Rearing without paternal help in the bolivian owl monkey *Aotus azarae boliviensis*: A case study 1." *Folia Primatol* 69 (2):115–120.

Jantschke, B., C. Welker, and A. KlaiberSchuh. 1995. "Notes on breeding of the titi monkey *Callicebus cupreus*." *Folia Primatol* 65 (4):210–213. doi:10.1159/000156890.

Khimji, S.N., and G. Donati. 2014. "Are rainforest owl monkeys cathemeral? Diurnal activity of black-headed owl monkeys, *Aotus nigriceps*, at Manu Biosphere Reserve, Peru." *Primates* 55 (1):19–24.

Koga, A., H. Tanabe, Y. Hirai, H. Imai, M. Imamura, T. Oishi, R. Stanyon, and H. Hirai. 2017. "Co-opted megasatellite dna drives evolution of secondary night vision in azara's owl monkey." *Genome Biol Evol* 9 (7):1963–1970. doi:10.1093/gbe/evx142.

Le Maho, Y., M. Goffart, A. Rochas, H. Felbabel, and J. Chatonnet. 1981. "Thermoregulation in the only nocturnal simian: The night monkey *Aotus trivirgatus*." *Am J Physiol* 240 (3):R156–R165. doi:10.1152/ajpregu.1981.240.3.R156.

Levenson, D. H., E. Fernandez-Duque, S. Evans, and G. H. Jacobs. 2007. "Mutational changes in S-cone opsin genes common to both nocturnal and cathemeral Aotus monkeys." *Am J Primatol* 69 (7):757–765. doi:10.1002/ajp.20402.

Macdonald, E. A., E. Fernandez-Duque, S. Evans, and L. R. Hagey. 2008. "Sex, age, and family differences in the chemical composition of owl monkey (*Aotus nancymaae*) subcaudal scent secretions." *Am J Primatol* 70 (1):12–8. doi:10.1002/ajp.20450.

Malaga, C. A., R. E. Weller, R. L. Buschbom, J. F. Baer, and B. B. Kimsey. 1997. "Reproduction of the owl monkey (*Aotus* spp.) in captivity." *J Med Primatol* 26 (3):147–52. doi:10.1111/j.1600-0684.1997.tb00046.x.

Marín-Gómez, O. H. 2008. "Consumo de néctar por *Aotus lemurinus* y su rol como posible polinizador de las flores de *Inga edulis* (Fabales: Mimosoideae)." *Neotrop Primates* 15 (1):30–32.

Meritt, J.R., and A. Dennis. 1980. "Captive reproduction and husbandry of the *Douroucouli Aotus trivirgatus* and the titi monkey *Callicebus* spp.". *Int Zoo Yearb* 20 (1):52–59.

Meyerson, N. R., A. Sharma, G. K. Wilkerson, J. Overbaugh, and S. L. Sawyer. 2015. "Identification of owl monkey CD4 receptors broadly compatible with early-stage HIV-1 isolates." *J Virol* 89 (16):8611–8622. doi:10.1128/JVI.00890-15.

Moreno-Perez, D. A., R. Garcia-Valiente, N. Ibarrola, A. Muro, and M. A. Patarroyo. 2017. "The Aotus nancymaae erythrocyte proteome and its importance for biomedical research." *J Proteomics* 152:131–137. doi: 10.1016/j.jprot.2016.10.018.

Moynihan, M. 1964. *Some Behavior Patterns of Platyrrhine Monkeys: I. The Night Monkey (Aotus Trivirgatus)*. Washington: Smithsonian Institution Collections.

Mundy, N. I., N. C. Morningstar, A. L. Baden, E. Fernandez-Duque, V. M. Davalos, and B. J. Bradley. 2016. "Can colour vision re-evolve? Variation in the X-linked opsin locus of cathemeral Azara's owl monkeys (*Aotus azarae azarae*)." *Front Zool* 13:9. doi:10.1186/s12983-016-0139-z.

Nevo, O., and E. W. Heymann. 2015. "Led by the nose: Olfaction in primate feeding ecology." *Evol Anthropol* 24 (4):137–148. doi:10.1002/evan.21458.

Nishihara, H., R. Stanyon, J. Kusumi, H. Hirai, and A. Koga. 2018. "Evolutionary origin of owlrep, a megasatellite dna associated with adaptation of owl monkeys to nocturnal lifestyle." *Genome Biol Evol* 10 (1):157–165. doi:10.1093/gbe/evx281.

Obaldia 3rd, N., E. Meibalan, J. M. Sa, S. Ma, M. A. Clark, P. Mejia, R. R. Moraes Barros, W. Otero, M. U. Ferreira, J. R. Mitchell, D. A. Milner, C. Huttenhower, D. F. Wirth, M. T. Duraisingh, T. E. Wellems, and M. Marti. 2018. "Bone marrow is a major parasite reservoir in plasmodium vivax infection." *mBio* 9 (3):e00625-18. doi:10.1128/mBio.00625-18.

Morrison, P., and J. Simoes. 1962. "Body temperatures in two Brazilian primate." *Zoologia* 24:167–178.

Paul, A. S., E. S. Egan, and M. T. Duraisingh. 2015. "Host-parasite interactions that guide red blood cell invasion by malaria parasites." *Curr Opin Hematol* 22 (3):220–226. doi:10.1097/MOH.0000000000000135.

Plavcan, J. M., and C. P. van Schaik. 1992. "Intrasexual competition and canine dimorphism in anthropoid primates." *Am J Phys Anthropol* 87 (4):461–477. doi:10.1002/ajpa.1330870407.

Puertas, P., R. Aquino, and F. Encarnacion. 1992. "Uso de alimentos y competición entre el mono nocturno Aotus vociferans y otros mamíferos, Loreto, Perú." *Folia Amazónica* 4 (2):151–160.

Rajendra, R. S., A. G. Brady, V. L. Parks, C. V. Massey, S. V. Gibson, and C. R. Abee. 2010. "The normal and abnormal owl monkey (*Aotus* sp.) heart: Looking at cardiomyopathy changes with echocardiography and electrocardiography." *J Med Primatol* 39 (3):143–150. doi:10.1111/j.1600-0684.2010.00403.x.

Rogge, J., K. Sherenco, R. Malling, E. Thiele, S. Lambeth, S. Schapiro, and L. Williams. 2013. "A comparison of positive reinforcement training techniques in owl and squirrel monkeys: Time required to train to reliability." *J Appl Anim Welf Sci* 16 (3):211–220. doi:10.1080/10888705.2013.798223.

Rotundo, M., E. Fernandez-Duque, and A. F. Dixson. 2005. "Infant development and parental care in free-ranging *Aotus azarai azarai* in Argentina." *Int J Primatol* 26 (6):1459–1473. doi:10.1007/s10764-005-5329-z.

Savagian, A., and E. Fernandez-Duque. 2017. "Do predators and thermoregulation influence choice of sleeping sites and sleeping behavior in Azara's Owl Monkeys (*Aotus azarae azarae*) in Northern Argentina?" *Int J Primatol* 38 (1):80–99.

Sereno, M. I., C. T. McDonald, and J. M. Allman. 2015. "Retinotopic organization of extrastriate cortex in the owl monkey–dorsal and lateral areas." *Vis Neurosci* 32:E021. doi:10.1017/S0952523815000206.

Shanee, S., N. Allgas, and N. Shanee. 2013. "Preliminary observations on the behavior and ecology of the Peruvian night monkey (*Aotus miconax*: Primates) in a remnant cloud forest patch, north eastern Peru." *Trop COnserv Sci* 6 (1):138–148. doi:10.1177/194008291300600104.

Shostak, Y., Y. Ding, and V. A. Casagrande. 2003. "Neurochemical comparison of synaptic arrangements of parvocellular, magnocellular, and koniocellular geniculate pathways in owl monkey (*Aotus trivirgatus*) visual cortex." *J Comp Neurol* 456 (1):12–28. doi:10.1002/cne.10436.

Silveira, L. C. L., E. S. Yamada, E. C. S. Franco, and B. L. Finlay. 2001. "The specialization of the owl monkey retina for night vision." *Color Research and Application* 26 (S1):S118–S122. doi:10.1002/1520-6378.

Smith, O. A., and C. A. Astley. 2007. "Naturally occurring hypertension in New World nonhuman primates: Potential role of the perifornical hypothalamus." *Am J Physiol Regul Integr Comp Physiol* 292 (2):R937–R945. doi:10.1152/ajpregu.00400.2006.

Spence-Aizenberg, A., B. A. Kimball, L. E. Williams, and E. Fernandez-Duque. 2018. "Chemical composition of glandular secretions from a pair-living monogamous primate: Sex, age, and gland differences in captive and wild owl monkeys (*Aotus* spp.)." *Am J Primatol* 80 (2):e22730. doi:10.1002/ajp.22730.

Stallings, J.R. 1989. "Status y conservación de primates en el Paraguay." *La Primatol en Latinoam. Washington, DC, World Wildlife Fund.* 133–151.

Suarez, C. F., L. Pabon, A. Barrera, J. Aza-Conde, M. A. Patarroyo, and M. E. Patarroyo. 2017. "Structural analysis of owl monkey MHC-DR shows that fully-protective malaria vaccine components can be readily used in humans." *Biochem Biophys Res Commun* 491 (4):1062–1069. doi:10.1016/j.bbrc.2017.08.012.

Svensson, M. S., R. Samudio, S. K. Bearder, and K. A. Nekaris. 2010. "Density estimates of Panamanian owl monkeys (*Aotus zonalis*) in three habitat types." *Am J Primatol* 72 (2):187–192. doi:10.1002/ajp.20758.

Tardif, S., K. Bales, L. Williams, E. L. Moeller, D. Abbott, N. Schultz-Darken, S. Mendoza, W. Mason, S. Bourgeois, and J. Ruiz. 2006. "Preparing new world monkeys for laboratory research." *ILAR J* 47 (4):307–315. doi:10.1093/ilar.47.4.307.

Van der Heide, G.W., D. Iriart, C.P. Juárez, and E. Fernandez-Duque. 2009. "Do forest composition and food availability predict demographic differences between owl monkey (*Aotus azarai*) groups inhabiting a gallery forest in Formosa, Argentina?" *Am J Primatol* 71: 59.

Villaseñor, J. L. 2003. "Diversidad y distribución de las Magnoliophyta de México." *Interciencia* 28 (3):160–167.

Wartmann, F. M., C. P. Juarez, and E. Fernandez-Duque. 2014. "Size, site fidelity, and overlap of home ranges and core areas in the socially monogamous owl monkey (*Aotus azarae*) of Northern Argentina." *Int J Primatol* 35 (5):919–939. doi:10.1007/s10764-014-9771-7.

Williams, L. personal obs. "personal obs.".

Williams, L. E., C. S. Coke, and J. L. Weed. 2017. "Socialization of adult owl monkeys (*Aotus* sp.) in Captivity." *Am J Primatol* 79 (1):1–7. doi:10.1002/ajp.22521.

Wolovich, C. K., and S. Evans. 2007. "Sociosexual behavior and chemical communication of *Aotus nancymaae*." *Int J Primat* 28 (6):1299–1213. doi:10.1007/s10764-007-9228-3.

Wolovich, C. K., S. Evans, and S. M. Green. 2010. "Mated pairs of owl monkeys (*Aotus nancymaae*) exhibit sex differences in response to unfamiliar male and female conspecifics." *Am J Primatol* 72 (11):942–950. doi:10.1002/ajp.20858.

Wolovich, C. K., A. Feged, S. Evans, and S. M. Green. 2006. "Social patterns of food sharing in monogamous owl monkeys." *Am J Primatol* 68 (7):663–674. doi:10.1002/ajp.20238.

Wolovich, C. K., J. P. Perea-Rodriguez, and E. Fernandez-Duque. 2008. "Food transfers to young and mates in wild owl monkeys (*Aotus azarai*)." *Am J Primatol* 70 (3):211–221. doi:10.1002/ajp.20477.

Wright, P. C. 1978. "Home range, activity pattern, and agonistic encounters of a group of night monkeys (*Aotus trivirgatus*) in Peru." *Folia Primatol (Basel)* 29 (1):43–55. doi:10.1159/000155825.

Wright, P. C. 1989. "The nocturnal primate niche in the new-world." *J Hum Evol* 18 (7):635–658. doi:10.1016/0047-2484(89)90098-5.

Wright, P. C. 1986. "Ecological correlates of monogamy in Aotus and Callicebus." *Prim Eco Conserv* 2:159–167.

Wright, P. C. 1994. "The behavior and ecology of the owl monkey." In *Aotus: The Owl Monkey*, edited by J.F. Baer, R.E. Weller and I. Kakoma, 97–112. San Diego, CA: Academic Press.

Wright, P. C. 1984. "Biparental care in Aotus and Callicebus". In *Female Primates: Studies by Women Primatologists*, edited by M. Small, 59–75. New York: Liss.

Wright, P. C. 1985. "Costs and benefits of nocturnality to the night monkey (Aotus)." PhD Dissertation, City University of New York, New York.

Wright, P.C., D.M. Haring, M.K. Izard, and E.L. Simons. 1989. "Psychological well-being of nocturnal primates in captivity." In *Housing, Care and Psychological Well-being of Captive and Laboratory Primates*, edited by E.F. Segal, 61–74. Park Ridge, NJ: Noyes Publications.

Yin, A., J. An, G. Lehew, M. A. Lebedev, and M. A. L. Nicolelis. 2016. "An automatic experimental apparatus to study arm reaching in New World monkeys." *J Neurosci Methods* 264:57–64. doi:10.1016/j.jneumeth.2016.02.017.

Zito, M., S. Evans, and P. J. Weldon. 2003. "Owl monkeys (*Aotus* spp.) self-anoint with plants and millipedes." *Folia Primatol (Basel)* 74 (3):159–161. doi:10.1159/000070649.

27
Behavioral Biology of Capuchin Monkeys

Marcela Eugenia Benítez
Emory University

Sarah F. Brosnan
Georgia State University

Dorothy Munkenbeck Fragaszy
University of Georgia

CONTENTS

Introduction 421
Typical Research Involvement 422
Behavioral Biology 423
    Ecology 423
    Social Organization and Behavior 424
    Feeding Behavior 424
    Communication 425
    Common Behaviors in Captivity 427
        Normal Behavior 427
        Abnormal Behavior in Captivity 429
Maintaining Behavioral Health 429
    Physical Environment 429
    Social Grouping 429
    Feeding Strategies 431
    Training 431
Special Situations 432
    Introductions 432
    Separations 432
Conclusions and Remarks 432
Appendix 432
References 433

Introduction

Capuchin monkeys are medium-sized (adults 2–5 kg) monkeys that can be found from Central America through much of South America, on both sides of the Andes, to northern Argentina. Currently, two sister genera are recognized, with 8 species in each (*Cebus*, gracile capuchins, and *Sapajus*, tufted, or robust, capuchins; Lima, Buckner, Silva-Júnior, et al., 2017). Physically, these species vary in morphology and coloration, but both are recognizable by a signature cap on their crown that resembles a friar's hood, which is how they got their English common name. Capuchins thrive in a variety of habitats, making them good subjects for field studies. Although less common in biomedical research than the smaller owl monkeys, marmosets, and squirrel monkeys from Central and South America, capuchin monkeys have been favored subjects for behavioral and cognitive research, due to their wide availability, ease of captive husbandry, high cognitive ability, and the complexity of their behavior in social and physical domains.

Capuchins, like all monkeys in the New World, last shared an evolutionary relative with Old World monkeys about 35 million years ago. Historically, resolving capuchins' taxonomy has been a challenge, because capuchins inhabit such a large geographic range and exhibit great variation in physical and behavioral traits across populations. However, molecular methods are now providing better resolution (Lima, Buckner, Silva-Júnior, et al., 2017). *Cebus*, the gracile capuchins, are often lighter in color, have longer limbs relative to body size, and less robust cranial anatomy (Masterson, 1997; Figure 27.1). *Sapajus*, the tufted, or robust, capuchins, are often darker, more sexually dimorphic, have more robust

FIGURE 27.1 Drawing of *Cebus* and *Sapajus*. Courtesy of Stephen Nash. (Modified from Fragaszy, Visalberghi, & Fedigan, 2004.)

mandibles, and take their name from the tufts of hair common on their crowns (Figure 27.1). One of the key challenges in our understanding of capuchins is that virtually all captive capuchins living in laboratories and zoos are tufted capuchins (*Sapajus*). These individuals are usually of unknown species, and probably many are hybrids, because prior to 2011, all were identified as *Cebus apella*. In contrast, many of the long-term field data are on gracile capuchins, *Cebus*, especially *C. imitator* and *C. capucinus*. In this chapter, we will synthesize findings from laboratory and field studies of both genera; however, we will focus our comments predominantly on tufted capuchins, as these are the type most often found in captive environments.

While capuchins share many similarities with other New World monkeys, they exhibit several anatomical and behavioral characteristics that set them apart (reviewed in Fragaszy, Visalberghi, & Fedigan, 2004). Second only to humans, capuchins have the largest brain in relation to body mass of any primate (Rilling & Insel, 1999). They have semiprehensile tails (Atelinae also have prehensile tails, but other New World monkeys do not) and elaborated manual dexterity compared to other New World monkeys. They have a variety of anatomical adaptations for feeding on hard foods, including thick tooth enamel and robust mandibles. They spend more time on the ground than other New World monkeys, using diverse tactics to find hidden foods, including sometimes using objects as tools to process foods protected by husks and spines. Capuchins maintain and keep track of multiple social relationships, form alliances and coalitions during conflict, and engage in unique social bonding traditions (Perry & Manson, 2011).

Typical Research Involvement

Field studies on capuchins have addressed community ecology, behavioral ecology, socioecology, communication, locomotor and positional behavior, reproduction, development, movement, decision-making, cooperation, tool-use, group structure, and social dynamics (reviewed in Fragaszy, Visalberghi, & Fedigan, 2004). Field experiments have documented impressive cognitive abilities, such as mental mapping (Janson, 1998, 2007), episodic-like memory (Janson, 2016), functional deception (Wheeler, 2009; Wheeler, Tiddi, & Heistermann, 2014), anticipatory actions while using tools (Fragaszy, Liu, Wright, et al., 2013; Fragaszy, Morrow, Baldree, et al., 2019), and perception of the functional characteristics of objects that could be used as tools (Visalberghi, Addessi, Truppa, et al., 2009). As extractive foragers, capuchins are experts in processing encased and otherwise protected foods. They acquire these skills slowly, with strong influence from their social partners (Perry & Jiménez, 2006; Eshchar, Izar, Visalberghi, et al., 2016; Fragaszy, Eshchar, Visalberghi, et al., 2017; Coelho, Falótico, Izar, et al., 2015). For example, white-faced capuchins (*Cebus*) in Panama extract fruit from spiny plants without getting stung (Perry, 2009), and bearded capuchins (*Sapajus*) in Brazil meticulously crack nuts and seeds with stone hammers (Mangalam & Fragaszy, 2015; Fragaszy, Greenberg, Visalberghi, et al., 2010; Figure 27.2). Capuchins also engage in a variety of unique social traditions (e.g., mutual hand-sniffing) that function to increase social bonds between individuals (Perry, 2011). Research with wild capuchins helps shed light on how social and ecological factors relate to heightened intelligence, behavioral innovation, and social transmission of behaviors.

Research with capuchins in captivity has certainly run the gamut of topics in animal behavior, as could be expected given their presence in laboratories since the 1920s. Captive capuchins have participated in such a wide variety of research studies that we cannot identify a "typical" research investigation. However, in the past few decades, capuchins have been popular subjects for studies of physical cognition (e.g., attention, memory, numerical reasoning, analogical reasoning, symbolic representations, spatial problem-solving, manual dexterity, and self-control; Addessi, Mancini, Crescimbene, et al., 2008; Anderson, Hattori, & Fujita, 2008; Anderson, Kuroshima, & Fujita, 2010; Beran, Evans, Leighty, et al., 2008; Beran, Perdue, & Smith, 2014; Fragaszy, Johnson-Pynn, Hirsh, et al., 2003; Fragaszy, Visalberghi, & Fedigan, 2004; LaCour, Stone, Hopkins, et al., 2014). They have also been popular in studies of social cognition, including deception, communication, social relationships, fairness, decision-making, and social traditions (e.g., Crast, Hardy, & Fragaszy, 2010; Drayton & Santos, 2014). One focus in laboratory studies has been the mechanisms underlying the cooperation and social traditions seen in the wild. A robust line of work using a cooperative pulling task, in which subjects must work together to acquire food, demonstrates that capuchins appear to understand the

FIGURE 27.2 Typical behavioral observations and research paradigms conducted with wild and captive capuchins. Top (left) *Sapajus* cracking a nut with a stone hammer (Photograph by Barth Wright); (right) nose poking tradition in *Cebus* (Photograph by Marcela Benítez). Bottom (left), trading paradigm in captive capuchins; (right), typical computer paradigm. (Photographs by Sarah Brosnan.)

contingencies of cooperation and their partner's role (reviewed in Brosnan, 2010). They are also quite good at finding the optimal decision, taking into account other group members' strategies in decision-making tasks (reviewed in Watzek, Smith, & Brosnan, 2017). Capuchins respond negatively to inequity (Brosnan & de Waal, 2003), and cooperate less when payoffs are unequal (Brosnan, Freeman, & de Waal, 2006; de Waal & Davis, 2003). Other research has shown that social partners have a strong influence on the initiation of feeding (Addessi & Visalberghi, 2001; Galloway, Addessi, Fragaszy, et al., 2005; handling of objects (Bonnie & De Waal, 2007), and efforts to solve mechanical problems (Crast, Hardy, & Fragaszy, 2010; Dindo, Thierry, & Whiten, 2008). In general, capuchin monkeys, especially youngsters, are facilitated by others' activities to do something similar.

In addition to behavioral and cognitive research, capuchins sometimes participate in biomedical research. Capuchins have been subjects in laboratory research in a range of biomedical fields, including immunology (de Palermo et al., 1988), reproductive biology (Nagle et al., 1989), pharmacological and metabolic studies (Bergeron et al., 1992), parasitology (Garcez et al., 1997), physiology (Terpstra et al., 1991), neurotoxicity (Lifshitz et al., 1999), and most frequently, neuroscience. Neuroscience studies have focused particularly on brain anatomy and function related to motor coordination and movements of the hands (Mayer, Lewenfus, Bittencourt-Navarrete, et al., 2019, Padberg, Franca, Cooke et al., 2007; Phillips, Subiaul, & Sherwood, 2012; Rathelot & Strick, 2009). Capuchins tend to do well in captivity, can be socially housed, and are easy to train, which makes them good models for research related to cognitive side effects of medication, behavioral models for identifying disease, and cognitive deficits associated with aging. For example, capuchins have participated extensively in schizophrenia research, because they can develop some of the extrapyramidal side effects seen in schizophrenic patients, and can be trained, via touch screen, to participate in visual tests commonly used as a tool for schizophrenia screening (Pessoa et al., 2008).

Behavioral Biology

Ecology

Capuchin monkeys occupy virtually every type of Neotropical forest, including humid tropical forests, swamp forests, seasonally flooded forests, mangrove forests, and gallery forests, as well as dry, deciduous forests, where rainfall is absent for 5–6 months of the year (Fragaszy, Visalberghi, & Fedigan, 2004). They range from sea level up to 2,700 m above sea level. As suggested by the wide variety of habitats in which they live, capuchin monkeys are fairly hardy with respect to temperature and humidity. They are found even where temperatures drop below freezing briefly at night, as in higher elevations in the southernmost parts of their range. They can survive in isolated forest patches, including forested urban parks, and in early secondary growth areas that require them to move in open areas between forest gaps. Some species of capuchins are sympatric in some parts of their ranges (e.g., *Cebus olivaceus* and *Sapajus apella* in Guyana, and *Sapajus apella* and *Cebus albifrons* in eastern Peru). In areas of sympatry, the species display somewhat different dietary preferences and habitat use. All capuchins locomote primarily quadrupedally, and favor horizontal or oblique tree limbs, although they readily ascend and descend trees and cliffs, leap gaps of several meters, and clamber through thickets and vegetation with small branches. The monkeys use their semiprehensile tail to grasp surfaces as they move through the forest, and use the tail as an anchor, bracing their feet against a support and leaning the body out into space when reaching for items otherwise beyond their grasp. Capuchins occasionally crouch and stand bipedally, and even (although rarely) locomote bipedally; sometimes while transporting item(s) in their hands (Duarte, Hanna, Liu, et al., 2012).

Wild capuchin monkeys devote one-third to one-half of their waking hours to foraging. Females and juveniles devote more time per day than males to foraging; males use this extra time to rest more than females and juveniles. The monkeys spend about one-quarter of their day traveling and one-quarter of their day resting, and the remainder in social activities, such as play or grooming. They adjust their activity budget according to season and weather, resting more in midday in hotter weather, for example, and in accord with the availability of food and water.

Capuchins spend most of their time in the middle layers of the forest, but will use all levels from the canopy to the understory, going to the ground to drink, forage, or travel. Capuchins in some regions readily travel and feed on the ground. At one site in the Cerrado of Brazil, bearded capuchin monkeys spend nearly 30% of their daylight hours on the ground (Wright, Biondi, Fragaszy, et al., 2019). In other regions, the monkeys rarely come to the ground. The monkeys' use of the ground is balanced between attraction to food resources and, perhaps, easy travel paths on the one hand, and the risk of predation on the other.

In natural settings, a group's home range may vary from less than 100 ha to about 400 ha. A study of one group of bearded capuchins revealed that they traveled 1.8 km/day, on average (Howard et al., 2012). Capuchins travel primarily along habitual routes in some regions; in others, they travel more randomly across the landscape (Presotto & Izar, 2010; Presotto, Verderane, Biondi, et al., 2018). Variations in travel patterns are thought to reflect the distribution and abundance of resources, which can vary widely. Capuchins are not territorial; groups encountering one another may mix relatively amicably for a while, or they may engage in noisy agonistic displays toward one another. One group may simply move away from another group to avoid an encounter. Neighboring groups may contain members that have emigrated from one group to the other; they are not necessarily "strangers" to one another.

Social Organization and Behavior

Capuchins generally live in mixed-sex groups of 5–40 individuals, with one alpha male and one alpha female. The alpha male is typically the most dominant individual, although the alpha female may be more dominant than other adult males in the group (Fragaszy, Visalberghi, et al., 2004; Izar, Stone, Carnegie, et al., 2009). The size and composition of social groups are flexible, depending on the species, local ecology, and season (e.g., Izar, Verderane, Peternelli-dos Santos, et al., 2012; Lynch Alfaro, 2007). Although there may be more than one male present, there is strong reproductive skew, such that the alpha male monopolizes the majority of mating (Carosi, Linn, & Visalberghi, 2005). Males leave their natal group between 6 and 9 years of age, sometimes emigrating with kin, and may spend time as part of a bachelor group before joining a new group. Upon becoming alpha, males undergo a physical and physiological transformation, gaining about 20% of their body mass (Fragaszy, Izar, Liu, et al., 2016) and exhibiting a drastic increase in testosterone levels (Benítez, Sosnowski, Tomeo, et al., in prep). Males are likely to commit infanticide when they take over a new group, whether they come in from the outside or successfully take over the alpha position of a group in which they already reside (Brasington, Wikberg, Kawamura, et al., 2017; Izar, Stone, Carnegie, et al., 2009).

Females typically stay in the group into which they were born (but see Tokuda, Martins, & Izar, 2018). Females form strong bonds with their mothers and develop close friendships with other females, as evidenced by increased grooming and sharing food with specific social partners. Males, on the other hand, are more indiscriminate (de Waal, 2000; di Bitetti, 1997). Closely related females may nurse one another's offspring (Baldovino & di Bitetti, 2008; Perry, 1996; Sargeant, Wikberg, Kawamura, & Fedigan, 2015). Although there is little evidence of adoption, one case has been reported in wild white-faced capuchins (*C. capucinus*; Perry & Manson, 2011). Female capuchins are highly proceptive, courting males vigorously and persistently (Carosi, Linn, & Visalberghi, 2005; Janson, 1984), but they lack any physical sexual signal, such as the swellings seen in some Old World monkeys and apes. Females in estrus spend several days soliciting the male with distinctive vocalizations and facial expressions, approaching him, and touching him. These behaviors escalate over the period of courtship (see *Communication*). Initially, the male will typically avoid or even aggress the female, but ultimately the two engage in courtship and mating. This courtship system may increase certainty of paternity, and indeed, male capuchins are particularly tolerant of infants and juveniles, including sharing food (Fragaszy, Visalberghi, & Fedigan, 2004). During the period in which a female is in estrus, it is difficult to engage either the female or the alpha male in other activities, including cognitive and behavioral testing. The female capuchin menstrual cycle averages 20.8 days (Nagle & Denari, 1982). Gestation lasts 154–162 days (Fragaszy, Visalberghi, & Fedigan, 2004) and both sexes attain reproductive maturity at approximately 4–5 years of age (Fragaszy & Adams-Curtis, 1998), although males do not typically achieve breeding status until several years later, individuals achieve full adult body mass between 7 and 9 years of age (Fragaszy, Izar, Liu, et al., 2016).

Feeding Behavior

Capuchin monkeys are frugivores/omnivores. The bulk of their diet is fruit and seeds, and to a lesser extent other plant parts, but capuchins also devote considerable time and effort to obtaining animal protein. While the great majority of their animal prey is invertebrates, they also take vertebrate prey opportunistically; they capture lizards, snakes, and other escaping prey by chasing them, and they pounce on larger insects (e.g., locusts) and birds. In all populations, animal protein is a significant feature of the diet for all animals post weaning, regardless of the abundance of plant foods.

Capuchins, even immature monkeys, eat tougher foods than other primates (Wright, Wright, Chalk, et al., 2009; Chalk, Wright, Lucas, et al., 2016). Thick dental enamel and robust cranial anatomy support this diet, and so do their foraging behaviors. Capuchin monkeys are formidable extractive foragers; they search for edible items encased in husks, wood, and soil, or hidden in leaf litter, holes, or rock crevices. This style of feeding requires active searching with hands, ears, nose, and eyes. They tap and sniff dead branches before breaking them to find embedded invertebrates. They rub open unripe cashew drupes to extract the kernel, avoiding the caustic latex in the rind. They break apart wasps' nests to eat the larvae, braving the stinging insects to do so, and raid the nests of birds and mammals, such as squirrels, bats, and coatis, often despite desperate defensive actions by the parents. Once an item has been procured, the monkeys pound, rub, tear, peel, and break it to subdue it, if necessary, to remove stinging or sharp elements, and to expose or extract the edible portion. They use their hands, mouth, and feet in these efforts. Some populations use sticks and stones to probe insect nests, to flush prey (such as lizards hiding in rock crevices), to excavate underground plant storage organs, and to pound open seeds.

Communication

Capuchin monkeys communicate using a wide range of visual, vocal, tactile, and olfactory signals that function to communicate internal state, social intent, and to inform others about the environment (Fragaszy, Visalberghi, & Fedigan, 2004). Capuchins exhibit a rich repertoire of facial expressions, gestures, and body positions that signal social and sexual intent (See Figure 27.3). Capuchins will bare their teeth in submission ("submissive grin") or as a threat ("open mouth threat face"); tilt their heads ("head tilt") or smack their lips together ("lip smack") as a sign of affiliation; and protrude their lips ("duck face") and raise their eyebrows ("scalp lift") during courtship (Visalberghi, Valenzano, & Preuschoft, 2006; Weigel, 1979; see Figure 27.4). While there is some variation in how visual signals are produced (for example, robust capuchins often "flirt" by raising their eyebrows and grinning, whereas gracile capuchins "flirt" by protruding their lips in "duck face"), the functions of many of these signals are similar across species (Weigel, 1979).

Capuchins engage in several tactile behaviors that function to reinforce social bonds, such as sitting in proximity or in contact while resting or feeding, grooming, and social fur rubbing, also called anointing (e.g., rubbing pungent materials over themselves and occasionally over each other; Benítez, et al., 2018; see Figure 27.3). Tactile cues are also important in reinforcing the social hierarchy, resolving tension, and establishing alliances. Capuchins will bother each other in every way imaginable (e.g., pinch, push, pull, smack, bite, supplant) to reinforce dominance or during courtship. They embrace or mount each other after altercations ("reunion display"), and ally against a rival by stacking on top of each other ("overlord" or "double threat"; see Figure 27.5).

To a lesser extent, capuchins rely on olfactory communication. Pheromones are certainly implicated in behaviors, such as urine washing, hand sniffing, and genital inspections. Urine washing is performed by both sexes, and consists of an individual urinating on the palm of its hand and rubbing that palm on the sole of the foot. Capuchins can discriminate between groups and recognize their species, based on odors created by urine washing (Ueno, 1994). In addition, female capuchins show differential neural responses associated with arousal when encountering urine from an adult male compared to urine from a juvenile male (Phillips, Buzzell, Holder, & Sherwood, 2011), and alpha males tend to urine wash more often than subordinate males (Campos & Fedigan, 2013). There is no clear evidence linking urine washing to female reproductive states, although females urine wash less while they are lactating (Campos & Fedigan, 2013).

Capuchins rely heavily on vocalizations to communicate. Capuchins emit a wide range of vocalizations that differ in context. We do not yet have a complete description of the vocal repertoire of capuchins, but a few studies have aimed to decode the form and function of their vocal communication. Vocalizations can range from soft calls, such as calls that may help group mates stay in contact or space themselves from one another while foraging, and trills, when individuals approach one another to groom or make affiliative contact, to louder, long distance calls, such as lost calls, which function to reunite a lost individual with its social group. Capuchins only respond to lost calls from members of their own group, and are more responsive to lost calls from dominant members (Digweed, Fedigan, & Rendall, 2007), suggesting that they can recognize each other's individual voices. Capuchins vocalize to indicate information about a resource, predator, or rival group. While foraging, capuchins produce a number of food-associated calls that provide information about food type, amount, preference, and divisibility (di Bitetti, 2003). Capuchins produce a variety of alarm calls depending on the predator and the degree of the perceived threat. These calls vary acoustically and result in different behavioral strategies tailored to different types of predator (Fichtel, Perry, & Gros-Louis, 2005). Fichtel, Perry, and Gros-Louis (2005) report that white-faced capuchins (*Cebus capucinus*) look up and/or move toward cover when they hear an aerial predator alarm call, often emitted by a single individual, and rarely behave aggressively toward birds. In contrast, many individuals give alarm calls to terrestrial predators, such as snakes, canids, and caimans, and they routinely behave aggressively toward those predators, usually joining with others to approach and "mob" the predators. Black capuchins (*Sapajus nigritus*), like white-faced capuchins, emitted the most specific calls to decoys of aerial predators, and responded to playbacks of these calls by looking up and seeking cover, but some calls given to decoys of terrestrial predators were also used in other stressful contexts (Wheeler, 2010). Calls given to terrestrial predators by black capuchins were situationally variable, and these calls seemed to elicit less specific responses from listeners.

There is evidence that some of capuchins' vocalizations are triggered involuntarily by specific stimuli. For example, male capuchins produced a guttural call ("wah wah" vocalization) in response to loud explosive noises, such as a tree falling or thunder, and only produced that call during these conditions (de Resende, Oliveira, & da Silva, 2007). However, the vast majority of capuchins' vocalizations appear to be under

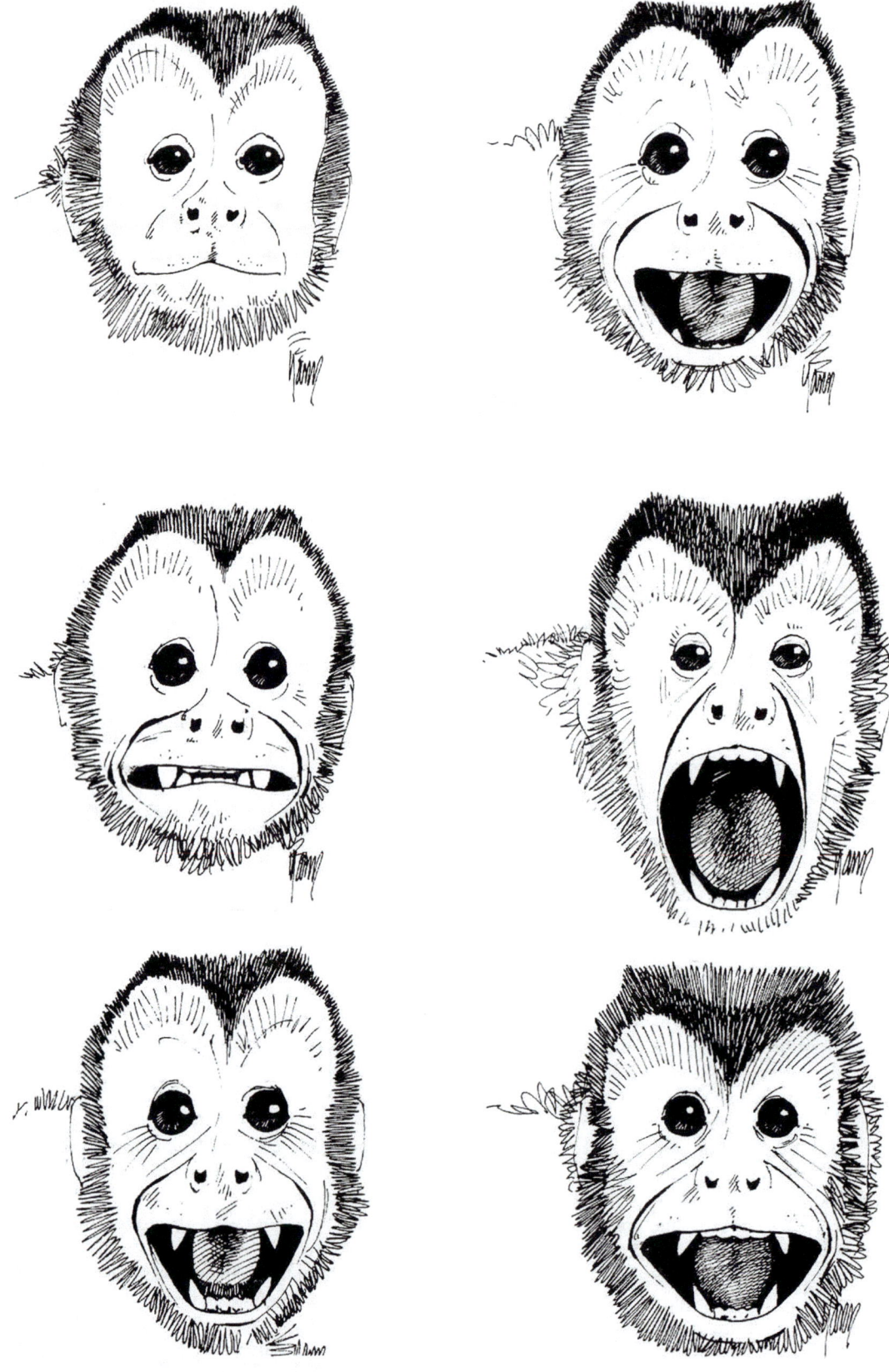

FIGURE 27.3 Examples of facial expressions in capuchins. Top: left, scalp lifting; right, relaxed open mouth. Both of these occur in affiliative or relaxed contexts. Middle: left, silent bared teeth display in a juvenile, associated with play; (right) open mouth threat. Bottom: left, relaxed open mouth display, associated with play; right, open mouth silent bared teeth display in an adult, associated with courtship. (Drawings by Lebrtrand Deputte and reproduced from Fragaszy, Visalberghi, & Fedigan, 2004.)

FIGURE 27.4 Examples of facial expressions and postures during sexual interactions. Top left, a female raising her eyebrows, grinning, and tilting her head. Top middle, a male responding to the female with the same facial expression, and top right, rubbing his chest. Bottom left, a female and a male gaze mutually at one another and touch each other's body. Bottom middle, the male and female engage in a ventro-ventral mount. Bottom right, the female turns and reaches back while the male mounts her dorsally; she is also grinning and vocalizing. (Drawn by Andy de Paoli from video images; reproduced from Fragaszy, Visalberghi, & Fedigan, 2004.)

voluntary control. Whether an individual produces food-associated calls, for example, depends on a variety of factors, such as the number and rank of individuals in close proximity (di Bitetti, 2003). Capuchin monkeys have been witnessed to alarm call deceptively, apparently to monopolize, or pull dominant individuals away from, a high value resource (Wheeler, 2009).

Capuchin communication, like human communication, is predominantly multimodal in nature. Visual, tactile, and vocal signals are often coupled together to communicate intent. In social interactions, a dominant individual may threaten a subordinate by emitting a vocal threat, while simultaneously showing an open mouth threat display, and the subordinate responds with a submissive grin coupled with a vocal scream. In courtship, females will often produce a variety of sexual facial expressions, such as duck face or scalp raise, hit the male and run away, and produce a series of courtship-specific vocalizations. While visual and tactile modes of communication function well in close contact, vocalizations can carry over longer distances and result in communication with a larger range of individuals. For example, as a female solicits sex from a specific male with visual and tactile cues, her vocalizations alert other males that she is in estrus.

Common Behaviors in Captivity

Normal Behavior

If socially housed with appropriate enrichment, capuchin monkeys should behave similarly in captivity as they do in the wild. This includes the communicative and social behaviors discussed above (affiliative behaviors, aggressive behaviors, dominance-seeking behaviors, sexual behaviors), and feeding behaviors (see Table 27.1). In captivity, capuchins show nearly constant environmental engagement (scanning the environment, manipulating objects, tearing and biting things open), which can result in destruction of materials that are not sufficiently sturdy and the production of noise (capuchins are fond of banging). As in the wild, they interact with one another frequently, including grooming, social play, and resting in proximity. It is important to monitor capuchins' behavior through regular (at least once per week) systematic observation to understand normal social dynamics, anticipate changes to those dynamics, monitor individuals' health and well-being, and determine compatible social partners for dyadic and group-level research. For example, capuchins often self-groom, shake the body, yawn, urine wash, or pace. While these are all common behaviors,

FIGURE 27.5 White-faced capuchins ally against a rival by stacking on top of each other (called overlord behavior, or double threat). (Photograph by Marcela Benítez.)

TABLE 27.1

Common Behaviors in Captivity

Category	Description
Self-scratch	Drag the nails/hand across one's own skin
Urine wash	Urinate on one palm and rub it on the sole of the foot
Self-groom	Pick through the hair, searching for and/or removing debris.
Body shake	Shake entire body or a body part. Often a sign of agitation.
Fur rub	Vigorously rub an object over fur repeatedly. Individuals may rub against another individual.
Play	Nonaggressive interactions involving two or more animals. Includes wrestle, tickle, chase. Usually between juveniles and infants.
Sexual	Any form of sexual behavior between two individuals. This includes solicit, penile erection (directed at a female), exploration behavior, mount, thrust, flirt (duck face, scalp raise, vocalize).
Groom	Pick through the hair, search for, and/or remove debris from another individual
Contact	Sit huddled or touching another individual
Proximity	Within arm's reach of another, but not touching
Aggression	Aggression can be noncontact (scream, threat, chase) or result in contact (hit, bite, pull tail)
Submissive behavior	Crouch, bob, flinch, avoid, flee, present (in a nonsexual context), or, during aggression received, fear grin
Dominant behavior	Threaten, mount, or displace another individual over a preferred space, food, or object.
Forage	Search for or extract food
Solo feed	Feed or forage alone
Proximity or contact feed	Feed in proximity or in contact with another individual
Beg	One individual begs for food from another, by sticking out the hand or placing the face close to that of the eating monkey.
Manipulate	Handle or play with a non-edible object
Locomote	Actively move around enclosure
Rest	Inactive, sleep, sit for long periods of time

TABLE 27.2

Abnormal Behavior Sometimes Seen in Captivity

Category	Description
Suck digits	Suck on any digit, usually thumb or index finger or repeated mouthing and/or licking any extremity
Hold body	Hold on to any part of the body (often the genital regions) for an extended period or in an awkward way (i.e., walking while holding one's genitals).
Coprophagy	Eat or play with feces or urine.
Exaggerated head turn	Throw head back in a circular motion
Pull hair out	Forcefully bite or pull clumps of hair out of own or other's body. Note the recipient if it is pulling out another's hair.

an increase in the rate of these behaviors can be indicative of stress or a decline in health. Regular observations can employ several sampling techniques, including focal observations of a single individual for a period of time or scans of the whole group in which each individual's behavior is recorded at regular frequent time points. Finally, capuchins respond well to humans, and to cognitive and behavioral testing; approaching experimenters and actively engaging with them and with experimental tasks.

Abnormal Behavior in Captivity

In captivity, some capuchins engage in a few stereotypical behaviors that are not seen in the wild (see Table 27.2). These behaviors can be indicative of stress, including forcefully biting or pulling out clumps of hair, while others just seem to be benign behaviors, like head tossing (an exaggerated head turn in which individuals throw their heads back in a circular motion when they change direction while locomoting). An abnormal behavior that may be of concern is coprophagy (eating or playing with feces). Coprophagy can be indicative of nutritional deficiency, boredom, social stress, or medical problems. As with other captive primates, capuchins can display locomotor stereotypies, such as pacing, and occasional emesis/regurgitation, although the latter is quite rare in our experience.

Maintaining Behavioral Health

Physical Environment

Many of the environmental factors important to capuchin monkeys' well-being are the same as for any gregarious nonhuman primate. Capuchins do best living with conspecifics (groups of mixed ages and sexes) in indoor-outdoor areas, with multiple points of connection to avoid individuals getting trapped, and ideally, with a clear or open doorway so that they can see who is on the other side. Although typically there is not much aggression between individuals, subordinates often will not approach dominants and will hesitate to enter an area where there has been an aggressive encounter in the past, unless they have clear visual access. With this in mind, it is best to have plenty of space and a variety of visual barriers, so that individuals can separate from one another voluntarily. Capuchins adapt to brief (30 min) crowding (e.g., for husbandry reasons) with a decrease in all social behavior and an increase in self-grooming (van Wolkenten et al., 2007). They also do well when they have visual access to neighboring groups, and in our experience, seek opportunities to interact with those groups.

Capuchins are primarily arboreal; when given the option they prefer to be higher in their enclosure, including for seated activities, such as eating. Thus, it is important to include shelves or platforms higher up on the walls of enclosures to give them a place to rest and engage in grooming or foraging. It is also important to provide aerial pathways from one platform to the next (such as hoses, trapezes, or branches). They will also use raised platforms on the ground, such as barrels and boxes (Figure 27.6). Although capuchins will readily forage on the ground, they will typically take food to an elevated platform to eat. Capuchins also like to climb, so vertical space is essential. Given their dexterous hands, feet, and semiprehensile tails, it is ideal to construct the walls and ceiling of the enclosure from chain link or large mesh that they can use for climbing.

Capuchins take every opportunity available to them to engage with the physical environment (Anderson, 1996; Fragaszy & Adams-Curtis, 1991; Fragaszy, Visalberghi, & Fedigan, 2004). Wild capuchins devote up to one-half of their day to foraging, with much of that time devoted to searching for hidden prey and extracting edible parts of foods, so giving them ample opportunities to search for food and to process encased foods promotes species-typical behavior. To model natural foraging and to encourage physical activity more generally, when possible, capuchins' foods should be scattered in some substrate that requires them to search for it (i.e., pine bark, wood wool, or hay). This can be used for daily fruits and vegetables to encourage activity. Dispersing small seeds among straw or other bedding is particularly effective. Capuchins can be given a number of feeding enrichments, including unshelled nuts (they can crack the shells of most nuts with their teeth), bags or boxes baited with treats, or puzzle boxes that require them to manipulate something in order to access food. Although capuchins typically will not eat leaves, browse is effective enrichment. The monkeys devote persistent effort to destroying it. Finally, capuchins consume substantial quantities of animal protein in the wild, and in captivity are particularly fond of mealworms.

Social Grouping

In the wild, capuchins live in mixed-sex groups, except when transitioning between groups. The social adaptability evident in the wild allows for some flexibility in captive housing as well. Social groups can contain one or more

FIGURE 27.6 Photo collage of an outdoor enclosure housing a group of tufted capuchins. Raised platforms, both rigid and flexible structures for climbing, options for going out of view of others, and loose objects on the ground are important features of the living space for captive capuchins. (Photograph by Kelly Leverett, courtesy of the Language Research Center, Georgia State University.)

adult males, depending on the temperament of the males. Anecdotal experience suggests that in captivity, groups with multiple adult males are more likely to be peaceful when they are at the larger end of group size (>15 individuals). In addition, they may be more stable with more than two adult males, possibly because the shifting alliances deter males from attacking each other. Even in smaller groups, however, two adult males may coexist peacefully and even exhibit affiliative behavior. This can occur when a juvenile male grows up in a group and never challenges the alpha, or when the alpha submits to a younger male without aggression (Brosnan, pers. obs.). On the other hand, adult males can fight each other viciously, and it can be hard to predict which males will not get along. Females more often harass other females in groups, biting the tail or hands, or pulling hair. Subordinate animals can become scapegoats, experiencing simultaneous targeted aggression from several others. Since it is difficult to predict these outcomes, it is important to make frequent standardized observations of the group to try to identify problems before they become severe, and so that there is a record of how relationships have changed over time when problems do emerge.

One key feature for stable capuchin groups is strong female bonds (Perry, 1996). In general, females should be kept together as "family", unless it is impossible to do so. On the other hand, males tend to leave their natal groups upon reaching adulthood. Allowing dispersal reduces the risk of inbreeding and instability in the male hierarchy. Facility managers should remove natal males from their captive social group when they are about 6 years of age or, if there is a reason to leave them, managers should be prepared to move them rapidly if fighting breaks out, as the risk of aggression and injury increases as the young male ages. When possible, it is good to swap males between groups (see Introductions, below) to mimic natural behavior and increase genetic diversity, if breeding is desired (but see below in relation to managing the potential for infanticide). If male swapping is not possible, small bachelor groups can be formed successfully. In our experience, males from a larger group will reduce or cease aggression if they are moved away from females, although anecdotal reports suggest that not all males can live together peacefully in pair-housed situations. Males in the wild reportedly sometimes emigrate with others from their group, including maternal brothers (Fragaszy, pers. obs.), but it is our experience that even brothers who have lived together their entire lives will not necessarily continue to live together peacefully once they reach puberty. A solution to inbreeding, if there is no option to remove a natal male at puberty, is to vasectomize the male. Note, however, that while this does remove the possibility of inbreeding, it will not reduce aggression and fighting among the males, nor will it prevent the male from sexual interest in close female relatives.

In captivity, there have been observations of males killing infants, even in groups in which they are well established, if they have never been seen mating with the female that is the mother of the infant (Brosnan, pers. obs.). To reduce the risk of infanticide upon the introduction of a new male, care should be taken to introduce new males only when there are no infants or pregnant females present. In cases in which there is a takeover or a female becomes pregnant, and there are no observations of her mating with the alpha male, the male should be separated from the female and her infant until the infant is about 6 months of age (it is not weaned at this time, but is fairly independent). This highlights the need for regular observations and recordkeeping of the social group, by every individual who works with the monkeys, including husbandry and research staff, to note mating, aggression, and other potentially important behaviors. This can be extremely useful in anticipating and working out problems with social group dynamics.

Feeding Strategies

Capuchins require a varied diet with sufficient vitamins (including D_3) and a comparatively high protein content. They can be fed high-protein monkey chow supplemented with nuts, peanut butter, cheese, yogurt, hard boiled eggs, or other foods containing protein and B vitamins. This is particularly important for females, if breeding is a priority. Capuchins particularly enjoy nuts and peanuts, fruits, *nonleafy* green vegetables (bell peppers, particularly red bell peppers, cucumbers), and cooked starches (sweet potatoes and potatoes).

New World monkeys have a tendency toward insulin resistance (i.e., type II diabetes; Harwood, Listrani, & Wagner, 2012), so sugar intake should be limited. Most diabetic capuchins' symptoms can be reasonably well controlled with diet, including switching the group to a low-sugar monkey chow and removing high glycemic index fruits and sugary treats from their diet. If necessary, a drug that increases insulin sensitivity, such as Metformin, can be given orally. Diabetes can be successfully managed while the individual remains in the social group, including changing the group's diet and providing oral medication. Anecdotally, diabetes appears to run in families, and the diabetic status of monkeys can be checked by looking at sugar and ketones in the urine, and a blood test for HbA1c during their annual physical examination. Generally, diabetic monkeys can live in large social groups with their diabetes under control.

Training

Capuchins are easy to train for husbandry and research tasks using positive reinforcement. They can be relatively easily trained to separate from their group into test boxes for small positive reinforcement, and for enriching cognitive and behavioral tasks, they will learn to separate without the need for the additional reinforcer. When training, always ensure that every person interacting with them follows the same training plan, and make use of a training log; capuchins are smart and will easily learn to manipulate humans into providing additional reinforcers. We also note that they do not learn well using negative reinforcement techniques (i.e., an unpleasant situation resolves when they cooperate). They learn most effectively with short, but intensive, training sessions, with positive reinforcement and judiciously applied "jackpot" rewards.

They can easily be trained to participate in tasks that require interacting with manual apparatus or tokens. For instance, capuchins have been extensively tested on cooperation using a cooperative "bar-pull" in which two or more individuals must work together to pull in a counterweighted tray that contains food rewards (Brosnan, 2010), and they can easily be trained to trade tokens back, for instance, to indicate choices or to study decision-making behavior (Brosnan & de Waal, 2004) or spatial reasoning (Johnson-Pynn, Fragaszy, & Hirsh, 1999; LaCour, Stone, Hopkins et al., 2014). They can be trained to use touchscreens or joysticks to participate in computerized tasks (Drayton & Santos, 2014; Evans, Beran, Chan, et al., 2008; Leighty & Fragaszy, 2003). On the other hand, capuchins do poorly when restrained or closely confined. They do not readily learn to sit in a "primate chair", and whereas macaques can readily be trained to present an arm or leg for venipuncture, for example, capuchin monkeys do not accept this training (Dettmer, Phillips, Rager, et al., 1996). In general, capuchins are most likely to cooperate if they are given some control over their movements and participation.

Several aspects of capuchins' behavior are worth keeping in mind when considering behavioral testing. First, capuchins routinely process foods by ripping, tearing, and banging. In captivity, they will destroy anything that they can. It is important to protect computer cables, microphones, cameras, etc., so that the monkeys cannot reach them. We have found that for tasks in which we need them not to grab, it is effective first to train them to target to an object (i.e., touch or handle an object separate from the task, such as a Kong toy) to keep their hands busy, and then train them in the task. This procedure was successfully used to train them to receive intranasal oxytocin from a nebulizer without restraint (Brosnan, Talbot, Essler, et al., 2015). As mentioned above, they enjoy banging objects against substrates and will often do so, which can be noisy and distracting (to humans and other monkeys), albeit not necessarily destructive.

Positive reinforcement training can be used very effectively for husbandry and veterinary purposes. Capuchins can be trained for tasks to help caretakers, such as touching a lixit to ensure water is flowing. They can also be desensitized to trays placed beneath them or taught to chew on dental tape to aid in the collection of urine, fecal, and salivary samples for hormone analyses. They can be trained to present body parts for veterinary inspection. Any person participating in training should be familiar with the monkeys' social dynamics, so that they can safely shift individuals, and the person should be trained in proper positive reinforcement techniques to maintain the monkeys' participation (see Prescott & Buchanan-Smith,

2016). Indeed, when the cooperation of the monkey has been attained using positive reinforcement techniques, training and testing can be quite enriching, and a sudden change in the monkey's willingness to participate or work output can be indicative of an illness or social issue that needs addressing.

Special Situations

Introductions

New individuals can be successfully introduced to stable social groups of capuchins (Cooper, Bernstein, Fragaszy, et al., 2001; Fragaszy, Baer, & Adams-Curtis, 1994). However, introductions are risky and often do not succeed, so they should be undertaken only when necessary and under continuous close observation. This could be due to husbandry concerns, including the acquisition of new animals or a pressing need to move an animal from its existing social group due to extreme and unremitting circumstances, or introducing a new male in order to bring in genetic diversity. Adding and removing animals may change social dynamics in unpredictable ways.

Females are relatively easy to integrate into an existing group (Fragaszy, Baer, & Adams-Curtis, 1994; Brosnan, pers. obs.), although that is attempted considerably less frequently than the introduction of males, therefore female introductions should be attempted with appropriate caution. Introducing males, however, is associated with variable outcomes. While some reports indicate that males may respond benignly to strangers (Becker & Berkson, 1979), in one case, introductions led to severe aggression, including the death of one of the males (Cooper, Bernstein, Fragaszy, et al., 2001). In general, introductions work best when a single male is introduced into a group in which there are no other adult males present. In some cases, it has been possible to introduce a group of males who knew each other into a group of females (Cooper, Bernstein, Fragaszy, & de Waal, 2001). However, in other cases even brothers who had lived together their whole lives were unable to establish a stable hierarchy (Brosnan, pers. obs.). It is also important to note that females may exert a substantial impact on which male becomes alpha by supporting a male (and different females may support different males), making it difficult to predict who will ultimately become the alpha male. We suspect that introducing multiple males works best if the males being introduced are substantially different in age and/or size, and have a strong established hierarchy before the introduction. It is important to note that exposure to novel females induces dramatic increases in testosterone in males (by at least an order of magnitude; Benítez, Sosnowski, Tomeo, et al., unpublished data), so even previously stable groups may show substantially altered dynamics. Thus, it is best to be prepared with an alternate plan, either to remove males to a bachelor group or split the females between two males. This may be logistically challenging and should be taken into consideration prior to attempted introductions. We do not recommend introducing males into a group in which one or more adult males is already present. We recommend a gradual introduction procedure moving from less to more contact (visual contact only, followed by partial contact, followed by supervised full contact, followed by unsupervised full contact, with progressive daily sessions spread out over at least a week). Details on an introduction procedure successfully used in Brosnan's laboratory are included in the Appendix.

Separations

There are situations in which an individual must be temporarily separated from its group for short periods (e.g., a few days) for medical or husbandry reasons. If this is unavoidable, it is best to leave the separated individual(s) in visual and vocal contact with the rest of the group, such as in a separate cage within the same room, and to separate for as short a period as possible. This allows for a relatively smooth reintroduction at the end of the separation period. If at all uncertain about the status of the group's relationships following the separation, it is best to follow at least a modified version of the introduction protocol to minimize the chance of a negative outcome and maximize the chance of spotting problems before they become severe.

Conclusions and Remarks

Capuchins are engaging and intelligent. Working with them can be a remarkably positive experience for both parties, as long as you recognize that they are independent and that to secure their voluntary participation, they require control of their environment. Capuchins are very adaptable, inhabiting a wide range of natural environments, and perhaps, because of this, they adapt well to captivity.

Appendix

Our lab (Brosnan) has developed a protocol for capuchin introductions based on the procedure used to introduce chimpanzees into large, mixed-sex social groups. In short, subjects are initially given noncontact, visual exposure over a period of 2–4 days. If there are no signs of overt stress when in proximity to one another through a clear divider, such as a Plexiglas panel (i.e., fear grimacing, fur plucking), then subjects may be given very limited direct contact (i.e., a small hole in a clear partition) under continuous observation for a period of one to several hours, then removed back to non-contact exposure. This is repeated for at least two days, until subjects show no signs of stress. Next, subjects are given extensive direct contact, but still with a barrier like mesh or a clear partition with multiple holes that are large enough for animals to make contact and groom. Again, this should take place for several hours, under continuous observation, and then subjects should be removed back to non-contact exposure. Not only does this approach reduce the chance of an injury, but also gives the subjects a chance to relax alone, which may reduce stress. Finally, subjects are given an hour or two of full contact interaction. If this goes well, subjects can spend the entire day together, although it is generally a good idea to separate subjects back out for the night for a few days, until it is clear that serious aggression is highly unlikely. All interactions should be continuously monitored by an experienced observer who is familiar with capuchin behavior. During all of this, there should be ample enrichment (scattered small food items or browse work well) provided to the group and the individuals being introduced, both to reduce stress and to give the animals a distraction from social tensions. In addition, some mild aggression is likely as they work out their new relationships, but fights and aggression should be monitored closely, in particular to ascertain if this is a situation in which they are not establishing a stable dominance hierarchy (i.e., there is no clear winner across fights). These are the most dangerous situations, as subjects will continue to fight until it is clarified.

REFERENCES

Addessi, E., & Visalberghi, E. (2001). Social facilitation of eating novel food in tufted capuchin monkeys (*Cebus apella*): input provided by group members and responses affected in the observer. *Animal Cognition* 4, 297–303. doi:10.1007/s100710100113.

Addessi, E., Mancini, A., Crescimbene, L., Padoa-Schioppa, C., & Visalberghi, E. (2008). Preference transitivity and symbolic representation in capuchin monkeys (*Cebus apella*). *PLoS ONE*, 3(6), e2414. doi:10.1371/journal.pone.0002414.

Anderson, J. R. (1996). Chimpanzees and capuchin monkeys: Comparative cognition. In A. Russon, K. Bard, & S. Parker (Eds.), *Reaching into Thought: The Minds of the Great Apes* (pp. 2–47). Cambridge, UK: Cambridge University Press.

Anderson, J. R., Hattori, Y., & Fujita, K. (2008). Quality before quantity: Rapid learning of reverse-reward contingency by capuchin monkeys (*Cebus apella*). *Journal of Comparative Psychology*, 122(4), 445–448. doi:10.1037/a0012624.

Anderson, J. R., Kuroshima, H., & Fujita, K. (2010). Delay of gratification in capuchin monkeys (*Cebus apella*) and squirrel monkeys (*Saimiri sciureus*). *Journal of Comparative Psychology*, 124(2), 205–210. doi:10.1037/a0018240

Baldovino, M. C., & di Bitetti, M. S. (2008). Allonursing in Tufted Capuchin Monkeys (*Cebus nigritus*): Milk or Pacifier? *Folia Primatologica*, 79(2), 79–92. doi:10.1159/000108780.

Becker, J. D., & Berkson, G. (1979). Response to neighbors and strangers by capuchin monkeys (*Cebus apella*). *Primates*, 20(4), 547–551. doi:10.1007/BF02373436.

Benítez, M. E., Sosnowski, M. J., Tomeo, O. B., & Brosnan, S. F. (2018). Urinary oxytocin in capuchin monkeys: Validation and the influence of social behavior. *American Journal of Primatology*, 80(10), e22877. doi:10.1002/ajp.22877.

Benítez, M. E., Sosnowski, M., Tomeo, O., & Brosnan, S. (2018, April). Testing the Challenge Hypothesis in male capuchin monkeys: hormonal, physical, and behavioral responses to novel females. [Abstract]. *American Journal of Physical Anthropology*, 165, 24.

Beran, M. J., Evans, T. A., Leighty, K. A., Harris, E. H., & Rice, D. (2008). Summation and quantity judgments of sequentially presented sets by capuchin monkeys (*Cebus apella*). *American Journal of Primatology*, 70, 191–194. doi:10.1002/ajp.20474.

Beran, Michael J., Perdue, B. M., & Smith, J. D. (2014). What are my chances? Closing the gap in uncertainty monitoring between rhesus monkeys (*Macaca mulatta*) and capuchin monkeys (*Cebus apella*). *Journal of Experimental Psychology: Animal Learning and Cognition*, 40(3), 303–316. doi:10.1037/xan0000020.

Bergeron, R. J., Liu, Z. R., McManis, J. S., & Wiegand, J. (1992). Structural alterations in desferrioxamine compatible with iron clearance in animals. *Journal of Medicinal Chemistry*, 35(25), 4739–4744. doi:10.1021/jm00103a012.

Bonnie, K., & De Waal, F. B. M. (2007). Copying without rewards: Socially influenced foraging decisions among brown capuchin monkeys. *Animal Cognition*, 10(3), 283–292. doi:10.1007/s10071-006-0069-9.

Brasington, L. F., Wikberg, E. C., Kawamura, S., Fedigan, L. M., & Jack, K. M. (2017). Infant mortality in white-faced capuchins: The impact of alpha male replacements. *American Journal of Primatology*, 79(12), e22725. doi:10.1002/ajp.22725.

Brosnan, S. F. (2010). What do capuchin monkeys tell us about cooperation? In D. R. Forsyth & C. L. Hoyt (Eds.), *For the Greater Good of All: Perspectives on Individualism, Society, and Leadership* (pp. 11–28). New York: Palgrave Macmillan Publishers.

Brosnan, S., & de Waal, F. B. M. (2003). Monkeys reject unequal pay. *Nature*, 425(6955), 297–299. doi:10.1038/nature01963.

Brosnan, S. F., & de Waal, F. B. M. (2004). Socially learned preferences for differentially rewarded tokens in the brown capuchin monkey (*Cebus apella*). *Journal of Comparative Psychology*, 118 (2), 133–139. doi:10.1037/0735-7036.118.2.133.

Brosnan, S. F., Freeman, C., & De Waal, F. B. M. (2006). Partner's behavior, not reward distribution, determines success in an unequal cooperative task in capuchin monkeys. *American Journal of Primatology*, 68(7), 713–724. doi:10.1002/ajp.20261.

Brosnan, S. F., Talbot, C. F., Essler, J. L., Leverett, K., Flemming, T., Dougall, P., Heyler, C., & Zak, P. J. (2015). Oxytocin reduces food sharing in capuchin monkeys by modulating social distance. *Behavior*. 152 (7-8), 941–961. doi:https://doi.org/10.1163/1568539X-00003268

Campos, F. A., & Fedigan, L. M. (2013). Urine-washing in white-faced capuchins: A new look at an old puzzle. *Behaviour*, 150(7), 763–798. doi:10.1163/1568539X-00003080.

Carosi, M., Linn, G. S., & Visalberghi, E. (2005). The sexual behavior and breeding system of tufted capuchin monkeys (*Cebus apella*). In *Advances in the Study of Behavior* (Vol. 35, pp. 105–149). Elsevier. doi:10.1016/S0065-3454(05)35003-0.

Chalk, J., Wright, B., Lucas, P., Schumacher, K., Vogel, E., Fragaszy, D.M., Visalberghi, E., Izar, P., & Richmond, B. (2016). Age-related variation in the mechanical properties of foods processed *by Sapajus libidinosus*. *American Journal of Physical Anthropology*, 159, 199–209. doi:10.1002/ajpa.22865.

Coelho, C.G., Falótico, T., Izar, P., Mannu, M., Resende, B. D., Siqueira, J. O., & Ottoni, E. B. (2015). Social learning strategies for nut-cracking by tufted capuchin monkeys (*Sapajus* spp.) *Animal Cognition*, 18: 911–919. doi:10.1007/s10071-015-0861-5.

Cooper, M., Bernstein, I., Fragaszy, D., & de Waal, F. (2001). Integration of new males into four social groups of tufted capuchins (*Cebus apella*). *International Journal of Primatology*, 22 (4), 663–683. doi:10.1023/A:1010745803740.

Crast, J., Hardy, J., & Fragaszy, D. (2010). Inducing traditions in captive capuchin monkeys (*Cebus apella*). *Animal Behaviour*, 80, 955–964. doi:10.1016/j.anbehav.2010.08.023.

de Resende, B. D., Oliveira, D. A. G., & da Silva, E. D. R. (2007). Capuchin monkey (*Cebus apella*) vocalizations in response to loud explosive noises. *Neotropical Primates*, 14 (1), 25–28. doi:10.1896/044.014.0105.

De Palmero, K. E., Carbonetto, C. H., Malchiodi, E. L., Margni, R. A., & Falasca, C. A. (1988). Humoral and cellular parameters of the immune system of *Cebus apella* monkeys. Cross reactivity between monkey and human immunoglobulins. *Veterinary Immunology and Immunopathology*, 19(3), 341–349. doi:10.1016/0165-2427(88)90119-5.

de Waal, F. B. M. (2000). Attitudinal reciprocity in food sharing among brown capuchin monkeys. *Animal Behavior*, 60, 253–261. doi:10.1006/anbe.2000.1471.

de Waal, F., & Davis, J. M. (2003). Capuchin cognitive ecology: Cooperation based on projected returns. *Neuropsychologia*, 41(2), 221–228. doi:10.1016/S0028-3932(02)00152-5.

Dettmer, E. L., Phillips, K. A., Rager, D. R., Bernstein, I. S., & Fragaszy, D. M. (1996). Behavioral and cortisol responses to repeated capture and venipuncture in *Cebus apella*. *American Journal of Primatology*, 38, 357–362. doi:10.1002/(SICI)1098-2345(1996)38:4<357::AID-AJP6>3.0.CO;2-Y.

di Bitetti, M. S. (2003). Food-associated calls of tufted capuchin monkeys (*Cebus apella nigritus*) are functionally referential signals. *Behaviour*, 140 (5), 565–592. doi:10.1163/156853903322149441.

di Bitetti, M. S. (1997). Evidence for an important social role of allogrooming in a platyrrhine primate. *Animal Behaviour*, 54, 199–211. doi:10.1006/anbe.1996.0416.

Digweed, S. M., Fedigan, L. M., & Rendall, D. (2007). Who cares who calls? Selective responses to the lost calls of socially dominant group members in the white-faced capuchin (*Cebus capucinus*). *American Journal of Primatology*, 69 (7), 829–835. doi:10.1002/ajp.20398.

Dindo, M., Theirry, B., & Whiten, A. (2008). Social diffusion of novel foraging methods in brown capuchin monkeys (*Cebus apella*). *Proceedings Royal Society London B*, 275, 187–193. doi:10.1098/rspb.2007.1318.

Drayton, L., & Santos, L. (2014). Insights into intraspecies variation in primate prosocial behavior: Capuchins (Cebus apella) fail to show prosociality on a touchscreen task. *Behavioral Sciences*, 4(2), 87–101. doi:10.3390/bs4020087.

Duarte, M., Hanna, J., Liu, Q., & Fragaszy, D. (2012). Kinematics of bipedal locomotion while carrying a load in the arms in beaded capuchin monkeys (*Sapajus libidinosus*). *Journal of Human Evolution*, 63 (6), 851–868. doi:10.1016/j.jhevol.2012.10.002.

Eshchar, Y., Izar, P., Visalberghi, E., Resende, B., & Fragaszy, D. (2016). When and where to practice: social influences on the development of nut-cracking in bearded capuchins (*Sapajus libidinosus*). *Animal Cognition*, 19, 605–618. doi:10.1007/s10071-016-0965-6.

Evans, T. A., Beran, M. J., Chan, B., Klein, E. D., & Menzel, C. R. (2008). An efficient computerized testing method for the capuchin monkey (*Cebus apella*): Adaptation of the LRC-CTS to a socially housed nonhuman primate species. *Behavior Research Methods*, 40(2), 590–596. doi:10.3758/BRM.40.2.590.

Fichtel, C., Perry, S., & Gros-Louis, J. (2005). Alarm calls of white-faced capuchin monkeys: An acoustic analysis. *Animal Behaviour*, 70(1), 165–176. doi:10.1016/j.anbehav.2004.09.020.

Fragaszy, D.M. & Adams-Curtis, L.E. 1998. Growth and reproduction in captive tufted capuchins (*Cebus apella*). *American Journal of Primatology*, 44, 197–203. doi:10.1002/(SICI)1098-2345(1998)44:3<197::AID-AJP2>3.0.CO;2-R.

Fragaszy, D. M., Baer, J., & Adams-Curtis, L. E. (1994). Introduction and integration of strangers into social groups of tufted capuchins. *International Journal of Primatology*, 15, 399–420. doi:10.1023/A:1010745803740.

Fragaszy, D. M., Morrow, K. S., Baldree, R., Unholz, E., Izar, P., Visalberghi, E., & Haslam, M. (2019). How bearded capuchin monkeys (*Sapajus libidinosus*) prepare to use a stone to crack nuts. *American Journal of Primatology*, 81, e22958. doi:10.1002/ajp.22958.

Fragaszy, D. M., Visalberghi, E., & Fedigan, L. M. (2004). *The Complete Capuchin: The Biology of the Genus Cebus*. Cambridge, UK: Cambridge University Press.

Fragaszy, D., Greenberg, R., Visalberghi, E., Ottoni, E. B., Izar, P., & Liu, Q. (2010). How wild bearded capuchin monkeys select stones and nuts to minimize the number of strikes per nut cracked. *Animal Behaviour*, 80 (2), 205–214. doi:10.1016/j.anbehav.2010.04.018.

Fragaszy, D., Johnson-Pynn, J., Hirsh, E., & Brakke, K. (2003). Strategic navigation of two-dimensional alley mazes: Comparing capuchin monkeys and chimpanzees. *Animal Cognition*, 6, 149–160. doi:10.1007/s10071-002-0137-8.

Fragaszy, D. M., & Adams-Curtis, L. E. (1991). Generative aspects of manipulation in tufted capuchin monkeys (*Cebus apella*). *Journal of Comparative Psychology*, 105, 387–397. doi:10.1037/0735-7036.105.4.387.

Fragaszy, D.M., Eshchar, Y., Visalberghi, E., Resende, B., Laity, K., & Izar, P. 2017. Synchronized practice helps bearded capuchin monkeys learn to extend attention while learning a tradition. *Proceedings of the National Academics of Science*, 114 (30), 7798–7805. doi:10.1073/pnas.1621071114.

Fragaszy, D.M., Izar, P., Liu, Q., Eshchar, Y., Young, L.A., & Visalberghi, E. (2016). Body mass in wild bearded capuchins (*Sapjaus libidinosus*). Ontogeny and sexual dimorphism. *American Journal of Primatology*, 78, 389–484. doi:10.1002/ajp.22509.

Fragaszy, D.M., Liu, Q., Wright, B.W., Allen, A., & Brown, C.W. (2013). Wild bearded capuchin monkeys (*Sapajus libidinosus*) strategically place nuts in a stable position during nut-cracking. *PLoS One*, 8(2): E56182. doi:10.1371/JOURNAL.PONE.0056182.

Galloway, A., Addessi, E., Fragaszy, D., & Visalberghi, E. (2005). Social facilitation of eating familiar food in tufted capuchin monkeys (*Cebus apella*): Does it involve behavioral coordination? *International Journal of Primatology*, 26, 181–189. doi:10.1007/s10764-005-0729-7.

Garcez, L. M., Silveira, F. T., El Harith, A., Lainson, R., & Shaw, J. J. (1997). Experimental cutaneous leishmaniasis: IV. The humoral response of Cebus apella (Primates: Cebidae) to infections of *Leishmania (Leishmania) amazonensis, L. (Viannia) lainsoni* and *L.(V.) braziliensis* using the direct agglutination test. *Acta Tropica*, 68(1), 65–76. doi:10.1016/S0001-706X(97)00078-8.

Harwood, H. J., Listrani, P., & Wagner, J. D. (2012). Nonhuman primates and other animal models in diabetes research. *Journal of Diabetes Science and Technology*, 6 (3), 503–514. doi:10.1177/193229681200600304.

Howard, A. M., Bernardes, S., Nibbelink, N., Biondi, L., Presotto, A., Fragaszy, D. M., & Madden, M. (2012). A maximum entropy model of the bearded capuchin monkey habitat incorporating topography and spectral unmixing analysis. *ISPRS Annals of the Photogrammetry, Remote Sensing and Spatial Information Sciences*, 1–2, 7–11. doi:10.5194/isprsannals-I-2-7-2012

Izar, P., Stone, A., Carnegie, S., & Nakai, É. S. (2009). Sexual selection, female choice and mating systems. In P. A. Garber, A. Estrada, J. C. Bicca-Marques, E. W. Heymann, & K. B. Strier (Eds.), *South American Primates* (pp. 157–189). New York: Springer. doi:10.1007/978-0-387-78705-3_7

Izar, P., Verderane, M., Peternelli-dos-Santos, L., Mendonça-Furtado, P., Presotto, A., Tokuda, M., Visalberghi, E., & Fragaszy, D. (2012). Flexible and conservative features of social systems in tufted capuchin moneys: Comparing the socioecology of *Sapajus libidinosus* and *Sapajus nigritus*. *American Journal of Primatology*, 74, 315–331. doi:10.10002/ajp.2096.

Janson, C. H. (1984). Female choice and mating system of the brown capuchin monkey *Cebus apella* (Primates: Cebidae). *Zeitschrift Für Tierpsychologie*, 65 (3), 177–200. doi:10.1111/j.1439-0310.1984.tb00098.x.

Janson, C. H. (1998). Experimental evidence for spatial memory in foraging wild capuchin monkeys, *Cebus apella*. *Animal Behaviour*, 55 (5), 1229–1242. doi:10.1006/anbe.1997.0688.

Janson, C. H. (2007). Experimental evidence for route integration and strategic planning in wild capuchin monkeys. *Animal Cognition*, 10 (3), 341–356. doi:10.1007/s10071-007-0079-2.

Janson, C. H. (2016). Capuchins, space, time and memory: an experimental test of what-where-when memory in wild monkeys. *Proceedings of the Royal Society B*, 283 (1840), 20161432. doi:10.1098/rspb.2016.1432.

Johnson-Pynn, J., Fragaszy, D. M., Hirsh, E., Brakke, K., & Greenfield, P. (1999). Strategies used to combine seriated cups by chimpanzees (*Pan troglodytes*), bonobos (*Pan paniscus*), and capuchins (*Cebus apella*). *Journal of Comparative Psychology*, 113, 137–148. doi:10.1037/0735-7036.113.2.137.

LaCour, L., Stone, B., Hopkins, W., Menzel, C., & Fragaszy, D. (2014). What limits tool use in nonhuman primates? Insights from tufted capuchin monkeys (*Sapajus* spp.) and chimpanzees (*Pan troglodytes*) aligning three-dimensional objects to a surface. *Animal Cognition*, 17, 113–125. doi:10.1007/s10071-013-0643-x.

Leighty, K., & Fragaszy, D. (2003). Joystick acquisition in tufted capuchin (*Cebus paella*). *Animal Cognition*, 6, 141–148.

Lifshitz, K., O'Keeffe, R. T., Linn, G. S., Lee, K. L., Camp-Bruno, J. A., & Suckow, R. F. (1997). Effects of dopamine agonists on Cebus apella monkeys with previous long-term exposure to fluphenazine. *Biological Psychiatry*, 41(6), 657–667. doi:10.1016/S0006-3223(96)00169-2.

Lima, M. G., Buckner, J. C., de Silva-Júnior, J., Aleixo, A., Martins, A. B., Boubli, J. P., Link, A., Farias, I. P., Silva, M. N., Röhe, F., Queiroz, H., Chiou, K. L., Di Fiore, A., Alfaro, M. E., & Lynch Alfaro, J. W. (2017). Capuchin monkey biogeography: Understanding *Sapajus* Pleistocene range expansion and the current sympatry between *Cebus* and *Sapajus*. *Journal of Biogeography*, 44, 810–820. doi:10.1111/jbi.12945.

Lynch Alfaro, J. W. (2007). Subgrouping patterns in a group of wild *Cebus apella nigritus*. *International Journal of Primatology*, 28 (2), 271–289. doi:10.1007/s10764-007-9121-0.

Mangalam, M., & Fragaszy, D.M. (2015). Wild bearded capuchin monkeys crack nuts dexterously. *Current Biology*, 25 (10), 1334–1339. doi:10.1016/j.cub.2015.03.035.

Masterson T. J. (1997). Sexual dimorphism and interspecific cranial form in two capuchin species: *Cebus albifrons* and *C. apella*. *American Journal of Physical Anthropology*, 104:487–511 doi:10.1002/(SICI)1096-8644(199712)104:4<487::AID-AJPA5>3.0.CO;2-P.

Mayer, A., Lewenfus, G., Bittencourt-Navarrete, R. E., Clasca, F., & Franca, J. G. (2019). Thalamic inputs to posterior parietal cortical areas involved in skilled forelimb movement and tool use in the capuchin monkey. *Cerebral Cortex*. doi:10.1093/cercor/bhz051.

Nagle, C. A., & Denari, J. H. (1982). The reproductive biology of capuchin monkeys. In *International Zoo Yearbook*, 22 (1), 143–150. doi:10.1111/j.1748-1090.1982.tb02023.x.

Nagle, C. A., Paul, N., Mazzoni, I., Quiroga, S., Torres, M., Mendizabal, A. F., & Farinati, Z. (1989). Interovarian relationship in the secretion of progesterone during the luteal phase of the capuchin monkey (*Cebus apella*). *Journal of Reproduction and Fertility*, 85 (2), 389–396. doi:10.1530/jrf.0.0850389.

Padberg, J., Franca, J. G., Cooke, D. F., Soares, J. G. Rosa, M. G. P., Fiorani, M., Gattass, R., & Krubitzer, L. (2007). Parallel evolution of cortical areas involved in skilled hand use. *Journal of Neuroscience*, 27 (38), 10106–10115. doi:10.1523/JNEUROSCI.2632-07.2007.

Perry, S. E. (1996). Female-female social relationships in wild white-faced capuchin monkeys, *Cebus capucinus*. *American Journal of Primatology*, 40, 167–182. doi:10.1002/(SICI)1098-2345(1996)40:2<167::AID-AJP4>3.0.CO;2-W.

Perry, S. (2011). Social traditions and social learning in capuchin monkeys (Cebus). *Philosophical Transactions Royal Society B*, 366, 0317. doi:10.1098/rstb.2010.0317.

Perry, S., & Manson, J. H. (2011). *Manipulative Monkeys: The Capuchins of Lomas Barbudal*. Cambridge, MA: Harvard Univ Press.

Perry, S. & Jiménez, J. C. O. (2006). The effects of food size, rarity, and processing complexity on white-faced capuchins' visual attention to foraging conspecifics. In G. Hohmann, M. Robbins & C. Boesch (Eds.), *Feeding ecology in apes and other primates. Ecological, physical and behavioral aspects* (pp. 203–234). Cambridge: Cambridge University Press.

Perry, S., & Jimenez, J. C. O. (2012). The effects of food size, rarity, and processing complexity on white-faced capuchins' visual attention to foraging conspecifics. *Feeding Ecology in Apes and Other Primates*, 48, 203.

Perry, S. (2009). Conformism in the food processing techniques of white-faced capuchin monkeys (*Cebus capucinus*). *Animal Cognition*, 12 (5), 705–716. doi:10.1007/s10071-009-0230-3.

Pessoa, V. F., Monge-Fuentes, V., Simon, C. Y., Suganuma, E., & Tavares, M. C. H. (2008). The Müller-Lyer illusion as a tool for schizophrenia screening. *Reviews in the Neurosciences*, 19 (2–3), 91–100. doi:10.1515/revneuro.2008.19.2-3.91.

Phillips, K. A., Buzzell, C. A., Holder, N., & Sherwood, C. C. (2011). Why do capuchin monkeys urine wash? An experimental test of the sexual communication hypothesis using fMRI. *American Journal of Primatology*, 73(6), 578–584. doi: 10.1002/ajp.20931.

Phillips, K. A., Subiaul, F., & Sherwood, C. (2012). Curious monkeys have increased gray matter density in the precuneus. *Neuroscience Letters*, 518 (2):172–175. doi:10.1016/j.neulet.2012.05.004.

Prescott, M., & Buchanan-Smith, H. (2016). *Training Nonhuman Primates Using Positive Reinforcement Techniques*. New York: Psychology Press.

Presotto, A., & Izar, P. (2010). Spatial reference of black capuchin monkeys in Brazilian Atlantic forest: Egocentric or allocentric? *Animal Behaviour*, 80 (1), 125–132. doi:10.1016/j.anbehav.2010.04.009.

Presotto, A., Verderane, M.P., Biondi, L., Mendonça-Furtado, O., Spagnoletti, N., Madden, M., & Izar, P. (2018). Intersection as key locations for bearded capuchin monkeys (*Sapajus libidinosus*) traveling within a route network. *Animal Cognition*, 21 (3), 393–405. doi:10.1007/s10071-018-1176-0.

Rathelot, J. A., & Strick, P. L. (2009). Subdivisions of primary motor cortex based on cortico-motoneuronal cells. *Proceedings of the National Academy of the United States of America*, 106 (3), 918–923. doi:10.1073/pnas.0808362106.

Rilling, J. K., & Insel, T. (1999). The primate neocortex in comparative perspective using magnetic resonance imaging. *Journal of Human Evolution*, 16, 191–233. doi:10.1006/jhev.1999.0313.

Sargeant, E. J., Wikberg, E. C., Kawamura, S., & Fedigan, L. M. (2015). Allonursing in white-faced capuchins (*Cebus capucinus*) provides evidence for cooperative care of infants. *Behaviour*, 152 (12–13), 1841–1869. doi:10.1163/1568539X-00003308.

Terpstra, A. H., Stucchi, A. F., & Nicolosi, R. J. (1991). Estimation of HDL cholesteryl ester kinetic parameters in the cebus monkey, an animal species with high plasma cholesteryl ester transfer activity. *Atherosclerosis*, 88 (2–3), 243–248. doi:10.1016/0021-9150(91)90087-j.

Tokuda, M., Martins, M. M., & Izar, P. (2018). Socio-genetic correlates of unbiased sex dispersal in a population of black capuchin monkeys (*Sapajus nigritus*). *Acta Ethologica*, 21, 1–11. doi:10.1007/s10211-017-0277-0.

Ueno, Y. (1994). Responses to urine odor in the tufted capuchin (*Cebus apella*). *Journal of Ethology*, 12 (2), 81–87. doi:10.1007/BF02350052.

van Wolkenten, M., Brosnan, S. F., & de Waal, F. B. M. (2007). Inequity responses of monkeys modified by effort. *Proceedings of the National Academy of Sciences of the United States of America*, 104 (47), 18854–18859. doi:10.1073/pnas.0707182104.

Visalberghi, E., Addessi, E., Truppa, V., Spagnoletti, N., Ottoni, E., Izar, P., & Fragaszy, D. (2009). Selection of effective stone tools by wild bearded capuchin monkeys. *Current Biology*, 19 (3), 213–217. doi:10.1016/j.cub.2008.11.064.

Visalberghi, E., Valenzano, D. R., & Preuschoft, S. (2006). Facial displays in *Cebus apella*. *International Journal of Primatology*, 27 (6), 1689–1707. doi:10.1007/s10764-006-9084-6.

Watzek, J., Smith, M. F., & Brosnan, S. F. (2017). Comparative economics: Using experimental economics paradigms to understand primate social decision-making. In L. Di Paolo, F. Di Vincenzo, & A. d'Almeida (Eds.), *The Evolution of Primate Social Cognition (pp. 129-141)*. Springer, Berlin.

Weigel, R. M. (1979). The facial expressions of thebrown capuchin monkey (*Cebus apella*). *Behaviour*, 68 (3–4), 250–276. doi:10.1163/156853979X00331.

Wheeler, B. C. (2009). Monkeys crying wolf? Tufted capuchin monkeys use anti-predator calls to usurp resources from conspecifics. *Proceedings of the Royal Society of London B: Biological Sciences*, 276 (1669), 3013–3018. doi:10.1098/rspb.2009.0544.

Wheeler, B. C. (2010). Production and perception of situationally variable alarm calls in wild tufted capuchin monkeys (*Cebus apella nigritus*). *Behavioral Ecology Sociobiology*, 64, 989–1000. doi:10.1007/s00265-010-0914-3.

Wheeler, Brandon C., Tiddi, B., & Heistermann, M. (2014). Competition-induced stress does not explain deceptive alarm calling in tufted capuchin monkeys. *Animal Behaviour*, 93, 49–58. doi:10.1016/j.anbehav.2014.04.016.

Wright, K. A., Biondi, L., Fragaszy, D. M., Visalberghi, E., & Izar, P. (2019). Positional behavior and substrate use in wild adult bearded capuchin monkeys (*Sapajus libidinosus*). *American Journal of Primatology*, 81 (12), e23067. doi:10.1002/ajp.23067.

Wright, B. W., Wright, K. A., Chalk, J., Verderane, M., Fragaszy, D., Visalberghi, E., Izar, P., Ottoni, E., Constatino, P., & Vinyard, C. (2009). Fallback foraging as a way of life: using dietary toughness to compare the fallback signal among capuchins and implications for interpreting morphological variation. *American Journal of Physical Anthropology*, 140 (4), 687–699. doi:10.1002/ajpa.21116.

28

Behavioral Biology of Macaques

Paul E. Honess
University of Nottingham

CONTENTS

Introduction 438
   Typical Research Involvement 438
Taxonomy, Distribution, and Conservation 439
Natural History 440
   Ecology 440
   Activity Patterns, Diet, and Habitat Use 440
      Activity Patterns 440
      Diet and Feeding Behavior 443
      Habitat Use 443
      Predation 444
   Social Organization and Behavior 444
      Social Organization 444
      Communication 445
      Affiliation 446
      Aggression and Reconciliation 446
   Reproduction 447
      Mating 447
      Births 447
      Maternal Care 447
Common Captive Behaviors 448
   Normal Behavior 448
      Grooming 448
      Locomotion 448
      Aggression 448
   Sources of Variation 449
      Variation due to Species and Source 449
      Variation due to Individual Temperamental Differences 449
   Abnormal Behavior 449
      Self-Injurious Behavior and Self-Directed Behavior 450
      Abnormal Repetitive Behavior 450
      Hair-Pulling/-Plucking 451
      Saluting 451
Ways to Maintain Behavioral Health 452
   Captive Considerations to Promote Species-Typical Behavior 452
      Environment 453
   Training 458
   Special Situations 458
Conclusions/Recommendations 459
Additional Resources 459
References 459

Introduction

This chapter describes the natural history and wild behavior of macaques (*Macaca* spp.) and aims to show their importance for understanding the captive behavior of these species when involved in research. Developing an understanding of natural behavior is essential for the informed interpretation of captive behavior. This is particularly important for designing environments and husbandry regimens that meet the animals' behavioral needs, reduce stress and anxiety, facilitate species-typical behavior, and prevent or treat abnormal behavior. These aspects are fundamental for ensuring that research models are as valid as possible, given that stress and maladaptation disrupt biological systems, producing atypical and more variable responses to research procedures, thereby threatening the reproducibility and translation of results (Garner 2005).

There are more macaque species in the wild than are involved in laboratory-type research (see Table 28.1). In many sections of this chapter, discussion will focus primarily on rhesus (*Macaca mulatta*) and long-tailed macaques (*M. fascicularis*), as these are the two major species in research (see Figure 28.1) and unless otherwise specified, the results/findings/statements included here apply to these species. Given the number of macaque species, it is not surprising that there is considerable diversity in their behavioral ecology. Therefore, in the natural history sections of this chapter, a number of interesting behaviors of a range of species, in addition to rhesus and long-tailed macaques, are presented. This enables cross-species comparisons and contrasts to be made and appreciation gained of patterns and exceptions in the behavioral biology of key macaque species. In recent years, there have been some excellent treatments of captive macaque behavior and welfare, and these should be consulted alongside this chapter (see: Winnicker et al. 2013; Schapiro 2017).

Typical Research Involvement

Historically, macaques have participated in a wide range of studies, including those related to human diseases and health conditions, such as polio (Jones-Engel 2020), tuberculosis (Peña and Ho 2015), HIV/SIV (Evans and Silvestri 2013), malaria (Joyner et al. 2015), neurodegeneration (Emborg 2017), genetics (Vinson et al. 2013), substance abuse (Platt et al. 2011), diet impact (Kaplan et al. 1982), and depression (Shively 2017). While research in many of these areas is ongoing, there has been growth in the involvement of macaques in other areas due to emerging diseases (e.g., Ebola, Zika, SARS/MERS/COVID) or enabling technological advances (e.g., Parkinson's disease deep brain stimulation (Kringelbach et al. 2007), neurologically linked prosthetics (Musallam et al. 2004), genetic alteration (Park and Silva 2019). In addition, macaques have, for many years, played a critical role in toxicology/safety testing, ensuring medical advances are safe for use in humans (Phillips et al. 2014).

Among publications in early 2020 are those that involve macaques in research on infectious disease (e.g., tuberculosis (Darrah et al. 2020), measles (Nelson et al. 2020), MERS (van Doremalen et al. 2020), COVID-19 (Munster et al. 2020), SIV/HIV (Virnik et al. 2020), Zika (Block et al. 2020), STLV (Murata et al. 2020), cancer (Sato et al. 2020), stem cells (Matsumoto et al. 2020), cardiology (Metzger et al. 2020), ricin toxin vaccines (Roy et al. 2020), microbiome (Jones-Engel 2020),

TABLE 28.1

Common, Scientific Names, and IUCN Threat Status of Macaques (Groves 2001; Li et al. 2015; IUCN Red List 2020), Including Scientific Name Abbreviations as Used in This Chapter

Common Name(s)	Scientific Name	IUCN Threat Status	Common Name(s)	Scientific Name	IUCN Threat Status
Rhesus[a] (5)	*M. mulatta* (*Mm*)	Least concern (decreasing)	Barbary (1)	*M. sylvanus*	Endangered
Long-tailed, cynomolgus or crab-eating[a] (4)	*M. fascicularis* (*Mfa*)	Least concern (decreasing)	Arunachal (6)	*M. munzala*	Endangered
Northern pig-tailed[a] (2)	*M. leonina* (*Mnl*)	Vulnerable	White-cheeked (6)	*M. leucogenys*	Unclassified but threatened by humans (crop raiding, development)
Southern or Sunda pig-tailed[a] (2)	*M. nemestrina* (*Mnn*)	Vulnerable	Mentawai (2)	*M. pagensis*	Critically endangered
Japanese[a] (5)	*M. fuscata* (*Mfu*)	Least concern (stable)	Siberut (2)	*M. siberu*	Vulnerable
Stump-tailed or bear[a] (4)	*M. arctoides* (*Ma*)	Vulnerable	Celebes crested or black (3)	*M. nigra*	Critically endangered
Bonnet[a] (6)	*M. radiata* (*Mr*)	Least Concern (decreasing)	Tonkean (3)	*M. tonkeana*	Vulnerable
Formosan rock (5)	*M. cyclopis*	Least Concern (stable)	Gorontalo (3)	*M. nigrescens*	Vulnerable
Lion-tailed (2)	*M. silenus*	Endangered	Booted (3)	*M. ochreata*	Vulnerable
Toque (6)	*M. sinica*	Endangered	Heck's (3)	*M. hecki*	Vulnerable
Tibetan or Milne-Edwards's (6)	*M. thibetana*	Near threatened	Moor (3)	*M. maura*	Endangered
Assam (6)	*M. assamensis*	Near threatened			

Species groups (Groves 2001): 1, *M. sylvanus* group; 2, *M. nemestrina* group; 3, Sulawesi group; 4, *M. fascicularis* group; 5, *M. mulatta* group; 6. *M. sinica* group.

[a] Species used in research.

FIGURE 28.1 The two most commonly used macaque species in research: Left: Rhesus. (Photo: William Sutton.) Right: Long-tailed. (Photo: P. Honess.)

neuroscience (Vanni et al. 2020), cognition (Gray et al. 2020), anesthesia (Brownlee et al. 2020), radiation (Endo et al. 2020), toxicology (Luetjens et al. 2020), alcohol/drug impact (Wang et al. 2020), and applied genetics (Zhang et al. 2020). The majority of these studies included long-tailed, rhesus, or Japanese (*M. fuscata*) macaques, while other macaque species (e.g., pig-tailed (*M. nemestrina*), stump-tailed (*M. arctoides*), bonnet (*M. radiata*)) feature only occasionally. Mauritian long-tailed macaques are notably free of the simian herpes B virus (*Macacine herpesvirus 1*), which makes them a sought-after research model (Stanley 2003). Simian herpes B virus produces few clinical signs in macaques, but has very serious consequences for humans (Engel et al. 2002). It is common in wild and captive macaques, and presents a serious hazard to human health, as well as a confounding variable for research (Wolfensohn and Honess 2005).

In 2014, only three species of macaques were reported in surveyed North American facilities: long-tailed, rhesus, and pig-tailed macaques (Lankau et al. 2014). While a similar number of facilities held long-tailed and rhesus macaques, the number of individuals of each species was markedly different (2010–2012: 45,541 vs 3,042, respectively). The majority of these animals were imported (99.07% vs 79.3%), though it is not stated whether those that were imported were captive-bred in sustainable breeding facilities or were wild-sourced. Only 985 pig-tailed macaques were acquired over the same period, mostly from domestic sources (69.5%). Long-tailed macaques are the model of choice for biomedical and drug-safety (toxicology) research (SCHER 2009; Heckel et al. 2015), which typically involves larger sample sizes and shorter studies than academic research, which includes more rhesus.

Taxonomy, Distribution, and Conservation

Macaques are Old World Monkeys (Superfamily: Cercopithecoidae). They comprise the genus *Macaca* and are closely related to baboons (Papionini), guenons, and vervets (Cercopithecini) (Groves 2001). Just over 20 species of macaque are recognized (Groves 2005; IUCN Red List 2020), with an additional species; the white-cheeked macaque (*M. leucogenys*), described in 2015 (Li et al. 2015). A list of currently recognized species and their scientific and common names is found in Table 28.1. Macaques may be classified into species-groups based on shared characteristics (Fooden 1976; Groves 2001) (see Table 28.1). Those species not recognized at the time of Groves' analysis are allocated here based on published affinities (*M. munzala*: Chakraborty et al. 2007, *M. leucogenys*: Li et al. 2015, *M. siberu*: Groves 2001). *M. fascicularis* has several common names including cynomolgus ("cyno") or crab-eating macaque, but will be referred to here as the long-tailed macaque. For brevity, in places throughout the chapter, macaque species will be abbreviated as: rhesus (*Mm*), long-tailed (*Mfa*), Japanese (*Mfu*), southern pig-tailed

(*Mnn*), northern pig-tailed (*Mnl*), unspecified pig-tailed (*Mn*), stump-tailed (*Ma*), and bonnet (*Mr*).

Macaques have the widest distribution of any primate genus (Melnick and Pearl 1987; Thierry et al. 2004); and are found predominantly in temperate and tropical forest, stretching from North Africa and southern Europe (Gibraltar) in the West to Japan in the East. Individual species' range countries are given in Table 28.2 and distribution maps are published elsewhere (Fooden 1980; Thierry et al. 2004). Macaques have been introduced outside of their natural range, including in Mauritius, Papua New Guinea, Hong Kong (*Mfa*: Fooden 1995), Cayo Santiago (Puerto Rico, *Mm*: Altmann 1962), and Mexico (*Mn*: Fooden 1990).

There is significant variation in rhesus macaque populations across their ranges, with some taxonomists recognizing multiple subspecies (Groves 2001). However, the major work on rhesus systematics notes that unambiguous diagnosis of rhesus subspecies is "virtually impossible" and therefore "it is unlikely that taxonomically useful subspecies can be defined in this species" (Fooden 2000, p. 79). Nevertheless, there is evidence that Indian- and Chinese-origin animals are sufficiently different, across a range of biological characteristics (Smith 2005), that the origin of research subjects needs to be accounted for in studies, and that captive breeding should be managed to minimize hybridization (Kanthaswamy et al. 2016). For this reason, data for Indian- and Chinese-origin rhesus macaques are presented separately in Table 28.2.

In parts of Indochina, macaque species appear to readily hybridize (e.g., Hong Kong: Burton and Chan 1996). Japanese macaques also hybridize with introduced nonnative macaque species (e.g., *Mfa*, *Mm*, Formosan rock: Fooden and Aimi 2005). Of particular importance is wild hybridization between long-tailed and rhesus macaques (Bunlungsup et al. 2017), which may have important consequences for the selection of research models where origin and/or genetics are not defined (Bunlungsup et al. 2017). Stump-tailed macaques are considered a hybrid species originating from hybridization of *M. assamensis-/thibetana*-like males with *M. fascicularis*-like females (Brandon-Jones et al. 2004).

Table 28.1 indicates the conservation status of macaque species: two are Critically Endangered, five Endangered, ten Vulnerable or Near Threatened, and only five are of Least Concern, and three of those populations are decreasing (IUCN Red List 2020). The major threat is from habitat loss and hunting, but some populations are at risk from capture for research (e.g., *Mfa*: Eudey 2008; *Mr*: Sinha 2001). Human population expansion and habitat loss have brought macaques into conflict with humans as aggression and disease risks (Engel et al. 2002; Fuentes and Gamerl 2005), and as agricultural pests (Lee and Priston 2005). However, crop damage may be outweighed by efficient predation of more significant pests (e.g., *Mnn*: Holzner et al. 2019). Macaques also pose a significant biodiversity threat, for example, as nest predators of rare endemic or threatened bird species (*Mfa*: Carter and Bright 2011; *Mnl*: Kaisin et al. 2018). In India, bonnet macaques' range has contracted due to competition from southwardly expanding rhesus macaques (Erinjery et al. 2017b).

Natural History

Ecology

Selected data, where possible from minimally disturbed sites, on ecological parameters are presented in Table 28.2. There is no "typical" habitat for many macaques, as they are found in a range of habitat types reflecting a degree of ecological flexibility often greater than that of other primates with which they are sympatric (Erinjery et al. 2017a).

Macaques primarily inhabit tropical, typically evergreen or semideciduous, forests. Some species are also found in temperate forest (e.g., *Mfu*, *Mm*, Formosan rock, Tibetan macaques), bamboo forest (e.g., *Mm*, Assamese, Formosan rock macaques), grassland (e.g., *Mm*, Formosan rock macaques), or mangrove swamps (e.g., *Mfa*, *Mm*, *Mn*) (Rowe 1996). Rhesus macaques occupy the broadest range of natural habitats and are even found in semidesert (Seth and Seth 1986). Expanding human populations, loss of natural forest, and the ecological flexibility of many macaques mean that many species can be found in human-altered, even urban environments (e.g., *Mfa*, *Mm*, *Mr*, Barbary macaques) (Richard et al. 1989; Erinjery et al. 2017b). Several species can be found at temple sites where they may be provisioned by local people and tourists (e.g., *Ma*, *Mfa*, *Mm*, *Mr*, toque, Tibetan: Rowe 1996).

Variation in habitat, often as a result of climate variation, tends to be reflected in temporal and spatial variation in food availability. Food distribution, as well as predation pressure, has a fundamental effect on group size, population density, and home range area (Wrangham et al. 1993; Chapman and Teichroeb 2012). Macaques are subject to these ecological factors, especially as seasonally available fruit typically makes up a large part of their diet (Table 28.2). Macaques are found across a wide range of climates varying in temperature from −20°C (*Mfu*, *Mm*) to +48°C (*Mm*), rainfall from 518 mm (*Mm*) to 8,600 mm/year (*Mfu*), and altitude from sea-level (several species) to over 3,000 m asl (*Mfu*, *Mm*) (Table 28.2). In areas with extremely cold winters, individuals may be heavy-bodied with thick fur (e.g., Chinese *Mm*, *Mfu*: Southwick et al. 1996).

Activity Patterns, Diet, and Habitat Use

Activity Patterns

As with other primates, macaques spend most of their time searching for and consuming food (Table 28.2) (Oates 1987). The specific nature of the animal's habitat and food type, availability, and distribution dictates the amount of time allocated for traveling and feeding. The need to share knowledge about the location of patchy food (e.g., ripe fruit vs mature leaves) has been seen as a driver of sociality and advanced mental abilities in primates (Milton 1981). The amount of dietary foliage is generally correlated positively with feeding time, and negatively with moving time (Clutton-Brock and Harvey 1977). For example, both Japanese and Chinese rhesus macaques eat more foliage (35% and 56%, respectively) and less fruit (30% and 27%, respectively) than long-tailed macaques (87% fruit) which travel further (Table 28.2). In fact, variation in food availability, temperature, and day length has a notable effect

TABLE 28.2

Selected Ecological and Reproductive Parameters of Macaque Species Used in Research

	Rhesus Macaque (Indian)	Rhesus Macaque (Chinese)	Long-Tailed Macaque	Japanese Macaque	Northern Pig-Tailed Macaque	Southern Pig-Tailed Macaque	Stump-Tailed Macaque	Bonnet Macaque
Scientific name[1]	*M. mulatta mulatta*, *M.m. villosa*	*M. m. lasiota*, *M.m. vestita*, *M.m. sanctijohannis*, *M.m. brevicauda*	*M. fascicularis*	*M. fuscata*	*M. leonina*	*M. nemestrina*	*M. arctoides*	*M. radiata*
Range[1]	Afghanistan, Pakistan, India, Nepal, Bangladesh, Myanmar, Thailand, Vietnam	China	Myanmar, Thailand, Cambodia, Vietnam, Malaysia, Indonesia, Borneo, Philippines, Timor	Japan	Thailand, Myanmar, Bangladesh, India, China	Malay peninsula, Sumatra, Borneo	N.E. India, Myanmar, W. Malaysia, S. China	S. India
Habitat	Semidesert to tropical and subalpine forest[2]	Tropical and temperate forest[3]	Tropical forest[4]	Subtropical temperate and sub-alpine forest[4,5]	Tropical forest[6]	Tropical forest[7]	Tropical forest[4]	Tropical forest[9]
Climate								
Rainfall (mm)	430–2,500[10,25]	641–1,575[11]	2,376[27]	2,500–8,600, 7,985[13]	Up to 3,000[6]	1,941[14], 5,570[15]	Up to 2,000[4]	958–3,526[16-18]
Temperature (°C)	−4 to 48[10,25]	−20 to 40[11]	25.5[27]	−20 to 28.3[13,19]	7–35[6]	27.2[14], 14–33[15]	-	Up to 34[16]
Altitude (masl)	400–1,000[10], 0–3,658[20]	0–1,960[11], up to 4,100m[a]	0–2,000[4]	0–3,180[21]	50–2,000[6,22]	0–1,700[23,15]	0–2,800[30]	Up to 26,000[24]
Activity Budget (%)								
Feeding	40.08[25], 27[20]	18[26], 11.6[12]	13[27]	38[28]	23.5[6]	28[14]	-	28[17]
Locomotion	26.21[25], 25[20]	11.3[26], 27.8[12]	45[27]	16[28]	17–19.4[6]	40[14]	-	18.5[17]
Resting	27.7[25], 28[20]	53.4[26], 41[12]	42[27]	32[28]	45[6]	20[14]	-	40[17]
Social	4.2[25], 22[20]	-, 19.5[12]		14[28]	12[6]	10[14]	-	14[17]
Of which grooming	1.79[25], 15[20]	-, 7[12]			6–8[6]			
Of which play	2.88[25], 3[20]	-, 12.5[12]						11.6[17]
Other	1.34[25]	17.3[26], -				2[14]	-	
Diet (%)								
Fruit	1.7–70[2]	27.3[3]	87[27]	30.2[29]	65.9[6]	32[14]	+[30]	41[16]
Leaves	1–79[2]	56.5[3]	4[27]	35[29]	7.7[6]	11[14]		25.6[16]
Flowers	5.5–9.2[2]	1.8[3]	3[27]	5.5[29]	2.1[6]	2[14]		8.7[16]
Roots/stems/Bark	1.2[2]	13.3[3]			5[6]	15[14]		5.9[16]
Seeds	Incl in Fruit	0.8[3]		13.2[29]	6.1[6]		+[30]	5.1[16]
Animal	+[2]		4[27]	10.3[29]	11.7[6]	32[14]	+[30]	14.5[16]
Other	2[2]	0.3[3]	2[27]	4.6[29]	1.5[6]	6[14]		
Arboreality (%)	34[31], 20[32]	35[12]	97[27]	60–32[33]	98.5[6]	43.7[14]	Occasionally[30]	Predominantly[18]
Group size	40–68[10]	20–110[11]	22–42[27]	40.8[33]	16–33[6]	20–81[7,14,15]	10–60[30]	17–35[9,16]

(*Continued*)

TABLE 28.2 (Continued)
Selected Ecological and Reproductive Parameters of Macaque Species Used in Research

	Rhesus Macaque (Indian)	Rhesus Macaque (Chinese)	Long-Tailed Macaque	Japanese Macaque	Northern Pig-Tailed Macaque	Southern Pig-Tailed Macaque	Stump-Tailed Macaque	Bonnet Macaque
Sex ratio (m:f)	1:3[34]	1:2.3–1:3.5[3]	1:2[27]	1:1.3[33], 1:3.4[35]	1:5.5[6]	1:2.2–1:7[7,14,15]	1:0.6[36]	1:1.8[9], 1:2[16]
Group home range (ha)	6–1,500[10,20]	37–2,200[11]	125[27]	101–530[35]	83–347[6,8]	62–828[7,23], 149[14]	22–several hundred[36,38]	240[18]
Day range (m)	830–1,895[10]	1,000–23,000[a]	1,900[27]	500–2,000[37]	690–2,240[6], 4,500[8]	825–2,964[23]	400–3,000[38]	1,000–2,500[18]
Breeding	Seasonal (April–November)[2]	Seasonal (March–August)[2]	Nonseasonal[4]	Seasonal (March–October)[21] Seasonal Spring–summer (Apr–Jul)[21]	Seasonal (January–May)[6] or weakly seasonal[40]	Nonseasonal[7]	Weakly seasonal or non-seasonal[30]	Seasonal (February–March)[39]
Sexual skin/swellings[41]	Present	Present	Present	Present	Present[40]	Present	Absent	Absent
Gestation (days)	164 (146–180)[42]	166–177[a]	162[43]	157–189[33]	Approx. 160[40]	171[7]	168–183[30,44]	150[45]
Wean (months)	10–14[34]	7–14[a]	14[43]	6–30[33]	–	12[4]	10–20[44]	12[45]

+, Unquantified component.

1. Groves (2001). 2. Fooden (2000). 3. Tang et al. (2015). 4. Rowe (1996). 5. Tsuji et al. (2015). 6. Choudhury (2008). 7. Oi (1996). 8. Bottin et al. (2008). 9. Coss and Ramakrishnan (2000). 10. Lindburg (1977). 11. Southwick et al. (1996). 12. Li et al. (2012). 13. Hanya et al. (2018). 14. Ruppert et al. (2018). 15. Oi (1990). 16. Ramanathan (1994). 17. Veeramani et al. (2019). 18. Kuruvilla (1978). 19. Hori et al. (1977). 20. Teas et al. (1980). 21. Fooden and Aimi (2003). 22. Choudhury (2003). 23. Caldecott (1986). 24. Molur et al. (2003). 25. Seth and Seth (1986). 26. Cui et al. (2015). 27. Wheatley et al. (1996). 28. Hanya (2004). 29. Hill (1997). 30. Fooden (1990). 31. Goldstein and Richard (1989). 32. Malik and Southwick (1987). 33. Fooden and Aimi (2005). 34. Lindburg (1971). 35. Melnick and Pearl (1987). 36. Syamil et al. (2019). 37. Maruhashi (1981). 38. Bertrand (1969). 39. Sinha (2001). 40. Feeroz (2003). 41. Nunn (1999). 42. Melnick and Pearl (1987). 43. van Schaik and van Noordwijk (1985). 44. Trollope and Blurton-Jones (1970). 45. Rahaman and Parthasarathy (1969).

* Guo Yongman, pers com, June 2020.

on macaque activity budgets (Li et al. 2020). Some species eat considerable amounts of poorly digestible bamboo, increasing feeding and reducing travel (e.g., Assamese, Chinese *Mm*: Zhou et al. 2011). Eating high-calorie foods, such as those acquired from humans, also reduces travelling time compared to natural diets (*Mfa*: Wheatley et al. 1996, Indian *Mm*: Seth and Seth 1986, *Mn*: Ruppert et al. 2018, Assamese: Koirala et al. 2017).

Macaques are intensely social animals and allocate considerable time to social behavior (e.g., Indian *Mm*: 22%), which increases, along with resting, in provisioned groups (Koirala et al. 2017). Social grooming often makes up a significant proportion of social activity time, reflecting its importance in forming and maintaining social bonds (Dunbar 2010). However, grooming may decrease in situations where animals need to spend more time being vigilant against threats (e.g., *Mr*: Balasubramaniam et al. 2020).

Diet and Feeding Behavior

Macaques are generalist consumers with diets dominated by fruit and other plant matter (Hanya and Chapman 2013) (Table 28.2). They are effective seed dispersers, often discarding seeds after eating the fruit (Kuruvilla 1978; Sengupta et al. 2020). Food is generally hand-picked, manipulated or processed, and then placed in the mouth for chewing and swallowing, or for storage in cheek pouches (Kurland 1973). Generally, macaques have eclectic diets, including many food types, such as fruit, flowers, leaves, seeds, gums, buds, grass, clover, roots, bark, mushrooms, termites, grasshoppers, ants, beetles, bird eggs, shellfish, worms, caterpillars, crabs, frogs, octopus, fish, and clay (*Mfa*: Wheatley et al. 1996; Yeager 1996; Son 2003; Gumert and Malaivijitnond 2012, *Mm*: Southwick et al. 1996; Fooden 2000; Majumder et al. 2012; Sengupta and Radhakrishna 2016).

Diets are often highly seasonal, reflecting seasonal availability of favored and most nutritious foods (e.g., fruit), with less nutritious items (e.g., mature leaves, buds, bark, roots) appearing in diets as fallback foods, especially in the winter (Southwick et al. 1996). When fallback foods are of particularly poor quality, there can be an impact on population size (e.g., *Mfu*: Hanya and Chapman 2013). Fruit consumption correlates positively with availability, with clear preferences for specific items (Indian *Mm*: Sengupta and Radhakrishna 2016, *Mr*: Ramanathan 1994), and makes up much of the diet when seasonally abundant and less when scarce. For example, in rhesus macaques in Pakistan, fruit availability and consumption are highly seasonal, dropping from 32% to 9% of the diet with concurrent increases in fallback foods, such as grass, clover, and other herbs (35.1% of diet), as well as bark and sap (*Mm*: Goldstein and Richard 1989). In the limestone forests of China, macaques have been reported to eat considerable amounts of high crude protein-containing young bamboo leaves year round, and when fruit has low seasonal availability, they increase their intake of less digestible, toxin-bearing, mature bamboo leaves, as a dietary fallback (up to 32% of the diet, Assamese: Zhou et al. 2011, *Mm*: Tang et al. 2015).

Many macaques actively hunt animal food, for example, insects on the wing or under bark and stones (*Mr*: Kuruvilla 1978), rats in oil palms (*Mnn*: Ruppert et al. 2018), and frogs and lizards (*Mfu*: Suzuki et al. 1990). They are also extractive foragers, accessing nutritionally rich foods by using stone tools (*Mfa* shellfish: Gumert and Malaivijitnond 2012; oil palm nuts: Luncz et al. 2017), or through force and skill (*Mfu* walnuts: Tamura 2020). Famously, Japanese macaques developed food and nonfood-related behavior, the spread of which through populations indicates cultural transmission (food-washing: Kawai 1965, stone handling: Huffman and Quiatt 1986). Other macaques also practice food rubbing of fruit, caterpillars (*Mr*: Kuruvilla 1978), and vegetables (*Mfa*: Wheatley 1988). Several species even self-medicate; eating clay, soil, and medicinal plants to address gastrointestinal problems (e.g., *Mfa, Mm, Mr*: Wheatley et al. 1996; Fooden 2000; Dileep and Boby 2014, *Mfu*: Tasdemir et al. 2020; Tibetan: Huffman et al. 2020).

Macaques may access human food through provisioning or crop-raiding, and it may comprise a significant proportion of their diet (e.g., *Mfa* 45%: Wheatley et al. 1996, *Mm* 30%: Gogoi and Das 2018). Consumption of human food can have a direct negative impact on the animals' general health (*Mm*: Gogoi and Das 2018, Barbary: Maréchal et al. 2016) and gut microbiome (*Mfu*: Lee et al. 2019). In human areas, rhesus macaques were found to be in poorer condition than those in natural forest, with diffuse hair loss and frequent regurgitation (Lindburg 1971). Crop-raiding may also have health impacts, for example, glucose intolerance linked to diabetes from a diet high in sugar cane (Mauritian *Mfa*: Dunaif and Tattersall 1987).

Feeding generally takes place during cooler periods of the day; early in the morning, and in the mid to late afternoon, just before roosting. Different food types may be sought in each bout, e.g., fruit in the morning, and insects, followed by foliage in the afternoon (*Mfa*: van Schaik et al. 1996). During feeding, groups may split into foraging parties that re-gather for travel between major food patches (e.g., *Mm*: Lindburg 1971; *Mnn*: Oi 1990). Dominant animals may be found feeding on richer food patches than subordinates (Indian *Mm*: Seth and Seth 1986).

Habitat Use

Macaques vary considerably in arboreality (Table 28.2). Some are almost exclusively arboreal (e.g., *Mfa, Mnl*), while others spend most of their time on the ground (e.g., *Ma, Mm, Mnn*). However, this can vary with habitat; for example, in mangrove forests, rhesus are almost exclusively arboreal, but only spend 20% of foraging time above 10 m in drier areas (Indian *Mm*: Mukherjee and Gupta 1965; Lindburg 1977). In other areas, they spend 20% of foraging time above 5 m, 15% at 0.1–5 m, and 60% on the ground (Chinese *Mm*: Li et al. 2012).

Macaque arboreality is reflected in differences in anatomy. Long-tailed macaques are highly adapted for arboreal locomotion; they are light, with relatively short legs and long tails for stabilization during climbing and leaping (Cant 1988; Fleagle 1999). *Mfa* also leap frequently (11% of locomotion) and jump gaps of up to 5 m (average 2.2 m) (Kurland 1973). More terrestrial species, on the other hand, typically have longer limbs and shorter tails (e.g., *Ma, Mnn, Mm*: Rodman 1979).

Many macaque species are capable swimmers (e.g., *Mfa, Mfu, Mm, Mnn, Mr*, toque, Tibetan: Fooden 1990; Rowe

1996), although stump-tailed macaques are not reported to swim (Fooden 1990). Macaques swim for recreation, to cool off, to cross rivers, and to access specific foods (Kawai 1965; Mukherjee and Gupta 1965; Lindburg 1971; Fittinghoff and Lindburg 1980; van Schaik et al. 1996) and may swim considerable distances (e.g., *Mfa* 100 m: Kurland 1973, *Mfu* 600 m: Fooden and Aimi 2005).

Home range size and day range length (Table 28.2) are largely determined by resource distribution (e.g., food and refuges) as well as group size (McNab 1963; Oates 1987). Home ranges must contain all necessary food and water, as well as suitable roosting sites (Milton and May 1976), and may change seasonally to ensure sufficient resources year-round (Lindburg 1971). Where macaques are provisioned, groups may be very large and sedentary, with small home ranges (Lindburg 1971; Wheatley et al. 1996). Richer habitats, and those with less seasonality, may support higher densities, as is the case for rhesus macaques in tropical ($120/km^2$) versus temperate ($7.2/km^2$) habitats (Southwick et al. 1996).

Predation

A range of predators are known to threaten macaques in their natural habitats, including: big cats (e.g., leopards, tigers: *Mfa*, *Mm*), large birds of prey (e.g., eagles and other raptors: *Mfa*, *Mfu*, *Mm*, *Mnn*), large reptiles (e.g., pythons: *Mfa*, *Mm*, *Mnn*; fresh and saltwater crocodiles: *Mfa*, *Mm*; monitor lizards: *Mnn*; Komodo dragons: *Mfa*), canids (e.g., feral dogs: *Mfa*, *Mfu*, *Mnn*; jackals: *Mm*; raccoon dogs: *Mfu*), bears (e.g., sun bears: *Mfa*), and sharks (*Mm*, *Mfa*) (Kurland 1973; van Schaik and Mitrasetia 1990; Fooden 1995; Wheatley et al. 1996; Imron et al. 2018; *Mfu*: Fooden and Aimi 2005; Indian *Mm*: Lindburg 1977; Mukherjee and Gupta 1965; *Mnn*: Ruppert et al. 2018). Many macaques are killed or captured by humans as agricultural pests (e.g., *Mfa*, *Mfu*, *Mnn*: Kurland 1973; Fooden and Aimi 2005; Ruppert et al. 2018) as are some for biomedical research (e.g., *Mfa*, *Mr*: Sinha 2001; Eudey 2008; Honess et al. 2010, and until recently *Mfu*: Isa et al. 2009).

Macaques vary in their response to threats. Being predominantly arboreal, if predators are detected while macaques are on the ground, most species give alarm calls and flee into trees (e.g., *Mm*, *Mr*: Lindburg 1971; Coss and Ramakrishnan 2000). An adult male may then position himself between the threat and the rest of his group (Lindburg 1971). There appears to be a difference in response between the two species of pig-tailed macaques, with southern pig-tailed macaques descending from as high as 10 m to the ground to flee from threats (Bernstein 1967), whereas northern pig-tailed macaques are reported to flee from threats arboreally (Choudhury 2008). The choice of strategy is likely influenced by the type of predator, the structure of the environment, and the relative agility of the macaque compared to the predator in an arboreal environment.

Macaques will also mob predators, often approaching quite near the predators to harass and vocalize at them (Lindburg 1977; van Schaik and Mitrasetia 1990; Wheatley et al. 1996). They may avoid sites where predators have been seen, for up to 12 days (van Schaik and Mitrasetia 1990). Macaques in the wild will even give antipredator responses to models of predators (e.g., snakes *Mfa*: van Schaik and Mitrasetia 1990, leopards *Mr*: Coss and Ramakrishnan 2000).

Social Organization and Behavior

Social Organization

Most macaques live in multimale-multifemale groups (Melnick and Pearl 1987), although uni-male groups are found in some species (e.g., *Mn*, *Mr*, lion-tailed macaques: Sinha 2001) and solitary males may often be found (e.g., *Mfa*: van Noordwijk and van Schaik 1988, *Mm*: Lindburg 1971, *Mnn*: Oi 1990). Data on group size and sex ratio can be found in Table 28.2. At the heart of macaque groups are the matrilines; groups of related females and their offspring bound together with especially close affiliative, cooperative, and contest-support relationships (Wrangham 1980).

Macaque social groups are characterized by dominance hierarchies (Flack and De Waal 2004). Dominance rank reduces the need for continual costly contests over resources; more dominant animals get preferential access, which results in reproductive benefits (Dixson 1998a; van Noordwijk and van Schaik 1999b; Liu et al. 2018). Typically, females inherit rank from their mothers, who reinforce their daughters' rank through intervention during aggressive interactions (Hill and Okayasu 1996), whereas males attain rank through contest and alliances (Lindburg 1971; van Noordwijk and van Schaik 1999b). A daughter will rarely rise above her mother's rank, but will out-rank all those who are subordinate to her mother, including any older sisters (Hill and Okayasu 1996). Each matriline has its own hierarchy, with an additional hierarchy across matrilines (Gouzoules and Gouzoules 1987).

Macaques exhibit sex-biased dispersal; typically females are philopatric, remaining in their natal group (*Mfa*: de Ruiter and Geffen 1998, *Mfu*: Fooden and Aimi 2005, *Mm*: Lindburg 1969), while males leave as they become sexually mature (*Mfa* 4.5–7 years: De Ruiter et al. 1992, *Mfu* 5 years: Fooden and Aimi 2005, *Mm* 4–5 years: Melnick et al. 1984, *Mnn* 4 years: Oi 1990). Sons of high-ranking females may disperse later than those of low-ranking females (Weiß et al. 2016). Bonnet macaques (*Mr*) are unusual, as maturing males occasionally remain in their natal group (Sinha 2001). Sex-biased dispersal acts to prevent inbreeding and is reflected in the genetic relatedness among adult females (high) and males (low) within groups (Melnick et al. 1984; de Ruiter and Geffen 1998). These aspects of genetic relatedness explain many patterns of behavior seen in macaque groups, especially in alliance formation, grooming, agonism, and reconciliation (Aureli et al. 1997; Thierry 2007).

In seasonally breeding macaque species (see Table 28.2), maturing males may emigrate just in time for the breeding season (e.g., *Mm*: Lindburg 1969). Once they have left their natal group, typically in peer groups, it may be some time before they become integrated into a new group (Pusey and Packer 1987; van Noordwijk and van Schaik 1988). Males may enter a new group either low in the hierarchy, eventually increasing their dominance rank or moving to a different group, or enter higher up, through direct confrontation with high-ranking resident males (Wheatley et al. 1996).

Once accepted into a group, males may eventually move on to another group. Acceptance by the females is generally key for a male to rise to and retain alpha status (Lindburg 1971). An alpha male may retain his status for as little as a few months to as long as 5 years (e.g., *Mfa*: De Ruiter et al. 1994; Wheatley et al. 1996). In *Mfu*, alpha males may remain dominant for up to 10 years (Fooden and Aimi 2005), which presents inbreeding risks. Tenure length may be dependent on group size, as well as internal social factors. For example, in long-tailed macaques, alpha males' tenures may last 2–3 years in large groups, but up to 5 years in smaller groups. Once the male loses his status, he may remain with the group for a further 1–2 years before migrating to another group, but he is unlikely to ever achieve alpha status again (*Mfa*: van Noordwijk and van Schaik 1988; De Ruiter et al. 1992).

Communication

Macaques communicate through olfactory, visual, tactile, and auditory channels. They have a strong visual bias, reflecting an evolutionary emphasis for close range social communication, and for the detection and assessment of food, especially fruit (Lucas et al. 1998; Osorio et al. 2004). As well as conveying information about the environment (e.g., presence of a predator, location of food), many forms of macaque communication transmit information about the sender, including his/her identity, size, social and reproductive status, and emotional state. These communication signals may be discrete or graded (Seyfarth 1987). Some actions may reflect the internal state of the animal, such as tension or anxiety (e.g., self-scratching, body shakes: Maestripieri et al. 1992). Visual, tactile, olfactory, and quiet vocal communication (unlike long-range loud calls) are typically used when in close proximity to conspecifics, emphasizing the importance of intragroup communication in the highly gregarious macaque species.

Visual communication is achieved through facial, gestural, and postural expression (van Hooff 1981; Maestripieri 2005), changes in color and/or size of female sexual swellings (e.g., *Mm*: Dubuc et al. 2009), and male facial coloration (e.g., *Mm*: Waitt et al. 2003). Species vary in the frequency of use of gestures. For example, stump-tailed macaques use these gestures more than pig-tailed macaques, which use them more than rhesus (Maestripieri 2005). Some postures typical of adult males that are related to contest interactions and escalating aggression may function to enlarge their appearance, and include a rigid pose with a bristling coat (Wolfensohn and Honess 2005) or a more relaxed, head raised, arms spread, and chest pushed out pose when seated (*Mfa*: P. Honess, Pers Obs.). Social yawns, which are wide and display the canines, are typically given by adult males as an early warning threat, but should be distinguished from yawns associated with fatigue or rest (Deputte 1994). As a sign of submission, macaques may present their rear ends, exposing the genitals (Maestripieri 2005), whereas high tail carriage, sometimes in a crook-shape, may indicate dominance (e.g., *Mm*: Ojha 1974; Vessey 1984; Partan 2002).

There is limited interspecies diversity in facial expressions that may be indicative of more aggressive/threatening or more affiliative/submissive/fearful states. These may be paired with accompanying vocalizations and include those listed in Table 28.3 (van Hooff 1981; for photographic illustrations, see "Expressions" on the NC3Rs website at https://www.nc3rs.org.uk/macaques/macaques/behaviour-and-communication/).

Pig-tailed macaques exhibit a species-specific facial expression: the "lips forward – ears back – neck extended"' (LEN) or pucker gesture (Oettinger et al. 2007). It is largely affiliative, or for appeasement, and is given in agonistic, sexual, or infant care contexts (Lee and Oettinger 2013).

A range of vocalizations are used primarily for intragroup communication (Lindburg 1971). However, there are some, like the long-tailed macaque "Ho!", that are given by dominant males during intergroup encounters (Wheatley et al. 1996). Most macaques produce calls across a range of functional contexts, including contact/affiliation, "coo" contact, group movement, food discovery, sexual interactions, submission/fear, agonistic recruitment, dominance aggression, intergroup aggression, and alarm situations (Hauser 1996). There is evidence of dialectic differences between populations, and even matrilines, particularly in the "coo" vocalization (e.g., *Mfu*, *Mm*: Hauser 1996). An individual's emotional and physiological state can influence vocalization; for example, macaques with suppressed cortisol levels produce lower intensity alarm calls (*Mm*: Bercovitch et al. 1995).

Macaques have a well-developed olfactory sense that is used, for example, in the assessment of reproductive state (Dixson 1998b), major histocompatibility complex (MHC)-based mate choice (Grob et al. 1998), and food quality determinations (Dominy 2004).

TABLE 28.3

Some Common Facial Expressions Used by Macaques

Expression	Description
Bared-teeth scream	Indicative of frustration and even fear.
Staring open-mouth face	A threat sometimes paired with a "huh" vocalization. Holding eye-contact is a key element and is typically seen as threatening in macaques. When made with a relaxed face and no stare, this expression may be used during, or as an invitation to, play.
Bared-teeth face (with/without vocalization) or fear grimace	Indicates submission in the face of potential aggression/contest. It is intensified when accompanied by vocalization.
Lip-smacking (+/- protruding tongue) and teeth-chattering face	These are affiliative, appeasing, and tension-reducing behaviors and are often associated with social grooming and other situations that involve close proximity.
Pout face	Given by infants separated from their mothers and may be accompanied by a "coo" call.

Affiliation

Macaques exhibit a range of affiliative behaviors, including allogrooming, play, sitting in contact, mounting, muzzle-to-muzzle contact, genital inspection, embracing, hand-touching, lip-smacking, and eyebrow-raising (*Mfa*: Aureli 1992). Allogrooming is one of the most important affiliative behaviors, indicated by the amount of time dedicated to it (Table 28.2). Macaques groom using both their hands and their mouths, and often focus on areas that the receiver finds hardest to reach (Barton 1985). Grooming techniques often appear highly stereotyped, systematic, and intense. Apart from removing detritus and parasites, allogrooming reduces tension for both parties (Terry 1970; Schino et al. 1988; Shutt et al. 2007), resulting in physiological changes that are indicative of "feel-good" effects (Dunbar 2010; Kiser et al. 2011).

Allogrooming assists in forming, strengthening, and maintaining social bonds, especially among the females in female-bonded social groups (Henzi and Barrett 1999; Dunbar 2010; Shimada and Sueur 2018). Females typically give more grooming than they receive; a pattern reversed in males (Thompson 1967; Mitchell and Tokunaga 1976). In despotic species (see discussion of levels of tolerance in macaques, below), grooming is asymmetric and usually directed upward in rank, whereas in more tolerant species, high-ranking individuals disproportionately groom lower-ranked animals (Butovskaya 2004). Unusually, in bonnet macaques, dominant females will groom subordinates that do not necessarily reciprocate (Sinha 2001). Allogrooming may function as "currency" in biological "markets" to influence sexual activity and access to infants (Henzi and Barrett 1999; Gumert 2007a, b).

Macaques, especially adult females, engage in close physical contact during periods of rest and at night while roosting. Females huddle in clusters that reflect kinship, and adult males may roost separately; often with the alpha male sitting above other group members (Kuruvilla 1978; van Schaik et al. 1996).

Play is predominantly an activity of younger monkeys performing an important function in developing social bonds and enhancing affiliative relationships (Shimada and Sueur 2018). It can be more or less aggressive and involve chasing, wrestling, pinning down, pulling, and mock-biting (Janus 1989). Aggressive play in male rhesus is rougher and up to 3.5 times more frequent than in females, and is thought to develop behavioral innovation and fighting skills that assist rank determination and, ultimately, reproductive success (Symons 1978).

Aggression and Reconciliation

Comparative studies of aggression in wild macaques are rare, but it has been noted that rhesus macaques are particularly aggressive (Southwick 1967). Other species, such as stump-tailed and Sulawesi macaques, are much less aggressive, and are more inclined to reconcile following aggressive interactions (Thierry 2000). Reconciliation is a vital social process that repairs relationships and ensures that winners and losers continue to benefit from harmonious group living (Cords 1992). Reconciliation most commonly takes place almost immediately after the conflict (e.g., within 3–4 min, *Mfa*: Aureli and van Schaik 1991). Postconflict reconciliation is more likely between kin than nonkin, and is less common following feeding conflicts, although these patterns are less marked in species with more egalitarian dominance styles (Aureli 1992; Aureli et al. 1997). Victims of aggression may redirect aggression onto other group-mates (Aureli and van Schaik 1991).

Macaques may be classified based on their tolerance and tendency to reconcile after conflict (Thierry 2000). Those species in Grade 1 (e.g., *Mfu, Mm*) are more "despotic"; being less tolerant and less inclined to reconcile than those in Grade 2 (e.g., *Mfa, Mn*), Grade 3 (e.g., *Ma, Mr*, Barbary, lion-tailed), or Grade 4 (Sulawesi), which are the most "egalitarian". It is noteworthy that those species that are most common in laboratory settings are those in Grades 1 and 2, which are the least tolerant or likely to reconcile. This fact presents important challenges for the management of captive macaques.

Indications of submission vary in their use among species. The silent bared-teeth expression or grimace face (van Hooff 1981) may serve as a clear sign of submission in more despotic species (e.g., *Mfa, Mm*) or to promote affiliative behavior in less despotic species (e.g., Barbary), and may signal play and affiliation in the least despotic (Sulawesi) (Aureli et al. 1997).

Branch shaking is used as a threat display by several species, especially during intergroup encounters (e.g., *Mfa*: Wheatley et al. 1996, *Mm*: Lindburg 1971). Such displays help maintain some exclusivity of core areas of home ranges; however, despite general antagonism and threats, fighting during such encounters is rare, with groups seeming to avoid each other based on established group dominance (Lindburg 1971). Head bobs, stares, and bluff charges may also be used during intergroup encounters (*Mm*: Lindburg 1971). Higher levels of inter- and intragroup aggressive behavior are noted in anthropogenic (competing for human-origin food), compared to forest, environments (Southwick 1967; Lindburg 1971; Southwick et al. 1976; Wheatley et al. 1996). In contrast to rhesus macaques, where adult males lead the majority of aggressive interactions (Lindburg 1971), female long-tailed macaques play a more significant part in such encounters (Wheatley et al. 1996). Intragroup aggression may be more significant than intergroup, and may be fatal, as has been observed as a result of male–male aggression during fission of a large group (Lindburg 1971). Females coming into estrus may become more aggressive and, as a result, attract more male aggression (*Mm*: Southwick 1967). Adult females may be more aggressive toward young females than young males, as females will remain in the group, becoming future competitors (Silk 1983).

Intragroup aggression is typically associated with the establishment or reinforcement of dominance status related to feeding or reproductive opportunities (de Waal 1989). Establishing mutually recognized dominance relationships reduces the likelihood of serious fighting (with associated risks of injury), social exclusion, or even death. In situations where ritualized dominance displays break down, a graded system of escalating aggressive behaviors gives the receiver opportunities to disengage, retreat, or submit. Aggression can escalate from yawns, bristling coat, and body shakes that indicate tension; to threat displays, using vocalizations, threat postures, and facial expressions; to noncontact aggression, such as lunging and chasing; and ultimately, to full-contact aggression, including hair-pulling, tail-pulling, slapping, hitting, and biting.

Reproduction

Mating

Most macaques mature at approximately 3–4 years of age, but may not attain adult body size or social maturity until they are 6–9 years old (Fooden 1990; De Ruiter et al. 1992; Oi 1996; Sinha 2001). Sexually mature females of most species exhibit swelling and/or changes in coloration of their perineal skin around the time of ovulation (Dixson 1998b) (Table 28.2). Sexual swellings, originally thought to be indicators of fertility, are not associated with mating competition and may be more pronounced in adolescent females or those least likely to successfully raise offspring (Nunn 1999; Nunn et al. 2001), and may instead help bias and confuse paternity, thereby reducing the risk of infanticide (Setchell and Kappeler 2003). Males may, in fact, make assessments of mate quality and fertility using cues related to estrogen levels (Engelhardt et al. 2004).

Females solicit mating by presenting their rear and looking back at the male, whereas males use a "puckering" facial expression (*Mnn*: Oi 1996). Mounting patterns vary between species with single (e.g., *Mr*), multiple (e.g., *Mm*), or single and multiple (e.g., *Mfa*) mounts to ejaculation being used depending on degree of intermale tolerance (Shively et al. 1982). Mounts last about 9 s with approximately 14 thrusts, again varying with species. The male may grimace during copulation and both sexes may vocalize, with females often giving specific calls associated with male ejaculation (e.g., *Mfa*: Deputte and Goustard 1980). At the end of the encounter, the female may also call as she rushes away (Oi 1996). In some species, copulations may involve male violence toward the female (e.g., biting, scratching, pressing down, *Mnl*: Feeroz 2003). Stump-tailed macaque males leave a solid copulatory plug in the vagina following copulation, to prevent sperm backflow and obstruct insemination by another male (Dixson 1998c; Feeroz 2003). Several other macaque species (e.g., *Mfu*, *Mm*, *M. nigra*, *M. silenus*) produce a seminal coagulate that is not as solid as a plug (Dixson and Anderson 2002).

Females generally copulate with several males, often out of the gaze of more dominant males (Lindburg 1971; Berard et al. 1994) who may try to monopolize mating through consortship (up to several days long) or disrupting subordinates' copulations (*Mfa*: De Ruiter et al. 1992, *Mnn*: Oi 1996). Attempts to monopolize females may be costly and not totally effective, and therefore, males may even cooperate to secure matings (*Ma*: Zothansiama et al. 2014; Toyoda et al. 2020). While unsuccessful in monopolizing matings, the alpha male may still secure parentage of the majority of a group's offspring (*Mfa*: alpha 52%–92%, beta 8%–33%, others 2%–8% each: Cowlishaw and Dunbar 1991; de Ruiter and van Hooff 1993; De Ruiter et al. 1994), particularly those born to higher ranking females (de Ruiter and Geffen 1998). Alpha male reproductive success may vary with group size, being higher in smaller groups (de Ruiter and van Hooff 1993), yet there is limited evidence of rank-related reproductive benefits in some species (*Mfu*: Fooden and Aimi 2005).

Births

Several species are strongly or weakly seasonal breeders (Table 28.2), though they may still breed across the year in captivity using timed mating techniques (*Mm*: Beck et al. 2020). Some generally nonseasonal breeders may breed seasonally in part of their range (e.g., *Mfa*: van Schaik and van Noordwijk 1985; De Ruiter et al. 1992). Seasonality of breeding is largely determined by the seasonality of the environment, particularly food availability (Janson and Verdolin 2004). For example, bonnet macaques live in variable habitats, with strongly seasonal fruit availability; and approximately 93% of births occur in February and March (Sinha 2001).

Birth rates often vary with species and environmental parameters. In some populations, it may be as high as 91% (*Mm:* Lindburg 1971) or as low as 53% (*Mfa:* van Noordwijk and van Schaik 1999a), with relatively little difference reported between closely related species (*Mnl*, 62%; *Mnn*, 57%–73%: Oi 1996; Feeroz 2003). Birth rate can also vary markedly with resource availability. For example, macaques in more productive tropical habitats tend to have higher birth rates than those in temperate conditions (Chinese *Mm*, 85% vs 38%: Southwick et al. 1996; *Mfu* 86% vs 25%: Fooden and Aimi 2005). Birth, and death, rates may be higher in smaller groups that are less able to compete for resources and are more vulnerable to predators (e.g., Chinese *Mm*: Liu et al. 2018).

Macaques typically give birth to a single infant, though there are rare reports of twins in the wild (*Ma*: Toyoda and Malaivijitnond 2018, *Mfu*: Fooden and Aimi 2005) and neonates may be born with a darker coat that is replaced with adult coloration after 2–3 months (e.g., *Mfa*: Fooden 1995). Uniquely among macaques, newborn stump-tailed macaques are whitish-buff, changing to brown by 12 months of age (Fooden 1990).

Maternal Care

Infants play a key role in social groups and draw considerable attention (Silk 1999). Alloparenting is relatively common among related females, reducing maternal load and providing parenting experience for young females (Quiatt 1979; Silk 1999). Females may trade grooming of the mother for contact opportunities with the infant (Gumert 2007b). In extreme cases, infants may be kidnapped, usually by a higher-ranking female, with potentially fatal consequences for the infant (Schino et al. 1993; Silk 1999). Infants will be carried ventrally when very young, but once they become larger, they begin to ride on their mother's back, though some may continue to be carried ventrally even when clearly too large (e.g., *Mm*: Altmann 1962). In contrast to daughters, developing sons may spend more time away from their mothers; forming relationships with peers, and spending less time suckling, though receiving richer milk (Hinde 2009). The microbiological content of milk varies across the suckling period, with positive implications for the gut microbiome (Muletz-Wolz et al. 2019). The duration and distance of mother–infant separation may reflect security perceptions, perhaps related to predation. In larger groups that tend to be better at detecting and deterring predators, there may be more separation than in smaller groups (e.g., Chinese *Mm*: Liu et al. 2018). Reduced anxiety, with increased vertical flight opportunities, may have the same effect (e.g., *Mfa*: Waitt et al. 2008). Weaning is achieved via gradual nutritional independence, with complete cessation of suckling by 6–30 months of age (Table 28.2) or by the time a new infant is born.

Common Captive Behaviors

Excellent coverage of the captive behavior of some macaques is provided in volumes edited by Schapiro (2017) and Winnicker et al. (2013). This section will examine key aspects of captive behavior, including abnormalities and ways in which they may be prevented or treated. It is always better to design environments and management systems that *prevent* maladaptation and the onset of behavioral abnormality, as once abnormal behaviors have been incorporated into the repertoire, they may be reduced with therapy, but they can rarely be eliminated (Bayne and Novak 1998; Turner and Grantham 2002).

Compared to wild conspecifics, captive macaques, especially when housed in restricted caging, exhibit behavioral repertoires that are typically impoverished in behavioral diversity and skewed in relation to the time allocated to different behaviors (Honess et al. 2004). In large part, this is due to the lack of behavioral opportunities, through reductions in social partners, lack of environmental complexity, restricted space, and loss of control. As a result, animals frequently become bored and inactive. Combine this with the imposition of other factors, such as husbandry, veterinary, and research interventions, it can often result in stress and even distress (Honess and Marin 2006a). This causes significant changes across almost all biological processes and functions, challenging the animals' coping abilities and their validity as research models. When animals are not able to use species-typical adaptations, fine-tuned over their evolutionary past, to deal with their needs, discomfort, anxiety, fear, and stress will persist with lasting consequences, including the development of abnormal behavior and other pathologies (see Novak et al. 2017a; Novak and Meyer this volume; Schapiro this volume). The modified behavior and physiology of captive macaques might be seen as "normal" and, therefore, sufficient for the species. However, there is general acceptance that providing opportunities for the expression of natural behavior is key for animal welfare provision, as represented in the Five Welfare Needs found in some national animal welfare legislation/regulation, e.g., in the UK (Animal Welfare Act, 2006) and in China (MacArthur Clark and Sun 2020).

In captivity, behavior can range from being wholly, or partly, replicative of wild behavior (e.g., grooming, mounting), analogous to wild behavior (e.g., cage shaking/branch shaking), or rare/absent in the wild (e.g., stereotypies, self-harming). In other cases, behaviors may be natural, but occur at unnatural frequencies, durations, or intensities (e.g., aggression). Both unnatural repertoires and/or unnatural time budgets are indicative of underlying issues, and efforts should be made to prevent these, by skillful planning that understands and incorporates key adapted aspects of natural coping into the design of environments (both social and physical) and care practices (husbandry and veterinary). Such *evolutionary accommodation* can enable animals to cope better with captive stressors and achieve better welfare. In research environments, where a justification has been made for using animals, there is often also a requirement to achieve research goals by minimizing both the number of animals used (Reduction) and the impact of research and husbandry procedures on the animals (Refinement); two of the 3Rs (Russell and Burch 1959) that are widely accepted in animal research.

Behavior varies due to a range of factors, both extrinsic, such as how animals are housed and managed, and intrinsic, such as species identity, age-sex class, temperament, experience, and genetics. Generally, observable behavior is the result of interactions among some, or even all, of these factors.

Normal Behavior

Grooming

In captive and wild macaques, allogrooming is crucial for managing stress and social relationships (Silk 1982; Schino et al. 1988; Matheson and Bernstein 2000), with the time allocated to it exceeding that for autogrooming. In groups, allogrooming may occupy approximately 20% of time budgets (Post and Baulu 1978), and is higher in females than males, but levels do not differ appreciably across species (*Mn, Mr*: Rosenblum et al. 1966). The presence of allogrooming is often used as an indicator of compatibility in animals that have recently been introduced. However, assessing levels of allogrooming may not be a meaningful measure, as grooming rates may vary considerably between pairs judged to be successfully socialized (Baker et al. 2012a).

Locomotion

Patterns of locomotion are largely dictated by a combination of (1) cage size and shape; (2) the availability of structures, feeding and watering sites, and enrichment devices; and (3) the activity of other animals, caregivers, and research staff. Well-placed furniture allows easy transitions between cage levels and promotes species-typical locomotion (e.g., running, walking, leaping, swinging, climbing). Although some species are more arboreal than others, all macaques should be provided with opportunities to climb. Younger animals are generally more active than older age groups, and swinging devices, often avoided by older animals, help youngsters develop strength and balance (Reinhardt 1992b; Waitt et al. 2008).

Appropriate spacing of structures will promote species-typical leaping; with sufficient space, even a typical leap distance of 2.2 m can be accommodated (*Mfa*: Honess 2017). Well-designed environments promote otherwise rare behavior (e.g., leaping), but also safely accommodate more common locomotion, including rapid chasing and vertical flight. Careful placement of furniture enables animals to run without colliding with others or cage structures, or being hit by swinging furniture.

Most macaques are naturally capable swimmers and benefit from opportunities to play in pools (Anderson et al. 1994; Robins and Waitt 2010). Adding food items to the water that sink (e.g., raisins, banana pieces in a box) provides additional enrichment (Anderson et al. 1992).

Aggression

Aggressive interactions are a natural part of the behavioral biology of wild macaques. The more, or less, aggressive dominance styles of different species are highly evolved, and therefore, it is reasonable to expect some aggression in captivity, even in the absence of relevant ecological drivers. Aggression, in captivity, is a major animal health and welfare issue, potentially resulting in stress, injury, and occasionally, death, and this is amplified when individuals behave hyperaggressively toward others or themselves (see below) (Anderson and Chamove 1980). Aggressive interactions may also vary with reproductive state (e.g., breeding season) and the number of matrilines in a group, as the availability of more allies may help mediate intermatriline aggression (Samuels and Henrickson 1983; Beisner and Isbell 2011; Stavisky et al. 2018).

Rhesus macaques, noted for their aggression and poor tendency to reconcile, nevertheless lead relatively peaceful lives in groups in expansive caging systems. They rely on aggression to establish and reinforce dominance relationships, and social instability (e.g., removing animals for research, veterinary, or management reasons) can dangerously elevate aggression. When adult group members are removed (permanently or for more than a few days), careful planning is necessary, such as removing lowest ranked animals first (e.g., for terminal procedures), or adding additional enrichment intended to reduce aggression by providing more hiding places and enabling visual separation (Maninger et al. 1998; Honess and Marin 2006b; Waitt et al. 2008).

Sources of Variation

Variation due to Species and Source

There are notable behavioral differences between species in captivity that broadly reflect those found in the wild, including aggressiveness and readiness to reconcile. Clarke and Boinski (1995) reviewed a range of studies that examined temperamental differences at species, population, and individual levels. Among macaques: rhesus are characterized as aggressive, exploratory, and weakly conciliatory; long-tailed as less aggressive, but readily disturbed; pig-tailed as highly social; lion-tailed as exploratory and manipulative; and bonnets as passive, with these tendencies relating to differences in natural ecology. Japanese macaques are considered to be gentle, compared to both rhesus and long-tailed macaques (Isa et al. 2009).

These differences are often reflected in responses in a range of contexts. For example, of three species examined (*Mfa*, *Mm*, and *Mr*), rhesus macaques were quicker to habituate to a transport cage than either long-tailed or bonnets (slowest), and of the three, rhesus showed the lowest stress and behavioral disturbance in response to training, novelty, and restraint. Bonnets were intermediate between rhesus and long-tailed (Clarke et al. 1988, 1994). Species also differ in their responses to humans, with rhesus being hostile, long-tailed fearful, bonnets appeasing, and pig-tailed sociable and unaggressive (Clarke and Mason 1988; Sussman et al. 2013). Macaque species also differ in response to conspecifics. For example, previously singly housed long-tailed macaques adapted better to subsequent group living, with lower stress and more affiliative hierarchy formation, than lion-tailed macaques (Clarke et al. 1995a). Differences observed among adult macaques are often replicated in infants, where, for example, long-tailed macaques are bolder in the presence of humans than pig-tailed macaques (Heath-Lange et al. 1999). However, species-typical temperament may not be fixed; cohoused rhesus can learn less aggressive conflict resolution techniques from their stumptailed macaque partners (de Waal and Johanowicz 1993).

In addition to species-level differences in behavior, there is also variation at population or subspecies level. For example, long-tailed macaques from different geographical origins exhibit different levels of social behavior, with animals from Indochina exhibiting the most aggression and least affiliation (including grooming), compared to conspecifics from the Philippines, Indonesia, and Mauritius (Brent and Veira 2002). It should be noted, however, that these intraspecific differences may be a result of local differences in rearing and welfare. There are also temperamental differences between Chinese- and Indian-origin rhesus macaques, with the former reported to have more irritable, reactive, and aggressive traits that emerge even in infancy (Champoux et al. 1994). However, similar patterns were not found in their response to humans (Baker et al. 2009).

Variation due to Individual Temperamental Differences

Individual temperament differences are a product of life experiences and epigenetic factors (Higley and Suomi 1989; Clarke and Boinski 1995; Itoh 2002; Rogers 2018) and are directly relevant to behavioral management (Capitanio 2017; Coleman 2017). Infant rearing conditions have a profound impact on temperament and behavior as macaques grow and develop. This is particularly true for early maternal deprivation (Harlow 1958), but is also apparent in less extreme examples that deviate from natural rearing, including nursery rearing (Shannon et al. 1998; Heath-Lange et al. 1999), peer rearing (Timmermans and Vossen 1996; Goin and Gust 1998), and early weaning (Reinhardt 2002b; Prescott et al. 2012).

While accounting for some behavioral variation, temperament is also linked to health outcomes, for example, rhesus macaques of more nervous, gentle, and vigilant personality types are more likely to have chronic diarrhea in stressful environments (Gottlieb et al. 2018). The BioBehavioral Assessment (BBA) program developed at the California National Primate Research Center uses behavioral and physiological data to aid animal management, subject selection for studies, and interpretation of results. These data have been used both predictively and retrospectively to illuminate individually variable responses in a range of circumstances (Capitanio 2017). Other facilities also perform temperament assessments to facilitate behavioral management (e.g., Coleman 2017), but few integrate behavior and physiology as systematically or extensively as the BBA program (Capitanio 2017).

Abnormal Behavior

The identification, assessment, prevention, and treatment of abnormal behavior form an important part of welfare monitoring and the provision of behavioral management for captive macaques. There is an established relationship between stress and abnormal behavior that validates its monitoring as an index of welfare (Lutz et al. 2003b; Novak et al. 2017a).

Animals with significantly challenged welfare may spend as much as 73% of their time in abnormal behavior (Crockett et al. 1995). When assessing behavioral abnormality, it is vital to thoroughly understand natural behavior, including its function and adaptive significance. Only then is it possible to understand the animal's motivation for expressing a behavior, and to design ethologically appropriate strategies to prevent or treat the observed abnormalities.

There is considerable diversity in the range of abnormality of behavior in macaques; natural behavior exhibited in inappropriate contexts, natural behavior exhibited at atypical levels (rate, duration, intensity), and potentially harmful, unnatural behaviors that serve no function. Some of these may have underlying neuropathologies (Garner 2006). Additionally, whole repertoires and time budgets may be more or less abnormal, being either impoverished (lacking behavioral diversity) or skewed (imbalanced time allocation); for example, allocating an abnormally high 43% of daily time budget to rest and inactivity (*Mfa*: Crockett et al. 1995). Some abnormal behavior may be infrequent, brief, and easily overlooked, but of high impact (e.g., self-biting). Therefore, assessments must be designed appropriately to capture these behaviors when seeking to determine their significance or to demonstrate the therapeutic effectiveness of behavioral management strategies.

Extensive lists of abnormal behaviors found, and remarkably consistently expressed, across macaques can be obtained elsewhere (e.g., Erwin and Deni 1979; Honess 2013). Aspects of abnormal behavior, such as frequency, duration, and intensity, may vary with factors, including species and individual temperament, as well as the nature and intensity of the triggering stressor. Generally, the most significant extrinsic factor in the development of abnormal behavior is environmental deficit (Mason 2006). This is typically due to insufficient (1) space, (2) environmental complexity, and (3) social stimulation (Erwin and Deni 1979; Novak 2003; Honess and Marin 2006a). In particular, early-life or prolonged single-housing in small caging, nursery-rearing, and number of procedures (blood draws) are known risk factors for the development of abnormal behaviors, with stronger effects on males than females (Lutz et al. 2003b; Novak 2003; Rommeck et al. 2009).

Abnormal behavior, especially when repetitive, is often labeled "stereotypic", but Mason (2006) cautions against this when causal factors remain unknown, and proposes the alternative, broad term: "*abnormal repetitive behavior*", which will be used here. Abnormal behaviors are also frequently classified descriptively as being "whole body" (e.g., bouncing, pacing, rocking, somersaulting), "self-directed" (e.g., saluting/eye poking, hair-plucking, self-grasping, digit sucking), "appetitive" (e.g., coprophagy, regurgitating, urine-drinking) or "self-injurious" (e.g., self-biting, self-injury) (Lutz et al. 2003b; Honess et al. 2013b). This categorization is theoretically problematic, mixing directionality, motivation, and consequence, but nevertheless has some practical utility. It is also worth noting that a number of abnormal behaviors observed in macaques have analogous abnormalities in humans, such as hair-plucking/trichotillomania (Honess et al. 2005; Novak and Meyer 2009; Coleman et al. 2017) and self-harming (Bayne and Novak 1998; Novak 2003).

Self-Injurious Behavior and Self-Directed Behavior

Abnormal behavior directed by an animal to itself (self-directed behavior: SDB) includes behaviors that may be physically harmful (self-injurious behavior: SIB), such as self-biting and self-hitting. SDB, with less trauma risk, includes behaviors such as self-sucking (e.g., a digit), self-clasp/grasp/hug, floating limb (sometimes with attacking of the limb), eye-poking/saluting, and self-plucking (Honess et al. 2013b). While SIB occurs in singly housed monkeys (e.g., 15%: Bayne et al. 1992b; Novak 2003), it is rare in pair- or group-housed animals (Reinhardt et al. 2004). Self-biting can vary from sham biting, through minor abrasions and bruising, to severe laceration, which may expose or even break the bone (Reinhardt and Rossell 2001). Self-biting can be repeatedly focused on the same site on the body (Bayne and Novak 1998; Reinhardt et al. 2004). Using an objective scoring system, like that devised by the Behavioral Management Consortium of the National Primate Research Centers (see Honess et al. 2013b; Baker et al. 2017), is important for the objective assessment of the traumatic impact, and management, of self-biting.

SIB is associated with high levels of emotional excitement in response to stressful events (e.g., feeding time: Novak 2003), rather than aggression (Reinhardt and Rossell 2001; Lutz et al. 2003a). Animals that exhibit it may also threaten and vocalize more, be less affiliative, have lower cortisol levels (Novak 2003), and show less anxiety during the human intruder test (Peterson et al. 2017). SIB may even represent a coping strategy to reduce excitement, as self-biting appears to significantly reduce an elevated heart rate (Novak 2003).

Rehousing animals with SIB with compatible conspecifics can be therapeutic (e.g., *Mfa*: Line et al. 1990a; Statz and Borde 2001; *Mm*: Reinhardt 1999). Although SIB can be resistant to treatment with feeding or grooming enrichment, drug therapy can be successful (Reinhardt and Rossell 2001). Treatment with guanfacine has been shown to reduce SIB in rhesus macaques, with the strongest effects seen in those animals with the most pronounced SIB, and with some positive effects even continuing after the end of the drug treatment (Freeman et al. 2015). However, the use of diazepam to treat SIB in rhesus macaques only reduced SIB in some animals, while appearing to increase it in others, depending on the individual's housing and procedural history (Tiefenbacher et al. 2005). Self-injury may have significant welfare consequences, such as surgery, amputation (Reinhardt et al. 2004), and even euthanasia, and may result in an animal having to be removed from study.

Abnormal Repetitive Behavior

There is a range of abnormal locomotion exhibited by macaques, including spinning/pirouetting, flipping/somersaulting, head tossing/weaving, and bouncing, which in general, is less common in older animals (Lutz et al. 2003b). Pacing is frequently the most common abnormal behavior seen in laboratory macaques (e.g., *Mm* 78%: Lutz et al. 2003a).

It is important to distinguish normally elevated locomotion from patterns that are abnormal. For example, macaques may elevate natural pacing in advance of feeding, while patrolling,

(Honess et al. 2013b) or in response to agonistic events (Poirier et al. 2019). During abnormal repetitive pacing, or route tracing, each circuit is typically identical with hands/feet placed in exactly the same place, particularly at turning points, as may be indicated by dirty marks on the cage/enclosure. Pacing is typically rhythmically ritualized and it can be difficult to distract the animal from this behavior. Abnormal repetitive pacing may also be embellished with other abnormalities, such as head flagging/tossing, often at turning points. It is not necessarily associated with all stressful events as, unlike agitated locomotion, it does not increase in response to others' aggressive interactions (Poirier et al. 2019). Accurate definitions and classification of pacing behavior are important, if it is to be used as a meaningful indicator of welfare.

Abnormal locomotion responds well to enrichment, while it remains present (e.g., Bryant et al. 1988), but having a sufficiently large cage with furniture that can periodically be reconfigured helps prevent the onset of route-tracing and provides cognitive stimulation (Young 2003b; Honess 2017).

Hair-Pulling/-Plucking

It has already been highlighted that allogrooming plays a key role in the hygienic and social life of macaques. In certain circumstances, autogrooming, similar to self-scratching, can be seen as a displacement activity associated with anxiety (Troisi and Gabriele 1987; Maestripieri et al. 1992), but is also clearly hygienic. Autogrooming can be abnormally high when macaques are singly housed, but may decrease significantly once the animals are paired (*Mfa* 24%–10%: Line et al. 1990a; *Mfa* 26%–17%, *Mm* 10%–5%: Baker et al. 2012b) or given foraging mats or grooming boards (*Mm*: Schapiro et al. 1996).

Some propose that normal grooming may be pathologically intensified under stress and inappropriate housing conditions, resulting in hair-plucking with hair loss, which may be indicative of challenged welfare (Reinhardt 2005). Hair-plucking may be self-inflicted or imposed on social partners, with hairs being removed, singly or in clumps, by the mouth or fingers, and typically ingested (Reinhardt et al. 1986; Honess et al. 2005; Heagerty et al. 2017) with potential health consequences (e.g., trichobezoars: Mook 2002; Wilson et al. 2020).

Hair-plucking may result in diffuse, patchy, or extensive to total alopecia (Honess et al. 2005). It is important to distinguish between patchy hair coats resulting from hair-plucking, as opposed to other causes. There are many potential causes of baldness in monkeys, some of which may indicate other clinical challenges, and some of which may actually be natural (e.g., reproductive state, seasonal molt) (Novak and Meyer 2009). However, where plucking is involved, specific patterns of baldness may be indicative of self-plucking (arm, leg, and lower back patches), while other patterns may be indicative of plucking by another animal (patches in areas not reachable by the animal itself, e.g., middle and upper back, and the back of the head (Honess et al. 2005)). It may be difficult to identify the plucking culprit; however, by feeding each animal a different marker (e.g., nontoxic dye) and looking for hair in marked feces, identification may be possible (Honess et al. 2013b).

Alopecia can reach high levels in some facilities with, for example, 34%–86% of rhesus having hair loss, although only 0.6%–20% of animals were reported to hair-pluck (Lutz et al. 2013). Hair loss may be exhibited more in females (Honess et al. 2005; Lutz et al. 2013; Coleman et al. 2017), but hair-plucking may be performed more by males (Lutz et al. 2003b, 2013). It also appears to correlate with dominance rank, with lower ranked females being plucked more and having more alopecia (Reinhardt et al. 1986; Heagerty et al. 2017). Mothers may groom their infants to baldness, which resolves rapidly after weaning (Honess et al. 2013b).

Singly-housed macaques are at significant risk of developing hair-plucking, although those with more anxious temperaments may have the least alopecia (*Mm*: Lutz et al. 2003b; Coleman et al. 2017). There is evidence of species and age variation in alopecia, with pig-tailed macaques exhibiting more than either rhesus or long-tailed macaques, and hair-plucking more common in older animals (Crockett et al. 2007). There may also be an effect related to the animal supply source (Kroeker et al. 2017). Rhesus with extensive alopecia have been shown to have increased hair cortisol, which reduces when hair regrows (Novak et al. 2017b). While this may indicate changes in stress levels, alopecia has not been found to be linked with a range of other abnormal behaviors, including self-harm, abnormal repetitive locomotion, and floating limb (Rommeck et al. 2009).

Oversimplification of the relationship between hair-plucking and alopecia can lead to the misuse of alopecia as an index of welfare (Honess et al. 2005; Novak et al. 2017a). Complexities related to conclusively identifying the source of hair-loss and its apparent resistance to treatment in some contexts (*Mn*: Crockett et al. 2007) have led some to question the dedication of significant resources to attempts to resolve alopecia in the face of higher priorities for animal welfare (Crockett et al. 2007; Coleman et al. 2017). However, grooming, foraging, or visual barrier enrichment have been shown to reduce, or even eliminate, hair-plucking (Bayne et al. 1991; Beisner and Isbell 2008; Honess et al., in prep).

A variety of tools are available for the objective quantification of alopecia to facilitate colony-wide surveys and individual monitoring. These range in complexity from scoring just the dorsum (Honess et al. 2005), to whole-body scoring using a modified Rule of Nines (Baker et al. 2017; Luchins et al. 2011b), through to digital scoring of images (Novak et al. 2017b).

Saluting

Saluting can be defined as "Pressing a digit against the eye, usually on one constant side but occasionally bilaterally" (Erwin and Deni 1979, Honess 2013). A significant proportion (75%) of rhesus macaques that exhibited this behavior have ophthalmic lesions (Thomasy et al. 2020) and therefore, this behavior, if not causative of the lesions, may in part function as an attempt to improve vision. As with other abnormal behaviors, saluting can be performed in a ritualized, repetitive fashion and is considered common among socially deprived rhesus macaques, but has also been noted in group-reared long-tailed macaques (Erwin and Deni 1979). There does not appear to be a sex difference in the incidence of saluting but, like hair-plucking and SIB, it is more common in older animals (Lutz et al. 2003b).

Ways to Maintain Behavioral Health

Behavior, including changes in its form, frequency, duration, and intensity, is readily observable. It is a good reflector of internal state, and is therefore, useful for assessing the impact of housing, husbandry, and research protocols, and for demonstrating the welfare benefits of refinements (Young 2003b; Honess and Wolfensohn 2010; Winnicker and Honess 2014). There are natural behaviors that are generally undesirable in captivity; for example, it would be stressful to provoke antipredator behavior using a predator. On the other hand, providing opportunities for the naturalistic use of the environment, such as cage height and visual barriers for species-typical, vertical flight, and concealment responses, enables better coping with the stress of perceived and real threats (Waitt et al. 2008). Any opportunities that promote natural behavior must, of course, also be safe for animals, people, and research outcomes (Bayne 2005; Wolfensohn and Honess 2005). Not all devices will be used safely by all animals, all the time, and appropriate risk and harm:benefit assessments should be conducted during planning that involve care staff, behavioral managers, and veterinarians. It is not possible to keep all animals free from injury, all of the time. To do so would mean retreating to single housing in small metal cages with grid floors and no furniture or devices. Such accommodations might be safe, but animal welfare would be very poor, science would suffer, and it would not be ethically acceptable today.

There is a close association between behavioral and physical health (e.g., Schapiro and Bushong 1994; Bayne and Novak 1998; Novak et al. 2017a; Shively 2017). It follows that working to ensure both psychological and physical health is important for securing overall welfare (Hutchinson 2017). An example is provided by chronic diarrhea in macaques, which can be debilitating, dramatically retard the growth of infants (Haertel et al. 2018), and can even result in death. However, while diarrhea is often associated with a range of pathogens, including viruses and bacteria (Wang et al. 2019), in many cases pathogens are not evident (idiopathic diarrhea) (Russell et al. 1987; Wolfensohn 1998), and stress may be the cause (Schapiro and Bushong 1994; Brady and Morton 1998). Stress is further implicated in the occurrence of diarrhea, as there appear to be both strong temperament effects (more nervous/anxious animals being more prone to chronic diarrhea) and stressful event-related effects (e.g., multiple housing changes (Elfenbein et al. 2016; Gottlieb et al. 2018)). Interestingly, infants born to mothers that were housed outdoors or experienced a range of minor stressors during pregnancy, appear to be protected against the risk of diarrhea, with the gut microbiome likely playing an important role (Elfenbein et al. 2016).

Diarrhea represents just one area where there is an interaction between behavioral and physical health, but it highlights the need for a holistic approach to welfare (Hutchinson 2017). Veterinarians and behaviorists should work closely to achieve overall welfare, both because it is ethically important and because it aims to produce animals that are fit for the purpose for which they are being kept; as productive breeders or as valid research models. This section discusses the main areas in which, through more finely-tuned environments and better occupied animals, behavioral health can be improved and maintained.

Captive Considerations to Promote Species-Typical Behavior

There are a range of options available to the practitioner in the design and equipping of captive macaque housing that have proven ability to promote species-typical, and reduce abnormal, behavior, and both functions should be given equal consideration. Whenever behavioral management strategies to enrich or refine the environment are being considered, the right question must be asked: "What behavior is exhibited that we do not want, and/or what behavior is missing that we do want?" Importantly, enrichment strategies must be assessed for safety and practicality prior to implementation, and must then be demonstrated to be effective utilizing objective assessments involving appropriate hypotheses, study designs, and research tools (Young 2003a; Winnicker and Honess 2014).

Apart from the refinements covered below, a wide variety of enrichment options is reviewed extensively elsewhere (e.g., Reinhardt et al. 1988; Chamove 1989; Fajzi et al. 1989; Reinhardt 1993a; Reinhardt and Roberts 1997; Honess and Marin 2006b). Many enrichment options, including some off-the-shelf devices, lack empirical evidence that they conclusively promote natural behavior or reduce diagnosed abnormal behavior (Line et al. 1991; Winnicker and Honess 2014). Without evidence of efficacy, attaching such devices to the outside of cages or scattering them inside may address human objectives, but may result in little more than cage decoration and be of minimal benefit to the animals. Enrichment devices may have an initial novelty effect that rapidly diminishes. Novelty may stimulate initial interest in some animals, but may be stressful for those that are neophobic or inhibited. In addition to individual temperamental differences, there are also species, experience, age, and origin differences in macaques' willingness to approach and explore novel objects (e.g., Line et al. 1991; Lloyd et al. 2005; Honess and Marin 2006b; Coleman 2017; Fischer 2017). As macaques readily habituate to enrichment devices, especially manipulable objects (toys) (Line et al. 1991), imaginative ways need to be found to make them enduringly appealing, such as by turning them into foraging devices (e.g., frozen juice in Kong toys), or by regularly rotating different devices among animals (Crockett et al. 1989). In addition to limited effectiveness, commercially available enrichment devices can be expensive, and are often designed to meet the behavioral needs of other species (e.g., Kong toys, without macaque-appropriate food fillings, to promote species-typical chewing in dogs). Effective enrichment devices can be produced cheaply in-house (Schapiro et al. 1995), allowing fine-tuning to the specific needs of individuals, husbandry procedures, and/or research contexts.

Other forms of enrichment, such as videos (e.g., cartoons, natural history documentaries), music, and computer games, have been shown to have some positive effects on behavior, but, overall, do not compare well to other forms of enrichment for reducing stress, aggression, and abnormal behavior (Schapiro et al. 1996; Honess and Marin 2006b). However, it is likely that the beneficial effects vary with the engagement, temperament, and experience of the individual, as well as the specifics of their content. For example, singly housed male rhesus macaques that were temperamentally inhibited did not interact with tablet devices (e.g., iPads) (Coleman 2017).

Music may have its strongest influence on staff welfare, and as a means to mask environmental noise generated by husbandry activities (e.g., cage cleaning), which can adversely affect animals (Turner et al. 2007). Music and other sounds may also have an effect on cognitive functions, which may confound results in some studies (Carlson et al. 1997; Zarei et al. 2019). While macaques may be attentive to visual and auditory enrichment, evidence indicates that where they have control of the equipment, engagement and behavioral benefits are likely to increase (Honess and Marin 2006b). However, it is likely that the ability of the animals to exert some control over part of their environment is the primary enriching aspect of such systems (Bassett and Buchanan-Smith 2007).

Materials for cage structures need not be restricted to plastics and metals. Wood is natural, thermoneutral, cheaply replaceable, sanitizable, and is suitable for the construction of manipulanda and structures (Eckert et al. 2000; Waitt et al. 2010; Luchins et al. 2011a); however, the use of wood can result in the ingestion of splinters (Eckert 1999; Hahn et al. 2000). Sticks are more attractive toys than some other options (e.g., nylon balls), although most toys have few beneficial effects on abnormal behavior (Line and Morgan 1991; Line et al. 1991). Types of wood that are toxic or that may confound study results must be avoided (Eckert et al. 2000).

Many varyingly successful enrichment options have been tried, but the most important enrichment option for macaques is at least one compatible conspecific (Schapiro et al. 1996).

Environment

Cage Size

Perhaps the two most significant factors that influence captive behavior are cage size and access to social partners. Cage size determines the space in which the animal can move, including the ability to retreat from perceived threats, as well as influencing behavioral managers' ability to include conspecifics, appropriate structures, and devices intended to elicit natural behavior, allow concealment, and occupy the animal. Providing compatible social partners is only part of creating a behaviorally healthy environment. Historical aspects of macaque housing design and the high cost of replacing existing caging systems mean that many facilities still house their macaques in caging that was (a) not designed for social housing, and (b) engineered to be structurally robust, rather than functionally flexible. Most facilities are able to provide a reasonable level of welfare for their animals *in spite* of the caging they have available. Traditional macaque cages were little more than small metal box cages designed to house, and provide ready access to, a single animal that could be sanitized using industrialized, staff-efficient processes. Minimum cage dimensions (according to regulations) around the world appear to be divisible into "European" and "US" sizes, with European dimensions considered "expansive" and US dimensions considered "conservative". A number of reviews of cage size and space provision for primates have been published (e.g., Reinhardt et al. 1996; Buchanan-Smith et al. 2004; Honess and Marin 2006b).

Increasing space, on its own, does little, other than create a bigger empty space, but may improve locomotion, and potentially decrease abnormal behavior (Crockett et al. 1995). However, unless the opportunity is taken to increase complexity, in addition to increasing space, the benefits are often ambiguous (Line et al. 1990b; Honess and Marin 2006b). Despite financial and practical challenges, there are examples of expansive modern laboratory caging systems that have been designed from scratch, rather than through modification of traditional cages. These systems aim to maximize safe social housing, facilitate research and veterinary access to the animals, and provide suitable space for devices and structures that elicit more species-typical behavior and promote coping. These may include, perches, visual barriers, swings, and platforms (see Figure 28.2). Enlarging cages and modifying infrastructure (e.g., drainage systems that allow the use of foraging substrates), contrary to conventional wisdom, can actually increase the number of animals that can be housed at a facility, while reducing cleaning time and costs (Wolfensohn and Honess 2005; Bennett et al. 2010). In expansive outdoor enclosures, enrichment structures can be arranged imaginatively to mimic aspects of the structural and functional complexity of a macaque's natural habitat (e.g., Waitt et al. 2008; Fernandez et al. 2017; see Neal Webb et al. 2018 for a discussion of Functionally Appropriate Captive Environments).

Most macaques, except some pig-tailed and stump-tailed macaque populations, spend the majority of their time in trees, at heights above those provided in many laboratory or breeding environments. Wherever height is provided, macaques will take advantage of it in species-appropriate ways. For example, rhesus macaque groups housed in a 2 m high cage spent 90% of the time off the floor, with height use reflecting natural patterns; higher levels were used more by dominant animals, and vertical flight was used in response to fearful stimuli (Reinhardt 1992b). A similar pattern of preference for height is reported for pig-tailed, bonnet, and long-tailed macaques (Rosenblum et al. 1964; Waitt et al. 2008). Appropriately positioned perches also facilitate dominance spacing and reduce aggression (*Mfu*: Nakamichi and Asanuma 1998).

Larger cages also allow for the inclusion of more visual barriers, which have benefits (e.g., aggression reduction) for macaques housed in groups (*Mfa*: Waitt et al. 2008; *Mm*: Westergaard et al. 1999; McCormack and Megna 2001, *Mn*: Maninger et al. 1998) or pairs (Reinhardt and Reinhardt 1991; Basile et al. 2007). Additional visual barriers help reduce aggression when introducing animals or forming new groups (Westergaard et al. 1999), and may increase breeding contributions from lower ranked males who can mate out of sight of dominant males that would otherwise disrupt their mating attempts (Baker et al. 1987; Waitt et al. 2008).

Simply reconfiguring, not enlarging, group cages, by halving floor area and doubling height, paired with additional visual barriers, can dramatically reduce aggression and anxiety (*Mfa*: Waitt et al. 2008), by promoting more species-typical affiliation, vertical flight reactions, vertical dominance spacing, and subordinate male reproduction.

Social Groupings

In 2003, approximately 75% of laboratory primates housed in 22 facilities in the USA were kept socially, with most housed outdoors, and approximately 50% (about 2500) of those housed inside remained singly housed (Baker et al. 2007). Single housing is a major risk factor for abnormal behavior ("Sources of Variation" section), deprives animals of important buffering benefits (Vogt et al. 1981; Smith and Peffer-Smith 1982; Gilbert and Baker 2011), and impacts a range of physiological systems (e.g., immune: Schapiro et al. 2000), with even temporary separation from a cage mate causing stress (Hannibal et al. 2018). Regulations requiring default social housing have resulted in renewed efforts to provide compatible partners for animals, especially through pairing (see McGrew 2017; Truelove et al. 2017). An increasingly small number of facilities remain reluctant to socially house their macaques, particularly male rhesus, largely due to an aversion to dealing with the consequences of aggression as a result of social incompatibility. Separating animals and treating wounds from fights may mean that subjects must be temporarily or permanently withdrawn from study. The "benefits" of single housing, in terms of reduced wounding and staff/study convenience, must be balanced against the psychological health and model validity "costs" of social isolation (Honess and Marin 2006a; Baker et al. 2012a; Novak et al. 2017a).

FIGURE 28.2 (a) An expansive indoor, experimental housing system incorporating vertical height, forage substrate, and a variety of wooden perches to minimize stress and promote species-typical behavior. Animals can use horizontal bars for climbing and easier lateral visual tracking. (Photos: Andrew Winterborn, Queen's University, Ontario, Canada.)

(Continued)

FIGURE 28.2 (CONTINUED) (b) An expansive outdoor, breeding housing system incorporating vertical height, high social diversity, and variety of internal structures that mimic the complexity of the natural environment to provide cognitive stimulation, minimize stress, and promote species-typical behavior. (Photo: Bioculture, Mauritius.)

Species-appropriate social housing remains a critical component of good welfare for macaques (Honess and Marin 2006b). Social housing with a compatible partner(s) improves welfare, regardless of the sex (Eaton et al. 1994; Doyle et al. 2008) or dominance status (Abbott et al. 2003; Baker et al. 2012a) of the animals. Wild macaques make considerable investments in forging and maintaining high-quality, mutually beneficial social relationships, through engagement in affiliative behaviors and the resolution of conflicts. Opportunities to perform these important natural behaviors should be fostered in captivity.

Living in a compatible pair, especially with full, rather than just protected, contact, meets some of the basic social needs of macaques and can reduce abnormal behavior (Weed et al. 2003; Baker et al. 2012b). However, living in a pair does not provide the same level of social diversity, cognitive stimulation, and emotional support that is present in a macaque social group. In groups, any given animal is likely to be able to find at least one "friend", with such complex, multigenerational social environments producing more socially competent offspring (Wooddell et al. 2019b). Group housing also provides more species-typical play opportunities and results in less SDB in young macaques (Schapiro et al. 1996).

One of the most successful pairing strategies is to form mixed-sex pairs of adults, although contraception or sterilization may be required (Statz and Borde 2001; Weed et al. 2003). When creating same-sex (isosexual) pairs, simply familiarizing potential partners before pairing may not promote success (Bernstein 1991). A variety of systematic approaches seek to determine compatibility prior to pairing, with the goal to increase successful outcomes (Truelove et al. 2017). Approaches to assess probable compatibility prior to actual pairing include examinations of aggression levels (Line et al. 1990a), partner preferences (Crockett et al. 1994), unidirectional displays of dominance (Reinhardt et al. 1995b), and assessments of temperament (Coleman 2012; Capitanio et al. 2017; McGrew 2017). Retrospective analyses of compatibility in rhesus in conjunction with temperament assessments conducted when the animals were 3 months of age, found that females that were similar in temperament were more likely to be compatible with one another, while the most compatible pairs of males included animals that differed in body weight and had specific temperamental characteristics (Capitanio et al. 2017).

Reported success rates for enduring pair compatibility in macaques are high. In rhesus macaques, there is a high success in isosexual adult pairings of both females (80–92%) and males (83–87%) (Reinhardt 1992a; Eaton et al. 1994; Reinhardt 2002a; Coleman 2017; MacAllister et al. 2020). Pairing either sex with infants increases success rates still further (female-infant: 94%, male-infant: 92%: Reinhardt 2002a). In long-tailed macaques, isosexual pairs of both adult females (83–100%) and juveniles (87%) are very successful, though adult male pairing appears more challenging (sometimes as low as a 50% success rate) (Line et al. 1990a; Crockett et al. 1994; McGrew 2017). Even when pairs "fail" (i.e., do not meet criteria of successful pairing (e.g., Baker et al. 2017)), rates of serious wounding can be as low as 0.8% (Reinhardt 1994).

Success, when defined as a pair remaining together without significant aggression, may not prove a sufficient indicator of relationship quality and welfare value. Where 92% of 37 adult female rhesus pairings were judged successful, closer examination found four successful pairs to be incompatible based on observed levels of separation and aggression similar to that seen in failed pairings (Coleman 2017). Behavior should continue to be monitored even after apparently successful pairing, as displays of submissive, rather than just affiliative, behavior predict lower levels of wounding (*Mm:* Pomerantz and Baker 2017). Physiological measures may, however, provide less ambiguous indications of compatibility; for example, levels of the hormone oxytocin, known to promote prosocial behavior, are significantly positively correlated within compatible pairs of male rhesus macaques (Berg et al. 2019).

Securing the social welfare of macaques starts where they are bred. Management systems that mimic those in the wild ensure the best welfare for mothers and infants, as unnatural practices, such as nursery-rearing and hand-rearing, can be problematic for welfare and future social behavior (e.g., Timmermans and Vossen 1996; Bayne and Novak 1998; Goin and Gust 1998; Bastian et al. 2003; Rommeck et al. 2011). Nursery-reared infants also differ in gut microbiome compared with those reared in social groups (*Mm:* Dettmer et al. 2019). Lone infants resulting from maternal death or infant rejection can be successfully managed using proactive strategies for adoption by nursing females that may cope with two infants or that have recently lost their own infant (Honess et al. 2013a). "Weaning" in a captive management context typically refers to the physical removal of the infant from its mother and out of her cage/enclosure. Weaning infant macaques too early has negative effects on the welfare of both mothers and infants, as well as on productivity (Champoux et al. 1989; Reinhardt 2002b; Prescott et al. 2012). Infants should be removed from their mothers no earlier than at about 10–14 months of age (and at an appropriate body weight), as this is the point at which nutritional weaning is generally achieved in the wild (Wolfensohn and Honess 2005) (see Table 28.2). Where possible, infants from the same natal group should be kept together upon weaning, as they have established dominance relationships, alliances, and compatibility (Honess et al. 2015). Serious consideration should be given to retaining some daughters in the breeding group, to serve as future breeders; again in an attempt to mimic natural patterns. Allowing daughters to remain in their natal group also provides insurance against natural group attrition, provides daughters with opportunities to learn parenting skills and to assist female relatives in dominance and caregiving interactions (Levallois and Desvaux de Marigny 2015). Successful group building with retained daughters requires considerable planning and monitoring. Where possible, daughters of higher ranking females should be preferentially retained, as they are likely to receive considerable support from their powerful mothers. Also, breeding males should be rotated out of groups every few years to avoid inbreeding. New males should only be introduced to breeding groups when they are at least 3.5 years of age and when there are as few infants in the group as possible, as introductions can result in infanticide (Clarke et al. 1995b; Rox et al. 2019). Such introductions are more successful when there are a small number of large matrilines in the social group (Rox et al. 2019).

Compatible animals should be kept together at the breeder as they are prepared for selection for client supply, notwithstanding other constraints, such as body weight, sexual maturity, and study-related genetic, physiological or behavioral characteristics. Responsible breeders should always supply compatibility information with their animals, and such information must reach those involved in pairing or grouping the animals prior to quarantine, and preferably upon arrival. Good communication between end-users and breeders also allows animals to be selected prior to shipment for inclusion in specific experimental groups and to be compatibly housed accordingly (Fernandez et al. 2017). Reforming pairs or groups in each new context makes no sense for welfare, and is costly and time-consuming.

Feeding Strategies

Feeding can take up to 60% of a wild macaques' time budget. Finding imaginative ways to encourage captive animals to work for their food and to extend feeding is important for naturalizing activity budgets and preventing boredom and obesity (Honess and Marin 2006b). Presenting food in easy to access containers means that animals can rapidly acquire their daily food ration, leaving considerable "empty" time that is likely to contribute to boredom and create opportunities for abnormal behavior (Reinhardt 1993a; Honess and Marin 2006b, Schapiro this volume). All too often, those macaques that are housed in impoverished environments have little else to occupy themselves, other than to eat. While manufactured diets are balanced and nutritionally complete, they are also quite uniform and not particularly palatable. Nevertheless, bored animals, with *ad lib* access to food, may have little else to do and can just sit at a food hopper and eat; the associated reduced activity is likely to result in unhealthy increases in weight (Bauer et al. 2012).

In the wild, unprovisioned populations of macaques are generally lean, with low percentages of body fat compared to their single-housed, captive conspecifics (2% v 8%, *Mfa, Mm*: Sterck et al. 2019). In a survey of 13 facilities, up to 20% of macaques were overweight (*Mfa:* Bauer et al. 2010). This was not necessarily related to housing condition, as up to 23% of group-housed animals were above normal body weight in another study (*Mfa, Mm*: Sterck et al. 2019). Dominance, however, does appear to be a factor in maintaining normal body weights (Bauer et al. 2012). While most facilities implement weight-monitoring and/or calorie-control strategies to reduce or prevent obesity (Bauer et al. 2010), some do feed *ad libitum*, and there is considerable variation in the strategies used to increase activity and in the use of high-calorie or fatty foods (e.g., peanut butter) as supplements, treats, or training rewards. All food items provided and/or consumed should be accounted for in the animal's daily ration. Particular care needs to be taken with training treats and off-the-shelf forage mixes, which often contain high calorie/fat items, such as nuts and sugar-coated dried fruit (Honess 2017).

Time-consuming food processing and extractive foraging tasks can be created that promote species-typical behavior and fill unoccupied time that may otherwise be available for engaging in abnormal behavior (Reinhardt 1993b; Murchison 1994; Reinhardt and Roberts 1997; Honess and Marin 2006b). Simple options include food that needs processing (e.g., unhusked corn: Beirise and Reinhardt 1992) or is challenging to access, such as food provided on the mesh of the cage roof (Reinhardt 1993b), or in puzzle feeders (Murchison 1994), forage substrate (Anderson and Chamove 1984; Bryant et al. 1988; Boccia and Hijazi 1998), forage boards/mats/racks (Bayne et al. 1991; Lutz and Novak 1995; Schapiro et al. 1996), and artificial turf (Bayne et al. 1992a). Different foraging enrichments may have different effects on rates of abnormal behavior. Some (e.g., supertubes) may result in decreases in abnormal behavior, whereas others (e.g., shakers) may increase it, and still others (e.g. puzzle balls) may initially yield decreases, then ultimately, increases in abnormal behavior (Gottlieb et al. 2011). These differences may be related to frustration arising from puzzle difficulty or the exhaustion of treats in less challenging tasks. Some beneficial options may nevertheless have potential practical or health-related challenges. For example, forage substrate has resolvable drainage (Bennett et al. 2010) and rare ingestion (Seier et al. 2005) consequences, but clear benefits related to reductions in abnormal behavior (Lam et al. 1991; Bayne et al. 1992a; Beisner and Isbell 2008).

Fresh produce, when appropriately distributed, spatially and temporally, can be sensorially stimulating and nutritionally valuable, and can encourage species-typical food quality assessment and processing, and reduce aggression trauma in macaque groups (Honess and Marin 2006b; Honess 2017; Wooddell et al. 2019a). It should be noted that most produce is selectively cultivated for human tastes, and often has higher caloric and lower fiber content than wild equivalents (Honess 2017). If browse is to be used, care needs to be exercised to avoid plants/species that may be toxic or confound experimental results (Gilbert and Wrenshall 1989; Shumaker and Woods 1995; Eckert et al. 2000; Bayne 2003; Schrier 2004). Some colony managers feel that feeding produce can be problematic, fearing that macaques will consume it preferentially over a balanced pelleted formulation, potentially resulting in nutritional deficiencies (Plesker and Schuhmacher 2006; Schultz 2017). Many facilities feed the pelleted diet in the morning, when animals are most hungry, and therefore, most likely to eat it, and then feed produce in the afternoon.

It is important to consider species, group, and individual temperaments when designing feeding regimens. Clumped food distribution can cause aggressive competition and food monopolization by dominant individuals (*Mfa*: Bauer et al. 2010; Waitt et al. In prep; *Mm*: Southwick 1967; Gore 1993). If dominant individuals exhibit excessive aggression in feeding contexts, then food should be scattered and sufficient feeding enrichment devices physically spaced to reduce the potential for aggressive monopolization (Honess and Marin 2006b). Surprisingly, reducing food availability may not automatically elevate aggressive behavior, as even a 50% reduction of food may halve aggressive events, with access to the scarce food being determined by established dominance hierarchies, with lowest ranked animals getting little to eat (Southwick 1967). While not a recommendation to reduce food provision, this reflects how macaques are able to use established dominance hierarchies to manage situations that otherwise might result in the elevation in aggressive competition.

Training

Macaques are cognitively sophisticated and highly social, making them ideal for training for a range of activities and tasks. Outside of laboratories, macaques are trained to harvest coconuts (*Mnn*: Bertrand 1967) and for use in entertainment (Agoramoorthy and Hsu 2005). In laboratories, humans are an inevitable part of macaques' environment and they will readily form relationships with care and research staff. It is important for animal welfare and research progress that these are positive relationships (Reinhardt 1997; Waitt et al. 2002). Positive relationships, created by actions, such as using hand-feeding to familiarize the animals to humans, will reduce aggression to staff, reduce time taken for research procedures, minimize model-challenging stress, improve data quality, and form a foundation for further training (Bayne et al. 1993; Fernandez et al. 2017; Graham 2017). If macaques are to be trained, systematic approaches using operant conditioning principles, such as positive reinforcement training (PRT), are likely to prove most successful (Perlman et al. 2012).

In many countries, PRT is promoted as best practice; both the NRC *Guide for the Care and Use of Laboratory Animals* & EU Directive (EU 2010; National Research Council 2010) recommend or require its use to train animals for cooperation with husbandry and research. Training is an important tool for gaining animal cooperation in their care and management (Luttrell et al. 1994; Knowles et al. 1995; Laule et al. 2003; Perlman et al. 2012), including for veterinary procedures (Magden 2017) and research protocols (Schapiro et al. 2005; Graham 2017; Mason et al. 2019). PRT can be highly successful: 90% of study animals were trained for cooperative handling, thereby reducing the need for the use of stressful physical and chemical restraint techniques (*Mfa:* Graham et al. 2012), which can produce changes that confound research results (Reinhardt et al. 1995a; Iliff et al. 2004; Lambeth et al. 2006; Graham 2017).

As PRT is generally positive and stimulating for those animals that readily engage, it is also enriching, reduces aggression, and enables close monitoring of the animals for health/behavioral problems (Heath 1989; Schapiro et al. 1995; Laule and Desmond 1998; Perlman et al. 2012; Westlund 2014; Westlund 2015). PRT can also be used to improve affiliation and cooperation (Schapiro et al. 2001; Laule and Whittaker 2007), and to reduce abnormal or unwanted behavior (Bloomsmith et al. 2007; Coleman and Maier 2010), including the aggressive exclusion of subordinates from feeding opportunities (Laule and Desmond 1998; Perlman et al. 2012).

While PRT remains the "gold standard" in animal training, it is often the case that negative reinforcement, used minimally, may be required to achieve certain challenging goals or to train animals of certain temperaments (Perlman et al. 2012; Wergård et al. 2015; Westlund 2015; Graham 2017). It has been shown that animal temperament correlates with training success (Coleman et al. 2005; Wergård et al. 2016). For example, less than half of female rhesus classed as "inhibited" trained to the simplest goal (a clicker) and very few to a more challenging goal (a target), whereas 82% of "exploratory" animals reached the more challenging goal (Coleman et al. 2005). Furthermore, only 45% of "inhibited" female long-tailed macaques successfully trained to present for vaginal swabbing, compared to 98% of "exploratory" animals (Coleman 2017). Additional factors, such as animal experience, age, and time in the facility, may also influence training success (Mason et al. 2019).

Training is likely to be most successful when the right motivating reward is identified (Graham 2017). While this is commonly a food or drink treat, other rewarding and valuable experiences can be effective reinforcers, including opportunities to view a females' rear-end (male *Mm*: Deaner et al. 2005) and to engage in grooming behavior with a human (*Mm*: Taira and Rolls 1996). The selected reward may need to be varied, in quantity, nature, or schedule, depending on the difficulty, motivation, or technical issues associated with the required task (Deaner et al. 2005; Prescott et al. 2010). All training-related preferences, resource use, and success, as well as positive behavioral and physiological effects on the animals should be monitored and recorded for evaluation and refinement (Laule et al. 2003; Schapiro et al. 2003; Perlman et al. 2012).

Special Situations

There are a range of circumstances that represent particular challenges to the welfare of macaques in laboratories. Among the most important are:

- Single housing associated with social incompatibility, veterinary care, or research constraints
- Water control paradigms often related to neurophysiological research
- High-level biological containment for some infectious disease models
- Housing of animals with special behavioral needs, e.g., geriatric animals in aging studies.

Given the negative impact of single housing and the importance of positive social behavior for psychological well-being and coping, all efforts should be made to ensure that no macaque is singly housed if it need not be. If single housing is unavoidable, it must be performed with appropriate regulatory consent (e.g., approval from Animal Care and Use Committees), closely monitored, of minimum duration, involve appropriate enrichment, and reintegration to a social setting must occur at the earliest opportunity.

The case is often made that water control paradigms are essential for certain protocols, especially in neuroscience. It may be based on the large number of trials (c. 700–1,000 per day) required in single-cell recording, where standard food rewards would result in rapid satiation. Some researchers feel that controlling access to water (i.e., starting experimental sessions when animals have not had access to water for a while) is required to sufficiently motivate the animal to do the work (Prescott et al. 2010). Such regimens are considered controversial due to their potential negative impact on animal welfare and the claim that more positive motivational tools are available (Westlund 2012). Refining research protocols to minimize adverse impacts on welfare, while still meeting research goals, is a key part of implementing the 3Rs (Russell and Burch 1959).

The avoidance of cross-contamination or pathogen escape may require that macaques be housed in containment facilities that inevitably restrict enrichment. This may also be the case in studies involving the use of genetically modified primate models (Park and Silva 2019). However, it has been shown that many of the social and other behavioral needs of macaques can be satisfied in both toxicological (Bayne 2003), and high containment environments (Dennis 2009).

Thanks to modern medical progress, to which macaques have made critical contributions, humans live longer. This has resulted in an increasing interest in studies of aging-related conditions that require older macaques. Due to their restricted mobility, older, especially geriatric macaques, need to be provided with specialized environments. Furniture should be fixed, rather than mobile, and height transition structures (ladder rungs, distance between runners and platforms) should cover shorter distances, to enable easier movement (Waitt et al. 2010). Consideration should also be given to other comfort-related measures, such as the use of more thermoneutral materials and the provision of resting sites with additional and/or softer bedding. While it is important to keep such animals active and in supportive social settings, enrichment that makes food unnecessarily physically difficult to attain should be avoided.

Conclusions/Recommendations

There is a significant body of published evidence and considerable accumulated expertise that can successfully guide attempts to maintain high standards of welfare for macaques in laboratories. Considerable in-roads have been made, with regulatory support, toward addressing single housing, which has been the root of many behavioral and stress-related problems in captive macaques. In some regions (e.g., Europe), and increasingly in the USA, caging systems have been developed that can accommodate a range of behavior-enhancing enrichment options that provide additional space and opportunities to engage in natural macaque behavior. While social housing and cage size are important factors in providing for the animals' well-being, the animals' welfare must be considered holistically; while aiming to further improve all aspects of their health and environment, in terms of compatible conspecifics, space, furniture, substrate, diet, and sensory stimuli (Honess 2017). With more focus on the development, use, and scientific assessment of enrichment for its therapeutic value and its promotion of species-typical natural behavior, significant additional progress will be possible. Such progress would improve animal welfare, the quality of scientific data, and the ethical case for the responsible participation of macaques in research.

For macaques to be most valuable in biomedical research, they need to be as biologically natural as possible; living in an environment and under management systems where inappropriate stress is minimized and evolved coping mechanisms are enabled (Wechsler 1995; Fernandez et al. 2017). Naturalistically functional environments paired with compassionate, attentive, skillful, and informed management enable animals to be behaviorally healthy and natural, as well as making them easier to manage (Novak et al. 1994; Snowdon 1994; Fernandez et al. 2017; Honess 2017; Neal Webb et al. 2018). The essence, or *telos*, of a macaque is the sum of all its evolved life processes that enable it, through adaptation, to be successful in its natural environment (Rollin 2015). Preserving this essence, the "monkey-ness of the monkey", will maximize the value of the animal as a research model. Maximizing the evolutionary "fit" of the environment to the animal, by designing-in elements that mimic the function of natural features (*evolutionary accommodation*) will help reduce dysfunction, maladaptation, and the collapse of the animal's ability to cope. An expert primate behavioral scientist's advanced understanding of natural behavior and its adaptive function is critical in this process. Some research and veterinary procedures directly challenge the implementation of the concept of evolutionary accommodation and it is important for multidisciplinary teams, including primate behavioral scientists, care staff, veterinarians, researchers, and managers, to work together, making the best use of internal and external expertise to plan the best care for the animals. This will support safe, naturalistic environments and humane management regimens, while still enabling the cost-effective and ethically acceptable achievement of research goals (Bayne 2005; Weed and Raber 2005; Honess 2017).

ADDITIONAL RESOURCES

American Society of Primatologists: The Welfare of Primates: https://asp.org/welfare/.

Animal Welfare Institute, Refinement Database: Database on refinement of housing, husbandry, care, and use of animals in research. https://awionline.org/content/refinement-database.

National Centre for the Replacement, Refinement and Reduction of Animals in Research (NC3Rs), The Macaque website: covers many aspects of the behavior and welfare of macaques in research settings. https://www.nc3rs.org.uk/macaques/.

Primate Info Net (PIN): Species fact sheets covering taxonomy, morphology, ecology, behavior, conservation of macaques: http://pin.primate.wisc.edu/factsheets/links/macaca.

US Department of Agriculture, National Agriculture Library, Environmental enrichment for nonhuman primates resource guide: macaques. https://www.nal.usda.gov/awic/environmental-enrichment-nonhuman-primates-resource-guide-macaques.

REFERENCES

Abbott, D. H., E. B. Keverne, F. B. Bercovitch, C. A. Shively, S. P. Mendoza, W. Saltzman, C. T. Snowdon, T. E. Ziegler, M. Banjevic, T. Garland and R. M. Sapolsky. 2003. Are subordinates always stressed? A comparative analysis of rank differences in cortisol levels among primates. *Hormones and Behavior* 43(1): 67–82.

Agoramoorthy, G. and M. J. Hsu. 2005. Use of nonhuman primates in entertainment in Southeast Asia. *Journal of Applied Animal Welfare Science* 8(2): 141–149.

Altmann, S. A. 1962. A field study of the sociobiology of rhesus monkeys, *Macaca mulatta*. *Annals of the New York Academy of Sciences* 102(2): 338–435.

Anderson, J. R. and A. S. Chamove. 1980. Self-aggression and social aggression in laboratory-reared macaques. *Journal of Abnormal Psychology* 89(4): 539–550.

Anderson, J. R. and A. S. Chamove. 1984. Allowing captive primates to forage. *Standards in Laboratory Animal Management, Part 2*. Potters Bar, UK, Universities Federation for Animal Welfare, UFAW: 253–256.

Anderson, J. R., P. Peignot and C. Adelbrecht. 1992. Task-directed and recreational underwater swimming in captive rhesus monkeys (*Macaca mulatta*). *Laboratory Primate Newsletter* 31(4): 1–5.

Anderson, J. R., A. Rortais and S. Guillemein. 1994. Diving and underwater swimming as enrichment activities for captive rhesus macaques (*Macaca mulatta*). *Animal Welfare* 3(4): 275–283.

Animal Welfare Act 2006. Legislation.gov.uk. 2015. [online] Available at: http://www.legislation.gov.uk/ukpga/2006/45/section/9.

Aureli, F. 1992. Post-conflict behavior among wild long-tailed macaques (*Macaca fascicularis*). *Behavioral Ecology and Sociobiology* 31(5): 329–337.

Aureli, F., M. Das and H. C. Veenema. 1997. Differential kinship effect on reconciliation in three species of macaques (*Macaca fascicularis*, *M. fuscata*, and *M. sylvanus*). *Journal of Comparative Psychology* 111(1): 91–99.

Aureli, F. and C. P. van Schaik. 1991. Post-conflict behavior in long-tailed macaques (*Macaca fascicularis*). 1. The social events. *Ethology* 89(2): 89–100.

Baker, K., R. Bellanca, M. Bloomsmith, K. Coleman, C. Crockett, A. Maier, B. McCowan and J. Perlman. 2009. Behavioral contrasts between Chinese-origin and Indian-origin rhesus macaques (*Macaca mulatta*) in a caged laboratory setting. *American Journal of Primatology* 71: 32.

Baker, K. C., M. A. Bloomsmith, K. Coleman, C. M. Crockett, J Worlein, C. K. Lutz, B. McCowan, P. Pierre and J. Weed. 2017. *Handbook of Primate Behavioral Management*, S. Schapiro. London, CRC Press, Taylor & Francis, 9–23.

Baker, K. C., M. A. Bloomsmith, B. Oettinger, K. Neu, C. Griffis, V. Schoof and M. Maloney. 2012a. Benefits of pair housing are consistent across a diverse population of rhesus macaques. *Applied Animal Behaviour Science* 137(3–4): 148–156.

Baker, K. C., C. M. Crockett, G. H. Lee, B. C. Oettinger, V. Schoof and J. P. Thom. 2012b. Pair housing for female longtailed and rhesus macaques in the laboratory: behavior in protected contact versus full contact. *Journal of Applied Animal Welfare Science* 15(2): 126–143.

Baker, K. C., J. L. Weed, C. M. Crockett and M. A. Bloomsmith. 2007. Survey of environmental enhancement programs for laboratory primates. *American Journal of Primatology* 69: 377–394.

Baker, S., D. Estep and P. Walters. 1987. The effect of visual barriers on sexual behavior of male stumptail macaques. *American Journal of Primatology* 12(3): 328–328.

Balasubramaniam, K. N., P. R. Marty, M. E. Arlet, B. A. Beisner, S. S. K. Kaburu, E. Bliss-Moreau, U. Kodandaramaiah and B. McCowan. 2020. Impact of anthropogenic factors on affiliative behaviors among bonnet macaques. *American Journal of Physical Anthropology* 171(4): 704–717.

Barton, R. 1985. Grooming site preferences in primates and their functional implications. *International Journal of Primatology* 6(5): 519–532.

Basile, B. M., R. R. Hampton, A. M. Chaudhry and E. A. Murray. 2007. Presence of a privacy divider increases proximity in pair-housed rhesus monkeys. *Animal Welfare* 16(1): 37–39.

Bassett, L. and H. M. Buchanan-Smith. 2007. Effects of predictability on the welfare of captive animals. *Applied Animal Behaviour Science* 102: 223–245.

Bastian, M. L., A. C. Sponberg, S. J. Suomi and J. D. Higley. 2003. Long-term effects of infant rearing condition on the acquisition of dominance rank in juvenile and adult rhesus macaques (*Macaca mulatta*). *Developmental Psychobiology* 42: 44–51.

Bauer, S. A., K. E. Leslie, D. L. Pearl, J. Fournier and P. V. Turner. 2010. Survey of prevalence of overweight body condition in laboratory-housed cynomolgus macaques (*Macaca fascicularis*). *Journal of the American Association for Laboratory Animal Science* 49(4): 407–414.

Bauer, S. A., D. A. Pearl, K. E. Leslie, J. Fournier and P. V. Turner. 2012. Causes of obesity in captive cynomolgus macaques: influence of body condition, social and management factors on behaviour around feeding. *Laboratory Animals* 46: 193–199.

Bayne, K. 2003. Environmental enrichment of nonhuman primates, dogs and rabbits used in toxicology studies. *Toxicologic Pathology* 31(Suppl.): 132–137.

Bayne, K. 2005. Potential for unintended consequences of environmental enrichment for laboratory animals and research results. *ILAR Journal* 46(2): 129–139.

Bayne, K., S. Dexter, H. Mainzer, C. McCully, G. Campbell and F. Yamada. 1992a. The use of artificial turf as a foraging substrate for individually housed rhesus monkeys (*Macaca mulatta*). *Animal Welfare* 1: 39–53.

Bayne, K., S. Dexter and G. M. Strange. 1993. The effects of food treat provisioning and human interaction on the behavioral well-being of rhesus macaques (*Macaca mulatta*). *Contemporary Topics in Laboratory Animal Science* 32(2): 6–9.

Bayne, K., S. Dexter and S. Suomi. 1992b. A preliminary survey of the incidence of abnormal behavior in rhesus monkeys (*Macaca mulatta*) relative to housing condition. *Laboratory Animals* 21: 38–46.

Bayne, K., H. Mainzer, S. Dexter, G. Campbell, F. Yamada and S. Suomi. 1991. The reduction of abnormal behaviors in individually housed rhesus monkeys (*Macaca mulatta*) with a foraging/grooming board. *American Journal of Primatology* 23: 23–35.

Bayne, K. and M. Novak. 1998. Behavioral disorders. *Nonhuman Primates in Biomedical Research: Diseases*, B. T. Bennett, C. R. Abee and R. Hendrickson. New York, Academic Press: 485–500.

Beck, R. T., G. R. Lubach and C. L. Coe. 2020. Feasibility of successfully breeding rhesus macaques (*Macaca mulatta*) to obtain healthy infants year-round. *American Journal of Primatology* 82(1): e23085.

Beirise, J. and V. Reinhardt. 1992. Three inexpensive environmental enrichment options for group-housed *Macaca mulatta*. *Laboratory Primate Newsletter* 31(1): 7–8.

Beisner, B. A. and L. A. Isbell. 2008. Ground substrate affects activity budgets and hair loss in outdoor captive groups of rhesus macaques (*Macaca mulatta*). *American Journal of Primatology* 70(12): 1160–1168.

Beisner, B. A. and L. A. Isbell. 2011. Factors affecting aggression among females in captive groups of rhesus macaques (*Macaca mulatta*). *American Journal of Primatology* 73(11): 1152–1159.

Bennett, A. J., C. A. Corcoran, V. A. Hardy, L. R. Miller and P. J. Pierre. 2010. Multidimensional cost–benefit analysis to guide evidence-based environmental enrichment: providing bedding and foraging substrate to pen-housed monkeys. *Journal of the American Association for Laboratory Animal Science* 49(5): 571–577.

Berard, J. D., P. Nurnberg, J. T. Epplen and J. Schmidtke. 1994. Alternative reproductive tactics and reproductive success in male rhesus macaques. *Behaviour* 129: 177–201.

Bercovitch, F. B., M. D. Hauser and J. H. Jones. 1995. The endocrine stress response and alarm vocalizations in rhesus macaques. *Animal Behaviour* 49(6): 1703–1706.

Berg, M. R., A. Heagerty and K. Coleman. 2019. Oxytocin and pair compatibility in adult male rhesus macaques (*Macaca mulatta*). *American Journal of Primatology* 81(8): e23031.

Bernstein, I. S. 1967. A field study of the pigtail monkey (*Macaca nemestrina*). *Primates* 8(3): 217–228.

Bernstein, I. S. 1991. Social housing of monkeys and apes: group formations. *Laboratory Animal Science* 41(4): 323–328.

Bertrand, M. 1967. Training without reward: traditional training of pig-tailed macaques as coconut harvesters. *Science* 155: 484–486.

Bertrand, M. 1969. *The Behavioral Repertoire of the Stumptailed Macaque: A Descriptive and Comparative Study*. Basel, Karger.

Block, L. N., M. T. Aliota, T. C. Friedrich, M. L. Schotzko, K. D. Mean, G. J. Wiepz, T. G. Golos and J. K. Schmidt. 2020. Embryotoxic impact of Zika virus in a rhesus macaque in vitro implantation model. *Biology of Reproduction* 102(4): 806–816.

Bloomsmith, M. A., M. J. Marr and T. L. Maple. 2007. Addressing nonhuman primate behavioral problems through the application of operant conditioning: is the human treatment approach a useful model? *Applied Animal Behaviour Science* 102(3–4): 205–222.

Boccia, M. L. and A. S. Hijazi. 1998. A foraging task reduces agonistic and stereotypic behaviours in pigtail macaque social groups. *Laboratory Primate Newsletter* 37(3): 1–5.

Bottin, G., M. C. Huynen and T. Savini. 2008. Preliminary results on activity budget, feeding ecology and ranging behaviour of wild pig-tailed macaques (*Macaca nemestrina leonina*) in Khao Yai National Park, Thailand. *Folia Primatologica* 79: 313.

Brady, A. G. and D. G. Morton. 1998. Digestive system. *Nonhuman Primates in Biomedical Research: Diseases*, E. L. Bennett, C. R. Abee and R. Hendrickson. London, Academic Press: 377–414.

Brandon-Jones, D., A. A. Eudey, T. Geissmann, C. P. Groves, D. J. Melnick, J. C. Morales, M. Shekelle and C. B. Stewart. 2004. Asian primate classification. *International Journal of Primatology* 25(1): 97–164.

Brent, L. and Y. Veira. 2002. Social behavior of captive Indochinese and insular long-tailed macaques (*Macaca fascicularis*) following transfer to a new facility. *International Journal of Primatology* 23(1): 147–159.

Brownlee, R. D., P. H. Kass and R. L. Sammak. 2020. Blood pressure reference intervals for ketamine-sedated rhesus macaques (*Macaca mulatta*). *Journal of the American Association for Laboratory Animal Science* 59(1): 24–29.

Bryant, C. E., N. M. Rupniak and S. D. Iversen. 1988. Effects of different environmental enrichment devices on cage stereotypies and autoaggression in captive cynomolgus monkeys. *Journal of Medical Primatology* 17: 257–269.

Buchanan-Smith, H. M., M. Prescott and N. J. Cross. 2004. What factors should determine cage sizes for primates in the laboratory? *Animal Welfare* 13(S): 197–201.

Bunlungsup, S., S. Kanthaswamy, R. F. Oldt, D. G. Smith, P. Houghton, Y. Hamada and S. Malaivijitnond. 2017. Genetic analysis of samples from wild populations opens new perspectives on hybridization between long-tailed (*Macaca fascicularis*) and rhesus macaques (*Macaca mulatta*). *American Journal of Primatology* 79(12): e22726.

Burton, F. D. and L. Chan. 1996. Behavior of mixed species groups of macaques. *Evolution and Ecology of Macaque Societies*, J. E. Fa and D. G. Lindburg. Cambridge, Cambridge University Press: 389–412.

Butovskaya, M. 2004. Social space and degrees of freedom. *Macaque Societies: A Model for the Study of Social Organization*, B. Thierry, M. Singh and W. Kaumanns. Cambridge, Cambridge University Press: 182–185.

Caldecott, J. 1986. *An Ecological and Behavioural Study of the Pig-tailed Macaque*. Basel, Karger.

Cant, J. G. H. 1988. Positional behavior of long-tailed Macaques (*Macaca fascicularis*) in Northern Sumatra. *American Journal of Physical Anthropology* 76(1): 29–37.

Capitanio, J. P. 2017. Variation in biobehavioral organization. *Handbook of Primate Behavioral Management*, S. Schapiro. Boca Raton, FL, CRC Press: 55–73.

Capitanio, J. P., S. A. Blozis, J. Snarr, A. Steward and B. J. McCowan. 2017. Do "birds of a feather flock together" or do "opposites attract"? Behavioral responses and temperament predict success in pairings of rhesus monkeys in a laboratory setting. *American Journal of Primatology* 79(1): 1–11.

Carlson, S., P. Rama, D. Artchakov and I. Linnankoski. 1997. Effects of music and white noise on working memory performance in monkeys. *Neuroreport* 8(13): 2853–2856.

Carter, S. P. and P. W. Bright. 2011. Habitat refuges as alternatives to predator control for the conservation of endangered Mauritian birds. *Turning the Tide: The Eradication of Invasive Species*, C. R. Veitch and M. N. Clout. Gland, IUCN: 71–78.

Chakraborty, D., U. Ramakrishnan, J. Panor, C. Mishra and A. Sinha. 2007. Phylogenetic relationships and morphometric affinities of the Arunachal macaque *Macaca munzala*, a newly described primate from Arunachal Pradesh, northeastern India. *Molecular Phylogenetics and Evolution* 44: 838–849.

Chamove, A. S. 1989. Environmental enrichment: a review. *Animal Technology* 40(3): 155–178.

Champoux, M., C. L. Coe, S. M. Schanberg, C. M. Kuhn and S. J. Suomi. 1989. Hormonal effects of early rearing conditions in the infant rhesus monkey. *American Journal of Primatology* 19: 111–117.

Champoux, M., S. J. Suomi and M. L. Schneider. 1994. Temperament differences between captive Indian and Chinese-Indian hybrid rhesus macaque neonates. *Laboratory Animal Science* 44(4): 351–357.

Chapman, C. A. and J. A. Teichroeb. 2012. What influences the size of groups in which primates choose to live? *Nature Education Knowledge* 3(10): 9.

Choudhury, A. 2003. The pig-tailed macaque *Macaca nemestrina* in India - status and conservation. *Primate Conservation* 19: 91–98.

Choudhury, A. 2008. Ecology and behaviour of the pig tailed macaque *Macaca nemestrina leonina* in some forests of Assam in North East India. *The Journal of the Bombay Natural History Society* 105: 279–291.

Clarke, A. S. and S. Boinski. 1995. Temperament in nonhuman primates. *American Journal of Primatology* 37(2): 103–125.

Clarke, A. S., N. M. Czekala and D. G. Lindburg. 1995a. Behavioral and adrenocortical responses of male cynomolgus and lion-tailed macaques to social stimulation and group formation. *Primates* 36(1): 41–56.

Clarke, A. S. and W. A. Mason. 1988. Differences among 3 macaque species in responsiveness to an observer. *International Journal of Primatology* 9(4): 347–364.

Clarke, A. S., W. A. Mason and S. P. Mendoza. 1994. Heart rate patterns under stress in three species of macaques. *American Journal of Primatology* 33(2): 133–148.

Clarke, A. S., W. A. Mason and G. P. Moberg. 1988. Interspecific contrasts in responses of macaques to transport cage training. *Laboratory Animal Science* 38(3): 305–309.

Clarke, M. R., J. L. Blanchard and J. A. Snyder. 1995b. Infant-killing in pigtailed monkeys: a colony management concern. *Laboratory Primate Newsletter* 34(4): 1–4.

Clutton-Brock, T. H. and P. H. Harvey. 1977. Species differences in feeding and ranging behaviour in primates. *Primate Ecology: Studies of Feeding and Ranging Behaviour in Lemurs, Monkeys and Apes*, T. H. Clutton-Brock. London, Academic Press: 557–579.

Coleman, K. 2012. Individual differences in temperament and behavioral management practices for nonhuman primates. *Applied Animal Behaviour Science* 137(3–4): 106–113.

Coleman, K. 2017. Individual differences in temperament and behavioral management. *Handbook of Primate Behavioral Management*, S. Schapiro. Boca Raton, FL, CRC Press: 95–113.

Coleman, K., C. K. Lutz, J. M. Worlein, D. H. Gottlieb, E. Peterson, G. H. Lee, N. D. Robertson, K. Rosenberg, M. T. Menard and M. A. Novak. 2017. The correlation between alopecia and temperament in rhesus macaques (*Macaca mulatta*) at four primate facilities. *American Journal of Primatology* 79(1): 1–10.

Coleman, K. and A. Maier. 2010. The use of positive reinforcement training to reduce stereotypic behavior in rhesus macaques. *Applied Animal Behaviour Science* 124(3–4): 142–148.

Coleman, K., L. A. Tully and J. L. McMillan. 2005. Temperament correlates with training success in adult rhesus macaques. *American Journal of Primatology* 65(1): 63–71.

Cords, M. 1992. Post-conflict reunions and reconciliation in long-tailed macaques. *Animal Behaviour* 44(1): 57–61.

Coss, R. G. and U. Ramakrishnan. 2000. Perceptual aspects of leopard recognition by wild bonnet macaques (*Macaca radiata*). *Behaviour* 137: 315–335.

Cowlishaw, G. and R. I. M. Dunbar. 1991. Dominance rank and mating success in male primates. *Animal Behaviour* 41: 1045–1056.

Crockett, C. M., K. L. Bentson and R. U. Bellanca. 2007. Alopecia and overgrooming in laboratory monkeys vary by species but not sex, suggesting a different etiology than self-biting. *American Journal of Primatology* 69(S1): 87–88.

Crockett, C. M., J. Bielitzki, A. Carey and A. Velez. 1989. Kong toys as enrichment devices for singly-caged macaques. *Laboratory Primate Newsletter* 28(2): 21–22.

Crockett, C. M., C. L. Bowers, D. M. Bowden and G. P. Sackett. 1994. Sex differences in compatibility of pair-housed adult longtailed macaques. *American Journal of Primatology* 32(2): 73–94.

Crockett, C. M., C. L. Bowers, M. Shimoji, M. Leu, D. M. Bowden and G. P. Sackett. 1995. Behavioral responses of longtailed macaques to different cage sizes and common laboratory experiences. *Journal of Comparative Psychology* 109(4): 368–383.

Cui, Z., Z. Wang, G. Zhao and J. Lu. 2015. Seasonal variations in activity budget of adult female rhesus macaques (*Macaca mulatta*) at Mt. Taihangshan area, Jiyuan, China: effects of diet and temperature. *Acta Theriologica Sinica* 35: 138–146.

Darrah, P. A., J. J. Zeppa, P. Maiello, J. A. Hackney, M. H. Wadsworth, T. K. Hughes, S. Pokkali, P. A. Swanson, N. L. Grant, M. A. Rodgers, M. Kamath, C. M. Causgrove, D. J. Laddy, A. Bonavia, D. Casimiro, P. L. Lin, E. Klein, A. G. White, C. A. Scanga, A. K. Shalek, M. Roederer, J. L. Flynn and R. A. Seder. 2020. Prevention of tuberculosis in macaques after intravenous BCG immunization. *Nature* 577(7788): 95–102.

de Ruiter, J. R. and E. Geffen. 1998. Relatedness of matrilines, dispersing males and social groups in long-tailed macaques (*Macaca fascicularis*). *Proceedings of the Royal Society B: Biological Sciences* 265(1391): 79–87.

De Ruiter, J. R., W. Scheffrahn, G. J. J. M. Trommelen, A. G. Uitterlinden, R. D. Martin and J. A. R. A. M. van Hooff. 1992. Male social rank and reproductive success in wild long-tailed macaques. *Paternity in Primates: Genetic Tests and Theories*, R. D. Martin, A. Dixson and E. J. Wickings. Basel, Karger: 175–191.

de Ruiter, J. R. and J. A. R. A. M. van Hooff. 1993. Male dominance rank and reproductive success in primate groups. *Primates* 34(4): 513–523.

De Ruiter, J. R., J. A. R. A. M. van Hooff and W. Scheffrahn. 1994. Social and genetic aspects of paternity in wild long-tailed macaques (*Macaca fascicularis*). *Behaviour* 129: 203–224.

de Waal, F. B. 1989. Dominance 'style' and primate social organisation. *Comparative Socioecology: The Behavioural Ecology of Humans and Other Mammals*, V. Standen and R. A. Foley. Oxford, Blackwell Scientific Publications: 243–263.

de Waal, F. B. M. and D. L. Johanowicz. 1993. Modification of reconciliation behavior through social experience: an experiment with two macaque species. *Child Development* 64(3): 897–908.

Deaner, R. O., A. V. Khera and M. L. Platt. 2005. Monkeys pay per view: adaptive valuation of social images by rhesus macaques. *Current Biology* 15(6): 543–548.

Dennis, M. 2009. An improved containment system for experimentally infected macaques. *NC3Rs Occassional Articles* 18: 7.

Deputte, B. L. 1994. Ethological study of yawning in primates. I. Quantitative analysis and study of causation in two species of Old World Monkeys (*Cercocebus albigena* and *Macaca fascicularis*). *Ethology* 98(3–4): 221–245.

Deputte, B. L. and M. Goustard. 1980. Copulatory vocalizations of female macaques (*Macaca fascicularis*): variability factors analysis. *Primates* 21(1): 83–99.

Dettmer, A. M., J. M. Allen, R. M. Jaggers and M. T. Bailey. 2019. A descriptive analysis of gut microbiota composition in differentially reared infant rhesus monkeys (*Macaca mulatta*) across the first 6 months of life. *American Journal of Primatology* 81(10–11): e22969.

Dileep, K. and J. Boby. 2014. Food preference and feeding habit of bonnet macaque (*Macaca radiata*). *International Journal of Pure and Applied Zoology* 2(3): 256–260.

Dixson, A. F. 1998a. Mating tactics and reproductive success. *Primate Sexuality: Comparative Studies of the Prosimians, Monkeys, Apes and Human Beings*. Oxford, Oxford University Press: 51–92.

Dixson, A. F. 1998b. *Primate Sexuality: Comparative Studies of the Prosimians, Monkeys, Apes, and Human Beings*. Oxford, Oxford University Press: 546.

Dixson, A. F. 1998c. Sperm competition. *Primate Sexuality: Comparative Studies of the Prosimians, Monkeys, Apes, and Human Beings*. Oxford, Oxford University Press: 217–243.

Dixson, A. and M. Anderson, 2002. Sexual selection, seminal coagulation and copulatory plug formation in primates. *Folia Primatologica* 73: 63–69.

Dominy, N. J. 2004. Fruits, fingers, and fermentation: the sensory cues available to foraging primates. *Integrative and Comparative Biology* 44(4): 295–303.

Doyle, L. A., K. C. Baker and L. D. Cox. 2008. Physiological and behavioral effects of social introduction on adult male rhesus macaques. *American Journal of Primatology* 70: 542–550.

Dubuc, C., L. Brent, A. K. Accamando, M. S. Gerald, A. MacLarnon, S. Semple, M. Heistermann and A. Engelhardt. 2009. Sexual skin color contains information about the timing of the fertile phase in free-ranging rhesus macaques. *International Journal of Primatology* 30: 777–789.

Dunaif, A. and I. Tattersall. 1987. Prevalence of glucose intolerance in free-ranging *Macaca fascicularis* of Mauritius. *American Journal of Primatology* 13: 435–442.

Dunbar, R. I. M. 2010. The social role of touch in humans and primates: behavioural function and neurobiological mechanisms. *Neuroscience & Biobehavioral Reviews* 34(2): 260–268.

Eaton, G. G., S. T. Kelley, M. K. Axthelm, S. A. Iliff-Sizemore and S. M. Shigi. 1994. Psychological well-being in paired adult female rhesus (*Macaca mulatta*). *American Journal of Primatology* 33: 89–99.

Eckert, K. 1999. Warning: rope in environmental enrichment. *Laboratory Primate Newsletter* 38(4): 3.

Eckert, K., C. Niemeyer, Anonymous, R. W. Rogers, J. Seier, B. Ingersoll, L. Barklay, C. Brinkman, S. Oliver, C. Buckmaster, L. Knowles, S. Pyle and V. Reinhardt. 2000. Wooden objects for enrichment: a discussion. *Laboratory Primate Newsletter* 39(3): 1–4.

Elfenbein, H. A., L. D. Rosso, B. McCowan and J. P. Capitanio. 2016. Effect of indoor compared with outdoor location during gestation on the incidence of diarrhea in indoor-reared rhesus macaques (*Macaca mulatta*). *Journal of the American Association for Laboratory Animal Science* 55(3): 277–290.

Emborg, M. E. 2017. Nonhuman primate models of neurodegenerative disorders. *ILAR Journal* 58(2): 190–201.

Endo, S., K. Ishii, M. Suzuki, T. Kajimoto, K. Tanaka and M. Fukumoto. 2020. Dose estimation of external and internal exposure in Japanese macaques after the Fukushima nuclear power plant accident. *Low-Dose Radiation Effects on Animals and Ecosystems: Long-Term Study on the Fukushima Nuclear Accident*, M. Fukumoto. Singapore, Springer Singapore: 179–191.

Engel, G. A., L. Jones-Engel, M. A. Schillaci, K. G. Suaryana, A. Putra, A. Fuentes and R. Henkel. 2002. Human exposure to herpesvirus B-seropositive macaques, Bali, Indonesia. *Emerging Infectious Diseases* 8(8): 789–795.

Engelhardt, A., J.-B. Pfeifer, M. Heistermann, C. Niemitz, J. A. R. A. M. van Hooff and J. K. Hodges. 2004. Assessment of female reproductive status by male longtailed macaques, *Macaca fascicularis*, under natural conditions. *Animal Behaviour* 67(5): 915–924.

Erinjery, J. J., T. S. Kavana and M. Singh. 2017a. Behavioural variability in macaques and langurs of the Western Ghats, India. *Folia Primatologica* 88(3): 293–306.

Erinjery, J. J., S. Kumar, H. N. Kumara, K. Mohan, T. Dhananjaya, P. Sundararaj, R. Kent and M. Singh. 2017b. Losing its ground: a case study of fast declining populations of a 'least-concern' species, the bonnet macaque (*Macaca radiata*). *PLoS One* 12(8): e0182140.

Erwin, J. and R. Deni. 1979. Strangers in a strange land: abnormal behaviors or abnormal environments? *Captivity and Behavior: Primates in Breeding Colonies, Laboratories and Zoos*, J. Erwin, T. L. Maple and G. Mitchell. New York, Van Norstrand Reinhold: 1–28.

EU. 2010. Directive 2010/63/EU of the European Parliament and of the Council of 22 September 2010 on the Protection of Animals Used for Scientific Purposes. *Official Journal of the European Union* L276: 33–79.

Eudey, A. A. 2008. The crab-eating macaque (*Macaca fascicularis*): widespread and rapidly declining. *Primate Conservation* 23: 129–132.

Evans, D. T. and G. Silvestri. 2013. Nonhuman primate models in AIDS research. *Current Opinion in HIV and AIDS* 8(4): 255–261.

Fajzi, K., V. Reinhardt and M. D. Smith. 1989. A review of environmental enrichment strategies for singly caged nonhuman primates. *Laboratory Animals* 18(2): 23–35.

Feeroz, M. 2003. Breeding activities of the pig-tailed macaque (*Macaca leonina*) in Bangladesh. *Zoos' Print Journal* 18: 1175–1179.

Fernandez, L., M.-A. Griffiths and P. Honess. 2017. Providing behaviorally manageable primates for research. *Handbook of Primate Behavioral Management*, S. Schapiro. Boca Raton, FL, CRC Press: 481–494.

Fischer, J. 2017. On the social life and motivational changes of aging monkeys. *Gerontology* 63(6): 572–579.

Fittinghoff, N. A. and D. G. Lindburg. 1980. Riverine refuging in east Bornean *Macaca fascicularis*. *The Macaques: Studies in Ecology*, Behavior, and Evolution, D. G. Lindburg. New York, Van Nostrand Rheinhold: 182–214.

Flack, J. C. and F. B. M. De Waal. 2004. Dominance style, social power, and conflict management in macaque societies: a conceptual framework. *Macaque Societies: A Model of the Study of Social Organization*, B. Thierry, M. Singh and W. Kaumanns. Cambridge, Cambridge University Press: 157–182.

Fleagle, J. G. 1999. Locomotor adaptations. *Primate Adaptation and Evolution*. London, Academic Press: 297–306.

Fooden, J. 1976. Provisional classification and key to living species of macaques (Primates: Macaca). *Folia Primatol* 25: 225–236.

Fooden, J. 1980. Classification and distribution of living macaques (*Macaca* Lacepede, 1799). *The Macaques: Studies in Ecology, Behavior, and Evolution*, D. G. Lindburg. New York, Van Nostrand Reinhold: 1–9.

Fooden, J. 1990. The bear macaque, *Macaca arctoides*: a systematic review. *Journal of Human Evolution* 19(6): 607–686.

Fooden, J. 1995. Systematic review of southeast Asian longtail macaques, *Macaca fascicularis* (Raffles, 1821). *Fieldiana Zoology* 81: 1–206.

Fooden, J. 2000. Systematic review of the rhesus macaque, *Macaca mulatta* (Zimmermann, 1780). *Fieldiana Zoology* 96: 1–180.

Fooden, J. and M. Aimi. 2003. Birth-season variation in Japanese macaques, *Macaca fuscata*. *Primates* 44(2): 109–117.

Fooden, J. and M. Aimi. 2005. Systematic review of Japanese macaques, *Macaca fuscata* (Gray, 1870). *Fieldiana Zoology* 104: 1–198.

Freeman, Z. T., K. A. Rice, P. L. Soto, K. A. Pate, M. R. Weed, N. A. Ator, I. G. DeLeon, D. F. Wong, Y. Zhou, J. L. Mankowski, M. C. Zink, R. J. Adams and E. K. Hutchinson. 2015. Neurocognitive dysfunction and pharmacological intervention using guanfacine in a rhesus macaque model of self-injurious behavior. *Translational Psychiatry* 5(5): e567.

Fuentes, A. and S. Gamerl. 2005. Disproportionate participation by age/sex classes in aggressive interactions between long-tailed macaques (*Macaca fascicularis*) and human tourists at Padangtegal monkey forest, Bali, Indonesia. *American Journal of Primatology* 66(2): 197–204.

Garner, J. P. 2005. Stereotypies and other abnormal repetitive behaviors: potential impact on validity, reliability, and replicability of scientific outcomes. *ILAR Journal* 46(2): 106–117.

Garner, J. P. 2006. Perseveration and stereotypy: systems-level insights from clinical psychology. *Stereotypic Animal Behaviour: Fundamentals and Applications to Welfare*, G. Mason and J. Rushen. Wallingford, UK, CABI: 121–152.

Gilbert, M. H. and K. C. Baker. 2011. Social buffering in adult male rhesus macaques (Macaca mulatta): effects of stressful events in single vs. pair housing. *Journal of Medical Primatology* 40(2): 71–78.

Gilbert, S. G. and E. Wrenshall. 1989. Environmental enrichment for monkeys used in behavioral toxicology studies. *Housing, Care and Psychological Wellbeing of Captive and Laboratory Primates*, E. F. Segal. Park Ridge, NJ, Noyes Publications: 244–254.

Gogoi, S. and A. N. Das. 2018. Diet and feeding behaviour of rhesus macaque at Navagraha Temple, Kamrup District of Assam. *Periodic Research* 7(2): 93–98.

Goin, D. L. and D. A. Gust. 1998. Peer-rearing influences subsequent maternal behavior and infant survival in a newly formed herpes B-virus negative rhesus macaque group. *Primates* 39(4): 539–543.

Goldstein, S. and A. Richard. 1989. Ecology of rhesus macaques (*Macaca mulatta*) in northwest Pakistan. *International Journal of Primatology* 10(6): 531–567.

Gore, M. A. 1993. Effects of food distribution on foraging competition in rhesus monkeys, *Macaca mulatta*, and Hamadryas baboons, *Papio hamadryas*. *Animal Behaviour* 45(4): 773–786.

Gottlieb, D. H., L. Del Rosso, F. Sheikhi, A. Gottlieb, B. McCowan and J. P. Capitanio. 2018. Personality, environmental stressors, and diarrhea in rhesus macaques: an interactionist perspective. *American Journal of Primatology* 80(12): e22908.

Gottlieb, D. H., S. Ghirardo, D. E. Minier, N. Sharpe, L. Tatum and B. McCowan. 2011. Efficacy of 3 types of foraging enrichment for rhesus macaques (*Macaca mulatta*). *Journal of the American Association for Laboratory Animal Science* 50(6): 888–894.

Gouzoules, S. and H. Gouzoules. 1987. Kinship. *Primate Societies*, B. Smuts, D. Cheney, R. Seyfarth, R. Wrangham and T. Struhsaker. Chicago, IL, University of Chicago Press: 299–305.

Graham, M. L. 2017. Positive reinforcement training and research. *Handbook of Primate Behavioral Management*, S. Schapiro. Boca Raton, FL, CRC Press: 187–200.

Graham, M. L., E. F. Rieke, L. A. Mutch, E. K. Zolondek, A. W. Faig, T. A. Dufour, J. W. Munson, J. A. Kittredge and H. J. Schuurman. 2012. Successful implementation of cooperative handling eliminates the need for restraint in a complex non-human primate disease model. *Journal of Medical Primatology* 41(2): 89–106.

Gray, D. T., L. Umapathy, N. M. De La Peña, S. N. Burke, J. R. Engle, T. P. Trouard and C. A. Barnes. 2020. Auditory processing deficits are selectively associated with medial temporal lobe mnemonic function and white matter integrity in aging macaques. *Cerebral Cortex* 30(5): 2789–2803.

Grob, B., L. A. Knapp, R. D. Martin and G. Anzenberger. 1998. The major histocompatibility complex and mate choice: inbreeding avoidance and selection of good genes. *Experimental and Clinical Immunogenetics* 15(3): 119–129.

Groves, C. P. 2001. *Primate Taxonomy*. Washington, DC, Smithsonian Institution Press.

Groves, C. P. 2005. Primates. *Mammal Species of the World: A Taxonomic and Geographic Reference*, D. E. Wilson and D. M. Reeder. Baltimore, MD, The Johns Hopkins University Press, Vol 1: 111–184.

Gumert, M. 2007a. Payment for sex in a macaque mating market. *Animal Behaviour* 74: 1655–1667.

Gumert, M. D. 2007b. Grooming and infant handling interchange in *Macaca fascicularis*: the relationship between infant supply and grooming payment. *International Journal of Primatology* 28(5): 1059–1074.

Gumert, M. D. and S. Malaivijitnond. 2012. Marine prey processed with stone tools by Burmese long-tailed macaques (*Macaca fascicularis aurea*) in intertidal habitats. *American Journal of Physical Anthropology* 149(3): 447–457.

Haertel, A. J., K. Prongay, L. Gao, D. H. Gottlieb and B. Park. 2018. Standard growth and diarrhea-associated growth faltering in captive infant rhesus macaques (*Macaca mulatta*). *American Journal of Primatology* 80(9): e22923.

Hahn, N. E., D. Lau, K. Eckert and H. Markowitz. 2000. Environmental enrichment-related injury in a macaque (*Macaca fascicularis*): intestinal linear foreign body. *Comparative Medicine* 50(5): 556–558.

Hannibal, D. L., L. C. Cassidy, J. Vandeleest, S. Semple, A. Barnard, K. Chun, S. Winkler and B. McCowan. 2018. Intermittent pair-housing, pair relationship qualities, and HPA activity in adult female rhesus macaques. *American Journal of Primatology* 80(5): e22762.

Hanya, G. 2004. Seasonal variations in the activity budget of Japanese macaques in the coniferous forest of Yakushima: effects of food and temperature. *American Journal of Primatology* 63(3): 165–177.

Hanya, G. and C. A. Chapman. 2013. Linking feeding ecology and population abundance: a review of food resource limitation on primates. *Ecological Research* 28: 183–190.

Hanya, G., Y. Otani, S. Hongo, T. Honda, H. Okamura and Y. Higo. 2018. Activity of wild Japanese macaques in Yakushima revealed by camera trapping: patterns with respect to season, daily period and rainfall. *PLoS One* 13(1): e0190631.

Harlow, H. F. 1958. The nature of love. *American Psychologist* 13(12): 673–685.

Hauser, M. D. 1996. Vocal communication in macaques: causes of variation. *Evolution and Ecology of Macaque Societies*, J. E. Fa and D. G. Lindburg. Cambridge, Cambridge University Press: 551–577.

Heagerty, A., R. A. Wales, K. Prongay, D. H. Gottlieb and K. Coleman. 2017. Social hair pulling in captive rhesus macaques (*Macaca mulatta*). *American Journal of Primatology* 79(12): e22720.

Heath-Lange, S., J. C. Ha and G. P. Sackett. 1999. Behavioral measurement of temperament in male nursery-raised infant macaques and baboons. *American Journal of Primatology* 47: 43–50.

Heath, M. 1989. The training of cynomolgus monkeys and how the human/animal relationship improves with environmental and mental enrichment. *Animal Technology* 40(1): 11–22.

Heckel, T., A. Singh, A. Gschwind, A. Reymond and U. Certa. 2015. Genetic variations in the *Macaca fascicularis* genome related to biomedical research. *The Nonhuman Primate in Nonclinical Drug Development and Safety Assessment*, J. Bluemel, S. Korte, E. Schenck and G. F. Weinbauer. San Diego, CA, Academic Press: 53–64.

Henzi, S. and L. Barrett. 1999. The value of grooming to female primates. *Primates* 40(1): 47–59.

Higley, J. and S. Suomi. 1989. Temperamental reactivity in non-human primates. *Temperament in Childhood*, G. Kohnstamm, J. Bates and M. Rothbart. New York, Wiley: 153–167.

Hill, D. A. 1997. Seasonal variation in the feeding behavior and diet of Japanese macaques (*Macaca fuscata yakui*) in lowland forest of Yakushima. *American Journal of Primatology* 43(4): 305–320.

Hill, D. A. and N. Okayasu. 1996. Determinants of dominance among female macaques: nepotism, demography and danger. *Evolution and Ecology of Macaque Societies*, J. E. Fa and D. G. Lindburg. Cambridge, Cambridge University Press: 459–472.

Hinde, K. 2009. Richer milk for sons but more milk for daughters: sex-biased investment during lactation varies with maternal life history in rhesus macaques. *American Journal of Human Biology* 21(4): 512–519.

Holzner, A., N. Ruppert, F. Swat, M. Schmidt, B. M. Weiß, G. Villa, A. Mansor, S. A. Mohd Sah, A. Engelhardt, H. Kühl and A. Widdig. 2019. Macaques can contribute to greener practices in oil palm plantations when used as biological pest control. *Current Biology* 29(20): R1066–R1067.

Honess, P. 2013. Behavior and enrichment of long-tailed (cynomolgus) macaques (*Macaca fascicularis*). *A Guide to the Behavior and Enrichment of Laboratory Macaques*, C. Winnicker. Wilmington, MA, Charles River Publications: 4–87.

Honess, P. 2017. Behavioral management of long-tailed macaques (*Macaca fascicularis*). *Handbook of Primate Behavioral Management*, S. Schapiro. London, CRC Press, Taylor & Francis: 305–337.

Honess, P., T. Andrianjazalahatra, P. Matai and S. Naiken. 2013a. Primate adoption: an essential alternative to hand-rearing. *Journal of the American Association for Laboratory Animal Science* 52(3): 253.

Honess, P., T. Andrianjazalahatra, S. Naiken and M.-A. Griffiths. 2015. Factors influencing aggression in peer groups of weaned long-tailed macaques (*Macaca fascicularis*). *Archives of Medical and Biomedical Research* 2(3): 88.

Honess, P., J. Gimpel, S. Wolfensohn and G. Mason. 2005. Alopecia scoring: the quantitative assessment of hair loss in captive macaques. *ATLA* 33(3): 193–206.

Honess, P., Y. Jiang and J. McDonnell. In prep. The influence of social factors and environmental enrichment on hair pulling behaviour in rhesus macaques (*Macaca mulatta*).

Honess, P., S. Schapiro, M. A. Bloomsmith, D. R. Lee, B. McCowan and B. Oettinger. 2013b. Abnormal behavior in macaques. *A Guide to the Behavior and Enrichment of Laboratory Macaques*, C. Winnicker. Wilmington, MA, Charles River Publications: 138–165.

Honess, P., M. A. Stanley-Griffiths, S. Narainapoulle, S. Naiken and T. Andrianjazalahatra. 2010. Selective breeding of primates for use in research: consequences and challenges. *Animal Welfare* 19: 57–65.

Honess, P. and S. Wolfensohn. 2010. A matrix for the assessment of welfare in experimental animals. *ATLA* 38: 205–212.

Honess, P. E., P. J. Johnson and S. E. Wolfensohn. 2004. A study of behavioural responses of non-human primates to air transport and re-housing. *Laboratory Animals* 38(2): 119–132.

Honess, P. E. and C. M. Marin. 2006a. Behavioural and physiological aspects of stress and aggression in nonhuman primates. *Neuroscience and Biobehavioral Reviews* 30(3): 390–412.

Honess, P. E. and C. M. Marin. 2006b. Enrichment and aggression in primates. *Neuroscience and Biobehavioral Reviews* 30(3): 413–436.

Hori, T., T. Nakayama, H. Tokura, F. Hara and M. Suzuki. 1977. Thermoregulation of the Japanese macaque living in a snowy mountain area. *Japanese Journal of Physiology* 27(3): 305–319.

Huffman, M. and D. Quiatt. 1986. Stone handling by Japanese macaques (*Macaca fuscata*): implications for tool use of stone. *Primates* 27: 413–423.

Huffman, M. A., B.-H. Sun and J.-H. Li. 2020. Medicinal properties in the diet of Tibetan macaques at Mt. Huangshan: a case for self-medication. *The Behavioral Ecology of the Tibetan Macaque*, J.-H. Li, L. Sun and P. M. Kappeler. Cham, Springer International Publishing: 223–248.

Hutchinson, E. 2017. The veterinarian-behavioral management interface. *Handbook of Primate Behavioral Management*, S. Schapiro. Boca Raton, FL, CRC Press: 217–224.

Iliff, S. A., B. H. Friscino and L. C. Anderson. 2004. Refinements of study design using positive reinforcement training in macaques. *Folia Primatologica* 75: 282–283.

Imron, M. A., R. A. Satria and M. F. P. Ramlan. 2018. Komodo dragon predation on crab-eating macaques at the Rinca Island's Visitor Centre, Indonesia. *Folia Primatologica* 89(5): 335–340.

Isa, T., I. Yamane, M. Hamai and H. Inagaki. 2009. Japanese macaques as laboratory animals. *Experimental Animals* 58(5): 451–457.

Itoh, K. 2002. Personality research with non-human primates: theoretical formulation and methods. *Primates* 43(3): 249–261.

IUCN Red List. 2020. Macaque Species Threat Status. Retrieved 20/04/2020, 2020, from https://www.iucnredlist.org/search?taxonomies=130129&searchType=species.

Janson, C. and J. Verdolin. 2004. Seasonality of primate births in relation to climate. *Seasonality in Primates: Studies of Living and Extinct Human and Non-human Primates*, D. K. Brockman and C. P. van Schaik. Cambridge, Cambridge University Press: 307–350.

Janus, M. 1989. Reciprocity in play, grooming, and proximity in sibling and nonsibling young rhesus monkeys. *International Journal of Primatology* 10(3): 243–261.

Jones-Engel, L. 2020. Commentary: trust but verify. *Cambridge Quarterly of Healthcare Ethics* 29(1): 42–45.

Joyner, C., J. W. Barnwell and M. R. Galinski. 2015. No more monkeying around: primate malaria model systems are key to understanding *Plasmodium vivax* liver-stage biology, hypnozoites, and relapses. *Frontiers in Microbiology* 6: 145–145.

Kaisin, O., E. Gazagne, T. Savini, M.-C. Huynen and F. Brotcorne. 2018. Foraging strategies underlying bird egg predation by macaques: a study using artificial nests. *American Journal of Primatology* 80(11): e22916.

Kanthaswamy, S., J. Ng, J. Broatch, J. Short and J. Roberts. 2016. Mitigating Chinese–Indian rhesus macaque (*Macaca mulatta*) hybridity at the California National Primate Research Center (CNPRC). *Journal of Medical Primatology* 45(6): 333–335.

Kaplan, J., S. Manuck, T. Clarkson, F. Lusso and D. Taub. 1982. Social status, environment, and atherosclerosis in cynomolgus monkeys. *Arteriosclerosis, Thrombosis, and Vascular Biology* 2(5): 359–368.

Kawai, M. 1965. Newly-acquired precultural behavior of the natural troop of Japanese monkeys on Koshima Island. *Primates* 6: 1–30.

Kiser, D., B. Steemers, I. Branchi and J. Homberg. 2011. The reciprocal interaction between serotonin and social behaviour. *Neuroscience and Biobehavioral Reviews* 36: 786–798.

Knowles, L., M. Fourrier and S. Eisele. 1995. Behavioral training of group-housed rhesus macaques (*Macaca mulatta*) for handling purposes. *Laboratory Primate Newsletter* 34(2): 1–4.

Koirala, S., M. K. Chalise, H. B. Katuwal, R. Gaire, B. Pandey and H. Ogawa. 2017. Diet and activity of *Macaca assamensis* in wild and semi-provisioned groups in Shivapuri Nagarjun National Park, Nepal. *Folia Primatologica* 88(2): 57–74.

Kringelbach, M. L., N. Jenkinson, S. L. F. Owen and T. Z. Aziz. 2007. Translational principles of deep brain stimulation. *Nature Reviews Neuroscience* 8(8): 623–635.

Kroeker, R., G. H. Lee, R. U. Bellanca, J. P. Thom and J. M. Worlein. 2017. Prior facility affects alopecia in adulthood for rhesus macaques. *American Journal of Primatology* 79(1): 1–9.

Kurland, J. A. 1973. A natural history of kra macaques (*Macaca fascicularis* Raffles, 1821) at the Kutai Reserve, Kalimantan Timur, Indonesia. *Primates* 14(2): 245–262.

Kuruvilla, G. P. 1978. Ecology of the bonnet macaque (*Macaca radiata* Geoffroy) with special reference to feeding habits. *Journal of the Bombay Natural History Society* 75: 976–988.

Lam, K., N. M. Rupniak and S. D. Iversen. 1991. Use of a grooming and foraging substrate to reduce cage stereotypies in macaques. *Journal of Medical Primatology* 20(3): 104–109.

Lambeth, S. P., J. Hau, J. E. Perlman, M. Martino and S. J. Schapiro. 2006. Positive reinforcement training affects hematologic and serum chemistry values in captive chimpanzees (*Pan troglodytes*). *American Journal of Primatology* 68(3): 245–256.

Lankau, E. W., P. V. Turner, R. J. Mullan and G. G. Galland. 2014. Use of nonhuman primates in research in North America. *Journal of the American Association for Laboratory Animal Science* 53(3): 278–282.

Laule, G. and T. Desmond. 1998. Positive reinforcement training as an enrichment strategy. *Second Nature: Environmental Enrichment for Captive Animals*, D. J. Shepherdson, J. D. Mellen and M. Hutchins. Washington, DC, Smithsonian Institution Press: 31–46.

Laule, G. and M. Whittaker. 2007. Enhancing nonhuman primate care and welfare through the use of positive reinforcement training. *Journal of Applied Animal Welfare Science* 10(1): 31–38.

Laule, G. E., M. A. Bloomsmith and S. J. Schapiro. 2003. The use of positive reinforcement training techniques to enhance the care, management, and welfare of primates in the laboratory. *Journal of Applied Animal Welfare Science* 6(3): 163–173.

Lee, D. R. and B. Oettinger. 2013. Behavior and enrichment of pig-tailed macaques. *A Guide to the Behaviour and Enrichment of Laboratory Macaques*, C. Winnicker. Wilmington, MA, Charles River Publications: 116–137.

Lee, P. C. and N. E. C. Priston. 2005. Perceptions of pests: Human attitudes to primates, conflict and consequences for conservation. *Primate – Human Interaction and Conservation*, J. D. Patterson and J. Wallace. Alberta, American Society of Primatologists Publications.

Lee, W., T. Hayakawa, M. Kiyono, N. Yamabata and G. Hanya. 2019. Gut microbiota composition of Japanese macaques associates with extent of human encroachment. *American Journal of Primatology* 81(12): e23072.

Levallois, L. and S. Desvaux de Marigny. 2015. Reproductive success of wild-caught and captive-bred cynomolgus macaques at a breeding facility. *Lab Animal Europe* 15(11): 12–24.

Li, C., C. Zhao and P.-F. Fan. 2015. White-cheeked macaque (*Macaca leucogenys*): a new macaque species from Medog, southeastern Tibet. *American Journal of Primatology* 77(7): 753–766.

Li, D., Q. Zhou, H. Tang and C. Huang. 2012. Sex-age differences in activity budget and position behavior of rhesus macaques (*Macaca mulatta*). *ACTA Theriologica Sinica* 32(1): 25–32.

Li, Y., G. Ma, Q. Zhou and Z. Huang. 2020. Seasonal variation in activity budget of Assamese macaques in limestone forest of southwest Guangxi, China. *Folia Primatologica* 91(5): 495–511. DOI: 10.1159/000506593.

Lindburg, D. 1971. The rhesus monkey in North India: an ecological and behavioral study. *Primate Behavior: Developments in Field and Laboratory Research*, L. A. Rosenblum. New York, Academic Press, Vol 2: 1–106.

Lindburg, D. 1977. Feeding behaviour and diet of rhesus monkeys (*Macaca mulatta*) in a Siwalik forest in North India. *Primate Ecology: Studies of Feeding and Ranging Behaviour in Lemurs, Monkeys and Apes*, T. H. Clutton-Brock. London, Academic Press: 223–249.

Lindburg, D. G. 1969. Rhesus monkeys: mating season mobility of adult males. *Science* 166(3909): 1176–1178.

Line, S. and K. Morgan. 1991. The effect of two novel objects on the behavior of singly caged adult rhesus macaques. *Laboratory Animal Science* 41(4): 365–369.

Line, S., K. Morgan, H. Markowitz, J. A. Roberts and M. Riddel. 1990a. Behavioral responses of female long-tailed macaques (*Macaca fascicularis*) to pair formation. *Laboratory Primate Newsletter* 29(4): 1–5.

Line, S., K. Morgan, H. Markowitz and S. Strong. 1990b. Increased cage size does not alter heart rate or behavior in female rhesus monkeys. *American Journal of Primatology* 20: 107–113.

Line, S. W., K. N. Morgan and H. Markowitz. 1991. Simple toys do not alter the behaviour of aged rhesus monkeys. *Zoo Biology* 10: 473–484.

Liu, B.-J., C.-F. Wu, P. A. Garber, P. Zhang and M. Li. 2018. Effects of group size and rank on mother–infant relationships and reproductive success in rhesus macaques (*Macaca mulatta*). *American Journal of Primatology* 80(7): e22881.

Lloyd, C. R., G. H. Lee and C. M. Crockett. 2005. Puzzle-ball foraging by laboratory monkeys improves with experience. *Laboratory Primate Newsletter* 44(1): 1–3.

Lucas, P. W., B. W. Darvell, P. K. D. Lee, T. D. B. Yuen and M. F. Choong. 1998. Colour cues for leaf food selection by long-tailed macaques (*Macaca fascicularis*) with a new suggestion for the evolution of trichromatic colour vision. *Folia Primatologica* 69(3): 139–154.

Luchins, K. R., K. C. Baker, M. H. Gilbert, J. L. Blanchard and R. P. Bohm. 2011a. Manzanita wood: a sanitizable enrichment option for nonhuman primates. *Journal of the American Association for Laboratory Animal Science* 50(6): 884–887.

Luchins, K. R., K. C. Baker, M. H. Gilbert, J. L. Blanchard, D. X. Liu, L. Myers and R. P. Bohm. 2011b. Application of the diagnostic evaluation for alopecia in traditional veterinary species to laboratory rhesus macaques (*Macaca mulatta*). *Journal of the American Association for Laboratory Animal Science* 50(6): 926–938.

Luetjens, C. M., A. Fuchs, A. Baker and G. F. Weinbauer. 2020. Group size experiences with enhanced pre- and postnatal development studies in the long-tailed macaque (*Macaca fascicularis*). *Primate Biology* 7(1): 1–4.

Luncz, L. V., M. S. Svensson, M. Haslam, S. Malaivijitnond, T. Proffitt and M. Gumert. 2017. Technological response of wild macaques (*Macaca fascicularis*) to anthropogenic change. *International Journal of Primatology* 38(5): 872–880.

Luttrell, L., L. Acker, M. Urben and V. Reinhardt. 1994. Training a large group of rhesus macaques to co-operate during catching: analysis of time investment. *Animal Welfare* 3: 135–140.

Lutz, C., L. Marinus, W. Chase, J. Meyer and M. Novak. 2003a. Self-injurious behavior in male rhesus macaques does not reflect externally directed aggression. *Physiology & Behavior* 78(1): 33–39.

Lutz, C., A. Well and M. Novak. 2003b. Stereotypic and self-injurious behaviour in rhesus macaques: a survey and retrospective analysis of environment and early experience. *American Journal of Primatology* 60: 1–15.

Lutz, C. K., K. Coleman, J. Worlein and M. A. Novak. 2013. Hair loss and hair-pulling in rhesus macaques (*Macaca mulatta*). *Journal of the American Association for Laboratory Animal Science* 52(4): 454–457.

Lutz, C. K. and M. A. Novak. 1995. Use of foraging racks and shavings as enrichment tools for groups of rhesus monkeys (*Macaca mulatta*). *Zoo Biology* 14(5): 463–474.

MacAllister, R. P., A. Heagerty and K. Coleman. 2020. Behavioral predictors of pairing success in rhesus macaques (*Macaca mulatta*). *American Journal of Primatology* 82(1): e23081.

MacArthur Clark, J. A. and D. Sun. 2020. Guidelines for the ethical review of laboratory animal welfare People's Republic of China National Standard GB/T 35892-2018 [Issued 6 February 2018 Effective from 1 September 2018]. *Animal Models and Experimental Medicine* 3(1): 103–113.

Maestripieri, D. 2005. Gestural communication in three species of macaques (*Macaca mulatta*, *M. nemestrina*, *M. arctoides*): use of signals in relation to dominance and social context. *Gesture* 5(1–2): 57–73.

Maestripieri, D., G. Schino, F. Aureli and A. Troisi. 1992. A modest proposal: displacement activities as an indicator of emotions in primates. *Animal Behaviour* 44(5): 967–979.

Magden, E. R. 2017. Positive reinforcement training and health care. *Handbook of Primate Behavioral Management*, S. Schapiro. Boca Raton, FL, CRC Press: 201–215.

Majumder, J., R. Lodh and B. Agarwala. 2012. Fish feeding adaptation by rhesus macaque *Macaca mulatta* (Cercopithecidae) in the Sundarban mangrove swamps, India. *Journal of Threatened Taxa* 4: 2539–2540.

Malik, I. and C. H. Southwick. 1987. Feeding behavior of free-ranging rhesus of Tughlaqabad. *Journal of Bombay Natural History Society* 84(2): 336–349.

Maninger, N., J. H. Kim and G. C. Ruppenthal. 1998. The presence of visual barriers decreases agonism in group housed pigtail macaques (*Macaca nemestrina*). *American Journal of Primatology* 45(2): 193–194.

Maréchal, L., S. Semple, B. Majolo and A. MacLarnon. 2016. Assessing the effects of tourist provisioning on the health of wild Barbary macaques in Morocco. *PLoS One* 11(5): e0155920.

Maruhashi, T. 1981. Activity patterns of a troop of Japanese monkeys (*Macaca fuscata yakui*) on Yakushima Island, Japan. *Primates* 22(1): 1–14.

Mason, G. 2006. Stereotypic behaviour in captive animals: fundamentals and implications for welfare and beyond. *Stereotypic Animal Behaviour: Fundamentals and Applications to Welfare*, G. Mason and J. Rushen. Wallingford, UK, CABI: 325–356.

Mason, S., E. Premereur, V. Pelekanos, A. Emberton, P. Honess and A. S. Mitchell. 2019. Effective chair training methods for neuroscience research involving rhesus macaques (*Macaca mulatta*). *Journal of Neuroscience Methods* 317: 82–93.

Matheson, M. D. and I. S. Bernstein. 2000. Grooming, social bonding, and agonistic aiding in rhesus monkeys. *American Journal of Primatology* 51: 177–186.

Matsumoto, S., C. J. Porter, N. Ogasawara, C. Iwatani, H. Tsuchiya, Y. Seita, Y.-W. Chang, I. Okamoto, M. Saitou, M. Ema, T. J. Perkins, W. L. Stanford and S. Tanaka. 2020. Establishment of macaque trophoblast stem cell lines derived from cynomolgus monkey blastocysts. *Scientific Reports* 10(1): 6827.

McCormack, K. M. and N. L. Megna. 2001. The effects of privacy walls on aggression in a captive group of rhesus macaques (*Macaca mulatta*). *American Journal of Primatology* 54(Suppl.): 50–51.

McGrew, K. 2017. Pairing strategies for cynomolgus macaques. *Handbook of Primate Behavioral Management*, S. Schapiro. Boca Raton, FL, CRC Press: 255–264.

McNab, B. K. 1963. Bioenergetics and the determination of home range size. *The American Naturalist* 97(894): 133–140.

Melnick, D. and M. Pearl. 1987. Cercopithecines in multimale groups: genetic diversity and population structure. *Primate Societies*, B. Smuts, D. Cheney, R. Seyfarth, R. Wrangham and T. Struhsaker. Chicago, IL, University of Chicago Press: 121–134.

Melnick, D., M. Pearl and A. Richard. 1984. Male migration and inbreeding avoidance in wild rhesus monkeys. *American Journal of Primatology* 7: 229–243.

Metzger, J. M., M. S. Lopez, J. K. Schmidt, M. E. Murphy, R. Vemuganti and M. E. Emborg. 2020. Effects of cardiac sympathetic neurodegeneration and PPARγ activation on rhesus macaque whole blood miRNA and mRNA eProfiles. *BioMed Research International* 2020: 9426204.

Milton, K. 1981. Distribution patterns of tropical plant foods as an evolutionary stimulus to primate mental development. *American Anthropologist* 83: 534–548.

Milton, K. and M. L. May. 1976. Body weight, diet and home range area in primates. *Nature* 259(5543): 459–462.

Mitchell, G. and D. H. Tokunaga. 1976. Sex differences in nonhuman primate grooming. *Behavioural Processes* 1: 335–345.

Molur, S., D. Brandon-Jones, W. Dittus, A. Eudey, A. Kumar, M. Singh, M. M. Feeroz, M. Chalise, P. Priya and S. Walker. 2003. *Status of South Asian Primates: Conservation Assessment and Management Plan (CAMP). Workshop Report, 2003*. Coimbatore, India, Zoo Outreach Organization/CBSG-South Asia.

Mook, D. M. 2002. Gastric trichobezoars in a rhesus macaque (*Macaca mulatta*). *Comparative Medicine* 52(6): 560–562.

Mukherjee, A. K. and T. S. Gupta. 1965. Habits of the rhesus macaque *Macaca mulatta* (Zimmermann) in the Sunderbans, 24-Parganas, West Bengal. *Journal of the Bombay Natural History Society* 62: 145–146.

Muletz-Wolz, C. R., N. P. Kurata, E. A. Himschoot, E. S. Wenker, E. A. Quinn, K. Hinde, M. L. Power and R. C. Fleischer. 2019. Diversity and temporal dynamics of primate milk microbiomes. *American Journal of Primatology* 81(10–11): e22994.

Munster, V. J., F. Feldmann, B. N. Williamson, N. van Doremalen, L. Pérez-Pérez, J. Schulz, K. Meade-White, A. Okumura, J. Callison, B. Brumbaugh, V. A. Avanzato, R. Rosenke, P. W. Hanley, G. Saturday, D. Scott, E. R. Fischer and E. de Wit. 2020. Respiratory disease in rhesus macaques inoculated with SARS-CoV-2. *Nature*. DOI: 10.1038/s41586-020-2324-7.

Murata, M., J.-i. Yasunaga, A. Washizaki, Y. Seki, M. Kuramitsu, T. W. Keat, A. Hu, K. Okuma, I. Hamaguchi, T. Mizukami, M. Matsuoka and H. Akari. 2020. Frequent horizontal and mother-to-child transmission may contribute to high prevalence of STLV-1 infection in Japanese macaques, *Retrovirology* 17: 15. DOI: 10.1186/s12977-020-00525-1.

Murchison, M. A. 1994. Primary forage feeder for singly-caged macaques. *Laboratory Primate Newsletter* 33(1): 7–9.

Musallam, S., B. D. Corneil, B. Greger, H. Scherberger and R. A. Andersen. 2004. Cognitive control signals for neural prosthetics. *Science* 305(5681): 258–262.

Nakamichi, M. and K. Asanuma. 1998. Behavioral effects of perches on group-housed adult female Japanese monkeys. *Perceptual and Motor Skills* 87: 707–714.

National Research Council. 2010. *Guide for the Care and Use of Laboratory Animals*. Washington DC, National Academies Press.

Neal Webb, S. J., Hau, J., Schapiro, S. J. 2018. Captive chimpanzee (*Pan troglodytes*) behavior as a function of space per animal and enclosure type. *American Journal of Primatology* 80(3): e22749. DOI: 10.1002/ajp.22749.

Nelson, A. N., W.-H. W. Lin, R. Shivakoti, N. E. Putnam, L. Mangus, R. J. Adams, D. Hauer, V. K. Baxter and D. E. Griffin. 2020. Association of persistent wild-type measles virus RNA with long-term humoral immunity in rhesus macaques. *JCI insight* 5(3): e134992.

Novak, M. A. 2003. Self-injurious behavior in rhesus monkeys: new insights into its etiology, physiology, and treatment. *American Journal of Primatology* 59(1): 3–19.

Novak, M. A., A. F. Hamel, A. M. Ryan, M. T. Menard and J. S. Meyer. 2017a. The role of stress in abnormal behavior and other abnormal conditions such as hair loss. *Handbook of Primate Behavioral Management*. S. Schapiro. Boca Raton, FL, CRC Press: 75–94.

Novak, M. A., M. T. Menard, S. N. El-Mallah, K. Rosenberg, C. K. Lutz, J. Worlein, K. Coleman and J. S. Meyer. 2017b. Assessing significant (>30%) alopecia as a possible biomarker for stress in captive rhesus monkeys (*Macaca mulatta*). *American Journal of Primatology* 79(1): 1–8.

Novak, M. A. and J. S. Meyer. 2009. Alopecia: possible causes and treatments, particularly in captive nonhuman primates. *Comparative Medicine* 59(1): 18–26.

Novak, M. A., P. O'Neill, S. A. Beckley and S. Suomi. 1994. Naturalistic environments for captive primates. *Captive Environments for Animal Behavior Research*, E. F. Gibbons, E. J. Wyers, E. Waters and E. W. Menzel. Albany, NY, State University of New York Press: 236–258.

Nunn, C. L. 1999. The evolution of exaggerated sexual swellings in primates and the graded-signal hypothesis. *Animal Behaviour* 58(2): 229–246.

Nunn, C. L., C. P. van Schaik and D. Zinner. 2001. Do exaggerated sexual swellings function in female mating competition in primates? A comparative test of the reliable indicator hypothesis. *Behavioral Ecology* 12(5): 646–654.

Oates, J. F. 1987. Food distribution and foraging behavior. *Primate Societies*, B. B. Smuts, D. L. Cheney, R. M. Seyfarth, R. W. Wrangham and T. T. Struhsaker. Chicago, IL, University of Chicago Press: 197–209.

Oettinger, B. C., C. M. Crockett and R. U. Bellanca. 2007. Communicative contexts of the LEN facial expression of pigtailed macaques (*Macaca nemestrina*). *Primates* 48(4): 293–302.

Oi, T. 1990. Population organization of wild pig-tailed macaques (*Macaca nemestrina nemestrina*) in West Sumatra. *Primates* 31(1): 15–31.

Oi, T. 1996. Sexual behaviour and mating system of the wild pig-tailed macaque in West Sumatra. *Evolution and Ecology of Macaque Societies*, J. Fa and D. Lindburg. Cambridge, Cambridge University Press: 342–368.

Ojha, P. R. 1974. Tail carriage and dominance in the rhesus monkey, *Macaca mulatta*. *Mammalia* 38(2): 163–170.

Osorio, D., A. C. Smith, M. Vorobyev and H. M. Buchanan-Smith. 2004. Detection of fruit and the selection of primate visual pigments for color vision. *The American Naturalist* 164(6): 696–708.

Park, J. E. and A. C. Silva. 2019. Generation of genetically engineered non-human primate models of brain function and neurological disorders. *American Journal of Primatology* 81(2): e22931.

Partan, S. 2002. Single and multichannel signal composition: facial expressions and vocalizations of Rhesus macaques (*Macaca mulatta*). *Behaviour* 139: 993–1027.

Peña, J. C. and W.-Z. Ho. 2015. Monkey models of tuberculosis: lessons learned. *Infection and Immunity* 83(3): 852–862.

Perlman, J. E., M. A. Bloomsmith, M. A. Whittaker, J. L. McMillan, D. E. Minier and B. McCowan. 2012. Implementing positive reinforcement animal training programs at primate laboratories. *Applied Animal Behaviour Science* 137(3–4): 114–126.

Peterson, E. J., J. M. Worlein, G. H. Lee, A. M. Dettmer, E. K. Varner and M. A. Novak. 2017. Rhesus macaques (*Macaca mulatta*) with self-injurious behavior show less behavioral anxiety during the human intruder test. *American Journal of Primatology* 79(1): 1–8.

Phillips, K. A., K. L. Bales, J. P. Capitanio, A. Conley, P. W. Czoty, B. A. 't Hart, W. D. Hopkins, S. L. Hu, L. A. Miller, M. A. Nader, P. W. Nathanielsz, J. Rogers, C. A. Shively and M. L. Voytko. 2014. Why primate models matter. *American Journal of Primatology* 76(9): 801–827.

Platt, D. M., G. Carey and R. D. Spealman. 2011. Models of neurological disease (substance abuse): self-administration in monkeys. *Current Protocols in Pharmacology* 56: 10.5.1–10.5.17.

Plesker, R. and A. Schuhmacher. 2006. Feeding fruits and vegetables to nonhuman primates can lead to nutritional deficiencies. *Laboratory Primate Newsletter* 45(4): 1–5.

Poirier, C., C. J. Oliver, J. Castellano Bueno, P. Flecknell and M. Bateson. 2019. Pacing behaviour in laboratory macaques is an unreliable indicator of acute stress. *Scientific Reports* 9(1): 7476.

Pomerantz, O. and K. C. Baker. 2017. Higher levels of submissive behaviors at the onset of the pairing process of rhesus macaques (*Macaca mulatta*) are associated with lower risk of wounding following introduction. *American Journal of Primatology* 79(8): e22671.

Post, W. and J. Baulu. 1978. Time budgets of *Macaca mulatta*. *Primates* 19(1): 125–140.

Prescott, M. J., V. J. Brown, P. A. Flecknell, D. Gaffan, K. Garrod, R. N. Lemon, A. J. Parker, K. Ryder, W. Schultz, L. Scott, J. Watson and L. Whitfield. 2010. Refinement of the use of food and fluid control as motivational tools for macaques used in behavioural neuroscience research: report of a Working Group of the NC3Rs. *Journal of Neuroscience Methods* 193(2): 167–188.

Prescott, M. J., M. E. Nixon, D. A. H. Farningham, S. Naiken and M.-A. Griffiths. 2012. Laboratory macaques: when to wean? *Applied Animal Behaviour Science* 137(3–4): 194–207.

Pusey, A. E. and C. Packer. 1987. Dispersal and philopatry. *Primate Societies*, B. B. Smuts, D. L. Cheney, R. M. Seyfarth, R. W. Wrangham and T. T. Struhsaker. Chicago, IL, University of Chicago Press: 250–266.

Quiatt, D. 1979. Aunts and mothers: adaptive implications of allomaternal behavior in nonhuman primates. *American Anthropologist* 81: 310–319.

Rahaman, H. and M. D. Parthasarathy. 1969. Studies on the social behaviour of bonnet monkeys. *Primates* 10(2): 149–162.

Ramanathan, K. 1994. Diet composition of the Bonnet macaque (*Macaca radiata*) in a tropical dry evergreen forest of Southern India. *Tropical Biodiversity* 2(2): 285–302.

Reinhardt, V. 1992a. Avoiding aggression during and after pair formation of adult rhesus macaques. *Laboratory Primate Newsletter* 31(3): 10–12.

Reinhardt, V. 1992b. Space utilization by captive rhesus macaques. *Animal Technology* 43(11–17): 11.

Reinhardt, V. 1993a. Foraging enrichment for caged macaques: a review. *Laboratory Primate Newsletter* 32(4): 1–5.

Reinhardt, V. 1993b. Using the mesh ceiling as a food puzzle to encourage foraging behaviour in caged rhesus macaques (*Macaca mulatta*). *Animal Welfare* 2: 165–172.

Reinhardt, V. 1994. Pair-housing rather than single-housing for laboratory rhesus macaques. *J Med Primatol* 23(8): 426–431.

Reinhardt, V. 1997. Training nonhuman primates to cooperate during blood collection: a review. *Laboratory Primate Newsletter* 36(4): 1–4.

Reinhardt, V. 1999. Pair-housing overcomes self-biting in macaques. *Laboratory Primate Newsletter* 38(1): 4–6.

Reinhardt, V. 2002a. Addressing the social needs of macaques used for research. *Laboratory Primate Newsletter* 41(3): 7–11.

Reinhardt, V. 2002b. Artificial weaning of Old World monkeys: benefits and costs. *Journal of Applied Animal Welfare Science* 5(2): 151–156.

Reinhardt, V. 2005. Hair pulling: a review. *Laboratory Animals* 39(4): 361–369.

Reinhardt, V., K. Baker, A. Lablans, E. Davis, J. P. Garner, C. Sherwin, S. Banjanin, L. Bell, J. Barley and J. Schrier. 2004. Self-injurious biting in laboratory animals: a discussion. *Laboratory Primate Newsletter* 43(2): 11–13.

Reinhardt, V., S. Eisele and D. Houser. 1988. Environmental enrichment program for caged macaques: a review. *Laboratory Primate Newsletter* 27(2): 5–6.

Reinhardt, V., C. Liss and C. Stevens. 1995a. Restraint methods of laboratory non-human primates: a critical review. *Animal Welfare* 4(3): 221–238.

Reinhardt, V., C. Liss and C. Stevens. 1995b. Social housing of previously single-caged macaques: what are the options and the risks? *Animal Welfare* 4: 307–328.

Reinhardt, V., C. Liss and C. Stevens. 1996. Space requirement stipulations for caged non-human primates in the United States: a critical review. *Animal Welfare* 5(4): 361–372.

Reinhardt, V. and A. Reinhardt. 1991. Impact of a privacy panel on the behavior of caged female rhesus monkeys living in pairs. *Journal of Experimental Animal Science* 34(2): 55–58.

Reinhardt, V., A. Reinhardt and D. Houser. 1986. Hair pulling and eating in captive rhesus monkey troops. *Folia Primatologica* 47: 158–164.

Reinhardt, V. and A. Roberts. 1997. Effective feeding enrichment for non-human primates: a brief review. *Animal Welfare* 6(3): 265–272.

Reinhardt, V. and M. Rossell. 2001. Self-biting in caged macaques: cause, effect and treatment. *Journal of Applied Animal Welfare Science* 4(4): 285–294.

Richard, A., S. Goldstein and R. Dewar. 1989. Weed macaques: the evolutionary implications of macaque feeding ecology. *International Journal of Primatology* 10(6): 569–594.

Robins, J. G. and C. D. Waitt. 2010. Improving the welfare of captive macaques (*Macaca sp.*) through the use of water as enrichment. *Journal of Applied Animal Welfare Science* 14(1): 75–84.

Rodman, P. S. 1979. Skeletal differentiation of *Macaca fascicularis* and *Macaca nemestrina* in relation to arboreal and terrestrial quadrupedalism. *American Journal of Physical Anthropology* 51(1): 51–62.

Rogers, J. 2018. The behavioral genetics of nonhuman primates: status and prospects. *American Journal of Physical Anthropology* 165: 23–36.

Rollin, B. E. 2015. *Telos*, conservation of welfare, and ethical issues in genetic engineering of animals. *Current Topics in Behavioral Neuroscience* 19: 99–116.

Rommeck, I., K. Anderson, A. Heagerty, A. Cameron and B. McCowan. 2009. Risk factors and remediation of self-injurious and self-abuse behavior in rhesus macaques. *Journal of Applied Animal Welfare Science* 12(1): 61–72.

Rommeck, I., J. P. Capitanio, S. C. Strand and B. McCowan. 2011. Early social experience affects behavioral and physiological responsiveness to stressful conditions in infant rhesus macaques (*Macaca mulatta*). *American Journal of Primatology* 73(7): 692–701.

Rosenblum, L. A., I. C. Kaufman and A. J. Stynes. 1964. Individual distance in two species of macaque. *Animal Behaviour* 12(2–3): 338–342.

Rosenblum, L. A., I. C. Kaufman and A. J. Stynes. 1966. Some characteristics of adult social and autogrooming patterns in two species of macaque. *Folia Primatologica* 4: 438–451.

Rowe, N. 1996. *The Pictorial Guide to the Living Primates*. East Hampton, NY, Pogonias Press.

Rox, A., A. H. van Vliet, E. H. M. Sterck, J. A. M. Langermans and A. L. Louwerse. 2019. Factors determining male introduction success and long-term stability in captive rhesus macaques. *PLoS One* 14(7): e0219972.

Roy, C. J., G. Van Slyke, D. Ehrbar, Z. A. Bornholdt, M. B. Brennan, L. Campbell, M. Chen, D. Kim, N. Mlakar, K. J. Whaley, J. W. Froude, F. J. Torres-Velez, E. Vitetta, P. J. Didier, L. Doyle-Meyers, L. Zeitlin and N. J. Mantis. 2020. Passive immunization with an extended half-life monoclonal antibody protects Rhesus macaques against aerosolized ricin toxin. *NPJ Vaccines* 5(1): 13.

Ruppert, N., A. Holzner, K. W. See, A. Gisbrecht and A. Beck. 2018. Activity budgets and habitat use of wild southern pig-tailed macaques (*Macaca nemestrina*) in oil palm plantation and forest. *International Journal of Primatology* 39(2): 237–251.

Russell, R. G., S. L. Rosenkranz, L. A. Lee, H. Howard, R. F. DiGiacomo, M. A. Bronsdon, G. A. Blakley, C. C. Tsai and W. R. Morton. 1987. Epidemiology and etiology of diarrhea in colony-born *Macaca nemestrina*. *Laboratory Animal Science* 37: 309–316.

Russell, W. M. S. and R. L. Burch. 1959. *The Principles of Humane Experimental Technique*. London, Methuen.

Samuels, A. and R. V. Henrickson. 1983. Brief report: outbreak of severe aggression in captive *Macaca mulatta*. *American Journal of Primatology* 5(3): 277–281.

Sato, N., K. Stringaris, J. K. Davidson-Moncada, R. Reger, S. S. Adler, C. Dunbar, P. L. Choyke and R. W. Childs. 2020. In Vivo tracking of adoptively transferred natural killer cells in rhesus macaques using ^{89}zirconium-oxine cell labeling and PET imaging. *Clinical Cancer Research* 26(11): 2573–2581. DOI: 10.1158/1078-0432.CCR-19-2897.

Schapiro, S., Ed. 2017. *Handbook of Primate Behavioral Management*. Boca Raton, FL, CRC Press, Taylor and Francis Group.

Schapiro, S., P. Nehete, J. Perlman and J. Sastry. 2000. A comparison of cell-mediated immune responses in rhesus macaques housed singly, in pairs, or in groups. *Applied Animal Behaviour Science* 68: 67–84.

Schapiro, S. J., M. Bloomsmith and G. Laule. 2003. Positive reinforcement training as a technique to alter nonhuman primate behaviour: quantitative assessments of effectiveness. *Journal of Applied Animal Welfare Science* 6(3): 175–187.

Schapiro, S. J., M. A. Bloomsmith, S. A. Suarez and L. M. Porter. 1996. Effects of social and inanimate enrichment on the behavior of yearling rhesus monkeys. *American Journal of Primatology* 40: 247–260.

Schapiro, S. J. and D. Bushong. 1994. Effects of enrichment on veterinary treatment of laboratory rhesus macaques (*Macaca mulatta*). *Animal Behaviour* 3: 25–36.

Schapiro, S. J., G. E. Laule, M. A. Bloomsmith and T. J. Desmond. 1995. Exploring and advancing environmental enrichment: a primate training and enrichment workshop. *Lab Animal* 24(4): 35–39.

Schapiro, S. J., J. E. Perlman and B. A. Boudreau. 2001. Manipulating the affiliative interactions of group-housed rhesus macaques using positive reinforcement training techniques. *American Journal of Primatology* 55(3): 137–149.

Schapiro, S. J., J. E. Perlman, E. Thiele and S. P. Lambeth. 2005. Training nonhuman primates to perform behaviors useful in biomedical research. *Lab Animal Europe* 5(5): 19–26.

SCHER. 2009. *The Need for Non-human Primates in Biomedical Research, Production and Testing of Products and Devices*. Brussels, Scientific Committee on Health and Environmental Risks, European Commission: 38.

Schino, G., F. Aureli, F. R. Damato, M. Dantoni, N. Pandolfi and A. Troisi. 1993. Infant kidnapping and co-mothering in Japanese macaques. *American Journal of Primatology* 30(3): 257–262.

Schino, G., S. Scucchi, D. Maestripieri and P. G. Turillazzi. 1988. Allogrooming as a tension-reduction mechanism: a behavioral approach. *American Journal of Primatology* 16(1): 43–50.

Schrier, J. 2004. Plants for browse, plants *not* for browse. *Laboratory Primate Newsletter* 43(1): 34.

Schultz, C. L. 2017. Nutrition, feeding, and behavioral management. *Handbook of Primate Behavioral Management*. S. Schapiro. Boca Raton, FL, CRC Press: 473–480.

Seier, J. V., M. A. Dhansay and A. Davids. 2005. Risks associated with environmental enrichment: intestinal obstruction caused by foraging substrate. *Journal of Medical Primatology* 34(3): 154–155.

Sengupta, A., E. Gazagne, A. Albert-Daviaud, Y. Tsuji and S. Radhakrishna. 2020. Reliability of macaques as seed dispersers. *American Journal of Primatology* 82(5): e23115.

Sengupta, A. and S. Radhakrishna. 2016. Influence of fruit availability on fruit consumption in a generalist primate, the rhesus macaque *Macaca mulatta*. *International Journal of Primatology* 37(6): 703–717.

Setchell, J. M. and P. M. Kappeler. 2003. Selection in relation to sex in primates. *Advances in the Study of Behavior*, P. J. B. Slater, J. S. Rosenblatt, T. J. Roper, C. T. Snowdon and M. Naguib. New York, Academic Press, 33: 87–173.

Seth, P. K. and S. Seth. 1986. Ecology and behaviour of rhesus monkeys in India. *Primate Ecology and Conservation*, J. G. Else and P. C. Lee. Cambridge, Cambridge University Press: 89–103.

Seyfarth, R. 1987. Vocal communication and its relation to language. *Primate Societies*, B. B. Smuts, D. L. Cheney, R. M. Seyfarth, R. W. Wrangham and T. T. Struhsaker. Chicago, IL, University of Chicago Press: 440–451.

Shannon, C., M. Champoux and S. J. Suomi. 1998. Rearing condition and plasma cortisol in rhesus monkey infants. *American Journal of Primatology* 46(4): 311–321.

Shimada, M. and C. Sueur. 2018. Social play among juvenile wild Japanese macaques (*Macaca fuscata*) strengthens their social bonds. *American Journal of Primatology* 80(1): e22728.

Shively, C. A. 2017. Depression in captive nonhuman primates: theoretical underpinnings, methods, and application to behavioral management. *Handbook of Primate Behavioral Management*. S. Schapiro. Boca Raton, FL, CRC Press: 115–125.

Shively, C., S. Clarke, N. King, S. Schapiro and G. Mitchell. 1982. Patterns of sexual behavior in male macaques. *American Journal of Primatology* 2(4): 373–384.

Shumaker, R. and S. Woods. 1995. Browse for nonhuman primates in captivity. *Laboratory Primate Newsletter* 34(4): 28.

Shutt, K., A. MacLarnon, M. Heistermann and S. Semple. 2007. Grooming in Barbary macaques: better to give than to receive? *Biology Letters* 3(3): 231–233.

Silk, J. B. 1982. Altruism among female *Macaca radiata*: explanations and analysis of patterns of grooming and coalition formation. *Behaviour* 79: 162–188.

Silk, J. B. 1983. Local resource competition and facultative adjustment of sex ratios in relation to competitive abilities. *American Naturalist* 121: 56–66.

Silk, J. B. 1999. Why are infants so attractive to others? The form and function of infant handling in bonnet macaques. *Animal Behaviour* 57(5): 1021–1032.

Sinha, A. 2001. The bonnet macaque revisited: ecology, demography and behaviour. *ENVIS Bulletin: Wildlife and Protected Areas* 1: 30–39.

Smith, D. G. 2005. Genetic characterization of Indian-origin and Chinese-origin rhesus macaques (*Macaca mulatta*). *Comparative Medicine* 55(3): 227–230.

Smith, E. O. and P. G. Peffer-Smith. 1982. Triadic interactions in captive Barbary macaques (*Macaca sylvanus*, Linnaeus, 1758): "Agonistic buffering"? *American Journal of Primatology* 2: 99–107.

Snowdon, C. T. 1994. The significance of naturalistic environments for primate behavioural research. *Naturalistic Environments in Captivity for Animal Behavioral Research*, E. F. Gibbons, E. J. Wyers, E. Waters and E. W. Menzel. Albany, State University of New York Press: 217–235.

Son, V. D. 2003. Diet of *Macaca fascicularis* in a mangrove forest, Vietnam. *Laboratory Primate Newsletter* 42(4): 1–5.

Southwick, C. H. 1967. An experimental study of intragroup agonistic behavior in rhesus monkeys (*Macaca mulatta*). *Behaviour* 28(1/2): 182–209.

Southwick, C. H., M. F. Siddioi, M. Y. Farooqui and B. C. Pal. 1976. Effects of artificial feeding on aggressive behaviour of rhesus monkeys in India. *Animal Behaviour* 24(1): 11–15.

Southwick, C. H., Z. Yongzu, J. Haisheng, L. Zhenhe and Q. Wenyuan. 1996. Population ecology of rhesus macaques in tropical and temperate habitats in China. *Evolution and Ecology of Macaque Societies*, J. Fa and D. Lindburg. Cambridge, Cambridge University Press: 95–105.

Stanley, M. A. 2003. The breeding of naturally occurring B virus-free cynomolgus monkeys (Macaca fascicularis) on the island of Mauritius. *International Perspectives: The Future of Nonhuman Primate Resources, Proceedings of the Workshop Held April 17–19, 2002*. Washington, DC, National Academies Press: 46–48.

Statz, L. M. and M. Borde. 2001. Pairing successes with male cynomolgus macaques after vasectomy. *Contemporary Topics in Laboratory Animal Science* 40(4): 91.

Stavisky, R. C., J. K. Ramsey, T. Meeker, M. Stovall and M. M. Crane. 2018. Trauma rates and patterns in specific pathogen free (SPF) rhesus macaque (*Macaca mulatta*) groups. *American Journal of Primatology* 80(3): e22742.

Sterck, E. H. M., D. G. M. Zijlmans, H. de Vries, L. M. van den Berg, C. P. van Schaik and J. A. M. Langermans. 2019. Determining overweight and underweight with a new weight-for-height index in captive group-housed macaques. *American Journal of Primatology* 81(6): e22996.

Sussman, A. F., J. C. Ha, K. L. Bentson and C. M. Crockett. 2013. Temperament in rhesus, long-tailed, and pigtailed macaques varies by species and sex. *American Journal of Primatology* 75(4): 303–313.

Suzuki, S., D. Hill, T. Maruhashi and T. Tsukahara. 1990. Frog- and lizard-eating behaviour of wild Japanese macaques in Yakushima, Japan. *Primates* 31: 421–426.

Syamil, A. R., A. R. Mohd-Ridwan, M. A. Amsah, M. A. B. Abdul-Latiff and B. M. Md-Zain. 2019. Population census and age category character of stump-tailed macaque, *Macaca arctoides*, in Northern Peninsular Malaysia. *Biodiversitas Journal of Biological Diversity* 20: 2446–2452.

Symons, D. 1978. *Play and Aggression: A Study of Rhesus Monkeys*. New York, Columbia University Press.

Taira, K. and E. T. Rolls. 1996. Receiving grooming as a reinforcer for the monkey. *Physiology & Behavior* 59(6): 1189–1192.

Tamura, M. 2020. Extractive foraging on hard-shelled walnuts and variation of feeding techniques in wild Japanese macaques (*Macaca fuscata*). *American Journal of Primatology* 82(6): e23130.

Tang, C., L. Huang, Z. Huang, A. Krzton, C.-h. Lu and Q. Zhou. 2015. Forest seasonality shapes diet of limestone-living rhesus macaques at Nonggang, China. *Primates* 57: 83–92.

Tasdemir, D., A. J. J. MacIntosh, P. Stergiou, M. Kaiser, N. R. Mansour, Q. Bickle and M. A. Huffman. 2020. Antiprotozoal and antihelminthic properties of plants ingested by wild Japanese macaques (Macaca fuscata yakui) in Yakushima Island. *Journal of Ethnopharmacology* 247: 112270.

Teas, J., T. Richie, H. Taylor and C. H. Southwick. 1980. Population patterns and behavioral ecology of rhesus monkeys (*Macaca mulatta*) in Nepal. *The Macaques: Studies in Ecology, Behavior and Evolution*, D. G. Lindburg. New York, Van Norstrand Reinhold: 247–262.

Terry, R. L. 1970. Primate grooming as a tension reduction mechanism. *Journal of Psychology: Interdisciplinary and Applied* 76(1): 129–136.

Thierry, B. 2000. Covariation of conflict management patterns across macaque species. *Natural Conflict Resolution*, F. Aureli and F. B. M. de Waal. Berkeley, University of California Press: 106–128.

Thierry, B. 2007. Unity in diversity: lessons from macaque societies. *Evolutionary Anthropology* 16: 224–238.

Thierry, B., M. Singh and W. Kaumanns. 2004. Why macaque societies? *Macaque Societies: A Model for the Study of Social Organization*, B. Thierry, M. Singh and W. Kaumanns. Cambridge, Cambridge University Press: 3–10.

Thomasy, S. M., S. Kim, S. Park, O. Pomerantz, I. Casanova, B. Gates, L. J. Young, A. N. Cameron, L. A. Tatum, J. A. Roberts, T. Stout, R. Chen, J. Rogers and A. Moshiri. 2020. Spontaneous ocular lesions in rhesus macaques with eye poking behavior. *Investigative Ophthalmology & Visual Science* 61(7): 187–187.

Thompson, N. S. 1967. Some variable affecting the behaviour of irus macaques in dyadic encounters. *Animal Behaviour* 15: 307–311.

Tiefenbacher, S., M. A. Fahey, J. K. Rowlett, J. S. Meyer, A. L. Pouliot, B. M. Jones and M. A. Novak. 2005. The efficacy of diazepam treatment for the management of acute wounding episodes in captive rhesus macaques. *Comparative Medicine* 55(4): 387–392.

Timmermans, P. J. A. and J. M. H. Vossen. 1996. The influence of rearing conditions on maternal behavior in cynomolgus macaques (*Macaca fascicularis*). *International Journal of Primatology* 17(2): 259–276.

Toyoda, A. and S. Malaivijitnond. 2018. The first record of dizygotic twins in semi-wild stump-tailed macaques (*Macaca arctoides*) tested using microsatellite markers and the occurrence of supernumerary nipples. *Mammal Study* 43(3): 207–212, 206.

Toyoda, A., T. Maruhashi, S. Malaivijitnond, H. Koda and Y. Ihara. 2020. Cooperation for copulation: a novel ecological mechanism underlying the evolution of coalition for sharing mating opportunities. *bioRxiv*. DOI: 10.1101/2020.01.30.927772.

Troisi, A. and S. Gabriele. 1987. Environmental and social influences on autogrooming behaviour in a captive group of Java monkeys. *Behaviour* 100(1/4): 292–302.

Trollope, J. and N. G. Blurton-Jones. 1970. Breeding the stump-tailed macaque, *Macaca arctoides*. *Laboratory Animals* 4: 161–169.

Truelove, M. A., A. L. Martin, J. E. Perlman, J. S. Wood and M. A. Bloomsmith. 2017. Pair housing of macaques: a review of partner selection, introduction techniques, monitoring for compatibility, and methods for long-term maintenance of pairs. *American Journal of Primatology* 79(1): 1–15.

Tsuji, Y., T. Y. Ito, K. Wada and K. Watanabe. 2015. Spatial patterns in the diet of the Japanese macaque *Macaca fuscata* and their environmental determinants. *Mammal Review* 45(4): 227–238.

Turner, J. G., C. A. Bauer and L. P. Rybak. 2007. Noise in animal facilities: why it matters. *Journal of the American Association for Laboratory Animal Science* 46(1): 10–13.

Turner, P. V. and L. E. Grantham. 2002. Short-term effects of an environmental enrichment program for adult cynomolgus monkeys. *Contemporary Topics in Laboratory Animal Science* 41(5): 13–17.

van Doremalen, N., E. Haddock, F. Feldmann, K. Meade-White, T. Bushmaker, R. J. Fischer, A. Okumura, P. W. Hanley, G. Saturday, N. J. Edwards, M. H. A. Clark, T. Lambe, S. C. Gilbert and V. J. Munster. 2020. A single dose of ChAdOx1 MERS provides protective immunity in rhesus macaques. *Science Advances*: eaba8399.

van Hooff, J. A. R. A. M. 1981. Facial expressions. *The Oxford Companion to Animal Behaviour*, D. McFarland. Oxford, Oxford University Press: 165–176.

van Noordwijk, M. and C. van Schaik. 1999a. The effects of dominance rank and group size on female lifetime reproductive success in wild long-tailed macaques, *Macaca fascicularis*. *Primates* 40(1): 105–130.

van Noordwijk, M. A. and C. P. van Schaik. 1988. Male careers in Sumatran long-tailed macaques (*Macaca fascicularis*). *Behaviour* 107(1–2): 24–43.

van Noordwijk, M. A. and C. P. van Schaik. 1999b. The effects of dominance rank and group size on female lifetime reproductive success in wild long-tailed macaques, *Macaca fascicularis*. *Primates* 40(1): 105–130.

van Schaik, C. and T. Mitrasetia. 1990. Changes in the behaviour of wild long-tailed macaques (*Macaca fascicularis*) after encounters with a model python. *Folia Primatologica* 55(2): 104–108.

van Schaik, C. P., A. J. J. van Amerongen and M. A. van Noordwijk. 1996. Riverine refuging by wild Sumatran long-tailed macaques (*Macaca fascicularis*). *Evolution and Ecology of Macaque Societies*, J. E. Fa and D. G. Lindburg. Cambridge, Cambridge University Press: 160–181.

van Schaik, C. P. and M. A. van Noordwijk. 1985. Interannual variability in fruit abundance and the reproductive seasonality in Sumatran Long-tailed macaques (*Macaca fascicularis*). *Journal of Zoology* 206(4): 533–549.

Vanni, S., H. Hokkanen, F. Werner and A. Angelucci. 2020. Anatomy and physiology of macaque visual cortical areas V1, V2, and V5/MT: bases for biologically realistic models. *Cerebral Cortex* 30(6): 3483–3517.

Veeramani, A., K. B. Pushpalatha, H. Mohanakrishnan, B. Ramakrishnan, P. Santhoshkumar, A. Samsan and S. Karthick. 2019. Demography, activity pattern, food and feeding habits of primates in Nilgiris, Tamil Nadu, India. *International Journal of Zoology and Animal Biology* 2(1): 1–11.

Vessey, S. H. 1984. Dominance among rhesus monkeys. *Political Psychology* 5(4): 623–628.

Vinson, A., K. Prongay and B. Ferguson. 2013. The value of extended pedigrees for next-generation analysis of complex disease in the rhesus macaque. *ILAR Journal* 54(2): 91–105.

Virnik, K., M. Rosati, A. Medvedev, A. Scanlan, G. Walsh, F. Dayton, K. E. Broderick, M. Lewis, Y. Bryson, J. D. Lifson, R. M. Ruprecht, B. K. Felber and I. Berkower. 2020. Immunotherapy with DNA vaccine and live attenuated rubella/SIV gag vectors plus early ART can prevent SIVmac251 viral rebound in acutely infected rhesus macaques. *PLoS One* 15(3): e0228163.

Vogt, J. L., C. L. Coe and S. Levine. 1981. Behavioural and adrenocorticoid responsiveness of squirrel monkeys to a live snake: is flight necessarily stressful? *Behavioural and Neural Biology* 32(4): 391–405.

Waitt, C., H. M. Buchanan-Smith and K. Morris. 2002. The effects of caretaker-primate relationships on primates in the laboratory. *Journal of Applied Animal Welfare Science* 5(4): 309–319.

Waitt, C., M. Bushmitz and P. Honess. In prep. Refining feeding routines for macaques to decrease feeding competition and improve welfare.

Waitt, C., P. Honess and M. Bushmitz. 2008. Creating housing to meet the behavioural needs of long-tailed macaques. *Laboratory Primate Newsletter* 47(4): 1–5.

Waitt, C., A. C. Little, S. Wolfensohn, P. Honess, A. P. Brown, H. M. Buchanan-Smith and D. I. Perrett. 2003. Evidence from rhesus macaques suggests that male coloration plays a role in female primate mate choice. *Proceedings of the Royal Society of London Series B-Biological Sciences* 270: S144–S146.

Waitt, C. D., M. Bushmitz and P. E. Honess. 2010. Designing environments for aged primates. *Laboratory Primate Newsletter* 49(3): 5–9.

Wang, K.-Y., K. L. Christe, J. Yee, J. A. Roberts and A. Ardeshir. 2019. Rotavirus is associated with decompensated diarrhea among young rhesus macaques (*Macaca mulatta*). *American Journal of Primatology* 81(1): e22948.

Wang, X., V. C. Cuzon Carlson, C. Studholme, N. Newman, M. M. Ford, K. A. Grant and C. D. Kroenke. 2020. In utero MRI identifies consequences of early-gestation alcohol drinking on fetal brain development in rhesus macaques. *Proceedings of the National Academy of Sciences* 117(18): 10035–10044.

Wechsler, B. 1995. Coping and coping strategies: a behavioural view. *Applied Animal Behaviour Science* 43(2): 123–134.

Weed, J. L. and J. M. Raber. 2005. Balancing animal research with animal well-being: establishment of goals and harmonization of approaches. *ILAR Journal* 46(2): 118–128.

Weed, J. L., P. O. Wagner, R. Byrum, S. Parrish, M. Knezevich and D. A. Powell. 2003. Treatment of persistent self-injurious behavior in rhesus monkeys through socialization: a preliminary report. *Contemporary Topics in Laboratory Animal Science* 42(5): 21–23.

Weiß, B. M., L. Kulik, A. V. Ruiz-Lambides and A. Widdig. 2016. Individual dispersal decisions affect fitness via maternal rank effects in male rhesus macaques. *Scientific Reports* 6(1): 32212.

Wergård, E.-M., H. Temrin, B. Forkman, M. Spångberg, H. Fredlund and K. Westlund. 2015. Training pair-housed rhesus macaques (*Macaca mulatta*) using a combination of negative and positive reinforcement. *Behavioural Processes* 113: 51–59.

Wergård, E.-M., K. Westlund, M. Spångberg, H. Fredlund and B. Forkman. 2016. Training success in group-housed long-tailed macaques (*Macaca fascicularis*) is better explained by personality than by social rank. *Applied Animal Behaviour Science* 177: 52–58.

Westergaard, G. C., M. K. Izard, J. H. Drake, S. J. Suomi and J. D. Higley. 1999. Rhesus macaque (*Macaca mulatta*) group formation and housing: wounding and reproduction in a specific pathogen free (SPF) colony. *American Journal of Primatology* 49: 339–347.

Westlund, K. 2012. Can conditioned reinforcers and variable-ratio schedules make food- and fluid control redundant? A comment on the NC3Rs Working Group's report. *Journal of Neuroscience Methods* 204(1): 202–205.

Westlund, K. 2014. Training is enrichment—and beyond. *Applied Animal Behaviour Science* 152: 1–6.

Westlund, K. 2015. Training laboratory primates - benefits and techniques. *Primate Biology* 2(1): 119–132.

Wheatley, B. P. 1988. Cultural behavior and extractive foraging in *Macaca fascicularis*. *Current Anthropology* 29(3): 516–519.

Wheatley, B. P., D. K. Harvya Putra and M. K. Gonder. 1996. A comparison of wild and food provisioned long-tailed macaques (*Macaca fascicularis*). *Evolution and Ecology of Macaque Societies*, J. E. Fa and D. G. Lindburg. Cambridge, Cambridge University Press: 182–206.

Wilson, J. M., C. K. Wallace, A. K. Brice and L. Makaron. 2020. Mineralized trichobezoars in a rhesus macaque (*Macaca mulatta*). *Journal of Medical Primatology* 49(3): 158–161.

Winnicker, C. and P. Honess. 2014. Evaluating the effectiveness of environmental enrichment. *Laboratory Animal Science Professional* 2014: 16–20.

Winnicker, C., P. Honess, S. Schapiro, M. A. Bloomsmith, D. R. Lee, B. McCowan, B. Oettinger and J. H. Simmons. 2013. *A Guide to the Behavior and Enrichment of Laboratory Macaques*. Wilmington, MA, Charles River Publications.

Wolfensohn, S. E. 1998. *Shigella* infection in macaque colonies: case report of an eradication and control program. *Laboratory Animal Science* 48: 330–333.

Wolfensohn, S. E. and P. E. Honess. 2005. *Handbook of Primate Husbandry and Welfare*. Oxford, UK, Blackwell Publications.

Wooddell, L. J., B. Beisner, D. L. Hannibal, A. C. Nathman and B. McCowan. 2019a. Increased produce enrichment reduces trauma in socially-housed captive rhesus macaques (*Macaca mulatta*). *American Journal of Primatology* 81(12): e23073.

Wooddell, L. J., S. S. K. Kaburu and A. M. Dettmer. 2019b. Dominance rank predicts social network position across developmental stages in rhesus monkeys. *American Journal of Primatology* 82: e23024.

Wrangham, R. W. 1980. An ecological model of female-bonded primate groups. *Behaviour* 75(3/4): 262–300.

Wrangham, R. W., J. L. Gittleman and C. Chapman. 1993. Constraints on group size in primates and carnivores: population density and day-range as assays of exploitation competition. *Behavioral Ecology and Sociobiology* 32: 199–209.

Yeager, C. P. 1996. Feeding ecology of the long-tailed macaque (*Macaca fascicularis*) in Kalimantan Tengah, Indonesia. *International Journal of Primatology* 17(1): 51–62.

Young, R. J. 2003a. Designing and analysing enrichment studies. *Environmental Enrichment for Captive Animals*. Oxford, Blackwell Science Ltd: Xii, 228.

Young, R. J. 2003b. *Environmental Enrichment for Captive Animals*. Oxford, UK, Blackwell Science Ltd.

Zarei, S. A., V. Sheibani and F. A. Mansouri. 2019. Interaction of music and emotional stimuli in modulating working memory in macaque monkeys. *American Journal of Primatology* 81(7): e22999.

Zhang, X.-L., M.-T. Luo, J.-H. Song, W. Pang and Y.-T. Zheng. 2020. An Alu element insertion in intron 1 results in aberrant alternative splicing of APOBEC3G pre-mRNA in northern pig-tailed macaques (*Macaca leonina*) that may reduce APOBEC3G-mediated hypermutation pressure on HIV-1. *Journal of Virology* 94(4): e01722-19.

Zhou, Q., H. Wei, Z. Huang and C. Huang. 2011. Diet of the Assamese macaque *Macaca assamensis* in limestone habitats of Nonggang, China. *Current Zoology* 57: 18–25.

Zothansiama, G. Solanki and C. L. Varte. 2014. Monopolizing females and the cost incurred in male Stump-tailed macaques (*Macaca arctoides*). *Issues and Trends of Wildlife Conservation in Northeast India*, Lalnuntluanga, J. Zothanzama, Lalramliana, Lalduhthlana and H. Lalremsanga. Mizoram, India, Mizo Academy of Sciences: 170–179.

29
Behavioral Biology of Vervets/African Green Monkeys

Matthew J. Jorgensen
Wake Forest School of Medicine

CONTENTS

Introduction	475
Note on Common Names and Taxonomy	476
General Characteristics	476
Typical Research Areas	477
Behavioral Biology	477
Natural History	477
Ecology, Habitat Use, Predator/Prey Relations	477
Social Organization and Behavior	478
Feeding Behavior	480
Communication	480
Common Captive Behaviors	481
Normal	481
Abnormal	481
Ways to Maintain Behavioral Health	482
Captive Considerations to Promote Species-Typical Behavior	482
Environment	482
Social Groupings	482
Feeding Strategies	483
Training	483
Special Situations	484
Conclusions/Recommendations	484
Additional Resources	484
Acknowledgments	485
References	485

Introduction

Vervets, also known as African green monkeys (*Chlorocebus* spp.), are Old World monkeys that have frequently been subjects in laboratory and biomedical research. Despite their repeated employment in laboratory settings, there are relatively few resources available describing the behavioral biology of this species (though see Jorgensen 2017, Turner, Schmitt, and Cramer 2019c). The purpose of this chapter is to provide researchers, veterinarians, enrichment coordinators, and animal care staff with information on the behavior of this species, in both the wild and captivity. Better knowledge of their natural history and behavioral patterns should help enhance the care and well-being of this species in captivity. I have had nearly 20 years of experience working with a breeding colony of Caribbean-origin vervets, the Vervet Research Colony (VRC; Jasinska et al. 2013), which has served as a National Institutes of Health-supported national biomedical research resource facilitating the involvement of vervets in a wide range of research areas.

Vervets are nonendangered and are widely distributed across savanna, riverine woodland, and forested grassland areas of sub-Saharan Africa (for maps of distribution, see Haus et al. 2013; Pfeifer 2017; Svardal et al. 2017; Warren et al. 2015). A subpopulation also exists on three Caribbean islands, having most likely been accidentally introduced there as part of the slave trade (Denham 1987, Horrocks 1986, McGuire 1974, Sade and Hildrech 1965). While much of the field work on vervets has focused on the African populations (e.g., Cheney and Seyfarth 1986, Seyfarth, Cheney, and Marler 1980b), the majority of vervets involved in biomedical research within the U.S. are derived from the Caribbean population (Jasinska et al. 2013, Smith 2012).

Note on Common Names and Taxonomy

For simplicity, I will use the term "vervet" to describe the entire genus *Chlorocebus*. However, there must be some acknowledgement of the inconsistent and often confusing common name usage and taxonomy associated with these animals (Groves and Kingdon 2013, Turner, Schmitt, and Cramer 2019b). The term "vervet" has typically been used by primatologists to refer to the entire genus *Chlorocebus*, while the term "African green monkey" has often been used by immunologists and virologists (Jasinska et al. 2013). Other common names include savanna monkeys, grivets, tantalus monkeys, callithrix monkeys, and Malbrouck monkeys, among others (Groves and Kingdon 2013).

Vervets have commonly been separated into five or six different taxonomic groupings, though whether these represent separate species or subspecies is still actively debated (Leffler 2017, Turner, Schmitt, and Cramer 2019b). The taxonomic identification is further confused by the fact that the genus had previously been known as *Cercopithecus*, and all the subgroups had previously been described as the single species, *Cercopithecus aethiops* (Lernould 1988, Smith 2012).

Groves and Kingdon (2013) present one taxonomic viewpoint in which they refer to each taxa as *separate species*: *Chlorocebus pygerythrus* (vervet monkey), *C. sabaeus* (green monkey, Callithrix), *C. aethiops* (grivet monkey), *C. tantalus* (Tantalus monkey), and *C. cynosuros* (Malbrouck monkey). In contrast, Warren et al. (2015), in the publication of the vervet genome, divided vervets into five *subspecies*: *Chlorocebus aethiops pygerythrus*, *C. a. sabaeus*, *C. a. aethiops*, *C. a. tantalus*, and *C. a. cynosuros*. In a more recent genetic analysis, using whole genome sequencing data from 163 animals across Africa and the Caribbean, Svardal et al. (2017) determined that *Chlorocebus* should be separated into "six African and two Caribbean taxonomic groups: *sabaeus* (West Africa), *aethiops, tantalus, hilgerti, cynosuros, pygerythrus*; and *sabaeus* (St. Kitts and Nevis) and *sabaeus* (Barbados)". They further conclude that *sabaeus, aethiops, tantalus*, and *pygerythrus* represent "genetically well-separated taxa" while *pygerythrus, cynosuros*, and *hilgerti* are more closely related. For an alternative analysis, one much disputed by Svardal et al., see Pfeifer (2017).

Whenever possible, I will try to clarify which taxa were discussed in specific studies, though this is not always possible. Again, it is usually safe to assume that laboratory work involved the Caribbean-origin *sabaeus* subgroup, while field research most often involved the *pygerythrus* subgroup.

General Characteristics

In the wild, *pygerythrus* females reach sexual maturity between 4 and 5 years of age, while males reach full body size around 6-years old (Cheney and Seyfarth 1990). Captive *sabaeus* females typically give birth for the first time at 3–4 years old (Fairbanks and McGuire 1986), with females reaching full body weight/size by age 4–5, while males reach full body size by age 5–6 (Schmitt et al. 2018). Growth patterns and overall body weight/size obviously differ between wild and captive populations (Jarrett et al. 2019). Vervets in the wild rarely live beyond 11–13 years old, while in captivity, *sabaeus* have been reported to live up to 30 years of age (Cramer et al. 2018, Latimer et al. 2019), with reproductive senescence typically occurring around age 20 (Atkins et al. 2014).

In captivity, adult female vervets typically range in body weight from 4 to 7 kg (mean 5.33 kg), with males being slightly larger, ranging from 6 to 9 kg (mean 7.22 kg). Crown-rump length averages 43.9 cm for females and 49.8 cm for males (Kavanagh, Fairbanks, et al. 2007). Body weights and lengths from wild populations are generally lower than captive populations (females: 2.57–3.44 kg; males: 4.13–4.43 kg; from Turner, Anapol, and Jolly (1997)), and often show significant variation across different locations in both absolute measures and the extent of sexual dimorphism (Turner et al. 2016, 2018). In a comparison of captive and wild *sabaeus* populations on St. Kitts, Turner et al. (2016) found that weight and measurements of arm, leg, and chest girth were higher in captive populations, while body, arm, and leg length measures did not differ between wild and captive animals.

Vervets are primarily quadrupedal, though they occasionally stand upright. They have long, nonprehensile tails (Bernstein et al. 1978). Like macaques, vervets have cheek pouches that can be used to store food. Unlike macaques, both sexes have long, sharp canines (Estes 1991, Fedigan 1992, Rowell 1971).

While often called "green" monkeys, vervet color patterns are geographically variable and typically range from grayish-brown to olive, yellow, and reddish green (Figure 29.1). The hair on their abdomens and part of their faces is white to yellowish. Adults have black pigmentation on their face, hands, and feet, and their face is framed with a white ruff and brow band, with pale pink eyelids. Infants have a lighter, pink pigmentation at birth that grows darker after about 5–6 months of age (Estes 1991, Groves and Kingdon 2013, Skinner and Chimimba 2005). Males have a bright blue scrotum, a red penis, and white perineal skin, though the blue scrotal color is less evident in the Caribbean and captive *sabaeus* populations. This scrotal coloration becomes bluer with age in *pygerythrus*, while it fades with age in *sabaeus* (Cramer et al. 2013). Despite their relatively muted coloration, experimental manipulation of scrotal color in captive *sabaeus* was found to influence dominance interactions among males (Gerald 2001).

FIGURE 29.1 Photo of captive-born *sabaeus* from the VRC, depicting coloration and facial pigmentation patterns.

Typical Research Areas

Early field studies of vervets included the pioneering work of Sade and Hildrech (1965), Hall and Gartlan (1965), Struhsaker (1967c), Gartlan (1969), and Poirier (1972) (see review by Fedigan and Fedigan 1988). Perhaps the most well-known field studies involved seminal work on the semantics of alarm calls (Seyfarth, Cheney, and Marler 1980a, b) and social behavior (Seyfarth and Cheney 1984).

Some examples of more recent field studies include investigations of social learning and cultural conformity (van de Waal, Borgeaud, and Whiten 2013, van de Waal, Claidiere, and Whiten 2013), studies of third-party rank relationships and social network analysis (Borgeaud, Schnider, et al. 2017, Borgeaud, Sosa, et al. 2017, Borgeaud, van de Waal, and Bshary 2013), studies of sex differences during intergroup aggression (Arseneau-Robar et al. 2016, Arseneau-Robar et al. 2018, Arseneau-Robar et al. 2017), and investigations of coalition formation and social network integration (Freeman et al. 2016, Jarrett et al. 2018, Young et al. 2017). For an overview of field studies of communication, cooperation and trade, and social learning, see Mertz et al. (2019).

In addition to field work, vervets have been subjects in a wide variety of laboratory research (Jasinska et al. 2013, Magden et al. 2015, Smith 2012). According to a survey by Carlsson et al. (2004), vervets are one of the most common nonhuman primate species in biomedical research. The relatively recent growth in the involvement of vervets may be due to the fact that they pose fewer biosafety concerns than macaques, since they are not carriers of herpes B virus (Baulu, Evans, and Sutton 2002), are typically less expensive (Freimer et al. 2008, Smith 2012), and have been recognized as a promising model for Alzheimer's disease (NIH-ORIP 2018).

One of the primary areas of laboratory work with vervets has involved studies of immunology, vaccine development, and infectious disease. Vervets have been studied in numerous investigations of HIV/AIDS (Goldstein et al. 2006, Pandrea et al. 2012, Zahn et al. 2010). They are of particular interest since African vervets are natural hosts of simian immunodeficiency virus (SIV), though they seldom show signs of immunodeficiency (Broussard et al. 2001, Ma et al. 2013, 2014). Caribbean-origin vervets, in contrast, are typically SIV-free (Jasinska et al. 2013, Pandrea et al. 2006). Recent work has shown that a high-fat diet exacerbates SIV disease progression (He et al. 2019) and that evolutionarily conserved wound-healing mechanisms in vervets may help explain why vervets remain nonpathogenic compared to macaques (Barrenas et al. 2019). Other infectious disease work with vervets has included respiratory syncytial virus (Eyles et al. 2013), vesicular stomatitis virus (Westcott et al. 2018), Nipah virus (Lee et al. 2019, Lo et al. 2019), and dengue virus (Briggs et al. 2014). Vervets have also been involved in testing influenza vaccine candidates (Matsuoka et al. 2014) and developing better influenza vaccine adjuvants for neonates (Holbrook et al. 2016, 2018).

There is an extensive literature involving vervets as models of lipid biology, metabolism, obesity, and diabetes. For example, Rudel and colleagues characterized lipid profiles of vervets (Koritnik et al. 1984, Parks and Rudel 1979) and studied the impact of age and dietary fat on cholesterol concentrations, atherosclerosis, and cardiovascular risk factors (Jorgensen et al. 2012, Rudel et al. 2002, Scobey, Wolfe, and Rudel 1992, Wolfe et al. 1993, 1994). Kavanagh and colleagues have studied the heritability of obesity (Kavanagh, Fairbanks, et al. 2007), the characterization of naturally occurring diabetes mellitus (Cann et al. 2010), the influence of mitochondrial quality on metabolic health (Kavanagh et al. 2017), the effects of dietary trans fats and fructose on obesity, insulin sensitivity, endotoxemia, and hepatic injury (Kavanagh, Jones, et al. 2007, Kavanagh et al. 2013), the impact of microbial translocation on metabolic disease in elderly animals (Wilson et al. 2018), and the effects of hydrotherapy on heat shock proteins and glucose metabolism (Kavanagh et al. 2016). Other groups have also studied vervets in investigations of glucose tolerance and diabetes (Liddie et al. 2018), as well as microRNA expression profiles associated with lipid metabolism (Zhou et al. 2019).

As previously noted, vervets have recently become an emerging model of Alzheimer's disease (Cramer et al. 2018, Kalinin et al. 2013, Latimer et al. 2019). Vervets experience age-related increases in amyloid burden (Kalinin et al. 2013), neurodegenerative disease biomarkers (Chen et al. 2018, Latimer et al. 2019), and synaptic degeneration similar to that of humans (Postupna et al. 2017). They also exhibit age-related changes in tau and amyloid beta (Aβ) in cerebrospinal fluid, histological evidence of tauopathies, and age-related cognitive and motor deficits (Cramer et al. 2018, Latimer et al. 2019). Like humans, vervets also show similar changes in cerebrospinal fluid markers of Alzheimer's disease with development of type 2 diabetes (Kavanagh et al. 2019).

Neuroscience research has included studies of brain imaging (Fears et al. 2009, Fedorov et al. 2011, Maldjian et al. 2016, Menzel et al. 2019, Woods et al. 2011), Parkinson's disease (Emborg 2007), substance and alcohol abuse (Groman et al. 2013, Jentsch et al. 1997, Melega et al. 2008, Rowland et al. 2017, Thomsen et al. 2018), retinal structure and ophthalmology (Bouskila et al. 2013, Bouskila et al. 2018), the endocannabinoid system (Bouskila et al. 2016, Kucera et al. 2018), and sleep (Mizrahi-Kliger et al. 2018).

Additional areas of research have included reproduction (Atkins et al. 2014, Kavanagh et al. 2011), behavior/temperament (Fairbanks 2001, Fairbanks and McGuire 1986, Laudenslager et al. 2011, Raleigh et al. 1991), the microbiome (Becker, Gavins, et al. 2019, Petrullo et al. 2019), hypertension (Rhoads et al. 2017), and genetics (Freimer et al. 2007, Huang et al. 2015, Jasinska et al. 2009, 2012, 2017, Warren et al. 2015). Investigators have also developed the vervet as a model of lymphatic obstruction and intestinal inflammation (Becker, Romero, et al. 2019). Jasinska et al. (2013) provide an overview of the variety of biomedical research that has been done with this species.

Behavioral Biology

Natural History

Ecology, Habitat Use, Predator/Prey Relations

Vervets are one of the most numerous and widespread African monkeys, found in many regions of sub-Saharan Africa

(Estes 1991, Groves and Kingdon 2013, Skinner and Chimimba 2005), in addition to the sub-population of animals found on the Caribbean islands of St. Kitts, Nevis, and Barbados (Chapman and Fedigan 1984, Poirier 1972). Within Africa, *sabaeus* are found as far west as Senegal; *tantalus* in north central Africa, from Ghana to Sudan; *aethiops* in Ethiopia; and *pygerythrus* in eastern Africa, from Kenya to the Cape region of South Africa.

They occupy a wide range of habitats including semi-desert, savannas, swamps and forests, typically preferring riverine woodlands (Estes 1991, Groves and Kingdon 2013). They are highly adaptable and are able to survive in marginal and ecologically rich environments (Fedigan and Fedigan 1988, Turner, Schmitt, and Cramer 2019a). Vervets can also adapt to a wide range of temperatures and can even survive during periods with no standing water (Lubbe et al. 2014, McFarland et al. 2014). They are considered agricultural pests in both Africa and the Caribbean, often raiding crops (Brennan, Else, and Altmann 1985, Saj, Sicotte, and Paterson 1999). Vervets are diurnal and typically sleep in trees at night. They are semiterrestrial and semiarboreal, able to forage on the ground, as well as in trees (Fedigan and Fedigan 1988).

Territoriality. Unlike baboons and macaques, vervets are territorial, with distinct home ranges that are defended from other troops (Cheney 1981). Territorial behavior has been reported in *pygerythrus* populations (Cheney and Seyfarth 1987), as well as *sabaeus* populations in Africa (Harrison 1983b) and the Caribbean (Chapman and Fedigan 1984). The strength of the territoriality ranges from mutual tolerance to intense aggression and has been shown to vary across populations, depending, in part, on such factors as seasonal variation in resources and the extent of previous group associations (Chapman and Fedigan 1984, Harrison 1983b). Vervets also adjust their home ranges in response to demographic and ecological factors (Isbell, Cheney, and Seyfarth 1990). For a more recent examination of the variability in communal range defense in different vervet populations, and in nonhuman primates in general, see Willems et al. (2015).

Predation and Mortality. Predation is a significant risk for vervets in Africa, with leopards, eagles, pythons, crocodiles, and baboons being common predators (Baldellou and Henzi 1992, Cheney, Lee, and Seyfarth 1981). Given their status as agricultural pests, humans are also significant predators in both Africa and the Caribbean (Turner, Schmitt, and Cramer 2019a). In a well-studied *pygerythrus* population in Amboseli (Kenya), predation accounted for 69% of mortality over a 7-year period (Cheney et al. 1988). Isbell (1990) documented a significant increase in leopard predation within the same population in just a single year. As an adaptation to predation, vervets have evolved a system of alarm calls to both alert group members to the presence of predators (Seyfarth, Cheney, and Marler 1980a, b) and as a predator deterrent (Isbell and Bidner 2016). See the "Communication" section for additional detail. In some locations, recent changes in climate have also had an impact on food availability and increased mortality (Young, Bonnell, et al. 2019).

Cheney, Lee, and Seyfarth (1981) found that causes of mortality varied by dominance rank. Low-ranking individuals were more likely to die from illness, presumably due to restricted access to food and water, while high-ranking animals were more likely to die from predation. They suggested that the increased predation risk in high-ranking animals may have been due to their higher frequency of alarm calling to predators and their higher rates of aggression during intergroup encounters. Baldellou and Henzi (1992), studying a different *pygerythrus* population in South Africa, found that males were more vigilant than females, though they were no more effective at detecting predators. Males were more vigilant during the breeding season, and the highest-ranking male showed the greatest vigilance, suggesting that increased male vigilance may have been directed more toward other males than just for predator defense.

Social Organization and Behavior

Vervets females, like most Old World monkeys, are philopatric and remain in their natal groups for their whole lives, while males emigrate into neighboring groups upon sexual maturity (Cheney and Seyfarth 1983). Social groups thus contain matrilines consisting of multiple generations of closely related females and their offspring, in addition to adult males that have immigrated from other groups. Male emigration typically occurs during the breeding season, which is also when male-male aggression is highest (Cheney 1983b, Henzi and Lucas 1980). Younger males tend to migrate to neighboring groups along with brothers or peers, while older males usually transfer alone and to more distant groups (Cheney and Seyfarth 1983).

Groups typically consist of multiple adult males, which is in contrast to the single-male groups seen in most guenons (Isbell, Cheney, and Seyfarth 2002). Group sizes typically range between 10 and 40 individuals, but can be as low as 5 and as high as 76 or more (Fedigan and Fedigan 1988; Isbell, Pruetz, and Young 1998; Melnick and Pearl 1987; Struhsaker 1967e). West African *sabaeus* group sizes are typically smaller (Dunbar 1974). Group fusions can occur when a minimum number of adults cannot be maintained (Isbell, Cheney, and Seyfarth 1991).

Dominance. Like other Old World monkeys, vervets form linear dominance hierarchies that are relatively stable over time (Cheney and Seyfarth 1990). Infants typically inherit the rank just below that of their mothers (Horrocks and Hunte 1983). Within captive *sabaeus*, dominance hierarchies of females have been considered to be separate from those of males (Raleigh and McGuire 1989), while recent work in wild *pygerythrus* described more integrated hierarchies (Young et al. 2017).

Higher ranking *pygerythrus* females at Amboseli received more grooming (Seyfarth 1980) and received more social support during aggressive interactions than lower ranking females (Cheney 1983a). These findings were not replicated in a *pygerythrus* population in South Africa (Henzi et al. 2013), possibly due to increased group size and ecological factors. Whitten (1983) reported that dominant females had greater access to some feeding sites, and in some groups, dominant females also had reproductive benefits (see "Feeding Behavior" section). Fairbanks and McGuire (1984) found that, in captive *sabaeus*, higher ranking females had better infant survival and shorter interbirth intervals than lower ranking females. In contrast, findings from wild *pygerythrus* in Amboseli found no correlation between dominance rank and measures of reproductive success (Cheney, Lee, and Seyfarth 1981, Cheney et al. 1988). As mentioned previously, Cheney, Lee, and Seyfarth (1981) found that higher ranking *pygerythrus* females alarm called more frequently and were more aggressive during intergroup conflicts, but were also at higher risk of death due to predation. Lower ranking females, in contrast, were more likely to die from illness. In captive *sabaeus*, Raleigh and McGuire (1989) found that adult females can influence the rank of adult males, based on which males establish affiliative relationships with the dominant female. Attainment of rank by adult males and the length of their residency within a group is highly contingent upon their integration into the social network of females (Young, McFarland, et al. 2019).

Unlike baboons, female vervets regularly form coalitions to control potential male aggression, particularly when new males have joined the groups. Cheney (1983a) reported that in wild *pygerythrus*, an alliance of two adult females or juveniles was sufficient to drive away an adult male. Fairbanks and McGuire (1987) report similar female alliances in captive *sabaeus*, such that the threat of female coalitions makes newly introduced adult males very cautious, with new males avoiding initiation of social interactions, especially with younger animals. Rates of female alliance formation are highest among kin; however, grooming between unrelated animals increases the chance that they will form coalitions in the future (Cheney and Seyfarth 1990). In contrast, recent grooming between kin does not lead to an increased likelihood of responding to calls for support (Seyfarth and Cheney 1984). Borgeaud and Bshary (2015) found that vervets will exchange grooming for coalition support, but only if the grooming partner is of higher rank than the target of the coalition.

Some have suggested that dominance hierarchies in vervets are less rigid than that of baboons or macaques (Bloomstrand and Maple 1987, Fedigan 1992, Rowell 1971), and that dominance displays are often quite subtle (Fairbanks and McGuire 1986). Unlike macaques, vervets do not use stylized mounting as a dominance gesture. Fairbanks (1980), for example, noted that while determination of the dominant female is typically unambiguous, it can be difficult to assign the rank of the nondominant females. Rank assignments in captive *sabaeus* have either been categorized as high, medium, or low (Fairbanks and McGuire 1984), or dominant vs. nondominant (Fairbanks et al. 2004), rather than as an ordinal list. As Rowell (1971) stated, it may be clearer to describe "top" and "bottom" animals, rather than using a linear ranking method. Male and female dominance hierarchies are relatively independent of each other. Maintenance of rank among females is subtle, since rates of aggression are low and rank reversals are rare (Cheney and Seyfarth 1990). Dominance rank in males, in contrast, depends on age, size, strength, and behavioral characteristics (Fairbanks et al. 2004, Raleigh and McGuire 1989). In a series of playback experiments in wild South African *pygerythrus*, Borgeaud and colleagues found evidence that adult females recognized rank relationships within their own hierarchy, as well as that of adult males, while males and juveniles failed to show recognition of the female hierarchy (Borgeaud et al. 2015, Borgeaud, van de Waal, and Bshary 2013). Not surprisingly, juveniles focus most of their social attention on matrilineal kin, and juvenile females show a stronger bias toward higher ranking animals than males (Grampp et al. 2019). Rowell (1971) speculated that the existence of relatively long canines in adult female vervets, along with the lack of pronounced sexual dimorphism, may help explain some of the differences in vervet dominance interactions compared to baboons or macaques.

Reproduction and Parental Care. Reproduction in vervets is highly seasonal in the wild (Andelman 1987), with births occurring during a 2–3 month period at the peak of food availability (Eley, Hutchinson, and Else 1984, Horrocks 1986, Poirier 1972, Struhsaker 1971). The extent of this seasonality has been less pronounced in captive populations (Else 1985, Hess, Hendrickx, and Stabenfeldt 1979, Seier 1986). Females typically give birth once per year, but can give birth twice in one season if the first infant does not survive (Fairbanks and McGuire 1984). Females usually give birth for the first time at 3–4 years of age, but first births can occur when females are as young as 2.5 years old. Weaning can start as early as when the infant is 12 weeks of age and is typically completed by 8.5 months (Fedigan and Fedigan 1988).

Infant mortality in the wild and captivity can be quite high. In Amboseli, 60% of *pygerythrus* infants die in the first year and only 27% reach sexual maturity (Cheney and Seyfarth 1990). Another study of wild *pygerythrus* in Kenya found variability in annual infant survivorship and a high rate of miscarriage (Turner et al. 1987). Captive populations have also reported high rates of infant mortality (Eley 1992, Johnson, Valerio, and Thompson 1973, Kushner et al. 1982, Seier 1986). In captive *sabaeus*, Fairbanks and McGuire (1984) reported that infant mortality was higher in younger females and that high-ranking females had shorter interbirth intervals and higher fecundity. Kavanagh et al. (2011), in a later study within the same captive *sabaeus* population, found that infant mortality was higher in females that were lower ranking, primiparous, and had poorer metabolic health. A recent study found that fetal cortisol exposure, as measured from neonatal hair, was positively associated with neonatal body mass in captive vervets, while maternal gestational hormones were not (Petrullo and Lu 2019).

Unlike baboons and macaques, female vervets do not develop swelling of the perineal skin and typically do not show any overt signs of menstruation or ovulation (Andelman 1987, Carroll et al. 2007, Eley et al. 1989, Rapkin et al. 1995, Rowell 1970, Seier et al. 1991, Tarara, Else, and Eley 1984). There is generally no consort formation and the bonds between males and females are relatively weak (Cheney and Seyfarth 1990). Mounting and other sexual behavior are rare and are usually only seen during the mating season and only for reproduction (not for dominance). Females generally prefer higher ranking males (Andelman 1987), though only high-ranking females are able to reject copulation attempts by low-ranking males (Keddy 1986). The lack of facial pigmentation makes young infants very attractive to other members of the group (Struhsaker 1967b, 1971).

In contrast to macaques and baboons (Chism 2000), allomothering is common in vervets, with young infants often carried by other group members, including juvenile and adolescent females (Fairbanks 1990, Johnson et al. 1980, Struhsaker 1971). Grandmothers often assist with infant care, and females with a living mother are often less protective of their infants, with those infants becoming independent of their mothers at a younger age (Fairbanks 1988b). There have been numerous studies of different factors that influence maternal care in captive *sabaeus* (Fairbanks 1988a, 1996, Fairbanks, Blau, and Jorgensen 2010, Fairbanks and McGuire 1993). Males are generally indifferent to infants, though Keddy-Hector, Seyfarth, and Raleigh (1989) did report evidence of male parental care in captive *sabaeus* under experimental situations. Additional details regarding a range of reproductive parameters have been reported for numerous wild (Cheney and Seyfarth 1987, Gartlan 1969, Turner et al. 1987) and captive populations (Bramblett, Pejaver, and Drickman 1975, Eley et al. 1989, Fairbanks and McGuire 1984, 1986, Hess, Hendrickx, and Stabenfeldt 1979, Kavanagh et al. 2011, Kushner et al. 1982, Rowell 1970, Seier 1986, 2005).

Feeding Behavior

Vervets are often described as "opportunistic omnivores", predominantly eating fruits, seeds, pods, gum, leaves, and flowers, while occasionally eating insects, small vertebrates, and bird's eggs (Estes 1991, Struhsaker 1967d). Fedigan and Fedigan (1988) describe vervets' flexibility and adaptability to fluctuating resources and their ability to exploit new or marginal habitats. Harrison (1984), in a study of foraging strategies in an African *sabaeus* population, found that their diet was quite diverse, consisting of over 65 species of plants, invertebrates, eggs, and meat. Food selection varied from month to month, and specific food selection often matched availability, rather than nutrient content. Food selection also varied by age and sex, with adult males eating more fruit and flowers in the dry season, and adult females eating greater amounts of foliage in the wet season (Harrison 1983a). A study of two social groups of *pygerythrus* in northern Kenya (Whitten 1983) found that dominant females had greater access to preferred feeding sites and this manifested itself in a difference in food intake. However, this relationship was only true when preferred food sites were clumped and dominant females could monopolize preferred feeding sites. Lee and Hauser (1998), in a study of *pygerythrus* in Amboseli (Kenya) during a period of habitat deterioration, also found that vervets selected foods based on availability, rather than macronutrients. As food availability declined, territory size increased, as did feeding time, but food selection remained relatively stable. Lee (1984) found that seasonal variability in dietary quality and distribution also affected rates of social interactions in immature animals. In a study comparing food distribution and movement patterns of sympatric vervets and patas monkeys, Isbell, Pruetz, and Young (1998) found that vervets travelled less and spent more time at each foraging site than patas. However, when vervets and patas foraged in similar habitats, vervets adjusted their feeding and movement patterns to more closely match that of patas. Observational studies have highlighted how social and ecological factors can impact time budgets for foraging and other activities (Isbell and Young 1993, Canteloup et al. 2019), and experimental studies have shown that vervets can adjust foraging strategies in response to changes in value and distance of food sources (Teichroeb and Aguado 2016).

In an innovative study of food preference and conformity to cultural norms, van de Waal, Borgeaud, and Whiten (2013) presented groups of *pygerythrus* in South Africa with two bins of differently colored corn. One color was made to taste bitter, while the other was palatable, with the palatable color alternated across different groups. Animals learned to avoid the bitter colored corn. When the bins were re-presented months later, with neither color made to taste bitter, infants that had been naïve to the original exposure adopted the color preferences of their mothers. In addition, when males migrated to new groups that had been exposed to the alternate color pattern, the males adopted the preference of their new group.

Communication

Struhsaker (Struhsaker 1967b, c) described a range of gestural communication seen in the wild *pygerythrus* population in Amboseli, including a list of 60 physically distinct behavioral patterns. Many of these behavior patterns are similar to those seen in rhesus macaques, baboons and patas, including an "eyelid display", a threat gesture in which the eyebrows are retracted to expose the lighter-colored eyelids along with a stare. This is considered to be a defensive behavior when the animal is in a crouching position, and aggressive when in an upright posture. "Head-popping" or "head-jerking" is another aggressive behavior in which the head is jerked back-and-forth. "Branch shaking" occurs during intergroup interactions, in which animals, typically males, bounce from branch to branch, often causing a large commotion with the group. Struhsaker (1967c) indicated that yawning did not occur in Amboseli vervets, though this behavior is often seen in captive *sabaeus*, and is typically associated with anxiety during stressful situations (Fairbanks 2001). Like macaques and baboons, grooming is the most common affiliative behavior pattern seen in vervets (Seyfarth 1980).

Struhsaker (1967c) also described gestural behavior patterns seen in vervets that are not seen in other Old World monkeys. He described a "penile display" in which a male approaches another male, stands bipedally in front of the other male, displays his inguinal area, and places his hands on the other male's head, shoulder, or back. Fairbanks and McGuire (1986), in describing dominance behavior in captive *sabaeus*, described this hands-on-head or hands-on-shoulder behavior as "manipulation." They indicate that "manipulation" is often the best predictor of dominance in captive animals (Rowell 1971) and is frequently observed during dominance disputes between closely ranked individuals. Struhsaker also described a "red, white, and blue" display in which an adult male holds his tail erect, presenting his red penis, blue scrotum, and white fur. This display is not seen in all African populations and is absent in Caribbean *sabaeus* populations, probably due to their reduced coloration.

In a series of elegant and inventive playback experiments, Cheney and Seyfarth described the vocal communication patterns of wild *pygerythrus* groups in Amboseli (Cheney and Seyfarth 1980, 1982a, 1988, Seyfarth, Cheney, and Marler 1980a, b). Struhsaker (1967a) had initially observed that vervets produced acoustically distinct alarm calls in response to different predators: leopards, eagles, and snakes. Different alarm calls elicited different behavioral responses. Leopard calls caused animals to run into trees, eagle calls caused animals to look up, and snake calls caused them to look at the ground. When recorded alarm calls were played back in the absence of any predator, the animals responded as if the specific predator was present, which suggested that the animals had a semantic understanding of each call type (Seyfarth, Cheney, and Marler 1980a, b). Subsequent playback experiments found that vervets were also able to distinguish the calls of individual animals within their own social groups, as well as those of neighboring groups (Cheney and Seyfarth 1980, 1982b). For more recent assessments of vervet alarm calls and vocal recognition of rank relationships, see Ducheminsky, Henzi, and Barrett (2014), Price and Fischer (2014), Borgeaud et al. (2015), Mercier et al. (2017), and Mercier et al. (2019).

Common Captive Behaviors

Normal

There are only a handful of publications that include detailed ethograms describing normative behavioral patterns in captive vervets (Dillon et al. 1992, Fairbanks, McGuire, and Page 1978, Raleigh, Flannery, and Ervin 1979, Raleigh et al. 1991). Two additional papers that focused more exclusively on affiliative and agonistic behavior are Fairbanks (1980) and Fairbanks and McGuire (1986). A more recent paper that focused on positive reinforcement training in a zoo setting also included an ethogram (Spiezio et al. 2015). All but one of these publications involved work done at the VRC, a captive population of Caribbean-origin *sabaeus* that was founded in the late 1970s in southern California and was moved to the Wake Forest School of Medicine in 2008 (see Jasinska et al. 2013). Table 29.1 attempts to consolidate these published ethograms along with an unpublished ethogram developed by Lynn Fairbanks from work done in the same colony.

Abnormal

As with other Old World monkeys, vervets can display abnormal and stereotypical behavior in captivity. Seier et al. (2011), in one of the few reports on stereotypical behavior in vervets, examined the effects of different enrichment devices, cage sizes, and social conditions on stereotypical behavior in captive-born, individually housed *pygerythrus*. They found that females engaged in stereotypical behavior more than males and that stereotypical behavior was highest when monkeys were housed in small, single cages. Rates of stereotypy were lowest

TABLE 29.1

Ethogram of Normative Behavior Patterns in Captive Vervets

Behavior	Description
Aggressive	
Manipulate	Stylized behavior involving placing the hands on the head or shoulders of another, embracing.
Display	Ritualized behavior including standing broadside to another animal with tail up, sideways prancing, circling another animal with tail-up, showing the hindquarters, or showing an erection to another.
Aggression	Stare, threaten (open mouth, chin forward, ears back, eye lids up), head jerk (rapid sideways jerking of the head), slap, push away, chase (not in play) attack.
Branch shake	Vigorous whole-body jumping onto cage or play structures.
Yawn	Any occurrence of yawning with teeth showing.
Submissive	
Submissive	Pawing at ground, backwards shuffling, ambivalent lunges from crouching position, screaming, defensive aggression.
Lipsmack	Rapid opening and closing of the lips directed to another.
Affiliative	
Groom	Using fingers to manipulate hair of another monkey; may remove parasites.
Muzzle	Bringing muzzle to any part of another animal's body except the genitals, usually muzzle to muzzle.
Social play	Any form of social play including wrestling, jumping, chasing, mouthing.
Social contact	Sitting within 1-m of another animal (initiating or receiving).
Sexual behavior	Sniffing, touching or manipulating the genital area of another animal, grabbing hips, mounting or copulating.
Self-Directed	
Scratch	Vigorous scratching of self.
Self-directed behavior	Any other self-directed behavior including self-grooming, brief touches to face or body, hand or foot rubbing, licking hand or body, gentle hair picking, masturbation, shrugging.
Abnormal	
Pacing	Pacing back and forth, at least three turns.
Other abnormal	Any other abnormal behavior, e.g., hair plucking, self-biting, nonfunctional repetitive behavior, spinning, crouching.
Miscellaneous	
Vocalization	Any vocalization.
Locomotion	Whole body movement of one or more steps.
Feeding	Any feeding behavior including eating chow, foraging on the ground, drinking and chewing on any edible object.
Object manipulation	Any object manipulation, exploring the environment, handling nonfood object or trying to see inside or under hidden object.

Source: Adapted from Fairbanks and McGuire (1986), Fairbanks, McGuire, and Page (1978), Jorgensen (2017), Struhsaker (1967b, c).

when animals had access to foraging logs or exercise cages, and when housed in larger cages. The same lab also tested the impact of administering fluoxetine to individually housed *pygerythrus* (Hugo et al. 2003). They found that 6 weeks of daily oral fluoxetine administration (mixed in diet) significantly decreased stereotypical behavior compared to controls. Fairbanks et al. (2001) found that 9 weeks of daily intramuscular injections of fluoxetine reduced impulsive behavior in an experimental setting (e.g., lower latency to approach and increased touching, sniffing, displaying to or threatening an unfamiliar animal in an intruder challenge test); abnormal and stereotypical behavior were not recorded as part of this study. Daniel, dos Santos, and Vincente (2008) reported that the rates of self-directed behavior in a group of vervets at the Lisbon Zoo increased after agonistic interactions, suggesting that self-directed behavior was indicative of anxiety. Fairbanks (2001) also found that some animals showed increases in anxiety, including high rates of scratching, yawning, and locomotion, when confronted with an unfamiliar animal as part of an experimental assessment of temperament.

The captive-born *sabaeus* within the VRC have historically had low rates of stereotypical behavior, and self-injurious behavior is extremely rare. This is likely due to the fact that all animals are mother-reared within multi-generational, species-typical social groups. In addition, infants typically remain in their natal groups with their mothers and close kin for at least 3–4 years, if not longer. Even within the limited number of pair- or individually housed animals, rates of stereotypy are quite low. In contrast, a recent cohort of captive-born males imported from St. Kitts have shown higher rates of stereotypy, suggesting that rearing and housing history may play an important role in the manifestation of stereotypical behavior in vervets.

Ways to Maintain Behavioral Health

Captive Considerations to Promote Species-Typical Behavior

Taking into account the natural history of vervets, in particular how they differ from macaques and baboons, may help inform behavioral management practices in captivity. To reiterate some of the key differences mentioned thus far: vervets are territorial; vervets are equally adept in trees and on the ground; female vervets have relatively long canines, have no overt signs of menstruation or ovulation, and are tolerant of allomothering; vervets have less rigid dominance hierarchies; and have a few unique patterns of behavior and vocalization not seen in other Old World monkeys.

Environment

As Bloomstrand and Maple (1987) report in their overview of management and husbandry practices for African primates, the use of vertical space is very important to vervets, given that they are more arboreal than macaques and baboons. They emphasize the importance of "volumetric space [as] a means to provide cover, privacy and social distance" (p. 228). Therefore, the use of barrels, visual barriers, hanging devices, sleeping platforms/shelves, and elevated cage furniture within social group housing is strongly encouraged. Kaplan (1987) speculated that vervets and patas, in contrast to rhesus macaques, may resolve aggressive conflicts using spatial factors more often than using facial expressions and postures. Therefore, the need for sufficient physical space to allow for separation/avoidance may be critical for vervets. This can obviously be problematic when animals are housed in standard, indoor pair caging. Some species-typical behavior patterns commonly observed in social groups are infrequent, nonexistent, or even physically impossible within smaller caging. The use of a mobile cage to allow increased vertical movement, like that used by Seier and deLange (1996), could have a major impact on behavioral health.

Social Groupings

Vervets can successfully be housed in heterosexual and isosexual pairs, harems, and in multimale, multifemale social groups. Despite the fact that a 2003 survey reported 0% of 179 vervets across six facilities within the U.S. were socially housed (Baker et al. 2007), an updated survey in 2014 reported 59%–86% of 797 vervets across two facilities were socially housed (Baker 2016). Harem groups (Else 1985) and short-term heterosexual pairing (Seier 1986) have been previously reported in Africa. Multimale, multifemale groups are most common in wild populations and have also been maintained in captivity (Bramblett et al. 1982). While the majority of social groups in the VRC currently contain one adult male per group, historically, multimale groups were the norm (Fairbanks and McGuire 1986, 1987). The introduction of new breeding males, even though mimicking animal movements seen in the wild, should be avoided when young infants are in the groups, since the presence of young infants can lead to increased aggression toward the males (Morland, Suleman, and Tarara 1992). Introduction of new males can also alter maternal behavior (Fairbanks and McGuire 1987).

In a detailed survey of vervet pair housing, Jorgensen et al. (2017) examined 271 isosexual pair-housing attempts at Wake Forest School of Medicine, the Tulane National Primate Research Center, and the New Iberia Research Center. We examined pairing success rates at 14, 30, and 60 days postintroduction. We found that success rates, defined as pairs still being housed together at least 14 days postintroduction, were >95% for both male pairs and female pairs in two cohorts of animals at Wake Forest and New Iberia. Success rates for a smaller cohort of males at Tulane and a smaller cohort of imported males at Wake Forest were 50% and 28%, respectively. Mean age of the pair and body weight differences were significant predictors of pairing success in males, but not in females. The odds of pairing success in males were higher in younger animals and in pairs with greater weight differences between pair-mates. Within one cohort, we also found that prior familiarity of pair-mates within group housing did not result in increased pairing success for either males or females.

To summarize my previous recommendations for pair housing techniques with vervets (Jorgensen 2017, Jorgensen et al. 2017), I would discourage the use of a period of protected contact and gradual introduction methods (Truelove et al. 2017), avoiding what my colleagues and I have termed a "macaque-centric" approach. In our experience, vervets rarely show overt aggression immediately after introduction. We have theorized that, in contrast to macaques, physical interactions (e.g., the "manipulate" embracing behavior; see Table 29.1) may be an essential component of vervet social interactions, suggesting that a period of familiarization without such contact may simply increase anxiety and have little beneficial effect. Of course, the need for controlled studies testing the effectiveness of different pairing techniques in vervets is needed before this notion can be fully supported. Another important factor to consider is the territorial nature of vervets. Forming new pairs in rooms that are unfamiliar to each animal will likely reduce territorial behavior, since neither animal is resident. Moving established pairs into new caging or even new positions within the same room should be avoided. Introducing a new animal to a pair-mate already established in a given housing location will likely have a lower chance of success than introducing the same two animals in a location unfamiliar to both animals.

Feeding Strategies

It has been reported in captive rhesus macaques that dominant animals can monopolize access to feeding locations, with this situation being most pronounced when available food amounts are lowest, even though food intake does not differ between dominant and subordinate animals (Deutsch and Lee 1991). While Whitten (1983) described rank-related differences in access to preferred food sites under certain ecological conditions in wild vervets, this situation is rarely seen in captive vervets. As Fairbanks and McGuire (1984) reported for captive *sabaeus*, there were no differences between high- and low-ranking females in access to diet, with no differences in body weight between dominant and subordinate females over a 5-year period. Anecdotally, my only experience of noticeable competition for food and restricted access to food has come in rare situations of limited food availability or briefly during the integration of adult males to new social groups. Rowell (1971) also reported that it was rare for one animal to take food from another.

Wilson et al. (2008) used an automated feeder system to measure food intake in socially housed rhesus macaques. Each animal was implanted with a subcutaneous ID chip that enabled the recording of individual food removal 24 h/day. They found that subordinate animals consumed more food and ate more at night than dominant animals. We used a similar system (BioDAQ, Research Diets) in group-housed vervets and found individual differences in food removal. As shown in Figure 29.2, older animals tended to consume less food than younger animals. Figure 29.3 shows the hourly percent of food intake for individual animals in two different social groups. Group B (Figure 29.3b) shows the standard pattern in which animals eat the majority of their diet at the beginning and end of the daylight periods (feeders were temporarily removed for a variable period from 7 to 9 AM each morning for cleaning, which shifted AM feeding times). Group A (Figure 29.3a) shows some examples of individual differences, in which one animal consumes >5% of its diet at 2 AM, another peaks at 9 AM, while a third peaks at 6 PM (right after lights are turned off).

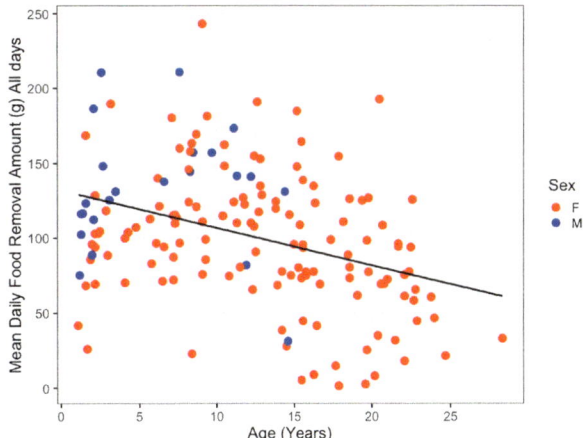

FIGURE 29.2 Automated feeder data depicting individual differences in mean daily food removal (g) by age and sex.

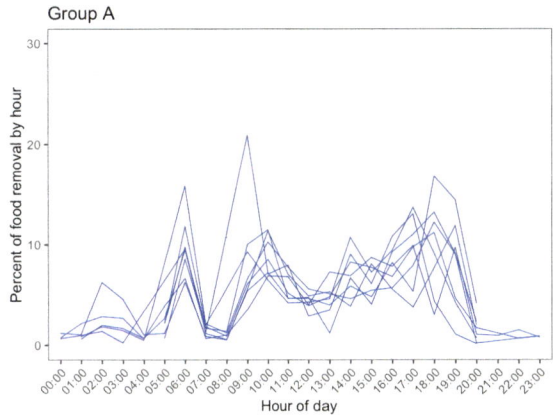

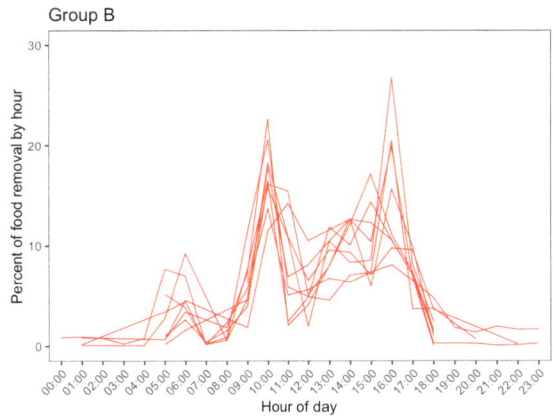

FIGURE 29.3 (a and b) Percent of food removal by hour for individual animals in two groups (A and B). Each graph represents a different social group and each line represents an individual animal within that group. Note: Feeders were temporarily removed for a variable period of time between 7 AM and 9 AM for daily cleaning.

Training

Unfortunately, there are only a handful of publications that describe the use of positive reinforcement training (PRT) in vervets. As I have summarized previously (Jorgensen 2017), most published reports include either training for awake sample collections or habituation/training necessary for cognitive testing or other research procedures. I am not aware of any publications describing the use of PRT methods for daily husbandry, animal care, or colony management procedures, though I suspect at least some work is being implemented without formal reporting (see Jorgensen 2017 for examples of unreported husbandry/veterinary procedures).

Spiezio et al. (2015) used food rewards to train a social group of ten animals within a zoo setting to individually enter a separate testing area in order to participate in a problem-solving study. Unlike most of the other studies listed below, this group described both the training procedures and the behavioral effects of the training. They found that the training program reduced aggressive behavior and increased social interactions compared to baseline, suggesting that the use of such procedures could also improve husbandry.

Other examples include descriptions of training vervets for awake bleeding procedures (Suleman, Njugana, and Anderson 1988, Wall, Worthman, and Else 1985) or the use of conditioning procedures for the collection of urine without requiring animals to be removed from their social groups (Kelley and Bramblett 1981). Molskness et al. (2007) described the training of 39 wild-caught, imported, adult female *sabaeus* in which animals were individually housed and trained to move into a sample collection cage for daily awake vaginal swabbing and saphenous venipuncture. Stephens et al. (2013) also described habituation/training of 12 pair-housed adult female *sabaeus* for daily vaginal swabbing.

Vervets have also been trained to participate in cognitive testing or other experimental procedures. James et al. (2007) described transferring adolescent male *sabaeus* into a capture tunnel attached to their home cage and the training/habituation procedures as part of daily cognitive testing tasks. Myers and Hamilton (2011) described cognitive testing of individually housed adult *sabaeus* that were trained to use a touch screen testing panel attached to their home cage. Mizrahi-Kliger et al. (2018) habituated two young adult female *sabaeus* to sleep in a primate chair within a sound-attenuating chamber in order to measure sleep stages via electroencephalogram, electromyogram, and electrooculogram, but provided no methodological details.

Special Situations

There are a few clinical risk factors that should be taken into account when working with vervets. Amory et al. (2013) noted that gastrointestinal tract disease is common in vervets. They found that bloat (gastric dilation) was associated with stress, including dietary changes and antibiotic treatment. Bloat is a serious concern in vervets, as it can often be fatal. Avoiding or limiting the feeding of preferred, denser food items after anesthesia, especially after fasting or during dietary changes, is recommended to reduce the risk of bloat. Gastric ulceration was also found to occur in stressful situations and was most common in wild-caught animals. Tarara, Tarara, and Suleman (1995) reported that nearly one-third (n=260) of captive *pygerythrus* had evidence of gastric ulcers at necropsy. They also found that gastric ulcers were associated with individual housing at rates that were much higher than reported rates in other nonhuman primates. There have also been reports that gastric ulcerations are related to capture and restraint (Suleman et al. 2000, 2004). Finally, Fincham, Seier, and Lombard (1992) reported that adult male vervets fed an atherogenic Western diet (high-fat, high-cholesterol) were at greater risk of torsion of the colon. The increased risk was not seen in females. Neither bloat nor colon torsion were observed during previous diet studies in the VRC using Western or high-fiber diets (Fairbanks, Blau, and Jorgensen 2010, Jorgensen et al. 2013, Voruganti et al. 2013).

Conclusions/Recommendations

While vervets share many similarities with macaques, baboons, and other Old World monkeys, I hope that this summary of their natural history highlights a few critical differences that should be taken into consideration when working with this unique species. Crafting species-specific behavioral management and enrichment plans is essential if we hope to maintain the psychological well-being of our nonhuman primate subjects.

Being aware that vervets are territorial will hopefully help avoid detrimental housing and husbandry decisions. Understanding that vervets are equally adept in trees, as well as on the ground, will hopefully lead to more utilization of vertical space and climbing structures, when applicable. Being attentive to the unique characteristics of female vervets; that they have relatively long canines, have no overt signs of menstruation or ovulation, and are tolerant of allomothering; should help clarify some of the behavioral differences seen in these animals. Knowing that vervets have less rigid dominance hierarchies and recognizing the unique patterns of vocal and gestural behavior seen in this species should help those working with vervets to have a better understanding of their social interactions.

As I have stated in previous publications, vervets are not macaques. While many of the behavioral management techniques, enrichment procedures, and housing options used with macaques can be successfully used with vervets, it is important to be aware of situations in which the more commonly used techniques and available facilities may be inappropriate for vervets. Understanding the important differences between vervets and other Old World monkeys in terms of their biology, behavioral repertoire, and natural history should hopefully benefit the researchers, veterinarians, enrichment coordinators, and animal care staff that work with these animals, as well as the animals themselves.

Additional Resources

For the last few years, I have tried to maintain a PubMed bibliography of research articles involving vervets. Due to the focus of PubMed, this reference list is obviously biased toward biomedical research and contains only a subset of the field work done with this species. The bibliography can be found here: https://www.ncbi.nlm.nih.gov/myncbi/browse/collection/44734281/.

An excellent starting point for learning about vervets is Cheney and Seyfarth's popular book *How Monkeys See the World* (Cheney and Seyfarth 1990), which describes their field studies and also provides an excellent introduction to vervet social behavior. A recent edited volume, *Savanna Monkeys: The Genus Chlorocebus* (Turner, Schmitt, and Cramer 2019c) also provides a summary of a wide range of recent field work. Struhsaker's *Science* paper on vervet behavior is also an accessible and useful starting point (Struhsaker 1967c). I hope that this current chapter will serve as a useful companion to my previous publications on pair housing and behavioral management of vervets.

Acknowledgments

The Vervet Research Colony is supported by P40-OD010965, UL1-TR001420. I wish to thank the research, veterinary, enrichment, and animal care staff that work with the colony. I also wish to thank Elizabeth Hilber for editorial corrections and assistance with literature review and Lynn Fairbanks for years of mentorship and comments on the chapter.

REFERENCES

Amory, J. T., W. M. du Plessis, A. Beierschmitt, J. Beeler-Marfisi, R. M. Palmour, and T. Beths. 2013. "Abdominal ultrasonography of the normal St. Kitts vervet monkey (*Chlorocebus sabaeus*)." *Journal of Medical Primatology* 42 (1):28–38. doi: 10.1111/jmp.12028.

Andelman, S. J. 1987. "Evolution of concealed ovulation in vervet monkeys (*Cercopithecus aethiops*)." *American Naturalist* 129 (6):785–799. doi: 10.1086/284675.

Arseneau-Robar, T. J. M., E. Muller, A. L. Taucher, C. P. van Schaik, R. Bshary, and E. P. Willems. 2018. "Male monkeys use punishment and coercion to de-escalate costly intergroup fights." *Proceedings of the Royal Society B: Biological Sciences* 285 (1880). doi: 10.1098/rspb.2017.2323.

Arseneau-Robar, T. J. M., A. L. Taucher, A. B. Schnider, C. P. van Schaik, and E. P. Willems. 2017. "Intra- and interindividual differences in the costs and benefits of intergroup aggression in female vervet monkeys." *Animal Behaviour* 123:129–137. doi: 10.1016/j.anbehav.2016.10.034.

Arseneau-Robar, T. J., A. L. Taucher, E. Muller, C. van Schaik, R. Bshary, and E. P. Willems. 2016. "Female monkeys use both the carrot and the stick to promote male participation in intergroup fights." *Proceedings of the Royal Society B: Biological Sciences* 283 (1843). doi: 10.1098/rspb.2016.1817.

Atkins, H. M., C. J. Willson, M. Silverstein, M. Jorgensen, E. Floyd, J. R. Kaplan, and S. E. Appt. 2014. "Characterization of ovarian aging and reproductive senescence in vervet monkeys (*Chlorocebus aethiops sabaeus*)." *Comparative Medicine* 64 (1):55–62.

Baker, K. C. 2016. "Survey of 2014 behavioral management programs for laboratory primates in the United States." *American Journal of Primatology* 78 (7):780–796. doi: 10.1002/ajp.22543.

Baker, K. C., J. L. Weed, C. M. Crockett, and M. A. Bloomsmith. 2007. "Survey of environmental enhancement programs for laboratory primates." *American Journal of Primatology* 69 (4):377–394. doi: 10.1002/ajp.20347.

Baldellou, M., and S. P. Henzi. 1992. "Vigilance, predator detection and the presence of supernumerary males in vervet monkey troops." *Animal Behaviour* 43 (3):451–461. doi: 10.1016/S0003-3472(05)80104-6.

Barrenas, F., K. Raehtz, C. Xu, L. Law, R. R. Green, G. Silvestri, S. E. Bosinger, A. Nishida, Q. Li, W. Lu, J. Zhang, M. J. Thomas, J. Chang, E. Smith, J. M. Weiss, R. A. Dawoud, G. H. Richter, A. Trichel, D. Ma, X. Peng, J. Komorowski, C. Apetrei, I. Pandrea, and M. Gale, Jr. 2019. "Macrophage-associated wound healing contributes to African green monkey SIV pathogenesis control." *Nature Communications* 10 (1):5101. doi: 10.1038/s41467-019-12987-9.

Baulu, J., G. Evans, and C. Sutton. 2002. "Pathogenic agents found in Barbados *Chlorocebus aethiops sabaeus* and in old world monkeys commonly used in biomedical research." *Laboratory Primate Newsletter* 41:4–6.

Becker, F., F. N. E. Gavins, J. Fontenot, P. Jordan, J. Y. Yun, R. Scott, P. R. Polk, R. E. Friday, M. Boktor, M. Musso, E. Romero, S. Boudreaux, J. Simmons, D. L. Hasselschwert, J. E. Goetzmann, J. Vanchiere, U. Cvek, M. Trutschl, P. Kilgore, and J. S. Alexander. 2019. "Dynamic gut microbiome changes following regional intestinal lymphatic obstruction in primates." *Pathophysiology*. doi: 10.1016/j.pathophys.2019.06.004.

Becker, F., E. Romero, J. Goetzmann, D. L. Hasselschwert, B. Dray, J. Vanchiere, J. Fontenot, J. W. Yun, P. C. Norris, L. White, M. Musso, C. N. Serhan, J. S. Alexander, and F. N. E. Gavins. 2019. "Endogenous specialized proresolving mediator profiles in a novel experimental model of lymphatic obstruction and intestinal inflammation in African green monkeys." *American Journal of Pathology* 189 (10):1953–1972. doi: 10.1016/j.ajpath.2019.05.013.

Bernstein, P. L., J. W. Smith, A. Krensky, and K. Rosene. 1978. "Tail positions of *Cercopithecus aethiops*." *Zeitschrift Tierpsychologie* 46:268–278.

Bloomstrand, M. A., and T. L. Maple. 1987. "Management and husbandry of African monkeys in captivity." In *Comparative Behavior of African Monkeys*, edited by E. L. Zucker, 197–234. New York: Alan R. Liss.

Borgeaud, C., M. Alvino, K. van Leeuwen, S. W. Townsend, and R. Bshary. 2015. "Age/sex differences in third-party rank relationship knowledge in wild vervet monkeys, *Chlorocebus aethiops pygerythrus*." *Animal Behaviour* 102:277–284. doi: 10.1016/j.anbehav.2015.02.006.

Borgeaud, C., A. Schnider, M. Krutzen, and R. Bshary. 2017. "Female vervet monkeys fine-tune decisions on tolerance versus conflict in a communication network." *Proceedings of the Royal Society B: Biological Sciences* 284 (1867). doi: 10.1098/rspb.2017.1922.

Borgeaud, C., E. van de Waal, and R. Bshary. 2013. "Third-party ranks knowledge in wild vervet monkeys (*Chlorocebus aethiops pygerythrus*)." *PLoS One* 8 (3):e58562. doi: 10.1371/journal.pone.0058562.

Borgeaud, Christele, and Redouan Bshary. 2015. "Wild vervet monkeys trade tolerance and specific coalitionary support for grooming in experimentally induced conflicts." *Current Biology* 25 (22):3011–3016.

Borgeaud, Christele, Sebastian Sosa, Cedric Sueur, and Redouan Bshary. 2017. "The influence of demographic variation on social network stability in wild vervet monkeys." *Animal Behaviour* 134:155–165.

Bouskila, J., V. Harrar, P. Javadi, A. Beierschmitt, R. Palmour, C. Casanova, J. F. Bouchard, and M. Ptito. 2016. "Cannabinoid receptors CB1 and CB2 modulate the electroretinographic waves in vervet monkeys." *Neural Plasticity* 2016:1253245. doi: 10.1155/2016/1253245.

Bouskila, J., P. Javadi, C. Casanova, M. Ptito, and J. F. Bouchard. 2013. "Rod photoreceptors express GPR55 in the adult vervet monkey retina." *PLoS One* 8 (11):e81080. doi: 10.1371/journal.pone.0081080.

Bouskila, J., R. M. Palmour, J. F. Bouchard, and M. Ptito. 2018. "Retinal structure and function in monkeys with fetal alcohol exposure." *Experimental Eye Research* 177:55–64. doi: 10.1016/j.exer.2018.07.027.

Bramblett, C. A., S. S. Bramblett, D. A. Bishop, and A. M. Coelho. 1982. "Longitudinal stability in adult status hierarchies among vervet monkeys (*Cercopithecus aethiops*)." *American Journal of Primatology* 2 (1):43–51. doi: 10.1002/ajp.1350020107.

Bramblett, C. A., L. D. Pejaver, and D. J. Drickman. 1975. "Reproduction in captive vervet and Sykes' monkeys." *Journal of Mammalogy* 56 (4):940–946.

Brennan, E. J., J. G. Else, and J. Altmann. 1985. "Ecology and behavior of a pest primate - Vervet monkeys in a tourist-lodge habitat." *African Journal of Ecology* 23 (1):35–44. doi: 10.1111/j.1365-2028.1985.tb00710.x.

Briggs, C. M., K. M. Smith, A. Piper, E. Huitt, C. J. Spears, M. Quiles, M. Ribeiro, M. E. Thomas, D. T. Brown, and R. Hernandez. 2014. "Live attenuated tetravalent dengue virus host range vaccine is immunogenic in African green monkeys following a single vaccination." *Journal of Virology* 88 (12):6729–6742. doi: 10.1128/JVI.00541-14.

Broussard, S. R., S. I. Staprans, R. White, E. M. Whitehead, M. B. Feinberg, and J. S. Allan. 2001. "Simian immunodeficiency virus replicates to high levels in naturally infected African green monkeys without inducing immunologic or neurologic disease." *Journal of Virology* 75 (5):2262–2275. doi: 10.1128/JVI.75.5.2262-2275.2001.

Cann, J. A., K. Kavanagh, M. J. Jorgensen, S. Mohanan, T. D. Howard, S. B. Gray, G. A. Hawkins, L. A. Fairbanks, and J. D. Wagner. 2010. "Clinicopathologic characterization of naturally occurring diabetes mellitus in vervet monkeys." *Veterinary Pathology* 47 (4):713–718. doi: 10.1177/0300985810370011.

Canteloup, C., C. Borgeaud, M. Wubs, and E. van de Waal. 2019. "The effect of social and ecological factors on the time budget of wild vervet monkeys." *Ethology* 125 (12):902–913. doi: 10.1111/eth.12946.

Carlsson, H. E., S. J. Schapiro, I. Farah, and J. Hau. 2004. "Use of primates in research: A global overview." *American Journal of Primatology* 63 (4):225–237. doi: 10.1002/Ajp.20054.

Carroll, R. L., K. Mah, J. W. Fanton, G. N. Maginnis, R. M. Brenner, and O. D. Slayden. 2007. "Assessment of menstruation in the vervet (*Cercopithecus aethiops*)." *American Journal of Primatology* 69 (8):901–916. doi: 10.1002/ajp.20396.

Chapman, C., and L. M. Fedigan. 1984. "Territoriality in the St Kitts vervet, *Cercopithecus aethiops*." *Journal of Human Evolution* 13 (8):677–686. doi: 10.1016/S0047-2484(84)80019-6.

Chen, J. A., S. C. Fears, A. J. Jasinska, A. Huang, N. B. Al-Sharif, K. E. Scheibel, T. D. Dyer, A. M. Fagan, J. Blangero, R. Woods, M. J. Jorgensen, J. R. Kaplan, N. B. Freimer, and G. Coppola. 2018. "Neurodegenerative disease biomarkers Abeta1-40, Abeta1-42, tau, and p-tau181 in the vervet monkey cerebrospinal fluid: Relation to normal aging, genetic influences, and cerebral amyloid angiopathy." *Brain and Behavior* 8 (2):e00903. doi: 10.1002/brb3.903.

Cheney, D. L. 1981. "Inter-group encounters among free-ranging vervet monkeys." *Folia Primatologica* 35 (2–3):124–146. doi: 10.1159/000155970.

Cheney, D. L. 1983a. "Extra-familial alliances among vervet monkeys." In *Primate Social Relationships: An Integrated Approach*, edited by R. A. Hinde, 278–286. Oxford: Blackwell Scientific.

Cheney, D. L. 1983b. "Proximate and ultimate factors related to the distribution of male migration." In *Primate Social Relationships: An Integrated Approach*, edited by R. A. Hinde, 241–249. Oxford: Blackwell Scientific.

Cheney, D. L., P. C. Lee, and R. M. Seyfarth. 1981. "Behavioral-correlates of nonrandom mortality among free-ranging female vervet monkeys." *Behavioral Ecology and Sociobiology* 9 (2):153–161. doi: 10.1007/Bf00293587.

Cheney, D. L., and R. M. Seyfarth. 1980. "Vocal recognition in free-ranging vervet monkeys." *Animal Behaviour* 28 (May):362–364. doi: 10.1016/S0003-3472(80)80044-3.

Cheney, D. L., and R. M. Seyfarth. 1982a. "How vervet monkeys perceive their grunts - field playback experiments." *Animal Behaviour* 30 (Aug):739–751. doi: 10.1016/S0003-3472(82)80146-2.

Cheney, D. L., and R. M. Seyfarth. 1982b. "Recognition of individuals within and between groups of free-ranging vervet monkeys." *American Zoologist* 22 (3):519–529.

Cheney, D. L., and R. M. Seyfarth. 1983. "Nonrandom dispersal in free-ranging vervet monkeys - social and genetic consequences." *American Naturalist* 122 (3):392–412. doi: 10.1086/284142.

Cheney, D. L., and R. M. Seyfarth. 1986. "The recognition of social alliances by vervet monkeys." *Animal Behaviour* 34:1722–1731. doi: 10.1016/S0003-3472(86)80259-7.

Cheney, D. L., and R. M. Seyfarth. 1987. "The influence of intergroup competition on the survival and reproduction of female vervet monkeys." *Behavioral Ecology and Sociobiology* 21 (6):375–386. doi: 10.1007/Bf00299932.

Cheney, D. L., and R. M. Seyfarth. 1988. "Assessment of meaning and the detection of unreliable signals by vervet monkeys." *Animal Behaviour* 36:477–486. doi: 10.1016/S0003-3472(88)80018-6.

Cheney, D. L., and R. M. Seyfarth. 1990. *How Monkeys See the World*. Chicago: University of Chicago Press.

Cheney, D. L., R. M. Seyfarth, S. J. Andleman, and P. E. Lee. 1988. "Reproductive success in vervet monkeys." In *Reproductive Success*, edited by T. H. Clutton-Brock, 384–402. Chicago: University of Chicago Press.

Chism, J. 2000. "Allocare patterns among cercopithecines." *Folia Primatologica* 71 (1–2):55–66. doi: 10.1159/000021730.

Cramer, J. D., T. Gaetano, J. P. Gray, P. Grobler, J. G. Lorenz, N. B. Freimer, C. A. Schmitt, and T. R. Turner. 2013. "Variation in scrotal color among widely distributed vervet monkey populations (*Chlorocebus aethiops pygerythrus* and *Chlorocebus aethiops sabaeus*)." *American Journal of Primatology* 75 (7):752–762. doi: 10.1002/ajp.22156.

Cramer, P. E., R. C. Gentzel, K. Q. Tanis, J. Vardigan, Y. Wang, B. Connolly, P. Manfre, K. Lodge, J. J. Renger, C. Zerbinatti, and J. M. Uslaner. 2018. "Aging African green monkeys manifest transcriptional, pathological, and cognitive hallmarks of human Alzheimer's disease." *Neurobiology of Aging* 64:92–106. doi: 10.1016/j.neurobiolaging.2017.12.011.

Daniel, J. R., A. J. dos Santos, and L. Vincente. 2008. "Correlates of self-directed behaviors in captive *Cercopithecus aethiops*." *International Journal of Primatology* 29 (5):1219–1226.

Denham, W. W. 1987. *West Indian Green Monkeys: Problems in Historical Biogeography*. Basel: Karger.

Deutsch, J. C., and P. C. Lee. 1991. "Dominance and feeding competition in captive rhesus monkeys." *International Journal of Primatology* 12 (6):615–628.

Dillon, J. E., M. J. Raleigh, M. T. McGuire, and D. B. Pollack. 1992. "Acute changes in social composition and agonistic behavior in male vervet monkeys (*Cercopithecus aethiops sabaeus*)." *American Journal of Primatology* 27:225–230.

Ducheminsky, N., S. P. Henzi, and L. Barrett. 2014. "Responses of vervet monkeys in large troops to terrestrial and aerial predator alarm calls." *Behavioral Ecology* 25 (6):1474–1484.

Dunbar, R. I. M. 1974. "Observations on the ecology and social organization of the green monkey, *Cercopithecus sabaeus*, in Senegal." *Primates* 15 (4):341–350.

Eley, R. M. 1992. "Reproductive biology of the vervet monkey (*Cercopithecus aethiops*)." *Utafiti* 4:1–33.

Eley, R. M., A. S. Hutchinson, and J. G. Else. 1984. "Reproductive seasonality in captive vervet monkeys, *Cercopithecus aethiops*." *International Journal of Primatology* 5 (4):335–335.

Eley, R. M., R. P. Tarara, C. M. Worthman, and J. G. Else. 1989. "Reproduction in the vervet monkey (*Cercopithecus aethiops*) 3. The menstrual-cycle." *American Journal of Primatology* 17 (1):1–10. doi: 10.1002/ajp.1350170102.

Else, J. G. 1985. "Captive propagation of vervet monkeys (*Cercopithecus aethiops*) in harems." *Laboratory Animal Science* 35 (4):373–375.

Emborg, M. E. 2007. "Nonhuman primate models of Parkinson's disease." *ILAR Journal* 48 (4):339–55.

Estes, R. D. 1991. *The Behavior Guide to African Mammals*. Berkeley and Los Angeles: University of California Press.

Eyles, J. E., J. E. Johnson, S. Megati, V. Roopchand, P. J. Cockle, R. Weeratna, S. Makinen, T. P. Brown, S. Lang, S. E. Witko, C. S. Kotash, J. Li, K. West, O. Maldonado, D. J. Falconer, C. Lees, G. J. Smith, P. White, P. Wright, P. T. Loudon, J. R. Merson, K. U. Jansen, and M. K. Sidhu. 2013. "Nonreplicating vaccines can protect African green monkeys from the memphis 37 strain of respiratory syncytial virus." *Journal of Infectious Disease* 208 (2):319–29. doi: 10.1093/infdis/jit169.

Fairbanks, L. A. 1980. "Relationships among adult females in captive vervet monkeys - Testing a model of rank-related attractiveness." *Animal Behaviour* 28 (Aug):853–859. doi: 10.1016/S0003-3472(80)80145-X.

Fairbanks, L. A. 1988a. "Mother-infant behavior in vervet monkeys - Response to failure of last pregnancy." *Behavioral Ecology and Sociobiology* 23 (3):157–165. doi: 10.1007/Bf00300350.

Fairbanks, L. A. 1988b. "Vervet monkey grandmothers - Effects on mother-infant relationships." *Behaviour* 104:176–188. doi: 10.1163/156853988x00665.

Fairbanks, L. A. 1990. "Reciprocal benefits of allomothering for female vervet monkeys." *Animal Behaviour* 40:553–562. doi: 10.1016/S0003-3472(05)80536-6.

Fairbanks, L. A. 1996. "Individual differences in matternal style: Causes and consequences for mothers and offspring." *Advances in the Study of Behavior* 25:579–611.

Fairbanks, L. A. 2001. "Individual differences in response to a stranger: Social impulsivity as a dimension of temperament in vervet monkeys (*Cercopithecus aethiops sabaeus*)." *Journal of Comparative Psychology* 115 (1):22–28. doi: 10.1037/0735-7036.115.1.22.

Fairbanks, L. A., K. Blau, and M. J. Jorgensen. 2010. "High-fiber diet promotes weight loss and affects maternal behavior in vervet monkeys." *American Journal of Primatology* 72 (3):234–241. doi: 10.1002/Ajp.20772.

Fairbanks, L. A., M. J. Jorgensen, A. Huff, K. Blau, Y. Y. Hung, and J. J. Mann. 2004. "Adolescent impulsivity predicts adult dominance attainment in male." *American Journal of Primatology* 64 (1):1–17. doi: 10.1002/Ajp.20057.

Fairbanks, L. A., and M. T. McGuire. 1984. "Determinants of fecundity and reproductive success in captive vervet monkeys." *American Journal of Primatology* 7 (1):27–38. doi: 10.1002/ajp.1350070106.

Fairbanks, L. A., and M. T. McGuire. 1986. "Age, reproductive value, and dominance-related behavior in vervet monkey females - cross-generational influences on social relationships and reproduction." *Animal Behaviour* 34:1710–1721. doi: 10.1016/S0003-3472(86)80258-5.

Fairbanks, L. A., and M. T. McGuire. 1987. "Mother infant relationships in vervet monkeys - response to new adult males." *International Journal of Primatology* 8 (4):351–366. doi: 10.1007/Bf02737388.

Fairbanks, L. A., and M. T. McGuire. 1993. "Maternal protectiveness and response to the unfamiliar in vervet monkeys." *American Journal of Primatology* 30 (2):119–129. doi: 10.1002/ajp.1350300204.

Fairbanks, L. A., M. T. McGuire, and N. Page. 1978. "Social roles in captive vervet monkeys (*Cercopithecus aethiops sabaeus*)." *Behavioural Processes* 3 (4):335–352. doi: 10.1016/0376-6357(78)90006-2.

Fairbanks, L. A., W. P. Melega, M. J. Jorgensen, J. R. Kaplan, and M. T. McGuire. 2001. "Social impulsivity inversely associated with CSF5-HIAA and fluoxetine exposure in vervet monkeys." *Neuropsychopharmacology* 24 (4):370–378. doi: 10.1016/S0893-133x(00)00211-6.

Fears, S. C., W. P. Melega, S. K. Service, C. Lee, K. Chen, Z. Tu, M. J. Jorgensen, L. A. Fairbanks, R. M. Cantor, N. B. Freimer, and R. P. Woods. 2009. "Identifying heritable brain phenotypes in an extended pedigree of vervet monkeys." *Journal of Neuroscience* 29 (9):2867–75. doi: 10.1523/JNEUROSCI.5153-08.2009.

Fedigan, L., and L. M. Fedigan. 1988. "*Cercopithecus aethiops*: A review of field studies." In *A Primate Radiation: Evolutionary Biology of the African Guenons*, edited by A. Gautier-Hion, F. Bourliere, J. P. Gautier and J. Kingdon, 389–411. New York: Cambridge University Press.

Fedigan, L. M. 1992. *Primate Paradigms: Sex Roles and Social Bonds*. Chicago, IL: University of Chicago Press.

Fedorov, A., X. Li, K. M. Pohl, S. Bouix, M. Styner, M. Addicott, C. Wyatt, J. B. Daunais, W. M. Wells, and R. Kikinis. 2011. "Atlas-guided segmentation of vervet monkey brain MRI." *Open Neuroimaging Journal* 5:186–197. doi: 10.2174/1874440001105010186.

Fincham, J. E., J. V. Seier, and C. J. Lombard. 1992. "Torsion of the colon in vervet monkeys: Association with an atherogenic Western-type of diet." *Journal of Medical Primatology* 21 (1):44–46.

Freeman, N. J., C. Young, L. Barrett, and S. Peter Henzi. 2016. "Coalition formation by male vervet monkeys (*Chlorocebus pygerythrus*) in South Africa." *Ethology* 122 (1):45–52.

Freimer, N. B., S. K. Service, R. A. Ophoff, A. J. Jasinska, K. McKee, A. Villeneuve, A. Belisle, J. N. Bailey, S. E. Breidenthal, M. J. Jorgensen, J. J. Mann, R. M. Cantor, K. Dewar, and L. A. Fairbanks. 2007. "A quantitative trait locus for variation in dopamine metabolism mapped in a primate model using reference sequences from related species." *Proceedings of the National Academy of Sciences USA* 104 (40):15811–15816. doi: 10.1073/pnas.0707640104.

Freimer, N., K. Dewar, J. Kaplan, and L. Fairbanks. 2008. "The importance of the vervet (African green monkey) as a biomedical model." Accessed 11/23/2018. http://www.genome.gov/Pages/Research/Sequencing/SeqProposals/TheVervetMonkeyBiomedicalModel.pdf.

Gartlan, J. S. 1969. "Sexual and maternal behavior of the vervet monkey, *Cercopithecus aethiops*." *Journal of Reproduction and Fertility Supplements* 6:137–150.

Gerald, M. S. 2001. "Primate colour predicts social status and aggressive outcome." *Animal Behaviour* 61:559–566.

Goldstein, S., C. R. Brown, I. Ourmanov, I. Pandrea, A. Buckler-White, C. Erb, J. S. Nandi, G. J. Foster, P. Autissier, J. E. Schmitz, and V. M. Hirsch. 2006. "Comparison of simian immunodeficiency virus SIVagmVer replication and CD4+ T-cell dynamics in vervet and sabaeus African green monkeys." *Journal of Virology* 80 (10):4868–4877. doi: 10.1128/JVI.80.10.4868-4877.2006.

Grampp, M., C. Sueur, E. van de Waal, and J. Botting. 2019. "Social attention biases in juvenile wild vervet monkeys: Implications for socialisation and social learning processes." *Primates* 60 (3):261–275. doi: 10.1007/s10329-019-00721-4.

Groman, S. M., A. M. Morales, B. Lee, E. D. London, and J. D. Jentsch. 2013. "Methamphetamine-induced increases in putamen gray matter associate with inhibitory control." *Psychopharmacology (Berl)* 229 (3):527–538. doi: 10.1007/s00213-013-3159-9.

Groves, C. P., and J. Kingdon. 2013. "Genus *Chlorocebus* savanna monkeys." In *The Mammals of Africa: Volume II Primates*, edited by T. M. Butynski, J. S. Kingdon and J. Kalina, 264–266. London: Bloomsbury Publishing.

Hall, K. R. L., and J. S. Gartlan. 1965. "Ecology and behaviour of the vervet monkey, *Cercopithecus aethiops*, Lolui Island, Lake Victoria." *Proceedings of the Royal Society of London* 145:37–56.

Harrison, M. J. S. 1983a. "Age and sex differences in the diet and feeding strategies of the green monkey *Cercopithecus sabaeus*." *Animal Behaviour* 31 (4):969–977.

Harrison, M. J. S. 1983b. "Territorial behaviour in the green monkey, *Cercopithecus sabaeus*: Seasonal defense of local food supplies." *Behavioral Ecology and Sociobiology* 12:85–94.

Harrison, M. J. S. 1984. "Optimal foraging strategies in the diet of the green monkey *Cercopithecus sabaeus* at Mount Assirik Senegal." *International Journal of Primatology* 5 (5):435–472.

Haus, T., E. Akom, B. Agwanda, M. Hofreiter, C. Roos, and D. Zinner. 2013. "Mitochondrial diversity and distribution of African green monkeys (chlorocebus gray, 1870)." *American Journal of Primatology* 75 (4):350–360. doi: 10.1002/ajp.22113.

He, T., C. Xu, N. Krampe, S. M. Dillon, P. Sette, E. Falwell, G. S. Haret-Richter, T. Butterfield, T. L. Dunsmore, W. M. M. Jr, K. J. Martin, B. B. Policicchio, K. D. Raehtz, E. P. Penn, R. P. Tracy, R. M. Ribeiro, D. N. Frank, C. C. Wilson, A. L. Landay, C. Apetrei, and I. Pandrea. 2019. "High-fat diet exacerbates SIV pathogenesis and accelerates disease progression." *Journal of Clinical Investigation* 129 (12):5474–5488. doi: 10.1172/JCI121208.

Henzi, S. P., N. Forshaw, R. Boner, L. Barrett, and D. Lusseau. 2013. "Scalar social dynamics in female vervet monkey cohorts." *Philosophical Transactions of the Royal Society of London Series B - Biological Sciences* 368 (1618):20120351. doi: 10.1098/rstb.2012.0351.

Henzi, S. P., and J. W. Lucas. 1980. "Observations on the inter troop movement of adult vervet monkeys *Cercopithecus aethiops*." *Folia Primatologica* 33 (3):220–235.

Hess, D. L., A. G. Hendrickx, and G. H. Stabenfeldt. 1979. "Reproductive and hormonal patterns in the African green monkey (*Cercopithecus aethiops*)." *Journal of Medical Primatology* 8 (5):273–281.

Holbrook, B. C., S. T. Aycock, E. Machiele, E. Clemens, D. Gries, M. J. Jorgensen, M. B. Hadimani, S. B. King, and M. A. Alexander-Miller. 2018. "An R848 adjuvanted influenza vaccine promotes early activation of B cells in the draining lymph nodes of non-human primate neonates." *Immunology* 153 (3):357–367. doi: 10.1111/imm.12845.

Holbrook, B. C., J. R. Kim, L. K. Blevins, M. J. Jorgensen, N. D. Kock, R. B. D'Agostino, Jr., S. T. Aycock, M. B. Hadimani, S. B. King, G. D. Parks, and M. A. Alexander-Miller. 2016. "A novel R848-conjugated inactivated influenza virus vaccine Is efficacious and safe in a neonate nonhuman primate model." *Journal of Immunology* 197 (2):555–564. doi: 10.4049/jimmunol.1600497.

Horrocks, J. A. 1986. "Life-history characteristics of a wild population of vervets (*Cercopithecus aethiops sabaeus*) in Barbados, West Indies." *International Journal of Primatology* 7 (1):31–47.

Horrocks, J., and W. Hunte. 1983. "Maternal rank and offspring rank in vervet monkeys - an appraisal of the mechanisms of rank acquisition." *Animal Behaviour* 31 (Aug):772–782. doi: i 10.1016/S0003-3472(83)80234-6.

Huang, Y. S., V. Ramensky, S. K. Service, A. J. Jasinska, Y. Jung, O. W. Choi, R. M. Cantor, N. Juretic, J. Wasserscheid, J. R. Kaplan, M. J. Jorgensen, T. D. Dyer, K. Dewar, J. Blangero, R. K. Wilson, W. Warren, G. M. Weinstock, and N. B. Freimer. 2015. "Sequencing strategies and characterization of 721 vervet monkey genomes for future genetic analyses of medically relevant traits." *BMC Biology* 13:10. doi: 10.1186/s12915-015-0152-2.

Hugo, C., J. Seier, C. Mdhluli, W. Daniels, B. H. Harvey, D. Du Toit, S. Wolfe-Coote, D. Nel, and D. J. Stein. 2003. "Fluoxetine decreases stereotypic behavior in primates." *Progress in Neuro-Psychopharmacology & Biological Psychiatry* 27 (4):639–643. doi: 10.1016/S0278-5846(03)00073-3.

Isbell, L. A. 1990. "Sudden short-term increase in mortality of vervet monkeys *Cercopithecus aethiops* due to leopard predation in Amboseli National Park Kenya." *American Journal of Primatology* 21 (1):41–52.

Isbell, L. A., and L. R. Bidner. 2016. "Vervet monkey (*Chlorocebus pygerythrus*) alarm calls to leopards (*Panthera pardus*) function as a predator deterrent." *Behaviour* 153:591–606.

Isbell, L. A., D. L. Cheney, and R. M. Seyfarth. 1990. "Costs and benefits of home range shifts among vervet monkeys (*Cercopithecus aethiops*) in Amboseli National Park, Kenya." *Behavioral Ecology and Sociobiology* 27 (5):351–358. doi: 10.1007/Bf00164006.

Isbell, L. A., D. L. Cheney, and R. M. Seyfarth. 1991. "Group fusions and minimum group sizes in vervet monkeys (*Cercopithecus aethiops*)." *American Journal of Primatology* 25 (1):57–65. doi: 10.1002/ajp.1350250106.

Isbell, L. A., J. D. Pruetz, and T. P. Young. 1998. "Movements of vervets (*Cercopithecus aethiops*) and patas monkeys (*Erythrocebus patas*) as estimators of food resource size, density, and distribution." *Behavioral Ecology and Sociobiology* 42 (2):123–133.

Isbell, L. A., and T. P. Young. 1993. "Social and ecological influences on activity budgets of vervet monkeys and their implications for group living." *Behavioral Ecology and Sociobiology* 32 (6):377–385.

Isbell, L. A. Cheney, D. L., Seyfarth, R. M. 2002. "Why vervet monkeys (*Cercopithecus aethiops*) live in multimale groups." In *The Guenons: Diversity and Adaptation in African Monkeys*, edited by M. Cords Glenn and M. E. Glenn, 173–187. New York: Kluwer Academic/Plenum Publishers.

James, A. S., S. M. Groman, E. Seu, M. Jorgensen, L. A. Fairbanks, and J. D. Jentsch. 2007. "Dimensions of impulsivity are associated with poor spatial working memory performance in monkeys." *Journal of Neuroscience* 27 (52):14358–14364. doi: 10.1523/JNEUROSCI.4508-07.2007.

Jarrett, J. D., T. Bonnell, M. J. Jorgensen, C. A. Schmitt, C. Young, M. Dostie, L. Barrett, and S. P. Henzi. 2019. "Modeling variation in the growth of wild and captive juvenile vervet monkeys in relation to diet and resource availability." *American Journal of Physical Anthropology* 171 (1):89–99. doi: 10.1002/ajpa.23960.

Jarrett, J. D., T. R. Bonnell, C. Young, L. Barrett, and S. P. Henzi. 2018. "Network integration and limits to social inheritance in vervet monkeys." *Proceedings of the Royal Society B: Biological Sciences* 285 (1876). doi: 10.1098/rspb.2017.2668.

Jasinska, A. J., M. K. Lin, S. Service, O. W. Choi, J. DeYoung, O. Grujic, S. Y. Kong, Y. Jung, M. J. Jorgensen, L. A. Fairbanks, T. Turner, R. M. Cantor, J. Wasserscheid, K. Dewar, W. Warren, R. K. Wilson, G. Weinstock, J. D. Jentsch, and N. B. Freimer. 2012. "A non-human primate system for large-scale genetic studies of complex traits." *Human Molecular Genetics* 21 (15):3307–3316. doi: 10.1093/hmg/dds160.

Jasinska, A. J., C. A. Schmitt, S. K. Service, R. M. Cantor, K. Dewar, J. D. Jentsch, J. R. Kaplan, T. R. Turner, W. C. Warren, G. M. Weinstock, R. P. Woods, and N. B. Freimer. 2013. "Systems biology of the vervet monkey." *ILAR Journal* 54 (2):122–143. doi: 10.1093/ilar/ilt049.

Jasinska, A. J., S. Service, O. W. Choi, J. DeYoung, O. Grujic, S. Y. Kong, M. J. Jorgensen, J. Bailey, S. Breidenthal, L. A. Fairbanks, R. P. Woods, J. D. Jentsch, and N. B. Freimer. 2009. "Identification of brain transcriptional variation reproduced in peripheral blood: An approach for mapping brain expression traits." *Human Molecular Genetics* 18 (22):4415–4427. doi: 10.1093/hmg/ddp397.

Jasinska, A. J., I. Zelaya, S. K. Service, C. B. Peterson, R. M. Cantor, O. W. Choi, J. DeYoung, E. Eskin, L. A. Fairbanks, S. Fears, A. E. Furterer, Y. S. Huang, V. Ramensky, C. A. Schmitt, H. Svardal, M. J. Jorgensen, J. R. Kaplan, D. Villar, B. L. Aken, P. Flicek, R. Nag, E. S. Wong, J. Blangero, T. D. Dyer, M. Bogomolov, Y. Benjamini, G. M. Weinstock, K. Dewar, C. Sabatti, R. K. Wilson, J. D. Jentsch, W. Warren, G. Coppola, R. P. Woods, and N. B. Freimer. 2017. "Genetic variation and gene expression across multiple tissues and developmental stages in a nonhuman primate." *Nature Genetics* 49 (12):1714–1721. doi: 10.1038/ng.3959.

Jentsch, J. D., D. E. Redmond, J. D. Elsworth, J. R. Taylor, K. D. Youngren, and R. H. Roth. 1997. "Enduring cognitive deficits and cortical dopamine dysfunction in monkeys after long-term administration of phencyclidine." *Science* 277 (5328):953–955. doi: 10.1126/science.277.5328.953.

Johnson, C., C. Koerner, M. Estrin, and D. Duoos. 1980. "Alloparental care and kinship in captive social groups of vervet monkeys (*Cercopithecus aethiops sabaeus*)." *International Journal of Primatology* 21 (3):406–415.

Johnson, P. T., D. A. Valerio, and G. E. Thompson. 1973. "Breeding the African green monkey, *Cercopithecus aethiops*, in a laboratory environment." *Laboratory Animal Science* 23 (3):355–359.

Jorgensen, M. J. 2017. "Behavioral management of *Chlorocebus* spp." In *Handbook of Primate Behavioral Management*, edited by S. J. Schapiro. Boca Raton, FL: CRC Press.

Jorgensen, M. J., S. T. Aycock, T. B. Clarkson, and J. R. Kaplan. 2013. "Effects of a Western-type diet on plasma lipids and other cardiometabolic risk factors in African green monkeys (*Chlorocebus aethiops sabaeus*)." *Journal of the American Association of Laboratory Animal Science* 52 (4):448–453.

Jorgensen, M. J., K. R. Lambert, S. D. Breaux, K. C. Baker, B. M. Snively, and J. L. Weed. 2017. "Pair housing of vervets/African green monkeys for biomedical research." *American Journal of Primatology* 79 (1):1–10. doi: 10.1002/ajp.22501.

Jorgensen, M. J., L. L. Rudel, M. Nudy, J. R. Kaplan, T. B. Clarkson, N. M. Pajewski, and P. F. Schnatz. 2012. "25(OH)D3 and cardiovascular risk factors in female nonhuman primates." *Journal of Women's Health* 21 (9):959–965. doi: 10.1089/jwh.2011.3416.

Kalinin, S., S. L. Willard, C. A. Shively, J. R. Kaplan, T. C. Register, M. J. Jorgensen, P. E. Polak, I. Rubinstein, and D. L. Feinstein. 2013. "Development of amyloid burden in African green monkeys." *Neurobiology of Aging* 34 (10):2361–2369. doi: 10.1016/j.neurobiolaging.2013.03.023.

Kaplan, J. R. 1987. "Dominance and affiliation in the Cercopithecini and Papionini: A comparative examination." In *Comparative Behavior of African Monkeys*, edited by E. L. Zucker, 127–150. New York: Alan R. Liss.

Kavanagh, K., A. T. Davis, K. A. Jenkins, and D. M. Flynn. 2016. "Effects of heated hydrotherapy on muscle HSP70 and glucose metabolism in old and young vervet monkeys." *Cell Stress Chaperones* 21 (4):717–25. doi: 10.1007/s12192-016-0699-z.

Kavanagh, K., A. T. Davis, D. E. Peters, A. C. LeGrand, M. S. Bharadwaj, and A. J. Molina. 2017. "Regulators of mitochondrial quality control differ in subcutaneous fat of metabolically healthy and unhealthy obese monkeys." *Obesity (Silver Spring)* 25 (4):689–696. doi: 10.1002/oby.21762.

Kavanagh, K., S. M. Day, M. C. Pait, W. R. Mortiz, C. B. Newgard, O. Ilkayeva, D. A. McClain, and S. L. Macauley. 2019. "Type-2-diabetes alters CSF but not plasma metabolomic and AD risk profiles in vervet monkeys." *Frontiers in Neuroscience* 13:843. doi: 10.3389/fnins.2019.00843.

Kavanagh, K., B. L. Dozier, T. J. Chavanne, L. A. Fairbanks, M. J. Jorgensen, and J. R. Kaplan. 2011. "Fetal and maternal factors associated with infant mortality in vervet monkeys." *Journal of Medical Primatology* 40 (1):27–36. doi: 10.1111/j.1600-0684.2010.00441.x.

Kavanagh, K., L. A. Fairbanks, J. N. Bailey, M. J. Jorgensen, M. Wilson, L. Zhang, L. L. Rudel, and J. D. Wagner. 2007. "Characterization and heritability of obesity and associated risk factors in vervet monkeys." *Obesity (Silver Spring)* 15 (7):1666–1674. doi: 10.1038/oby.2007.199.

Kavanagh, K., K. L. Jones, J. Sawyer, K. Kelley, J. J. Carr, J. D. Wagner, and L. L. Rudel. 2007. "Trans fat diet induces abdominal obesity and changes in insulin sensitivity in monkeys." *Obesity (Silver Spring)* 15 (7):1675–1684. doi: 10.1038/oby.2007.200.

Kavanagh, K., A. T. Wylie, K. L. Tucker, T. J. Hamp, R. Z. Gharaibeh, A. A. Fodor, and J. M. Cullen. 2013. "Dietary fructose induces endotoxemia and hepatic injury in calorically controlled primates." *American Journal of Clinical Nutrition* 98 (2):349–357. doi: 10.3945/ajcn.112.057331.

Keddy-Hector, A. C., R. M. Seyfarth, and M. J. Raleigh. 1989. "Male parental care, female choice and the effect of an audience in vervet monkeys." *Animal Behaviour* 38:262–271.

Keddy, A. C. 1986. "Female mate choice in vervet monkeys." *American Journal of Primatology* 10:125–134.

Kelley, T. M., and C. A. Bramblett. 1981. "Urine collection from vervet monkeys by instrumental-conditioning." *American Journal of Primatology* 1 (1):95–97. doi: 10.1002/ajp.1350010112.

Koritnik, D. R., L. L. Wood, L. N. Shandilya, and L. L. Rudel. 1984. "Lipids, lipoproteins, and endocrine profiles during pregnancy in the African green monkey (*Cercopithecus aethiops*)." *Metabolism* 33 (9):840–844.

Kucera, R., J. Bouskila, L. Elkrief, A. Fink-Jensen, R. Palmour, J. F. Bouchard, and M. Ptito. 2018. "Expression and localization of CB1R, NAPE-PLD, and FAAH in the vervet monkey nucleus accumbens." *Scientific Reports* 8 (1):8689. doi: 10.1038/s41598-018-26826-2.

Kushner, H., N. Kraft-Schreyer, E. T. Angelakos, and E. M. Wudarski. 1982. "Analysis of reproductive data in a breeding colony of African green monkeys." *Journal of Medical Primatology* 11 (2):77–84.

Latimer, C. S., C. A. Shively, C. D. Keene, M. J. Jorgensen, R. N. Andrews, T. C. Register, T. J. Montine, A. M. Wilson, B. J. Neth, A. Mintz, J. A. Maldjian, C. T. Whitlow, J. R. Kaplan, and S. Craft. 2019. "A nonhuman primate model of early Alzheimer's disease pathologic change: Implications for disease pathogenesis." *Alzheimer's and Dementia* 15 (1):93–105. doi: 10.1016/j.jalz.2018.06.3057.

Laudenslager, M. L., M. J. Jorgensen, R. Grzywa, and L. A. Fairbanks. 2011. "A novelty seeking phenotype is related to chronic hypothalamic-pituitary-adrenal activity reflected by hair cortisol." *Physiology and Behavior* 104 (2):291–295. doi: 10.1016/j.physbeh.2011.03.003.

Lee, J. H., D. A. Hammoud, Y. Cong, L. M. Huzella, M. A. Castro, J. Solomon, J. Laux, M. Lackemeyer, J. K. Bohannon, O. Rojas, R. Byrum, R. Adams, D. Ragland, M. St Claire, V. Munster, and M. R. Holbrook. 2019. "The use of large-particle aerosol exposure to Nipah virus to mimic human neurological disease manifestations in the African green monkey." *Journal of Infectious Disease* 91:245–262. doi: 10.1093/infdis/jiz502.

Lee, P. C. 1984. "Ecological constraints on the social development of vervet monkeys." *Behaviour* 221:S419–S430. doi: 10.1163/156853984x00092.

Lee, P. C., and M. D. Hauser. 1998. "Long-term consequences of changes in territory quality on feeding and reproductive strategies of vervet monkeys." *Journal of Animal Ecology* 67 (3):347–358. doi: 10.1046/j.1365-2656.1998.00200.x.

Leffler, E. M. 2017. "Evolutionary insights from wild vervet genomes." *Nature Genetics* 49 (12):1671–1672. doi: 10.1038/ng.3992.

Lernould, J. M. 1988. "Classification and geographical distribution of guenons: a review." In *A Primate Radiation: Evolutionary Biology of the African Guenons*, edited by A. Gautier-Hion, F. Bourliere, J.-P. Gautier and J. Kingdon, 54–78. Cambridge: Cambridge University Press.

Liddie, S., H. Okamoto, J. Gromada, and M. Lawrence. 2018. "Characterization of glucose-stimulated insulin release protocols in african green monkeys (*Chlorocebus aethiops*)." *Journal of Medical Primatology*. doi: 10.1111/jmp.12374.

Lo, M. K., F. Feldmann, J. M. Gary, R. Jordan, R. Bannister, J. Cronin, N. R. Patel, J. D. Klena, S. T. Nichol, T. Cihlar, S. R. Zaki, H. Feldmann, C. F. Spiropoulou, and E. de Wit. 2019. "Remdesivir (GS-5734) protects African green monkeys from Nipah virus challenge." *Science Translational Medicine* 11 (494). doi: 10.1126/scitranslmed.aau9242.

Lubbe, A., R. S. Hetem, R. McFarland, L. Barrett, P. S. Henzi, D. Mitchell, L. C. R. Meyer, S. K. Maloney, and A. Fuller. 2014. "Thermoregulatory plasticity in free-ranging vervet monkeys, *Chlorocebus pygerythrus*." *Journal of Comparative Physiology B Biochemical Systemic and Environmental Physiology* 184 (6):799–809.

Ma, D., A. J. Jasinska, F. Feyertag, V. Wijewardana, J. Kristoff, T. He, K. Raehtz, C. A. Schmitt, Y. Jung, J. D. Cramer, M. Dione, M. Antonio, R. Tracy, T. Turner, D. L. Robertson, I. Pandrea, N. Freimer, C. Apetrei, and Consortium International Vervet Research. 2014. "Factors associated with siman immunodeficiency virus transmission in a natural African nonhuman primate host in the wild." *Journal of Virology* 88 (10):5687–5705. doi: 10.1128/JVI.03606-13.

Ma, D., A. Jasinska, J. Kristoff, J. P. Grobler, T. Turner, Y. Jung, C. Schmitt, K. Raehtz, F. Feyertag, N. Martinez Sosa, V. Wijewardana, D. S. Burke, D. L. Robertson, R. Tracy, I. Pandrea, N. Freimer, C. Apetrei, and Consortium International Vervet Research. 2013. "SIVagm infection in wild African green monkeys from South Africa: Epidemiology, natural history, and evolutionary considerations." *PLoS Pathology* 9 (1):e1003011. doi: 10.1371/journal.ppat.1003011.

Magden, E. R., K. G. Mansfield, J. H. Simmons, and C. R. Abee. 2015. Chapter 17 - Nonhuman primates. In: *Laboratory Animal Medicine*. 3rd ed., edited by J. G. Fox, L. C. Anderson, G. M. Otto, K. R. Pritchett-Corning, M. T. Whary, 771–930. London: Academic Press.

Maldjian, J. A., C. A. Shively, M. A. Nader, D. P. Friedman, and C. T. Whitlow. 2016. "Multi-atlas library for eliminating normalization failures in non-human primates." *Neuroinformatics* 14 (2):183–90. doi: 10.1007/s12021-015-9291-4.

Matsuoka, Y., A. Suguitan, Jr., M. Orandle, M. Paskel, K. Boonnak, D. J. Gardner, F. Feldmann, H. Feldmann, M. Marino, H. Jin, G. Kemble, and K. Subbarao. 2014. "African green monkeys recapitulate the clinical experience with replication of live attenuated pandemic influenza virus vaccine candidates." *Journal of Virology* 88 (14):8139–8152. doi: 10.1128/JVI.00425-14.

McFarland, R., L. Barrett, R. Boner, N. J. Freeman, and S. P. Henzi. 2014. "Behavioral flexibility of vervet monkeys in response to climatic and social variability." *American Journal of Physical Anthropology* 154 (3):357–364. doi: 10.1002/ajpa.22518.

McGuire, M. T. 1974. "St. Kitts vervet (*Cercopithecus aethiops*)." *Journal of Medical Primatology* 3 (5):285–297.

Melega, W. P., M. J. Jorgensen, G. Lacan, B. M. Way, J. Pham, G. Morton, A. K. Cho, and L. A. Fairbanks. 2008. "Long-term methamphetamine administration in the vervet monkey models aspects of a human exposure: Brain neurotoxicity and behavioral profiles." *Neuropsychopharmacology* 33 (6):1441–1452. doi: 10.1038/sj.npp.1301502.

Melnick, D. J., and M. C. Pearl. 1987. "Cercopithecine multimale groups." In *Primate Societies*, edited by B. B. Smuts, D. L. Cheney, R. M. Seyfarth, R. W. Wrangham and T. T. Struhsaker, 121–134. Chicago: University of Chicago Press.

Menzel, M., M. Axer, K. Amunts, H. De Raedt, and K. Michielsen. 2019. "Diattenuation Imaging reveals different brain tissue properties." *Scientific Reports* 9 (1):1939. doi: 10.1038/s41598-019-38506-w.

Mercier, S., E. C. Deaux, E. van de Waal, A. E. J. Bono, and K. Zuberbuhler. 2019. "Correlates of social role and conflict severity in wild vervet monkey agonistic screams." *PLoS One* 14 (5):e0214640. doi: 10.1371/journal.pone.0214640.

Mercier, S., C. Neumann, E. van de Waal, E. Chollet, J. M. de Bellefon, and K. Zuberbuhler. 2017. "Vervet monkeys greet adult males during high-risk situations." *Animal Behaviour* 132:229–245. doi: 10.1016/j.anbehav.2017.07.021.

Mertz, J., A. Surreault, E. van de Waal, and J. Botting. 2019. "Primates are living links to our past: The contribution of comparative studies with wild vervet monkeys to the field of social cognition." *Cortex* 118:65–81. doi: 10.1016/j.cortex.2019.03.007.

Mizrahi-Kliger, A. D., A. Kaplan, Z. Israel, and H. Bergman. 2018. "Desynchronization of slow oscillations in the basal ganglia during natural sleep." *Proceedings of the National Academy of Sciences USA* 115 (18):E4274–E4283. doi: 10.1073/pnas.1720795115.

Molskness, T. A., D. L. Hess, G. M. Maginnis, J. W. Wright, J. W. Fanton, and R. L. Stouffer. 2007. "Characteristics and regulation of the ovarian cycle in vervet monkeys (*Chlorocebus aethiops*)." *American Journal of Primatology* 69 (8):890–900. doi: 10.1002/ajp.20395.

Morland, H. S., M. A. Suleman, and E. B. Tarara. 1992. "Changes in male-female interactions after introduction of new adult males in vervt monkey (*Cercopithecus aethiops*) groups." *Laboratory Primate Newsletter* 31 (2):1–4.

Myers, T. M., and L. R. Hamilton. 2011. "Delayed match-to-sample performance in African green monkeys (*Chlorocebus aethiops sabaeus*): Effects of benzodiazepine, cholinergic, and anticholinergic drugs." *Behavioral Pharmacology* 22 (8):814–823. doi: 10.1097/FBP.0b013e32834d6292.

NIH-ORIP. 2018. Nonhuman Primate Evaluation and Analysis. Part 2: Report of the Expert Panel Forum on Challenges in Assessing Nonhuman Primate Needs and Resources for Biomedical Research. https://orip.nih.gov/about-orip/research-highlights/nonhuman-primate-evaluation-and-analysis-part-2-report-expert-panel

Pandrea, I., C. Apetrei, J. Dufour, N. Dillon, J. Barbercheck, M. Metzger, B. Jacquelin, R. Bohm, P. A. Marx, F. Barre-Sinoussi, V. M. Hirsch, M. C. Muller-Trutwin, A. A. Lackner, and R. S. Veazey. 2006. "Simian immunodeficiency virus SIVagm.sab infection of Caribbean African green monkeys: A new model for the study of SIV pathogenesis in natural hosts." *Journal of Virology* 80 (10):4858–4867. doi: 10.1128/JVI.80.10.4858-4867.2006.

Pandrea, I., N. F. Parrish, K. Raehtz, T. Gaufin, H. J. Barbian, D. Ma, J. Kristoff, R. Gautam, F. Zhong, G. S. Haret-Richter, A. Trichel, G. M. Shaw, B. H. Hahn, and C. Apetrei. 2012. "Mucosal simian immunodeficiency virus transmission in African green monkeys: Susceptibility to infection is proportional to target cell availability at mucosal sites." *Journal of Virology* 86 (8):4158–4168. doi: 10.1128/JVI.07141-11.

Parks, J. S., and L. L. Rudel. 1979. "Isolation and characterization of high-density lipoprotein apoproteins in the non-human primate (vervet)." *Journal of Biological Chemistry* 254 (14):6716–6723.

Petrullo, L., M. J. Jorgensen, N. Snyder-Mackler, and A. Lu. 2019. "Composition and stability of the vervet monkey milk microbiome." *American Journal of Primatology* 81 (10–11):e22982. doi: 10.1002/ajp.22982.

Petrullo, L., and A. Lu. 2019. "Natural variation in fetal cortisol exposure is associated with neonatal body mass in captive vervet monkeys (*Chlorocebus aethiops*)." *American Journal of Primatology* 81 (1):e22943. doi: 10.1002/ajp.22943.

Pfeifer, S. P. 2017. "The demographic and adaptive history of the African green monkey." *Molecular Biology and Evolution* 34 (5):1055–1065. doi: 10.1093/molbev/msx056.

Poirier, F. E. 1972. "The St. Kitts green monkey (*Cercopithecus aethiops sabaeus*): Ecology, population dynamics, and selected behavioral traits." *Folia Primatologica* 17:20–55.

Postupna, N., C. S. Latimer, E. B. Larson, E. Sherfield, J. Paladin, C. A. Shively, M. J. Jorgensen, R. N. Andrews, J. R. Kaplan, P. K. Crane, K. S. Montine, S. Craft, C. D. Keene, and T. J. Montine. 2017. "Human striatal dopaminergic and regional serotonergic synaptic degeneration with Lewy body disease and inheritance of APOE epsilon4." *American Journal of Pathology* 187 (4):884–895. doi: 10.1016/j.ajpath.2016.12.010.

Price, T., and J. Fischer. 2014. "Meaning attribution in the West African green monkey: Influence of call type and context." *Animal Cognition* 17 (2):277–286. doi: 10.1007/s10071-013-0660-9.

Raleigh, M. J., J. W. Flannery, and F. R. Ervin. 1979. "Sex differences in behavior among juvenile vervet monkeys (*Cercopithecus aethiops sabaeus*)." *Behavioral and Neural Biology* 26 (4):455–465.

Raleigh, M. J., and M. T. McGuire. 1989. "Female influences on male dominance acquisition in captive vervet monkeys *Cercopithecus aethiops sabaeus*." *Animal Behaviour* 38 (1):59–67.

Raleigh, M. J., M. T. McGuire, G. L. Brammer, D. B. Pollack, and A. Yuwiler. 1991. "Serotonergic mechanisms promote dominance acquisition in adult male vervet monkeys." *Brain Research* 559 (2):181–190.

Rapkin, A. J., D. B. Pollack, M. J. Raleigh, B. Stone, and M. T. McGuire. 1995. "Menstrual cycle and social behavior in vervet monkeys." *Psychoneuroendocrinology* 20 (3):289–297.

Rhoads, M. K., S. B. Goleva, W. H. Beierwaltes, and J. L. Osborn. 2017. "Renal vascular and glomerular pathologies associated with spontaneous hypertension in the nonhuman primate *Chlorocebus aethiops sabaeus*." *American Journal of Physiology-Regulatory, Integrative and Comparative Physiology* 313 (3):R211–R218. doi: 10.1152/ajpregu.00026.2017.

Rowell, T. E. 1970. "Reproductive cycles of two Cercopithecus monkeys." *Journal of Reproduction and Fertility* 22:321–338.

Rowell, T. E. 1971. "Organization of caged groups of cercopithecus monkeys." *Animal Behaviour* 19 (4):625-&. doi: 10.1016/S0003-3472(71)80165-3.

Rowland, J. A., J. R. Stapleton-Kotloski, G. E. Alberto, A. T. Davenport, R. J. Kotloski, D. P. Friedman, D. W. Godwin, and J. B. Daunais. 2017. "Changes in nonhuman primate brain function following chronic alcohol consumption in previously naive animals." *Drug and Alcohol Dependence* 177:244–248. doi: 10.1016/j.drugalcdep.2017.03.036.

Rudel, L. L., M. Davis, J. Sawyer, R. Shah, and J. Wallace. 2002. "Primates highly responsive to dietary cholesterol up-regulate hepatic ACAT2, and less responsive primates do not." *Journal of Biological Chemistry* 277 (35):31401–31406. doi: 10.1074/jbc.M204106200.

Sade, D. S., and R. W. Hildrech. 1965. "Notes on the green monkey (*Cercothithecus aethiops sabaeus*) on St. Kitts, West Indies." *Caribbean Journal of Science* 5 (1–2):67–81.

Saj, T., P. Sicotte, and J. D. Paterson. 1999. "Influence of human food consumption on the time budget of vervets." *International Journal of Primatology* 20 (6):977–994. doi: 10.1023/A:1020886820759.

Schmitt, C. A., S. K. Service, A. J. Jasinska, T. D. Dyer, M. J. Jorgensen, R. M. Cantor, G. M. Weinstock, J. Blangero, J. R. Kaplan, and N. B. Freimer. 2018. "Obesity and obesogenic growth are both highly heritable and modified by diet in a nonhuman primate model, the African green monkey (*Chlorocebus aethiops sabaeus*)." *International Journal of Obesity (London)* 42 (4):765–774. doi: 10.1038/ijo.2017.301.

Scobey, M. W., M. S. Wolfe, and L. L. Rudel. 1992. "Age- and dietary fat-related effects on biliary lipids and cholesterol gallstone formation in African green monkeys." *Journal of Nutrition* 122 (4):917–23.

Seier, J. 2005. "Vervet monkey breeding." In *The Laboratory Primate*, edited by S. Wolfe-Coote, 175–180. Amsterdam: Elsevier: Associated Press.

Seier, J., C. de Villiers, J. van Heerden, and R. Laubscher. 2011. "The effect of housing and environmental enrichment on stereotyped behavior of adult vervet monkeys (*Chlorocebus aethiops*)." *Lab Animal* 40 (7):218–224. doi: 10.1038/laban0711-218.

Seier, J. V. 1986. "Breeding vervet monkeys in a closed environment." *Journal of Medical Primatology* 15 (5):339–349.

Seier, J. V., and P. W. deLange. 1996. "A mobile cage facilitates periodic social contact and exercise for singly caged adult Vervet monkeys." *Journal of Medical Primatology* 25 (1):64–68.

Seier, J. V., F. S. Venter, J. E. Fincham, and J. J. Taljaard. 1991. "Hormonal vaginal cytology of vervet monkeys." *Journal of Medical Primatology* 20 (1):1–5.

Seyfarth, R. M. 1980. "The distribution of grooming and related behaviors among adult female vervet monkeys." *Animal Behaviour* 28 (Aug):798–813. doi: 10.1016/S0003-3472(80)80140-0.

Seyfarth, R. M., and D. L. Cheney. 1984. "Grooming, alliances and reciprocal altruism in vervet monkeys." *Nature* 308 (5959):541–543. doi: 10.1038/308541a0.

Seyfarth, R. M., D. L. Cheney, and P. Marler. 1980a. "Monkey responses to 3 different alarm calls - Evidence of predator classification and semantic communication." *Science* 210 (4471):801–803. doi: 10.1126/science.7433999.

Seyfarth, R. M., D. L. Cheney, and P. Marler. 1980b. "Vervet monkey alarm calls - semantic communication in a free-ranging primate." *Animal Behaviour* 28 (Nov):1070–1094. doi: 10.1016/S0003-3472(80)80097-2.

Skinner, J. D., and C. T. Chimimba. 2005. *The Mammals of the Southern African Subregion*. Cambridge: Cambridge University Press.

Smith, D. G. 2012. "Taxonomy of nonhuman primates used in biomedical research." In *Nonhuman Primates in Biomedical Research: Biology and Management*, edited by C. R. Abee, K. Mansfield, S. D. Tardiff and T. Morris, 57–85. London: Academic Press.

Spiezio, C., F. Piva, B. Regaiolli, and S. Vaglio. 2015. "Positive reinforcement training: A tool for care and management of captive vervet monkeys (*Chlorocebus aethiops*)." *Animal Welfare* 24:283–290.

Stephens, S. M., F. K. Pau, T. M. Yalcinkaya, M. C. May, S. L. Berga, M. D. Post, S. E. Appt, and A. J. Polotsky. 2013. "Assessing the pulsatility of luteinizing hormone in female vervet monkeys (*Chlorocebus aethiops sabaeus*)." *Comparative Medicine* 63 (5):432–8.

Struhsaker, T. T. 1967a. "Auditory communication among vervet monkeys (*Cercopithecus aethiops*)." In *Social Communication among Primates*, edited by S. A. Altmann, 281–324. Chicago: University of Chicago Press.

Struhsaker, T. T. 1967b. *Behavior of vervet monkeys (Cercopithecus aethiops)*. Vol. 82, *University of California Publications in Zoology*. Berkeley and Los Angeles: University of California Press.

Struhsaker, T. T. 1967c. "Behavior of vervet monkeys and other cercopithecines." *Science* 156 (3779):1197–203.

Struhsaker, T. T. 1967d. "Ecology of vervet monkeys (*Cercopithecus aethiops*) in the Masai-Amboseli Game Reserve, Kenya." *Ecology* 48 (6):891–904.

Struhsaker, T. T. 1967e. "Social structure among vervet monkeys (*Cercopithecus aethiops*)." *Behaviour* 29 (2):6–121.

Struhsaker, T. T. 1971. "Social behaviour of mother and infant vervet monkeys (*Cercopithecus aethiops*)." *Animal Behaviour* 19 (2):235–250. doi: 10.1016/S0003-3472(71)80004-0.

Suleman, M. A., J. Njugana, and J. Anderson. 1988. "Training of vervet monkeys, Skyes monkeys and baboons for collection of biological samples." *Proceedings of the XIIth Congress of the Internatinoal Primatological Society*, p. 12. Brasilia, Brazil.

Suleman, M. A., E. Wango, I. O. Farah, and J. Hau. 2000. "Adrenal cortex and stomach lesions associated with stress in wild male African green monkeys (*Cercopithecus aethiops*) in the post-capture period." *Journal of Medical Primatology* 29 (5):338–342.

Suleman, M. A., E. Wango, R. M. Sapolsky, H. Odongo, and J. Hau. 2004. "Physiologic manifestations of stress from capture and restraint of free-ranging male African green monkeys (*Cercopithecus aethiops*)." *Journal of Zoo and Wildlife Medicine* 35 (1):20–24. doi: 10.1638/01-025.

Svardal, H., A. J. Jasinska, C. Apetrei, G. Coppola, Y. Huang, C. A. Schmitt, B. Jacquelin, V. Ramensky, M. Muller-Trutwin, M. Antonio, G. Weinstock, J. P. Grobler, K. Dewar, R. K. Wilson, T. R. Turner, W. C. Warren, N. B. Freimer, and M. Nordborg. 2017. "Ancient hybridization and strong adaptation to viruses across African vervet monkey populations." *Nature Genetics* 49 (12):1705–1713. doi: 10.1038/ng.3980.

Tarara, E. B., R. P. Tarara, and M. A. Suleman. 1995. "Stress-induced gastric-ulcers in vervet monkeys (*Cercopithecus aethiops*) - the influence of life-history factors." *Journal of Zoo and Wildlife Medicine* 26 (1):72–75.

Tarara, R. P., J. G. Else, and R. M. Eley. 1984. "The menstrual-cycle of the vervet monkey, *Cercopithecus aethiops*." *International Journal of Primatology* 5 (4):384–384.

Teichroeb, J. A., and W. D. Aguado. 2016. "Foraging vervet monkeys optimize travel distance when alone but prioritize high-reward food sites when in competition." *Animal Behaviour* 115:1–10. doi: 10.1016/j.anbehav.2016.02.020.

Thomsen, M., J. J. Holst, A. Molander, K. Linnet, M. Ptito, and A. Fink-Jensen. 2018. "Effects of glucagon-like peptide 1 analogs on alcohol intake in alcohol-preferring vervet monkeys." *Psychopharmacology (Berl)* 236:603–611. doi: 10.1007/s00213-018-5089-z.

Truelove, M. A., A. L. Martin, J. E. Perlman, J. S. Wood, and M. A. Bloomsmith. 2017. "Pair housing of macaques: A review of partner selection, introduction techniques, monitoring for compatibility, and methods for long-term maintenance of pairs." *American Journal of Primatology* 79 (1):1–15. doi: 10.1002/ajp.22485.

Turner, T. R., F. Anapol, and C. J. Jolly. 1997. "Growth, development, and sexual dimorphism in vervet monkeys (*Cercopithecus aethiops*) at four sites in Kenya." *American Journal of Physical Anthropology* 103:19–35.

Turner, T. R., J. D. Cramer, A. Nisbett, and J. Patrick Gray. 2016. "A comparison of adult body size between captive and wild vervet monkeys (*Chlorocebus aethiops sabaeus*) on the island of St. Kitts." *Primates* 57 (2):211–220. doi: 10.1007/s10329-015-0509-8.

Turner, T. R., C. A. Schmitt, and J. D. Cramer. 2019a. "Behavioral ecology of savanna monkeys." In *Savanna Monkeys: The Genus Chlorocebus*, edited by T. R. Turner, C. A. Schmitt and J. D. Cramer, 109–126. Cambridge: Cambridge University Press.

Turner, T. R., C. A. Schmitt, and J. D. Cramer. 2019b. "Savanna monkey taxonomy." In *Savanna Monkeys: The Genus Chlorocebus*, edited by T. R. Turner, C. A. Schmitt and J. D. Cramer, 31–54. Cambridge: Cambridge University Press.

Turner, T. R., C. A. Schmitt, and J. D. Cramer. 2019c. *Savanna Monkeys: The Genus Chlorocebus*. Cambridge: Cambridge University Press.

Turner, T. R., C. A. Schmitt, J. D. Cramer, J. Lorenz, J. P. Grobler, C. J. Jolly, and N. B. Freimer. 2018. "Morphological variation in the genus *Chlorocebus*: Ecogeographic and anthropogenically mediated variation in body mass, postcranial morphology, and growth." *American Journal of Physical Anthropology* 166 (3):682–707. doi: 10.1002/ajpa.23459.

Turner, T. R., P. L. Whitten, C. J. Jolly, and J. G. Else. 1987. "Pregnancy outcome in free-ranging vervet monkeys (*Cercopithecus aethiops*)." *American Journal of Primatology* 12 (2):197–203. doi: 10.1002/ajp.1350120207.

van de Waal, E., C. Borgeaud, and A. Whiten. 2013. "Potent social learning and conformity shape a wild primate's foraging decisions." *Science* 340 (6131):483–485. doi: 10.1126/science.1232769.

van de Waal, E., N. Claidière, and A. Whiten. 2013. "Social learning and spread of alternative means of opening an artificial fruit in four groups of vervet monkeys." *Animal Behaviour* 85 (1):71–76. doi: 10.1016/j.anbehav.2012.10.008.

Voruganti, V. S., M. J. Jorgensen, J. R. Kaplan, K. Kavanagh, L. L. Rudel, R. Temel, L. A. Fairbanks, and A. G. Comuzzie. 2013. "Significant genotype by diet (G x D) interaction effects on cardiometabolic responses to a pedigree-wide, dietary challenge in vervet monkeys (*Chlorocebus aethiops sabaeus*)." *American Journal of Primatology* 75 (5):491–499. doi: 10.1002/ajp.22125.

Wall, H. S., C. Worthman, and J. G. Else. 1985. "Effects of ketamine anesthesia, stress and repeated bleeding on the hematology of vervet monkeys." *Laboratory Animals* 19 (2):138–144. doi: 10.1258/002367785780942633.

Warren, W. C., A. J. Jasinska, R. Garcia-Perez, H. Svardal, C. Tomlinson, M. Rocchi, N. Archidiacono, O. Capozzi, P. Minx, M. J. Montague, K. Kyung, L. W. Hillier, M. Kremitzki, T. Graves, C. Chiang, J. Hughes, N. Tran, Y. Huang, V. Ramensky, O. W. Choi, Y. J. Jung, C. A. Schmitt, N. Juretic, J. Wasserscheid, T. R. Turner, R. W. Wiseman, J. J. Tuscher, J. A. Karl, J. E. Schmitz, R. Zahn, D. H. O'Connor, E. Redmond, A. Nisbett, B. Jacquelin, M. C. Muller-Trutwin, J. M. Brenchley, M. Dione, M. Antonio, G. P. Schroth, J. R. Kaplan, M. J. Jorgensen, G. W. Thomas, M. W. Hahn, B. J. Raney, B. Aken, R. Nag, J. Schmitz, G. Churakov, A. Noll, R. Stanyon, D. Webb, F. Thibaud-Nissen, M. Nordborg, T. Marques-Bonet, K. Dewar, G. M. Weinstock, R. K. Wilson, and N. B. Freimer. 2015. "The genome of the vervet (*Chlorocebus aethiops sabaeus*)." *Genome Research* 25 (12):1921–1933. doi: 10.1101/gr.192922.115.

Westcott, M. M., J. Smedberg, M. J. Jorgensen, S. Puckett, and D. S. Lyles. 2018. "Immunogenicity in African green monkeys of M protein mutant vesicular stomatitis virus vectors and contribution of vector-encoded flagellin." *Vaccines (Basel)* 6 (1):16. doi: 10.3390/vaccines6010016.

Whitten, P. L. 1983. "Diet and dominance among female vervet monkeys (*Cercopithecus aethiops*)." *American Journal of Primatology* 5 (2):139–159. doi: 10.1002/ajp.1350050205.

Willems, E. P., T. J. Arseneau, X. Schleuning, and C. P. van Schaik. 2015. "Communal range defence in primates as a public goods dilemma." *Philosophical Transactions of the Royal Society of London Series B - Biological Sciences* 370 (1683):20150003. doi: 10.1098/rstb.2015.0003.

Wilson, M. E., J. Fisher, A. Fischer, V. Lee, R. B. Harris, and T. J. Bartness. 2008. "Quantifying food intake in socially housed monkeys: Social status effects on caloric consumption." *Physiology and Behavior* 94 (4):586–594. doi: 10.1016/j.physbeh.2008.03.019.

Wilson, Q. N., M. Wells, A. T. Davis, C. Sherrill, M. C. B. Tsilimigras, R. B. Jones, A. A. Fodor, and K. Kavanagh. 2018. "Greater microbial translocation and vulnerability to metabolic disease in healthy aged female monkeys." *Scientific Reports* 8 (1):11373. doi: 10.1038/s41598-018-29473-9.

Wolfe, M. S., J. S. Parks, T. M. Morgan, and L. L. Rudel. 1993. "Childhood consumption of dietary polyunsaturated fat lowers risk for coronary artery atherosclerosis in African green monkeys." *Arteriosclerosis, Thrombosis, and Vascular Biology* 13 (6):863–875.

Wolfe, M. S., J. K. Sawyer, T. M. Morgan, B. C. Bullock, and L. L. Rudel. 1994. "Dietary polyunsaturated fat decreases coronary artery atherosclerosis in a pediatric-aged population of African green monkeys." *Arteriosclerosis, Thrombosis, and Vascular Biology* 14 (4):587–597.

Woods, R. P., S. C. Fears, M. J. Jorgensen, L. A. Fairbanks, A. W. Toga, and N. B. Freimer. 2011. "A web-based brain atlas of the vervet monkey, *Chlorocebus aethiops*." *Neuroimage* 54 (3):1872–1880. doi: 10.1016/j.neuroimage.2010.09.070.

Young, C., T. R. Bonnell, L. R. Brown, M. J. Dostie, A. Ganswindt, S. Kienzle, R. McFarland, S. P. Henzi, and L. Barrett. 2019. "Climate induced stress and mortality in vervet monkeys." *Royal Society Open Science* 6 (11):191078. doi: 10.1098/rsos.191078.

Young, C., R. McFarland, A. Ganswindt, M. M. I. Young, L. Barrett, and S. P. Henzi. 2019. "Male residency and dispersal triggers in a seasonal breeder with influential females." *Animal Behaviour* 154:29–37. doi: 10.1016/j.anbehav.2019.06.010.

Young, C., R. McFarland, L. Barrett, and S. P. Henzi. 2017. "Formidable females and the power trajectories of socially integrated male vervet monkeys." *Animal Behaviour* 125:61–67.

Zahn, R. C., M. D. Rett, M. Li, H. Tang, B. Korioth-Schmitz, H. Balachandran, R. White, S. Pryputniewicz, N. L. Letvin, A. Kaur, D. C. Montefiori, A. Carville, V. M. Hirsch, J. S. Allan, and J. E. Schmitz. 2010. "Suppression of adaptive immune responses during primary SIV infection of *sabaeus* African green monkeys delays partial containment of viremia but does not induce disease." *Blood* 115 (15):3070–3078. doi: 10.1182/blood-2009-10-245225.

Zhou, X. J., J. Wang, H. H. Ye, and Y. Z. Fa. 2019. "Signature microRNA expression profile is associated with lipid metabolism in African green monkey." *Lipids in Health and Disease* 18 (1):55. doi: 10.1186/s12944-019-0999-2.

30
Behavioral Biology of Baboons

Corrine K. Lutz
Animal Behavior Scientist

CONTENTS

Introduction	495
Research Contributions	495
Reproductive Research	496
Fetal and Infant Studies	496
Complex Diseases	496
Transplantation	496
Behavioral Biology	496
Natural History	496
Ecology	496
Social Organization	497
Mating and Reproduction	497
Common Social Behaviors	498
Feeding Behavior	500
Communication	500
Captive Behavior	501
Normal Behavior	501
Abnormal Behavior	501
Ways to Maintain Behavioral Health in Captivity	502
Environment	502
Social Groupings	503
Feeding Strategies	503
Training	505
Conclusions/Recommendations	505
References	506

Introduction

The baboon is an important model in biomedical and behavioral research due to the number of physical, biochemical, genetic, and physiological traits that it shares with humans. Baboons are Old World monkeys that are part of the taxonomic family Cercopithecidae. Depending on the source, baboons may be listed either as separate species: olive (*Papio anubis*), yellow (*Papio cynocephalus*), hamadryas (*Papio hamadryas*), guinea (*Papio papio*), and chacma (*Papio ursinus*; Nowak, 1999; Napier and Napier, 1994), or as multiple subspecies under the single species *Papio hamadryas* (e.g., *Papio hamadryas anubis* or *Papio hamadryas cynocephalus;* Williams-Blangero et al., 1990). There are naturally occurring hybrid zones between ranges in the wild (Charpentier et al., 2012; Jolly et al., 2011; Nagel, 1973; Phillips-Conroy and Jolly, 1981) and the subspecies readily hybridize in captivity (Williams-Blangero et al., 1990). Baboons are also highly adaptable and can be easily maintained in a variety of physical and social environments (Goodwin and Coelho, 1982; Maclean et al., 1987; Roder and Timmermans, 2002; Schlabritz-Loutsevitch et al., 2004).

Research Contributions

Because of their similarities to humans and their relatively large size in comparison to other monkeys, baboons are an excellent model for a wide range of biomedical research. For example, they have been involved as models in the study of infectious diseases, such as parasitic (e.g., *Plasmodium knowlesi*; Ozwara et al., 2003) and viral infections (e.g., Zika virus, Gurung et al., 2018; pertussis, Warfel et al., 2014), as well as the study of epilepsy (Szabo et al., 2009). Below are additional examples in which baboons have successfully contributed as models in biomedical research.

Reproductive Research

The baboon is an excellent model for reproductive research. Its menstrual cycle is similar to, but slightly longer than, that of humans, lasting approximately 32–36 days (range: 24–47 days), and the hormonal patterns during the cycle are also similar (Wildt et al., 1977; Goncharov et al., 1976; Kling and Westfahl, 1978; Stevens, 1997). The baboon is a continuous breeder that cycles throughout the year, and the menstrual cycle can be easily followed externally (D'Hooghe, 1997). By monitoring the perineal sexual skin swelling that changes accordingly throughout the cycle, researchers can avoid the need for blood sampling to determine hormone levels (Bauer, 2015; D'Hooghe et al., 2009). In contrast to some other monkey species, the baboon also has a relatively straight cervical canal, which facilitates gynecological manipulations (Bauer, 2015; Chai et al., 2007). Because of these unique characteristics, the baboon has participated extensively in reproductive research. For example, baboons have been subjects in studies of endometriosis; they are known to have spontaneous endometriosis and in situations when endometriosis is experimentally induced, the resulting lesions are similar to those seen in human disease (D'Hooghe et al., 1991, 1995, 2009). Additional reproductive research conducted with baboons includes the testing of birth control methods (Goldberg et al., 1981; Jensen et al., 2016; Stevens, 1997) and uterine transplantation (Johannesson et al., 2012).

Fetal and Infant Studies

Baboons have also been subjects in studies of fetal and infant development, in part because they show similar risk factors to those of women for fetal loss (Schlabritz-Loutsevitch et al., 2008). For example, maternal nutrient restriction studies have demonstrated that when baboon mothers were fed a restricted diet, placental weight decreased (Schlabritz-Loutsevitch et al., 2007). Teratology and pharmacokinetic studies with baboons have included the testing of agents such as Depo-Provera (Tarara, 1984), Thalidomide (Hendrickx et al., 1966), Fluoxetine (Shoulson et al., 2014), and Bendectin (Hendrickx et al., 1985). Because fetal development of the baboon lung is similar to that of the human (Yoder et al., 2009), baboons have also been models for the study of neonatal lung disease. Early model development has demonstrated that baboons delivered preterm via cesarean section develop hyaline membrane disease (HMD) with similarities to that of the human disease (Escobedo et al., 1982). Treatments for HMD, such as high-frequency ventilation (Kinsella et al., 1991) and surfactant therapy (Vidyasagar et al., 1985), were successfully tested as treatment options with baboons.

Complex Diseases

The baboon is an excellent model for the study of complex diseases that can develop due to defects in genetically controlled systems in the face of environmental challenges (Cox et al., 2013). Obesity, along with its associated comorbidities, is one example of a complex condition that is studied in baboons. For example, studies have tested the impact of early nutrition on obesity and found that maternal weight, sire, and overnutrition during infancy have an impact on fat levels later in life (Lewis et al., 1986, 1991). The baboon has also been shown to be susceptible to arterial lipid deposition and is an excellent model for the study of atherosclerosis (McGill et al., 1960). In comparison to rhesus and squirrel monkeys, cholesterol metabolism in baboons is more similar to that of humans (Eggen, 1974), and there is evidence of heritability for high-density lipoprotein (HDL) cholesterol (MacCluer et al., 1988). Increasing levels of dietary cholesterol and fat in the baboon diet also impact levels of serum cholesterol (MacCluer et al., 1988; McGill et al., 1981).

Transplantation

Because of their large size, and because they are susceptible to many of the same, or similar, infectious agents that affect humans, baboons have been routinely employed in transplantation studies. Discordant models of xenotransplantation involving baboons typically focus on pig-to-baboon transplantations, including renal (Fishman et al., 2018; Rivard et al., 2018), cardiac (Chen et al., 2000; Langin et al., 2018), and lung (Watanabe et al., 2018) transplantations. Concordant models of xenotransplantation involving baboons typically focused on rhesus to baboon transplantations, including lung (Matsumiya et al., 1996b) and heart (Asano et al., 2003; Matsumiya et al., 1996a) transplantations.

Behavioral Biology

Natural History

Ecology

Baboons are large, primarily terrestrial Old World monkeys (Altmann, 1980). They have a dog-like muzzle and a tail that is held erect, but with a downward bend (Nagel, 1973). Baboons typically walk quadrupedally (Hall, 1962a), but they also climb trees for sleeping and foraging (Altmann and Altmann, 1970; DeVore and Hall, 1965; Rose, 1977). Baboons have physical characteristics similar to other cercopithecines, such as ischial callosities and cheek pouches (Hall, 1962a; Lambert and Whitham, 2001; Vilensky, 1978; Warren, 2008). Ischial callosities are sitting pads of dense epidermal thickening that provide stable seating on tree branches (Hall, 1962a; Vilensky, 1978), while cheek pouches are utilized for food storage, and allow baboons to avoid aggression over limited resources and to move to safer, or less crowded, locations to feed (Lambert and Whitham, 2001; Rose, 1977; Warren, 2008). However, some differences in anatomical features do exist across subspecies, such as the amount of bending in the tail, the color of the face, the shape of the anal patches, as well as coat color and coat length (Nagel, 1973; Nowak, 1999). Common names for the subspecies reflect differences in coat coloration, such as the olive baboon (*Papio hamadryas anubis*), which has a greenish-gray coat; the yellow baboon (*Papio hamadryas cynocephalus*), which has a yellow-brown coat; and the Guinea, or red baboon (*Papio hamadryas papio*), which has a reddish-brown coat (Nowak, 1999). Hamadryas male baboons are known for their long silvery-gray coats (Nowak, 1999). Baboons reach sexual maturity at around 6 years of age for females and 9 years of age for males (Cheney and Seyfarth, 2007). They are also sexually dimorphic, with females typically weighing 30%–42% less than males (Coelho, 1985). Depending on the subspecies, weights can range from 13.2 to 33.5 kg for males and 7.3–17 kg for females (Altmann et al., 1993; Coelho, 1985; Leigh, 2009; Phillips-Conroy and Jolly, 1981; Strum, 1991). Additionally, males possess large canines that can be linked to male–male competition for access to females (Galbany et al., 2015).

The baboon is the most widely distributed and abundant species of nonhuman primate in Africa, ranging across sub-Saharan Africa, including southern and western Africa (Altmann and Altmann, 1970; Hall, 1968; DeVore and Hall, 1965; Wolfheim, 1983). The hamadryas baboon's range also includes northeast Africa and the southern part of the Arabian Peninsula, while the Guinea baboon is found in extreme West Africa (Wolfheim, 1983). Baboons are adaptable to a wide range of habitats, including semidesert steppe, open grassland, rocky hills, woodlands, and forests (Hall, 1968; DeVore and Hall, 1965; Wolfheim, 1983). Day range length and home range size are associated with troop size, habitat, food availability, and season (Altmann, 1980; Altmann and Altmann, 1970; DeVore and Hall, 1965; Hoffman and O'Riain, 2012). Average day ranges can vary from 1.2 to 8.8 km (Altmann and Altmann, 1970; Dunbar and Dunbar, 1974; Hoffman and O'Riain, 2012; Rose, 1977), while home ranges can vary from 1.5 to 37.7 km^2 (Altmann and Altmann, 1970; Hoffman and O'Riain, 2012; Stacey, 1986). A core area within the home range is an area of more frequent use, often containing sleeping trees, water, resting places, and food sources (DeVore and Hall, 1965). Although home ranges of neighboring groups may overlap, the core areas overlap little, if any, and the daily routines also tend to keep the troops apart (DeVore and Hall, 1965). In the evening, baboons congregate on sleeping cliffs (Kummer, 1968; Schreier and Swedell, 2009) or they climb into trees to sleep; the choice of location within the tree depends, in part, on the weight of the animal (Altmann, 1980; Altmann and Altmann, 1970).

Social Organization

Most baboon subspecies are organized in multi-male, multi-female troops ranging in size from 8 to almost 200 individuals, with a modal troop size closer to 20–50 animals (Altmann and Altmann, 1970; DeVore and Hall, 1965; Rose, 1977). Group size may depend on environmental constraints, such as food availability, local terrain, as well as predation risks (Altmann and Altmann, 1970; DeVore and Hall, 1965). Detection or avoidance of predators, such as leopards, eagles, jackals, hyenas, and lions, may be a primary benefit of living in larger groups (Altmann and Altmann, 1970; DeVore and Hall, 1965; Jooste et al., 2012; Stacey, 1986). Within a troop, adult females often outnumber adult males, which could be due in part to sex differences in maturation rates or to male migrations (Altmann and Altmann, 1970; DeVore and Hall, 1965). In contrast, the hamadryas baboon has a unique social structure. They live in one-male units (OMUs) consisting of a male and approximately two to three females (range: 1–10 females; Kummer, 1968). These OMUs are sometimes accompanied by one or more subadult "follower" males, who seldom copulate with the females, but may confer added fitness benefits to the leader male (Chowdhury et al., 2015; Kummer, 1968). The OMU is the smallest social unit of hamadryas society. The next largest level is the clan, consisting of two or more OMUs. Two or more clans join together to form a band, or foraging unit, which is similar to the multimale, multifemale social groups of other baboon subspecies. Finally, several bands may congregate on the same sleeping rock to form a troop, consisting of approximately 50 to well over 100 individuals (Abegglen, 1984; Kummer, 1968; Schreier and Swedell, 2009, 2012). These social units can cleave and coalesce in response to food availability and predation risk (Schreier and Swedell, 2009, 2012). Guinea baboons appear to have a similar social structure to that of hamadryas baboons, although there may be some slight variations (Fischer et al., 2017; Kopp et al., 2015; Patzelt et al., 2014).

Most baboon troops are organized around female-bonded and ranked matrilines (Bentley-Condit and Smith, 1999). In this type of social organization, females are philopatric; they remain in the troop and attain rank positions similar to those of their mothers. Therefore, female dominance ranks show a high degree of stability (Cheney and Seyfarth, 2007; Hausfater, 1975, 1982) and high-ranking females enjoy priority of access to resources (Silk, 1993). The hamadryas females, on the other hand, are forcibly taken by males and typically do not form matrilines (Polo and Colmenares, 2012; Sigg et al., 1982). Male baboons have an alternative strategy, emigrating from their natal troop and attempting to join a new troop (Beehner et al., 2006; Cheney and Seyfarth, 2007; Packer, 1979a). Male dominance hierarchies are less stable than those of females, due to the entry of immigrant males into the social group (Hausfater, 1975; MacCormick et al., 2012), but adult males are consistently dominant to adult females (Cheney and Seyfarth, 2007; Hausfater, 1975). The arrival of a new male into a troop can cause anxiety for the troop members, due to risks of infanticide and hierarchy takeover by that male (Cheney and Seyfarth, 2007). Males compete with each other for dominance rank (Kitchen et al., 2005), which is strongly correlated with testosterone and age (Beehner et al., 2006; Gesquiere et al., 2011; Kalbitzer et al., 2015). Dominance tends to peak in young adulthood and then decreases as the animal ages (Beehner et al., 2006; Packer, 1979a). Although male dominance is also correlated with mating success and quicker recovery time from illness (Alberts et al., 2003; Archie et al., 2012; Beehner et al., 2006; Packer, 1979b), dominant males may also experience greater energetic and physiological demands due to high levels of agonistic and mating behaviors (Akinyi et al., 2017; Gesquiere et al., 2011).

Mating and Reproduction

Baboons do not typically have a specified breeding season; instead they produce offspring throughout the year (Altmann, 1980; Bercovitch and Harding, 1993). However, birth peaks sometimes occur with seasonal fluctuations and may be associated with resource availability (Bercovitch and Harding, 1993; Barrett et al., 2006; Cheney et al., 2006; Kummer, 1968). Baboon females have anogenital skin swellings that change in tumescence according to their ovulatory cycles (Daspre et al., 2009). Mean cycle length can vary, but is approximately 32–36 days (Garcia et al., 2008; Hausfater, 1975; Maxim and Buettner-Janusch, 1963). Swellings are largest during the time that the female is most likely to be fertile (Higham et al., 2008), and swelling size may advertise the period during which males should consort with females (Higham et al., 2009; Rigaill et al., 2013). The majority of copulations occur during full tumescence (Daspre et al., 2009). Swelling size has been positively correlated with the proportion of a female's offspring that survive and therefore may be an advertisement of female quality (Domb and Pagel, 2001). However, Fitzpatrick et al. (2015) did not find a relationship between larger sex skin swellings and male preference or reproductive success. Larger swelling size was also reported to be associated with higher rank (Garcia et al., 2008), but this association has not been reported in all studies (Domb and Pagel, 2001).

Although social interaction between males and females may persist throughout all reproductive states (Seyfarth, 1978b), consortships between a male and female are typically formed when the sex skin of the female is maximally swollen (Cheney and Seyfarth, 2007; Saayman, 1971a). Typically, the dominant male is favored by the females, is involved in the majority of consortships, has consortships of longer duration, and copulates with the females during maximum tumescence (Beehner et al., 2006; Bulger, 1993; Hall and DeVore, 1965; Seyfarth, 1978a). In contrast, juvenile, subadult, and less dominant males are more likely to copulate with the female during the initial stages of her swelling (Cheney and Seyfarth, 2007; Hall and DeVore, 1965). Although dominant males tend to father a disproportionately large number of offspring, they do not father all offspring (Packer, 1979b). The ability of the dominant male to monopolize matings is, in part, associated with the number of females in the troop (Bulger, 1993). Baboon females can copulate with as many as four to five different males during a single cycle, but the average is closer to one to two males (Bercovitch, 1987; Bulger, 1993; Hall, 1962b; Hausfater, 1975). Similarly, hamadryas baboon females typically copulate only with their leader male, but they sometimes copulate with nonleaders (Swedell and Saunders, 2006).

Sexual behaviors can include genital inspect, sniffing the anogenital region, present, hip hold, ankle clasping, mounting, and thrusting (Coelho and Bramblett, 1989; Packer, 1979a). Although behaviors that females employ to initiate copulation can vary, the majority of mounts and copulations are initiated by the female presenting her hindquarters to the male (Hausfater, 1975; Packer, 1979a). Less frequently, males will manipulate females to present to them (Packer, 1979a). However, when a consort relationship is formed, either the male or the female may initiate the copulation (Hall and DeVore, 1965). During copulation, the male mounts the female as she stands, holds her back, and grips her hind legs with his feet so both of his feet are off the ground (Hall, 1962b; Hall and DeVore, 1965; Hausfater, 1975). Copulations can occur in a series of mounts, with ejaculation occurring at the end of the series, or ejaculation can occur at the end of a single mount (Hall and DeVore, 1965; Hausfater, 1975).

Females are between approximately 5 and 8 years of age when they first become pregnant (Bercovitch and Strum, 1993; Cheney et al., 2006). Physical changes occur to the anogenital area as a result of pregnancy; the paracollosal skin becomes depigmented, and changes from black to pinkish-red or deep scarlet. After parturition, the skin gradually turns black again (Altmann, 1973). The gestation period lasts approximately 6 months (Altmann and Alberts, 2003; Cheney and Seyfarth, 2007; Hall and DeVore, 1965; Wasser and Norton, 1993). Baboon infants are born with bright pink skin and a black hair coat. During the third month of life, infants develop some gray skin pigmentation and gold coloration in their hair coat, reaching adult coloration by age 4–6 months (Altmann, 1980; Hall and DeVore, 1965; Kummer, 1968). Infants initially ride ventrally; however, at about 5 weeks of age, they begin to ride dorsally (Altmann, 1980; Hall and DeVore, 1965). Infants are still dependent on their mothers for virtually all food and transportation during the third and fourth month of life (Altmann, 1980). By 10 months of age, they nurse only rarely and are weaned by approximately 15 months of age, depending on food availability (Barrett et al., 2006; Hall and DeVore, 1965). Contact time between mother and infant decreases with infant's age and infants are independent by 1 year of age (Altmann, 1980; Cheney and Seyfarth, 2007).

Common Social Behaviors

Affiliative behavior is a category of friendly behaviors that includes social contact, grooming, touch, present, social approach, and lipsmack (the rapid, repetitive opening and closing of the lips; Coelho and Bramblett, 1989; Easley and Coelho, 1991; Hall and DeVore, 1965). Lipsmacking is a behavior that provides a form of amicable social communication (Easley and Coelho, 1991); it is positively associated with several affiliative behaviors (e.g., attempt to touch, muzzle-muzzle, groom, social approach), and negatively associated with most agonistic behaviors (Easley and Coelho, 1991). Grooming (see below) and proximity have also been used as measures of affinity between individuals. For example, frequency of grooming and/or proximity (see Figure 30.1) is related to kinship, consort relationships, and dominance status (Hall and DeVore, 1965). Based on grooming and proximity measures, older, higher-ranking females tend to have more friendships with older, long-term resident males (Smuts, 1985). Additional variables that can impact social behavior in baboons include temperament or personality traits. Personality traits may have a genetic component and can be categorized on axes ranging from bold to shy and anxious to calm (Carter et al., 2014; Johnson et al., 2015). These traits can also include descriptors such as "nice", "loner", and "aloof" (Seyfarth et al., 2012, 2014). Although personality traits remain relatively stable over time (Seyfarth et al., 2012, 2014), they are not fixed.

Social grooming is a very prominent affiliative behavior in baboons (Hall and DeVore, 1965). Social grooming occurs when an individual parts the hair of another animal and picks or scrapes at the hair or skin, removing ectoparasites and detritus (Hausfater, 1975; Packer, 1979a). One function

FIGURE 30.1 Proximity can be used as a measure of affinity between individuals. (Photo courtesy of Kathy West Studios.)

of grooming is skin care and ectoparasite removal (Akinyi et al., 2013). In this case, the groomer often covers areas that would be difficult for the groomee to reach, and those that receive more grooming tend to have lower ectoparasite loads (Akinyi et al., 2013). Grooming plays a role in social communication (Saunders and Hausfater, 1988), is an important aspect of female bonding (Henzi et al., 1997), and may also play a role in stress reduction. For example, female chacma baboons with focused, predictable grooming partners showed less of a stress increase during social instability (Wittig et al., 2008), and captive female hamadryas baboons were reported to groom their harem male more during crowded than during less crowded conditions (Judge et al., 2006). Baboons often solicit grooming by presenting a portion of their body toward another individual (Packer, 1979a). The main recipients of grooming tend to be younger animals, females, and high-ranking individuals (Akinyi et al., 2013), but grooming of females by females is the most common type (Hall and DeVore, 1965). The majority of male–female interactions also involve grooming, and males tend to be the recipients (Harding, 1980). Adult males typically only groom female consorts that have a swollen sex skin (Saayman, 1971a).

Play is another important affiliative behavior that is believed to promote the development of social and/or motor skills in infant and juvenile baboons (Chalmers, 1980). Play can involve contact, such as wrestling, grappling, pulling, pushing, jumping on, hitting, mouthing, and mock-biting between individuals, but it can also include noncontact behaviors such as chasing and being chased (Coelho and Bramblett, 1989; Owens, 1975). These behaviors are often accompanied by a play face, which is a relaxed, open-mouth face with eyelids lowered (Coelho and Bramblett, 1989). Play usually occurs in infants by 6 weeks of age (Owens, 1975). The amount of play peaks anywhere from 5 to 14 months, but then declines with age, and is significantly reduced after 3 years of age (Chalmers, 1980; Owens, 1975). Infants tend to play more than juveniles (Coelho and Bramblett, 1982), while play is infrequent in adults (Owens, 1975). Males typically play more than females, engaging in both more frequent and longer bouts (Coelho and Bramblett, 1982; Owens, 1975; Young et al., 1982). However, sex differences in play behavior were not observed in all studies (Young and Bramblett, 1977; Young and Hankins, 1979).

Not all social behaviors are affiliative. Aggressive behavior can range from threats, such as aggressive display, branch shake, brow raise, canine display (yawn), head bob, stare, and slapping the ground, to actual attacks, which can include charging, hitting, and biting (Coelho and Bramblett, 1989; Hall and DeVore, 1965; Hausfater, 1975; Saayman, 1971a). However, aggression accounts for only a small percentage of time spent in social behavior (Harding, 1980). The intensity of the threat or attack can vary greatly with the stressors and the social situation. For example, a male may simply stare at another individual or he may immediately initiate an attacking charge (Hall and DeVore, 1965). In the wild, levels of aggression are based, in part, on the animal's sex. Females are almost always the recipients, rather than the instigators, of aggressive behavior (Harding, 1980), but males are wounded more often (Drews, 1996; MacCormick et al., 2012), and are typically wounded by another male (MacCormick et al., 2012). When females are the instigators of aggression, high-ranking females tend to be more aggressive than low-ranking females (Ramirez et al., 2004; Silk, 1987b), but they are injured less often (Silk, 1987b).

Postconflict or reconciliation behaviors are friendly or peaceful behaviors performed by former opponents soon after an aggressive encounter (Castles and Whiten, 1998a; Silk et al., 1996). During reconciliation, the rates of approaches and grunts are elevated over baseline rates (Silk et al., 1996), while displacement behaviors and incidence of further aggression are reduced (Castles and Whiten, 1998b; Judge and Bachmann, 2013). These behaviors help to reduce aggression between former opponents postconflict (Silk et al., 1996), and they reduce postconflict arousal in the opponents, as well as in the bystanders (Judge and Bachmann, 2013). Those with higher quality relationships show higher levels of reconciliation (Romero et al., 2008). For example, closely related and closely ranked animals were more likely to reconcile conflicts than those that were more distantly related or ranked (Castles and Whiten, 1998a).

Self-directed or displacement behaviors are characterized by being seemingly irrelevant to the situation and may occur in times of uncertainty (Hall and DeVore, 1965; Maestripieri et al., 1992). They include behaviors, such as self-scratching, self-grooming, self-touching, yawning, brow or muzzle wipe, and body shaking (Castles et al., 1999; Coelho and Bramblett, 1989; Hall and DeVore, 1965; Maestripieri et al., 1992). The actions tend to be jerky and of short duration (Hall and DeVore, 1965). These behaviors may be used as an indicator of stress or anxiety (Maestripieri et al., 1992) and can vary according to the dominance status of the nearest baboon (Castles et al., 1999).

Submissive behaviors are usually performed by lower ranking animals in response to more dominant individuals and can include behaviors such as avoid, crouch, flee, fear grimace, and present (Coelho and Bramblett, 1989; Hall and DeVore, 1965; Hausfater, 1975). A fear grimace involves the retraction of the lips and exposing the teeth, but the teeth are not separated. This is a gesture often performed by a subordinate animal when near, or when attacked by, a dominant individual (Hall and DeVore, 1965). Adult males rarely exhibit a fear grimace (Hall and DeVore, 1965). Although rump presents may occur in other contexts (e.g., sexual, affiliative, grooming), almost half of presents can be considered submissive; typically, the animal presenting is subordinate to the recipient (Hausfater and Takacs, 1987). These presents may help diffuse potential threats from the recipient of the present (Hausfater and Takacs, 1987). Males that are in their prime present significantly less and receive significantly more presents than do old or young males (Fraser and Plowman, 2007). Similarly, high-ranking females elicit submissive behavior, such as lean away, grimace, crouch, and cower, particularly from lower ranking adult females (Silk, 1987b). Subordinate mothers also give a fear bark when more dominant females interact with their infants (Cheney et al., 1995).

Feeding Behavior

In the wild, foraging and feeding account for more than 40% of the baboon's daylight time budget and, for some, as much as 80% (Altmann, 1980; Hall, 1962a; Kunz and Linsenmair, 2008; Rose, 1977). However, time budgets can vary seasonally (Altmann, 1980; Silk, 1987a). The baboon foraging strategy can be characterized as flexible and adaptable, allowing them to cope with the severe seasonality of plant growth (DeVore and Hall, 1965; Whiten et al., 1991). This lack of specialized adaptations allows baboons to exploit a wide variety of foods in diverse habitats (Altmann and Altmann, 1970). The baboon food niche is typically characterized by high levels of fiber and, depending on the habitat and season, low levels of protein (Whiten et al., 1991). The majority of the baboon diet is plant material; approximately 90% of their diet includes bulbs, flowers, seeds, and leaves from more than 180 species of plants (DeVore and Hall, 1965; Hall, 1962a; Kunz and Linsenmair, 2008; Norton et al., 1987; Stacey, 1986). Although baboons have been observed to eat vertebrate meat, such as birds, eggs, lizards, or hares, the amount is very small in comparison to insect- or plant-eating (DeVore and Hall, 1965; Kunz and Linsenmair, 2008; Rhine et al., 1986), and predation on mammals by baboons is quite rare (Sommer et al., 2016). Geophagy, such as the consumption of clay, has also been documented in baboons (Hall, 1962a). This is performed primarily by pregnant females and may be used to alleviate gastrointestinal distress and/or to supplement minerals in the diet (Pebsworth et al., 2012). As agriculture and farming expand into their ranges, baboons have adapted by raiding crops and adding human food, such as maize and sweet potatoes, to their diet (Strum, 2010; Warren, 2008). Baboons that crop-raid spend less time feeding and more time resting, have shorter inter-birth intervals, and tend to have smaller home range sizes (Strum, 2010). Those living in areas with more abundant food (e.g., near a garbage dump) also tend to weigh more and have more body fat than those living in the more typical wild feeding environment (Altmann et al., 1993).

Communication

Vocalizations are one form of communication utilized by baboons. Baboons can vary their vocalizations in intensity and complexity, and these vocalizations can also vary with respect to habitat type (e.g., dense forest vs. open range; Ey et al., 2009). Although not exhaustive, below are examples of some of the more common baboon vocalizations.

- *Grunt*: Grunts are short, quiet, low-pitched calls (Ey et al., 2009) that are used as contact calls for short-distance communication, such as during group movement (Cheney and Seyfarth, 2007; Ey et al., 2009; Rendall et al., 1999), or in an affiliative context, such as friendly approaches, infant handling, resting, feeding, and before group movements (Ey et al., 2009; Maciej et al., 2013; Rendall et al., 1999). Social grunts are often performed in bouts of a number of calls (Maciej et al., 2013) and are typically the most common vocalizations (Cheney and Seyfarth, 2007; Hall and DeVore, 1965; Maciej et al., 2013). Grunts performed in an affiliative context tend to increase the probability of friendly interactions (Cheney and Seyfarth, 2007; Silk et al., 2016).
- *Bark*: Barks are short, one-syllable explosive vocalizations (Coelho and Bramblett, 1989). They can constitute a graded continuum, ranging from

"contact barks", which are more tonal calls indicating that the signaler is at risk of separation from the group or when a mother or infant have been separated (Fischer et al., 2001; Hall and DeVore, 1965; Rendall et al., 2000), to harsher alarm barks in response to predators (Fischer et al., 2001; Hall and DeVore, 1965; Maciej et al., 2013). The majority of barks produced during a travel/forage context occur when visibility is poor (e.g., in dense vegetation; Maciej et al., 2013).

- *Loud or wahoo call*: These two-syllable calls are the loudest of the baboon vocalizations and may be adapted for long-range communication (Fischer et al., 2004). Wahoo calls can be used as an alarm to predators (Fischer et al., 2002), as an indicator of male competitive ability (Cheney and Seyfarth, 2007; Fischer et al., 2002, 2004; Kitchen et al., 2003), or when a male has been separated from a group (Fischer et al., 2002). Similar to barks, the majority of wahoo calls produced in the forage/travel context occur when visibility is poor (Maciej et al., 2013).
- *Scream or screech*: This is a shrill, high-pitched sound that is produced when an individual is being attacked or threatened (Coelho and Bramblett, 1989). Screams or screeches can be emitted by males during agonistic interactions (Maciej et al., 2013) or by females when harassed or chased by a male (Hall and DeVore, 1965; Maciej et al., 2013).
- *Roar grunt*: Roar grunts are produced in calling bouts and are often associated with aggressive displays, such as yawning or branch shaking (Maciej et al., 2013). Roar grunts also occur during fights between adult males (Hall and DeVore, 1965).
- *Ooer (or Ick-Ooer)*: This is a soft nasal two-phase vocalization (Coelho and Bramblett, 1989) typically performed by infants when frustrated, such as when separated from the mother (Hall and DeVore, 1965).

Captive Behavior

Normal Behavior

Most baboon behavior in captivity is similar to what is observed in the wild. However, there are some differences in the frequency or duration of behaviors due to differences in group size, space, and food availability. In the wild, baboons spend a significant amount of time feeding, moving, and resting, with social behavior comprising a smaller portion of the activity budget (Harding, 1980; Stacey, 1986). In captivity, food is typically provided in an easily accessible form and can be consumed quickly, so less time is devoted to feeding or foraging behavior. In addition, captive baboons typically have higher levels of social interaction than do their wild counterparts (Rowell, 1967), which may be due, in part, to a reduction in available space (Judge et al., 2006).

Abnormal Behavior

Abnormal behavior has been reported to occur in a number of captive nonhuman primate species, ranging from prosimians to New and Old World monkeys and apes (Birkett and Newton-Fisher, 2011; Camus et al., 2013; McGrogan and King, 1982; Tarou et al., 2005). Baboons are no exception and have been observed to display abnormal behavior in captivity (Lutz, 2018; Lutz et al., 2014). Behaviors can be considered abnormal if they are qualitatively different (i.e., occur in captivity, but not typically in the wild) or if they are quantitatively different (i.e., occur significantly more or significantly less than what is typically observed in the wild; Erwin and Deni, 1979). Although differences in abnormal behavior exist across even closely related species of nonhuman primates (Lutz, 2018), these often uniquely individualistic abnormal behaviors can be categorized for ease of comparison. Some examples of abnormal behaviors noted in captive baboons are listed below (categories and definitions taken from Lutz et al., 2014).

Motor Stereotypy

- *Head toss*: Repeated circular movement of the head at the neck can be performed rapidly or slowly
- *Pace*: Repeated walking in the same pattern (e.g., back and forth) for at least three revolutions
- *Rock*: Repeated back-and-forth or side-to-side movement of the body, occurring at least three times
- *Swing*: Repetitive back-and-forth movement when hanging from the cage side or ceiling

Self-Directed Behavior

- *Eye poke*: Placement of fingers or toes into, or right next to, the eye for an extended period of time; often appears as if the animal is "saluting"
- *Hair pull*: Plucking own hair from body
- *Self-bite*: Mouth-to-self contact where teeth contact the skin

Abnormal Appetitive

- *Abnormal mouth movements*: Repeated movement of mouth, lips, or tongue not associated with eating or manipulating an object in the mouth.
- *Coprophagy*: Ingesting or manipulating feces in the mouth
- *Hair-eat*: Chewing or ingestion of hair
- *Regurgitate*: The backward flow of already swallowed food
- *Wiggle digits*: Repeated movement of fingers or toes usually at, in, or around the mouth, often associated with regurgitation.

Other Abnormal

- Idiosyncratic abnormal behaviors not included in the above list

Although baboons tend to exhibit overall lower levels of abnormal behavior than what is typically noted in macaques (Lutz, 2018), the presence of abnormal behavior can still point to past or present environmental deficits (Mason, 1991). Risk factors for abnormal behavior in baboons can include single housing at an early age, increased numbers of clinical procedures (e.g., sedation and blood draw), increased number of days singly housed, and nursery rearing (Brent and Hughes, 1997; Lutz et al., 2014; Veira and Brent, 2000). Sex (being male) may also be a risk factor for some specific abnormal behaviors, such as abnormal appetitive behavior, abnormal body movements, and self-directed behaviors (Brent and Hughes, 1997; Lutz et al., 2014). Abnormal behavior can be triggered by environmental disturbances, such as other animals being sedated in the room (Crockett and Gough, 2002) or maintenance being conducted in the room (Macy et al., 2000). Because baboons with reported abnormal behavior are more likely to be euthanized for humane/management reasons than those of the general population (Veira and Brent, 2000), these behaviors need to be addressed. Heightened concern occurs in situations of self-inflicted injury, which can range from lesions caused by self-biting (De Villiers and Seier, 2010; Macy et al., 2000) to gastrointestinal inflammation caused by the accumulation of ingested hair (Mejido et al., 2009). If the behavior causes injury, encompasses a large portion of the day, or significantly interferes with species-typical behavior, the behavior and associated environmental factors should be assessed and, if warranted, an intervention should be implemented.

Various interventions have been implemented in the treatment of abnormal behavior in baboons. For example, the provisioning of food enrichment (e.g., novel food items or foraging devices), inanimate enrichment (e.g., play cage, toys, visual stimuli, auditory stimuli), positive reinforcement training (PRT), and social enrichment have all helped to reduce abnormal behavior (Bourgeois and Brent, 2005; Brent and Belik, 1997; De Villiers and Seier, 2010; Kessel and Brent, 1995; Pyle et al., 1996). Of these treatments, animate enrichment (e.g., interaction with a human or conspecific) showed the most promising outcomes (Bourgeois and Brent, 2005). However, interventions are not successful in all cases (Nevill and Lutz, 2015). For example, when treating a baboon with self-injurious behavior, periods of social contact were less successful than the provisioning of a Kong toy. In this case, the animal redirected self-biting from its knee toward the toy (Crockett and Gough, 2002). Drug therapy, such as the use of guanfacine, may also be used as a tool for treatment of self-injurious behavior (Macy et al., 2000). However, because long-term effects of drug therapy on the animal or on research outcomes are not well known, it may be best used as a way to provide time for assessing and addressing the issue via other routes. Because most interventions reduce, rather than eliminate, abnormal behavior in captive nonhuman primates, it is important to understand the impact of the captive environment on an animal's behavior in order to implement behavioral management strategies that are likely to prevent the behavior from occurring in the first place.

Ways to Maintain Behavioral Health in Captivity

Environment

The environment in which nonhuman primates are maintained has a significant impact on their well-being. However, in addition to a focus on animal welfare, the relatively large size and strength of baboons need to be addressed when developing captive enclosures. Therefore, when designing captive environments for baboons, one should ensure the durability of the caging materials, in addition to making accommodations for husbandry procedures, safety (animal and personnel), and accessibility.

Indoor single caging for baboons is typically made of stainless steel. For adult baboons, the minimum cage size recommended by the *Guide for the Care and Use of Laboratory Animals* (the *Guide*) can range from 0.56 to more than 2.32 m^2 in floor space and from 0.81 to 1.52 m high, depending on the animal's weight (NRC, 2011). European cage size standards are larger, ranging from 4.0 to 7.0 m^2, with a minimum height of 1.8 m (European Union, 2010). The cage height should be sufficient for the animal to comfortably stand erect, but professional judgment also plays an important role when determining appropriate cage dimensions (NRC, 2011). Caging should also have perches or platforms to allow the baboon to rest above the cage floor, and containers for food, and either water bottles or spigots for water. To allow for better accessibility, the cages typically contain a squeeze-back mechanism that allows for physical restraint when treating or sedating the animal. Alternatively, the animal can be trained to enter a transfer box that is connected to the cage for an additional means of access or for transferring the animal to another location. This type of caging accommodates husbandry procedures, safety, and accessibility. However, single housing negatively affects the animal's welfare and should only be utilized for Institutional Animal Care and Use Committee (IACUC)-approved research protocols or for clinical care.

Social housing for baboons can include indoor pens, indoor/outdoor runs, or large outdoor corrals. These enclosures typically exceed recommendations for square footage noted in the *Guide* (NRC, 2011). Because crowding can increase physiological stress (Pearson et al., 2015), such larger structures may help to improve welfare. For example, baboons moved from a smaller indoor space to larger indoor/outdoor enclosures showed an increase in reproductive efficiency (Cary et al., 2003). Although housing baboons in large outdoor corrals more closely replicates wild conditions, smaller spaces can also be designed with baboon social structure in mind. For example, a circular building with radially arranged outdoor pens attached to central indoor caging was designed for the social structure of hamadryas baboons that separate into smaller OMUs while foraging during the day, and then congregate into larger groups at night (Maclean et al., 1987). However, although wellbeing may increase with cage size in these types of structures, access to the animals and safety may decrease. For example, observation of the animals may be more difficult and accessing individual animals may be more challenging, when they are free to range in a large corral. To improve access to socially housed baboons, chute systems can be integrated into the caging system and the animals can be trained to enter the chutes (Holmes et al., 1996). Dividers in the chutes allow for safe separation of the animals. The chutes can also be outfitted with electronic scales for weighing individuals, cages for physical restraint, or cages for the individualized feeding of group-housed animals (Holmes et al., 1996; Schlabritz-Loutsevitch et al., 2004). The use of the chutes helps to reduce contact between the animals and the handlers, reduces stress, and increases safety.

Regardless of size, a barren enclosure is inadequate for ensuring the welfare of nonhuman primates. In order to encourage cage use and promote well-being, the interior and the contents of the cage must also be considered. Structural enrichment, or "cage furniture", is one aspect of environmental enrichment and enclosure design. Structural enrichment can include perches, hammocks, and other structures that add a vertical dimension to the cage. Although wild baboons are mainly terrestrial, they do utilize trees or cliffs as a refuge from predators or as a safe sleeping site at night (Altmann and Altmann, 1970; DeVore and Hall, 1965). Cage furniture gives captive baboons similar opportunities, by providing better access to vertical space, increased options for privacy, and better views of their surroundings (Kessel and Brent, 1996). The addition of climbing structures, such as ladders and plastic barrels, has been shown to decrease the amount of time group-housed baboons spend on the floor (Kessel and Brent, 1996). Visual barriers, such as solid panels, can also be used to provide privacy, giving subordinate animals opportunities to hide from view. These structures can be made out of metal, PVC, old firehose, or recycled plastic barrels, and they should be arranged in a way that does not interfere with cage cleaning, animal capture, or animal observation (see Figure 30.2).

In the wild, baboons, especially juveniles, have been observed to manipulate nonedible objects, such as stones (Kummer, 1968). In captivity, sturdy chew toys and other enrichment items that baboons can manipulate both manually and orally can be utilized to promote this species-typical behavior. When three types of chew toys were provided to group-housed baboons, approximately one-quarter of the animals were observed to use the toys at any given time (Brent and Belik, 1997). The provisioning of various types of chew toys has also resulted in a decrease in abnormal behavior in group-housed baboons (Brent and Belik, 1997). However, because some baboons show preferences for certain toys and because interest in toys can wane with time (Hienz et al., 1998), it is beneficial to routinely rotate different types of manipulable enrichment. In addition, as with any items provided to nonhuman primates, the toys need to be safe and not cause harm to the animals. They need to be designed so that small parts cannot be ingested, and they must be routinely inspected and discarded if damaged (Matz-Rensing et al., 2004).

Social Groupings

Baboons are highly adaptable animals, but to ensure enhanced wellbeing, they should be housed in social groups. The *Guide* emphasizes the importance of social housing by stating that nonhuman primates should be socially housed and that single housing should be the exception (NRC, 2011). Social housing allows for increased species-specific behaviors and improved welfare. In the wild, baboons often live in large multimale, multifemale groups, which can easily be replicated in captivity given enough space. For example, a large breeding group of baboons (over 400 individuals) was successfully housed in a 6-acre corral, producing approximately 200 infants per year (Goodwin and Coelho, 1982). However, small- to medium-sized social groups of 10–15 individuals (e.g., same sex, harem, or juvenile groups) can also be successfully maintained when space is limited or better access to the animals is needed. For example, a comparison of bachelor groups, single-male groups, and multimale, multifemale groups of hamadryas baboons housed in zoos showed no difference in wounding rates across group types (Wiley et al., 2018). In some cases, an animal will need to be singly housed due to clinical or approved research purposes, which can have a negative impact on wellbeing. For example, single housing at an early age or for an extended period of time has been associated with abnormal behaviors, such as pacing, hair pulling, and self-biting (Brent and Hughes, 1997; Lutz et al., 2014), and abnormal behavior was shown to decrease when animals were moved from single to group housing (Kessel and Brent, 2001). Because of these negative impacts, single housing should be limited to the shortest possible duration (NRC, 2011).

Feeding Strategies

Captive baboons are typically fed a nutritionally complete biscuit diet as their main source of nutrition. Ideally, baboons should be fed two or more times per day. In one study of group-housed baboons, aggression peaked just prior to feeding and then remained low for several hours afterwards. Multiple feedings per day were proposed to reduce the risk of aggression (Wasserman and Cruikshank, 1983). In singly housed baboons, abnormal behavior was reduced during the time the animals ate their regular chow diet, also demonstrating that providing food more often throughout the day may be beneficial (Brent and Long, 1995). However, increasing feedings from once to twice per day surprisingly resulted in an increase in hair eating in a study of group-housed baboons (Nevill and Lutz, 2015).

Although a biscuit diet may provide for the nutritional needs of the animal, it does not provide the diversity of food items that baboons obtain in the wild (Altmann and Altmann, 1970; DeVore and Hall, 1965; Whiten et al., 1991). Therefore, nutritional enrichment, such as fruit, vegetables, and grains, should routinely be provided to the baboons. Nutritional enrichment includes not only a variety of food items but also variety in its form and presentation. Depending on the size of the group and the amount of food being provided, produce items may be left whole to promote foraging behaviors and increase processing time, as long as the food is not clumped in a way that increases feeding competition (Gil-Burmann et al., 1998). An alternative to providing whole produce is to chop the food items or provide smaller items (e.g., grains) that can be scattered throughout the enclosure for more equitable access. Scattering a grain mix (see Figure 30.3) to increase foraging was reported to reduce

FIGURE 30.2 Structural enrichment adds a vertical dimension to an enclosure. (Photo courtesy of Texas Biomedical Research Institute.)

hair eating in some group-housed baboons, but it did not result in a significant reduction overall (Nevill and Lutz, 2015).

In the wild, baboons spend much of their waking hours foraging for food (Hall, 1962a; Kunz and Linsenmair, 2008). To encourage and simulate such foraging behavior in captivity, a variety of foraging devices can be installed on cages. Foraging devices can simulate foraging in the wild by requiring a baboon to engage in some manipulation to retrieve a food item. These devices can vary in design and can be made in-house or purchased from a manufacturer. Examples of foraging devices include polyvinyl chloride (PVC) tubes, with holes drilled on the side, that contain various food items (Brent and Long, 1995), a wooden log with drilled holes that contain a honey/seed mix (De Villiers and Seier, 2010), and a grooming board covered in fleece and coated with corn syrup and foraging crumbles (Pyle et al., 1996). The provisioning of

FIGURE 30.3 Scattering corn in an enclosure promotes natural foraging behavior. (Photo courtesy of Kathy West Studios.)

foraging devices has been demonstrated to reduce levels of abnormal and self-injurious behavior in singly housed baboons (Bourgeois and Brent, 2005; Brent and Long, 1995; De Villiers and Seier, 2010; Pyle et al., 1996). However, if foraging devices are to be provided to group-housed animals, enough devices should be provided to reduce the possibility of aggression or monopolization by the dominant individuals.

Training

PRT is a form of operant conditioning that involves rewarding an animal for responding appropriately to the trainer's cues or commands, increasing the chance that the behavior will occur again (Laule et al., 2003). PRT has been shown to reduce stress (Lambeth et al., 2006) and has been used as a form of environmental enrichment to reduce abnormal behavior (Bourgeois and Brent, 2005) in nonhuman primates. PRT also promotes cooperation between the animals and personnel, improves clinical care and husbandry procedures, and increases psychological well-being. PRT has been successfully used to train baboons to perform a variety of behaviors to enhance husbandry, clinical, and research procedures. For example, shift training utilizes positive reinforcement to encourage baboons to move from their home cage to another cage, to a testing area, to a physical restraint area, or to separate feeding cages for research projects requiring strict monitoring of the diet (Holmes et al., 1996; Schlabritz-Loutsevitch et al., 2004). Shift training reduces the time needed to access the animals, allows for easy separation of animals, and allows for individual testing and monitoring. In addition, shift training reduces or eliminates the need for personnel to enter the cage or to dart the animal for capture. Target training, another form of PRT, utilizes positive reinforcement to train an animal to hold a target or a position within a cage (O'Brien et al., 2008). Target training can be used to position an animal for assessment and/or treatment, thus avoiding the need for sedation or the separation of animals to prevent the interference of social partners with procedures. The utilization of PRT can significantly enhance research and clinical procedures, allowing veterinarians or researchers to obtain samples (e.g., saliva; Pearson et al., 2008) or measurements (e.g., blood pressure; Mitchell et al., 1980; Turkkan, 1990) from awake animals. PRT has also been invaluable in studies utilizing touch screen tasks for cognitive testing (Fagot et al., 2014; Maugard et al., 2014; Rodriguez et al., 2011; Zurcher et al., 2010).

Conclusions/Recommendations

Baboons are useful models in biomedical research. Because they are similar to humans in both size and physiology, they can contribute to studies ranging from reproductive research to complex diseases. Although they are highly adaptable to multiple environments and are relatively easy to socialize, their large size can make providing adequate housing in captivity difficult. However, appropriate planning for their physical, behavioral, and social needs will help promote baboon welfare and their value as a biomedical research animal.

REFERENCES

Abegglen, J.J. 1984. *On Socialization in Hamadryas Baboons.* Toronto: Bucknell University Press.

Akinyi, M.Y., L.R. Gesquiere, M. Franz, P.O. Onyango, J. Altmann, and S.C. Alberts. 2017. Hormonal correlates of natal dispersal and rank attainment in wild male baboons. *Hormones and Behavior* 94:153–161.

Akinyi, M.Y., J. Tung, M. Jeneby, N.B. Patel, J. Altmann, and S.C. Alberts. 2013. Role of grooming in reducing tick load in wild baboons (*Papio cynocephalus*). *Animal Behaviour* 85:559–568.

Alberts, S.C., H.E. Watts, and J. Altmann. 2003. Queuing and queue-jumping: long-term patterns of reproductive skew in male savannah baboons, *Papio cynocephalus*. *Animal Behaviour* 65:821–840.

Altmann, J. 1980. *Baboon Mothers and Infants.*, Cambridge, MA: Harvard University Press.

Altmann, J. and S.C. Alberts. 2003. Variability in reproductive success viewed from a life-history perspective in baboons. *American Journal of Human Biology* 15:401–409.

Altmann, J., D. Schoeller, S.A. Altmann, P. Muruthi, and R.M. Sapolsky. 1993. Body size and fatness of free-living baboons reflect food availability and activity levels. *American Journal of Primatology* 30:149–161.

Altmann, S.A. 1973. The pregnancy sign in savannah baboons. *The Journal of Zoo Animal Medicine* 4:8–12.

Altmann, S.A. and J. Altmann. 1970. *Baboon Ecology African Field Research.* New York: S. Karger.

Archie, E.A., J. Altmann, and S.C. Alberts. 2012. Social status predicts wound healing in wild baboons. *PNAS* 109:9017–9022.

Asano, M., S.R. Gundry, H. Izutani, S.N. Cannarella, O. Fagoaga, and L.L. Bailey. 2003. Baboons undergoing orthotopic concordant cardiac xenotransplantation surviving more than 300 days: effect of immunosuppressive regimen. *Journal of Thoracic and Cardiovascular Surgery* 125:60–70.

Barrett, L., S.P. Henzi, and J.E. Lycett. 2006. Whose life is it anyway? maternal investment, developmental trajectories, and life history strategies in baboons. In *Reproduction and Fitness in Baboons: Behavioral, Ecological, and Life History Perspectives*, ed. L. Swedell and S.R Leigh, 199–224. New York: Springer.

Bauer, C. 2015. The baboon (*Papio sp.*) as a model for female reproduction studies. *Contraception* 92:120–123.

Beehner, J.C., T.J. Bergman, D.L. Cheney, R.M. Seyfarth, and P.L. Whitten. 2006. Testosterone predicts future dominance rank and mating activity among male chacma baboons. *Behavioral Ecology and Sociobiology* 59:469–479.

Bentley-Condit, V.K. and E.O. Smith. 1999. Female dominance and female social relationships among yellow baboons (*Papio hamadryas cynocephalus*). *American Journal of Primatology* 47:321–334.

Bercovitch, F.B. 1987. Reproductive success in male savanna baboons. *Behavioral Ecology and Sociobiology* 21:163–172.

Bercovitch, F.B. and R.S.O. Harding. 1993. Annual birth patterns of savanna baboons (*Papio cynocephalus anubis*) over a ten-year period at Gilgil, Kenya. *Folia Primatologica* 61:115–122.

Bercovitch, F.B. and S.C. Strum. 1993. Dominance rank, resource availability, and reproductive maturation in female savanna baboons. *Behavioral Ecology and Sociobiology* 33:313–318.

Birkett, L.P. and N.E Newton-Fisher. 2011. How abnormal is the behaviour of captive, zoo-living chimpanzees? *PLos One* 6(6): e20101. doi:10.1371/journal.pone.0020101.

Bourgeois, S.R. and L. Brent. 2005. Modifying the behaviour of singly caged baboons: evaluating the effectiveness of four enrichment techniques. *Animal Welfare* 14:71–81.

Brent, L. and M. Belik. 1997. The response of group-housed baboons to three enrichment toys. *Laboratory Animals* 31:81–85.

Brent, L. and A. Hughes. 1997. The occurrence of abnormal behavior in group-housed baboons. *American Journal of Primatology* 42:96–97.

Brent, L. and K.E. Long. 1995. The behavioral response of individually caged baboons to feeding enrichment and the standard diet: a preliminary report. *Contemporary Topics* 34:65–69.

Bulger, J.B. 1993. Dominance rank and access to estrous females in male savanna baboons. *Behaviour* 127:67–103.

Camus, S.M.J., C. Blois-Heulin, Q. Li, M. Hausberger, and E. Bezard. 2013. Behavioural profiles in captive-bred cynomolgus macaques: towards monkey models of mental disorders? *PLos One* 8(4):e62141.doi:10.1371/journal.pone.0062141.

Carter, A.J., H.H. Marshall, R. Heinsohn, and G. Cowlishaw. 2014. Personality predicts the propensity for social learning in a wild primate. *PeerJ* 2:e283; doi: 10.7717/peerj.283.

Cary, M.E., B. Valentine, and G.L. White. 2003. The effects of confinement environment on reproductive efficiency in the baboon. *Contemporary Topics in Laboratory Animal Science* 42:35–39.

Castles, D.L. and A. Whiten. 1998a. Post-conflict behaviour of wild olive baboons. I. Reconciliation, redirection and consolation. *Ethology* 104:126–147.

Castles, D.L. and A. Whiten. 1998b. Post-conflict behaviour of wild olive baboons. II. Stress and self-directed behaviour. *Ethology* 104:148–160.

Castles, D.L., A. Whiten, and F. Aureli. 1999. Social anxiety, relationships and self-directed behaviour among wild female olive baboons. *Animal Behaviour* 58:1207–1215.

Chai, D., S. Cuneo, H. Falconer, J.M. Mwenda, and T. D'Hooghe. 2007. Olive baboon (*Papio anubis anubis*) as a model for intrauterine research. *Journal of Medical Primatology* 36:365–369.

Chalmers, N.R. 1980. The ontongeny of play in feral olive baboons (*Papio anubis*). *Animal Behaviour* 28:570–585.

Charpentier, M.J.E., M.C. Fontaine, E. Cherel, et al. 2012. Genetic structure in a dynamic baboon hybrid zone corroborates behavioural observations in a hybrid population. *Molecular Ecology* 21:715–731.

Chen, R.H., A. Kadner, R.N. Mitchell, and D.H. Adams. 2000. Mechanism of delayed rejection in transgenic pig-to-primate cardiac xenotransplantation. *Journal of Surgical Research* 90:119–125.

Cheney, D.L. and R.M. Seyfarth. 2007. *Baboon Metaphysics.* Chicago: The University of Chicago Press.

Cheney, D.L., R.M. Seyfarth, J. Fischer, et al. 2006. Reproduction, mortality, and female reproductive success in chacma baboons of the Okavango Delta, Botswana. In *Reproduction and Fitness in Baboons: Behavioral, Ecological, and Life History Perspectives*, ed. L. Swedell and S.R. Leigh, 147–176. New York: Springer.

Cheney, D.L., R.M. Seyfarth, and J.B. Silk. 1995. The responses of female baboons (*Papio cynocephalus ursinus*) to anomalous social interactions: evidence for causal reasoning? *Journal of Comparative Psychology* 109:134–141.

Chowdhury, S., M. Pines, J. Saunders, and L. Swedell. 2015. The adaptive value of secondary males in the polygynous multi-level society of hamadryas baboons. *American Journal of Physical Anthropology* 158:501–513.

Coelho, A.M. Jr. 1985. Baboon dimorphism: growth in weight, length and adiposity from birth to 8 years of age. In *Monographs in Primatology*, Vol. 6, ed. E.S. Watts, 125–159. New York: Alan R. Liss, Inc.

Coelho, A.M. Jr. and C.A. Bramblett. 1982. Social play in differentially reared infant and juvenile baboons (*Papio sp*). *American Journal of Primatology* 3:153–160.

Coelho, A.M. Jr. and C.A. Bramblett. 1989. Behaviour of the genus *Papio*: ethogram, taxonomy, methods, and comparative measures. In *Perspectives in Primate Biology*, Vol. 3, ed. P.K. Seth and S. Seth, 117–140. New Delhi, India: Today & Tomorrow's Printers and Publishers.

Cox L.A., A.G. Comuzzie, L.M. Havill, et al. 2013. Baboons as a model to study genetics and epigenetics of human disease. *ILAR Journal* 54:106–121.

Crockett, C.M. and G.M. Gough. 2002. Onset of aggressive toy biting by a laboratory baboon coincides with cessation of self-injurious behavior. *American Journal of Primatology* 57 (supplement 1):39.

Daspre, A., M. Heistermann, J.K. Hodges, P.C. Lee, and L. Rosetta. 2009. Signals of female reproductive quality and fertility in colony-living baboons (*Papio h. anubis*) in relation to ensuring paternal investment. *American Journal of Primatology* 71:529–538.

De Villiers, C. and J.V. Seier. 2010. Stopping self injurious behaviour of a young male chacma baboon (*Papio ursinus*). *Animal Technology and Welfare* 9:77–80.

DeVore, I. and K.R.L. Hall. 1965. Baboon ecology. In *Primate Behavior Field Studies of Monkeys and Apes*, ed. I. DeVore, 20–52. New York: Holt, Rinehart and Winston.

D'Hooghe, T.M. 1997. Clinical relevance of the baboon as a model for the study of endometriosis. *Fertility and Sterility* 68:613–625.

D'Hooghe, T.M., C.S. Bambra, F.J. Cornillie, M. Isahakia, and P.R. Koninckx. 1991. Prevalence and laparoscopic appearance of spontaneous endometriosis in the baboon (*Papio anubis, Papio cynocephalus*). *Biology of Reproduction* 45:411–416.

D'Hooghe, T.M., C.S. Bambra, B.M. Raeymaekers, I. De Jonge, J.M. Lauweryns, and P.R. Koninckx. 1995. Intrapelvic injection of menstrual endometrium causes endometriosis in baboons (*Papio cynocephalus* and *Papio anubis*). *American Journal of Obstetrical Gynecology* 173:125–134.

D'Hooghe, T.M., C.K. Kyama, and J.M. Mwenda. 2009. Baboon model for endometriosis. In *The Baboon in Biomedical Research*, ed J.L. VandeBerg, S. Williams-Blangero, and S.D. Tardif, 139–156. New York: Springer Science and Business Media.

Domb, L.G. and M. Pagel. 2001. Sexual swellings advertise female quality in wild baboons. *Nature* 410:204–206.

Drews, C. 1996. Contexts and patterns of injuries in free-ranging male baboons (*Papio cynocephalus*). *Behaviour* 133:443–474.

Dunbar, R.I.M. and E.P. Dunbar. 1974. Ecological relations and niche separation between sympatric terrestrial primates in Ethiopia. *Folia primatologica* 21:36–60.

Easley, S.P. and A.M. Coelho, Jr. 1991. Is lipsmacking an indicator of social status in baboons? *Folia Primatologica* 56:190–201.

Eggen, D.A. 1974. Cholesterol metabolism in rhesus monkey, squirrel monkey, and baboon. *Journal of Lipid Research* 15:139–145.

Erwin, J. and R. Deni. 1979. Strangers in a strange land: abnormal behaviors or abnormal environments? In *Captivity and Behavior: Primates in Breeding Colonies, Laboratories, and Zoos*, ed. J. Erwin, T.L. Maple, and G. Mitchell, 1–28. New York: Van Nostrand Reinhold.

Escobedo, M.B., J.L. Hilliard, F. Smith, et al. 1982. A baboon model of bronchopulmonary dysplasia I. clinical features. *Experimental and Molecular Pathology* 37:323–334.

European Union. 2010. Directive 2010/63/EU of the European Parliament and of the Council of 22 September 2010 on the protection of animals used for scientific purposes. *Official Journal of the European Union* L276/33.

Ey, E., C. Rahn, K. Hammerschmidt, and J. Fischer. 2009. Wild female olive baboons adapt their grunt vocalizations to environmental conditions. *Ethology* 115:493–503.

Fagot, J., J. Gullstrand, C. Kemp, C. Defilles, and M. Mekaouche. 2014. Effects of freely accessible computerized test systems on the spontaneous behaviors and stress level of guinea baboons (*Papio papio*). *American Journal of Primatology* 76:56–64.

Fischer, J., K. Hammerschmidt, D.L. Cheney, and R.M. Seyfarth. 2002. Acoustic features of male baboon loud calls: influences of context, age, and individuality. *Journal of the Acoustical Society of America* 111:465. doi: 10.1121/1.1433807.

Fischer, J., D.M. Kitchen, R.M. Seyfarth, and D.L. Cheney. 2004. Baboon loud calls advertise male quality: acoustic features and their relation to rank, age, and exhaustion. *Behavioral Ecology and Sociobiology* 56:140–148.

Fischer, J., G.H. Kopp, F.D. Pesco, et al. 2017. Charting the neglected West: the social system of Guinea baboons. *American Journal of Physical Anthropology* 162:15–31.

Fischer, J., M. Metz, D.L. Cheney, and R.M. Seyfarth. 2001. Baboon responses to graded bark variants. *Animal Behaviour* 61:925–931.

Fishman, J.A., D.H. Sachs, K. Yamada, and R.A. Wilkinson. 2018. Absence of interaction between porcine endogenous retrovirus and porcine cytomegalovirus in pig-to-baboon renal xenotransplantation in vivo. *Xenotransplantation* 25:e12395.

Fitzpatrick, C.L., J. Altmann, and S.C. Alberts. 2015. Exaggerated sexual swellings and male mate choice in primates: testing the reliable indicator hypothesis in the Amboseli baboons. *Animal Behaviour* 104:175–185.

Fraser, O. and A.B. Plowman. 2007. Function of notification in *Papio hamadryas*. *International Journal of Primatology* 28:1439–1448.

Galbany, J., J. Tung, J. Altmann, and S.C. Alberts. 2015. Canine length in wild male baboons: maturation, aging and social dominance rank. *PLos One* 10(5):e0126415. doi:10.1371/journal.pone.0126415.

Garcia, C., P.C. Lee, and L. Rosetta. 2008. Impact of social environment on variation in menstrual cycle length in captive female olive baboons (*Papio anubis*). *Reproduction* 135:89–97.

Gesquiere, L.R., N.H. Learn, M.C.M. Simao, P.O. Onyango, S.C. Alberts, and J. Altmann. 2011. Life at the top: rank and stress in wild male baboons. *Science* 333:357–360.

Gil-Burmann, C., F. Pelaez, and S. Sanchez. 1998. Variations in competitive mechanisms of captive male hamadryas-like baboons in two feeding situations. *Primates* 39:473–484.

Goldberg, E., T.E. Wheat, J.E. Powell, and V.C. Stevens. 1981. Reduction of fertility in female baboons immunized with lactate dehydrogenase C_4. *Fertility and Sterility* 35:214–217.

Goncharov, N., T. Aso, Z. Cekan, N. Pachalia, and E. Diczfalusy. 1976. Hormonal changes during the menstrual cycle of the baboon (*Papio hamadryas*). *European Journal Obstet Gynec Reprod Biol* 6(4):209–217.

Goodwin, W.J. and A.M. Coelho Jr. 1982. Development of a large scale baboon breeding program. *Laboratory Animal Science* 32:672–676.

Gurung, S., A.N. Preno, J.P. Dubaut, et al. 2018. Translational model of Zika virus disease in baboons. *Journal of Virology* 92:e00186-18.

Hall, K.R.L. 1962a. Numerical data, maintenance activities and locomotion of the wild chacma baboon *Papio ursinus*. *Proceedings of the Zoological Society of London* 139:181–220.

Hall, K.R.L. 1962b. The sexual, agonistic and derived social behaviour patterns of the wild chacma baboon, *Papio ursinus*. *Proceedings of the Zoological Society of London* 139:283–327.

Hall, K.R.L. 1968. Experiment and quantification in the study of baboon behavior in its natural habitat. In *Primates Studies in Adaptation and Variability*, ed. P.C. Jay, 120–130. New York: Holt, Rinehart and Winston.

Hall, K.R.L. and I. DeVore. 1965. Baboon social behavior. In *Primate Behavior Field Studies of Monkeys and Apes*, ed. I. DeVore, 53–110. New York: Holt, Rinehart and Winston.

Harding, R.S.O. 1980. Agonism, ranking, and the social behavior of adult male baboons. *American Journal of Physical Anthropology* 53:203–216.

Hausfater, G. 1975. Dominance and reproduction in baboons (*Papio cynocephalus*). In *Contributions to Primatology*, Vol. 7, ed. H. Kuhn, W.P. Luckett, C.R. Noback, A.H. Schultz, D. Starck, and F.S. Szalay, 1–150. New York: S. Karger.

Hausfater, G., J. Altmann, and S. Altmann. 1982. Long-term consistency of dominance relations among female baboons (*Papio cynocephalus*). *Science* 217:752–755.

Hausfater, G. and D. Takacs. 1987. Structure and function of hindquarter presentations in yellow baboons (*Papio cynocephalus*). *Ethology* 74:297–319.

Hendrickx, A.G., L.R. Axelrod, and L.D. Clayborn. 1966. 'Thalidomide' syndrome in baboons. *Nature* 210:958–959.

Hendrickx, A.G., M. Cukierski, S. Prahalada, G. Janos, and J. Rowland. 1985. Evaluation of Bendectin embryotoxicity in nonhuman primates: I. Ventricular septal defects in prenatal macaques and baboon. *Teratology* 32:179–189.

Henzi, S.P., J.E. Lycett, and T. Weingrill. 1997. Cohort size and the allocation of social effort by female mountain baboons. *Animal Behaviour* 54:1235–1243.

Hienz, R.D., T.J. Zarcone, J.S. Turkkan, D.A. Pyle, and R.J. Adams. 1998. Measurement of enrichment device use and preference in singly caged baboons. *Laboratory Primate Newsletter* 37:6–10.

Higham, J.P., A.M. MacLarnon, C. Ross, M. Heistermann, and S. Semple. 2008. Baboon sexual swellings: information content of size and color. *Hormones and Behavior* 53:452–462.

Higham, J.P., S. Semple, A. MacLarnon, M. Heistermann, and C. Ross. 2009. Female reproductive signaling, and male mating behavior, in the olive baboon. *Hormones and Behavior* 55:60–67.

Hoffman, T.S., and M.J. O'Riain. 2012. Troop size and human-modified habitat affect the ranging patterns of a chacma baboon population in the Cape Peninsula, South Africa. *American Journal of Primatology* 74:853–863.

Holmes, K.A., M.D. Paull, A.M. Birrell, A. Hennessy, A.G. Gillin, and J.S. Horvath. 1996. A unique design for ease of access and movement of captive *Papio hamadryas*. *Laboratory Animals* 30:327–331.

Jensen, J.T., C. Hanna, S. Yao, E. Thompson, C. Bauer, O.D. Slayden. 2016. Transcervical administration of polidocanol foam prevents pregnancy in female baboons. *Contraception* 94:527–533.

Johannesson, L., A. Enskog, P. Dahm-Kahler, et al. 2012. Uterus transplantation in a non-human primate: long-term follow-up after autologous transplantation. *Human Reproduction* 27:1640–1648.

Johnson, Z., L. Brent, J.C. Alvarenga, et al. 2015. Genetic influences on response to novel objects and dimensions of personality in *Papio* baboons. Behavioral Genetics 45:215–227.

Jolly C.J., A.S. Burrell, J.E. Phillips-Conroy, C. Bergey, and J. Rogers. 2011. Kinda baboons (*Papio kindae*) and grayfoot chacma baboons (*P. ursinus griseipes*) hybridize in the Kafue River Valley, Zambia. *American Journal of Primatology* 73:291–303.

Jooste, E., R.T. Pitman, W. van Hoven, and L.H. Swanepoel. 2012. Unusually high predation on chacma baboons (*Papio ursinus*) by female leopards (*Panthera pardus*) in the Waterberg Mountains, South Africa. *Folia Primatologica* 83:353–360.

Judge, P.G. and K.A. Bachmann. 2013. Witnessing reconciliation reduces arousal of bystanders in a baboon group (*Papio hamadryas hamadryas*). *Animal Behaviour* 85:881–889.

Judge, P.G., N.S. Griffaton, and A.M. Fincke. 2006. Conflict management by hamadryas baboons (*Papio hamadryas hamadryas*) during crowding: a tension-reduction strategy. *American Journal of Primatology* 68:993–1006.

Kalbitzer, U., M. Heistermann, D. Cheney, R. Seyfarth, and J. Fischer. 2015. Social behavior and patterns of testosterone and glucocorticoid levels differ between male chacma and Guinea baboons. *Hormones and Behavior* 75:100–110.

Kessel, A.L. and L. Brent. 1995. An activity cage for baboons, part II: long-term effects and management issues. *Contemporary Topics* 34:80–83.

Kessel, A.L. and L. Brent. 1996. Space utilization by captive-born baboons (*Papio sp.*) before and after provision of structural enrichment. *Animal Welfare* 5:37–44.

Kessel, A. and L. Brent. 2001. The rehabilitation of captive baboons. *Journal of Medical Primatology* 30:71–80.

Kinsella, J.P., D.R. Gerstmann, R.H. Clark, et al. 1991. High-frequency oscillatory ventilation versus intermittent mandatory ventilation: early hemodynamic effects in the premature baboon with hyaline membrane disease. *Pediatric Research* 29:160–166.

Kitchen, D.M., D.L. Cheney, and R.M. Seyfarth. 2005. Contextual factors mediating contests between male chacma baboons in Botswana: effects of food, friends, and females. *International Journal of Primatology* 26:105–125.

Kitchen, D.M., R.M. Seyfarth, J. Fischer, and D.L. Cheney. 2003. Loud calls as indicators of dominance in male baboons (*Papio cynocephalus ursinus*). *Behavioral Ecology and Sociobiology* 53:374–384.

Kling, O.R. and P.K. Westfahl. 1978. Steroid changes during the menstrual cycle of the baboon (*Papio cynocephalus*) and human. *Biology of Reproduction* 18:392–400.

Kopp, G.H., J. Fischer, A. Patzelt, C. Roos, and D. Zinner. 2015. Population genetic insights into the social organization of Guinea baboons (*Papio papio*): evidence for female-biased dispersal. *American Journal of Primatology* 77:878–889.

Kummer, H. 1968. *Social Organization of Hamadryas Baboons*. New York: S. Karger.

Kunz, B.K. and K.E. Linsenmair. 2008. The disregarded West: diet and behavioural ecology of olive baboons in the Ivory Coast. *Folia Primatologica* 79:31–51.

Lambert, J.E. and J.C. Whitham. 2001. Cheek pouch use in *Papio cynocephalus*. *Folia Primatol* 72:89–91.

Lambeth, S.P., J. Hau, J.E. Perlman, M. Martino, and S.J. Schapiro. 2006. Positive reinforcement training affects hematologic and serum chemistry values in captive chimpanzees (*Pan troglodytes*). *American Journal of Primatology* 68:245–256.

Langin, M., A. Panelli, B. Reichart, et al. 2018. Perioperative telemetric monitoring in pig-to-baboon heterotopic thoracic cardiac xenotransplantation. *Annals of Transplantation* 23:491–499.

Laule, G.E., M.A. Bloomsmith, and S.J. Schapiro. 2003. The use of positive reinforcement training techniques to enhance the care, management, and welfare of primates in the laboratory. *Journal of Applied Animal Welfare Science* 6:163–173.

Leigh, S.R. 2009. Growth and development of baboons. In *The Baboon in Biomedical Research*, ed J.L. VandeBerg, S. Williams-Blangero, and S.D. Tardif, 57–88. New York: Springer Science and Business Media.

Lewis, D.S., H.A. Bertrand, C.A. McMahan, H.C. McGill Jr., K.D. Carey, and E.J. Masoro. 1986. Preweaning food intake influences the adiposity of young adult baboons. *Journal of Clinical Investigation* 78:899–905.

Lewis, D.S., A.M. Coelho Jr., and E.M. Jackson. 1991. Maternal weight and sire group, not caloric intake, influence adipocyte volume in infant female baboons. *Pediatric Research* 30:534–540.

Lutz, C.K. 2018. A cross-species comparison of abnormal behavior in three species of singly-housed old world monkeys. *Applied Animal Behaviour Science* 199:52–58.

Lutz, C.K., P.C. Williams, and R.M. Sharp. 2014. Abnormal behavior and associated risk factors in captive baboons (*Papio hamadryas spp.*). *American Journal of Primatology* 76:355–361.

MacCluer, J.W., C.M. Kammerer, J. Blangero, et al. 1988. Pedigree analysis of HDL cholesterol concentration in baboons on two diets. *American Journal of Human Genetics* 43:401–413.

MacCormick, H.A., D.R. MacNulty, A.L. Bosacker, et al. 2012. Male and female aggression: lessons from sex, rank, age, and injury in olive baboons. *Behavioral Ecology* 23:684–691.

Maciej, P., I. Ndao, K. Hammerschmidt, and J. Fischer. 2013. Vocal communication in a complex multi-level society: constrained acoustic structure and flexible call usage in Guinea baboons. *Frontiers in Zoology* 10:58.

Maclean, J.M., A.F. Phippard, M.G. Garner, G.G. Duggin, J.S. Horvath, and D.J. Tiller. 1987. Group housing of hamadryas baboons: a new cage design based upon field studies of social organization. *Laboratory Animal Science* 37:89–93.

Macy, J.D. Jr., T.A. Beattie, S.E. Morgenstern, and A.F.T. Arnsten. 2000. Use of guanfacine to control self-injurious behavior in two rhesus macaques (*Macaca mulatta*) and one baboon (*Papio anubis*). *Comparative Medicine* 50:419–425.

Maestripieri, D., G. Schino, F. Aureli, and A. Troisi. 1992. A modest proposal: displacement activities as an indicator of emotions in primates. *Animal Behaviour* 44:967–979.

Mason, G.J. 1991. Stereotypies and suffering. *Behavioural Processes* 25:103–115.

Matsumiya, G., S.R. Gundry, N. Fukushima, M. Kawauchi, C.W. Zuppan, and L.L. Baily. 1996a. Pediatric cardiac zenograft growth in a rhesus monkey-to-baboon transplantation model. *Xenotransplantation* 3:76–80.

Matsumiya, G., S.R. Gundry, S. Nehlsen-Cannarella, et al. 1996b. Successful long-term concordant xenografts in primates: alteration of the immune response with methotrexate. *Transplantation Proceedings* 28:751–753.

Matz-Rensing, K., A. Floto, and F-J. Kaup. 2004. Intraperitoneal foreign body disease in a baboon (*Papio hamadryas*). *Journal of Medical Primatology* 33:113–116.

Maugard, A., E.A. Wasserman, L. Castro, and J. Fagot. 2014. Effects of training condition on the contribution of specific items to relational processing in baboons (*Papio papio*). *Animal Cognition* 17:911–924.

Maxim, P.E. and J. Buettner-Janusch. 1963. A field study of the Kenya baboon. *American Journal of Physical Anthropology* 21:165–180.

McGill, H.C. Jr., C.A. McMahan, A.W. Kruski, J.L. Kelley, and G.E. Mott. 1981. Responses of serum lipoproteins to dietary cholestrerol and type of fat in the baboon. *Arteriosclerosis* 1:337–344.

McGill, H.C. Jr., J.P. Strong, R.L. Holman, and N.T. Werthessen. 1960. Arterial lesions in the Kenya baboon. *Circulation Research* 8:670–679.

McGrogan, H.J. and J.E. King. 1982. Repeated separations of 2-year-old squirrel monkeys from familiar mother surrogates. *American Journal of Primatology* 3:285–290.

Mejido, D.C.P., E.J. Dick Jr., P.C. Williams, et al. 2009. Trichobezoars in baboons. *Journal of Medical Primatology* 38:302–309.

Mitchell, D.S., H.S. Wigodsky, H.H. Peel, and T.A. McCaffrey. 1980. Operant conditioning permits voluntary, noninvasive measurement of blood pressure in conscious, unrestrained baboons (*Papio cynocephalus*). *Behavior Research Methods & Instrumentation* 12:492–498.

Nagel, U. 1973. A comparison of anubis baboons, hamadryas baboons and their hybrids at a species border in Ethiopia. *Folia Primatologica* 19:104–165.

Napier, J.R. and P.H. Napier. 1994. *The Natural History of the Primates*. Cambridge, MA: The MIT Press.

National Research Council (NRC). 2011. *The Guide for the Care and Use of Laboratory Animals* (8th edition). Washington, DC: National Academies Press.

Nevill, C.H. and C.K. Lutz. 2015. The effect of a feeding schedule change and the provision of forage material on hair eating in a group of captive baboons (*Papio hamadryas sp.*). *Journal of Applied Animal Welfare Science* 18:319–331.

Norton, G.W., R.J. Rhine, G.W. Wynn, and R.D. Wynn. 1987. Baboon diet: a five-year study of stability and variability in the plant feeding and habitat of the yellow baboons (*Papio cynocephalus*) of Mikumi National Park, Tanzania. *Folia Primatologica* 48:78–120.

Nowak, R.M. 1999. *Walker's Primates of the World*. Baltimore, MD: The Johns Hopkins University Press.

O'Brien, J.K., S. Heffernan, P.C. Thomson, and P.D. McGreevy. 2008. Effect of positive reinforcement training on physiological and behavioural stress responses in the hamadryas baboon (*Papio hamadryas*). *Animal Welfare* 17:125–138.

Owens, N.W. 1975. Social play behaviour in free-living baboons, *Papio anubis*. *Animal Behaviour* 23:387–408.

Ozwara, H., J.A.M. Langermans, J. Maamun, et al. 2003. Experimental infection of the olive baboon (*Papio anubis*) with *Plasmodium knowlesi*: severe disease accompanied by cerebral involvement. *American Journal of Tropical Medicine and Hygiene* 69:188–194.

Packer, C. 1979a. Inter-troop transfer and inbreeding avoidance in *Papio anubis*. *Animal Behaviour* 27:1–36.

Packer, C. 1979b. Male dominance and reproductive activity in *Papio anubis*. *Animal Behaviour* 27:37–45.

Patzelt, A., G.H. Kopp, I. Ndao, U. Kalbitzer, D. Zinner, and J. Fischer. 2014. Male tolerance and male-male bonds in a multilevel primate society. *PNAS* 111:14740–14745.

Pearson, B.L., P.G. Judge, and D.M. Reeder. 2008. Effectiveness of saliva collection and enzyme-immunoassay for the quantification of cortisol in socially housed baboons. *American Journal of Primatology* 70:1145–1151.

Pearson, B.L., D.M. Reeder, and P.G. Judge. 2015. Crowding increases salivary cortisol but not self-directed behavior in captive baboons. *American Journal of Primatology* 77:462–467.

Pebsworth, P.A., M. Bardi, and M.A. Huffman. 2012. Geophagy in chacma baboons: patterns of soil consumption by age class, sex, and reproductive state. *American Journal of Primatology* 74:48–57.

Phillips-Conroy, J.E. and C.J. Jolly. 1981. Sexual dimorphism in two subspecies of Ethiopian baboons (*Papio hamadryas*) and their hybrids. *American Journal of Physical Anthropology* 56:115–129.

Polo, P. and F. Colmenares. 2012. Behavioural processes in social context: female abductions, male herding and female grooming in hamadryas baboons. *Behavioural Processes* 90:238–245.

Pyle, D.A., A.L. Bennett, T.J. Zarcone, J.S. Turkkan, R.J. Adams, and R.D. Hienz. 1996. Use of two food foraging devices by singly housed baboons. *Lab Primate Newsletter* 35:10–15.

Ramirez, S.M., M. Bardi, J.A. French, and L. Brent. 2004. Hormonal correlates of changes in interest in unrelated infants across the peripartum period in female baboons (*Papio hamadryas anubis sp.*). *Hormones and Behavior* 46:520–528.

Rendall, D., D.L. Cheney, and R.M. Seyfarth. 2000. Proximate factors mediating "contact" calls in adult female baboons (*Papio cynocephalus ursinus*) and their infants. *Journal of Comparative Psychology* 114:36–46.

Rendall, D., R.M. Seyfarth, D.L. Cheney, and M.J. Owren. 1999. The meaning and function of grunt variants in baboons. *Animal Behaviour* 57:583–592.

Rhine, R.J., G.W. Norton, G.M. Wynn, R.D. Wynn, and H.B. Rhine. 1986. Insect and meat eating among infant and adult baboons (*Papio cynocephalus*) of Mikumi National Park, Tanzania. *American Journal of Physical Anthropology* 70:105–118.

Rigaill, L., J.P. Higham, P.C. Lee, A. Blin, and C. Garcia. 2013. Multimodal sexual signaling and mating behavior in olive baboons (*Papio anubis*). *American Journal of Primatology* 75:774–787.

Rivard, C.J., T. Tanabe, M.A. Lanaspa, et al. 2018. Upregulation of CD80 on glomerular podocytes plays an important role in development of proteinuria following pig-to-baboon xenorenal transplantation - an experimental study. *Transplant International* 31:1164–1177.

Roder, E.L. and P.J.A. Timmermans. 2002. Housing and care of monkeys and apes in laboratories: adaptations allowing essential species-specific behaviour. *Laboratory Animals* 36:221–242.

Rodriguez, J.S., N.R. Zurcher, T.Q. Bartlett, P.W. Nathanielsz, and M.J. Nijland. 2011. CANTAB delayed matching to sample task performance in juvenile baboons. *Journal of Neuroscience Methods* 196:258–263.

Romero, T., F. Colmenares, and F. Aureli. 2008. Postconflict affiliation of aggressors in *Papio hamadryas*. *International Journal of Primatology* 29:1591–1606.

Rose, M.D. 1977. Positional behaviour of olive baboons (*Papio anubis*) and its relationship to maintenance and social activities. *Primates* 18:59–116.

Rowell, T.E. 1967. A quantitative comparison of the behaviour of a wild and a caged baboon group. *Animal Behaviour* 15:499–509.

Saayman, G.S. 1971a. Behaviour of the adult males in a troop of free-ranging chacma baboons (*Papio ursinus*). *Folia Primatologica* 15:36–57.

Saayman, G.S. 1971b. Grooming behaviour in a troop of free-ranging chacma baboons (*Papio ursinus*). *Folia Primatologica* 16:161–178.

Saunders, C.D. and G. Hausfater. 1988. The functional significance of baboon grooming behavior. *Annals New York Academy of Sciences* 525:430–432.

Schlabritz-Loutsevitch, N.E., B. Ballesteros, C. Dudley, et al. 2007. Moderate maternal nutrient restriction, but not glucocorticoid administration, leads to placental morphological changes in the baboon (*Papio sp.*). *Placenta* 28:783–793.

Schlabritz-Loutsevitch, N.E., K. Howell, K. Rice, et al. 2004. Development of a system for individual feeding of baboons maintained in an outdoor group social environment. *Journal of Medical Primatology* 33:117–126.

Schlabritz-Loutsevitch, N.E., C.M. Moore, J.C. Lopez-Alvarenga, B.G. Dunn, D. Dudley, and G.B. Hubbard. 2008. The baboon model (*Papio hamadryas*) of fetal loss: maternal weight, age, reproductive history and pregnancy outcome. *Journal of Medical Primatology* 37:337–345.

Schreier, A.L. and L. Swedell. 2009. The fourth level of social structure in a multi-level society: ecological and social functions of clans in hamadryas baboons. *American Journal of Primatology* 71:948–955.

Schreier, A.L. and L. Swedell. 2012. Ecology and sociality in a multilevel society: ecological determinants of spatial cohesion in hamadryas baboons. *American Journal of Physical Anthropology* 148:580–588.

Seyfarth, R.M. 1978a. Social relationships among adult male and female baboons. I. Behaviour during sexual consortship. *Behaviour* 64:204–226.

Seyfarth, R.M. 1978b. Social relationships among adult male and female baboons. II. Behaviour throughout the female reproductive cycle. *Behaviour* 64:227–247.

Seyfarth, R.M., J.B. Silk, and D.L. Cheney. 2012. Variation in personality and fitness in wild female baboons. *PNAS* 109:16980–16985.

Seyfarth, R.M., J.B. Silk, and D.L. Cheney. 2014. Social bonds in female baboons: the interaction between personality, kinship and rank. *Animal Behaviour* 87:23–29.

Shoulson, R.L., R.L. Stark, and M. Garland. 2014. Pharmacokinetics of fluoxetine in pregnant baboons (*Papio spp.*). *JAALAS* 53:708–716.

Sigg, H., A. Stolba, J.J. Abegglen, and V. Dasser. 1982. Life history of hamadryas baboons: physical development, infant mortality, reproductive parameters and family relationships. *Primates* 23:473–487.

Silk, J.B. 1987a. Activities and feeding behavior of free-ranging pregnant baboons. *International Journal of Primatology* 8:593–613.

Silk, J.B. 1987b. Correlates of agonistic and competitive interactions in pregnant baboons. *American Journal of Primatology* 12:479–495.

Silk, J.B. 1993. The evolution of social conflict among female primates. In *Primate Social Conflict*, ed W.A. Mason and S.P. Mendoza, 49–83. Albany: State University of New York Press.

Silk, J.B., D.L. Cheney, and R.M. Seyfarth. 1996. The form and function of post-conflict interactions between female baboons. *Animal Behaviour* 52:259–268.

Silk, J.B., R.M. Seyfarth, and D.L. Cheney. 2016. Strategic use of affiliative vocalizations by wild female baboons. *PLoS One* 11(10):e0163978. doi: 10.1371/journal.pone.0163978.

Smuts, B.B. 1985. *Sex and Friendship in Baboons*. New York: Aldine Publishing Company.

Sommer, V., A. Lowe, G. Jesus, et al. 2016. Antelope predation by Nigerian forest baboons: ecological and behavioural correlates. *Folia Primatologica* 87:67–90.

Stacey, P.B. 1986. Group size and foraging efficiency in yellow baboons. *Behavioral Ecology and Sociobiology* 18:175–187.

Stevens, V.C. 1997. Some reproductive studies in the baboon. *Human Reproduction Update* 3:533–540.

Strum, S.C. 1991. Weight and age in wild olive baboons. *American Journal of Primatology* 25:219–237.

Strum, S.C. 2010. The development of primate raiding: implications for management and conservation. *International Journal of Primatology* 31:133–156.

Swedell, L. and J. Saunders. 2006. Infant mortality, paternity certainty, and female reproductive strategies in hamadryas baboons. In *Reproduction and Fitness in Baboons: Behavioral, Ecological, and Life History Perspectives*, ed. L. Swedell and S.R. Leigh, 19–51. New York: Springer.

Szabo, C.A., M.M. Leland, K.D. Knape, and J.T. Williams. 2009. The baboon model of epilepsy: current applications in biomedical research. In *The Baboon in Biomedical Research*, ed. J.L. VandeBerg, S. Williams-Blangero, and S.D. Tardif, 351–370. New York: Springer Science and Business Media.

Tarara, R. 1984. The effect of medroxyprogesterone acetate (Depo-Provera) on prenatal development in the baboon (*Papio anubis*): a preliminary study. *Teratology* 30:181–185.

Tarou, L.R., M.A. Bloomsmith, and T.L. Maple. 2005. Survey of stereotypic behavior in prosimians. *American Journal of Primatology* 65:181–196.

Turkkan, J.S. 1990. New methodology for measuring blood pressure in awake baboons with use of behavioral training techniques. *Journal of Medical Primatology* 19:455–466.

Veira, Y. and L. Brent. 2000. Behavioral intervention program: enriching the lives of captive nonhuman primates. *American Journal of Primatology* 51:97.

Vidyasagar, D., H. Maeta, T.N.K. Raju, et al. 1985. Bovine surfactant (surfactant TA) therapy in immature baboons with hyaline membrane disease. *Pediatrics* 75:1132–1142.

Vilensky, J.A. 1978. The function of ischial callosities. *Primates* 19:363–369.

Warfel, J.M., J.F. Papin, R.F. Wolf, L.I. Zimmerman, and T.J. Merkel. 2014. Maternal and neonatal vaccination protects newborn baboons from pertussis infection. *Journal of Infectious Diseases* 210:604–610.

Warren, Y. 2008. Crop-raiding baboons (*Papio anubis*) and defensive farmers: a West African perspective. *West African Journal of Applied Ecology* 14:1–11.

Wasser, S.K. and G. Norton. 1993. Baboons adjust secondary sex ratio in response to predictors of sex-specific offspring survival. *Behavioral Ecology and Sociobiology* 32:273–281.

Wasserman, F.E. and W.W. Cruikshank. 1983. The relationship between time of feeding and aggression in a group of captive hamadryas baboons. *Primates* 24:432–435.

Watanabe, H., H. Sahara, S. Nomura, et al. 2018. GalT-KO pig lungs are highly susceptible to acute vascular rejection in baboons, which may be mitigated by transgenic expression of hCD47 on porcine blood vessels. *Xenotransplantation* 25:e12391.

Whiten, A., R.W. Byrne, R.A. Barton, P.G. Waterman, and S.P. Henzi. 1991. Dietary and foraging strategies of baboons. *Philosophical Transactions of the Royal Society of London B* 334:187–197.

Wildt, D.E., L.L. Doyle, S.C. Stone, and R.M. Harrison. 1977. Correlation of perineal swelling with serum ovarian hormone levels, vaginal cytology, and ovarian follicular development during the baboon reproductive cycle. *Primates* 18:261–270.

Wiley, J.N., A. Leeds, K.D. Carpenter, and C.J. Kendall. 2018. Patterns of wounding in hamadryas baboons (*Papio hamadryas*) in North American zoos. *Zoo Biology* 37:74–79.

Williams-Blangero, S., J.L. Vandeberg, J. Blangero, L. Konigsberg, and B. Dyke. 1990. Genetic differentiation between baboon subspecies: relevance for biomedical research. *American Journal of Primatology* 20:67–81.

Wittig, R.M., C. Crockford, J. Lehmann, P.L. Whitten, R.M. Seyfarth, and D.L. Cheney. 2008. Focused grooming networks and stress alleviation in wild female baboons. *Hormones and Behavior* 54:170–177.

Wolfheim, J.H. 1983. *Primates of the World Distribution, Abundance, and Conservation*. Seattle: University of Washington Press.

Yoder, B.A., D.C. McCurnin, and J.J. Coalson. 2009. Baboon models for neonatal lung disease. In *The Baboon in Biomedical Research*, ed J.L. VandeBerg, S. Williams-Blangero, and S.D. Tardif, 179–205. New York: Springer Science and Business Media.

Young, G.H. and C.A. Bramblett. 1977. Gender and environment as determinants of behavior in infant common baboons (*Papio cynocephalus*). *Archives of Sexual Behavior* 6:365–385.

Young, G.H., A.M. Coelho Jr., and C.A. Bramblett. 1982. The development of grooming, sociosexual behavior, play and aggression in captive baboons in their first two years. *Primates* 23:511–519.

Young, G.H. and R.J. Hankins. 1979. Infant behaviors in mother-reared and harem-reared baboons (*Papio cynocephalus*). *Primates* 20:87–93.

Zurcher, N.R., J.S. Rodriguez, S.L. Jenkins, et al. 2010. Performance of juvenile baboons on neuropsychological tests assessing associative learning, motivation and attention. *Journal of Neuroscience Methods* 188:219–225.

Part 3

Selected Ethograms

Part 3 of this book contains ethograms for 16 of the species addressed in Part 2. In addition, the links to online ethograms for mice and macaques are provided below. As you know, an ethogram is a list of behaviors that can be observed and scored while watching animals (including humans), along with the operational definitions for those behaviors. Ethograms facilitate a process whereby different observers can observe animal behavior and record their observations (data) in the same way. Good ethograms facilitate high levels of interobserver reliability, a critical requirement for the effective observation and analysis of behavior. If behavioral measures are intended to be used to assess the welfare, and/or changes in welfare, of laboratory animals, a functional ethogram is absolutely essential. Ethograms can either be quite specific, with many operationally defined behaviors, or they can be quite general, with only a few operationally defined behavioral categories. Ethograms with many behaviors typically yield more detailed data, while ethograms with general behavioral categories are typically easier for observers to use. It must be noted that individual behaviors can always be combined into categories for the purposes of analysis, but categories, once collected, can never be split into individual behaviors. This fact will have to be considered when determining the level of detail to include in the ethogram that you choose.

While the ethograms in Part 3 have all been presented in a similar format, contain many of the same behaviors, and are, overall, reasonably similar, there are a number of differences across ethograms. There is no such thing as a single ethogram that is appropriate for the study of ALL animals.

All of the ethograms in Part 3 contain species-typical behaviors, and many contain abnormal behaviors as well. Some of the ethograms contain behaviors that are specific for responses in particular situations (e.g., during behavioral tests), and some include operational definitions of postures and/or vocalizations, in addition to behaviors. The purpose of putting all of these ethograms in a single location is to provide those interested in laboratory animal behavior with a centralized resource for basic ethograms, which can then be adapted and optimized to facilitate behavioral observations for a variety of situations.

A comprehensive mouse ethogram can be found at: https://mousebehavior.org/ethogram/.

A comprehensive macaque ethogram can be found at: https://www.nc3rs.org.uk/macaques/macaques/behaviour-and-communication/.

A comprehensive dog ethogram can be found in Tables A1-A5 (Chapter 13) of this volume.

Ethogram: Rats (Chapter 7)

(Adapted from Blanchard & Blanchard 1977; Blanchard et al. 1975; Panskeep 1981; Pellis & Pellis 1987; Pellis 1988; Vanderschuren et al. 1997)

Behavior	Description
Active Behaviors	
Anogenital investigation/sniff	Rat extends his nose toward and within a few millimeters of the anogenital area of another animal. It may also involve licking of the region.
Burrowing	Rat is displacing soil using fore legs and/or kicking out with the hind legs.
Climb	Rat is suspended with all paws in contact with a vertical surface or the cage ceiling.
Curved back approach	Normal walking posture of nonfrightened rats.
Gnaw	Rat pulls his lower jaw forward using the jaw muscles, such that his incisors touch each other and its molars do not. The upper incisors hold the object, and the lower incisors are pulled powerfully upward to cut against it. Gnawing is a rodent's specialty, and their specialized jaw muscles and jaw articulation give rodents a very effective, powerful gnawing action.
Jump	While rearing, rat jumps up with all four paws off the floor. Jump starts when first back leg leaves the floor.
Hop	Rat moves forward with small hops or big jumps. Hop starts when the first back leg leaves the ground and ends when the first back leg touches the ground.
Nibble	Light bites (no breaking of skin) and licks to a conspecific body or to a human hand or finger. When directed at a human hand, it is used to solicit playful interactions such as tickling.
Sniffing	Rat's nose and head are moving up and down, and whiskers are moving sideways, or rat is moving slowly with nose close or in contact with a surface such as cage floor, lid, or nesting material.
Stand	Rat is standing immobile with all four feet on the floor, but is not in a freezing posture. The rat is standing, immobile, in a crouching posture, nose is up, head/nose may move as he is sniffing the air and looking around. The front paw may be picked up and put down without locomotion (i.e., movement forward or sideways).
Stretch	*Lateral*: Rat is parallel to the ground with the body elongated and back slightly arched; head and tail often angled upward; hind legs and sometimes one fore leg are outstretched; rat is often yawning. *Vertical (or Rear)*: Rat stands on hind feet on floor and front feet in contact or not with a vertical surface (e.g., wall, objects). Rear starts when first forepaw leaves the ground and ends when first forepaw touches the ground. A new rear is initiated whenever the rat touches the floor before going up again.
Play	
Social (rough-and-tumble) play[a]	Refers to play directed at conspecifics.
Dorsal contact	Rat pounces upon his/her partner's back and nuzzles the nape with the forepaws and snout. However, because the "target" rat is moving, other parts of the body (head, back rump) in addition to the nape may be touched. The paws must return to the floor before a second contact can be counted.
Pin	One rat lying on its back in a supine position, so the nape is against the ground and all four paws are removed from the cage floor; at the same time, the second rat is standing over the bottom rat and has its front paws (and sometimes also the back paws) on top of the bottom rat. A pin can be brief or last up to a few seconds.
Locomotor play	Refers to apparently spontaneous movements that carry the individual about its environment.
Scampering	Running rapidly alone, not toward a cage mate, and performing at least three hops.
Object play	Refers to play directed at inanimate objects; object manipulation using the paws or mouth.
Inactive Behaviors	
Immobile	Rat is standing or lying. If standing, rat may move one paw sideways, but movement is not linear or directional (no locomotor activity).
Lying	Rat is lying in lateral or ventral recumbency, motionless, may be sleeping.
Maintenance Behaviors	
Eat/drink	Rat snout is touching sipper tube or feeder or can be seen holding food in paws or chewing.
Groom	Rat is moving both forepaws on fur and/or face in rapid strokes, which starts when paws first reach for the head. Also includes scratching body with hind paws and licking/nibbling various parts of the body. Starts when foot is first lifted or when the head is first tucked before licking/nibbling.
Shake	Rat shakes head or entire body (like dogs do).
Aggressive/Defensive Behaviors[b]	
Bite	Rat's mouth opens and closes on conspecific body part or experimenter finger or other parts of the hand scratching or breaking the skin. During fights between rats, offensive bites tend to be directed at the lower back and flanks, while defensive bites tend to be directed at the face. Defensive bites are sometimes delivered in a lunge-and-bite sequence. The bite to the head is characteristic of hurt, frightened, or defensive rats.

(Continued)

Flat back approach	Rat is moving toward a stimulus with the top of the back lower than the ears and the ventral surface of the snout held parallel to the ground.
Stretch attend posture	Rat is standing with the body elongated, the back is lower than the ears, and the hind paws are extended back. From this position, it may lean forward, but without locomotion. Locomotion in this posture is called the Flat Back Approach.
Freeze	Rat is standing, immobile, in a crouching posture with no visible movement other than respiration.
Boxing	Two rats stand on their hind legs face to face and nearly nose to nose, and push or paw at each other with their front legs and paws, usually around the head, neck, shoulders and front legs of their opponent. In high intensity boxing, the rats stand erect on their hind feet and rapidly push, paw, and grab at each other. In low intensity boxing, the rats squat and paw at each other gently. Boxing is a defensive strategy. Thus, as long as the subordinate rat maintains whisker-to-whisker contact, the dominant rat cannot bite his rump. The dominant rat may respond to the boxing tactic with a lateral display.
Lateral display	Threatening posture in which one rat (usually but not always the dominant one) approaches another rat sideways or broadside, with his back strongly arched, and crowds the second rat. It may be a successful strategy to counter boxing and achieve a rump bite. If physical contact is made, the lateral display may become a push. The rat may also kick.
Mounting Behaviors	
Ear wiggle	Female vibrates her ears rapidly. Ear wiggling is part of a suite of solicitation behaviors in which the female initiates and maintains mounting behavior by the male. Ear wiggling occurs when the female is in behavioral estrus, about every 4–5 days.
Lordosis (female arched back)	Female mating posture. Female stands immobile, with her back arched downward toward the floor, her rump pushed upward and tail deflected to the side. Her vulva, which normally faces the floor, rotates almost 90° to the vertical, backward-facing position. Without lordosis, copulation would be impossible. Lordosis is a reflexive behavior that is triggered by a touch on the lower back, flanks, or genital region. The female may also solicit mounting behavior by the male, which in turn triggers lordosis.
Mount	One rat places its forequarters on another rat's rump from behind. Mounting is the male copulatory position, and is seen when a male mounts a female prior to mating. Mounting is also sometimes seen between rats of the same sex, usually in an aggressive context.
Maternal Behaviors	
Nest building	Any manipulation using paws and mouth on the nesting or bedding materials. It can include pushing and pulling in materials around nest area or pups.
Moving pups	Dam is carrying pups in her mouth to a different location in the cage.
Nursing	Dam is huddled, crouching over a majority of the litter in the typical nursing position. This posture allows pups to nurse while staying warm. The dam may also be lying outstretched with pups attached to nipples.
Grooming pups	Dam licks and grooms the pups with the forepaws.
Press posture	Adult female (and male if present) is seen to wedge the ventral surface of their body (at least half) in the corner of the cage for at least 5 s. Eyes may be open or closed.
Behaviors Specific to Some Tests: Elevated Plus Maze	
Entries in closed arm	When a rat enters a closed arm. An entry is defined as when the four paws cross the line separating the neutral (also called central) zone from one arm of the maze. The rat is considered to be out of the arm when two paws have left the arm.
Entries in open arm	When a rat enters an opened arm. An entry is defined as when the four paws cross the line separating the neutral zone from one arm of the maze. The rat is considered to be out of the arm when two paws have left the arm.
Closed arm returns	When a rat exits a closed arm with only the forepaws, and returns back (i.e., moving backward or turning around) into the same arm.
Open end exploring	When a rat reaches the end of an open arm, i.e., nose reaches, touches, or passes the end of the arm.
Head dipping	When a rat moves his head/shoulder over the sides of an open arm of the maze. Record a dip when the nose (up to the eyes) passes or is equal to the side of the maze. Record a new dip every time the head moves up to horizontal, or higher, and then back down.
Head scan	The rat points his head (with the face pass the eyes out in the open arm) in the open arm but his body is still in the neutral zone of the maze.
Open Field	
Squares crossing	A square is entered when the four paws cross the line separating it from another square.
Emergence Test	
Head out	When the tip of the nose to level of eyes emerges into the bright box from the dark box out of the dark box. It is considered a risk assessment behavior.
Emerge	Rat full body is out from the dark box, i.e., to the base of the tail or all four paws are out in the bright box.

[a] Dorsal contact and pin are the two main components of rat social play. For more details about the other components and differences between rough and tumble play and aggressive behaviors, see Panksepp (1981), Pellis and Pellis (1987), and Pellis (1988).

[b] For a more detailed description of rat aggressive behaviors, see Blanchard and Blanchard (1977) and Blanchard et al. (1975).

Ethogram: Guinea Pig (Chapter 8)

(Adapted from Berryman 1976; Brewer et al. 2014; Corat et al. 2012; Hennessy & Jenkins 1994; Rood 1972; Young 1969)

Behavior	Description
Activity	
Nonspecific locomotion	Quadrupedal movement other than social interaction or consumption behavior.
Jump (popcorn)	All four limbs simultaneously leave the ground while twisting the body (not a startle response).
Startle response	A rapid generalized motor response often followed by either rapid locomotion or freezing in response to a loud noise or other unexpected sensory stimulation.
Rear	Rising up on the hindlimbs.
Facial wipe	Grooming action pattern of alternating and/or simultaneous wiping of the face with the forelimbs by bringing the forelimbs from behind the ears down over the head to the nose.
Head shake	Rapid side to side shaking motion of the head.
Self-groom	Starts with facial wiping and culminates in grooming of the fur using teeth, tongue and a back and forth motion of the nose, accompanied by paw raking of the fur.
Scratch	Rapid rubbing of the hind paws over the body or head, followed by raking the nails through the teeth.
Freeze	Immobility and motionless with full body tone, can last from seconds to hours.
Resting	Cessation of movement with a relaxed body posture, either sitting or lying.
Marking	Dragging the perineal region on the surface to scent mark with the anal glands or side to side movements of the hindquarters against an object to scent mark with the supracaudal glands.
Feeding	
Natural grass	Grasping one or more grass stalks with the mouth and thrusting the head upwards to tear off the grass. The stalk is then consumed by drawing it into the mouth a small segment at a time until the whole stalk is consumed.
Hay rack	Similar to natural grass feeding except no head thrust is observed.
Hay taking	A second guinea pig chews on the opposite end of a hay blade, often taking the hay from the first.
Crowding	Hording of the space in front of the food source (hopper/rack/clump of grass), using the body to push against another guinea pig, moving it away from the food source.
Pellet food	Lips and incisors are used to draw the pellet into the mouth with no paw manipulation.
Water dish	Standing on the edge of the dish or bowl, water is scooped up with the mouth and the head tilted back to swallow.
Water bottle	The nipple of the water bottle is grasped with the teeth and shaken in an up-and-down motion to release water into the mouth.
Water, automatic system	The needle of the water valve is pushed with the tongue or nose, allowing water to flow into the mouth, then the head is tilted back to swallow.
Coprophagy	Bending forward to retrieve a fecal pellet by mouth from the anus (only rarely from the ground).
Prosocial	
Approach	Movement toward another guinea pig, with the nose coming within 2–4 cm.
Follow	Close movement behind another guinea pig, typically within a distance of 2–4 cm.
Contact jumping (popcorning)	A jump that occurs immediately after another guinea pig has made contact (not a startle response).
Simultaneous jumping (popcorning)	Two or more guinea pigs jumping at the same time. Thought to indicate play behavior.
Anogenital Exploration	Sniffing or licking the anogenital region of another guinea pig.
Nose-to-Nose	Nose is moved to within 1 cm of the nose of another guinea pig with both remaining motionless for at least 3 s.
Aggression	
Chase	Quick movement toward another guinea pig, typically from behind.
Nose Up	An upward head toss. Often used to determine dominance.
Kick	Kicking of a hind limb toward another guinea pig.
Offensive stance	Orientation toward an adversary in a crouched stance with mouth open and forelimbs somewhat extended. Teeth chattering and piloerection are often observed as well.
Head thrust	The head is raised then thrust downward toward the adversary, typically without contact.
Bite	Biting another guinea pig.

(Continued)

Urine spray	Flattening of the back and lifting of the hind quarters (lordosis position in females) and spraying urine toward another guinea pig (typically undesirable males). In males, hip waggling (the Rumba) during urine spray occurs during mating and may serve the purpose of scent marking the female to identify her as part of his harem.
Reproduction	
Rumblestrut (Rumba)	Swaying of the hindquarters during mating or dominance. Often accompanied by a purr vocalization.
Circling	Movement by males in a circular pattern around a female during mating.
Nuzzling	The male runs its chin up on the hindquarters of a female during mating.
Mounting	Approaching from behind, forelimbs are placed on the upper back of another guinea pig. Hip thrusting may or may not occur. Often used by either sex in dominance situations, as well as male–female mating.
Lordosis	The rump is raised with the hips tilted anteriorly, hindlimbs extended, and forelimbs lowered. This movement makes the genital region prominent from behind.
Crouch nursing	The female's head and body are held in an immobile posture with the hindlimbs rigid and her back arched allowing offspring to nurse.
Passive lift nursing	Offspring burrow under the lactating female to gain access to her nipples.
Leg out nursing	The nursing female lifts a hindlimb, allowing offspring access to the adjacent nipple.
Side nursing	A lactating female lays on her side for pups to nurse.
Ambulatory nursing	A lactating female nurses one or two offspring while locomoting.
Grooming offspring	Licking anywhere on the body of offspring by adults.
Anogenital licking of offspring	Females lick the anogenital region of pups to help stimulate elimination.
Boxing of offspring	A forceful push of the head into the side flanks of offspring.
Snapping at offspring	Snapping motion of jaw while opening the mouth. Often directed at offspring.
Vocalization	
Chut	Brief low short sounds of 0–3 kHz uttered during exploration of the environment or another guinea pig, or by mothers during nursing or grooming of the young. Thought to be sounds of contentment.
Chudder	Chains of brief sounds slightly higher than the chut and wavering up and down in tone over many minutes. Often accompanied by the whine. Emitted during discomfort, distress, or flight.
Whine	A fluctuating quavering sound often started and punctuated with chudders and may last 10–20 min. Occurs during defense, evasion, and flight.
Low whistle	A single component sound at the low frequency of the chut in bouts of 10–20. Commonly followed by whistles and emitted when separated from others or in anticipation of care-taking activities such as feeding.
Whistle/Wheek	A two-part call with the first component a similar frequency to the chut, the second part rising to as high as 20 kHz. The mouth is open and respiratory movements are pronounced. Often interspersed in a bout of low whistling. This call is issued when first separated from others or in anticipation of care taking activities, such as feeding.
Squeal	A brief single sound with a frequency of 1–16 kHz emitted during any kind of injury. The mouth is open and respiratory movements are pronounced.
Scream	Similar to the squeal, but exhibiting an ascending frequency and a wider range of up to 1–32 kHz. Heard when a guinea pig is cornered or loses a fight to another. This call also shares the rising frequency of the whistle, but is much more piercing in intensity. The mouth is open and respiratory movements are pronounced.
Purr	A low burst of sounds at 0–3.5 kHz, often emitted when contact is solicited. Body may appear to make vibrating movements. This vocalization is observed in both sexes during mating, and in young approaching to nurse.
Drrr	A short Purr that is exhibited during sudden changes to the environment. May be a warning call that starts with orientation toward the stimulus, then emission of the drrr call. Body may appear to make vibrating movements during the call.
Chirrup	A rare call with downturned frequencies falling from 3 to 1.5 kHz. During the call the body remains motionless. The exact circumstances of the call are unknown.
Tweet	Calls emitted from offspring during stimulation of the anogenital region by mothers. Chirrup calls have a rising frequency from 0.5 to 4 kHz.

Ethogram: Hamster (Chapter 10)

(Provided by C Winniker)

Behavior	Description
Active	
Climbing	Vertical movement: up walls or onto wire, hind legs lifted off the floor.
Digging	Scraping with forepaws, ejecting material with hind legs.
Locomotion	Horizontal movement: walking, running.
Rearing	Head raised, looking around, erect on hind legs.
Resting	Standing still >1 s, sitting, lying.
Stretching	Back stretched, forelegs forward, hind legs backward.
Maintenance	
Drinking	Mouth interacting with water spiggot.
Feeding	Eating food, filling cheek pouches.
Grooming	Licking of the fur, grooming with forepaws and scratching with any limb.
Aggression	
Attack	One hamster jumps on or bites another.
Chase	One hamster pursuing another.
Threat	One hamster displays upright or sideways offensive postures, usually followed by behaviors listed below.
Tumble/fight	Hamsters engage and roll repeatedly, often accompanied by vocalizations and biting.
Submission	
Defensive posture	Orienting with head turned away from another hamster, forepaws raised to cover chest.
Evade/retreat	One hamster attempts escape or runs from pursuer.
Freeze	Cessation of movement, frequently with abdomen exposed.
Reproduction	
Flank marking	Vigorous rubbing of the flank gland against objects.
Food hoarding	Carrying of food and piling in distinct location(s).
Lordosis	Standing posture with the back flattened horizontally and tail elevated.
Mount	One animal mounts another, irrespective of orientation or intromission.
Nest building	Carrying of nesting substrate to and piling in distinct location(s).

Ethogram: Rabbit (Chapter 11)

(Adapted from Morton et al. 1993; Gunn & Morton 1995; Leach et al. 2009)

Behavior	Description
Aggressive/Submissive Behavior	
Aggressive circling	Slow or rapid chasing around and around in one spot; participants may have rear end of opponent gripped between their teeth.
Biting	Seen during inter- or intrasexual chasing/fighting.
Bowing	Head lowered, neck outstretched (sometimes with eyes partly or fully closed) toward approaching rabbit.
Chasing	Rabbit rapidly pursues another, often with tail erect.
Crouching	A submissive behavior; animal "freezes" and presses head and shoulders against ground with ears flattened.
Fighting	Involves aggressive attack with limbs or teeth, often with combatants leaping into the air/past each other.
Courtship/Mating Behavior	
Courtship circling	Male runs semicircles, alternating around stationary or slow-moving female. Also occurs between females in single-sex groups.
Copulation	Sexual mounting and thrusting followed by ejaculation. Successful ejaculation accompanied by male vocalization as he falls backward/sideways off the female.
Lordosis	Female crouching still with curved, convex spine, tail elevated, and vulva visible.

(Continued)

Mounting	Sexual mounting of hindquarters (or head) of conspecific. May see homosexual mounting by males or females grouped in the absence of opposite sex individuals in captivity.
Sexual following	Male approaches female from behind and she moves forward with male repeatedly following. He may sniff under her tail and put his chin on her rump as a prelude to mounting, before she moves away. An unreceptive female will usually keep her tail erect as he moves away. Also occurs between does in single-sex groups.
Sexual submission	Female crouches with tail fully erect so that a male may mate with her (see also "Lordosis").
Other Social Behavior	
Allogroom	Rabbits lick the fur of another rabbit (usually around the head, particularly the ears), typically of the opposite sex.
Nose to nose approach	Two rabbits approach head on with necks outstretched to sniff nose to nose and/or nose to chin.
Nose to tail approach	One rabbit approaches another from the rear and sniffs under his/her tail (typically in the context of a male sniffing to determine the estrous status of a female).
Nudging	Rabbit pushes nose into body or rump of conspecific – may be in a sexual or nonsexual context.
Locomotion	
Hopping	Forward movement achieved by alternate extension of fore and hindlimbs. Distinguished from running by its slower speed and shorter distance covered per forward jump.
Jump	Vertical movement by rabbit either onto an elevated surface or into the air.
Leaping	Jumping forward while hopping.
Parallel running	Two rabbits run in parallel, with elevated gait, tail erect and at a slow pace along a mutual territorial boundary. May be interspersed with bouts of jump-fighting and/or parallel paw-scraping and scent-marking displays.
Play gambolling ('frisky hop')	Forward hopping/jumping accompanied by sideways tossing of the head/ears, shaking/twisting the body or kicking out with the feet. Young rabbits may also run back/forth at some speed during this activity.
Running	Rapid forward movement achieved by alternate, fully stretched extension of fore and hindlimbs.
Shuffling	Walking at a very slow pace.
General	
Air-boxing	Fast-forward flicking of forelimbs while rabbit sits upright on haunches. Usually precedes body grooming.
Chin marking	Rubbing the chin over an object or conspecific, releasing secretion from the chin gland.
Coprophagy/re-ingestion	Rabbit removes, chews, and swallows soft, mucus-covered coprophagy pellets directly from anus.
Digging	Prolonged paw-scraping at deep substrate, usually associated with burrow excavation.
Dozing	Lying or sitting with eyes slightly open to half open and one or both ears erect; aware of the environment and responsive to sound and movement.
Drinking	Drink from water bottle.
Eating	Eating from the food hopper or hay from the floor.
Exploring	Search in the pen substrate with nose.
Fur-pulling	Only performed by females. Rabbit pulls mouthfuls of hair from her body in order to line her nest prior to parturition.
Gnawing	Gnawing cage bars and/or litter tray.
Head shaking	Shaking the head from side-to-side.
Olfactory investigation	Close sniffing of object/conspecific.
Pawscraping	Rapid scratching at the ground with the forepaws in the context of (1) foraging (typically for roots); (2) aggressive encounters between two rabbits; or (3) a scent-marking session by a rabbit alone. Scent products in the form of urine, feces and/or chin-gland secretions are typically deposited during (2) and (3).
Rearing	Sitting up on hindlimbs with both forepaws off the ground; ears partly or fully down (or, if rearing alert – ears erect).
Resting/inactive	*Sitting* – in upright stationary position, with rear end and forepaws on ground and ears down (or if sitting alert – with ears erect). *Lying, limbs tucked under* – resting with trunk on ground, hindlimbs tucked under the forelimbs, which are lying under, or forward stretched from, the body. *Lying, limbs outstretched* – resting with body trunk on ground, all four limbs outstretched and belly exposed. Rabbits often sunbathe in this position.
Scratching	Scratching at own body with a hindfoot.
Sleep	Lying or sitting with both eyes closed, ears usually flat against the back. Facial twitching and rapid eye movements may be seen.
Thumping	Loud thumping of the ground with the hindfoot (feet), usually when alarmed but males may also foot-thump after mating.
Vocalization	Includes low-pitched grunting (heard in pursued does, sexually aroused does, sexually pursued but unreceptive does) and screaming (very high pitched screeching when rabbit is injured or frightened).

Ethogram: Ferret (Chapter 12)

(Adapted from Vinke et al. 2008)

Behavior	Description
Offensive Behavior	
Positions for attack	Orientation to reach a position from which an attack may be launched.
Attack	Movement into a position from which the neck or ear regions of the opponent may be bitten. Attacker crosses over its opponent's back to reach the far side of the neck so that the attack is generally from above.
Neck bite	Bites opponent in the neck and holds.
Bite other	Bites to other places (for example: flank, belly, anal).
Drag	Grips opponent by the neck with teeth and pulls it around (neck bite is implied).
Shake	Animal bites in a body part of another animal and shakes violently with the head up and down and from right to left (neck bite is implied).
Oblique attack with physical contact	Approach to the opponent from the side or behind. The attacker's head is turned away from its adversary and its back is arched; attacker pushes its flank against its adversary and may roll on to it.
Position for lunge attack	Orientation to a position from which a lunge attack can be launched.
Lunge attack with contact	Lunges toward the opponent directing the open mouth toward the near side of the opponent's neck and touches opponent's fur, sometimes results in a bite.
Defensive Behavior	
Ward off	Flank is directed toward an adversary, the head turned away so that the convex side of the neck is presented to the opponent. Generally in response to an attack.
Extricate	Any movements made by an animal to free itself from its opponent's bite. May include rolling, kicking, scratching with forepaws, brief attacks.
Extricate with vocalization	As for extricate, but the animal makes high-pitch vocalizations that sound like screaming.
Defensive	Faces the opponent with back arched.
Defensive threat	As for defensive, but accompanied with vocalizations (vocalizations accompanying different levels of intensity are hiss, scream).
Snap	Biting movements made without any attempt actually to bite the opponent.
Passive	The animal that is being bitten makes no attempts to free itself from the opponent's grip.
Scream	Animal vocalizes with high pitch tones.
Social Behavior (Other)	
Investigate ano-genital	Sniffing another animal in anogenital region.
Investigate neck-nose	Sniffing another animal in the neck/ear region.
Investigate nose-nose	Sniffing another animal on the nose.
Nose push	Animal pushes with nose into the flank of the opponent.
Play Behavior	
Locomotory play	Play behavior consisting of gallop forward (bouncing jerky gait), gallop away, leap over the opponent, jump on the opponent.
Rough and tumble play	Play behavior by which the animal may roll over, pushes with paws, ward off with paw, inhibited chin biting.
Weasel war dance	Fast movements with jerking, galloping, play fighting (with a partner or object) and back and forth movements, which makes the behavior look like a dance; often displayed in conjunction with "dook" vocalizations.
Other Behavior	
Back away	Movement away from the opponent while remaining facing it.
Face	Facing the opponent.
Evade	Movements away from the opponent (in step), while not facing opponent.
Flee	Movement away from the opponent (in gallop) while not facing opponent.
Follow	Follows opponent while it backs away/evades (walking).
Chase	Animal follows opponent while it backs away/evades (galloping).
Approach	Animal increases proximity to opponent.
Explore	Walking around, sniffing, and scratching at the walls or floor.
Prostration	Flattens itself on the ground with head down, front paws pointing forward and hind legs splayed out on the ground.
Rubbing	Animal rubs the ground or wall with face or neck and may roll over.
Fur shake	Animal shakes off water from fur.

(Continued)

Grooming	Animal licks or rubs its pelt with forepaws and tongue.
Coughing	Animal coughs/sneezes.
Alert	Animal actively observes surrounding.
General	
Defecating	Animal defecates – defecating typically starts by turning the body 180 degrees around, subsequently stepping backward toward a fixed defecating place in the cage (defecating place mostly as far as possible from the food and resting places).
Urinating	Animal urinates – urinating typically with a hind-paw lift or in a squatted position.
Anal dragging	Dragging/sledging the perineum across surfaces to release scents from the peri-anal sebaceous gland.
Vocalizations	
Barking	A loud, scratchy chirp that occurs as a result of (over) excitement, surprise, or as part of protection of the territory.
Dooking	Low, clucking/chattering vocalizations that may be heard in contexts of excitement and comfort – positively/rewarding perceived inanimate and animate situations. Will often be displayed in conjunction with the "weasel war dance".
Hissing	Sissing sounds (similar to cats, but not as scratchy), which are usually heard in contexts of fear, anger, pain, and/or other forms of irritation or discomfort; occasionally may occur during rough play.
Honking	Sound that usually is made by kits when nervous or calling out for their mother.
Huffing	Blow out air loudly.
Screaming/screeching/squealing/crying	High-impact and loud vocalizations, consisting of high-pitched sounds accompanied or replaced by rapid chattering that may be heard in contexts of danger, such as extreme pain, fright, anger, or other forms of discomfort.
Squeaking	Short, high-pitched sounds (similar to the sound of a squeaky toy) that are usually made by a submissive ferret in contexts of discomfort (e.g., to communicate to a conspecific to let go).
Whimpering/whining	A low or high complaining, crying sound that may be heard in contexts of pain, fear, and/or discomfort. May also be heard when a (young) ferret is calling out for its mother or wanting attention.

Ethogram: Domestic Cat (Chapter 14)

(Provided by J Stella)

Behavior	Description
Affiliative Behaviors	
Approach	Approaches the observer in a friendly manner.
Rub	Rubs its body along the ground or an object. This behavior can be subdivided according to the part of the body used: head (cheek, forehead, ears, lips, chin), body (neck, flank), tail.
Eye contact: head/body turn toward	Oriented toward the observer and the eyes are looking at the observer.
Tail-up	Raises its tail to a vertical position.
Play	Plays in its environment. May be any of the following types of play:
Object	Manipulates an object with its paws in an apparently playful manner; may pat, throw, pounce, or wrestle with the object.
Self	Plays with its own body, usually the tail.
Solicit	Directs play at another cat or the observer.
Vocalization (purr, meow)	Purr – makes a low rhythmical tone from its chest and throat, produced during both exhalation and inhalation. Meow – makes a distinct sound, usually when it is trying to obtain something from another cat, but it is often directed toward human caretakers.
Agonistic Behaviors	
Lunge, swat	Lunge in a threatening manner at the observer, with or without swatting with the paw, hissing, or growling.
Crouch	Positioned with its ventrum and legs in contact with the ground, the paws are folded.
Stare	Gazes fixedly at another cat or human and is not easily distracted. It is often directed at another's eyes.
Lip lick	Licks its lips briefly. This is not in relation to feeding behavior.
Vocalization (hiss, growl)	Hiss – makes a drawn-out SSSS sound, which is unvoiced. Growl – makes a low-pitched rumbling noise.

(*Continued*)

Avoidance Behaviors

Stay, freeze, feign sleep	Stay – does not move from its position. Freeze-immobile with body tensed. Feign sleep-pretends to be asleep; it is motionless except for ear flicking, eye movement beneath partially closed eyelids and an increased respiratory rate.
Head, body turn away	Turns its head and/or body away from the observer.
Attempt to hide	Part or the entire body is behind or under something in the cage.
Avert gaze	Avoids looking at another cat or the observer for an extended period, but it may keep it in its peripheral vision.

Vigilance Behaviors

Startle-Start	Starts or jumps involuntarily, as by surprise or alarm.
Eyes fixed on observer; pupils dilated; no blink	The eyes are open fully, the pupils are fully dilated and the eyes do not blink while being observed.
Ears erect	The ears are pointed upward.
Body tense	Entire body is visibly tense, usually in a crouched position.
Increased RR	Breathing fast or at an increased rate with or without being active.
Vocalization (none, hiss, growl)	None – makes no sound. Hiss – a drawn-out SSSS sound, which is unvoiced. Growl1 – a low-pitched rumbling noise.

Maintenance Behaviors

Eat	Eats while being observed.
Drink	Drinks while being observed.
Groom	Grooms itself by licking its body or by licking its paw and passing the paw over its head, while being observed.
Rest	Remains generally inactive with eyes closed, but occasionally opens them to scan the area; ears flicking regularly.
Stretch	Bows its front end and then alternates with its hind end, often seen after waking.
Yawn	Opens its mouth wide and yawns.

Ethogram: Pig (Chapter 15)

Selected pig behaviors. (Adapted from Machado et al. 2017)

Behavior	Description
Stance	
Sleeping or lying	Animal lying with the body touching the floor or stretched on it with eyes closed or open.
Moving around or sitting	The slow movement is walking in the pen. Sitting.
Interaction with the Environment	
Eating or drinking	Pig with its head by the drinking trough or feeding trough.
Nuzzling or exploring the environment	Exploratory function, investigating the environment, nuzzling, sniffing or biting some element of the pen.
Interacting with the object	Sniffing, biting, or nuzzling the enrichment object.
Social Interactions	
Sexual behavior	Mounting or letting a partner mount,
Agonistic behavior	Confrontation, headbutts, fights, and chasing a partner in the pen,
Interaction with a partner	Nuzzling some part of another pig, playing.

Ethogram: Horse (Chapter 18)

Selected social horse behaviors (Christensen et al. 2011)

Behavior	Description
Noncontact Agonistic Interactions	
Displacement	Approach of one horse causes another to move away so that distance is maintained or increased, without overt aggression (also termed retreat/avoidance, depending on whether receiver or sender is noted).
Threat to bite	Bite intention movement with ears back and neck extended, with no actual contact.

(Continued)

Threat to kick	Kick intention movement, performed by swinging rump or backing up, and by waving or stamping hind leg toward another horse, without making contact.
Chase	One horse pursuing another usually at the gallop in an apparent attempt to direct the movement of the other. The chaser typically pins the ears, exposes the teeth and bites at the rump and tail of the pursued horse.
Mouth clapping	Submissive behavior. Opening and closing mouth with lips retracted. Typically, the head and neck are extended, and the ears oriented back or laterally (also termed snapping).
Contact Agonistic Interactions	
Bite	Opening and rapid closing of the jaws with actual contact to another horse's body. The ears are back and lips retracted.
Kick	One or both hind legs lift off the ground and rapidly extend backwards toward another horse, with apparent intent to make contact.
Push	Pressing of the head, neck, shoulder, chest or body against another horse, causing it to move one or more legs to retain balance.
Play Interactions	
Low intensity play	Playful interactions directed at another individual, which may or may not reciprocate; includes low intensity play movements such as nipping, grasping and pulling mane or tail.
Play fight	High intensity play, which is reciprocated by one or more partners; includes vigorous play movements such as rearing, boxing, circling, kneeling and chasing.
Friendly Interactions	
Social grooming	Reciprocal coat care in which the partners stand beside one another, usually head-to-shoulder or head-to-tail, grooming each other's neck, mane, rump or tail by gently nipping, nuzzling or rubbing.
Head rest	One horse rests its chin or entire head on the neck, body or rump of another horse.
Greeting Behaviors	
Nasal sniff	Olfactory investigation. Two or more horses sniff mutually head to head.
Body sniff	Olfactory investigation. A horse sniffs the neck, withers, flank or tail of another horse, which may or may not reciprocate.
Genital sniff	Olfactory investigation. A horse sniffs the genital region of another horse, which may or may not reciprocate.
Strike	One foreleg is rapidly extended forward toward another horse. The strike is typically associated with arched neck threat and posturing. The strike is often accompanied by a snort or squeal.

Ethogram: Zebra Finch (Chapter 20)

Selected zebra finch behaviors (Adapted from Figueredo et al. 1992). See Chapter 21 for vocalizations.

Behavior	**Description**
Allofeeds	Bird regurgitates food for another.
Allopreens	Bird preens another.
Approaches	Bird comes within a few inches of another.
Autopreens	Bird preens, grooms, scratches self, wipes beak, bathes, dries body, etc.
Beak fences	Bird exchanges facial pecks with another.
Broods (nest/eggs/chicks)	Bird sits upon nest, eggs, or nestlings.
Builds/rearranges nest	Bird manipulates materials within nest.
Charges	Bird rapidly approaches another.
Chases	Bird sustains rapid pursuit of another.
Clumps	Extensive body contact without preening.
Drinks water	Bird imbibes water from any source.
Eats	Bird consumes food or grit.
Fights (escalated)	Bird bites or grapples with another.
Flees (no pursuit)	Bird escapes another without chase.
Flies	Bird moves on the wing.
Gathers nesting materials	Bird collects typical nesting materials and transports them to nest site.
Hops	Bird moves on foot in a hopping motion.
Investigates nonfood/nonnest	Bird closely examines nonfood item, which is not typical nesting material.
Manipulates nonfood/nonnest	Bird pushes, pulls, or transports nonfood item, not directed toward nest.

(Continued)

Mounts	Bird sexually mounts another.
Plucks feathers	Bird removes feathers from another.
Pulls tail	Bird vigorously pulls tail of another.
Remains near (passive)	Bird remains within a few inches of another, mostly stationary.
Remains near (active)	Bird follows movements of another to keep within a few inches.
Rests	Bird remains still with eyes mostly closed.
Solicits/courts	Bird directs lateral darting movements with rapid tail fluttering toward another.
Stays	Bird remains still with eyes mostly open.
Supplants (attack)	Bird rapidly flies directly at and attempts to displace another.
Is supplanted	Bird quickly moves away to avoid contact with the oncoming bird.
Threatens	Bird, with head up and beak agape, tilts body toward and visually tracks another.

Ethogram: Zebrafish (Chapter 21)

Selected common locomotor behaviors in larval and adult zebrafish (according to Zebrafish Behavioral Catalog, ZBC; Kalueff et al. 2013).

Behavior	Description
Alarm reaction (both)	An adaptive escape reaction which serves as an antipredatory response exhibited in the context of fear-inducing stimulation characterized by increased speed of movement and rapid directional changes.
Avoidance (both)	Increased movement away and/or time spent away from an object or a stimulus (e.g., predator, bright light)
Attraction (larval)	Increased time spent nearby or movement toward an object (visual) or chemical stimulus (e.g., food extract).
Backward swimming (adult)	Aberrant motor behavior observed under some circumstances, such as following exposure to selected hallucinogenic drugs (e.g., lysergic acid diethylamide).
Beat-and-glide (larval)	An intermittent form of swimming characterized by tail beating followed by gliding; appears at 4 dpf in larvae.
Bend/bending (larval)	Aberrant neurological phenotype involving swimming with the body in a laterally bent position; can be observed as part of seizure behavior.
Burst-and-coast behavior (larval)	Darting pattern specific to larval fish not yet able to perform continuous swimming. Fish move forward (burst) in a single motion and glide (coast) to a slow speed, or stop from which they burst forward again.
Burst swim (larval)	Fast forward swim with large bend angles, maximally at mid-body of larval zebrafish, appears at 2 dpf in larvae. Includes larger amplitude bending (large bend angles), faster speeds, and greater yaw than during slow swimming; often associated with escape behaviors; pectoral fins are tucked against the body and not active.
C-start (C-bend/turn, Mauthner reflex) (larval)	Quick escape/startle response in which the fish body first curves to form a C-shape, and then the fish propels itself away at an angle from its previous position using a fast swim.
Circling (adult)	Repetitive swimming in a circular direction (usually seen during seizures, neurological impairments, and following the selected drug's action).
Coasting (adult)	Passive sliding without body/fin movements.
Coil/coiling (larval)	Embryonic movement describing a full body contraction that brings the tip of the tail to the head (coils); can be spontaneous or evoked by touch.
Corkscrew swimming (adult)	Spiral swimming with an increased speed and in an uncoordinated direction; commonly observed as part of seizure phenotype.
Creeping (adult)	Very slow swimming during which only the pectoral fins propel the fish forward.
Darting (adult)	A single fast acceleration in one direction (e.g., as part of escape behavior) with the use of caudal fin.
Dashing (adult)	A series of directed (propulsive) darting movements; commonly seen as an escape response.
Droopy tail (adult)	Motor phenotype associated with neurological deficits, akinesia, and global hypolocomotion. Can be evoked by aging, motor impairments, or genetic and pharmacological modulations.
Erratic movement (adult)	Complex behavior characterized by sharp changes in direction or velocity and repeated rapid darting.
Fast turn (larval)	Escape-like turns in larval zebrafish, characterized by fast, large-angle turns that involve bending of the entire body with high angular velocity.
"Figure eight" swimming (adult)	A specific swimming pattern observed in zebrafish following selected drug treatments (e.g., nicotine or ketamine).
Flee/fleeing (both)	Accelerating movement away from another fish or stimulus.
Floating (both)	Passive swimming (typically near the water surface).
Freeze/freezing (adult)	A complete cessation of movement (except for gills and eyes) by the fish while at the bottom of the tank.
Hyperlocomotion/ hyperactivity (both)	Abnormally fast swimming endured for an extended period of time; typically related to psychostimulant/convulsant action or anxiety-like behavior.
Hypolocomotion/ hypoactivity (both)	Abnormally slow swimming for an extended period of time; typically related to sedation, neuromotor deficits, and akinesia.

(Continued)

Selected Ethograms

Immobility (both)	A complete cessation of movement (except for gills and eyes) at the bottom of the tank; differs from freezing and resting as it is not always associated with altered (respectively, increased or reduced) opercular movements.
Inclined swimming (larval)	An aberrant phenotype (swimming with an angle relative to the water surface; tilting), commonly induced by neuroactive/neurotoxic substances.
"Jittery" swimming (adult)	A specific pattern of swimming characterized by multiple short "jerky" movements with reduced smoothness of swimming trajectories, common for some seizure behavior.
Loop/looping (larval)	Distinct circular swimming behavior in larvae around a virtual point outside of the larva's body.
Meandering (adult)	Movement without a fixed direction or path.
O-bend (larval)	Orientation movement in which the larval zebrafish body curves to change the orientation of swimming.
Optomotor response (larval)	Locomotion induced by a repetitive moving stimulus presentation (e.g., rotating drum), as zebrafish will generally swim in the same direction as the moving pattern.
Photomotor response (larval)	Stereotypic series of motor behaviors in embryonic zebrafish in response to light stimulation, as zebrafish show motor excitation.
Slow swim (larval)	Larval zebrafish slow swimming (scoots) characterized by small bend angles with bend location near the tail.
S-start/bend (larval)	Quick escape/startle response where the body curves to form an S-shaped body bend with simultaneous activity rostrally on one and caudally on the other side.
Startle response (both)	An evolutionarily conserved, adaptive behavior in response to sudden, usually aversive, stimuli, such as vibration, light, sound, or touching.
Stereotypic locomotion (both)	A pattern of rigid, repetitive behaviors (e.g., swimming from corner to corner) evoked in zebrafish under some conditions (e.g., treatment with psychostimulants like nicotine and caffeine, or hallucinogens like ibogaine).
Swim/swimming (both)	Simple zebrafish locomotion; can be categorized by its duration as "prolonged", which may be maintained for minutes, or as "sustained", which may be maintained for hours.
Tremor (both)	Specific shivering-like behavior, most typically evoked in adult or larval zebrafish by selected neurotoxic/convulsant drugs.
Undulating movements (adult)	A wavelike or snakelike motion; part of aggression-related behaviors.
Vertical drift (adult)	An aberrant phenotype that involves passive floating vertically.
Vertical swim (adult)	An aberrant phenotype that involves swimming vertically (typically heads up at the surface).
Weavering (adult)	Aberrant tremor-/shaking-like phenotype, typically evoked by neurotoxic or convulsant agents.

Ethogram: Marmoset (Chapter 24)

(Adapted from https://primate.wisc.edu/wp-content/uploads/sites/1409/2020/05/Obs-Ethogram-Marmoset-worksheet.pdf)

Behavior	**Description**
Walking or climbing	Walk quadrupedally, that is, with all four limbs. Walking is simply normal movement.
Leaping	Like walking, a way to get around by jumping from place to place.
Drinking	Consuming water from pipes, lixits, bottles, or dishes.
Eating or foraging	A monkey sifts through its surroundings with hands or mouth in search of food items.
Tree gnawing	Tree gnawing is a type of foraging in which the animals gnaw on the tree to extract sap/gum.
Grooming (self or others)	Picking at or licking the fur of oneself or other individuals with hands or mouth. A monkey may be receiving or initiating grooming. Grooming can be either hygienic or for forming and strengthening social bonds.
Infant comfort	Infant approaches another monkey with physical contact, usually in response to conflict or being alone.
Play	Social interactions that are characterized by relaxed-looking huddling or other close physical contact and apparent low tension; may include relaxed-looking wrestling, jumping on, jumping over, chasing, fleeing, hiding, playful "nips".
Vocalizing	Marmosets are chirping, calling or otherwise vocalizing
Resting	Monkey is neither moving nor engaged in any activity.
Sleep	Monkey is unmoving with eyes closed.
Scent marking	Rubbing scent glands (located in chest and near genitals) on branches or other areas to mark them to communicate with other marmosets through scent.
Dominant interaction (includes threats)	Marmosets bite, scratch, or chase another animal. Other dominant behaviors include displacement, facial, vocal, or physical threats, such as ear tuft flicking, tongue flicking, teeth baring, and head shaking. May be accompanied by piloerection (hair standing on end).
Submissive interaction	Marmoset flees from another monkey that is aiming dominance interactions at it.
Allo-parenting	Care of infants by individuals other than the parents.

Ethogram: Capuchin (Chapter 27)

(Modified from Benitez et al. Chapter 27)

Behavior	Description
Normal	
Aggression	Aggression can be noncontact (scream, threat, chase) or result in contact (hit, bite, pull tail).
Beg	One individual begs for food from another, by sticking out the hand or placing the face close to that of the eating monkey.
Body shake	Shake entire body or a body part. Often a sign of agitation.
Contact	Sit huddled or touching another individual.
Dominant behavior	Threaten, mount, or displace another individual over a preferred space, food, or object.
Forage	Search for or extract food.
Fur rub	Vigorously rub an object over fur repeatedly. Individuals may rub against another individual.
Groom	Pick through the hair, search for and/or remove debris from another individual.
Locomote	Actively move around enclosure.
Manipulate	Handle or play with a nonedible object.
Play	Nonaggressive interactions involving two or more animals. Includes wrestle, tickle, chase. Usually between juveniles and infants.
Proximity	Within arm's reach of another, but not touching.
Proximity or contact feed	Feed in proximity or in contact with another individual.
Rest	Inactive, sleep, sit for long periods of time.
Self-groom	Pick through the hair, searching for and/or removing debris.
Self-scratch	Drag the nails/hand across one's own skin.
Sexual	Any form of sexual behavior between two individuals. This includes solicit, penile erection (directed at a female), exploration behavior, mount, thrust, flirt (duck face, scalp raise, vocalize).
Solo feed	Feed or forage alone.
Submissive behavior	Crouch, bob, flinch, avoid, flee, present (in a nonsexual context), or, during aggression received, fear grin.
Urine wash	Urinate on one palm and rub it on the sole of the foot.
Abnormal	
Coprophagy	Eat or play with feces or urine.
Exaggerated head turn	Throw head back in a circular motion.
Hold body	Hold on to any part of the body (often the genital regions) for an extended period or in an awkward way (i.e., walking while holding one's genitals).
Pull hair out	Forcefully bite or pull clumps of hair out of own or other's body. Note the recipient if it is pulling out another's hair.
Suck digits	Suck on any digit, usually thumb or index finger or repeated mouthing and/or licking any extremity.

Ethogram: Vervet Monkey (Chapter 29)

Normative behavior patterns in captive vervets. (Adapted from Fairbanks & McGuire 1986; Fairbanks et al. 1978; Jorgensen 2017; Struhsaker 1967a, b; see also Jorgensen Chapter 29)

Behavior	Description
Aggressive	
Manipulate	Stylized behavior involving placing the hands on the head or shoulders of another, embracing.
Display	Ritualized behavior including standing broadside to another animal with tail up, sideways prancing, circling another animal with tail up, showing the hindquarters, or showing an erection to another.
Aggression	Stare, threaten (open mouth, chin forward, ears back, eye lids up), head jerk (rapid sideways jerking of the head), slap, push away, chase (not in play) attack.
Branch shake	Vigorous whole-body jumping onto cage or play structures.
Yawn	Any occurrence of yawning with teeth showing.
Submissive	
Submissive	Pawing at ground, backward shuffling, ambivalent lunges from crouching position, screaming, defensive aggression.
Lipsmack	Rapid opening and closing of the lips directed to another.
Affiliative	
Groom	Using fingers to manipulate hair of another monkey; may remove parasites.

(*Continued*)

Selected Ethograms

Muzzle	Bringing muzzle to any part of another animal's body except the genitals, usually muzzle to muzzle.
Social play	Any form of social play including wrestling, jumping, chasing, mouthing.
Social contact	Sitting within 1-meter of another animal (initiating or receiving).
Sexual behavior	Sniffing, touching or manipulating the genital area of another animal, grabbing hips, mounting or copulating.
Self-Directed	
Scratch	Vigorous scratching of self.
Self-directed behavior	Any other self-directed behavior including self-grooming, brief touches to face or body, hand or foot rubbing, licking hand or body, gentle hair picking, masturbation, shrugging.
Abnormal	
Pacing	Pacing back and forth, at least three turns.
Other abnormal	Any other abnormal behavior, e.g., hair plucking, self-biting, nonfunctional repetitive behavior, spinning, crouching.
Miscellaneous	
Vocalization	Any vocalization.
Locomotion	Whole body movement of one or more steps.
Feeding	Any feeding behavior including eating chow, foraging on the ground, drinking and chewing on any edible object.
Object manipulation	Any object manipulation, exploring the environment, handling nonfood object or trying to see inside or under hidden object.

Ethogram: Baboon (Chapter 30)

(Adapted from Coelho & Bramblett 1989)

Behavior	Description
Self-Directed Behavior	
Body shake	A quick side-to-side shaking movement of the head and upper body.
Self-groom	Manipulating, brushing or licking of hair of oneself.
Self-scratch	Movement of the hand or foot during which the fingers/fingernails and/or toes/toenails are drawn across the hair or skin. A new bout is counted if a different body part is scratched or a different limb is used to scratch.
Self-touch	Making directed contact with one's own body with the hands, feet, or mouth that does not qualify as self-grooming or self-scratching.
Swipe	Dragging hand across nongravel substrate in a sweeping/digging motion.
Yawn	Slowly opening the mouth wide with a downward thrust of the lower mandible. Most often includes narrowing the eyes and exposing the teeth.
Affiliative Behavior	
Contact	Touching another individual. Bouts end if partners move more than 1 m away before 5 s pass. Not scored during grooming.
Groom	Manipulating, brushing or licking of hair of another individual. Bouts end if partners move more than 1 m away before 5 s pass.
Play	Social interactions characterized by apparent low tension, usually accompanied by a "play face" (mouth is open and facial features are relatively relaxed). May include: grunting, wresting, sham-biting, jumping on or over, chasing, fleeing, hiding.
Proximity	Approaching to within 1 m of another animal. Not scored during more intimate affiliative states (contact, groom).
Aggressive Behavior	
Bite	Intense agonistic interaction during which the skin/limb of another animal is grasped with the teeth. Does not include "nips" which consists of a brief pinch of the skin with the incisors (classified as "rough behavior").
Chase	Strong agonistic interaction involving pursuit past the location the recipient maintained at the start of the interaction.
Rough behavior	Mild agonistic interaction involving slight physical contact, usually no facial signal. May include: nipping, grabbing, kicking, pulling, pushing, poking, slapping, pulling hair, butting, shoving.
Threat	Moderate agonistic interaction containing any of the following components: *Head thrust* – Jabbing head toward another individual, often accompanied by raised eyebrows and staring at an individual. *Open-mouth threat* – Opening mouth wide quickly, usually accompanied by thrusting of the head toward target and staring at an individual. *Raised eyebrows* – Glare with display of white portion of skin around eyes. *Teeth gnashing* – Conspicuous, exaggerated grinding of teeth. *Lunge* – Charging toward another animal that does not exceed the recipient's location at the time the action begins. *Pin* – Holding another animal down.

(Continued)

Submissive Behavior

Avoid	Moving more than one step from another animal upon its approach.
Bare-teeth	Pulling back the face muscles to display the teeth.
Bark	Emitting a high-pitched yapping vocalization (usually repeated) during which the mouth may remain open and in which no component noises are longer than 1 s.
Flee	Rapid withdrawal from another animal, usually occurring in response to aggressive behavior or an approach.
Present	Orientating the hindquarters toward another animal, usually accompanied by lowering of the forelimbs, lifting of the tail, or looking back over the shoulder.
Scream	Emitting a loud high-pitched vocalization occurring in a defensive or retreating context in which one of the component noises is sustained for longer than 1 s.

REFERENCES

Berryman JC. Guinea-pig vocalizations: Their structure, causation and function. *Zeitschrift für Tierpsychologie* 41:80–106, 1976.

Blanchard RJ and Blanchard DC. Aggressive behavior in the rat. *Behavioral Biology* 21:157–161, 1977

Blanchard RJ, Fukunaga K, Blanchard DC, and Kelley MJ. Conspecific aggression in the laboratory rat. *Journal of Comparative and Physiological Psychology* 89:1204–1209, 1975.

Brewer JS, Bellinger SA, Joshi, P., and Kleven GA. Enriched open field facilitates exercise and social interaction in 2 strains of guinea pigs (*Cavia porcellus*). *Journal of the American Association for Laboratory Animal Science* 53:344–355, 2014.

Christensen JW, Sondergaard E, Thodberg K, and Halekoh U. Effects of repeated regrouping on horse behavior and injuries. *Applied Animal Behaviour Science* 133:199–206, 2011.

Coelho AM Jr. and Bramblett CA. Behaviour of the genus *Papio*: Ethogram, taxonomy, methods, and comparative measures. In PK Seth and S Seth (eds), *Perspectives in Primate Biology Vol. 3*, Today & Tomorrow's Printers and Publishers: New Delhi, pp. 117–140, 1989.

Corat C, Tarallo R, Branco CR, Savalli C, Tokumaru RS, et al. The whistles of the guinea pig: An evo-devo proposal. *Revista de Etologia* 11:46–55, 2012.

Fairbanks LA and McGuire MT. Age, reproductive value, and dominance-related behavior in vervet monkey females: Cross-generational influences on social relationships and reproduction. *Animal Behaviour* 34:1710–172, 1986.

Fairbanks LA, McGuire MT, and Page N. Social roles in captive vervet monkeys (*Cercopithecus aethiops sabaeus*). *Behavioral Processes* 3:335–352, 1978.

Figueredo AJ, Ross DM, and Petrinovich L. The quantitative ethology of the zebra finch: A study in comparative psychometrics. *Multivariate Behavioral Research* 27:435–458, 1992.

Gunn D and Morton DB. Inventory of the behavior of New Zealand White rabbits in laboratory cages. *Applied Animal Behaviour Science* 45:277–292, 1995.

Hennessy MB and Jenkins R. A descriptive analysis of nursing behavior in the guinea pig (*Cavia porcellus*). *Journal of Comparative Psychology* 108:23–28, 1994.

Jorgensen MJ. Behavioral management of *Chlorocebus spp*. In SJ Schapiro (ed), *Handbook of Primate* Behavioral *Management*, Boca Raton, FL: CRC Press, pp. 339–366, 2017.

Kalueff AV, Gebhardt M, Stewart AM, Cachat JM, Brimmer M, et al. Towards a comprehensive catalog of zebrafish behavior 1.0 and beyond. *Zebrafish* 10:70–86, 2013.

Leach MC, Allweiler S, Richardson C, Roughan JV, Narbe R, et al. Behavioural effects of ovariohysterectomy and oral administration of meloxicam in laboratory housed rabbits. *Research in Veterinary Science* 87:336–347, 2009.

Machado SP, Caldara FR, Foppa L, de Moura R, Goncalves LMP, et al. Behavior of pigs reared in enriched environment: Alternatives to extend pigs attention. *PLOS ONE* 12:e0168427, 2017.

Morton D, Jennings M, Batchelor GR, et al. Refinements in rabbit husbandry. Second report of the BVAAWF/FRAME/RSPCA/UFAW, joint working group on refinement. *Laboratory Animals* 27:301–329, 1993.

Panskeep J. The ontogeny of play in rats. *Developmental Psychobiology* 14:327-332, 1981.

Pellis SM and Pellis VC. Play-fighting differs from serious fighting in both target of attack and tactics of fighting in the laboratory rat *Rattus norvegicus*. *Aggressive Behavior* 13:227–242, 1987.

Pellis SM. Agonistic versus amicable targets of attack and defense: Consequences for the origin, function, and descriptive classification of play-fighting. *Aggressive Behavior* 14:85–104, 1988.

Rood JP. Ecological and behavioural comparisons of three genera of Argentine cavies. *Animal Behaviour Monographs* 5:3–83, 1972.

Struhsaker TT. *Behavior of vervet monkeys (Cercopithecus aethiops)*. Vol 82. Berkeley and Los Angeles: University of California Press. 1967a.

Struhsaker TT. Behavior of vervet monkeys and other cercopithecines. *Science* 156:1197–1967b.

Vanderschuren LJMJ, Niesink RJM, and VanRee JM. The neurobiology of social play behavior in rats. *Neuroscience and Biobehavioral Reviews* 21:309–326, 1997.

Vinke CM, van Dejik R, Houx BB, and Shoemaker NJ. The effects of surgical and chemical castration on intermale aggression, sexual behavior and play behavior in the male ferret (*Mustela putorius furo*). *Applied Animal Behaviour Science* 115:104–121, 2008.

Young WC. Psychobiology of sexual behavior in the guinea pig. *Advances in the Study of Behavior* 2:1–110, 1969.

Index

abnormal affective states 29
abnormal appetitive behavior, baboons 501
abnormal behaviors 4, 74
    abnormal affective states 29
    baboons 501–502
    baseline standard 28
    capuchin monkeys 429
    cattle 278
    causes
        correlated factors 35–37
        fixed and causal risk factors 33–34
        retrospective and correlational approaches 33
        variable risk factors 34–35
    chickens and quail 304–305
    deer mice 154
    definition 27–28
    displacement activities 29–30
    dogs 209–210
    ethological standard 28
    ferrets 196–198
    fowl 304–305
    gerbils 154–155
    guinea pig 138–139
    hamsters 167
    macaques 449–451, 457
    medical conditions 32
        arthritic and painful disorders 32–33
        gastrointestinal disease 32
        neurological and retinal disorders 33
    mice 95–97
    owl monkeys 413
    pigs and mini pigs 248
    persistence 37–38
        hypotheses 37–38
    prevalence 30
        elephants 32
        mice 32
        nonhuman primates 30–31
        sheep 32
    rabbits 179, 180
    rats 119
    reptiles 370–371
    self-injurious behavior 29
    sheep 266
    stereotypic behaviors 28
    therapeutic interventions 38
        environmental approaches 38–40
        pharmacotherapy 40–41
    treatment approaches 27
    types 28–30
    voles 155
    welfare 55–56
    white-footed mice 154
    zebra finch 322–323
    zebrafish 336–337
abnormal repetitive behaviors (ARBs) 32, 95; *see also* stereotypic behaviors
    definition 55–56
acoustic signals 17

activity patterns, of ferrets 195
ad libitum feeding 213
adrenergic system 41
adult neurogenesis 316
affective disorders 34
African green monkeys *see* vervets (*Chlorocebus* spp.)
aggression
    baboons 499–500
    capuchin monkeys 430–431
    cattle 276
    cats 228–229
    chickens and quail 301, 305, 307
    dogs 213
    ferrets 196, 198–199, 201
    guinea pig 134–135, 138
    hamsters 168
    macaques 446, 448–449
    marmosets 388
    mice 95–97, 100
    pigs and mini pigs 245, 247, 248
    rabbits 178, 181
    rats 116
    sheep 266
    squirrel monkeys 401
    zebrafish 335, 336
agonistic behaviors, of zebrafish 335
allopreening 318
alopecia
    in cats 230
    in ferrets 196
    in macaques 451
    in squirrel monkeys 399
α2A-receptor agonist 41
Ambystoma mexicanum 346
amphibians 5
    characteristics 345
    common captive behaviors 351
    communication 351
    courtship and mating behavior 350–351
    ecology 348–349
    feeding behavior 349
    natural history 347–348
    range of species 346
    in research 346–347
    social organization and behavior 349–350
    ways to maintain behavioral health 352–354
animal behavior 3
    biologically relevant questions 10–11
        bird song 11–12
        proximate question 11
        ultimate question 11
    challenges 8
    definition 8
    description phase 8–9
    explanation phase 10
    quantification phase 9–10
    scientific method 12–13
    sensory motor capabilities 21
animal cognition 19, 20

animal model
    definition 68
    effects of 68–69
anthropomorphism 21
antibody production, in rabbits 185
antipredator behavior
    of chickens and quail 302
    of guinea pig 134, 139
    of horses 287–288
    of macaques 452
    of marmosets 383
    of rabbits 175
    of rats 114
    of sheep 262
anura 345
anxiety-like behaviors 36
    of dogs 215
    of macaques 447
    of marmosets 383, 384
    of mice 100, 101
    of rats 117
    of vervets 480, 482
    of zebrafish 336, 337
Aotus see owl monkeys (*Aotus*)
ark calls 320
arousal/tension reduction hypothesis, of abnormal behavior 37

baboons (*Papio* sp.) 5, 495
    captive behavior 501–502
    complex diseases research 496
    ethogram 527–528
    fetal and infant studies 496
    natural history
        common social behaviors 498–500
        communication 500–501
        ecology 496–497
        feeding behavior 500
        mating and reproduction 497–498
        social organization 497
    reproductive research 496
    in research 495
    transplantation research 496
    ways to maintain behavioral health
        environment 502–503
        feeding strategies 503–505
        social groupings 503
        training 505
barbering 55, 74
    in deer and white-footed mice 154
    in mice 95, 100
    in rats 118, 119
bark, vocalizations 500–501
beagles 4, 206, 209, 214
begging calls 320–321
behavioral management 65–66
    components 75
        empirical assessments 78
        environmental enrichment 75–77
        positive reinforcement training 77–78
        socialization 75

behavioral management (*cont.*)
    data and research 78
        costs and benefits 78–79
    definition 66
    future directions 81
    implementation 79–81
    principles 66–67
        animal model 68–69
        enhancing welfare 67–68
        functionally simulating natural
            conditions 69–74
        natural behavioral biology 67
        separate from captive management 74–75
        well-being and wellness 67–68
    strategies 66
"benign" noise 77
Benítez, M.E. 5
biological functioning approach, of assessing welfare 52
    behavioral correlates 59–60
    behavioral indicators 60
bird song 11–12
blood sampling, V-shaped restraint benches 253–254
body-directed activities, in ARBs 28
Bos Taurus see Cattle (*Bos* sp.)
Breeding 16
    amphibians 348
    chickens and quail 301, 302
    deer mice and white-footed mice 151
    dogs 207, 208
    ferrets 193
    guinea pigs 135
    hamsters 169
    macaques 447
    marmosets 381, 386
    mice 90, 92, 94
    pigs and minipigs 245, 246
    rabbits 176, 181–182
    reptiles 368
    squirrel monkeys 401
    vervets 478
    zebra finch 322, 324, 325
    zebrafish 334
the Brown rat 114
    cache 7, 20

cage biting, of guinea pig 138
Callithrix penicillata see marmosets (*Callithrix* sp.)
Canis lupus familiaris see dogs (*Canis* sp.)
captive behaviors, of zebrafish 336–337
captive management strategies 66
capuchin monkeys 5, 421–422
    common behaviors in captivity 427–429
    communication 425–427
    ecology 423–424
    ethogram 526–527
    feeding behavior 424–425
    maintaining behavioral health
        feeding strategies 431
        physical environment 429
        social grouping 429–431
        training 431–432
    in research 422–423
    social organization and behavior 424
    special situations 432
Caro, T. 12

Cas9 editing 11
cattle (*Bos* sp.) 4
    characteristics 275
    common captive behavior 277–279
    communication 277
    ecology and habitat use 275–276
    feeding behavior 276–277
    in research 274–275
    social organization and behavior 276
    special situations 281
    species-typical behavior
        environmental enrichment 279
        feeding strategies 280
        social groupings 280
        training 280
    use of 273
    ways to maintain behavioral health 279
Cavia porcellus see guinea pigs (*Cavia* sp.)
chemical cues 18
chemical signals 17–18
chickens 4; *see also* fowl
    archaeological evidence 299
    natural history
        communication 303
        ecology 302
        feeding behavior 303
        social organization 302–303
    in research 300
    ways to maintain behavioral health
        environment 305–306
        social groupings 306–308
chimpanzees
    prevalence rates of abnormal behavior 30
    sex differences in abnormal behavior 30, 31
    stereotypic behavior 30, 31
chinchillas, fur chewing in 36
Chlorocebus pygerythrus see vervets (*Chlorocebus* spp.)
Christensen, J.W. 293
cicadas 412
Cloutier, S. 4
C58 mice, stereotypic behavior 32, 33
cognitive bias paradigm 58–59
cognitive enrichment 76
Coleman, K. 3
combined valence and arousal testing paradigms 58
common captive behaviors
    amphibians 351
    cattle 277–279
    deer mice 153–154
    dogs 209–210
    domestic cat 229–235
    ferrets 193–198
    fowl 303–305
    gerbils 153–155
    guinea pigs 137–139
    hamsters 167–168
    horses 291–292
    macaques 448–451
    marmosets 382–384
    owl monkeys 413
    pigs 247–249
    rabbits 178–179
    rats 118–119
    reptiles 368–371
    sheep 265–266

squirrel monkeys 399
vervets 481–482
voles 154–155
white-footed mice 153–154
zebra finch 321–323
communication 17–18
    abnormal behavior 38
    amphibians 351
    baboons 500–501
    capuchin monkeys 425–427
    cattle 277
    chickens 303
    dogs 208–209
    domestic cat 228–229
    ferrets 193
    guinea pig 137
    hamsters 167
    horses 291
    macaques 445
    marmosets 382
    mice 91–92
    owl monkeys 413
    rabbits 177–178
    rats 117–118
    reptiles 368
    sheep 265
    squirrel monkeys 399, 401–402
    vervets 480–481
    zebrafish 320–321, 335–336
comorbidities
    anxious and impulsive behavior 36
    perseveration 36
    sleep disturbance 35–36
    stress and HPA axis dysregulation 35
conditioned place preference/aversion 58
consumer demand 57–58
copulation calls 320
corticosterone 352
cortisol 35–36
    cat 234
    dogs 209
    guinea pigs 137
    macaques 445, 450
    rabbits 185
    vervets 479
    zebrafish 338
costs and benefits, of behavioral management programs 78
    ethical implications 79
    exceptions 79
    expanded capabilities 79
    subject selection 78
    validity 78
    variability 78
Coturnix japonica see quail (*Coturnix* sp.)
crib-biting, in horses 36–37
cross-fostering, of pigs 255
cytomegalovirus (CMV) 33

Danio rerio see zebrafish (*Danio rerio*)
deer mice (*Peromyscus* sp.) 4, 148
    common captive behaviors 153–154
    feeding behavior and substrates 153
    habitat use, predator/prey relations 150
    natural history and ecology 150
    in North America 148
    in research 150
    social organization and behavior 151

Index

special situations 156
stereotypic behavior 32
ways to maintain behavioral health
    environment 155
    feeding strategies 156
    social groupings 155–156
    training 156
demes 92
DeNardo, D.F. 5
depression 29
description phase, in animal behaviors 8–9
desired behaviors 60
diabetes, of guinea pig 133
differential reinforcement of alternative behavior (DRA) 38
displays 15
  in baboons 499, 501
  in capuchin monkeys 425, 426, 427
  in cattle 277
  in chicken and quail 301–302, 306–308
  in gerbil 154
  in macaques 446
  in marmosets 382, 384
  reptiles 364, 365, 368
  in sheep 263
  in vervets 480–481
  in zebra finches 319, 321
distance calls 320
distress calls 320
Dixon, L. 4
"Do as I Do" 20
dogs (*Canis* sp.) 4, 205–206
  common captive behaviors 209–210
  ethogram 217–218
  natural history
    ecology 207
    feeding behavior 208
    senses and communication 208–209
    social organization and behavior 208
  research use 206–207
  special situations 216
  ways to maintain behavioral health
    environment 210–211
    feeding strategies 213–214
    social groupings 211–213
    training 214–216
dog's sense of smell 8
domestic cat (*Felis* sp.)
  common captive behaviors
    environment 231–233
    feeding strategies 235
    human–cat interactions 235
    normal behavior 229
    social groupings 233–235
    stress-related and sickness behavior 229–230
    training 235
    ways to maintain behavioral health 230–231
  communication 228–229
  ethogram 521–522
  feeding behavior 227–228
  mating behavior 226–227
  natural history 224–225
  in research 223–224
  social organization and behavior 225–226
  in United States 223
domestic fowl (*Gallus gallus domesticus*) 299

dominance hierarchies 15
  baboons 497
  chickens and quail 304
  deer mice and white-footed mice 153
  guinea pigs 134, 135, 138
  hamsters 166, 167
  macaques 444
  pigs and minipigs 245
  rabbits 177
  rats 116
  sheep 263
  squirrel monkeys 397
  vervets 478–479
  zebra finch 318
  zebrafish 335
dosing, of minipigs 252
  dermal application 253
  injections and sprays/drops 252–253
  oral dosing 252
  special dosing procedures 253
drinking behavior, reptiles 367–368
Dwyer, C.M. 4

early life stress exposure, of abnormal behavior 34
ecology 13–14
  amphibians 348–349
  baboons 496–497
  capuchin monkeys 423–424
  chickens 302
  chickens and quail 302
  dogs 207
  fowl 300–301
  guinea pigs 134
  horses 287–288
  macaques 440
  marmosets 379–380
  mice 90–91
  owl monkeys 410–411
  pigs 245
  rabbits 175
  rats 114–115
  reptiles 362–363
  sheep 262–263
  squirrel monkeys 396–397
  vervets 477–478
  zebra finch 317
  zebrafish 332–334
Edwards, S. 4
electrocardiograms, of minipigs 254
elephants, stereotypic behavior 32
environmental enrichment (EE) 21, 39, 75
  baboons 503
  cattle 279
  chickens and quail 306
  cognitive 76
  dogs 211
  feeding 76
  ferrets 198–200
  marmosets 383
  mice 94, 99–100
  in microenvironment considerations 99–100
  occupational 76
  physical 76
  rabbits 179
  rats 122
  sensory 76–77

error variance 78
estrus
  in cattle 276, 277
  in cats 226
  in dogs 208
  in ferrets 193
  in guinea pigs 135, 140
  in horses 289, 293
  in mice 92
  in pigs 245, 246
  in rabbits 176
  in rats 116
  in sheep 263
ethical implications, in behavioral management programs 79
ethogram 5, 9, 217–218, 513
  baboon 527–528
  capuchin 526–527
  domestic cat 521–522
  ferret 520–521
  guinea pig 516–517
  hamster 518
  horse 523
  marmoset 526
  pigs 522
  rabbit 518–519
  rats 513–515
  vervet 527
  zebra finch 524
  zebrafish 524–525
explanation phase, in animal behaviors 10
exploratory behavior
  of marmosets 383
  of pigs 247–248
Equus ferus przewalskii see horse (*Equus* sp.)

FACEs *see* Functionally Appropriate Captive Environments (FACEs)
feeding behaviors 17
  amphibians 349, 354
  baboons 500, 503–505
  capuchin monkeys 424–425
  cattle 276–277
  chickens and quail 303
  deer mice 153, 156
  dogs 208, 213–214
  domestic cat 227–228
  ferrets 193
  fowl 302
  gerbils 153, 156
  guinea pig 137, 140
  hamsters 167, 169
  horses 290–291
  macaques 457
  marmosets 380–381, 383, 387–388
  owl monkeys 412–413
  pigs 245
  rabbits 177, 182–184
  rats 116–117, 122–123
  reptiles 367
  sheep 264–265
  squirrel monkeys 399
  vervets 480, 483
  voles 153, 156
  white-footed mice 153, 156
  zebra finch 320
  zebrafish 335

feeding enrichment, in behavioral
        management 76
feelings approach, of animal welfare 52, 54–55
    affective state behaviors 55–57
    preference tests 57–59
feline idiopathic cystitis (FIC) 224
feline immunodeficiency virus (FIV) 224
Felis catus see domestic cat (*Felis* sp.)
Ferret Grimace Scale 197
ferrets (*Mustela* sp.) 3
    characteristics 191
    common captive behaviors
        abnormal behavior and signs of
            discomfort 196–198
        normal behavior 193–196
    ethogram 520–521
    natural history 192
        breeding and parental care 193
        communication 193
        feeding behavior and predator–prey
            relationships 193
        habitat and habitat use 192
        home ranges and territoriality 192
        social organization 192
    research participation 191–192
    special situations 201
    ways to maintain behavioral health
        early life considerations 198
        environmental enrichment 198–200
        social structure preferences 198
        training 200–201
Five Freedoms 53
flehmen response 228, 277–278, 289
fluoxetine 41
foraging 17, 69
    amphibians 349
    baboons 500, 504
    capuchin monkeys 424
    chickens and quail 302, 303
    deer mice and white-footed mice 156
    dogs 213
    ferrets 193, 195
    gerbils 153, 156
    hamsters 169
    horses 290
    macaques 443, 457
    marmosets 379, 380, 383, 387
    mice 101
    owl monkeys 412, 415
    pigs 136, 247–248
    rabbits 177
    rats 123
    reptiles 367
    sheep 262
    squirrel monkeys 397, 399
    vervets 480
    voles 157
    zebrafish 335
    zebra finches 323
Forced Swim Test 36
fowl
    common captive behaviors
        abnormal behaviors 304–305
        normal behaviors 303–304
    natural history
        communication 302
        ecology 300–301
        feeding behavior 302

social organization and behavior
    301–302
potential benefits 307–308
ways to maintain behavioral health
    environment 305–306
    social groupings 306–308
free-living house mice 94
free-ranging animals 9, 10
Friedrich, S.R. 4
Functionally Appropriate Captive
        Environments (FACEs) 69, 70
    appropriate behavior 73–74
    for cats 71
    control and choice 70
    decreasing abnormal behavior 74
    for dogs 71
    individual differences *vs.* group norms 74
    for macaques 73
    for marmosets 73
    for pigs 71
    predictability 72–73
    providing opportunities 69–70
    for rats 70
    voluntary participation 71–72
    for zebra finches 72
    for zebrafish 72
functionally simulating natural conditions
        69–74
fur chewing, in chinchillas 36

GABAergic system 41
Gallus gallus domesticus see domestic fowl
        (*Gallus gallus domesticus*)
Gallus gallus gallus see Junglefowl (*Gallus
        gallus gallus*)
Garcia de la Chica, A. 5
gastrointestinal disease, abnormal behavior 32
genetically altered animals 7
genome editing technologies 11
gerbils (*Meriones unguiculatus*) 4, 148–149
    common captive behaviors 153–155
    feeding behavior and substrates 153
    natural history and ecology 150–151
    in research 150
    social organization and behavior 151–152
    special situations 156
    ways to maintain behavioral health
        environment 155
        feeding strategies 156
        social groupings 156
Gottlieb, D.H. 3, 4, 56
Grand, N. 4
"green" monkeys 476
grooming 29, 53
    baboons 498–499
    capuchin monkeys 424, 427
    cats 229, 230
    ferrets 196
    deer mice and white-footed mice 151,
        153, 154
    gerbils 153–154
    guinea pigs 135, 136
    horses 286
    macaques 448, 450
    marmosets 383
    mice 91, 92
    rabbits 178
    rats 116, 119

sheep 264
squirrel monkeys 400
vervets 479
grunt, vocalizations 500
guanfacine 41
guinea pigs (*Cavia* sp.) 3, 4
    classification 132
    common captive behaviors 137–139
    domestication 131
    ethogram 516–517
    hairless 141
    history 133
    natural history 134–137
    research application 133–134
    special situations 141
    strains 131–132
    ways to maintain behavioral health
        environment 139
        feeding strategies 140
        husbandry 139–140
        social groupings 139
        training 140

habituation 18, 77, 101, 124–125, 198, 215,
        268, 387, 484
hairless guinea pigs 141
hamsters (*Mesocricetus* sp.) 4
    biomedical research 165
    common captive behaviors 167–168
    ethogram 518
    natural history
        communication 167
        ecology, habitat use and predator/prey
            relations 166
        feeding behavior 167
        social organization and behavior
            166–167
    research contributions 165–166
    ways to maintain behavioral health
        environment 168
        feeding strategies 169
        social groupings 168–169
        training 169
home cage physical features, in mice 99
home ranges 8, 13
    baboons 497
    cat 225
    cattle 276
    chickens and quail 300, 301
    deer mice and white-footed mice
        148, 151
    dogs 208
    ferrets 192
    gerbils 150
    guinea pigs 134
    macaques 444
    marmosets 379, 383
    mice 90
    owl monkeys 411
    pigs and minipigs 245
    rabbits 175, 177
    rats 114
    sheep 263
    squirrel monkey 396
    vervets 478
    zebra finches 318
    zebrafish 333
horse (*Equus* sp.) 4

common captive behaviors 291–292
crib-biting in 36–37
ecology, habitat use, and antipredator behavior 287–288
ethogram 523
Middle Ages 285
in research 286–287
social organization and behavior 288–289
    communication 291
    feeding behavior 290–291
    foals 290
    mare-foal interactions 289–290
    mares, sexual behavior of 289
    separation and weaning 290
    stallions, sexual behavior of 289
special situations 294
ways to maintain behavioral health
    environment 292–293
    feeding strategies 293–294
    social groupings 293
    training 294
Hosie, C.A. 5
Human Intruder Test 36
humidity 98
hyper-aggressiveness, abnormal behavior 29
hypothalamic-pituitary-adrenocortical (HPA) axis, of dysregulation 35

ick-ooer, vocalizations 501
impoverishment hypothesis, of abnormal behavior 37–38
individual variation, in animal behaviors 20–21
infectious disease, of guinea pig 133
Institute Armand Frappier (IAF) hairless guinea 141
interindividual variability, in data 78
International Mouse Strain Resource (IMSR) 90
"isolation syndrome" 100

Jorgensen, M.J. 5, 482
Junglefowl (*Gallus gallus gallus*) 299

kackle calls 320
Kasuga, D.C. 151
Kleven, G.A. 4

Lambton, S. 4
landscape of fear 13–14
learning 18–20, 77–78
    amphibians 350, 354
    cat 226, 228
    in chickens and quail 302, 304
    in ferrets 199
    in chickens and quail 303–304
    complex forms 19–20
    habituation 18
    in horses 287, 291, 294
    in marmosets 385
    in mice 91, 100
    operant conditioning 18–19
    in rats 114, 121, 123–124
    in sheep 268
    social learning 20
    as training 77, 123–124, 294
    vervets 477
    zebra finch 321

zebrafish 334, 339
Lidfors, L. 4
Lithobates catesbeianus 346
locomotor stereotypies, in ARBs 28
long tonal calls 321
lordosis
    cat 226
    guinea pigs 135, 136
    hamster 167
loud, vocalizations 501
lower urinary tract (LUT) disease 224
Lutz, C.K. 31

macaques (*Macaca* spp.) 5, 19
    affiliation 446
    aggression and reconciliation 446
    common captive behaviors 448
        abnormal behavior 449–451
        normal behavior 448–449
        sources of variation 449
    Common, Scientific Names, and IUCN Threat Status 438
    communication 445
    natural history
        activity patterns 440–443
        diet and feeding behavior 443
        ecology 440
        habitat use 443–444
        predation 444
    reproduction 447
    in research 438–439
    sex differences in 31
    social organization 444–445
    stereotypic behavior 31
    taxonomy, distribution, and conservation 439–440
    ways to maintain behavioral health 452
        special situations 458–459
        species-typical behavior 452–457
        training 458
Manciocco, A. 5
mares, sexual behavior of 289
marmosets (*Callithrix* sp.) 5, 14
    care of aged animals 388
    common captive behaviors 382–384
    ecology 379–380
    ethogram 526
    feeding behavior 380–381
    mating and reproduction 381–382
    in research 377–379
    social organization and behavior 380
    social problems 388
    types 377
    wasting syndrome 388
    ways to maintain behavioral health
        environmental complexity 384–385
        feeding 387–388
        social groupings 385–387
        training 388
Marmoset Wasting Syndrome (MWS) 388
maternal care
    guinea pigs 136
    mice 101
    sheep 263
    macaques 447
mating behavior

amphibians 350–351
baboons 497–498
breeding periods 16
capuchin monkeys 431
chickens and quail 302–303
dog 208
domestic cat 226–227
female choice 16
guinea pig 135
hamsters 167
horses 288
macaques 447
male competition 15–16
marmosets 381–382
mating strategies 15
mating system 16–17
mice 92–94
owl monkeys 411, 412
parental behavior 17
pigs 254
rabbits 176
reptiles 366, 370
rats 115–116, 119
sheep 263
squirrel monkeys 398
voles 152, 156
zebra finch 318–320
mating system 16–17
meadow voles 10
Mello, C.V. 4
Meriones unguiculatus see gerbils (*Meriones* sp.)
Mesocricetus auratus see hamsters (*Mesocricetus* sp.)
Meyer, J.S. 3, 4
mice (*Mus* sp.) 4
    common captive behaviors 93
        abnormal behaviors 95–97
        normal behaviors 93–94
    natural history
        ecology, diets, and threats 90–91
        mating and mate choice 92–93
        reproduction and maternal behavior 93
        senses and communication 91–92
        social organization and behavior 92
    special situations 102
    stereotypic behavior 32
    typical research involvement 89–90
    ways to maintain behavioral health
        environment 97–100
        feeding strategies 101
        handling and reduce fear of humans 101
        maternal care and optimal weaning 101
        social groupings 100–101
        training 101–102
Microtus 4; *see also* voles (*Microtus* sp.)
minipigs 4; *see also* pigs
Mongolian gerbil *see* gerbils
monogamy 14
mood congruent bias 59
motor stereotypy, baboons 501
mouthing behavior
    in ARBs 28
    ferrets 196
Mus musculus see mice (*Mus* sp.)
Mustela putorius furo see ferrets (*Mustela* sp.)

naltrexone 41
natural behavioral biology 67
natural history
    amphibians 347–348
    baboons
        common social behaviors 498–500
        communication 500–501
        ecology 496–497
        feeding behavior 500
        mating and reproduction 497–498
        social organization 497
    chickens and quail 302–303
    deer mice 150
    dogs 207–209
    domestic cat 224–225
    ferrets 192–193
    fowl 300–302
    gerbils 150–151
    guinea pigs 134–137
    hamsters 166–167
    macaques
        activity patterns 440–443
        diet and feeding behavior 443
        ecology 440
        habitat use 443–444
        predation 444
    mice 90–92
    pigs 245–247
    rabbits 174–175
    rats 114–117
    reptiles
        communication 368
        drinking behavior 367–368
        ecology 362–363
        environment 363
        feeding behavior 367
        parental care 365–367
        social organization and behavior 364–365
        thermoregulation 364
    sheep 262–265
    vervets
        communication 480–481
        ecology, habitat use, predator/prey relations 477–478
        feeding behavior 480
        social organization and behavior 478–480
    voles 151
    white-footed mice 150
    zebra finch 317–321
    zebrafish 332–336
naturalistic approach, of animal welfare 52
    species-typical behaviors 52–53
    use of 54
natural odors 77
negative affective state behaviors 55
negative parental care 97
neurological disease, abnormal behavior 33
neuronal gene expression 317
New Zealand White (NZW) rabbits 3
nonbeagle dogs, use of 216
nonhuman primates, abnormal behavior in 40–41
normal behaviors
    baboons 501
    deer mice 153
    dogs 209
    fowl 303–304
    gerbils 153–154
    hamsters 167
    macaques 448–449
    marmosets 383
    in mice
        homeostasis 93
        interaction with environment 93–94
        mating and parental care 94
        self-maintenance 93
        social interactions 94
    owl monkeys 413
    reptiles 368
    sheep 266
    squirrel monkeys 399
    voles 154
    white-footed mice 153
    zebra finch 321–322
    zebrafish 336
Norway rat 114
Novak, M.A. 3, 4
nutcrackers 20
nutrition, of guinea pig 133

observational learning *see* social learning
occupational enrichment 76
one-male units (OMUs) 497
ooer, vocalizations 501
operant conditioning 18–19
ophthalmoscopy, of minipigs 254
opioidergic system 41
orangutans 19
Oryctolagus cuniculus see Rabbits (*Oryctolagus* sp.)
otology, of guinea pig 133
Ovis aries see sheep (*Ovis* sp.)
owl monkeys (*Aotus*) 5
    characteristics 409
    common captive behavior 413
    communication 413
    ecology 410–411
    feeding behavior 412–413
    in research 410
    social organization and behavior 411–412
    ways to maintain behavioral health
        enrichment 415
        environment/housing 413–414
        feeding strategies 415
        social groupings 414–415
        training 415

painful disorders, abnormal behavior 32–33
parental behavior
    guinea pig 136
    horses 293
    rats 115–116
    voles 149, 151
parental care 17
    ferrets 193
    reptiles 365–367
    vervets 479
paternal care 10
    guinea pigs 136
    owl monkeys 410
    mice 94
peptide arsenal 348
Peromyscus 4; *see also* deer mice (*Peromyscus* sp.); white-footed mice (*Peromyscus* sp.)
perseveration 36
personality 20–21, 59
    amphibians 350, 354
    cats 233, 235
    dogs 209
    macaques 449
    baboons 498
Phillips, C. 4
physical enrichment 76
pigs (*Sus* sp.) 4
    common captive behaviors 247
        farrowing and maternal behavior 248–249
        foraging and exploratory behavior 247–248
        maintenance behaviors 249
        reproductive behavior 248
        social behavior 247
    in disease models 244–245
    domestication 243
    ethogram 522
    natural history
        ecology and feeding behavior 245
        reproductive behaviors 246–247
        social organization and behavior 245–246
    in pharmaceutical research 244
    special situations 254–255
    use of 243–244
    ways to maintain behavioral health 249–250
    commonly applied study procedures 252–254
    enclosure environment and enrichment 250–251
    feeding 251
    housing 250
    human–animal interaction and training 251–252
place conditioning *see* conditioned place preference/aversion
play behavior 53, 56
    baboons 499
    cattle 278
    ferrets 195–196
    guinea pigs 130
    hamsters 169
    horses 290
    pigs and minipigs 247
    rats 120
    squirrel monkeys 397
play fighting 116
Pleurodeles waltl 346
polyandry 14
polygynandry 14
polygyny 14
Pomerantz, O. 3, 4
popcorning 4, 138
positive affective state behaviors 56
positive reinforcement training (PRT) 40, 66, 79
    macaques 458
    marmosets 388
    pigs and minipigs 251–252
    training approaches 77–78
    voluntary participation 71–73, 77
    zebrafish 339
poultry *see* chickens

Powell, C. 5
prairie voles 10
predation 13–14
   macaques 444
   vervets 478
   zebrafish 334
preference tests 57
   cognitive bias paradigm 58–59
   confounds 59
   degree evaluation 57–58
   measurement 57
Prescott, M. 4
Pritchett-Corning, K.R. 4
PRT *see* positive reinforcement training (PRT)
Przewalski horses 285
psychological well being 30, 39, 54, 99, 458

quail (*Coturnix* sp.)
   natural history
      communication 303
      ecology 302
      feeding behavior 303
      social organization 302–303
   in research 300
   ways to maintain behavioral health
      environment 305–306
      social groupings 306–308
quantification phase, in animal behaviors 9–10

Rabbit Grimace Scale (RbtGS) 185
Rabbits (*Oryctolagus* sp.)
   antibody production in 185
   common captive behaviors 178–179
   communication 177–178
   ecology, habitat use, predator/prey
      relations 175
   ethogram 518–519
   feeding behavior 177
   natural history 174–175
   reproductive behavior 176–177
   research participation 174
   social organization and behavior 175–176
   training 184
   in United States 173
   ways to maintain behavioral health
      environment 179–181
      feeding strategies 182–184
      social groupings 181–182
   welfare of 184–185
rats (*Rattus norvegicus*) 4
   animal-based research 113–114
   in Canada 113
   common captive behaviors 118–119
   communication 117–118
   ethogram 513–515
   natural history
      ecology 114–115
      feeding behavior 116–117
      predator/prey relations 114–115
      social organization and behavior
         115–116
   ways to maintain behavioral health
      environment 119–122
      experimental testing and sampling 123
      feeding strategies 122–123
      social groupings 122
      training and handling 123–125
Rattus norvegicus see rats (*Rattus norvegicus*)

repetitive behaviors 8
reproduction behaviors
   baboons 497–498
   guinea pig 133, 135
   macaques 447
   marmosets 381–382
   mice 93
   pigs 246–248
   rabbits 176–177
   reptiles 370
   squirrel monkeys 398
   vervets 479
   zebra finch 318–320
   zebrafish 334
reptiles 5, 361–362
   common captive behaviors
      abnormal 370–371
      feeding and drinking behavior 370
      normal 368
      reproductive behavior 370
      thermoregulation 368–370
   natural history
      communication 368
      drinking behavior 367–368
      ecology 362–363
      environment 363
      feeding behavior 367
      parental care 365–367
      social organization and behavior
         364–365
      thermoregulation 364
   in research 362
   special situations 372
   ways to maintain behavioral health
      371–372
respiratory disease, of guinea pig 133
restricted feeding 213
retinal disorders, abnormal behavior 33
rhesus monkeys 19–20
   abnormal repetitive behavior 56
   aggression 446, 449
   alopecia 451
   cognition 19
   cognitive bias paradigm. 58
   diet 443
   dominance 15
   habitat use 443
   pair housing 8, 15, 456
   play 446
   positive reinforcement training 19, 458
   reproduction 15
   self-injurious behavior 450
   systematics 440
   temperament 449, 452, 456, 458
roar grunt, vocalizations 501
rough-and-tumble play 116
3Rs 66, 81, 192, 210, 249, 331, 448, 458
rumblestrut 4, 135

Saimiri spp. *see* squirrel monkeys
   (*Saimiri* spp.)
satellite males 16
scent marking 18
Schapiro, S.J. 3, 4
scream/screech, vocalizations 501
Seier, J.V. 481, 484
selective serotonin reuptake inhibitors (SSRIs)
   40, 41

self-directed behavior, baboons 501
self-groom 29
   baboons 500
   capuchin monkeys 429
   cattle 279
   guinea pigs 135
   marmosets 383
   mice 93
   rabbit 180
   rats 116, 119
self-injurious behavior (SIB) 29
   macaques 33, 36, 450
   prevalence 31
Selye, H. 54
sensory enrichment 76–77
sensory motor capabilities 21
serotonergic system 40–41
serotonin transporter gene (SERT) 33–34
sex differences, in stereotypic behavior 31, 33
sex dimorphism, of zebra finch 316–317
sexual behavior
   of mares 289
   of stallions 289
sexual selection
   breeding periods 16
   female choice 16
   male competition 15–16
   mating strategies 15
   mating system 16–17
   parental behavior 17
sham chew
   horses 291
   rabbits 179
sheep (*Ovis* sp.) 4, 261
   common captive behaviors 265–266
   natural history
      communication 265
      ecology and habitat use 262–263
      feeding behavior 264–265
      social organization and behavior
         263–264
   in research 261–262
   special situations 269
   ways to maintain behavioral health
      environment 266–267
      feeding strategies 268
      social groupings 267–268
      training 268–269
sheep, stereotypic behavior 32
shoaling, zebrafish 334–335
SIB *see* self-injurious behavior (SIB)
sickness behavior, of mice 97
simian immunodeficiency virus (SIV) 33
single housing, in dogs 216
sleep disturbance 35–36
Smith, T.E. 5
social behaviors 69
   amphibians 349–350
   capuchin monkeys 424
   cattle 276
   deer mice 151
   dogs 208
   domestic cat 225–226
   fowl 301–302
   gerbils 151–152
   guinea pigs 134–137
   hamsters 166–167
   horses 288–291

social behaviors (*cont.*)
    marmosets 380
    mice 92
    owl monkeys 411–412
    pigs 245–247
    rabbits 175–176
    rats 115–116
    reptiles 364–365
    sheep 263–264
    squirrel monkeys 397
    vervets 478–480
    voles 152–153
    white-footed mice 151
    zebra finch 317–318
    zebrafish 334–335
social grouping
    amphibians 353–354
    baboons 503
    capuchin monkeys 429–431
    cattle 280
    chickens 306–308
    chickens and quail 306–308
    deer mice 155–156
    dogs 211–213
    domestic cat 233–235
    fowl 306–307
    gerbils 156
    guinea pig 139
    hamsters 168–169
    horses 293
    macaques 453–457
    marmosets 385–387
    mice 100–101
    owl monkeys 414–415
    rabbits 181–182
    rats 122
    reptiles 371
    squirrel monkeys 400–401
    vervets 482–483
    voles 156
    white-footed mice 155–156
    zebra finch 324
social housing
    and abnormal behavior 38–39
    baboons 502, 503
    chickens and quail 306–308
    dogs 211–213
    guinea pigs 138
    hamsters 169
    macaques 453, 456
    marmosets 387
    mice 94, 100
    rabbits 181
    rats 122
    squirrel monkeys 402
social learning 20
    cats 228
    chickens and quail 302, 303
    marmosets 381, 388
    rats 123
    vervets 477
    zebrafish 334
social organization 14–15
    amphibians 349–350
    baboons 497
    capuchin monkeys 424
    cattle 276
    chickens and quail 302–303

    deer mice 151
    dogs 208
    domestic cat 225–226
    ferrets 192
    fowl 301–302
    gerbils 151–152
    guinea pigs 134–137
    hamsters 166–167
    horses 288–291
    macaques 444–445
    marmosets 380
    mice 92
    owl monkeys 411–412
    pigs 245–246
    rabbits 175–176
    rats 115–116
    reptiles 364–365
    sheep 263–264
    squirrel monkeys 397
    vervets 478–480
    voles 152–153
    white-footed mice 151
    zebra finch 317–318
    zebrafish 334–335
social play 116
solitary living 14
song 11–12
spatial distributions 13
species-typical behaviors (STBs)
    amphibians 352–353
    cattle 279–280
    dogs 210–214
    ferrets 198–201
    guinea pig 139–140
    hamsters 168–169
    horses 292–294
    macaques 452–457
    mice 97–101
    rabbits 179–182
    rats 119–125
    reptiles 371–372
    and welfare 52–53
    zebrafish 337–339
species-typical constraints hypothesis, of abnormal behavior 37, 38
squirrel monkeys (*Saimiri* spp.) 5
    common captive behavior 399
    communication 399
    ecology 396–397
    feeding behavior 399
    life history and reproduction 398
    physical characteristics 395
    in research 395–396
    social organization and social behaviors 397
    ways to maintain behavioral health
        communication 401–402
        enrichment and feeding strategies 401
        environment 400
        social groupings 400–401
        special situations 402
        training 402
stack calls 320
stallions, sexual behavior of 289
stampede 4
Stella, J.L. 4
stereotypic behaviors 17, 28, 30–31
    in elephants 32

    enrichment effects on 39–40, 100
    in hamsters 168
    in horses 287, 288, 292
    in macaques 31
    in mice 33, 95
        sex differences 33
    in mink 37
    pair housing 39, 42
    pharmacotherapy 41
    positive reinforcement training 40
    in rats 119
    in sheep 32, 266
    single cage housing 34
    stress 34, 292
    in zebrafish 337
stereotypies *see* stereotypic behavior
steroid action, in zebra finch 316–317
Stone, A.I. 5
stotting 10
stress 35
stressors, in adolescents and adults 34–35
Struhsaker, T.T. 477, 480
subjective well-being (SWB) assessments 56–57
submissive behaviors, in baboons 500
subordinate animals 67
Sucrose Preference Test 36
Summer, F. 150
Sus scrofa see pigs (*Sus* sp.)

tactile communication 18
Taeniopygia guttata castanotis see zebra finch (*Taeniopygia guttata castanotis*)
Tail Suspension Test 36
taste aversion learning 19
temperament 20–21, 74, 77, 78, 206, 235, 430, 449–453, 456–458, 498
temperatures, housing mice at 98
teratology, of guinea pig 133
territorial behavior
    ferrets 193
    vervets 478
territory 8, 13, 15–16
    ferrets 193
    cat 224, 225, 228, 233
    cattle 276
    chickens and quail 301
    deer mice and white-footed mice 151, 155
    gerbils 152
    hamsters 167
    marmosets 380
    mice 100
    rabbits 175–177
    rats 115–116
    reptiles 364, 365
    vervets 480
    voles 152, 156
tet calls 320
thermoregulation, of reptiles 364, 368–370
thigmotaxis 337
thuk calls 320
Tinbergen, N. 10, 11
toxicology, of guinea pig 133
tryptophan hydroxylase-2 gene (TPH2) 33–34

umwelt 8
undesired behaviors 60
urodela 345

Index

vervets (*Chlorocebus* spp.) 5, 8, 475
   characteristics 476
   common captive behaviors 481–482
   common names and taxonomy 476
   ethogram 527
   natural history
      communication 480–481
      ecology, habitat use, predator/prey relations 477–478
      feeding behavior 480
      social organization and behavior 478–480
   in research 477
   special situations 484
   ways to maintain behavioral health
      environment 482
      feeding strategies 483
      social groupings 482–483
      training 483–484
visual signals 17
voles (*Microtus* sp.) 4
   common captive behaviors 154–155
   feeding behavior and substrates 153
   natural history and ecology 151
   in research 150
   social organization and behavior 152–153
   special situations 156
   ways to maintain behavioral health
      environment 155
      feeding strategies 156
      social groupings 156
voluntary participation, in behavioral management 77

wahoo call, in vocalizations 501
welfare 4, 51
   amphibians 352
   biological functioning approach 52
      behavioral correlates 59–60
      behavioral indicators 60

feelings approach 52, 54–55
   affective state behaviors 55–57
   preference tests 57–59
measuring 52
naturalistic approach 52
   species-typical behaviors 52–53
   use of 54
practical applications 60
   behavioral biology 61
   environmental evaluations 60
   individual assessments 60
of rabbits 184–185
whine calls 320
white-crowned sparrows 11–12
white-footed mice (*Peromyscus* sp.) 148
   common captive behaviors 153–154
   feeding behavior and substrates 153
   habitat use, predator/prey relations 150
   natural history and ecology 150
   in North America 148
   in research 150
   social organization and behavior 151
   special situations 156
   ways to maintain behavioral health
      environment 155
      feeding strategies 156
      social groupings 155–156
      training 156
"Whitten effect" 94
Whitten, P.L. 478, 483
Williams, L. 5
Winnicker, C. 4, 448
wsst calls 320

Xenopus laevis 346

zebra finch (*Taeniopygia guttata castanotis*) 3, 4
   common captive behaviors
      abnormal behaviors 322–323

      normal behaviors 321–322
   domestication 315–316
   ethogram 524
   natural history
      ecology 317
      feeding behavior 320
      mating and reproduction 318–320
      social organization and behavior 317–318
      vocal communication 320–321
   in research 316–317
   subspecies 315
   ways to maintain behavioral health
      breeding 325
      environment 323–324
      feeding strategies 324
      social groupings 324
      training 324–325
zebrafish (*Danio rerio*) 4–5, 331–332
   captive behaviors 336–337
   communication 335–336
   diet 333
   ecology 332–334
   ethogram 524–525
   feeding behaviors 335
   geography 332–333
   growth and mortality 333–334
   natural history 332–336
   predation 334
   in research 332
   social organization and behavior 334–335
   special situations 339
   standard laboratory housing conditions for 331, 332
   ways to maintain behavioral health
      environment 337–338
      feeding strategies 338–339
      training 339